ENCYCLOPEDIA

OF

SPECTROSCOPY

AND

SPECTROMETRY

ENCYCLOPEDIA
OF
SPECTROSCOPY
AND
SPECTROMETRY

Editor-in-Chief

JOHN C. LINDON

Editors

GEORGE E. TRANTER

JOHN L. HOLMES

ACADEMIC PRESS

A Harcourt Science and Technology Company

San Diego   San Francisco   New York   Boston
London   Sydney   Tokyo

This book is printed on acid-free paper

ACADEMIC PRESS
A Harcourt Science and Technology Company
24–28 Oval Road
London NW1 7DX, UK
http.hbuk.co.uk/ap/

ACADEMIC PRESS
525 B Street, Suite 1900,
San Diego, CA 92101–4495, USA
http://www.apnet.com

ISBN 0-12-226680-3   1001865621

A catalogue record for this Encyclopedia is available from the British Library

Library of Congress Catalog Card Number: 98–87952

Access for a limited period to an on-line version of the Encyclopedia of Spectroscopy and Spectrometry
is included in the purchase price of the print edition.
This on-line version has been uniquely and persistently identified by the Digital Object Identifier (DOI)

10.1006/rwsp.2000

By following the link

http://dx.doi.org/10.1006/rwsp.2000

from any Web Browser, buyers of the Encyclopedia of Spectroscopy and Spectrometry
will find instructions on how to register for access.

Typeset by Macmillan India Limited, Bangalore, India
Printed and bound in Great Britian by The University Printing House, Cambridge, UK.
00 01 02 04 05 CU 9 8 7 6 5 4 3 2 1

Editors

EDITOR-IN-CHIEF

John C. Lindon
Biological Chemistry
Division of Biomedical Sciences
Imperial College of Science, Technology and Medicine
Sir Alexander Fleming Building
South Kensington
London SW7 2AZ, UK

EDITORS

George E. Tranter
Glaxo Wellcome Medicines Research
Physical Sciences Research Unit
Gunnells Wood Road
Stevenage
Hertfordshire SG1 2NY, UK

John L. Holmes
University of Ottawa
Department of Chemistry
PO Box 450
Stn 4, Ottawa, Canada KIN 6N5

Editorial Advisory Board

Preface

This encyclopedia provides, we believe, a comprehensive and up-to-date explanation of the most important spectroscopic and related techniques together with their applications.

The *Encyclopedia of Spectroscopy and Spectrometry* is a cumbersome title but is necessary to avoid misleading readers who would comment that a simplified title such as the "Encyclopedia of Spectroscopy" was a misnomer because it included articles on subjects other than spectroscopy. Early in the planning stage, the editors realized that the boundaries of spectroscopy are blurred. Even the expanded title is not strictly accurate because we have also deliberately included other articles which broaden the content by being concerned with techniques which provide localized information and images. Consequently, we have tried to take a wider ranging view on what to include by thinking about the topics that a professional spectroscopist would conveniently expect to find in such a work as this. For example, many professionals use spectroscopic techniques, such as nuclear magnetic resonance, in conjunction with chromatographic separations and also make use of mass spectrometry as a key method for molecular structure determination. Thus, to have an encyclopedia of spectroscopy without mass spectrometry would leave a large gap. Therefore, mass spectrometry has been included. Likewise, the thought of excluding magnetic resonance imaging (MRI) seemed decidedly odd. The technique has much overlap with magnetic resonance spectroscopy, it uses very similar equipment and the experimental techniques and theory have much in common. Indeed, today, there are a number of experiments which produce multidimensional data sets of which one dimension might be spectroscopic and the others are image planes. Again the subject has been included.

This led to the general principle that we should include a number of so-called spatially-resolved methods. Some of these, like MRI, are very closely allied to spectroscopy but others such as diffraction experiments or scanning probe microscopy are less so, but have features in common and are frequently used in close conjunction with spectroscopy. The more peripheral subjects have, by design, not been treated in the same level of detail as the core topics. We have tried to provide an overview of as many as possible techniques and applications which are allied to spectroscopy and spectrometry or are used in association with them. We have endeavoured to ensure that the core subjects have been treated in substantial depth. No doubt there are omissions and if the reader feels we got it wrong, the editors take the blame.

The encyclopedia is organized conventionally in alphabetic order of the articles but we recognize that many readers would like to see articles grouped by spectroscopic area. We have achieved this by providing separate contents lists, one listing the articles in an intuitive alphabetical form, and the other grouping the articles within specialities such as mass spectrometry, atomic spectroscopy, magnetic resonance, etc. In addition each article is flagged as either a "Theory", "Methods and Instrumentation" or "Applications" article. However, inevitably, there will be some overlap of all of these categories in some articles. In order to emphasize the substantial overlap which exists among the spectroscopic and spectrometric approaches, a list has been included at the end of each article suggesting other articles in this encyclopedia which are related and which may provide relevant information for the reader. Each article also comes with a "Further Reading" section which provides a source of books and major reviews on the topic of the article and in some cases also provides details of seminal research papers. There are a number of colour plates in each volume as we consider that the use of colour can add greatly to the information content in many cases, for example for imaging studies. We have also included extensive Appendices of tables of useful reference data and a contact list of manufacturers of relevant equipment.

We have attracted a wide range of authors for these articles and many are world recognized authorities in their fields. Some of the subjects covered are relatively static, and their articles provide a distillation of the established knowledge, whilst others are very fast moving areas and for these we have aimed at presenting up-to-date summaries. In addition, we have included a number of entries which are retrospective in nature, being historical reviews of particular types of spectroscopy. As with any work of this magnitude some of the articles which we desired and commissioned to include did not make it for various reasons. A selection of these will appear in a separate section in the on-line version of the encyclopedia, which will be available to all purchasers of the print version and will have extensive hypertext links and advanced search tools. In this print version there are 281 articles contributed by more than 500 authors from 24 countries. We have persuaded authors from Australia, Belgium, Canada, Denmark, Finland, France, Germany, Hungary, India,

Israel, Italy, Japan, Mexico, New Zealand, Norway, Peru, Russia, South Africa, Spain, Sweden, Switzerland, The Netherlands, the UK and the USA to contribute.

The encyclopedia is aimed at a professional scientific readership, for both spectroscopists and non-spectroscopists. We intend that the articles provide authoritative information for experts within a field, enable spectroscopists working in one particular field to understand the scope and limitations of other spectroscopic areas and allow scientists who may not primarily be spectroscopists to grasp what the various techniques comprise in considering whether they would be applicable in their own research. In other words we tried to provide something for everone, but hope that in doing so, we have not made it too simple for the expert or too obscure for the non-specialist. We leave the reader to judge.

John Lindon
John Holmes
George Tranter

Acknowledgements

Without a whole host of dedicated people, this encyclopedia would never have come to completion. In these few words I, on behalf of my co-editors, can hope to mention the contributions of only some of those hard working individuals.

Without the active co-operation of the hundreds of scientists who acted as authors for the articles, this encyclopedia would not have been born. We are very grateful to them for endeavouring to write material suitable for an encyclopedia rather than a research paper, which has produced such high-quality entries. We know that all of the people who contributed articles are very busy scientists, many being leaders in their fields, and we thank them. We, as editors, have been ably supported by the members of the Editorial Advisory Board. They made many valuable suggestions for content and authorship in the early planning stages and provided a strong first line of scientific review after the completed articles were received. This encyclopedia covers such a wide range of scientific topics and types of technology that the very varied expertise of the Editorial Advisory Board was particularly necessary.

Next, this work would not have been possible without the vision of Carey Chapman at Academic Press who approached me about 4 years ago with the excellent idea for such an encyclopedia. Four years later, am I still so sure of the usefulness of the encyclopedia? Of course I am, despite the hard work and I am further bolstered by the thought that I might not ever have to see another e-mail from Academic Press. For their work during the commissioning stage and for handling the receipt of manuscripts and dealing with all the authorship problems, we are truly indebted to Lorraine Parry, Colin McNeil and Laura O'Neill who never failed to be considerate, courteous and helpful even under the strongest pressure. I suspect that they are now probably quite expert in spectroscopy. In addition we need to thank Sutapas Bhattacharya who oversaw the project through the production stages and we acknowledge the hard work put in by the copy-editors, the picture researcher and all the other production staff coping with very tight deadlines.

Finally, on a personal note, I should like to acknowledge the close co-operation I have received from my co-editors George Tranter and John Holmes. I think that we made a good team, even if I say it myself.

John Lindon
Imperial College of Science, Technology and Medicine
London
22 April 1999

Guide to Use of the Encyclopedia

Structure of the Encyclopedia

The material in the Encyclopedia is arranged as a series of entries in alphabetical order.

There are 4 categories of entry:

- Historical Overview
- Theory
- Methods and Instrumentation
- Applications

To help you realize the full potential of the material in the Encyclopedia we have provided the following features to help you find the topic of your choice.

1. Contents lists

Your first point of reference will probably be the main alphabetical contents list. The complete contents list appearing in each volume will provide you with both the volume number and the page number of the entry.

Alternatively you may choose to browse through a volume using the alphabetical order of the entries as your guide. To assist you in identifying your location within the Encyclopedia a running headline indicates the current entry. Furthermore, a "reference box" is provided on the opening page of each entry so that it is immediately clear whether it is a theory, methods & instrumentation, applications, or historical entry, and which of the following areas of spectroscopy it covers.

- Atomic Spectroscopy
- Electronic Spectroscopy
- Fundamentals in Spectroscopy
- High Energy Spectroscopy
- Magnetic Resonance
- Mass Spectrometry
- Spatially Resolved Spectroscopic Analysis
- Vibrational, Rotational, & Raman Spectroscopies

Example:

Nuclear Overhauser Effect

Anil Kumar and **R Christy Rani Grace**,
Indian Institute of Science, Bangalore, India

MAGNETIC RESONANCE
Theory

You will find "dummy entries" in the following instances:

1. where obvious synonyms exist for entries. For example, a dummy entry appears for ESR Imaging which directs you to EPR Imaging where the material is located.
2. where we have grouped together related topics. For example, a dummy entry appears for Arsenic, NMR Applications which leads you to **Heteronuclear NMR Applications (As, Sb, Bi)** where the material is located.

3. where there is debate over whether an entry title begins with the application of a technique, or with the technique itself. For example, a dummy entry appears for Raman Spectroscopy in Biochemistry which directs you to Biochemical Applications of Raman Spectroscopy where the material is located.

Dummy entries appear in both the contents list and the body of the text.

Example:

If you were attempting to locate material on the application of spectroscopic techniques in astronomy via the contents list the following information would be provided.

Astronomy, Applications of Spectroscopy   *See* Interstellar Molecules, Spectroscopy of; Stars, Spectroscopy of.

The page numbers for these entries are given at the appropriate location in the contents list.
If you were trying to locate the material by browsing through the text and you looked up Astronomy then the following would be provided.

Astronomy, Applications of Spectroscopy

See   **Interstellar Molecules, Spectroscopy of; Stars, Spectroscopy of.**

Alternatively if you looked up Stars the following information would be provided.

STARS, SPECTROSCOPY OF   2199

Stars, Spectroscopy of

AGGM Tielens, Rijks Universiteit, Groningen, The Netherlands

Copyright © 1999 Academic Press

| ELECTRONIC SPECTROSCOPY |
| Applications |

Further to aid the reader to locate material the main alphabetical Contents list is followed by a list of the entries grouped within their relevant subject area. (Subject areas follow each other alphabetically). Within each subject area the entries are further broken down into those covering historical aspects, theory, methods & instrumentation, or applications. The entries are listed alphabetically within these categories, and their relevant page numbers are given.

2. Cross References

To direct the reader to other entries on related topics a "see also" section is provided at the end of each entry

Example:

The entry Nuclear Overhauser Effect includes the following cross-references:

See also: **Chemical Exchange Effects in NMR; Macromolecule–Ligand Interactions Studied By NMR; Magnetic Resonance, Historical Perspective; NMR Pulse Sequences; NMR Relaxation Rates; Nucleic Acids Studied Using NMR; Proteins Studied Using NMR Spectroscopy; Structural Chemistry Using NMR Spectroscopy, Organic Molecules; Structural Chemistry Using NMR Spectroscopy, Peptides; Structural Chemistry Using NMR Spectroscopy, Pharmaceuticals; Two-Dimensional NMR Methods.**

3. Index

The index appears in each volume. Any topic not found through the Contents list can be located by referring to the index. On the opening page of the index detailed notes on its use are provided.

4. Colour plates

The colour figures for each volume have been grouped together in a plate section. The location of this section is cited at the end of the contents list.

5. Appendices

The appendices appear in volume 3.

6. Contributors

A full list of contributors appears at the beginning of each volume.

Contributors

Adams, Fred
Department of Chemistry
University of Instelling Antwerp
University Pleim 1, B-2610, Antwerp, Belgium

Aime, S
University of Torino
Department of Chemistry
via Giuria 7, 10125, Torino, Italy

Andersson, L A
Vassar College
Poughkeepsie, Box 589, NY 12604-0589, USA

Ando, Isao
Department of Polymer Chemistry
Tokyo Institute of Technology
Meguro Ku, Tokyo, 152, Japan

Andrenyak, David M
University of Utah
Center for Human Toxicology
Salt Lake City, Utah 84112, USA

Andrews, David L
School of Chemical Sciences
University of East Anglia
Norwich, NR4 7TJ, UK

Andrews, Lester
Chemistry Department
University of Virginia
McCormick Road,
Charlottesville, VA 22901, USA

Appleton, T G
Department of Chemistry
The University of Queensland
Brisbane, Queensland 4072, Australia

Arroyo, C M
USA Medical Research Institute for Chemical
Defense, Drug Assessment Division
Advanced Assessment Branch
3100 Ricketts Point Road, Aberdeen Proving
Ground, Maryland, MD 21010, USA

Artioli, Gilberto
Dipartimento di Scienze della Terra
Universita degli Studi di Milano
via Botticelli 23, I-20133, Milan, Italy

Ashfold, Michael N R
University of Bristol
School of Chemistry
Bristol, BS8 1TS, UK

Aubery, M
Laboratorie de Glycobiologie et Reconnaissance
Cellulaire
Université Paris V - UFR Biomédicale
45, rue des Saints-Péres, 75006, Paris, France

Baer, Tom
Department of Chemistry
University of North Carolina
Chapel Hill, NC 27599-3290, USA

Bain, A D
Department of Chemistry
McMaster University
1280 Main Street W., Hamilton,
Ontario L8S 4M1, Canada

Baker, S A
Beltsville Human Nutrition Research Center
U.S. Department of Agriculture
Food Composition Lab
Beltsville, MD 21054, USA

Baldwin, Mike
Mass Spectrometry Facility
University of California
San Francisco, CA 94143-0446, USA

Bateman, R
Micromass LTD
Manchester, M23 9LZ, UK

Batsanov, Andrei
Department of Chemistry
University of Durham
South Road, Durham, DH1 3LE, UK

Beauchemin, Diane
Department of Chemistry
Queen's University
Kingston, ONT K7L 3N6, Canada

Bell, Jimmy D
The Robert Steiner MR Unit, MRC Clinical
Sciences Centre
Imperial College School of Medicine
Hammersmith Hospital
Du Cane Road, London, W12 0HS, UK

Belozerski, G N
Post Box 544, B-155, 199155, St. Petersburg,
Russia

Bernasek, S L
Princeton University
Department of Chemistry
Princeton, NJ 05844, USA

Berova, Nina
Columbia University
Department of Chemistry
New York, NY 10027, USA

Berthezene, Y
Hospital L. PradelUMR CNRS 5515
Dept de Imagerie Diagnostique et Therapeutique
BP Lyon Montchat, F-69394,
Lyon 03, France

Boesl, Ulrich
Institut für Physikalische und Theoretische
Chemie
Technische Universität München
Lichtenbergstrasse 4,
D-85748, München, Germany

Bogaerts, Annemie
Department of Chemistry
University of Antwerp
Universiteitsplein 1, B-2610, Wilrijk, Belgium

Bohme, D
Department of Chemistry & Centre for Research in
Earth & Space Science
York University
North York, Ontario M3J 1P3, Canada

Bonchin, Sandra L
Los Alamos National Laboratory
Nuclear Materials Technology-Analytical
Chemistry
NMT-1, MS G740, Los Alamos,
NM 87545, USA

Bowie, John H
The University of Adelaide
Department of Organic Chemistry
South Australia 5005, Australia

Brand, Willi A
Max-Planck-Institute for Biochemistry
P.O. Box 100164, 07701, Jena, Germany

Braslavsky, Silvia E
Max Planck-Institut für Strahlenchemie
Postfach 101365,
D-45470, Mülheim an der Ruhr, Germany

Braut-Boucher, F
Laboratorie de Glycobiologie et Reconnaissance
Cellulaire
Université Paris V - UFR Biomédicale
45, rue des Saints-Peres,
75006, Paris, France

Brittain, H G
Center for Pharmaceutical Physics
10 Charles Road, Milford,
NJ 08848, USA

Brumley, W C
National Exposure Research Laboratory
US EPA, Division of Environmental Science
PO Box 93478, Las Vegas,
Nevada 89193, USA

Bryce, David L
Dalhousie University
Department of Chemistry
Halifax, Nova ScotiaCanada

Bunker, Grant
Department of Physics
Illinois Institute of Technology
3101 S. Dearborn, Chicago, IL 60616, USA

Burgess, C
Rose Rae
Startforth
Barnard Castle, Durham, DL12 9AB, UK

Buss, Volker
University of Duisberg
Department of Theoretical Chemistry
D-47048, Duisberg, Germany

Callaghan, P T
Department of Physics
Massey University
Palmerston North, New Zealand

Calucci, Lucia
Dipartimento di Chimica e Chimica Industriale
via Risorgimento 35, 56126, Pisa, Italy

Cammack, Richard
Division of Life Sciences, Kings College
University of London
Campden Hill Road, London, W8 7AH, UK

Canè, E
Universita di Bologna
Dipartimento di Chimica Fisica e Inorganica
Viale Risorgimento 4,
40136, Bologna, Italy

Canet, D
Laboratorie Methode RMN
Universite de Nancy 1
FU CNRS E008, INCM,
F-S4506, Vandoeuvre, Nancy, France

Carter, E A
University of Bradford
Chemical and Forensic Sciences
Bradford, BD7 1DP, UK

Caruso, Joseph A
University of Cincinnati
Department of Chemistry
Cincinnati, OH 45221-0037, USA

Cerdan, Sebastian
Instituto de Investigaciones Biomedicas,
C.S.I.C
c/ Arturo Duperier 4, 28029,
Madrid, Spain

Chakrabarti, C L
Chemistry Department
Carlton University
Ottawa, Ontario K1S 5B6, Canada

Chen, Peter C
Department of Chemistry
Spelman College
Spelman Lane, Atlanta, Georgia 30314-4399,
USA

Cheng, H N
University of Delaware
Department of Chemistry
Newark, DE 19176, USA

Chichinin, A I
Institute of Chemical Kinetics and Combustion
Institutskaya 3, 630090, Novosibirsk, Russia

Claereboudt, Jan
University of Antwerp
Department of Pharmaceutical Sciences
Universiteitsplein 1, B-2610, Antwerp, Belgium

Claeys, Magda M
University of Antwerp
Department of Pharmaceutical Sciences
Universiteitsplein 1, B-2610, Antwerp, Belgium

Colarusso, Pina
National Instiues of Digestive and Diabetes and
Kidney Diseases, National Institutes of Health
Laboratory of Chemical Physics
Bethesda, MD 20892, USA

Conrad, Horst
Fritz Haber Institute of the
Max Planck Gessellschaft
Faradayweg 4-6, D14195, Berlin, Germany

Cory, D G
Department of Nuclear Engineering
MIT
Cambridge, MA 02139, USA

Crouch, Dennis J
University of Utah
Center for Human Toxicology
Salt Lake City, Utah 84112, USA

Cruz, Fatima
Instituto de Investigaciones Biomedicas, C.S.I.C
c/ Arturo Duperier 4, 28029,
Madrid, Spain

Curbelo, Raul
Digilab Division
Bio-Rad Laboratories
237 Putnam Avenue, Cambridge, MA 02139, USA

Dåbakk, Eigil
Foss Sverige
Turebergs Torg 1, Box 974, SE 191 92,
Sollentuna, Sweden

Davies, M C
The University of Nottingham
Laboratory of Biophysics and Surface Analysis,
School of Pharmaceutical Siences
University Park, Nottingham,
NG7 2RD, UK

Dawson, P H
Iridian Spectral Technologies Ltd
Industry Partnership Facility [M5O]
1200, Montreal Road,
Ottawa, K1A 0R6, Ontario Canada

Demtroder, W
Fachbereich Physik
Universtität Kaiserslautern
D-6750, Kaiserslautern, Germany

Di, Qiao Qing
University of Florida
College of Pharmacy
P.O. Box 100485, Gainesville,
FL 32610, USA

Dirl, Rainer
Centre for Computational Material Science
Institut für Theoretische Physik
Tu Wien, Wiedner Hauptstrabe 8-10,
A-1040, Vienna, Austria

Dixon, Ruth M
MRC Biochemical and Clinical Magnetic
Resonance Unit
Department of Biochemistry
South Parks Road, Oxford, OX1 3QU, UK

Docherty, John
Institute of Biodiagnostics
436 Ellice Avenue, Winnipeg,
Manitoba R3B 1Y6, Canada

Dong, R Y
Department of Physics and Astronomy
Brandon University
Brandon, Manitoba R7A 6A9, Canada

Douglas, D
University of British Columbia
Department of Chemistry
2036 Main, Mall,
Vancouver, BC, V6T 1Z1, Canada

Dua, Suresh
The University of Adelaide
Department of Chemistry
South Australia 5005, Australia

Dugal, Robert
The Canadian Pharmaceutical Manufacturers
Association
Doping Control Laboratory
Ottawa, Canada

Durig, J R
University of Missouri-Kansas City
5100 Rockhill Road, Kansas City,
Missouri 64110-2499, USA

Dworzanski, Jacek
Center for Micro Analysis and Reaction Chemistry
University of Utah
110 South Central Campus Drive, Room 214,
Salt Lake City, Utah 84112, USA

Dybowski, Cecil R
University of Delaware
Department of Chemistry
Newark, DE 19716, USA

Eastwood, DeLyle
Air Force Institution of Technology
MS AFIT/ENP
Wright-Patterson AFB,
OH 45433-7765, USA

Edwards, H G M
Chemistry and Forensic Sciences
University of Bradford
Bradford,
West Yorkshire BD7 1DP, UK

Eggers, L
Gerhard-Mercator-Universität
Institut für Physikalische und Theoretische
Chemie
D-47048, Duisburg, Germany

Emsley, James W
Department of Chemistry
University of Southampton
Highfield, Southampton, UK

Endo, I
Department of Physics
Hiroshima University
1-3-1 Kagamiyama,
Higashi Hiroshima, 739, Japan

Ens, W
Department of Physics
University of Manitoba
Winnipeg,
Manitoba R3T2N2, Canada

Farley, J W
University of Nevada
Department of Physics
Las Vegas, NV 89154, USA

Farrant, R D
Physical Sciences
GlaxoWellcome R & D
Gunnels Wood Road, Stevenage, SG1 2NY, UK

Feeney, J
National Institute for Medical Research
Medical Research Council
The Ridgeway, Mill Hill, London, NW7 1AA, UK

Fennell, Timothy R
Chemical Industry Institute of Toxicology
6 Davis Drive, PO Box 12137, Research Triangle
Park, North Carolina NC 27709-2137, USA

Ferrer, N
Serv. Cientif. Tecn. University of Barcelona
Lluis Sole Sabaris 1, E-08028, Barcelona, Spain

Fisher, A J
Department of Physics and Astronomy
University College London
Gower Street, London, WC1E 6BT, UK

Flack, H D
University of Geneva
Laboratory of Crystallography
24 Quai Ernest Ansermet,
CH 1211, Geneva, Switzerland

Flytzanis, Chr.
Laboratorie d'Optique Quantique
CNRS - Ecole Polytechnique
F-91128, Palaiseau, Cedex France

Foltz, Rodger L
Center for Human Toxicology
University of Utah
20 S 2030 ERM 490,
Salt Lake City, UT 84112-9457, USA

Ford, Mark
University of York
Department of Chemistry
Heslington, York Y010 5DD, UK

Friedrich, J
Lehrstuhl für Physik, Weihenstephan
Technische Universität München
D-85350, Freisling, Germany

Fringeli, Urs
Insitute of Physical Chemistry
University of Vienna
Althanstrasse 14/UZA II,
A-1090, Vienna, Austria

Frost, T
GlaxoWellcome R & D
Temple Hill, Dartford, Kent DA1 5AH, UK

Fuller, Watson
Keele University
Department of Physics
Keele, Staffs ST5 5BG, UK

Futrell, J H
Department of Chemistry & Biochemistry
University of Delaware
Newark, Delaware 19716, USA

Geladi, Paul
Department of Chemistry
Umeå University
SE 901 87, Umeå, Sweden

Gensch, Thomas
Katholieke Universiteit of Leuven
Department of Organic Chemistry
Molecular Dynamics and Spectroscopy
Celestijnenlaan 200F, B-3001,
Heverlee, Belgium

Gerothanassis, I P
Department of Chemistry
University of Iannina
GR-45110, Iannina, Greece

Gilbert, A S
19 West Oak, Beckenham,
Kent BR3 5EZ, UK

Gilchrist, Alison
University of Leeds
Department of Colour Chemistry
Leeds, LS2 9JT, UK

Gilmutdinov, AKh
Kazan Lenin State University
Department of Physics
Kazan, 420008, Russia

Gorenstein, D G
Sealy Centre for Structural Biology
Medical Branch
University of Texas
Galveston, Texas 77555-1157, USA

Grace, R Christy Rani
Department of Physics
Indian Institute of Science
Bangalore, India

Green-Church, Kari B
Department of Chemistry
Louisiana State University
Baton Rouge, LA 70803, USA

Greenfield, Norma J
Department of Neuroscience and Cell Biology,
Robert Wood Johnson Medical School
University of Medicine and Dentistry of
New Jersey
675 Hoes Lane, Piscataway,
NJ 08854, USA

Grime, G
Department of Materials
University of Oxford
Parks Road, Oxford, UK

Grutzmacher, Hans
Universität Bielefeld
Fakultat für Chemie
Postfach 100131,
D-33501, Bielefeld, Germany

Guillot, G
Unite de Recherche en Resonance,
Magnetique Medicale, CNRS URS 2212
Bat.220 Universite Paris-Sud
91405, ORSAY, Cedex France

Hallett, F R
Department of Physics
University of Guelph
Guelph, Ontario N1G 2W1, Canada

Hannon, A C
ISIS Facility, Rutherford Appleton Laboratory
Didcot, Oxon OX11 0QX, UK

Harada, Noboyuki
Tohoku University
Institute of Chemical Reaction Science
Sendai, 980 77, Japan

Hare, John F
SmithKline Beecham Pharmaceuticals
The Frythe, Welwyn,
Herts, AL6 9AR, UK

Harmony, Marlin D
Department of Chemistry, Marlott Hall
University of Kansas
Lawrence, Kansas 66045, USA

Harrison, A G
Chemistry Department
University of Toronto
80 St George Street, Toronto,
Ontario M5S 3H6, Canada

Hawkes, G E
Department of Chemistry,
Queen Mary and Westfield College
University of London
Mile End Road, London, E1 4NS, UK

Hayes, Cathy
Department of Botany
Trinity College
Dublin 2, Eire

Heck, Albert J R
Bijvoet Center for Biomolecular Research,
Utrecht University
Department of Chemistry and Pharmacy
Sorbonnelaan 16, 3584 CA, Utrecht,
The Netherlands

Herzig, Peter
Institut für Physikalische Chemie
Universität Wien
Währingerstraße 42,
A-1090, Wien, Austria

Hess, Peter
Physikalisch-Chemisches Institut
Universität Heidelberg
Im Neuenheimer Feld 253,
D-69120, Heidelberg, Germany

Hicks, J M
Department of Chemistry
Georgetown University
Washington DC, 20057, USA

Hildebrandt, Peter
Max-Planck-Institut für Strahlenchemie
Postfach 101365,
D-45413, Mülheim/Ruhr, Germany

Hill, Steve J
Department of Environmental Science
University of Plymouth
Drake Circus, Plymouth PL4 8AA, UK

Hills, Brian P
Institute of Food Research
Norwich Laboratory
Norwich Research Park, Colney,
Norwich NR4 7UA, UK

Hockings, P D
SmithKline Beecham Pharmaceuticals
Analytical Sciences Department
The Frythe, Welwyn,
Herts, AL6 9AR, UK

Hofer, Tatiana
Universität Kaiserslautern
Fachbereich Chemie der
D-67663, Kaiserslautern, Germany

Hoffmann, G G
Hoffmann Datentechnik
Postfach 10 06 31,
D-46006, Oberhausen, Germany

Holcombe, James A
Department of Chemistry
University of Texas
Austin, Texas7871-1167, USA

Holliday, Keith
University of San Francisco
Department of Physics
2130 Fulton Street, San Francisco,
CA 94117, USA

Holmes, John L
Department of Chemistry
University of Ottawa
PO Box 450, Stn 4, Ottawa, K1N 6N5, Canada

Homer, J
Chemical Engineering and Applied Chemistry,
School of Engineering and Applied Science
Aston University
Aston Triangle, Birmingham, B4 7ET, UK

Hore, P J
Physical and Theoretical Chemistry Laboratory
University of Oxford
South Parks Road, Oxford, OX1 3QZ, UK

Huenges, Martin
Technische Universität Munchen
Institut für Organische Chemie and Biochemie -
Leharul II
Lichtenbergatrahe 4,
D-85747, Garching, Germany

Hug, W
Institut de Chimie Physique
Universite de Fribourg
CH-1700, Fribourg, Switzerland

Hunter, Edward P L
Physical and Chemistry Properties Division (838)
Physics Building (221), Room A 113
NIST, PHY A 111, Gaithersburg,
Maryland 20899, USA

Hurd, Ralph
GE Medical Systems
47697 Westinghouse Drive, Fremont,
California 94539, USA

Imhof, Robert E
Department of Physics and Applied Physics
Strathclyde University
Glasgow, G4 0NG, UK

Jackson, Michael
National Research Council Canada
Institute for Biodiagnostics
435 Ellice Avenue, Winnipeg,
Manitoba R3B 1Y6, Canada

Jalsovszky, G
Chemistry Research Centre, Institute of Chemistry
Hungarian Academy of Sciences
PO Box 17, H-1525, Budapest, Hungary

Jellison, G E
Oak Ridge National Laboratory
Solid State Division
POB 2008, Oak Ridge, Tennessee, TN 37831, USA

Jokisaari, J
Department of Physical Sciences
University of Oulu
P O Box 3000, Oulu, FIN-90Y01, Finland

Jonas, J
School of Chemical Sciences
University of Illinois
Urbana, Illinois, 61801, USA

Jones, J R
Department of Chemistry
University of Surrey
Guildford, Surrey GU2 5XH, UK

Juchum, John
University of Florida
College of Pharmacy
P.O.Box 100485, Gainsville, FL 32610, USA

Katoh, Etsuko
National Institute of Agrobiological Resources
2-1-2, Kannondai, Tsukuba, Ibaraki 305-0856,
Japan

Kauppinen, J
University of Turku
Department of Applied Physics
FIN-20014, Turku 50, Finland

Kessler, Horst
Institut für Organische Chemie und Biochemie
Technische Universität München
Lichtenbergstrabe 4,
D-85747, Garching, Germany

Kettle, S F
School of Chemical Sciences
University of East Anglia
Norwich, NR4 7TJ, UK

Kidder, Linda H
National Institues of Digestive and Diabetes and
Kidney Diseases, National Institutes of Health
Laboratory of Chemical Physics
Bethesda, MD 20892, USA

Kiefer, Wolfgang
Institut für Physikalische Chemie
Der Universität Wurzburg
Am Hubland,
D-97074, Wurzburg, Germany

Kiesewalter, Stefan
Universität Kaiserslautern
Fachbereich Chemie der
D-67663, Kaiserslautern, Germany

Kimmich, Rainer
Universität Ulm
Sektion Kernresonanzspektroskopie
D-89069, Ulm, Germany

Klinowski, J
Department of Chemistry
University of Cambridge
Lensfield Road,
Cambridge, CB2 1EW, UK

Koenig, J L
Case Western Reserve University
Department of Macromolecular Science
10900 Euclid Avenue, Cleveland,
Ohio 44106-7202, USA

Kolemainen, E
Department of Chemistry
University of Jyvaskyla
Jyvaskyla, FIN-40351, Finland

Kooyman, R P H
University of Twente
Department of Applied Physics
Enschede, NL 7500 AE, The Netherlands

Kordesch, Martin E
Department of Physics and Astronomy
Ohio University
Athens, Ohio 45701, USA

Kotlarchyk, M
Department of Physics
3242 Gosnell, Rochester Institute of Technology
85 Lomb Memorial Drive, Rochester,
NY 14623-5603, USA

Kramar, U
Institute of Petrography and Geochemistry
University of Karlsruhe
Kaiserstrasse 12,
D-76128, Karlsruhe, Germany

Kregsamer, P
Atominstitut der Osterreichischen Universitaten
Stadionallee 2, 1020, Wien, Austria

Kruppa, Alexander I
Institute of Chemical Kinetics and Combustion
Novosibirsk-90, 630090, Russia

Kuball, Hans-Georg
Universität Kaiserslautern
Fachbereich Chemie der
D-67653, Kaiserslautern, Germany

Kumar, A
Department of Physics and Sophisticated
Instruments Facility
Indian Institute of Science
Bangalore, 560012,
Karnataka, India

Kushmerick, J G
The Pennsylvania State University
Department of Chemistry
University Park, PA 16802-6300, USA

Kvick, Ake
European Synchrotron Radiation Facility
BP 220, Avenue des Martyrs, F-38043,
Grenoble, France

Laeter, J Rde
Curtin University of Technology
Bentley, Western Australia 6102, Australia

Latosińska, Jolanta N
Insitute of Physics
Adam Mickiewicz University
Umultowska 85, 61-614, Poznań, Poland

Leach, M O
Clinical Magnetic Resonance Research Group
Institute of Cancer Research
Royal Marsden Hospital
Sutton, Surrey SM2 5PT, UK

Lecomte, S
CNRS-Université Paris VI
Thiais, France

Leshina, T V
Russian Academy of Sciences
Institute of Chemical Kinetics and Combustion
Novosibirsk-90, Russia

Levin, Ira W
National Institues of Digestive and Diabetes and
Kidney Diseases, National Institutes of Health
Laboratory of Chemical Physics
Bethesda, MD 20892, USA

Lewen, Nancy S
Bristol Myers Squibb, 1 Squibb Dr., Bldg. 101 Rm
B18, New Brunswick, NJ 08903, USA

Lewiński, J
Department of Chemistry
Warsaw University of Technology
Noakowskiego 3, PL-00664, Warsaw, Poland

Lewis, Neil
National Institutes of Digestive and Diabetes and
Kidney Diseases, National Institute of Health
Laboratory of Chemical Physics
Bethesda, MD 20892, USA

Leyh, Bernard
F.N.R.S. and University of Leige
Department of Chemistry (B6)
B.4000, Sart Tilman, Belgium

Lias, S
Physical and Chemical Properties Division (838)
Physics Building (221), Room A 113
NIST, PHY A 111, Gaithersburg,
Maryland 20899, USA

Lifshitz, Chava
Department of Physical Chemistry
The Farkas Centre for Light-induced Processes
The Hebrew University of Jerusalem
Jerusalem, 91904, Israel

Limbach, Patrick A
Louisiana State University
Department of Chemistry
Baton Rouge, LA 70803, USA

Lindon, John C
Biological Chemistry
Division of Biomedical Sciences
Imperial College School of Science, Technology
and Medicine
Sir Alexander Fleming Building
South Kensington, London SW7 2AZ, UK

Linuma, Masataka
Hiroshima University
Department of Physics
1-3-1 Kagamiyama, Higashi,
Hiroshima, 739, Japan

Liu, Maili
The Chinese Academy of Sciences
Wuhan Institute of Physics and Mathematics
Laboratory of Magnetic Resonance and Atomic
and Molecular Physics
Wuhan, 430071,
Peoples' Republic of China

Lorquet, J C
Departement of Chemie
Universite de Liege
Sart-Tilman B6 (Batiment. B6), B-4000, Liege 1,
Belgium

Louer, D
Groupe Crystallochimique, Chemical Solids and
Inorganic Molecules Laboratory
University of Rennes
CNRS UMR 6511 Ave. Gen. Leclerc,
F-35042, Rennes, France

Luxon, Bruce A
Sealy Center for Structural Biology
University of Texas Medical Branch
Galveston, Texas 77555, USA

Maccoll, Allan
10, The Avenue, Claygate, Surrey KT10 0RY, UK

Macfarlane, R D
Chemistry Department
Texas A & M University
College Station, TX 77843-3255, USA

Maerk, T D
Institut fuer Ionenphysik
Leopold Franzens Universitaet
Technikerstr. 25, A-6020, Innsbruck, Austria

Magnusson, Robert
Department of Electronic Engineering
University of Texas
Arlington, Texas TX 76019, USA

Mahendrasingam, A
Keele University
Physics Department
Staffordshire, ST5 5BG, UK

Maier, J P
Institut für Physikalische Chemie
Universitat Basel
Klingelbergstrasse, CH 4056, Basel, Switzerland

Makriyannis, A
School of Pharmacy
University of Connecticut
Storrs, CT 06269, USA

Malet-Martino, Myriam
Universite Paul Sabatier
Groupe de RMN Biomedicale, Laboratorie des
IMRCP (UMR CNRS 5623)
31062, Toulouse, Cedex, France

Mamer, O
Mass Spectrometry Unit
McGill University
1130 Pine Avenue West, Montreal,
Quebec H3A 1A3, Canada

Mandelbaum, Asher
Technion-Israel Institute of Technology
Department of Chemistry
Technion, Haifa 32000, Israel

Mantsch, H H
National Research Council of Canada
Institute for Biodiagnostics
435 Ellice Avenue, Winnipeg, R3B 1Y6, Canada

Mao, Xi-an
The Chinese Academy of Sciences
Wuhan Institute of Physics and Mathematics,
Laboratory of Magnetic Resonance and Atomic
and Molecular Physics
Wuhan, 430071, Peoples' Republic of China

March, Raymond E
Department of Chemistry
Trent University
Peterborough, Ontario K9J 7B8, Canada

Marchetti, Fabio
Universita di Camerino
Dipartmento di Scienze Chemiche
via S. Agostino 1, 62032, Camerino MC, Italy

Mark, Tilmann D
Loepold Franzens Universitat
Institut für Ionenphysik
Technikerstrasse 25, A-6020, Innsbruck, Austria

Marsmann, H C
University Gesemthsch. Paderborn
Fachbereich Chem.
Warburger Str. 100,
D-33095, Paderborn, Germany

Martino, Robert
Universite Paul Sabatier
Groupe de RMN Biomedicale
Laboratore des IMRCP (UMR CNRS 5623)
31062, Toulouse Cedex, France

Maupin, Christine L
Michigan Technological University
Department of Chemistry
Houghton, MI 4993, USA

McClure, C K
Department of Chemistry
Montana State University
Bozeman, MT 59171, USA

McLaughlin, D
Kodak Research Laboratories
Eastman Kodak Co.
Rochester, New York NY 14650, USA

McNab, Iain
Department of Physics
University of Newcastle
Newcastle-upon-Tyne, NE1 7RU, UK

McNesby, K L
6735 Indian River Drive, Citrus Heights,
CA 95621, USA

Meuzelaar, H L C
Center for Micro Analysis & Reaction Chemistry
University of Utah
110 South Central Campus Drive, Room 214,
Salt Lake City, Utah 84112, USA

Michl, Josef
Department of Chemistry & Biochemistry
University of Colorado
Boulder, CO 80309-0215, USA

Miklos, Andras
Institute of Physical Chemistry
University of Heidelberg
Im Neuenheimer Feld 253,
D-69120, Heidelberg, Germany

Miller, S A
Princeton University
Department of Chemistry
Princeton, NJ 08544, USA

Miller-Ihli, Nancy J
U.S. Department of Agriculture
Food Composition Laboratory
Building 161, Rm. 1, BARC-East, Beltsville,
MD 20705, USA

Morris, G A
Department of Chemistry
University of Manchester
Oxford Road, Manchester, M13 9PL, UK

Mortimer, R J
Department of Chemistry
Loughborough University
Loughborough, Leics LE11 3TU, UK

Morton, Thomas H
Department of Chemistry
University of California
Riverside, CA 92521-0403, USA

Muller-Dethlefs, K
University of York
Department of Chemistry
Heslington, York YO1 5DD, UK

Mullins, Paul G M
SmithKline Beecham Pharmaceuticals
Analytical Sciences Department
The Frythe, Welwyn, Herts, AL6 9AR, UK

Murphy, Damien M
Cardiff University
National ENDOR Centre, Department of Chemistry
Cardiff, CF1 3TB, UK

Nafie, L A
Department of Chemistry
Syracuse University
Syracuse, New York 13244-4100, USA

Naik, Prasad A
Room #201, R & D Block "D"
Centre for Advanced Technology
Indore, 452013, Madhya Pradesh, India

Nakanishi, Koji
Department of Chemistry
Columbia University
New York NY 10027, USA

Nicholson, J K
Biological Chemistry, Division of Biomedical Sciences
Imperial College of Science, Technology & Medicine
Sir Alexander Fleming Building,
South Kensington, London,
SW7 2AZ, UK

Nibbering, N M M
Institute of Mass Spectrometry
University of Amsterdam
Nieuwe Achtergracht 129, 1018 WS,
Amsterdam, The Netherlands

Niessen, W M A
hyphen MassSpec Consultancy
De Wetstraat 8, 2332 XT, Leiden,
The Netherlands

Nobbs, Jim
University of Leeds
Department of Colour Chemistry
Leeds, LS2 9JT, UK

Norden, B
Department of Physical Chemistry
Chalmers University of Technology
S-41296, Gothenburg, Sweden

Norwood, T
Department of Chemistry
University of Leicester
Leicester, LE1 7RH, UK

Olivieri, A C
Facultad de Ciencias Biochimicas y Farmaceuticas, Departamento Quimica Analitica
Universita Nacional Rosario
Suipacha 531,
RA-2000, Rosario, Santa Fe, Argentina

Omary, Mohammed A
University of Maine
Department of Chemistry
Orono, Maine ME 04469, USA

Parker, S F
Rutherford Appleton Laboratory, ISIS Facility
Oxon, Didcot OX11 0QX, UK

Partanen, Jari O
University of Turku
Department of Applied Phyics
FIN 20014,
Turku, Finland

Patterson, Howard H
Department of Chemistry
University of Maine
Orono, Maine ME 04469, USA

Pavlopoulos, Spiro
University of Connecticut
Institute of Materials Science and School of
Pharmacy
Storrs, Connecticut 06269, USA

Pettinari, C
Università degli Studi di Camerino
Scienze Chimiche
Via S. Agostino 1, 62032, Camerino, MC, Italy

Poleshchuk, O K
Department of Inorganic Chemistry
Tomsk Pedagogical University
Komsomolskii 75, 634041, Tomsk,
Russian Federation

Rafaiani, Giovanni
Università di Camerino
Dipartimento di Scienze Chemie
Via S Agostino 1, 62032, Camerino MC, Italy

Ramsey, Michael H
Imperial College of Science Technology and
Medicine
TH Huxley School of Environmental, Earth
Science and Engineering
London, SW7 2BP, UK

Randall, Edward W
Department of Chemistry, Queen Mary &
Westfield College
University of London
Mile End Road,
London, E1 4NS, UK

Rehder, D
Institute of Inorganic Chemistry
University of Hamburg
Martin-Luther-King Platz 6,
D-20146, Hamburg, Germany

Reid, David G
Smithkline Beecham Pharmaceuticals
Analytical Sciences Department
The Frythe, Welwyn, Herts, AL6 9AR, UK

Reid, Ivan D
Paul Scherrer Institute
CH-5232, Villigen PSI, Switzerland

Reynolds, William F
Lash Miller Chemical Laboratories
University of Toronto
80 George Street, Toronto,
Ontario M5S 1A1, Canada

Richards-Kortum, Rebecca
Dept. of Elec. and Computer Engin.
University of Texas at Austin
Austin, TX 78712, Inter-Office C0803, USA

Riddell, Frank G
Department of Chemistry
University of St Andrews
The Purdie Building, St Andrews, Fife KY16 9ST,
Scotland

Riehl, J P
Michigan Technological University
Department of Chemistry
Houghton, MI 49931, USA

Rinaldi, Peter L
Department of Chemistry
University of Akron
Akron, Ohio OH 44325-3061, USA

Roberts, C J
The University of Nottingham
Department of Pharmaceutical Sciences
University Park, Nottingham, NG7 2RD, UK

Rodger, Alison
University of Warwick
Department of Chemistry
Coventry, CV4 7AL, UK

Rodger, C
University of Strathclyde
Department of Pure & Applied Chemistry
Glasgow, G1 1XL, UK

Roduner, Emil
Universität Stuttgart
Physikalisch-Chemisches Institut
Pfaffenwaldring 55, D-70550, Stuttgart, Germany

Rost, F W D
45 Charlotte Street, Ashfield, NSW 2131, Australia

Rowlands, C C
Cardiff University
National ENDOR Centre, Department of Chemistry
Cardiff, CF1 3TB, UK

Rudakov, Taras N
7 Reen Street, St James, WA 6102, Australia

Salamon, Z
University of Arizona
Department of Biochemistry
Tuscon, AZ 85721, USA

Salman, S R
Chemistry Department
University of Qatar
PO Box 120174, Doha, Qatar

Sanders, Karen
University of Warwick
Department of Chemistry
Coventry, CV4 7AL, UK

Sanderson, P N
Protein Science Unit
GlaxoWellcome Medicines Research Centre
Gunnels Wood Road, Stevenage,
Herts, SG1 2NY, UK

Santini, Carlo
Università di Camerino
Dipartimento di Scienze Chemiche
Via s. Agostino 1, 62032, Camerino MC, Italy

Santos Gómez, J
Instituto de Estructura de la Materia, CSIC
28006, Madrid, Spain

Schafer, Stefan
Institute of Physical Chemistry
University of Heidelberg
Im Neuenheimer Feld 253,
D-69120, Heidelberg, Germany

Schenkenberger, Martha M
Bristol-Myers Squibb
Pharmaceutical Research Institute
1 Squibb Dr., New Brunswick,
NJ 08903, USA

Schrader, Bernhard
Institut für Physikalische und Theoretisch Chemie
Universität Essen
Fachbereich 8, D-45117, Essen, Germany

Schulman, Stephen G
Department of Medicinal Chemistry
College of Pharmacy
University of Florida
Gainesville, FL 32610-0485, USA

Schwarzenbach, D
University of Lausanne
Institute of Crystallography
BSP Dorigny,
CH-1015, Lausanne, Switzerland

Seitter, Ralf-Oliver
Universität Ulm
Sektion Kernresonanzspektroskopie
89069, Ulm Germany

Seliger, J
Department of Physics
Faculty of Mathematics and Physics
University of Ljubljana
Jadranska, 19, 1000,
Ljubljana Slovenia

Shaw, R Anthony
National Research Council Canada
Institute for Biodiagnostics
435 Ellice Avenue, Winnipeg,
Manitoba R3B 1Y6, Canada

Shear, Jason B
Department of Chemistry
University of Texas
Austin, TX 78712, USA

Sheppard, Norman
School of Chemical Sciences
University of East Anglia
Norwich, NR4 7TJ, UK

Shluger, A L
Department of Physics and Astronomy
University College London
Gower Street, London, WC1E 6BT, UK

Shockcor, J P
Bioanalysis and Drug Metabolism
Stine-Haskell Research Center
Dupont-Merck, P O Box 30, Elkton Road, Newark,
Delaware DE 19714, USA

Shukla, Anil K
University of Delaware
Department of Chemistry and Biochemistry
Newark,
DE 19176, USA

Shulman, R G
MR Center, Department of Molecular Biophysics
Yale University
New Haven, Connecticut 06520, USA

Sidorov, Lev N
Physical Chemistry Division
Chemistry Department
Moscow State University
119899, Moscow, Russia

Sigrist, Markus
Swiss Federal Insitute of Technology (ETH)
Insitute of Quantum Electronics, Laboratory for
Laser Spectroscopy and Environmental Sensing
Hoenggerberg,
CH-8093, Zurich, Switzerland

Simmons, Tracey A
Department of Chemistry
Louisiana State Universty
Baton Rouge, LA 70803, USA

Smith, W E
University of Strathclyde
Department of Pure & Applied Chemistry
Glasgow, G1 1XL, UK

Smith, David
The University of Keele
Department of Biomedical Engineering and
Medical Physics, Hospital Centre
Thornburrow Drive, Hartshill,
Stoke-on-Trent ST4 7QB, UK

Snively, C M
Department of Macromolecular Science
Case Western Reserve University
Cleveland, OH 44106, USA

Somorjai, J
Institute for Biodiagnostics
435 Ellice Avenue, Winnipeg,
R3B 1Y6, Canada

Španel, Patrick
Keele University
Department of Biomedical Engineering and
Medical Physics
Thornburrow Drive, Hartshill,
Stoke-on-Trent ST4 7QB, UK

Spanget-Larsen, Jens
Department of Life Sciences and Chemistry
Roskilde University
POB 260, DK-4000, Roskilde, Denmark

Spiess, H W
Max Plank Institute of Polymer Research
Postfach 3148, D-55021, Mainz, Germany

Spinks, Terence
PET Methodology Group, MRC Clinical Sciences
Centre, Royal Postgraduate Medical School
Hammersmith Hospital
Du Cane Road,
London, W12 0NN, UK

Spragg, R A
Perkin-Elmer Analytical Instruments
Post Office Lane, Beaconsfield,
Bucks HP9 1QA, UK

Standing, K G
Department of Physics
University of Manitoba
Winnipeg, Manitoba R3T 2N2, Canada

Steele, Derek
The Centre for Chemical Sciences
Royal Holloway
University of London
Egham, Surrey TW40 0EX, UK

Stephens, Philip J
University of Southern California
Department of Chemistry
Los Angeles, CA 90089-0482, USA

Stilbs, Peter
Royal Institute of Technology Physical Chemistry
S-10044, Stockholm, Sweden

Streli, C
Atominstitut of the Austrian Universities
Stadionallee 2, 1020, Wien, Austria

Styles, Peter
John Radcliffe Hospital
MRL Biochemistry & Chemical Magnetic
Resonance Unit
Headington, Oxford,
OX3 9DU, UK

Sumner, Susan C J
Chemical Industry Institute of Toxicology
6 Davis Drive, PO Box 12137, Research Triangle
Park, North Carolina
NC 27709-2137, USA

Sutcliffe, L H
Institute of Food Research
Norwich NR4 7UA, UK

Szepes, Laszlo
Eotvos Lorand University
Department of General and Inorganic Chemistry
Pazmany Peter Satany 2, 1117,
Budapest Hungary

Taraban, Marc B
Institute of Chemical Kinetics and Combustion
Novosibirsk-90, 630090, Russia

Tarczay, Gyorgy
Eotvos University
Department of General and Inorganic Chemistry
Pazmany Peter S. 2,
Budapest H-1117, Hungary

Taylor, A
Robens Institute of Health and Safety
Trace Elements Laboratory
University of Surrey
Guildford, Surrey GU2 5XH, UK

Tendler, S J B
University of Nottingham
Department of Pharmaceutical Sciences
University Park, Nottingham, NG7 2RD, UK

Terlouw, J
McMaster University
Department of Chemistry
ABB-455, 1280 Main Street West,
Hamilton, ON L8S 4M1, Canada

Thompson, Michael
University of London
Birkbeck College, Department of Chemistry
Gordon House, 29 Gordon Square,
London, WC1H 0PP, UK

Thulstrup, Erik W
Department of Chemistry & Life Sciences
Roskilde University (RUC)
Building 17.2, PO Box 260,
DK-4000, Roskilde, Denmark

Tielens, A G G M
Kapteyn Astronomical Institute
PO Box 800, 9700 AV, Groningen, The Netherlands

Tollin, Gordon
University of Arizona
Department of Biochemistry
Tucson, AZ 85721, USA

Traeger, John C
Department of Chemistry
La Trobe University
Bundoora, Victoria 3083, Australia

Tranter, George E
Glaxo Wellcome Medicines Research Centre
Gunnels Wood Road, Stevenage,
Herts, SG1 2NY, UK

Trombetti, A
Università di Bologna
Dipartimento Chimica Fisica e Inorganica
Viale Risorgimento 4, I-40136, Bologna, Italy

True, N S
Department of Chemistry
University of California Davis
Davis, CA 95616, USA

Ulrich, Anne S
Institute of Molecular Biology
University of Jena
Winzerlaer Strasse 10, D-07745, Jena, Germany

Utzinger, Urs
The University of Texas at Austin
Texas USA

Van Vaeck, Luc
Department of Chemistry
University Instelling Antwerp
University Pleim 1, B-2610, Antwerp, Belgium

Vandell, Victor E
Louisiana State University
Department of Chemistry
Baton Rouge, LA 708023, USA

Varmuza, Klaus
Department of Chemometrics
Technical University of Vienna
A-1060, Vienna Austria

Veracini, C A
Dipartimento Chemica
University of Pisa
I-56126, Pisa, Italy

Viappiani, Christiano
Dipartimento di Scienze Ambientali
Universita degli Studi di Parma
viale delle Scienze, I-43100, Parma, Italy

Vickery, Kymberley
Chalmers University of Technology
Department of Physical Chemistry
S-412 96, Gothenburg, Sweden

Wagnière, Georges H
Physikalisch-Chemisches Institut der Universitat
Zurich
Winterhurerstrasse 190,
CH-8057, Zurich, Switzerland

Waluk, Jacek
Institute of Physical Chemistry
Polish Academy of Sciences
01-224 Warszawa, Kasprzaka, 44/52, Poland

Wasylishen, Roderick E
Department of Chemistry
Dalhousie University
Halifax, Nova Scotia B3H 4J3, Canada

Watts, Anthony
Department of Biochemistry
University of Oxford
Oxford, OX1 3QU, UK

Webb, G A
University of Surrey
Department of Chemistry
Guildford, GU2 5XH, UK

Weiss, P S
Penn State University
Department of Chemistry
University Park,
Pennsylvania PA 16802-6300, USA

Weller, C T
School of Biomedical Sciences
University of St Andrews
St Andrews, KY16 9ST, UK

Wenzel, Thomas J
Department of Chemistry
Bates College
Lewiston, Maine ME 04240, USA

Wesdemiotis, Chrys
Chemistry Department
University of Akron
Akron, OH 44325-3601, USA

Western, Colin M
University of Bristol
School of Chemistry
Bristol, BS8 1TS, UK

White, R L
Department of Chemistry
University of Oklahoma
Norman, Oklahoma OK 73019-0390, USA

Wieser, Michael E
University of Calgary
2500 University Drive NW, Calgary,
Alberta T2N 1N4, Canada

Wilkins, John
Unilever Research
Colworth, Sharnbrook Beds. MK44 1LQ, UK

Williams, P M
The University of Nottingham
Laboratory of Biophysics and Surface Analysis,
School of Pharmaceutical Sciences
University Park, Nottingham, NG7 2RD, UK

Williams, Antony J
Wobrauschek P Atominstitut der Osterreichischen
Universitaten
Stadionallee 2, 1020, Wien, Austria

Wlodarczak, G
Laboratorie de Spectroscopie Hertzienne
URA CNRS 249, Univ. de Lille 1,
F59655, Villeneured'Aacg Cedese, France

Woźniak, Stanisław
Mickiewicz University
Warsaw, Poland

Young, Ian
Robert Steiner MR Unit
Royal Postgraduate Medical School
Hammersmith Hospital
Ducane Road, London, W12 0HS, UK

Zagorevskii, Dimitri
Department of Chemistry
University of Missouri-Columbia
Columbia MO 65201, USA

Zoorob, Grace K
Biosouth Research Laboratories Inc.
5701 Crawford Street, Harahan,
LA 70123, USA

Zwanziger, J W
Department of Chemistry
Indiana University
Bloomington, Indiana 47405, USA

Contents

Volume 1

G

H

Volume 2

I

Volume 3

O

Q

R

S

X

Y

Z

Entry Listing by Subject Area

Subject areas follow each other alphabetically in this list. Within each subject area entries are categorised into those covering historical aspects, theory, methods and instrumentation, or applications. The entries are listed alphabetically within these categories.

Atomic Spectroscopy

Electronic Spectroscopy

Fundamentals of Spectroscopy

Theory

Methods & Instrumentation

High Energy Spectroscopy

Theory

Methods & Instrumentation

Applications

Mass Spectrometry

Historical Overview

Theory

Methods & Instrumentation

Applications

Agricultural Applications of Atomic Spectroscopy

See Environmental and Agricultural Applications of Atomic Spectroscopy.

Alkali Metal Nuclei Studied By NMR

See NMR Spectroscopy of Alkali Metal Nuclei in Solution.

Aluminium NMR, Applications

See Heteronuclear NMR Applications (B, Al, Ga, In, Tl).

Angiography Applications of MRI

See MRI Applications, Clinical Flow Studies.

Anisotropic Systems Studied By Chiroptical Spectroscopy

See Chiroptical Spectroscopy, Oriented Molecules and Anisotropic Systems.

Antimony NMR, Applications

See **Heteronuclear NMR Applications (As, Sb, Bi).**

Arsenic NMR, Applications

See **Heteronuclear NMR Applications (As, Sb, Bi).**

Art Works Studied Using IR and Raman Spectroscopy

Howell GM Edwards, University of Bradford, UK

> **VIBRATIONAL ROTATIONAL &
> RAMAN SPECTROSCOPIES**
> **Applications**

Introduction

The scientific study of works of art and the materials that have been used in their creation has received impetus with the development of nondestructive, microsampling analytical techniques and an increasing awareness on the part of art historians, museum conservators and scientists of the importance of characterization for the attribution of the historical period and genuiness of an artefact. Useful information about ancient technologies and methods used in the construction of works of art is now forthcoming; in particular, spectroscopists are realizing the challenge that is being provided by the analytical characterization of ancient materials.

In this article, the applications of vibrational spectroscopic techniques to art are addressed and examples are taken to illustrate the following:

- identification of pigments, nondestructively, in preserved illuminated manuscripts, paintings and watercolours;
- characterization of genuine and fake artefacts, e.g. ivories;
- rock art and frescoes – effects of environmental and climatic degradation on exposed artwork;
- provision of information about archaeological and historical trade routes from the spectroscopic analysis of dyes, pigments and resins;

- identification of biomaterials in carved artwork, e.g. horn, hoof and tortoiseshell;
- information of critical importance to art restorers and museum conservation scientists – ill-restored items and the preservation of deteriorating material.

Comparison of infrared and Raman spectroscopies

Although both Fourier transform-infrared (FT-IR) and Raman spectroscopies have been applied to the study of art and museum artefacts, it is possible to identify several indicators that will dictate the preferential use of either technique. Infrared studies have been reported from a much earlier date than Raman studies; hence there now exists much more comprehensive infrared database for pigments and natural or synthetic materials. Specific points also need to be considered relating to the sampling and composition of the specimens being studied. The presence of highly fluorescent coatings or impurities in ancient specimens dictated the use of infrared spectroscopy when the alternative was visible excitation for Raman spectroscopic studies; fluorescence often swamped the lower-intensity Raman signals from such samples.

However, the smaller scattering intensity of water and hydroxyl groups in the Raman effect, and their strong absorption in the infrared, generally favours the use of the Raman technique for the characterization of ancient, hydrated biomaterials such as linens and cottons. In the area of paintings and manuscripts, Raman spectroscopy has had extensive application in recent years because of the accessibility of the low-wavenumber regions of the vibrational spectrum (< 500 cm^{-1}), which are vitally important for the characterization of inorganic and mineral pigments. This region is observed only with difficulty in the infrared, and then with the use of special instruments with far-infrared capability.

The shape and reflectivity of surfaces are also important for specimens that cannot be subjected to mechanical or chemical pretreatment – the taking of small samples by drilling, scraping or excision is often prohibited. Curved surfaces are notoriously poor for infrared examination and specimens are better analysed after being subjected to a flattening or crushing process. The use of fibreoptic probes for 'remote' analysis is now advocated for many infrared and Raman applications, and art objects are no exception. However, the interpretation of the spectral data from remote-probe scanning experiments is dependent on accurate subtraction of probe background; this is particularly relevant for features below 1000 cm^{-1}, e.g. artists' pigments. However, the obvious advantage of having the capability of taking the Raman or infrared spectrometer into a museum environment for *in situ* examination of an artefact or painting has dictated the construction of several portable units that might well provide a new dimension to these applications.

Finally, the advent of FT-Raman spectroscopy with near-infrared excitation from a Nd^{3+}/YAG laser at 1064 nm has given a new dimension to art applications, especially for the study of naturally fluorescent biomaterials such as horn, hoof, tortoiseshell and ivory. Improvements in the generation of Raman spectra and their detection using CCD-visible Raman spectroscopy are now also providing valuable new ways of analysing often difficult museum specimens. The role of Raman microscopy (and also infrared microscopy) in the characterization of micro-sized specimens will be illustrated in this article. Generally, it is now possible to achieve good-quality Raman microscopy spectra from sample areas as small as 1–2 µm using visible excitation (~ 0.5 mm has been claimed in some instances) and about 5–15 µm using infrared and Raman FT techniques. The latest advances in confocal microscopy using visible excitation and CCD detection now

means that it is possible to depth-profile suitable specimens with a resolution of about 1 µm or better; this is important for the analysis of coatings applied by artists to paintings.

Applications of IR and Raman spectroscopies

Plastics

Since the first commercial plastic, parkesine, in 1862, a huge range of functional articles have been produced in these media; many advantages were appreciated, including the ability to incorporate pigments and to mould large objects. Museums now have exhibitions devoted to plastic articles, but in recent years the problems in their conservation and storage and the control of degradation in damaged plastic exhibits has become acute. In particular, the acid ester plastics such as cellulose nitrate and cellulose acetate are prone to degradation, which has often resulted in major destruction of the articles concerned. Examples include early motion picture films, Victorian substitute 'tortoiseshell' articles, Bakelite and children's toys, such as dolls. FT-IR and Raman spectroscopy have been instrumental in discovering the methods of degradation and the source of possible 'triggers'; suggestions have thus been made for the proper storage and conservation of plastic articles for future generations, such as dolls from the 1940s, space suits from the 1960s and 'art nouveau' objects from the 1920s.

Glass

Degradation processes in medieval stained glass windows have been studied using FT-IR spectroscopy and the pigments applied to early specimens have been characterized using the FT-Raman technique. Similar methods have been used to study early enamels and cloisonné specimens.

Faience

Faience was produced in Egypt over 5000 years ago and may truly be considered as the first 'high-technology ceramic'. The faience body, formed from sand with additives such as lime and natron (a hydrated sodium carbonate found naturally as a mineral in dried lake beds), was shaped and modelled, and a glaze was applied and fired.

Sixteen pieces of Egyptian faience from Tell-el-Amarna have been analysed using Raman microscopy and the red and yellow pigments identified as red ochre and lead antimonate yellow, respectively.

A major problem in the characterization of the blue, white and green pigments resulted from a swamping of the pigment Raman signals by the silica in the intact glazes. Results for the red and yellow specimens were obtained from fractured samples.

Biomaterials

The preservation and restoration of objects made from biomaterials is particularly challenging as their degradation products are complex and diverse. Examples of problems facing museum conservators include art objects made from ivory, horn, natural resins and textiles; these objects have often become fragile and restoration is of the utmost significance and importance. Often, earlier restorative procedures, which may have been incompletely documented, are no longer satisfactory and the results of chemical deterioration of applied restoratives under the influence of solar radiation and humidity changes in storage are sadly too often plain to see.

There are several good examples in the recent literature of the successful application of vibrational infrared and Raman spectroscopy to the characterization of art objects composed of biomaterials. Ivory, a generic name for the exoskeletal dental growths of certain mammalian species, has been appreciated as an art medium for thousands of years; it is soft enough to be carved and polished, yet hard enough to resist superficial weathering damage. However, the identification and attribution of archaeological ivories to species of elephant, sperm whale, narwhal, hog and hippopotamus, for example, is often fraught with difficulty, particularly where the ivory object is small and may only be a part of a larger specimen. The observation of Schreger lines and surface morphology is extremely difficult in such cases. Very recently, Raman spectroscopy has been used to provide a suggested protocol for ivory identification and characterization and has been used successfully to determine the animal origin (sperm whale) of a Roman die excavated at Frocester Roman villa (third century AD) in the UK (**Figure 1**).

There have also been several examples of the use of Raman spectroscopy to identify genuine and fake ivory articles. **Figures 2** and **3** shows some bangles, an Egyptian necklace and a carved cat, all of which were assumed to be genuine ivory dating from the seventeenth century or later; in some cases, the articles were found to be modern imitations. The case of the carved cat is very significant, as the Raman spectra show the presence of calcite that has been added to a polymer composite of poly(methyl methacrylate) and polystyrene to simulate the density of true elephant ivory. This specimen could not therefore be three hundred years old as these polymers have only been known in the last fifty years! The FT-Raman spectra of true ivories from different mammal sources are shown in **Figure 4** and of the fake specimens in **Figure 5**; the spectroscopic differences are clearly discernible and provide a means of identification between fake and real specimens.

'Scrimshaw' is a special name for carved whalebone and teeth of the sperm whale; many objects in museum collections date from the eighteenth-century production of carved decorative artwork in this material by whaling sailors for their families ashore. Genuine scrimshaw is now extremely valuable and items have been created to deceive the unwary.

Figure 1 Roman die, ca. AD 300, from archaeological excavations at Frocester Villa, Gloucester, UK. Raman spectroscopy has suggested the origin of the die as sperm whale ivory. (See Colour Plate 1).

Figure 2 Selection of ornamental jewellery consisting of three bangles assumed to be ivory but which were shown spectroscopically to be composed of modern resins, and a genuine ivory necklace. (See Colour Plate 2).

Figure 3 'Ivory' cat, which was identified spectroscopically as a modern limitation composed of poly(methyl methacrylate) and polystyrene resins with added calcite to give the texture and density of ivory. Reproduced with permission from Edwards HGM and Farwell DW, Ivory and simulated ivory artefacts: Fourier-transform Raman diagnostic study, *Spectrochimica Acta, Part A*, **51**: 2073–2081 © 1995, Elsevier Science B. V. (See Colour Plate 3).

Spectroscopy has played a role in the characterization of genuine scrimshaw. **Figure 6** shows a stackplot of Raman spectra of carved solid and hollow sperm whale-tooth scrimshaw specimens, a staybusk,

and a spill vase/quill pen holder; from these spectra the whale teeth and staybusk are confirmed to be genuine, but the spill vase/quill pen holder is identified as a modern fake made from polymer resin.

Similarly, infrared and Raman spectroscopic studies of ancient textiles are being undertaken to derive information about the possible processes of their degradation under various burial environments, e.g. Egyptian mummy wrappings, silk battle banners, Roman woollen clothing and artistic linens from funerary depositions. A classic, topical and ongoing controversy covers the spectroscopic studies associated with the Shroud of Turin and the nature of the pigmented areas on the ancient linen.

Ancient technologies and cultures often used naturally occurring biomaterials as decorative pigments and as functional repairing agents on pottery and glass. Recent studies have centred on the provision of a Raman spectroscopic database for native waxes and resins, which has been used for the nondestructive evaluation of archaeological items, e.g. 'Dragon's blood', and it is possible to identify different sources of this material from the spectra. FT-IR spectroscopy, too, has been successful in the characterization of conifer resins used in ancient amphorae and has been particularly useful for attribution of the Baltic geographical origins of amber jewellery through its succinic acid content and for the detection of modern fakes made from phenol–formaldehyde resins that are often passed off as examples of ancient ethnic jewellery.

The identification of dammar and mastic resins that have been used as spirit-soluble varnishes on

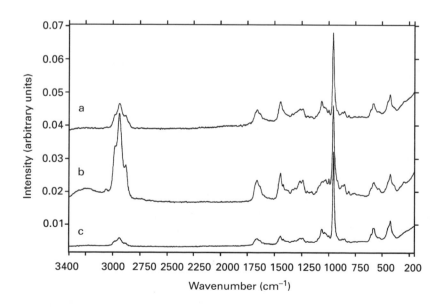

Figure 4 FT-Raman spectra of true ivory; 1064 nm excitation, 500 spectral scans accumulated, 4 cm⁻¹ spectral resolution: (a) sperm whale ivory, (b) elephant ivory, (c) walrus ivory.

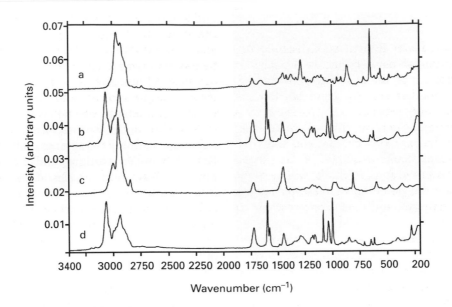

Figure 5   FT-Raman spectra of fake ivory specimens; conditions as for **Figure 4**: (a) carved Victorian bangle, (b) large bangle, (c) small bangle, (d) cat. The absence of the characteristic proteinaceous features in true ivory near 1650 and 1450 cm⁻¹ and the strong phosphate mode near 960 cm⁻¹ should be noted. Also, the presence of the aromatic ring bands at 3060, 1600 and 1000 cm⁻¹ in (b) and (d) indicate a polystyrene resin content, while the carbonyl stretching band at 1725 cm⁻¹ in all fake specimens indicates the presence of poly(methyl methacrylate). In the cat specimen, the band at 1086 cm⁻¹ uniquely identifies a calcite additive in the specimens of imitation ivory studied. Reproduced with permission from Edwards HGM and Farwell DW, Ivory and simulated ivory artefacts: Fourier-transform Raman diagnostic study, *Spectrochimica Acta, Part A*, **51**: 2073–2081 © 1995, Elsevier Science B.V.

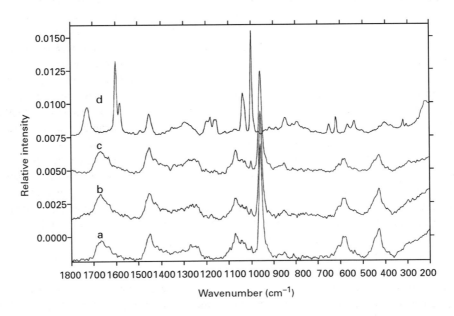

Figure 6   FT-Raman stack-plot spectra of scrimshaw specimens: (a) hollow sperm whale tooth, (b) solid sperm whale tooth, (c) whalebone staybusk and (d) spill vase/quill pen holder. Minor spectroscopic differences confirm the whalebone origin of the staybusk. The modern resin composition of the spill vase/quill pen holder is also unambiguously identified from the aromatic ring stretching bands at 3060 cm⁻¹ and 1600 cm⁻¹. Reproduced with permission from Edwards HGM, Farwell DW, Sedder T and Tait JKF, Scrimshaw: real or fake? An FT-Raman diagnostic study, *Journal of Raman Spectroscopy*, **26**: 623–628 © 1995, John Wiley and Sons Ltd.

paintings and on early photographic prints, and the conservation of the latter, in particular, is very important because of the embrittlement of the substrate due to exposure to light and humidity changes.

Raman spectroscopy has been used to identify wax coatings on early photographs from the American Civil War (ca. 1865) that are showing evidence of deterioration in museum collections.

Paintings

The use of infrared and Raman spectroscopies for the characterization of paint pigments depends on two critical factors: first, the ability to record good-quality spectra nondestructively from small paint flakes or chips, and, second, the provision of a database of mineral, natural and synthetic pigments on which the basis of a comparison and attribution can be made. Some examples of the success of vibrational spectroscopic methods will now be given to illustrate the potential of these techniques for pigment analysis, and the conservation of degraded media.

Several databases now exist for the comparison of unknown pigments with contemporary specimens. A most useful basis for the attribution of specimens is provided by synthetic pigments, for which the dates of first manufacture or use are well established. For example, titanium(IV) oxide (TiO_2) is the most important white pigment in use today; its two major naturally occurring forms are rutile and anatase, which have been in use as paint media since 1923 and 1947, respectively. Their identification by vibrational spectroscopy is unambiguous; hence, if either is found on a disputed Renaissance painting, it could indicate a forgery or at least a recently restored work. It is interesting that the distinction between these two forms of TiO_2 is not achievable with X-ray fluorescence techniques normally used for pigment analysis in art.

A rather different example is provided by the work of Guys, who painted social life in France between 1843 and 1860. Seventy samples taken from 43 of Guys' works revealed a heavy dependence on newly synthesized and experimental pigments at that time, including Prussian blue, Cobalt blue and French ultramarine, sometimes in admixture and unique to an artist of his time. Restoration, therefore, needs careful attention to unusual combinations of experimental pigments. A similar study emerges from the FT-IR study of Morisot's work, *Bois de Boulogne*; this also showed the presence of novel combinations of pigments, but ones that were more stable than others that were in use at that time. Analysis of paint flakes from the *Virgin and Child*, a suspected fifteenth-century work, revealed the presence of expected organic binders and varnishes but also suggested that a very heavy reworking had taken place more recently, and that the work could be a forgery. On the 'Jónsbók' Icelandic manuscript, six pigments were identified by Raman microscopy, only bone white being indigenous to Iceland; the others, vermilion, orpiment, realgar, red ochre and azurite would all have been imported.

The most important naturally occurring and synthetic inorganic pigments used over periods of time and which feature in paintings from medieval to contemporary ages are given in **Tables 1** to **7** along with the year of manufacture or documented use; the tables are constructed according to colour or type, viz. blue, black, brown/orange, green, red, white and yellow. These tables form the basis of dating of paintings and manuscripts by vibrational spectroscopy. The stability of inorganic mineral pigments relative to 'fugitive' organic dyes was realized in mediaeval times.

The principal dyes used by medieval dyers were indigo from woad for blue, alizarin and purpurin from madder for red, and luteolin from weld or crocetin from saffron for yellow; some had been used long before the Middle Ages and weld was known in the Stone Age. Organic pigments have also been used on manuscripts, notably saffron, weld, indigo, woad, Tyrian purple, madder and carmine.

Many natural products were extracted from lichens in the past and used to dye textiles. Notable among these were orchil for purple and crottle for brown. Other dye plants can yield greens, browns and blacks (in the last case, for example, marble gall from oak *Quercus* trees with added iron sulfate).

Organic pigments are prone to both fluorescence and photochemical degradation. Moreover, they often scatter only very weakly, perhaps owing to the fact that they may have been made into Colourfast lakes with a mordant such as alum; hence they are difficult to identify uniquely on a manuscript owing to the lack of concentration of pigment at the sampling point. The better known organic dyes and pigments are listed in **Table 8**.

The Raman and infrared spectra of organic pigments and dyes of relevance to artwork have now been recorded, including those of modern dyes such as methyl blue (a synthetic triarylmethane dye), methyl violet and perylene reds.

Perhaps the greatest advances in recent years in the field of nondestructive pigment identification in art works have come from historiated manuscripts that have been studied using Raman microscopy. *In situ* analyses of the brightly coloured initials and borders of the fourteenth-century Icelandic 'Jónsbók', Chinese manuscripts, illuminated Latin texts and early medieval Bibles have demonstrated the power of the technique. From samples that often represented only a 20 μm fragment that had fallen into the bindings of early manuscripts, FT-IR and Raman microscopy have made considerable advances in the knowledge of colour hue technology in medieval times; for example, the different blues in an historiated initial could be attributed to a finer particle size

Table 1 Blue inorganic pigments

Pigment	Chemical name	Formula	Date / Source[a]
Azurite	Basic copper(II) carbonate	$2CuCO_3 \cdot Cu(OH)_2$	Mineral
Cerulean blue	Cobalt(II) stannate	$CoO \cdot nSnO_2$	1821
Cobalt blue	Cobalt(II)-doped alumina glass	$CoO \cdot Al_2O_3$	1775
Egyptian blue	Calcium copper(II) silicate	$CaCuSi_4O_{10}$	3rd millennium BC/Mineral
Lazurite (from lapis lazuli)	Sulfur radical anions in a sodium aluminosilicate matrix	$Na_8[Al_6Si_6O_{24}]S_n$	Mineral/1828
Manganese blue	Barium manganate(VII) sulfate	$Ba(MnO_4)_2 + BaSO_4$	1907
Phthalocyanine blue (Winsor blue)	Copper(II) phthalocyanine	$Cu(C_{32}H_{16}N_6)$	1936
Posnjakite	Basic copper(II) sulfate	$CuSO_4 \cdot 3Cu(OH)_2 \cdot H_2O$	Mineral
Prussian blue	Iron(III) hexacyanoferrate(II)	$Fe_4[Fe(CN)_6]_3 \cdot 14–16H_2O$	1704
Smalt	Cobalt(II) silicate	$CoO \cdot nSiO_2 (+K_2O + Al_2O_3)$	ca. 1500
Verdigris	Basic copper(II)	$Cu(O_2CCH_3)_2 \cdot 2Cu(OH)_2$	Mineral

[a] The pigment is either specified to be a mineral and/or the date of its manufacture is listed.

Table 2 Black inorganic pigments

Pigment	Chemical name	Formula	Date / Source
Ivory black[a]	Calcium phosphate + carbon	$Ca_3(PO_4)_2 + C + MgSo_4$	4th century BC ?
Lamp black	Amorphous carbon	C	~ 3000 BC
Magnetite	Iron(II,III) oxide	Fe_3O_4	Mineral
Mineral black	Aluminium silicate + carbon (30%)	$Al_2O_3 \cdot nSiO_2 + C$	Mineral
Vine black	Carbon	C	Roman

[a] Bone black is similar to ivory black.

Table 3 Brown/orange inorganic pigments

Pigment	Chemical name	Formula	Date / Source
Cadmium orange	Cadmium selenosulfide	$Cd(S,Se)$ or CdS (>5 µm)	Late 19th century
Ochre (goethite)	Iron(III) oxide hydrate	$Fe_2O_3 \cdot H_2O$ + clay, etc.	Mineral
Sienna (burnt)	Iron(III) oxide	Fe_2O_3 + clay, etc.	Antiquity?

[a] Bone black is similar to ivory black.

Table 4 Green inorganic pigments

Pigment	Chemical name	Formula	Date / Source
Atacamite	Basic copper(II) chloride	$CuCl_2 \cdot 3Cu(OH)_2$	Mineral
Chromium oxide	Chromium(III) oxide	Cr_2O_3	Early 19th century
Cobalt green	Cobalt(II) zincate	$CoO \cdot nZnO$	1780
Emerald green	Copper(II) arsenoacetate	$Cu(C_2H_3O_2)_2 \cdot 3Cu(AsO_2)_2$	1814
Green earth – a mix of celadonite and glauconite	Hydrous aluminosilicate of magnesium, iron and potassium	Variations on $K[(Al^{III} \cdot Fe^{III})(Fe^{II} \cdot Mg^{II})](AlSi_3 \cdot Si_4)O_{10}(OH)_2$	Mineral
Malachite	Basic copper(II) carbonate	$CuCO_3 \cdot Cu(OH)$	Mineral
Permanent green deep	Hydrated chromium(III) oxide + barium sulfate	$Cr_2O_3 \cdot 2H_2O + BaSO_4$	Latter half of 19th century
Phthalocyanine green	Copper(II) chlorophthalocyanine	$Cu(C_{32}H_{15}ClN_8)$	1938
Pseudo-malachite	Basic copper(II) phosphate	$Cu_3(PO_4)_2 \cdot 2Cu(OH)_2$	Mineral
Verdigris (basic)	Basic copper(II) acetate	$Cu(C_2(C_2H_3O_2)_2 \cdot 2Cu(OH)_2$	Mineral and synthetic (BC)
Viridian	Hydrated chromium(III) oxide	$Cr_2O_3 \cdot 2H_2O$	1838 (?1850)

Table 5 Red inorganic pigments

Pigment	Chemical name	Formula	Date / Source
Cadmium red	Cadmium selenide	$CdSe$	ca. 1910
Chrome red	Basic lead(II) chromate	$PbCrO_4 \cdot Pb(OH)_2$	Early 19th century
Litharge	Lead(II) oxide	PbO	Antiquity
Realgar	Arsenic(li) sulfide	As_2S_2	Mineral
Red lead (minimum)	Lead(II,IV) oxide	Pb_3O_4	Antiquity
Red ochre	Iron(III) oxide + clay + silica	$Fe_2O_3 \cdot H_2O$ + clay + silica	Mineral
Vermilion (cinnabar)[a]	Mercury(II) sulfide	HgS	Mineral and synthetic (13th century)

[a] Limited lightfastness (→ black form).

Table 6   White inorganic pigments

Pigment	Chemical name	Formula	Date / Source
Anatase	Titanium(IV) oxide	TiO_2	1923
Barytes	Barium sulfate	$BaSO_4$	Mineral
Bone white	Calcium phosphate	$Ca_3(PO_4)_2$	Antiquity
Chalk (whiting)	Calcium carbonate	$CaCO_3$	Mineral
Gypsum	Calcium sulfate	$CaSO_4 \cdot 2H_2O$	Mineral
Kaolin	Layer aluminosilicate	$Al_2(OH)_4Si_2O_5$	Mineral
Lead white	Lead(II) carbonate (basic)	$2PbCO_3 \cdot Pb(OH)_2$	Mineral and synthetic (500–1500 BC)
Lithopone	Zinc sulfide and barium sulfate	$ZnS + BaSO_4$	1874
Rutile	Titanium(I) oxide	TiO_2	1947
Zinc white	Zinc oxide	ZnO	1834

Table 7   Yellow inorganic pigments

Pigment	Chemical name	Formula	Date / Source
Barium yellow	Barium chromate	$BaCrO_4$	Early 19th century
Cadmium yellow	Cadmium sulfide	CdS	Mineral (greenockite) + synthetic ca. 1845
Chrome yellow	Lead(II) chromate	$PbCrO_4$ or $PbCrO_4 \cdot 2PbSO_4$	1809
Cobalt yellow (aureolin)	Potassium cobaltinitrite	$K_3[Co(NO_2)_6]$	1861
Lead antimonate yellow	Lead(II) antimonate	$Pb_2Sb_2O_7$ or $Pb_3(SbO_4)_2$	Antiquity
Lead tin yellow	Lead(II) stannate	[1] Pb_2SnO_4	Antiquity ?
		[1] $PbSn_{0.76}Si_{0.24}O_3$	Antiquity ?
Massicot	Lead(II) oxide	PbO	Antiquity
Ochre	Geothite + clay + silica	$Fe_2O_3 \cdot H_2O$ + clay + silica	Mineral (and synthetic)
Orpiment	Arsenic(II) sulfide	As_2S_3	Mineral
Strontium yellow	Strontium chromate	$SrCrO_4$	Early 19th century
Zinc yellow	Zinc chromate	$ZnCrO_4$	Early 19th century

Table 8   Organic pigments and dyes

Colour	Pigment	Formula / Composition	Origin (Date)
Blue	Indigo	Indigotin $C_{16}H_{10}N_2O_2$	Plant leaf (BC), synthetic (1878)
Black	Bitumen	Mixture of hydrocarbons	(BC)
Brown	Sepia	Melanin	Ink of cuttlefish (ca. 1880)
	Van Dyck brown	Humic acids	Lignite containing manganese (16th century?)
		Allomelanins	
Green	Sap green	Organic dye	Buckthorn berry (14th century?), coal-tar dye
Purple	Tyrian purple	6,6′-Dibromoindigotin, $C_{16}H_8Br_2N_2O_2$	Marine mollusc (1400 BC), synthetic (1903)
Red	Carmine	Carminic acid, $C_{22}H_{20}O_{13}$	Scale insect, cochineal (Aztec)
		Kermesic acid, $C_{16}H_{10}O_8$	Scale insect, kermes (antiquity)
	Madder	{ Alizarin, $C_{14}H_8O_4$	Madder root (3000 BC)
		Purpurin, $C_{14}H_8O_5$	Synthetic alizarin (1868)
	Permanent red	Various azo dyes	Synthetic (after 1856)
Yellow	Gamboge	α- and β-Gambogic acids $C_{38}H_{44}O_8$ and $C_{29}H_{36}O_6$	Gum-resin (before 1640)
	Hansa yellow	Various azo dyes	Synthetic (1900)
	Indian yellow	Magnesium salt of euxanthic acid $MgC_{19}H_{16}O_{11} \cdot 5H_2O$	Cow urine (15th century)
	Quercitron	Quercitrin $C_{21}H_{20}O_{11}$	Inner bark of *Quercus* oak
	Saffron	Crocetin $C_{20}H_{24}O_4$	Crocus flower stigma (antiquity)
	Weld	Luteolin $C_{15}H_{10}O_6$	Plant foliage (Stone Age)

Tables 1–8 are reproduced with some modifications from Clark RJH, Raman microscopy: application to the identification of pigments on medieval manuscripts, *Chemical Reviews*, 187–196. 1995 © The Royal Society of Chemistry.

and not to a dilution with other materials or colours. Useful information has also been provided about additives and binding media.

On some paintings and manuscripts, an inappropriate blend of adjacent colours or mixtures of pigments and binders was achieved; Raman microscopy has identified two examples of these in cadmium sulfide and copper arsenoacetate (which yields black copper sulfide) and egg tempera with lead white (which yields black lead sulfide). The term 'inappropriate' here refers to the instability of pigments and pigment mixtures resulting in a chemical reaction over periods of time and through aerial or substratal influences.

Where the pigment is coloured, the choice of exciting line for Raman spectroscopy is extremely important because absorption of the scattered light by the sample may affect the spectrum. In such cases, the exciting line is chosen to fall outside the contour of the electronic absorption bands of the pigment. Vermilion (HgS) and red ochre (Fe_2O_3) give poor-quality Raman spectra using green excitation but give strong Raman spectra when excited with red radiation. An interesting example of the effects due to absorption of laser radiation is provided by the Raman spectra of red lead (Pb_3O_4) obtained using 514.5 and 632.8 nm excitation (**Figure 7**). In each case, well-defined Raman spectra are obtained, but only with 632.8 nm excitation is the spectrum that of the genuine material; that obtained with 514.5 nm excitation matches that of massicot (PbO). These observations may be explained because red lead can be converted into massicot by heat. Red lead absorbs 514.5 nm radiation strongly, leading to localized heating that results in conversion of the irradiated particle or particles into massicot. Since 632.8 nm radiation is not absorbed by the red lead, there is minimal local heating with this exciting line and thus no decomposition.

For coloured pigments, the absorption of exciting radiation can be used to advantage to produce enhanced Raman scattering through the resonance Raman effect. This is particularly useful for weakly scattering species, and selection of excitation within the contour of an electronic absorption band of a chromophore can produce a spectral intensity several orders of magnitude larger than that obtained in conventional Raman spectra. A classic example of this is provided by the deep blue mineral lapis lazuli, $Na_8[Al_6Si_6O_{24}]S_n$, where the species S_2^- and S_3^- are responsible for the colour. The species S_3^-, although present to an extent of < 1% in the pigment, gives such an intense Raman spectrum with green-red excitation that no bands due to the host lattice are observed. Other techniques fail to discriminate the presence of S_3^- in the aluminosilicate lattice.

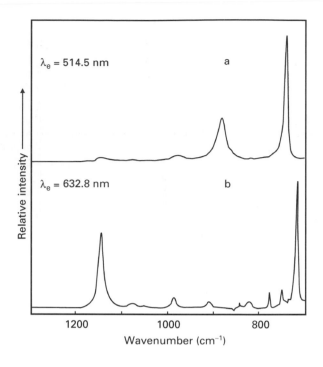

Figure 7 Illustration of the effect of wavelength of laser excitation on the Raman spectra of a pigment, red lead (Pb_3O_4), excited with (a) 514.5 nm and (b) 632.8 nm radiation. The genuine spectrum is (b); the spectrum excited by green radiation in (a) corresponds to massicot (PbO), converted from Pb_3O_4 by localized heating in the laser beam. Reproduced with permission from Bert SP, Clark RJH and Withnall R, Non-destructive pigment analysis of artefacts by Raman microscopy, *Endeavour, New Series*, **16**: 66–73 © 1992, Elsevier Science.

The Raman spectrum is sensitive to both composition and crystal form, as is demonstrated by titanium(IV) dioxide, the most important white pigment in use today. White pigments normally present considerable problems in pigment identification, comprising sulfates, carbonates and phosphates, which are easily distinguished in Raman spectroscopy (**Figure 8**). The symmetric vibration of the anion gives rise to an intense band at ~1000 cm⁻¹ for sulfates, ~1050 cm⁻¹ for carbonates and ~960 cm⁻¹ for phosphates. The exact wavenumbers of these bands are also sensitive to the cation (1050 cm⁻¹ for $PbCO_3$ and 1085 cm⁻¹ for $CaCO_3$). This factor becomes extremely important in database construction since artistic vocabulary generally describes the colour rather than a precise mineral origin, e.g. cadmium yellow, although strictly CdS has also been designated for organic substitutes of similar hue. Old recipes for obtaining pigment colours are often vague and employ unidentifiable materials; the same chemical compound or mixture can even have different names according to geographical locality or historical period. For example, three contemporary samples designated Naples Yellow, assumed to be

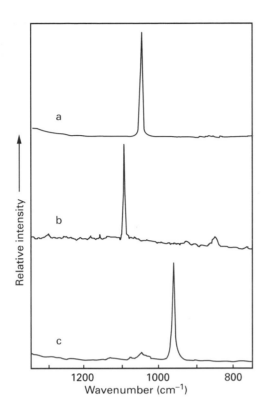

Figure 8 Raman spectra of white pigments; differentiation between (a) lead white ($PbCO_3$) (b) chalk ($CaCO_3$) and (c) bone white ($Ca_3(PO_4)_2$). Reproduced with permission from Best SP, Clark RJH and Withnall R. Nondestructive pigment analysis of artefacts by Raman microscopy, *Endeavour, New Series* **16**: 66–73 © 1992, Elsevier Science.

$Pb_2(SbO_4)_2$ – a lead antimonate – have been shown by Raman spectroscopy to be a lead antimony oxide, $Pb_2Sb_2O_6$, another oxide, $Pb_2Sb_2O_3$, and the third sample mainly $Pb_2(CrO_4)_2$, but also containing $PbCO_3$. None of the samples tested actually proved to be Naples Yellow of the assumed formula!

Medieval manuscripts

Most pigments in medieval manuscript illuminations are inorganic. They are coated with a small quantity of binding material and deposited on parchment that has previously been rubbed with pumice. The small size of the illuminations limits the number of spectroscopic samplings.

The blue colour in a set of six French manuscripts from the twelfth century was studied in different kinds of initial – historiated (with a descriptive scene), decorated, or simply coloured – which showed variable hues of pure blue, grey and sometimes violaceous. Raman microspectra obtained from all of the samples show common features: they are dominated by a strong band at 548 cm^{-1}, with

Table 9 The Raman bands of inorganic pigments in the Lucka Bible

Pigment	Raman bands (cm^{-1})
Azurite	248w, 404vw, 770m, 838vw, 1098m, 1424w, 1578w
Lapis lazuli	259w, 549vw, 807w, 1096m, 1355vw, 1641w
White lead	1054s
Malachite	225w, 274w, 355w, 437m, 516vw, 540w, 724w, 757vw, 1064w, 110w, 1372w, 1498m
Orpiment	136m, 154m, 179w, 203m, 293s, 311s, 355s, 384m, 587vw
Realgar	124vw, 143m, 166w, 172w, 183s, 193s, 214w, 222s, 329w, 345m, 355s, 370w, 376w
Red lead	121vs, 152m, 223w, 232w, 313w, 391w, 477w, 549s
Vermilion	254s, 281w, 344m

Reproduced with permission from Best SP, Clark RJH, Daniels MAM and Withnall R, A bible laid open, *Chemistry in Britain*, **2**: 118–122 © 1993, The Royal Society of Chemistry.

weaker bands at 260, 585 and 1098 cm^{-1}. This group of four bands is characteristic of ultramarine blue and reveals the presence of only this pigment. In the blue-grey samples, microscopic observation showed tiny black particles that were identified by their Raman spectra as graphite.

Ultramarine has been identified in a manuscript commentary on *Ezekiel*, ca. AD 1000, in the abbey of St Germain, Auxerre, France. The pigment ultramarine is first mentioned in written sources in the thirteenth century. This discovery established for the first time that the pigment was in fact introduced into Europe almost two centuries earlier. The manuscript was to reveal more fascinating details. Part of the dedication image at the beginning of the manuscript is a figure representing the Abbot, Heldric, kneeling before St Germain, who died about 448 and is buried in the abbey. From the spectrum of the pigments, it was established that the Abbot's blue garments were painted with indigo (woad) while only the patron saint's more glorious garment was covered with the rare and expensive lapis lazuli, thus emphasizing his importance.

The letter 'I' in the Lucka Bible at the beginning of the Book of Genesis displays seven scenes representing the seven days of Creation. The varied palette (**Table 9**) has been well characterized using Raman microscopy (**Figure 9**).

The Skard copy of the Jónsbók Icelandic Law Code dates from ca. 1360 and is one of the most outstanding examples of an Icelandic medieval manuscript. The pigments have been analysed by Raman microscopy; three historiated initials and about 15 illuminated initials with associated background paintings and embellishments have been examined. Six pigments were identified unambiguously by

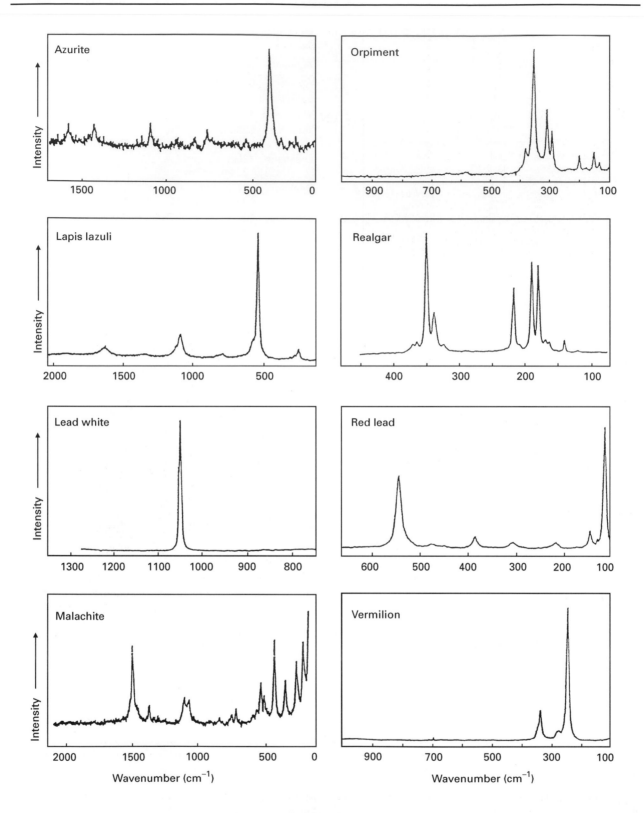

Figure 9 Raman spectra of the historiated letter 'I' from the Book of Genesis, Lucka Bible, from which the data and identification of the mineral pigments in **Table 9** have been derived. Reproduced with permission of Clark RJH, Raman microscopy: application to the identification of pigments on medieval manuscripts, *Chemical Reviews*, 187–196 © 1995, The Royal Society of Chemistry.

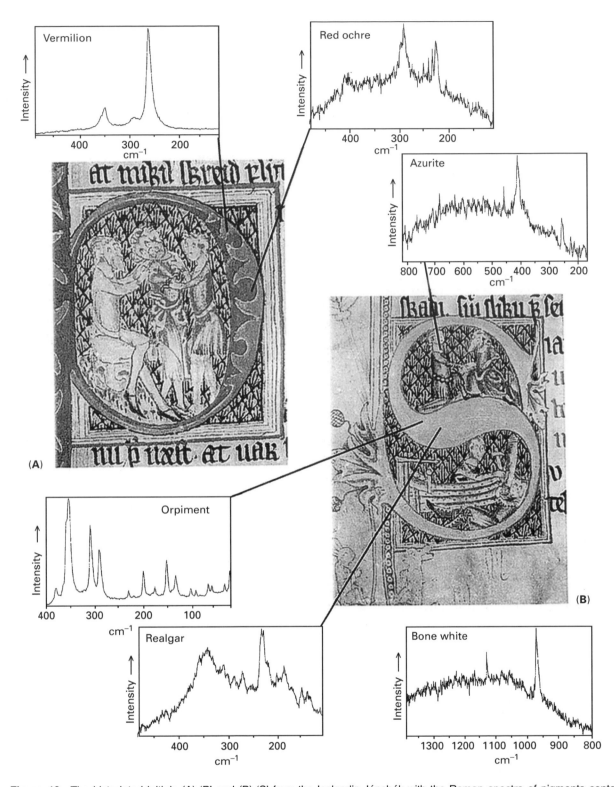

Figure 10 The historiated initials (A) 'P' and (B) 'S' from the Icelandic Jónsbók with the Raman spectra of pigments contained therein. The combination of vermilion and red ochre in the 'P' should be noted. The spectrum of bone white has been obtained from the letter 'H' (not shown). Reproduced with permission from Best SP, Clark RJH, Daniels MAM, Porter CA and Withnall R (1995). Identification by Raman microscopy and visible reflectance spectroscopy of pigments on an Icelandic manuscript. *Studies in Conservation* **40**: 31–40. (See Colour Plate 4.)

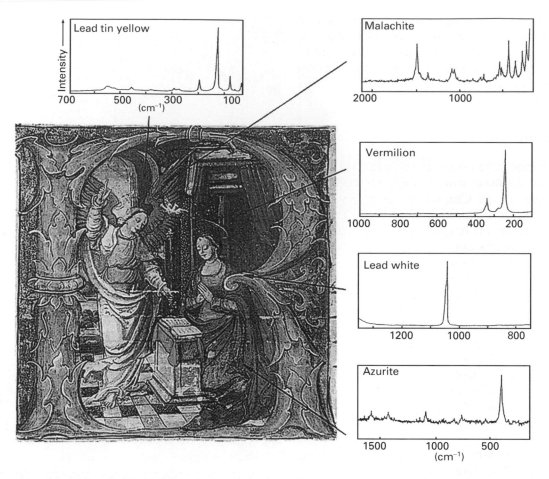

Figure 11 Elaborately historiated initial 'R' in sixteenth-century German choir book, with Raman microscopy spectra of selected pigmented regions. Reproduced with permission of Clark RJH (1995) Raman microscopy: application to the identification of pigments on medieval manuscripts. *Chemical Society Reviews* 187–196 © The Royal Society of Chemistry. (See Colour Plate 5a).

Raman microscopy: vermilion, orpiment, realgar, red ochre, azurite and bone white. Neither red lead nor lead white, pigments commonly used in Northern Europe, was identified on the manuscript: this raises questions as to the availability of these pigments in Iceland, to which they are not native, and of the effectiveness of the trading routes. Examples of two historiated initials and the Raman spectra of the pigments used are shown in **Figure 10**.

An elaborately historiated initial 'R' on a sixteenth-century German choir book has been studied by Raman microscopy and no fewer than eight pigments have been identified, namely, azurite, lead tin yellow, malachite, vermilion, white lead, red lead, carbon and massicot (**Figure 11**).

Raman microscopy demonstrated that in this case the two shades of blue on the garments of the right-hand figure, although appearing to be due to different pigments, arise from the same pigment, azurite, the illuminator having used less azurite and proportionately more binder to produce the lighter

shade without choosing to add any white pigment, such as lead white, to gain the same effect. The azurite used in the deeper blue robe arises from coarse grains of pigment (~30 μm diameter) whereas the lighter blue used in the undergarment arises from fine grains (~3 μm diameter). The effect of the particle size on the depth of colour of a powder is a common feature of powdered materials, whose colour is determined by diffuse reflectance. As the particle size is reduced, the average depth to which radiation penetrates before being scattered is also reduced and hence the depth of colour is reduced.

The angel's wing and podium are painted in malachite, which is a basic copper carbonate similar to azurite. Both azurite and malachite have a similar provenance in that they are both associated with secondary copper ore deposits, but their spectra are very different.

It is of particular interest that the dark grey colour of the pillar top is not obtained via a single pigment but by colour subtraction of a mixture of at least seven pigments: white lead, carbon, azurite, as well as small amounts of vermilion, red lead, massicot and lead tin yellow type I. The mixture is evident at ×100 magnification by Raman microscopy, whereby each individual pigment grain may be identified (**Figure 12**).

Wall paintings

The application of FT-Raman spectroscopy to the study of red pigments from English early medieval wall paintings in the Chapter House of Sherborne Abbey and the Holy Sepulchre Chapel in Winchester Cathedral (**Figure 13**) has been reported. The complexity of the spectra can be observed in **Figure 14**, where the pigment and substratal features are clearly differentiated. The operations of the two monastic houses were very different, as the Sherborne Abbey pigment consisted of pure vermilion, whereas that of Winchester Abbey was a 3:1 mixture of red ochre and vermilion. In both cases, spectral features due to sandstone and marble could be seen, but no identifying feature could be ascribed to a plaster substrate.

The biodeterioration of the Renaissance frescoes in the Palazzo Farnese, Caprarola, Italy, arising from aggressive colonization by the lichen *Dirina massiliensis* forma *sorediata* has been studied using Raman microscopy. Over 80% of the masterpieces painted by Zuccari in 1560 have now been destroyed. The role of the lichen metabolic products and the ability

Figure 13 Holy Sepulchre Chapel, Winchester Cathedral. Wall painting of ca. 1175–85 on the east well depicting the *Deposition, Entombment, Maries at the Sepulchre and the Harrowing of Hell*. Reproduced with permission from Edwards HGM, Brooke C and Tait JKF, An FT-Raman spectroscopic study of pigments or medieval English wall paintings, *Journal of Raman Spectroscopy* **28**: 95–98, © 1997 John Wiley and Sons Ltd. (See Colour Plate 6).

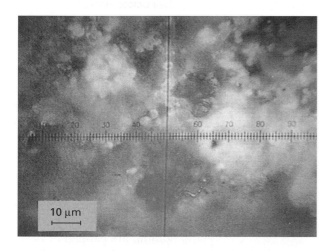

Figure 12 Portion of top of column of historiated initial 'R' shown in **Figure 11**; the individual pigment grains can be clearly seen under the ×100 magnification and can be separately identified using Raman microscopy. Reproduced with permission of Clark RJH (1995) Raman microscopy: application to the identification of pigments on medieval manuscripts. *Chemical Society Reviews* 187–196 © The Royal Society of Chemistry. (See Colour Plate 5b).

of primitive organisms to survive hostile environments generated by large concentrations of heavy metals such as mercury, lead and antimony are now being understood as a result of these studies, which will assist conservators in future projects.

The red pigment on the Carlovingian frescoes in the crypt of the Abbey of St Germain at Auxerre shows up to 13 layers of paint identified by Raman microscopy, and vermilion and iron(II) oxide have both been found.

The North American palaeo-Indian shelters at Seminole Canyon, at the confluence of the Rio Grande and Devils Rivers, have provided the Raman spectra of the oldest cave paintings yet studied. From a limited palette, these 3500-year-old pictographs have yielded red, white and black colours identified

(A)

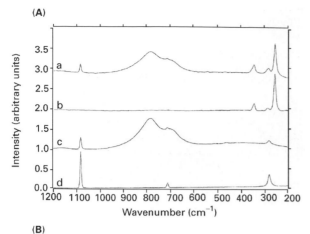

(B)

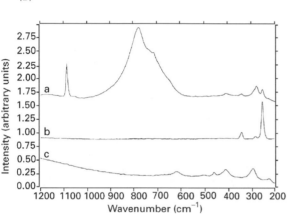

Figure 14   FT-Raman spectra of twelfth-century wall paintings in (A) Sherborne Abbey ((a) red pigment, (b) vermilion, (c) unpainted stone, (d) calcite) and (B) Winchester Cathedral ((a) red pigment, (b) vermilion, (c) red ochre). Reproduced with permission from Edwards HGM, Brooke C and Tait JKF, An FT-Raman spectroscopic study of pigments on medieval English wall paintings, *Journal of Raman Spectroscopy* **28**: 95–98, © 1997 John Wiley and Sons Ltd.

as red ochre, calcium oxalate monohydrate (whewellite) and manganese(IV) oxide, respectively. In addition, the black pigment showed traces of an organic additive recently identified from DNA profiling as bison or deer bone marrow, presumably added in a symbolic ritual.

FT-IR spectroscopy, too, has had considerable success in the identification of pigments and associated materials in paintings and manuscripts. In particular, the use of infrared microspectroscopy has proved to be an essential prerequisite for the scientific examination of complex pigment mixtures. Generally, however, inorganic mineral pigments are not well characterized in the infrared because of their low band wavenumbers; hydroxyl groups in hydrated minerals cause problems in absorption and

the substrates themselves, e.g. cellulose, linen or starch, often give rise to broad backgrounds. IR spectroscopy, therefore, is much more applicable to organic pigments, binders and mixtures.

One project dealing with medieval manuscripts has had as its objective the application of small-particle-analysis techniques to the study of pigments in medieval Armenian and Byzantine manuscripts. At the University of Chicago, 10 decorated manuscripts were sampled (**Table 10**). These manuscripts represent a broad chronological span ranging from the tenth century to the post-Byzantine era (sixteenth century or later).

Interpretation of IR spectra was complicated by the fact that the pigment samples often contained several components in addition to the pigment. These additives often served as binding media – thickening agents or extenders, for example. Because the amount of pigment is small relative to the amounts of these nonpigment components, the spectral bands of the pigment are often difficult to distinguish from interfering spectral bands of nonpigment components. Chromatographic separation prior to spectroscopic analysis was not feasible because of the insolubility and limited quantity of the samples. Spectral subtraction has been used in many cases as a means to 'separate' the components.

The infrared spectrum of a yellow pigment sample from manuscript (MS) 46 (Haskell gospels) shows several bands that can be attributed to egg yolk, a typical medium used in mediaeval times, as well as calcium carbonate and kaolin. The absorption bands at 3382, 3298 and 1600 cm^{-1} suggested that a primary amine is present. Putrescine (1,4-diaminobutane), an amine found in decaying proteinaceous matter, has similar spectral features. Because parchment (being of animal skin origin) would be subject to putrefaction if not properly preserved, this amine could be a product of bacteriological degradation.

Table 10   University of Chicago Special Collections manuscripts analysed by FT-IR spectroscopy

Manuscript number	Name
972	Archaic Mark
1054	Elfleda Bond Goodspeed *Gospels*
965	Rockefeller-McCormick *New Testament*
131	Chrysanthus gospels
232	Greek gospels
46	Haskell gospels
129	Nicolaus gospels
727	Georgius gospels
879	Lectionary of Constantine the reader
948	Lectionary of St Menas the wonder worker

Although the pigment was not identified in this case, the information that egg tempera is present as the binding medium is valuable; the binding medium in MS 965 is hide glue.

Pigment components have also been identified from their IR spectra. The infrared spectrum obtained from a purple pigment removed from MS 965 showed bands due to ultramarine blue and white lead. A red pigment (cochineal?) must also be present with the blue to produce the purple colour.

In MS 972 (Archaic Mark), the presence of Prussian blue from the infrared spectrum indicates possible restoration because Prussian blue, a synthetic dye, was not available until 500 years later.

Another example of the use of FT-IR spectroscopy in the authentication of paintings is provided by the analysis of the *Virgin and Child*, a fifteenth-century Italian painting on panel. There were already stylistic doubts and, although IR spectroscopy confirmed the presence of egg tempera as a binding medium, the blue pigment was not azurite or lapis lazuli. A prominent infrared absorption at $2091\,cm^{-1}$ identified the blue pigment on the Virgin's robe as Prussian blue, unknown before 1704. Hence, the painting has either been very heavily reworked or is possibly a nineteenth-century copy of an earlier work.

See also: **Colorimetry, Theory; Dyes and Indicators, Use of UV-Visible Absorption Spectroscopy; Fourier Transformation and Sampling Theory; Rayleigh Scattering and Raman Spectroscopy, Theory; Vibrational CD Spectrometers; Vibrational CD, Applications; Vibrational CD, Theory.**

Further reading

Adelantado JVG, Carbo MTD, Martinez VP and Reig BF (1996) FTIR spectroscopy and the analytical study of works of art for purposes of diagnosis and conservation. *Analytica Chimica Acta* **330**: 207–215.

Best SP, Clark RJH and Withnall R (1992) Non-destructive pigment analysis of artefacts by Raman microscopy. *Endeavour* **16**: 66–73.

Clark RJH (1995) Raman microscopy: application to the identification of pigments on mediaeval manuscripts. *Chemical Society Reviews* 187–196.

Coupry C and Brissaud D (1996) Applications in art, jewellery and forensic science. In Corset J and Turrel G. (eds) *Raman Microscopy: Developments and Applications*, pp 421–453. London: Academic Press.

Edwards HGM and Farwell DW (1995) FT-Raman spectroscopic study of ivory: a nondestructive diagnostic technique. *Spectrochimica Acta Part A* **51**: 2073–2079.

Hoepfner G, Newton T, Peters DC and Shearer JC (1983) FTIR in the service of art conservation. *Analytical Chemistry* **55**: 874A–880A.

Katon JE, Lang PI, Mathews TF, Nelson RS and Orna MV (1989) Applications of IR microspectroscopy to art historical questions about mediaeval manuscripts. *Advances in Chemistry Series* **220**: 265–288.

Learner T (1996) The use of FTIR in the conservation of 12th century paintings. *Spectroscopy Europe* 14–19.

Mills JS and White R (1994) *The Organic Chemistry of Museum Objects*, 2nd edn. London: Butterworth-Heinemann.

Astronomy, Applications of Spectroscopy

See **Interstellar Molecules, Spectroscopy of; Stars, Spectroscopy of.**

Atmospheric Pressure Ionization in Mass Spectrometry

WMA Niessen, hyphen MassSpec Consultancy,
Leiden, The Netherlands

MASS SPECTROMETRY
Methods & Instrumentation

Introduction

Ion sources for atmospheric-pressure ionization (API) in mass spectrometry (MS) were first described in 1958 by Knewstubb and Sugden. The work of the research group of Horning and Carroll in the late 1960s and early 1970s resulted in a system commercially available from Franklin GNO Corp. These types of instruments were mainly used in the study of ion–molecule reactions in gases in the atmosphere. Furthermore, in 1974 this type of instrument was applied with an atmospheric-pressure corona discharge ion source in the coupling of liquid chromatography to MS (LC-MS). Despite the promising results, the technique did not attract much attention at that time.

Subsequently, several other API instruments were described and built. The most successful commercial instrument was the Sciex TAGA (trace atmospheric gas analyser), while API systems were also available from Extranuclear and Hitachi. In the late 1980s, the TAGA API-MS system was reengineered by Sciex for online LC-MS and was called the API-III. This system was applied by a growing group of scientists, especially within pharmaceutical industries. In the early 1980s, Yamashita and Fenn fundamentally investigated the potential of electrospray ionization in an API source. This research led to another approach to LC-MS using an API system. In 1988, Fenn and co-workers demonstrated that with this electrospray system it was possible to perform the MS analysis of high molecular-mass proteins via multiple charging. As a result of this observation, API-MS went through a period of tremendous growth, the end of which is certainly not yet here. In the mid 1980s, similar API technology was also applied in the coupling of inductively coupled plasmas (ICP) to MS.

At present, there are three major application areas of API-MS:

- air or gas analysis,
- online LC-MS, and
- ICP-MS.

In this paper, technical and instrumental aspects of API-MS are discussed and a number of typical applications are briefly reviewed.

API ion source design

An API ion source generally consists of four parts:

1. the sample introduction device, the design of which is highly dependent on the type of application,
2. the actual ion source region, i.e. the region where the ions are generated,
3. an ion sampling aperture, where the ions are sampled from atmospheric pressure into the vacuum system of the MS, and
4. an ion transfer system, where the pressure difference between the atmospheric-pressure source and the high-vacuum mass analyser is bridged and the ions entering through the ion sampling aperture are preferentially transferred and focused into the mass analyser region.

The proper design of parts 3 and 4 is of utmost importance, as it determines to what extent the very high ionization efficiency that can be achieved in an API source actually becomes available to the analytical applications. The inevitable ion losses in the sampling and transfer regions should be kept to a minimum.

Ion sampling aperture and ion transfer system

An API source system consists of a region where the ions are generated. This region can have a relatively large volume of several litres, although in practice only the ions that come close to the ion sampling aperture are efficiently sampled into the MS. Therefore, in many systems the ions are generated just in front of the ion sampling aperture. A general scheme of an API source is shown in **Figure 1**.

In the early API systems, the ion sampling aperture consisted of a 25–200 μm i.d. pinhole in a metal plate. The size of the pinhole is limited by the pumping capacity of the vacuum system of the MS.

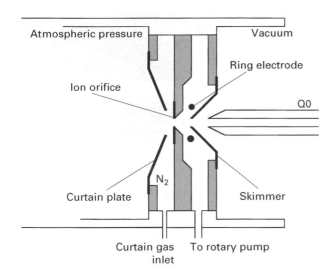

Figure 1 Schematic diagram of a typical API source, as for instance used in LC-MS.

In the first prototypes, the pinhole inner diameter was only 25 μm, while in the Sciex TAGA a larger pinhole could be applied, because of the high pumping efficiency of the cryogenic vacuum-pumping system available in this instrument.

The sampling of ions through an orifice in a plate or in the tip of a cone is used in many commercial API systems. However, two alternative ion sampling devices based on the transport of ions through capillary tubes have been developed and commercialized. The first, designed by Fenn and co-workers and subsequently commercialized by Analytica of Branford, incorporates a 100–150 mm × 0.5 mm ID glass capillary. Both ends of the glass capillary are coated with metal, e.g. silver, gold or platinum, in order to define the potential at both tips. The use of an insulating glass capillary enables independent control over the potentials applied in the atmospheric-pressure ion source region and in the ion acceleration region inside the vacuum. The second alternative to the ion sampling orifice is a heated stainless steel capillary, initially proposed by the group of Chait and subsequently commercialized by Finnigan.

In the strong cooling that takes place upon the expansion of a gas into a vacuum chamber, as is taking place in an electrospray source, the ions act as condensation nuclei for water and solvent molecules. As a result, clusters of analyte ions and numerous solvent molecules are formed. This cluster formation can be counteracted or avoided in two ways: either by preventing the solvent ions from reaching the ion sampling orifice, or by increasing the temperature of the gas, vapour and ion mixture entering the vacuum. By flushing the area just in front of the sampling orifice with a stream of dry nitrogen, the solvent vapour can be swept away, while the ions are forced to pass through this nitrogen 'curtain' by the application of an electric field between the curtain plate and the orifice plate in **Figure 1**. An additional benefit of a curtain gas is the prevention of other neutral contaminants, such as particulate material, from entering the ion sampling orifice or capillary. This is important for the ruggedness required for day-to-day use of LC-MS, e.g. in the analysis of large series of biological samples. A curtain or counter-current gas is applied in the sources built by Sciex, Analytica of Branford and Hewlett-Packard. Alternatively, the temperature of (part of) the ion source or of the ion sampling capillary can be increased to avoid the formation of clusters between analyte ions and solvent molecules. This obviates the need for a curtain gas, but also leaves the ion sampling device unprotected from contamination by particulate material.

Ions are sampled from the atmospheric-pressure region and together with nitrogen, which is applied as nebulizing gas as well as curtain gas, entering the vacuum system. The low-pressure side of the ion sampling pinhole or the low-pressure end of the ion sampling capillary acts as a nozzle, where the mixture of gas, solvent vapour and ions expands. The expanding jet is subsequently sampled by means of a skimmer into a second vacuum stage. In most systems, the region between the nozzle and the skimmer, which is pumped by a high-capacity rotary pump, contains an additional lens, e.g. a ring or tube electrode, to preferentially have the ions sampled by the skimmer. From a gas dynamics point of view, it might be important to position the skimmer just in front of the Mach disk of the expansion. However, because ions are preferentially sampled, the positioning of the skimmer relative to the Mach disk does not appear to be very important. In some systems, nozzle and skimmer are actually not precisely aligned in order to reduce the number of neutral molecules in the highly directed flow from the nozzle entering the region with higher vacuum. Optimum ion transfer is achieved by means of a tube lens between nozzle and skimmer.

At the lower pressure side of the skimmer a second expansion step takes place. In older systems, the ions entering through the skimmer were extracted, transferred and focused into the mass analyser region by means of a set of conventional flat lenses. Later, it was determined that a more efficient ion transfer and focusing can be achieved by means of an RF-only quadrupole, hexapole or octapole device (cf. **Figure 1**). These types of ion transfer and focusing devices are especially in use in systems for LC-MS. In other systems, as used for ICP-MS for instance, ion

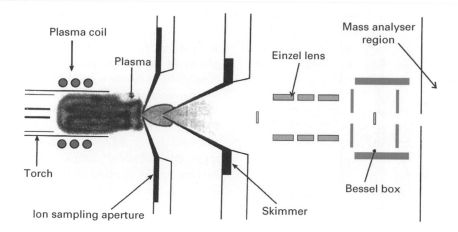

Figure 2 Schematic diagram of a typical API source for ICP-MS.

focusing is achieved by means of a combination of Einzel lenses and a Bessel box (cf. **Figure 2**).

The source designs described so far are used in combination with a (triple) quadrupole instrument. Modifications of this design are required in the coupling to other types of mass analysers, e.g. the implementation of a gate or ion pulse electrode for use in combination with ion-trap and time-of-flight mass analysers.

Sample introduction devices

A wide variety of sample introduction devices are available for gas analysis by API-MS. In the study of atmospheric processes, direct sampling of ionic species and clusters in various atmospheric layers may be performed. For these *in situ* studies, air is directly sampled into the system, eventually through specially designed sampling tubes. Alternatively, gases may be sampled and mass analysed after ionization by electrons from a corona discharge (similar to APCI).

For use in combination with LC-MS and other liquid-introduction techniques, two main types of sample introduction devices are available, i.e. one for electrospray ionization and one for APCI.

The devices for APCI consist of a heated nebulizer, which is a combination of a concentric pneumatic nebulizer and a heated quartz tube for droplet evaporation. Additional auxiliary gas is used to flush the evaporating droplets through the vaporizer into the actual ionization region. A schematic diagram of such a device is shown in **Figure 3**. Reactant ions for APCI are generated in a point-to-plane corona discharge in the atmospheric-pressure ion source. The discharge needle is operated at 3–5 kV and provides a corona emission current of a few μA.

For electrospray ionization, initially, 100 μm i.d. stainless-steel capillaries, i.e. hypodermic needles, were used for sample introduction. With such a device, the flow-rate is limited to ~10 μL min^{-1}, which is too high for many biochemical applications and too low for effective LC-MS coupling. Therefore, the initial system was modified.

For biochemical applications, the dimensions of the introduction capillary were decreased to a few μm i.d. capillaries, mostly a glass or fused silica. This approach is called nanoelectrospray as it allows the continuous introduction of between 5 and 100 nL min^{-1} of a liquid into the API source. Nano-electrospray is especially useful for applications where the sample amount is limited. In this way, it is possible to perform a series of MS experiments with a minute amount of sample, e.g. 30 min of various MS experiments with only 1 μL of sample.

For LC-MS applications, the electrospray nebulization process was assisted by pneumatic nebulization, i.e. a nebulization gas at a high linear velocity is forced around the electrospray needle. This approach, introduced by Sciex as ionspray in 1986, was subsequently adopted by other instrument manufacturers and is now the most widely applied

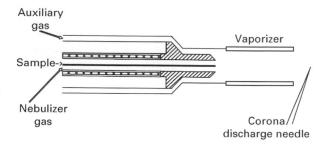

Figure 3 Schematic diagram of a sample introduction device for APCI: a heated nebulizer.

approach to LC-MS via electrospray interfacing. A schematic diagram of a typical sample introduction device for LC-MS via pneumatically-assisted electrospray is shown in **Figure 4**.

Significant research has taken place in the past few years concerning the optimum geometry of the spray device in the atmospheric-pressure ion source. While initially the spray device was positioned axially relative to the ion sampling orifice, an approach which is still in use, other instrument manufacturers use a different setup, e.g. the spray device positioned orthogonally to the ion sampling orifice, or at an angle of 45°, eventually with a perpendicular heated gas probe to assist in droplet evaporation. The two main perspectives in this type of research are the ability to introduce higher liquid flow-rate while still avoiding the risk of clogging the ion sampling orifice, and the ability to introduce samples containing significant amounts of nonvolatile material, e.g. samples from biological fluids after minimum sample pretreatment, or LC mobile phases containing phosphate buffers. While the advantages of the alternative positioning of the spray device is well documented for biological fluids, the evidence for the possibilities of the use of phosphate buffers is less convincingly demonstrated.

For use in ICP-MS, ions are sampled directly from the inductively coupled plasma. Special precautions are required related to the high temperatures in the plasma. A schematic diagram of an ICP-MS system is shown in **Figure 2**. In most cases, the liquid sample, which can be introduced directly from a vial, of a liquid-phase separation technique such as LC is nebulized into a separate spray chamber. The aerosol, containing the smaller droplets, is transported by argon to the torch, where the ICP is generated and sustained. In this so-called ICP flame, the analytes are atomized and ionized. The ions are directly sampled from the flame by the ion sampling aperture for mass analysis.

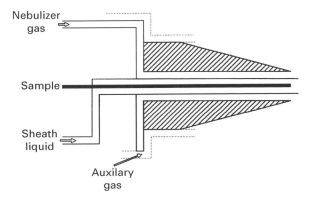

Figure 4 Schematic diagram of a pneumatically-assisted electrospray sample introduction device.

Ionization in API conditions

Ionization in an API source can take place in a variety of ways, depending on the type of applications. The various way are briefly reviewed here.

Gas-phase ionization

In initial applications of API sources for gas analysis and monitoring of atmospheric ionic processes, either direct sampling of the ions present in flames or in the atmosphere was applied, or ions were generated by means of electrons from an electron-emitting ^{63}Ni foil or by means of a corona discharge. Currently, the corona discharge is the most widely applied approach for generating ions, in both gas analysis and APCI in LC-MS applications.

In an API source equipped with a ^{63}Ni foil, energetic electrons ionize the nitrogen present in a series of consecutive steps:

$$N_2 + e^- \rightarrow N_2^{\bullet+} + 2\,e^-$$

$$N_2^{\bullet+} + 2N_2 \rightarrow N_1^{\bullet+} + N_2$$

Similarly, other gases may be ionized. In the presence of traces of water in the ion source, these ionic species react with the water molecules:

$$N_4^{\bullet+} + H_2O \rightarrow H_2O^{\bullet+} + 2N_2$$

$$H_2O^{\bullet+} + H_2O \rightarrow H_3O^+ + HO^{\bullet+}$$

$$H_3O^+ nH_2O \rightarrow H_3O^+(H_2O)_n$$

The main ionic species generated, i.e. $N_4^{\bullet+}$ and $H_3O^+.(H_2O)_n$, may enter in ion–molecule reactions, leading to charge exchange of proton transfer reactions, respectively. This is valid for positive-ion acquisition, while negative ions may be generated either by electron-capture processes with electrons or via proton-transfer ion-molecule reactions starting from $OH^-.(H_2O)_n$ for instance.

The same reactant ions are formed in positive-ion mode in a point-to-plane corona discharge. Approximately 3–9 kV is applied at the corona discharge, creating a corona current of 1–5 μA, depending on the point-to-plane distance and the composition of the vapours in the ion source. Corona discharges are very complicated processes that involve avalanches of ionization reactions by ion–molecule and electron–molecule collisions together with quenching processes, as discussed in detail by Meek and Craggs.

In the positive-ion mode, the ionization of analyte molecules follows the general rules of chemical ionization, i.e. proton transfer may take place when the proton affinity of the analyte molecule exceeds that of the reagent gas, leading to protonated molecules:

$$M + H_3O^+ \cdot (H_2O)_n \rightarrow [M+H]^+ + nH_2O$$

while, in addition, various adduct ions and cluster ions with solvent molecules may be generated. In the negative-ion mode, either electron-capture products $M^{-\bullet}$ or deprotonated molecules $[M-H]^-$ may be generated, although the electron-capture products may further react in ion–molecule reactions to form different ionic species.

The positive-ion mass spectra obtained from APCI and conventional medium-pressure CI might show differences due to the fact that the mass spectrum in medium-pressure CI is more determined by the reaction kinetics, while in APCI, due to the longer residences times of ions in the sources as well as the more frequent ion–molecule collisions, the mass spectrum more reflects the thermodynamic equilibrium conditions.

Liquid-based ionization

The liquid-based ionization technique in API-MS is applicable in electrospray interfacing as well as with a heated nebulizer introduction under certain conditions, i.e. without switching the corona discharge electrode on, and for certain compounds. Here, the discussion is restricted to electrospray interfacing.

In electrospray ionization and interfacing, a high potential, typically 3 kV, is applied to a liquid emerging from a capillary. This causes the liquid to break into fine threads emerging from the so-called Taylor cone formed at the capillary tip. The threads subsequently disintegrate into small droplets. The electrohydrodynamic or Rayleigh disintegration results from autorepulsion of the electrostatically charged surface which exceeds the cohesive forces of surface tension (the Rayleigh limit). In this electrospray nebulization process uniform droplets in the 1-μm diameter range are formed. These charged droplets shrink by solvent evaporation, whereas electrohydrodynamic droplet disintegration again takes place as soon as the Coulomb repulsion of the surface charges exceeds the surface tension forces. The offspring droplets generated in the field-induced droplet disintegration process carry only 2% of the mass of the parent droplet and 15% of its charge.

Analyte ions in these offspring droplets, present as preformed ions, desorb or evaporate into the gas phase and become available for mass analysis. The ion evaporation process, described first by Iribarne and Thomson and generally considered as the most important ionization mechanism in electrospray ionization, is a process which thermodynamically corresponds to bringing a solvated ion from the surface of a charged droplet to infinity. Below a certain size/charge ratio for a droplet, the Gibbs free energy of solvation of an ionic species present in the droplet will exceed the energy required to bring this species as a solvated ion from infinity to the surface of the droplet. Under these conditions, ion evaporation is possible. According to Iribarne and Thomson, ion evaporation will take place instead of electrohydrodynamic disintegration when the critical ion evaporation droplet radius exceeds the Rayleigh limit.

Electrospray ionization is especially effective for analytes that are present as preformed ions in solution, e.g. organic acids or bases that are present as deprotonated or protonated molecules, respectively, at suitable pH values, quaternary ammonium compounds, etc. For biomacromolecules like proteins, which exist in solution as a distribution of ionic species carrying different numbers of charge, electrospray ionization results in a spectrum containing an ion envelope of multiply charged ions (see **Figure 5**). To a first approximation, this mass spectrum reflects the charge-state distribution of the protein in solution. As the ion separation in a mass analyser is based on mass-to-charge ratio, the multiple charging allows the mass analysis of large molecules within the limited m/z-range of, for instance, a quadrupole mass spectrometer.

The observation of multiple charging of proteins in electrospray ionization attracted much attention: all major instrument manufacturers introduced an API system to be used in combination with an electrospray interface. At the same time, it was found that electrospray ionization is not only suitable for the mass analysis of large molecules, but is also a very efficient ionization technique of small polar or ionic molecules, e.g. drugs and their metabolites. In the past few years, hundreds of dedicated API-MS systems equipped with electrospray and APCI interfaces have found their way into many different laboratories, especially within pharmaceutical companies and biochemistry/biotechnology laboratories.

Plasma-based ionization

A plasma is an ionized gas that is macroscopically neutral, i.e. an equal number of positive and negative particles are present. In ICP-MS, the gas used to

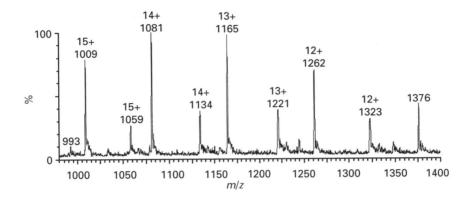

Figure 5 Ion envelope of multiply charged ions, obtained in the electrospray ionization of haemoglobin. In the spectrum, the *m/z* and charge state of the ions are indicated. The spectrum consists of two series: one due to a Hb-α chain (M_r 15126 Da) and one due to a Hb-β^A chain (M_r 15 867 Da).

generate the plasma is argon, which has a high ionization energy (15.76 eV). The hot argon plasma has the capability to atomize most samples, and excite and ionize most elements of the periodic table. The plasma is generated in a torch. A high-frequency generator, typically operated at 27 or 40 MHz and with a power of 1–2 kW, is used to produce a high-frequency field through an induction coil, which is positioned at the outside of the torch. In this way, the argon plasma is produced, i.e. the argon is ionized and the plasma is sustained. Liquid samples are nebulized and the aerosol is injected into the plasma for atomization and ionization. The ionized atoms generated in this way are sampled by the ion sampling aperture and mass analysed.

Applications

Gas analysis

There are a number of applications of API-MS in gas analysis, e.g. the study of environmental air contamination, human breath analysis, the study of (ionic) processes in the atmosphere, the detection of trace impurities in gases that are important in microelectronic manufacturing processes, or the identification of ions in flames. The power of this technique is perhaps best illustrated by the ability to achieve detection limits in the ppt range for oxygen, water, methane and carbon dioxide in high-purity gases like hydrogen, nitrogen, argon and helium. A proper characterization of the trace impurities in these gases is of utmost importance in semiconductor processing. Using a mobile API-MS system, such as the Sciex TAGA, real-time environmental monitoring of TNT, industrial emissions, PCBs and other environmental contaminants in the ground-level troposphere is possible.

LC-MS

The use of API-MS for the online LC-MS coupling is the largest commercial field of application of API-MS. From the early 1990s, when the potential of API-MS for LC-MS, and especially its robustness and 'user-friendliness', were substantially demonstrated, an astonishingly large number of API-MS systems for LC-MS were purchased. Particularly in various stages of drug development within pharmaceutical industries, the use of API-based LC-MS is preferred over the more conventional LC-UV system. In general, better reliability, confirmation, selectivity and sensitivity are achieved in LC-MS systems. In this way, high-throughput quantitative bioanalysis is possible. In addition, API-based LC-MS systems using electrospray or atmospheric-pressure chemical ionization (APCI) interfacing and ionization are applied for the quantitative and qualitative analysis of a wide variety of polar compounds in numerous matrices in a number of other fields, e.g. pesticides, herbicides and their metabolites and degradation products in environmental samples, natural products in plant extracts and cell cultures, peptides, proteins and other biomacromolecules in biochemical and biotechnological applications.

ICP-MS

ICP-MS is used in practically every discipline where inorganic analytical support is required. This includes environmental, geological, biological, medical, nuclear, metallurgical (semiconductor industry) and nutritional studies. An important advantage of ICP-MS in quantitative analysis, required in many fields of application, is the substantial gain in precision and accuracy that can be achieved by the use of isotope-dilution mass spectrometry, where stable isotopes or long-lived radio-isotopes of the elements to be analysed are added as internal standards, e.g. the use of ^{111}Cd in the analysis of ^{114}Cd. In addition, ICP-MS can be extremely useful in speciation. In this respect, there is a growing interest in the use of ICP-MS directly coupled to LC or capillary electrophoresis for speciation and analysis of elements of toxicological interest, like As, Se, Pb and Hg.

See also: **Chemical Ionization in Mass Spectrometry; Chromatography-MS, Methods; Cosmochemical Applications Using Mass Spectrometry; Inductively Coupled Plasma Mass Spectrometry, Methods; Inorganic Chemistry, Applications of Mass Spectrometry.**

Further reading

Bruins AP, Covey TR and Henion JD (1987) Ion spray interface for combined liquid chromatography/atmospheric pressure ionization mass spectrometry. *Analytical Chemistry* 59: 2642–2646.

Carroll DI, Dzidic I, Horning EC and Stillwell RN (1981) Atmospheric pressure ionization mass spectrometry. *Applied Spectroscopic Review* 17: 337–406.

Chowdhury SK, Katta V and Chait BT (1990) An electrospray-ionization mass spectrometer with new features. *Rapid Communications of Mass Spectrometry* 4: 81–87.

Cole RB (ed) (1997) *Electrospray Ionization Mass Spectrometry.* Chichester, UK: Wiley.

Fenn JB, Mann M, Meng CK, Wong SF and Whitehouse CM (1989) Electrospray ionization for mass spectrometry of large biomolecules. *Science* 246: 64–71.

Hieftje GM and Norman LA (1992) Plasma source mass spectrometry. *International Journal of Mass Spectrometry and Ion Processes* 118/119: 519–573.

Iribarne JV and Thomson BA (1976) On the evaporation of small ions from charged droplets. *Journal of Chemical Physics* 64: 2287 and (1979) Field induced ion evaporation from liquid surfaces at atmospheric pressure. *Journal of Chemical Physics* 71: 4451.

Knewstubb PF and Sugden TM (1958) Mass-spectrometric observation of ions in flames. *Nature* 181: 474–475.

Knewstubb PF and Sugden TM (1958) Mass-spectrometric observation of ions in hydrocarbon flames. *Nature* 181: 1261.

Meek JM and Craggs JD (1978) *Electrical Breakdown of Gases.* Chichester: Wiley.

Meng CK, Mann M and Fenn JB (1988) *Proceedings of the 36th ASMS Conference on Mass Spectrometry and Allied Topics,* June 5–10, San Francisco, CA, pp. 771–772.

Niessen WMA (1998) *Liquid Chromatography – Mass Specctrometry,* 2nd edn. New York: Marcel Dekker.

Niessen WMA (1998) Advances in instrumentation in liquid chromatography – mass spectrometry and related liquid-introduction techniques. *Journal of Chromatography A* 794: 407–435.

Vela NP, Olson LK and Caruso JA, Elemental speciation with plasma mass spectrometry. *Analytical Chemistry* 65: 585A–597A.

Yamashita M and Fenn JB (1984) Electrospray ion source. Another variation on the free-jet theme. *Journal of Physical Chemistry* 88: 4451–4459.

Yamashita M and Fenn JB (1984) Negative ion production with the electrospray ion source. *Journal of Physical Chemistry* 88: 4671–4675.

Atomic Absorption, Methods and Instrumentation

Steve J Hill and **Andy S Fisher**, University of Plymouth, UK

ATOMIC SPECTROSCOPY
Methods & Instrumentation

Introduction

Atomic absorption spectroscopy has become one of the most frequently used tools in analytical chemistry. This is because for the determination of most metals and metalloids the technique offers sufficient sensitivity for many applications and is relatively interference free. There are two basic atom cells (a means of turning the sample, usually a liquid, into free atoms) used in atomic absorption spectroscopy:

(i) the flame and (ii) electrothermal heating of a sample cell. It is generally acknowledged that if sufficient analyte is present in the sample, then it should be determined using a flame technique because this has the added advantages of being rapid (assuming only a few elements need be determined) and, in comparison with alternative techniques, very simple to use. Electrothermal atomic absorption spectroscopy (ETAAS) requires more operator skill and is less rapid, but yields substantially superior limits of detection when compared with flame atomic absorption spectroscopy (FAAS). This section describes some of the methods and instrumentation that have been developed for both flame and electrothermal techniques of atomic absorption spectroscopy.

Instrumentation

The basic principle of both FAAS and ETAAS is that sample is introduced into the atom cell, where it is desolvated and then atomized. The analyte atoms so formed then quantitatively absorb light in a way that is proportional to the concentration of the atoms of the analyte in the cell. The light, which is at a specific wavelength, is then isolated from other wavelengths that may be emitted by the atom cell and then detected. Thus, much of the instrumentation used for electrothermal and flame absorption spectroscopy is identical. Both techniques require a similar light source, background correction system, line isolation device (monochromator or polychromator), detector (photomultiplier or charge coupled device) and readout system. Each of these components is discussed below, together with details of the individual atom cells (flame and the electrothermal atomizer) and sample introduction systems.

Light source

The fundamental requirement of the light source is to provide a narrow line profile with little background. It should also have a stable and reproducible output with sufficient intensity to ensure that a high signal-to-noise ratio is obtained. Two basic types of light source are used for atomic absorption, of which the hollow-cathode lamp (HCL) is the more commonly used. This is a lamp in which the cathode is coated with the analyte metal of interest. Within the lamp, inert filler gas (neon or argon) is ionized by an electric current and these ions are then attracted by the cathode. The inert gas ions bombard the cathode and in so doing excite the metal ions coated on it. It is this excitation of the metal that produces the emission of radiation with wavelengths characteristic of the analyte. Hollow-cathode lamps are available for most metallic elements. A schematic diagram of a hollow-cathode lamp is shown in **Figure 1**.

Electrodeless discharge lamps are used less frequently than the hollow-cathode lamps except for analytes such as arsenic and selenium. These lamps may be excited using either microwave energy (although these tend to be less stable) or radiofrequency energy. The radiofrequency-excited lamps are less

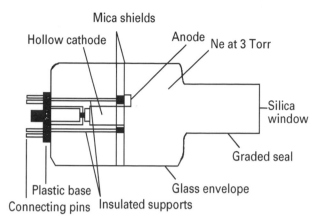

Figure 1 Schematic representation of a hollow-cathode lamp. From Ebdon L (1998) *Introduction to Analytical Atomic Spectrometry*. Reproduced by permission of John Wiley & Sons Limited.

intense than the microwave-excited ones, but are still 5–100 times more intense than a standard hollow-cathode lamp. In the electrodeless discharge lamp, a bulb contains the element of interest (or one of its salts) in an argon atmosphere. The radiofrequency energy ionizes the argon and this in turn excites the analyte element, causing it to produce its characteristic spectrum. For analytes such as arsenic and selenium, these lamps give a better signal-to-noise ratio than hollow-cathode lamps and have a longer useful lifetime. A schematic diagram of an electrodeless discharge lamp is shown in **Figure 2**.

Background correction systems

There are basically three types of automatic background correction system available for atomic absorption, although manual methods such as the use of nearby non-atomic absorbing lines to estimate background absorbance may also be used. The three main automatic methods are the deuterium or hydrogen lamp, the Zeeman effect, and Smith–Hieftje background correction. The deuterium lamp produces a continuum of radiation, some of which will be absorbed by molecular species within the atom cell. The amount of atomic absorption observed using the deuterium lamp is negligible and hence the atomic absorption signal is obtained by subtracting the absorbance from the continuum lamp from the total analyte absorbance from the hollow cathode lamp. The deuterium or hydrogen lamp is of most value at wavelengths in the UV region (< 350 nm). The Zeeman effect background correction system is more versatile. It relies on a strong magnetic field operating at approximately

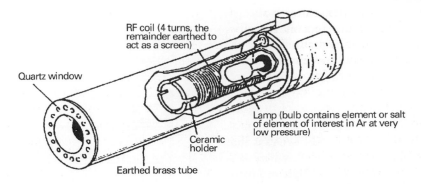

Figure 2 Electrodeless discharge lamp. From Ebdon L (1998) *Introduction to Analytical Atomic Spectrometry*. Reproduced by permission of John Wiley & Sons Limited.

1 T and 50–60 Hz, placed either around the light source or, more commonly, around the atom cell to split the signal into a number of components. The π component is at the normal analyte wavelength; the σ components are typically 0.01 nm (depending on the strength of the magnetic field) either side of the π component and hence lie outside the atomic absorption profile. Background can be corrected by subtracting the absorbance with the magnetic field 'on', from the signal with the magnet 'off'. It should be noted that this is a simplified description of the process, as different elements have different splitting patterns and the magnetic field may be applied longitudinally or transversely, which may also produce different splitting patterns.

When utilizing the Smith–Hieftje system, the hollow-cathode lamp is boosted periodically to a much higher current, causing the lamp to 'self-absorb'. In this state no atomic absorption occurs in the atom cell but the molecular absorption still remains. Background correction is achieved automatically by subtracting the signal obtained at high current from that obtained using the normal current. The process is particularly efficient at removing interferences such as that caused by phosphate or selenium determinations, although the high currents used may shorten the lifetime of the lamp. In modern instruments, modulation of the source is achieved electronically. This enables discrimination between absorption and emission signals. Previously, a rotating sector, often referred to as a 'chopper', placed between the source and the atom cell was used.

Single-beam and double-beam instruments

The vast majority of instruments used for atomic absorption measurements have a single-beam configuration, using the optical layout shown in **Figure 3A**. The double-beam arrangement (**Figure 3B**) is far more complex and offers far fewer advantages for atomic absorption when compared with double-beam spectrometers in molecular spectroscopy. This is because the reference beam does not pass through the atom cell. Despite this, double-beam instruments can compensate for source drift and for warm-up and source noise.

Line isolation devices

To ensure that only light of a wavelength specific to the analyte of interest is being measured, a line isolation device is required. Until recently, the line isolation device used for atomic absorption was a

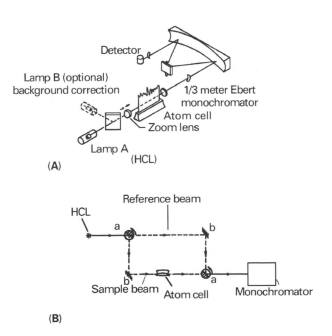

Figure 3 Schematics of (A) single-beam and (B) double-beam spectrometers. From Ebdon L (1998) *Introduction to Analytical Atomic Spectrometry*. Reproduced by permission of John Wiley & Sons Limited.

monochromator. There are numerous types of monochromator, but modern instruments typically use one of three designs: Ebert, Czerny–Turner or Littrow configuration. These are shown schematically in **Figure 4**.

Light of all wavelengths enters the monochromator through an entrance slit and is then split into specific wavelengths using either a prism, or more commonly, a diffraction grating. By altering the position of this dispersing element, light of only the desired wavelength passes through the exit slit to the detector. Since the method of atomic absorption is so specific, very highly resolving monochromators are not required. Thus, the focal length of an atomic absorption monochromator is often 0.25 or 0.5 m compared with a minimum of 0.75 m required for conventional optical emission spectroscopy.

A monochromator enables only one wavelength to be interrogated at any instant; this was something of a weakness of atomic absorption spectroscopy. New technology, however, has enabled the development of multielement spectrometers that use a bank of between 4 and 6 hollow-cathode lamps and an echelle-style polychromator employing orders of 100 or more and are thus capable of considerable dispersion. In addition, such instruments have no exit slit and so a huge number of wavelengths may be focused onto an array detector such as a charge-coupled device. A schematic diagram of such an instrument is shown in **Figure 5**. Research continues into producing a continuum light source that would enable 30–40 analytes to be determined simultaneously.

Detection systems

Traditionally, detection of the light isolated by the monochromator has been accomplished using a photomultiplier tube (PMT). Several configurations exist, e.g. end-on and side-on, and the construction may be of different materials, to increase the efficiency at different wavelengths; but basically they all work in a similar way. Light enters the multiplier through a quartz window and impacts with a photocathode that is usually made from one of a number of alloys (e.g. Cs–Sb, Na–K–Sb–Cs or Ga–As), which then emits electrons. These electrons are then accelerated down a series of dynodes, each being at a more positive potential than the previous one. As the electrons impact with successive dynodes, further electrons are ejected and hence a cascade effect occurs. In this way a single photon may cause the ejection of 10^6 electrons. The number of electrons is

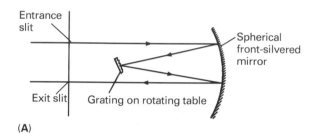

(A)

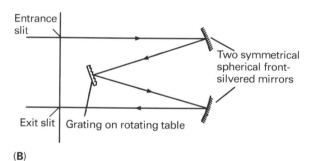

(B)

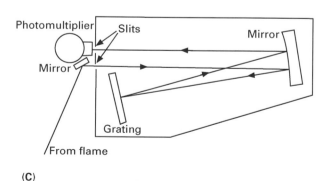

(C)

Figure 4 Schematics of different monochromator types: (A) Ebert, (B) Czerny–Turner, (C) Littrow. From Ebdon L (1998) *Introduction to Analytical Atomic Spectrometry*. Reproduced by permission of John Wiley & Sons Limited.

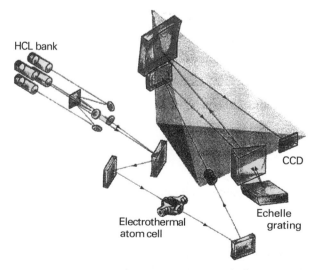

Figure 5 Echelle-based polychromator.

then measured at the anode and the resulting current is proportional to the radiation reaching the PMT.

As described previously, a number of different materials may be used to coat the photocathode in the PMT, and each has a different response curve. Some, for example the Cs–Sb tube, have very little sensitivity above 600 nm, but others, for example Ga–As, may be used up to almost 900 nm. The response profiles of some commonly used photomultipliers are shown in **Figure 6**.

Conventional spectrometers use a detector that is capable of measuring only one signal. Multielement instruments use state-of-the-art diode arrays or charge-coupled devices that can measure numerous spatially separated signals and can therefore simultaneously determine the signals arising from a bank of hollow-cathode lamps. A charge-coupled device may be considered to be similar to an electronic photographic plate. The device consists of several hundred linear photodetector arrays on a silicon chip with dimensions of typically 13×18 mm. The line isolation device (often an echelle grating) separates the analytical wavelengths and these may then be detected on different regions of the array. A detailed description of how the array works is beyond the scope of this text.

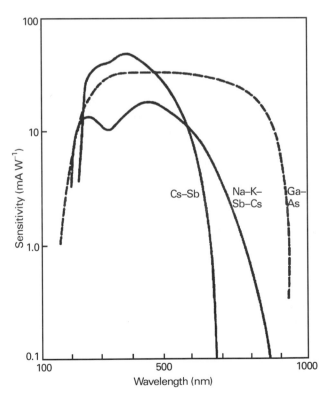

Figure 6 Response curves for several commercial photomultiplier tubes. From Ebdon L (1998) *Introduction to Analytical Atomic Spectrometry.* Reproduced by permission of John Wiley & Sons Limited.

Instrument control and output devices

Manual controls and dials have slowly disappeared, with the large majority of modern instruments being controlled by computer. In addition to controlling the instrumental parameters, the computer may be used to program autosamplers, save experimental parameters and data, calibrate with known standards, plot calibration curves, use scale expansion facilities (if necessary) and produce statistical data from the results. A continuous graphics mode enables the signal to be monitored over a period of time. This is especially useful when flow injection, chromatography or hydride-generation transients are to be detected and integrated. Older instrumentation relied on a chart recorder for the same purpose.

Sample introduction and atom cells

For flame systems, the sample is introduced via a nebulizer and spray chamber assembly. Sample is drawn up the intake capillary by the Venturi effect. The liquid sample column comes into contact with the fast moving flame gases and is shattered into small droplets. In many systems it is further smashed into still smaller droplets by an impact bead. The flame gases then carry the aerosol (the nebular) through a spray chamber containing a series of baffles or flow spoilers. These act as a droplet size filter, passing the larger droplets to waste (85–90% of the sample) and allowing the finer droplets to be transported to the flame. For electrothermal atomizers, the sample is placed into the atom cell either using a hand-held micropipette or by an autosampler. Since most instruments are supplied with an autosampler and the precision associated with their use is superior to that of the micropipette, most laboratories make use of this method of sample introduction.

As described previously, there are two basic types of atom cell; flame and electrothermal tube, but both can have numerous modifications. The flame cell most commonly used is the 10 cm length air–acetylene burner. This provides a flame that is at approximately 2300°C (although the actual temperature will depend on the fuel/air ratio). The flame may be used in a fuel-lean mode, which produces a hot, oxidizing blue flame; in a fuel-rich mode, which produces a cooler, reducing yellow flame; or in a stoichiometric mode whose properties are between the two. Different analytes give different sensitivities depending on the flame type. Chromium, for instance, gives better sensitivity in a reducing flame, whereas for magnesium it is better to use a lean, oxidizing flame.

Different areas of the flame have different temperatures and different chemical properties. The efficiency of atomization of analytes will therefore depend critically on the area within the flame. Hence, it is important that the light beam from the HCL passes through the region of the flame where atomization is optimal. Optimization of the burner height is therefore necessary to ensure that maximum sensitivity is achieved.

Although an air–acetylene flame is sufficient for the atomization of the majority of analytes, it is not sufficiently hot or reducing to atomize analytes that form refractory oxides (e.g. Al, Mo, Si and Ti). For analytes such as these a nitrous oxide–acetylene flame of length 5 cm is used. This flame is hotter (2900°C) and contains typically 2.5–3 times as much fuel as the air–acetylene flame. Other flame types, e.g. air–propane and air–hydrogen also exist, although they are less commonly used. The air–hydrogen flame (2200°C) is almost invisible and has the advantage of being transparent at lower wavelengths offering improved noise characteristics for elements such as lead or tin at around 220 nm. The inhomogeneity of flame chemistry and the need for burner height optimization hold true for all flame types.

One common modification to a flame cell is to place a quartz tube on the burner head so that the light beam passes through the length of the tube. This method is especially useful when using sample introduction by hydride generation or by gas chromatography. The tube is heated either by a flame or electrically using a wire winding. The analyte, which enters the tube via a hole or slit cut into the side, is atomized within the tube. The atoms then leave the tube but, because of their retention in the tube, spend longer in the light beam than in normal flame systems. An increase in sensitivity is therefore obtained. Another small modification of this system is that, instead of a quartz tube, a quartz T-piece may be used. However, the function is the same.

Electrothermal atom cells have changed radically since their inception in the late 1950s. The majority of electrothermal devices have been based on graphite tubes that are heated electrically (resistively) from either end. Modifications such as the West Rod Atomizer (a carbon filament) were also devised but were later abandoned. Tubes and filaments made from highly refractory metals such as tungsten and tantalum have also been made, but they tend to become brittle and distorted after extended use and have poor resistance to some acids. Their use continues, however, in some laboratories that need to determine carbide-forming elements. For example, silicon reacts with the graphite tube to form silicon carbide, which is both very refractory and very stable. The silicon is therefore not atomized and is lost analytically. Use of a metal vaporizer prevents this.

In electrothermal atomization the sample is introduced into the tube, which is then heated in a series of steps at increasing temperature. The sample is dried at a temperature just above the boiling point of the solvent (but not so hot as to cause frothing and spitting of the sample); ashed (charred at an intermediate temperature to remove as much of the concomitant matrix and potential interferences as possible without losing any analyte); and then atomized at a high temperature. During the atomization stage, the atoms leave the graphite surface and enter the light beam, where they absorb the incident radiation. The ashing and atomizing temperatures used will depend on the analyte of interest; for example, some analytes such as lead are relatively volatile and so cannot be ashed at temperatures above 450°C, otherwise volatile salts such as chlorides will be lost. Other analytes, for example, magnesium, are less volatile and can be ashed at temperatures close to 1000°C without analyte loss. Such elements require a much higher atomization temperature. The sensitivity of electrothermal atomization AAS is greater than for flame AAS because the atoms are formed within the confines of a tube and hence spend longer in the light beam. Also, since the sample is placed within the atom cell, 100% of it is available for analysis compared with the 10–15% available in flame systems. A comparison of characteristic concentrations (concentration that gives an absorbance of 0.0044) obtained for flame and electrothermal techniques for many analytes is shown in **Table 1**. It must be stressed that the figures for ETAAS will depend critically on the injection volume and so values tend to be given in absolute terms (i.e. a weight).

Overall, the Massmann design of electrothermal atomizer in which the tube is heated from either end is still the most common (**Figure 7B**), but more recently transversely heated tubes have been developed (**Figure 7A**). The longitudinally heated tubes have a temperature gradient along the tube, with the central portion being several hundred degrees hotter than the ends. This can lead to condensation of analyte at the cooler ends and subsequent re-atomization from the hot graphite surface. Several atomization peaks may therefore result. The transversely heated tubes do not have a temperature gradient and therefore do not suffer from this problem.

Interferences

Flame techniques are regarded as being relatively free from interferences, but some distinct classes of interference do exist. These include a few spectral interferences (e.g. Eu at 324.753 nm on Cu at 324.754 nm), ionization interferences and chemical

Table 1 Comparison of characteristic concentrations for flame and electrothermal AAS

Analyte	Flame AAS ($\mu g\ L^{-1}$)	Electrothermal AAS (pg)
Ag	30	5
Al	300	30
As	800[a],5[b]	42
Au	100	18
B	8 500	600
Ba	200	15
Be	16	2.5
Bi	200	67
Ca	13	1
Cd	11	1
Co	50	17
Cr	50	7
Cs	40	10
Cu	40	17
Fe	45	12
Ge	1 300	25
Hg	2 200[a],0.1[b]	220
K	10	2
La	48 000	7 400
Mg	3	0.4
Mn	20	6
Mo	280	12
Ni	50	20
Pb	100	30
Rb	30	10
Sb	300	60
Se	350[a],4[b]	45
Si	1 500	120
Sn	400	100
Sr	40	4
Te	300	50
Ti	1 400	70
Tl	300	50
U	110 000	40 000
V	750	42
Yb	700	3
Zn	10	1

[a] Under normal flame conditions.
[b] With vapour generation.

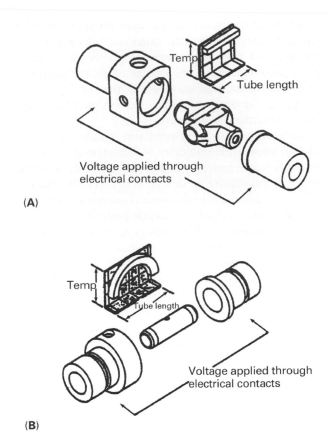

(A)

(B)

Figure 7 Electrothermal tubes available commercially: (A) transversely heated graphite atomizer (THGA), (B) longitudinally heated Massmann atomizer. From Ebdon L (1998) *Introduction to Analytical Atomic Spectrometry*. Reproduced by permission of John Wiley & Sons Limited.

interferences. Ionization interference is a vapour-phase interference (in the past often termed cation enhancement) that occurs when the sample contains large amounts of an easily ionized element. The presence of large concentrations of easily ionized elements will lead to a large concentration of electrons in the flame. These electrons prevent the ionization of the analyte and hence lead to higher atomic absorption signals. If the easily ionized element is not present in the standards, the analyte may be partially ionized (and lost analytically) and hence serious overestimates of the true concentration will be obtained. This type of interference may be overcome by adding an excess of easily ionized element to all standards and samples.

Chemical interferences may exist in several different forms:

- Formation of less volatile compounds, e.g. when phosphate is present during the determination of calcium. Calcium phosphate is refractory and hence atomization will be retarded in comparison with calcium in the standards.
- Formation of more volatile compounds, e.g. chlorides.
- Occlusion into refractory compounds. Small amounts of analyte may become trapped in a refractory substance and hence not be atomized efficiently.
- Occlusion into volatile compounds. Some compounds sublime explosively and hence atomization may be enhanced.

Most of these interferences may be overcome by using a hotter flame (e.g. nitrous oxide–acetylene); by adding a chelating reagent, e.g. EDTA, to complex preferentially with the analyte; by adding a releasing agent, e.g. lanthanum, that will combine preferentially with phosphate; or by optimizing the flame conditions and viewing height.

Electrothermal AAS was renowned for being highly prone to interferences. However, modern methods and instrumentation have decreased this problem substantially. Interferences include memory effects, chemical interferences (loss of analyte as a volatile salt, carbide formation, condensation and recombination), background absorption (smoke), and physical interferences such as those resulting from placing the sample on a different part of the tube. The stabilized temperature platform furnace (STPF) concept has gone a long way to eliminating the problem. The concept is that a 'matrix modifier' is required for interference-free determinations. This is a material that either decreases the volatility of the analyte, enabling higher ash temperatures to be achieved and hence boiling away more interference (a common example is a mix of palladium and magnesium nitrates) or increases the volatility of the matrix (for example, the introduction of air 'burns away' many interferences leaving the analytes in the atomizer). The matrix modifier may be placed in the autosampler, and added to all samples and standards. Other requirements for the STPF concept include a rapid heating rate during atomization, integrated signals (rather than peak height), a powerful background correction system (e.g. Zeeman), fast electronics to measure the transient signal, and isothermal operation. Isothermal operation means that the analyte is vaporized from the graphite surface into hot gas. In normal electrothermal AAS, the analyte leaves the hot tube wall and enters the cooler gas phase. Under these circumstances it may recombine with some other species to form a compound and thus be lost analytically. To overcome this, platforms that are only loosely in contact with the tube walls have been developed. The analytes in this case are atomized not by the resistively heated graphite tube but by the surrounding gas. The transversely heated tubes described earlier have a platform as an integral part of the tube. A modification to this technique, called probe atomization, has also been developed. Here the sample is placed on a probe and is then dried and ashed in the normal way. The probe is then removed from the tube, which is heated to the atomization temperature. The probe is then re-introduced into the hot environment.

Most analytical techniques for use with furnace work now utilize the advantages of the STPF

concept. Other interferences, e.g. formation of carbides, may be partially overcome by treating the atomizer with a carbide-forming element, for example by soaking in tantalum solution. The principle is that the tantalum occupies the active sites on the surface, thereby preventing the analyte from forming a carbide.

Methods

Numerous methods have been described for use with AAS, but the majority require the sample to be in a liquid form. Some ETAAS systems allow the analysis of solids directly (e.g. by weighing small amounts onto sampling boats that may be slotted into specialized tubes), but usually, if solids are to be analysed, acid decomposition methods are required. An alternative is the analysis of slurries, using finely ground material ($< 10~\mu m$) dispersed in a solvent. Manual agitation of the slurry ensures homogeneity of the sample, enabling a representative aliquot to be introduced. Some autosamplers come equipped with an ultrasonic agitator that will perform the same task. This method of analysis is frequently used in ETAAS.

If samples containing high dissolved (or suspended) solids are to be analysed by FAAS, a flow injection technique (in which discrete aliquots of sample are introduced) may be used. This has also been referred to as 'gulp sampling'. This prevents salting up of the nebulizer and blockage of the burner slot, which are obviously undesirable effects that have an adverse effect on sensitivity and signal stability. Flow injection techniques may also be used to preconcentrate analytes. If the sample flows through a column containing an ion-exchange resin, the analytes will be retained. Elution with a small volume of acid may yield very high preconcentration factors. This technique has been used for both FAAS and, more recently, for ETAAS, yielding limits of detection far superior to those achieved under normal conditions. A typical flow injection manifold suitable for this type of application is shown in **Figure 8**.

A more traditional method of preconcentration is solvent extraction. Flame AAS may be used as a detector for analytes in organic solvents (sensitivity

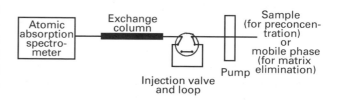

Figure 8 Typical flow injection manifold for matrix analyte preconcentration or matrix elimination.

may even be improved because of improved nebulization efficiency), but occasional blocking of the burner slot by carbon may occur. This may be removed by gentle rubbing with a noncombustible material (such as a spatula). Organic solvents may also be introduced to ETAAS, but the 'dry'-stage temperature has to be modified to prevent sample loss by spitting.

Hydride generation is a common method for the detection of metalloids such as As, Bi, Ge, Pb, Sb, Se, Sn and Te, although other vapours, e.g. Hg or alkylated Cd, may also be determined. This technique improves the sensitivity of the analysis substantially. Since the sample is in the gas phase, the sample transport efficiency is close to 100%. The hydrides atomize readily in the flame, although this approach is usually used in conjunction with a quartz T-piece in the atom cell. Methods have been developed that trap the hydrides on the surface of a graphite tube for use with ETAAS. This leads to preconcentration and further improvements in detection limit.

Chromatography has been coupled to AAS to effect speciation analysis. Here different chemical forms of an analyte are separated (either by high-performance liquid chromatography or, if they are sufficiently volatile, by gas chromatography) prior to introduction to the atomic absorption instrument. Gas chromatography is usually coupled directly with a T-piece interface arrangement, but HPLC couplings introduce the sample through the conventional nebulizer/spray chamber assembly. A small postcolumn air bleed may be necessary to compensate for the differences in flow rate between the chromatograph (1–2 mL min^{-1}) and the uptake rate of the nebulizer (5–10 mL min^{-1}). Chromatography is not frequently coupled with ETAAS because the atom cell is not well suited to continuous monitoring.

The limited linear range offered by the AAS technique may be partially overcome by using alternative, less sensitive lines. This is a useful technique that avoids the requirement to dilute samples. Care must be taken, however, to ensure that viscosity effects do not cause a difference in nebulization efficiency between samples and standards. Alternatively, burner head rotation may be used. This shortens the path in the flame through which the light beam travels and hence decreases sensitivity. However, there is often an increase in noise as a result of using this approach.

Conclusions

The sensitivity offered by the various techniques utilizing AAS for the determination of metals and metalloids is often sufficient for many applications. Direct aspiration of samples in solution into a flame atomic absorption instrument provides the analyst with rapid acquisition of data with good precision. The technique is remarkably free from interferences and those that do exist may usually be overcome by judicious choice of operating conditions. The instrumentation involved is relatively simple and, for flame work, analyses may easily be performed by non-expert operators. If increased sensitivity is required, the analyst has the option of preconcentrating the analyte using one of a number of methods prior to introduction to flame AAS, or alternatively may use electrothermal AAS. The latter technique, however, does require more experience if reliable results are to be obtained.

A potential disadvantage of using atomic absorption is that in the past instruments have been capable of determining only one analyte at a time. Modern instrumentation has improved this number to as many as six, although it must be stressed that compromise conditions (e.g. in ETAAS temperature programs) must be used. This may lead to a decrease in overall sensitivity compared with single-element determinations. However, the ongoing development of multielement AAS will certainly boost the use of this already popular and versatile technique.

See also: **Atomic Absorption, Theory; Atomic Fluorescence, Methods and Instrumentation; Atomic Spectroscopy, Historical Perspective; Light Sources and Optics.**

Further reading

Dedina J and Tsalev DL (1995) *Hydride Generation Atomic Absorption Spectrometry.* Chichester: Wiley.

Ebdon L, Evans EH, Fisher A and Hill SJ (1998) *An Introduction to Analytical Atomic Spectrometry.* Chichester: Wiley.

Harnly JM (1996) Instrumentation for simultaneous multielement atomic absorption spectrometry with graphite furnace atomization. *Fresenius Journal of Analytical Chemistry* 355: 501–509.

Haswell SJ (ed) (1991) *Atomic Absorption Spectrometry, Theory, Design and Applications,* Analytical Spectroscopy Library. Amsterdam: Elsevier.

Hill SJ, Dawson JB, Price WJ, Shuttler IL, Smith CMM and Tyson JF (1998) Atomic spectrometry update — advances in atomic absorption and fluorescence spectrometry and related techniques. *Journal of Analytical Atomic Spectrometry* 13: 131R–170R.

Vandecasteele C and Block CB (1993) *Modern Methods for Trace Element Determination.* Chichester: Wiley.

Atomic Absorption, Theory

Albert Kh Gilmutdinov, Kazan State University, Russia

ATOMIC SPECTROSCOPY
Theory

Atomic absorption spectroscopy (AAS) is a technique for quantitative determination of metals and metalloids by conversion of a sample to atomic vapour and measurement of absorption at a wavelength specific to the element of interest. Owing to high sensitivity and selectivity, the technique is widely used for fundamental studies in physics and physical chemistry: measurements of oscillator strengths, diffusion coefficients of gas phase species, partial pressure of vapours, rate constants of homogeneous and heterogeneous reactions, etc. The widest application of AAS, however, is in analytical chemistry. Nowadays it is one of the most popular techniques for trace analysis of over 65 elements in practically all types of samples (environmental, biological, industrial, etc.).

The vast majority of substances to be analysed by AAS are in the condensed phase. At the same time, atomic absorption, like any other atomic spectrometry technique, can only detect free atoms that are in the gas phase. Thus, as analyte initially present in solution at a concentration c(liq) must first be transferred into the gas phase via the atomization process to produce N(g) free atoms in an atomizer volume. The analyte atoms are then detected by absorption of radiation from a primary source at a wavelength characteristic of the element resulting in an absorption signal, A. Thus, the general scheme of AAS can be presented as follows:

$$c(\text{liq}) \xrightarrow{\text{atomization}} N(\text{g}) \xrightarrow{\text{absorption}} A$$

The primary goal of the theory of AAS is to establish the relationship between the measured analytical signal, atomic absorbance, A, and the analyte concentration, c, in the sample. Theoretical description of the two stages involves entirely different sciences: theory of atomization is based on thermodynamics, kinetics and molecular physics, while description of absorbances is based on optics and spectroscopy. Below, the two stages will be considered separately.

Atom production

In the first step of an AA determination the element of interest must be atomized. The ideal atomizer would provide complete atomization of the element irrespective of the sample matrix producing a well-defined and reproducible absorption layer of the analyte atoms. Excitation processes, however, should be minimal so that analyte and background emission noise is small. For achieving the best detection limits, the analyte vapour should not be highly diluted by the atomizer gas. Various types of electric discharges, laser radiation, electron bombardment and inductively coupled plasmas were tested as atomizers. Although not ideal, high-temperature flames and electrothermal atomizers have gained acceptance as the most common technique of atom production in AAS.

Flame atomization

The longest practiced technique for converting a sample into atoms is the spraying of a sample solution into a combustion flame. The most popular flames used in AAS are the air–C_2H_2 flame providing a maximum temperature of about 2500 K and the N_2O–C_2H_2 flame with a maximum temperature exceeding 3000 K. Atomization occurs because of the high enthalpy and temperature of the flame, and through chemical effects. Owing the generation of cyanogen radicals, which are known to be an efficient scavenger for oxygen, the nitrous oxide-acetylene flame provides not only a hotter but also a more reducing environment for atom production: the lack of oxygen moves equilibria such as MeO $\leftrightharpoons$ Me + O to the right. Therefore a N_2O–C_2H_2 flame provides greater atomization efficiencies and thus better detection limits for refractory elements, such as Al, Ta Ti, Zr, Si, V and the rare earths.

Atom production in flames is an extremely complex process consisting of many stages and involving numerous sides processes. A quantitative theory of analyte atomization in flames is absent. Qualitatively, the process can be described as follows: the solution droplets sprayed into the flame are first dried, the resulting solid microparticles become molten and vaporize or thermally decompose to produce gaseous molecules that are finally dissociated into free atoms. The atoms may further be excited and ionized and form new compounds. These processes are dependent upon the temperature and the reducing power of the flame and occur within a few milliseconds – the time required by the sample to pass through the flame. The processes taking place in flames are shown in more detail in **Figure 1**.

Electrothermal atomization

In the majority of cases this occurs in a small graphite tube where the sample to be analysed is introduced via a small aperture in the upper tube wall. Metals such as tantalum or tungsten are also used to construct electrothermal atomizers as well as nontubular geometry for the atomizer (graphite cups, rods, filaments, etc.). After a 5–50 μL droplet of the sample solution has been deposited into the graphite furnace, it is heated resistively through a series of preprogrammed temperature stages to provide: (i) drying of the droplet to evaporate the solvent (~ 370 K for about 30 s); (ii) thermal pretreatment (pyrolysis) to remove volatiles without loss of analyte (600–1500 K for 45 s) (this allows selective volatilization of matrix components that are purged from the atomizer by a flow of inert gas); (iii) analyte atomization (1500–3000 K achieved rapidly at a heating rate of about 2000 K s⁻¹). A major difference of electrothermal atomization from atomization in flames is that some matrix components are removed at the pyrolysis step and the atomization takes place within a confined geometry in an inert gas atmosphere. During atomization, the purge gas flow is shut off so that analyte atoms remain in the probing beam as long as possible.

The number of analyte atoms in the atomizer volume $N(t)$, generated during the atomization stage is described by the convolution:

$$N(t) = \int_0^t S(t') R(t, t')\, \mathrm{d}t' \qquad [1]$$

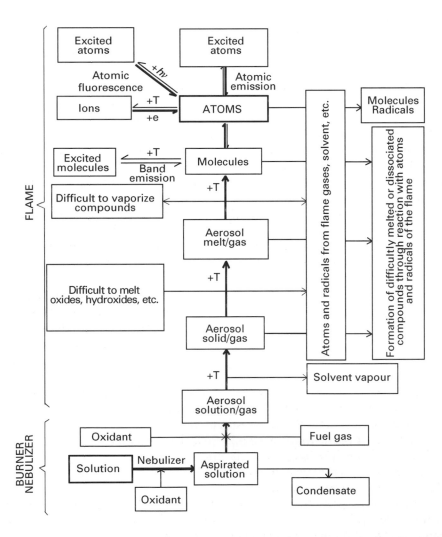

Figure 1 Schematic representation of the processes taking place in a flame. The bold arrows show the pathways for analyte atom production and thin arrows show side processes. Reproduced with permission from Welz B (1985) *Atomic Absorption Spectrometry*, 2nd edn, p. 167. Weinheim: VCH.

Here *the supply function* $S(t')$, gives the rate (atoms s^{-1}) of the primary generation of analyte atoms from the deposited sample, whereas *the removal function,* $R(t,t')$ characterizes the subsequent transfer of the vaporized atoms in the atomizer volume. In the simplest case of vaporization of a monolayer of atoms from the atomizer surface the supply function is given as:

$$S(t) = N_0 \, v \, \exp\left\{ -\frac{E}{RT(t)} - v \int_0^t \exp\left[-\frac{E}{RT(t')} \right] \mathrm{d}t' \right\}$$

[2]

where N_0 is the total number of atoms in the deposited sample, v and E are the frequency factor and the activation energy of the vaporization process, respectively, R is the gas constant and $T(t)$ is the time-dependent atomizer temperature. A typical example of the supply function in electrothermal AAS is given in **Figure 2**. The removal function $R(t,t')$ gives the probability of an analyte atom, vaporized at the time, t', to be present in the atomizer volume at a later instant, $t > t'$. Early in time, the probability is equal to unity and decreases with increasing time because of loss processes. Generally, the loss processes include concentration and thermodiffusion, and convection. The knowledge of the removal function allows calculation of the mean residence time, τ, of analyte in the atomizer volume. This parameter is defined as $\tau = \int_0^\infty R(t)\,\mathrm{d}t$ and gives the life-time of evaporated atoms in the atomizer volume. When the analyte loss is governed only by concentration diffusion through the tube ends, the residence time can be estimated by the following relationship:

$$\tau = \frac{1}{8D}\left(L^2 - \frac{a^2}{3} \right)$$

[3]

where L is the tube length, a is the linear dimension of the area occupied by the sample on the atomizer surface, D is the analyte diffusion coefficient. Typically, the analyte residence time in tube graphite atomizers is a few tenths of a second, which is about 100 times longer than in a flame. The long residence time, along with the high degree of analyte atomization caused by the reducing environment, leads to a 10- to 1000-fold increase in graphite furnace sensitivity compared with flame AAS.

The typical change in the total number of analyte atoms, $N(t)$, in the tube electrothermal atomizer obtained by convolution [1] of the supply function

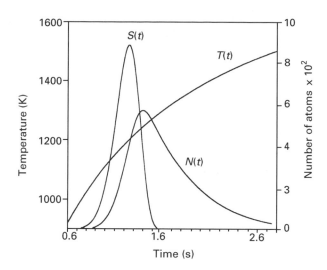

Figure 2 Supply function $S(t)$ of silver atoms computed for the change in the atomizer temperature $T(t)$ ($E = 280$ kJ mol^{-1}, $v = 10^{13}$ s^{-1}); the change in the total number of silver atoms within the atomizer volume $N(t)$.

with the diffusion removal function is given in **Figure 2**.

The actual atom production and dissipation processes in electrothermal atomizer are much more complex than that described above. In many cases the analyte is atomized as a result of a set of multistage processes taking place simultaneously. The basic processes that occur with analyte in a graphite furnace are shown schematically in **Figure 3** and listed in **Table 1**. The last column of the Table shows the elements for which the respective processes have been documented.

In addition to the above widely used atomization techniques, mention must also be made of 'the hydride generation technique'. The method is based on conversion of a hydride forming element (As, Bi, Ge, Sb, Se, Sn, Te) in an acidified sample to volatile hydride and transport of the released hydride to an atomizer (hydrogen diffuse flame, graphite furnace, heated quartz tube) where they are atomized to give free analyte atoms. Sodium borohydride is almost exclusively used as an agent for conversion of analyte to hydride:

$$NaBH_4 + 3H_2O + HCl \longrightarrow H_3BO_3 + NaCl + 8H$$
$$\xrightarrow{El^{m+}} ElH_n + H_2 \text{ (excess)}$$

where El, is the element of interest and m may or may not equal n. The advantage of sample volatilization as a gaseous hydride lies in the analyte's preconcentration and separation from the sample matrix. This results in an enhanced sensitivity and in a significant

Table 1 Selection of processes involving analyte in graphite tube atomizers

Process[a]	Ref. in Figure 3	Elements
Me (s,l) → Me (g)	1	Ag, Au, Co, Cu, Ni, Pb, Pd
MeO (s,l) → Me (g)	1	Al, Cd, In, Pb
MeO (s,l) + C(s) → Me (g)	1	Mn, Mg
Me C (s,l) → Me (g)	1	Al, Ba, Ca, Cu, Mo, Sr, V
MeOH (s,l) → Me (g)	1	Rb
MeO (s,l) → MeO (g)	1′	Al, As, Ga, In, Tl
Me (s,l) → Me$_2$ (g)	1′	Se, Co
Me (ad) → Me (g)	3	As, Cu
Me (g) → Me (s,l)	4	Ag, Au, Cu, Mg, Mn, Pd
Me (g) + O (g) → MeO (g)	5	Al, Si, Sn
MeO (g) → Me (g)	5′	Al, As, Cd, Ga, In, Mn, Pb, Zn
MeO (g) + C(s) → Me (g) + CO	6	Al, Ga, In, Tl
Me (g) + C(s) → MeC (g)	6′	Al

[a] s, l, g, ad stand for solid, liquid, gaseous and adsorbed species, respectively.

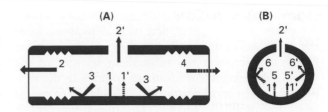

Figure 3 Schematic representation of the basic physical and chemical processes taking place in a tube electrothermal atomizer. Solid arrows denote pathways of free analyte atoms, dotted arrows show the pathways of the analytes that are bound into molecules. *Primary generation* of the analyte vapour from the site of sample deposition as an atomic (1) or a molecular (1′) species. Irreversible *loss of analyte* from the furnace through its ends (2) and through the sample dosing hole (2′) by diffusion and convection. *Physical adsorption/desorption* at the graphite surface (3). Gas phase *condensation* (4) at the cooler parts of the atomizer. *Gas phase reactions* (5) that bind free analyte atoms into stable molecules or those (5′) that increase the free atom density. *Heterogeneous reactions* of analyte vapour with the atomizer walls: includes both production (6) and loss (6′) of free atoms at the furnace wall.

reduction of interferences during atomization. The main disadvantages of the method are interferences by substances in the solution that reduce the efficiency of hydride generation.

Relative detection limits achieved by the described atomization techniques are presented in **Table 2**.

Table 2 Selected relative detection limits (µg L^{-1}) for different atomization techniques[a]

Element	Flame AA	GF AA	Hydride AA	Element	Flame AA	GF AA	Hydride AA
Ag	1.5	0.02		Mn	1.5	0.035	
Al	45	0.1		Mo	45	0.08	
As	150	0.2	0.03	Na	0.3	0.02	
Au	9	0.15		Ni	6	0.3	
B	1000	20		P	75 000	130	
Ba	15	0.35		Pb	15	0.06	
Be	1.5	0.008		Pd	30	0.8	
Bi	30	0.25	0.03	Pt	60	2	
Ca	1.5	0.01		Rb	3	0.03	
Cd	0.8	0.008		Ru	100	1	
Co	9	0.15		Sb	45	0.15	0.045
Cr	3	0.03		Se	100	0.3	0.03
Cu	1.5	0.1		Sn	150	0.2	0.03
Fe	5	0.1		Sr	3	0.025	
Ge	300	10	1	Te	30	0.4	0.03
Hg	300	0.6	0.009	Ti	75	0.35	
Ir	900	3		Tl	15	0.15	
K	3	0.008		V	60	0.1	
Li	0.8	0.06		Zn	1.5	0.1	
Mg	0.15	0.004					

[a] Detection limits are based on 98% confidence level (3 standard deviations). The values for the graphite furnace technique are referred to sample aliquots of 50 µL.
Adapted with permission from *The Guide to Techniques and Applications of Atomic Spectroscopy* (1997) p 5. Norwalk, CT, USA: Perkin-Elmer.

Atomic absorption

AAS is based on the interaction of the probing radiation beam from a primary source with gas phase analyte atoms. Two different types of primary source are used in AAS: line sources emitting narrow spectral lines of the element to be analysed and a continuum source (normally high-pressure xenon arc lamp) that produces radiation from below 180 nm to over 800 nm. In principle, the continuum source is best suited for AAS. It has a unique capability for multi-element determinations, and provides extended analytical range and inherent background correction. However, the lack of intensity found below 280 nm still limits the use of the source in research laboratories. In conventional AAS, the line radiation of the element of interest is generated in a hollow cathode lamp (HCL) or an electrodeless discharge lamp (EDL). The spectral profile, $J(\lambda)$, of the analytical line emitted by the source interacts with the absorption profile, $k(\lambda)$, of the analyte atoms in the atomizer (**Figure 4**). It is seen that radiation absorption is strongly dependent on the shape and relative position of the two spectral profiles. In order to describe quantitatively absorption, both emission and absorption profiles must first be well defined.

Emission and absorption line profiles

Three major broadening mechanisms determine emission and absorption profiles in AAS: the Doppler effect, interatomic collisions and hyperfine splitting. Other types of broadening (natural broadening, Stark broadening, etc.) are negligible. The Doppler broadening originates from the thermal agitation of emitting and absorbing atoms that results in a Gaussian-shaped spectral line (curve G in **Figure 5**). The full width, $\Delta\nu_D$, at one-half of the maximum intensity (FWHM) of the Doppler-broadened atomic line is given by:

$$\Delta\nu_D = \frac{7.16}{\lambda}\sqrt{\frac{T}{M}} \qquad [4]$$

where $\Delta\nu_D$ is in cm^{-1}, λ in nm, T in K and M in g mol^{-1}. The width is inversely proportional to the wavelength of the transition, to the square root of the absolute temperature, T, and to the reciprocal of the square root of the molecular weight of the analyte, M. The absorption line widths predicted from this equation range from 1 pm to 10 pm for flame and electrothermal atomizers. The Doppler widths of the lines emitted by HCLs and EDLs are 2–3 times lower than those for absorption profiles in atomizers.

The second major broadening mechanism in AAS is collision or pressure broadening which originates from deactivation of the excited state due to interatomic collisions. There are two types of collisions. Collisions between two analyte atoms lead to resonance broadening which is negligible in analytical

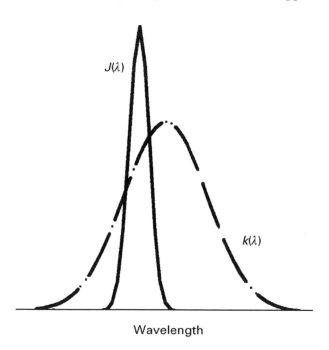

Figure 4 Schematic diagram of the emission $J(\lambda)$ and absorption $k(\lambda)$ spectral profiles typical in AAS.

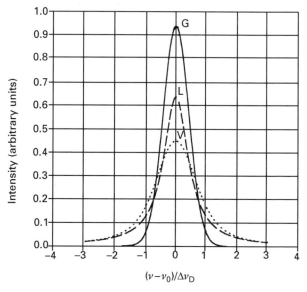

Figure 5 Normalized theoretical Gaussian (G), Lorentzian (L) and Voigt (V) line profiles versus dimensionless frequency. The FWHMs and areas of the Gaussian and Lorentzian peaks are equal; the Voigt profile is the convolution of the (G) and (L) profiles.

AAS because the analyte concentration is small. However, Lorentz broadening, which involves collisions between analyte atoms and foreign species (Ar atoms in electrothermal atomizers) called perturbers, are significant. Compared to a Doppler spectral profile, the Lorentzian profile is broader, has a lower peak height and is described by a Lorentzian function (curve L in **Figure 5**) with the FWHM given by:

$$\Delta v_\mathrm{L} = N_\mathrm{p} \sigma \sqrt{\frac{8RT}{\pi} \left(\frac{1}{M_\mathrm{A}} + \frac{1}{M_\mathrm{p}} \right)} \qquad [5]$$

Here N_p is the number density of perturbers (cm^{-3}), σ is the collisional cross-section (cm^2), M_A and M_p is the molecular weight (g mol^{-1}) of the analyte and the perturbers, respectively. The collision width predicted by Equation [5] for absorption profiles is about the same order of magnitude as the Doppler line width (1–10 pm). The Δv_L values for lines emitted by the HCLs and EDLs are much lower than those for the absorption profiles in atomizers because of low pressure in the primary sources. Therefore this part of the broadening is normally neglected in the sources. However, at high optical densities of an absorbing layer when the wings of the emission line play an important role, the Lorentzian component in the broadening of the emission line is also becoming important. In addition to broadening the lines, collisions also cause a small shift in the line centre (towards the longer wavelengths in an Ar environment) and an asymmetry in the far line wings. The shift is proportional to the collision width and theory predicts a value of 2.76 for the width-to-shift ratio (it is shown schematically in **Figure 4**).

The total spectral profile of an atomic line is given by the convolution of the Gaussian profile (due to Doppler broadening) and the Lorentzian profile (due to pressure broadening). The result is the Voigt function (curve V in **Figure 5**), $H(\omega,a)$:

$$H(\omega, a) = \frac{a}{\pi} \int_{-\infty}^{\infty} \frac{e^{-y^2}\,\mathrm{d}y}{a^2 + (\omega - y)^2} \qquad [6]$$

where $a = \sqrt{\ln 2}\,(\Delta v_\mathrm{L}/\Delta v_\mathrm{D})$ is the damping constant determining the shape of the Voigt profile: as the damping constant increases, the profile broadens and its peak height decreases. In the extreme cases of $a = 0$ and $a = \infty$, the Voigt function goes to the purely Gaussian function and purely Lorentzian function, respectively. Normally, a dimensionless frequency, ω, normalized by the Doppler width Δv_D

of the absorption profile is used for computations: $\omega = (v/\Delta v_\mathrm{D})2\sqrt{\ln 2}$. Within about 1% accuracy, the FWHM of the Voigt profile can be estimated from the following empirical equation:

$$\Delta v_\mathrm{V} = \frac{\Delta v_\mathrm{L}}{2} + \sqrt{\left[\left(\frac{\Delta v_\mathrm{L}}{2} \right)^2 + (\Delta v_\mathrm{D})^2 \right]} \qquad [7]$$

Finally, the third broadening mechanism is hyperfine splitting which is caused by isotope shifts and nuclear spin splitting. Generally a spectral line consists of n hyperfine components with relative intensities, b_j, ($\Sigma_{i=1}^{n} b_i = 1$) that are located at a distance $\Delta \omega_i$ from the component with the minimum frequency. Each component is Doppler and collision broadened and described by the Voigt function [6]. Thus in the general case the emission $J(\omega)$ and absorption $k(\omega)$ profiles are presented as follows:

$$J(\omega) = \sum_{i=1}^{n} b_i H_i \left(\frac{\omega - \Delta \omega_i}{\alpha}; a_\mathrm{e} \right),$$

$$k(\omega) = \sum_{j=1}^{n} b_j H_j (\omega - \Delta \omega_j + \Delta \omega_\mathrm{S}; a) \qquad [8]$$

where α is the ratio of the Doppler width of the emission line to the Doppler width of the absorption line. The equation takes into account that the emission profile, $J(\omega)$, is approximately $1/\alpha$ times narrower than the absorption profile, $k(\omega)$, in the scale of dimensionless frequency ω and that the absorption profile, is pressure-shifted relative to the emission profile to a value $\Delta \omega_\mathrm{S}$. For many analytical lines used in AAS the hyperfine splitting is greater than Doppler and pressure broadening and determines the total FWHM.

In graphite furnace AAS, the damping constant, a, varies from 0.17 (Be resonance line, 234.9 nm) to 3.27 (Cs, 852.1 nm) for absorption lines and its value for emission lines, a_e, in the primary sources varies from 0.01 to 0.05. A relatively small value of the a-parameter means that the emission profile is primarily determined by the Doppler broadening.

Figure 6 shows the emission and absorption spectral profiles of the lead 283.3 nm resonance line computed for the conditions of graphite furnace AAS using Equation [8]. The line consists of five hyperfine components that are clearly seen in the emission profile. Because of higher temperature and much higher pressure, the absorption profile is wider and shifted towards the longer wavelengths.

Calibration curve shapes

It has been demonstrated that in AAS the atomization efficiency is close to 100% for the majority of elements in aqueous solutions. Assuming complete atomization, the shape of calibration curves in AAS is determined by the dependence of detected atomic absorbance on the number of absorbing atoms. The

Absorbance, A, which is the analytical signal in AAS, is defined as the logarithm of the ratio of the incident radiant flux, Φ_0, to the radiant flux, Φ_{tr}, that is transmitted through the absorbing layer of the analyte atoms. Using the Beer–Lambert law [9] for expressing the transmitted radiant flux, the general relationship for the absorbance recorded by an AA spectrometer can be expressed as follows:

$$A = \lg \frac{\Phi_0}{\Phi_{\mathrm{tr}}} = \lg \left\{ \frac{\int\limits_0^h \int\limits_0^b \int\limits_{-\lambda^*}^{\lambda^*} J(x,y) J(\lambda)\, \mathrm{d}\lambda\, \mathrm{d}x\, \mathrm{d}y}{\int\limits_0^h \int\limits_0^b \int\limits_{-\lambda^*}^{\lambda^*} J(x,y)\, J(\lambda) \exp\left[-\int\limits_0^l k(\lambda;x,y,z)\, \mathrm{d}z\right] \mathrm{d}\lambda\, \mathrm{d}x\, \mathrm{d}y} \right\} \qquad [10]$$

relationship $A = f(N)$ is called the *concentration curve* and its description is one of the primary goals of theoretical AAS.

The phenomenon of radiation absorption had already been studied quantitatively early in the 18th century, mainly on liquids and crystals. The investigations resulted in the Beer–Lambert law relating the radiant flux, Φ_{tr}, transmitted through an absorbing layer to the optical properties of the layer. In modern formulation, the law applied to a spatially uniform absorbing layer of length l and a uniform parallel probing beam is written as follows:

$$\Phi_{\mathrm{tr}}(\lambda) = \Phi_0(\lambda) \exp[-\mu(\lambda)l] \qquad [9]$$

where $\Phi_0(\lambda)$ is the incident radiant flux incident at a wavelength λ, $\mu(\lambda)$ is a wavelength-dependent attenuation coefficient which is the sum of absorption, $k(\lambda)$, and scattering, $\varepsilon(\lambda)$, coefficients: $\mu(\lambda) = k(\lambda) + \varepsilon(\lambda)$. Curve $J_{\mathrm{tr}}(\lambda)$ in **Figure 6** shows the spectral profile of the lead 283.3 nm resonance line that is transmitted through a uniform layer of 3×10^{12} cm^{-2} lead atoms. It was assumed that the spatially uniform incident radiation with the spectral profile $J(\lambda)$ is absorbed by the absorption profile $k(\lambda)$, without scattering, following Equation [9]. Generally, a non-uniform radiation with intensity distribution $J(x,y)$ over the beam cross section is used to probe a non-uniform layer of absorbing species with the number density $n(x,y,z)$. With non-uniform probing beams, a radiant flux passing through a surface S within a spectral bandwidth $\{-\lambda^*, \lambda^*\}$ is expressed in terms of intensities as follows:

$$\Phi = \int\limits_S \int\limits_{-\lambda^*}^{\lambda^*} J(x,y) J(\lambda)\, \mathrm{d}x\, \mathrm{d}y\, \mathrm{d}\lambda$$

where $J(\lambda)$ is a function describing the spectral composition of the radiation.

Here, b and h are the width and the height of the monochromator entrance slit, l is the length of the absorbing layer; the limits of $-\lambda^*$ to λ^* are the spectral bandwidth isolated by the monochromator. It was assumed that radiation attenuation occurs only because of atomic absorption, i.e. $\varepsilon(\lambda) = 0$. The absorption coefficient, $k(\lambda;x,y,z)$, depends on the number density of absorbing atoms in the ground state, n_0 (cm^{-3}), and the spectral profile, $k(\lambda)$, of the absorption line, and is presented as:

$$k(\lambda;x,y,z) = \frac{2\sqrt{\pi \ln 2}\, e^2}{mc} \cdot \frac{f}{\Delta v_{\mathrm{D}}} \cdot n_o(x,y,z)\, k(\lambda)$$
$$= c(\lambda, T)\, n_o(x,y,z) \qquad [11]$$

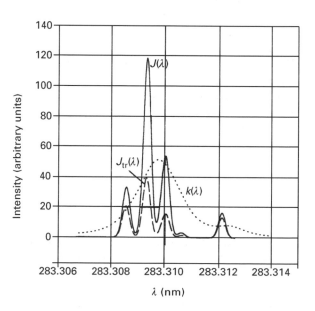

Figure 6 Emission $J(\lambda)$ and absorption $k(\lambda)$ spectral profiles of the 283.3 nm lead resonance line computed for the conditions of graphite furnace AAS ($a_{\mathrm{e}} = 0.01$, $a = 1.24$; $T = 2100$ K). $J_{\mathrm{tr}}(\lambda)$ is the emission profile that is transmitted through a uniform layer of 3×10^{12} cm^{-2} lead atoms.

Here e and m are the charge and mass of the electron, respectively, c is the velocity of light, f is the oscillator strength of the spectral transition, Δv_D is the Doppler width of the absorption line. The coefficient $c(\lambda, T)$ denotes the combination of values that depend only on the spectral features of the analysis lines.

The general relationship [10] is simplified significantly if the following assumptions are made: (i) radial distributions of the probing beam and the analyte atoms are uniform, i.e. $J(x,y) = \text{const}$, $n(x,y,z) = n(z)$; (ii) the atomizer is spatially isothermal, i.e. $T(x,y,z) = \text{const}$ and (iii) the emission line is a single and infinitely narrow line at λ_0 so that it is described by Dirac's delta-function: $J(\lambda) = J_0 \, \delta(\lambda - \lambda_0)$. Substitution of these simplifications into Equation [10] gives:

$$A = (\lg e) \int_0^l k(z; \lambda_0) \, \mathrm{d}z = (\lg e) c(\lambda_0, T) N \qquad [12]$$

where $c(\lambda_0, T)$ is a constant for given transition and for a given temperature, $N(\mathrm{cm}^{-2}) = \int_0^l n(z)\mathrm{d}z$ is the number of absorbing atoms per unit cross-sectional area along the radiation beam; in the case of uniform analyte distributions, this value is proportional to the total number of analyte atoms in the atomizer. Thus, if the above assumptions are valid, the recorded absorbance is proportional to the total number of absorbing atoms in the atomizer and the concentration curve $A = f(N)$ is a straight line. This is a basis for AAS to be an analytical technique, because the recorded signal, A, is directly proportional to the unknown number, N, of analyte atoms. Such a concentration curve computed for the case of a lead resonance line is presented by curve 1 in **Figure 7**.

The actual situation, however, differs substantially from that presented by simple Equation [12]. It follows from general relationship [10] that absorbance is dependent on three groups of factors: (1) the spectral characteristics of the analysis line (dependence of J and k on λ); (2) the cross-sectional distribution of intensity in the incident beam, $J(x,y)$; and (3) spatial distribution of the analyte and temperature within the atomizer (dependence of k on x,y,z). Thus detected absorbance, A, depends not only on the number, N, of absorbing atoms but also on the spectral features of the analysis lines used to probe the absorbing layer and the spatial distribution of analyte atoms and radiation intensity in the probing beam:

$$A = f(N; \text{Spectral, Spatial}) \qquad [13]$$

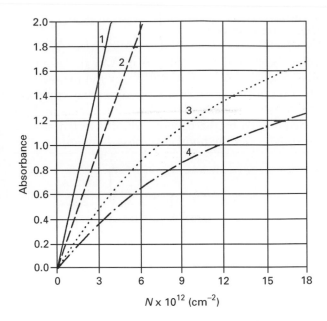

Figure 7 The evolution of the 283.3 nm lead concentration curve as the spectral and spatial features are successively taken into account. Emission line is: single and infinitely narrow (1), single and broadened (2); all the spectral features of the line are taken into account including hyperfine splitting and the pressure shift (3); (4) as (3) plus absorbing layer is assumed to be non-uniform in the radial cross section with the analyte number density at the atomizer bottom 4 times greater than that in the upper part of the absorbing layer.

The effect of spectral features of the analysis line and spatial nonuniformities of the absorbing layer is illustrated in **Figure** 7. Curve 2 is the lead concentration curve when the resonance line is assumed to be a single but broadened one; accounting for all the spectral features of the emission and absorption profiles, including hyperfine structure and pressure shift, results in curve 3. It can be seen that broadening of the analysis line lead to a bending of the concentration curve and decreasing of its slope, the major effect being the hyperfine broadening. Additional accounting for the fact that the analyte is distributed nonuniformly in the atomizer cross section leads to further curvature and a decrease in the slope of concentration curve (curve 4). **Figure** 7 illustrates clearly the importance of the *spectral* and *spatial* features in relationship [10]. Recent research shows, however, that these undesirable dependencies can be avoided if transmitted intensities are measured with spectral, spatial and temporal resolutions. The *multi-dimensional* absorbance detected in such a way is dependent only on the number of absorbing atoms irrespective of their spatial distribution and spectral features of the primary source.

Interferences

An interference effect is a change of the analytical signal by the sample matrix as compared to the reference standard, typically an acidified aqueous solution. There are two types of interference in spectrochemical analysis, *spectral* and *nonspectral*. Spectral interference originates from incomplete isolation of the radiation emitted or absorbed by the analyte from other radiation detected by the spectrometer. Any reasons other than specific atomic absorption that cause radiation attenuation in equation [9] result in spectral interferences. They arise from: absorption of radiation by overlapping molecular bands or atomic lines of concomitants; scattering of source radiation by nonvolatilized microparticles (smoke, salt particles, condensed microdrops); foreign line absorption if the corresponding radiation happens to be emitted by the radiation source, in addition to the analysis line, within the spectral bandwidth of the monochromator. The first two sources of spectral interferences are most common and described as background attenuation or background absorption. In general, they are more pronounced at shorter wavelengths. Most background attenuation can be distinguished from the analyte absorption because the element only absorbs in the very narrow spectral region while the background attenuation is less specific and extends over a considerably broader wavelength band. Instrumental methods of correction for background attenuation effects include compensation using a continuum source and the Zeeman effect.

For interferences other than spectral, the analyte itself is directly affected. The nonspectral interferences are best classified according to the stage at which the particular interference occurs, i.e. solute-volatilization and vapour-phase interferences. A nonspectral interference is found when the analyte exhibits a different sensitivity in the presence of sample concomitants as compared to the analyte in a reference solution. The difference in the signal may be due to: analyte loss during the thermal pretreatment stage in the electrothermal atomizer: analyte reaction with concomitants in the condensed phase to form compounds that are atomized to a lesser extent, analyte ionization or change the degree of ionization caused by concomitants.

Reactions of the analyte with the tube material (graphite) or the purge gas are not normally considered to be interferences because they influence the analyte in the sample and in the standard to the same degree. A nonspectral interference is in many instances best detected by the use of the analyte addition technique. An interference exists if the slope of the additions curve is different from that of the calibration curve. For most systems, nonspectral interferences in electrothermal AAS can be effectively eliminated by adding a proper matrix modifier ($Pd-Mg(NO_3)_2$ is the most common modifier) that delays analyte volatilization until the atomizer temperature is sufficiently high and steady for efficient atomization, and buffers the gas phase composition.

List of symbols

a = damping constant; a_e = damping constant for emission lines; a = the linear dimension of the area occupied by the sample on the atomizer surface; A = absorbance; b = width of monochromator entrance slit; $c(liq)$ = concentration of analyte in solution; c = velocity of light; D = analyte diffusion coefficient; e = charge of electron; E = activation energy; f = oscillator strength; h = height of monochromator entrance slit; $J(\lambda)$ = spectral profile; $k(\lambda)$ = absorption profile; l = length of absorbing layer; L = tube length; m = mass of electron; M = molecular weight of the analyte; M_A = molecular weight of analyte; M_p = molecular weight of perturbers; n_o = number density of absorbing atoms in the ground state; N = number of analyte atoms; $N(g)$ = number of free atoms in an atomizer volume; N_0 = total number of atoms in the deposited sample; N_P = number density of perturbers; $N(t)$ = number of analyte atoms in the atomizer volume; R = gas constant; $R(t,t')$ = removal function; $S(t')$ = supply function; t = time analyte atom present in atomizer volume; t' = time of vaporization; T = absolute temperature; $T(t)$ = time-dependent atomizer temperature; α = ratio of the Doppler width of the emission to the absorption line; Δv_D = width of Doppler-broadened atomic line; Δv_L = width of Lorentzian-broadened atomic line; $\varepsilon(\lambda)$ = scattering coefficient; λ = wavelength; λ = wavelength; $\mu(\lambda)$ = wavelength-dependent attenuation coefficient; v = frequency factor; σ = collisional cross section; τ = mean residence time; Φ = transmitted radiant flux; Φ_0 = incident radiant flux; ω = frequency (dimensionless).

See also: **Atomic Absorption, Methods and Instrumentation; Atomic Fluorescence, Methods and Instrumentation; Fluorescence and Emission Spectroscopy, Theory.**

Further reading

Alkemade CThJ, Hollander T, Snelleman W and Zeegers PJTh (1982) *Metal Vapours in Flames*. Oxford: Pergamon Press.

Chang SB and Chakrabarti CL (1985) Factors affecting atomization in graphite furnace atomic absorption spectrometry. *Progress in Analytical Atomic Spectroscopy* 8: 83–191.

Ebdon L (1982) *An Introduction to Atomic Absorption Spectroscopy*. London: Heyden.

Haswell SJ (Ed.) (1991) *Atomic Absorption Spectrometry*. Amsterdam: Elsevier.

Holcombe JA and Rayson GD (1983) Analyte distribution and reactions within a graphite furnace atomizer. *Progress in Analytical Atomic Spectroscopy* 6: 225–251.

L'vov BV (1970) *Atomic Absorption Spectrochemical Analysis*. London: Adam Hilger.

Mitchell ACG and Zemansky MW (1961) *Resonance Radiation and Excited Atoms*. Cambridge: Cambridge University Press.

Slavin W (1991) *Graphite Furnace AAS – A Source Book*, 2nd edn. Norwalk, CT: Perkin-Elmer Corporation.

Styris DL and Redfield DA (1993) Perspectives on mechanisms of electrothermal atomization. *Spectrochimica Acta Reviews* 15: 71–123.

Welz B (1985) *Atomic Absorption Spectrometry*, 2nd edn. Weinheim: VCH.

Winefordner JD (ed) (1976) *Spectroscopic Methods for Elements*. New York: Wiley-Interscience.

Atomic Emission, Methods and Instrumentation

Sandra L Bonchin, Los Alamos National Laboratory, NM, USA

Grace K Zoorob, Biosouth Research Laboratories, Inc., Harahan, LA, USA

Joseph A Caruso, University of Cincinnati, OH, USA

ATOMIC SPECTROSCOPY
Methods & Instrumentation

Atomic emission spectroscopy is one of the most useful and commonly used techniques for analyses of metals and nonmetals providing rapid, sensitive results for analytes in a wide variety of sample matrices. Elements in a sample are excited during their residence in an analytical plasma, and the light emitted from these excited atoms and ions is then collected, separated and detected to produce an emission spectrum. The instrumental components which comprise an atomic emission system include (1) an excitation source, (2) a spectrometer, (3) a detector, and (4) some form of signal and data processing. The methods discussed will include (1) sample introduction, (2) line selection, and (3) spectral interferences and correction techniques.

Atomic emission sources

The atomic emission source provides for sample vaporization, dissociation, and excitation. The ideal excitation source will allow the excitation of all lines of interest for the elements in the sample, and do this reproducibly over enough time to encompass full elemental excitation. Excitation sources include but are not limited to (1) inductively coupled plasma (ICP), (2) direct current plasmas (DCP), (3) microwave induced plasmas (MIP), and (4) capacitively coupled microwave plasmas (CMP). Glow discharges are utilized for direct solids analyses, but will not be discussed here. An analytical plasma is a high-energy, slightly ionized gas (about 0.01 to 0.1% ionized).

Inductively coupled plasmas

The most commonly used ion source for plasma spectrometry, the ICP, is produced by flowing an inert gas, typically argon, through a water-cooled induction coil which has a high-frequency field (typically 27 MHz) running through it (**Figure 1**). The alternating current in the coil has associated with it a changing magnetic field, which induces a changing electric field. The flowing gas is seeded with electrons by means of a Tesla coil. These electrons undergo acceleration by the electric field, and gain the energy necessary to excite and ionize the gaseous atoms by collision. This produces the plasma, self-sustaining as long as the RF and gas flows continue.

Sample particles entering the plasma undergo desolvation, dissociation, atomization, and excitation. The ICP has sufficiently long residence times and high enough temperatures so that the sample solvent is completely vaporized, and the analyte reduced to free atoms, which undergo excitation. This excitation results in the emission of light at specific frequencies for elements in the sample, which is

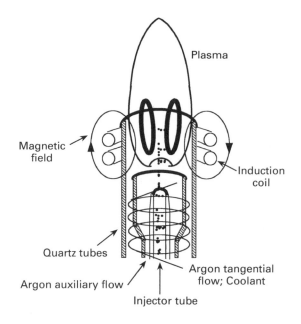

Figure 1   Inductively coupled plasma and torch schematic.

proportional to their concentration. An emission spectrometer separates the frequencies of light into discrete wavelengths, which are unique to given elements in the sample.

Direct current plasmas

In a direct current plasma, a dc current passing between two electrodes heats the plasma gas, again typically argon, and produces a discharge. The most common version is the 3-electrode system (**Figure 2**). This system has argon flowing around two graphite anodes and a tungsten cathode to produce the plasma. The sample is introduced between the anodes. Vaporization, atomization, ionization, and excitation occur. This technique is more tolerant of samples containing a high proportion of solids than the ICP method. However, it is less efficient due to lower plasma temperatures, and the electrodes need to be replaced frequently.

Microwave induced plasmas

A microwave induced plasma (MIP) is an electrodeless discharge generated in a glass or quartz capillary discharge tube, often in a resonant cavity. These tubes generally have an inner diameter of the order of a few millimetres, and the plasma gas is an inert gas, such as helium or argon. The resonant cavity, which is hollow and of the order of a few centimetres diameter, allows coupling of the microwave power into the plasma gas flowing through the capillary discharge tube. The microwave power supply operates at a frequency of 2.45 GHz. Microwave plasmas

can be produced at atmospheric pressure, if the design of the cavity allows. Helium plasmas typically require reduced pressure unless a 'Beenakker cavity' is employed (**Figure 3**). MIPs use lower power levels (hundreds of watts) than those required by the DCP and ICP. However, due to the decreased size of the MIP, the power densities produced are comparable. Argon has an ionization potential of 15.75 eV, which is not sufficient in the argon ICP to produce the excitation ionization energy needed to ionize some elements efficiently, such as the nonmetals. Helium plasmas are more efficient for these problem elements, as the ionization potential of helium is 24.6 eV. The MIP is not very adept at handling liquid samples, due to the low powers employed. However, at higher powers above 500 watts similar performance to the ICP may be obtained. The general lack of availability of these sources as part of commercially available instrumentation, has limited their application and use.

Capacitively coupled microwave plasmas

A capacitively coupled microwave plasma is formed using a magnetron to produce microwave energy at 2.45 GHz. This is brought to a hollow coaxial electrode via coaxial waveguides. The microwave power is capacitively coupled into the plasma gas, usually argon or nitrogen, via the electrode. This is an atmospheric pressure source with a small plasma volume. Power levels can range from 10 to 1000

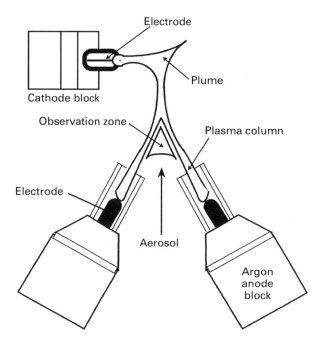

Figure 2   Direct current plasma schematic.

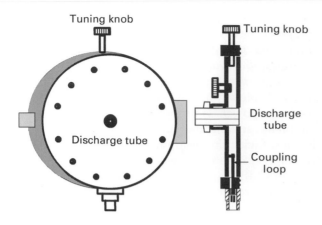

Figure 3 Microwave induced plasma schematic illustrating the 'Beenakker Cavity' design.

watts, with frequencies from 200 kHz to 30 MHz. Unlike other MIPs, the CMP is able to handle greater amounts of solvent. However, the CMP has been found to be easily contaminated due to the microwave energy being conducted through a coaxial waveguide to the electrode, which must be replaced regularly.

Spectrometers

The atomic emission source will produce unstable, excited atoms or ions, which spontaneously return to a lower energy state. The emission spectrum is produced when a photon of energy is generated during this transition. The basic assumption is that the emitted energy is proportional to the concentration of atoms or ions in the sample. The measurement of this energy is performed using the optics of the spectrometer to isolate the characteristic elemental emission wavelengths, and to separate this radiation from the plasma background. The spectrometer will need a high resolving power to be able to separate the lines of interest from adjacent spectral lines (at least 0.1 nm). The spectrometer consists of (1) a dispersive element, such as a grating, and (2) an image transfer assembly, which contains the entrance and exit slits, and mirrors or lenses.

The grating provides dispersion of the wavelength range of interest over a given angular range. Some commonly used grating spectrometers include (1) the Paschen–Runge spectrometer, which is used in both sequential and simultaneous instruments and has the advantage of extensive wavelength coverage, (2) the echelle grating spectrometer, with excellent dispersion and resolution with a small footprint, and (3) the Ebert and Czerny–Turner spectrometers, which are similar except that the latter has two mirrors to the Ebert's one.

The slit allows a narrow line of light to be isolated. The number of lines will represent the various wavelengths emitted by the plasma, with each line corresponding to the image produced by the spectrometer slit, and each wavelength corresponding to a specific element. There are two slits used: (1) the primary slit, which is where light enters the spectrometer, and (2) the secondary slit, where light exits, producing a line isolated from the rest of the spectrum. The imaging system consists either of lenses or of concave mirrors.

One of the most useful aspects of atomic emission spectrometry is its capability for multi-element analysis. This can be achieved using either the sequential monochromator, where elements in a sample are quickly read one at a time, or the polychromator, which allows the simultaneous measurement of the elements of interest.

Monochromators

Multi-element determinations using a monochromator must be sequential, as the monochromator can only observe one line at a time due to it only having one secondary slit. The slit can be set to scan the wavelengths, which is slower than in a simultaneous instrument, but allows for the selection of a wider range of wavelengths. A scanning monochromator user a movable grating to find known individual spectral lines at fixed positions, typically a Czerny–Turner configuration. Although this allows the observance of only one individual wavelength at a time, it also allows all wavelengths in the range of the spectrometer to be observed.

Polychromators

Polychromators have a permanently fixed secondary slit for certain individual wavelengths (individual elements). Each slit will have its own detector. This allows for truly simultaneous analyses of selected elements, resulting in much quicker analyses than the monochromator for elements that have a slit installed. However, the selection of possible analysis lines is limited. A simplified diagram of the beam path in a polychromator is shown in **Figure 4**. A polychromator can focus the emission lines on the circumference of the Rowland circle by using concave gratings; typically the Paschen–Runge configuration is used. In summary, the polychromator has a throughput advantage while the sequential scanning monochromator has flexibility as its advantage.

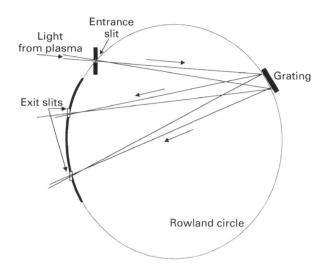

Figure 4 Beam path through a polychromator.

Detectors

Photomultiplier tubes

A photomultiplier tube (PMT) consists of a photosensitive cathode, several dynodes and a collection anode. The dynodes are responsible for the increase in signal by electron multiplication. PMTs see the elemental line intensity per unit time proportionally with current and have wide dynamic ranges. A PMT, housed in a suitable mechanical movement, can scan the range of wavelengths in a spectrum sequentially, which involves longer analysis times. Alternatively, there can be a series of PMTs, each collecting the signal at discrete wavelengths at its assigned exit slit. Unfortunately, it always takes one PMT for each wavelength of light observed, whether or not the PMT is fixed or roving. Modern instruments have taken advantage of the advent of multichannel solid-state detectors to provide more flexibility in multi-element analyses.

Charge-coupled and charge-injection devices

The charge-coupled and charge-injection devices are solid-state sensors with integrated silicon circuits. They are similar in that they both collect and store charge generated by the light from the emissions in metal-oxide semiconductor (MOS) capacitors. The amount of charge generated in a charge transfer detector is measured either by moving the charge from the detector element where it is collected to a charge-sensing amplifier (CCD), or by moving it within the detector element and measuring the voltage change induced by this movement (CID). The CCD is susceptible to 'blooming' in the presence of too much light, since there can be an overflow of charge from a full pixel to an adjacent one. CCDs are best used in very sensitive, low light level applications. The CID does not suffer from 'blooming' due to its method of nondestructively measuring the photon-generated charge. The CID allows the monitoring of any wavelength between 165 and 800 nm, whereas the range of the CCD is between 170 and 780 nm.

Methods

Sample introduction

Sample introduction into the plasma is a critical part of the analytical process in atomic emission spectroscopy (AES). Since the ICP is the most commonly used source, the sample introduction schemes described below will focus more on it than the other sources mentioned previously. Sample is carried into the plasma at the head of a torch by an inert gas, typically argon, flowing in the centre tube at 0.3–1.5 L min^{-1}. The sample may be an aerosol, a thermally or spark generated vapour, or a fine powder. Other approaches may also be taken to facilitate the way the analyte reaches the plasma. These procedures include hydride generation and electrothermal vaporization.

Torches One of the torch configurations more commonly used with the ICP is shown in **Figure 1**. The plasma torch consists of three concentric quartz tubes through which streams of argon flow. The nebulizer gas, which carries the analyte into the plasma, flows in the centre tube. The auxiliary gas flows around the centre tube and adjusts the position of the plasma relative to the torch. The coolant gas streams tangentially through the outer tube, serving to cool the inside walls and centre of the torch, and stabilize the plasma.

Nebulizers The great majority of analyses in ICP AES are carried out on liquid samples. The most convenient method for liquids to be introduced into the gas stream is as an aerosol from a nebulizer. The aerosol may be formed by the action of a high-speed jet across the tip of a small orifice or by the use of an ultrasonic transducer. A spray chamber is usually placed after the nebulizer to remove some of the larger droplets produced, and thereby improves the stability of the spectral emission.

The most commonly used nebulizer designs are pneumatic and ultrasonic, although other types (electrostatic, jet impact, and mono-dispersive generators) have been described. The selection of the appropriate nebulizer depends upon the characteristics of the sample: mainly density, viscosity, organic content, total dissolved solids, and total sample volume. Additionally, the performance of a particular nebulizer can be described using several attributes such as droplet size distribution, efficiency, stability, response time, tendency to clog, and memory effects.

Concentric nebulizers Solution sample introduction has been associated with pneumatic nebulizers (concentric and cross-flow) almost universally for routine analysis, due to their simplicity and low cost. However, they provide low analyte transport efficiency (< 5%) to the plasma, and may be prone to clogging.

The most widely used ICP nebulizer is the one-piece Meinhard concentric nebulizer (**Figure 5A**). It is a general-purpose nebulizer with low tolerance for total dissolved solids, used for applications requiring low nebulizer gas flows. It operates in the free running mode whereby the solutions are drawn up by the pressure drop generated as the nebulizer gas passes through the orifice. The viscosity of the solution and the vertical distance through which the liquid is lifted affect the rate of liquid transfer. Although the concentric nebulizer is easy to use, the transport efficiency is low, owing to the wide range of droplet sizes produced. In addition, nebulizer blockage may occur due to the presence of suspended solids becoming lodged in the narrow, central sample uptake capillary. Filtering or centrifuging the sample may minimize the risk of blockage. If a sample with a high dissolved-solids content dries in the nebulizer, it may also cause nebulizer blockage or interfere with the operation of the nebulizer by decreasing the signal. Frequent cleaning of the nebulizer solves this problem.

Cross-flow nebulizers The cross-flow nebulizer shows much of the same general behaviour of the concentric nebulizers. The cross-flow nebulizer is less prone to blockage and salting effects, although they may still occur as the sample solution passes through a capillary. The cross-flow nebulizer operates when a horizontal jet of gas passes across the top of a vertical tube (**Figure 5B**). The reduced pressure that is generated draws the liquid up the tube, where, at the top, it is disrupted into a cloud of fine droplets.

Babington type nebulizer The Babington nebulizer is designed to allow a film of liquid containing sample to flow over the surface of a sphere (**Figure 5C**). A gas, which is forced through an aperture beneath the film, produces the aerosol. This design features the liquid sample flowing freely over a small aperture, rather than passing through a fine capillary, and is

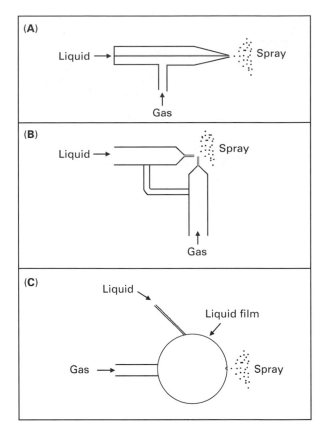

Figure 5 Schematic of the (A) concentric nebulizer, (B) cross-flow nebulizer and (C) Babington nebulizer.

therefore more tolerant of high dissolved-solids. This kind of nebulizer can be used to introduce slurries into the system since the delivery of the sample is not constrained by a capillary. However, the Babington-type nebulizer shows extensive memory effects since the solution is allowed to wet the entire face of the sphere. A modification of the design is also in use, whereby the liquid sample passes through a V-groove and a gas is introduced from a small hole in the bottom of the groove. The main advantage of the V-groove nebulizer is its resistance to blockage. However, the design produces aerosol less efficiently, and it is of coarser size distribution than that produced by other concentric nebulizers.

Frit-type nebulizer The concentric and cross-flow nebulizers are inefficient at the 1 mL min^{-1} flow rate at which they usually operate, producing only about 1% of droplets of the correct size to pass into the plasma. An alternative design is the frit nebulizer, which produces droplets with a mean size of 1 μm. The glass frit nebulizer is depicted schematically in **Figure 6**. The sample flows over the fritted glass disc as the nebulizing gas is passed through the frit. This

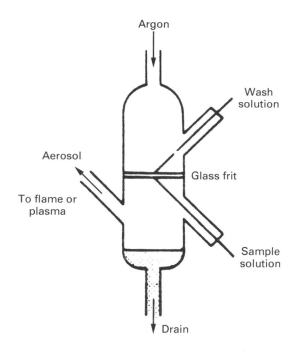

Figure 6 Glass frit nebulizer schematic. Reprint permission from Ingle JD and Crouch SR (1988) *Spectrochemical Analysis*. p. 193, Figure 7-5. Englewood Cliffs, NJ: Prentice Hall.

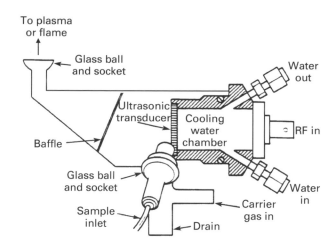

Figure 7 Ultrasonic nebulizer schematic. Reprint permission from Ingle JD and Crouch SR (1988) *Spectrochemical Analysis*. p. 194, Figure 7-6. Englewood Cliffs, NJ: Prentice Hall.

design provides an excellent fine aerosol, but at the expense of the fritted disc clogging over time. The transport efficiency is very high compared to the pneumatic-type nebulizers but the operation is at flows of 100 µL min^{-1} or less.

Ultrasonic nebulizer The ultrasonic nebulizer also produces fine aerosol and has a high sample-transport efficiency (**Figure 7**). The aerosol is generated when the solution is fed to the surface of a piezoelectric transducer which operates at a frequency between 0.2 and 10 MHz. Vibrations of the transducer cause the solution to break into small droplets which are then transported to the plasma. The ultrasonic nebulizer offers highly efficient aerosol production independent of the carrier-gas flow rate. With this nebulizer, more analyte can be transported into the plasma at a lower nebulizer gas flow rate than that seen with pneumatic nebulizers. This increases the sensitivity and improves detection limits since samples have a longer residence time in the plasma. However, the high efficiency also allows more water to enter the plasma, producing a cooling effect, which may decrease analyte ionization. Hence these nebulizers require desolvation to be realistically utilized. Apart from the analytical shortcomings of the ultrasonic nebulizer, the high cost of the commercial systems may be prohibitive.

Electrothermal vaporization Electrothermal vaporization can introduce liquid, solid, and gaseous (by trapping) samples to the plasma with high analyte transport efficiency (20–80%) and improved sensitivity over pneumatic nebulization. A small amount (5–100 µL) of sample is deposited into an electrically conductive vaporization cell (**Figure 8**). Initially, a low current is applied to the cell, which causes resistive heating to occur and dries the sample. A high current is then passed for a short time (typically 5 s) to completely vaporize the sample. An optional 'ashing' stage may be used to remove some of the matrix prior to the analyte vaporization stage. A stream of argon gas is passed through the unit and carries the sample vapour to the plasma. A variable current supply is required for controlling cell heating, and the sequence of heating steps is carried out using an electronic control system. Electrothermal vaporization allows sample matrix components, including the solvent, to be separated from the analytes of interest through judicious selection of temperature programming steps. This may reduce oxide formation and the number of spectroscopic interferences. Additionally, samples with high salt content and limited volume can be analysed with electrothermal vaporization.

Hydride generation The problems associated with liquid sample introduction, and the inefficiencies in analyte transport can be overcome by presenting the sample in a gaseous form to the plasma. Introducing the analyte in a gaseous form eliminates the use of a nebulizer and a spray chamber. It also provides nearly 100% analyte transport efficiency, which subsequently improves the detection limits. Additional

Drying stage

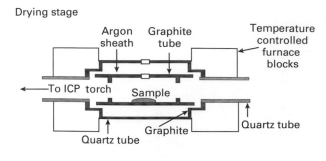

Atomization stage

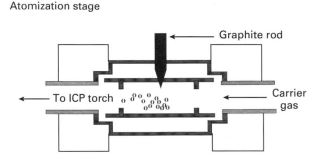

Figure 8 Illustration of electrothermal vaporization graphite furnace.

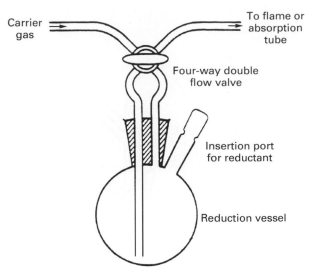

Figure 9 Hydride generation apparatus. Reprint permission from Ingle JD and Crouch SR (1988) *Spectrochemical Analysis*. p. 280, Figure 10-6. Englewood Cliffs, NJ: Prentice Hall.

benefits include no blockage problems, the possibility of matrix separation and analyte preconcentration, and the absence of water, which reduces the levels of many polyatomic ion species. The hydrides of the elements arsenic, bismuth, germanium, lead, antimony, selenium, and tin are easily formed in acidic solutions and are gaseous at room temperature. The most frequently used reaction is:

$$NaBH_4 + 3H_2O + HCl \rightarrow H_3BO_3 + NaCl + 8H$$
$$\rightarrow EH_n + H_2$$

where E is the hydride forming element of interest. **Figure 9** shows an apparatus for hydride generation. It allows the mixing of the sample and the reducing agent, followed by the transport of the volatile analyte species (hydride) by the carrier gas to the plasma.

The gaseous sample introduction line is connected directly to the central tube of the plasma torch, eliminating the need for the conventional nebulizer and spray chamber. Additional equipment to handle gas mixing and dilution at controlled flow rates may be required.

Direct solids introduction Direct insertion of solid samples into the ICP can be used for metal powders, salts, and geological samples. The sample is placed on a wire loop or cup made from graphite, molybdenum, tantalum, or tungsten. The probe is then moved along the axis of the torch and closer to the plasma where drying takes place. At the completion of this stage, the device is propelled rapidly into the core of the plasma and measurements of the analytes can be taken.

Line selection

The choice of which line to use for a given sample type is a difficult one, as most elements have many lines available for analysis. Not only should the intensity of the line be considered, but also whether the line is free of spectral interferences from both the plasma and other sample constituents. Additionally, the nature of the emission, atomic or ionic, should be considered. The emission could originate from an excited neutral atom, which is termed an atomic transition, or an excited ionized form of the element, which is called an ionic transition. While analysis of elements undergoing transitions of the ionic type tend to experience fewer consequences due to changes in the plasma operating conditions, there are elements that produce no ionic lines, such as aluminium and boron. There are very good reference books which list tables of spectral lines and are included in the Further reading section. Often modern instrumentation has sufficiently 'intelligent' software to assist the operator with line selection.

Spectral interferences and correction techniques

A spectrometer with good resolution will help greatly in the separation of adjacent lines from the

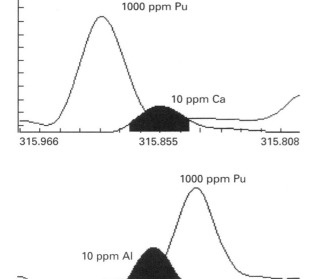

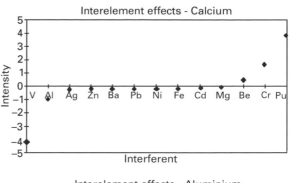

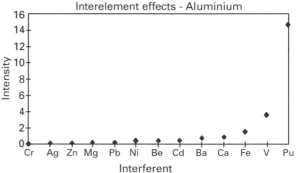

Figure 10 Interelement effects of plutonium on calcium (top) and aluminium (bottom).

Figure 11 Interelement equivalent concentration correction factors for several interfering elements on calcium (top) and aluminium (bottom) as calculated by a program written in-house by Gerth DJ.

spectral line of interest. The effect of partial overlap can be minimized in some cases with a high-resolution spectrometer, and in other cases, can be overcome using correction techniques.

When the interference is from the plasma emission background, there are background correction options available with most commercial instrumentation. The region adjacent to the line of interest can be monitored and subtracted from the overall intensity of the line. If direct spectral overlap is present, and there are no alternative suitable lines, the interelement equivalent concentration (IEC) correction technique can be employed. This is the intensity observed at an analyte wavelength in the presence of 1000 mg L^{-1} of an interfering species. It is expressed mathematically as:

$$\text{IEC} = (I_{\text{init}}/I_{\text{a}}) \times C_{\text{a}}$$

where the correction is in milligrams of analyte per litre of solution, and I_{init} is the intensity read at the analyte wavelength in the presence of 1000 mg L^{-1} of the interfering species, I_{a} is the intensity the instrument will produce for a certain analyte concentration at the analyte wavelength, and C_{a} is the concentration in mg L^{-1} used to give I_{a}. This correction can be applied to all the analytes in a method and

some instrument manufacturers' software will automatically perform the necessary corrections. Examples of the effect of 1000 ppm plutonium as the interfering element on calcium and aluminium are shown in **Figure 10**. This shows the subarrays obtained for separate solutions of plutonium, calcium, and aluminium. Graphical representations of interfering elements on calcium and aluminium are shown in **Figure 11**. These graphs are produced by a program written at Los Alamos National Laboratory by DJ Gerth, and illustrate which interfering elements have the strongest effect on the analytes. Those elements with the strongest deviation from the average are those needing the interelement correction.

List of symbols

C_{a} = analyte concentration used to give I_{a}; I_{a} = intensity at certain analyte concentration and wavelength; I_{init} = intensity in presence of 1000 mg L^{-1} of the interfering species.

See also: **Electronic Components, Applications of Atomic Spectroscopy; Fluorescence and Emission Spectroscopy, Theory; Inductively Coupled Plasma Mass Spectrometry, Methods.**

Further reading

Boumans PWJM (1987) *Inductively Coupled Plasma Emission Spectroscopy – Part 1, Chemical Analysis Series*, Vol. 90. Chichester: Wiley.

Edelson MC, DeKalb EL, Winge RK and Fassel VA (1986) *Atlas of Atomic Spectral Lines of Plutonium Emitted by an Inductively Coupled Plasma*. Ames Laboratory, US Department of Energy.

Ingle JD and Crouch SR (1988) *Spectrochemical Analysis*. Englewood Cliffs, NJ: Prentice-Hall.

Montaser A and Golightly DW (eds) (1992) *Inductively Coupled Plasmas in Analytical Atomic Spectrometry* Weinheim: VCH.

Slickers K (1993) *Automatic Atomic Emission Spectroscopy*. Brühlsche Universitätsdruckerei.

Sneddon J (ed) (1990) *Sample Introduction in Atomic Spectroscopy, Analytical Spectroscopy Library*, #4. Amsterdam: Elsevier.

Sweedler JV, Ratzlaff KL and Denton MB (eds) (1994) *Charge-Transfer Devices in Spectroscopy*. Weinheim: VCH.

Thompson M and Walsh JN (1989) *Handbook of Inductively Coupled Plasma Spectrometry*. Glasgow: Blackie.

Varma A (1991) *Handbook of Inductively Coupled Plasma Atomic Emission Spectroscopy*. Boca Raton, FL: CRC Press.

Winge RK, Fassel VA, Peterson VJ and Floyd MA (1985) *Inductively Coupled Plasma-Atomic Emission Spectroscopy, An Atlas of Spectral Information*. Amsterdam: Elsevier.

Atomic Fluorescence, Methods and Instrumentation

Steve J Hill and **Andy S Fisher**, University of Plymouth, UK

ATOMIC SPECTROSCOPY
Methods & Instrumentation

Introduction

Atomic fluorescence spectroscopy (AFS) has been used for elemental analysis for several decades. It has better sensitivity than many atomic absorption techniques and offers a substantially longer linear range. However, despite these advantages, it has not gained the widespread usage of atomic absorption or emission techniques. In recent years, the use of AFS has been boosted by the production of specialist equipment that is capable of determining individual analytes at very low concentrations (at the ng L^{-1} level). The analytes have tended to be introduced in a gaseous form and hence sample transport efficiency to the atom cell is very high. This article describes the instrumentation and methods available for atomic fluorescence spectroscopy, although it should be emphasized that much of the instrumentation associated with this technique is often very similar to that used for atomic absorption spectroscopy (AAS). A schematic diagram of the different parts of an AFS instrumentation is shown in **Figure 1**. It can be seen that the light source, atom cell, line isolation device, detector and readout system used for AFS are very similar to those used in AAS, although there may be subtle differences and the components may be less sophisticated, as described below.

Light sources

As can be seen from **Figure 1**, light sources for AFS are placed at right angles to the detector. This is to ensure that incidental radiation from the source is not detected. Since a more intense source will lead to greater sensitivity, in AFS a standard hollow-cathode lamp (HCL) is often insufficient for the majority of applications. A range of alternatives has therefore been developed. An electrodeless discharge lamp (EDL) is suitable because it has an intensity 200–2000 times greater than that of a hollow-cathode lamp. Boosted-discharge hollow-cathode lamps (BDHCL)

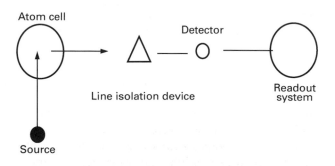

Figure 1   A schematic representation of an AFS instrument.

are similar to standard hollow-cathode lamps but can be operated at greatly increased current. The light output is consequently several times more intense. A similar effect may be obtained by pulsing a standard hollow-cathode lamp to 100–1000 mA, although this will reduce the lifetime of the lamp. Continuum sources such as 150–500 W xenon lamps have also been used, but there are several problems associated with their use. The most obvious drawback is the relatively poor intensity offered at each individual wavelength by the black body irradiator. The problem is particularly bad in the ultraviolet region of the spectrum. Scatter has also been reported as being problematic.

Vapour discharge lamps have been available for some elements (e.g. Cd, Ga, Hg, In, K, Na and Tl) since the inception of commercial instrumentation, although such devices are limited to the volatile elements and are therefore not universal.

Lasers have the greatest output intensity and would therefore be expected to yield the highest sensitivity. Unfortunately, they are expensive and have, in the past, been produced to give light of very specific wavelengths, which may not correspond to the frequency required to excite the analyte atoms. The development of the tunable dye laser has helped to overcome this problem. Frequency doubling (i.e., halving the wavelength of the light output) has enabled wavelengths well into the ultraviolet region to be obtained. A nitrogen laser pump can be used to obtain a wavelength of 220 nm while a Nd:YAG (neodymium:yttrium aluminium garnet) laser allows 180 nm to be reached. An added advantage of high-intensity lasers is that they may cause saturation fluorescence (population inversion), which nullifies the effects of quenching and self-absorption. Another source that has been used to initiate fluorescence is an inductively coupled plasma. This may be used either as a continuum source (if the actual fireball is used), or as a line source (if the tail-flame is used). The use of an ICP as a light source to initiate fluorescence is not yet a routine technique.

As with AAS, modulation of the light source enables an optimal signal-to-noise ratio to be obtained. The frequency of modulation depends on the light source, but may be up to 10 kHz for an EDL.

Atom cell

Several atom cells have been used for AFS. The majority of applications have used a circular flame atom cell (sometimes with a mirror placed to collect as much light as possible), but more recently inductively coupled plasmas and electrothermal atomizers have also been used. There is a risk of self-absorption at high analyte concentrations, i.e. light fluoresced by some atoms will be absorbed by others in different parts of the flame. A circular flame has a smaller path length compared with a slot burner, hence its popularity. A variety of fuels have been used for the flames. Air–acetylene, nitrous oxide–acetylene and argon–hydrogen diffusion flames have all been used successfully. Quenching (the nonradiative loss of energy) increases with increasing flame temperature and so the cooler air–acetylene and hydrogen flames are often used. Conversely, the hot nitrous oxide–acetylene flame aids the atomization of refractory compounds, The quenching crosssection of the colliding particles also effects the amount of quenching. The relative quenching cross-section of the different flame gases increases in the order argon (negligible) < hydrogen (low) < oxygen (high). It can therefore be seen that for easily atomized analytes an argon–hydrogen diffusion flame offers at the best properties for atomic fluorescence.

Inductively coupled plasmas atomize refractory analytes very efficiently because of their high temperature (8000 K) and therefore have a very high fluorescent yield. Care must be taken, however, to ensure that ionization of the analyte species does not occur. An added advantage is that the plasma is made up of argon and is therefore optically thin (i.e. it has a low quenching cross-section). The disadvantages of using a plasma as the atom cell are that it is more expensive and the instrumentation is more complex. When a plasma is used as an atom cell, the light source may be placed in one of two orientations. It may pass light through the side of the plasma although more recently it has been used axially, i.e. the source passes light through the entire length of the plasma. Since the second configuration has a longer path length, more analyte atoms may be excited and hence improved detection limits are obtained. The disadvantage with this configuration is that the light source must be protected in some way from the extreme temperatures of the plasma. This may be achieved by using either a water-cooled condenser containing a quartz window or by using a protective gas flow to protect the lens from the plasma tail flame. If mirrors or some other collection device are placed around the plasma, then all fluoresced light may be collected and directed to the detector. It has been noted that the optimum viewing height for fluorescence measurements using an ICP as the atom cell occurs substantially higher above the load coil than for emission measurements. Observation above 3 cm is routine and making measurements from the tail flame is common to ensure that a signal above the background and analyte emission noise is obtained.

Similarly, a decrease in plasma power has also been found to be beneficial.

Electrothermal atomizers for AFS share many of the advantages associated with their use for AAS. The argon inert gas that prevents oxidation of the graphite tube ensures that minimal quenching occurs, although a number of other interferences may occur. By and large, the interferences are very similar to those experienced in traditional electrothermal AAS, such as carbide formation (for some analytes), scatter by particulate matter, and losses during thermal pretreatment. A more comprehensive overview of interferences may be found in the sections on AAS. The use of matrix modifiers to assist in the separation of the analyte from the matrix is still often necessary.

Where a laser is used as light source, a technique termed laser-induced fluorescence electrothermal atomization (LIF-ETA) is possible. A schematic diagram of a typical LIF-ETA system is shown in **Figure 2A**. The light source passes through a pierced mirror and into the atomizer. Light fluoresced from the analytes within the atomizer then reflect off the mirror and onto the detector. An alternative configuration is shown in **Figure 2B**, in which the laser irradiates the sample through the injection port. The two configurations are known as longitudinal and transverse geometries, respectively. The technique is

capable of extremely low detection limits (below 1 pg g^{-1}) and has therefore found use in the nuclear, biomedical, electronics and semiconductor industries.

A heated atom cell in not required in the case of mercury determinations. Mercury produces an atomic vapour at room temperature and therefore does not require a flame to cause atomization. Instead, the mercury vapour may simply be transported to an atom cell at room temperature. The radiation from the light source excites the mercury atoms in the normal way.

Line isolation devices

A conventional monochromator (either Ebert, Czerny–Turner, Littrow or Echelle) may be used (see AAS sections for details), but some of the more basic instrumentation uses interference filters. These are optical filters that remove large bands of radiation in a nondispersive way. A dispersion element such as a prism or a grating is therefore not required. Only a relatively narrow band of radiation is allowed to pass to the detector. The disadvantage with such devices is that they are not particularly efficient and hence much of the fluoresced light is lost. An alternative development is the multi-reflectance filter. This is shown diagramatically in **Figure 3**, and has the

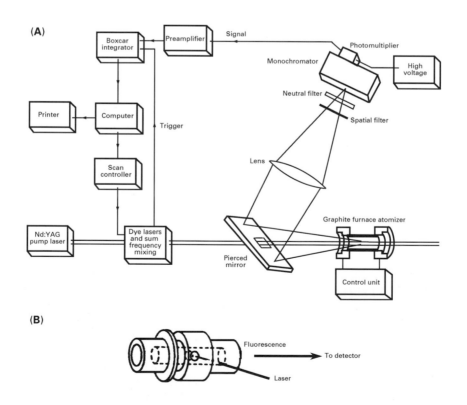

Figure 2 (A) Schematic diagram of longitudinal laser-induced fluorescence with electrothermal atomization. (B) Transverse laser-induced fluorescence. From Ebdon L (1998) *Introduction to Analytical Atomic Spectrometry*. Reproduced by permission of John Wiley & Sons Limited.

advantages of being 80% efficient at transmitting the wavelengths of interest while virtually eliminating background noise. Such a device therefore has a superior performance to other wavelength filters.

Interference and multi-reflectance filters may be used when the analyte of interest has been separated from the matrix and all the concomitant elements. In such a case the analyte atoms are the only elements that will fluoresce when light from a monochromatic light source is used to excite them. Powerful dispersion gratings are therefore not required because, theoretically, light specific to only one element is produced. A schematic diagram of a very basic instrument that uses a multi-reflectance filter is shown in **Figure 4**. If other analytes are present in the atom cell, then either a monochromator is required to separate the other wavelengths (produced by emission), or, the source must be modulated so that distinction between emission and fluorescence signals may be made.

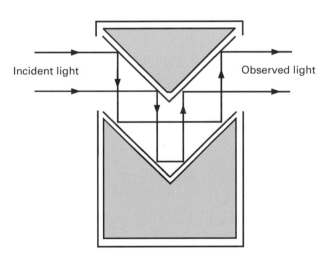

Figure 3 A multi-reflectance filter.

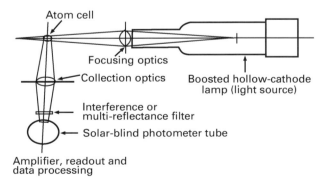

Figure 4 Schematic of a very basic nondispersive AF spectrometer.

Detectors

The light isolated by the line isolation device (monochromator, interference filter or multi-reflectance filter) is normally detected using a photomultiplier tube (PMT). A description of how these devices work may be found in the sections on AAS. It should be noted that some basic instruments (e.g. as shown in **Figure 4**), often use a solar-blind photomultiplier tube. This is a PMT that detects radiation in the UV region but does not detect light in the visible region of the spectrum.

Readout systems and data handling

The readout and data-handling systems used for AFS are similar to those found for AAS instrumentation. The signal from the photomultiplier tube passes through an amplifier and then on to the readout device. This may be a digital display, a chart recorder, an integrator, or a computer with software dedicated to the instrument. The last alternative is the most commonly used in modern instrumentation. The software is likely to be able to control calibration, plot graphs, perform quality assurance tests and calculate means and standard deviations of the results. One advantage of using a computer to control the analysis is that the use of an autosampler becomes a possibility. This enables unattended operation of the instrumentation.

Applications

As described previously, although techniques to utilize AFS were developed several decades ago, until recently they were not widely used. However, a new generation of simple AFS instruments have been developed to specifically detect the vapour-forming elements, such as those that form hydrides (As and Se) and mercury, which forms an atomic vapour. All of these analytes have a primary line below 260 nm and, since the analytes may be readily separated from the bulk matrix and concomitant elements, dispersion is not necessary. Instead these basic instruments originally used a simple interference filter, although these have now been superseded by the more efficient multi-reflectance filter.

The theory behind hydride generation is beyond the scope of this text, but in brief, if a sample is mixed with a reducing agent, such as sodium tetrahydroborate, then some elements, e.g. As, Bi, Ge, Pb, Se, Sb, Sn and Te, will form hydrides. It must be noted that the efficiency of hydride formation depends on the oxidation state of the analyte. Arsenic(III) forms a hydride more efficiently than As(V), and Se(IV) forms a hydride whereas Se(VI) does not. It is therefore necessary to ensure that all the analyte atoms are in the same oxidation state; this is normally achieved using one of a variety of reducing agents.

Once formed, the hydride leaves the reaction cell and is flushed towards the AFS instrument using an inert gas (usually argon). During this process, hydrogen is formed as a breakdown product of tetrahydroborate, and can be swept along with the hydride in the argon to the atom cell, where it can be used as the fuel for the hydrogen–argon diffusion flame. More hydrogen may be added as necessary from an external source. Since a hydrogen–argon diffusion flame is used, the amount of water entering the flame must be minimal (otherwise quenching and flame instability result). The hydride and water vapour are therefore separated, often by passing through a membrane drier tube. This is a tube containing a semi-permeable hydroscopic membrane surrounded by a counterflow of dry gas. Water vapour in the hydride passes through the membrane and is removed by the countergas flow. The dry hydride continues through the tube to be flushed toward the AFS instrument by the flow of argon.

Mercury may be determined using very similar instrumentation, but it may also be introduced in a number of other ways. Since mercury is volatile, solid samples may be placed on a small platform and then heated. The sample is thus vaporized and the vapours are passed through a trap (sand coated with gold). The mercury is trapped onto the gold while the smoke and matrix constituents pass through and are hence separated from the analyte. Heating of the gold trap to 900°C releases the mercury, which may then be swept in a flow of argon to the AFS instrument. This technique has the advantage that the mercury may be preconcentrated on the trap, since several repeat samples may be vaporized and the mercury collected before it is subsequently removed from the trap. The use of the gold trap preconcentrator is equally applicable to the analysis of liquid samples. When a gold trap preconcentrator is used, the limit of detection for mercury is substantially below $1 \ ng \ L^{-1}$, which is superior to that obtained using atomic absorption, emission or inductively coupled plasma–mass spectrometric (ICP-MS) techniques.

The instrumentation described above may also be used for speciation studies. In such cases a separation technique such as high-performance liquid chromatography (HPLC) or gas chromatography (GC) is coupled with the AFS instrument. The chromatography is used to separate different analyte species and some on-line chemistry is subsequently required to convert the species into a chemical form suitable for vapour generation. Although the chemistry behind these methods of speciation is beyond the scope of this text, it should be emphasized that the method used should be powerful enough to convert even very stable species, such as arsenobetaine or selenomethionine into atomic arsenic and selenium. The additional use of photolysis may therefore be necessary. Similarly, pyrolysis may be necessary to convert different mercury species into atomic mercury since this is the only species that will fluoresce.

As described previously, vapour introduction approaches are by far the most common application of atomic fluorescence. Despite this, mention of other methods should be made. If a conventional nebulizer and spray chamber assembly (see AAS section) is used, it is possible to introduce liquid samples directly to the atom cell. In circumstances such as these, it is necessary to use more robust air–acetylene or nitrous oxide–acetylene flame, or perhaps an ICP. The use of an ICP as an atom cell for AFS measurements has led to the development of a number of different techniques, e.g. ASIA, an acronym for atomiser, source, inductively coupled plasmas in AFS. This technique uses a high-powered ICP as a source and a low-powered ICP for the atom cell. It has been found that ICP-AFS yields linear calibrations over 4–6 orders of magnitude and is more sensitive than ICP-AES.

Several techniques have been developed that use an electrothermal atomizer as an atom cell. Laser-induced fluorescence (LIF) has been discussed previously, but other techniques such as laser-excited atomic fluorescence spectroscopy (LEAFS) also exist, although not used routinely.

Interferences

Atomic fluorescence has the advantage of being less prone to spectral interferences than either AES or AAS. Molecular fluorescence is less of a problem than molecular absorption is in AAS. Scatter from the light source and quenching from the gaseous species in the atom cell are often the major sources of interference. For many applications where the analyte is separated from the matrix (e.g. vapour generation) chemical interferences may exist; for example, the presence of high concentrations of some transition metals may interfere in the hydride formation process. This will inevitably lead to errors in the measurement unless preventative steps are taken.

Detection limits

A discussion of the detection limits for the different approaches utilizing atomic fluorescence is difficult because they depend so heavily on the type of source used. Similarly, samples introduced in the vapour phase tend to yield substantially improved limits of detection (LODs) when compared with those introduced as a liquid. In general, if an EDL is used as a source and a circular flame as an atom cell, LODs at the $\mu g\ L^{-1}$ level are obtainable. Improvement will be made if a laser is used as a source. If the sample is introduced as a vapour, LODs at the $ng\ L^{-1}$ level are obtained. The use of an ICP as an atom cell leads to LODs at the $\mu g\ L^{-1}$ level if a normal HCL is used as the source, improving by approximately an order of magnitude if a boosted discharge HCL is used. If an electrothermal atomizer is used in conjunction with a laser, the best LODs of all are obtained.

Conclusions

Atomic fluorescence spectroscopy is a popular technique for those analytes that readily form vapours, and specialized instrumentation is now available for individual elements. Such instruments are simple to operate, easily automated, and offer good sensitivity and freedom from interferences. The use of other flame-based fluorescence techniques has waned considerably over the years, but research continues into the use of electrothermal atomizers and inductively coupled plasmas as atom cells, reflecting the superior limits of detection that are potentially available. The availability of tuneable lasers may also encourage the resurgence of AFS as a routine analytical tool.

See also: **Atomic Absorption, Methods and Instrumentation; Light Sources and Optics; UV-Visible Absorption and Fluorescence Spectrometers.**

Further reading

Greenfield S (1994) Inductively coupled plasmas in atomic fluorescence spectrometry. A review. *Journal of Analytical Atomic Spectrometry* 9: 565–592.

Greenfield S (1995) Atomic fluorescence spectrometry — progress and future prospects. *Trends in Analytical Chemistry* 14: 435–442.

Hill SJ, Dawson JB, Price WJ, Shuttler IL, Smith CMM and Tyson JF (1998) Advances in atomic absorption and fluorescence spectrometry and related techniques. *Journal of Analytical Atomic Spectrometry* 13: 131R–170R.

Kirkbright GF and Sargent M (1974) *Atomic Absorption and Fluorescence Spectroscopy*. London: Academic Press.

Atomic Force Microscopes

See **Scanning Probe Microscopes.**

Atomic Force Microscopy, Theory

See **Scanning Probe Microscopy, Theory.**

Atomic Spectroscopy, Historical Perspective

CL Chakrabarti, Carleton University, Ottawa, Ontario, Canada

ATOMIC SPECTROSCOPY
Historical Overview

Professor B. V. L'vov, the inventor of the very powerful analytical technique graphite furnace atomic absorption spectroscopy, has rightly pointed out in the Introduction of his definitive book entitled *Atomic Absorption Spectrochemical Analysis* that, 'The discovery of atomic absorption and the history of research into it are integral parts of the entire history of spectroscopy and spectrochemical analysis'. Indeed, the early history of atomic spectroscopy, as far as spectrochemical analysis was concerned, consisted of the development of emission spectrochemical analysis, which was usually dependent on Fraunhofer lines (which are atomic absorption lines) for wavelength calibration.

Following Newton's study of the spectrum of the sun in 1666, there was a period of almost 136 years entirely concerned with emission spectroscopy. Only in 1802 did Wollaston report the presence of dark bands in the continuum emission spectrum of the sun and, after a more detailed study by Fraunhofer (1814), Brewster (1820) was able to ascribe them to absorption of radiation within the sun's atmosphere. Another 40 years passed before Kirchhoff and Bunsen showed that one of these dark bands in the emission spectrum of the sun corresponded exactly to the yellow emission band obtained when sodium vapour is heated in a flame. This led Kirchhoff to deduce that the Fraunhofer lines in the solar spectrum are absorption lines of elements whose flame emission spectra would contain lines at exactly the same position in the spectrum. Their work enabled Kirchhoff to develop the fundamental relationship between emission and absorption spectra: any species that can be excited to emit radiation at a particular wavelength will also absorb radiation at that wavelength. Thus Kirchhoff not only laid the foundations of atomic absorption methods of chemical analysis but also gave a striking example of their power. Indeed, it is difficult to imagine a more convincing and dramatic demonstration. Bunsen and Kirchhoff are thus rightfully considered to be founders of spectrochemical analysis. However, it is surprising that almost a century after the work of Bunsen and Kirchhoff the potentialities of atomic absorption measurements remained unexplored and unsuspected. Why?

As Alan Walsh has presented in his perceptive analysis of the reasons, it seems likely that one reason for neglecting atomic absorption methods was that Bunsen and Kirchhoff's work was restricted to visual observation of the spectra. In such visual methods the sensitivity of emission methods was probably better than that of absorption. Not only was the photographic recording of the spectra more tedious, but also the theory seemed to indicate they would only prove useful for quantitative analysis if observed under very high resolution. But possibly a more fundamental reason for neglecting atomic absorption is related to Kirchhoff's law (1859), which states that the ratio of the emissive power E and absorptive power A of a body depends only on the temperature of the body and not on its nature. Otherwise radiative equilibrium could not exist within a cavity containing substances of different kinds. The law is usually expressed as

$$E/A = K(\lambda T)$$

where $K(\lambda T)$ is a function of wavelength and temperature.

This law is perfectly correct but unfortunately much and possibly most of the subsequent teaching concerning it has been misleading. In most textbooks and lessons on this subject the enunciation of the law is followed by a statement to the effect that 'this means good radiators are good absorbers, poor radiators are poor absorbers'. This statement is patently absurd without any reference to temperature or wavelength.

This erroneous conclusion has been made, presumably, because Kirchhoff's constant, K, as he so clearly pointed out, is a function of wavelength and temperature. He was unable to find an analytical expression for it and it was not deduced until 1900 by Kirchhoff's successor at the University of Berlin, Max Planck, as Planck's distribution function. Many spectroscopists were misled by the widely held but misleading assumptions regarding the implications of Kirchhoff's law. What is even more surprising is that numerous spectroscopists wrote papers on atomic emission methods and referred to the problems caused by the effects of self-absorption and self-reversal, and yet failed to make the small-step connection between atomic emission and atomic absorption! It was left to Alan Walsh, whose research experience in atomic emission spectroscopy and molecular absorption spectroscopy virtually compelled him to see the obvious connection. Walsh expressed this experience in his inimitable way in the following words: 'It appears to be true that "having an idea" is not necessarily the result of some mental leap: it is often the result of merely being able, for one sublime moment, to avoid being stupid!'.

Walsh in 1953 and Alkemade and Milatz in 1955 independently published papers indicating the substantial advantage of atomic absorption methods over emission methods for quantitative spectrochemical analysis. Alkemade and Milatz considered only the selectivity in atomic absorption methods, but Walsh discussed general problems of development of absolute methods of analysis. During his early experiments with flame atomizer–burners, Walsh encountered the problems which arise when measuring atomic absorption using a continuum source, which might have been responsible for the neglect of atomic absorption methods. The use of a sharp-line source such as a hollow-cathode lamp not only solved this problem, it also provided the atomic absorption method with one of its important advantages, that is, the ease and certainty with which one can isolate the analysis line. The concept of 'putting the resolution in the source' also permitted the use of a simple monochromator since the function of the monochromator, in such a case, is only to isolate the analysis line from the neighbouring lines and the background.

Graphite furnace atomic absorption spectroscopy (GFAAS) has been widely used as a spectrochemical trace-analytical technique during the last 30 years. L'vov pioneered most of the theoretical and experimental developments in GFAAS and provided a masterly treatment of the possibilities of GFAAS in a paper entitled 'Electrothermal atomization—the way towards absolute methods of atomic absorption analysis'. RE Sturgeon, who has been responsible for much of the development of GFAAS, presented a critical analysis of what it does and what it cannot do. AKh Gilmutdinov, who did much of the theoretical development of GFAAS in the 1990s, also elucidated the complex processes and reactions that occur in an electrothermal atomizer by digital imaging of atomization processes using a charge-coupled device camera in my research laboratories.

Atomic spectroscopy in the three variations that are most commonly used in spectrochemical analysis, atomic absorption, atomic emission and atomic fluorescence, are all mature techniques, with their particular areas of strengths and weaknesses now well recognized. Many a battle has been fought and won to establish the superiority of one or the other technique over the rival techniques. It is sometimes claimed that Inductively Coupled Plasma mass spectrometry (ICP-MS), which has the winning combination of high-temperature ionization of elements in a plasma with the detection of the ions by mass spectrometry, is the ultimate trace-analytical technique that will triumph over the rival techniques. However, such claims are based on the limited applications of these techniques by practitioners whose allegiance to their own techniques gives more credit to their loyalty than to their scientific objectivity, as elaborated in the following paragraph. Some have also predicted the imminent demise of graphite furnace atomic absorption spectroscopy as a trace-analytical technique. Such a prediction, based as it is on inadequate comprehension of the enormous complexities and extreme diversities of real-life situations requiring a variety of trace-analytical techniques which possesses some special capabilities not possessed by many analytical techniques, is destined to be false, as is shown below.

In my research laboratories in recent years, my 10 PhD students and three adjunct research professors have been doing research on the chemical speciation of potentially toxic elements in the aquatic, the atmospheric and the terrestrial environment. Because freshwaters and soils are systems that are usually far removed from equilibrium, our research is mostly directed to the development of kinetic schemes of chemical speciation which require routine use of inductively coupled plasma mass spectrometry and graphite furnace atomic absorption spectroscopy to measure the kinetics of metal uptake from aqueous environmental samples. As such, we routinely make hundreds of determinations of trace and ultratrace elements every week. For these determinations, we have in our research laboratories the latest models of two ICP-MS and several GFAAS systems. We have made objective studies of the performance characteristics of ICP-MS and GFAAS for kinetic studies of real-life, aqueous, environmental samples over a period covering 6 years from 1993 to 1998, and have come to the following inescapable conclusions. The analytical sensitivities of both ICP-MS and GFAAS for toxic metals are comparable. For kinetic runs on aqueous, environmental samples, ICP-MS has the great advantage of providing continuous sampling with numerous data points having adequate time resolution, whereas GFAAS provides discrete sampling with fewer data points having inadequate time resolution. However, in the most important and decisive criterion of sample compatibility with the technique, ICP-MS fails completely on account of its inability to handle any solutions containing significant amounts of solute materials and/or corrosive chemicals such as hydrofluoric acid and strong inorganic acids and bases, whereas GFAAS can readily handle such samples. Since the aqueous, environmental samples for kinetic runs are always pretreated with solutes in the form of acid–base buffers, and since such solutions cannot be diluted in order to make them acceptable to ICP-MS, it is not possible to use ICP-MS for such kinetic studies without fouling the interior of the ICP-MS equipment with encrustations of the solute materials, which do serious damage. The alternative then is GFAAS, in spite of all its limitations. Because of the easy compatibility of GFAAS with the difficult sample type described above, GFAAS will continue to be used, as it is used now, until another new analytical technique which has all the advantages of ICP-MS without its fatal deficiencies mentioned above, is invented, developed and tested using real-life samples, i.e. for a long time.

See also: **Atomic Absorption, Theory; Atomic Emission, Methods and Instrumentation; Atomic Fluorescence, Methods and Instrumentation; Inductively Coupled Plasma Mass Spectrometry, Methods.**

Further reading

Chakrabarti CL, Gilmutdinov AKh and Hutton JC (1993). Digital imaging of atomization processes in electrothermal atomizer for atomic absorption spectrometry. *Analytical Chemistry* **65**: 716–723.

L'vov BV (1970) *Atomic Absorption Spectrochemical Analysis*. London: Adam Hilger.

L'vov BV (1978) Electrothermal atomization—the way toward absolute methods of atomic absorption analysis. *Spectrochimica Acta, Part B* **33**: 153–193.

Sturgeon R (1986) Graphite furnace atomic absorption spectrometry: fact and fiction. *Fresenius' Zeitschrift für Analytische Chemie* **324**: 807–818.

Walsh A (1980) Atomic absorption spectroscopy—some personal reflections and speculations. *Spectrochimica Actya, Part B* **35**: 639–642.

ATR and Reflectance IR Spectroscopy, Applications

UP Fringeli, University of Vienna, Austria

In Memory of N Jim Harrick.

VIBRATIONAL, ROTATIONAL & RAMAN SPECTROSCOPIES
Applications

Introduction

Chemical reactions that occur at gas–solid and liquid–solid interfaces are of central importance to a variety of research and technological areas, including biomembranes, drug design, drug–membrane interaction, biosensors, chemical sensors, heterogeneous catalysis, thin film growth, semiconductor processing, corrosion and lubrication. Many methods are used for interface studies, ranging from most simple ones like the octanol–water two-phase system for mimicking the partition of a drug between a biomembrane and the surrounding water, to most specialized and expensive techniques such as low-energy electron diffraction (LEED), Auger electron spectroscopy (AES), X-ray photoelectron spectroscopy (XPS/ESCA) and ion scattering spectroscopy (ISS).

Among this palette of techniques, optical reflection spectroscopy in the mid- and near-IR range occupies an important complementary position. The basic equipment consists of a commercial IR spectrometer and a suitable reflection accessory that usually fits into the sample compartment of the spectrometer. Many reflection techniques permit *in situ* applications, and if applied in the mid IR, result in quantitative and structural information on a molecular level. Moreover, IR reflection spectroscopy features a very high performance-to-price ratio.

There is a wide range of different spectroscopic reflection techniques. First one should distinguish between internal (total) and external reflection. Attenuated total reflection (ATR) belongs to the first group. It makes use of the evanescent wave existing at the interface of the IR waveguide and the sample. Commercial ATR attachments differ mainly in shape and mounting of the internal reflection element (IRE) in the light path. Most IREs enable multiple internal reflections, a prerequisite for monolayer and sub-monolayer spectroscopy, and a referred to as MIRE.

A variety of external reflection techniques are in use. In specular reflection (SR) the radiation reflected from the front surface of a bulk sample is collected. SR is often measured at or near normal incidence. Reflected spectral energy depends on the absorption behaviour of the sample. In regions of strong absorption the reflected energy is enhanced with respect to non-absorbing spectral regions, moreover, the reflection spectrum is usually very different from a corresponding absorption (AB) spectrum obtained by a transmission experiment. AB spectra may, however, be calculated from SR spectra by means of the Kramers–Kronig transformation (KKT). Corresponding software for SR data processing is supplied with most commercial IR instruments. While specular reflectance is measured at or near normal incidence, IR reflection–absorption spectroscopy (IRRAS) works from about 10° to grazing incidence. In this case the sample is placed on a reflecting substrate, usually a metal. The portion of reflected light from the sample surface is generally small compared with the energy reflected off the metal surface. Therefore, IRRAS data and transmission (T) data are analogous. From Fresnel's equations (see below), it follows that parallel (∥) and perpendicular (⊥) polarized electromagnetic waves undergo different phase shifts upon reflection. This phase shift is 180° for ⊥-polarized light at non-absorbing interfaces. As a consequence, incoming and reflected beams cancel at the interface (node). On the other hand, at large angles of incidence ∥-polarized incident light results in an enhanced electric field component perpendicular to the reflecting interface (z-axis). For thin samples, i.e. sample thickness (d) much smaller than the quarter wavelength ($\lambda/4$) of the reflected light, RA spectra report only partial information on orientation. It should be noted, however, that for a complete orientation analysis spectra obtained with light polarized in the plane of the interface (x,y-plane) is also necessary. ATR fulfils this requirement, in contrast to RA.

Diffuse reflectance (DR) is successfully applied to obtain IR spectra of rough (scattering) or dull surfaces, i.e. of media intractable by other reflection techniques. The interpretation of DR spectra, however, is sometimes handicapped by the fact that they may be a mixture of AB and SR spectra. DR spectroscopy is a sensitive tool, especially when used with an IR Fourier transform (FT) spectrometer (DRIFT).

Elucidation of structure–activity relationships is the aim of many applications of reflection spectroscopy to thin layers at interfaces. In this context, polarization measurements are of considerable importance, since molecules at interfaces exhibit often induced ordering.

Low signal intensities are common in different kinds of interface spectroscopy, especially when the sample consists of a monolayer or even submonolayer as usual in heterogeneous catalysis and substrate–biomembrane interaction. Although modern FTIR spectrometers exhibit very high stability, signal-to-noise (S/N) ratio enhancement by data accumulation is limited by environmental and instrumental instabilities. The fact that most commercial FT instruments are operated in a single-beam mode is disadvantageous in this respect, because the longer an experiment lasts, the greater is the time lag between sample and reference data, which facilitates the intrusion of instabilities. Several optional extras are available in order to reduce the time lag between acquisition of sample and reference spectra.

One possibility is the conversion of the single-beam instrument into a pseudo-double-beam instrument by means of a shuttle which moves alternately the sample and reference into the IR beam. Such attachments were first developed for transmission experiments, and were later adapted for ATR measurements. In the latter case the sample and the reference are placed on top of one another on the same trapezoidal MIRE. A parallel beam of half the height of the MIRE is directed alternately through the upper and lower half of the MIRE by computer-controlled vertical displacement of the ATR cuvette. This method is referred to as single-beam sample reference (SBSR) technique and is described in more detail below.

Polarization modulation (PM) in combination with IRRAS is a further possibility to enhance instrumental stability and background compensation when working at grazing incidence to a thin sample on a metal substrate. PM at about 50 kHz is achieved by means of a photoelastic modulator (PEM). Since under these experimental conditions the sample will only absorb light in the ∥-polarized half-wave, the ⊥-polarized half-wave of the signal is representative of the background, i.e. of the reference. Subtraction is performed by lock-in technique within each PM cycle, i.e. 50 000 times per second. As a consequence, environmental and instrumental contributions are largely compensated.

Finally, it should be noted that a more general application of modulation spectroscopy can be used to obtain selective information on an excitable sample. Modulated excitation (ME) spectroscopy can always be applied with samples allowing periodic stimulation via a periodic variation of any external thermodynamic parameter, e.g. temperature, pressure, concentration, electric field, light flux. ME causes a periodic variation of the absorbance at those wavelengths that are typical for the molecules involved in the stimulated process. Phase-sensitive detection (PSD) by digital lock-in technique adapted for FTIR instruments permits spectral registration of the modulated, i.e. affected, part of the sample. A typical feature of ME with PSD is the comparison of sample and reference within each period of stimulation. Within this time interval environmental and instrumental parameters are usually stable so that a very good baseline is achieved. Moreover, if one or more relaxation times τ_i of the kinetic response of the stimulated sample fulfil the condition $0.1 < \omega\tau_i < 10$, where ω denotes the angular frequency of stimulation, significant phase lags ϕ_i between stimulation and sample responses will occur which are related to the reaction scheme and the rate constants of the stimulated process.

Theory of reflectance spectroscopy

For a comprehensive description of the theory of reflectance the reader is referred to the Further reading section. In this article, theory will only be presented when necessary for a general understanding.

Fresnel's equations

The theory of reflection and transmission of an electromagnetic wave by a plane boundary was first derived by Fresnel. The geometry of specular reflection and transmission is depicted in **Figure 1**. The incident (i) plane wave consists of the parallel (∥) and perpendicular polarized (⊥) electric field components $E_{i\parallel}$ and $E_{i\perp}$, respectively. The corresponding components of the reflected (r) and refracted (transmitted t) field components are denoted by $E_{r\parallel}$, $E_{r\perp}$, $E_{t\parallel}$, and $E_{t\perp}$. Fresnel's equations relate the reflected and transmitted components to the corresponding incident components.

For a nonabsorbing medium, i.e. the absorption indices κ_1 and κ_2 equal to zero, one obtains for the ratio r between reflected and incident electric field

$$r_{\parallel} = \frac{E_{r\parallel}}{E_{i\parallel}} = \frac{n_2 \cos\theta_i - n_1 \cos\theta_t}{n_2 \cos\theta_i + n_1 \cos\theta_t}$$

$$r_{\perp} = \frac{E_{r\perp}}{E_{i\perp}} = \frac{n_1 \cos\theta_i - n_2 \cos\theta_t}{n_1 \cos\theta_i + n_2 \cos\theta_t} \qquad [1]$$

where ∥ and ⊥ denote parallel and perpendicular polarization, according to **Figure 1**. Equation [1] may be modified by introducing Snell's law of refraction:

$$n_1 \sin\theta_i = n_2 \sin\theta_t \qquad [2]$$

resulting in

$$r_{\parallel} = \frac{E_{r\parallel}}{E_{i\parallel}} = \frac{\tan(\theta_i - \theta_t)}{\tan(\theta_i + \theta_t)}$$

$$r_{\perp} = \frac{E_{r\perp}}{E_{i\perp}} = -\frac{\sin(\theta_i - \theta_t)}{\sin(\theta_i + \theta_t)} \qquad [3]$$

As concluded from Equation [3], perpendicular polarized incident light undergoes a phase shift of 180° upon reflection, i.e. there is a node at the reflecting interface resulting in zero electric field strength at this point. On the other hand, parallel polarized components remain in-phase. However, this conclusion holds no longer in the case of absorbing media.

The corresponding equations for the ratio t between transmitted and incident electric fields are

$$t_{\parallel} = \frac{E_{t\parallel}}{E_{i\parallel}} = \frac{2n_1 \cos\theta_i}{n_2 \cos\theta_i + n_1 \cos\theta_t}$$

$$t_{\perp} = \frac{E_{t\perp}}{E_{i\perp}} = \frac{2n_1 \cos\theta_i}{n_1 \cos\theta_i + n_2 \cos\theta_t} \qquad [4]$$

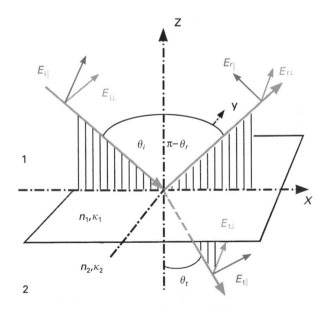

Figure 1 Specular reflection and transmission. The angles of incidence (i), reflection (r) and refraction (t) are denoted by θ_i, θ_r and θ_t, respectively. The corresponding electric field components are denoted by E. They are split into orthogonal portions, one parallel to the plane of incidence (x,z-plane) and the other perpendicular to this plane (parallel to the y-axis). Accordingly, electric fields are referred to as parallel (∥) and perpendicular (⊥) polarized, n_1, n_2, κ_1 and κ_2 denote the refractive and absorption indices in the two media.

In order to modify Equations [1], [3] and [4] for absorbing media one has to introduce the complex refractive index. For an incoming plane-wave it is given by

$$\hat{n} = n + i\kappa \qquad [5]$$

where n is the refractive index and κ is the absorption index. As a consequence, r and t become complex and the resulting phase shifts differ from $0°$ and $180°$ as mentioned above.

Energy flux density

The flux density of an electromagnetic wave is described by the Poynting vector. For the case of the plane wave field one obtains for the time average in the direction of propagation

$$\bar{S} = \frac{1}{2}\sqrt{\varepsilon/\mu}\,E_0^2 \qquad [6]$$

where μ is the permeability, i.e. the product of the permeability of vacuum μ_0 and the relative permeability μ_r which is unity for nonconductive materials, and ε denotes the permittivity which is the product of the permittivity of vacuum ε_0 and the relative permittivity (dielectric constant) ε_r which is complex for absorbing media, according to

$$\begin{aligned}
\hat{\varepsilon}_r &= \hat{n}^2 = n^2 - \kappa^2 + i \cdot 2n\kappa \\
&= (n^2 + \kappa^2) \cdot \exp\left[i \cdot \mathrm{atan}\left(\frac{2n\kappa}{n^2 - \kappa^2}\right)\right] \\
&= |\hat{\varepsilon}_r| \cdot \exp\left(i \cdot \varphi\right)
\end{aligned} \qquad [7]$$

Introducing the absolute value of Equation [7] into Equation [6] results in, for nonconducting media,

$$\bar{S} = \frac{1}{2}\sqrt{\varepsilon_0/\mu_0}\,\sqrt{n^2 + \kappa^2}\,E_0^2 \qquad [8]$$

Reflectance ρ and transmittance τ are defined as the ratios of the corresponding energy fluxes $\bar{J}$. According to Equation [8] they are proportional to the square of the electric field, i.e.

$$\rho = \frac{\bar{J}_r}{\bar{J}_i} = rr^* \quad \text{and} \quad \tau = \frac{\bar{J}_r}{\bar{J}_i} = \frac{\sqrt{(n_2^2 + \kappa_2^2)}}{\sqrt{(n_1^2 + \kappa_1^2)}}\frac{\cos\theta_t}{\cos\theta_i}tt^* \qquad [9]$$

The factor $\sqrt{(n_2^2 + \kappa_2^2)}/\sqrt{(n_1^2 + \kappa_1^2)}$, which reduces to n_{21} for nonabsorbing media, results from the change of dielectrica (see Equation [8]), and $\cos\theta_t/\cos\theta_i$ takes account of the different cross-sections of the beam in media 1 and 2, respectively.

rr^* and tt^*, become r^2 and t^2 for nonabsorbing media. In this case, Equations [1], [2] and [8] result in

$$\rho_\parallel = \left(\frac{n_{21}^2 \cos\theta_i - \sqrt{n_{21}^2 - \sin^2\theta_i}}{n_{21}^2 \cos\theta_i + \sqrt{n_{21}^2 - \sin^2\theta_i}}\right)^2$$
$$\rho_\perp = \left(\frac{\cos\theta_i - \sqrt{n_{21}^2 - \sin^2\theta_i}}{\cos\theta_i + \sqrt{n_{21}^2 - \sin^2\theta_i}}\right)^2 \qquad [10]$$

where n_{21} denotes the ratio of refractive indices of media 2 and 1, respectively (see **Figure 1**). For normal incidence, i.e. $\theta_i = 0$, Equation [10] reduces to

$$\rho_\parallel = \rho_\perp = \left(\frac{n_{21} - 1}{n_{21} + 1}\right)^2 \qquad [11]$$

In order to obtain the reflectance of an absorbing medium one may introduce Equation [5] into Equation [10] or [11]. The result for normal incidence is

$$\rho_\parallel = \rho_\perp = \frac{(n_2 - n_1)^2 + (\kappa_2 - \kappa_1)^2}{(n_2 + n_1)^2 + (\kappa_2 + \kappa_1)^2} \qquad [12]$$

It should be noted that in many applications medium 1 is air or a nonabsorbing crystal, i.e. $\kappa_1 = 0$. It follows from Equation [12] that the reflectance increases with increasing absorption index of medium 2 (κ_2). In the limiting case of $\kappa_2 \to \infty$ one obtains $\rho \to 1$, i.e. a perfect mirror. Expressions for the more complicated case of oblique incidence to absorbing media have been derived (see Further reading).

The Kramers–Kronig relations

For normal incidence ($\theta_i = 0$) and nonabsorbing medium 1 ($\kappa_1 = 0$, see **Figure 1**), one obtains from Equations [1] and [5] the following expression for the complex ratio $\hat{r}(\theta_i = 0)$ between reflected and incident electric fields:

$$\hat{r}_{\parallel}(\theta_i = 0) = \frac{E_{r\parallel}}{E_{i\parallel}} = \frac{n_2 - n_1 + i\kappa_2}{n_2 + n_1 + i\kappa_2} = \eta \exp(i\phi)$$

$$\hat{r}_{\perp}(\theta_i = 0) = \frac{E_{r\perp}}{E_{i\perp}} = -\frac{n_2 - n_1 + i\kappa_2}{n_2 + n_1 + i\kappa_2} = \eta \exp[i(\phi + \pi)]$$

[13]

where η and ϕ are the amplitude and phase of $\hat{r}(\theta_i = 0)$. They are functions of the wavenumber $\tilde{\nu}$ and related to each other by the Kramers–Kronig equations:

$$\ln \eta(\tilde{\nu}) = \frac{2}{\pi} \int_0^\infty \frac{\nu\,\phi(\nu)}{\nu^2 - \tilde{\nu}^2}\,d\nu$$

$$\phi(\tilde{\nu}) = \frac{2\tilde{\nu}}{\pi} \int_0^\infty \frac{\ln \eta(\nu)}{\nu^2 - \tilde{\nu}^2}\,d\nu$$

[14]

Experimentally, $\eta(\tilde{\nu})$ can be determined, since it is related to the reflectance at normal incidence by $\rho(\theta_i = 0) = \eta(\tilde{\nu})\eta(\tilde{\nu})^*$, i.e. $\eta(\tilde{\nu}) = \sqrt{\rho(\theta_i = 0)}$; see Equation [12]. The Drude model may be used to extrapolate the measurement to $\tilde{\nu} = 0$ and $\tilde{\nu} = \infty$. From $\eta(\tilde{\nu})$ and $\phi(\tilde{\nu})$ the components of the refractive index can be calculated according to

$$n = \frac{1 - \eta^2}{1 - 2\eta\cos\phi + \eta^2}$$

$$\kappa = \frac{2\eta\sin\phi}{1 - 2\eta\cos\phi + \eta^2}$$

[15]

For a detailed discussion, see the Further reading section.

Internal reflection spectroscopy (ATR)

Internal reflection can only occur when the angle of the refracted beam θ_t is larger than the angle of incidence θ_i. This means, according to Snell's law (Equation [2]), that the refractive index of medium 2 must be smaller than that of medium 1 ($n_2 < n_1$). This is contrary to the situation in **Figure 1** where $n_2 > n_1$ was assumed. The region of total reflection begins when θ_t reaches 90°, i.e. at the critical angle of incidence θ_c. It follows from Equation [2] that

$$\sin\theta_c = \frac{n_2}{n_1} = n_{21}$$

[16]

It follows from Equations [2] and [16] for $\theta_i > \theta_c$ that $\sin\theta_t = \sin\theta_i/\sin\theta_c > 1$, resulting in a complex value for the corresponding cosine:

$$\cos\theta_t = \pm i n_{12}\sqrt{\sin^2\theta_i - n_{21}^2}$$

[17]

The ratio r between internally reflected and incident electric field components is then obtained by introducing Equation [17] into Equation [1], resulting in

$$r_{\parallel} = \frac{E_{r\parallel}}{E_{i\parallel}} = \frac{n_{21}^2\cos\theta_i - i\sqrt{\sin^2\theta_i - n_{21}^2}}{n_{21}^2\cos\theta_i + i\sqrt{\sin^2\theta_i - n_{21}^2}}$$

$$r_{\perp} = \frac{E_{r\perp}}{E_{i\perp}} = \frac{\cos\theta_i - i\sqrt{\sin^2\theta_i - n_{21}^2}}{\cos\theta_i + i\sqrt{\sin^2\theta_i - n_{21}^2}}$$

[18]

The corresponding equations for medium 2 are obtained by introducing Equation [17] into Equation [4], resulting in

$$t_{\parallel} = \frac{E_{t\parallel}}{E_{i\parallel}} = \frac{2\cos\theta_i}{n_{21}(\cos\theta_i + i\sqrt{\sin^2\theta_i - n_{21}^2})}$$

$$t_{\perp} = \frac{E_{t\perp}}{E_{i\perp}} = \frac{2\cos\theta_i}{\cos\theta_i + in_{21}^2\sqrt{\sin^2\theta_i - n_{21}^2}}$$

[19]

Finally, it should be noted that incident electric fields undergo phase shifts in the ATR mode even if medium 2 is nonabsorbing. It follows from Equation [18] that

$$\tan\frac{\delta_{\parallel}}{2} = -\frac{\sqrt{\sin^2\theta_i - n_{21}^2}}{n_{21}^2\cos\theta_i}$$

$$\tan\frac{\delta_{\perp}}{2} = -\frac{\sqrt{\sin^2\theta_i - n_{21}^2}}{\cos\theta_i}$$

[20]

where $\delta_{\parallel}$ and $\delta_{\perp}$ are the phase shifts per internal reflection (no absorption) of $\parallel$-polarized and $\perp$-polarized incident light. Since the phase shifts and amplitudes are polarization dependent, linearly polarized incident light is elliptically polarized after an internal reflection. This phenomenon, however, does not hinder polarization measurement in the ATR mode.

Applications

Diffuse reflectance

The geometry of a diffuse reflection experiment is shown in **Figure 2**. The incident beam (I) is collimated to the sample S by means of the ellipsoidal mirror M_i. Two reflection mechanisms must be considered, specular reflection, R_s, and diffuse reflection, R_d. The former occurs at the surface and is governed by the Fresnel equations (Equations [1], [3] and [10–12]). As a consequence of anomalous dispersion, specular reflected light exhibits S-shaped intensity changes at the wavelengths of sample absorption. In contrast, diffuse reflected light exhibits absorption bands at frequencies observed also with transmitted light, but with intensities deviating significantly from those measured in a transmission experiment. The intensity of the diffuse reflection spectrum may be described by the Bouguer–Lambert law (Eqn [21]), the analogous expression to the Lambert–Beer law in transmission spectroscopy.

$$R_d = I_0 \exp(-\alpha \bar{d}) \qquad [21]$$

where $\bar{d}$ is the mean penetrated layer thickness, i.e. the depth of light penetration into the surface layer which results in an intensity decrease by a factor of 1/e, and α denotes the napierian absorption coefficient.

Diffuse reflectance infrared Fourier transform spectroscopy (DRIFT) has become a frequently used technique to obtain IR spectra from materials intractable by transmission spectroscopy. A number of high-performance reflection accessories are available from different manufacturers (see below), allowing

the detection of quantities down to the nanogram region. Nevertheless, DRIFT spectroscopy is confronted with two intrinsic problems: (i) the superposition of diffuse and specular reflected light (see **Figure 2**), which may lead to distorted line shapes, and (ii) the dependence of the mean penetration depth $\bar{d}$ on the absorption coefficient. $\bar{d}$ is found to be inversely proportional to the absorption coefficient α, thus leading to a certain leveling of the band intensities.

The disturbance by specular reflection may be reduced considerably by technical means (trapping) on the reflection attachment. The resulting diffuse reflection spectrum then has to be corrected in order to correspond to the absorbance of a transmission spectrum. This mathematical procedure is generally performed according to the Kubelka–Munk theory. For a comprehensive and critical discussion of this theory the reader is referred to the Further reading section.

Specular reflection spectroscopy (SRS)

In specular reflectance, only light reflected off the front surface is collected (see **Figure 2**). The reflected energy is generally small (<10%) for non-absorbing regions at normal or near-normal incidence. However, according to Equation [12], enhanced reflectance is observed in regions of sample absorption. As illustrated by **Figure 3**, radiation intensity is different to transmission intensity, since S-shaped bands result as a consequence of anomalous dispersion of the refractive index in the region of an absorption band. As a typical example, the specular reflectance spectrum of a black plastic is shown in **Figure 3**.

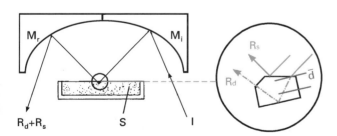

Figure 2 Diffuse reflection experiment. M_i, M_r = ellipsoidal mirrors for incident and reflected light; S = sample; I, R_d, R_s = incident diffuse, and specular reflected beams, respectively. In the magnified circle a possible ray tracing through a surface particle is shown, demonstrating the formation of mixed diffuse and specular reflected light. $\bar{d}$ is the mean penetrated layer thickness according to the Bouguer–Lambert law (Equation [21]).

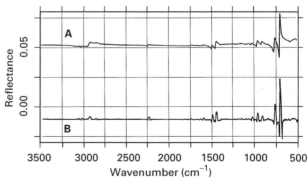

Figure 3 (A) Specular reflectance (SR) spectrum of a black acrylonitrile–butadiene–styrene polymer film measured at near normal incidence, (B) Absorbance spectrum after data treatment of SR spectrum by a Kramers–Kronig transformation. Reproduced in part with permission of Elsevier Science from Zachman G (1995) *Journal of Molecular Structure* **348**: 453–456.

Reflection absorption spectroscopy (RAS)

RAS at near normal incidence This is one of the most common and straightforward external reflection techniques. The IR beam is directed to the sample in the angular range 10–50°. The sample film must be on a reflective support. Under these conditions the RA spectrum is dominated by absorption since specular reflectance from the outer sample surface results in only 4–10% as shown by **Figure 4**. For this reason RA spectra resemble transmission spectra very closely. Accordingly, typical sample thicknesses are between 0.5 and 20 μm.

RAS at grazing angle Sensitivity at grazing angle is significantly enhanced with respect to near-normal reflectance and to transmission. The enhancement is explained by a polarization effect a the reflective surface. This effect is greatest for metal substrates and angles of incidence above 80°. As a general effect, ⊥-polarized incident light has a node at the interface (see above). The resulting reflectance will be very weak as long as the layer thickness is significantly smaller than λ/4. Therefore, ⊥-polarized reflectance may be used as reference spectrum. ∥-Polarized incident light produces electric field components in the x- and z-directions. On metal surfaces, however, the x- and y-components vanish, but, the z-component is significantly enhanced owing to interference of incident and reflected beams. Compared with near-normal incidence, RAS magnification may be by more than one order of magnitude, thus permitting monolayer spectroscopy by a single reflection. Detailed information on RAS techniques applied to study carbon monoxide (CO) adsorbed on metal surfaces can be found in the Further reading section.

More recently, RAS at grazing angle has been applied successfully for *in situ* spectroscopy of lipid monolayers and proteins at the air–water interface of a Langmuir trough, using the water surface as reflector (see Further reading section).

Polarization modulation RAS This technique makes use of the fact that ⊥-polarized incident light has a node at the reflecting interface resulting in zero absorbance at this point and nearly no absorbance of films significantly thinner than the quarter wavelength. Under these conditions, the ⊥-reflectance spectrum may be used as a reference spectrum for the ∥-reflectance spectrum. Since monolayer and submonolayer quantities of organic molecules result in low-intensity spectra, such measurements are susceptible to instrumental and environmental instabilities. This problem may be overcome by polarization modulation (PM). For this purpose, a photoelastic modulator (PEM) is placed in the light path, leading to very fast periodic polarization changes of the incident light. The frequency range is 40–100 kHz, i.e. high enough to avoid interference's with the interferometer frequencies. Phase-sensitive demodulation of the PEM signal results in the interferogram of the difference between ∥- and ⊥-RA spectra, which is then normalized by division by the stationary response featuring the sum of ∥- and ⊥-mean reflectance according to

$$\frac{\Delta R}{R} = \frac{R_{\parallel} - R_{\perp}}{R_{\parallel} + R_{\perp}} \qquad [22]$$

A description of an experimental setup and of the relevant equations for PM-IRRAS can be found in the Further reading section.

Instabilities are largely compensated because sample and reference spectra are measured and evaluated within one period, i.e. within 10–25 μs. If applied in the IR region, this technique is referred to as PM-IRRAS.

It should be noted, however, that significant baseline problems may occur owing to different transmittance of ∥- and ⊥-polarized light by the spectrometer. This is demonstrated by the IRRAS data of a cadmium arachidate monolayer on a gold surface in **Figure 5**.

One should note that the usually intense absorption bands of CH$_2$ stretching, $\nu(CH_2)$ in the 2800–2950 cm^{-1} region and of CH$_2$ bending near 1470 cm^{-1} are very weak in the IRRAS, spectrum in

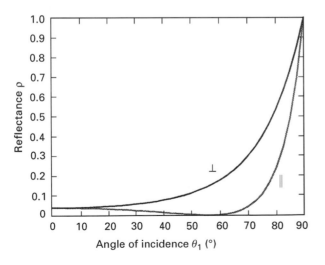

Figure 4 Specular reflectance calculated according to Equation [10]. Refractive indices: $n_1 = 1$ and $n_2 = 1.5$, ∥, ⊥ denote parallel and perpendicular polarized incident light. The Brewster angle, where $\rho_{\parallel} = 0$ for a nonabsorbing medium, is calculated as $\theta_i = \theta_B = 56.3°$, according to $\tan \theta_B = n_{21}$.

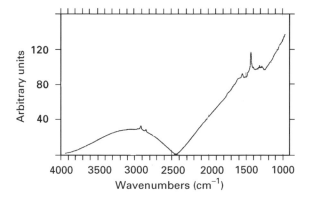

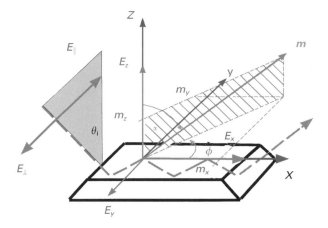

Figure 5 PM-IRRAS spectrum of an arachidic acid monolayer on a gold surface. The spectrum was normalized according to Equation [22]. It should be noted that only molecular vibration can be detected by IRRAS when the corresponding transition moment exhibits a component normal to interface (z-direction). Reproduced in part with permission from Beccard B and Mapanowicz R (1995) *Nicolet Application Note AN-9542*. Madison, WI: Nicolet.

Figure 6 ATR setup. Optical and structural features are related to the IRE fixed-coordinate system x,y,z. $E_{\parallel}$ and $E_{\perp}$ denote the parallel and perpendicular polarized electric field components of the light incident to the IRE under the angle θ_i. $E_{\parallel}$ results in the E_x and E_z components of the evanescent wave, while $E_{\perp}$ results in the E_y component. $\boldsymbol{m}$ denotes the unit vector in direction of the transition dipole moment vector of a given vibrational mode, and m_x, m_y, m_z are the corresponding components in the IRE coordinate system. $\boldsymbol{m}$ goes off at an angle α with respect to the z-axis and the projection of $\boldsymbol{m}$ to the xy-plane goes off at an angle ϕ with respect to the x-axis. Reproduced with permission of the American Institute of Physics from Fringeli UP *et al.* (1998) *AIP Conference Proceedings* **430**: 729–747.

contrast to a corresponding DPPA monolayer ATR spectrum (see **Figure 9**). The reason is that IRRAS of thin layers on metallic surfaces offers only the electric field component normal to the interface, i.e. E_z, as already explained above. The intensities of CH_2 stretching and CH_2 bending in **Figure 5** are consistent with hydrocarbon chains of arachidic acid aligned along the normal to the interface. The lack of a E_x and E_y components, i.e. of an electric field component parallel to the interface, disables the determination of a tilt angle. In this respect, IRRAS is at a disadvantage with respect to ATR, the technique presented in the following section. Moreover, ATR allows significantly better baseline control, especially if special techniques such as SBSR (single-beam sample reference technique) or ME (modulated excitation) are applied.

Attenuated total reflection (ATR) spectroscopy

A number of interesting conclusions may be drawn from Equations [18] and [19]. First, calculation of the reflectance according to Equation [9] results in $\rho_{\parallel} = \rho_{\perp} = 1$, which means *total reflection*. However, if medium 2 is absorbing, one has to insert the complex refractive index (Equation [5]) into Equation [18], resulting in $\rho_{\parallel} \neq \rho_{\perp} < 1$, which means *attenuated total reflection (ATR)*.

Second, in order to obtain more information on the nature of this process, one may calculate the propagation of a plane-wave in medium 2 under the conditions of total reflection. The result for a nonabsorbing medium 2 is that there is no transmittance normal to the interface, i.e. $\tau_z = 0$, however, there is an energy

flux in the x,y-plane near the interface, i.e. τ_x, $\tau_y \neq 0$. Hence, there is a electromagnetic wave beyond the interface, although the whole energy is totally reflected. This wave is referred to as *evanescent wave*. Straightforward calculation results in that the electric field strength of this wave decreases exponentially with distance z from the interface, according to

$$E_{x,y,z} = E_{0x,y,z}\ \exp\left[-\frac{z}{d_p}\right] \qquad [23]$$

The subscripts x, y, z stand for the electric field components of the evanescent wave in the corresponding directions. Subscript 0 denotes the value at $z = 0$ (interface in medium 2) and d_p is the so-called *penetration depth*, which results in

$$d_p = \frac{\lambda/n_1}{2\pi\sqrt{\sin^2\theta_i - n_{21}^2}} \qquad [24]$$

where λ denotes the wavelength in vacuum and $\lambda/n_1 = \lambda_1$ is the wavelength in medium 1, i.e. in the internal reflection element (IRE). A typical ATR setup is shown in **Figure 6**.

Hence the penetration depth is the distance in the z-direction within which the electric field is decreased by a factor of 1/e. This distance varies between a fraction of a micron and a few microns depending on the refractive indices of the IRE (e.g. $n_{Ge} = 4.0$, $n_{ZnSe} = 2.4$) and the medium 2($n_2 \approx 1.5$ for organic materials), as well as the angle of incidence. Harrick has given further practical details (see Further reading section). As a consequence of Equations [23] and [24], the ATR spectrum features information on materials within a distance of one or a few d_p from the reflecting interface, resulting the highest sensitivity at the interface. Therefore, ATR spectroscopy is an optimum tool for *in situ* thin immobilized layer analysis. For reviews on membrane spectroscopy the reader is referred to the Further reading section.

Quantitative analysis of ATR spectra

The concept of effective thickness The concept of effective thickness was introduced by Harrick. The quantity d_e indicates the thickness of a sample that would result in the same absorbance in a hypothetical transmission experiment, as obtained with the genuine ATR experiment. This concept enables the straightforward application of the Lambert–Beer's law on ATR data according to

$$T = 10^{-N\varepsilon c d_e} = 10^{-A} \qquad [25]$$

where $A = N\varepsilon c d_e$ denotes the absorbance resulting from N internal reflections. For an isotropic layer extending from $z = z_i$ to $z = z_f$ one obtains

$$d_e^{iso} = \frac{1}{\cos\theta_i} \frac{n_2}{n_1} \frac{d_p}{2} E_{02}^{r^2}$$
$$\times \left[\exp\left(-\frac{2z_i}{d_p}\right) - \exp\left(-\frac{2z_f}{d_p}\right) \right] \qquad [26]$$

According to Equation [26] d_e turns out to be wavelength dependent via d_p. As a consequence, ATR spectra of bulk media generally show increasing intensity with increasing wavelength. However, if the thickness of the layer $d = z_f - z_i$ is small compared with d_p then Equation [26] reduces to Equation [27], which is independent of the wavelength. E_{02}^r denotes the relative electric field component at the reflecting interface of medium 2. For $z_i = 0$ one obtains

$$d_e^{iso} = \frac{1}{\cos\theta_i} \frac{n_2}{n_1} d\, E_{02}^{r^2} \qquad [27]$$

For a bulk medium extending from $z_i = 0$ to $z_f = \infty$ Equation [26] results in

$$d_e^{iso} = \frac{1}{\cos\theta_i} \frac{n_2}{n_1} \frac{d_p}{2} E_{02}^{r^2} \qquad [28]$$

A more detailed presentation including an approximate calculation of the effective thickness for intermediate layer thickness, i.e. $d \approx d_p$, and references for a rigorous application of the general formalism are given in the Further reading section.

Relative electric field components The relative electric field components E_{02}^r are obtained from Equations [19] according to

$$E_{02,\parallel}^r = \sqrt{(t_\parallel t_\parallel^*)}; \quad E_{02,\perp}^r = \sqrt{(t_\perp t_\perp^*)}$$
$$E_{02,x}^r = E_{02,\parallel}^r \cdot \cos\theta_t; \quad E_{02,z}^r = E_{02,\parallel}^r \cdot \sin\theta_t \qquad [29]$$
$$E_{02,y}^r = E_{02,\perp}^r$$

Explicit expressions of Equations [29] for thin films and bulk media can be found in references in the Further reading section.

Validity of the effective thickness concept Since the effective thickness concept permits the application of Lambert–Beer's law to ATR data, experimental validation may be performed easily by comparing spectra of the same sample measured by both, ATR and transmission (T). As long as the results do not differ significantly from each other the formalism described above is considered to be applicable. ATR and T measurements with aqueous solutions of Na_2SO_4 have shown that at a 1 M concentration Lambert–Beer's law is still fulfilled for the very intense SO_4^{2-} stretching band at 1100 cm^{-1}. Even for the strong H_2O bending [$\delta(H_2O)$] band of liquid water at 1640 cm^{-1} the integral molar absorption coefficients determined by ATR with a germanium IRE at an angle of incidence of $\theta_i = 45°$ was found to be equal to T data within the experimental error. However, a few per cent deviations were found when peak values of the absorbance were used to determine the molar absorption coefficient. The latter indicates the onset of band distortion, a phenomenon well known in ATR spectroscopy under conditions of strong absorption. This finding is in accordance with calculations by Harrick using Fresnel's equations with complex refractive indices. For Ge in contact with liquid water and $\theta_i = 45°$ the analysis resulted in an upper limit of the absorption coefficient $\alpha_{max} \approx 1000$ cm^{-1}. The concept of effective thickness as described above may be considered to be valid for $\alpha < \alpha_{max}$. For organic compounds this condition is generally fulfilled. In case of $\delta(H_2O)$ of liquid water, however, the absorption coefficient results in $\alpha = \varepsilon(1640 \text{ cm}^{-1})c = 1.82 \times 10^4 \text{ cm}^2 \text{ mol}^{-1} \times 5.56 \times 10^{-2} \text{ mol cm}^{-3} = 1011.9 \text{ cm}^{-1}$, which indicates that the limit of validity of the approach is reached, in complete accordance with experimental data mentioned above. Furthermore, it turned out that the anomalous dispersion in the range of strong water absorption bands should be taken into account if sample absorption within this range is analysed quantitatively.

Quantitative analysis taking sample absorption and thickness into account Recently, the validity of Harrick's weak absorber approximations has been checked by comparison with the general thickness- and absorption-dependent model. It was found that the formalism depicted in the section above may be used for film thicknesses up to 20 nm. Especially if the film is in contact with a third bulk medium, e.g. water, the deviation between accurate and approximate calculation of relative electric field components according to Equation [29] was found to be below 3%, i.e. within the error of most experiments. A comprehensive description of ATR spectroscopy of polymers using the general formalism can be found in the Further reading section.

Since quantitative analysis of ATR data by the general formalism is very cumbersome, the use of more tractable approximations is recommended if possible. One possibility is the weak absorber approximations described above; another approach was derived for the study of electrochemical reactions by ATR spectroscopy.

Further more general information on quantitative methods and applications of ATR spectroscopy, see the Further reading section.

Orientation measurements Considering a transition dipole moment M associated with a vibrational mode of a given molecule and the electric field E, responsible for vibrational excitation, the magnitude of light absorption depends on the mutual orientation of these two vectors according to Equation [30] which is the basis for orientation measurements. M_x, M_y and M_z denote the components of the transition dipole moment in the IRE fixed coordinate system shown in **Figure 6**.

$$A = (E \cdot M)^2$$
$$= |E|^2 \cdot |M|^2 \cdot \cos^2(E, M) \quad [30]$$
$$= (E_x M_x + E_y M_y + E_z M_z)^2$$

It is usual to work with dimensionless relative intensities instead of absolute intensities in order to get rid of physical and molecular constants, e.g. the magnitude of the transition moment. Introducing the so-called dichroic ratio R, the absorbance ratio obtained from spectra measured with parallel and perpendicular polarized incident light, i.e.

$$R = \frac{A_\parallel}{A_\perp} = \frac{d_{e,\parallel}}{d_{e,\perp}}$$
$$= \frac{E_x^2 \langle m_x^2 \rangle + E_z^2 \langle m_z^2 \rangle + 2 E_x E_z \langle m_x m_z \rangle}{E_y^2 \langle m_y^2 \rangle} \quad [31]$$

where A and d_e denote absorbance and effective thickness relative to $\parallel$- and $\perp$-polarized incident light, respectively. E_x, E_y and E_z denote the relative electric field components according Equation [29]. $\langle m_x^2 \rangle$, $\langle m_y^2 \rangle$, $\langle m_z^2 \rangle$ and $\langle m_x m_z \rangle$ are ensemble mean values of the components of the unit vector in the direction of the transition dipole moment, see **Figure 6**. $\langle m_x m_z \rangle = 0$ for uniaxial orientation, e.g. isotropic distribution around all relevant orientation axes. On substitution of the unit vector components according to the geometry depicted in **Figure 6**, one obtains for the dichroic ratio

$$R^{ATR} = \frac{E_x^2}{E_y^2} + 2 \frac{E_z^2}{E_y^2} \frac{\langle \cos^2 \alpha \rangle}{1 - \langle \cos^2 \alpha \rangle} \quad [32]$$

where $\langle \cos^2 \alpha \rangle$ denotes the mean square cosine of the angle between the transition moment and the normal to the interface. This quantity is accessible via two measurements, one with $\parallel$-polarization and the other with $\perp$-polarized incident light, resulting in R. The relative electric field components are available from Fresnel's Equations [19] with Equation [29]. For an isotropic sample, $\langle \cos^2 \alpha \rangle_{iso} = 1/3$. Insertion of this value into Equation [32] result in

$$R_{iso}^{ATR} = \frac{E_x^2 + E_z^2}{E_y^2} \quad [33]$$

It should be mentioned that an isotropic sample results in $R = 1$ in transmission but, according to Equation [33], not in ATR, where $R_{iso}^{ATR} \neq 1$. As an example, for total reflection with a bulk material 2 and $\theta_i = 45°$, one obtains $R_{iso}^{ATR} = 2.0$ irrespective of n_1 and n_2, except $n_1 > n_2$.

The segmental order parameter S_{seg} is frequently used to characterize molecular ordering, e.g. To describe the fluctuation of a functional group in a molecule via the polarized absorption bands of a typical group vibration. For uniaxial orientation the order parameter is defined according to

$$S_{\text{seg}} = \frac{3}{2} \langle \cos^2 \alpha \rangle - \frac{1}{2} \qquad [34]$$

Thus, if the ATR geometry and the optical constants of the system are known, then S_{seg} may be determined measuring the dichroic ratio R of a given absorption band, followed by the calculation of the mean square cosine $\langle \cos^2 \alpha \rangle$ and inserting this value into Equation [34]. A typical example is discussed later.

For more details and examples of application the reader is referred to the Further reading section.

Determination of surface concentration The effective thickness as indicated by Equations [26]–[28] holds for isotropic samples. Modification for oriented samples results

$$d_{ex} = 3 \langle m_x^2 \rangle d_{ex}^{\text{iso}} = \frac{3}{2}(1 - \langle \cos^2 \alpha \rangle) d_{ex}^{\text{iso}}$$

$$d_{ey} = 3 \langle m_y^2 \rangle d_{ey}^{\text{iso}} = \frac{3}{2}(1 - \langle \cos^2 \alpha \rangle) d_{ey}^{\text{iso}} \qquad [35]$$

$$d_{ez} = 3 \langle m_z^2 \rangle d_{ez}^{\text{iso}} = 3 \langle \cos^2 \alpha \rangle d_{ez}^{\text{iso}}$$

From Equation [25] and the relation $d_{e\parallel} = d_{ex} + d_{ez}$ and $d_{e\perp} = d_{ey}$ one obtains for the surface concentration Γ.

$$\begin{aligned}\Gamma = c \cdot d &= \frac{d \cdot A_{\parallel}}{\nu \cdot N \cdot d_{e\parallel} \cdot \varepsilon} = \frac{d \cdot \int A_{\parallel} d\tilde{\nu}}{\nu \cdot N \cdot d_{e\parallel} \cdot \int \varepsilon d\tilde{\nu}} \\[2mm] &= \frac{d \cdot A_{\perp}}{\nu \cdot N \cdot d_{e\perp} \cdot \varepsilon} = \frac{d \cdot \int A_{\perp} d\tilde{\nu}}{\nu \cdot N \cdot d_{e\perp} \cdot \int \varepsilon d\tilde{\nu}}\end{aligned} \qquad [36]$$

where $A_{\parallel}$ and $A_{\perp}$ denote the absorbance measured with parallel and perpendicular polarized incident light, respectively, ε is the molar absorption coefficient, ν denotes the number of equal functional groups per molecule and N is the number of active internal reflections. It should be noted that Equation [36] holds for peak absorbance and integrated absorbance, provided that the corresponding molar absorption coefficients are used.

Special experimental ATR techniques

Single-beam sample reference (SBSR) technique Most FTIR spectrometers are working in the

single-beam (SB) mode. As a consequence a single-channel reference spectrum has to be stored for later conversion of single-channel sample spectra into transmittance and absorbance spectra. This technique suffers inaccuracy owing to drifts resulting from the instrument, the sample or atmospheric absorption. In order to eliminate these unwanted effects to a great extent, a new ATR attachment has been constructed, converting a single-beam instrument into a pseudo-double-beam instrument. The principle features of this attachment are depicted in **Figure 7**. As usual, a convergent IR beam enters the sample compartment. However, the focal point is now displaced from the centre of the sample compartment by means of the planar mirrors M1 and M2 to the new position F. The off-axis

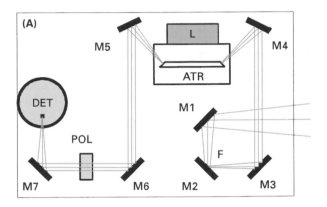

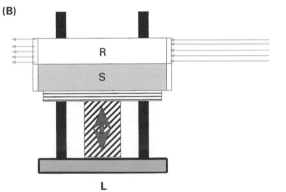

Figure 7 Single-beam sample reference (SBSR) ATR attachment. (A) The focus in the sample compartment is displaced to the position F by the planar mirrors M1 and M2. The off-axis parabolic mirror M3 produces a parallel beam with a diameter of 1 cm, i.e. half of the height of the MIRE. The cylindrical mirror M4 focuses the light to the entrance face of the MIRE. M5, which has the same shape as M4, reconverts to parallel light directing it via the planar mirror M6 through the polarizer POL and it is then being focused to the detector DET by the off-axis parabolic mirror M7. (B) Alternating change from sample to reference is performed by computer-controlled lifting and lowering of the ATR cell body. Reproduced with permission of the American Institute of Physics from Fringeli UP *et al.* (1998) *AIP Conference Proceedings* **430**: 729–747.

parabolic mirror M3 performs a conversion of the divergent beam into a parallel beam with fourfold reduced cross-section. This beam is focused to the entrance face of a trapezoidal MIRE by a cylindrical mirror M4. Therefore, the ray propagation in the MIRE is still parallel to the direction of light propagation (x-axis), enabling subdivision of the two reflective faces (x,y-planes) of the MIRE alongside at half-height. One half of the MIRE is then used for the sample (S) and the other one for the reference (R). Both S and R were encapsulated by flow-through cuvettes, independently accessible by liquids or gases. This principle is referred to as the Single beam sample reference (SBSR) technique.

SBSR absorbance spectra are calculated from sample and reference single-channel spectra which have been measured with very short mutual time delay. A most favourable benefit of SBSR technique is that no waisted time for purging is required before starting a measurement after closing the sample compartment. Moreover, the whole sequence of single-channel spectra in the S and R channels is also available, allowing reconstruction of the history of each channel at any time by conventional data handling.

Modulated excitation (ME) spectroscopy and 2D IR spectroscopy Change of any external thermodynamic parameter generally exerts a specific influence on the state of a chemical system. The system response will be relaxation from the initial state (e.g. an equilibrium) to the final state (a new equilibrium state or a stationary state). In the case of a periodic change (modulation) of the parameter, the system response will also be periodic with angular frequency ω, relaxing from the initial state to a stationary state. All absorption bands of the spectrum which result from stimulated molecules or parts of them will be labelled by the frequency ω. As a consequence, it is possible to separate the modulated response of a system from the stationary response, resulting from parts of the system that were not affected by modulated excitation (ME) as well as from the background. Moreover, if the kinetics of the stimulated process is in the same time range as the period of external excitation, phase lags and damped amplitudes will result. Both depend characteristically on the stimulation frequency, and therefore one can derive relevant information on the reaction scheme and the kinetics of the stimulated process (see also caption to **Figure 8**).

A variety of ME experiments have been reported. (i) Temperature ME of poly-L-lysine was used to study induced periodic secondary structural changes as well as the sequence of transients. (ii) The classical ATR setup (see **Figure 6**) facilitates the application of electric fields to immobilized thin films, such as biomembrane assemblies or to bulk materials such as liquid crystals, since a Ge ATR plate, supporting the membrane, may be used as one electrode, and the back-wall of the cuvette as counter electrode. (iii) Hydration modulation was used to detect hydration sites of model membranes, and (iv) ME by UV radiation permitted kinetic studies of photoinduced chemical reactions. (v) ME by chemical substrates is a further versatile method to study chemically induced conformational changes of a sample immobilized to the MIRE. For that purpose, two computer-controlled pumps are used for periodic exchange the liquid (water) environment of the sample in a flow-through cell. An example demonstrating the sensitivity and high quality of background compensation of ME techniques is presented later. The principles of ME spectroscopy are depicted schematically in **Figure 8**.

2D FTIR spectroscopy Absorption bands in a set of modulation spectra that exhibit equal phase shifts with respect to the external stimulation are considered to be correlated. 2D correlation analysis is a statistical graphical means to visualize such a correlation in a 2D plot. Consequently, phase-resolved modulation spectra are data of a higher level and unambiguously allow a more direct and accurate evaluation. 2D plots look attractive, but, one should be aware that the information content is lower than that of the underlying modulation spectra, first because band overlapping may result in inadequate phase information, and second because 2D spectra are affected much more by baseline errors than the original modulation spectra. A comprehensive discussion can be found in the Further reading section.

Sensitivity of ATR spectroscopy

Sensitivity of stationary ATR measurements Commercial multiple internal reflection elements MIRE permit up to 50 internal reflections. This is generally enough for thin-layer spectroscopy in the nanometre or even subnanometre region. As an example, **Figure 9** shows a dipalmitoylphosphatidic acid monolayer, i.e. a lipid monolayer of about 2 nm thickness, which has been transferred from the air–water interface to a germanium MIRE by means of the Langmuir–Blodgett technique.

The dominant bands in **Figure 9** result from the stretching vibrations of 28 CH_2 groups of the two saturated hydrocarbon chains of the DPPA molecule. Looking at three resolved weaker bands gives an impression of the absorbance to be expected from a monomolecular coverage by functional groups of medium or weak molar absorption. The first is the terminal methyl group of the hydrocarbon chains. The

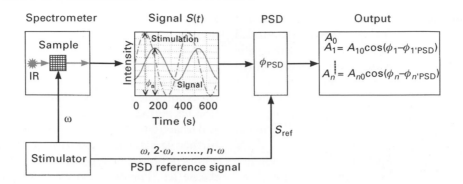

Figure 8 Schematic setup for modulated excitation (ME) experiments. A periodic excitation is exerted on the sample with frequency ω. The sample response $S(t)$, as sensed by IR radiation, contains the frequency ω and higher harmonics at wavelengths that are significant for those parts of the sample that have been affected by the stimulation. Selective detection of the periodic sample responses is performed by phase-sensitive detection (PSD), resulting in the DC output and A_n of fundamental ω ($n = 1$) and their harmonics $n\omega$ ($n = 2, 3, ...$), as well as the phase shifts ϕ_n between the nth harmonic and the stimulation. This phase shift is indicative of the kinetics of the stimulated process and of the underlying chemical reaction scheme. Since the PSD output A_n ($n = 1, 2, ...,n$; frequency $n\omega$) is proportional to $\cos(\phi_n - \phi_{n,PSD})$, absorption bands featuring the same phase shift ϕ_n are considered to be correlated, i.e. to be representative of a population consisting of distinct molecules or molecular parts. $\phi_{n,PSD}$ is the operator-controlled PSD phase setting. Because of the cosine dependence, different populations will have their absorbance maxima at different $\phi_{n,PSD}$ settings, thus allowing selective detection. Moreover, since in the case that $0.1 < \omega\tau_i < 10$ (τ_i denotes the ith relaxation time of the system), ϕ_n becomes ω dependent, $\phi_n = \phi_n(\omega)$. The spectral information can then be spread in the $\phi_{n,PSD} - \omega$ plane, resulting in a significant enhancement of resolution with respect to standard difference spectroscopy and time-resolved spectroscopy.

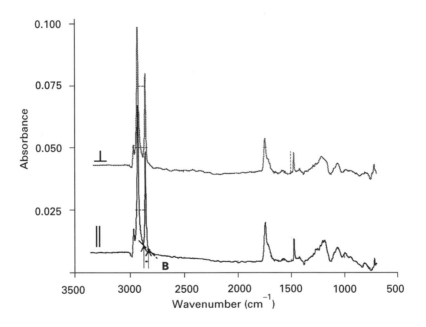

Figure 9 Parallel (∥) and perpendicular (⊥) polarized ATR absorbance spectra of a dipalmitoylphosphatidic acid (DPPA) monolayer transferred at 30 mN m^{-1} from the aqueous subphase (10^{-4} M CaCl$_2$) to a germanium multiple internal reflection element (MIRE). Spectra were obtained from the dry monolayer in contact with dry air. A surface concentration of $\Gamma = 3.93 \times 10^{-10}$ mol cm^{-2} was calculated by means of Equation [36] using the dichroic ratio of the symmetric CH$_2$ stretching vibration at 2850 cm^{-1} with respect to a linear baseline (B), resulting in R^{ATR} [ν^s(CH$_2$)] = 0.923. Angle of incidence $\theta_i = 45°$; number of equal functional groups $\nu = 28$; number of active internal reflections $N = 39$.

antisymmetric stretching vibration, ν_{as}(CH$_3$) absorbs at ~2960 cm^{-1}. As concluded from **Figure 9**, this monolayer results in a peak absorbance of about 6 mAU. A weaker band is observed near 1420 cm^{-1} and may be assigned to the bending vibration of the α-methylene groups of the hydrocarbon chains,

$\delta(\alpha\text{-CH}_2)$. Thus an approximate monolayer of α-CH$_2$ groups results in an absorbance of only about 1 mAU. Third, a monolayer of phosphate head groups results in more intense absorption bands because of the larger transition dipolemoment of the polar group. The corresponding absorbances of PO$_3$ stretching vibrations in the range 1000–1250 cm^{-1} are within 5 and 10 mAU. It is concluded that conventional ATR measurements may allow significant access to bands of about 0.2–0.5 mAU, which corresponds to 20–50% of a monolayer of weak absorbers.

Quantitative analysis of stationary ATR spectra The DPPA monolayer spectra shown in **Figure 9** are now used to demonstrate the ease of application of the formalism for quantitative analysis of ATR spectra presented earlier.

Dichroic ratio of symmetric CH$_2$ stretching The dichroic Ratio according to Equation [32] was calculated from the integrated absorbances of the symmetric CH$_2$ stretching bands, $\nu_s(CH_2)$, using linear baselines as marked in **Figure 9** with lower and upper limits at 2828 and 2871 cm^{-1}, respectively. The corresponding integrals were found to be $\int A_\parallel d\tilde{\nu} = 0.381$ cm^{-1}, and $\int A_\perp d\tilde{\nu} = 0.413$ cm^{-1} resulting in $R^{ATR} = 0.923$. This is the relevant experimental quantity.

Mean orientation of hydrocarbon chains Uniaxial orientation, i.e. isotropic distribution of DPPA around the z-axis, is assumed. The mean square cosine of the angle between the transition dipole moments of $\nu_s(CH_2)$ of the whole population of CH$_2$ groups of the molecule (28 groups in hydrocarbon chains, 1 in the glycerol part, slightly shifted in frequency) can be calculated from Equation [32], resulting in Equation [37].

$$\langle \cos^2 \alpha \rangle = \frac{(R^{ATR} - E_x^2/E_y^2)(E_y^2/E_z^2)}{2 + (R^{ATR} - E_x^2/E_y^2)(E_y^2/E_z^2)} \quad [37]$$

The squares of relative electric field components at the interface $(z = 0)$ in medium 2 as calculated from Equation [29] for $\theta_i = 45°$, $n_1 = 4.0$ (germanium), $n_2 = 1.5$ (DPPA monolayer) and $n_3 = 1.0$ (dry air) result in $E_{0x,2}^2 = 1.991$, $E_{0y,2}^2 = 2.133$, and $E_{0z,2}^2 = 0.450$. It follows that $E_{0\parallel,2}^2 = E_{0x,2}^2 + E_{0z,2}^2 = 2.441$ and $E_{0\perp,2}^2 = E_{0y,2}^2 = 2.133$. The dichroic ratio for an isotropic film under these conditions would result in, according to Equation [33], $R_{iso,2}^{ATR} = 1.144$. Explicit equations for relative electric components calculated by means of Harrick's weak absorber approximation can be found in the Further reading section.

Introducing the experimental value of R^{ATR} and the calculated squares of relative electric field components into Equation [37], one obtains for the mean square cosine of the angle between the transition moment of $\nu_s(CH_2)$ and the z-axis $\langle \cos^2 \alpha \rangle = -0.025$. This value should not be negative because its minimum is zero; however, since it is

small, we consider it to be within experimental and predominantly systematic errors. Therefore, we set $\langle \cos^2 \alpha \rangle = 0$, resulting in $\alpha = 90°$. This result requires that all methylene groups of the hydrocarbon chains assume an all-*trans* conformation and, moreover, all hydrocarbon chains are aligned normal to the MIRE, i.e. parallel to the z-axis (tilt angle 0°). The exact wavenumber of the symmetric stretching vibration of the CH$_2$ group in glycerol is not known. However, overlapping with $\nu_s(CH_2)$ of the hydrocarbon chains is probable. Consequently, the bisectrice of the glycerol CH$_2$ group may also be concluded to be predominately parallel to the x,y-plane.

Mean order parameter of CH$_2$ groups The mean segmental order parameter resulting from Equation [34] is found to be $\bar{S}_{seg}[\nu_s(CH_2)] = -\frac{1}{2}$. This value is representative of a perfectly ordered molecular entity with isotropic arrangement of transition dipole moments around the z-axis and perfect parallel alignment to the interface (x,y-plane). It should be noted that for $\langle \cos^2 \alpha \rangle = 1$, i.e. transition moments perfectly aligned normal to the interface (z-axis), Equation [34] results in the upper limit $\bar{S}_{seg} = 1$. Lipids in natural biomembranes consist of a considerable amount of unsaturated hydrocarbon chains. Since double bonds cause unavoidably *gauche* defects in elongated hydrocarbon chains, which leads to a reduced chain ordering, $\bar{S}_{seg}[\nu_s(CH_2)]$ is increased, reaching zero for an isotropic chain arrangement, since $\langle \cos^2 \alpha \rangle = 1/3$ in this case.

It should be noted that the determination of order parameters of individual methylene groups in the hydrocarbon chains requires generally selective deuteration. In this respect, comprehensive deuterium NMR work should be mentioned (see Further reading section).

A more general case of sample geometry is that of a transition moment being inclined by an angle Θ with respect to the molecular axis a and isotropically distributed around a. Furthermore, the molecular axis a forms an angle γ with respect to the tilt axis t, and is isotropically distributed around it, and finally, the axis t forming a tilt angle δ with the z-axis and is isotropically distributed around it. In this case, the segmental order parameter, e.g. $S_{seg}[\nu_s(CH_2)]$, may be expressed as superposition of three uniaxial orientations according to

$$S_{seg} = \left(\frac{3}{2}\langle \cos^2 \delta \rangle - \frac{1}{2}\right) \cdot \left(\frac{3}{2}\langle \cos^2 \gamma \rangle - \frac{1}{2}\right)$$
$$\times \left(\frac{3}{2}\langle \cos^2 \Theta \rangle - \frac{1}{2}\right)$$
$$= S_\delta \cdot S_\gamma \cdot S_\Theta \quad [38]$$

The angles δ, γ and Θ may be distinct or fluctuating (partly or all), describing a microcrystalline ultrastructure (MCU) and a liquid crystalline ultrastructure (LCU), respectively. S_γ is referred to as the molecular order parameter S_{mol}.

Applying Equation [38] to the DPPA monolayer under discussion, one obtains: $S_\delta = 1$, $S_\gamma = 1$, $S_\Theta = -\frac{1}{2}$, meaning no tilt ($\delta = 0$), molecular axis (hydrocarbon chain) normal to the interface ($\gamma = 0$), and transition dipole moment normal to the molecular axis ($\Theta = 90$).

Surface concentration and area per molecule The surface concentration may be calculated using Equation [36]. The following additional information is required: (i) the integrated molar absorption coefficient related to a linear baseline from 2828 to 2871 cm^{-1} (see **Figure 9**) was $\int \varepsilon d\tilde{v} = 5.7 \times 10^5$ cm mol^{-1}, (ii) the real thickness of the layer was assumed to be $d = 2.5$ nm, (iii) the number of equal functional groups $v = 28$, and (iv) the effective thicknesses d_e for parallel or perpendicular polarized incident light, which were calculated from Equations [27] and [35], resulted in $d_{e,\parallel} = 3.97$ nm and $d_{e,\perp} = 4.30$ nm. The mean surface concentration was found to be 3.93×10^{-10} mol cm^{-2}, corresponding to a molecular cross-section of 0.427 nm^2 per molecule (42.3 Å^2 per molecule). This value leads to the conclusion that the two hydrocarbon chains of a DPPA molecule predominantly determine the area per molecule, since the cross-section of an elongated hydrocarbon chain is 20–21 Å^2.

Conclusions Quantitative analysis, including orientation measurements, has been shown to be straightforward when the formalism based on Harrick's weak absorber approximation is applied. For thin adsorbed layers, such as the DPPA monolayer under discussion, the results are fairly good. Application to bulk materials may introduce systematic errors as discussed above. If the weak absorber approximation is still to be applied, one should take care to work with an angle of incidence which is at least 15° larger than the critical angle, in order to avoid significant band distortions. In many cases it is possible to use quantitative data from transmission experiments to check the validity of the formalism applied to ATR data.

A general critical aspect concerning the baseline selection should be mentioned. A linear tangential baseline has been used for quantitative analysis of the symmetric CH$_2$ stretching vibration of DPPA (see **Figure 9**). Obviously the correct baseline is lower, i.e. the integrated absorbances used for analysis are systematically too small. The reason for this procedure is only to permit good reproducibility. While the determination of the dichroic ratio is indifferent with respect to the choice of the baseline, it is mandatory to use integrated or peak molar absorption coefficients which have been determined under the same conditions. Even then deviations in the range of several per cent may occur among different operators.

Finally, it should be noted that ATR spectroscopy allows very good background compensation, when adequate equipment is used.

Sensitivity of modulated excitation (ME) ATR spectroscopy An impression of the sensitivity of stationary measurements was given the last but one section. A limit of 0.2 mAU is suggested. This limit is beaten by one order or magnitude when the ME technique is applied. As mentioned above, the sample must fulfil the condition of a reversible stimulation by a periodically altered external thermodynamic parameter. Here, the excellent sensitivity and instrumental stability will be demonstrated, for example, with a chemical modulation experiment performed in liquid water, a very strong absorber in the 3400 and 1640 cm^{-1} region.

In order to study the influence of immobilized charges on a lipid model membrane, an arachidic acid (ArAc) bilayer was prepared on a germanium MIRE by means of the Langmuir–Blodgett (LB) technique. The MIRE was transferred in the hydrated state from the LB trough into a flow-through cell and kept in permanent contact with an aqueous buffer solution. Since the carboxylic acid groups of the second monolayer were facing the aqueous phase, the degree of protonation could be controlled via the environmental pH. A periodic pH modulation between pH 3 and 10 induced a periodic protonation and deprotonation of the carboxylic acid group. It should be noted that the first ArAc LB layer was attached by head to the Ge MIRE. Obviously, this binding was so special that typical absorption bands of the carboxylic acid groups were not visible in the spectrum. Therefore, one may assume that the head group signals shown in **Figure 10** result predominately from the outer monolayer of ArAc. The stationary spectral intensity is comparable to that of the DPPA monolayer shown in **Figure 9**. Moreover, one should note that the experiments were performed in H$_2$O, where in the 1640 and 3400 cm^{-1} regions there is very low spectral energy available, favouring perturbations by incomplete background compensation.

In this context, only the sensitivity and selectivity of ME techniques will be discussed. A comprehensive presentation and analysis of polarized pH modulation spectra will be given elsewhere.

Sensitivity of ME-spectroscopy Taking the pH-modulated spectrum shown in **Figure 10** as a typical example, one may estimate the sensitivity by comparing the most intense ME spectra ($\phi_{PSD} = 60°/\phi_{PSD} = 90°$, the maximum is expected at $\phi_{PSD} \approx 75°$, consequently, $\phi_{PSD} \approx 165°$ should result intensity zero) with the lowest intensity ME spectrum at $\phi_{PSD} = 0°$. In order to check the S/N ratio, the $\phi_{PSD} = 0°$ spectrum was expanded 25 times in the CH_2 wagging region and is plotted as a dashed inset in **Figure 10**. The ordinate scaling factor for the zoomed spectrum is 4.0×10^{-5}. Comparing it with the other ME spectra (scaling factor 1.0×10^{-3}) one can conclude that bands as weak as 1.0×10^{-5} AU are still detectable.

Selectivity of pH ME The highest selectivity of ME spectroscopy is achieved if the stimulation frequency ω and the kinetics of the stimulated process are matched, i.e. if $0.1 < \omega\tau_i < 10$, where τ_i denotes the *i*th relaxation time of the system. τ_i is a function of the rate constants involved in the stimulated process. Under these conditions, significant ω-dependent phase shifts are expected, resulting in $\phi_i = \arctan(-\omega\tau_i)$ for a linear system. Consequently, a molecular or confor-

mational population represented by the relaxation time τ_i exhibits maximum absorbance in the ME spectrum at a PSD phase setting $\phi_i = \phi_{PSD}$, thus allowing selective detection and kinetic analysis by means of phase-resolved ME spectra. Moreover, ω acts as an additional experimental degree of freedom in this context, since information on selectivity and kinetics can be spread in the ω/ϕ_{PSD} plane which is more selective than the unidirectional information resulting from conventional relaxation measurements.

In the actual case of pH modulation exerted on a monolayer of ArAc, there is no phase resolution observed, owing to the long modulation period of $\tau_m = 16$ min, i.e. no kinetic information is available. However, unambiguous discrimination between the protonated and deprotonated populations is possible. Only one characteristic example will be given here. The most prominent band from the protonated state is the C=O stretching vibration ν(COOH) of the carboxylic acid group near 1700 cm^{-1}. All other bands in the ME spectrum that have the same phase belong to the protonated population, whereas the remaining bands featuring opposite sign are members of the deprotonated population. Consequently, if no

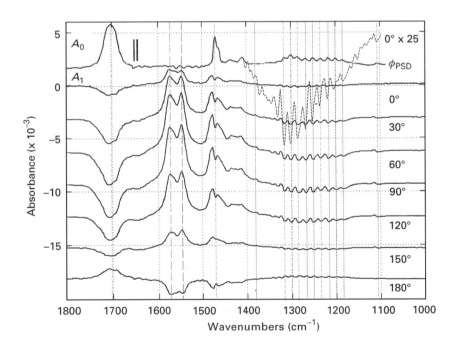

Figure 10 pH-modulated excitation (ME) of an arachidic acid (ArAc) bilayer attached to a germanium multiple internal reflection element (MIRE). ME was performed by pumping alternatively two buffer solutions (100 mM NaCl, pH 3 and 100 mM NaCl, pH 10) through the ATR cuvette with a modulation period of $\tau = 16$ min. T = 10°C. Upper trace A_0; stationary spectrum of a protonated ArAc layer for comparison with modulation spectra. Traces A_1; modulation spectra at PSD phase settings $\phi_{PSD} = 0, 30,..., 180°$. The 180° spectrum corresponds to the 0° spectrum with opposite sign, because the PSD output is proportional to cos $(\phi - \phi_{PSD})$, see also **Figure 8**. ϕ denotes the phase difference between a given band and the stimulation. Owing to the long period of $\tau_m = 16$ min, the observed bands in the modulation spectra exhibit only two resolved ϕ values, which are 180° apart, as a consequence of the fact that the chemical relaxation time of protonation/deprotonation of ArAc is much shorter than the stimulation period. In order to demonstrate the excellent S/N ratio, the ordinate of the weakest modulation spectrum has been expanded in the CH_2 wagging region by a factor of 25, i.e. the ordinate scaling factor for the dashed spectrum results in 4.0×10^{-5} (see text).

phase resolution is achieved, ME spectra reduce to difference spectra, which, however, have a considerably better background and instability compensation than conventional difference spectra, since corresponding sample and reference spectra are measured and evaluated/accumulated within each period of stimulation.

Consider now the wagging region $\gamma_w(CH_2)$ of the spectra shown in **Figure 10**. The wagging motion $\gamma_w(CH_2)$ is described as in-phase displacement of both H atoms through the H–C–H plane of a methylene group, where the C atom remains predominately in place. In an all-*trans* hydrocarbon chain the transition dipole moment of $\gamma_w(CH_2)$ is expected to be parallel to the chain direction. Deviations may occur, however, from coupling with a polar end group. In the stationary absorbance spectrum A_0, one can observe nine weak bands between about 1180 and 1320 cm^{-1}. This sequence results from concerted wagging vibrations of all methylene groups in a hydrocarbon chain with an all-*trans* conformation. According to IR selection rules one has to expect $n/2$ IR-active vibrations for an even number n of CH_2 groups in an all-*trans* conformation. ArAc has 18 CH_2 groups per chain, resulting in the above-mentioned sequence of nine bands in accordance with theory. Since these bands are found to be in phase with $\nu(COOH)$, one can conclude that deprotonation of COOH is paralleled by reversible disordering of the chain structure, most probably by introducing *gauche* defects.

Finally, it should be mentioned that $\gamma_w(CH_2)$ belongs to the group of weak absorption bands. One can conclude, therefore, that ME IR ATR spectroscopy allows significant quantitative studies on a molecular level with submonolayer quantities of weak absorbers.

Manufacturers of reflection accessories

Standard equipment for reflection spectroscopy

ASI Sense IR Technologies, 15 Great Pasture Road, Danbury, CT 06810, USA.

Bruker Optics, Wikingerstrasse 13, D-76189 Karlsruhe, Germany.

Graseby Specac Inc., 301 Commerce Drive, Fairfield, CT 06432, USA.

Harrick Scientific Corporation, 88 Broadway, Ossining, NY 10562, USA.

International Crystal Laboratories, 11 Erie Street, Garfield, NJ 07026, USA.

Spectra-Tech, Inc., Warrington WA3 7BH, UK.

Special equipment for SBSR-ATR and ME-ATR spectroscopy

Optispec, Rigistrasse 5, CH-8173 Neerach, Switzerland.

List of symbols

A = absorbance (decadic); d = sample thickness; d_e = *effective thickness* (Harrick); d_p = penetration depth; $\bar{d}$ = mean penetrated layer thickness; E = electric field; E_x = electric field component in x-direction; E_y = electric field component in y-direction; E_z = electric field component in z-direction; $\bar{J}$ = energy flux (time average); m_x = x-component of the unit vector in the direction of M; m_y = y-component of the unit vector in the direction of M; m_z = z-component of the unit vector in the direction of M; M = transition dipole moment; n = refractive index; $\hat{n}$ = complex refractive index; N = number of active internal reflections; r = ratio of reflected to incident field; R^{ATR} = dichroic ratio related to ATR spectra; R = dichroic ratio; R = reflectance; $S_\gamma = S_{mol}$ = molecular order parameter; S_θ = tilt order parameter; S_θ = order parameter with respect to the molecular axis; S_{seg} = segmental order parameter; S = Poynting vector (time average); t = ratio of transmitted to incident field; T = transmittance; z = distance from surface; $\parallel$, $\perp$ = parallel and perpendicular polarized light, respectively; $*$ = conjugate complex; α = absorption coefficient; α = angle between transition dipole moment and z-axis; δ = phase shift; ε = permittivity; ε = molar absorption coefficient; ε_0 = permittivity of vacuum; ε_r = relative permittivity (dielectric constant); η = amplitude; θ_c = critical angle; θ_i = angle of incidence; θ_r = angle of reflection; θ_t = angle of refraction (transmission); κ = absorption index; λ = wavelength; μ = permeability; μ_0 = permeability of vacuum; μ_r = relative permeability; ν = number of equal functional groups per molecule; $\tilde{\nu}$ = wavenumber; ρ = reflectance; τ = transmittance; τ = relaxation time; Γ = surface concentration; ϕ = phase lag; ω = angular frequency.

See also: **Electromagnetic Radiation; Industrial Applications of IR and Raman Spectroscopy; Polymer Applications of IR and Raman Spectroscopy; Raman and IR Microspectroscopy; Surface Studies by IR spectroscopy.**

Further reading

Baurecht D and Fringeli UP (2000) Surface charge induced conformational changes in an arachidic acid Langmuir-Blodgett bilayer observed by pH-modulated excitation FTIR ATR spectroscopy. *Langmuir* (in preparation).

Beccard B and Hapanowicz R (1995) Polarization Modulation FT-IR spectroscopy. *Nicolet Application Note AN-9592*. Madison, WI: Nicolet.

Blaudez D, Turlet J-M, Dufourcq D, Bard D, Buffeteau T and Desbat B (1996) Investigation at the air-water interface using polarization modulation IR spectroscopy. *Journal of the Chemical Society, Faraday Transactions* **92**: 525–530.

Born M and Wolf E (1983) *Principles of Optics*, Chapter I. Oxford: Pergamon Press.

Fringeli UP (1992) *In situ* infrared attenuated total reflection membrane spectroscopy. In: Mirabella FM (ed.) *Internal Reflection Spectroscopy, Theory and Applications*, Chapter 10, pp 255–324. New York: Marcel Dekker.

Fringeli UP (1997) Simulatneous phase-sensitive digital detection process for time-resolved, quasi-simultaneously captured data arrays of a periodically stimulated system. *PCT International Patent Application*, WO97/08598.

Fringeli UP, Goette J, Reiter G, Siam M and Baurecht D (1998) Structural investigation of oriented membrane assemblies by FTIR-ATR spectroscopy. In: deHaseth JA (ed.) *Fourier Transform Spectroscopy*; 11th International Conference, AIP Conference Proceedings 430, pp. 729–747. Woodbury, New York: American Institute of Physics.

Galant J, Desbat B, Vaknin D and Salesse Ch (1998) Polarization-modulated infrared spectroscopy and X-ray reflectivity of photosystem II core complex at the gas-water interface. *Biophysical Journal* **75**: 2888–2899.

Greenler RG (1966) Infrared study of adsorbed molecules on metal surfaces by reflection techniques. *Journal of Chemical Physics* **44**: 310–315.

Hansen WH (1973) Internal reflection spectroscopy in electrochemistry. In: Delahay P and Tobias ChW (eds) *Advances in Electrochemistry and Electrochemical Engineering*, Vol 9, Muller RH (ed.) *Optical Techniques in Electrochemistry*, pp 1–60. New York: John Wiley & Sons.

Hapke B (1993) *Theory of Reflectance and Emittance Spectroscopy*. New York: Cambridge Univeristy Press.

Harrick NJ (1967) *Internal Reflection Spectroscopy*, New York: Interscience; 2nd edn (1979) Ossining, NY: Harrick, Scientific.

Hoffmann FM (1983) Infrared reflection-absorption spectroscopy of adsorbed molecules. *Surface Science Reports* **3**: 107–192.

Kortüm G (1969) *Reflectance Spectroscopy*. New York: Springer.

Mendelsohn R, Brauner JW and Gericke A (1995) External infrared reflection absorption spectroscopy of monolayer films at the air-water interface. *Annual Review of Physical Chemistry* **46**: 305–334.

Mirabella FM (ed) (1992) *Internal Reflection Spectroscopy, Theory and Application*. New York: Marcel Dekker.

Müller M, Buchet R and Fringeli UP (1996). 2D-FTIR ATR spectroscopy of thermo-induced periodic secondary structural changes of poly-(L)-lysine: A cross-correlation analysis of phase-resolved temperature modulation spectra. *Journal of Physical Chemistry* **100**: 10810–10825.

Picard F, Buffeteau T, Desbat B, Auger M and Pézolet M (1999) Quantitative orientation measurements in thin films by attenuated total reflection spectroscopy. *Biophysical Journal* **76**: 539–551.

Seelig J and Seelig A (1980) Lipid confromation in model membranes and biological membranes. *Quarterly Review of Biophysics* **13**: 19–61.

Tamm L and Tatulian S (1997) Infrared spectroscopy of proteins and peptides in lipid bilayers. *Quarterly Review of Biophysics* **30**: 365–429.

Urban MW (1996) *Attenuated Total Reflectance Spectroscopy of Polymers*. Washington, DC: American Chemical Society.

Wendlandt WWM and Hecht HG (1996) *Reflectance Spectroscopy*, New York: Interscience.

Wenzl P, Fringeli M, Goette J and Fringeli UP (1994) Supported phospholipid bilayers prepared by the "LB/Vesicle Method": A Fourier transform infrared attenuated total reflection spectroscopic study on structure and stability. *Langmuir* **10**: 4253–4264.

Wooten F (1972) *Optical Properties of Solids*. New York: Academic Press.

Zachman G (1995) A rapid and dependable identification system for black polymeric materials. *Journal of Molecular Structure* **348**: 453–456.

Zbinden R (1964) *Infrared Spectroscopy of High Polymers*, New York: Academic Press.

Biochemical Applications of Fluorescence Spectroscopy

Jason B Shear, University of Texas at Austin, TX, USA

ELECTRONIC SPECTROSCOPY
Applications

Introduction

A remarkable range of fluorescence spectroscopy techniques has been developed in the last forty years to characterize fundamental properties of biological systems. The rapid progress over this period has led to strategies for monitoring events that transpire on the femtosecond time scale, for discerning features smaller than the Rayleigh resolution limit of light, and for detecting individual molecules. More routinely, fluorescence spectroscopy has provided invaluable insights into the structure, function, and concentrations of macromolecules, small molecules, lipids, and inorganic ions. Moreover, fluorescence analysis of living systems has begun to reveal directly how these biochemicals function and interact *in situ*. The initial section of this article presents an overview of the measurement formats used in fluorescence studies of biochemical systems. In the sections that follow, the application of diverse spectroscopic techniques to the study of biomolecular systems is examined, considering both the properties and uses of intrinsic biological fluorophores and also of fluorescent probes that have been designed to report on the presence or condition of biochemicals. In the Conclusion, several fluorescence approaches that may offer new capabilities in future biochemical studies are discussed. A fully comprehensive treatment of the diverse applications of fluorescence in biological studies is not possible in this article. For example, fluorescence studies of biological porphyrins – a vast field of spectroscopy – is mentioned in the most cursory fashion. Readers are encouraged to refer to other articles as listed later, for additional information, as well as to the sources in the Further reading list at the end of this article.

Overview of measurement formats

In situ measurements

In general, when information is sought regarding the spatial distribution of chemicals within cells or tissue, some variant of fluorescence microscopy is used – often in combination with transmission or Nomarski imaging. From an optical standpoint, fluorescence microscopy measurements either use a *wide-field* geometry, in which an entire field-of-view is illuminated simultaneously, or rely on raster scanning of a small resolution element to sequentially generate a fluorescence image. In the first approach, an arc lamp is commonly used as the excitation source, and large-format images can be acquired at video rate using an array detector. In scanning fluorescence microscopy, a laser typically provides the excitation light, which is focused critically to diffraction spot sizes as small as ~0.2 μm using a high numerical aperture (NA) microscope objective. The sequential nature of image formation in scanning microscopy makes this procedure relatively slow. Although large format two-dimensional (2D) images typically require half a second or more to acquire, small 2D images or line scans can be produced much more rapidly. A fundamental advantage of the two scanning techniques most commonly used – confocal and multiphoton microscopies – is an ability to acquire 3D images of various biological specimens by scanning the laser focal spot in a given plane, then shifting the fine focus of the microscope to repeat the process in a new plane. In confocal microscopy, out-of-plane fluorescence is prevented from reaching a single-element detector by placing a small aperture in a plane conjugate to the focal spot; multiphoton microscopy achieves a similar degree of axial resolution through an inherent 3D localization of the excitation focal volume.

Measurements with chemical separations

Though microscopy can be used to quantify simultaneously up to several chemical species by acquiring detailed spectroscopic information, when analysis of *many*, potentially unidentified, cellular components is desired, a separation procedure typically is used in combination with fluorescence detection. Commonly used techniques include high-pressure liquid chromatography (HPLC), capillary electrophoresis (CE) and related capillary chromatography techniques, and gel electrophoresis. In HPLC and CE, trace analysis frequently is performed by labelling nonfluorescent components either pre- or post-separation using fluorogenic reagents. Capillary techniques frequently offer several advantages over HPLC, including faster and more efficient separations, lower sample volume requirements, and alleviation of the need for high-pressure equipment. In DNA gel electrophoresis, fluorescent intercalating dyes can be used to stain oligonucleotide bands in gels after electrophoresis. For DNA sequencing in agarose gels, the Sanger method can be used to generate fluorescent fragments by labelling primers with fluorophores whose emission spectra encode a specific terminating dideoxyribo-nucleotide. Analysis of proteins using SDS-polyacrylamide slab gel electrophoresis or various capillary techniques can be accomplished by fluorescently labelling analytes before separation.

Cuvette, flow, and 'chip' measurements

A large number of studies on isolated biochemical systems can be performed in cuvettes or deep-well slides. In particular, characterizations of macromolecular structure frequently are conducted using thermostatted cells with capabilities for stopped-flow exchange of various solution modifiers (e.g. ligands, denaturants) on a millisecond time scale. In many cases, instruments used for such studies also can perform nanosecond or microsecond time-resolved measurements of emission intensity or polarization in addition to standard steady-state detection after polarization.

In many instances, it is desirable to rapidly count or isolate a subpopulation of cells exhibiting a particular chemical characteristic. In flow cytometry, this property is linked to a corresponding difference in cellular fluorescence (e.g. by selectively labelling a relevant gene product), which provides a means for measuring relatively large numbers of cells in a flowing stream using automated instrumentation.

Fluorescence-activated cell sorting (FACS) uses a flow cytometer in combination with a shunting system to isolate cells exhibiting characteristic fluorescence.

The presence of various oligonucleotides, peptides, and small molecules in solution can be analysed using solid state 'chip' based devices, in which receptor molecules (often antibodies or complementary oligonucleotides) immobilized on the chip surface bind the analyte(s) of interest. As a result of analyte binding, a characteristic fluorescence signal is generated from the receptor, analyte, or some other interacting species. In ELISA (enzyme-linked immunosorbent assay), the binding of analyte is ultimately linked to the enzymatic production of a fluorescent solution-phase molecule.

Analyses based on intrinsic biological fluorescence

Ideally, all biological chemicals would carry highly specific spectroscopic signatures that define not only their identities but also every aspect of their physical states and environments, and they would yield optical signals intense enough to reveal the smallest changes relevant to a particular experiment. Biological analyses, of course, are not so straightforward, and generally provide very limited information in the absence of exogenous 'probe' chemicals. Nevertheless, there are several classes of intrinsically fluorescent biological molecules, and a host of elegant spectroscopic techniques are used to investigate the properties of these species.

The aromatic amino acids – tryptophan (Trp), tyrosine (Tyr) and phenylalanine (Phe) – have strong deep-UV absorption bands ($\lambda_{ex} < 230$ nm) corresponding to $S_0 \rightarrow S_2$ transitions, but commonly are excited to the S_1 state in fluorescence studies ($\lambda_{ex} \approx 260–280$ nm) to minimize photoreaction and enhance fluorescence quantum yields (Φ_f). At these longer wavelengths, Trp has the largest molar extinction coefficient ($\varepsilon_{max} \approx 5600$) and quantum yield ($\Phi_f \approx 0.2$) of the three amino acids; for Phe, the values of ε and Φ_f are so poor that this species is rarely useful in fluorescence studies. When subjected to UV irradiation, proteins with both Trp and Tyr (**Figure 1**) typically exhibit emission spectra whose shape is characteristic of Trp residues ($\lambda_{max} \approx 350$ nm) because of nonradiative energy transfer from Tyr to Trp.

Despite the fact that Trp and Tyr are not highly fluorescent, and undergo intersystem crossing and myriad photoreactions with large quantum yields, these species can be useful probes of protein structure. Emission from both Trp and Tyr is

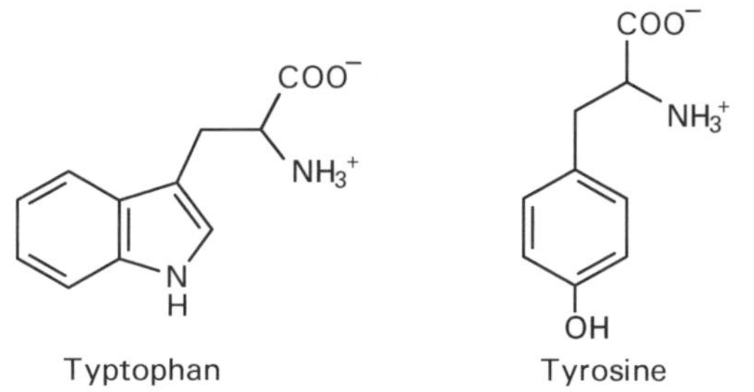

Figure 1 Fluorescent amino acids.

quenched to different extents by a variety of intramolecular polar moieties (including carboxylic acids, amino groups, and imidazole groups), bound cofactors (e.g. NAD^+), prosthetic groups (e.g. heme) and small molecules in solution, making determination of protein structural changes possible through measurement of excited-state lifetimes or steady-state fluorescence intensities. Unlike Tyr, Trp fluorescence often is found to be higher in proteins than in free solution. In molecules containing multiple fluorescent residues, the overall modulation of fluorescence associated with protein conformational changes can be small due to an averaging of advantageous and deleterious quantum yield changes. For this reason, time-resolved measurements of multiexponential excited state decay times sometimes can provide information unavailable from steady-state intensity measurements.

The Stokes shift for Trp also can depend significantly on environment. When Trp is sequestered in the hydrophobic core of a protein, little or no reorientation of solvent takes place after excitation of the chromophore, and emission is blue-shifted relative to a Trp residue on the exterior of a protein. Extreme examples include azurin, in which Trp appears to be completely isolated from H_2O ($\lambda_{max} < 310$ nm), and denatured parvalbumin, in which the Trp emission spectrum is essentially identical to free Trp. For proteins that contain Tyr but not Trp, the shapes of the emission spectra closely match free Tyr ($\lambda_{max} \approx 305$ nm), regardless of local chemical environment.

A number of macromolecular diffusion and conformational properties can be studied using fluorescence anisotropy, fluorescence correlation spectroscopy (FCS), and fluorescence recovery after photobleaching (FRAP). These techniques most commonly are applied to proteins labelled with highly fluorescent probes, but can exploit intrinsic fluorescence in some instances. In fluorescence anisotropy studies, polarized light is used to selectively excite molecules whose transition dipole moments are aligned with the electric field vector. Steady-state measurements of fluorescence anisotropy can be accomplished using a continuous excitation source; provided that anisotropy decays monoexponentially after excitation, information can be obtained on properties such as rotational diffusion times in membranes and the existence of ligand–host interactions. Time-resolved measurements, in which anisotropy is measured as a function of time following pulsed excitation, can provide more detailed information on phenomena such as anisotropic rotational diffusion, and can be used to study complex processes such as electron transfer in light-harvesting chlorophyll complexes. FCS is a steady-state technique for measuring the concentration and diffusion coefficient of a fluorescent molecule based on the magnitude of signal fluctuations as molecules diffuse through a finite probe volume. Diffusion coefficients can also be determined with FRAP, a technique in which fluorescent molecules within the probe volume are photobleached with a high-intensity pulse of light. After the bleaching event, the kinetics of fluorescence recovery (representing the diffusion of 'fresh' molecules into the probe volume) are monitored with low-intensity irradiation. Depending on the circumstances, FCS or FRAP can be useful for measuring the local viscosity of cellular or isolated biochemical environments, for monitoring ligand/host interactions, and for determining changes in diffusion coefficients associated with macromolecular conformational changes.

Identification and measurement of UV fluorescent proteins in complex biological samples requires mixtures to be fractionated into individual components, often with CE or HPLC. CE with on-column UV fluorescence detection has been shown to be useful in measuring attomole quantities of protein in individual red blood cells. Peptide mapping can be accomplished using HPLC with post-column UV fluorescence detection by purifying individual proteins, then subjecting various fractions to proteolytic digests before separating the resulting peptide segments with chromatography.

Metabolic derivatives of Trp and Tyr also produce significant UV fluorescence. In some secretory cells, Trp is converted to the indolamine neurotransmitter, serotonin, which in turn can be processed into the pineal hormone, melatonin. Metabolic conversion of Tyr yields the catecholamines (dopamine, norepinephrine, and epinephrine). Although the cellular concentration of transmitters is typically low, their concentrations within secretory granules can approach 1 molar. The fluorescence properties of these neurotransmitters are analogous to the parent amino acids, with indolamines exhibiting stronger emission and at longer excitation and emission wavelengths than the catecholamines.

Intrinsic UV fluorescence has been used to perform *in situ* measurements of serotonin. Loss of serotonin from astrocytes can be monitored using wide-field UV fluorescence microscopy after various chemical treatments. In addition, three-photon scanning fluorescence microscopy has been shown to be useful for tracking secretion of serotonin from granules in cultured cells in response to cross-linking of cell surface IgE receptors. Fluorescence from indolamines and catecholamines separated with CE has provided a means to measure these species in the low- to mid-attomole range.

In some instances, protein fluorescence does not derive directly from the aromatic amino acids. Two such proteins native to the jellyfish genus *Aequorea* – aequorin and the *green fluorescent protein* (GFP) – have been particularly useful tools in biochemical studies. Aequorin emits light ($\lambda_{max} \approx 470$ nm) from the bound luminophore coelenterazine as the result of a Ca^{2+}-promoted oxidation reaction, and hence has been useful as an intracellular Ca^{2+} probe both through microinjection and recombinant gene expression. The discovery and cloning of GFP has attracted much attention recently in part because of its utility as a reporter for various gene products when incorporated into fusion proteins. In GFP, the chromophoric unit has been identified as an imidazolone anion formed by cyclization and oxidation of tripeptide sequence (-Ser-Tyr-Gly-); fluorescence of this species is characterized by excitation maxima at ~400 and ~475 nm and peak emission at ~510 nm. The fluorescence from GFP is intense enough to detect individual molecules of this species immobilized in aqueous gels. Both aequorin and GFP can be genetically targetted to accumulate in specific organelles to measure localized properties, such as $[Ca^{2+}]$ and protein diffusion coefficients in mitochondria.

A variety of protein cofactors have significant intrinsic fluorescence, including the reduced nicotinamide nucleotides (NADH and NADPH) and the oxidized flavins (**Figure 2**). The nicotinamides are excited in the near-UV and emit maximally at ~470 nm; flavins exhibit both near-UV and visible excitation maxima, and have a peak emission at ~515 nm. In both groups of cofactors, the adenine group partially quench fluorescence through collisions; in flavins, adenine also can quench fluorescence by forming an intramolecular complex with the fluorescent ring system. The significance of adenine quenching is underscored by the fact that flavin mononucleotide (FMN), a highly fluorescent cofactor that lacks adenine, has a fluorescence quantum yield tenfold greater than that of flavin adenine dinucleotide (FAD). Flavins that are covalently or non-

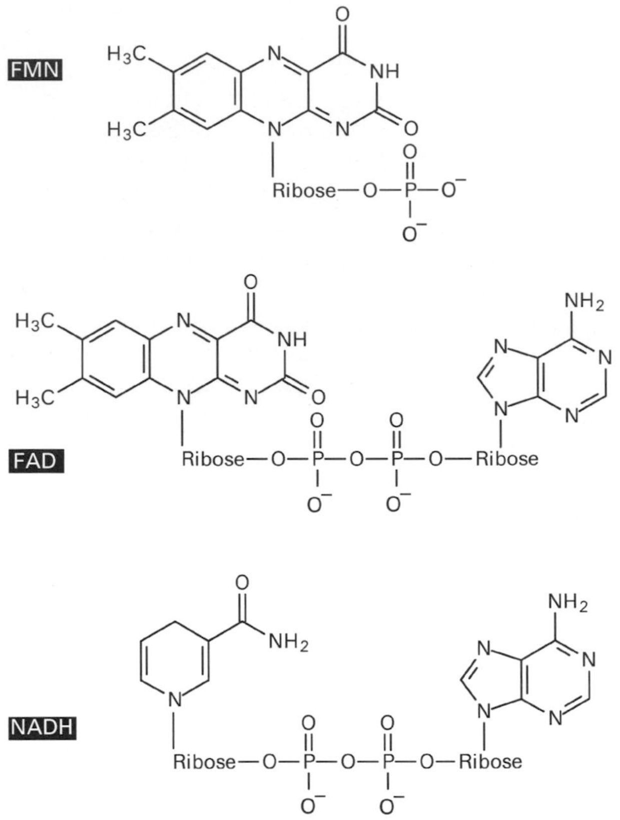

Figure 2   Three fluorescent cofactors.

covalently bound to proteins display widely varying differences in fluorescence quantum yields, although flavoproteins often are relatively nonfluorescent. In contrast, NADH bound to proteins sometimes has substantially greater fluorescence than free NADH because of a decreased ability of adenine to quench the nicotinamide group. In dehydrogenases, for example, crystallographic studies indicate that nicotinamide cofactors are bound in an extended fashion, with the adenine and nicotinamide groups bound to different pairs of β-sheets. Moreover, conformational changes in lactate dehydrogenase induced by effector molecules can be monitored as changes in NADH fluorescence. As a rule, the redox cofactors – like the fluorescent amino acids – have spectroscopic properties that are sensitive to environment, and can be used to probe conformational changes in proteins.

NADH and oxidized flavin (flavoprotein) fluorescence can be used as an indicator of the redox states of cells challenged with a variety of chemical effectors, and has been used to correlate the oxidative status of cells *in situ* and *in vivo* to functions such as electrical activity in brain tissue. The fluorescence of NADH also has been used to characterize the *in vitro* activity of individual lactate dehydrogenase molecules, which reduce the nonfluorescent NAD+

during oxidative production of pyruvate. This general strategy, in which single-enzyme molecules are characterized through the generation of a fluorescent product, also can be applied to reactions using engineered (nonbiological) fluorogenic substrates.

Most nucleic acids have poor fluorescence properties because of extremely short excited state lifetimes, and are most often studied using nonbiological intercalating or groove-binding dyes. One notable exception is the Y base, which has been used as an indicator of tRNA folding induced by divalent cation binding. In addition, some nucleotides exhibit moderate fluorescence under highly acidic conditions, and can be measured in attomole to femtomole amounts using capillary electrophoresis.

Analyses using nonbiological probes

Although fluorimetric analysis of biological systems is experimentally simpler and less disruptive to inherent chemical properties when the use of synthetic fluorescent molecules can be avoided, in many instances the intrinsic properties of biological chemicals do not provide sufficient means for characterizing or quantifying desired properties. In the case of protein analysis, for example, many species lack Trp or Tyr residues altogether; in protein molecules that do contain these amino acids, the fluorescence characteristics may be inadequate to yield the needed information.

Numerous probes have been developed for investigating biological molecules *in vitro* and in living cells, with greatly differing strategies for uncovering desired information. No common photophysical or photochemical characteristics can be ascribed to this entire assortment of nonbiological fluorophores. Many probes have been engineered to have very large absorption cross sections and fluorescence quantum yields, and to be resistant to photoreaction. The probes based on laser dyes such as fluorescein ($\varepsilon_{max} > 50\,000$; $\Phi_f > 0.9$) or one of the rhodamines are good examples of this class of molecules, and often are incorporated in a number of the calcium probes, fluorescent antibodies (see below) and reagents used to label proteins for conformational studies involving fluorescence anisotropy, FCS and FRAP. In addition, by conjugating two spectroscopically distinct probes to different sites on a molecule, Förster energy transfer between the fluorophores can be used to measure intramolecular distances. Despite the advantages of these highly fluorescent molecules, probes having *less* optimal fluorescence properties are deliberately selected for some applications – often as a compromise to achieve a greater *change* in optical characteristics under different chemical

environments, or to avoid toxicity to living cells. For example, a dye whose fluorescence properties depend sensitively on the polarity and rigidity of its local environment may be more useful in many instances than a rhodamine for characterizing protein conformation. One such compound is NBD chloride, an amine-reactive reagent whose product has a markedly greater Φ_f value when occupying a hydrophobic site.

Reactive fluorogenic reagents (e.g. NBD chloride, fluorescamine and CBQCA) are essentially nonfluorescent until they react with analyte molecules, and are commonly used to tag biological species that share a particular reactive group (e.g. a thiol or amine). This low-specificity labelling approach is extremely useful for identification and measurement of many low-concentration species in a mixture when some means exist for fractionating the multiple components (e.g. HPLC). Because of the low level of reagent fluorescence, a large excess of these species typically can be present in the reaction mixture without interfering with analysis.

When characterizing samples containing many biomolecules, a probe that reacts with a particular chemical moiety sometimes does not provide adequate specificity, even when used with a highly efficient separation procedure. Analysis of the protein distribution in cellular specimens is perhaps the quintessential high-specificity requirement. Because of the value of such measurements in characterizing cellular composition, immunohistochemistry is one of the most ubiquitous biological applications of fluorescence. In routine determinations of gene products, cultured cellular samples or sectioned tissue are chemically fixed and subjected to sequential application of primary and secondary antibodies, followed by imaging with fluorescence microscopy. Primary antibodies can be raised against a tremendous diversity of cellular species, ranging from membrane proteins to small diffusable species such as neurotransmitters. Several species can be analysed in a single specimen by covalently attaching dyes with different excitation/emission spectra to different secondary antibodies. Most commonly, fixed and labelled samples are analysed using either wide-field or confocal laser scanning microscopy. When one seeks to analyse a solution sample for the presence of a known antigen, ELISA can be used. In one format, an immobilized antibody binds the antigen, removing it from solution. A second antibody which is conjugated to an enzyme capable of converting a nonfluorescent substrate into a fluorescent product then is applied, and binds to the immobilized antigen. After a washing step, the enzymatic reaction is initiated. In this way, large signal amplifications provide extremely sensitive assays for analytes such as hormones and drug metabolites. Although immunological techniques can be useful for measuring analyte concentrations, uncertainties in cross-reactivity and matrix effects on the antibody-antigen conjugation efficiency often limit this approach to semi-quantitative determinations.

Diffusable cytosolic species (e.g. second messengers) can be measured in living cells using highly specific fluorescent probes. The calcium-sensitive dyes, a broad class of molecules whose excitation or emission properties are dependent on chelation of Ca^{2+}, are ubiquitous in cellular biology studies. For such compounds, it is important that affinity for Ca^{2+} is much greater than for other cationic species that may be present in the cytosol at much higher concentrations (e.g. Mg^{2+}). In general, cytosolic free Ca^{2+} concentrations vary over the approximate range 100 to 1000 nM, although much higher levels are sometimes reached transiently. Common examples of Ca^{2+}-sensitive dyes include fluo-3 (100-fold intensity increase when bound to Ca^{2+}), fura-2 (excitation λ_{max} changes when bound to Ca^{2+}), and indo-1 (emission λ_{max} changes when bound to Ca^{2+}) (**Figure 3**). For these species, the Ca^{2+} dissociation constant has been designed to approximately match free cytosolic levels, thus maximizing the modulation in fluorescence for a given change in $[Ca^{2+}]$. Like tryptophan, indo-1 has an indole-based chromophore, but in this case the π-system is further delocalized to electronically interact with carboxylate groups responsible for Ca^{2+} chelation. In many instances, fura-2 and indo-1 may offer advantages over fluo-3, as ratiometric measurements of excitation or emission intensities at two different wavelengths can help avoid errors associated with uncertainties in intracellular dye concentrations. Calcium probes have played an invaluable role in identifying a range of cytosolic phenomena, such as $[Ca^{2+}]$ spiking after hormone binding to cell surface receptors and Ca^{2+} waves in oocytes following a fertilization event. Probes exist for other important cytosolic effectors, such as cyclic AMP (cAMP), which is monitored by the level of Förster energy transfer between fluorescein and rhodamine attached to a cAMP-binding enzyme.

Although it is possible to microinject dyes for probing cytosolic milieus through patch or intracellular pipettes, it is more common to incubate many cultured cells in a medium containing low molecular mass dyes in an esterified form (e.g. acetoxymethyl, or 'AM', ester). This parent species is hydrophobic, and thus can diffuse freely through cell membranes. Once the compound is localized to the cytosol, nonspecific esterases hydrolyse the hydrophobic group, generating a charged form of the probe that

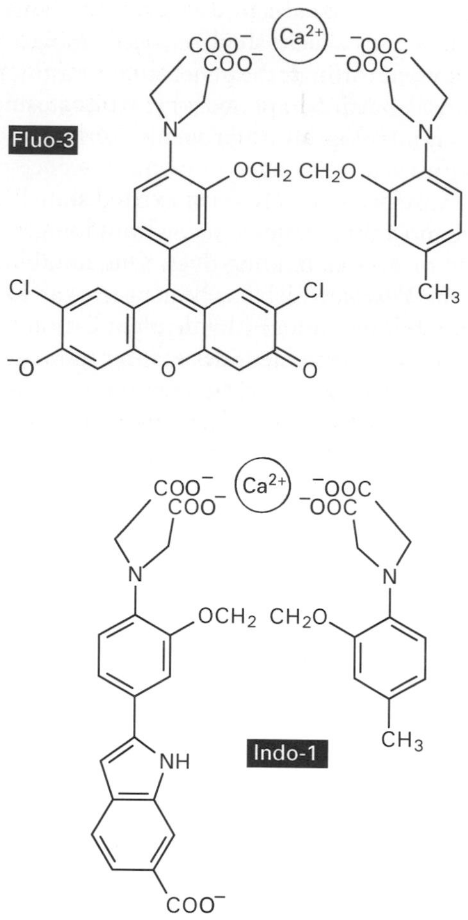

Figure 3 Two common calcium probes.

can no longer diffuse through the membrane. Large numbers of cells also can be loaded rapidly using electroporation.

Structures in living cells often can be characterized using probes with specificity for *classes* of molecules rather than for particular target analytes. Hydrophobic or amphipathic dyes, for example, can be used to gauge cell membrane fluidity and diffusion, as well as exocytosis of secretory vesicles and subsequent membrane recycling. Examples of this class of dyes are 8-anilinonaphathalene sulfonate (ANS), which displays significant fluorescence only when bound to membranes (or protected hydrophobic regions of proteins), and FM-143, a compound used extensively to study regulated secretion from neurons. Some membrane dyes have been engineered (or serendipitously found) to alter their fluorescence properties based on the electric field across the phospholipid bilayer of cell or organelle membranes. Typically, such potentiometric dyes either respond with fast kinetics to electric field changes but display small modulations in fluorescence (e.g. the styrylpyridinium probes), or undergo large fluorescence changes but only very slowly. Because large, rapid changes in fluorescence generally are not obtained with existing probes, it is difficult to perform voltage measurements on individual neurons and other excitable cells.

Other fluorogenic dyes, such as DAPI, ethidium bromide and the cyanine dyes, bind to DNA and/or RNA and undergo changes in fluorescence intensity of up to 1000-fold. Depending on the probe, these compounds can be used to image fixed cells or track mitotic events. Combinations of probes with different specificities for single-stranded, double-stranded and ribosomal RNA have been used to characterize the quantity and conformation of nucleic acids throughout the cell cycle.

Conclusion

In the forty years since the intrinsic fluorescence of amino acids was first characterized, fluorescence spectroscopy has developed into one of the most versatile and powerful tools for investigating biochemical systems. In addition to the enormous range of applications that now exist, a variety of emerging analysis strategies provide evidence of a continued phase of rapid growth for fluorescence applications. Developments in instrumentation – laser sources, optics, and detectors – have made possible the characterization of individual enzyme molecules and highly fluorescent proteins such as GFP and β-phycoerythrin, a light-harvesting protein. Solid-state femtosecond sources recently have made it feasible to generate multiphoton-excited fluorescence of biological molecules in solution using relatively low integrated irradiation, and have provided reproducibility necessary for experiments involving living specimens.

Various technologies that rely on fluorescence for high sensitivity analysis have been made possible through advances in microfabrication and robotics. Examples include chip-based microarray sensors for detecting a multitude of ligands simultaneously, and chips containing arrays of microscopic electrophoresis channels for performing high-throughput, rapid separations of DNA fragments.

Other fluorescence techniques exploit the evanescent, or near-field, properties of light to excite fluorescence in highly restricted regions of space, thereby improving measurement sensitivity or the spatial resolution of fluorescence imaging. Near-field scanning optical microscopy (NSOM) can image structures smaller than the diffraction limit of light using fluorescence generated by an evanescent field that escapes from the aperture at the end of a drawn,

metallized fibre. The resolution obtained with this approach is determined by the dimensions of the aperture, which is usually limited by losses in throughput as the fibre tip diameter is reduced. Thus far, only a few biological applications of this powerful technique have been reported.

New developments in the chemistry of fluorescent probes undoubtedly will open new applications for fluorescence in biochemistry. A growing number of companies and academic researchers are devising new biological probes, and in some cases are using nontraditional strategies – such as combinatorial chemistry – to generate hosts with high specificity and large binding constants for desired ligands. Genetic manipulation of cultured cells offers particularly exciting opportunities for *in situ* biochemical measurements. Technology now exists for using an enzyme as a reporter for neurotransmitter-activated transcription in individual mammalian cells, with sensitivities capable of measuring fewer than 50 gene product molecules. In an initial demonstration of this strategy, large amplifications of gene product expression were obtained by localizing an engineered substrate to the cytosol whose fluorescence properties change when it is degraded by the reporter enzyme. Through the judicious use of molecular and cellular biology, chemistry, and fluorescence spectroscopy, eventually it may be feasible to track the production and breakdown of individual molecules in living cells.

List of symbols

ε = molar extinction coefficient; λ = wavelength; Φ = quantum yield.

See also: **Fluorescence Microscopy, Applications; Fluorescence Polarization and Anisotropy; Fluorescent Molecular Probes; Inorganic Condensed Matter, Applications of Luminescence Spectroscopy; Luminescence, Theory; Organic Chemistry Applications of Fluorescence Spectroscopy; UV-Visible Absorption and Fluorescence Spectrometers; X-ray Fluorescence Spectrometers; X-ray Fluorescence Spectroscopy, Applications.**

Further reading

Cantor CR and Schimmel PR (1980) *Biophysical Chemistry*, Parts I–III. New York: Freeman.

Chalfie M, Tu Y, Euskirchen G, Ward WW and Prasher DC (1994) Green fluorescent protein as a marker for gene expression. *Science* **263**: 802–805.

Cobbold PH and Rink TJ (1987) Fluorescence and bio-luminescence measurement of cytoplasmic free calcium. *Biochemical Journal* 248: 313–328.

Craig DB, Arriaga E, Wong JCY, Lu H and Dovichi NJ (1998) Life and death of a single enzyme molecule. *Analytical Chemistry* 70: 39A–43A.

Creed D (1984) The photophysics and photochemistry of the near-UV absorbing amino acids – I. Tryptophan and its simple derivatives. *Photochem. Photobiol.* 39: 537–562.

Creed D (1984) The photophysics and photochemistry of the near-UV absorbing amino acids – II. Tyrosine and its simple derivatives. *Photochem. Photobiol.* 39: 563–575.

Everse J, Anderson B and You K-S (1982) *The Pyridine Nucleotide Coenzymes*. New York: Academic Press.

Freifelder D (1982) *Physical Biochemistry*, 2nd edn. New York: W.H. Freeman.

Hoagland RP (1996) *Handbook of Fluorescent Probes and Research Chemicals*, 6th edn. Eugene, OR: Molecular Probes.

Lakowicz JR (ed) (1991) *Topics in Fluorescence Spectroscopy*, Vol 1, *Techniques*. New York: Plenum Press.

Lakowicz JR (ed) (1992) *Topics in Fluorescence Spectroscopy*, Vol 3, *Biochemical Applications*. New York: Plenum Press.

Lakowicz JR (1983) *Principles of Fluorescence Spectroscopy*. New York: Plenum Press.

Lillard SJ, Yeung ES, Lautamo RMA and Mao DT (1995) Separation of hemoglobin variants in single human erythrocytes by capillary electrophoresis with laser-induced native fluorescence detection. *J. Chromatogr. A* 718: 397–404.

Permyakov EA (1993) *Luminescent Spectroscopy of Proteins*. Boca Raton, FA: CRC Press.

Rutter GA, Burnett P, Rizzuto R et al (1996) Subcellular imaging of intramitochondrial Ca^{2+} with recombinant targeted aequorin: significance for the regulation of pyruvate dehydrogenase activity. *Proceedings of the National Academy of Science, USA* 93: 5489–5494.

Stryer L (1995) *Biochemistry*, 4th edn. New York: W.H. Freeman.

Teale FWJ and Weber G (1957) Ultraviolet fluorescence of the aromatic amino acids. *Biochemical Journal* 65: 476–482.

Tsien RY (1994) Fluorescence imaging creates a window on the cell. *Chemical Engineering News* 72: 34–44.

Zlokarnik G, Negulescu PA, Knapp TE et al (1998) Quantitation of transcription and clonal selection of single living cells with beta-lactamase as reporter. *Science* 279: 84–88.

Biochemical Applications of Mass Spectrometry

Victor E Vandell and **Patrick A Limbach**,
Louisiana State University, Baton Rouge, LA, USA

MASS SPECTROMETRY
Applications

Mass spectrometry is a powerful tool for the characterization of various biomolecules including proteins, nucleic acids and carbohydrates. The advantages of mass spectrometry are high sensitivity, high mass accuracy, and more importantly, structural information. Historically, biomolecules have proven difficult to characterize using mass spectrometry. Problems often arise with impure samples, low ion abundance for analysis due to inefficient ionization processes, and low mass accuracy for higher molecular weight compounds. Recent advances in the development of electrospray ionization (ESI) and matrix-assisted laser desorption/ionization (MALDI) now permit the analysis of biomolecules with high sensitivity and good mass accuracy. These improvements now allow the use of mass spectrometry for the identification of unknown structures and are sui for applications focused on acquiring sequence information on the samples of interest.

This article provides a brief introduction to the various applications of mass spectrometry to biomolecule analysis. More detailed discussions of particular applications of mass spectrometry to such analyses can be found in other articles in this encyclopedia.

Ionization and mass analysis

Mass spectrometric measurements of biomolecules involve the determination of molecular mass, or of the masses of various components of the original molecule, which can then be related to various structural properties, such as sequence. In either case, the crucial step lies in the conversion of liquid- or solid-phase solutions of the analytes into gaseous ions. Common ionization sources used in mass spectrometry for biomolecule analysis experiments are: fast-atom bombardment (FAB), ESI and MALDI. FAB was historically the ionization method of choice, but has been largely replaced by ESI or MALDI at the present time.

ESI- and MALDI-based methods have particularly benefited the analysis of biomolecules because these techniques accomplish the otherwise experimentally difficult task of producing gas-phase ions from solution species that are both thermally labile and polar. **Table 1** summarizes the different instrument configurations used in the analysis of various classes of biomolecules.

MALDI has extremely high sensitivities with reports of detection limits at the sub-femtomolar level. A higher efficiency for protonated molecular ion production with MALDI has been observed relative to FAB. MALDI typically yields intact protonated molecular ions with minimal fragmentation and is therefore commonly referred to as a 'soft' ionization process. MALDI sources are generally coupled to time-of-flight (TOF) mass analysers. TOF mass analysers are characterized by high upper mass limits with reduced resolution at the higher masses. The production of high molecular weight ions and subsequent analysis of these ions makes the MALDI-TOF combination a powerful tool for biomolecule analysis.

ESI can be considered a complementary method to MALDI. As with MALDI, electrospray ionization of biomolecules yields protonated or cationized molecular ions with little or no fragmentation, and it is also referred to as a 'soft' ionization source. A particular advantage of ESI compared to MALDI is that the analyte is sampled from the solution phase. Under these conditions, ESI is readily coupled to high performance liquid chromatography (HPLC) or capillary electrophoresis (CE) separation systems. Such combinations permit online LC-MS or CE-MS experiments.

Table 1 Summary of mass spectrometry techniques used in biomolecule analysis

Analyte	Ionization source	Mass analyser
Proteins/peptides	FAB	Quadrupole, sector
	MALDI	TOF, FTICR
	ESI	Quadrupole, sector, FTICR, TOF
Oligonucleotides	FAB	Quadrupole, sector
	MALDI	TOF, FTICR
	ESI	Sector, FTICR, TOF
Oligosaccharides	FAB	Quadrupole, sector
	MALDI	TOF
	ESI	Quadrupole, sector, TOF
Lipids	FAB	Quadrupole
	MALDI	TOF
	ESI	Quadrupole, FTICR, TOF

The overriding feature of an ESI-generated mass spectrum is the appearance of multiply charged ions. Multiple charging results from the loss or addition of multiple hydrogen ions or metal ions (e.g. potassium or sodium) to the biomolecule. The multiple charging effect is advantageous because it allows for the analysis of high molecular weight biomolecules using mass analysers with low m/z limits. The disadvantage of multiple charging is that a spectrum has to be deconvoluted and thus spectral interpretation can be complicated, especially during the analysis of mixtures.

Molecular weight determinations of biomolecules

Relative mass is an intrinsic molecular property which, when measured with high accuracy, becomes a unique and unusually effective parameter for characterization of synthetic or natural biomolecules. Mass spectrometry based methods can be broadly applied not only to unmodified synthetic biomolecules, but also to modified synthetic and natural biomolecules (e.g. glycosylated proteins). The level of mass accuracy one obtains during the measurement will depend on the capabilities of the mass analyser used. Quadrupole and TOF instruments yield lower mass accuracies than sector or Fourier transform ion cyclotron resonance (FTICR) instruments. High mass accuracy is not only necessary for qualitative analysis of biomolecules present in a sample, but is necessary to provide unambiguous peak identification in a mass spectrum.

The primary challenge to accurate molecular weight measurements of biomolecules is reduction or complete removal of salt adducts. In the majority of situations, the buffers used to prepare or isolate the analyte of interest contains Group 1 or Group 2 metal salts. These metal salts can potentially interfere with the accurate mass analysis of the analyte due to the gas-phase adduction of one to several metal cations to the analyte. Optimal results are obtained only after substantive (and in some cases, exhaustive) removal of these contaminants. Recent developments for sample purification involve the use of solution additives or online purification cartridges which reduce the presence of interfering salts while retaining the ability to characterize minimal amounts of analyte.

Measurement of molecular mass of biomolecules is now a suitable replacement for prior methods based on the use of gel electrophoresis. Molecular weight measurement for all classes of biomolecules is a relatively routine procedure, and the results obtained are typically of greater accuracy than those previously obtained by gel electrophoresis. A common application of mass spectrometry and molecular weight measurement is for the identification of unknown proteins. The experimentally obtained molecular weight value can be searched against the available protein databases (e.g. Swiss–Prot or PIR) to potentially identify the protein. This application becomes particularly useful for the identification of unknown proteins when used in conjunction with enzymatically generated peptide fragments (see below).

Determination of the primary sequence of biomolecules

One of the most popular and productive uses of mass spectrometry for biomolecule analysis is the sequence or structure determination through analysis of smaller constituents of the original molecule. Two different approaches for generating the smaller, sequence informative constituents from the original molecule are available. An indirect approach is to generate sequence-specific information by solution-based chemical or enzymatic reactions. Chemical or enzymatic digestion of an intact biomolecule results in a number of smaller constituents that are amenable to mass spectrometric analysis. The so-called direct approach involves fragmentation of the analyte in the gas phase. Fragmentation can be induced by the desorption or ionization process, or can be induced by collisions with neutral target molecules or by collisions with surfaces. The analysis of fragment ions initiated by gas-phase dissociation resulting from collisions with neutral molecules or surfaces are broadly referred to as tandem mass spectrometry (MS/MS) experiments. With both approaches, the sequence of the original molecule is obtained by interpreting the resulting fragments.

Enzymatic approaches

Enzymatic digestion followed by mass spectrometric analysis of the resulting products is a popular and powerful approach to sequence determination. A number of enzymes are available which either are specific for a particular substituent of the biomolecule or are nonspecific but sequentially hydrolyse the analyte of interest. Trypsin is an enzyme commonly used to digest proteins into smaller peptides. Trypsin selectively cleaves proteins at the C-terminal side of lysine and arginine amino acid residues. Digestion of a protein using trypsin will generate a tryptic digest whose components are amenable to mass spectrometric analysis. In many cases, the masses of the tryptic fragments can be measured more accurately than the mass of the original molecule, thereby

improving the identification of unknown proteins. As mentioned earlier, the masses of tryptic peptides, used in conjunction with molecular weight measurements of intact proteins, can be used to search known protein databases for efficient and accurate identification of unknown proteins.

Sequential digestion of biomolecules is an alternative approach to determining sequence information. The utility of any mass spectrometric sequencing method that relies on consecutive backbone cleavages depends on the formation of a mass ladder. The sequence information is obtained by determining the mass difference between successive peaks in the mass spectrum. For example, phosphodiesterases are enzymes which sequentially hydrolyse the phosphodiester linkage between oligonucleotides and nucleic acids. In the case of oligodeoxynucleotides, the expected mass differences between successive peaks will correspond to the loss of: dC = 289.5, dT = 304.26, dA = 313.27, and dG = 329.27 Da. Mass ladder methods have a distinct advantage for sequence determination, because it is the difference in two mass measurements that results in the desired information. A drawback to this approach is the limited size of the analyte that is amenable to sequential digestion.

Tandem mass spectrometry approaches

The gas-phase approach to determining structural information about biomolecules is through tandem mass spectrometry (MS/MS). Tandem mass spectrometry involves isolation of the ion of interest (commonly referred to as the parent ion in MS/MS) and then dissociating this ion via collisions with neutrals or surfaces to produce fragment ions (commonly referred to as product ions in MS/MS) from which the primary sequence of the molecular ion of interest can be determined. Generally, the fragmentation process in MS/MS experiments generates product ions which contain sequence specific information from throughout the molecular ion of interest.

Tandem mass spectrometry is applicable to all types of biomolecules but typically requires specialized mass spectrometry instrumentation for implementation. The most common tandem mass spectrometer is a triple quadrupole instrument, wherein the first and third quadrupoles are used as mass analysers and the middle quadrupole is used as a collision chamber. MS/MS can be performed with double focusing sector instruments and in some cases with specialized TOF mass analysers. Quadrupole ion traps and FTICR mass spectrometers are ideally suited for MS/MS experiments, and due to the operational characteristics of these mass analysers additional MS/MS experiments can be performed, allowing for MS^n studies of biomolecules.

Higher order gas-phase structure and non-covalent interactions studied using mass spectrometry

Similar to studies performed using NMR, hydrogen/deuterium (H/D) exchange experiments on biomolecules are feasible with mass spectrometry. Labile hydrogens can be exchanged in solution prior to mass spectrometric analysis, or alternatively H/D exchange experiments can be performed in the gas phase. The former studies have been used to localize sites of H/D exchange on biomolecules of interest and are used for mechanistic studies of biomolecule function. The latter studies typically involve ESI-MS and permit investigations into the gas-phase conformation of biomolecules in the absence of the solvent. The majority of H/D exchange experiments have focused on peptide and protein analysis, although the method is amenable to other classes of biomolecules.

A particular advantage of ESI-MS for biomolecule analysis is realized by generating the analyte ions from solution conditions that retain the secondary, tertiary and even quaternary structure of the biomolecules. Noncovalent binding of biomolecules has been observed in the ESI mass spectrum and, when operated under the appropriate conditions, the mass spectral data are a direct probe of the solution-phase biomolecule assembly. Protein assemblies, protein-nucleic acid complexes, duplex DNA and other non-covalently bound biomolecule assemblies have been studied using mass spectrometry.

See also: **Chemical Structure Information from Mass Spectrometry; Chromatography–MS, Methods; Fast Atom Bombardment Ionization in Mass Spectrometry; Fragmentation in Mass Spectrometry; Hyphenated Techniques, Applications of in Mass Spectrometry; Medical Applications of Mass Spectrometry; MS-MS and MS^n; Nucleic Acids and Nucleotides Studied Using Mass Spectrometry; Peptides and Proteins Studied Using Mass Spectrometry; Quadrupoles, Use of in Mass Spectrometry; Sector Mass Spectrometers; Surface Induced Dissociation in Mass Spectrometry; Time of Flight Mass Spectrometers.**

Further reading

Fenn JB, Mann M, Meng CK, Wong SF and Whitehouse CM (1989) Electrospray ionization for mass spectrometry of large biomolecules. *Science* **246**: 64–70.

Karas M, Bahr U and Hillenkamp F (1989) UV laser matrix desorption/ionization mass spectrometry of proteins in the 100,000 Dalton range. *International Journal of Mass Spectrometry and Ion Processes* **92**: 231–242.

Loo JA (1995) Bioanalytical mass spectrometry: many flavors to choose. *Bioconjugate Chemistry* **6**: 644–665.

McCloskey JA (ed) (1990) *Methods in Enzymology*, Vol. 193 San Diego: Academic Press, Inc.

Senko MW and McLafferty FW (1994) Mass spectrometry of macromolecules: Has its time now come? *Annual Review of Biophysics and Biomolecular Structure* **23**: 763–785.

Smith RD, Loo JA, Edmonds CG, Barinaga CJ and Udseth HR (1990) New developments in biochemical mass spectrometry: Electrospray ionization. *Analytical Chemistry* **62**: 882–899.

Biochemical Applications of Raman Spectroscopy

Peter Hildebrandt, Max-Planck-Institut für Strahlenchemie, Mülheim, Germany
Sophie Lecomte, CNRS-Université Paris VI, Thiais, France

VIBRATIONAL, ROTATIONAL & RAMAN SPECTROSCOPIES
Applications

Until the late 1960s, Raman spectroscopy was largely restricted to small inorganic and organic molecules. Concomitant to the substantial improvement of excitation sources and detection systems beginning in the early 1970s, biological molecules, which in most cases are weak Raman scatterers, became more and more accessible to this technique. Now, Raman spectroscopy has been developed to a versatile tool for studying biological systems on various levels of complexity, i.e. ranging from small amino acids to biopolymers and even to living cells.

Raman spectra – like IR spectra – include detailed information about the molecular structure and, hence, may provide a key for elucidating structure–function relationships. Unfortunately, the ability to extract structural data from the spectra is much less advanced than, for example, in NMR spectroscopy, since a sound vibrational analysis based on normal mode calculations is not straightforward. Empirical force fields only provide meaningful results in the case of small and/or symmetric molecules for which a large set of isotopomers are available. Alternatively, quantum chemical methods can be employed to calculate the force constants. Despite the recent progress in hard- and software development, this approach is as yet restricted to prosthetic groups and building blocks of biopolymers. Thus, in most cases the interpretation of the Raman spectra of biomolecules is based on empirical relationships derived from comparative studies of model compounds.

The large number of normal modes of biomolecules is also associated with an experimental difficulty inasmuch as individual Raman-active modes may be closely spaced so that the observed peaks include several unresolved bands. This is particularly true for those modes which originate from chemically identical building blocks of the biopolymers such as the amide vibrations of proteins. In these cases, however, the analysis of these bands provides valuable information about the structure of the ensemble of these building blocks, e.g. the secondary structure of proteins.

The selectivity of Raman spectroscopy can be substantially increased by the choice of an appropriate excitation wavelength. When the excitation energy approaches that of an electronic transition of a molecule, exclusively the Raman intensities of the modes originating from this chromophore are enhanced by several orders of magnitude (resonance Raman, RR). In this way, it is possible to probe the vibrational band pattern of this specific part of the macromolecule regardless of the size of the remaining optically transparent matrix.

Specific experimental considerations

Under non-resonant conditions, quantum yields for the Raman effect are in the order of 10^{-9}, implying that sample concentrations in the millimolar range are required for obtaining Raman spectra of satisfactory quality. Resonance enhancement substantially improves the sensitivity, but even RR intensities are too weak to obtain spectra of samples which exhibit a strong fluorescence or which contain strongly fluorescent impurities. In those cases, Fourier-transform near-infrared Raman spectroscopy may be a useful alternative approach.

Possible thermal denaturation or photochemical processes which may be induced by laser irradiation may impose serious constraints on Raman (RR) spectroscopic experiments of biomolecules. Such effects can be reduced by using appropriate sample arrangements such as flowing or rotating devices, or by lowering the temperature.

In any case, the photon flux through a volume element of a sample should be kept as small as possible to ensure the integrity of the biomolecule during the Raman experiment. These requirements are best fulfilled using CW-lasers as excitation sources. A second parameter which is essential for measuring the spectra of sensitive biomaterial is a short duration of the Raman experiment. Scanning monochromatic detection systems (photon counting devices) may require several hours to obtain Raman spectra of sufficient quality. Polychromatic detection systems can strongly reduce the total signal accumulation time and, in the case of CCD detectors, yield a spectral resolution and signal-to-noise ratio comparable to scanning systems.

Structure of biomolecules

Raman spectroscopic studies which are directed to gain information about the structure of biomolecules are generally performed in the stationary mode. Such experiments have been carried out with all kinds of biopolymers and their building blocks. However, up to now, the level of spectral analysis and interpretation is such that reliable structural information can be extracted from the spectra only for proteins, nucleic acids and membranes.

Proteins and amino acids

Non-resonant Raman spectra of proteins display bands originating from the protein backbone, i.e. the amide modes and the amino acid side chains, in particular, the aromatic amino acids tryptophan (Trp), tyrosine (Tyr) and phenylalanine (Phe). The strongest amide band in the non-resonant Raman spectra originates from the amide I mode which is essentially a pure C=O stretching of the peptide bond. Its frequency depends on the type of hydrogen bonding and dipole–dipole interactions associated with the peptide bond. As these interactions are different for the various elements of secondary structure of proteins, systematic Raman and IR spectroscopic studies of peptides and proteins with known three-dimensional structure allow the determination of amide I frequencies which are characteristic not only for the main secondary structure elements α-helix (1657 cm^{-1}) and β-sheet (1626 and 1635 cm^{-1}) but also for turns (1666 cm^{-1}) and β-sheet/turns (1679 cm^{-1}). For irregular or random coil peptide bonds the amide I is found at 1644 cm^{-1}. These findings constitute the basis for the secondary structure determination of proteins and peptides of unknown structure. In the most general way, the measured amide I envelope is fitted by a set of these bands with constant frequencies and half widths. The relative intensities, which are the only adjustable parameters, represent a measure for the relative contributions of the various secondary structure elements. This approach is also frequently used in IR spectroscopy which offers the advantage of a better signal-to-noise ratio but is restricted to measurements in D$_2$O due to the strong absorption of water. With an increased set of reference proteins, the method has gained a considerable accuracy which can well compete with CD measurements.

Complementary information about the secondary structure can be obtained by using RR spectroscopy. Upon excitation in resonance with the π → π* transitions of the peptide bonds at ~190 nm, the amide modes II, III and II′, which include the C–N stretching

vibrations, are preferentially enhanced so that they appear with substantially higher intensity than the amide I mode (C=O stretching) which in turn gains intensity only via a higher lying electronic transition at ~160 nm. Like the amide I mode, the amide modes II, III and II′ are at different frequencies in the various structural elements. More important for secondary structure determination, however, are the RR intensities of these modes. The qualitatively different dipole–dipole interactions in α-helical and β-sheet structures lead to a respective decrease (hypochromic effect) and increase (hyperchromic effect) of the oscillator strength which in turn affects the Raman cross-sections of these amide modes. Thus, it was possible to derive (linear) relations between the RR intensities of these modes (determined relative to an internal standard) and the secondary structure. Due to the high energy of the deep UV laser pulses required as the excitation source, special attention has to be paid to avoiding photodegradation processes of the biomolecule.

Such effects are also a restriction on probing the aromatic amino acid side chains of proteins using excitation lines between 200 and 240 nm. Under these conditions, several ring vibrational modes are selectively enhanced via the L$_a$ (Phe, Tyr) and B$_{a,b}$ transitions (Trp). The frequencies and relative intensities of these modes can be used to study the local environment of these amino acids, i.e. to obtain information about this specific part of the tertiary structure. In particular, the modes of Tyr and Trp may provide more detailed information about the local protein structure. The p-hydroxy substituent of Tyr is generally involved in hydrogen bonding interactions with hydrogen bond acceptors or donors. These interactions are sensitively reflected by the modes ν_{7a}, $\nu_{7a'}$ and ν_{9a} since their frequencies show linear relations with the solvent donor number. Monitoring these modes in proteins allows conclusions about the kind and the strength of hydrogen bonding interactions of Tyr. In the extreme case, when Tyr is deprotonated, the entire vibrational band pattern drastically changes and, due to the red-shift of the electronic transition, the resonance enhancement is altered. Also Trp can undergo hydrogen bonding interactions via the indole N–H group. In that case, a band at ~880 cm^{-1} (ν_{10a}, W17) is an appropriate marker in the RR spectra of proteins. The second important spectral marker of Trp, a Fermi doublet at ~1345 cm^{-1}, exhibits a moderate resonance enhancement and can only be identified in the spectra of relatively small proteins. This band is an indicator for the hydrophobicity of the local environment of the indole ring. Based on systematic studies of Trp and model compounds, an empirical relationship has been derived between the torsional angle of the C_2–C_3–C_β–C_α entity and the frequency of the band at ~1550 cm^{-1} (W3). This marker band, however, does not appear to be applicable for analysing the Trp structure in proteins inasmuch as it appears in a spectrally crowded region.

Although *a priori* the size of the protein is not a limitation for UV RR spectroscopic analysis of aromatic amino acids, increasing numbers of Phe, Tyr and Trp residues strongly complicate the interpretation of the spectra. Nevertheless, working with high quality spectra it is possible to probe subtle structural changes of a single amino acid even in relatively large proteins as demonstrated in the studies on haemoglobin by Spiro and co-workers. This protein consists of two ($\alpha\beta$) dimers, each of them including 15 Phe, 6 Tyr and 3 Trp residues. It had been assumed that the $T \rightarrow R$ transition of haemoglobin was associated with a change of hydrogen bonding interactions of one Tyr (out of 12) which in fact could be confirmed by UV RR (difference) spectra.

The application of Raman spectroscopy for probing amino acid side-chain structures in proteins is not restricted to excitation in the deep UV. Some ring vibrations of the aromatic amino acid side chains can also be identified in the non-resonance Raman spectra. These are, for instance, the Fermi doublet of Trp at ~1345 cm^{-1} and a band pair of Tyr at 850 and 830 cm^{-1}, assigned to the fundamental mode ν_1 and the overtone $2\nu_{16a}$, which together form a Fermi doublet as well. This doublet, which is only weakly resonance enhanced upon UV excitation, appears with considerable intensity under non-resonance conditions and is known to be a sensitive marker for hydrogen bonding interactions as derived from systematic studies on model compounds. In fact, the intensity distribution of this doublet can be used to monitor changes of hydrogen bonding interactions of Tyr in proteins.

Among the aliphatic amino acid side chains, cysteine (Cys) and methionine (Met) give rise to particular strong Raman bands which appear in less cogent spectral regions. The C–S stretching is found between 620 and 750 cm^{-1} depending on the conformation of the side chain, e.g. *trans* or *gauche* conformation of Met. In addition, the C–S stretching vibrations of disulfide bridges, which constitute an important stabilizing factor for the tertiary structure of proteins, are observed at 630–650, 700 and 720 cm^{-1} when the *trans* substituent X of the X–C_α–C_β–S–S entity is a hydrogen, nitrogen and carbon atom, respectively. However, the most characteristic feature of disulfide bridges is the S–S stretching (500–525 cm^{-1}). There is a large body of experimental data for proteins and model compounds and evidence has been provided for a linear correlation between the frequency and the C–S–S–C dihedral angle.

Chromoproteins

Chromoproteins are characterized by an electronic absorption band in the near-UV, visible or near-IR spectral range. These bands may arise from $\pi \rightarrow \pi^*$ transitions of prosthetic groups or from charge-transfer transitions of specifically bound transition metal ions. Thus, chromoproteins which may serve as electron transferring proteins, enzymes or photoreceptors, are particularly attractive systems to be studied by RR spectroscopy since an appropriate choice of the excitation wavelength readily leads to a selective enhancement of the Raman bands of the chromophoric site. Moreover, these chromophores generally constitute the active sites of these biomolecules so that RR spectroscopic studies are of utmost importance for elucidating structure–function relationships.

Photoreceptors may use light either for energy conversion (photosynthetic pigments, bacterial rhodopsins) or as a source of information to trigger a physiological response (rhodopsins, phytochromes). Typical chromophores are retinal Schiff bases which are covalently attached to the apoprotein [(bacterial) rhodopsins] and cyclic or linear tetrapyrroles (photosynthetic pigments, phytochromes). In the case of (bacterial) rhodopsins and phytochromes, light absorption initiates a photochemical reaction which is followed by a series of thermal relaxation processes. Thus, RR spectroscopic studies of these biomolecules encounter the difficulty that the exciting laser beam induces the reaction sequence. In order to avoid accumulation of various intermediate states, low temperature spectroscopy is frequently employed so that thermally activated reaction steps can be blocked. Alternatively, time-resolved RR spectroscopy may be an appropriate approach (e.g. bacteriorhodopsin) which, in addition, provides information about the chromophore–protein dynamics.

Primary interest of RR spectroscopic investigations of the visual pigment rhodopsin was directed to the analysis of the photochemical reaction. Single-beam experiments carried out at 77 K yielded a photostationary equilibrium of the parent state and the primary photoproduct. Using an additional laser beam of a wavelength in resonance with the electronic transition of the product, it was possible to shift the equilibrium largely to the parent state. Thus, subtracting the dual-beam- from the single-beam-excited spectrum yielded a pure spectrum of the parent state. The interpretation of these spectra was greatly facilitated by the large body of experimental data of retinal model compounds including a variety of isotopomers as well as of the detailed investigations of bacteriorhodopsin (see below). Moreover, normal mode analyses based on empirical force field calculations support a reliable vibrational assignment allowing for the determination of the configurational and conformational state of the retinal chromophore. In this way, it could be shown that the primary photoprocess of the parent state in which the retinal is in the 11-*cis* configuration leads to a distorted all-*trans* configuration. Unusually strong intensities of the C–H out-of-plane vibrations of the retinal chain were taken as a measure of the torsion of the C–C single bonds of the chain. These intensities decrease in the intermediates formed during the subsequent thermal reactions indicating a relaxation of the chromophore geometry and the immediate protein environment. In the final step, the Schiff base, i.e. the covalent linkage to the apoprotein, was deprotonated as indicated by the shift of the C–N stretching vibration and its insensitivity towards H/D exchange. Furthermore, analysis of the RR spectra in terms of retinal–protein interactions contributed essentially to the understanding of colour regulation mechanism in vision.

In addition to its photolability, the plant photoreceptor phytochrome, whose chromophore is a linear tetrapyrrole, exhibits a relatively strong fluorescence which severely hinders RR spectroscopic studies under rigorous resonance conditions. Thus, Mathies and co-workers employed shifted-excitation Raman difference spectroscopy in which two spectra, measured with excitation lines separated by $10 \, cm^{-1}$, were subtracted from each other. In this way, the fluorescence background was removed yielding only difference signals which were used to calculate the absolute RR spectrum. Similar spectra were obtained by an alternative approach using Fourier-transform Raman spectroscopy at low temperature (77 K). In that case, the excitation line in the near-infrared is shifted from the electronic transition of the chromophore to circumvent fluorescence (and unwanted photochemical processes) but yields a sufficient preresonance enhancement so that the spectra are dominated by the RR bands of the tetrapyrrole. In addition to the spectra of the parent states, those of the intermediates of the photoinduced reaction cycle were obtained by controlled illumination and increase of the temperature. As shown in **Figure 1**, the spectra of the various states of phytochrome (of the $P_r \rightarrow P_{fr}$ transformation) reveal substantial differences reflecting the changes of the chromophore configuration, conformation and interactions with the surrounding protein. In contrast to retinal proteins, however, analysis of the vibrational spectra of linear tetrapyrroles is much less advanced so that up to now only a little structural information can be extracted from the spectra. Progress in spectral interpretation can be expected by extending the studies to recombinant phytochromes including chemically modified and isotopically labelled chromophores as well as by normal analyses based on quantum chemical force field calculations.

For RR spectroscopic studies of photosynthetic pigments, the intrinsic chromophore fluorescence is again an annoyance. In addition to the techniques employed in phytochrome studies, chlorophyll-containing biomolecules allow excitation in resonance with the second (non-fluorescing) electronically excited state (~350–400 nm). In the RR spectra of chlorophylls several bands have been identified which are known to be sensitive markers for the coordination state of the central Mg ion. These bands as well as those originating from the C=O stretchings of the formyl and keto substituents provide information about chlorophyll–chlorophyll and chlorophyll–protein interactions. This information was particularly helpful for elucidating the structural motifs of the chromophore assemblies in light-harvesting pigments. RR spectroscopic studies of genetically engineered protein variants of reaction centres have contributed to the understanding of those parameters controlling redox potentials and electron-transfer processes.

In contrast to photoreceptors, haem proteins are readily accessible for RR spectroscopy inasmuch as they exhibit strong electronic transitions in the visible and near UV. Moreover, they are photostable and do not reveal any interfering fluorescence. There are numerous review articles on the RR spectroscopic results obtained for various representatives of this versatile class of chromoproteins which includes redox enzymes, electron transferring proteins and oxygen transport proteins. Interpretation of the RR spectra is backed by extensive empirical material accumulated from studies on metalloporphyrins in which spectral and structural properties were correlated. Thus, the RR spectra of haem proteins can be interpreted in terms of the kind of the porphyrin, the coordination sphere of the central iron, distortions of the haem geometry, and haem–protein interactions via the porphyrin substituents and the axial ligands. As an example, **Figure 2** shows the RR spectra of cytochrome c_{552} in the oxidized and reduced state, displaying the RR bands in the region of those marker bands whose frequencies are characteristic for the oxidation, spin and coordination state of the haem iron.

In addition to those investigations which aim to characterize the structure of the haem pocket, many

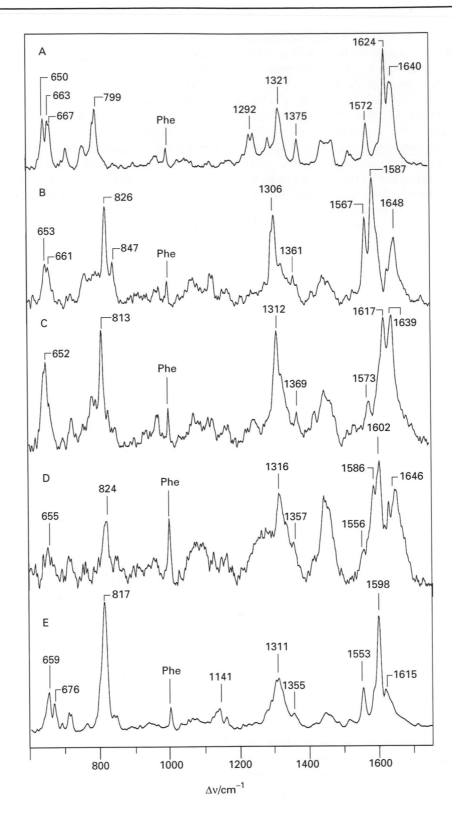

Figure 1   Fourier-transform RR spectra (1064 nm) of various states of phytochrome trapped at low temperature. (A) P_r; (B) Lumi-R; (C) Meta-R_a; (D) Meta-R_c; (E) P_{fr}.

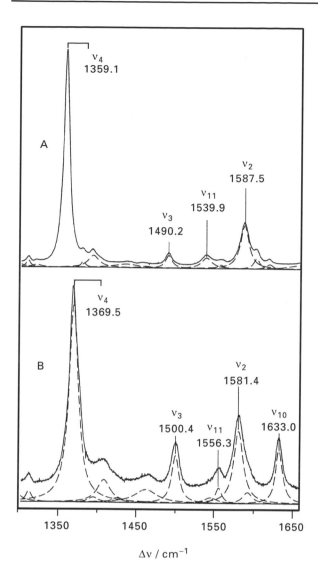

Figure 2 RR spectra (413 nm) of cytochrome c_{552} of *Thermus thermophilus* in the reduced state (A) and the oxidized state (B).

were directed to elucidate substrate-binding and substrate-specificity on a molecular level.

An interesting approach for probing molecular details of the haem pocket is the use of carbon monoxide which may occupy a vacant coordination site of the haem. The vibrations of the Fe–CO entity can be analysed in terms of the orientation of the CO ligand with respect to the haem plane, which, in turn, depends on the molecular interactions in the haem pocket. Such studies are particularly relevant for oxygen binding proteins inasmuch as CO is an appropriate analogue for molecular oxygen.

Whereas those chromoproteins discussed so far include prosthetic groups with delocalized π-electron systems, the chromophoric site in metalloproteins is constituted by a transition metal ion coordinated by various amino acid side chains. Excitation in resonance with charge-transfer transitions leads to a predominant enhancement of the metal–ligand vibrations and, to some extent, of internal ligand vibrations. Thus, the corresponding RR spectra are less complex although the interpretation is not necessarily unambiguous and straightforward in each case. Among these proteins, iron–sulfur proteins, blue-copper proteins and non-haem iron proteins have been studied extensively by RR spectroscopy.

Iron–sulfur proteins reveal a characteristic vibrational band pattern in the low frequency region which, in many cases, can be used to identify the type of the iron–sulfur cluster based on a comparison with the spectra of model compounds and related proteins of known structure. However, such studies are restricted to the oxidized form since the reduced complexes do not exhibit a charge-transfer transition in the visible region. This is also true for the blue-copper proteins for which the RR spectrum is dominated by modes including the Cu–S(Cys) stretching coordinate as well as internal modes of the Cys ligand. A substantial improvement of the understanding of these spectra was achieved by using protein variants with isotopically labelled amino acids.

The key issue for RR spectroscopists dealing with non-haem iron proteins is to determine the geometry of the oxygen-bridged binuclear complexes as well as to identify the nature of the ligands. In these studies, the RR spectra of ^{18}O-labelled samples significantly supports the interpretation.

Nucleic acids

Raman bands of nucleic acids originate from in-plane vibrations of the nucleic acid bases (adenine, guanine, cytosine, thymine and uracil) and from the furanose-phosphate backbone. In general, Raman spectra of DNA or RNA reveal structural

studies were directed to solve specific problems related to the function of the haem protein under consideration. Detailed studies were carried out on the cytochrome *c*–cytochrome *c* oxidase redox couple to understand the interprotein electron transfer mechanism. Specific emphasis was laid on the analysis of the protein complexes formed prior to the electron transfer in order to assess the effect of protein–protein interaction on the structure of the individual haem sites. In these experiments, dual channel Raman spectroscopy, which permits the quasi-simultaneous measurement of two spectra, is particularly useful for the detection of subtle spectral changes. In this way, RR spectroscopy provided evidence for a considerable conformational flexibility of both proteins which may be of functional relevance. Many studies of haem enzymes (e.g. cytochrome P-450) information about base stacking and interbase hydrogen bonding interactions. Whereas base stacking is reflected solely by intensity variations of purine and pyrimidine bands, changes of interbase hydrogen interactions, for instance, due to helix formation or conformational transitions of the helix, are indicated by both intensity changes and frequency shifts. Raman spectroscopy offers the possibility to determine the conformation of the nucleic acid inasmuch as there are marker bands indicative for the A-, B- and Z-forms of DNA. The most characteristic marker bands originate from the backbone since the various helical structures differ with respect to the sugar–phosphate conformation. Based on the comparison of various DNAs of known crystal structure, it was found that the phosphodiester symmetric stretching vibration is at 811 ± 3 cm^{-1} in A-DNA whereas it is upshifted to 835 ± 5 cm^{-1} in the B-form. Furthermore, some ring vibrations of nucleic acid bases, particularly of guanine and cytosine, also serve to distinguish between the various DNA structures (**Table 1**). For quantitative analysis of the DNA structure, the relative intensities of these specific marker bands can be used to assess the percentage of A-, B- and Z-conformation using the conformation-insensitive band at 1100 cm^{-1} (symmetric stretching of the PO$_2$ group) as a reference.

In general, structural changes of DNA in biological processes such as transcription, replication or DNA packaging, are not global but are restricted only to a few nucleotides along the chain. Such changes can induce local melting of the secondary structure, reorientations and/or disruptions of the base stacking. Melting of RNA and DNA double helices always leads to the disappearance of the 814 cm^{-1} and 835 cm^{-1} bands, indicating a decrease in furanose conformational order and an increase of the backbone chain flexibility. Precise thermal melting profiles of both RNA and DNA double helical complexes have been determined based on the intensity variations of these bands. A careful analysis of the Raman spectra of nucleic acids allows the detection of subtle alterations of the helical organization. This has been demonstrated in studies of DNA packaging, which involves formation of DNA–protein complexes with a series of histone and non-histone proteins. A backbone conformation of the B-type family was consistently observed for the nucleosome DNA complexes irrespective of the histone and non-histone content. Furthermore, complex formation is reflected by the intensity decrease of the adenine and guanine ring modes at 1490 and 1580 cm^{-1}, respectively. Whereas the intensity attenuation of the 1580 cm^{-1} band was attributed to the binding of histone proteins in the small grooves of double helical

Table 1 Conformation-sensitive non-resonant Raman bands of nucleic acid

Frequency (cm^{-1})	Description
805–815	phosphodiester symmetric stretching: C-3′-endo conformation (A-DNA)
835 (weak)	phosphodiester symmetric stretching: C-2′-endo conformation (B-DNA)
870–880	phosphodiester symmetric stretching: C-DNA
682	guanine ring breathing: C-2′-endo-anti (B-DNA)
665	guanine ring breathing: C-3′ endo-anti (A-DNA)
625	guanine ring breathing: C-3′ endo-syn (Z-DNA)
1260 (weak)	cytosine band: B-DNA
1265 (strong)	cytosine band: Z-DNA
1318 (moderate)	guanine band: B-DNA
1318 (very strong)	cytosine band: Z-DNA
1334 (moderate)	guanine band: B-DNA
1355 (moderate)	guanine band: Z-DNA
1362 (moderate)	guanine band: B-DNA
1418 (weak)	guanine band: Z-DNA
1420 (moderate)	guanine band: B-DNA
1426 (weak)	guanine band: Z-DNA

B-form DNA, the spectral changes of the 1490 cm^{-1} band is related to the binding of non-histone proteins in the large grooves.

Raman spectroscopy has also been applied to the analysis of the B–Z transformation in synthetic polymers induced by drug binding. For example, it was shown that binding of *trans*-dichlorodiamine platinium to poly(dG-dC)·poly(dG-dC) appeared to inhibit the right-to-left handed transition whereas the anti-tumour drug *cis*-dichlorodiamine platinium was found to reduce the salt requirement for adopting a left-handed modified Z-like conformation and rendered the transition essentially non-cooperative.

A Raman spectroscopic method to obtain information about dynamic aspects of nucleic acid secondary structure monitors the deuterium exchange of the C-8 purine hydrogen. The exchange rates can be determined from the time-dependent intensity increase of the bands characteristic for the deuterated bases. As the exchange process depends on the chemical environment and the solvent accessibility for the individual bases, these data allow differentiation of helical structures.

Pyrimidine and purine bases of nucleic acids have electronic absorptions in the range between 200 and 280 nm. Using excitation lines in this spectral region, RR spectra of nucleic acids exclusively display bands of in-plane ring modes of the bases without any interference of bands from the backbone. However, a selective enhancement of modes originating from purine or pyrimidine bases is not possible so that the practical use of UV RR spectroscopy for structural investigations of nucleic acids is limited. On the other hand, RR spectroscopy can be employed to probe drug–DNA interactions if the electronic transition of the drug is shifted towards the near-UV and visible region. In this way, intercalation of adriamycin by DNA was studied using different excitation lines to probe either the DNA Raman spectrum (364 nm) or the RR spectrum of the bound drug (457 nm).

Membranes

Natural biological membranes represent multicomponent systems as they include a large variety of different lipids, proteins and carbohydrates. This heterogeneity is a serious obstacle to the interpretation of Raman spectra. Thus, the majority of spectroscopic studies have focused on liposomes which are taken as models for biological membranes.

Studying the fluidity and phase transition of the lipid hydrocarbon chain, a sensitive marker band originates from the longitudinal acoustic mode (LAM) at 100–200 cm^{-1}, which corresponds to an accordion-like stretching of the entire molecule along its long axis. The frequency of this mode depends on the all-*trans* chain length and is sensitive to *gauche* rotations of the chain. The LAM is quite strong in pure hydrocarbons. However, it is difficult to observe due to its proximity to the laser line so that in the Raman spectra of phospholipid dispersions, it could be detected only in a few cases, i.e. at low water content and low temperature.

The skeletal optical modes between 1050 and 1150 cm^{-1} and the ratio of the C–H stretching bands at 2850 and 2885 cm^{-1}, were used to quantify membrane in order to detect membrane order perturbations induced by the incorporation of small hydrophobic molecules in the bilayer. The intensity at 1130 cm^{-1} is thought to be a measure for the number of all-*trans* sections involving more than three carbon atoms. For phosphatidylcholine vesicles, it was found that the intensity of this band

relative to either the 1660 cm^{-1} band or the C–N choline stretching at 718 cm^{-1} could be used to monitor both phase transitions and gradual melting of the lipids. Changes of a stiffening of hydrocarbon chains, e.g. by binding of divalent cations to the phosphate head groups of phosphatidylcholines, is inferred from alterations of the intensity ratio of the 1064 and 1089 cm^{-1} bands.

Interactions of membranes with larger molecules are difficult to probe by Raman spectroscopy unless selectivity can be increased by taking advantage of the RR effect. This has been demonstrated in studies on drug–membrane interactions by tuning the excitation line in resonance with the electronic transition of the drug molecule, e.g. amphotericin.

Dynamics of biomolecules

Operating in the time-resolved domain, RR spectroscopy can be employed to probe the dynamics of biological systems. Hence, such studies can provide simultaneously kinetic and structural data. In many cases, time-resolved RR spectroscopy, albeit technically demanding, may represent the method of choice for elucidating the molecular mechanisms of biological reactions.

Photoinduced processes

Bacterial retinal proteins such as bacteriorhodopsin and halorhodopsin act as light-driven ion pumps. The ion gradient that is generated across the membrane is then converted into chemical energy (i.e. synthesis of adenosine triphosphate). The ion translocation is linked to a photoinduced reaction cycle of the retinal chromophore. The photophysical and kinetic properties as well as the stability of this class of proteins make them an ideal system for time-resolved RR spectroscopic studies. Moreover, these proteins represent suitable objects for developing and optimizing time-resolved spectroscopic techniques. The main advantage is the reversibility of the photoinduced reaction cycle, i.e. the system comes back to the parent state in a few ms after the primary photochemical event. Thus, time-resolved RR spectra can be accumulated continuously since the 'fresh sample' condition is readily established. This condition ensures that the protein is always in the same state when irradiated by the exciting laser beam.

There are two approaches for time-resolved RR spectroscopy of these retinal proteins. Using cw-excitation a time-resolution down to 100 ns can be achieved by rapidly moving the sample through the laser focus (rotating cell, capillar flow system). Pulsed laser excitation can provide a time-resolution even in the subpicosecond range depending on the pulse width of the laser. In both methods, intermediate states are probed in dual-beam experiments with the probe beam irradiating the sample after a delay time δ with respect to the photolysis beam which initiates the reaction cycle. The systematic variation of δ allows the determination of the photocycle kinetics which, along with the structural data derived from the RR spectra, can provide a comprehensive picture of the dynamics of the protein. **Figure 3** shows a selection of time-resolved RR of halorhodopsin. The bands displayed in these spectra are diagnostic for the retinal configuration (C=C stretching) and confirm that the photoreaction includes the isomerization from the all-*trans* to the 13-*cis* configuration.

Redox processes

Cytochrome *c* oxidase, a membrane-bound enzyme in the respiratory chain of aerobic organisms, reduc-

es oxygen to water. This process which takes place at the binuclear metal centre constituted by a haem a_3 and a Cu ion runs via several intermediate states with life times in the micro- and millisecond range. Technical improvements now make it possible to monitor this reaction sequence by time-resolved RR spectroscopy using rapid flow systems. The starting point of the experiments is the thermally stable carbon monoxide complex of the reduced enzyme in oxygen-saturated solution. This complex is photodecomposed by a laser beam so that oxygen can bind to the catalytic centre constituting the time 'zero' of the reaction sequence. The various intermediates are probed by a second laser beam irradiating the (photolysed) sample volume after a delay time. The oxygen-sensitive vibrational modes which can readily be identified based on the $^{18}O/^{16}O$ isotopic shifts give insight into the nature of the various intermediates and, hence, into the molecular mechanism of the oxygen reduction. A particularly interesting technical approach for these studies is based on an artificial cardiovascular device designed to maintain a continuous enzymatic reaction. Thus, it is possible to accumulate the RR signals during a sufficiently long period of time using a minimum amount of sample.

Living cells and biomedical applications

Recent progress in laser technology and signal detection has substantially increased the sensitivity of Raman spectroscopy. This is a prerequisite for the development of Raman microspectroscopy and Raman imaging which has opened a new field for the application of Raman spectroscopy in biological and medical research. A particularly powerful method has been introduced by Greve and co-workers who combined Raman spectroscopy with confocal microscopy which allows the measurement of spatially resolved Raman spectra with a lateral resolution of less than 0.5 μm. The technique makes use of sensitive CCD detection systems by probing either the intensity of a marker band in two (spatial) dimensions (global imaging) or a complete spectrum in one dimension (linescan imaging). In these experiments, low energy excitation (>600 nm) has to be employed and the power of the tightly focused laser beam must be reduced as far as possible to avoid photoinduced degradation of the biological objects. In this way, it is possible to study single living cells and to obtain Raman spectra of the cytoplasm and the nucleus separately. The spectral information allows determination of the DNA conformation, the relative content of the individual nucleic acid bases and the

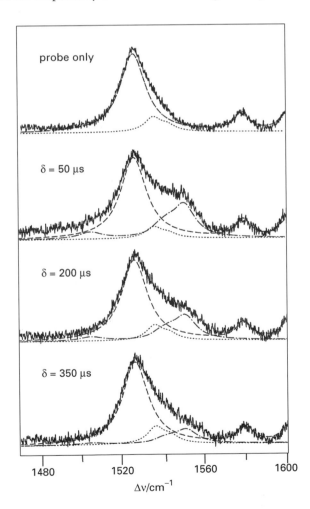

Figure 3 Time-resolved RR spectra of halorhodopsin from *Natronobacterium pharaonis* obtained in single (probe only, 514 nm) and dual beam experiments with variable delay times δ of the probe laser (514 nm) relative to the pump laser (600 nm).

DNA/protein ratio. Furthermore, less abundant biomolecules such as carotenoids or haem enzymes or non-fluorescent drugs (e.g. cobalt octacarboxyphthalocyanine) can be detected by taking advantage of the RR effect. These studies are particularly promising for elucidating metabolic processes in cells.

This method has also been applied to medical problems related to disease diagnosis (e.g. artherosclerotic plaque) or optimization of medical treatment (e.g. bone implants). However, medical applications of Raman spectroscopy do not necessarily require the combination with microscopy. Numerous studies have indicated that based on the analysis of Raman spectra by statistical methods it is possible to differentiate between normal and pathological tissues. Despite the substantial technical improvements, the intrinsically low sensitivity of Raman spectroscopy constitutes a limit for general applicability. Fluorescent samples may impose an additional constraint, which, however, can be overcome by near-infrared (1064 nm) Fourier-transform Raman spectroscopy and microscopy. In summary, it appears that for special medical applications Raman spectroscopic techniques may become a powerful diagnostic tool in clinical situations.

See also: **Biofluids Studied By NMR; Chiroptical Spectroscopy, General Theory; Forensic Science, Applications of IR Spectroscopy; Industrial Applications of IR and Raman Spectroscopy; IR Spectrometers; Medical Science Applications of IR; Membranes Studied By NMR Spectroscopy; Nucleic Acids and Nucleotides Studied Using Mass Spectrometry; Nucleic Acids Studied By NMR; Peptides and Proteins Studied Using Mass Spectrometry; Raman Spectrometers; Surface-Enhanced Raman Scattering (SERS), Applications.**

Further reading

Althaus T, Eisfeld W, Lohrmann R and Stockburger M (1995) Application of Raman spectroscopy to retinal proteins. *Israel Journal of Chemistry* 35: 227–251.

Andel F, Lagarias JC and Mathies RA (1996) Resonance Raman analysis of chromophore structure in the lumi-R photoproduct of phytochrome. *Biochemistry* 35: 15997–16008.

Andrew CR, Han J, den Blaauwen T *et al* (1997) Cysteine ligand vibrations are responsible for the complex resonance Raman spectrum of azurin. *Journal of Bioinorganic Chemistry* 2: 98–107.

Callender R, Deng H and Gilmanshin R (1998) Raman difference studies of protein structure and folding, enzymatic catalysis and ligand binding. *Journal of Raman Spectroscopy* 29: 15–21.

Carey PR (1998) Raman spectroscopy in enzymology: the first 25 years. *Journal of Raman Spectroscopy* 29: 7–14.

Clark RJH and Hester RE (eds) (1986) Spectroscopy of biological systems. *Advances in Spectroscopy* 13.

Clark RJH and Hester RE (eds) (1993) Biomolecular spectroscopy, part A. *Advances in Spectroscopy* 20.

Clark RJH and Hester RE (eds) (1993) Biomolecular spectroscopy, part B. *Advances in Spectroscopy* 21.

Hildebrandt P (1992) Resonance Raman spectroscopy of cytochrome P-450. In: Ruckpaul K and Rein H (eds) *Frontiers in Biotransformation*, Vol. 7, pp 166–215. Berlin/Weinheim: Akademie-Verlag/VCH.

Hildebrandt P (1995) Resonance Raman spectroscopy of cytochrome c. In: Scott RA and Mauk AG (eds) *Cytochrome c. A Multidisciplinary Approach*, pp 285–314. Mill Valley: University Science.

Kneip C, Mozley D, Hildebrandt P, Gärtner W, Braslavsky SE and Schaffner K (1997) Effect of chromophore exchange on the resonance Raman spectra of recombinant phytochromes. *FEBS Letters* 414: 23–26.

Kochendoerfer GG, Wang Z, Oprian DD and Mathies RA (1997) Resonance Raman examination of wavelength regulation mechanism in human visual pigments. *Biochemistry* 36: 6577–6587.

Manoharan R, Wang Y and Feld MS (1996) Histochemical analysis of biological tissues using Raman spectroscopy. *Spectrochimica Acta A* 52: 215–249.

Otto C, de Grauw CJ, Duindam JJ, Sijtsema NM and Greve J (1997) Applications of micro-Raman imaging in biomedical research. *Journal of Raman Spectroscopy* 28: 143–150.

Spiro TG (ed) (1987) *Biological Application of Raman Spectroscopy*, Vol 1: Raman spectra and the conformations of biological macromolecules. New York: John Wiley & Sons.

Spiro TG (ed) (1987) *Biological Application of Raman Spectroscopy*, Vol 2: Resonance Raman spectra of polyenes and aromatics. New York: John Wiley & Sons.

Spiro TG (ed) (1988) *Biological Application of Raman Spectroscopy*, Vol 3: Resonance Raman spectra of haem and metalloproteins. New York: John Wiley & Sons.

Tensmeyer LG and Kauffmann EW (1996) Protein structure as revealed by nonresonance Raman spectroscopy. In: Havel HA (ed) *Spectroscopic Methods for Determining Protein Structure in Solution*, pp 69–95. New York: VCH.

Thamann TJ (1996) Probing local protein structure with ultraviolet resonance Raman spectroscopy. In: Havel HA (ed) *Spectroscopic Methods for Determining Protein Structure in Solution*, pp 96–134. New York: VCH.

Twardowski J and Anzenbacher P (1994) *Raman and IR Spectroscopy in Biology and Biochemistry*. New York: Ellis Horwood.

Verma SP and Wallach DFH (1984) Raman spectroscopy of lipids and biomembranes. In: Chapman D (ed) *Biomembrane Structure and Function*, pp 167–198. Weinheim: Verlag Chemie.

Zhao X and Spiro TG (1998) Ultraviolet resonance Raman spectroscopy of hemoglobin with 200 and 212 nm excitation: H-bonds of tyrosines and prolines. *Journal of Raman Spectroscopy* 29: 49–55.

Biofluids Studied By NMR

John C Lindon and **Jeremy K Nicholson**, Imperial College of Science, Technology and Medicine, London, UK

MAGNETIC RESONANCE
Applications

Introduction

Investigation of biofluid composition provides insight into the status of a living organism in that the composition of a particular fluid carries biochemical information on many of the modes and severity of organ dysfunction. One of the most successful approaches to biofluid analysis has been the application of NMR spectroscopy.

The complete assignment of the ^{1}H NMR spectrum of most biofluids is not possible (even by use of 900 MHz NMR spectroscopy) owing to the enormous complexity of the matrix. However, the assignment problems vary considerably between biofluid types. For instance, seminal fluid and blood plasma are highly regulated with respect to metabolite composition and concentrations and the majority of the NMR signals have been assigned at 600 and 750 MHz for normal human individuals. Urine composition is much more variable because its composition is normally adjusted by the body in order to maintain homoeostasis and hence complete analysis is much more difficult. There is also enormous variation in the concentration range of NMR-detectable metabolites in urine samples. With every new increase in available spectrometer frequency the number of resonances that can be resolved in a biofluid increases and although this has the effect of solving some assignment problems, it also poses new ones. Furthermore, problems of spectral interpretation arise due to compartmentation and binding of small molecules in the organized macromolecular domains that exist in some biofluids such as blood plasma and bile.

All biological fluids have their own characteristic physicochemical properties and a summary of some of these is given in **Table 1** for normal biofluids. These partly dictate the types of NMR experiment that may be employed to extract the biochemical information from each fluid type. An illustration of the complexity of biofluid spectra, and hence the need for ultrahigh field measurements, is given in **Figure 1** which shows 800 MHz ^{1}H NMR spectra of normal human urine, bile and blood plasma.

It is clear that at even the present level of technology in NMR, it is not yet possible to detect many important biochemical substances, e.g. hormones, in body fluids because of problems with sensitivity, dispersion and dynamic range and this area of research will continue to be technology limited. With this in mind, it would seem prudent to interpret quantitative ^{1}H NMR measurements of intact biological materials and assignment of resonances in 1D spectra with considerable caution even when measured at ultrahigh field.

Resonance assignment in NMR spectra of biofluids

Usually in order to assign ^{1}H NMR spectra of biofluids, comparison is made with spectra of authentic materials and by standard addition to the biofluid sample. Additional confirmation of assignments is usually sought from the application of 2-dimensional (2D) NMR methods, particularly COSY and TOCSY and, increasingly, inverse-detected heteronuclear correlation methods such as HMQC and HSQC. In addition, the application of the 2D J-resolved (JRES) pulse sequence is useful for spreading out the coupling patterns from the multitude of small molecules in a biofluid. Even 2D correlation NMR spectra of complex biofluids can show much overlap of cross-peaks and further editing is often desirable. Thus simplification of 1D and 2D NMR spectra of biofluids can also be achieved using (i) spin–echo methods particularly for fluids containing high levels of macromolecules (T_2 editing), (ii) editing based on T_1 relaxation time differences, (iii) editing based on diffusion coefficient differences, (iv) multiple quantum filtering.

One major advantage of using NMR spectroscopy to study complex biomixtures is that measurements can often be made with minimal sample preparation (usually with only the addition of 5–10% D_2O) and a detailed analytical profile can be obtained on the whole biological sample. This in turn requires good methods for suppressing solvent resonances.

Detailed ^{1}H NMR spectroscopic data for a wide range of low molecular weight metabolites found in biofluids are given in **Table 2**.

Table 1 Normal biofluids and their physicochemical properties

Biofluid	Function	Water content[a]	Viscosity	Protein content[b]	Lipid content[b]	Peak overlap[c]
Urine	Excretion Homoeostasis	+++	e	e	e	e
Bile	Excretion Digestion	++	++	+	+	+++
Blood plasma	Transport Homoeostasis Mechanical	+++	++	+++	+++	++
Whole blood[d]	Transport Oxygenation	+++	+++	+++	++	+++
Cerebrospinal fluid	Transport Homoeostasis Mechanical	+++	+	+	+	++
Milk	Nutrition	++	+	++	+++	+++
Saliva	Excretion Digestion	+++	++	++	++	+
Gastric juice	Digestion	+++	+++	+++	e	++
Pancreatic juice	Digestion	++	++	+++	+	+++
Seminal fluid	Support for spermatozoa	+	+++	+	+	+++
Prostatic fluid	Support for spermatozoa	+	++	+	+	++
Seminal vesicle fluid	Support for spermatozoa	+	+++	+++	+	+++
Amniotic fluid	Protection of fetus	+++	+	+	+	+
Follicular fluid	Reproduction	+++	+	+	+	+
Synovial fluid	Joint protection	+++	+++	++	+	+
Aqueous humour	Eye function	+++	+	++	e	+

+++, ++, + indicate high, medium, low degree of constraint for NMR studies.
[a] Relative water intensity when compared with concentrations of metabolities of interest.
[b] In biofluids with high protein or lipid contents, spin-echo spectra must normally be employed to eliminate broad resonances.
[c] A subjective indication of spectral crowding (at 600 MHz) due to abundance of endogenous metabolites with a wide range of shifts.
[d] Presence of cells and process of cell sedimentation gives rise to magnetic field inhomogeneity problems.
[e] Not a limiting factor.

NMR studies of dynamic interactions

Although NMR spectroscopy of biofluids is now well established for probing a wide range of biochemical problems, there are still many poorly understood physicochemical phenomena occurring in biofluids, particularly the subtle interactions occurring between small molecules and macromolecules or between organized multiphasic compartments. The understanding of these dynamic processes is of considerable importance if the full diagnostic potential of biofluid NMR spectroscopy is to be realized. Typically, it is now possible to study enzymatic reactions, chemical reactions and biofluid instability, microbiological activity in biofluids, macromolecular binding of small molecules, membrane-based compartmentation, metal complexation and chemical exchange processes.

NMR spectroscopy of blood plasma and whole blood

NMR peak assignments

The physicochemical complexity of plasma shows up in its ^{1}H NMR spectra by the range of linewidths of the signals. This means that a number of different multiple pulse NMR experiments and/or physicochemical interventions must be applied to extract useful biochemical information. Numerous high resolution ^{1}H NMR studies have been performed on the biochemistry of blood and its various cellular components and plasma. The physical properties of whole blood pose serious limitations on direct NMR investigations, but packed erythrocytes yield more useful information on cell biochemistry. Well resolved spectra are given by plasma, and ^{1}H NMR

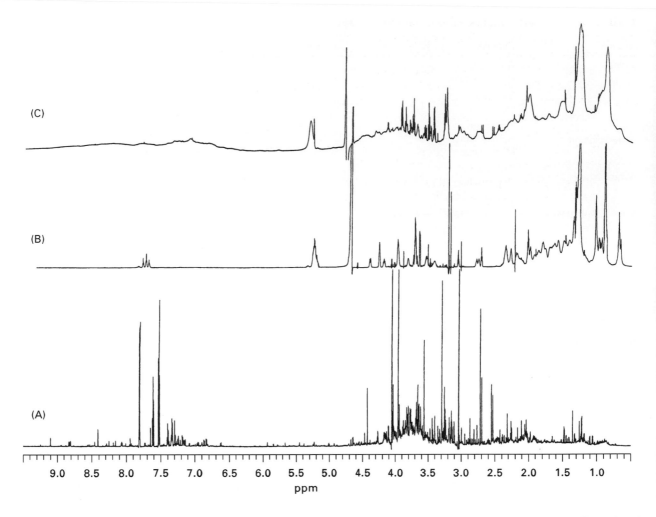

Figure 1 800 MHz ¹H NMR spectra of control human biofluids; (A) urine; (B) gall bladder bile and (C) blood plasma. Reproduced with permission of Academic Press from Lindon JC, Nicholson JK and Everett JR (1999) NMR spectroscopy of biofluids. *Annual Reports on NMR Spectroscopy* **38**: in the press.

measurements on blood serum and plasma can provide much useful biochemical information on both low molecular weight metabolites and macromolecular structure and organization.

In blood plasma, the ¹H NMR peaks of metabolites, proteins, lipids and lipoproteins are heavily overlapped even at 800 MHz (**Figure 1**). Most blood plasma samples are quite viscous and this gives rise to relatively short T_1 relaxation times for small molecules and this allows relatively short pulse repetition cycles without signal saturation. The spectral profile can be simplified by use of spin–echo experiments with an appropriate T_2 relaxation delay to allow signals from broad macromolecular components and compounds bound to proteins to be attenuated. By the early 1980s many metabolites had been detected in normal blood plasma although assignments were, in general, based on the observation of only one or two resonances for each metabolite. In addition, peaks from certain macromolecules

such as α_1-acid glycoprotein (*N*-acetyl neuraminic acid and related sialic acid fragments) have been assigned and used diagnostically, in particular their *N*-acetyl groups which give rise to relatively sharp resonances presumably due to less restricted molecular motion. The signals from some lipid and lipoprotein components, e.g. very low density lipoprotein (VLDL), low density lipoprotein (LDL), high density lipoprotein (HDL) and chylomicrons, have also been partially characterized.

For normal plasma at pH 7, the largest peak in the spectral region to high frequency of water is that of the α-anomeric H1 resonance of glucose at $\delta 5.223$ (which provides a useful internal chemical shift reference). In spectra of normal human and animal plasma, there are few resonances in the chemical shift range to high frequency of $\delta 5.3$ when measured in the pH range 3 to 8.5. However, on acidification of the plasma to pH < 2.5, resonances from histidine and phenylalanine become detectable. Experiments with model solutions suggested that serum albumin has a high capacity for binding aromatic amino acids and histidine at neutral pH and this is responsible for their NMR-invisibility in normal human blood plasma. Serum albumin also binds a large number of other species of both endogenous and xenobiotic origin. Even commercial 'purified' bovine serum albumin (BSA) can be shown to contain a significant amount of bound citrate and acetate which become NMR-detectable in BSA solutions at pH 2. Acidification of human plasma also renders citrate NMR-detectable in spin–echo spectra as it becomes mobilized from the protein binding sites. Through the increased spectral dispersion available from the use of 600, 750 and 800 MHz ^{1}H NMR measurements and through the use of a variety of 2D methods, the assignment of resonances in blood plasma spectra in normal individuals is now extensive (see **Table 2**).

Blood plasma also has intrinsic enzymatic activities although many of these are not stable (particularly if the sample is not frozen immediately on collection). It has been noted that under certain pathological conditions, such as those following liver or kidney damage, enzymes that are present at elevated levels in the plasma because of leakage from the damaged tissue can cause NMR-detectable alterations to spin–echo spectra of plasma.

Molecular diffusion coefficients are parameters that are not related directly to NMR spectral intensities under normal conditions. However, molecular diffusion can cause NMR signal intensity changes when pulsed field gradients are applied during the NMR experiment. A number of pulse sequence developments have meant that measurement of diffusion coefficients is relatively routine. The editing of ^{1}H NMR spectra of biofluids based on diffusion alone or on a combination of spin relaxation and diffusion has been demonstrated recently. This approach is complementary to the editing of ^{1}H NMR spectra based on differences in T_1 and T_2. New methods for editing TOCSY NMR spectra of biofluids have been proposed based on differences in molecular diffusion coefficients and this has been termed Diffusion Edited TOCSY (DETOCSY).

Much study has been devoted to the problem of lipoprotein analysis in blood plasma using ^{1}H NMR spectroscopy. This has been comprehensively reviewed recently by Ala-Korpela. Lipoproteins are complex particles that transport molecules normally insoluble in water. They are spherical with a core region of triglyceride and cholesterol ester lipids surrounded by phospholipids in which are embedded various proteins known as apolipoproteins. In addition, free cholesterol is found in both the core and surface regions. The lipoproteins are in a dynamic

equilibrium with metabolic changes going on *in vivo*. Lipoproteins are usually classified into five main groups, chylomicrons, very low density lipoprotein (VLDL), low density lipoprotein (LDL), intermediate density lipoprotein (IDL) and high density lipoprotein (HDL) based on physical separation using centrifugation. Based on the measurement of ^{1}H NMR spectra of the individual fractions and using line-shape fitting programs, it has been possible to identify the chemical shifts of the CH_2 and CH_3 groups of the fatty acyl side chains. Quantification can be carried out using either time-domain or frequency-domain NMR data. The usefulness of ^{1}H NMR spectra for lipoprotein analysis and ^{31}P NMR spectroscopy for phospholipid analysis in blood plasma has been explored. More recently a neural network software approach has been used to provide rapid lipoprotein analyses.

NMR spectra of blood plasma in pathological states

A good deal of excitement was generated by a publication which reported that ^{1}H NMR spectroscopy of human blood plasma could be used to discriminate between patients with malignant tumours and other groups, namely normals, patients with non-tumour disease and a group of patients with certain benign tumours. This paper stimulated many other research groups around the world to investigate this approach to cancer detection.

The test as originally published involved the measurement of the averaged line width at half height of the two composite signals at δ1.2 and δ0.8 in the single pulse ^{1}H NMR spectrum of human blood plasma. On comparing the averaged signal width, W, for normal subjects with those for patients with a variety of diseases and with those for pregnant women, it was reported that several statistically significant differences existed between the groups. It was proposed that the lowering of W observed in the malignant group was due to an increase in the T_2^* of the lipoproteins. A mass of literature now exists on this subject and it was immediately apparent that the perfect 100% sensitivity and specificity reported in the early publications has never been repeated by any group. In addition to the presence of cancer, a number of other factors have been found to cause changes in the linewidth index, W. These include diet, age and sex, pregnancy and trauma as well as hyperlipidaemia. In all cases the observed changes in W are caused by alterations in the plasma lipoprotein composition, especially the VLDL/HDL ratio.

The success of heart transplantation has improved recently with the chronic use of cyclosporin A to

Table 2 ¹H NMR assignments and chemical shifts for metabolities found in biofluids

Metabolite	Assignment	$\delta(^1H)$	Multiplicity	$\delta(^{13}C)$	Biofluid[a]
α-Hydroxyisovalerate	CH_3	0.81	d		C,U
α-Hydroxy-n-butyrate	CH_3	0.90	t		C,U
n-Butyrate	CH_3	0.90	t		U
α-Hydroxy-n-valerate	CH_3	0.92	t		U
Isoleucine	δ-CH_3	0.94	t		A,C,P,S,U
Leucine	δ-CH_3	0.96	t		A,C,P,U
Leucine	δ-CH_3	0.97	t		A,C,P,U
α-Hydroxyisovalerate	CH_3	0.97	d		C,U
Valine	CH_3	0.99	d	19.6	C,P,U
Isoleucine	β-CH_3	1.01	d	14.6	A,C,P,S,U
Valine	CH_3	1.04	d		A,C,P,U
Ethanol	CH_3	1.11	t		C,P,U
Propionate	CH_3	1.12	t		U
Isobutyrate	CH_3	1.13	d		P
β-Hydroxybutyrate	CH_3	1.20	d		C,P,S,U
Isoleucine	γ-CH_2	1.26	m		A,C,P,S,U
Fucose	CH_3	1.31	d		P
Lactate	CH_3	1.33	d	20.9	A,C,P,S,U
Threonine	CH_3	1.34	d		A,C,P,S,U
α-Hydroxyisobutyrate	CH_3	1.36	s		U
α-Hydroxy-n-valerate	γ-CH_2	1.37	m		U
Alanine	CH_3	1.48	d	16.8	A,C,P,S,U
Lysine	γ-CH_2	1.48	m		C,P,S,U
Isoleucine	γ-CH_2	1.48	m		C,P,S,U
n-Butyrate	β-CH_2	1.56	d		P,U
Adipate	CH_2	1.56	m		P,U
Citrulline	γ-CH_2	1.58	m		C,U
α-Hydroxy-n-valerate	β-CH_2	1.64	m		U
α-Hydroxybutyrate	CH_2	1.70	m		U
Arginine	γ-CH_2	1.70	m		U
Leucine	CH_2	1.71	m	40.7	C,P,S,U
Lysine	δ-CH_2	1.73	m		C,P,S,U
Ornithine	γ-CH_2	1.81	m		C,P,S,U
Citrulline	β-CH_2	1.88	m		C,P,S,U
N-Acetylglutamate	β-CH_2	1.89	m		C,P,S,U
γ-Amino-n-butyrate	β-CH_2	1.91	m		C,P,S,U
Lysine	β-CH_2	1.91	m	30.3	C,P,S,U
Arginine	β-CH_2	1.93	m		C,P,S,U
Ornithine	β-CH_2	1.95	m		A,S,U
Acetate	CH_3	1.95	s		A,C,P,S,U
Isoleucine	β-CH	1.98	m		C,P,U
Acetamide	CH_3	2.01	s		U
N-Acetyl groups(glycoproteins)	CH_3	2.02	s	23.0	P,U
α-Hydroxyisovalerate	β-CH	2.02	m		U

Table 2 *continued*

Metabolite	Assignment	$\delta(^1H)$	Multiplicity	$\delta(^{13}C)$	Biofluid[a]
N-Acetylaspartate	CH_3	2.03	s		C,U
Proline	γ-CH_2	2.01	m		C,S,U
N-Acetyl groups(glycoproteins)	CH_3	2.05	s		P,U
Proline	β-CH_2	2.07	m		C,S,U
N-Acetylglutamate	CH_3	2.04	s		U
N-Acetylglutamate	β-CH_2	2.06	m		U
Glutamate	β-CH_2	2.10[b]	m	30.1	A,C,P,S,U
Glutamine	β-CH_2	2.14[b]	m		C,P,S,U
Methionine	S-CH_3	2.14	s		A,C,P,S,U
n-Butyrate	α-CH_2	2.16	t		C,P,S,U
Methionine	β-CH_2	2.16	m		A,C,P,S,U
Adipate	CH_2COOH	2.22	m		U
N-Acetylglutamate	γ-CH_2	2.23	t		U
Acetone	CH_3	2.23	s		P,U
Valine	β-CH	2.28	m		C,P,S,U
Acetoacetate	CH_3	2.29	s		C,P,S,U
γ-Amino-n-butyrate	α-CH_2	2.30	t		C
β-Hydroxybutyrate	CH_2	2.31	ABX		C,P,S,U
Proline	β-CH_2	2.35	m		A,C,S,U
Glutamate[d]	γ-CH_2	2.36	m	34.5	C,U
Oxalacetate	CH_2	2.38	s		C,P,U
Pyruvate	CH_3	2.38	s		C,P,U
Malate	CH_2	2.39	dd		U
β-Hydroxybutyrate	CH_2	2.41	ABX		C,P,S,U
Succinate	CH_2	2.43	s		C,P,S,U
Carnitine	$CH_2(COOH)$	2.44	dd		C,P,S,U
α-Ketoglutarate	γ-CH_2	2.45	t		C,P,S,U
Glutamine	γ-CH_2	2.46[b]	m	31.9	C,P,S,U
Glutamate[d]	γ-CH_2	2.50	m		A,C,P,S,U
N-Acetylaspartate	CH_2	2.51	ABX		C,U
Methylamine	CH_3	2.54	s		P
Citrate	$\frac{1}{2}CH_2$	2.67	AB		A,C,P,S,U
Methionine	S-CH_2	2.65	t		C,P,S,U
Aspartate	β-CH_2	2.68[c]	ABX		C,P,S,U
Malate	CH_2	2.69	ABX		U
N-Acetylaspartate	CH_2	2.70	ABX		C,U
Dimethylamine	CH_3	2.72	s		C,P,S,U
Sarcosine	CH_3	2.74	s		U
Dimethylglycine	CH_3	2.78	s		P,U
Citrate	$\frac{1}{2}CH_2$	2.80	AB		A,C,P,S,U
Aspartate	β-CH_2	2.82	ABX		C,P,S,U
Methylguanidine	CH_3	2.83	s		U
Asparagine	β-CH_2	2.86	m		C,P,S,U
Trimethylamine	CH_3	2.88	s		U

Table 2 *continued*

Metabolite	Assignment	$\delta(^1H)$	Multiplicity	$\delta(^{13}C)$	Biofluid[a]
Asparagine	β-CH$_2$	2.96	m		C,P,S,U
α-Ketoglutarate	β-CH$_2$	3.01	t		C,P,S,U
γ-Amino-*n*-butyrate	γ-CH$_2$	3.02	t		C
Lysine	ε-CH$_2$	3.03	t	40.3	C,P,S,U
Cysteine	CH$_2$	3.04	m		C,U
Creatine	CH$_3$	3.04	s		A,C,P,S,U
Phosphocreatine	CH$_3$	3.05	s		S
Creatinine	CH$_3$	3.05	s		A,C,P,S,U
Tyrosine	CH$_2$	3.06	ABX		C,P,S,U
Ornithine	δ-CH$_2$	3.06	t		C,S,U
Cysteine	CH$_2$	3.12	ABX		C,U
Malonate	CH$_2$	3.13	s		U
Phenylalanine	β-CH$_2$	3.13	m		C,P,S,U
Histidine	β-CH$_2$	3.14	ABX		C,P,S,U
Citrulline	α-CH$_2$	3.15	m		C,U
cis-Aconitate	CH$_2$	3.17	s		U
Tyrosine	CH$_2$	3.20	ABX		C,P,S,U
Choline	N(CH$_3$)$_3$	3.21	s	55.0	C,P,S,U
Phosphorylethanolamine	NCH$_2$	3.23	t		S
β-Glucose	C-H2	3.24	dd		A,C,P,S,U
Histidine	β-CH$_2$	3.25	ABX		C,P,S,U
Arginine	δ-CH$_2$	3.25	t	41.3	C,S,U
Trimethylamine-*N*-oxide	N(CH$_3$)$_3$	3.27	s		C,P,S,U
Taurine	CH$_2$SO$_3$	3.25	t		A,C,P,S,U
Betaine	N(CH$_3$)$_3$	3.27	s		C,P,S,U
Phenylalanine	β-CH$_2$	3.28	m		C,P,S,U
Myo-inositol	H5	3.28	t		P
Tryptophan	CH$_2$	3.31	ABX		P,S,U
Glycerophosphorylcholine	N(CH$_3$)$_3$	3.35	s		S
Proline	δ-CH$_2$	3.33	m		A,P,U
β-Glucose	C-H4	3.40	t	70.6	A,C,P,S,U
α-Glucose	C-H4	3.41	t	70.6	A,C,P,S,U
Proline	δ-CH$_2$	3.42	m		A,P,U
Carnitine	NCH$_2$	3.43	m		C,P,S,U
Taurine	NCH$_2$	3.43	t		A,C,P,S,U
Acetoacetate	CH$_2$	3.45	s		C,P,S,U
β-Glucose	C-H5	3.47	ddd	76.7	A,C,P,S,U
trans-Aconitate	CH$_2$	3.47	s		U
α-Glucose	C-H3	3.49	t	76.7	A,C,P,S,U
Tryptophan	CH$_2$	3.49	ABX		P,U
Choline	NCH$_2$	3.52	m		C,P,S,U
Glycerophosphorylcholine	NCH$_2$	3.52	m		S
α-Glucose	C-H2	3.53	dd	72.3	A,C,P,S,U
Glycerol	CH$_2$	3.56	ABX	63.5	C,P
Myo-inositol	H1/H3	3.56	dd		P

Table 2 *continued*

Metabolite	Assignment	$\delta(^1H)$	Multiplicity	$\delta(^{13}C)$	Biofluid[a]
Glycine[d]	CH_2	3.57	s		C,P,S,U
Threonine	α-CH	3.59	d		U
Fructose (β-furanose)	H1	3.59	d		S
Fructose (β-furanose)	H1	3.59	d		S
Fructose (β-pyranose)	H1	3.60	d		S
Sarcosine	CH_2	3.61	s		C,P,S,U
Ethanol	CH_2	3.61	d		C,P,S,U
Valine	α-CH	3.62	d	64.2	C,P,S,U
Myo-inositol	H4/H6	3.63	dd		P
Fructose (β-pyranose)	H6'	3.63	m		S
Fructose (β-pyranose)	H3	3.64	m		S
Glycerol	CH_2	3.65	ABX	63.5	C,P
Isoleucine	α-CH	3.68	d		A,C,P,S,U
Fructose (β-furanose)	H6'	3.70	m		S
Fructose (β-furanose)	H1	3.70	m		S
α-Glucose	C-H3	3.71	t	73.6	C,P,S,U
β-Glucose	C-H6'	3.72	dd	61.6	A,C,P,S,U
Leucine	α-CH	3.73	t	55.1	A,C,P,S,U
α-Glucose	C-H6'	3.74	m	61.4	A,C,P,S,U
Ascorbate	CH_2	3.74	d		U
Ascorbate	CH_2	3.76	d		U
Citrulline	α-CH	3.76	t		C,U
Lysine	α-CH	3.76	t		C,P,S,U
Glutamine	α-CH	3.77	t	55.4	C,P,S,U
Glutamate	α-CH	3.77	t		A,C,P,S,U
Arginine	α-CH	3.77	t		C,U
Glycerol	CH	3.79	ABX	72.6	C,P
Alanine	CH	3.79	q		A,C,P,S,U
Ornithine	α-CH	3.79	t		C,P,S,U
Guanidoacetate	CH_2	3.80	s		U
Mannitol	CH3	3.82	d		C
Fructose (β-furanose)	H4	3.82	m		S
Fructose (β-pyranose)	H3	3.84	m		S
α-Glucose	C-H6	3.84	m	61.4	A,C,P,S,U
α-Glucose	C-H5	3.84	ddd	72.3	A,C,P,S,U
Fructose (β-furanose)	H6	3.85	m		S
α-Hydroxyisovalerate	α-CH	3.85	d		U
Serine	α-CH	3.85	ABX		U
Methionine	α-CH	3.86	t		A,C,P,S,U
β-Glucose	C-H6	3.90	dd	61.6	A,C,P,S,U
Fructose (β-pyranose)	H4	3.90	m		S
Betaine	CH_2	3.90	s		C,P,S,U
Aspartate	α-CH	3.91	ABX		C,U
4-Aminohippurate	CH_2	3.93	d		U
Creatine	CH_2	3.93	s		P

Table 2 *continued*

Metabolite	Assignment	$\delta(^1H)$	Multiplicity	$\delta(^{13}C)$	Biofluid [a]
Glycolate	CH$_2$	3.94	s		U
Tyrosine	CH	3.94	ABX		C,P,S,U
Phosphocreatine	CH$_2$	3.95	s		[e]
Serine	β-CH$_2$	3.95	ABX		U
Hippurate	CH$_2$	3.97	d		C,P,S,U
Histidine	α-CH	3.99	ABX		C,P,S,U
α-Hydroxybutyrate	CH	4.00	ABX		U
Asparagine	α-CH	4.00	ABX		C,S,U
Cysteine	CH	4.00	ABX		U
Serine	β-CH$_2$	4.00	ABX		C,U
Phosphorylethanolamine	OCH$_2$	4.00	m		S
Phenylalanine	α-CH	4.00	m		C,P,S,U
Fructose (β-pyranose)	H5	4.01	m		S
Ascorbate	CH	4.03	m		U
Fructose (β-pyranose)	H6	4.01	m		S
α-Hydroxy-*n*-valerate	α-CH	4.05	m		U
Creatinine	CH$_2$	4.06	s		A,C,P,S,U
Tryptophan	CH	4.06	ABX		S,U
Myo-Inositol	H2	4.06	t		C,S,U
Choline	OCH$_2$	4.07	m		C,P,S,U
Lactate	CH	4.12	q	69.2	A,C,P,S,U
Fructose (β-furanose)	H5	4.13	m		S
Proline	α-CH	4.14	m		A,C,P,S,U
β-Hydroxybutyrate	CH	4.16	ABX		C,P,S,U
Threonine	β-CH	4.26	ABX		A,C,P,S,U
Malate	CH	4.31	dd		U
N-methylnicotinamide	CH$_3$	4.48	s		U
Glycerophosphorylcholine	NCH$_2$	3.52	m		S
N-Acetylaspartate	CH	4.40	ABX		C,U
β-Galactose	C-H1	4.52	d		C
Ascorbate	CH	4.52	d		U
β-Galactose	C-H1	4.53	d		P
β-Glucose	C-H1	4.64	d		A,C,P,S,U
Water	H$_2$O	4.79	s		all fluids
Phospho(enol)pyruvate	CH	5.19	t		[e]
α-Glucose	C-H1	5.23	d	92.9	A,C,P,S,U
Allantoate	CH	5.26	s		U
Phospho(enol)pyruvate	CH	5.37	t		[e]
Allantoin	CH	5.40	d		U
Urea	NH$_2$	5.78	s		P,U
Uridine	H5	5.80	d		C,S,U
Uridine	H1'	5.82	d		C,S,U
cis-Aconitate	CH	5.92	s		U
Urocanate	CH(COOH)	6.40	d		U
Fumarate	CH	6.53	s		U

Table 2 *continued*

Metabolite	Assignment	$\delta(^1H)$	Multiplicity	$\delta(^{13}C)$	Biofluid[a]
trans-Aconitate	CH	6.62	s		U
4-Aminohippurate	C-H3/H5	6.87	d		U
3,4-Dihydroxymandelate	cyclic H	6.87	d		U
3,4-Dihydroxymandelate	cyclic H	6.90	s		U
Tyrosine	C-H3/H5	6.91	d′	116.7	C,P,S,U
3,4-Dihydroxymandelate	cyclic H	6.94	d		U
3-Methylhistidine	C-H4	7.01	s		P
Histidine	C-H4	7.08	s		C,P,S,U
1-Methylhistidine	C-H4	7.05	s		P
Tyrosine	C-H2/H6	7.20	d		C,P,S,U
Tryptophan	C-H5/H6	7.21	t		S,U
Indoxyl sulphate	C-H5	7.20	m		U
Indoxyl sulphate	C-H6	7.28	m		C,U
Tryptophan	C-H5/H6	7.29	t		S,U
Urocanate	CH(ring)	7.31	d		U
Tryptophan	C-H2	7.33	s		S,U
Phenylalanine	C-H2/H6	7.33	m		C,P,S,U
Phenylalanine	C-H4	7.38	m		C,P,S,U
Urocanate	C-H5	7.41	s		U
Phenylalanine	C-H3/H5	7.43	m		C,P,S,U
Nicotinate	cyclic H	7.53	dd		U
Hippurate	C-H3/H5	7.55	t		U
Tryptophan	C-H7	7.55	d		S,U
3-Methylhistidine	C-H2	7.61	s		P
Hippurate	C-H4	7.64	t		U
4-Aminohippurate	C-H2/H6	7.68	d		U
Tryptophan	C-H4	7.74	d		S,U
1-Methylhistidine	C-H2	7.77	s		P
Uridine	C-H6	7.81	d		C,S,U
Histidine	C-H2	7.83	s		C,P,S,U
Hippurate	C-H2/H6	7.84	d		U
Urocanate	cyclic C-H3	7.89	s		U
N-methylnicotinamide	C-H5	8.19	t		U
Nicotinate	cyclic H	8.26	dt		U
Formate	CH	8.46	s		C,P,S,U
Nicotinate	cyclic H	8.62	dd		U
N-methylnicotinamide	C-H4	8.90	d		U
Nicotinate	cyclic H	8.95	d		U
N-methylnicotinamide	C-H6	8.97	d		U
N-methylnicotinamide	C-H2	9.28	s		U

[a] Main biofluid in which compounds are found or have been observed; A, C, P, S and U refer to observation in amniotic fluid, CSF, Serum, Seminal fluid and urine, respectively.

[b] Varies according to the presence of divalent metal ions (especially Ca, Mg and Zn).

[c] Highly variable over physiological pH range.

[d] Significant shift differences due to the formation of fast exchanging complexes with divalent metal ions in biofluids.

[e] Not normally observed in biofluids because of low stability.

suppress rejection, and the better treatment of acute rejection episodes when they do occur. However, the early detection of acute cardiac graft rejection still relies on invasive and iterative right ventricular endomyocardial biopsies. This led to a new application of the measurement of the linewidth index, W, for the assessment of heart graft rejection after transplantation. Patients in moderate and severe rejection showed W values that were significantly different from those of the light rejection patients. However, as occurred in the test for cancer, the wide overlap between the values observed for each group meant that the W value alone could not be used to classify the patients into the four rejection grades. It has also been reported that the areas of two glycoprotein signals in the spin–echo ^{1}H NMR spectra of blood plasma from heart transplant patients correlated with a standard echocardiography parameter used to monitor rejection. Although a good correlation was found in 5 patients and an acceptable correlation in 3, only a poor correlation was seen in 5 further patients. It was thought that infections and inflammatory states unrelated to rejection interfered with the correlation.

In human beings, diabetes is a relatively common condition which can have serious, complex and far-reaching effects if not treated. It is characterized by polyuria, weight loss in spite of increased appetite, high plasma and urinary levels of glucose, metabolic acidosis, ketosis and coma. The muscles and other tissues become starved of glucose, whilst highly elevated levels of glucose are found in the urine and plasma. Based on NMR spectra, there are marked elevations in the plasma levels of the ketone bodies and glucose, post-insulin withdrawal. In general, the NMR results were in good agreement with conventional assay results. The CH_3 and CH_2 resonances of the lipoproteins VLDL and chylomicrons also decreased significantly in intensity relative to the CH_3 signal of HDL and LDL, indicating the rapid metabolism of the mobile pool of triglycerides in VLDL and chylomicrons.

The ^{1}H NMR spectra of the blood plasma from patients with chronic renal failure during dialysis, patients in the early stages of renal failure and normals have also been analysed. For patients on acetate dialysis, the method clearly showed how the acetate was accumulated and metabolized during the course of the dialysis, as well as allowing changes in the relative concentrations of endogenous plasma components to be monitored. A subsequent ^{1}H, ^{13}C and ^{14}N NMR study of the plasma and urine from chronic renal failure patients showed that the plasma levels of trimethylamine-N-oxide (TMAO) correlated with those of urea and creatinine, suggesting

that the presence of TMAO is closely related to the degree of renal failure.

The uraemic syndrome is associated with a complex set of biochemical and pathophysiological changes that remain poorly understood. The first application of 750 MHz NMR for studying the biochemical composition of plasma has been reported from patients on haemodialysis (HD) and peritoneal dialysis (PD). Increased plasma levels of low MW metabolites including methylhistidine, glycerol, choline, TMAO, dimethylamine and formic acid were found. The concentrations of these metabolites, and ratios to others, varied with the type of dialysis therapy. For example, the biochemical composition of plasma from patients on PD was remarkably consistent whereas pre- and post-HD significant fluctuations in the levels of TMAO, glucose, lactate, glycerol, formate and lipoproteins (VLDL, LDL and HDL) were observed. Elevated concentrations of glycoprotein fragments were also observed and may relate to the presence of high levels of N-acetyl glucosaminidase (NAG), and other glycoprotein cleaving enzymes, in the plasma originating from damaged kidney cells.

^{1}H NMR spectroscopy of whole blood and red blood cells

Single pulse ^{1}H NMR measurements on whole blood give very little biochemical information owing to the presence of a broad envelope of resonances from haemoglobin and plasma proteins. Spin–echo spectroscopy of whole blood can give rise to moderately well resolved signals from plasma metabolites and those present inside erythrocytes, notably glutathione, but whole blood spectra are not easily reproducible because of erythrocyte sedimentation which progressively (within a few minutes) degrades the sample field homogeneity during the course of data collection. Furthermore, there are substantial intracellular/extracellular field gradients, which give a major contribution to the T_2 relaxation processes for the nuclei of molecules diffusing through those gradients, and very weak spectra are obtained. Spin–echo NMR measurements on packed erythrocyte samples do give rise to well resolved signals from intracellular metabolites and a variety of transport and cellular biochemical functions can be followed by this method.

One major parameter that can be obtained from NMR is the intracellular pH, and this has been measured for erythrocytes by ^{1}H NMR spectroscopy using a suitable ^{1}H NMR pH indicator that has an NMR chemical shift which varies with pH in the range desired. Candidates include the C2-H protons from histidines in haemoglobin, but an exogenous compound can be added as an indicator.

The transport of substances between the inside and outside of red cells can be monitored using NMR if the resonances from the two environments have different chemical shifts or intensities. Also, resonances outside the cell can be selectively broadened by the addition of paramagnetic species that do not cross the red cell membrane. Those used include the ferric complex of desferrioxamine, dysprosium-DTPA and the copper-cyclohexanediaminetetraacetic acid complex. To measure the rate of influx of a compound into red cells, the compound is added with the paramagnetic agent to a red cell suspension and the intensity of the resonance from the intracellular component is monitored as a function of time. This approach has been used to study the transport of glycerol, alanine lactate, choline and glycylglycine. The time scale that can be addressed covers the range of ms to hours.

Disease processes studied include the use of NMR spectroscopy of erythrocytes to investigate the effect of lithium treatment in manic depressive patients showing an increase in choline. In heavy metal poisoning a considerable proportion of the metal is found in the blood and the binding of heavy metals inside erythrocytes has received much attention. Rabenstein has reviewed much of the work in the area of red blood cell NMR spectroscopy. More recently, diffusion coefficient measurement has been used to study the binding between diphosphoglycerate and haemoglobin inside intact red blood cells.

NMR spectroscopy of human and animal urine

Sample preparation and NMR assignments

The composition and physical chemistry of urine is highly variable both between species and within species according to lifestyle. A wide range of organic acids and bases, simple sugars and polysaccharides, heterocycles, polyols, low molecular weight proteins and polypeptides are present together with inorganic species. Many of these moieties also interact, forming complexes some of which are amenable to NMR study, that may undergo chemical exchange reactions on a variety of different time scales. The ionic strength of urine varies considerably and may be high enough to adversely affect the tuning and matching of the RF circuits of a spectrometer probe, particularly at high field strengths. The presence of high concentrations of protein in the urine, e.g. due

to renal glomerular or tubular damage, can result in the broadening of resonances from low MW compounds that may bind to these urinary proteins.

Urinary pH values may vary from 5 to 8, according to the physiological condition in the individual, but usually lie between 6.5 and 7.5. Urine samples should, therefore, be frozen as soon as possible after collection if NMR measurements cannot be made immediately. When experiments involve collections from laboratory animals housed in metabolic cages, urine samples should be collected into receptacles that are either cooled with dry ice or have a small amount of sodium azide present as a bacteriocide. However, both these procedures may inhibit or destroy urinary enzymes that may frequently be assayed by conventional biochemical methods for assessment of kidney tubular integrity in toxicological experiments. Urinary pH values are unstable because of the progressive and variable precipitation of calcium phosphates which may be present in the urine close to their solubility limits. One solution to this problem is the addition of 100–200 mM phosphate buffer in the D_2O added for the lock signal followed by centrifugation to remove precipitated salts. This has the effect of normalizing the pH to a range of 6.7–7.6 which is stable for many hours during which NMR measurements can be made. Few metabolites (except for histidine and citrate) show major chemical shift variations over this pH range.

The vast majority of urinary metabolites have ^1H T_1 relaxation times of 1 to 4 s. Thus, the general rule of applying a $5 \times T_1$ relaxation delay between successive 90° transients to obtain >99% relaxation and hence quantitative accuracy cannot be applied routinely. Instead, by using 30–45° pulse angles and leaving a total T_1 relaxation delay of 5 s between successive pulses, spectra with good signal-to-noise ratios can usually be obtained for most metabolites in 5–10 minutes at high field with a generally low level of quantitative signal distortion. It has become standard practice to replace simple solvent resonance presaturation with pulse methods, which either induce such saturation or which leave the solvent water resonance unexcited, and one popular method involves using simply the first increment of the 2-dimensional NOESY pulse sequence.

A vast number of metabolites may appear in urine samples and problems due to signal overlap can occur in single pulse experiments. The magnitude of the signal assignment problem in urine is also apparent at ultrahigh field. There are probably >5000 resolved lines in single pulse 750 or 800 MHz ^1H NMR spectra of normal human urine (**Figure 1**), but there is still extensive peak overlap in certain chemical shift ranges and even at

800 MHz many signals remain to be assigned. The dispersion gain at 800 MHz even over 600 MHz spectra is particularly apparent in 2- (or 3-) dimensional experiments that can be used to aid signal assignment and simplify overlapped spectra. Many of the endogenous compounds that have been detected in NMR spectra of human and animal urines are shown in **Table 2**.

The biochemical composition of urine varies considerably from species to species and also with age as almost all species have age-related changes in renal function. Rats and other rodents have much higher levels of taurine, citrate, succinate, 2-oxoglutarate and allantoin than humans. Rat urine (and that of other rodents) is generally much more concentrated than human urine, and so NMR signal-to-noise ratios may be better for many metabolites. All animals have physiological processes that are modulated by biological rhythms. These include excretory processes and the urinary composition of an animal may vary considerably according to the time it is collected. Given these types of variation it is obviously of paramount importance to have closely matched (and in many cases timed) control samples where toxicological or disease processes are being studied. Dietary composition also affects the urinary metabolite profiles of man and animals, and it is important to distinguish these from disease-related processes in clinical or toxicological studies. For example, persons consuming large quantities of meat/poultry before urine collection may have NMR detectable levels of carnosine and anserine in their urine, consumption of cherries is associated with elevated urinary fructose and consumption of shellfish and fish are associated with high levels of betaine and trimethylamine in the urine. The 500 MHz and 600 MHz ^{1}H NMR spectra of neonate urines are characterized by strong signals from amino acids, organic acids, amines, sugars and polyols. In particular, high concentrations of *myo*-inositol have been detected in both pre- and full-term urines. This is an important intracellular organic osmolyte in renal cells, present in high concentrations in the renal inner medulla.

NMR spectra of urine in disease

Given the serious outcome of many inborn errors of metabolism, if not diagnosed and treated at an early stage, there has been a search for a rapid, sensitive and general method for the detection and diagnosis of inborn errors of metabolism in neonates. Conventional methods including specific enzyme assays and gas chromatography/mass spectrometry are sensitive but time consuming, involve considerable sample preparation and are not general. However, NMR spectroscopy of biofluids has been shown to be a very powerful and general method for the detection of inborn errors of metabolism and many disorders have been studied by the use of biofluid NMR.

A good example is provided by the spectra from individuals with a deficiency in methylacetoacetyl CoA thiolase (MACT). The conversion of 2-methylacetoacetyl CoA into acetyl-CoA and propionyl-CoA is inhibited and an accumulation of abnormal catabolic products is observed in the urine. Both 1D and 2D ^{1}H NMR spectroscopy have been used to investigate the urinary metabolites of patients with this disorder. The urine spectra from the patients clearly showed the presence of both 2-methyl-3-hydroxybutyrate and tiglylglycine, which is characteristic for MACT deficiency due to the build up of metabolites close to the position of enzyme deficiency.

A similar success was observed for studies of branched chain ketoaciduria in which the second stage of the catabolism of leucine, valine and isoleucine involves an oxidative decarboxylation. In patients with branched chain ketoaciduria, this step is blocked for all three of these amino acids. The urine of these patients takes on the odour of maple syrup and hence this condition is also known as maple syrup urine disease. ^{1}H NMR spectroscopy was used to study the urines of patients with such branched chain ketoaciduria. The spectra showed several abnormal metabolites including the amino acids leucine, isoleucine and valine and their corresponding transamination products. It was noted that 2-hydroxyisovalerate levels were very high in the urines of all the patients studied and that, as in other inborn errors of metabolism, the levels of urinary glycine were elevated.

High field ^{1}H NMR spectroscopy has shown itself to be a very powerful tool with which to diagnose inborn errors of metabolism and to monitor the clinical response of patients to therapeutic interventions. Characteristic patterns of abnormal metabolites are observed in the patient's urine for each disease. The main advantages of using ^{1}H NMR spectroscopy in this area are its speed, lack of sample preparation and the provision of non-selective detection for all the abnormal metabolites in the biofluid regardless of their structural type, providing only that they are present above the detection limit of the NMR experiment.

There are a number of studies of other disease processes using NMR spectroscopy of urine. These include glomerularnephritis, kidney transplant patients over a period of 14 days post-operation to investigate the biochemical effects of rejection, phenol poisoning and metabolic acidoses caused by alcohol ingestion.

Evaluation of toxic effects of xenobiotics using NMR spectroscopy of urine

The successful application of ^{1}H NMR spectroscopy of biofluids to study a variety of metabolic diseases and toxic processes has now been well established and many novel metabolic markers of organ-specific toxicity have been discovered. The method is based on the fact that the biochemical composition of a biofluid is altered when organ damage occurs. This is particularly true for NMR spectra of urine in situations where damage has occurred to the kidney or liver. It has been shown that specific and identifiable changes can be observed which distinguish the organ that is the site of a toxic lesion. Also it is possible to focus in on particular parts of an organ such as the cortex of the kidney and even in favourable cases to very localized parts of the cortex. Finally it is possible to deduce the biochemical mechanism of the xenobiotic toxicity, based on a biochemical interpretation of the changes in the urine.

A wide range of toxins has now been investigated including the kidney cortical toxins mercury chloride, p-aminophenol, ifosfamide, the kidney medullary toxins propylene imine and 2-bromoethanamine hydrochloride and the liver toxins hydrazine, allyl alcohol, thioacetamide and carbon tetrachloride. The testicular toxin cadmium chloride has been investigated in detail and the aldose reductase inhibitor HOE-843 has also been studied.

Use of combined NMR spectroscopy–pattern recognition (PR) to evaluate biochemical changes in urine

The first studies using PR to classify biofluid samples used a simple scoring system to describe the levels of 18 endogenous metabolites measured by ^{1}H NMR spectroscopy of urine from rats which were either in a control group or had received a specific organ toxin which affected the liver, the testes, the renal cortex or the renal medulla. The data were used to construct non-linear maps and various types of principal components (PC) scores plots in 2 or 3 dimensions. The samples were divided into two subsets, a training set and a test set. This study showed that samples corresponding to different organ toxins mapped into distinctly different regions and that in particular the renal cortical and renal medullary toxins showed good separation (**Figure 2**).

In addition, the NMR-PR approach has been used to investigate the time course of metabolic urinary changes induced by two renal toxins. In this case, toxic lesions were induced in rats by a single acute dose of the renal cortical toxin mercury(II) chloride and the medullary toxin 2-bromoethanamine. The

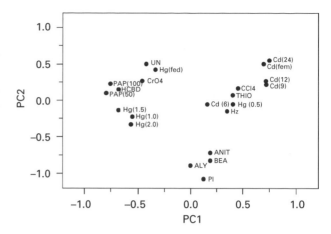

Figure 2 A plot of PC1 versus PC2 for a set of rat urine samples using simple scored levels of 16 metabolites as the descriptor set. A cluster of renal cortical toxins is seen at the left, whilst the testicular and liver toxins are to the right. At the bottom the two renal papillary toxins (BEA and PI) appear to cluster with liver toxins (ALY and ANIT) but are well separated in the third principal component. Reproduced with permission of Academic Press from Lindon JC, Nicholson JK and Everett JR (1999) NMR spectroscopy of biofluids. *Annual Reports on NMR Spectroscopy* **38**: 1–88.

onset, progression and recovery of the lesions were also followed using histopathology to provide a definitive classification of the toxic state relating to each urine sample. The concentrations of 20 endogenous urinary metabolites were measured at eight time points after dosing. A number of ways of presenting the data were investigated and it was possible to construct PC scores plots which showed metabolic trajectories which were quite distinct for the two toxins. These showed that the points on the plot can be related to the development of, and recovery from, the lesions. These trajectories allowed the time points at which there were maximal metabolic differences to be determined and provided visualization of the treated groups of animals.

A toxicological assessment approach based on neural network software has been tested by analysing the toxin-induced changes in endogenous biochemicals in urine as measured using ^{1}H NMR spectroscopy to ascertain whether the methods provide a robust approach which could lead to automatic toxin classification. The neural network approach to sample classification was in general predictive of the sample class and once the network is trained, the prediction of new samples is rapid and automatic. The principal disadvantage, common to most neural network studies, in that it is difficult to ascertain from the network which of the original sample descriptors are responsible for the classification.

Recently, studies have been published using pattern recognition to predict and classify drug toxicity effects including lesions in the liver and kidney and using supervized methods as an approach to an expert system. In this case the NMR spectra were segmented into integrated regions automatically, thereby providing standard intensity descriptors for input to the PR.

The usefulness of the NMR-PR approach to the classification of urine samples from patients with inborn errors of metabolism has also been attempted. Urine samples from adult patients with inborn errors of metabolism comprising cystinuria, oxalic aciduria, Fanconi syndrome, porphyria and 5-oxoprolinuria were studied. PCA mapping produced a reasonable separation of control urines from the clusters derived from the different types of inborn error.

The rapid and unambiguous distinction of the various clinical situations that can occur following a kidney transplant remains a problem. For example, it is necessary to distinguish rejection, which requires increased immunosuppressive therapy such as cyclosporin A (CyA), from tubular necrosis, which can be caused by too high levels of CyA. At present, kidney function is generally characterized using the level of blood plasma creatinine as a monitor. However the balance between rejection and tubular necrosis remains a difficult distinction and the only clear diagnosis is an invasive kidney biopsy.

A study of patients after renal transplantation has shown that NMR can reveal new markers of abnormality and in particular that high levels of trimethylamine-N-oxide (TMAO) in the urine are correlated with rejection episodes. However, the use of TMAO as a single diagnostic marker of rejection is probably not sufficiently discriminatory for clinical use because of interpatient variability. It is likely that TMAO is only one of a number of potential markers of rejection and that a combination of other metabolites present in urine will provide a new differential diagnosis of rejection and CyA toxicity. A combination of high field ^{1}H NMR urinalysis, automatic data reduction and PR methods has been applied to investigate this.

NMR-PR was seen to separate the samples into three distinct groups with the greatest separation being between the CyA toxicity group and the rejection group. It was also notable that, in general, the greater the degree of allograft dysfunction (as assessed by conventional means) the further the samples plotted away from the control group. Identification of the compounds that gave rise to the significant descriptors has not been conclusively proven. However, close examination of spectra from samples from the various classes using both 600 MHz and

750 MHz ^{1}H NMR spectroscopy suggested that these were 3-hydroxyisovaleric acid, N-acetylated glycoproteins, TMAO, hippuric acid and a molecule related to N-methylnicotinamide.

NMR spectroscopy of cerebrospinal fluid (CSF)

Because of the low protein content of normal human CSF, it is possible to use single pulse ^{1}H NMR experiments to obtain biochemical information without recourse to the spin–echo techniques often required for studies on blood and plasma. However, where serious cerebral damage has occurred, or in the presence of an acute infection, spectra may become dominated by broad resonances from proteins. Problems may also arise from protein binding and metal binding of some metabolites, with a consequential broadening of their proton resonances.

There are a number of ^{1}H NMR studies of CSF in the literature. Also, it has been shown that high quality 2D ^{1}H NMR spectra can be obtained from human CSF, and that changes in NMR patterns can be roughly related to disease states in the donor. Examination of a number of *ex vivo* control samples showed high consistency in the aliphatic region of the 1D ^{1}H NMR spectra of CSF and many assignments could be made by inspection. Diseases studied using NMR spectra of CSF include lumbar disk herniations, cerebral tumours, drug overdose, diabetes, hepatic encephalitis, multiple sclerosis, AIDS dementia complex, Parkinson's disease, Creutzfeld–Jakob disease, Guillain–Barre syndrome and vitamin B$_{12}$ deficiency.

One study has reported 500 and 600 MHz ^{1}H NMR data on the post mortem CSF from Alzheimer's disease (AD) patients and controls. The main differences between the spectra of the two groups were found to be in the region $\delta 2.4$–2.9, and principal components analysis showed that separation of the two groups was possible based mainly on lower citrate levels in the AD patients. Non-matching in patient age and the time interval between death and autopsy caused a reduction in the inter-group differences but they were still significant ($p < 0.05$).

NMR spectroscopy of seminal fluids

The 750 MHz ^{1}H NMR spectra of seminal fluid obtained from a healthy individual is complex but many of the signals have been assigned (see **Table 2**). Fresh undiluted seminal fluid gives rise to NMR spectra with very broad and poorly resolved signals due to the presence of high concentrations of peptides (which are cleaved to amino acids by endogenous peptidase activity) and the high viscosity of the matrix. The complexity of the biochemical composition of seminal fluids together with their reactivity poses a number of assignment and quantitation problems. Some metabolite signals in seminal fluids also appear to have anomalous chemical shifts when compared to simple standard solutions measured at the same pH or even when compared to other body fluids. The application of 2D NMR methods has been shown be very useful in the case of seminal fluids. An investigation of dynamic molecular processes that occur in seminal fluid has been undertaken using ^{1}H NMR spectroscopy. Reactions that could be followed included hydrolysis of phosphorylcholine and nucleotides and zinc complexation.

In view of the importance of artificial insemination in farming, it is somewhat surprising that so few studies of animal seminal fluid have been reported. However there are studies on boar seminal plasma giving details of resonance assignments.

The comparison of ^{1}H NMR spectra of seminal fluid from normal controls with those from patients with vasal aplasia (obstruction of the vas deferens leading to blockage of the seminal vesicles) and those with non-obstructive infertility has been reported. The ^{1}H NMR spectra of the seminal fluid from patients with non-obstructive infertility were similar to those of normal subjects. However, the ^{1}H NMR spectra of the seminal fluid from patients with vasal aplasia were grossly different from those of normal subjects and corresponded closely to those of prostatic secretions from normals, owing to the lack of seminal vesicle secretion into the fluid. In the ^{1}H NMR spectra of the vasal aplasia patients, signals from amino acids were either absent or present at very low levels. Similarly, choline is at a low level or absent in the seminal fluid from vasal aplasia patients, as it derives (indirectly) from the seminal vesicle component. Significant differences were observed between the normal and vasal aplasia patient groups for the molar ratios of citrate:choline and spermine:choline.

Other studies of infertility include a procedure providing automatic diagnosis based on NMR spectroscopy and work on azoospermic subjects and prostate cancer. In addition ^{31}P NMR spectroscopy has also been used to distinguish semen from healthy and infertile men.

NMR spectroscopy of bile

NMR spectroscopy of bile and dynamic interactions of metabolites

The single pulse NMR spectra of bile are dominated by broad resonances that arise from bile acids that are present in mixed micelles with phospholipids and cholesterol (**Figure 1**). They are broad as a result of short spin–spin (T_2) relaxation times reflecting constrained molecular motions within micellar particles. On lyophilization and reconstitution with water, the molecular mobility of a number of biliary metabolites changes significantly because of disruption of the micellar compartments. In particular the T_2 relaxation times of the aliphatic side chains of lipid moieties are increased in lyophilized bile suggesting greater mobility of these molecules. Increases in signal intensities that occur on lyophilization reflect changes in compartmentation of molecules, which is related to the disruption/reorganization of the biliary micellar compartments. Signals from β-hydroxybutyrate, valine and other branched chain amino acids do not contribute significantly to the ^{1}H NMR spectra of non-lyophilized bile, but resonances from these components are clearly resolved after lyophilization. The sharper signals in bile give rise to well resolved 2D COSY spectra that allow a comprehensive assignment of the bile salt signals.

Variable temperature ^{1}H NMR studies of human bile show that considerable dynamic structural information is available, particularly at high fields. The micellar cholesteryl esters that are abundant in bile appear to show liquid crystal behaviour, and it is possible to use NMR measurements to map the phase diagram for the complex biliary matrix.

A number of studies have used both ^{1}H and ^{13}C NMR spectroscopy of bile to aid characterization of its composition and structure. Thus, ^{13}C spectra of bile from fish exposed to petroleum have been studied. ^{31}P NMR spectra of human bile have also been investigated. ^{1}H NMR spectroscopy of bile has been used to investigate the micellar cholesterol content and lipids, and both ^{1}H and ^{31}P NMR have been used to study the distribution of lecithin and cholesterol.

The use of ^{1}H NMR spectroscopy of bile as a means of monitoring liver function has been proposed. The ^{31}P NMR spectra of the bile from patients with primary biliary cirrhosis of the liver and from clinically healthy men have been compared. Also the level of lactate in bile from patients with hepatobiliary diseases including cancer has been investigated using ^{1}H NMR spectroscopy.

NMR spectroscopy of miscellaneous body fluids

Amniotic and follicular fluids

The first study of human amniotic fluid using ^{1}H NMR spectroscopy detected 18 small molecule metabolites including glucose, leucine, isoleucine, lactate and creatinine. Following this, other studies used a combination of 1D and 2D COSY spectroscopy to assign resonances, assess NMR methods of quantitation and investigate the effects of freezing and thawing. In addition NMR results have been correlated with other clinical chemical analyses. A total of 70 samples were measured using ^{1}H NMR at 600 MHz at different stages of gestation and with different clinical complications and significant correlations between the NMR spectral changes and maternal pre-eclampsia and foetal open spina bifida were observed. The effects of various pathological conditions in pregnancy have been investigated using ^{1}H NMR spectroscopy of human amniotic fluid. Several studies of amniotic fluid using ^{31}P NMR spectroscopy have also been carried out, principally to analyse the phospholipid content. One study of the metabolic profiling of ovarian follicular fluids from sheep, pigs and cows has been reported.

Milk

Surprisingly, very little has been published on the NMR spectroscopy of milk given that it is both a biofluid and a food substance. The studies generally focus on milk as a food and Belton has also reviewed the information content of NMR spectra of milk. Very recently, ^{19}F NMR spectroscopy has been used to detect trifluoroacetic acid in milk.

Synovial fluid

^{1}H NMR has been used to measure the levels of a variety of endogenous components in the synovial fluid (SF) aspirated from the knees of patients with osteoarthritis (OA), rheumatoid arthritis (RA) and traumatic effusions (TE). The spin–echo NMR spectrum of synovial fluid shows signals from a large number of endogenous components. Many potential markers of inflammation could not be monitored because of their low concentrations or because of their slow tumbling (e.g. hyaluronic acid, a linear polysaccharide that imparts a high viscosity to synovial fluid). The low molecular weight endogenous components showed a wide patient-to-patient variability and showed no statistically significant correlation with disease state. However, correlations were reported between the disease states and the synovial fluid levels of the N-acetyl signals from acute phase glycoproteins. Correlations between the disease state and the levels and type of triglyceride in the synovial fluid have also been reported.

The ^{1}H NMR spectra of the SF of a female patient with seronegative erosive RA and of another female patient with sarcoidosis and independent inflammatory OA were followed over the course of several months and standard clinical tests were performed on paired blood serum samples taken at the same time. It was found that the SF levels of triglyceride CH_3, CH_2 and CH, glycoprotein N-acetyl signals and creatinine all correlated well with one another, and with standard clinical measures of inflammation. The correlation of disease state with creatinine level is of particular interest, and the altered triglyceride composition and concentration in OA was suggested as a potential marker for the disease in SF.

Spin-echo ^{1}H NMR spectroscopy has been used to detect the production of formate and a low molecular weight, N-acetyl-containing oligosaccharide, derived from the oxygen radical-mediated depolymerization of hyaluronate, in the SF of patients with RA, during exercise of the inflamed joint. Gamma radiolysis of rheumatoid SF and of aqueous hyaluronate solutions was also shown to produce formate and the oligosaccharide species. It has been proposed that the hyaluronate-derived oligosaccharide and formate could be novel markers of reactive oxygen radical injury during hypoxic reperfusion injury in the inflamed rheumatoid joint.

^{13}C NMR can be used to monitor the synovial fluids from patients with arthritis. In contrast to ^{1}H NMR studies, signals are seen from hyaluronic acid, the main determinant of the viscoelasticity of the synovial fluid, even though the molecular weight is in the region 500 to 1600 kD. ^{13}C NMR spectra of synovial fluids from patients with RA, OA, TE and cadaver controls have been compared with one another and with spectra of authentic hyaluronic acid, both before and after the incubation of the latter with hyaluronidase, an enzyme which depolymerizes the hyaluronic acid. Depolymerization of the hyaluronic acid was accompanied by a decrease in the half-band widths of its ^{13}C resonances. The synovial fluid NMR spectra from the patients with rheumatoid arthritis had sharper signals for the C-1 and C-1′ carbons of hyaluronic acid than those from the osteoarthritic patients, which in turn exhibited sharper signals than those from the cadavers or the joint trauma patients. Thus the degree of polymerization of hyaluronic acid was deduced to decrease in the order: controls/joint trauma patients > osteoarthritic patients > rheumatoid arthritis patients. Since it is known that the consequence of hyaluronate depolymerization may be articular cartilage damage, it was concluded that ^{13}C NMR spectroscopy may be a valuable method for studying these clinical relevant biophysical changes in synovial fluid.

Aqueous humour and vitreous humour

The first study by NMR spectroscopy on aqueous humour was on 9 samples taken during surgery for other conditions, and NMR spectra were measured at 400 MHz. A number of metabolites were detected, including acetate, acetoacetate, alanine, ascorbate, citrate, creatine, formate, glucose, glutamine or glutamate, β-hydroxybutyrate, lactate, threonine and valine. Following this, there have been a number of other studies. These include ^{1}H NMR spectra from aqueous humour of rabbits and cod fish, ^{31}P NMR spectra of aqueous and vitreous humour from pigs and ^{23}Na NMR spectra of vitreous humour. Finally, the penetration of dexamethasone phosphate into the aqueous humour has been followed using ^{1}H and ^{19}F NMR spectroscopy.

Saliva

Only limited studies using NMR spectroscopy of saliva have been reported. ^{1}H NMR spectroscopy of human saliva was used in a forensic study and following this another group reported that only parotid gland saliva gave a well resolved ^{1}H NMR spectrum showing significant circadian effects. No age- or sex-related differences were observed for saliva from healthy subjects but marked differences were observed in cases of sialodentitis. Finally, the biochemical effects of an oral mouthwash preparation have been studied using ^{1}H NMR spectroscopy.

Digestive fluids

The analysis of pancreatic juice and small bowel secretions using ^{1}H NMR spectroscopy has been reported.

Pathological cyst fluid

A ^{1}H NMR study of the fluid from the cysts of 6 patients with autosomal dominant polycystic kidney disease (ADPKD) has been reported. ADPKD in adults is characterized by the slow progressive growth of cysts in the kidney, and when these cysts reach a large size they can significantly distort the kidney and disrupt both the blood supply and renal function. ADPKD is one of the commonest causes for renal transplantation in adults. The ^{1}H NMR spectra revealed a number of unusual features and showed the cyst fluids to be distinct from both blood plasma and urine. Unusually, the fluids from all patients contained high levels of ethanol, which was not related to consumption of alcoholic beverages or drug preparations. In general there was little variation in the composition of the cyst fluids as revealed by ^{1}H NMR, although the protein signal intensity

did vary somewhat. It was hypothesized that this constancy of composition reflected the chronic nature of the accumulation of the cyst fluid and a long turnover time of the cyst components, which thus has the effect of averaging the compositions. The unique biochemical composition of the cyst fluids was ascribed to abnormal transport processes occurring across the cyst epithelial wall.

Drug metabolites in biofluids

Several studies have used NMR spectroscopy to determine the number and identity of drug metabolites in bile. These include the use of ^{19}F NMR spectroscopy to study doxifluridine catabolites in human bile, ^{19}F and ^{13}C NMR spectroscopy of perfluorinated fatty acids in rat bile and ^{13}C NMR for monitoring the formation of formaldehyde from demethylation of antipyrine. ^{1}H NMR spectroscopy of rat bile has been used to monitor the excretion of paracetamol metabolites. A combination of ^{1}H and ^{1}H–^{13}C 2D NMR methods has allowed identification of 4-cyano-N,N-dimethylaniline, cefoperazone and benzyl chloride in rat bile.

NMR spectroscopy of urine combined with labelling with stable isotopes such as ^{13}C and ^{2}H has been used to monitor the silent process of deaceteylation and subsequent reacetylation (futile deacteylation) in the rat as this has implications for the toxicity of paracetamol and phenacetin. The metabolites of 5-fluorouracil in plasma and urine have been analysed using ^{19}F NMR spectroscopy in patients receiving chemotherapy. ^{2}H NMR has been employed to study the pharmacokinetics of benzoic acid in relation to liver function. The metabolism of hydrazine in rats has been probed using ^{15}N NMR spectroscopy of urine.

Often it is advantageous to carry out a partial extraction of metabolites using a solid phase separation cartridge. In this case, the urine containing the drug metabolites is loaded on to a C18 cartridge that is then washed with acidified water to remove very polar endogenous components. The metabolites of interest can be then be selectively washed off using methanol or water/methanol mixtures.

The direct coupling of HPLC with NMR spectroscopy has required a number of technological developments to make it a feasible routine technique and now on-line NMR detection of HPLC fractions is a useful adjunct to the armoury of analytical methods.

To date, HPLC-NMR spectroscopy has been applied to the profiling and identification of the metabolites of a number of drugs and xenobiotics present in biofluids such as plasma and urine and in bile samples from rats and humans. In general the simple stop-flow approach has predominated in these studies. The HPLC-NMR method is most useful for compounds, and their metabolites, with suitable NMR reporter groups. This includes the use of ^{19}F NMR spectroscopy for molecules containing fluorine and diagnostic ^1H NMR resonances in regions of the NMR spectrum where solvent signals do not cause interference. For example, the pattern of NMR resonances from glucuronide conjugates is particularly diagnostic.

HPLC-NMR of biofluids has been employed to identify the unusual endogenous metabolites found in rat urine after administration of compounds that induce liver enzymes leading to elevated urinary levels of a number of carbohydrates and related molecules.

A wide range of ^1H and ^{19}F HPLC-NMR studies has been carried out on human and animal biofluids, predominantly urine and bile, to characterize the metabolites of exploratory drugs and marketed pharmaceuticals. In some cases, the biofluid has been subjected to solid phase extraction to remove many of the very polar materials, which facilitates the HPLC separation by avoiding overloading the column. These studies include the identification and structuring of metabolites of antipyrine, ibuprofen, flurbiprofen, paracetamol, the anti-HIV compound GW524W91, iloperidone and tolfenamic acid. Other studies using HPLC-NMR on bile include the identification of a metabolite of 7-(4-chlorobenzyl)-7,8,13,13a-tetrahydroberberine chloride in rat bile and metabolites of a drug LY335979, under development to prevent multidrug resistance to cancer chemotherapy. In addition, ^{31}P HPLC-NMR has been used to study metabolites of ifosfamide, and ^2H HPLC-NMR has been employed to detect metabolites of dimethylformamide-d_7.

The further coupling of HPLC-NMR with mass spectroscopy has also been used to analyse drug metabolites in urine for the human excretion of paracetamol metabolites. This approach has also been used to identify metabolites of substituted anilines in rat urine.

Capillary electrophoresis (CE), and related techniques, is a relatively new technique which uses a length of fused silica capillary with an optical window to enable detection, a detector (UV, fluorescence or mass spectrometry), a high voltage source, two electrode assemblies and buffer solutions in suitable reservoirs. The technique has been shown to provide very high separation efficiencies, with hundreds of thousands of theoretical plates achievable.

However, the small injection volume available to CE (a few nl) means that high sensitivity can only be achieved if concentrations of the analyte in the sample are high. Nevertheless, some results have been reported using CE-NMR and the method has been applied to the identification of paracetamol metabolites in human urine. Finally the application of capillary electrochromatography (CEC) directly coupled to NMR has been explored using the same paracetamol metabolite samples.

See also: **Cells Studied By NMR; Chromatography–NMR, Applications; Diffusion Studied Using NMR Spectroscopy; Drug Metabolism Studied Using NMR Spectroscopy;** *In vivo* **NMR, Applications, Other Nuclei;** *In vivo* **NMR, Applications, ^{31}P;** *In vivo* **NMR, Methods; Perfused Organs Studied Using NMR Spectroscopy; Proteins Studied Using NMR Spectroscopy; Solvent Suppression Methods in NMR Spectroscopy; Structural Chemistry Using NMR Spectroscopy, Pharmaceuticals.**

Further reading

Ala-Korpela M (1997) ^1H NMR spectroscopy of human blood plasma. *Progress in NMR Spectroscopy* 27: 475–554.

Gartland KPR, Beddell CR, Lindon JC and Nicholson JK (1991) The application of pattern recognition methods to the analysis and classification of toxicological data derived from proton NMR spectroscopy of urine. *Molecular Pharmacology* 39: 629–642.

Lindon JC, Nicholson JK and Everett JR (1999) NMR spectroscopy of biofluids. In Webb GA (ed) *Annual Reports on NMR Spectroscopy.* Vol 38, in press. London, Academic Press.

Lindon JC, Nicholson JK, Sidelman UG and Wilson ID (1997) Directly-coupled HPLC–NMR and its application to drug metabolism. *Drug Metabolism Reviews* 29: 705–746.

Lindon JC, Nicholson JK and Wilson ID (1996) Direct coupling of chromatographic separations to NMR spectroscopy. *Progress in NMR Spectroscopy* 29: 1–49.

Nicholson JK and Wilson ID (1989) High resolution proton magnetic resonance spectroscopy of biofluids. *Progress in NMR Spectroscopy* 21: 449–501.

Nicholson JK, Foxall PJD, Spraul M, Farrant RD and Lindon JC (1995) 750 MHz ^1H and ^1H–^{13}C NMR spectroscopy of human blood plasma. *Analytical Chemistry* 67: 793–811.

Rabenstein DL (1984) ^1H NMR methods for the non-invasive study of metabolism and other processes involving small molecules in intact erythrocytes. *Journal of Biochemical and Biophysical Methods* 9: 277–306.

Biomacromolecular Applications of Circular Dichroism and ORD

Norma J Greenfield, University of Medicine and
Dentistry of New Jersey, Piscataway, NJ, USA

Most biological macromolecules, including proteins, nucleic acids and carbohydrates, are built of repeating units that are assembled into highly asymmetric structures. The specific conformations of the macromolecules cause the optical transitions of the chromophores in their repeating units (e.g. peptides, nucleotides, glycosides) to align with each other so that their electronic and magnetic transitions interact. The interactions may result in hypo or hyperchromism and in splitting of the transitions into multiple transitions. In addition the aligned transitions may exhibit a high degree of circular dichroism (CD) and optical rotatory dispersion (ORD). Because the ORD and CD spectra of macromolecules depend on their conformation, they are excellent probes for following changes in structure as a function of temperature, denaturants or ligand binding. In addition, non-optically active chromophores sometimes bind to macromolecules in highly asymmetric fashions, and generate large extrinsic CD and ORD bands in the region of absorption. These extrinsic bands can be used to follow interactions of the macromolecules with the chromophores, or to probe structural transitions of the macromolecules. The following article discusses how ORD and CD are used to study the conformation, folding and interactions of proteins, nucleic acids and carbohydrates. Specific examples are given to illustrate each application.

Proteins and polypeptides

Proteins and polypeptides have two major classes of chromophores, the amide groups of the peptide backbone, which absorb light in the far UV (below 250 nm) and the aromatic amino acid side chains and disulfide bonds, which absorb light in both the near (320–250 nm) and far-UV. Far-UV ORD and CD are useful for studying protein structure and folding because many conformations that are common in proteins, including α-helixes, β-pleated sheets, poly-L-proline II-like helices and turns, have characteristic spectra. **Figure 1** illustrates representative CD spectra of model polypeptides with different secondary structures. In addition, the chromophores of the aromatic amino acids of proteins are often in

asymmetric environments, leading to characteristic CD spectra in the near-UV, which may serve as useful probes of the tertiary structure of proteins.

Analysis of the secondary structure of proteins

Many methods have been developed to extract the conformation of proteins in solution from CD and ORD data. In early studies, the ORD spectra of proteins were analysed to yield structural data, but when commercial CD spectrometers became available in the late 1960s, CD became the method of choice because CD spectra have better resolution

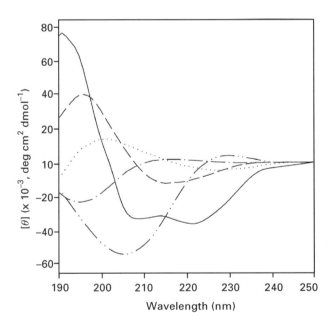

Figure 1 The circular dichroism of polypeptides with different conformations. (—) α-helix, poly-L-lysine in water, pH 11.1, 22°C; (---) β-sheet, poly(Lys-Leu-Lys-Leu) in 0.5 M NaF at pH 7; (·····) β-turn (Type III), N-acetyl-Pro-Gly-Leu-OH in trifluoroethanol at −60°C; (—·—·) random coil, Glu-Lys-Lys-Leu-Glu-Glu-Ala in 20 mM sodium phosphate, pH 2, 0°C; (—··—··) P$_\parallel$, poly-L-proline in trifluoroethanol. Reprinted from data in Greenfield NJ and Fasman GD (1969) *Biochemistry* **8**: 4108–4116, Brahms S and Brahms J (1980) *J. Mol. Biol.* **138**: 149–178, by permission of Academic Press, Venyaminov SY, Baikalov IA, Shen ZM, Wu C-SC and Yang JT (1993) *Anal. Biochem.* **214**: 17–24, and Bovey FA and Hood FP (1967) *Biopolymers* **5**: 325–326 by permission of John Wiley & Sons, Inc.

than ORD spectra. Basically, all of the analysis methods assume that the spectrum of a protein can be represented by a linear combination of the spectra of the secondary structural elements and a noise term. In constrained fits the sum of all the fractional weights of each structure must be equal to one. In the first attempts to elucidate protein conformation from CD spectra, the spectra of proteins were fitted by a combination of the spectra of polypeptides with known conformations using the method of least squares (multilinear regression). Once the conformation of many proteins had been determined from X-ray crystallographic analysis, the CD spectra of several well-characterized proteins were deconvoluted into reference (basis) spectra for the α-helix, antiparallel and parallel β-pleated sheets, β-turn and random conformations using multilinear-regression analysis. These extracted curves were then used in place of the polypeptide standard curves. Later, singular-value decomposition (SVD) and convex-constraint analysis (CCA) were used to extract the basis curves. The basis curves obtained from deconvoluting a set of protein CD spectra may vary greatly depending on the choice of reference proteins. This occurs because some proteins have unusual CD spectra in the far UV, due to aromatic amino acids, disulfide bridges, or rare conformations. To overcome these difficulties, some methods use selection procedures so that only proteins with spectral characteristics that are similar to those of the protein to be evaluated are used as standards. Methods that use selected reference data include ridge-regression, variable-selection and neural-network procedures.

The following computer programs for analysing CD data to yield the secondary structures of proteins and polypeptides are widely available: MLR is a non-constrained and G&F, LINCOMB and ESTIMATE are constrained least-squares-analysis programs; CONTIN uses the ridge-regression technique; VARSLC and SELCON use singular-value decomposition combined with variable selection to compare the structure of unknown proteins with standard; CCA uses the convex-constraint algorithm to deconvolute sets of CD spectra into a minimum number of basis curves and K2D is a neural-network recall program. **Table 1** compares the agreement between the secondary structures of 16 proteins and poly-L-glutamate, determined by analysis of the CD spectra using these computer programs, with the structures determined by X-ray crystallography.

Determination of the thermodynamics of protein folding

The change in the CD spectra as a function of temperature can be used to determine the thermodynamics

of unfolding or folding of a protein because the observed change in ellipticity, θ_{obs}, is directly proportional to the change in concentration of native and denatured forms. When the protein is fully folded $\theta_{obs} = \theta_F$ and when it is fully unfolded $\theta_{obs} = \theta_U$. For a monomeric protein, the equilibrium constant of folding, K = folded/unfolded. If we define α as the fraction folded at a given temperature, T, then

$$K = \frac{\alpha}{1 - \alpha} \quad [1]$$

$$\Delta G = nRT \ln K \quad [2]$$

where ΔG is the free energy of folding, R is the gas constant and n is the number of molecules.

$$\Delta G = \Delta H(1 - T/T_M) - \Delta C_p[(T_M - T) + T(\ln(T/T_M))] \quad [3]$$

where ΔH is the Van't Hoff enthalpy of folding, ΔS is the entropy of folding, T_M is the observed midpoint of the thermal transition and ΔC_p is the change in heat capacity for the transition. At T_M, $K = 1$; therefore $\Delta G = 0$ and $\Delta S = \Delta H/T_M$. Rearranging these equations, we obtain:

$$K = \exp(-\Delta G/RT) \quad [4]$$

$$\alpha = \frac{K}{1 + K} \quad [5]$$

$$\theta_{obs} = (\theta_F - \theta_U)\alpha + \theta_U \quad [6]$$

To calculate the values of ΔH and T_M that best describe the folding curve, initial values of ΔH, ΔC_p, T_M, θ_F and θ_U are estimated, and Equation [6] is fitted to the experimentally observed values of the change in ellipticity as a function of temperature, by a nonlinear least-squares curve-fitting routine such as the Levenberg–Marquardt algorithm. Similar equations can be used to estimate the thermodynamics of folding of proteins and peptides that undergo folded multimer to unfolded monomer transitions.

Figure 2 illustrates how changes in ellipticity as a function of temperature and concentration have been used to determine the enthalpy of folding of a peptide derived from the coiled-coil domain of the yeast GCN4 transcription factor, which undergoes

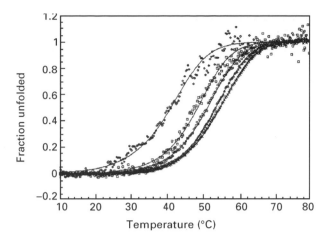

Figure 2 The change in fraction folded of a coiled-coil leucine zipper peptide, GCN4-P1, as a function of temperature and peptide concentration. The concentrations range from 1 μM with the lowest T_M to 20 μM with the highest T_M. The enthalpy of unfolding was 35.0 ± 1.1 kcal mol^{-1} (monomer) which compared to that of 34.7 ± 0.3 kcal mol^{-1} measured by calorimetry. Abstracted with permission from data in Thompson KS, Vinson CR and Freire E (1993) *Biochemistry* **32**: 5491–5496. Copyright 1993 American Chemical Society.

a two-state transition between a folded two-stranded α-helical coiled coil and a monomeric disordered state. The values for the enthalpy of the folding transition, determined from the changes in ellipticity as a function of temperature, agree with the values obtained from scanning calorimetry.

In some cases thermal folding and unfolding studies are impractical. The protein may have a very high T_M, or may aggregate and precipitate upon heating. In these cases, denaturants, such as urea or guanidine-HCl, may be added to a protein or peptide to induce unfolding. K, the equilibrium constant of folding, is determined from the ellipticity observed at each concentration of denaturant. It is assumed that the protein is native in the absence of denaturant, and fully unfolded when there is no further change in ellipticity upon addition of higher concentrations of perturbant. At each concentration of denaturant,

$$\alpha = (\theta_{obs} - \theta_U)/(\theta_F - \theta_U) \qquad [7]$$

$$K = \alpha/(n[C]^{n-1})(1 - \alpha)^n \qquad [8]$$

α, θ_F and θ_U have the same definitions as in thermal unfolding above and n is the number of identical subunits. $[C]$ is the total concentration of protein monomers. The free energy of folding is evaluated at every concentration of denaturant

using Equation [2] and is plotted as a function of denaturant. The curve is extrapolated to zero denaturant to obtain the free energy of folding of the native material. **Figure 3** depicts the determination of the free energy of folding of a coiled-coil DNA binding protein from a urea-denaturation study. The peptide forms a two-stranded coiled-coil α-helix when folded and is a single-stranded unordered peptide when dissociated. The association constant and free energy of folding, determined from the denaturation data (**Figures 3B** and **3C**), agree with those determined from the change in ellipticity as a function of peptide concentration (**Figure 3A**).

Analysis of conformational transitions

Some native proteins undergo conformational transitions that do not result in formation of a denatured state. Such transitions may involve changes from an α-helical to a β-pleated sheet conformation (or vice versa). Conformational changes of proteins are sometimes followed by oligomerization and/or aggregation and precipitation reactions that may have important clinical consequences. For example, transitions from a random or α-helical form to a β-sheet may be involved in the pathology of diseases such as scrapie, 'mad cow disease' and Alzheimer's disease. **Figure 4** illustrates how CD has been used to follow the thermal stability and conformational transitions of the scrapie amyloid (prion) protein, PrP27-30. The CD spectra of the solvent-exposed (**Figure 4A**) and rehydrated solid state PrP27-30, obtained under various conditions, were deconvoluted using the CCA algorithm and five common spectral components were identified (**Figure 4B**). Infectivity quantitatively correlated with an increasing proportion of a native, β-sheet-like secondary-structure component (**Figure 4C**), a decreasing amount of an α-helical component, and an increasingly ordered tertiary structure.

Detection of folding intermediates

Often when a protein or peptide folds or unfolds, the process is not a two-state transition between the native and totally disordered forms; intermediate states exist. These states may be transient, and may be observed during kinetic measurements of folding or unfolding. Alternatively, they may be stable intermediates, seen when the peptide is subject to denaturing conditions, such as high temperatures, or exposure to urea, guanidine or detergents. These states may be partially folded, with well-defined tertiary structures, or may be so called 'molten

Table 1 Comparisons of methods of analysing protein conformation from circular dichroism data[a]

Computer program	Standards	Wavelength range, nm	α-helix		β-sheet		β-turn	
			P	σ	P	σ	P	σ
Linear regression – unconstrained fit								
MLR (1)	4 peptides (1)	240–178	0.91	0.13	0.43	0.21	0.07	0.16
MLR (1)	4 peptides (1)	240–200	0.92	0.14	0.74	0.16	0.23	0.16
Linear regression – constrained fit								
G&F (2)	Poly-L-lysine (2)	240–208	0.92	0.13	0.61	0.18	ND	ND
LINCOMB (3)	4 peptides (1)	240–178	0.93	0.11	0.58	0.15	0.61	0.11
LINCOMB (3)	4 peptides (1)	240–200	0.94	0.11	0.71	0.13	0.53	0.14
LINCOMB (3)	17 proteins (4)	240–178	0.94	0.09	0.62	0.14	0.21	0.13
LINCOMB (3)	17 proteins (4)	240–200	0.92	0.10	0.09	0.28	0.52	0.12
Singular-value decomposition								
SVD (4)	17 proteins (4)	240–178	0.98	0.05	0.68	0.12	0.22	0.10
SVD (4)	17 proteins[b] (4)	240–200	0.96	0.07	0.43	0.14	0.04	0.13
Convex-constraint algorithm								
CCA (5)	17 proteins (4)	260–178	0.96	0.10	0.62	0.18	0.39	0.18
CCA (5)	17 proteins (4)	240–200	0.97	0.10	0.42	0.20	0.52	0.22
Ridge regression								
CONTIN (6)	17 proteins (4)	260–178	0.93	0.11	0.56	0.15	0.58	0.08
CONTIN (6)	17 proteins (4)	240–200	0.95	0.13	0.60	0.15	0.74	0.07
Variable selection								
VARSLC (7)	17 proteins (4)	260–178	0.97	0.07	0.81	0.10	0.60	0.07
Variable selection – self consistent method								
SELCON (8)	17 proteins (4)	260–178	0.95	0.09	0.84	0.08	0.77	0.05
SELCON (8)	17 proteins (4)	260–190	0.94	0.09	0.73	0.09	0.84	0.05
SELCON (8)	17 proteins (4)	240–200	0.93	0.10	0.73	0.11	0.71	0.06
SELCON (8)	33 proteins (9)	260–178	0.93	0.09	0.91	0.07	0.53	0.09
SELCON (8)	33 proteins (9)	260–200	0.88	0.12	0.86	0.09	0.46	0.09
Neural-network analysis								
K2D (10)	18 proteins (10)	240–200	0.95	0.09	0.77	0.10	ND	ND
			Mean	*STD*	*Mean*	*STD*	*Mean*	*STD*
			0.36	0.27	0.20	0.16	0.22	0.08

[a] The secondary structures of 16 proteins plus poly-L-glutamate were analysed using each computer program as described by Sreerama and Woody (8). P is the correlation coefficient between the CD-estimated and X-ray conformations and σ is the mean-square error between the CD-estimated and X-ray conformations. The secondary structures were assigned by the method of Kabsch and Sander (11). When protein databases were used as standards, each protein analysed was excluded from the data set used as the references. When the σ value is higher than the standard deviation (STD) of the mean value of each conformation found in the 17 samples which were analysed, the program does a relatively poor job of analysing that conformation.

[b] PGA was excluded from the calculation of P and σ because the fit was obviously impossible.

(1) Brahms S and Brahms J (1980) Determination of protein secondary structure in solution by vacuum ultraviolet circular dichroism. *Journal of Molecular Biology* **138**: 149–178.

Continued

Table 1 *continued*

(2) Greenfield N and Fasman GD (1969) Computed circular dichroism spectra for the evaluation of protein conformation. *Biochemistry* **8**: 4108–4116.

(3) Perczel A, Park K and Fasman GD (1992) Analysis of the circular dichroism spectrum of proteins using the convex constraint algorithm: a practical guide. *Analytical Biochemistry* **203**: 83–93.

(4) Hennessey JP and Johnson WC Jr (1981) Information content in the circular dichroism of proteins. *Biochemistry* **20**: 1085–1094.

(5) Perczel A, Hollósi M, Tusnady G and Fasman GD (1991) Convex constraint analysis: a natural deconvolution of circular dichroism curves of proteins. *Protein Engineering* **4**: 669–679.

(6) Provencher SW and Glöckner J (1981) Estimation of globular protein secondary structure from circular dichroism. *Biochemistry* **20**: 33–37.

(7) Manavalan P and Johnson WC Jr (1987) Variable selection method improves the prediction of protein secondary structure from circular dichroism spectra. *Analytical Biochemistry* **167**: 76–85.

(8) Sreerama N and Woody RW (1994) Poly(pro)II helices in globular proteins: identification and circular dichroic analysis. *Biochemistry* **33**: 10022–10025.

(9) Toumadje A, Alcorn SW and Johnson WC Jr (1992) Extending CD spectra of proteins to 168 nm improves the analysis for secondary structures. *Analytical Biochemistry* **200**: 321–331.

(10) Andrade MA, Chacón P, Merelo JJ and Morán F (1993) Evaluation of secondary structure of proteins from UV circular dichroism spectra using an unsupervised learning neural network. *Protein Engineering* **6**: 383–390.

(11) Kabsch W and Sander C (1983) Dictionary of protein secondary structure: pattern recognition of hydrogen-bonded and geometrical features. *Biopolymers* **22**: 2577–2637.

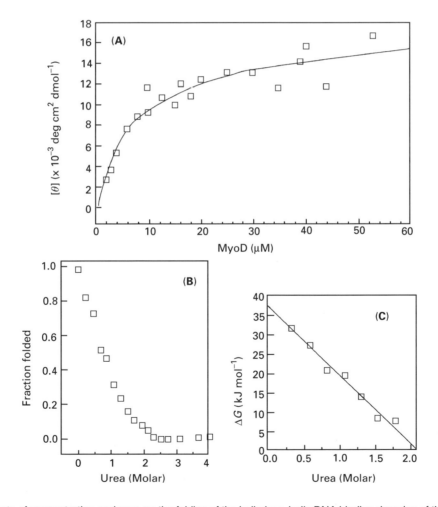

Figure 3 The effects of concentration and urea on the folding of the helix-loop-helix DNA binding domains of the muscle-regulatory DNA transcription factors, MyoD. (A) The change in ellipticity at 222 nm of Myo D as a function of concentration. The data were used to calculate the constant of dimerization $K_{dimer} = 9.4$ µM. (B) The change in fraction folded of MyoD as a function of urea concentration. (C) ΔG of folding calculated from the data in (B). The extrapolated free energy of folding calculated from the value at 0 urea was similar to those calculated from the dimerization constant. Redrawn with permission from data in Wendt H, Thomas RM and Ellenberger T (1998) *J. Biol. Chem.* **273**: 5735–5743. Copyright 1998. The American Society for Biochemistry & Molecular Biology.

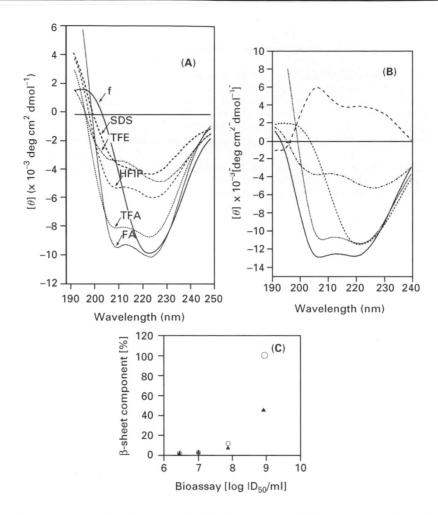

Figure 4 (A) Solvent-induced conformational transition of PrP27-30 scrapie amyloid (prion) protein in the solid state on quartz glass. Films were exposed for 20 to 30 min to: SDS, sodium dodecyl sulfate, TFE, trifluoroethanol; HFIP, hexafluorisopropanol; TFA, trifluoroacetic acid; FA, 99% formic acid. "f" is the spectrum with no treatment at 23°C. (B) Data on PrP27-30 obtained under various conditions, deconvoluted into five basis curves using the convex-constraint algorithm. (—) β-turn type I or III and/or 3_{10}-type helix; (·····) alpha helix; (----) β-sheet; (– – –) β-turn type II; (—·—) additional chiral contribution. (C) Correlation of the solvent-induced changes in the β-sheet component with infectivity. The data for dehydrated films are open circles and rehydrated films are closed circles. The data obtained with TFA and FA, where the proteins were highly helical and the infectivity was very low, are excluded from the fits. (o) dehydrated films (▲) rehydrated films. Redrawn with permission from data in Safar J, Roller PP, Gajdusek DC and Gibbs CJ Jr (1993) *Protein Sci.* **2**: 2206–2216. Copyright 1993 The Protein Society.

globules'. 'Molten globule' is a term describing a compact state with native-like secondary structure but slowly fluctuating tertiary structure. For example, methanol induces the formation of a molten-globule state in the globular protein, cytochrome C. **Figure 5** illustrates the effect of increasing quantities of methanol on the CD spectra of cytochrome C in the near and far UV. A loss of ellipticity occurs in the near UV, that is present in the native state arising from the tertiary interactions of the aromatic chromophores (**Figure 5A**). The far-UV spectra (**Figure 5B**), however, show that the helical content of cytochrome C actually increases upon the addition of methanol, as shown by the increasing ellipticity at 222 and 208 nm.

Determination of the kinetics of protein folding

Besides the study of protein folding under equilibrium conditions, CD can be used to measure rapid events in protein folding and unfolding by attaching a rapid mixing device to a CD spectropolarimeter. Stopped-flow CD can be used both to obtain the kinetic constants of folding reactions and to define the conformation of folding intermediates. Data can be collected in the far and near UV. The far-UV data give information about the secondary structure of the protein, while the near-UV data are windows that monitor the formation of tertiary structure.

Stopped-flow CD studies of highly helical two-stranded coiled coils and four-helix bundle proteins

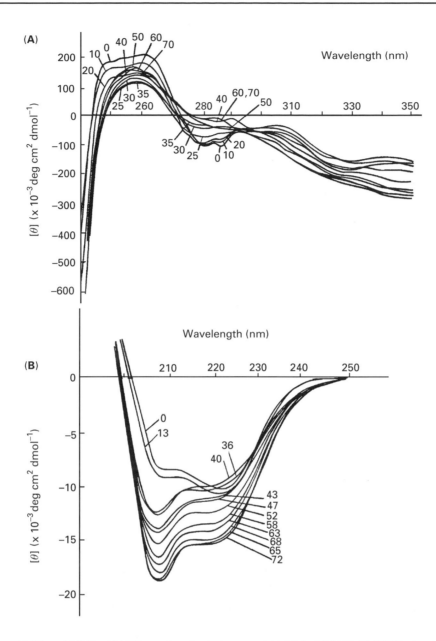

Figure 5 (A) Near-UV CD and (B) Far-UV CD spectra of cytochrome c (pH 4.0 in 0.5 M NaCl, 24°C) at methanol concentrations shown near the curves. The negative bands in the near-UV spectrum are attributed to the side chain of the single Trp59. They decrease and almost vanish at methanol concentration > 40%. Abstracted with permission from data in Bychkova VE, Dujsekina AE, Klenin SI, Elisaveta IT, Uversky VN and Ptitsyn OB (1996) *Biochemistry* **35**: 6058–6063. Copyright 1996 American Chemical Society.

have shown them to fold in essentially two-state transitions. The kinetics of folding of globular proteins, however, can be very complex. For example, the kinetics of the unfolding and refolding of bovine beta-lactoglobulin, a predominantly β-sheet protein in the native state, are illustrated in **Figure 6**. The kinetics of unfolding (**Figure 6A, B**) show a single-phase transition between the folded and denatured form, with the loss of secondary structure (**Figure 6A**) and tertiary structure (**Figure 6B**) being simultaneous. The refolding reaction, in contrast, is a complex process composed of different kinetic phases

(**Figure 6C**). In particular, a burst-phase intermediate is formed during the dead time of stopped-flow measurements that shows more intense ellipticity signals in the far UV than does the native state. This more-intense CD is due to the transient formation of a non-native alpha-helical structure (**Figure 6D**). The CD spectrum suggests the folding intermediate exists in a molten globule state, since there are no near-UV CD bands indicative of tertiary structure. Similarly, intermediate states have been shown to exist in the refolding of diverse proteins including cytochrome C, ribonuclease HI from *E. coli*, dihydrofolate

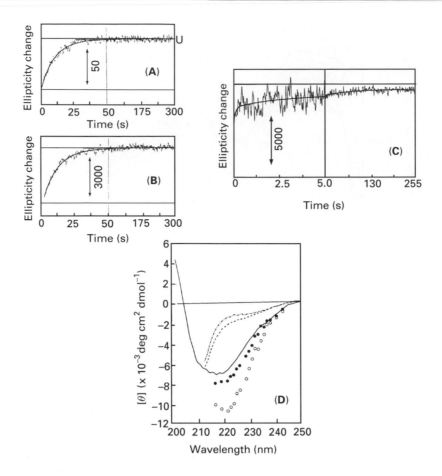

Figure 6   (A) Unfolding curve of β-lactoglobulin (β-LGA) measured by the CD changes at 293 nm at pH 3.2 at 4.5°C. The change was initiated by jumping the guanidine HCl (GdnHCl) concentration from 0 to 4.0 M. The vertical arrows show the change in ellipticity in deg.cm^2.dmol^{-1} (B) Same, measured at 220 nm. (C) Kinetic refolding curve of β-LGA measured by the CD change at 221 nm. Refolding was initiating by dropping the concentration of GdnHCl from 4 to 0.4 M. (D) CD spectrum of β-LGA in the far UV at pH 3.2 and 4.5°C. (—) native state; (---) unfolded state in 4 M GdnHCl; (—·—·) disulfide-cleaved and carboxymethylated material in 4 M GdnHCl; (o) ellipticity of β-LGA at 0 time after refolding; (•) ellipticity of β-LGA at 255 min after refolding. Reprinted from data in Kuwajima K, Yamaya H and Sugai S (1996) *J. Mol. Biol.* **264**: 806–822, by permission of Academic Press.

reductase, alcohol dehydrogenase, carbonic anhydrase, retinoic acid-binding protein and staphylococcal nuclease.

Analysis of protein–ligand interactions

Circular dichroism can be used to follow the binding of ligands to proteins, peptides, nucleic acids and carbohydrates provided one of two criteria is fulfilled. Either the ligand must bind in an asymmetric fashion that induces *extrinsic* optical activity in the chromophores of the bound ligand, or the binding must result in a conformational change in the macromolecule that results in a change in its *intrinsic* CD spectrum. In the case of equivalent binding sites, the change in ellipticity due to complex formation is directly proportional to how much ligand is bound. Thus, the change in ellipticity as a function of substrate concentration can be used to estimate the binding constants.

Extrinsic ellipticity bands (called Cotton effects), developed when chromophores bind to proteins, have been used to study protein–ligand interactions in many diverse systems. **Figure 7A** illustrates the induction of extrinsic CD bands when retinol binds to interphotoreceptor retinol-binding protein, and **Figure 7B** shows the change in the ellipticity at the wavelength of maximal ellipticity as a function of ligand concentration. Extrinsic Cotton effects have been used to study the binding of many other cofactors to proteins, including the binding of nucleotides to lactic dehydrogenase, folates and antimetabolites to dihydrofolate reductase, ATP to GroEL and pyridoxal phosphate to tryptophan synthase. They have also been used to study the interactions of substrates with enzymes including tryptophan synthase and N-acetylglucosamyl transferase, to study the interactions of drugs with bovine serum albumin, and to examine the interactions of DNA binding proteins with DNA.

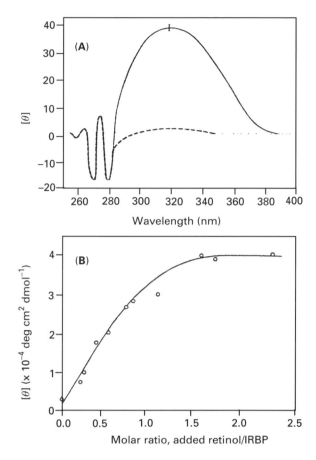

Figure 7 (A) Near-UV mean-residue circular-dichroism ellipticity, $[\theta]$, in deg.cm^2.dmol^{-1}, of interphotoreceptor retinol-binding protein (IRBP), 0.4–0.8 mg ml^{-1}, with and without bound retinol. (B) Circular-dichroism titration curve of apo-IRBP, 0.4 mg ml^{-1}, with retinol. The height of the extrinsic Cotton effect at 330 nm is calculated in ellipticity per mole of IRBP. Redrawn with permission from data in Adler AJ, Evans CD and Stafford WF (1985) *J. Biol. Chem.* **260**: 4850–4855. Copyright 1985 The American Society for Biochemistry & Molecular Biology.

The binding of ligands to proteins often results in conformational transitions. The resultant change in the intrinsic ellipticity of the backbone amides caused by the conformational change can be used to obtain the ligand-binding constants. For example, divalent cations are effectors in many biological systems, and their binding often induces large conformational changes in the target protein. Calcium-ion binding to calmodulin and troponin C induces an increase in the ellipticity of the proteins due to stabilization of α-helical regions. Many other broad classes of protein-ligand interactions result in conformational changes. For example, DNA binding to many different DNA transcription factors results in increases in the helical content of the factors. These DNA binding proteins often contain α-helical-coiled-coil or helix-loop-helix motifs, which serve as dimerization domains, and a relatively unstructured basic region which binds to the DNA. Upon DNA binding the basic region assumes a helical conformation.

Nucleic acids

Nucleic acids are polymers of nucleotides that consist of purine or pyrimidine bases attached to a phosphorylated sugar. Monomeric sugars are asymmetric and they induce a small amount of ellipticity in the attached planar-symmetric base. Polymerized nucleic acids, however, are rigid and highly asymmetric. In the polymerized nucleic acids, the bases stack with one another in precise orientations depending on the conformational state. Base stacking leads to characteristic splittings of the transitions of chromophores of the bases into multiple transitions, and the rigidity leads to CD bands with high rotational strength. CD spectra of polynucleotides are sensitive both to their sequence and conformation, and can provide a good deal of structural information.

Analysis of the secondary structure of polynucleotides

Homopolynucleotides   Even very simple polynucleotides show multiple conformational states and complex CD and ORD spectra. Compounds as small as homodimers of ribo and deoxyribonucleotides show evidence of base stacking in solution from measurements of optical activity. For example, **Figure 8A** illustrates the circular dichroism properties of several riboadenylate compounds. AMP shows a single absorption band and a single negative CD band at 260 nm. Diadenylic acid has a single UV band at 260 nm, but a split CD spectrum with a positive band at 270 nm and a negative band at 250 nm. Theoretical calculations of the CD spectrum agree with a model where the bases are stacked in a parallel fashion in the dinucleotide. The stacking leads to hypochromicity in the absorption spectrum, and splitting of the transitions into two transitions, one perpendicular and one parallel to the helix axis with equal strength and opposite rotatory strength. Poly(rA) has a spectrum similar to that of diadenylate but higher in magnitude.

The CD spectrum of poly(rA) changes with pH, chain length and temperature. At neutral pH, poly(rA) is unprotonated and forms a single-stranded helix. At acidic pH, poly(rA) forms dimers. Depending on the pH there are two conformational forms, one where poly(rA) is fully protonated, and one where it is half-protonated. **Figure 8B** shows the

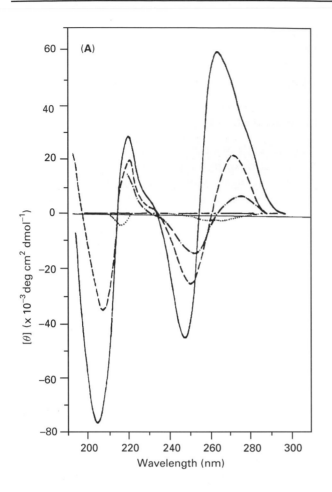
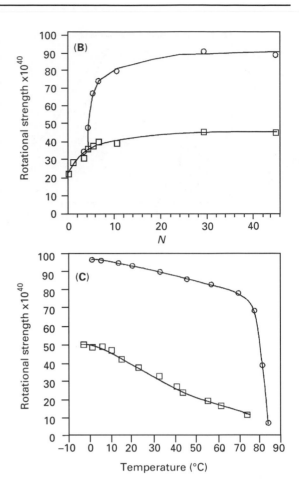

Figure 8 (A) Circular dichroism of adenylate compounds at pH 8.5 in unbuffered 0.1 M NaF at 22°C. (—) poly (rA); (—·—·) rA-2'O-methyl-prAp; (-----) rA(3'p5)rA; (·····) rAMP-5'. Abstracted with permission from data in Adler AJ, Grossman L and Fasman GD (1969) *Biochemistry* **8**: 3846–3859. Copyright 1969 American Chemical Society. (B) Rotational strength of the positive band of adenylate oligomers at 0°C as a function of the chain length, N. (□) pH 7.0; (○) pH 4.5. (C) Rotational strength of the positive band of poly(rA) as a function of temperature at (□) pH 7.4 and (○) pH 4.5. (B) and (C) are reprinted from data in Brahms J, Michelson AM and Van Holde KE (1966) *J. Mol. Biol.* **15**: 457–488, by permission of Academic Press.

chain-length dependence, and **Figure 8C** the temperature dependence, of the rotational strength of the positive-ellipticity band of the single-stranded form at pH 7.4 and the double-stranded half-protonated form at pH 4.5. Both the chain-length dependence and the unfolding of the double-stranded forms as a function of temperature, are more cooperative than that of the single-stranded form.

Heteropolynucleotides Circular dichroism spectra of nucleic acids are complex, compared to those of homopolynucleotides because four different bases contribute to the CD spectra, and the spectra are sequence dependent. In addition, nucleic acids display great conformational diversity and may form single-, double- or triple-stranded helices. For example, the spectra of poly(rA), poly(rU), the dimer poly(rA)·poly(rU) and the trimer poly(rU)·poly(rA)·poly(rU) are illustrated in **Figure 9**.

Complex formation results in shifts in wavelength and intensity of the bands, compared to the addition of the spectra of the unmixed components.

The various oligomerization states of the polynucleotides, moreover, can exist in multiple conformations. For example the A, B and Z forms of double-stranded nucleic acids have distinct CD spectra. **Figure 10A** illustrates the CD spectra of poly (dAdC)·poly(dGdT) in the A, B and Z conformations and **Figure 10B** illustrates CD spectra of a double-helical polynucleotide, poly(dGdC)·poly(dGdC) in the B form, the Z form and in a form in which the bases assume the Hoogsteen base-pairing conformation.

Analysis of conformational transitions of nucleic acids

Conformational transitions of nucleic acids can be followed by examining CD and ORD spectra as a

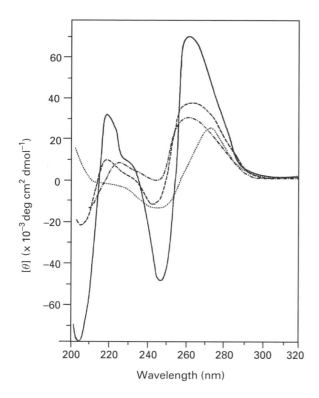

Figure 9 The CD spectra of (——) Poly(rA); (·····) Poly(rU); (----) poly(rU)·poly(rA); (—·—·) poly (rU)·poly (rA)·poly (rU) at pH 7 in sodium phosphate buffer. Redrawn with permission from data in Steely T, Gray DM and Ratiff RL (1986) *Nucleic Acid Research* **14**: 10 071–10 090. Copyright 1986 Oxford University Press.

function of temperature and denaturants, and the thermodynamic parameters of the transitions can be determined using the same equations used to determine the thermodynamics of folding of proteins. In addition, the spectra of nucleic acids, obtained under varying conditions, can be deconvoluted using the CCA and SVD methods, to determine

whether there are folding intermediates, as in the case of proteins.

Analysis of ligand binding to nucleic acids

Circular dichroism and ORD are excellent methods to follow the binding of ligands to nucleic acids. The same equations that are used to analyse protein–ligand interactions may be applied to the study of DNA–ligand interactions. For example, CD and ORD have been used to examine the binding of DNA to proteins, drugs and amines. **Figure 11** illustrates the effect of binding of spermidine to poly(dT)·poly(dA)·poly(dT). The addition of the amine has dramatic stabilizing effects. The conformation of the polynucleotide complex undergoes sequential changes from B-DNA to triplex DNA as the concentration of spermidine is increased from 0 to 50 μM. At 60 μM spermidine, the CD spectrum of triplex DNA is comparable to that of ψ-DNA, with a strong positive band centred around 260 nm. A negative band is also found at 295 nm. At higher concentrations of spermidine, however, the intensity of the positive band progressively decreases. The peak intensity is found at a 1:0.3 molar ratio of DNA phosphate-spermidine.

Figure 12 illustrates an example of DNA-protein binding study. The gene 32 coded protein of bacteriophage T4 is necessary for genetic recombination and DNA replication. It denatures poly-(dAdT)·poly(dAdT)] and T4 DNA at temperatures far below their regular melting temperatures. The protein interacts with native DNA and a variety of synthetic polynucleotides. Illustrated here is the interaction of the protein with poly(dA). The CD spectra suggest that the protein keeps poly(dA) in a conformation that is equivalent to that of a single-stranded form in high salt. Similar results were

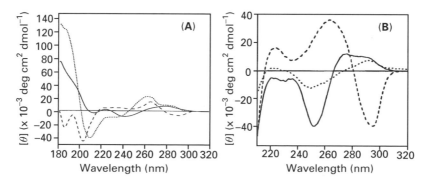

Figure 10 (A) The CD spectra of poly (dAdC)·poly(dGdT) in the (·····) A, (——) B and (---) Z conformations. Redrawn from data in Riazance-Lawrence JH and Johnson WC Jr (1992) *Biopolymers* **32**: 271–276 by permission of John Wiley and Sons, Inc. (B) CD spectra of poly(dGdC)·poly(dGdC) in the (——) B-form, (- - -) Z-form and (····) in a form in which the bases assume the Hoogsteen base-pairing conformation. Abstracted with permission from data of Seger-Nolten GMJ, Sijtsema NM and Otto C (1997) *Biochemistry* **36**: 13 241–13 247. Copyright 1997 American Chemical Society.

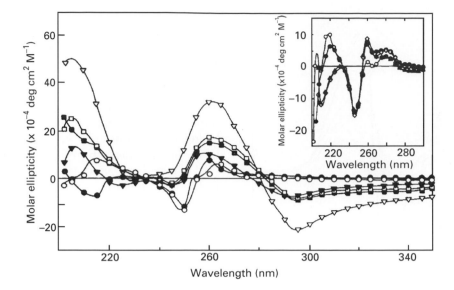

Figure 11 CD spectra of poly(dT)·poly(dA)·poly(dT) in the presence of (○) 0; (●) 60; (▽) 70; (■) 80; (▼) 90 and (□) 100 μM spermidine. The insert shows the effects of (○) 0, (●) 5, (▼) 10, and (♦) 25 μM spermidine. Data were recorded at 25°C in 10 mM sodium cacodylate and 0.5 mM EDTA at pH 7.2. Redrawn with permission from data in Thomas TJ, Kulkarni GD, Greenfield NJ, Shirahata A and Thomas T. (1996) *Biochem J.* **319**: 591–599. Copyright (1996) Portland Press.

obtained with other dimeric DNA analogues including poly(dA)·poly(dT) and poly(dT).

Carbohydrates

Carbohydrates are much more difficult to study using ORD and CD than proteins and nucleic acids and much less literature is available. First, carbohydrates mainly absorb below 190 nm, a region difficult to examine using commercial CD spectrometers. Second, unlike proteins and nucleic acids, there are no common sets of spectral characteristics which easily define the conformation of carbohydrates. However, ORD and CD have been used to obtain useful information about the structure of carbohydrates. In favourable circumstances they can give information about the structure and linkages in carbohydrates. In addition, they can distinguish helical from disordered conformations and be used to monitor conformational transitions induced by salts, solvents and temperature.

Determination of carbohydrate structure and linkages

While no simple basis spectra exist that can be used to determine the conformation of carbohydrates, analysis of the CD spectra of monomeric carbohydrates has provided empirical rules for extracting some conformational information. By comparing the

spectra of monomeric sugars, researchers have been able to assign the CD bands to specific transitions, and to determine quadrant rules for the effect of structure on the sign of the transitions. For example, the CD spectra of methyl-β-galactopyranoside, carrageenan and agarose are shown in **Figure 13**. X-ray films of carrageenan show that the conformation is a double helix. The CD spectrum of carrageenan is consistent with the conformation seen in the X-ray analysis. There are conflicting models for the structure of agarose in gels. In one model the chains are extended and in the other they are helical. The geometry-dependent linkage contributions to the CD of agarose were determined by subtracting the monomeric CD from the CD of the polymer. Application of the quadrant rules indicated that agarose was also helical. The difference in CD sign between carrageenan and agarose arises from a translocation of the beta-D-galactose O-2 atom from one quadrant to the neighbouring quadrant of the C-5-O-5-C-1 ether chromophore of the preceding anhydro sugar residue.

Analysing conformational transitions of carbohydrates using CD and ORD

While the analysis of the CD spectrum of carbohydrates to obtain detailed structural information is complex, CD and ORD serve as simple methods for following order–disorder transitions in carbohydrates. Because the absorption bands of

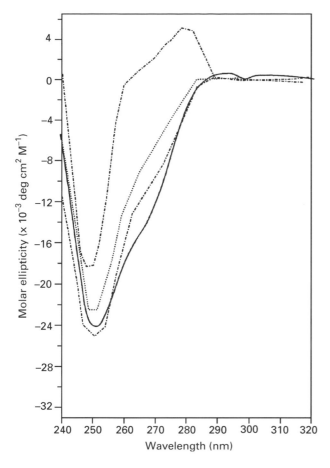

Figure 12 CD spectra of poly(dA) complexed with the gene 32 coded protein of bacteriophage T4 at (—) 1.0, (----) 29.9 and (·····) 47.7°C. The nucleotide to protein ratio is 8.5. Also illustrated (—··—) is the spectrum of free poly(dA) at 1°C. Redrawn with permission from Greve J, Maestre MF, Moise H and Hosoda J (1978) *Biochemistry* 17: 887–893 Copyright (1978) American Chemical Society.

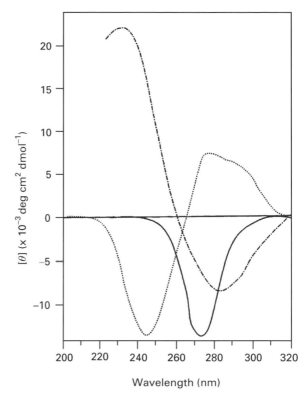

Figure 13 CD spectrum of (—) methyl β-D-galactopyranoside, (···) carrageenan and (—··—) agarose. Redrawn with permission from data in Arndt ER and Stevens ES (1993) *J. Am. Chem. Soc.* **115**: 7849–7853, Copyright (1993) American Chemical Society, Arndt ER and Stevens ES (1994) *Biopolymers* **34**: 1527–1534 and Stevens ES and Morris ER (1990) *Carbohydr. Polym.* **12**: 219–224.

carbohydrates are at very low wavelength, many of the studies of conformational transitions have been done using ORD, since ORD bands extend even into the visible region and are more easily accessible. For example, **Figure 14** shows how the effects of potassium-ion concentration and temperature on the order–disorder transition of xanthan, an extracellular polysaccharide produced by *Xanthomonas campestris*, have been followed by measuring the optical rotation at 365 nm. The data were used to find the enthalpy of folding of the polymer. The data could also be fitted by Zimm–Bragg helix-coil transition analysis. Similar studies of conformational transitions have been performed on other carbohydrates including acetan, carrageenan, succinoglycan and hyaluronate.

In addition to monitoring the folding of carbohydrates by examining changes in the intrinsic optical activity of the carbohydrate, structural information has been obtained by examining the CD and ORD spectra of bound dyes. For example when methylene blue binds to carrageenan, the binding induces optical activity in the dye which changes as a function of temperature. There is a progressive loss of ellipticity as a function of temperature, suggesting that the dye binds to the double-helical form of the polysaccharide, but not the coil form.

List of symbols

K = equilibrium folding constant; n = number of molecules; N = chain length; P = correlation coefficient; R = gas constant; T = temperature; ΔC_p = change in heat capacity; ΔG = free energy of folding; ΔH = Van't Hoff enthalpy of folding; ΔS = entropy of folding; θ = ellipticity; σ = mean-square error.

See also: **Biomacromolecular Applications of UV-Visible Absorption Spectroscopy; Carbohydrates Studied By NMR; Chiroptical Spectroscopy, General**

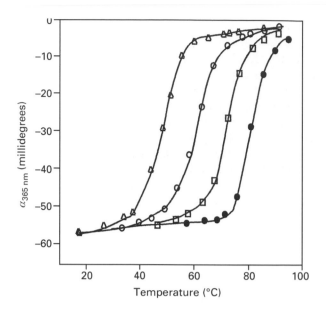

Figure 14   The unfolding of xanthan as a function of potassium ion concentration. (Δ) 4.3, ($\bigcirc$) 15, ($\square$) 30 and ($\bullet$) 500 mM. Redrawn with permission from data in Norton IT, Goodall DM, Frangou SA, Morris ER and Rees DA (1984) *J. Mol. Biol.* **175**: 371–394 Copyright (1984) Academic Press.

Theory; Induced Circular Dichroism; Macromolecule–Ligand Interactions Studied By NMR; Magnetic Circular Dichroism, Theory; Nucleic Acids and Nucleotides Studied Using Mass Spectrometry; Nucleic Acids Studied Using NMR; ORD and Polarimetry Instruments; Proteins Studied Using NMR Spectroscopy; Peptides and Proteins Studied Using Mass Spectrometry; Vibrational CD Spectrometers; Vibrational CD, Applications; Vibrational CD, Theory.

Further reading

Adler AJ, Greenfield NJ and Fasman GD (1973) Circular dichroism and optical rotatory dispersion of proteins and polypeptides. *Methods in Enzymology* **27** **part D**: 675–735.

Breslauer KJ (1987) Extracting thermodynamic data from equilibrium melting curves for oligonucleotide order-disorder transitions. *Methods in Enzymology* **259**: 221–245.

Eftink MR (1995) Use of multiple spectroscopic methods to monitor equilibrium unfolding of proteins. *Methods in Enzymology* **259**: 487–512.

Greenfield NJ (1996) Methods to estimate the conformation of proteins and polypeptides from circular dichroism data. *Analytical Biochemistry* **235**: 1–10.

Johnson WC Jr (1996) Determination of the conformation of nucleic acids by electronic CD. In: Fasman GD (ed) *Circular Dichroism and the Conformational Analysis of Biomolecules*, pp 433–468. New York and London: Plenum Press.

Perrin JH and Hart PA (1970) Small molecule-macromolecule interactions as studied by optical rotatory dispersion-circular dichroism. *Journal of Pharmaceutical Science* **59**: 431–448.

Stevens ES (1996) Carbohydrates. In: Fasman GD (ed) *Circular Dichroism and the Conformational Analysis of Biomolecules*, pp 501–530. New York and London: Plenum Press.

Woody RW (1995) Circular dichroism. *Methods in Enzymology* **246**: 34–71.

Biomacromolecular Applications of UV-Visible Absorption Spectroscopy

Alison Rodger and **Karen Sanders**, University of Warwick, Coventry, UK

ELECTRONIC SPECTROSCOPY
Applications

Introduction

Ultraviolet (UV) or visible radiation is absorbed by a molecule when the frequency of the light is at the correct energy to cause the electrons of the molecule to rearrange (or become excited) to another, higher-energy, state of the system. Frequency, ν (measured in σ^{-1}), wavelength, λ (usually measured in nm) and energy, E (measured in J) are related by

$$E = h\nu = hc/\lambda$$

where $h = 6.626\ 196 \times 10^{-34}$ J s is the Planck constant and $c = 2.997\ 925 \times 10^{17}$ nm s^{-1} is the speed of light in units to match the choice of nm for λ.

Absorption can be pictorially viewed as either the electric field or the magnetic field (or both) of the radiation pushing the electron density from a starting arrangement to a higher-energy final one. The direction of net linear displacement of charge is known as the polarization of the transition. The polarization and intensity of a transition are characterized by the so-called transition moment, which is a vectorial property having a well-defined direction (the transition polarization) within each molecule and a well-defined length (which is proportional to the square root of the absorbance). The transition moment may be regarded as an antenna by which the molecule absorbs light. Each transition thus has its own antenna and the maximum probability of absorbing light is obtained when the antenna and the electric field of the light are parallel.

Absorbance is defined in terms of the intensity of incident, I_0, and transmitted, I, light:

$$A = \log_{10}\left|\frac{I_0}{I}\right|$$

In most experiments we use the Beer–Lambert law to relate the absorption of light, A, to the sample concentration, C:

$$A = \varepsilon C l$$

where l is the length of the sample through which the light passes and ε is the molar absorption coefficient (extinction coefficient); if l is measured in cm and C in M = mol dm^{-3}, then ε has units of mol^{-1} dm^3 cm^{-1}. The Beer–Lambert law breaks down when the sample absorbs too high a percentage of the incident photons for the instrument to measure the emitted photons. An absorbance of 2, for example, means that 99% of the photons are absorbed. Biological samples present additional challenges to the Beer–Lambert law: if there are local high concentrations of sample (as in vesicles for example) or if molecules interact and perturb the spectroscopy of the isolated molecule, then the Beer–Lambert law becomes invalid.

The most common application of UV-visible absorption spectroscopy is to determine the concentration of a species in solution using the Beer–Lambert law. Other applications follow because the energy of UV-visible light is usually sufficient only to excite valence electrons which are the ones involved in bonding. Thus any UV-visible absorption spectrum is directly related to bonds and hence the structure of a molecule. The challenge is then to relate the plot of absorbance versus wavelength, which the spectrometer produces, to the structure of the molecules in the cuvette. With complicated systems, such as biological macromolecules and their complexes with small molecules, we usually interpret UV-visible spectra by considering changes in the spectrum as a function of a variable such as temperature, ionic strength, solvent, concentration, etc. Alternatively the absorption spectra data are used as input for interpreting other spectra such as fluorescence, circular dichroism (CD) or linear dichroism (LD).

Biological macromolecule structure and UV spectroscopy

Proteins and DNAs are linear polymers where a limited set of residues are joined together by, respectively, the amide or phosphodiester bonds. The situation is similar for carbohydrates though the linking options are more varied.

To a first approximation the absorbance spectrum expected for a biomolecule is therefore the sum of the spectra for the component parts.

In the case of nucleic acids (DNA and RNA) the UV absorbance from 200–300 nm is due exclusively to transitions of the planar purine and pyrimidine bases (**Figure 1**). (The backbone begins to contribute at about 190 nm.) The accessible region of the spectrum (nitrogen purging is required below ~200 nm as oxygen absorption interferes with the spectrum) is therefore dominated by $\pi \rightarrow \pi^*$ transitions of the bases. The UV spectra of the bases (**Figure 2**) look as if there are two simple bands; however, each 'simple' band observed is a composite of more than one transition. This makes detailed analysis of DNA absorption difficult, but usually ensures that the absorption spectrum changes when the system is perturbed. Thus absorption spectroscopy is a useful qualitative or empirical probe of structural changes. Typical DNA UV absorption spectra are illustrated in **Figure 3**. The base transitions are significantly perturbed by the so-called π–π stacking interactions and so both wavelength maxima and transition intensities vary depending on the base sequence and structure adopted (cf. **Table 1**).

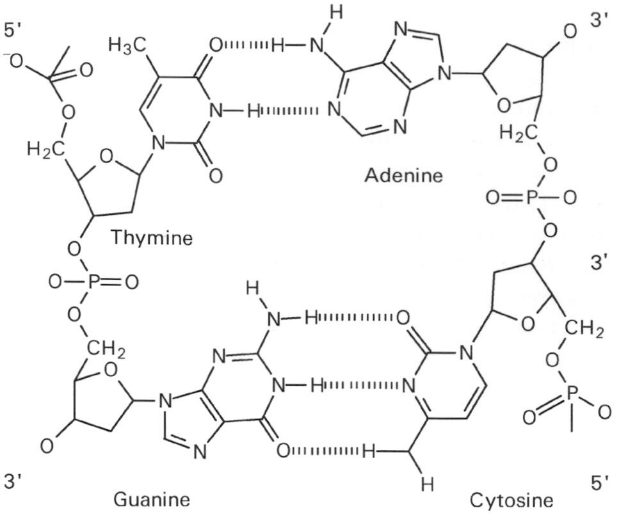

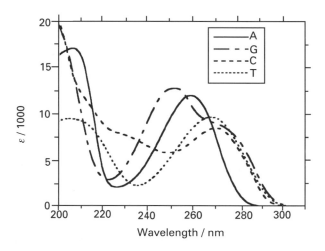

Figure 1 Structural formula of DNA.

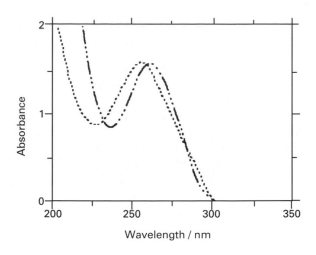

Figure 3 Absorbance spectra in 1 cm pathlength cuvettes of 190 μM poly[d(G-C)]₂ (dotted line) and 240 μM poly[d(A-T)₂] (dashed line).

Figure 2 UV spectra of the DNA nucleotides: deoxyadenosine 5′-monophosphate (A), deoxyguanosine 5′-monophosphate (G), deoxycytidine 5′-monophosphate (C) and thymidine 5′-monophosphate (T). The spectrum of uracil is almost indistinguishable from that of thymine.

Table 1 Long-wavelength absorbance maxima and extinction coefficients for some DNAs. Calf thymus DNA is ~60% A-T

DNA	Wavelength of A_{max}(nm)	ε_{max} (mol⁻¹ dm³ cm⁻¹)
Calf thymus	260	6 600
Poly[d(A-T)]₂	262	6 600
Poly(dA).poly(dT)	260	6 000
Poly[d(G-C)]₂	254	8 400
Poly(dG).poly(dC)	253	7 400

(A = adenine, T = thymine, G = guanine and C = cytosine)

In the case of peptides and proteins the spectroscopy of the amide bonds, the side chains and any prosthetic groups (such as haems) determines the observed UV-visible absorption spectrum. However, as with DNA, intensities and wavelengths can be perturbed by the local environment of the groups. UV spectra of proteins are usually divided into the 'near' and 'far' UV regions. The near-UV in this context means 250–300 nm and is also described as the aromatic region, though transitions of disulfide bonds (cystines) also contribute to the total absorption intensity in this region. The far-UV (< 250 nm) is dominated by transitions of the peptide backbone of the protein, but transitions from some side chains also contribute to the spectrum below 250 nm.

The aromatic side chains, phenylalanine, tyrosine and tryptophan all have transitions in the near-UV region (**Figure 4**). At neutral pH, the indole of tryptophan has two or more transitions in the 240–290 nm region with total maximum extinction coefficient ε_{max}(279 nm) ~5000 mol⁻¹ dm³ cm⁻¹; tyrosine has one transition with ε_{max}(274 nm) ~1400 mol⁻¹ dm³ cm⁻¹; phenylalanine also has one transition with ε_{max} ~190 mol⁻¹ dm³ cm⁻¹; and a cystine disulfide bond absorbs from 250–270 nm with ε_{max} ~300 mol⁻¹ dm³ cm⁻¹. Although tryptophans have by far the most intense transitions, many proteins have few tryptophans compared with the other aromatic groups, so the near UV is not necessarily dominated by tryptophan transitions.

The peptide chromophore (**Figure 5**) which gives rise to the transitions observed in the far-UV region (180–240 nm) has non-bonding electrons on the oxygen and also on the nitrogen atoms, π-electrons which are delocalized to some extent over the carbon, oxygen and nitrogen atoms, and σ bonding electrons. The lowest energy transition of the peptide

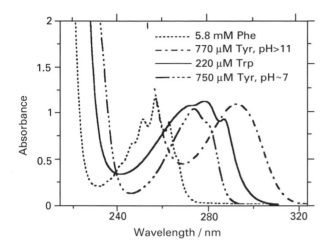

Figure 4 Aromatic absorption spectra of tryptophan, tyrosine and phenylalanine. Note the different concentrations required to ensure absorbance maxima of 1 absorbance unit.

chromophore is an $n \rightarrow \pi^*$ transition analogous to that in ketones, and the next transition is $\pi \rightarrow \pi^*$. As in the carbonyl case, the $n \rightarrow \pi^*$ transition is predominantly of magnetic transition dipole character and is thus of low intensity ($\varepsilon \sim 100 \ mol^{-1} \ dm^3 \ cm^{-1}$), though it is not as low as for a simple ketone; it occurs at about 210–230 nm (depending mainly upon the extent of hydrogen bonding of the oxygen lone pairs) and its small electric character is polarized more or less along the carbonyl bond. The $\pi \rightarrow \pi^*$ transition ($\varepsilon \sim 7000 \ mol^{-1} \ dm^3 \ cm^{-1}$) is dominated by the carbonyl π-bond and is also affected by the involvement of the nitrogen in the π orbitals; its electric dipole transition moment is polarized somewhere near the line between the oxygen and nitrogen atoms, and it is centred at 190 nm.

In an α-helix, the coupling of the $\pi \rightarrow \pi^*$ transition moments in each amide chromophore results in a component at about 208 nm which contributes to the characteristic α-helix CD spectrum. For UV-visible spectroscopy, however, the far-UV spectroscopy is usually of little use either for concentration determination or structural analysis as the accessible region (above 200 nm) is almost a linear plot of increasing intensity with decreasing wavelength.

Carbohydrate UV-visible spectroscopy is essentially that of any substituents that have spectroscopy; a simple sugar system such as starch has no spectroscopy above 200 nm. Carbohydrates are typically derivatized with thiols or aromatic chromophores for UV spectroscopy. The spectroscopy of these compounds is largely determined by that of the derivatives. These data may be found in UV-visible spectroscopy atlases.

Wavelength scanning

Simple wavelength scans of biological macromolecules

Nucleic acids (DNA and RNA), proteins and peptides absorb very little light above 300 nm in the absence of ligands or prosthetic groups with chromophores (absorbing units). However, it is usually wise to collect a simple UV scan of a sample from about 350 nm. If the spectrum is not flat between 350 and 310 nm then the sample has condensed into particles whose size is of the order of the wavelength of light; therefore what is being observed is scattering of the incident light rather than absorption. UV absorbance is most commonly used to determine the concentration of a sample and also to give an indication of its purity.

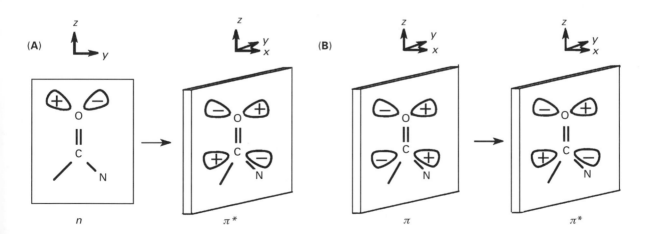

Figure 5 Schematic illustrations of (A) $n \rightarrow \pi^*$ and (B) $\pi \rightarrow \pi^*$ transitions of peptides.

To perform a wavelength scan proceeds as follows:

1. First choose the parameters.

   The wavelength range will usually be 350–200 nm, unless the buffer cuts out the low-wavelength end (see below).

   A data interval of 0.5–1 nm is adequate as the bands are all broad. Shorter data intervals will not improve the spectrum you plot but will fill up the computer memory. (If data is being collected to aid interpretation of LD spectra, ensure that the data interval is the same in both cases.)

   The bandwidth is the wavelength range of the radiation at any specified wavelength. If data is being collected every 0.5 nm, then a bandwidth of 0.5 nm is appropriate. A larger value will lead to a greater averaging of data than the data interval suggests. Do not choose a much smaller bandwidth as the incident light intensity will be reduced and the spectral noise will increase without improving the quality of your data.

   The data averaging time determines the signal-to-noise ratio and affects the scan speed required to produce undistorted spectra. 0.033–0.1 s is usually a good choice; longer times are required if the absorbance is very small (less than 0.1 absorbance units) or small differences in absorbance are being determined (as for DNA melting curves; see below).

2. Choose the solvent or buffer for your experiment and fill a pair of matched cuvettes of the desired pathlength with it. Either switch off the baseline correction or use an air/air baseline in the instrument. Place one cuvette in the sample holder (usually the front position) and leave the reference holder empty. Run a spectrum over the desired wavelength range to check that the solvent/buffer spectrum is not significant over the wavelength range of interest. If it is then you need to change the solvent/buffer.

   In choosing your buffer note phosphate buffers are essentially spectroscopically invisible over the wavelength range usually used; however, you need to ensure that phosphate does not interact in any way with your sample. Cacodylate and ammonium acetate have a window down to ~210 nm and have the added advantage of preventing bacterial growth (whereas phosphate promotes it). Chloride ions begin absorbing just above 200 nm so high salt spectra cannot be collected at the lower-wavelength end of the spectrum. Higher buffer concentrations can be accommodated in shorter pathlength cuvettes (see

Beer–Lambert law discussion above), so if it is not possible to dilute or change the buffer try a smaller pathlength cuvette. (5 mm cuvettes will stand in a normal sample holder, 1 mm cuvettes will need spacers to hold them vertical – if they are not held vertical then you will be working with a variable pathlength. Smaller pathlength cuvettes will need special holders.) UV cut-off wavelengths for solvents are readily available in liquid chromatography texts.

3. Place the matched solvent/buffer cuvettes in both the reference holder (usually the rear position) and the sample holder and perform a baseline accumulation. An alternative to having the solvent/buffer in the reference beam is to use a reference cell of water and collect a spectrum of the solvent/buffer which is subsequently subtracted from all spectra. This method may be preferable if the solvent/buffer has a significant absorbance as it is easier to determine whether an apparent peak or dip in the spectrum is due to the buffer in some way.

4. Place the sample in the sample cuvette in the sample holder and record the spectrum.

The absorbance at ~260 nm (or wherever the maximum is for a particular molecule) is generally used to determine nucleic acid concentrations using the Beer–Lambert law. As noted above, for DNA samples the linear relationship between concentration and absorbance seems to break down when the absorbance of a 1 cm pathlength solution exceeds ~1.5–2 absorbance units. Some DNA extinction coefficients are given in **Table 1**.

Protein absorbances will be dominated by tryptophan residues (if there are any) and will have a maximum at 280 nm. The other aromatic residues also absorb at 280 nm. Absorbance at 280 nm may therefore be used to give an estimate of protein concentrations. At 280 nm a 1 mg cm^{-3} protein solution in a 1 cm pathlength cell often has an absorbance of ~1 absorbance unit. This is because many proteins have a similar percentage of 'aromatic' amino acid residues. However, the A_{280} (1 mg cm^{-3}) can vary from 0.3 to 1.8. For example, the A_{280} (1 mg cm^{-3}) for bovine serum albumin is ~0.66. In cases where the protein amino acid content and molecular weight are known then a reasonably accurate estimate of ε can be made using the above ε values for the residues (instead of assuming $A_{280} = 1$ for 1 mg cm^{-3}) and then the Beer–Lambert law applied.

The Beer–Lambert method for concentration determination of nucleic acids and proteins is based on the assumption that the samples are pure. Nucleic acids, if present, will interfere with the protein concentration determination because they also absorb at 280 nm, and conversely. In these cases the following formula permits a rough estimate of protein concentration in the presence of nucleic acids (for a 1 cm pathlength cuvette):

$$1.55 \times A_{280} - 0.76 \times A_{260} = \text{mg protein cm}^{-3}$$

It is also common practice to report the A_{280}/A_{260} ratio as an indication of purity for a given sample.

Wavelength scans of derivatized protein samples for concentration determination

Although proteins have different percentages of the different amino acids, each residue has an amide bond linking it to the next residue in the chain. A number of concentration-determination methods have thus been developed that involve derivatizing the amides and spectroscopically determining the concentration of the derivatives. The three methods mentioned below all rely on a standard of known concentration to enable a calibration curve to be plotted. The calibration is not necessarily linear.

Biuret method This method is simple and reasonably specific as it depends on the reaction of copper(II) with four N atoms in the peptide bonds of proteins. Compounds containing peptide bonds give a characteristic purple colour when treated in alkaline solution with copper sulfate. This is termed the 'biuret' reaction because it is also given by the substance biuret $NH_2CONHCONH_2$. For a wide variety of proteins, 1.0 mg of protein in 2 cm³ of solution results in an absorbance at 540 nm of 0.1. Many haemoproteins, for example, give spurious results due to their intrinsic absorption at 540 nm, but modifications which overcome this difficulty are known (either removal of the haem before protein estimation or destruction of the haem by hydrogen peroxide treatment). The protein content of cell fractions such as nuclei and microsomes can be estimated by this method after solubilization by detergents such as deoxycholate or sodium dodecyl sulfate.

The biuret reagent may be made by placing $CuSO_4.5H_2O$ (1.5 g) and sodium potassium tartrate.$4H_2O$ (6.0 g) into a dry 1 dm³ volumetric flask and adding about 500 cm³ of water. With constant swirling, NaOH solution (300 cm³, 10% w/v) is added and the solution made to volume (1 dm³) with water. The reagent prepared in this manner is a deep blue colour. It may be stored indefinitely if KI

(1 g) is also added and the reagent is kept in a plastic container. The protein solution (x cm³, where $x < 1.5$) is then mixed with water [$(1.5 - x)$ cm³] to make a total volume of 1.5 cm³ to which 1.5 cm³ of biuret reagent is added. The purple colour is developed by incubating for 20 min at 37°C. The tubes must then be cooled rapidly to room temperature and the absorbance at 540 nm determined. The colour of the solution is stable for hours.

Folin–Ciocalteau or Lowry method While the biuret method is sensitive in the range 0.5 to 2.5 mg protein per assay, the Lowry method is 1 to 2 orders of magnitude more sensitive (5 to 150 μg). The main disadvantage of the Lowry method is the number of interfering substances; these include ammonium sulfate, thiol reagents, sucrose, EDTA, Tris, and Triton X-100.

The final colour in the Lowry method is a result of two reactions. The first is a small contribution from the biuret reaction of protein with copper ions in alkali solution. The second results from peptide-bound copper ions facilitating the reduction of the phosphomolybdic-tungstic acid (the Folin reagent) which gives rise to a number of reduced species with a characteristic blue colour. The amino acid residues which are involved in the reaction are tryptophan and tyrosine as well as cysteine, cystine and histidine. The amount of colour produced varies slightly with different proteins. In this respect it is a less-reliable assay than the biuret method, but it is more reliable than the absorbance method since A_{280} may include contribution from other species, and also the absorption of a given residue is dependent on its environment within the protein.

Two solutions are required for the Lowry method. For the alkaline copper solution, mix 50 cm³ Na_2CO_3 (2% w/v) in NaOH (0.1 M) with 1 cm³ of $CuSO_4.5H_2O$ (0.5% w/v) and 1 cm³ of sodium potassium tartrate (1% w/v). This solution must be discarded after 1 day. The Folin reagent (phosphomolybdic-tungstic acid) may be made by diluting the concentrated Folin reagent obtained from e.g. Sigma with an equal volume of water so that it is 1 N (i.e. 1 M H^+).

To perform an assay add x cm³ of sample (where $x < 0.6$) containing 5–100 μg of protein as required to $(0.6 - x)$ cm³ of water. Then add 3 cm³ of the alkaline copper solution. The solutions must then be mixed well and allowed to stand for 10 min at room temperature. 3.0 cm³ of Folin reagent is then added and after 30 min the absorbance at 600 nm is determined.

Coomassie blue dye binding assay This protein-determination method involves the binding of

Coomassie brilliant blue G-250 to protein. The protonated form of Coomassie blue is a pale orange-red colour whereas the unprotonated form (**Figure 6**) is blue. When proteins bind to Coomassie blue in acid solution their positive charges suppress the protonation and a blue colour results. The binding of the dye to a protein causes a shift in the absorption maximum of the dye from 465 to 595 nm and it is the increase in absorbance at 595 nm that is monitored. The assay is very reproducible and rapid with the dye binding process virtually complete in ~2 min with good colour stability.

The reagent is prepared as follows. Coomassie brilliant blue G-250 (100 mg) is dissolved in 50 cm^3 95% ethanol. To this solution phosphoric acid (100 cm^3, 85% w/v) is added and the solution diluted to 1 dm^3. To perform the assay, x cm^3 of the sample containing 5–100 µg of protein is placed in a clean, dry test tube. $(0.5 - x)$ cm^3 water and 5.0 cm^3 of diluted dye reagent are added and the solution mixed well. After a period of from 5–60 min, A_{595} is determined.

The only compounds found to give excess interfering colour in the assay are relatively large amounts of detergents such as sodium dodecyl sulfate, Triton X-100 and commercial glassware detergents. Interference by small amounts of detergent may be eliminated by the use of proper controls. The assay is non-linear and requires a standard curve.

Simple wavelength scans of macromolecules with bound ligands

When a ligand is added to, for example, a DNA solution, if it does not bind to the DNA then the UV-visible spectrum will simply be the sum of the DNA spectrum and the ligand spectrum. If the ligand binds to the DNA then the spectrum of the complex will be different from the sum spectrum. One should note that the observed spectrum is probably a complicated mixture of that due to bound and unbound ligand and free and complexed DNA.

When a planar aromatic molecule binds intercalatively (sandwiched between two base pairs) to DNA there is usually a characteristic decrease in the ligand-absorbance signal (this can be up to 50%) and a shift to the red (bathochromic shift) of between ~2 nm and 20 nm as illustrated in **Figure 7**. The DNA spectrum is also affected by any molecule such as an intercalator that causes a structural change. This makes such spectra a useful probe of DNA/drug interactions but renders absorbance useless for concentration determinations unless the perturbed extinction coefficients are known.

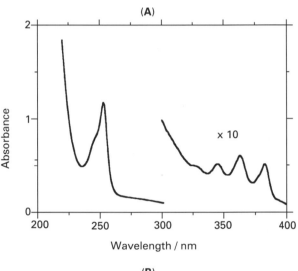

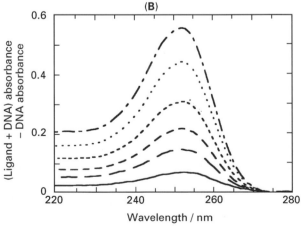

Figure 7 (A) 5 µM anthracene-9-carbonyl-N^1-spermine in water. (B) 2 µM, 4 µM, 5 µM, 7 µM, 10 µM and 13 µM anthracene-9-carbonyl-N^1-spermine in water with 200 µM calf thymus DNA. Note broadening and magnitude decrease of 250 nm band absorbance; this molecule intercalates between DNA base pairs.

Figure 6 Coomassie blue.

Titrations

The label 'titration' is used to cover experiments where spectra are collected as a function of concentration, ionic strength, pH, etc. To minimize macromolecule consumption and also (perhaps surprisingly) to minimize concentration errors, the best method is often to add solution to the cuvette. A simple way to avoid dilution effects is to proceed as follows. Consider a starting sample that has concentration x M of species X. Each time y cm^3 of Y is added, also add y cm^3 of a $2x$ M solution of X. The concentration of X remains constant at x M. Many variants on this theme may be derived.

If a titration series where ligand concentration is held constant while the macromolecule varies has a constant absorbance at a wavelength, λ_i, where the ligand absorbs this is called an isosbestic point. Isosbestic points only occur if two species are present in solution and those species have the same absorbance at λ_i.

Binding constants

If the spectral shape of a ligand spectrum remains unchanged during a titration experiment, but the magnitude changes in a manner that is proportional to the concentration of bound ligand then the spectral data can be used to determine the equilibrium binding constant, K. This is because in such a case the ligands are binding in one binding mode or in constant proportions in more than one mode (meaning site, orientation, sequence, etc). The data must be of very high quality for absorbance (or any other spectroscopic data) to be used to determine K. A simple plot of change in absorbance versus either macromolecule or ligand concentration (whichever is being varied) will probably enable the quality of the data set to be determined. It should be a smooth curve.

Consider the equilibrium

$$L_f + S_f \overset{K}{\rightleftharpoons} L_b$$

where L_f is a free ligand, L_b is a bound ligand and S_f is a free site. In the simple case where the macromolecule can be treated as a series of binding sites of n residues in size, then the total site concentration $S_{tot} = [M]/n$, where $[M]$ is the residue concentration of the macromolecule. (For proteins it is sometimes preferable to think in terms of the concentration of molecules rather than residues. In this case $S_{tot} = n'[M]$ where n' is the number of binding sites per protein.) Then

$$K = \frac{nc_b}{c_f[M]}$$

where c_b is the concentration of the bound ligand and c_f is the concentration of the free ligand.

There are a number of methods for determining K using absorbance data. The simplest is the enhancement method. This method is commonly used for fluorescence spectroscopy and may also be used to interpret absorbance data. We write

$$c_{tot}A = c_fA_f + c_bA_b$$
$$c_{tot}A = (c_{tot} - c_b)A_f + c_bA_b$$
$$c_b = \frac{c_{tot}(A - A_f)}{A_b - A_f}$$

Application of this equation requires knowledge of the absorbance of free and bound ligand. Determining the latter requires measuring an absorbance spectrum under conditions where it is known that all the ligand is bound to the macromolecules. K may then be determined directly. A more accurate value of K will be achieved if the data is used to perform a Scatchard plot.

The Scatchard plot is based on rewriting the equation for the equilibrium constant as:

$$\frac{r}{c_f} = \frac{KS_f}{[M]}$$
$$= \frac{K}{n} - rK$$

where

$$r = \frac{c_b}{[M]}$$

So, a plot of r/c_f versus r has slope $-K$ and y-intercept K/n. The x-intercept occurs where $r = n$.

Other methods more commonly used with CD or LD data may be used with normal absorption data if the change in absorbance (the absorbance of the DNA/ligand system minus the absorbance of a free-ligand solution of the same ligand concentration) is used in the analysis.

Macromolecule condensation

A UV spectrum may be used to follow the condensation of a macromolecule sample into particles, though, as discussed above, what is being probed is really scattering of the light rather than its absorbance. A monotonic increase in absorbance is observed above 300 nm as condensation takes place. In the case of DNA, the addition of a highly charged DNA binding ligand (such as spermine or $[Co(NH_3)_6]^{3+}$) will effect this change. Concomitantly with the increase in absorbance signal above 300 nm, a decrease in the 260 nm DNA absorbance is observed.

DNA melting curves: absorption as a function of temperature

If there were no residue–residue interactions in a biomolecule then the UV spectrum would be independent of its geometry and the absorption spectrum would simply be the sum of the contributions from the residues are discussed above. This is particularly true for DNA, where π–π stacking interactions lower the magnitude of the absorbance at 260 nm (the hypochromic effect) and change it at most other wavelengths. The extent of this change depends on the DNA structure. In principle if the DNA is heated enough to disrupt all base structure then the spectrum would become the sum of the base spectra in appropriate proportions. A high enough temperature to achieve this cannot usually be reached in water; however, we can use the change in the UV absorbance signal at a chosen wavelength (usually 260 nm, though at ~280 nm A-T base pairs show very little change in absorbance so this wavelength may be used to probe the role of G-C relative to A-T base pairs) to follow the disruption of base stacking and hence also base-pair hydrogen bonding.

The data from such an experiment is usually illustrated as a melting curve or a derivative melting curve and summarized by the so-called melting temperature, T_m (e.g. **Figure 8**). T_m is the temperature where the absorbance is the average of the duplex and single-stranded DNA absorbances where 50% of the DNA has melted. Thermodynamic data relating to the stability of the duplex may also be extracted from melting curves.

List of symbols

A = absorbance; $c_{b,f}$ = ligand concentration; I = transmitted light intensity; I_o = incident light

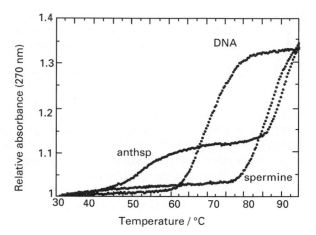

Figure 8 UV melting curves for 200 μM calf thymus DNA in 10 mM salt (denoted DNA), and also with 20 μM spermine and 20 μM anthracene-9-carbonyl-N^1-spermine (denoted anthsp). Note the premelting transition with anthracene-9-carbonyl-N^1-spermine.

intensity; l = sample light path length; ε = molar absorption coefficient; λ = wavelength of radiation (usually in nm); ν = frequency of radiation (s^{-1}).

See also: **Biomacromolecular Applications of Circular Dichroism and ORD; Dyes and Indicators, Uses of UV-Visible Absorption Spectroscopy; Macromolecule–Ligand Interactions Studied By NMR; Nucleic Acids and Nucleotides Studied Using Mass Spectrometry; Nucleic Acids Studied Using NMR; Peptides and Proteins Studied Using Mass Spectrometry; Proteins Studied Using NMR Spectroscopy.**

Further reading

Atkins PW (1983) *Molecular Quantum Mechanics.* Oxford: Oxford University Press.

Atkins PW (1991) *Physical Chemistry*, 4th edn. Oxford: Oxford University Press.

Bradford MM (1976) *Analytical Biochemistry* **72**: 248.

Craig DP and Thirunamachandran T (1984) *Molecular Quantum Electrodynamics: An Introduction to Radiation-Molecule Interaction.* London: Academic Press.

Eriksson S, Kim SK, Kubista M and Nordén B (1993) *Biochemistry* **32**: 2987.

Gornall AG, Bardawils CJ and David MM (1949) Determination of serum proteins by means of the biuret reagent. *Journal of Biological Chemistry* **177**: 751–766.

Hiort C, Nordén B and Rodger A (1990) Enantioselective DNA binding of [Ru(1,10-phenanthroline)$_3$]$^{2+}$ studied with linear dichroism. *Journal of the American Chemical Society* **112**: 1971.

Hollas JM (1992) *Modern Spectroscopy*, 2nd edn. Chichester: John Wiley and Sons.

Legler G *et al* (1985) *Analytical Biochemistry* **150**: 278.

Marky LA and Breslauer KJ (1987) Calculating thermo-dynamic data for transitions of any molecularity from equilibrium melting curves. *Biopolymers* **26**: 1601–1620.

Michl J and Thulstrup EW (1986) *Spectroscopy with Polarized Light*. New York: VCH.

Newbury SF, McClellan JA and Rodger A (1996) Spectroscopic and thermodynamic studies of conformational changes in long, natural mRNA molecules. *Analytical Communications* **33**: 117–122.

Puglisi JD and Tinoco IJ (1989) Absorbance melting curves of RNA. *Methods in Enzymology* **180**: 304–325.

Read SM and Northcliffe DH (1981) *Analytical Biochemistry* **96**: 53.

Rodger A (1993) Linear dichroism. *Methods in Enzymology* **226**: 232–258.

Rodger A and Nordén B (1997) *Circular Dichroism and Linear Dichroism*. Oxford: Oxford University Press.

Rodger A, Blagbrough IS, Adlam G and Carpenter ML (1994) DNA binding of a spermine derivative: spectroscopic studies of anthracene-9-carbonyl-N[1]-spermine with poly(dG-dC)$_2$ and poly(dA-dT)$_2$. *Biopolymers* **34**: 1583–1593.

Rodger A, Taylor S, Adlam G, Blagbrough IS and Haworth IS (1995) Multiple DNA binding modes of anthracene-9-carbonyl-N[1]-spermine. *Bioorganic and Medicinal Chemistry* **3**: 861–872.

UV-VIS Atlas of Organic Compounds (1992) 2nd edn. Weinheim: VCH.

Biomedical Applications of Atomic Spectroscopy

Andrew Taylor, Royal Surrrey County Hospital and University of Surrey, Guildford, UK

ATOMIC SPECTROSCOPY
Applications

The techniques of flame, electrothermal and vapour generation atomic absorption, flame and inductively coupled plasma atomic emission and inductively coupled plasma mass spectrometry for the measurement of minerals and trace elements in biological specimens are described. Situations in which each of the techniques might be employed for the analysis of biomedical samples are reviewed. Interferences associated with the types of samples typically examined are mentioned with accounts of how these are removed in regular practice, either during the sample preparation or by features of the instrumentation. The advantages and disadvantages of these techniques are given with particular reference to sensitivity, single to multielement measurements and special applications such as the determination of stable isotopes. While total analyte concentrations are typically determined, examples of how speciation may be of interest are included. It is seen that for biomedical measurements, atomic spectroscopic techniques are complementary and that each may be appropriate for particular sample types or applications. Except for a few very special purposes these are the techniques of choice for measurement of minerals and trace elements in biomedical specimens.

Quantitative analytical techniques included under the general heading of atomic spectroscopy are almost always employed for the direct determination of inorganic elements. For a few biomedical applications the specimen preparation gives indirect measurements of molecular compounds, generally by using a metal-based reagent in a separation or extraction step and measurement of the metal as a surrogate for the analyte of interest. The few examples of indirect measurements are included in the extensive reviews of atomic spectrometry published annually as the *Atomic Spectrometry Updates* (ASU).

As described in the articles on 'Theory' and 'Methods & Instrumentation' in this Encyclopedia, analytical atomic spectroscopy concerns measurements of about 60 elements, generally metals, although consideration of mass spectrometry extends the range to most other elements. Biomedical applications require the measurement of almost all these elements, which are often loosely termed the minerals and trace elements. As given in **Table 1**, biological specimens contain various bulk elements which fulfill essential structural and functional roles. The trace elements, which individually account for less than 0.01% of the dry weight of the organism, are represented by a series of elements essential to health, development and well-being (**Table 2**) and others such as lead, present as contaminants from the environment. With sufficiently sensitive analytical techniques almost all elements of the periodic table can be found and are included among the trace elements.

Figure 1 demonstrates that minerals and trace elements are relevant to a large number of disciplines which may be regarded as 'biomedical'. Some of

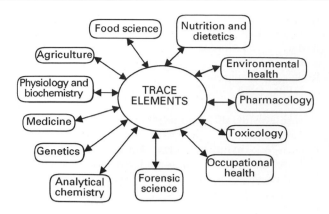

Figure 1 The relevance of minerals and trace elements to biomedical sciences.

these are described in detail in related articles and are not considered further here.

From the preceding discussion it is evident that measurements of minerals and trace elements in biomedical samples are appropriate to investigations related to:

- the biological importance of essential trace and bulk elements;
- undue effects, i.e. the harmful consequences of a sufficiently large exposure to any element, whether essential or nonessential;
- the pharmacology of certain metals which are administered as the active principal of a pharmacological agent.

Measurements are vital to situations such as understanding mechanisms of action at cellular or molecular levels, assessing the status of, e.g. zinc, within an individual subject and monitoring the effects of exposure.

Table 1 Bulk elements found in biological specimens

Minerals	Nonminerals
Calcium	Carbon
Iron	Chlorine
Magnesium	Hydrogen
Potassium	Nitrogen
Sodium	Oxygen
	Phosphorus
	Sulfur

Table 2 Essential trace elements (not all are proven to have essential roles in man)

Minerals	Chromium, cobalt, copper, iron, manganese, molybdenum, nickel, selenium, silicon, vanadium, zinc
Nonminerals	Fluorine, iodine

Many analytical techniques have been applied to the determination of minerals and trace elements but the most successful are those based on atomic spectroscopy. The techniques included in the subsequent sections afford the sensitivity required to measure concentrations below 1 ppm in specimens of just a few μL or mg with almost total specificity and relatively few interferences. As situations of deficiency and toxicity are investigated, analytes may be present at very low or very high concentrations within the same specimen type. Therefore, techniques with differing sensitivities may be used, as appropriate, for the same application. Many biomedical investigations require the measurement of just one or two metals. Consequently, multi-element techniques are not necessarily as important as in some other application areas. Nevertheless, there are situations in which this facility does become important.

Flame atomic absorption spectroscopy (FAAS)

Biomedical samples and analytes

Flame AAS is suitable for measurement of a limited range of elements present at concentrations greater than about 1 μg mL^{-1} in biological fluids, and for the analysis of solutions obtained from biological tissues at the completion of the sample preparation steps.

Typical biological fluids include blood and blood serum, blood plasma, urine and saliva. Measurement of calcium in serum was the first analysis to which the technique of AAS was applied and is an obvious example of how FAAS is useful for biomedical analysis. Other specimens e.g. dialysis fluids, intestinal contents, total parenteral nutrition solutions, may be analysed on rare occasions. Elements present at a sufficiently high concentration are lithium and gold when used to treat depression and rheumatoid arthritis respectively, and calcium, magnesium, iron, copper and zinc. Sodium and potassium can be determined by FAAS but are more usually measured by flame atomic emission spectroscopy or with ion selective electrodes. Other elements are present in fluids at too low a concentration to be measured by conventional FAAS with pneumatic nebulization. With other fluids, e.g. seminal plasma, cerebrospinal fluid, analysis may just be possible for a very few elements.

The concentrations of many metals in plant, animal or human tissues are usually much higher than in biological fluids and very often the weight of an available specimen is such that a relatively large mass of analyte is recovered into a small volume of solution, thus enhancing the concentration still further. For the analysis of tissues (these include specimens such as hair and the cellular fractions of blood) following sample dissolution steps, FAAS may be suitable for measurement of most of the biologically important elements.

Interferences and sample preparation

FAAS is subject to certain interferences associated with the nature of biological specimens. Mechanisms of the more important ionization, chemical and matrix interferences (**Table 3**) are discussed elsewhere.

Ionization interferences   Falsely high results may be obtained because of the high concentrations of sodium and potassium, which are among the more easily ionized elements. As with matrix interferences, they may not be evident with all nebulizer–burner systems and the presence of possible interferences should be investigated with any new instrument. If ionization interference is observed the calibration solutions should be prepared with sodium and potassium at approximately the same concentration as in the test specimens. This applies to biological fluids and to solutions prepared from tissues.

Chemical interferences   The only example of any consequence pertains to the measurement of calcium. The Ca^{2+} ion forms a refractory phosphate molecule and atomization is thereby impaired. Diluents containing La^{3+} or Sr^{2+}, preferentially bind to the phosphate and release the Ca^{2+}, or EDTA to complex the Ca^{2+} in solution allowing its release in the flame, effectively eliminate this interference. An alternative strategy is to use the hotter nitrous oxide–acetylene flame. A few other chemical interferences have been recorded in atypical settings and are not relevant to biomedical samples.

Matrix interferences   Two examples occur with biomedical specimens.

(a) Typical biological fluids contain protein and other macromolecules which increase the sample viscosity compared with simple aqueous solutions. The viscosity may reduce the aspiration rate through the narrow capillary tubing of the pneumatic nebulizer and absorbance signals will be attenuated compared with aqueous calibration solutions, giving falsely low results. Techniques to eliminate these effects include:

- Dilution of the sample with water (**Table 4**). A fivefold dilution is usually sufficient but some nebulizers may require a larger sample dilution to equalize the aspiration rates. Dilution may, however, reduce the analyte concentration to below the useful working range. If so, a lower dilution can be tolerated by matching the viscosity of the calibration and sample solutions by either addition of dilute glycerol to the calibrants or addition of agents such as butanol, propanol or Triton X-100 to the sample. Whatever measure is adopted it is important to determine the aspiration rates of calibration and test solutions to demonstrate that the corrective action has been effective.
- Precipitation of the protein with concentrated acid and removal by centrifugation. Trichloroacetic acid or nitric acid are used for this purpose.
- Use of matrix-matched calibration solutions. Either protein is added to the solutions or certified reference materials are used (if available).
- Standard additions for calibration.

(b) Samples with a high content of dissolved solids are liable to produce falsely high results because of nonatomic absorption or light scattering. If this occurs the analyte can be removed from the matrix by extraction into an organic solvent, or a background correction technique can be employed.

Flame composition

Elements in biomedical specimens which may be measured by FAAS are determined using the air–acetylene flame. The more refractory metals, requiring a higher temperature nitrous oxide–acetylene flame for

Table 3   Interferences in FAAS

Element	Interference	Remedy
Na, K, Fe, Li, Cu, Zn	Ionization	Add ionization suppressant (Cs, Na, K)
Ca	Chemical	La, Sr, EDTA
		Higher temperature (greater energy)
Fe, Au, Cu, Zn	Matrix	TCA precipitation
		Dilution with water
		Dilution with detergent, alcohol
		Add glycerol to standards

Table 4   Typical dilution factors for measurements of minerals in serum or urine

Gold	1+1
Iron	1+1
Copper, zinc	1+1 >> 1+4
Lithium	1+9
Potassium	1+19
Calcium	1+50
Magnesium	1+50 >> 1+100
Sodium	1+200

atomization are at concentrations too low to be determined by flame atomization (except in a few tissue specimens or in indirect methods). Other flames which have historically been used for special applications now have no real place in the analysis of biomedical samples.

For most elements the proportion of acetylene to air in the flame has little influence on formation of the ground state atomic vapour and a large variation in flow rates can be tolerated. The important exception is calcium which, as shown in **Figure 2A**, is more efficiently atomized in a reducing, fuel-rich flame. The position of the light path in the flame is also more critical for calcium than for other elements (**Figure 2B**). Modern computer-controlled instruments are preprogrammed by the manufacturers to operate under optimal conditions.

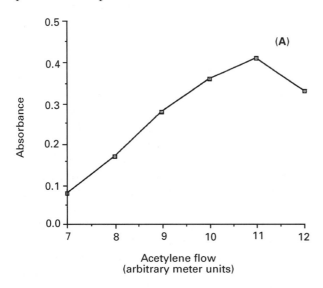

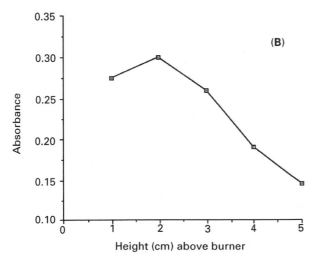

Figure 2 (A) The effect of increasing the proportion of fuel in an air–acetylene flame on the absorbance given by a solution of calcium. (B) The effect of raising the light path above the burner head on the absorbance given by a solution of calcium.

Nebulization and sensitivity enhancement

Pneumatic nebulizers were the original devices to introduce the sample as an aerosol into the flame gases. Despite the inefficiency of these nebulizers, coupled with depletion of the atom population in the flame, there are advantages of speed, precision and simplicity, and they have never been replaced by other devices. In practice there can be blockages associated with particulate matter in biological materials, precipitation of protein on the inner wall of the capillary tube and the viscosity effects referred to previously. However, daily cleaning and maintenance will usually obviate these difficulties. Techniques to improve sensitivity so as to allow the positive features of flame atomization to be retained, are widely applied to the analysis of biomedical specimens. These involve atom trapping, preventing dispersal of the atoms, preconcentration by liquid–liquid extraction and preconcentration using solid-phase traps.

Atom trapping Ground state atoms condense onto the surface of a water-cooled tube placed above the burner head and below the light path. Dilute solutions are nebulized for periods of up to several minutes to trap the analyte. Then the water is displaced by air with a rapid rise in temperature of the trap so that all the atoms are released together into the light path to give a strong atomic absorption signal. The technique has been applied to the measurement of cadmium and lead in urine.

Preventing dispersion of the atoms A hollow tube is mounted on the burner and the light path passes along the length of the tube. Samples are nebulized and atoms enter the tube where they are prevented from dispersing throughout and beyond the flame by the physical presence of the tube. Consequently, a more concentrated atom population is favoured with an improved atomization signal. Such devices were employed for the Delves' microcup technique for solid sample introduction directly into the flame – a procedure which is still used in a few laboratories for the measurement of lead in blood. More use is made of the slotted quartz tube and a number of different designs have been used. The tube is best suited for more volatile elements (e.g. Cd, Cu, Pb, Zn, etc.) and sensitivity enhancements of 3–10-fold are obtained. Furthermore, with the greater sensitivity, larger specimen dilutions can be made so that matrix interferences are eliminated or specimens of small size can be analysed.

Preconcentration methods are employed to enhance sensitivity for FAAS and for other atomic spectrometric techniques. These are described in a later section.

Electrothermal atomization atomic absorption spectroscopy (ETAAS)

Electrothermal heating for sample vaporization has proven to be an extremely powerful tool for atomic absorption, atomic emission and for introduction of atoms into the sampling torch for inductively coupled plasma (ICP) mass spectrometry (MS). With the exception of a few designs modelled on the graphite rod, all commercial systems are developments of the Massman furnace, and are constructed from graphite. As explained elsewhere, ETAAS affords considerable improvements in sensitivity compared with FAAS and the higher temperatures attained permit the measurement of elements such as Al, Cr and V which do not form an appreciable population of ground state atoms within the air–acetylene flame. Because of the small sample volumes required, less than 50 μL, the technique is ideal for biomedical applications which are often characterized by limited material (as in paediatric investigations, for example).

Biomedical specimens and analytes

A glimpse at any of the ASU reviews immediately shows that ETAAS is used for the analysis of all specimen types within the biomedical field and for the determination of virtually all applicable elements. With special furnace materials and/or linings, even the very refractory elements such as boron and the rare earths can be measured with useful sensitivity.

Although not extensively exploited, nonliquid samples can be accommodated. Tissue samples may be loaded into the furnace but this is a cumbersome procedure and because only a few mg of material can be taken it is essential that the original specimen is homogeneous – which rarely applies to biological tissues. However, it is possible to prepare suspensions of finely powdered solid samples in a thixotropic medium and the resultant slurry may then be handled as if it were a liquid and be injected into the furnace. Analyses of tissue specimens as slurries are regularly reported.

Interferences

As explained by L'Vov, who first introduced the technique of ETAAS, electrothermal atomization is ideally accomplished under stable, isothermal conditions. However, the essential design of the Massman furnace imposes dynamic conditions with complex temperature profiles both temporally and spatially within the furnace. While this allows for simplified operational designs it does however lead to more interferences. The topic of interferences is dealt with in more detail in other articles. Of special significance to biomedical specimens, which typically contain large amounts of sodium, chloride and carbon-based biomolecules, are the following.

Drying Protein-rich viscous specimens may dry unevenly with 'explosive sputtering' at the start of the ash phase, causing poor reproducibility.

Ashing Volatile halides, e.g. $AlCl_3$ may form, causing preatomization loss of the analyte.

Atomization Stable compounds such as carbides can be produced, giving low atomization rates (e.g. Mo); vapour phase reactions, especially molecular condensation at the cooler ends of the furnace (e.g. $Pb(g) + 2NaCl(g) \rightarrow PbCl_2(g) + 2Na(g)$), can result in nonatomic absorption of incident radiation and scattering of incident radiation by particulates (carbon, smoke, salts). The majority of developments since the introduction of ETAAS are concerned with the elimination of these interferences.

Furnace materials and design

Commercially available furnaces are all made from electrographite, electrographite with a pyrolytic coating or total pyrolytic graphite. Devices that delay atomization until the temperature of the gas in the furnace has reached a plateau are described elsewhere. The L'Vov platform is widely used with biomedical specimens, especially in larger furnaces, which are slow to heat to the atomization temperature. A second device, the graphite probe, produces a less sensitive analytical arrangement and is not often used. Transverse heating causes less of a temperature difference between the centre and ends of the furnace compared with longitudinal heating of the conventional Massman design and, therefore, vapour phase condensation is reduced. An authorative review of materials suitable for use in furnace construction, and of recent developments in design, has been prepared by Frech (see Further reading).

Chemical modifiers

Effective analysis of most biomedical materials requires addition of reagents that modify the behaviour of the specimen during the heating program so as to reduce the interferences described above. The chemical modifiers most commonly employed with these specimens are given in **Table 5**. Triton X-100 is used at a concentration of around 0.1% and is included in with the sample diluent. Gaseous oxygen or air are effective ashing aids but will cause rapid deterioration of the graphite furnace unless a desorption step is included before the temperature is increased for atomization. Other modifier solutions

Table 5 Chemical modifiers used in the analysis of biomedical specimens by ETAAS

Modifier	Purpose
Triton X-100	To ensure smooth, even drying of protein-rich specimens and avoid the formation of a dried crust around a liquid core
Gaseous oxygen	To promote destruction of the organic matrix and reduce formation of smoke and other particulates which give nonatomic absorption
Ni, Cu, Pd	To stabilize volatile elements, e.g. Se, As, during the dry and ash phases
Potassium dichromate	Stabilizes Hg up to a temperature of 200°C
HNO_3 or NH_4NO_3	To stabilize analyte atoms by removal of halides as HCl or NH_4Cl during the ash phase
$Mg(NO_3)_2$	Becomes reduced to MgO which traps the metals to reduce volatilization losses and also to delay atomization and separate the analyte signal from the background absorption
$NH_3H_2PO_4$ or $(NH_3)_2HPO_4$	Usually used together with magnesium nitrate; reduces volatilization losses and also delays atomization to separate the analyte signal from the background absorption

can be included with the sample diluent or separately added by the autosampler to the specimen inside the furnace. The choice of modifier often depends on the availability of a source material which is free from contamination.

Background correction

Background correction is essential for almost all biomedical applications in order to correct for nonatomic absorption of the incident radiation. Nonatomic absorption is much reduced at wavelengths above 300 nm but this is relevant to just a few elements. Each of the techniques for background correction (**Table 6**) have their advantages and disadvantages. The three systems used in commercial instruments are effective although the Smith Hieftje variation is included by very few manufacturers. The Zeeman-effect technique is ideal for dealing with structured background which is of particular significance in the measurement of cadmium in urine

Table 6 Background correction (BC) techniques

Measure absorbance at a nearby nonabsorbing line
Deuterium BC
Smith-Hieftje BC
Zeeman-effect BC;   (i) applied to the light source
(ii) applied to the furnace

and elements such as arsenic and selenium in iron-rich samples, i.e. blood. Without Zeeman-effect background correction these applications are difficult to carry out successfully by ETAAS.

Novel atomizers

Several research groups have designed atomizers with the purpose of separating the analyte from interfering species, to permit simple atomic absorption. While some appear to be effective, none are commercially available. It was shown some years ago that a 150 W tungsten filament from a light bulb could be used as an electrothermal atomizer. More recently, this concept has been used to develop very small portable instruments for onsite measurement of lead in blood. Excellent results have been reported but a commercial model is still awaited.

Other techniques for atomic absorption spectroscopy

Biomedical samples and analytes

The technical and instrumental features of cold vapour and hydride generation are described elsewhere. Mercury is used extensively in industry and there may be situations of environmental use, but the typical sources of exposure are from the diet, especially fish, and from dental amalgam. Depending on the nature of the exposure, e.g. inorganic salts, organomercury compounds, the metal or its vapour, it may be necessary to analyse specimens of urine, blood or tissues and foods. Many of the elements which form volatile hydrides have important biomedical properties. Selenium is an essential micronutrient, arsenic is especially recognized as toxic, while bismuth and antimony have valuable pharmacological properties. As with mercury, all types of biological specimens may require to be analysed.

Mercury: cold vapour atomic absorption spectrometry

The concentration of mercury in urine is an excellent indicator of recent exposure to mercury vapour or inorganic compounds, and the measurement is very simple following digestion with $KMnO_4$–H_2SO_4 to ensure that all carbon–mercury bonds are broken. With other samples a more aggressive approach is required to not only release the mercury but to also destroy the organic matrix which inhibits vaporization of elemental mercury. These more powerful reducing conditions and heating are necessary for analysis of blood and tissues. To prevent loss of volatile mercury either a closed digestion system or heating with a reflux condenser must be used.

Because the clinical and toxicological effects of alkylmercurials are so different from those of inorganic species it is useful to separately measure organomercury compounds. The essential reaction to the cold vapour technique is the reduction of Hg^{2+} to Hg^0. Thus, total mercury concentration in a specimen will be determined after the disruption of carbon–mercury bonds, as described above, while if this step is omitted only the inorganic mercury is measured.

Hydride forming elements

Considerable interest in methods to measure arsenic and other hydride forming elements has been evident in recent years. The basic procedure involves careful digestion of the specimens to convert all the different species to a single valency form, reduction with BH_4^-, and vaporization to the hydride, followed by atomization in a quartz tube heated in an air–acetylene flame or with electrical thermal wire. However, measurement of total arsenic is not always entirely helpful. Fish contain large amounts of organoarsenic species, which are absorbed and excreted without further metabolism and with no adverse health effects. These species will be included in a total arsenic determination and can mask attempts to measure toxic As^{3+} and metabolites. Thus, methods to measure the individual species, or related groups of compounds in urine or other samples, have been described to provide more meaningful results. These methods include separation by chromatography or solvent extraction, and pretreatment steps which transform only the species of interest into the reducible form.

Accessories from instrument manufacturers for hydride generation allow for either a 'total consumption' measurement when all the sample is reacted at once with the BH_4^- and the gaseous hydride taken to the heated cell as a single pulse, or by flow injection to give a continuous, steady-state signal. The former gives a lower detection limit but the latter is easier to automate. The chemical hydride reaction is impaired by other hydride-forming elements and by transition metals so that careful calibration is essential. It has been shown that the hydride may be taken to a graphite furnace where the gas is trapped onto the surface. Rapid electrothermal heating then gives a very sensitive signal. Trapping is more efficient when the graphite is coated with a metal salt, e.g. Ag, Pd, Ir. Electrolysis releases nascent hydrogen and this reaction has been employed as an alternative to chemical hydride generation. With this arrangement the interferences are much less.

With careful sample preparation hydride generation is an extremely valuable analytical technique appropriate to ICP-AES and ICP-MS as well as atomic absorption spectrometry. Within the context of biomedical analysis, measurements of arsenic and selenium are currently the more important.

Atomic emission spectroscopy

Flame atomic emission spectroscopy (FAES)

Of the different techniques for atomic emission spectroscopy (AES) only those which use a flame or an ICP are of any interest for analysis of biomedical specimens. Flame AES, also called flame photometry, has been an essential technique within clinical laboratories for measuring the major cations, sodium and potassium. This technique, usually with an air–propane flame, was also used to determine lithium in specimens from patients who were given this element to treat depression, and was employed by virtually all clinical laboratories throughout the world until the recent development of reliable, rapid-response ion selective electrodes. Biological fluids need only to be diluted with water and in modern equipment the diluter is an integral part of the instrument so that a specimen of plasma or urine can be introduced without any preliminary treatment.

Inductively coupled plasma atomic emission spectroscopy (ICP-AES)

At the temperature of the flame there is no useful emission of other biologically important elements. However, with the greater energy of the ICP, much lower detection limits, typically around $1\ \mu g\ mL^{-1}$, are obtained and many elements may be determined in solutions prepared from biological tissues. Recent developments with the optical systems and array detectors offer improvements in sensitivity and data collection. In consequence, elements such as copper, zinc and aluminium may be measured in blood plasma, while the expanded information caught by detectors is making it possible for powerful chemometric manipulation of individual signals to be undertaken.

The most valuable features of ICP-AES is the multi-element capability. In most clinical situations this is not an important requirement but for some research work and in other biomedical applications the comprehensive information derived can be of considerable interest. There are few interferences associated with ICP-AES but, as with FAAS, matrix interferences associated with uptake of sample via the nebulizer may be encountered with some systems.

Inductively coupled plasma mass spectrometry (ICP-MS)

Other instrumental approaches to inorganic MS are available which, together with X-ray fluorescence spectroscopy, are mainly applied to the analysis of very thin solid specimens such as tissue sections. For more general application, the ICP as the ion source has provided enormous improvements to inorganic MS. These improvements are: convenience to operate, low detection limits (μg L^{-1} or below), multielement analyses, and the ability to measure isotopes, thus allowing stable isotopes to be used as tracers for metabolic and other studies. However, there are important disadvantages associated with ICP-MS. Sampling times of several minutes at a nebulization rate of approximately 3 mL min^{-1} are required to ensure a sufficient number of counts at each mass number and, therefore, the volume of undiluted specimen necessary to give a result can be large. As a consequence of the long sampling times and the frequent measurement of blanks, calibration specimens and quality control samples, the total number of real specimens analysed in a working day is not high. With a slow rate of analysis, the requirement for experienced staff and the very high capital and servicing expenditure, the measurement costs are extremely high.

Interferences and sample preparation

Matrix interferences affecting the nebulizer are similar to those of FAAS and ICP-AES, and sample preparation generally involves dilution or acid digestion of liquid specimens and digestion or solubilization of tissues. In addition, some suppression of analyte signals may be caused by sodium and other dissolved inorganic components. To overcome this the salts may be removed by chelating resins or by using reference materials with a composition similar to the test specimens as calibrants. Of greater importance, ICP-MS is subject to a number of spectral interferences due to the formation of polyatomic species with the same mass number as the element being measured. For example, in the determination of zinc there are interferences due to sulfur ($^{32}S^{16}O_2$, $^{33}S^{16}O_2{}^{1}H$ and $^{32}S^{16}O^{18}O$), and chlorine ($^{35}Cl^{16}O_2$) on ^{64}Zn, ^{66}Zn and ^{67}Zn. Various techniques to reduce spectral interferences include the use of ion-exchange, either off-line or on-line (usually an HPLC column), or liquid–liquid extraction, to separate the analyte from the interferants or to count an isotope which is relatively free from interferences. The more recent generation of high-resolution mass spectrometers is reported to give good discrimination of analyte isotopes from other species but at a cost of higher detection limits. To improve the signal associated with isotopes of low abundance, techniques for sample introduction that avoid the use of nebulizer have been developed. These include mercury vapour and hydride generation, electrothermal vaporization and laser ablation techniques.

Biomedical applications

The technique of ICP-MS is rarely appropriate for measurement of a single element. Exceptions are where sensitivity is unmatched by other techniques, and for isotope work. The very low detection limits for rare earth elements and the actinides have permitted a number of studies relating to the biochemistry and unusual sources of exposure to these elements. Similarly, low levels of occupational and environmental exposure to platinum and other noble metals are now being investigated. Isotope work includes: fundamental biochemical studies using isotopes of elements such as H, C, N and O; absorption and distribution of essential trace minerals, e.g. Fe, Zn, Se in situations such as childhood, pregnancy and dietary deficiencies; and the identification of sources of exposure to lead and other nonessential elements by comparisons of isotopic compositions, i.e. a 'fingerprinting' approach. More complex investigations require a combination of these features: e.g. determination of intestinal absorption rates of different dietary selenium species involves feeding subjects with foods previously enriched with stable isotopes, and separation of the selenium species in specimens such as blood by HPLC coupled directly to the ICP-MS, which affords both very sensitive analysis and differentiation of the selenium isotopes. Indeed, speciation work such as this represents a major area of biomedical interest to which the technique of ICP-MS is applied. A further activity involves the characterization and certification of biomedical reference materials where the high degree of accuracy required is achieved by inclusion of isotope dilution analysis into the multielement measurement.

Sample preparation

The objectives for preparation of biomedical specimens are to (i) remove interfering components from the matrix and (ii) adjust the concentration of analyte to facilitate the actual measurement. These objectives may be realized by a number of approaches (**Table 7**) which in general are appropriate to all the techniques described in this article.

Methods for destruction of the organic matrix by simple heating or by acid digestion have been used extensively and are thoroughly validated. Microwave

Table 7 Approaches to sample preparation

Procedure	Remarks
Dilution, protein precipitation	Using simple off-line arrangements or flow-injection manifold
Dry ashing	Using a muffle furnace or a low-temperature asher
Acid digestion	(i) In open vessels with convection or microwave heating
	(ii) In sealed vessels to increase the reaction pressure
Base dissolution	Using quaternary ammonium hydroxides
Chelation and solvent extraction	For analyte enhancement and removal of interferences
Trapping onto solid-phase media	For analyte enhancement and removal of interferences

accomplished off-line, developments in flow-injection analysis provide for the assembly of simple on-line manifolds so that complete measurements may be carried through automatically. Developments in these applications involving a wide range of biomedical sample types and elements are regularly described in the ASU reviews. Biological reference materials with and without certified values are readily available and several interlaboratory comparison programmes are established for analysts to determine the accuracy of their methods.

See also: **Atomic Absorption, Methods and Instrumentation; Atomic Absorption, Theory; Atomic Emission, Methods and Instrumentation; Fluorescence and Emission Spectroscopy, Theory; Food and Dairy Products, Applications of Atomic Spectroscopy; Forensic Science, Applications of Atomic Spectroscopy; Inductively Coupled Plasma Mass Spectrometry, Methods; Pharmaceutical Applications of Atomic Spectroscopy; X-ray Fluorescence Spectrometers; X-ray Fluorescence Spectroscopy, Applications.**

heating is now well established for this purpose with specifically constructed apparatus to avoid dangers of excessive pressure within reaction vessels. Although the number of specimens which can be processed is not large, microwave heating affords rapid digestion and low reagent blanks. More recent developments include continuous flow systems for automated digestion linked directly to the instrument for measurement of the analyte(s).

Preconcentration by liquid–liquid partitioning is a widely used procedure. Analyte atoms in a large volume of aqueous specimen are complexed with an appropriate agent and then extracted into a smaller volume of organic solvent. This leads to enhancement of concentration and also removes the analyte from potential/real interferences in the original matrix. It is used to measure lead in blood and metals in urine and for other applications. Preconcentration by trapping onto solid-phase media represents the area where much of the recent interest in FAAS has been focussed but is relevant to all sample-preparation work. The original work involved adsorption onto material such as charcoal or alumina but newer phases include ion-exchange resins and novel support systems to which functional groups are added to confer increased selectivity and capacity. While trapping of an analyte from a dilute sample, and elution into a small volume of release solution may be

Further reading

Bacon J, Crain J, McMahon A and Williams J (1998) Atomic spectrometry update – Atomic mass spectrometry. *Journal of Analytical Atomic Spectrometry* 13: 171R–208R.

Ellis A, Holmes M, Kregsamer P, Potts P, Streli C, West M and Wobrauschek P (1998) Atomic spectroscopy update – X-ray fluorescence spectrometry. *Journal of Analytical Atomic Spectrometry* 13: 209R–232R.

Frech W (1996) Recent developments in atomizers for electrothermal atomic absorption spectrometry. *Fresenius Journal of Analytical Chemistry* 355: 475–486.

Roelandts I (1997) Biological and environmental reference materials: update 1996. *Spectrochimica Acta, Part B* 52B: 1073–1086.

Taylor A (1997) Applications of recent developments for trace element analysis. *Journal of Trace Elements in Medicine and Biology* 11: 185–187.

Taylor A (1998) Atomic absorption and emission spectrometry. In: Crocker J and Burnett D (eds) *The Science of Laboratory Diagnosis*. Oxford: Isis Medical Media.

Taylor A, Branch S, Halls DJ, Owen L and White M (1999) Atomic spectrometry update – Clinical and biological materials, foods and beverages. *Journal of Analytical Atomic Spectrometry* 14: 717–781.

Bismuth NMR, Applications

See **Heteronuclear NMR Applications (As, Sb, Bi).**

Boron NMR, Applications

See **Heteronuclear NMR Applications (B, Al, Ga, In, Tl).**

¹³C NMR, Methods

Cecil Dybowski, Alicia Glatfelter and HN Cheng,
University of Delaware Newark, DE, USA

MAGNETIC RESONANCE
Methods

The analysis of carbonaceous materials with spectroscopy takes on special importance because, without doubt, the study of the chemistry of carbon is more extensive than that of any other element. The strong sensitivity of NMR parameters to chemical state has long made NMR spectroscopy a favourite tool for analysis of organic materials. Early on, proton NMR spectroscopy was used to analyse materials, but with the development of ever-more-sensitive instrumentation, the NMR analysis of carbon in organic molecules has become the essential tool for identification of carbon-containing materials.

Several isotopes of carbon occur naturally. The most abundant isotope of naturally occurring carbon is ¹²C, which comprises 98.90% of all carbon. It has no magnetic moment and therefore is not active in NMR spectroscopy. Most people are familiar with the existence of ¹⁴C, a minute percentage of naturally occurring carbon, because, through measurement of its radioactive emissions, one can define the age of certain ancient objects. It is not NMR active because its spin is also zero. ¹³C, comprising only 1.10% of naturally occurring carbon, is the carbon isotope on which NMR spectroscopists focus strongly because it has a spin of $\frac{1}{2}$ like the proton or the ¹⁹F nucleus.

Properties of the ¹³C nucleus

The NMR properties of ¹³C (**Table 1**) give an indication of both the strengths and the weaknesses of analysis using it. The gyromagnetic ratio of ¹³C is approximately one-fourth that of the proton and, as a result, the resonance condition for ¹³C at a given magnetic field strength occurs at a radiofrequency approximately one-fourth that of the proton resonance. For a sample with equal numbers of spins and equal spin–lattice relaxation rates, the predicted signal-to-noise ratio of the ¹³C spectrum at a particular

Table 1 ¹³C NMR properties

Abundance	0.011
Spin	$\frac{1}{2}$
Gyromagnetic ratio (Hz T⁻¹)	1.0705×10^7
Typical shift range (ppm)	300
Receptivity relative to ¹H	1.59×10^{-2}

magnetic field strength would be over an order of magnitude lower than that of a proton spectrum. Additionally accounting for the concentration difference due to the natural abundance, one finds that typically the absolute receptivity of ¹³C relative to the proton is only 1.76×10^{-4}. Thus, there are difficulties with obtaining ¹³C NMR spectra compared to ¹H NMR spectra. However, the utility of carbon NMR in studies of organic molecules guarantees ¹³C will be one of the nuclei more heavily investigated by NMR spectroscopists.

Single-pulse Fourier transform NMR

The analysis of carbonaceous materials would indeed be difficult if it were not for the fact that Fourier transform NMR spectroscopy allows one to increase the signal-to-noise ratio by coadding experiments. The simplest experiment is the gathering of the response to a single pulse. The length of the pulse for maximum signal, the $\pi/2$ pulse, on most modern spectrometers is typically in the range of 5 to 20 μs. Data acquisition goes on for a time that depends on the spectrum width and resolution desired, often for times greater than a second.

In the single-pulse experiment, an important parameter is the rate at which the experiment can be repeated, determined by the time it takes to repolarize the system after the magnetization has been destroyed in the excitation and acquisition. This is a function of the spin–lattice relaxation time, T_1, of the carbons in the sample. For typical small organic molecules, T_1 can be of the order 10 to 60 s or more at room temperature. A rule of thumb for quantitative evaluation of NMR spectra is that the repolarization time is five times the longest T_1 in the sample. For some samples, particularly those containing quaternary carbon atoms or carbonyl groups, this time can be substantially longer than a few minutes. Thus, a compromise is often struck, in which the repolarization time is shortened to some few tens of seconds, with the knowledge that certain resonances may be attenuated, making the resultant spectrum nonquantitative.

Another compromise involved to obtain data faster is to use a pulse width different from $\pi/2$ and a repolarization time shorter than $5T_1$. For a resonance with a relaxation time, T_1, and a repolarization time, T_R, one may calculate the angle, α_E (the Ernst angle), that gives the largest signal-to-noise by coaddition in a given time:

$$\cos \alpha_E = \exp(-T_R/T_1) \qquad [1]$$

The variation of the Ernst angle with T_R/T_1 is shown in **Figure 1**. The optimal pulse angle may be substantially less than $\pi/2$ for typical conditions encountered for carbon NMR spectroscopy where the repolarization time is short compared to T_1. For a fixed repetition time, the Ernst angles of carbons in a sample will not be identical because T_1 varies from carbon to carbon. Thus, even in defining the pulse angle, some compromise must be made, depending on the range of T_1 values in the sample.

Heteronuclear decoupling

Detection of the ^{13}C NMR spectrum of an organic compound in solution with the single pulse experiment, as described above, is quite feasible. However, were the spectrum acquired in this manner, one would notice that the number of peaks in it exceeds the number of unique carbon environments. Most organics contain protons as well as carbon atoms. These protons are coupled to the carbons through the indirect (or scalar) J coupling, resulting in splitting of the carbon resonances of any carbon with attached protons.

Generally, carbons are coupled to one (AX), two (AX$_2$), or three (AX$_3$) protons. In these cases, one sees the appearance of a doublet (AX) a triplet (AX$_2$), or a quartet (AX$_3$). If the carbon is coupled to several chemically distinct protons, one may see patterns that arise from spin systems labelled AXY,

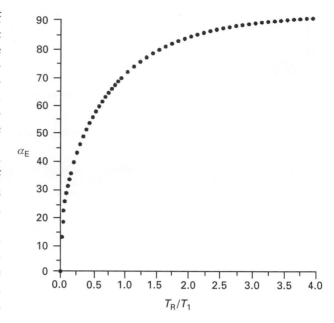

Figure 1 The Ernst angle as a function of the ratio of the repolarization time to the spin–lattice relaxation time.

AX$_2$Y, etc. These splittings contain information on the coupling constants between carbons and protons that can be of value in structure elucidation, but frequently the study of carbon NMR spectra focusses mainly on the unique carbon environments in a sample, and the splitting that the coupling brings only confuses the analysis. Thus, carbon NMR spectra are often obtained with simultaneous irradiation of the offending protons which suppresses the effects of coupling – the heteronuclear decoupled NMR experiment.

One question sometimes asked is why it is not necessary also to decouple the carbons from one another. Because of the low natural abundance of ^{13}C, it is not very likely that a ^{13}C will be near another spin-active ^{13}C. Coupling requires relatively close proximity of the spins that are coupled, and it is simply quite unlikely that the spectrum has contributions from molecules containing two or more ^{13}C atoms in a sample with the natural abundance of ^{13}C. For sample containing substantial enrichment of the carbon in ^{13}C (e.g. to increase signal in a single scan), such couplings may affect the spectrum, and consideration of them is an important point to keep in mind.

Another important feature of carbon spectroscopy is the fact that the spectrum in natural abundance is really a superposition of spectra from different molecules, each with substitution of ^{13}C at different sites in the molecule. The spectrum is not that of a single molecule with ^{13}C at each position. This fact is sometimes

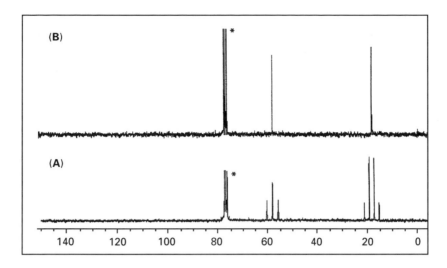

Figure 2 (A) Ethanol: the coupled ¹³C NMR spectrum; (B) the decoupled ¹³C NMR spectrum. The solvent resonance is indicated with an asterisk.

forgotten because the analysis of the spectrum is often presented as if one were analysing a single molecule.

Heteronuclear decoupling is effected by irradiation of the protons during the acquisition of the carbon signal. While this may sound simple, the process can be quite complex. For example, exposing the protons to a single-frequency excitation will only decouple protons whose resonances are at that frequency; others will be more or less affected, depending on how close their resonances are to the irradiation frequency. Thus, NMR spectroscopists have developed ways to irradiate over a band of frequencies that encompasses all of the protons coupled to the carbons. One means to spread the irradiation over the various protons is noise modulation of the proton excitation. This random modulation can be shown to result in excitation in a band that will excite all of the protons coupled to carbons. Such a technique is known as 'noise decoupling' or 'broadband decoupling'.

The use of decoupling with pure-frequency sources has been important in assigning resonances, since the observation of collapse of a coupling pattern indicates the existence of coupling between the carbon and the particular proton resonance irradiated. Experiments providing similar information are 'off-resonance decoupling' and 'spin ticklings'. In these cases, the centre frequency of the noise irradiation or the field strength is varied to observe the effects on the resonance lines of coupling partners. From these, one gathers information on the coupling partners, the sizes of couplings, and the resonance frequencies of the coupling partners. In former times, such experiments were frequently the method of choice to assign resonances; however, the current use of two-dimensional NMR methods (see below) to determine coupling partners probably presages the decline of the use of selective-decoupling techniques in the future.

Truly random noise decoupling deposits energy in the sample over a range of frequencies near the centre irradiation frequency. Unfortunately, most of this energy falls at frequencies where there are no proton resonances. This portion of the energy does no useful work in suppressing the coupling of protons to carbons, but it does affect the sample. The energy deposited ultimately results in heating of the sample. To limit heating and to extend the range of useful decoupling, in the early 1980s, spectroscopists focussed on the possibility of using coherent pulse sequences to produce the same effect as noise decoupling – the suppression of effects due to coupling of carbons to protons. The techniques go by various acronyms such as MLEV-16 and WALTZ-16, and the concept behind them is quite simple. Coherent fast switching of a proton spin between its two spin states produces a kind of short-term cancellation of evolution due to the J coupling, as observed by the detected carbon magnetization. Importantly, because of the coherent nature of the process, the energy deposited in the sample goes to spin inversion, which limits heating of the sample. Most modern spectrometers implement some version of this technique when the spectrometer is performing an experiment with broadband decoupling.

The importance of heteronuclear decoupling can be seen in **Figure 2**, where the carbon spectra of ethanol obtained with and without decoupling are shown. The decoupled spectrum in **Figure 2B** shows only lines for the various unique carbons sites. The simplicity of the spectrum in **Figure 2B** indicates the power of decoupling in unravelling the spectrum of a complex material.

Nuclear Overhauser enhancement

In the heteronuclear decoupling experiment, it has proved convenient to have noise decoupling active not just during acquisition, but also during the repolarization time between experiments. If the experiment is performed in this manner, the spectrum is modified by another effect, the nuclear Overhauser effect enhancement (NOE). The presence of the perturbing field during the repolarization period affects the return to equilibrium of the spin system by saturating the proton resonance. This saturation directly impacts the cross-relaxation between the carbons and protons, increasing the signal in the case of ¹³C. If the system is allowed to reach equilibrium under the perturbation during repolarization, one predicts an enhancement of the signal, I_{eq}, over the signal without the NOE, I_0:

$$I_{eq} = I_0(1 + \eta) = I_0 \left(1 + \frac{\gamma_H}{2\gamma_C}\right) \qquad [2]$$

This increase is about 2.998, making the detection of the carbon resonance easier. Of course, for a particular repolarization time, not all ¹³C spins in a sample may have reached an equilibrium. The NOE is maximal if the relaxation mechanism solely originates in the direct dipole–dipole coupling between protons and carbons. This mechanism seems to be dominant for carbons directly bonded to protons, but for quaternary carbons and carbonyl carbons, such is not the case, and reduced enhancements are observed for these carbons. For this reason, a spectrum obtained with the NOE present should not be considered quantitative.

The presence of the NOE can be avoided by turning proton irradiation off during the repolarization time between experiments. If the experiment includes the activation of the proton irradiation for decoupling during acquisition, one obtains a spectrum with no enhancement and with coupling suppressed. On the other hand, if the proton irradiation is off during acquisition and the repolarization time, the spectrum has no enhancement and displays all couplings. There are four possible kinds of experiment, depending on when proton irradiation is on, as indicated in Table 2. Spectroscopists often use the NOE in obtaining carbon spectra as it gives a significantly larger signal for many of the carbons in a sample.

Pulse techniques for analysis

The single-pulse NMR experiment is a powerful tool for analysing materials, especially carbon. However,

Table 2 Effects of decoupling and NOE in carbon spectra

Irradiation during repolarization time	Irradiation during acquisition	Character of the spectrum
No	No	Coupled, no enhancement
Yes	No	Coupled, with enhancement
No	Yes	Decoupled, no enhancement
Yes	Yes	Decoupled, with enhancement

there are questions and ambiguities that cannot be unravelled with this simple experiment alone. Some may be addressed with selective-decoupling techniques. However, the use of spectral-editing sequences, and ultimately 2D NMR, has aided assignment of resonances in carbon spectroscopy. As an example, knowing the number of protons to which a carbon is coupled brings additional information to the process of identifying the structure of some unknown carbon-containing material. Generally, the sequences for carrying out these investigations involve pulse excitation of the carbons and the protons in such a way that the magnetization from a particular carbon centre behaves in a predictable manner. The techniques are often referred to by spectroscopists in the acronymical shorthand so prevalent in many spectroscopies: DEPT, INEPT, APT, and many, many more. The techniques involve modulation of the spin evolution or outright transfer of polarization from one species to the other.

As an example, we discuss the so-called attached-proton test (APT). It involves the use of a spin echo, which is produced by a sequence of two pulses, as shown in **Figure 3**.

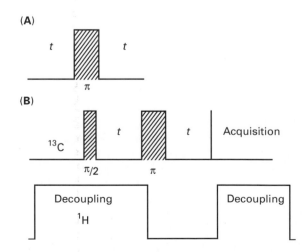

Figure 3 (A) The single π pluse. (B) The gated-decoupling attached-proton test, showing the sequence applied to the carbons (top) and the sequence applied to the protons (bottom).

First, we discuss the simple spin echo produced by a single π pulse (**Figure 3A**). Consider a magnetic moment, subject only to the chemical-shift interaction, that lies initially along the x axis of the rotating frame. After a time t, it will have moved ahead of the x axis of the rotating frame by an angle determined by the time t and the chemical shift. The π pulse inverts this magnetic moment so that immediately after the pulse it sits at that angle behind the x axis. During the subsequent period, however, it catches up to the x axis, so that at a time t after the π pulse it is aligned with the x axis as it was at the beginning. In effect, at that time, the position of the magnetic moment is as if it were not subject to a chemical shift.

On the other hand, consider a spin that has no chemical shift (i.e. it sits at the frequency of the excitation) but is subject to a J coupling to another spin. During the first period, the magnetic moment begins to deviate from the x axis of the rotating frame, depending on its spin state and that of its coupling partner. Let us now apply a π pulse only to the spin under observation. Once again, the magnetic moment returns to alignment with the x axis in the second period.

In a third experiment, we propose to carry out the experiment of **Figure 3B** on the spins. Once again, we assume that the chemical shift is zero, but the J coupling is not. During the first period, there is no evolution of the spin because the decoupling suppresses this interaction, so the magnetic moment remains along the x axis in the rotating frame. The π pulse does not change the direction of the spin, since it is along the x direction. However, in the second period, the J coupling causes the magnetic moment to deviate from the x axis, so that a time t after the π pulse, the magnetization is moved away from the x axis by an amount determined by the time t and the J coupling.

Were we to carry out the same experiment with a nonzero chemical shift and a nonzero J coupling, we would discover that the evolution due to chemical shift would be cancelled but the evolution due to J coupling would not. Thus it is possible with this experiment to label magnetic moments at time t after the π pulse according to the effects of J coupling. If, at that point, an acquisition is started, the peaks in the resulting spectrum will have magnitudes corresponding to the amplitude at that time, which will be determined by the J coupling.

Consider an AX system subject to this experiment. The spectrum will show no splittings, for it is taken with decoupling. The amplitude of the line will oscillate with t. In particular, for $t = 1/J_{CH}$, this signal will appear totally inverted. On the other hand, a carbon not coupled to an X spin (such as carbonyl or quaternary carbon) experiences no deviation and gives a positive signal under the same set of circumstances. A careful examination of the evolution of the components of the AX_2 and AX_3 species shows that, under this same experiment, the signal of the carbon in an AX_2 environment experiences no net deviation of the magnetic moment at a time t after the pulse, whereas the AX_3 system experiences a net inversion. Thus, the response to such an experiment is a spectrum with the signals from AX and AX_3 inverted relative to those of A and AX_2. From such a spectrum, one can easily discriminate these kinds of environment, as one can see from the spectrum of **Figure 4**.

The utility of the APT sequence hinges on the fact that one may choose a time t that is equal to π/J_{CH} for all carbons in the sample. This will only be true if the coupling constant is similar for carbons in various environments. As it turns out, the coupling constants are frequently in the range of 120 to 150 Hz, which is a sufficiently narrow range to allow the technique to work relatively well. A typical value of t that is used in these experiments is 7 ms. However, it must be emphasized that there are systems for which the coupling constants are very different from this nominal range, and lines from these carbons will not necessarily appear to be of the 'right phase'; such deviations should always be checked by additional experiments with different values of t.

The other commonly seen spectral editing techniques used in carbon NMR involve the transfer of polarization from protons to carbons in particular ways. A good example of this type of experiment is the so-called DEPT (distortionless enhancement by polarization transfer) sequence (**Figure 5**). By careful choice of experimental parameters, one can obtain spectra that represent only the quaternary, only the methines, only the methylenes and only the methyl groups in the sample. This is done in this case by use of the sequence shown in **Figure 5**. Four spectra are obtained with $\delta = \pi/4$, $\pi/2$, and $3\pi/4$ in the sequence

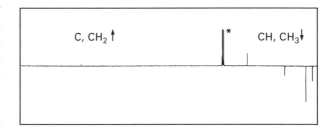

Figure 4 The APT spectrum of ethyl isobutyrate in deuterochloroform. The solvent resonance is indicated with an asterisk.

and a single-pulse experiment. **Figure 6** shows these spectra for ethyl isobutyrate. In **Figure 6A** all the lines appear. In **Figure 6B** all the lines except the quaternary carbons appear positive. In **Figure 6C** only resonances from CH groups appear. In **Figure 6D** the CH and CH$_3$ appear positive and the CH$_2$ groups appear negative. With these spectra, one can identify the various species present.

The INEPT (insensitive nuclei enhanced by polarization transfer) experiment is similar to the DEPT experiment, except that it uses an evolution period to produce the appropriate condition for spectral selection that depends critically on the size of the coupling constant. The technique is therefore somewhat more susceptible to artefacts resulting from the variability of coupling constants than is DEPT.

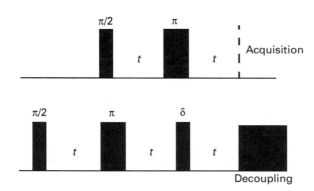

Figure 5 The DEPT experiment. The top line is the sequence applied to the carbons and bottom is applied to the protons. The time t is fixed as 1/(2J_{CH}).

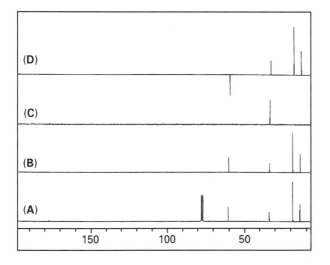

Figure 6 DEPT spectra of ethyl isobutyrate: (A) the single-pulse spectrum; (B) DEPT-π/4; (C) DEPT-π/2; (D) DEPT-3π/4. The triplet arises from the carbon of the solvent, deuterochloroform.

Two-dimensional carbon NMR spectroscopy

Two-dimensional (2D) NMR spectroscopy is now used routinely for many analytical problems. A convenient way to conceptualize 2D NMR spectroscopic studies is to divide the time of a 2D pulse experiment into four periods:

Preparation → Evolution → Mixing → Detection
Characteristic time : t_1 t_2

During the preparation period the spin system is 'prepared' for the 2D experiment. This may entail a simple π/2 pulse or a complex series of pulses and delay times. These preparation pulses and delays are applied in such a way, depending on the experiment, as to produce a particular behaviour of the spins during the evolution period (time = t_1). The mixing period is needed in some 2D experiments to permit transfer of magnetization from one spin to another. The signal is finally detected in the detection period (time = t_2).

In a typical 2D experiment the evolution time t_1 is varied systematically. For every value of t_1, the signal as a function of t_2 is collected in the detection period. The complete set of these signals (as a function of t_1 and t_2) constitutes the 2D data set, $S(t_1, t_2)$. Suitable Fourier transformation of $S(t_1, t_2)$ with respect to both t_1 and t_2 then produces the 2D spectrum, $S(f_1, f_2)$, where f_i denotes the frequency.

A large number of 2D experiments have been devised. A list of some simpler 2D experiments is shown in **Table 3**. These experiments can be broadly classified as either J-resolved or shift-correlated. In J-resolved experiments, the chemical shift δ and the coupling constant J are plotted to form a 2D representation with δ and J axes. In this way, the scalar coupling pattern for spins with different chemical shifts can be visualized clearly. In shift-correlated experiments, 2D plots with different chemical shifts can be obtained. These experiments are found to be very useful in resolving different spectral assignment problems.

An example is shown of the C–H HETCOR (heteronuclear shift correlations) experiment. The pulse sequence is given in **Figure 7**. In the preparation period, a π/2 pulse is applied to the ^{1}H nuclei (a), causing the ^{1}H magnetization to evolve with time t_1. Midway in the evolution period, a π pulse is applied to the ^{13}C nuclei (b). The π pulse produces refocussing of the ^{13}C–^{1}H coupling at the start of the mixing period (c). The mixing time Δ_1 is set to 1/(2J_{CH}) to polarize the ^{1}H magnetization. At (d), π/2 pulses at

Table 3   Some simple 2D experiments

Type	Name	Function
J–resolved	Homonuclear	Shows ¹H−¹H coupling patterns; permits precise J_{HH} measurements
	Heteronuclear	Shows ¹³C−¹H coupling patterns; permits precise J_{CH} measurements
Shift-correlated	COSY (HH-COSY)	Permits correlation of ¹H signals through ¹H−¹H couplings
	NOESY	Permits correlation of ¹H signals through NOE interactions
	HETCOR (CH-COSY)	Permits correlation of ¹H and ¹³C signals through ¹J_{CH}
	HMQC	Gives similar information as HETCOR; indirectly detects the less receptive ¹³C through the more receptive ¹H
	RELAY	Permits correlation of ¹H signals through relayed coherence; ¹H−¹H or ¹³C−¹H
	CH COLOC	Permits correlation of ¹³C−¹H signals through long-range ¹³C−¹H coupling
	HMBC	Gives similar information as COLOC; indirectly detects the less receptive ¹³C through the more receptive ¹H
	2D INADEQUATE	Gives correlation of ¹³C signals through direct ¹³C−¹³C coupling

COSY = homonuclear chemical shift correlation spectroscopy; NOESY = 2D nuclear Overhauser effect (NOE) spectroscopy; HMQC = heteronuclear mulltiple-quantum coherence; RELAY = relayed coherence transfer; COLOC = correlation via long-range coupling; HMBC = heteronuclear long-range coupling; INADEQUATE = incredible natural abundance double quantum transfer experiment.

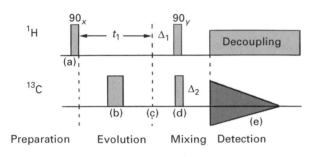

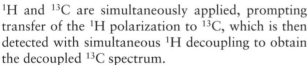

Figure 7   The C–H HETCOR pulse sequence.

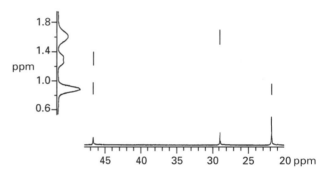

Figure 8   The C–H HETCOR spectrum of polypropylene.

¹H and ¹³C are simultaneously applied, prompting transfer of the ¹H polarization to ¹³C, which is then detected with simultaneous ¹H decoupling to obtain the decoupled ¹³C spectrum.

A C–H HETCOR plot for polypropylene is given in **Figure 8**. The projection along one axis gives the ¹³C spectrum, and the projection along the other axis gives the ¹H spectrum. Thus, if the ¹H and the ¹³C spectra are both partially assigned, this experiment sometimes permits the information to be combined in order to derive improved assignments of both spectra.

Spectral assignments

¹³C NMR spectroscopy is ideally suited for organic structure determination. As noted above, it is a rare spin (only 1.1% in natural abundance), which means ¹³C−¹³C couplings are not usually seen. Secondly, the ¹³C shift range is large (220 ppm), and neighbouring atoms (as far as ε position away) affect the chemical shift of a given carbon. Thirdly, the chemical shift is less sensitive than ¹H to inter- and intramolecular interactions, such as hydrogen bonding and self-association. As a result, in most cases a one-to-one correspondence can be made between an organic structure and the ¹³C NMR spectrum.

The ¹³C chemical shift is diagnostic of different functional groups. An abbreviated shift scale is shown in **Figure 9**. This shift scale is useful for spectral interpretation and is reminiscent of similar structure–spectral correlation charts used in infrared (IR) spectroscopy. Yet ¹³C NMR provides much more detailed structural information than IR. It has been known since the 1960s that the ¹³C shifts of alkanes are approximately additive with increasing number of substituents. Similar empirical additivity rules have been found for alkyl carbons with different functional groups and other additivity rules have been proposed for olefinic, aromatic, and carbonyl carbons. These linear additivity rules are a distinct feature of ¹³C NMR, and they simplify spectral interpretation.

There are literally thousands of papers using ¹³C NMR for structure determination of organic materials. To many NMR users, spectral assignment is an

Table 4   Substitutent chemical shift parameters (in ppm) for ^{13}C NMR ($k_0 = -2.3$ for tetramethylsilane reference)

Substituent		α^b	β	γ	δ
Paraffin[a]	$-C\lessgtr$	9.1	9.4	−2.5	0.3
Ether, alcohol[a]	−O−	49.0	10.1	−6.0	0.3
Amine[a]	$-N\lessgtr$	28.3	11.3	−5.1	0.3
Ammonium[a]	$-\overset{+}{N}\lessgtr$	30.7	5.4	−7.2	−1.4
Thioether, thiol[a]	−S−	11.0	12.0	−3.0	−0.5
Phenyl	$-C_6H_5$	22.1	9.3	−2.6	0.3
Fluoro	−F	70.1 (1,2)	7.8	−6.8	0.0
		69.0 (3)			
		66.0 (4)			
Chloro	−Cl	31.1 (1,2)	10.0	−5.1	−0.5
		35.0 (3)			
		43.0 (4)			
Bromo	−Br	18.9 (1,2)	11.0	−3.8	−0.7
		27.9 (3)			
		36.9 (4)			
Iodo	−I	−7.2 (1,2)	10.9	−1.5	−0.9
		3.8 (3)			
		20.8 (4)			
Ammonium	$-NH_3^+$	26.0	7.5	−4.6	−0.1
Nitrile	−CN	3.1	2.4	−3.3	−0.5
Nitrate	$-NO_3$	62.0	4.4	−4.0	0.0
Peroxy, hyd roperoxy	−OO−	55.0	2.7	−4.0	0.0
Oxime, *syn*	−C=NOH	11.7	0.6	−1.8	0.0
Oxime, *anti*		16.1	4.3	−1.5	0.0
Thiocyanate	−SCN	21.0	7.2	−4.0	0.3
Sulfoxide	−S(O)−	31.1	9.0	−3.5	0.0
Sulfonate	SO_3H	38.9	0.5	−3.7	0.2
Aldehyde	−CHO	29.9	−0.6	−2.7	0.0
Ketone	−C(O)−	22.5	3.0	−3.0	0.0
Acid	−COOH	20.1	2.0	−2.8	0.0
Carboxylate	$-COO^-$	24.5	3.5	−2.5	0.0
Acyl chloride	−COCl	33.1	2.3	−3.6	0.0

Table 4 Continued

Substituent		α^b	β	γ	δ
Ester	−C(O)O−	22.6	2.0	−2.8	0.0
	−OC(O)−	54.5 (1,2,3)	6.5	−6.0	0.0
		62.5 (4)			
Amide	−CONH−	22.0	2.6	−3.2	−0.4
	−NHCO−	28.0	6.8	−5.1	0.0
Olefin[a]	C=C	21.5	6.9	−2.1	0.4
Acetylene[a]	C≡C	4.4	5.6	−3.4	−0.6

[a] Steric correction parameters (**Table 5**) apply to these substitutents.
[b] Number(s) in parentheses denotes(s) the number of non-hydrogen substituents on the carbon in question.

Table 5   Steric correction parameters, $S(i,j)^a$

i	$j = 1$	$j = 2$	$j = 3$	$j = 4$
Primary	0.0	0.0	−1.1	−3.4
Secondary	0.0	0.0	−2.5	−7.5
Tertiary	0.0	−3.7	−9.5	−15
Quaternary	−1.5	−8.4	−15	−25

[a] Designation i = the carbon in question, and j = number of nonhydrogen substituents directly attached to the α-substituent (applicable only to α-substituents marked with footnote a in **Table 4**).

A sample calculation for 2-ethyl-1-butanol is as follows:

$(CH_3CH_2)_2CH–CH_2OH$

	C_1	C_2	C_3	C_4
Base	−2.3	−2.3	−2.3	−2.3
α	9.1	18.2	27.3	58.1
β	9.4	18.8	28.9	18.8
γ	−5.0	−8.5	0	−5.0
δ	+0.6	0	0	0
$S(2,3)$	0	−2.5	0	−2.5
$S(3,2)$	0	0	−11.1	0
Calculated	11.8	23.7	42.8	67.1
Observed	11.1	23.0	43.6	64.6

art that depends on experience and good memory. The shift scale in **Figure 9**, coupled with spectral intensities and other chemical information, often permits simple structures to be deciphered. For more complex molecules, one common method involves looking up in spectral libraries for either the compound in question or, if not available, compounds with similar structures. This process has been automated, and many computer-assisted structure-determination methods have been developed. Several research groups have been very active in advancing this important area.

Another commonly used method involves the empirical shift rules described earlier. For alkyl carbons, the ¹³C shifts can be approximated by a linear combination of additive terms related to the neighbouring functionalities:

$$\delta_{observed} = k_0 + n_\alpha S_\alpha + n_\beta S_\beta + n_\gamma S_\gamma + n_\delta S_\delta$$
$$+ n_\varepsilon S_\varepsilon + S_c \qquad [3]$$

where n_i refers to the number of i neighbouring carbons ($i = \alpha, \beta, \gamma, \delta$, and ε), S_i is a constant characteristic of the ith carbon, and S_c represents steric corrective terms to be used for contiguous secondary, tertiary, and quaternary carbons. One set of additive parameters is shown in **Table 4**. The term $n_\varepsilon S_\varepsilon$ is often neglected because it is usually insignificantly small. The steric correction parameters S_c are given in **Table 5**.

Computer programs have been written that automate the rather tedious arithmetic involved. A recent program provides predicted ¹³C shifts and spectra of organic compounds containing common functional groups by incorporating most of the reported activity rules.

In practice, spectral assignment of more complex structures or unknown compounds often requires a combination of assignment methodologies, with or without computer assistance and experimental techniques. Thus, for a given problem it is not unusual to obtain the single-pulse spectrum, the APT, INEPT, and/or DEPT spectrum of a sample. If needed, 2D spectra are also obtained. The information gathered is used, together with shift-prediction or spectral-search methods, to determine the structure of the sample in question.

Summary

Carbon NMR is a valuable tool for the study of organic materials. The NMR parameters are directly correlated with the structural properties of the various carbon sites. Modern spectrometers make carbon spectroscopy on samples containing ¹³C feasible at natural abundance. With the development of modern NMR technology, a host of techniques may be applied to obtain a wide variety of information about the molecule under study. In particular, spectral-editing experiments can provide information on carbon–proton connectivity that aid in the assignment of resonances and the elucidation of structure.

List of symbols

f_i = frequency; I_0 = signal intensity without NOE; I_{eq} = signal enhancement; J = coupling constant; k_0 = base chemical shift; n = number of neighbouring carbon atoms; S = signal; t = time; T_1 = spin–lattice relaxation time; T_R = repolarization time; α_E = Ernst angle; γ = magnetogyric ratio; δ = chemical shift.

See also: **¹³C NMR, Parameter Survey; NMR Pulse Sequences; NMR Relaxation Rates; NMR Spectrometers; NQR, Applications; Nuclear Overhauser Effect; Structural Chemistry Using NMR Spectroscopy, Organic Molecules; Structural Chemistry Using NMR Spectroscopy, Pharmaceuticals; Two-Dimensional NMR, Methods.**

Further reading

Bax A (1982) *Two-Dimensional Nuclear Magnetic Resonance in Liquids.* Dordrecht: Delft University Press.

Becker ED (1980) *High-Resolution NMR. Theory and Chemical Applications*, 2nd edn. New York: Academic Press.

Carabedian M, Dagane I, Dubois J-E (1988) Elucidation of progressive intersection of ordered substructures for carbon-13 nuclear magnetic resonance. *Analytical Chemistry* 60: 2186–2192, and references therein.

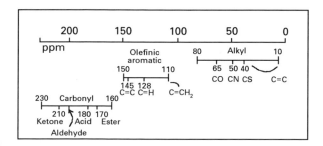

Figure 9 A simplified chemical-shift scale for carbon.

Carbon-13 Database, Bio-Rad Sadtler Division, 3316 Spring Garden Street, Philadelphia, PA 19104, USA.

Cheng HN and Bennett MA (1991) Trends in shift rules in carbon-13 nuclear magnetic resonance spectroscopy and computer-aided shift prediction. *Analytica Chimica Acta* 242: 43–53.

Cheng HN and Ellingsen SJ (1983) Carbon-13 nuclear magnetic resonance spectral interpretation by a computerized substituent chemical shift method. *Journal of Chemical Information and Computer Science* 23: 197.

Cheng HN and Kasehagen LJ (1994) Integrated approach for ^{13}C nuclear magnetic resonance shift prediction, spectral simulation and library search. *Analytica Chimica Acta* 285: 223–235.

Clerc JT and Sommerauer H (1977) A minicomputer program based on additivity rules for the estimation of ^{13}C NMR chemical shifts. *Analytica Chimica Acta* 95: 33–40.

Clerc JT, Pretsch E and Sternhell S (1973) *^{13}C-Kernresonanz-spektroskopie*. Frankfurt: Akademische Verlagsgesellschaft.

Dodrell DM, Pegg DT and Bendall MR (1983) Correspondence between INEPT and DEPT pulse sequences for coupled spin-half nuclei. *Journal of Magnetic Resonance* 51: 264–269.

Ernst RR (1966) Nuclear magnetic double resonance with an incoherent radio-frequency field. *Journal of Chemical Physics* 45: 3845.

Ernst RR, Bodenhausen G and Wokaun A (1987) *Principles of Nuclear Magnetic Resonance in One and Two Dimensions*. Oxford: Clarendon Press.

Fürst A and Pretsch E (1990) A computer program for the prediction of ^{13}C-NMR chemical shifts of organic compounds. *Analytica Chimica Acta* 229: 17–25.

Hearmon RA (1986) Microcomputer-based ^{13}C NMR spectral simulation of substituted alkanes. *Magnetic Resonance in Chemistry* 24: 995–998.

INKA Database, Scientific Information Service, Larchmont, NY.

Levitt M, Freeman R and Frenkiel T (1983) Broadband decoupling in high-resolution nuclear magnetic resonance spectroscopy. In: Waugh JS (ed) *Advances in Magnetic Resonance*, Vol 11, pp 48–110. New York: Academic Press.

Lindley MR, Gray NAB, Smith DH, Djerassi C (1982) Applications of artificial intelligence for chemical inference. 40. Computerized approach to the verification of carbon-13 nuclear magnetic resonance spectral assignments. *Journal of Organic Chemistry* 47: 1027–1035.

Martin GE and Zektzer AS (1988) *Two-Dimensional NMR Methods for Establishing Molecular Connectivity*. New York: VCH.

Morris GA and Freeman R (1979) Enhancement of nuclear magnetic resonance signals by polarization transfer. *Journal of the American Chemical Society* 101: 760–762.

Munk ME and Christie BD (1989) The characterization of structure by computer. *Analytica Chimica Acta* 216: 57–75, and references therein.

Nakanishi K (ed) (1990) *One-Dimensional and Two-Dimensional NMR Spectra by Modern Pulse Techniques*. Tokyo: Kodansha.

Noggle JH and Schirmer RS (1971) *The Nuclear Overhauser Effect*. New York: Academic Press.

Paul EG and Grant DM (1964) Carbon-13 magnetic resonance II. Chemical shift data for the alkanes. *Journal of the American Chemical Society* 86: 2984–2990.

Pretsch E, Clerc JT, Seibl J and Simon W (1989) *Tables of Spectral Data for Structure Elucidation of Organic Compounds*, 2nd edn. Berlin: Springer.

SPECINFO Database, STN International, 2540 Olentangy River Rd., Columbus, OH 43210, USA.

Takahashi Y, Maeda S and Sasaki S (1987) Automated recognition of common geometrical patterns among a variety of three-dimensional molecular structures. *Analytica Chimica Acta* 200: 363–377, and references therein.

^{13}C NMR, Parameter Survey

R Duncan Farrant, GlaxoWellcome R&D, Stevenage, UK

John C Lindon, Imperial College of Science, Technology and Medicine, London, UK

Copyright © 1999 Academic Press

MAGNETIC RESONANCE
Applications

Introduction

High-resolution ^{13}C NMR spectroscopy is in widespread and routine use in many laboratories as a tool for the identification of organic compounds. This article is intended to provide a compilation of the chemical shifts and coupling constants available from ^{13}C NMR spectra and to show how these are related to molecular structures. ^{13}C NMR of organic molecules in solution is particularly powerful because ^{13}C chemical shifts, and spin–spin coupling involving ^{13}C nuclei generally fall into well defined ranges. Comprehensive tabulations of such values have been compiled (see Further reading) and more recently a number of computer-based predictive schemes have become available.

Table 1 Estimation of the ^{13}C chemical shifts in aliphatic compounds

Substituent	Increment Z_i for substituents in position			
	α	β	γ	δ
–H	0.0	0.0	0.0	0.0
–C< *	9.1	9.4	–2.5	0.3
–C=C–*	19.5	6.9	–2.1	0.4
–C≡C–	4.4	5.6	–3.4	–0.6
–Ph	22.1	9.3	–2.6	0.3
–F	70.1	7.8	–6.8	0.0
–Cl	31.0	10.0	–5.1	–0.5
–Br	18.9	11.0	–3.8	–0.7
–I	–7.2	10.9	–1.5	–0.9
–O–*	49.0	10.1	–6.2	0.3
–O–CO–	56.5	6.5	–6.0	0.0
–O–NO–	54.3	6.1	–6.5	–0.5
–N< *	28.3	11.3	–5.1	0.0
–N$^+$< *	30.7	5.4	–7.2	–1.4
–NH$_3^+$	26.0	7.5	–4.6	0.0
–NO$_2$	61.6	3.1	–4.6	–1.0
–NC	31.5	7.6	–3.0	0.0
–S–*	10.6	11.4	–3.6	–0.4
–S–CO–	17.0	6.5	–3.1	0.0
–SO–*	31.1	7.0	–3.5	0.5
–SO$_2$–*	30.3	7.0	–3.7	0.3
–SO$_2$Cl	54.5	3.4	–3.0	0.0
–SCN	23.0	9.7	–3.0	0.0
–CHO	29.9	–0.6	–2.7	0.0
–CO–	22.5	3.0	–3.0	0.0
–COOH	20.1	2.0	–2.8	0.0
–COO$^-$	24.5	3.5	–2.5	0.0
–COO–	22.6	2.0	–2.8	0.0
–CON<	22.0	2.6	–3.2	–0.4
–COCl	33.1	2.3	–3.6	0.0
–CS–N<	33.1	7.7	–2.5	0.6
–C=NOH syn,	11.7	0.6	–1.8	0.0
–C=NOH anti	16.1	4.3	–1.5	0.0
–CN	3.1	2.4	–3.3	–0.5

Steric corrections S:

Observed ^{13}C centre	Number of substituents other than H at the α–atom[a]			
	1	2	3	4
Primary (CH$_3$)	0.0	0.0	–1.1	–3.4
Secondary (CH$_2$)	0.0	0.0	–2.5	–6.0
Tertiary (CH)	0.0	–3.7	–8.5	–10.0
Quaternary (C)	–1.5	–8.0	–10.0	–12.5

[a] For α-substituents marked with asterisks above.

Conformation corrections K for γ-substitutents:

Conformation		K
Synperiplanar		–4.0
Synclinal		–1.0
Anticlinal		0.0
Antiperiplanar		2.0
Not fixed	–	0.0

From Pretsch E, Simon W, Seibl J and Clerc T (1989) *Spectral Data for Structure Determination of Organic Compounds*, 2nd edn. Reproduced with permission from Springer-Verlag.

Table 2 ^{13}C Chemical shifts for methyl groups

Substituent X	δ_{CH_3-X}	Substituent X	δ_{CH_3-X}	Substituent X	δ_{CH_3-X}	Substituent X	δ_{CH_3-X}
–H	–2.3	–cyclopentyl	20.5	–CH$_2$COCH$_3$	7.0	–Ocy	55.1
–CH$_3$	7.3	–cy	23.1	–CH$_2$COOH	9.6	–OCH=CH$_2$	52.5
–CH$_2$CH$_3$	15.4	–CH=CH$_2$	18.7	–OPh	54.8	–SO$_2$CH$_2$CH$_3$	39.3
–CH(CH$_3$)$_2$	24.1	–C≡CH	3.7	–OCOOCH$_3$	54.9	–SO$_2$Cl	52.6
–C(CH$_3$)$_3$	31.3	–phenyl	21.4	–OCOCH$_3$	51.5	–SO$_3$H	39.6
–(CH$_2$)$_6$CH$_3$	14.1	– α–naphthyl	19.1	–OCOcy	51.2	–SO$_3$Na	41.1
–CH$_2$Ph	15.7	–β –naphthyl	21.5	–OCOCH=CH$_2$	51.5	–CHO	31.2
–CH$_2$F	15.8	–2–pyridyl	24.2	–OCOPh	51.8	–COCH$_3$	30.7
–CH$_2$Cl	18.7	–3–pyridyl	18.0	–OSO$_2$OCH$_3$	59.1	–COCH$_2$CH$_3$	27.5
–CH$_2$Br	19.1	–4–pyridyl	20.6	–NH$_2$	28.3	–COCCl$_3$	21.1
–CH$_2$I	20.4	–2–furyl	13.7	–NH$_3^+$	26.5	–COCH=CH$_2$	25.7
–CHCl$_2$	31.6	–2–thienyl	14.7	–NHCH$_3$	38.2	–COcy	27.6
–CHBr$_2$	31.8	–2–pyrrolyl	11.8	–NHcy	33.5	–COPh	25.7
–CCl$_3$	46.3	–2–indolyl	13.4	–NHPh	30.2	–COOH	21.7
–CBr$_3$	49.4	–3–indolyl	9.8	–N(CH$_3$)$_2$	47.5	–COO–	24.4
–CH$_2$OH	18.2	–4–indolyl	21.6	–N–pyrrolidinyl	42.7	–COOCH$_3$	20.6
–CH$_2$OCH$_3$	14.7	–5–indolyl	21.5	–N–piperidinyl	47.7	–COOCOCH$_3$	21.8
–CH$_2$OCH$_2$CH$_3$	15.4	–6–indolyl	21.7	–N(CH$_3$)Ph	39.9	–CONH$_2$	22.3
–CH$_2$OCH=CH$_2$	14.6	–7–indolyl	16.6	–1–pyrrolyl	35.9	–CON(CH$_3$)$_2$	21.5
–CH$_2$OPh	14.9	–F	71.6	–1–imidazolyl	32.2	–COSH	32.6
–CH$_2$OCOCH$_3$	14.4	–Cl	25.6	–1–pyrazolyl	38.4	–COSCH$_3$	30.2
–CH$_2$NH$_2$	19.0	–Br	9.6	–1–indolyl	32.1	–COCOCH$_3$	23.2
–CH$_2$NHCH$_3$	14.3	–I	–24.0	–NHCOCH$_3$	26.1	–COCl	33.6
–CH$_2$N(CH$_3$)$_2$	12.8	–OH	50.2	–N(CH$_3$)CHO	36.5; 31.5	–CN	1.7
–CH$_2$NO$_2$	12.3	–OCH$_3$	60.9	–NC	26.8	–SC$_8$H$_{17}$	15.5
–CH$_2$SH	19.7	–OCH$_2$CH$_3$	57.6	–NCS	29.1	–SPh	15.6
–CH$_2$SO$_2$CH$_3$	6.7	–OCH(CH$_3$)$_2$	54.9	–NO$_2$	61.2	–SSCH$_3$	22.0
–CH$_2$SO$_3$H	8.0	–OC(CH$_3$)$_3$	49.4	–SH	6.5	–SOCH$_3$	40.1
–CH$_2$CHO	5.2	–OCH$_2$CH=CH$_2$	57.4	–SCH$_3$	19.3	–SO$_2$CH$_3$	42.6

From Pretsch E, Simon W, Seibl J and Clerc T (1989) *Spectral Data for Structure Determination of Organic Compounds*, 2nd edn. Reproduced with permission from Springer-Verlag.

Table 3   Effect of a substituent on the ^{13}C chemical shifts in vinyl compounds

Substituent X	z_1	z_2	Substituent X	z_1	z_2
–H	0.0	0.0	–OCH$_3$	29.4	–38.9
–CH$_3$	12.9	–7.4	–OCH$_2$CH$_3$	28.8	–37.1
–CH$_2$CH$_3$	17.2	–9.8	–OCH$_2$CH$_2$CH$_2$CH$_3$	28.1	–40.4
–CH$_2$CH$_2$CH$_3$	15.7	–8.8	–OCOCH$_3$	18.4	–26.7
–C(CH$_3$)$_2$	22.7	–12.0	–N(CH$_3$)$_2$	28.0*	–32.0*
–CH$_2$CH$_2$CH$_2$CH$_3$	14.6	–8.9	–N$^+$(CH$_3$)$_3$	19.8	–10.6
–C(CH$_3$)$_3$	26.0	–14.8	–N–pyrrolidonyl	6.5	–29.2
–CH$_2$Cl	10.2	–6.0	–NO$_2$	22.3	–0.9
–CH$_2$Br	10.9	–4.5	–NC	–3.9	–2.7
–CH$_2$I	14.2	–4.0	–SCH$_2$Ph	18.5	–16.4
–CH$_2$OH	14.2	–8.4	–SO$_2$CH=CH$_2$	14.3	7.9
–CH$_2$OCH$_2$CH$_3$	12.3	–8.8	–CHO	15.3	14.5
–CH=CH$_2$	13.6	–7.0	–COCH$_3$	13.8	4.7
–C=CH	–6.0	5.9	–COOH	5.0	9.8
–Ph	12.5	–11.0	–COOCH$_2$CH$_3$	6.3	7.0
–F	24.9	–34.3	–COCl	8.1	14.0
–Cl	2.8	–6.1	–CN	–15.1	14.2
–Br	–8.6	–0.9	–Si(CH$_3$)$_3$	16.9	6.7
–I	–38.1	7.0	–SiCl$_3$	8.7	16.1

From Pretsch E, Simon W, Seibl J and Clerc T (1989) *Spectral Data for Structure Determination of Organic Compounds*, 2nd edn. Reproduced with permission from Springer-Verlag.

Chemical shifts

Alkanes

¹³C NMR chemical shifts are referenced to that of tetramethylsilane (TMS) added as an internal standard and taken as 0.0 ppm. Often secondary standards are used and one common approach is to use the ¹³C NMR resonance of the organic solvent. For the two commonest NMR solvents, these have values relative to TMS of CDCl₃ at 77.5 ppm and dimethyl sulfoxide-d_6 at 39.5 ppm.

The ¹³C chemical shift of methane is at −2.3 ppm relative to TMS and from this base value it is possible to calculate shifts for other alkanes (and even substituted alkanes provided that the appropriate base shifts for the substituted methane are available) using the increment data given in **Table 1** together with the equation

$$\delta C = -2.3 + \sum Z_i + \sum S_j + \sum K_k \qquad [1]$$

where Z denotes the substituent effects, s values are included to take into account steric effects and K allows for conformations of γ-substituents.

Methyl groups can have a fairly large range of shift values and these can also be predicted using substituent effects as shown in **Table 2**. Rules have been given by Grant and co-workers for calculating ¹³C chemical shifts of methyl and ring carbons in cyclohexanes.

The ¹³C NMR chemical shifts for cyclohexanes can be calculated using similar parameters to those for the aliphatic compounds above.

Much study has been devoted to halogenated alkanes including fluorinated compounds and predictive rules have been derived.

Alkenes

The ¹³C shifts for carbons of double bonds generally range from 80 to 160 ppm and they can again be estimated from empirical rules based on substituent effects. Thus the shifts of double bond carbons can be estimated from the equation

$$\delta = 123.3 + z_i \qquad [2]$$

where z_1 is an increment for substitution on the *ipso* carbon and z_2 is the effect of a substituent on the vicinal carbon. The substituent effects are given in **Table 3**.

Carbons involved in keto–enol tautomerism can experience large chemical shift differences according to the tautomer present. For example, for acetylacetone the carbonyl carbon resonates at 201.1 ppm in the keto form but is at 190.5 ppm in the enol form. Similarly, the carbon which is a methylene group in the keto form with a shift of 56.6 ppm moves to 99.0 ppm as an olefinic carbon in the enol form.

Ring strain can induce significant effects on ¹³C shifts. For example, the olefine carbons in cyclopropene are at 108.7 ppm, in cyclobutene they are at 137.2 ppm, in cyclopentene they appear at 130.8 ppm and in cyclohexene at 127.4 ppm.

Conjugation can also have large effects. The inner olefine carbon in butadiene is at 136.9 ppm, whilst the central carbon in allene is at 213.5 ppm with the outer carbon at 74.8 ppm.

Alkynes

Acetylene has a ¹³C shift of 71.9 ppm and the effects of substituents have been evaluated. These are given in **Table 4**.

Aromatic hydrocarbons

Benzene has a chemical shift of 128.5 ppm and the 1-, 2- and 4a-positions of naphthalene have shifts of 128.0, 126.0 and 133.7 ppm, respectively. Well

Table 4   ¹³C Chemical shifts in alkynes (H–C$_b$≡C$_a$–X)

X	a	b
–H	71.9	71.9
–CH₃	80.4	68.3
–CH₂CH₃	85.5	67.1
–CH₂CH₂CH₃	84.0	68.7
–CH₂CH₂CH₂CH₃	83.0	66.0
–CH(CH₃)₂	89.2	67.6
–C(CH₃)₃	92.6	66.8
–cy	88.7	68.3
–CH₂OH	83.0	73.8
–CH=CH₂	82.8	80.0
–C≡C–CH₃	68.8	64.7
–Ph	84.6	78.3
–OCH₂CH₃	88.2	22.0
–SCH₂CH₃	72.6	81.4
–CHO	81.8	83.1
–COCH₃	81.9	78.1
–COOH	74.0	78.6
–COOCH₃	74.8	75.6

From Pretsch E, Simon W, Seibl J and Clerc T (1989) *Spectral Data for Structure Determination of Organic Compounds*, 2nd edn. Reproduced with permission from Springer-Verlag.

Table 5   Effect of substituents on the ^{13}C chemical shifts in monosubstituted benzenes

Substituent X	z_1	z_2	z_3	z_4	Substituent X	z_1	z_2	z_3	z_4
–H	0.0	0.0	0.0	0.0	–N$^+$(CH$_3$)$_3$	19.5	–7.3	2.5	2.4
–CH$_3$	9.2	0.7	–0.1	–3.0	–NHCOCH$_3$	9.7	–8.1	0.2	–4.4
–CH$_2$CH$_3$	15.7	–0.6	–0.1	–2.8	–NHNH$_2$	22.8	–16.5	0.5	–9.6
–CH(CH$_3$)$_2$	20.2	–2.2	–0.3	–2.8	–N=N–Ph	24.0	–5.8	0.3	2.2
–CH$_2$CH$_2$CH$_2$CH$_3$	14.2	–0.2	–0.2	–2.8	–N$^+$≡N	–12.7	6.0	5.7	16.0
–C(CH$_3$)$_3$	22.4	–3.3	–0.4	–3.1	–NC	–1.8	–2.2	1.4	0.9
–cyclopropyl	15.1	–3.3	–0.6	–3.6	–NCO	5.1	–3.7	1.1	–2.8
–CH$_2$Cl	9.3	0.3	0.2	0.0	–NCS	3.0	–2.7	1.3	–1.0
–CH$_2$Br	9.5	0.7	0.3	0.2	–NO	37.4	–7.7	0.8	7.0
–CF$_3$	2.5	–3.2	0.3	3.3	–NO$_2$	19.9	–4.9	0.9	6.1
–CCl$_3$	16.3	–1.7	–0.1	1.8	–SH	2.1	0.7	0.3	–3.2
–CH$_2$OH	12.4	–1.2	0.2	–1.1	–SCH$_3$	10.0	–1.9	0.2	–3.6
–CHOCH$_3$	9.2	–3.1	–0.1	–0.5	–SC(CH$_3$)$_3$	4.5	9.0	–0.3	0.0
–CH$_2$NH$_2$	14.9	–1.4	–0.2	–2.0	–SPh	7.3	2.5	0.6	–1.5
–CH$_2$SCH$_3$	9.8	0.4	–0.1	–1.6	–SOCH$_3$	17.6	–5.0	1.1	2.4
–CH$_2$SOCH$_3$	0.8	1.5	0.4	–0.2	–SO$_2$CH$_3$	12.3	–1.4	0.8	5.1
–CH$_2$CN	1.6	0.5	–0.8	–0.7	–SO$_2$Cl	15.6	–1.7	1.2	6.8
–CH=CH$_2$	8.9	–2.3	–0.1	–0.8	–SO$_3$H	15.0	–2.2	1.3	3.8
–C≡CH	–6.2	3.6	–0.4	–0.3	–SO$_2$OCH$_3$	6.4	–0.6	1.5	5.9
–Ph	13.1	–1.1	0.5	–1.1	–SCN	–3.7	2.5	2.2	2.2
–F	34.8	–13.0	1.6	–4.4	–CHO	8.2	1.2	0.5	5.8
–Cl	6.3	0.4	1.4	–1.9	–COCH$_3$	8.9	0.1	–0.1	4.4
–Br	–5.8	3.2	1.6	–1.6	–COCF$_3$	–5.6	1.8	0.7	6.7
–I	–34.1	8.9	1.6	–1.1	–COPh	9.3	1.6	–0.3	3.7
–OH	26.9	–12.8	1.4	–7.4	–COOH	2.1	1.6	–0.1	5.2
–ONa	39.6	–8.2	1.9	–13.6	–COO$^-$	9.7	4.6	2.2	4.6
–OCH$_3$	31.4	–14.4	1.0	–7.7	–COOCH$_3$	2.0	1.2	–0.1	4.3
–OCH=CH$_2$	28.2	–11.5	0.7	–5.8	–CONH$_2$	5.0	–1.2	0.1	3.4
–OPh	27.6	–11.2	–0.3	–6.9	–CON(CH$_3$)$_2$	8.0	–1.5	–0.2	1.0
–OCOCH$_3$	22.4	–7.1	0.4	–3.2	–COCl	4.7	2.7	0.3	6.6
–OSi (CH$_3$)$_3$	26.8	–8.4	0.9	–7.1	–CSPh	18.7	1.0	–0.6	2.4
–OCN	25.0	–12.7	2.6	–1.0	–CN	–15.7	3.6	0.7	4.3
–NH$_2$	18.2	–13.4	0.8	–10.0	–P(CH$_3$)$_2$	13.6	1.6	–0.6	–1.0
–NHCH$_3$	21.4	–16.2	0.8	–11.6	–P(Ph)$_2$	8.9	5.2	0.0	0.1
–N(CH$_3$)$_2$	22.5	–15.4	0.9	–11.5	–PO(OCH$_2$CH$_3$)$_2$	1.6	3.6	–0.2	3.4
–NHPh	14.7	–10.6	0.9	–10.5	–SiH$_3$	–0.5	7.3	–0.4	1.3
–N(Ph)$_2$	19.8	–7.0	0.9	–5.6	–Si(CH$_3$)$_3$	11.6	4.9	–0.7	0.4
–NH$_3$$^+$	0.1	–5.8	2.2	2.2					

From Pretsch E, Simon W, Seibl J and Clerc T (1989) *Spectral Data for Structure Determination of Organic Compounds*, 2nd edn. Reproduced with permission from Springer-Verlag.

defined substituent effects on aromatic carbon shifts in benzene derivatives have been derived. These are calculated according to the equation

$$\delta = 128.5 + \sum z_i \qquad [3]$$

and the substituent effects for *ortho*, *meta* and *para* positions are listed in **Table 5**. Similar data compilations are available for naphthalene derivatives and for pyridines. The naphthalene substituent parameters can be used with a reasonable degree of success in other condensed ring hydrocarbons provided that the base values for a particular ring system are available. These have been given for a number of ring systems, including benzofuran, benzthiophen, benzimidazole, quinoline, isoquinoline and many others (see Further reading).

The ^{13}C shifts for purine and pyrimidine bases have also been measured and these form useful base values for studies of nucleosides, nucleotides and nucleic acids.

Oxygen-containing compounds

Alcohols show large chemical shifts according to the proximity of the oxygen to the measured carbon. Thus *n*-propanol has shifts of 64.2, 25.9 and 10.3 ppm for the carbons α, β and γ to the OH, respectively. Ethylene glycol has a shift of 63.4 ppm, whilst glycerol has shifts of 64.5 and 73.7 ppm for the CH_2 and CH carbons, respectively. Trifluoroethanol, often used in protein and peptide NMR studies, has carbon shifts of 61.4 and 125.1 ppm for the CH_2 and CF_3 carbons, respectively, with $^1J_{CF} = 278$ Hz and $^2J_{CCF} = 35$ Hz (see later). Ethers show similar substituent effects to alcohols. For cyclic ethers, ring strain again induces large chemical shift changes. Thus for ethylene oxide the CH_2 shift is 39.5 ppm whereas for the four-membered ring analogue the shift is 72.6 ppm.

Extensive data compilations are available for monosaccharides and other sugars, including the tabulation by Bock.

Amines

The protonation of amines causes a shielding of carbons adjacent to the nitrogen except for branched systems, where some deshielding is often seen. The effect of protonation can be about −2 ppm at an α-carbon, −3 ppm at a β-carbon and up to −1 ppm at a γ-carbon. For example, the ¹³C shifts of ethylamine are 36.9 ppm (CH_2) and 19.0 (CH_3) and these are shifted by −0.2 and −5.0 ppm, respectively, on protonation. Again, ring strain effects in cyclic amines are mirrored by large ¹³C shift effects.

Amino acids

¹³C NMR data for amino acids have been tabulated extensively as these provide base values for studies of NMR spectra of peptides and proteins. The data compilation by Pretsch and co-workers is comprehensive, as are the results shown in the book by Wuthrich. Values are very pH dependent.

Other functional groups

Nitrile carbon shifts are in the range of 115–125 ppm whereas in isonitriles the shifts are around 155–165 ppm. In oximes, the carbon in the C=N bond appears around 150–160 ppm and in isocyanates the carbon shift is in the region of 120–125 ppm. For hydrazones with the functionality C=N–N, the ¹³C shifts are usually around 165 ppm. Methylene carbons adjacent to nitro groups generally appear near 70–80 ppm, and for methylene groups in nitrosoamines these appear around 45–55 ppm.

The SH group induces much smaller shift changes. The CH_2 resonance of carbons adjacent to a thiol group appear around 24–35 ppm according to the presence of other substituents and chain branching. Similar effects are seen in thioethers. Sulfoxides of the type RR'SO are chiral because of the pyramidal structure of the sulfoxide group and hence nonequivalence can be seen in the NMR spectra. Methylene carbons adjacent to sulfone groups appear around 55–65 ppm.

Carbonyl carbons

For the fragment, Cβ–Cα–CO–Cα–Cβ, it is possible to estimate the carbonyl carbon chemical shift from the substituent effects given in **Table 6A** using the equation

$$\delta = 193.0 + \sum z_i \qquad [4]$$

For carboxylic acids and esters, Equation [4] and **Table 6B** apply with a base shift of 166.0 ppm. For amides, the base shift is 165.0 ppm and the substituent effects in **Table 6C** are used.

Table 6 Additivity rules for estimating the chemical shift for carbonyl groups

(A) Aldehydes and ketones:

Substituent i	z_α	z_β
–C≤	6.5	2.6
–CH=CH₂	−0.8	0.0
–CH=CH–CH₃	0.2	0.0
–Ph	−1.2	0.0

(B) Carboxylic acids and esters:

Substituent i	z_α	z_β	z_γ	$z_{\alpha'}$
–C≤	11.0	3.0	−1.0	−5.0
–CH=CH₂	5.0			−9.0
–Ph	6.0	1.0		−8.0

(C) Amides:

Substituent i	z_α	z_β	z_γ	$z_{\alpha'}$	$z_{\beta'}$
–C≤	7.7	4.5	−0.7	−1.5	−0.3
–CH=CH₂	3.3				
–Ph	4.7			−4.5	

From Pretsch E, Simon W, Seibl J and Clerc T (1989) *Spectral data for Structure Determination of Organic Compounds*, 2nd edn. Reproduced with permission from Springer-Verlag.

Aldehyde carbons have a similar range of chemical shifts as ketones; for example, the shift for cyclohexane-1-aldehyde is 204.7 ppm. Cyclic ketones tend to be more deshielded than alicyclic ketones. For example, methyl ethyl ketone has a carbonyl shift of 207.6 ppm but cyclohexanone is at 209.7 ppm. The carbonyl group of lactones is highly shielded relative to other carbonyl carbons, with an upfield shift of around 20 ppm. Similarly, amide and lactam carbons are in the region of 165–175 ppm. The carbonyl carbons of anhydrides are also similarly shielded.

Coupling constants

One-bond J_{CH}

An extensive review of these is now available (see Further reading). In general, these lie in the range 125–170 Hz but there are a number of values beyond this upper figure. The value can be estimated in substituted methanes of the type $CHZ_1Z_2Z_3$ using Equation [5] and the substituent effects in **Table 7**.

$$J_{CH} = 125.0 + \sum z_i \qquad [5]$$

Two-bond J_{CH}

These couplings can be small and vary considerably. For example, for the fragment ^{13}C–C–H, J_{CH} is 1–6 Hz, for ^{13}C=C–H it is 0–16 Hz and for ^{13}C–(C=O)–H it is 20–50 Hz.

Three-bond J_{CH}

These generally lie in the range 0–10 Hz but they depend on the dihedral angle in similar fashion to proton–proton coupling constants. Thus, when the ^{13}C and ^1H are in bonds which are eclipsed, the coupling is around 6 Hz, when they are antiperiplanar, the coupling is about 9 Hz and when the bonds are at right-angles, the coupling drops to close to zero.

In alkenes, the three-bond *trans* coupling is always larger than the *cis* coupling for a pair of *cis–trans* isomers. Typical values are 7–10 Hz for a *cis* coupling and 8–13 Hz for a *trans* coupling.

Three-bond couplings in aromatic molecules are generally around 7–10 Hz.

Other couplings involving ^{13}C

One-bond ^{13}C–^{13}C couplings have a typical value of around 30–40 Hz in unstrained, unsubstituted

Table 7 Additivity rule for estimating the ^{13}C–^1H coupling constants in aliphatic compounds

Substituent	Increments z_i
–H	0.0
–CH_3	1.0
–C(CH_3)_3	–3.0
–CH_2Cl	3.0
–CH_2Br	3.0
–CH_2I	7.0
–CHCl_2	6.0
–CCl_3	9.0
–C≡CH	7.0
–Ph	1.0
–F	24.0
–Cl	27.0
–Br	27.0
–I	26.0
–OH	18.0
–OPh	18.0
–NH_2	8.0
–NHCH_3	7.0
–N(CH_3)_2	6.0
–SOCH_3	13.0
–CHO	2.0
–COCH_3	–1.0
–COOH	5.5
–CN	11.0

From Pretsch E, Simon W, Seibl J and Clerc T (1989) *Spectral Data for Structure Determination of Organic Compounds*, 2nd edn. Reproduced with permission from Springer-Verlag.

systems. Ring strain causes a reduction in coupling constant, the value in cyclopropane, for example, being only 12.4 Hz. Couplings in conjugated systems tend to be higher, an extreme example being acetylene with a value of 171.5 Hz. ^{13}C–^{13}C coupling through two or three bonds can also be measured. Values are typically 1–5 Hz.

^{13}C–^{19}F coupling constants can also often be measured. The one-bond coupling is around 280–300 Hz depending on substituents, with the two-bond coupling also being of considerable magnitude, around 20 Hz. Longer range ^{13}C–^{19}F couplings are also observed, especially in aromatic and conjugated molecules. In fluorobenzene $^1J_{CH}$ is 245.1 Hz, $^2J_{CH}$ is 21.0 Hz, $^3J_{CH}$ is 7.8 Hz and $^4J_{CH}$ is 3.2 Hz.

In addition, a number of studies have made use of ^{13}C–^{15}N and ^{13}C–^{31}P coupling constants.

List of symbols

J = coupling constant; K = conformations of γ-substituents; s = steric effects; Z = substituent effects; δ = chemical shift.

See also: **Parameters in NMR Spectroscopy, Theory of; Structural Chemistry Using NMR Spectroscopy, Organic Molecules; Structural Chemistry Using NMR Spectroscopy, Peptides; Structural Chemistry Using NMR Spectroscopy, Pharmaceuticals.**

Further reading

Bock K and Pedersen C (1983) ¹³C NMR spectroscopy of monosaccharides. *Advances in Carbohydrate Chemistry and Biochemistry* 41: 27–66.

Bock K, Pedersen C and Pedersen H (1984) ¹³C NMR spectroscopy of oligosaccharides. *Advances in Carbohydrate Chemistry and Biochemistry* 42: 193–225.

Bruhl TS, Heilmann D and Kleinpeter E (1997) ¹³C NMR chemical shift calculations for some substituted pyridines. A comparative consideration. *Journal of Chemical Information and Computer Science* 37: 726–730.

Dalling DK and Grant DM (1972) Carbon-13 magnetic resonance. XXI. Steric interactions in the methylcyclohexanes. *Journal of the American Chemical Society* 94: 5318–5324.

Grant DM and Paul EG (1964) Carbon-13 magnetic resonance: chemical shift data for the alkanes. *Journal of the American Chemical Society* 86: 2984–2995.

Hansen PE (1981) Carbon-hydrogen spin–spin coupling constants. *Progress in Nuclear Magnetic Spectroscopy* 14: 175–296.

Kalinowski H-O, Berger S and Braun S (1988) *Carbon-13 NMR Spectroscopy*. New York: Wiley.

Ovenall DW and Chang JJ (1977) Carbon-13 NMR of fluorinated compounds using wide-band fluorine decoupling. *Journal of Magnetic Resonance* 25: 361–372.

Pathre S (1996) Drawing structures and calculating ¹³C NMR spectra-ACD/CNMR. *Analytical Chemistry* 68: A740–A741.

Pretsch E, Simon W, Seibl J and Clerc T (1989) *Spectral Data for Structure Determination of Organic Compounds*, 2nd edn. Berlin: Springer.

Sarneski JE, Surprenant HL, Molen FK and Reilley CN (1975) Chemical shifts and protonation shifts in carbon-13 nuclear magnetic resonance studies of aqueous amines. *Analytical Chemistry* 47: 2116–2124.

Van Bramer S (1997) ACD/CMR and ACD/HNMR spectrum prediction software. *Concepts in Magnetic Resonance* 9: 271–273.

Wuthrich K (1986) *NMR of Proteins and Nucleic Acids*. New York: Wiley.

Cadmium NMR, Applications

See **Heteronuclear NMR Applications (Y–Cd).**

Caesium NMR Spectroscopy

See **NMR Spectroscopy of Alkali Metal Nuclei in Solution.**

Calibration and Reference Systems (Regulatory Authorities)

C Burgess, Burgess Consultancy, Barnard Castle,
Co Durham, UK

FUNDAMENTALS OF SPECTROSCOPY

Methods & Instrumentation

Overview

Spectroscopic measurements are at the heart of many analytical procedures and processes. Proper and adequate calibration and qualification of spectrometers is an essential part of the process for ensuring data integrity. Wherever practicable, the standards used should be traceable to a recognized national or international standard. The International Organization for Standardization (ISO) 9000 series of guides are the international standard for quality systems. Quality of measurement is central to compliance: for example ISO 9001 Section 4.11 requires that:

> …the user shall identify, calibrate and adjust all inspection, measuring and test equipment and devices that can affect product quality at prescribed intervals, or prior to use, against certified equipment having a known valid relationship to nationally recognised standards.

More detailed information on the compliance requirements for the competence of calibration and testing laboratories is contained in ISO Guide 25. This guide has been interpreted to meet national requirements in some countries: for example the UK Accreditation Service refers to it as UKAS M10. Many laboratories seek accreditation to ISO Guide 25 and operate quality and testing systems which meet the requirements. The ISO Guide 25 approach heavily focuses on good analytical practices and adequate calibration of instruments with nationally or internationally traceable standards wherever possible.

In addition to the ISO approach, environmental measurements (for example by the US Environmental Protection Agency) and the pharmaceutical industry are subject to rigorous regulatory requirements. The pharmaceutical industry has placed great emphasis on method validation in, for example, HPLC (see Further Reading section for details). However, until recently, there has been little specific regulatory requirement for assuring that the analytical instruments are working properly.

The US Food and Drug Administration (FDA) specifically requires that:

> Laboratory controls shall include: …

The calibration of instruments, apparatus, gauges, and recording devices at suitable intervals in accordance with an established written program containing specific directions, schedules, limits for accuracy and precision, and provisions for remedial action in the event accuracy and/or precision limits are not met. Instruments, apparatus, gauges, and recording devices not meeting established specifications shall not be used.

The major regulatory guidelines for Good Manufacturing Practice (GMP) and Good Laboratory Practice (GLP) are similarly vague. 'Fitness for purpose' is the phrase that is commonly used, but what does this mean in practice?

Only the Pharmacopoeias (for example the US Pharmacopoeia (1995) and the British Pharmacopoeia (1998)), and the Australian Regulatory Authority have been sufficiently worried by instrumental factors to give written requirements for instrument performance. Whilst these guidelines are not entirely consistent at least they are attempting to ensure consistency of calibration practices between laboratories.

However, there has been a resurgence of interest in the underlying data quality by regulatory authorities particularly the FDA following a major legal ruling involving Barr Laboratories, Inc. in 1993. This legal case has changed the regulatory focus and put the laboratory firmly in the spotlight in terms of the assurance of the quality of the data it produces.

Up to this point we have focused on the spectrometer and the underlying data quality. However from an analytical science perspective, process is important as illustrated in **Figure 1**. Scientific knowledge is founded upon relevant and robust information generated from reliable data.

The overall analytical process is illustrated in **Figure 2**.

Quality has to be built in at all stages of the process, for failure to ensure integrity at any level invalidates results from further processing. One of the key areas not covered by this article is the validation of the software and systems involved in instrument control and data transformation. Clearly this must be addressed as part of the overall assessment for the 'fitness for purpose' of any computerized spectrometer.

Table 1 American Society for Testing and Materials (ASTM) standards relating to spectrometry and spectrometer performance

ASTM Standard	Title
E 131-95	Standard Terminology Relating to Molecular Spectroscopy
E 168-92	Standard Practices for General Techniques of Infrared Quantitative Analysis
E 169-93	Standard Practices for General Techniques of Ultraviolet-Visible Quantitative Analysis
E 275-93	Standard Practice for Describing and Measuring Performance of Ultraviolet, Visible and Near-Infrared Spectrophotometers
E 386-90	Standard Practice for Data Presentation Relating to High Resolution Nuclear Magnetic Resonance (NMR) Spectroscopy
E 387-84	Standard Test Method for Estimating Stray Radiant Power Ratio of Spectrophotometers by the Opaque Filter Method
E 573-96	Standard Practices for Internal Reflection Spectroscopy
E 578-83	Standard Test Method for Linearity of Fluorescence Measuring Systems
E 579-84	Standard Test Method for Limit of Detection of Fluorescence of Quinine Sulphate
E 925-94	Standard Practice for the Periodic Calibration of Narrow Band-pass Spectrophotometers
E 932-93	Standard Practice for Describing and Measuring Performance of Dispersive Infrared Spectrometers
E 958-93	Standard Practice for Measuring Practical Spectral Bandwidth of Ultraviolet-Visible Spectrophotometers
E 1252-94	Standard Practice for General Techniques of Qualitative Infrared Analysis
E 1421-94	Standard Practice for Describing and Measuring Performance of Fourier Transform-Infrared Spectrometers; Level Zero and Level One Tests
E 1642-94	Standard Practices for General Techniques of Gas Chromatography Infrared (GC-IR) Analysis
E 1655-94	Standard Practices for Infrared, Multivariate Quantitative Analysis
E 1683-95a	Standard Practice for Testing the Performance of Scanning Raman Spectrometers
E 1790-96	Standard Practice for Near Infrared Qualitative Analysis
E 1791-96	Standard Practice for Transfer Standards for Reflectance Factor for Near-Infrared Instruments Using Hemispherical Geometry
E 1840-96	Standard Guide for Raman Shift Standards for Spectrometer Calibration
E 1865-97	Standard Guide for Open-path Fourier Transform Infrared (OP/FT-IR) Monitoring of Gases and Vapours in Air
E 1866-97	Standard Guide for Establishing Spectrophotometer Performance Tests

Note: These standards are available from ASTM, 100 Barr Harbor Drive, West Conshohocken, PA 19428, USA

Table 2 Summary of pharmacopoeial requirements for the calibration of IR and NIR spectrometers

Monograph	Authority	Test	Standard	Measurement process	Acceptance criteria
IR	Ph. Eur. 1997 Method 2.2.24 BP 1998 IIA	Resolution	0.05 mm polystyrene film 0.038 mm polystyrene film	% transmittance differences at 2870 and 2871 cm^{-1} and at 1589 and 1583 cm^{-1}	Greater than 18 and 12 respectively
	Ph. Eur 1997. Method 2.2.24	Wavenumber accuracy	0.05 mm polystyrene film	Location of transmittance minima 3027.1 (± 0.3) 1583.1 (± 0.3) 2924 (± 2) 1181.4 (± 0.3) 2850.7 (± 0.3) 1154.3 (± 0.3) 1994 (± 1) 1069.1 (± 0.3) 1871.0 (± 0.3) 1028.0 (± 0.3) 1801.0 (± 0.3) 906.7 (± 0.3) 1601.4 (± 0.3) 698.9 (± 0.5)	Not specified. The uncertainties quoted in the table reflect the uncertainities in the quoted values and not the acceptance criteria for the spectrometer
	USP 23 (851) 1995	Wavenumber accuracy	Polystyrene, water vapour, carbon dioxide, ammonia gas	Location of transmittance minima	Not specified
NIR	Ph. Eur. 1997 Method 2.2.40	Wavelength scale accuracy	Polystyrene or rare earth oxides	Location of absorbance or reflectance minima	Not specified
		Wavelength scale repeatability	Polystyrene or rare earth oxides	Location of absorbance or reflectance minima	Standard deviation is consistent with spectrometer specification
		Response scale repeatability	Reflective thermoplastic resins doped with carbon black	Measurement of reflectance values	Standard deviation is consistent with spectrometer specification
		Photometric noise	Reflective thermoplastic resins doped with carbon black or reflective ceramic tile	Peak to peak or at a given wavelength	The photometric noise is consistent with spectrometer specification

Table 3 Summary of pharmacopoeial requirements for the calibration of UV-visible spectrometers and polarimeters

Monograph	Authority	Test	Standard	Measurement process	Acceptance criteria
UV-VIS	Ph. Eur. 1997 Method 2.2.25	Wavelength scale accuracy	4 % holmium oxide in 1.4 M perchloric acid (Ho) or atomic lines from deuterium (D), hydrogen (H) or mercury vapour arc (Hg)	Location of absorbance maxima 241.15 nm (Ho) 404.66 nm (Hg) 253.7 nm (Hg) 435.83 nm (Hg) 287.15 nm (Ho) 486.0 nm (Dβ) 302.25 nm (Hg) 486.1 nm (Hβ) 313.16 nm (Hg) 536.3 nm (Ho) 334.15 nm (Ho) 546.07 nm (Hg) 316.5 nm (Ho) 576.96 nm (Hg) 365.48 nm (Hg) 579.07 nm (Hg)	±1 nm 200 to 400 nm ±3 nm 401 to 700 nm
	USP 23 ⟨851⟩ 1995		NIST SRM[a] 2034, holmium oxide solution	As required by certificate	±1 nm
	Ph. Eur. 1997 Method 2.2.25	Absorbance scale accuracy	57.0 to 63.0 mg of potassium dichromate per litre of 0.005 M sulfuric acid solution	Wavelength nm / A (1%, 1 cm): 235 / 124.0 257 / 144.0 313 / 48.6 350 / 106.6	Wavelength nm / Tolerance: 235 / 122.9 to 126.2 257 / 142.4 to 145.7 313 / 47.0 to 50.3 350 / 104.9 to 109.2
	USP 23 ⟨851⟩	Absorbance scale accuracy	NIST SRM 931 liquid filters NIST SRM 930 glass filters	As required by certificate	Not specified but implied within NIST range of values
		Resolution	0.02 % v/v toluene in hexane	Absorbance ratio at the maximum at 269 nm to the minimum at 266 nm	Greater than 1.5
		Stray light	1.2 % w/v potassium chloride in water	Absorbance of 10 mm pathlength at 200 ± 2 nm	Greater than 2
	BP 1998 IIB	Second derivative	0.020% solution of toluene in methanol	Record the second derivative spectrum between 255 nm and 275 nm for a 10 mm path length	A small negative extremum located between two large negative extrema at about 261 and 268 nm should be clearly visible
Polarimeter	Ph. Eur. 1997 Method 2.2.7 USP 23 ⟨781⟩ 1995	Accuracy of rotation	Quartz plate	Not specified	Not specified
	USP 23 ⟨781⟩ 1995		NIST traceable quartz plate	Not specified	Not specified but implied within NIST range of values
	Ph. Eur. 1997 Method 2.2.7 USP 23 ⟨781⟩ 1995	Linearity of scale	Sucrose solutions	Not specified	Not specified
	USP 23 ⟨781⟩ 1995		NIST SRM 17 sucrose NIST SRM 41 dextrose	Not specified	Not specified but implied within NIST range of values

[a] National Institute for Science and Technology, Standard Reference Material (USA).

Table 4 Summary of pharmacopoeial requirements for the calibration of NMR and CD spectrometer

Monograph	Authority	Test	Standard	Measurement process	Acceptance criteria
NMR	Ph. Eur. 1997 Method 2.2.33	Resolution	20% v/v 1,2 dichlorobenzene in deuterated acetone or 5% v/v tetramethylsilane in deuterochloroform	Peak width at half height of the band at δ7.33 ppm or δ7.51 ppm of the symmetrical multiplet for dichlorobenzene or Peak width at half height of the band at δ0.00 ppm for tetramethylsilane.	Less than or equal to 0.5 Hz
		Signal to noise ratio	1% v/v ethylbenzene in carbon tetrachloride	Measure the spectrum over range δ2 to 5 ppm. Measure the peak amplitude, A, of the largest peak of the methylene quartet centred at δ2.65 ppm and the peak to peak amplitude of the baseline noise between δ4 and δ5 ppm. The amplitude is measured from a baseline constructed from the centre of the noise on either side of the quartet and at a distance of at least 1 ppm from its centre.	The S/N ratio is given by 2.5 (A/H) which has to be greater than 25:1 for a mean of five measurements
		Side band amplitude			Amplitude of spining side bands is not greater than 2% of the sample amplitude at the rotational speed used
		Integrator response repeatability	5% v/v ethylbenzene in carbon tetrachloride	For quantitative measurements, carry out five successive scans of the protons of the phenyl and ethyl groups and determine the mean values.	No Individual value shall differ by more than 2.5% from the mean
CD	Ph. Eur. 1997 Method 2.2.41	Absorbance accuracy	1.0 mg ml^{-1} isoandrosterone	Record the spectrum between 280 and 360 nm and determine the minimum at 304 nm	$\Delta\varepsilon$ is +3.3
		Linearity of modulation	1.0 mg ml^{-1} (1S)-(+)-camphorsulfonic acid	Record the spectrum between 185 and 340 nm and determine the minima at 192.5 and 290.5 nm	At 192.5 nm, $\Delta\varepsilon$ is −4.3 to −5 At 290.5 nm, $\Delta\varepsilon$ is +2.2 to +2.5

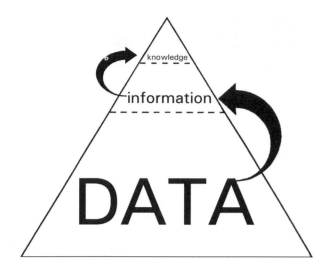

Figure 1 Data, information and knowledge triangle.

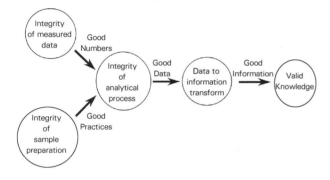

Figure 2 Analytical process integrity.

This article is restricted to the calibration requirements for spectrometers within the regulatory context which from a practical standpoint means the major pharmacopoeias.

Spectroscopic coverage

Current spectrometric monographs include calibration requirements for: infrared (IR), near infrared (NIR), ultraviolet and visible (UV-visible), nuclear magnetic resonance (NMR), circular dichroism and polarimetry. Monographs for atomic spectrophotometry, emission and absorption, fluorescence and X-ray fluorescence are available in the European Pharmacopoeia but do not specify calibration requirements other than that they are to be operated in accordance with the manufacturer's instructions. A series of ASTM standards are available and are listed in **Table 1**. Whilst primarily intended for use by instrument vendors and for instrumental fitness for purpose for ASTM analytical procedures, they are sometimes useful in supplementing pharmacopoeial guidance.

Tables 2–4 summarize the current requirements of the European Pharmacopoeia (Ph. Eur.), the British Pharmacopoeia (BP) and the United States Pharmacopoeia (USP).

See also: **Food and Dairy Products, Applications of Atomic Spectroscopy; Food Science, Applications of Mass Spectrometry; Food Science, Applications of NMR Spectroscopy; Laboratory Information Management Systems (LIMS); Pharmaceutical Applications of Atomic Spectroscopy.**

Further reading

NIS 45, Edition 2, May 1996, Accreditation for Chemical Laboratories.

Environmental Protection Agency, 2185-Good Automated Laboratory Practices, 1995.

International Conference on Harmonisation Note for guidance on validation of analytical procedures: methodology, 1995, CPMP/ICH/281/95.

FDA Center for Drug Evaluation and Research, Reviewer Guidance; Validation of Chromatographic Methods, November 1994.

21 CFR 211 SUBPART I LABORATORY CONTROLS §211.160 (b) (4).

e.g. United States Pharmacopoeia 23, 1995, <621> chromatography, <711> dissolution testing, <831> Refractive index, <851> spectrophotometry and light scattering.

e.g. BP 1998, Appendix II; Infrared spectrophotometry, Ultraviolet and Visible spectrophotometry and Pharm. Eur. V.6.19 Absorption Spectrophotometry.

Australian Code of GMP for Therapeutic Goods—Medicinal Products; Appendix D, November 1991; Guidelines for Laboratory Instrumentation.

DEPARTMENT OF HEALTH AND HUMAN SERVICES Food and Drug Administration [Docket No. 93N-0184] Barr Laboratories, Inc.; Refusal to Approve Certain Abbreviated Applications; Opportunity for a Hearing, Vol. 58 No. 102 Friday, May 28, 1993 p 31035 (Notice) 1/1061.

Carbohydrates Studied By NMR

Charles T Weller, School of Biomedical Sciences,
University of St Andrews, UK

Introduction

Carbohydrates, the most widely abundant biological molecules, are key components within a wide variety of biological phenomena. Polysaccharides such as glycogen, cellulose and starch have important structural or nutritional roles, but it is the mediation of specific recognition events by carbohydrates that has sparked detailed analyses of structure, conformation and function. The non-invasive, non-destructive nature of the technique is commonly mentioned as the most important reason for analysing carbohydrates by NMR, and is of considerable advantage when dealing with small amounts of precious material. This advantage is, however, offset by the relative insensitivity of the technique when compared with other methods such as mass spectrometry or enzymatic approaches that may be more appropriate for specific problems of sequence determination or compositional analysis. The value of NMR analysis is apparent in the additional data obtainable regarding composition, structure, conformation and mobility necessary for understanding recognition processes. The analysis of carbohydrates by NMR is characterized by a modification of existing techniques to address the particular problems of oligosaccharide spectra, namely poor dispersion, heterogeneity, a relatively high degree of interresidue mobility and a lack of conformational restraints.

This article will attempt to provide an introduction to a large field of study, and readers are encouraged to look to the Further reading section for additional information.

Sample preparation

In high-field NMR, the use of high quality sample tubes can have a significant impact upon the quality of the spectra; 5 mm tubes are suitable for most work, with 10 mm sample tubes often being used for ^{13}C analysis.

For biologically relevant NMR analysis, experiments are generally carried out in aqueous solution. A strong water signal prevents direct analysis of proton signals lying beneath it and, owing to the problems of dynamic range, would reduce the sensitivity with which weak solute signals are detected. If, as is often the case, the exchangeable OH or NH protons are not of interest, then experiments are usually carried out in ^2H_2O. Repeated dissolution and evaporation or lyophilization is recommended to reduce the proportion of residual protons remaining. In most cases, two dissolutions into >99.8% ^2H_2O followed by a final dissolution into >99.95% ^2H_2O, preferably from a sealed ampoule, should be sufficient. Other precautions, such as pre-wetting pipettes, further rounds of dissolution and drying, or the use of a dry box may be used for greater sensitivity or quality.

Although extremes of pH or pD (outside the range δ 5–8) are to be avoided, buffer solutions are only necessary in two cases: if the molecule is charged, and the charge state is relevant to the study; and in the study of acid-labile carbohydrates, such as sialylated oligosaccharides.

The use of ^1H_2O as solvent is necessary in some cases, and requires appropriate water suppression techniques. Care must be taken to ensure that the water is free from impurities, as well as dissolved gases or paramagnetic species. A deuterated solvent to provide an appropriate lock signal should be added, usually 5–10% ^2H_2O.

Oligosaccharides are not normally soluble in non-aqueous solvents, except for dimethyl sulfoxide (DMSO). Despite the biologically irrelevant environment, this permits the observation of exchangeable protons, and there is also some increase in the proton spectral dispersion. Similar precautions to those noted for the use of ^2H_2O should be observed, such as the use of high isotopic purity (>99.96%) solvent, and repeated dissolution to remove exchangeable protons.

To approach optimal line widths, it is often advisable to remove soluble paramagnetic components by passage through a suitable chelator. Degassing to remove dissolved oxygen is also recommended. For acceptable spectra within a reasonable time using a 500 MHz spectrometer, 10 nmol is an approximate lower limit for 1D ^1H spectra; for 2D ^1H spectra, amounts of 1 μmol are preferable, although spectra with as little as 100 nmol are possible with care.

Sample volumes vary between 0.4–0.7 mL, depending on the size of the spectrometer RF coils. Temperature should be maintained at a constant level, preferably one or two degrees above room temperature for stability. Sample spinning is not necessary, except in cases where resolution is a particular problem. T_1 for protons in carbohydrates is of the order of 0.1 to 0.5 s, and recycle delays of 1–1.5 s are usually sufficient.

Referencing of samples is commonly to acetone at δ 2.225 relative to DSS (5,5-dimethylsilapentanesulfonate) at 25°C, or DMSO at 2.5 ppm.

Analysis of carbohydrate structure by NMR

The problems of carbohydrate structure addressed by NMR can be divided into four parts:

1. What is the composition of the molecule: what monosaccharides are present, are they in furanose or pyranose configurations, and α or β anomeric forms?
2. What is the nature of the linkages between the monosaccharides: at what positions do the substituents occur, and in what sequence?
3. What is the conformation of the molecule: what conformation or family of conformations are populated about the O-glycosidic bonds and hydroxymethyl rotamers?
4. What are the dynamic properties of the molecule?

Whilst the first and second questions can be satisfactorily answered by NMR, they are perhaps better approached in tandem with chemical or enzymatic methods, such as hydrolysis, followed by reduction and analysis of alditol acetates by GC-MS, or digestion with specific glycosidases. Such a concerted approach has been used to successfully determine the composition and configuration of many oligosaccharides and glycans.

The unique advantages of NMR in the analysis of carbohydrate structure are only fully apparent in consideration of points 3 and 4. In contrast to the unbranched polymeric nature of polypeptide and nucleic acid chains, carbohydrates may be branched structures, capable of substitution at several points. The monosaccharide constituents are polymerized in nature by a non-template directed, enzymatic process; the resulting oligosaccharides are often heterogeneous, differing in detail from a consensus structure. NMR is particularly efficient at investigating the solution conformations, and the dynamic properties of such molecules.

¹H NMR of carbohydrates

Primary analysis and assignment: ¹H 1D spectra

The ¹H NMR spectrum of the disaccharide galactopyranose β 1–4 linked to glucopyranose (Galpβ1-4Glcp, lactose, shown in **Figure 1**) is shown in **Figure 2**. Despite the relative simplicity of the molecule, with only 14 proton signals observable, the spectrum is surprisingly complex. This is due to the relatively small chemical shift dispersion of the non-exchangeable ring protons, resonating between δ 3–4. The small chemical shift dispersal combined with homonuclear spin–spin coupling gives rise in many cases to non-first-order spectra, complicating the measurement of spin-coupling values. This makes the assignment and analysis of even quite small saccharides quite difficult, and as a consequence the use of spectrometers of field strength 400 MHz or above is recommended.

Primarily owing to the electron-withdrawing effect of the ring oxygen, the anomeric (H1) protons give rise to signals lying outside this envelope of ring protons, between δ 4.25–5.5. Since equatorial protons experience a shift of approximately δ –0.5 relative to axial proton signals, α anomeric protons of D-sugars tend to be found between δ 4.9–5.5, and β protons between δ 4.3–4.7. The ring protons of each monosaccharide type give rise to characteristic patterns within the envelope of overlapping ring proton resonances. However, incorporation of a monosaccharide residue into an oligosaccharide will cause changes in these chemical shift patterns. For example, substitution at a given carbon causes the signal arising from the attached proton to change by δ 0.2–0.5.

The glucose residue of the sample shown in **Figure 2** is free to mutarotate between α and β forms. The signals arising from these are easily distinguished from those owing to the galactose H1 by the difference in area. The two glucose resonances have a combined area approximately equal to that of the galactose H1 resonance.

Figure 1 Structure of the disaccharide Galpβ1-4Glcp, showing the numbering of protons within the sugar rings.

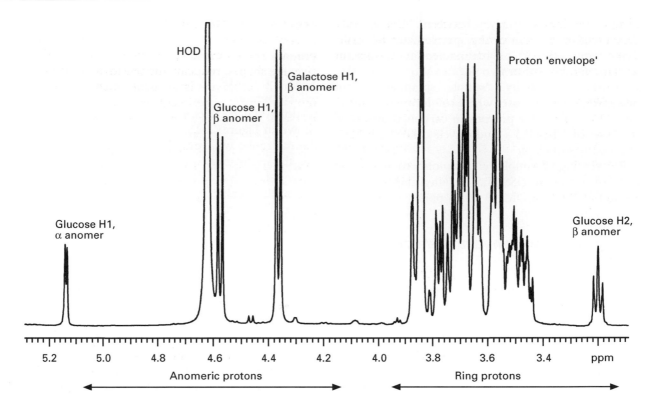

Figure 2 ^1H 1D spectrum of the disaccharide galactopyranose β1-4glucopyranose at 500 MHz and 303 K in ^2H_2O. Peak assignments are given for well-resolved peaks, following the numbering of **Figure 1**. The unresolved proton 'envelope' consists of the remaining ring proton resonances. The peak marked 'HOD' arises from residual water within the sample.

Anomeric signals typically exhibit characteristic doublets, arising from the $^3J_{H-H}$ H1–H2 coupling, whereas ring protons show more complex multiplet patterns. These couplings are proportional to the dihedral angle between the two protons. Typical values for $^3J_{H-H}$ in carbohydrates are: axial–axial 7–8 Hz, axial–equatorial and equatorial–equatorial 3–4 Hz.

During the early 1980s, Vliegenthart and co-workers identified a number of 'structural reporter groups', outlined in **Table 1**. Measurement of the chemical shift values, coupling patterns and line widths of signals arising from these elements can be interpreted to aid in the assignment and interpretation of 1D ^1H NMR spectra.

Sugar composition can thus be identified, and signals assigned on the basis of chemical shift and characteristic coupling values. The type and number of each can be established by consideration of chemical shifts, coupling patterns and relative integrals. Owing to the poor dispersion of the majority of resonances, this is usually limited to resolved anomeric protons or other structural reporter groups. Despite the development of highly efficient experiments based upon the use of selective excitation to produce edited one-dimensional spectra with more manageable information content, full proton assignment usually relies upon two- or, in some cases, three-dimensional techniques.

Two-dimensional homonuclear analysis

The poor dispersion and strong coupling present in carbohydrate proton spectra present particular problems for the assignment process. In addition, the

Table 1   Structural reporter groups as described by Vliegenthart and co-workers

Sugar type	Proton(s)	Parameter	Information content
All	H1	δ, J	Residue and linkage type
Mannose	H2, H3	δ	Substitution within core region of branched glycan
Sialic acid	H3	δ	Type and configuration of linkage
Fucose	H5, CH3	δ	Type and configuration of linkage: structural environment
Galactose	H4, H4	δ	Type and configuration of linkage
Amino sugars	N-acetyl CH₃	δ	Sensitive to small structural variations

characteristics of a given spin system differ for each oligosaccharide and are highly dependant on structure. The use of two-dimensional experiments alleviates these problems.

Coherence transfer methods   The correlated spectroscopy (COSY) experiment reduces the degree of resonance overlap by separating resonances into two orthogonal proton dimensions. The 1D spectrum lies along the diagonal, with cross-peaks joining pairs of J-coupled spins. For carbohydrates, assignment can then start with the well-resolved anomeric protons, and continue by stepwise J-correlation to the remaining nuclei. This simple process is complicated in many cases by overlap and strong coupling between signals, even in two dimensions. To reduce these difficulties, high quality spectra are needed, with optimal digital resolution.

Sometimes absolute value mode, as opposed to pure phase spectra, can be preferable for easy assignment, despite the theoretical disadvantages of a phase-twist line shape. Carbohydrate line widths are generally narrower than those of proteins and nucleic acids, and the use of absolute value spectra does not significantly diminish the quality of spectra, as seen in **Figure 3**. The COSY spectrum requires a pseudo- echo weighting function to remove dispersive line shapes, reducing the sensitivity of the experiment. If sensitivity is an issue, then an inherently more sensitive experiment, such as the absorption mode DQ-COSY (see below) should be used.

If necessary, linkage positions can be identified using a COSY spectrum of a sample dissolved in DMSO-d_6. The hydroxyl protons do not exchange with solvent, and give rise to observable signals within the spectrum. Substitution removes the hydroxyl proton at that position, which can then be identified by the absence of a hydroxyl proton signal.

The COSY experiment may fail to generate the expected cross-peaks, for several reasons:

1. For strongly coupled neighbours, the cross-peak lies close to the diagonal, and may not be seen.
2. Pairs of nuclei with small scalar coupling values ($^3J_{\text{H-H}}$) will produce low intensity cross-peaks. Since the active coupling is antiphase in COSY cross-peaks, they will cancel if the value of J is less than the line width. This is particularly

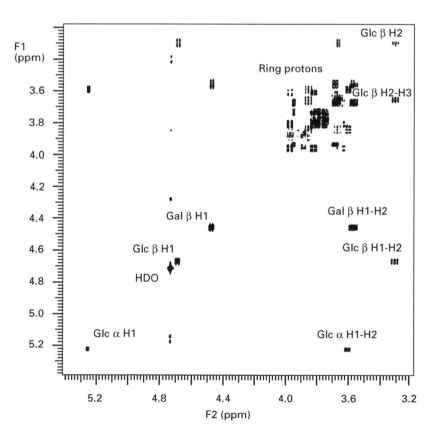

Figure 3 ^{1}H–^{1}H 2D COSY spectrum of galactopyranose β 1-4glucopyranose at 500 MHz and 303 K in ^{2}H$_2$O. A full proton assignment is usually possible from a spectrum like this, and peak identities are shown where space permits. The well-resolved anomeric protons provide a useful starting point for assignment, and subsequent pairs of spin-coupled nuclei are correlated through off-diagonal peaks.

noticeable in larger molecules. Common examples are: D-mannopyranose β H1–H2 ($J < 1$ Hz) and D-galactopyranose H4–H5 ($J < 1$ Hz).

3. Unrelated resonances may simply occur at the same position, and cannot be distinguished.

Experiments incorporating isotropic mixing sequences The homonuclear Hartmann–Hahn (HOHAHA) and the closely related total correlation spectroscopy (TOCSY) experiments are used to overcome problems of overlap and strong coupling. These experiments give correlations to all other spins within the same coupling network. The second pulse in the COSY sequence is replaced by an isotropic mixing period, such as a spin–lock sequence (TOCSY) or an efficient decoupling sequence, such as MLEV-17 (HOHAHA). The spin system becomes strongly coupled and coherence transfer occurs between all nuclei within the coupling network. This illustrates an advantage over experiments such as RELAY-COSY: values of $^3J_{H-H}$ between adjacent nuclei vary, and the use of an isotropic mixing period produces an increased efficiency of transfer. The presence of small couplings between nuclei, reducing the efficiency of transfer, still presents a problem.

Resolution in this type of experiment is better than for a COSY spectrum; however, more cross-peaks are generated, producing a more complicated spectrum. This is at first sight hard to interpret; however, proton assignments in each ring can be determined by inspection of a line lying through the frequency of the anomeric proton of the sugar residue. The cross-peaks found along this line can be distinguished by reference to the COSY spectrum, and by consideration of the multiplet structure. Pyranose rings have an effectively rigid ring geometry, and it is possible to estimate the couplings between adjacent protons based upon the dihedral angle between them. Judicious variation of the length of the isotropic mixing period can also be used to give stepwise correlations along the spin-system, since shorter periods correlate proportionately smaller fragments. One-dimensional analogues of this experiment, using selective pulses to specifically excite anomeric protons, may be quicker and more efficient, especially when the anomeric protons are well resolved.

Multiple-quantum homonuclear experiment Incorporation of a multiple-quantum filter into a COSY-type sequence reduces the complexity of the spectrum, and aids in the assignment process.

A DQF-COSY, incorporating a double-quantum filter transparent to two or more coupled spins, therefore passes all signals except singlets. Such a sequence has a slightly reduced sensitivity (cross-peak intensity is approximately half that of a comparable COSY), but there is no need for the sensitivity-reducing weighting functions that are necessary with COSY. One noticeable feature is the absence of a diagonal, thus cross-peaks lying close to the diagonal can be more easily identified.

Incorporation of a triple-quantum filter produces a spectrum with signals due only to three or more coupled spins with two resolvable couplings. There are no singlets or doublets, including anomeric resonances, although suppression is not absolute, due to the presence of small couplings along four or more bonds.

Under ideal conditions, one would expect to see resonances from only H5 and H6 protons, and this experiment is best used in conjunction with HOHAHA for full assignment. Sensitivity is much reduced relative to COSY spectra.

Through-space dipolar interaction methods Although primarily used for conformational analysis, the nuclear Overhauser effect spectroscopy (NOESY) experiment is also helpful in the assignment process. The resulting spectra show correlations between protons coupled by through-space dipolar interactions. Both inter- and intraresidue NOEs help to confirm the consistency of assignments made by the techniques described above. Long-range NOEs are not often seen in carbohydrates, and the observation of NOEs between protons on adjacent residues is often evidence of the linkage positions between the two sugars.

Through-space dipolar interactions are of most use in the conformational analysis of biomolecules, as described in the section on conformational analysis below.

At 500 MHz, moderate-sized (more than six residues) oligosaccharides lie within the spin-diffusion limit. However, for smaller molecules, as the value of the function $\omega_0 \tau_c$ (where ω_0 is the Larmor frequency of protons, and τ_c is the correlation time of the molecule) approaches 1 then the value of the NOE tends towards 0. Cross-peak intensities of NOESY spectra of smaller oligosaccharides (2–5 residues) may thus become too small to measure accurately. In such cases, the rotating frame Overhauser effect spectroscopy (ROESY, originally referred to as CAMELSPIN) experiment is commonly used to measure NOE values. To reduce the appearance of HOHAHA-like cross-peaks, a low power spin–lock field should be used, and the transmitter carrier offset to the low-field end of spectrum. The offset dependency of cross-peak intensities should also be removed by 90° pulses at either end of spin–lock period.

Three-dimensional experiments, often essentially hybrids of simpler sequences such as HOHAHA-COSY, have been used successfully to resolve cases of particularly difficult resonance overlap.

Hydroxyl protons

The quest for additional conformational information has led to the investigation of hydroxyl protons in aqueous solution. Samples are dissolved in mixed methanol–water or acetone–water solvents, and analysed in capillary NMR tubes at low (–5 to –15°C) temperatures. Chemical exchange of hydroxyl protons is reduced to the point that it is possible to use them as probes of hydration and hydrogen bonding. Distance information can also be extracted from NOESY or ROESY spectra under these conditions.

^{13}C NMR of carbohydrates

Primary analysis and assignment

Initial investigations of the ^{13}C spectra of carbohydrates were of natural abundance samples, although spectra of isotopically-enriched samples have become more common in recent years.

^{13}C spectra are usually acquired with broad band proton decoupling. The resulting spectra are not very complex with, at natural abundance, a single sharp peak for each nucleus, there being no visible carbon–carbon coupling. The value of the carbon–proton coupling is approximately 150–180 Hz, and without proton decoupling this splitting, along with further two- and three-bond couplings, can render the spectrum quite complex.

Monosaccharides exhibit characteristic carbon chemical shift patterns. Anomeric carbons typically occur between δ 90–110 in pyranose rings, C6 (hydroxymethyl) signals between δ 60–75, and the remaining carbons resonate between δ 65–110. Substitution at a given carbon causes an approximate shift of up to $\delta + 10$. The usefulness of this shift in identifying the position of substitution is complicated by an additional $\delta -1$ to -2 shift of the surrounding carbon signals. In addition, the local structure of the saccharide can have a considerable effect upon the chemical shift values. Residue type and substitution patterns can thus be discovered, with care, from measurements of the carbon chemical shifts of a simple sugar. Databases of ^{13}C chemical shift information for large numbers of simple carbohydrates are available for comparison in the literature.

Carbon–proton and carbon–carbon spin couplings

A consistent estimate of the anomeric configuration is given by the value of the one-bond $^1J_{C-H}$ coupling; β anomers show a value of ~160 Hz, and α anomers ~170 Hz.

Three-bond carbon–proton scalar coupling values, such as $^3J_{CCCH}$, or $^3J_{COCH}$, provide useful indicators of conformation. These couplings can be used to confirm the internal conformation of a saccharide residue, using appropriate Karplus curves. Across the glycosidic link, values of the couplings H1–C1–O–Cx (where x is the aglyconic carbon) and C1–O–Cx–Hx are proportional to the glycosidic dihedral angles ϕ and ψ respectively (see section on conformational analysis below). Selective incorporation of ^{13}C has been successfully used to simplify and enhance the measurement of these parameters. Linkage positions can be unambiguously determined by the identification of these couplings using experiments such as the 2D ^{13}C–^1H multiple-bond correlation experiment (HMBC). This experiment correlates carbon and proton resonances by means of $^3J_{C-H}$ couplings; $^1J_{C-H}$ correlations are suppressed, leaving only long-range couplings. Those couplings that span the glycosidic bond can thus be identified and measured.

Multidimensional, heteronuclear analysis

Enrichment with ^{13}C allows the use of a wide range of useful experiments that greatly facilitate the investigation of carbohydrate structure. Proton detected versions of carbon–proton correlation experiments, such as that shown in **Figure 4**, overcome the sensitivity limitations of observing carbon signals. If the proton assignments are known, assignment of carbon resonances from such spectra is a relatively simple process. The use of carbon nuclei to edit spectra into an orthogonal frequency dimension to overcome overlap has led to the development of a series of useful hybrid 3D or pseudo-3D experiments, such as HCCH-COSY or HMQC-NOESY. Editing spectra in this way has helped to assign the spectra not only of complex glycans in free solution, but also of the attached carbohydrate chains of glycoproteins, both at natural abundance as well as enriched with ^{13}C and ^{15}N. The use of these heteronuclei allows the carbohydrate resonances to be observed separately from those due to the protein portion of the molecule.

Conformational analysis

Measurement of $^3J_{H-H}$ couplings shows that pyranose rings do not show any large degree of internal flexibility, except for pendant groups such as the hydroxymethyl in hexopyranoses. Interpretation of these uses the Haasnoot parametrization of the Karplus equation for $^3J_{H-H}$ couplings in HCCH

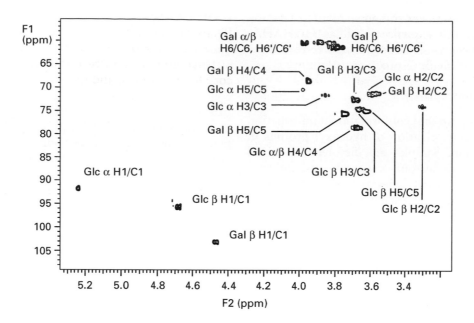

Figure 4 HSQC ^{1}H–^{13}C correlation spectrum of galactopyranose β1-4glucopyranose at 500 MHz and 303 K in ^{2}H$_2$O. Resonances are labelled with the appropriate assignment for each correlation. In practice, these can be relatively easily identified by comparison with previously assigned proton or carbon chemical shifts. Reproduced by courtesy of Homans SW and Kiddle GR, University of St. Andrews.

portions of the saccharide ring:

$$^3J_{H-H'} = 13.22\cos^2\theta - 0.99\cos\theta$$
$$+ \Sigma_\iota\Delta\chi_\iota[0.87 - 2.46\cos^2(\xi_\iota\theta)] \quad [1]$$

where θ is the HCCH′ torsion angle, $\Delta\chi_\iota$ are the Huggins' electronegativities of the substituents relative to protons, and ξ is either +1 or −1 depending upon the orientation of the substituent. If stereo-specific assignments have been made, orientation about hydroxymethyl rotamers can be determined by similar parametrizations.

Conformational variability in pyranoside oligosaccharides is mostly owing to variations about the glycosidic torsion angles ϕ and ψ and, for $1 \rightarrow 6$ linkages, the ω angle (**Figure 5**). NMR analysis of carbohydrate conformation thus concentrates upon determining the orientations about these dihedral angles. This is made difficult by the lack of conformational information available: even in the most favourable conditions, a maximum of around three NOEs are seen between each pair of residues in oligosaccharides, and observation of long-range correlations between different parts of the molecule is extremely unlikely.

Interresidue correlations in NOESY or ROESY spectra can be used to determine distance information by making use of the fact that the intensity of the cross-peak is proportional to the inverse sixth

power of the distance between the two nuclei. The relatively inflexible ring geometry of pyranose rings allows intraresidue NOEs to be used to calibrate adjacent intraresidue correlations. When implementing such experiments it is important to ensure that the initial rate approximation holds. The motion of the molecule, both internal and overall, also affects the intensity of the NOE; thus interpretation of NOE values must take into account the motional model

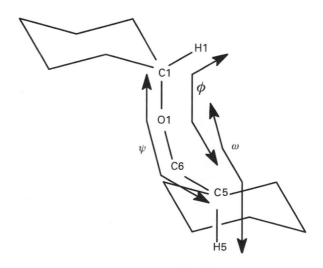

Figure 5 A $1 \rightarrow 6$ linked disaccharide showing the dihedral angles ϕ, ψ and ω, about which conformational variation is greatest.

used. Since it is now clear that the majority of carbohydrate structures are flexible to some degree, oscillating through an ensemble of related conformations, this can be a problem. Flexibility is understood to be particularly marked about $1 \rightarrow 6$ linkages owing to rotation about the additional dihedral angle, ω.

The use of three-bond $^3J_{\text{C–H}}$ couplings as conformational probes has been described, using the Tvaroska parametrization of the Karplus relationship:

$$^3J_{(\text{C,H})} = 5.7 \cos^2 \theta - 0.6 \cos \theta + 0.5 \qquad [2]$$

where θ is the HCOC torsion angle. However, in a similar situation to that described for NOE restraints, these analyses have been hampered by the extent to which these values are subject to conformational averaging. Because of these problems, angular information derived from $^3J_{\text{C–H}}$ values is perhaps better used for evaluating the quality of conformational ensembles.

Recent work has addressed the use of ^{13}C within isotopically enriched compounds to derive additional conformational parameters, such as ^{13}C–^{13}C scalar couplings, and ^{13}C–^{1}H or ^{13}C–^{13}C NOEs. These approaches have yielded useful results, but have been hampered by the lack of a well-established Karplus-type relationship for ^{13}C–^{13}C scalar coupling.

Investigators have thus made considerable use of theoretical methods such as molecular mechanics to complement the limited experimental data available.

Dynamics

The time-averaged nature of many NMR derived parameters, and the degree to which it affects oligosaccharide conformations has made it necessary to attempt to measure the degree of mobility within oligosaccharides. Since the relaxation of protonated carbons is almost entirely due to the directly attached proton it is possible to measure relaxation parameters from them that are not affected by variations in internuclear distance. Investigation of carbohydrate dynamics has been approached by molecular modelling in combination with the measurement of relaxation rates, including the T_1, T_2 of ^{13}C and ^{1}H, as well as ^{1}H–^{1}H and ^{13}C–^{1}H NOE values, to define the spectral density, given by

$$J(\omega) = (2/5)\tau_c/(1 + \omega\tau_c)^2 \qquad [3]$$

where τ_c denotes the rotational correlation time.

These measurements have been interpreted by the Lipari–Szabo model-free approach, in which molecular motion is separated into overall and internal motions, related to the overall and internal correlation times τ_O and τ_i:

$$J(\omega) = S^2(2/5)\tau_O/(1 + \omega\tau_O)^2 \\ + (1 - S^2)(2/5)\tau/(1 + \omega\tau)^2 \qquad [4]$$

where $1/\tau = 1/\tau_O + 1/\tau_O$. S^2 is a measure of the degree of internal reorientation, varying in value from 0 for isotropic reorientation to 1 for no internal motion.

These measurements can then be used to derive dynamic models of carbohydrates and oligosaccharides. It is now generally accepted that carbohydrates are dynamic molecules with internal motions that occur on a time-scale faster than the overall motion of the molecules.

List of symbols

J = coupling constant; $J(\omega)$ = spectral density function; T_1, T_2 = relaxation constants; $\Delta\chi_l$ = Huggins electronegativities; θ = torsion angle; τ_c, τ_i and τ_O = rotational, internal and overall correlation times respectively; ϕ, ψ and ω = glycosidic torsion angles.

See also: **^{13}C NMR, Methods; Labelling Studies in Biochemistry Using NMR; Nuclear Overhauser Effect; Nucleic Acids Studied By NMR; Proteins Studied Using NMR Spectroscopy; Two-Dimensional NMR, Methods.**

Further reading

Bush CA (1996) Polysaccharides and complex oligosaccharides. In: Grant DM and Harris RK (eds) *Encyclopaedia of Nuclear Magnetic Resonance*, Vol 6, pp 3746–3750. Chichester: Wiley.

Homans SW (1993) ^{1}H NMR studies of oligosaccharides. In: Roberts GCK (ed) *NMR of Macromolecules: a Practical Approach*, pp 289–314. Vol 134 of Rickwood D and Hames BD (series eds) *Practical Approach Series*. Oxford: Oxford University Press.

Homans SW (1993) Conformation and dynamics of oligosaccharides in solution. *Glycobiology*, 3: 551–555.

Hounsell EF (1995) ^{1}H NMR in the structural and conformational analysis of oligosaccharides and glycoconjugates. *Progress in Nuclear Magnetic Resonance Spectroscopy* 27: 445–474.

Serianni AS (1992) Nuclear magnetic resonance approaches to oligosaccharide structure elucidation. In: Allen HJ and Kisalius EC (eds) *Glycoconjugates: Composition, Structure and Function*, pp 71–102. New York: Marcel Dekker.

Tvaroska I (1990) Dependence on saccharide conformation of the one-bond and three-bond coupling constants. *Carbohydrate Research*, **206**: 55–64.

van Halbeek H (1994) NMR developments in structural studies of carbohydrates and their complexes. *Current Opinion in Structural Biology* 4: 697–709.

van Halbeek H (1996) Carbohydrates and glycoconjugates. In: Grant DM and Harris RK (eds) *Encyclopaedia of Nuclear Magnetic Resonance*, Vol 2, pp 1107–1137. Chichester: Wiley.

Vleigenthart JFG, Dorland L and van Halbeek H (1983) High resolution spectroscopy as a tool in the structural analysis of carbohydrates related to glycoproteins. *Advances in Carbohydrate Chemistry and Biochemistry* **41**: 209–374.

CD Spectroscopy of Biomacromolecules

See **Biomacromolecular Applications of Circular Dichroism and ORD.**

Cells Studied By NMR

Fátima Cruz and **Sebastián Cerdán**, Instituto de Investigaciones Biomédicas, Madrid, Spain

MAGNETIC RESONANCE
Applications

Introduction

Nuclear magnetic resonance (NMR) methods have become available recently to study metabolism and morphology at the cellular level. The NMR methods may be classified as magnetic resonance spectroscopy (MRS), magnetic resonance imaging (MRI) and combinations of both, magnetic resonance spectroscopic imaging (MRSI). MRS provides information on the chemical composition of cells and its changes under specific circumstances, MRI yields X-ray like images of cellular anatomy and physiology and MRSI can study the spatial distribution of some metabolites within large cells or cellular aggregates. Notably, most of the nuclei participating in cellular reactions or some of their isotopes are NMR active. This allows a large variety of cellular functions to be monitored by different NMR methods. **Table 1** summarizes the magnetic properties of the nuclei most commonly used in NMR studies of cells as well as the biological information which can be obtained.

NMR studies of cells probably started in the mid 1950s analysing the dynamics of water in blood using ^1H NMR. However, modern NMR studies of cellular metabolism began later with the introduction of commercially available high-field spectrometers and Fourier transform NMR techniques. Improve-ments in signal-to-noise ratios allowed, in the early 1970s, a study by ^{13}C NMR of the metabolism of glucose in a suspension of yeast and to determine by ^{31}P NMR the intracellular pH in erythrocyte suspensions. Since then studies of cellular metabolism by NMR methods have been used routinely in many laboratories.

The development of cellular NMR is supported by some inherent advantages of the NMR method with respect to more classical approaches. First, the noninvasive character of NMR allows repetitive, noninvasive measurements of metabolic processes as they occur in their own intracellular environment. Second, the magnetic properties of nuclei like the relaxation times T_1 and T_2 or the homonuclear and

Table 1 NMR properties and type of biological information provided by various nuclei

Nuclei	Frequency (MHz)	Spin	Natural abundance (%)	Relative sensitivity	Information
^1H	400	1/2	99.98	1.0	Metabolite concentration. Cell fingerprinting. Flow through some pathways. Intra and extracellular pH. Cellular volume. Microscopic imaging. MRI
^{31}P	161.9	1/2	100	6.6×10^{-2}	Concentration of phosphorylated metabolites. Bioenergetic status. Intra- and extracellular pH. Cellular volume. Phospholipid metabolism
^{13}C	100.6	1/2	1.11	1.6×10^{-2}	Quantitative measurements of metabolic flow through specific pathways
^{23}Na	105.8	3/2	100	9.3×10^{-2}	Membrane potential. Cellular volume. Na$^+$/H exchange
^{19}F	376.3	1/2	100	0.83	Intra- and extracellular pH. Oxygenation state. Divalent metal ion concentration
^2H	61.8	1	0.016	9.7×10^{-3}	Lipid structure. Rates of hydration–dehydration reactions. Water flow and perfusion
^{39}K	18.7	3/2	93.1	5.1×10^{-4}	Membrane potential
^{17}O	54.2	5/2	0.037	2.9×10^{-2}	Water transport and metabolism
^{15}N	40.5	1/2	0.37	1.0×10^{-3}	Nitrogen metabolism
^{14}N	28.9	1	99.63	1.0×10^{-3}	Nitrogen metabolism

heteronuclear spin coupling patterns, contain unique information on the physiological or pathological status of the cells and on the flux through specific metabolic pathways. Third, NMR methods allow the acquisition of images of the spatial distribution of water in sufficiently large single cells or in cellular aggregates, and it seems likely that this approach will be extended to other metabolites in the near future. Despite these advantages, the NMR method is not devoid of drawbacks. In particular, NMR is a relatively insensitive technique with a metabolite detection threshold of around 10^{-1} mM for *in situ* cells and 50 μM in extracts for moderate-field spectrometers. Thus, to be able to obtain useful NMR spectra with an adequate signal-to-noise ratio, the cell cultures need to be grown to densities similar to those found in tissues. Even if cell extracts are used, these must be prepared from a sufficiently large number of cells to generate, in the NMR tube, metabolite concentrations in excess of the lower threshold for detection. These demanding conditions require the use of specialized perifusion systems for *in situ* NMR studies or facilities for large-scale cell culture in work with cell extracts.

In this article, we describe general procedures for cell culture compatible with NMR studies and illustrate the information that NMR methods can provide on cellular metabolism and morphology, with examples involving mainly the use of ^{31}P, ^{13}C and ^1H NMR. Several reviews, some of them quoted in the Further reading section, have covered this topic previously.

Large-scale and high-density cell culture procedures compatible with NMR studies

The use of cell cultures provides a powerful tool to understand the cellular and molecular mechanisms occurring *in vivo*. Different types of cell culture exist depending on the origin of the cells used in the preparation. Primary cultures are obtained after enzymatic, chemical or mechanical disaggregation of the original tissue. Cell lines, transformed cells or cell strains are obtained from primary cultures after several passages from transplantable tumours or by cellular cloning. The *in vivo* behaviour is more closely resembled by primary cultures, but these usually require complex preparation resulting in relatively small cellular yields. Cellular lines or strains, and cellular clones have the advantage of being homogenous cell populations that are easy to grow and maintain in large scale, but their behaviour is not as comparable to the *in vivo* situation.

Figure 1 illustrates the general procedure used to obtain a primary culture. The tissue of interest is dissected from the appropriate organ obtained from embryonic, newborn or adult donors. The dissected tissue is then chopped and disaggregated, digested enzymatically and the resulting cells are washed and collected by sedimentation. Cell counting and assessment of viability are easily accomplished later by optical microscopy, using a haemocytometer for cell counting and the dye exclusion criteria for viability assessment. The next step consists of cell seeding.

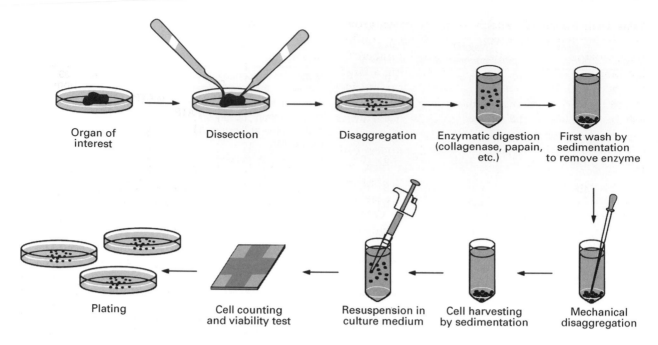

Figure 1 General steps in the preparation of primary cell cultures.

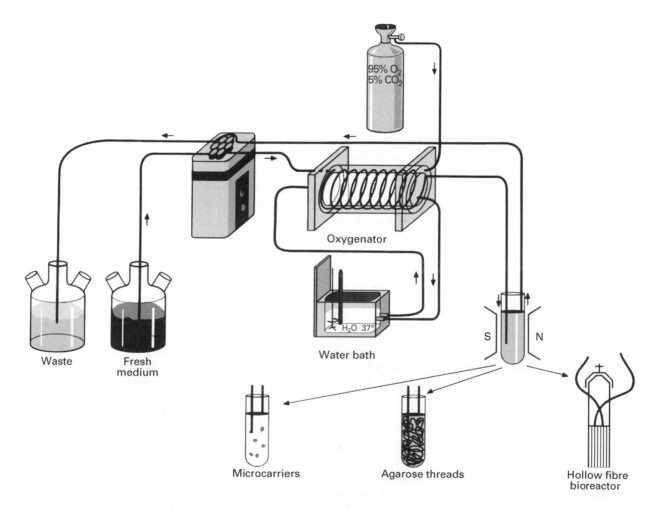

Figure 2 High density cell culture devices compatible with *in situ* NMR spectroscopy. The tubes containing the entrapped cells can normally be accommodated in commercial high-resolution probes and placed inside the magnet (indicated by N and S).

Cells seeded normally attach to flat surfaces, grow, proliferate and differentiate until reaching confluence. Usually, NMR studies demand cell numbers in the range $10^7–10^8$ cells, a quantity requiring a large number of culture dishes or flasks. To further increase the surface of the culture it is possible to use rollers. These devices allow cells to grow in the inner surface of a rolling cylinder which contains a moderate amount of culture medium. Several cylinders can be maintained rolling simultaneously in a rack, increasing enormously the surface area of the culture and thus the number of confluent cells. The excellent monograph of Freshney provides a more detailed description of every cell culture procedure and optimized techniques for particular cells.

There are two different kinds of experimental design in NMR studies of cells: (i) the study can be performed inside the magnet in real time, monitoring the metabolism of cells under *in situ* conditions; or (ii) extracts can be prepared from the cells after different incubation times and examined by NMR under high-resolution conditions (**Figure 2**). In the first case special precautions need to be taken to guarantee adequate nutrient supply, good oxygenation conditions and efficient removal of waste products from the NMR tube containing the cells. Different methods have been developed to fulfil these requirements. All of them involve a perifusion system adapted to the NMR tube which contains the cells immobilized or grown in different matrices. The perifusion system consists basically of a peristaltic pump, a thermostatically stabilized gas-exchange chamber or oxygenator and the NMR tube containing the cells. The peristaltic pump delivers fresh medium to the oxygenator and NMR tube and removes the used medium to a waste reservoir. The oxygenator is constructed of gas-permeable Silastic tubing equilibrated with an atmosphere of 95% O_2/5% CO_2. The NMR tube contains the entrapped cell suspension. There are basically three types of cell entrapment protocol compatible with NMR studies: (i) polymer threads, (ii) microcarrier beads or (iii) hollow fibre bioreactors.

Cells are entrapped in polymer threads (agarose or alginate) by mixing the cell suspension with a solution of the liquid polymer (above the gelling temperature) and extruding the mixture through a Teflon tube immersed in ice. The polymer solidifies at low temperature entrapping the cells in filaments which can be easily perifused. The advantage of this method lies in its simplicity, but on the other hand, cells are placed in a nonphysiological environment. Alternatively, anchorage dependent cells can be grown on the surface of solid or porous microcarrier beads of approximately 100 µM diameter. The beads can be made of different materials and these include

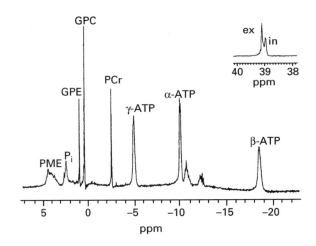

Figure 3 ^{31}P NMR spectrum (161.1 MHz) of C6 glioma cells grown in a hollow fibre bioreactor. Note the important contribution of the GPC and GPE resonances. Inset: intra- (in) and extracellular (ex) resonances of DMMP. From Gillies RJ, Galons JP, McGovern KA, Scherer PG, Lien YH, Job C, Ratcliff R *et al. NMR in Biomedicine* **6**: 95–104. Copyright 1993 John Wiley & Sons Limited. Reproduced with permission.

collagen, polystyrene, polyacrylamide, gelatine, and Sephadex. This method has the advantage that cells can be treated as a suspension. However the density of cells may not be sufficiently high since most of the volume in the NMR tube is filled by the beads. The hollow fibre bioreactor approach consists of growing cells on the surface of capillary fibres which are gas and nutrient permeable. Cells are grown directly in a bioreactor adapted to the NMR tube where the experiment is performed, reaching higher density than with other methods. However, there are some technical difficulties in setting up this model properly and it is relatively expensive. For a detailed description of the bioreactor approach see the literature given in the Further reading section.

Applications

^{31}P NMR of cells

^{31}P NMR spectra of cells show a small number of resonances but contain crucial information on cellular biochemistry and physiology. **Figure 3** depicts a representative ^{31}P NMR spectrum obtained *in situ* from a culture of C6 glioma cells grown in a bioreactor. It is possible to distinguish clearly resonances from the β, α and γ phosphates of nucleotide triphosphates, phosphocreatine (PCr), phosphodiesters (glycerolphosphorylcholine, GPC, and glycerolphosphorylethanolamine, GPE), inorganic phosphate (P_i) and phosphomonoesters (PME). The resonances from the nucleotide triphosphates contain primarily (>90%) the contribution of MgATP and are normally referred to as ATP resonances. The PME peak is a composite resonance containing mainly contributions from phosphocholine (PCho) and phosphoethanolamine (PE), as well as from sugar phosphates in a smaller proportion.

^{31}P NMR spectra provide quantitative information on the relative concentrations of phosphorylated metabolites. For spectra acquired under fully relaxed conditions, the area of the ^{31}P NMR resonance is proportional to the concentration of the metabolite. Absolute quantifications are more difficult to perform but can also be obtained if the concentration of one of the phosphorylated metabolites in the sample (normally ATP) is measured by an independent method, such as enzymatic analysis. Unfortunately, the resonances from the α and β phosphates of ADP overlap under *in situ* conditions with those from the α and γ phosphates of ATP, thus precluding direct quantification of ADP *in situ*. The concentration of ADP *in situ* can be estimated indirectly from the difference in area between the α ATP or γ ATP peaks and the area of the β ATP peak, which is almost exclusively derived from ATP.

Probably, the main interest in ^{31}P NMR spectroscopy lies in its ability to assess the bioenergetic status of the cell. This is determined by the relative rates of ATP-producing and -consuming processes according to Equation [1]

$$MgATP^{2-} + H_2O$$
$$\rightleftharpoons MgADP^- + P_i^{2-} + H^+ + energy \qquad [1]$$

In most mammalian cells, ATP is produced aerobically by oxidative phosphorylation and hydrolysed to provide the energy needed to support biosynthetic activities, maintain transmembrane ion gradients and perform cellular work. The balance between production and consumption of ATP determines its steady-state level and therefore the net availability for energy. If oxygen and nutrients become limiting, ATP synthesis may proceed for a short time using PCr as a phosphate donor in reaction [2]

$$PCr^{2-} + MgADP^- + H^+ \rightleftharpoons MgATP^{2-} + Cr^0 \qquad [2]$$

Under these conditions no significant change occurs in the ATP resonances but a net decrease in the intensity of PCr is observed. When the ATP-buffering capacity of the PCr-creatine system is exceeded, net ATP hydrolysis occurs, resulting in a net increase in P_i and acidification, as described by Equation [1].

Another important aspect of ^{31}P NMR spectroscopy is that it allows the determination of intracellular and extracellular pH (in P_i-containing media) from the chemical shift of the P_i resonance. This is because P_i has an apparent pK_a of 6.7 and its chemical shift is pH-dependent in the physiological range. Normally the chemical shift of the P_i resonance is measured with respect to internal references like PCr or α-ATP, which are nontitratable in the physiological pH range, ($pK_a \approx 4.5$). The following expression can be used to measure intracellular pH from the chemical shift of $P_i(\delta)$ measured with respect to the α-ATP resonance:

$$pH_i = 6.7 + \log{(\delta - 11.26)/(13.38 - \delta)}$$

The values 11.26 and 13.38 represent the acidic and alkaline limits of the P_i titration curve chemical shifts with respect to the α-ATP resonance. Phosphonic acid derivatives, like methylene diphosphonic acid (MDPA) or dimethyl methylphosphonate (DMMP), resonating far from physiological phosphates (*ca.* 20–30 ppm) can be used as an external reference for chemical shift and concentration calibrations. Sometimes these references cross the cell membrane and two different resonances are observed for the intracellular and extracellular compounds. In these cases it is possible to determine the relative cellular volume from the relative areas of the intra- and extracellular resonances (**Figure 3** inset).

^{31}P NMR spectra also contain resonances from crucial phospholipid metabolites. These include the phosphodiesters, glycerolphosphocholine and glycerolphosphoethanolamine and the phosphomonoesters, PCho and PE. Tumoural cells and *in vivo* tumours show increased PME resonances, often caused by increased PCho content. This characteristic of tumour cells has been proposed to contain diagnostic or even prognostic information on tumours. The metabolic reasons underlying the increase in PCho in tumour cells are not completely understood but seem to be caused by an increase in its synthesis rather than to an increase in the degradation of phosphatidylcholine.

^{13}C NMR of cells

^{13}C NMR spectroscopy allows the detection of resonances from ^{13}C, the stable isotope of carbon with a magnetic moment. The natural abundance for ^{13}C is only 1.1% of the total carbon and its magnetogyric ratio is approximately one-fourth of that of the proton. These two circumstances make ^{13}C NMR a relatively insensitive technique. However, this sensitivity problem can be improved by using ^{13}C–enriched substrates. The combination of ^{13}C NMR detection

and substrates selectively enriched in ^{13}C have made it possible to follow *in vivo* and *in vitro* the activity of a large variety of metabolic pathways in cells and subcellular organelles. These include glycolysis and the pentose phosphate pathway, glycogen synthesis and degradation, gluconeogenesis, the tricarboxylic acid cycle, ketogenesis, ureogenesis and the glutamate, glutamine, γ-aminobutyric acid cycle in the brain among others.

The design of ^{13}C NMR experiments with selectively ^{13}C–enriched substrates is similar to the classical radiolabelling experiments using radioactive ^{14}C. A relevant difference is that ^{13}C precursors are used in substrate amounts, while ^{14}C substrates are normally used in tracer amounts. However, ^{13}C NMR presents important advantages over ^{14}C. First, the metabolism of the ^{13}C-labelled substrate can be followed in real time, *in situ* and noninvasively. Second, even if tissue extracts are prepared, the detection of ^{13}C labelling in the different carbon atoms of a metabolite does not require separation and carbon-by-carbon degradation of the compound, two normal requirements in experiments with ^{14}C. Finally, the analysis by ^{13}C NMR of homonuclear spin-coupling patterns and isotope effects allows the determination of whether two or more ^{13}C atoms occupy contiguous positions *in the same metabolite molecule*, a possibility only available to ^{13}C NMR methods. As a counterpart to these advantages, ^{13}C NMR is significantly less sensitive than other conventional metabolic techniques like radioactive counting, mass spectrometry or optical methods. In particular, ^{13}C NMR studies with cells *in situ* are difficult to find and most of the ^{13}C NMR work has been performed with cell extracts. This allows operation under high-resolution conditions and increases significantly both the sensitivity and the resolution.

Figures 4 and **5** illustrates the use of ^{13}C NMR spectroscopy to study the metabolism of (1-^{13}C)glucose in primary cultures of neurons and astrocytes. A simplified scheme of the metabolism of (1-^{13}C)glucose in neural cells is given in **Figure 4**. Briefly, (1-^{13}C)glucose is metabolized to (3-^{13}C)pyruvate through the Embden Meyerhoff glycolytic pathway. The (3-^{13}C)pyruvate produced can be transaminated to (3-^{13}C)alanine, reduced to (3-^{13}C)lactate or enter the tricarboxylic acid (TCA) cycle through the pyruvate dehydrogenase (PDH) or pyruvate carboxylase (PC) activities. A net increase in (3-^{13}C)lactate reveals increased aerobic glycolysis and is normally observed under hypoxic conditions in normal cells. If (3-^{13}C)pyruvate enters the TCA cycle though PDH it produces (2-^{13}C)acetyl-coenzyme A first, and subsequently (4-^{13}C)α-ketoglutarate. In contrast, if (3-^{13}C)pyruvate is carboxylated to (3-^{13}C)oxalacetate by

PC, it will produce (2-^{13}C)α-ketoglutarate in the next turn of the cycle. The α-ketoglutarate–glutamate exchange is many times faster than the TCA cycle flux and therefore, ^{13}C labelling in the glutamate carbon atoms reflects accurately the labelling in the α-ketoglutarate precursor. Thus, (4-^{13}C)glutamate or (2-^{13}C)glutamate are produced from (1-^{13}C)glucose depending on the route of entry of the (3-^{13}C) pyruvate into the TCA cycle. Therefore, a comparison of intensities from the C4 and C2 glutamate resonances provides a qualitative estimation of the relative contribution of PDH and PC to the TCA cycle.

Figure 5 shows proton decoupled ^{13}C NMR spectra of perchloric acid extracts obtained from primary cultures of neurons (upper panel) and astrocytes (lower panel) incubated for 24 h with 5 mM (1-^{13}C)glucose. The most relevant resonances are the ones derived from the C3, C4 and C2 carbons of glutamate; C3, C4 and C2 carbons of glutamine (or glutathione); C3 carbon of alanine; C3 carbon of lactate; C6 carbon of *N*-acetylaspartic acid and C2 carbon of acetate (see **Figure 5** legend for resonance assignments). The spectra of **Figure 5** reveal important differences in the metabolism of (1-^{13}C)glucose by neurons and astrocytes. Briefly, the lactate C3 resonance is higher in the spectrum of astrocytes than in that of the neurons. The glutamate C4 resonance is significantly larger in neurons than in astrocytes, while the C2 glutamine resonance in astrocytes is higher than in neurons. Taken together these results suggest that astrocytes metabolize glucose mainly through aerobic glycolysis having an important contribution from PC, while neurons metabolize glucose mainly through the TCA cycle, using PDH as the main route.

It is possible to obtain in cells quantitative determinations of flux through specific metabolic pathways or through specific steps of a pathway using ^{13}C NMR. There are two possible strategies, both based on the principles of isotopic dilution and both requiring the use of mathematical models of metabolism. The first strategy allows the determination of fluxes by solving linear systems of equations relating flux to fractional ^{13}C enrichments in specific carbons of precursors and products. The second strategy involves the determination of relative isotopomer populations from an analysis of the ^{13}C–^{13}C spin coupling patterns. For a more precise description of both methods the reader is referred to the article of Portais and co-workers and Künnecke and co-workers in the Further reading section.

^{1}H NMR of cells

^{1}H NMR has an inherently higher sensitivity than ^{31}P or ^{13}C NMR and thus a larger number of metabolites

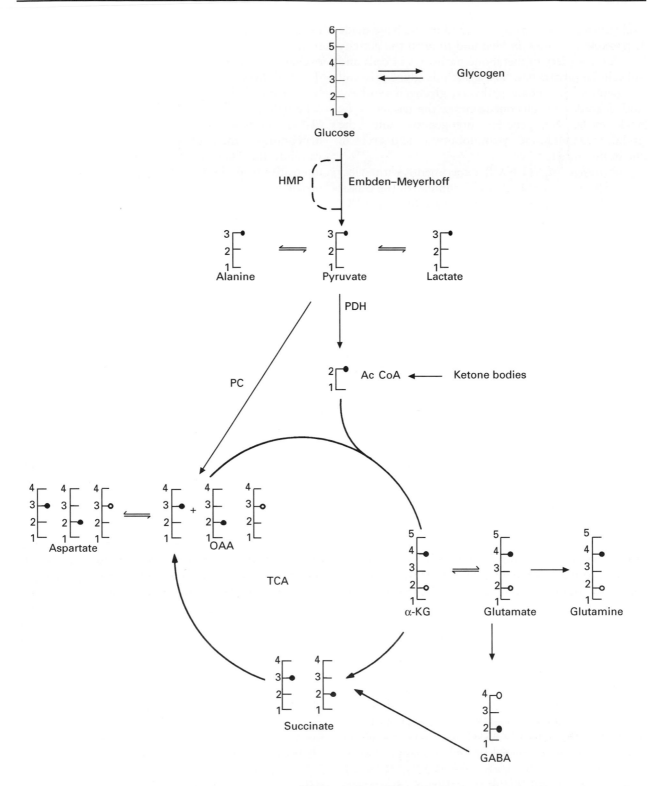

Figure 4 Metabolism of (1-^{13}C) glucose in neural cells. HMP: hexose monophosphate shunt. TCA: tricarboxylic acid cycle; OAA: oxalacetate; α-KG: 2-oxoglutarate; GABA: γ–aminobutyric acid; Ac CoA: acetyl coenzyme A. Only one turn of the TCA is considered for clarity. Filled or open circles in TCA cycle intermediates indicate labelling from PDH or PC, respectively.

is potentially detectable. However, two limitations need to be considered: (i) the water signals from the cells need to be eliminated and (ii) the complex homonuclear spin coupling patterns due to overlapping proton resonances have to be resolved. Both drawbacks have been overcome in part by the

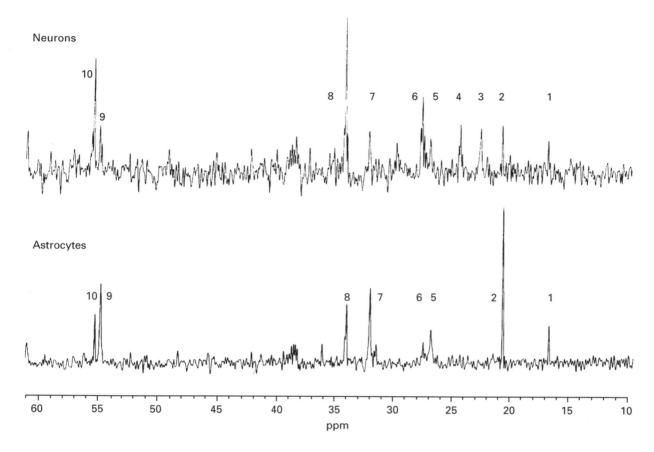

Figure 5 Proton decoupled ^{13}C NMR spectrum (125.7 MHz) of neutralized perchloric acid extracts obtained after the incubation of primary cultures of neurons (upper trace) or astrocytes (lower trace) with 5 mM (1-^{13}C) glucose during 24 h. 1: alanine C3, 2: lactate C3, 3: *N*-acetylaspartic C6, 4: acetate C2, 5: glutamine C3, 6: glutamate C3, 7: glutamine C4, 8: glutamate C4, 9: glutamine C2, 10: glutamate C2. An artificial line broadening of 1 Hz was used in both cases.

development of water suppression techniques and different one-, two- or *n*-dimensional editing methods.

The information provided by ^{1}H NMR spectra of cell suspensions is illustrated in **Figure 6** which shows a representative ^{1}H NMR spectrum of a suspension of rat erythrocytes acquired using a simple one-pulse sequence (**Figure 6B**). The most prominent resonances derive from the H3 proton of lactate, the choline protons of ergothioneine and the glutamate and glutamine resonances from glutathione. All these resonances are located on top of a much broader resonance arising from low-mobility lipids. Therefore ^{1}H NMR spectroscopy allows lactate production to be followed easily. This is normally done using spin echo sequences which eliminate the broad lipid component because of its short T_2 (**Figure 6A**). In addition to lactate, ^{1}H NMR normally allows the direct quantification of creatine singlets, singlets from choline-containing compounds and many other metabolites. Sometimes, quantification of these compounds requires very high resolution conditions, high spectrometer observation frequencies and computerized deconvolution

algorithms. Indeed the ^{1}H NMR spectrum provides a true fingerprint of the metabolites present in the cell and it appears to be characteristic for every cell type. This has been clearly shown in studies of cell typing with neural cells. Furthermore, it has been shown recently that fully transformed cells present an increase in the ^{1}H resonance of the PCho peak, and that the ratio PCho to GPC may serve as an indicator of multiple oncogene lesions.

A more recently developed application of ^{1}H NMR to cellular studies relies on the noninvasive determination of intra- and extracellular pH. The procedure is based in the use of a probe containing a reporter proton with a chemical shift sensitive to pH in the physiological range. This method is inherently more sensitive than previous ^{31}P NMR approaches. **Figure 6A** illustrates more clearly this procedure, by showing the spectrum of an erythrocyte suspension after the addition of the extrinsic pH probe imidazol-1-yl acetic acid methyl ester. The imidazole protons, H2 (*ca.* 8.2 ppm) and H4, H5 (*ca.* 7.25 ppm) are clearly observed in the aromatic region of the spectrum between 6 and 9 ppm (inset to **Figure 6A**). Two

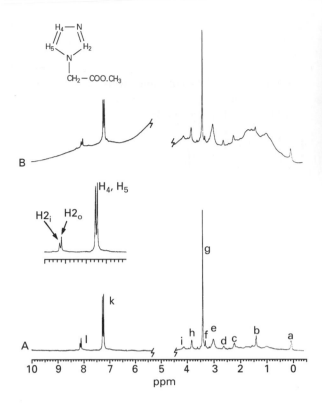

Figure 6 ^{1}H NMR spectra (360.13 MHz) of a rat erythrocyte suspension in the presence of a ^{1}H NMR pH probe. The water resonance (not shown) was attenuated with a 2 s presaturating pulse. (A) Spin echo acquisition with 5 ms interpulse delay. Inset: expansion of the aromatic region. (B) Conventional one-pulse acquisition. Inset: Chemical structure of imidazole-1-yl acetic acid methyl ester, the pH probe used. a: TSP, b: lactate H3, c: glutamate H3,H3′ in glutathione, d: aspartate H3,H3′, e: creatine (methyl), f: trimethyl groups in choline, carnitine and ergothioneine, g: methanol, h: H2 of amino acids, i: lactate H2, k: H4 and H5 of the pH probe, l: H2 of the pH probe. H2$_i$: intracellular probe, H2$_o$: extracellular probe. Methanol is produced by endogenous esterases immediately after the addition of the probe.

signals for the imidazole H2 proton (H2$_i$ and H2$_o$) with different chemical shifts are observed. These signals have been shown to be derived from molecules of the probe located in the intra- (H2$_i$) and extracellular (H2$_o$) space. Thus the different chemical shift of intra- and extracellular probes allows the determination of the intra- and extracellular pH values as well as the transmembrane pH gradient.

Finally, impressive progress in NMR microscopy methods has allowed the acquisition of ^{1}H NMR images of single cells. T_2-weighted NMR images from single cells normally show a dark cytoplasm and relatively bright nucleus. **Figure 7** provides an illustrative example of cellular NMR imaging, showing the morphology of a neuron from *Aplysia californica* before (**Figure 7A**) and after (**Figure 7B**) hypotonic shock. The hypotonic shock increased the relaxation times T_1 and T_2 of cytoplasm and nucleus,

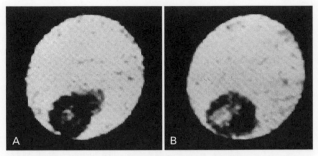

Figure 7 T_2-weighted spin echo microscopic ^{1}H NMR images (360.13 MHz) of an isolated neuron from *Aplysia californica* before (A) and after (B) hypotonic shock. Resolution (x,y,z) is 20 μm × 20 μm × 150 μm. Artificial sea water appears as a bright background surrounding the cell. T_2 differences between nucleus and cytoplasm cause these structures to appear as bright and dark, respectively. Reproduced with permission of the American Physiological Society from Hsu EW, Aiken NR, Blackband SJ (1996) *American Journal of Physiology* **271**: C1895–C1900.

but did not change the apparent diffusion coefficient of water. In conclusion, ^{1}H NMR methods provide information on glycolysis, fingerprint cell typing, state of proliferation, intra- and extracellular pH and cellular micromorphology. Representative articles describing these ^{1}H NMR applications are given in the Further reading section.

Other nuclei

In addition to the most commonly used nuclei, such as ^{31}P, ^{13}C and ^{1}H, other nuclei can be used in NMR studies of cellular metabolism (**Table 1**). The ^{19}F nucleus has a magnetic moment slightly smaller than ^{1}H, whereas its chemical shift range is larger by a factor of approximately 20. No contributions from tissue background signals are observed, which makes ^{19}F a useful magnetic label for metabolic studies. Using difluoromethylalanine or fluorinated derivatives of BAPTA, ^{19}F NMR spectroscopy has been extensively used to measure intracellular pH as well as cytosolic free Ca^{2+} and free Mg^{2+} in peripheral blood lymphocytes and other cells.

The ^{23}Na nucleus is a quadrupolar spin 3/2 nucleus with approximately 100% natural abundance. Because of its quadrupolar nature, some of the ^{23}Na is invisible to the NMR experiment. However, it has been possible to resolve intra- and extracellular ^{23}Na resonances in yeast cells and erythrocytes and to monitor Na^+–H^+ exchange in the latter cells.

The ^{17}O nucleus has an unfavourable gyromagnetic ratio but a large chemical shift range, close to 1000 ppm. ^{17}O NMR has been used extensively to study water metabolism in erythrocytes and subcellular organelles.

Finally, ^{2}H is one of the most recent additions to the repertoire of cellular NMR methods. ^{2}H has a natural abundance of only 0.02% and therefore the use of ^{2}H NMR requires the use of selectively enriched substrates. ^{2}H has been mainly used in cells to investigate the structure *in situ* of choline lipids in the cell membrane.

List of symbols

ADP = adenosine 5′-diphosphate; ATP = adenosine 5′-triphosphate; BAPTA = 1,2-bis(2-aminophenoxy) ethane-N,N,N′,N′-tetraacetic acid; CoA = coenzyme-A; CDP = cytidine 5′-diphosphate; DMMP = dimethyl methylphosphonate; GPC = glycerolphosphocholine; GPE = glycerolphosphoethanolamine; MDPA = methylene diphosphonic acid; MRI = magnetic resonance imaging; MRS = magnetic resonance spectroscopy; MRSI = magnetic resonance spectroscopic imaging; PCho = phosphocholine; PE = phosphoethanolamine; PC = pyruvate carboxylase; PCr = phosphocreatine, PDH = pyruvate dehydrogenase; P$_i$ = inorganic phosphate; PME = phosphomonoesters; TCA = tricarboxylic acid cycle; TSP = 2,2′,3,3′-tetradeutero trimethylsilyl propionate; T_1 and T_2 = relaxation times.

See also: **Biofluids Studied By NMR; Carbohydrates Studied By NMR;** *In Vivo* **NMR, Applications, ^{31}P;** *In Vivo* **NMR, Methods; NMR Microscopy; Nucleic Acids Studied Using NMR; Perfused Organs Studied Using NMR Spectroscopy; Solvent Supression Methods in NMR Spectroscopy; Two-Dimensional NMR Methods.**

Further reading

Bhakoo KK, Williams SR, Florian CL, Land H and Noble MD (1996) Immortalization and transformation are associated with specific alterations in choline metabolism. *Cancer Research* 56: 4630–4635.

Freshney RI (1994) In: Freshney RI (ed) *Culture of Animal Cells: a Manual of Basic Technique.* New York: Wiley-Liss.

Gil S, Zaderenko P, Cruz F, Cerdan S and Ballesteros P (1994) Imidazol-1-ylalkanoic acids as extrinsic ^{1}H NMR probes for the determination of intracellular pH, extracellular pH and cell volume. *BioOrganic and Medicinal Chemistry* 2: 305–314.

Gillies RJ, Galons JP, McGovern KA, Scherer PG, Lien YH, Job C, Ratcliff R, Chapa F, Cerdan S, Dale BE (1993) Design and application of NMR-compatible bioreactor circuits for extended perifusion of high-density mammalian cell cultures. *NMR in Biomedicine* 6: 95–104.

Hsu EW, Aiken NR, Blackband SJ (1996) Nuclear magnetic resonance microscopy of single neurons under hypotonic perturbation. *American Journal of Physiology* 271 (Cell Physiol 40): C1895–C1900.

King GF and Kuchel PW (1994) Theoretical and practical aspects of NMR studies of cells. *Immunomethods* 4: 85–97.

Künnecke B, Cerdan S and Seelig J (1993) Cerebral metabolism of (1,2-^{13}C$_2$)glucose and (U-^{13}C$_4$)3-hydroxybutyrate in rat brain, as detected by ^{13}C spectroscopy. *NMR in Biomedicine* 6: 264–277.

Lundberg P, Harmsen E, Ho C, and Vogel HJ (1990) Nuclear magnetic resonance studies of cellular metabolism. *Analytical Biochemistry* 191: 193–222.

Portais JC, Schuster R, Merle R, Canioni P (1993) Metabolic flux determination in C6 glioma cells using carbon-13 distribution upon (1-^{13}C)glucose incubation. *European Journal of Biochemistry* 217: 457–468.

Szwergold BS (1992) NMR spectroscopy of cells. *Annual Review of Physiology* 54: 775–798.

Urenjak J, Williams SR, Gadian DG and Noble M (1993) Proton nuclear magnetic resonance spectroscopy unambiguously identifies different neural cell types. *Journal of Neuroscience* 13: 981–989.

Vogel HJ, Brodelius P, Lilja H and Lohmeier-Vogel EM (1987) Nuclear magnetic resonance studies of immobilized cells. *In* Mosbach K (eds.), *Methods in Enzymology*, vol. 135, B: pp. 512–536. New York: Academic Press.

Chemical Applications of EPR

Christopher C Rowlands and **Damien M Murphy**,
Cardiff University, UK

Introduction

EPR spectroscopy is a technique that affords a means for the detection and quantification of paramagnetism, i.e. the presence of unpaired electrons. It is applicable to solids, liquids or even gases. A compound or molecular fragment containing an odd number of electrons is paramagnetic. In this article we give examples to show the power and breadth of the technique. The reader is advised, however, to read the Encyclopedia article on the theoretical aspects of EPR to appreciate fully the strength of the technique and also the excellent series *Specialist Periodical Reports on Electron Spin Resonance*, published by the Royal Society of Chemistry, which has now reached volume 16.

The technique can be applied to obtain information on the following four areas.

1. How atoms are bonded together in molecules (structure).
2. The route by which compounds are transformed from one to another, i.e. which bonds break and which are formed (mechanism).
3. The rate at which these processes occur (kinetics).
4. The molecular interactions that exist, for example, between solvent and solute (environment).

Knowledge of the *g* values and the detailed hyperfine interactions (*A* values) allow us to help identify radical species, and these parameters contain information about the electron distribution within the molecule. Radicals are often present as intermediates during a reaction; consequently their identification will give information concerning the reaction mechanism and measurement of how their concentration changes with time will give kinetic data.

Let us now look in greater detail at several major applications of the technique.

Organic and organometallic free radicals; structure, kinetics and mechanism

There are many many examples of the power of the technique in this area and space does not permit us to cover fully all examples. Articles by Tabner and by Rhodes fully cover the basic principles and practical aspects for the detection and evaluation of radical anions and cations (see Further reading) while more recently Davies and Gescheidt have reviewed the literature up to 1996 in the Specialist Reports referred to earlier. In the same publication, Alberti and Hudson cover the topics of organic and organometallic free radicals up to 1995. Consequently we will select just one or two examples from these areas.

An early method, widely used for radical generation is based on the redox system Ti^{3+}–H_2O_2. This was first used by Dixon and Norman to generate $HO^{\bullet}$ and $HOO^{\bullet}$ radicals obtained by rapid mixing of aqueous solutions of $TiCl_3$ and H_2O_2 with a substrate immediately before the EPR cavity. It has been suggested that the main reacting species in such a system is Ti–O–$O^{\bullet 3+}$. The EPR signals obtained from such a system depend upon such factors as flow rate, temperature, pH and the ratio of the two components. Flow systems such as this have been used quite successfully to study polymerization reactions that are initiated by redox reactions. Analysis of the hyperfine interactions of the reacting species enabled the structure of the radicals to be determined.

The use of such flow and/or stopped flow systems has been used quite extensively to investigate the kinetics of radical reactions. These include the use of a stopped flow EPR system to investigate the one-electron oxidation of a range of phenols by a dimeric manganese(IV/IV) triazacyclononane complex with and without hydrogen peroxide.

Another use has been the development of a technique of photolytic time-resolved EPR that allows kinetic data to be obtained with relative ease. The method involves flowing oxygen-free solutions of the materials under investigation through a flat reaction cell placed within the EPR cavity. The flow rates of the solutions could be adjusted so that signal intensities were not dependent on their cavity dwell time. The solutions could be irradiated either continuously or intermittently and thermostatted by the use of a stream of dinitrogen. Signal averaging is used to overcome signal-to-noise difficulties.

Spin trapping, spin labels and spin probes in polymers and biological systems

The role of free radicals in biological and medicinal systems is of considerable interest as they have been implicated in a wide range of diseases such as cancer. In such systems, the direct detection of free radicals is often not possible because of their low steady-state concentration arising from their high reactivity and transient nature. In this instance, the technique of spin trapping is used. Spin trapping involves the addition of a diamagnetic molecule, usually a nitrone or nitroso compound, which reacts with a free radical to give a stable paramagnetic spin adduct that accumulates until it becomes observable by EPR (Eqn [1]).

$$X^\bullet \quad + \quad ST \quad \longrightarrow \quad XST^\bullet \qquad [1]$$

Transient free radical	Spin trap	Long-lived spin adduct
Paramagnetic	*Diamagnetic*	*Paramagnetic*

By determination of the parameters of the spin adduct spectrum it is often possible to identify the nature of the primary trapped radical, or at least to determine the type of radical trapped. Table 1 gives examples of the range of hyperfine coupling constants obtained for a variety of trapped species for the two most commonly used spin traps DMPO (dimethylpyridine N-oxide) and PBN (α-phenyl-N-t-butyl nitrone). Care has to be taken, however, because the coupling constants vary with solvent polarity (for example, in water the DMPO adduct of the t-butoxy radical has $a(N) = 1.48$ and $a(H) = 1.60$ mT while in toluene $a(N) = 1.3$ and $a(H) = 0.75$ mT).

The spin trap must react at relatively fast trapping rates and the radical adduct must have a reasonable half-life; for example, DMPO has a $k_{trapping}$ for OH$^\bullet$ of 2×10^9 M^{-1} s^{-1} and half-life of about 15 min, whereas for O$_2^-$, $k_{trapping}$ is 1×10^1 M^{-1} s^{-1} and the half-life is about 90 s. Consequently the DMPO–O$_2^-$ adduct is rarely seen. This difficulty has been overcome by the use of phosphorus-substituted DMPO, in particular where one of the methyl groups in the 5 position has been replaced by a diethoxyphosphoryl group. This increases the half-life of the OOH adduct considerably in both organic and aqueous solvents, with the difference being more pronounced in aqueous solvents. One other advantage of the use of phosphorus-substituted DMPO spin traps is that the β-^{31}P hyperfine coupling constant has a much larger variation (2.5–5.5 mT) than the β-H hyperfine

Table 1 Hyperfine coupling constants (mT) for a variety of spin-trapped adducts

	PBN		DMPO	
Free radical	^{14}N	H	^{14}N	H
$\bullet$H	1.53	0.82(2H)		
$\bullet$Phenyl	1.60	0.44	1.57	2.5
$\bullet$CCl$_3$	1.39	0.175		
$\bullet$OH	1.54	0.27	1.50	1.50
O$_2^-$	1.43	0.225	1.45	1.16
Bu^{t}O$\bullet$			1.48	1.60
SO$_3^-$			1.47	1.60
PhS$\bullet$			1.29	1.415

PBN, *N*–*t*–butyl–α–phenylnitrone TEMPOL, 4–hydroxy–2,2,6,6–tetramethylpiperidinooxy

coupling constant (0.6–2.5 mT), thus allowing an easier identification of the trapped radical.

Table 2 gives a range of bimolecular rate constants with both PBN and DMPO for a variety of radical species.

In a study of heterogeneous catalysis, the spin trapping technique has been used to prove the presence of radical species on a catalyst surface. In a study of a palladium metal catalyst supported on alumina, it was shown that hydrogen is dissociatively chemisorbed by trapping hydrogen atoms with PBN. Alkyl and aromatic free radicals were also shown to be present when other feed stocks were used. In a separate study of the photodecomposition of acetaldehyde, it was shown that high-power irradiation (600 W mercury/xenon lamp) generated a different PBN adduct from that given with irradiation under direct sunlight and that the presence of oxygen played a major role in the photochemistry.

Table 2 Second-order rate constants for radical spin trapping for the traps PBN and DMPO

Spin trap	R$^\bullet$	$T(^\circ C)$	Rate constant (M^{-1}s^{-1})
PBN	Me$^\bullet$	25	4×10^6
	RC$^\bullet$H$_2$	40	1.3×10^5
	Ph$^\bullet$	25	2×10^7
	HO$^\bullet$	25	2×10^9
	ROO$^\bullet$	60	1×10^2
DMPO	HO$^\bullet$	25	2×10^9
	O$_2^-$	25	1×10^1
	RC$^\bullet$H$_2$	40	2.6×10^6

Since the first reported study on nitroxide spin labelling in 1965, the technique has found widespread application to biological and chemical problems. The introduction of a small free radical, the spin label, onto the molecule of interest gives environmental information at that point. It has the ability to measure very rapid molecular motion and is usually free from interference, thus making it a very powerful technique that gives useful details about dynamic processes at the molecular level.

The types of free radical mostly used as spin probes or spin labels are based on nitroxides, such as the label known as TEMPOL (4-hydroxy-2,2,6,6-tetramethylpiperidinooxy). **Figure 1** shows the change in the EPR spectrum of a spin-labelled system going from a fully mobile isotropic system (**Figure 1C**) to a fully immobilized glass type spectrum (**Figure 1A**). Measurement of the hyperfine parameters enables information to be obtained about the mobility or rigidity of the system under study.

The spin probe technique has been used to investigate the interactions of the anionic surfactant SDS and gelatin. It was found that for simple SDS solutions the rotational correlation time of the spin probe, 16-doxylstearic acid methyl ester, increased slightly with increased surfactant concentration. In

contrast, in the presence of gelatin these properties varied markedly as a function of the stoichiometric ratio of the surfactant to gelatin concentration, with the correlation time decreasing as the hyperfine coupling constant increased with increasing surfactant concentration. Such behaviour was attributed to the characteristics of the various amino acids present in the gelatin.

The use of spin labelling in polymer systems has enabled information to be obtained about polymer chain size and also their conformational state, with three main states being distinguished. The ability of some nitroxide radicals to diffuse easily into various polymer systems, coupled with the fact that their EPR spectra showing three lines caused by coupling to ^{14}N show a high degree of rotational motion in the amorphous parts of the polymer, means that studies varying the temperature through the glass transition temperature T_g will enable changes to be measured in the rotational correlation time. Since the rotational correlation time obeys the Arrhenius law (Eqn [2]),

$$\tau = \tau_0 \ \exp(E_a/RT) \qquad [2]$$

which is valid for $T > T_g$, values for the rotational activation energy of the nitroxide free radical can be measured. Since this does not depend on radical size but only on the polymer matrix, a measure of polymer mobility is obtained.

There are several excellent reviews on the application of the method to studies in polymer systems and also to investigations of biological membranes. These are listed in the 'Further reading' section.

Study of transition metal compounds

EPR spectroscopy is widely used to study the structure and geometry of d-transition metal ions (TMI) in inorganic complexes, in biological systems and in catalysts. Information obtained through analysis of the spectrum varies from simple identification of the metal centre to a thorough description on the electronic structure of the complex. At a simple level, the main interactions experienced by the TMI that consequently influence the EPR spectrum are (i) the electronic Zeeman effect and (ii) interaction between the electron and the nuclear spin. In the first case, the interaction is expressed via the g tensor which carries information on electronic structure; in the second case, the metal hyperfine interaction is expressed via the A tensor.

Although the information regarding the transition metal compound can be comprehensive, we shall

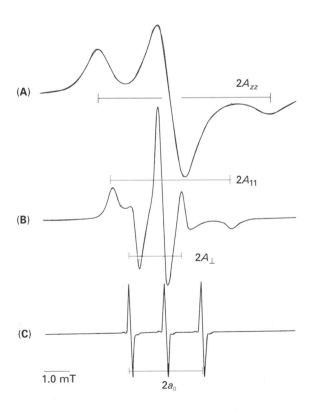

Figure 1 EPR spectra arising from a ^{14}N ($I = 1$) showing (A) randomly and rigidly oriented in a frozen matrix; (B) antisotropic motion in a randomly oriented environment; (C) isotropic spectrum with no restriction on movement.

confine ourselves here to a simple illustration on the power of EPR for investigating the symmetry of the metal centre in the solid state. The simplest case arises for TMI with one d electron. For a metal in a tetragonally distorted octahedral field experiencing compression along the 4-fold axis (D_{4h}), the five d orbitals have the splittings shown in **Figure 2** with the nondegenerate d_{yz} state lowest in energy. For a slight distortion, the low-lying 2-fold degenerate excited states (d_{xy}, d_{xz}) are close to the ground state, as depicted in **Figure 2A**. The small energy separation Δ_1 results in a short relaxation time with the result that the EPR signal can only be observed at low temperatures. Greater distortions lead to longer relaxation times, so that the EPR signals become observable at increasingly higher temperatures with $g_\perp > g_\parallel$ as shown in **Figure 2A**. In the case of elongation along the 4-fold axis, the unpaired electron now

occupies a doubly degenerate ground state (d_{xz}, d_{yz}) and no EPR signal is observed. However, if the ion experiences a trigonal distortion with a resulting point symmetry of D_{3d} as shown in **Figure 2B**, then the single electron occupies the d_{z^2} orbital so the order of the g-values changes or becomes 'inverted' (i.e. $g_\parallel > g_\perp$) as depicted in **Figure 2B**. This simple example illustrates the effects of the ground state of the TMI on the resulting EPR spectrum.

As shown above, the point symmetry at the metal can therefore be reflected in the **g** values, but the **A** values can also be influenced in a similar manner. Furthermore, depending on the point symmetry, the principal axes of **g** and **A** may or may not be coincident. The relationship between these tensors, the EPR spectral characteristics and the point symmetry of the metal are shown in **Table 3**. The importance of these relationships is that each type of EPR

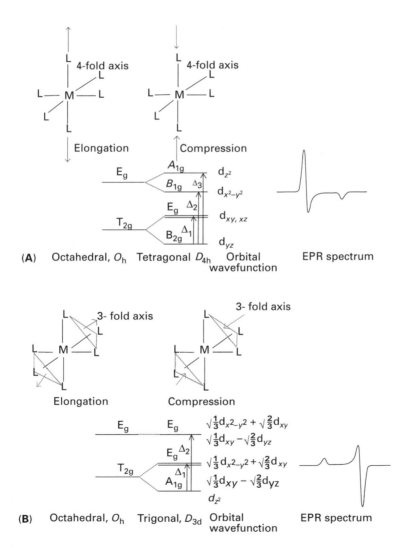

Figure 2   Orbitals and corresponding EPR spectrum for an octahedral complex with (A) tetragonal and (B) trigonal distortion.

Table 3 Relationship between **g** and **A** tensor, EPR symmetry and the point symmetry of the paramagnets

EPR symmetry	g and A tensors	Coincidence of tensor axis	Molecular point symmetry
Isotropic	$g_{xx} = g_{yy} = g_{zz}$ $A_{xx} = A_{yy} = A_{zz}$	All coincident	O_h T_d O, T_h, T
Axial	$g_{xx} = g_{yy} \neq g_{zz}$ $A_{xx} = A_{yy} \neq A_{zz}$	All coincident	D_{4h}, C_{4v}, D_4, D_{2d}, D_{6h}, C_{6v}, D_6, D_{3h}, D_{3d}, C_{3v}, D_3
Rhombic	$g_{xx} \neq g_{yy} \neq g_{zz}$ $A_{xx} \neq A_{yy} \neq A_{zz}$	All coincident	D_{2h}, C_{2v}, D_2
Monoclinic	$g_{xx} \neq g_{yy} \neq g_{zz}$ $A_{xx} \neq A_{yy} \neq A_{zz}$	One axis of g and A coincident	C_{2h}, C_s, C_2
Triclinic	$g_{xx} \neq g_{yy} \neq g_{zz}$ $A_{xx} \neq A_{yy} \neq A_{zz}$	Complete non-coincidence	C_1, C_1
Axial non-collinear	$g_{xx} \neq g_{yy} \neq g_{zz}$ $A_{xx} = A_{yy} \neq A_{zz}$	Only g_{zz} and A_{zz} coincident	C_3, S_6, C_4, S_4, C_{4h}, C_6, C_{3h}, C_{6h}

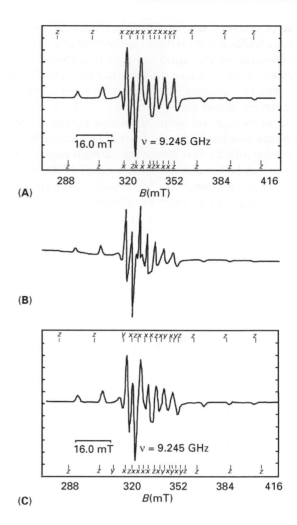

(A)

(B)

(C)

Figure 3 Experimental and simulated X-band powder EPR spectra of $H_2[V(R,R\text{-HIDPA})_2]$ diluted in the corresponding zirconium compound, where HIDPA = hydroxyiminodipropionate. (A) Simulated spectrum with axial symmetry and $g_{\parallel} = 1.9195$, $g_{\perp} = 1.9839$, $A_{\parallel} = 17.5$ mT and $A_{\perp} = 4.9$ mT; (B) experimental spectrum; and (C) simulated spectrum with rhombic symmetry $g_{zz} = 1.9195$, $g_{xx} = 1.9848$, $g_{yy} = 1.9829$, $A_{zz} = 17.15$ mT, $A_{xx} = 4.6$ mT and $A_{yy} = 5.2$ mT. Reproduced with permission of the Royal Society of Chemistry from Mabbs FE (1993) Some aspects of the electron paramagnetic resonance spectroscopy of d-transition metal compounds. *Chemical Society Reviews* **22**: 313–324.

behaviour is associated with a restricted number of point symmetries, which places constraints upon the geometrical structures of the paramagnet. For example, if the paramagnet is known to have, say, rhombic symmetry, then the associated geometry must belong to one of the point groups D_{2h}, C_{2v} or D_2 (**Table 3**). It would be incorrect to assign a structure that belongs to a more symmetric arrangement, e.g. D_{4h} which is strictly axial. Therefore, for a system with unknown structure, if the EPR symmetry can be determined (**Table 3**) we have invaluable structural information since the paramagnet can only belong to a restricted range of point symmetries. As an example, consider the EPR spectrum of $H_2[V(R,R\text{-}HIDPA)_2]$ diluted in the corresponding zirconium compound (**Figure 3B**). At first glance the spectrum appears to have axial symmetry, since the features associated with the x and y directions are not split. Simulation of a spectrum based purely on axial symmetry is not entirely satisfactory (**Figure 3A**). The introduction of a small anisotropy in both g_{xx}, g_{yy} and A_{xx}, A_{yy}, however, produces an improved simulation (**Figure 3C**). Although the rhombicity is small, it is consistent with the known structure of the compound. From the crystallographic data, the highest possible point symmetry at the vanadium is C_2. In more complex cases, other techniques such as variable-frequency CW-EPR, ENDOR and ESEEM are required to provide a more detailed description on the electronic structure of transition metal complexes. Nevertheless, one can start to appreciate the power of CW-EPR for studies of paramagnetic transition metal complexes.

Surface chemistry and catalysis

EPR has been widely applied to surface science and catalysis in order to examine a variety of surface paramagnetic species important in catalytic processes including adsorbed atoms, ions or molecules that may be intermediates in chemical reactions, intrinsic surface defects, transition metal ions supported on an oxide and spin labels interacting with a surface. Information regarding the important physicochemical characteristics of the surface can be gained through analysis of the EPR spectrum. Properties such as surface crystal field, surface redox properties, surface group morphology, mobility of adsorbed species, coordination of surface metal ions and identification of the active site have been studied on a variety of surfaces.

In many cases, the surface itself is diamagnetic and investigations by EPR rely on indirect methods. A large amount of information regarding the state of a surface can be gained by adsorption of probe molecules (i.e. a molecule whose properties in the adsorbed state can be monitored and evaluated by EPR). These molecules may be divided roughly into two classes. The first is paramagnetic probes (or probes that become paramagnetic upon adsorption). The spectroscopic features of simple paramagnetic probes such as NO and O_2^- can provide detailed information on surface electrostatic fields. For example, the g values of the adsorbed superoxide anion (O_2^-) are known to depend not only on the nominal charge of the surface cation where adsorption occurs but also on the coordination environment of the cation itself, as shown in **Figure 4**. Paramagnetic probes such as VCl_4 or $MoCl_5$ retain their paramagnetism during reaction with surface hydroxyl groups, allowing one to study the distribution of the surface groups themselves. The other type are diamagnetic probes that remain diamagnetic upon adsorption. These probes allow one to investigate the properties of a paramagnetic adsorption site, such as the changes in the coordination sphere of a surface transition metal ion.

Metal oxides are frequently involved in catalysis, either directly or indirectly as supports. By studying the charge transfer reactions at the gas–solid or liquid–solid interface, the nature, number and strength of the acidic or basic catalyst can be investigated. The oxidizing properties of the oxide can be studied by formation of the radical cations of aromatic molecules such as anthracene and perylene, while the reducing properties of the surface can be monitored through the formation of the radical anions of molecules with large electron affinity such as tetracyanoethylene (TCNE) or nitroaromatic compounds. The EPR spectra of the corresponding radical cations and anions yield both qualitative and quantitative information on the redox nature of the surface and the subject has been reviewed.

Finally, a molecular approach to catalysis requires a deep understanding of the coordination processes occurring at the catalyst surface. The use of EPR to investigate the various types of coordination chemistry that occur on catalytic systems involving transition metal ions and oxide matrices have been

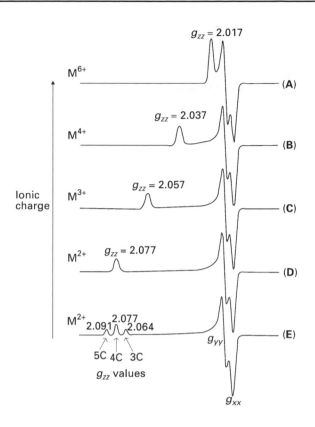

Figure 4 Simulated EPR spectra of the superoxide (O_2^-) anion adsorbed on differently charged cationic sites. The g_{zz} values of the radical have the form $g_{zz} = g_e + 2\lambda/\Delta$ where λ is the spin–orbit coupling constant for oxygen and Δ is the extent of the energy splitting between the two π^* orbitals, caused by the local electrostatic field of the cation. As the charge varies, Δ will vary and different g_{zz} values are observed (A–D). The anion is also sensitive enough to distinguish between cations with the same nominal charge which, because of different coordinative environments (e.g. 5-coordinated, 4C or 3C surface cations), exert slightly different electrostatic fields (E).

reviewed. It has been shown that the reactivity and properties of the central transition metal ion in the heterogeneous system are heavily influenced by the coordinating ligands (much more than in the homogeneous counterparts) so that a variety of TMI complexes can be created and stabilized at the surface. Changes to the EPR spectroscopic features as a function of variation in the coordination sphere of a surface-supported TMI can easily be followed. An example of this is shown in **Figure 5** for a Ni$^+$ supported catalyst. Changes in the spectroscopic features of the signal as a function of differing CO pressures illustrate the sensitivity of the technique to perturbations in the local environment. This coordination chemistry approach to EPR studies of the mechanism of catalytic reactions is illustrated in the dimerization of olefins and the oxidation of methanol on the surface of oxide catalysts.

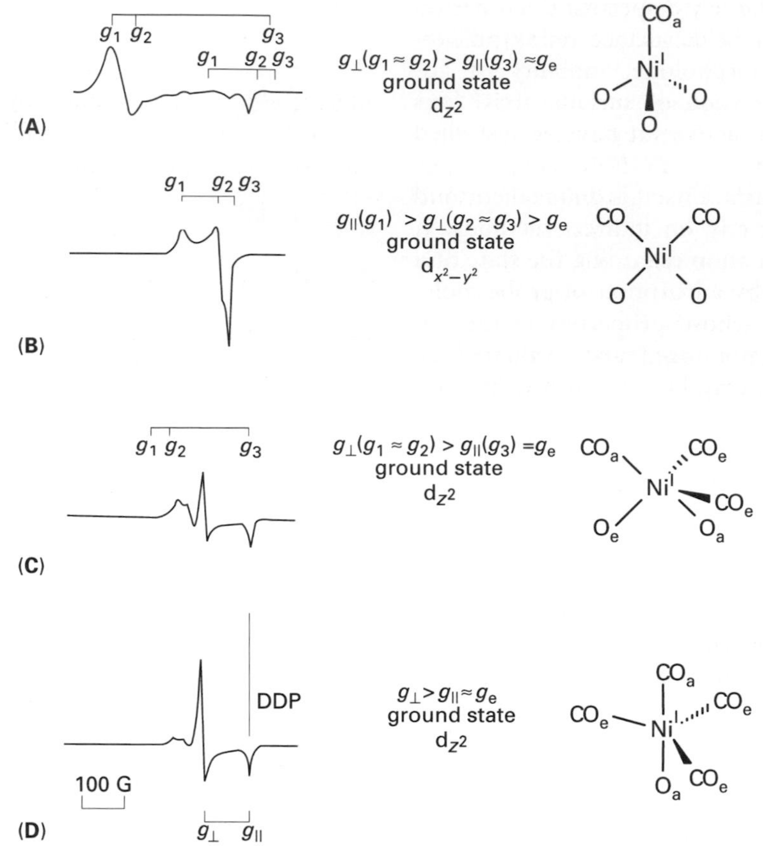

Figure 5 X-band EPR spectra of Ni$^+$ complexes formed upon CO adsorption on reduced Ni/SiO$_2$ catalysts under a pressure of (A) 10 torr CO followed by evacuation at 340 K, (B) 10 torr, (C) 100 torr, (D) 400 torr. Subscript a represents axial and subscript e represents equitorial positions. Reprinted with permission from Dyrek K and Che M (1997) EPR as a tool to investigate the transition metal chemistry on oxide surfaces. *Chemical Reviews* **97**: 303–331 Copyright 1997 American Chemical Society.

Applications of pulsed techniques and high-frequency EPR

The applications discussed above were performed using conventional CW (continuous wave) methods, predominantly at X-band (~9 GHz) frequency. However, in recent years EPR spectroscopy has experienced an extraordinarily rapid development in two noted areas, namely pulsed techniques and high-frequency EPR. Already both developments have proved to be invaluable tools in important areas of physics, chemistry and biology, and no article on the applications of EPR spectroscopy is complete without an explanation of the powerful attributes of these techniques.

The difference between high-frequency and low-frequency EPR is usually defined by the magnet system. Superconducting or Bitter magnets are employed for high-frequency EPR (such as those operating at 94 GHz), whereas traditional electromagnets are used for lower frequencies (e.g. Q-band

(~35 GHz) to L-band (~1 GHz)). High-frequency EPR offers numerous advantages over the low-frequency spectrometers, including increased resolution of *g* factors (which is particularly important in complex or disordered systems where spectral lines are not resolved at lower fields) and increased absolute sensitivity (which is important in studies of small single crystals or any system where the number of spins is inherently low). High-frequency EPR also allows measurement of a number of integer-spin systems (such as Ni(II) species) which often have very high zero-field splittings. These systems only start to become resolved at high frequency where the energy of the photon is above the zero-field energy. The number of applications is growing rapidly, but the most compelling areas of development are in complex disordered systems prevalent in structural biology, including new insights into the detailed mechanisms associated with photosynthesis and many metalloproteins. Because EPR is only sensitive to the sites with unpaired spins, it targets only the most important bonds that largely determine the chemical and physical properties of the metalloprotein. High-frequency EPR is also particularly appropriate for studying these complex enzymes since the active sites often involve metal ions or metal clusters with high zero-field splittings, which are EPR-silent at low frequency. Furthermore, most disordered systems are better studied at high frequencies because the improved spectral resolution allows the local structure to be inferred, especially in conjunction with other techniques such as ENDOR.

Exploitation of pulsed techniques in chemical research has been hampered in the past by expensive instrumentation and the lack of sufficiently fast digital electronics. The situation has undoubtedly changed in recent years, and the various pulsed EPR methods have found increased or exclusive use in areas such as time-resolved EPR, as tools for studying molecular motion, in electron–nuclear double resonance spectroscopy and in EPR imaging. Fourier transform EPR has proved particularly useful for studying the various spin polarization mechanisms involved in free radical chemistry, while the very high time resolution involved in echo-detected EPR makes it an ideal technique for measuring rates of chemical reactions and determining relaxation times.

The underlying basis of pulsed EPR, like NMR, requires the generation of a spin echo by a short intense microwave pulse. However, in some cases the coupling between the electron spin and the nuclear spins can cause a modulation of the echo intensity. As this electron spin echo envelope modulation (ESEEM) can only occur if anisotropic interactions are present, it is only found in the solid state. Analysis of the ESEEM spectra allows determination of very small hyperfine couplings which are too small to be measured by CW-EPR, so that information on the number, identity and distance of weakly inter-acting nuclei may be obtained. A wide variety of applications of the nuclear modulation effect have been demonstrated, including determination of ligand structure in metalloproteins, magnetic properties of triplet states and location and coordination of surface complexes. In fact, the ESEEM method ranks with EXAFS as one of the most important tools for determining the structures of active centres in metalloproteins.

Conclusions

It is intended in this article to demonstrate the diversity of applications in which CW-EPR is traditionally used, while briefly describing some current uses of more advanced EPR techniques. However, with recent technological developments in FT-EPR, pulsed

ENDOR, EPR imaging and high-frequency EPR, new and previously undreamed of prospects are likely to emerge. In many cases the spectroscopic data of interest cannot be obtained by any other method. This fact alone will ensure EPR a place as a dominant spectroscopic technique of major importance well into the future.

List of symbols

a = hyperfine splitting constant, units are mTesla; a_0 = isotropic hyperfine coupling constant, units are mTesla; $A_{\parallel}$ = hyperfine coupling constant parallel to a unique symmetry axis, units are mTesla; $A_{\perp}$ = hyperfine coupling constant perpendicular to a unique symmetry axis, units are mTesla; E_a = rotational activation energy, in kJ mol^{-1}; g = g factor; $g_{\parallel}$ = parallel to a unique symmetry axis; $g_{\perp}$ = perpendicular to a unique symmetry axis; τ = free radical tumbling correlational time (in seconds).

See also: **EPR Imaging; EPR, Methods; EPR Spectroscopy, Theory; Inorganic Chemistry, Applications of Mass Spectrometry; Spin Trapping and Spin Labelling Studied Using EPR Spectroscopy; Surface Studies By IR Spectroscopy.**

Further reading

Bales B, Griffiths PC, Goyffon P, Howe AM and Rowlands CC (1997) EPR insights into aqueous solutions of gelatin and sodium dodecylsulphate. *Journal of the Chemical Society, Perkin Transactions 2* 2473–2477.

Berliner LJ and Reubens J (eds) (1989) *Biological Magnetic Resonance*, vol. 8, *Spin Labelling Theory and Applications*. New York: Plenum Press.

Carley AF, Edwards HA, Hancock FE, *et al* (1994) Application of ESR to study the hydrogenation of alkenes and benzene over a supported Pd catalyst. *Journal of the Chemical Society, Faraday Transactions*, **90**: 3341–3346.

Che M and Giamello E (1987) In: Fierro JLG (ed) 'Spectroscopic Characterisation of Heterogeneous Catalysts', vol. 57, Part B, p B265. Amsterdam: Elsevier Science Publishers.

Dixon WT and Norman ROC (1962) Free radicals formed during the oxidation and reduction of peroxides. *Nature*, **196**, 891.

Dyrek K and Che M. (1997) *Chemical Review* 97: 305–331.

Egerton TA, Murphy DM, Jenkins CA and Rowlands CC (1997) An EPR study of spin trapped free radical intermediates formed in the heterogeneously assisted photodecomposition of acetaldehyde. *Journal of the Chemical Society, Perkin Transactions 2* 2479–2485.

Marsh D (1994) Spin labelling in biological systems. *Specialist Periodical Reports, Electron Spin Resonance* **14**: 166–202.

Möbius K (1994) EPR studies of photosynthesis. *Specialist Periodical Reports, Electron Spin Resonance* **14**: 203–245.

Ranby B and Rabek JF (1977) *ESR Spectroscopy in Polymer Research*. Berlin: Springer-Verlag (and references therein).

Rhodes CJ (1991) Organic radicals in solid matrices. *Specialist Periodical Reports, Electron Spin Resonance* **13a**: 56–103.

Schweiger A (1991) Pulsed electron spin resonance spectroscopy: basic principles, techniques and examples of applications. *Angewandte Chemie, International Edition, English* **30**: 265–292.

Tabner BJ (1991) Organic radicals in solution. *Specialist Periodical Reports, Electron Spin Resonance* **13a**: 1–55.

Wasserman AM (1996) Spin labels and spin probes in polymers. *Specialist Periodical Reports, Electron Spin Resonance* **15**: 112–151.

Chemical Exchange Effects in NMR

Alex D Bain, McMaster University, Hamilton, ON, Canada

MAGNETIC RESONANCE
Theory

Introduction

Chemical exchange, in NMR terms, means that a nucleus moves among a set of magnetic environments. The sample is macroscopically at equilibrium, but an individual nucleus exchanges among a number of sites, so the magnetic properties of the nucleus are modulated by the exchange. Chemical exchange effects were recognized early in the development of NMR: at about the same time (and in the same laboratory) as the scalar coupling. In N,N'-dimethyl formamide and many other molecules with N,N'-dimethyl groups (**Figure 1**), the two methyl groups have different chemical shifts if the molecule is rigid. At low temperature this is effectively the case, because of restricted rotation about the C–N bond. However, as the sample is warmed, the two signals broaden, coalesce, and eventually start to sharpen up to a single line at the average chemical shift, as shown in **Figure 1**.

This behaviour is usually interpreted in terms of an 'NMR timescale'. For rotation about the C–N bond, which is slow on this timescale at low temperature, the two signals are distinct and relatively sharp. At higher temperature, the rotation is faster and all that we can detect is the average resonance frequency. The broad coalescence line shape is characteristic of the intermediate timescale, and is perhaps the most familiar and obvious manifestation of chemical exchange. In this article we will concentrate on intermediate exchange, because a thorough

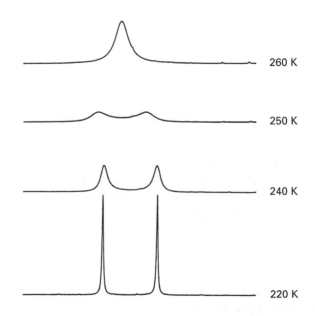

3-Dimethylamino-7-methyl-1,2,4-benzotriazine

260 K

250 K

240 K

220 K

Figure 1 ^{1}H NMR spectra of the two *N*-methyl groups in 3-dimethylamino-7-methyl-1,2,4-benzotriazine, as a function of temperature.

understanding of this case clarifies most other aspects of chemical exchange.

Chemical exchange in all three regimes – slow, intermediate and fast – has an effect on the NMR spectrum. Typical values of the parameters mean that processes with rates in the range of $1–10^4$ s^{-1} can be studied most easily. In other words, this means activation energies for the reaction in the range of approximately 40–80 kJ mol^{-1} (10–20 kcal mol^{-1}). This includes conformational changes in ring structures, restricted rotations about chemical bonds, ligand rearrangements in coordination complexes, and some intermolecular transfer processes. Molecules are dynamic entities, so the effects of chemical exchange are apparent in many NMR spectra.

The theory of chemical exchange in simple; uncoupled spin systems have often been couched in terms of the Bloch equations, and this is the approach used in most of the literature. However, we feel that it is simpler and easier to consider the time domain. This time-domain method allows us to treat all exchanging systems – slow, intermediate or fast, coupled or uncoupled – in a consistent and simple way. There is also a simple physical picture of the spectroscopic transition probability that helps in this interpretation of the theory.

The Bloch equations approach

The Bloch equations for the motion of the x and y magnetizations (usually called the u- and v-mode signals), in the presence of a weak radiofrequency field, B_1, are given in Equation [1].

$$\frac{du}{dt} + \frac{u}{T_2} - (\omega_0 - \omega_1)v = 0$$
$$\frac{dv}{dt} + \frac{v}{T_2} - (\omega_0 - \omega_1)u = \gamma B_1 M_z \quad [1]$$

In this equation, ω_1 is the frequency of the RF irradiation, ω_0 is the Larmor frequency of the spin, T_2 is the spin–spin relaxation time and M_z is the z magnetization of the spin system. We can simplify the notation somewhat by defining a complex magnetization, M, as in Equation [2].

$$M = u + iv \quad [2]$$

With this definition, the Bloch equations can be written as in Equation [3].

$$\frac{dM}{dt} + i(\omega_0 - \omega_1)M + \frac{1}{T_2}M = i\gamma B_1 M_z \quad [3]$$

In chemical exchange, the two exchanging sites, A and B will have different Larmor frequencies, ω_A and ω_B. The exchange will carry magnetization from A to B and vice versa. If we assume equal populations in the two sites, and the rate of exchange to be k, then we can set up two coupled Bloch equations for the two sites, as in Equation [4].

$$\frac{dM_A}{dt} + i(\omega_A - \omega_1)M_A + \frac{1}{T_2}M_A$$
$$- kM_B + kM_A = i\gamma B_1 M_{zA}$$
$$\frac{dM_B}{dt} + i(\omega_B - \omega_1)M_B + \frac{1}{T_2}M_B$$
$$- kM_A + kM_B = i\gamma B_1 M_{zB} \quad [4]$$

The observable NMR signal is the imaginary part of the sum of the two steady-state magnetizations, M_A and M_B. The steady state implies that the time derivatives are zero, and a little further calculation (and neglect of T_2 terms) gives the NMR spectrum of an exchanging system as in Equation [5].

$$\nu = \frac{1}{2}\gamma B_1 M_z$$
$$\times \frac{k(\omega_A - \omega_B)^2}{(\omega_A - \omega_1)^2(\omega_B - \omega_1)^2 + 4k^2[\omega_1 - (\omega_A + \omega_B)/2]^2} \quad [5]$$

This equation can be further extended to systems with unequal populations and more sites, using the same techniques.

Chemical exchange in the time domain

If we create (by a pulse) the magnetization, M_A and M_B, at time zero, and then turn off the B_1 magnetic field, Equation [4] can be simplified as in Equation [6].

$$\frac{d}{dt}\begin{pmatrix} M_A \\ M_B \end{pmatrix} = -L\begin{pmatrix} M_A \\ M_B \end{pmatrix} \quad [6]$$

$$L = \begin{pmatrix} i\delta + \frac{1}{T_2} + k & -k \\ -k & -i\delta + \frac{1}{T_2} + k \end{pmatrix} \quad [7]$$

In this equation, the matrix L is given by Equation [7], where we have made $\omega_1 = (\omega_A + \omega_B)/2$ and $\delta = (\omega_A - \omega_B)/2$.

$$\begin{pmatrix} M_A(t) \\ M_B(t) \end{pmatrix} = \exp(-Lt) \begin{pmatrix} M_A(0) \\ M_B(0) \end{pmatrix} \qquad [8]$$

Equation [6] is a set of first-order differential equations, so its formal solution is given by Equation [8], in which exp( ) means the exponential of the matrix. In practice, we diagonalize the matrix L with a matrix of eigenvectors, U, as in Equation [9] to give a diagonal matrix, Λ, with the eigenvalues of L down the diagonal.

$$\Lambda = U^{-1} L U \qquad [9]$$

Equation [8] becomes Equation [10]. The exponential of a diagonal matrix is again a diagonal matrix with exponentials of the diagonal elements, as in Equation [11].

$$\begin{pmatrix} M_A(t) \\ M_B(t) \end{pmatrix} = U \exp(-\Lambda t) U^{-1} \begin{pmatrix} M_A(0) \\ M_B(0) \end{pmatrix} \qquad [10]$$

$$\begin{pmatrix} M_A(t) \\ M_B(t) \end{pmatrix} = U \begin{pmatrix} e^{-\lambda_1 t} & 0 \\ 0 & e^{-\lambda_2 t} \end{pmatrix} U^{-1} \begin{pmatrix} M_A(0) \\ M_B(0) \end{pmatrix} \qquad [11]$$

As was mentioned above, the observed signal is the imaginary part of the sum of M_A and M_B, so Equation [11] predicts that the observed signal will be the sum of two exponentials, evolving at the frequencies λ_1 and λ_2. This is the free induction decay (FID). In the limit of no exchange, the two frequencies are simply ω_A and ω_B, as we would expect. When k is nonzero, the mathematics becomes slightly more complicated.

If we ignore relaxation and exchange, then L is a Hermitian matrix with real eigenvalues and eigenvectors. However, when the exchange is important, the Hermitian character is lost and the eigenvalues and eigenvectors have both real and imaginary parts. The eigenvalues are given by the roots of the characteristic equation (Eqn [12]).

$$\begin{vmatrix} i\delta + \dfrac{1}{T_2} + k - \lambda & -k \\ -k & -i\delta + \dfrac{1}{T_2} + k - \lambda \end{vmatrix} = 0 \qquad [12]$$

The eigenvalues of Equation [7] are given in Equation [13]

$$\lambda = \left(\frac{1}{T_2} + k \right) \pm \sqrt{k^2 - \delta^2} \qquad [13]$$

These eigenvalues are the (complex) frequencies of the lines in the spectrum, as in Equation [11]: the imaginary part gives the oscillation frequency and the real part gives the rate of decay. If $k < \delta$ (slow exchange) then there are two different imaginary frequencies, which become $\pm \delta$ in the limit of small k (see **Figure 2**). In fast exchange, when k exceeds the shift difference, δ, the quantity in the square root in [13] becomes positive, so the roots are pure real. This means that the spectrum is still two lines, but they are both at the average chemical shift, and have different widths. The full expressions for these line shapes are given in the Appendix.

Because of the role of the eigenvectors in Equation [11], the factor (amplitude) multiplying the complex exponential is itself complex. The magnitude of the complex amplitude gives the intensity of the line and its phase gives the phase of the line (the mixture of absorption and dispersion). In slow exchange, the

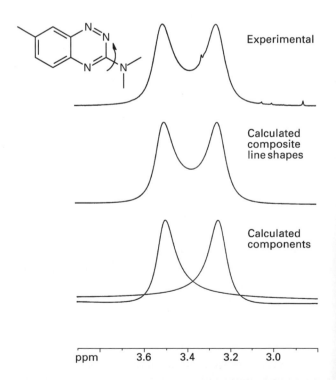

Figure 2 Decomposition of the coalescence line shape into individual lines. Top: experimental spectrum from **Figure 1**; middle: calculated line shape to match experimental spectrum; bottom: individual lines calculated as in the Appendix.

two lines have the same real part, but the imaginary parts have opposite signs, so the phase distortion is opposite. The sum of these distorted line shapes gives the familiar coalescence spectrum, as in **Figure 2**. In fast exchange, the two lines are both in phase, but one line is negative. This negative line is very broad, and decreases in absolute intensity as the rate increases, leaving only the single, positive, in-phase line for fast exchange.

Therefore, the real and imaginary parts of the frequency, and the real and imaginary parts of the amplitude, provide the four parameters that define a line in an NMR spectrum: its intensity, its phase, its position, and its width. This gives us a time-domain picture of the chemical exchange, which can easily be converted to a spectrum by using the Fourier transform.

Systems with scalar coupling

Scalar coupled systems are more complicated, but fundamentally no different from the uncoupled systems described in the time domain. If there is scalar coupling in the spectrum, the line shape becomes more complex. For instance, **Figure 3** shows the line shape of the diacylpyridine ligand as part of a rhenium complex. The rhenium bonds to the pyridine nitrogen and to only one of the carbonyls, lifting the symmetry of protons 3 and 5. However, the rhenium can break this latter coordination and bond to the other carbonyl, which effectively interchanges protons 3 and 5 on the pyridine ring. During this exchange, they retain their coupling to the central proton 4, and the shift of 4 does not change. Therefore, proton 4 remains as a sharp triplet even when protons 3 and 5 are very broad.

In any strongly coupled spin system, each line in the spectrum is a mixture of transitions of various nuclei, which depends on the chemical shifts and coupling constants. When a chemical exchange happens, all the spectral parameters change with it. Therefore, a magnetization (or a coherence) that was associated with a single spectral line in one site may be spread among several lines in the other site. In order to deal with these complexities, we must use the density matrix.

One way of thinking of the density matrix is that it is a list of all the observables of a spin system. For example, the magnetizations of the two exchanging sites form part of the density matrix for that system. At equilibrium, the density matrix is just the z magnetization created by the static magnetic field. In a pulse Fourier transform NMR experiment, this z magnetization is flipped into the xy plane, and

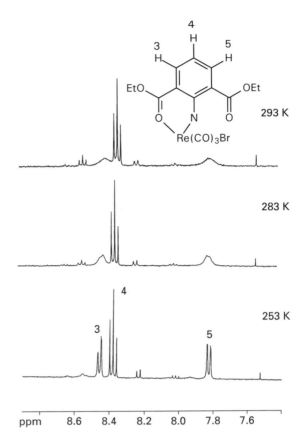

Figure 3 Example of chemical exchange in a coupled spin system. The rhenium can bond to one or other of the carbonyl oxygen atoms, so that the symmetry of the molecule is lifted. Exchange of the rhenium between carbonyl groups broadens the signals of protons 3 and 5 (at 8.46 ppm and 7.82 ppm), but leaves 4 (8.38 ppm) unaffected.

divided among the individual lines in the spectrum, which then precess around the z axis.

In order to calculate the spectrum of the system (or any other observable), we need to be able to follow the density matrix as a function of time. The equation of motion of the density matrix (ρ) is given in Equation [14], where H is the Hamiltonian of the spin system.

$$i\frac{h}{2\pi}\frac{\partial}{\partial t}\rho = [H, \rho] \qquad [14]$$

Some manipulation allows us to reformulate Equation [14] as Equation [15]. Mathematically, this means using superoperators in Liouville space L, but the details need not concern us here. The important point is that the density matrix becomes a vector of all possible observables of the system – in this case we only deal with the xy magnetizations. Anything we do to the system, pulses, free precession, relaxation, exchange, is represented by a matrix.

$$\mathrm{i}\frac{h}{2\pi}\frac{\partial}{\partial t}\rho = \boldsymbol{L}^{\mathrm{so}}\,\rho \qquad [15]$$

If we use frequency units ($h/2\pi = 1$), then the solution to Equation [15] is given in Equation [16], which is identical to Equation [8].

$$\rho(t) = \exp(-\mathrm{i}\,\boldsymbol{L}^{\mathrm{so}}t)\,\rho(0) \qquad [16]$$

Relaxation or chemical exchange can be easily added in Liouville space, by including a Redfield matrix, $\boldsymbol{R}^{\mathrm{so}}$ (so = superoperator), for relaxation, or a kinetic matrix, $\boldsymbol{K}^{\mathrm{so}}$, to describe exchange. Both relaxation and exchange are described very conveniently by superoperators. The complete equation of motion becomes Equation [17]. Note the similarity to the equations for two uncoupled sites, but now we derived them rigorously from the equations of motion of the density matrix.

$$\begin{aligned}\rho(t) &= \exp(-\mathrm{i}\boldsymbol{L}^{\mathrm{so}} - \boldsymbol{R}^{\mathrm{so}} - \boldsymbol{K}^{\mathrm{so}})t\,\rho(0)\\ &= U\exp(-\Lambda t)U^{-1}\,\rho(0)\end{aligned} \qquad [17]$$

The details of what the density matrix is at time zero, what we observe in the receiver, and how we construct the superoperators for a general spin system are all beyond the scope of this article. However, the basic idea remains: even the most complex exchange line shape is a sum of individual lines, with complex frequencies and complex intensities, as in Equation [17]. Each of those lines has a transition probability.

Physical interpretation of the transition probability

The standard expression for the probability of a transition from an initial state, ψ, to a final state, ϕ, induced by a perturbation I_x (the RF field along x), is given in Equation [18].

$$\text{Transition probability} \propto |\langle\,\phi\,|\,I_x\,|\,\psi\,\rangle|^2 \qquad [18]$$

We can write the matrix element in Equation [18] in a rather more complicated way, as in Equation [19].

$$\langle\,\phi\,|\,I_x\,|\,\psi\,\rangle =$$
$$\mathrm{trace}\left\{\left(\begin{array}{cc}\langle\,\phi|I_x|\phi\,\rangle & \langle\,\psi|I_x|\phi\,\rangle \\ \\ \langle\,\phi|I_x|\psi\,\rangle & \langle\,\psi|I_x|\psi\,\rangle\end{array}\right)\times\left(\begin{array}{cc}0 & 1 \\ 0 & 0\end{array}\right)\right\}$$
$$[19]$$

The reason for rewriting the equation is that the right hand side can be interpreted as the dot product of the operator I_x with the operator $|\phi\rangle\langle\psi|$. We have moved to Liouville space, in which all the operators are vectors. The dot product in Liouville space is defined as the trace of the product of the operators in the original spin space. One important point is that Equation [19] is independent of basis set. The matrix element in Equation [18] becomes a simple dot product of two Liouville space vectors, as in Equation [20].

$$\langle\,\phi\,|\,I_x\,|\,\psi\,\rangle = \left(\begin{array}{c}\langle\,\phi\,|\,I_x\,|\,\phi\,\rangle \\ \langle\,\psi\,|\,I_x\,|\,\phi\,\rangle \\ \langle\,\phi\,|\,I_x\,|\,\psi\,\rangle \\ \langle\,\psi\,|\,I_x\,|\,\psi\,\rangle\end{array}\right)\cdot\left(\begin{array}{c}0 \\ 0 \\ 1 \\ 0\end{array}\right) \qquad [20]$$

A dot product of two vectors gives the projection of one vector on the other, so Equation [20] defines the projection of $|\phi\rangle\langle\psi|$ onto I_x. In other words, this is the projection of the transition onto the total xy magnetization. The standard transition probability is the square of this projection.

This leads to a physical interpretation of the transition probability. The spin system at equilibrium is represented by the total z magnetization, F_z. This is the sum of the z magnetizations of the sites, weighted by the equilibrium populations of each site. To distinguish the weighted sum, we give this spin operator a symbol different from I_z. A pulse flips this into the xy plane, so that immediately after the pulse, the spin system is represented by F_x. Now each transition receives a share of the total x magnetization. Its share is given by the projection of the transition (as an operator) onto the operator F_x. The transitions evolve independently as a function of time. However, we do not observe the transition directly, but rather the total xy magnetization. The detector is simply a coil of wire, and we measure the xy magnetization as a function of time to give us an FID. Fourier transforming the FID gives us a spectrum. Therefore, each transition contributes to the total signal according to its projection along I_x. The intensity of a transition is the product of how much coherence it received from F_x at the start, times how visible it is to the receiver.

Note that the Liouville matrix, $iL+R+K$ may not be Hermitian, but it can still be diagonalized. Its eigenvalues and eigenvectors are not necessarily real, however, and the inverse of U may not be its complex-conjugate transpose. If we allow complex numbers in it, Equation [17] is a general result. Since Λ is a diagonal matrix we can expand in terms of the individual eigenvalues, λ_j. We can also apply U ('backwards') to I_x, and obtain Equation [21].

$$\text{NMR signal} = \sum_j (UI_x)_j^* \, (U^{-1}F_x)_j \exp(-i\lambda_j t) \quad [21]$$

Each of the terms in the sum is a transition. The NMR signal is always a sum of decaying sine waves, whose frequency and decay rate are given by the imaginary and the real parts of λ_j. The intensity is governed by the coefficients in Equation [21] of this term. This coefficient is the product of two terms. The first, $(UI_x)_j$, tells us how much the coherence overlaps with the receiver. The second term, $(U^{-1} F_x)_j$ tells us how much the coherence received from the equilibrium z magnetization. The product of these terms is the generalization of the transition probability. If we do not have relaxation, the two terms are complex conjugates, and we recover the

$$
\begin{pmatrix}
i\omega_A - k & iJ/2 & \cdot & -iJ/2 & k & \cdot & \cdot & \cdot \\
iJ/2 & i\omega_A - k & -iJ/2 & \cdot & \cdot & k & \cdot & \cdot \\
\cdot & -iJ/2 & i\omega_B - k & iJ/2 & \cdot & \cdot & k & \cdot \\
-iJ/2 & \cdot & iJ/2 & i\omega_B - k & \cdot & \cdot & \cdot & k \\
k & \cdot & \cdot & \cdot & i\omega_B - k & iJ/2 & \cdot & -iJ/2 \\
\cdot & k & \cdot & \cdot & iJ/2 & i\omega_B - k & -iJ/2 & \cdot \\
\cdot & \cdot & k & \cdot & \cdot & -iJ/2 & i\omega_A - k & iJ/2 \\
\cdot & \cdot & \cdot & k & -iJ/2 & \cdot & iJ/2 & i\omega_A - k
\end{pmatrix}
\quad [23]
$$

normal transition probability. However, the transition probability for a general system may have an imaginary part. If we cling to idea that each term in Equation [21] represents a transition, then an individual transition may be out of phase at $t = 0$. Immediately after the excitation pulse, the total magnetization will lie on the y axis, but the individual components may not. However, our true observable is still only the FID. Whether we want to decompose it into individual transitions of individual spins depends very much on the system we are studying.

Example of exchange of coupled spins

Provided we can construct the matrices, the description of exchange is quite simple. Again, the details are beyond our scope, but an example of the Liouvillian (no exchange) for a coupled spin system is given in Equation [22] for an AB spin system, with coupling constant J. The eigenvalues of this matrix are the familiar line positions of an AB system.

$$
\begin{pmatrix}
i\omega_A & iJ/2 & 0 & -iJ/2 \\
iJ/2 & i\omega_A & -iJ/2 & 0 \\
0 & -iJ/2 & i\omega_B & iJ/2 \\
-iJ/2 & 0 & iJ/2 & i\omega_B
\end{pmatrix}
\quad [22]
$$

If there is an exchange between A and B given by a rate k, then we set up the two blocks, as we set up the two exchanging spins in the Bloch equations. Note that we have made spin A in one block, exchange with spin B in the other. The full Liouvillian, including exchange, is given by Equation [23], in which dots replace zeroes to emphasize the form of the matrix.

In this particular case, called mutual exchange, the spins simply permute themselves. We can simplify this matrix to half its original size, as in Equation [24], but

for nonmutual exchange we must retain Equation [23].

$$
\begin{pmatrix}
i\omega_A - k & iJ/2 & k & -iJ/2 \\
iJ/2 & i\omega_A - k & -iJ/2 & k \\
k & -iJ/2 & i\omega_B - k & iJ/2 \\
-iJ/2 & k & iJ/2 & i\omega_B - k
\end{pmatrix}
\quad [24]
$$

Except for its size, this is exactly the same form as in our previous cases. The spectrum can therefore be expressed as a sum of four lines, as in **Figure 4**.

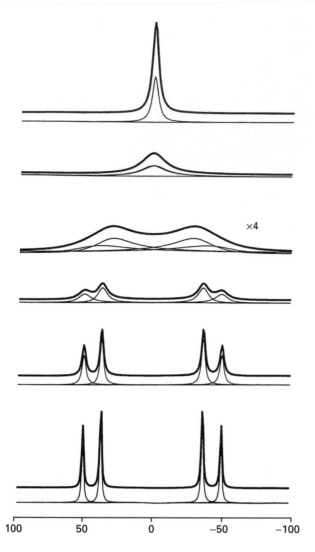

Figure 4 Calculated spectra for mutual exchange in an AB spin system, as a function of exchange rate. The heavy traces are the total line shape and the lighter lines show the individual components. The bottom spectrum shows the typical static AB spectrum, which broadens and coalesces at higher rates of exchange.

Slow chemical exchange

The term 'slow' in this case means that the exchange rate is much smaller than the frequency differences in the spectrum, so the lines in the spectrum are not significantly broadened. However, the exchange rate is still comparable with the spin–lattice relaxation times in the system. In this case, we can measure the rates by doing a modified spin–lattice relaxation experiment, pioneered by Hoffman and Forsen.

In the absence of exchange (and let us ignore dipolar relaxation), each z magnetization will relax back to equilibrium at a rate governed by its own T_1 as in

Equation [25].

$$\frac{\mathrm{d}}{\mathrm{d}t}[M(t) - M(\infty)] = -\frac{1}{T_1}[M(t) - M(\infty)] \quad [25]$$

If we have two sites, A and B, then we can write the analogous equation for two sites as in Equation [26].

$$\frac{\mathrm{d}}{\mathrm{d}t}\begin{bmatrix} M^{\mathrm{A}}(t) - M^{\mathrm{A}}(\infty) \\ M^{\mathrm{B}}(t) - M^{\mathrm{B}}(\infty) \end{bmatrix}$$

$$= \begin{pmatrix} \dfrac{1}{T_1^{\mathrm{A}}} & 0 \\ 0 & \dfrac{1}{T_1^{\mathrm{B}}} \end{pmatrix}\begin{bmatrix} M^{\mathrm{A}}(t) - M^{\mathrm{A}}(\infty) \\ M^{\mathrm{B}}(t) - M^{\mathrm{B}}(\infty) \end{bmatrix} \quad [26]$$

If the two sites exchange with rate k during the relaxation, then a spin can relax either through normal spin–lattice relaxation processes, or by exchanging with the other site. Equation [26] becomes Equation [27].

$$\frac{\mathrm{d}}{\mathrm{d}t}\begin{bmatrix} M^{\mathrm{A}}(t) - M^{\mathrm{A}}(\infty) \\ M^{\mathrm{B}}(t) - M^{\mathrm{B}}(\infty) \end{bmatrix}$$

$$= \begin{pmatrix} \dfrac{1}{T_1^{\mathrm{A}}} - k & k \\ k & \dfrac{1}{T_1^{\mathrm{B}}} - k \end{pmatrix}\begin{bmatrix} M^{\mathrm{A}}(t) - M^{\mathrm{A}}(\infty) \\ M^{\mathrm{B}}(t) - M^{\mathrm{B}}(\infty) \end{bmatrix} \quad [27]$$

This equation is very similar to Equations [6] and [7]. The basic situation is just as in intermediate exchange, except that now we are dealing with z magnetizations rather than xy. The frequencies are zero, and the matrix now has pure real eigenvalues, but the approach is the same. We also stay in the time domain, since a relaxation experiment follows the z magnetizations as a function of time. As before, the time dependence is obtained by diagonalizing the relaxation–exchange matrix, and calculating the magnetizations for each time at which they are sampled.

There are two main applications of slow chemical exchange: one is to determine the qualitative mechanism, and the other is to measure the rates of the processes as accurately as possible. For the first case, in which we have a spectrum in slow exchange, we need to establish the mechanism: which site is exchanging with which. For this purpose, the homonuclear two-dimensional (2D) experiment EXSY (exchange spectroscopy) (the same pulse sequence as NOESY (2D nuclear Overhauser effect spectroscopy), but involving exchange) is by far the best technique to use. If there is no exchange, then the spectrum lies along the diagonal and there are no cross-peaks. Exchange between sites leads to a pair of symmetrical cross-peaks joining the diagonal peaks of the same site, so the mechanism is very obvious.

The EXSY pulse sequence starts with two $\pi/2$ pulses separated by the incrementable delay, t_1. This modulates the z magnetizations, so that the relaxation that occurs during the mixing time which follows, t_m, is frequency-labelled. Finally the z magnetizations are sampled with a third $\pi/2$ pulse. Magnetization from a different site that enters via exchange will have a different frequency label. A two-dimensional Fourier transform then produces the spectrum. However, care must be taken in choosing the mixing time if there are multiple exchange processes. If the mixing time is too long, there is a substantial probability that a spin may have exchanged twice in that time, leading to spurious cross-peaks.

For careful rate measurements, once the mechanism is established, it is our opinion that one-dimensional (1D) methods are superior to quantitative 2D ones. Apart from the fact that 1D spectra can be integrated more easily, we also have more control over the experiment. Modern spectrometers can create almost any type of selective excitation, so that we can control the conditions at the start of the relaxa-

tion. For two sites, a nonselective inversion that inverts both sites equally will mask most of the exchange effects and the relaxation will be dominated by T_1. However, if one site is inverted selectively, then that site can regain equilibrium by either T_1 processes or by exchanging with the other site that was left at equilibrium. The inverted signal will relax at roughly the sum of the exchange and spin–lattice relaxation rate, while the signal that was unperturbed at the start of the experiment shows a characteristic transient, as in **Figure 5**. For multiple sites, a wide range of initial conditions is available. Standard nonlinear least-squares methods allow us to fit these curves and derive values for the rates involved.

Fast exchange

In fast exchange the spectrum has coalesced, usually to a single line, so the information in the spectrum is not so rich. For two equally-populated sites, for instance, we get a single Lorentzian line with a width proportional to δ^2/k. We can measure the line width (or better, T_2) via a CPMG (Carr–Purcell pulse sequence, Meiboom–Gill modification) experiment or a $T_{1\rho}$ measurement. However, if we do not know the frequency difference, δ, we cannot obtain the rate. If the exchange rate is not too fast ($< 10^4$ roughly) modifications of the T_2 experiments can help. Both experiments have an inherent timescale: in CPMG, it is the timing of the refocussing pulse; in $T_{1\rho}$, it is the precession frequency about the spin-locking field. If the exchange is fast with respect to the experiment,

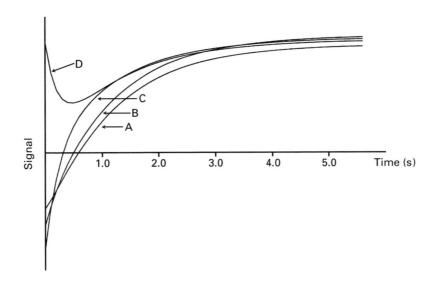

Figure 5   Inversion–recovery curves for slow exchange between two sites. Lines A and B are the results from one experiment. They show the recovery after a nonselective inversion of both sites, showing that the two sites have slightly different T_1 values. C and D are obtained from a different experiment. Line C shows the recovery of a site that has been selectively inverted, and line D shows the behaviour of the line that was not perturbed in this experiment. The inverted line relaxes faster, due to a combination of spin–lattice, relaxation and exchange, and the unperturbed line shows the characteristic transient.

we measure a T_2 appropriate to the coalesced spectrum. If, however, the exchange is slower than the experimental timescale (but still fast with respect to the frequency difference), the apparent T_2 reflects the individual sites. As we change the timescale of the experiment, the apparent T_2 changes, and so we can obtain a value for the rate itself.

Conclusions

All of the effects of chemical exchange can be calculated by following the appropriate magnetizations as a function of time. For slow exchange, this is the complex coupling of relaxation and exchange that leads to transient behaviour in modified spin–lattice relaxation experiments. For intermediate exchange, the rather complex line shapes can always be decomposed into a sum of individual transitions, even though the transitions are distorted in phase, intensity, position, and width by the dynamics of the system. For two uncoupled sites, this can be calculated quite easily, but for larger coupled systems the construction of the matrices may get complicated. However, the resulting lines are always governed by a transition probability, provided we extend our definition to allow probabilities with both real and imaginary parts. Working in the time domain is already familiar from Fourier transform spectroscopy, and it provides a complete and simple approach to chemical exchange.

Appendix

Eigenvalues and eigenvectors for two-site exchange

For two-site equally populated chemical exchange, the two magnetizations at time zero are equal, so that Equation [11] can be written as in Equation [28], where the eigenvalues are complex numbers.

$$\begin{pmatrix} M_A(t) \\ M_B(t) \end{pmatrix} = U \begin{pmatrix} e^{\lambda_1 t} & 0 \\ 0 & e^{\lambda_2 t} \end{pmatrix} U^{-1} \begin{pmatrix} 1 \\ 1 \end{pmatrix} \quad [28]$$

Since we are dealing with nonHermitian matrices, the matrix formed by the eigenvectors will not be unitary, and will have four independent complex elements. Let us simply call them a, b, c, and d, so that U is given by:

$$U = \begin{pmatrix} a & b \\ c & d \end{pmatrix} \quad [29]$$

Regardless of whether U is unitary, its inverse is given by Equation [30], where Δ is the determinant of Equation [29].

$$U^{-1} = \frac{1}{\Delta} \begin{pmatrix} d & -b \\ -c & a \end{pmatrix} \quad [30]$$

Equation [28] then says that the signal is given by the following, regardless of slow or fast exchange.

$$\text{Signal} = \frac{(a+c)(d-b)}{\Delta} \exp(\lambda_1 t) + \frac{(b+d)(-c+a)}{\Delta} \exp(\lambda_2 t) \quad [31]$$

The values of the eigenvectors have two forms, depending on whether $\delta > k$ (slow exchange), or $\delta < k$ (after coalescence). In the first case, the eigenvalues are given as:

$$\text{Eigenvalues} = -\left(\frac{1}{T_2} + k\right) \pm i\sqrt{\delta^2 - k^2} \quad [32]$$

and a convenient matrix of eigenvectors is given by:

$$\begin{pmatrix} k & i(\sqrt{\delta^2 - k^2} + \delta) \\ -i(\sqrt{\delta^2 - k^2} + \delta) & k \end{pmatrix} \quad [33]$$

After coalescence, when the rate is greater than the frequency difference, the two transitions are both at zero frequency (i.e. the average chemical shift), but have different widths and intensities. The eigenvalues are pure real, as in Equation [34],

$$\text{Eigenvalues} = -\left(\frac{1}{T_2} + k\right) \pm \sqrt{k^2 - \delta^2} \quad [34]$$

and the eigenvectors are similar, but reflect the fact that $k^2 - \delta^2$ is now positive.

$$\begin{pmatrix} \sqrt{k^2 - \delta^2} - i\delta & -\sqrt{k^2 - \delta^2} - i\delta \\ k & k \end{pmatrix} \quad [35]$$

List of symbols

B_1 = radiofrequency field; F_z = total z magnetization; h = Planck's constant; I_x = perturbation RF field along x; $i = \sqrt{-1}$; k = rate of exchange; K^{so} = kinetic superoperator matrix; L = Liouville superoperator; L^{so} = superoperator matrix; M_z = z magnetization of the spin system; R^{so} = Redfield superoperator matrix; T_2 = spin–spin relaxation time; $T_{1\rho} = T_1$ in rotating frame; $u = x$ magnetization of spin system; $v = y$ magnetization of spin system; γ = magnetogyric ratio; δ = chemical shift difference; λ = eigenvalue; Λ = diagonal matrix; ϕ = final state; ψ = initial state; ω_1 = frequency of RF irradiation; ω_0 = Larmor frequency of the spin.

See also: **NMR Principles; NMR Pulse Sequences; NMR Relaxation Rates; Two-Dimensional NMR, Methods.**

Further reading

Bain AD (1988) The superspin formalism for pulse NMR. *Progress in Nuclear Magnetic Resonance Spectroscopy* 20: 295–315.

Bain AD and Duns GJ (1996) A unified approach to dynamic nmr based on a physical interpretation of the transition probability. *Canadian Journal of Chemistry* 74: 819–824.

Binsch G (1969) A unified theory of exchange effects on nuclear magnetic resonance lineshapes. *Journal of the American Chemical Society* 91: 1304–1309.

Binsch G and Kessler H (1980) The kinetic and mechanistic evaluation of NMR spectra. *Angewandte Chemie, International Edition in English* 19: 411–494.

Gutowsky HS and Holm CH (1956) Rate processes and nuclear magnetic resonance spectra. II. Hindered internal rotation of amides. *Journal of Chemical Physics* 25: 1228–1234.

Jackman LM and Cotton FA (1975) *Dynamic Nuclear Magnetic Resonance Spectroscopy.* New York: Academic Press.

Johnson CS (1965) Chemical rate processes and magnetic resonance. *Advances in Magnetic Resonance* 1: 33–102.

Kaplan JI and Fraenkel G (1980) *NMR of Chemically Exchanging Systems.* New York: Academic Press.

Orrell KG, Sik V and Stephenson D (1990) Quantitative investigation of molecular stereodynamics by 1D and 2D NMR methods. *Progress in Nuclear Magnetic Resonance Spectroscopy* 22: 141–208.

Perrin CL and Dwyer T (1990) Application of two-dimensional NMR to kinetics of chemical exchange. *Chemical Review* 90: 935–967.

Sandstrom J (1982) *Dynamic NMR Spectroscopy.* London: Academic Press.

Sorensen OW, Eich GW, Levitt MH, Bodenhausen G and Ernst RR (1983) Product operator formalism for the description of NMR pulse experiments. *Progress in Nuclear Magnetic Resonance Spectroscopy* 20: 163–192.

Chemical Ionization in Mass Spectrometry

Alex G Harrison, University of Toronto, Ontario, Canada

MASS SPECTROMETRY
Methods & instrumentation

Introduction

In chemical ionization mass spectrometry (CIMS), ionization of the gaseous analyte occurs via gas-phase ion–molecule reactions rather than by direct electron impact (EI), photon impact or field ionization. EI ionization of a reagent gas (present in large excess) is usually followed by ion–molecule reactions involving the initially formed ions and the reagent gas neutrals to produce the chemical ionization (CI) reagent ion or reagent ion array. Collision of the reagent ion(s) with the analyte (usually present at ~1% of the reagent gas pressure) produces one or more ions characteristic of the analyte. These initial analyte ions may undergo fragmentation or, infrequently, react further with the reagent gas to produce a final array of ions representing the CI mass spectrum of the analyte as produced by the specific reagent gas.

To a considerable extent, the usefulness of CIMS arises from the fact that a wide variety of reagent gases and, hence, reagent ions can be used to ionize the analyte; often the reagent system can be tailored to the problem to be solved. Problems amenable to solution by CI approaches include (a) molecular mass determination, (b) structure elucidation and (c) identification and quantification. In many instances, CI provides information that is complementary rather than supplementary to that obtained by EI, and often both approaches are used. After a brief discussion of instrumentation, the major approaches, in both positive ion and negative ion CI, to the solution of the above problems are discussed. The focus is primarily on the use of medium-pressure mass spectrometry; CI at atmospheric pressure or at much lower pressures in ion trapping instruments are discussed elsewhere.

Instrumentation

Most commonly, CI studies have been performed at total ion source pressures of 0.1 to 1 torr using conventional sector, quadrupole or time-of-flight mass spectrometers with ion sources only slightly modified from those used for EI ionization. The essential change is that the ion source is made more gas-tight by using smaller electron beam entrance and ion beam exit slits; the latter is accomplished by having either exchangeable source volumes or a moveable ion exit slit assembly to change from EI to CI. Most manufacturers now supply instruments capable of both EI and CI and incorporating enhanced pumping to maintain adequately low pressures in the remainder of the instrument when the source is operated at elevated pressures. Provision is made for introduction of the reagent gas and for introduction of the analyte by a heated inlet system, by a solids probe, by a direct exposure probe or by interfacing a gas chromatograph to the ion source. Capillary gas chromatography–CI mass spectrometry is a particularly powerful and sensitive approach to identification and quantification of the components of complex mixtures.

Brønsted acid chemical ionization

Gaseous Brønsted acids, BH^+, react with the analyte M primarily by the proton transfer reaction

$$BH^+ + M \rightarrow MH^+ + B \qquad [1]$$

Other, usually minor, reactions that are possible include hydride ion abstraction (Eqn [2]) and charge exchange (Eqn [3])

$$BH^+ + M \rightarrow [M-H]^+ + BH_2(B + H_2) \qquad [2]$$

$$BH^+ + M \rightarrow M^{+\bullet} + BH^\bullet \ (B + H^\bullet) \qquad [3]$$

The enthalpy change for Equation [1] is given by

$$\Delta H^\circ = PA(B) - PA(M) \qquad [4]$$

where the proton affinity of X, PA(X), is given by

$$X_{(g)} + H^+_{(g)} \rightarrow XH^+_{(g)}, \quad -\Delta H^\circ = PA(X) \qquad [5]$$

If $PA(M) > PA(B)$, the reaction is exothermic and normally occurs at the ion–neutral collision rate determined by ion-induced dipole and ion–dipole interactions. Conversely, if the reaction is endothermic, the rate decreases exponentially by the enthalpy of activation and the sensitivity of the ionization process decreases.

Most of the exothermicity of Equation [1] resides in MH^+ and is available as internal energy to promote fragmentation of MH^+. To maximize the probability of MH^+ formation, which provides molecular mass information, one uses the least exothermic reaction possible. On the other hand, structural information is derived from the fragment ions observed in the CI mass spectrum and more exothermic protonation may be desirable to promote such fragmentation. The fragmentation of the even-electron MH^+ ions usually involves elimination of even-electron stable molecules to form an even-electron fragment ion. Thus if the molecule contains a functional group Y, the fragmentation of MH^+ frequently involves elimination of the stable molecule HY. For RYH^+, the tendency to eliminate HY is roughly inversely related to the PA of HY, with the result that the ease of loss of Y as HY is approximately

$$NH_2 < CH_3C(=O)O < CH_3S < CH_3O < C_6H_5$$
$$\approx HC(=O)O < CN < SH < OH < I \approx Cl \approx Br$$

although the stability of the resulting ion also plays a role in competing loss of different functional groups from a common MH^+. Obviously, the more readily fragmentation occurs, the less likely it is that MH^+ ions, giving molecular mass information, will be observed, although with suitable choice of reagent ions, MH^+ ions are frequently formed when no $M^{+\bullet}$ ions are observed in the EI mass spectrum. In other cases, such as hydroxylic compounds, the use of Brønsted base CI (see below) may prove useful in obtaining molecular mass information.

Table 1 lists Brønsted acid reagent systems which have seen significant use, along with the major reactant ions and the proton affinities of the conjugate bases. Most organic molecules have PAs in the range

760–1000 kJ mol⁻¹. Thus protonation by H_3^+ will be strongly exothermic, while protonation in *i*-butane or ammonia CI will only be mildly exothermic or even endothermic. CH_4 as a reagent gas is unique in that it forms two reactant species, CH_5^+ and $C_2H_5^+$, in almost equal abundance; the polar reagent gases H_2O, CH_3OH and NH_3 produce solvated protons with the extent of solvation depending on the partial pressure of the reagent gas. The usefulness of Brønsted acid CI in establishing molecular masses is illustrated by the comparison of the EI, CH_4 CI and NH_3 CI mass spectra of ephedrine in **Figure 1**. No molecular ion is observed in the EI mass spectra. By contrast, both the CH_4 CI and NH_3 CI mass spectra show abundant MH⁺ (*m/z* 166) ions, clearly establishing the molecular mass. In line with the relative PAs of **Table 1**, the less exothermic protonation in NH_3 CI leads to less extensive fragmentation of the MH⁺ ion. As is often the case, CH_4 CI provides not only molecular mass information (MH⁺) but also structurally informative fragment ions; the fragmentation of protonated ephedrine is rationalized in **Scheme 1** and is seen to involve the formation of even-electron fragment ions by the elimination of small, stable, even-electron neutral molecules (the numbers in **Scheme 1** indicate the *m/z* of the ionic fragment). In this case the CH_4 CI mass spectrum provides more structural information than the EI mass spectrum.

The effect of protonation exothermicity on the extent of fragmentation is shown more dramatically by the Brønsted acid CI mass spectra of the tripeptide H-Val-Pro-Leu-OH shown in **Figure 2** and obtained using mass-selected reactant ions. With NH_4^+ as the reactant ion [PA(NH_3) = 853.5 kJ mol⁻¹] only MH⁺ is observed, while the more exothermic protonation by $C_2H_5^+$ [PA(C_2H_4) = 680.3 kJ mol⁻¹] leads to significant fragmentation of MH⁺, and the very exothermic protonation by N_2OH^+ [PA(N_2O) = 580.7 kJ mol⁻¹] leads to essentially complete fragmentation of MH⁺.

Table 1 Brønsted acid reagent systems

Reagent gas	Reactant ion (BH⁺)	PA(B) (kJ mol⁻¹)
H_2	H_3^+	423.4
$N_2O–H_2$	N_2OH^+	580.7
CH_4	CH_5^+	550.6
	$C_2H_5^+$	680.3
H_2O	$H^+(H_2O)_n$[a]	696.0
CH_3OH	$H^+(CH_3OH)_n$[a]	773.6
i-C_4H_{10}	$C_4H_9^+$	819.6
NH_3	$H^+(NH_3)_n$[a]	853.5

[a] Degree of solvation depends on partial pressure of reagent gas; proton affinity given for monosolvated proton.

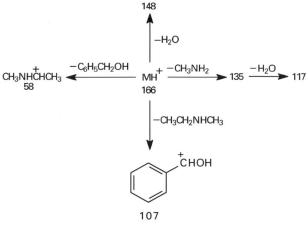

Scheme 1

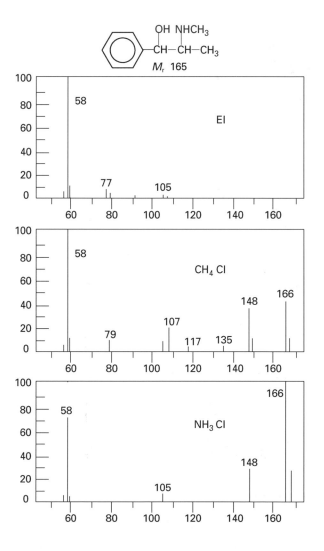

Figure 1 EI, CH_4 CI and NH_3 CI mass spectra of ephedrine (relative intensity as a function of *m/z*).

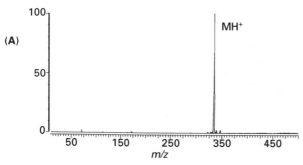

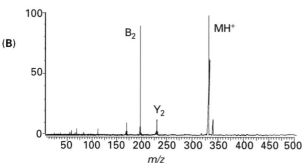

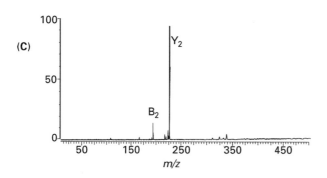

Figure 2 Brønsted acid CI mass spectra of H-Val-Pro-Leu-OH using (A) NH_4^+, (B) $C_2H_5^+$ and (C) N_2OH^+ as reagent ions. Reprinted with permission from Speir JP, Gorman GS, Cornett DS and Amster IJ (1991) Controlling the dissociation of peptide ions using laser desorption/chemical ionization Fourier transform mass spectrometry. *Analytical Chemistry* **63**: 65–69. Copyright (1991) American Chemical Society.

Ammonia has a very high PA, with the result that NH_4^+ (and solvated forms thereof) will efficiently protonate only those analytes with a PA greater than about 860 kJ mol^{-1}; in effect, this means that only nitrogen-containing analytes (such as **Figures 1** and **2**) are efficiently protonated. Analytes with lower PAs, which contain a lone pair of electrons, frequently give $M \cdot NH_4^+$ adducts.

Charge exchange chemical ionization

In charge exchange CI, the reagent gas is chosen to produce an odd-electron species $R^{+\bullet}$ on EI ionization. This ion reacts with the analyte M by charge exchange to produce, initially, the odd-electron molecular ion (Eqn [6]), which may undergo further fragmentation (Eqn [7]).

$$R^{+\bullet} + M \rightarrow M^{+\bullet} + R \qquad [6]$$

$$M^{+\bullet} \rightarrow \text{fragments} \qquad [7]$$

Since the $M^{+\bullet}$ ion formed initially is the same as that formed initially in EI ionization, the fragmentation reactions observed will be similar to those observed in EI mass spectra. The essential difference is that, while the $M^{+\bullet}$ ions formed by electron ionization have a wide range of internal energies, the $M^{+\bullet}$ ions formed by charge exchange have an internal energy, $E(M^{+\bullet})$, given approximately by

$$E(M^{+\bullet}) = RE(R^{+\bullet}) - IE(M) \qquad [8]$$

where $RE(R^{+\bullet})$, the recombination of $R^{+\bullet}$, is defined by

$$R^{+\bullet} + e \rightarrow R, \quad -\Delta H^\circ = RE(R^{+\bullet}) \qquad [9]$$

and IE(M) is the adiabatic ionization energy of M.

Reagent systems have been developed (**Table 2**) giving reactant ions with recombination energies over the range 9.3–16 eV. By comparison, most organic molecules have ionization energies in the range 8–12 eV, so that, by suitable choice of the charge exchange reagent, a wide range of internal energies can be imparted to the molecular ion in Equation [6]. A potential use of charge exchange ionization is to enhance differences in the mass spectra of isomeric molecules compared to EI ionization; the use of charge exchange has the advantage, compared to low-energy EI, that the sensitivity of ionization is maintained. This use of charge exchange is illustrated in **Figure 3**

Table 2 Charge exchange reagent systems

Reagent gas	Reactant ion (R^+)	$RE(R^+)(eV)$
C_6H_6	$C_6H_6^{+\bullet}$	9.3
N_2-CS_2	$CS_2^{+\bullet}$	~10.0
$CO_2-C_6F_6$	$C_6F_6^{+\bullet}$	10.2
$CO-COS$	$COS^{+\bullet}$	11.2
Xe	$Xe^{+\bullet}$	12.1, 13.4
CO_2	$CO_2^{+\bullet}$	13.8
Kr	$Kr^{+\bullet}$	14.0, 14.7
N_2	$N_2^{+\bullet}$	15.3
Ar	$Ar^{+\bullet}$	15.8, 15.9

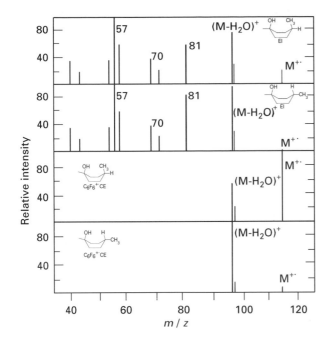

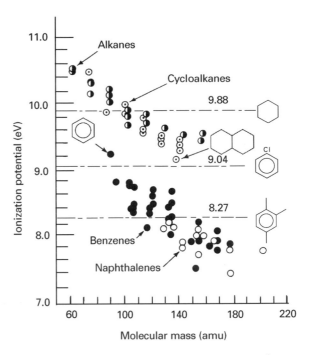

Figure 3 EI and $C_6F_6^{+\bullet}$ charge exchange (CE) mass spectra of *cis*- and *trans*-4-methylcyclohexanol. Harrison AG and Lin MS, (1984) Stereochemical applications of mass spectrometry. 3. Energy dependence of the fragmentation of stereoisomeric methyl cyclohexanols. *Organic Mass Spectrometry* 19: 67–71. Copyright John Wiley & Sons Limited. Reproduced with permission.

Figure 4 Ionization potentials of hydrocarbons as a function of molecular mass. Reprinted with permission from Sieck LW (1983) Determination of molecular weight distributions of aromatic components in petroleum products with chlorobenzene as reagent gas. *Analytical Chemistry* 55: 38–41. Copyright (1983) American Chemical Society.

which compares the $C_6F_6^{+\bullet}$ charge exchange mass spectra of *cis*- and *trans*-4-methylcyclohexanol with 70 eV mass spectra. The low-energy charge exchange spectra are much simpler; more importantly, the differences in the spectra of the two isomers are much enhanced following charge exchange ionization. A particularly useful application of charge exchange ionization is for the selective ionization of specific components of mixtures. The principle of the method is illustrated in **Figure 4** which shows the ionization energies of alkanes, cycloalkanes, alkylbenzenes and alkylnaphthalenes (components of petroleum products) plotted as a function of molecular mass. Also shown, as horizontal lines, are recombination energies of $c\text{-}C_6H_{12}^{+\bullet}$, $C_6H_5Cl^{+\bullet}$ and $(CH_3)_3C_6H_3^{+\bullet}$. These species should ionize only those hydrocarbons with ionization below the appropriate line. Clearly, $C_6H_5Cl^{+\bullet}$ will ionize only the benzene and naphthalene components, while $(CH_3)_3C_6H_3^{+\bullet}$ will be even more selective for the naphthalenes and higher molecular mass benzenes.

Brønsted base chemical ionization

Gaseous Brønsted bases, B^-, usually react with organic analytes M by proton abstraction to produce

$[M–H]^-$ as the initial product.

$$B^- + M \rightarrow [M - H]^- + BH \qquad [10]$$

Equation [10] is normally rapid provided it is exothermic, i.e. provided $\Delta H^\circ_{acid}(BH) > \Delta H^\circ_{acid}(M)$ or, alternatively, provided $PA(B^-) > PA([M-H]^-)$, where

$$BH_{(g)} \rightarrow B_{(g)}^- + H_{(g)}^+$$
$$\Delta H^\circ = \Delta H^\circ_{acid}(BH) = PA(B^-) \qquad [11]$$

In general, little fragmentation of $[M - H]^-$ is observed, with the result that Brønsted base CI often provides molecular mass information through formation of the $[M - H]^-$ ion that is 1 amu lower than the molecular mass.

Table 3 lists possible Brønsted base reagents along with the ΔH°_{acid} of the conjugate acid BH. For comparison, phenols have gas-phase acidities in the range of 1370–1465 kJ mol⁻¹, carboxylic acids have acidities in the range of 1350–1422 kJ mol⁻¹, alkynes in the range of 1520–1580 kJ mol⁻¹ and alkylbenzenes about 1590 kJ mol⁻¹. Also listed in the Table are the electron affinities of the neutral species

B; these are relevant to the tendency of B^- to react by electron transfer to form $M^{-\bullet}$. These electron affinities are quite high with the exception of $O_2^{-\bullet}$ which is the only reactant likely to react by electron transfer to M.

Of the bases listed OH^- has seen by far the greatest use. It can be prepared by electron bombardment of a CH_4–N_2O (~ 90:10) mixture (at usual CI source pressures) through the reaction sequence

$$e_{therm} + N_2O \rightarrow O^{-\bullet} + N_2 \qquad [12]$$

$$O^{-\bullet} + CH_4 \rightarrow OH^- + CH_3^{\bullet} \qquad [13]$$

where the CH_4 also serves to thermalize the electrons. Alternatively, the addition of CH_3ONO to the methane bath gas leads to formation of CH_3O^-. The Cl^- and Br^- ions can be prepared by dissociative electron capture by halogen-containing compounds such as CH_2Cl_2 or CH_2Br_2. The F^- ion is usually prepared by dissociative electron capture by NF_3. The OH^- ion is a strong base and will abstract a proton from most organic analytes to form $[M - H]^-$ Thus, OH^- CI can be usefully employed to obtain molecular mass information. For example, the OH^- CI mass spectra of ephedrine shows $[M - H]^-$ (*m/z* 164) as the dominant ion, and OH^- CI could be used as readily as CH_4 CI or NH_3 CI (**Figure 1**) to establish the molecular mass. More importantly, Brønsted base CI can often be used to provide molecular mass information in cases where Brønsted acid CI fails. This is illustrated in **Figure 5** which compares the CH_4 CI and OH^- CI mass spectra of 2,3-cyclopropyl-1,4-undecanediol. While the CH_4 CI mass spectrum is quite uninformative (as is the EI mass spectrum), the OH^- CI mass spectrum shows an abundant $[M - H]^-$ (*m/z* 199) ion, as well as fragmentation by sequential loss of two water molecules from $[M - H]^-$. The hydroxyl ion can react with some RX molecules by an S_N2 reaction to give X^- and does react with esters, R' COOR, in part, by addition/elimination to give $R'COO^-$

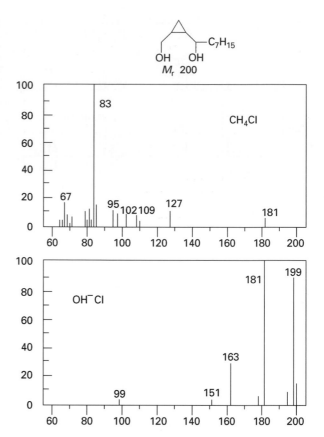

Figure 5　CH_4 CI and OH^- CI mass spectra of 2,3-cyclopropyl-1,4-undecanediol (relative intensity as a function of *m/z*).

+ ROH. The reactions of CH_3O^- have not been extensively studied; since it has a similar PA to that of OH^- one would expect the reactions to be similar. The F^- ion is a somewhat weaker base than OH^- or CH_3O^- and thus is somewhat more selective in its proton abstraction reactions. Cl^- is quite a weak base and will abstract only very acidic protons; in other cases $[M + Cl]^-$ adducts have been observed, which has proven useful in providing molecular mass information.

The $O^{-\bullet}$ ion, which can be produced by electron bombardment of a N_2–N_2O (~ 9:1) mixture, is unique in that it is both a strong Brønsted base and a radical. As a result, it reacts with many organic analytes not only by proton abstraction but also by H-atom abstraction, $H_2^{+\bullet}$-abstraction and H-atom and alkyl group displacement. This variety of reactions is illustrated in **Figure 6** for C_5 ketones. The $[M - H_2]^-$ · in ketones results to a significant extent by $H_2^{+\bullet}$ abstraction from one α-carbon; the resultant ion undergoes further fragmentation by loss of the opposite alkyl group (see *m/z* 41, **Figure 6**, for example). Alkyl group displacement results in formation of carboxylate ions for ketones. Clearly $O^{-\bullet}$ CI provides

Table 3　Brønsted base reagents

B^-	ΔH°_{acid} (BH) (kJ mol^{-1})	EA(B) (kJ mol^{-1})
NH_2^-	1690	75.3
OH^-	1636	176.6
$O^{-\bullet}$	1598	141.0
CH_3O^-	1594	151.5
F^-	1552	328.0
$O_2^{-\bullet}$	1477	42.3
Cl^-	1393	348.9
Br^-	1356	324.7

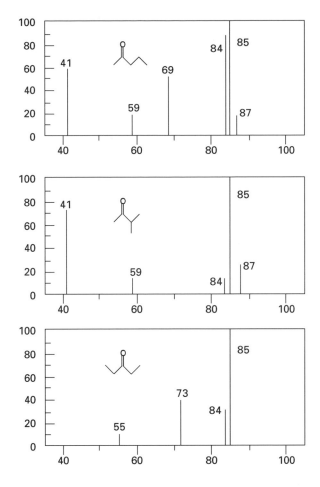

Figure 6   O⁻˙ CI mass spectra of C₅ ketones (Mᵣ 86). Data from Marshall A, Tkaczyk M and Harrison AG (1991) O⁻˙ chemical ionization of carbonyl compounds. *Journal of the American Society of Mass Spectrometry* 2: 292–298.

a greater possibility of isomer distinction than OH⁻ CI which produces only [M – H]⁻. However, O⁻˙ CI usually has a considerably lower sensitivity than OH⁻ CI or Brønsted acid CI.

Electron capture chemical ionization

In electron capture CI, ionization of the analyte M occurs by electron capture (Eqn [14]) or by dissociative electron capture (Eqn [15]).

$$e_{\text{therm}} + M \rightarrow M^{-\bullet} \qquad [14]$$

$$e_{\text{therm}} + M \rightarrow F^{-} + N \qquad [15]$$

Both reactions are resonance processes that require electrons of near-thermal energy to occur efficiently.

Typically, a high-pressure (0.1–1.0 torr) buffer gas is used to 'thermalize' the electrons (emitted from a heated filament) by inelastic scattering and positive ion ionization processes. Ideally, the buffer gas should not form stable anions, and buffer gases that have been used include He, H_2, N_2, CH_4, NH_3, CO_2 and $i\text{-}C_4H_{10}$. Polyatomic gases such as CH_4, NH_3 $i\text{-}C_4H_{10}$ or CO_2 have proven to be most efficient in promoting electron capture. The first three have seen the most use, in part because they are often used in Brønsted acid CI and are readily available.

The major attraction of electron capture chemical ionization (ECCI) is the prospect of enhanced sensitivity compared to other forms of CI or electron ionization. Whereas the maximum rate constants for the ion–molecule reactions involved in Brønsted acid or Brønsted base CI are in the range $(2–4) \times 10^{-9}$ cm³ molecule⁻¹ s⁻¹, electron capture rate constants can be as high as 10^{-7} cm³ molecule⁻¹ s⁻¹ because of the high mobility of the electron. Since the sensitivity of the ionization process depends, in part, on the rates of the ionization processes, in favourable circumstances ECCI can be up to 100 times more sensitive than other forms of ionization; this increase can be important when detection and quantification at trace levels is desired. At the same time the rate constants for electron capture are very compound-sensitive and can be very low, with the result that ECCI does show variable and, sometimes, very low sensitivity. It appears that only analytes with appreciable electron affinities (perhaps 0.5 eV) provide reasonable sensitivity in ECCI. The presence of halogen or nitro groups (particularly when attached to π-bonded systems), a conjugated carbonyl system or highly conjugated carbon systems leads to reasonably high electron affinities. Thus, as examples, dinitrophenols, dinitroanilines and polychlorinated aromatic compounds show good sensitivity in ECCI.

Analytes that are not suitable candidates for electron capture can often be made so by suitable derivatization. Perfluoracyl, pentafluorobenzyl, pentafluorobenzoyl and nitrobenzyl derivatives are often used not only to increase the electron capture rate constants but also to enhance chromatographic properties. However, in contrast to electron capture gas chromatography, it is not sufficient that electron capture be facile; in addition, ions characteristic of the analyte must be formed since ions characteristic only of the derivatizing agent leaves the identity of the analyte in question. This is illustrated by the data in **Table 4** for derivatization of phenols, where it can be seen that the perfluoroacyl derivatives yield ions characteristic only of the derivatizing agent, whereas the pentafluorobenzyl and pentafluorobenzoyl derivatives yield ions characteristic of the analyte. Also

Table 4 Derivatization of phenols[a]

Derivative	Relative intensity		Other ions (relative intensity)
	M⁻	ArO⁻	
–COCF₃	0	0.2	CF₃COO⁻ (100)
–COC₂F₅	0	0	C₂F₅COO⁻(100), C₂F₄CO⁻ (23)
–COC₃F₇	0	0	C₃F₇CO⁻ (100), C₃F₆CO⁻ (6)
–CH₂C₆F₅ (1)[b]	0	100	
–COC₆F₅ (0.4)[b]	100	5	C₆F₅⁻(8)
–COC₆H₃(CF₃)₂ (0.08)[b]	100	2	

[a] Data from Trainor TM and Vouros P (1987) Electron capture negative ion chemical ionization mass spectrometry of derivatized chlorophenols and chloroanilines. *Analytical Chemistry* **59**: 601–610.

[b] Relative molar response.

included in the Table are the relative sensitivities (relative molar response) for the three derivatives which give ions identifying the analyte; clearly, the pentafluorobenzyl or pentafluorobenzoyl derivatives are to be preferred in this case.

In addition to variable sensitivity, two further problems that can be encountered in ECCI are non-reproducibility of the spectra and the observation of artifact peaks, i.e. ions that are not explainable simply on the basis of electron capture by the analyte. ECCI mass spectra tend to depend rather strongly on experimental conditions such as buffer gas identity and pressure, source temperature, purity of the buffer gas, amount of analyte introduced, instrument used and instrument focusing conditions. The variability of the spectra obtained is illustrated by the example in **Figure 7**. At 100°C source temperature with CH₄ buffer gas, nordiazepam shows M⁻• (*m/z* 170) as the only significant ion, while at 150°C source temperature Cl⁻ is the dominant ion, and with 1 part in 2000 of oxygen in the methane, predominant formation of [M – H]⁻ (*m/z* 169) is observed. This amount of oxygen could easily be in the system if there is a small air leak. The occurrence of artifact peaks is illustrated in **Figure 8** which shows the ECCI spectra of tetracyanoquinodimethane with CH₄ and CO₂ as buffer gases. The major peaks in the spectrum with CH₄ buffer gas represent artifact peaks arising from reaction of the analyte with free radicals produced on electron bombardment of methane; these altered species are subsequently ionized by electron capture. By contrast, the spectrum with CO₂ buffer gas shows only M⁻•. CO₂ clearly is preferred in this case although when oxidation of the analyte occurs (such as formation of fluorenone from fluorene) the use of CO₂ as a buffer gas leads to substantial artifact peaks. It appears that the most

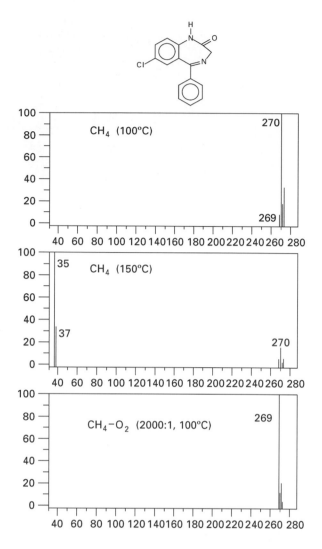

Figure 7 Electron capture CI mass spectra of nordiazepam under different conditions (relative intensity as a function of *m/z*). Data from Garland WA and Miwa BJ (1983) *Biomedical and Environmental Mass Spectrometry* **10**: 126–129.

prevalent pathway to these artifacts involves a surface-catalysed reaction of the analyte with some component of the reagent gas plasma, which produces one or more species with a higher cross section for electron capture than the original analyte. Alteration of the analyte presumably occurs when other ionization methods are used; the difference is that in other modes of CI all species are ionized with similar efficiencies, while in ECCI the altered species may be preferentially ionized.

Despite these difficulties with ECCI it is often the method of choice when identification and quantification at very low concentrations and/or involving complex mixtures is necessary. In such work it is necessary to control the experimental parameters carefully since they can affect both spectra and sensitivity.

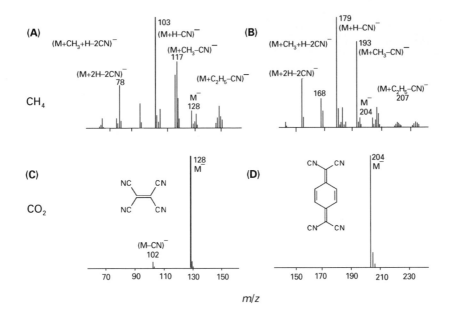

Figure 8 Electron capture CI mass spectra of tetracyanoethylene and tetracyanoquinodimethane using CH_4 and CO_2 as buffer gases. Reprinted with permission from Sears LJ and Grimsrud EP (1989) Elimination of unexpected ions in electron capture mass spectrometry using carbon dioxide buffer gas. *Analytical Chemistry*, **61**: 2523–2528. Copyright (1989) American Chemical Society.

List of symbols

e_{therm} = electron of near-thermal energy; E = energy; m/z = mass to charge ratio; $\Delta H°$ = enthalpy change.

See also: **Ion Energetics in Mass Spectrometry; Ion Molecule Reactions in Mass Spectrometry.**

Further reading

Budzikiewicz H (1986) Negative chemical ionization (NCI) of organic compounds. *Mass Spectrometry Reviews* 5: 345–380.

Burrows EP (1995) Dimethyl ether chemical ionization mass spectrometry. *Mass Spectrometry Reviews* 14: 107–115.

Chapman JS (1993) *Practical Organic Mass Spectrometry*, 2nd edn, Chapters 3 and 4. Chichester, UK: Wiley.

Creaser CS (1995) Chemical ionization in ion trap mass spectrometry. *In*: March RE and Todd JFJ (eds) *Practical Aspects of Ion Trap Mass Spectrometry*, Vol III, Chapter 7. Boca Raton, FL: CRC Press.

Harrison AG (1992) *Chemical Ionization Mass Spectrometry*, 2nd edn. Boca Raton, FL: CRC Press.

Munson MSB and Field FH (1996) Chemical ionization mass spectrometry I. General introduction. *Journal of the American Chemical Society* 88: 2621–2630.

Vairamani M, Mirza UA and Srinivas R (1990) Unusual positive ion reagents in chemical ionization mass spectrometry. *Mass Spectrometry Reviews* 9: 235–258.

Westmore JB and Alauddin MM (1986) Ammonia chemical ionization mass spectrometry. *Mass Spectrometry Reviews* 5: 381–465.

Winkler FJ and Splitter JS (1994) Sterochemical effects in the positive- and negative-ion chemical ionization mass spectra of stereoisomeric molecules. *In*: Splitter JS and Tureček F (eds) *Applications of Mass Spectrometry to Organic Stereochemistry*, Chapter 16. New York: VCH.

Chemical Reactions Studied By Electronic Spectroscopy

Salman R Salman, University of Qatar, Doha, Qatar

Introduction

This article covers applications of UV and visible electronic spectroscopy to studies of chemical reactivity. This encompasses both studies of reaction kinetics and of equilibria and chemical exchange phenomena. Chemical reactions can occur in different phases. Many of the reactions occur in the gas phase and the reactions which take place in this phase will be covered in the section on reaction kinetics.

In solution, the volume occupied by the molecules will be larger than that in the gas phase, thus the molecules will come close together and the freedom for translational motion is restricted. There is another important difference between the solution and the gas phase, which is the great proximity between the solvent molecules and the solute molecules. The solvent can act as a heat sink and energy can pass from the solvent to the solute and this affects the vibrational modes and causes the reactants to overcome activation barriers. Solvent polarity will affect reactions in solutions containing ions. Polar solvents lower the energy required for the formation of ions. In solution many processes and reactions will take place such as diffusion-controlled reactions, reactions of hydrogen and hydroxyl ions, the formation of solvated molecules and electron transfer reactions. The rate coefficients depend on many factors such as the solvent and the formation of a solvent cage, the pressure and the ionic strength of the solution.

Many reactions take place on solid surfaces, and solid materials catalyse such reactions. Surface catalysis is very important in industry, and makes important contributions to some atmospheric reactions. Molecules will be adsorbed on solid surfaces through physicosorption in which the attractive force between the reactant molecules and the surface is of the van der Waals type. Enthalpy of adsorption, ΔH_{ad}, of physicosorption is relatively small, about -40 kJ mol^{-1}. With chemical sorption the value for the enthalpy of adsorption is large because a strong bond is formed between the chemical molecule and the surface. Catalytic reactions can be classified as homogeneous catalysis or heterogeneous catalysis.

UV-visible spectroscopy for kinetics studies

The problem of variable response factors

UV-visible spectroscopy has proved useful in biochemical analysis, environmental studies, in forensic science, drug kinetics, food quality, identification and quantification of chemical and biological substances but is limited as a tool for the investigation of molecular structure compared with infrared and nuclear magnetic resonance. However, it is quite useful for the study of reaction kinetics and equilibrium.

There is a problem facing the use of a UV-visible spectrum for the calculation of the concentration of different species if two species absorb at the same region, and in this case there is a need to change the reagents to obtain different colours for different species. Also, it is important to carry out a close inspection for the two bands which have the same wavelength. An example of such a system is the absorption of potassium permanganate along with the absorption of a soluble azo dye, which appears as if it is absorbing at the same region, this being near 530 nm. A close inspection and calculations show that λ_{max} of $KMnO_4$ and the dye are 533 and 542 nm, respectively, while the extinction coefficients for $KMnO_4$ and the dye are 1.01 and 1.88×10^{14} dm^3 mol^{-1} cm^{-1}, respectively.

Theory of UV spectroscopy

In UV-visible spectroscopy we are interested in studying the electronic energy. This energy can be calculated from the observed bands in the UV-visible region of the spectrum (195–750 nm). The observed bands in this region are due to the excitation of the electrons by light of a certain wavelength, from the ground state to the excited state. This excitation of the energy of electrons between energy states is in accord with the Frank–Condon principle which states that an electronic transition in a molecule takes place so rapidly compared with the vibrational motion of the nuclei that the internuclear distance can be regarded as fixed during the transition.

The types of electronic transitions are given in Table 1. In organometallic compounds a very intense colour can arise due to the d → d, π → π* and n → π* transitions. This is true especially in transition metals connected to an organic molecule. Several factors, such as substitution (electronic and steric effect), solvent, temperature, and pH will affect electronic transitions and band intensity and shape. In the following paragraphs a brief discussion is given.

Sample preparation

To obtain useful and accurate information from the UV-visible spectrum it is important to fix appropriate conditions for recording the spectrum. The studied material must be stable and not undergo decomposition by itself under the experimental conditions, which include concentration, solvent and temperature. The solute must dissolve completely in the solvent. It must not be colloidal, because this will given an error in the calculation of the different species present in solution. Also, the solution must not undergo hydrolysis or decomposition in this solvent, otherwise a new species will form which may not be soluble in the solvent.

Cells for the UV region below 330 nm are made of fused silica. In the case of the visible region glass cells may be used. Cells for special measurements can be used and they are classified as sampling cells, flow cells and rectangular cells.

In all measurements it is important to work with concentrations which are within the Beer–Lambert law. There is a need to carry out calibration so that by preparing different concentrations of the sample in the solvent and plotting the concentration against the recorded absorbance we have a straight line. It is necessary to use a concentration of the sample within this calibrated curve. The usual concentration, which is used in recording a UV-visible spectrum, is 10^{-3}–10^{-5} M. The solute must not change its concentration due to decomposition or precipitation.

Solvent effects

It is very important to choose the appropriate solvent. The best solvent is that which is able to solubilize the solute completely, which is inert and will not interact with the solute and thus cause a perturbation of its electronic structure, and which will not absorb in the region where the sample will absorb. Table 2 gives the cutoff wavelength for a selective range of solvents.

Sometimes it is important to use a solvent with a high solubilizing ability like water, ethanol, etc. Solvents must be of high purity because of the presence of impurities will affect the spectrum and some of those impurities have absorptions in the same region as that of the solute, while others will interact with the solute and this will change its structure. It is important to take into consideration the solvent effect on the solute. So if we take a solute dissolved in an inert solvent such as cyclohexane as our reference absorption, then some solvents will cause the absorption of the solute to shift more to the red region and this shift is denoted as bathochromic or red-shift. Other solvents may cause the band to shift to the blue region and this is called blue-shift or hypsochromic. This phenomenon is important in UV-visible spectroscopy because it is possible to study the solute–solvent interactions and hydrogen bond strength. Some solute molecules can be used as a probe for measuring the solvent properties.

Table 1 UV-visible absorption of some chromophores

Chromo-phore	Compound	Electronic transition	λ_{max} (nm)	ε [a] (m^2 mol^{-1})
C$_6$H$_6$	Benzene	π → π*	200, 255	800
C=C	Ethylene	π → π*	180	1300
C=O	Acetone	π → π*	185	95
		n → π*	277	2
N=N	Azomethane	n → π*	347	1
N=O	Nitrosobutane	n → π*	665	2

[a] Molar absorptivity.

Table 2 λ_{max} for different solvents

Solvent	λ_{max} (nm)
Water	165
Acetonitrile (spectroscopic grade)	190
Hexane	199
Heptane	200
Isoctane	202
Diethyl ether	205
Ethanol	207
Propan-2-ol	209
Methanol	210
Cyclohexane	212
Acetonitrile (chemical grade)	213
Dioxan	216
Dichloromethane	233
Tetrahydrofuran	238
Trichloromethane	247
Tetrachloromethane	257
Dimethyl sulfoxide	270
Dimethylformamide	271
Benzene	280
Pyridine	306
Propanone	331

Light absorption and colour

Colour change is very important for the determination of different species present in the media whether they are the reacting species or the species formed during the reaction. From the spectra it is easy to calculate the concentration of all species and therefore to study the kinetics of the reaction.

Solution conditions and analysis

To obtain accurate data and concentrations for the different species present in solutions at specific times, it is important to obtain correct conditions for monitoring the species by the change of the absorbance of those species. It is important to determine the following conditions and their variation.

Temperature and its variation It is possible to change the temperature and this will lead to a change of absorbance, thus, it is possible to calculate the activation energy of the reaction.

pH This is very important for aqueous solutions. The pH can have a large affect on the UV-visible spectrum, rates of complexation or change in reaction kinetics. pH can have a pronounced effect on shifting the equilibrium in enol–keto tautomerism. A good example is the effect of pH on the equilibrium in the system dichromate/chromate.

Ionic strength The ionic strength of the solution is important for some reactions, especially for solutions which contain ions other than the neutral reactant molecules; interaction will occur between ions other than the reactants and this will affect the rate constant.

Effect of pressure For the majority of chemical reactions it is difficult to study the pressure effect. The reaction rate is insensitive to changes in pressure and a large pressure must be applied before we observe any important change. Sometimes we need to study the effect of pressure because it provides a useful insight into the mechanism of the reaction.

Solvent polarity In general, one should use solvents with low polarity so there will be no interference in the formation and decomposition of different species. On occasions, however, it is necessary to use solvents with high polarity to increase the solubility of the solutes or to affect the activation energy of a certain reaction.

Other factors that are important are the order of addition of different reagents, mixing and stirring to obtain solution homogeneity and choice of the proper wavelength which will be used to monitor the change of concentration during the reaction.

Specialized techniques

Stopped flow method This method can be applied for the study of gases and liquids. The usual detection techniques are absorption spectrometry. High flow rates are required, to promote mixing for fast reactions.

Flash photolysis and pulse radiolysis In the flash photolysis technique a reactant is irradiated with an intense flash of visible or ultraviolet light. The intensity must be sufficient to produce a measurable change in chemical composition, but of short duration compared with that of the ensuing reactions, which are to be studied.

Multidimensional spectroscopy and derivative spectroscopy When bands of reactants and reaction products overlap in the fundamental UV-visible absorption spectra the reaction kinetics cannot be followed by the classic UV method. In many cases, the second derivative UV-visible spectrophotometry (D-2) provides an alternative method to solve the problem. Even-order derivatives are suitable to follow kinetics because the maxima in the UV-visible derivative spectrum can be associated with the minima and a low-noise online spectra is obtained which can be computed up to the 6th order derivative and even up to the 10th order with the newly developed computers. On the other hand, the first derivative does not provide the above association; and other higher odd-order derivatives are less precise, though in practice it has proved valuable to work with spectra of the 3rd and 5th order.

Applications to chemical reactions

Many applications of UV-visible spectroscopy to the study of chemical reactions have been published and some illustrative examples which have been reported in the literature during the last two years will be covered.

Complex formation

The kinetics of transformation of complexes in organic media can be studied. Such studies will indicate whether the complexes will form in a one-step reaction mechanism or a multistep reaction mechanism. It is possible to determine the rate constants and the stability constant of the complex. UV-visible spectroscopy can be used to study the protonation of certain complexes and it is possible to obtain information on the position of protonation.

Reduction of complexes can be studied by UV-visible spectroscopy. Reduction of $trans\text{-}[Pt(CN)_4X_2]^{2-}$

(X = Cl or Br) [as model compounds for antitumour-active platinum(IV) pro-drugs] to $[Pt(CN)_4]^{2-}$ by L-methionine, MeSR, has been studied at 25°C in the range $0 < pH < 12$ (X = Cl) and $0 < pH < 6$ (X = Br) by use of stopped-flow spectrophotometry. It was concluded that methionine-containing biomolecules may compete with thiol compounds for reduction of platinum pro-drugs under acidic conditions, and also in neutral solutions with low concentrations of thiol-containing biomolecules.

The formation kinetics and the thermodynamic stability of iron(III) complexes with a new tetradentate ligand, N(alpha)-salicyl-L-alaninehydroxamic acid (H-2slalh), have been investigated by the use of UV-visible and stopped-flow spectrophotometric methods. By comparing the pH-dependent absorbance change, the mechanism of the complex formation was explained.

Photochemical reactions

Photochemical behaviour of full-aromatic β-carbolines, nor-harmane, harmane and harmine in chloroform medium and the corresponding UV absorption and fluorescence emission spectra have been discussed. Irreversible electron transfer from the singlet excited β-carboline molecule to the chloroform molecule in the transient excited charge transfer complex has been proposed as the primary photochemical process initiating the mechanism of secondary reactions in this system.

In deoxygenated water, methanol, and ethanol, 4-hydroxybenzonitrile (4-HBN) is photoisomerized into 4-hydroxybenzoisonitrile (4-HBIN), which is then hydrolysed into 4-hydroxyformanilide in acidic medium. The triplet–triplet absorption of 4-HBN ($\lambda_{max} = 300$ nm) is detected by transient absorption spectroscopy. The triplet is converted into long-lived transients absorbing in the far-UV. The analysis of the kinetics of 4-HBIN formation as a function of irradiating photon flux shows that the photoisomerization of 4-HBN is a two-stage photoprocess. According to triplet-quenching studies, the first stage proceeds via the 4-HBN triplet to yield an intermediate capable of absorbing a second UV photon, which then gives 4-HBIN in the second stage. Mechanistic considerations indicate that this intermediate is likely to be an arizine.

Tautomerism and isomerization

Tautomerism is an important phenomenon and an example of this process is the equilibrium in Schiff bases. These bases can exist as an equilibrium between two species, keto ⇌ enol [**Scheme 1**]. This process is affected by a number of factors such as solvent polarity, temperature and pH. The kinetics of tautomerism can be studied by UV-visible spectroscopy and we will deal with a few examples.

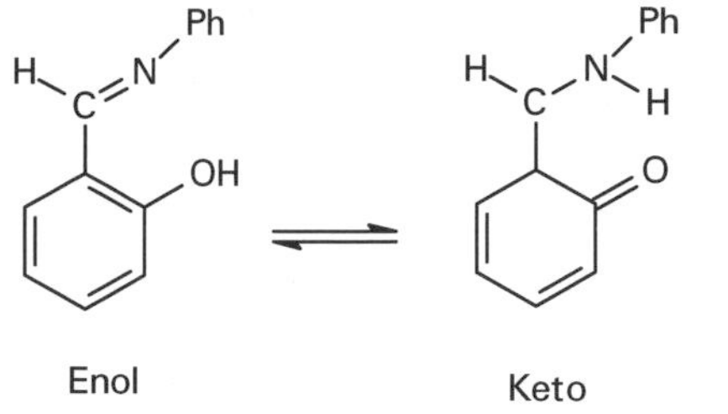

Enol   Keto

Scheme 1

The tautomerism between a hydroxy Schiff base and the corresponding ring-closed oxazolidine was kinetically studied in chloroform. This method indicated that this reaction is pseudo-first-order. UV examination was used to deduce the molecular species in various pH buffers. In an acid solution (e.g. pH 3.0) the Schiff base existed as the protonated Schiff base at the imine nitrogen atom, and in the alkaline region (e.g. pH 9.0) as the oxazolidine form.

The keto–enol equilibrium constants of acetylacetone, ethyl acetoacetate and ethyl benzoylacetate in water at 25°C were determined by studying the influence of surfactants on their UV-visible spectra. These measured equilibrium constants were used to obtain the reactivity of the ketones towards several nitrosating agents.

Atmospheric, environmental reactions, gas phase and free radical kinetics

The UV absorption cross-sections of peroxyacetyl nitrate (PAN), $CH_3C(O)O_2NO_2$, have been measured as a function of temperature (298, 273 and 250 K) between 195 and 345 nm. Photolysis becomes the most important atmospheric loss process for PAN, and the OH reaction is found to unimportant throughout the troposphere.

The atmospheric chemistry of benzene oxide/oxepin, a possible intermediate in the atmospheric oxidation of aromatic hydrocarbons, has been investigated by visible and UV photolysis. The results indicated that the major atmospheric sinks of benzene oxide/oxepin are the reaction with OH radicals and photolysis and under smog chamber conditions, with high NO_2 concentrations, also reaction with NO_3.

A flash photolysis-resonance fluorescence technique was used to study the rate constant for the reaction of OH radicals dimethyl carbonate over the temperature range 252–370 K. Pulse radiolysis/transient UV absorption techniques were used to study the ultraviolet absorption spectra and kinetics of $CH_3OC(O)OCH_2$ and $CH_3OC(O)CH_2O_2$ radicals at 296 K.

The codisposal of trace metals (e.g. Co), synthetic chelates (e.g. ethylenediaminetetraacetic acid, H(4) EDTA) and water-miscible organic solvents has occurred at some contamination sites. The reactions of Co(II)-EDTA with a redox reactive naturally occurring solid, goethite, in aqueous and semiaqueous (methanol–water, acetone–water) suspensions was studied. UV-visible spectroscopy indicated that goethite catalysed oxidation of Co(II)-EDTA to Co(III)-EDTA by dissolved O_2. These reactions have important implications on the fate of the redox-sensitive metal in complex, mixed waste environments.

Photocatalytic degradation

A number of organic pollutants present in industrial and domestic wastewater resist biodegradation and they are poisonous even at low concentrations. A new method for removing toxicants from wastewater is based on the use of photocatalysis. It was found that titanium dioxide (TiO_2) is the best photocatalyst for the detoxification of water because it is cheap, nontoxic and easy to use and handle. The energy gap in TiO_2 is 3.2 eV and thus it can be activated at >400 nm.

The primary oxidant responsible for most advanced oxidation processes, i.e. the use of photochemical methods, is the hydroxyl radical, which is formed by the reduction reactions of electrophiles with water or hydroxide ions. The mechanism of the formation of the hydroxyl radical is well discussed in the literature. Heterogeneous photocatalytic oxidation of organic compounds in aqueous solution is achieved by the reactive hydroxyl radical. The photocatalytic process can remove a large number of organic hazardous compounds in water. Phenols, carboxylic acids and herbicides are among a large number of pollutants, which are destroyed in air and in water using photocatalysis.

Photodegradation of 1,2,3,4-tetrachlorodibenzo-p-dioxin (1,2,3,4-TCDD) in hexane solution was studied under controlled near-UV light exposure in the spectral region from 325 to 269 nm. Irradiation experiments carried out at a constant light energy (700 mJ) showed that the percentage of 1,2,3,4-TCDD left in the solution after irradiation changed from about 55 to 75%, with a minimum of 55% at 310 nm. Further irradiation experiments carried out at two wavelengths, namely 310 and 269 nm, and light energy ranging from 0 to 4000 mJ, showed that the photodegradation reaction of the TCDD always followed pseudo-first-order kinetics.

A semitransparent TiO_2 film with extraordinarily high photocatalytic activity was prepared on a glass substrate by sintering a TiO_2 sol at 450 °C. The photocatalytic properties of the film were investigated by measuring the photodegradative oxidation of gaseous acetaldehyde at various concentrations under strong and weak UV light irradiation conditions. The kinetics of acetaldehyde degradation as catalysed by the TiO_2 film as well as by P-25 powder were analysed in terms of the Langmuir–Hinshelwood model. It is shown that the number of adsorption sites per unit true surface area is larger with the TiO_2 film, as analysed in powder form, than with P-25 powder. Meanwhile, the first-order reaction rate constant is also much larger with the film than with P-25 powder. Moreover, under most experimental conditions, particularly with high concentrations of acetaldehyde and weak UV illumination intensity, the quantum efficiency was found to exceed 100% on an absorbed-photon basis, assuming that only photogenerated holes play a major role in the reaction. This leads to the conclusion that the photodegradative oxidation of acetaldehyde is not mediated solely by hydroxyl radicals, generated via hole capture by surface hydroxyl ions or water molecules, but also by photocatalytically generated superoxide ion, which can be generated by the reduction of adsorbed oxygen with photogenerated electrons.

Solvents effects and solvolysis

Oxidations of arylalkanes by (Bu_4NMnO_4)-Bu^n have been studied, e.g. toluene, ethylbenzene, diphenylmethane, triphenylmethane, 9,10-dihydroanthracene, xanthene and fluorene. Toluene is oxidized to benzoic acid and a small amount of benzaldehyde; other substrates give oxygenated and/or dehydrogenated products. The manganese product of all of the reactions is colloidal MnO_2. The kinetics of the reactions, monitored by UV-visible spectrometry, show that the initial reactions are first order in the concentrations of both (Bu_4NMnO_4)-Bu^n and substrate. No induction periods are observed. The same rate constants for toluene oxidation are observed in neat toluene and in o-dichlorobenzene solvent, within experimental errors. The presence of O_2 increases the rate of (Bu_4NMnO_4)-Bu^n disappearance. The reactions of toluand dihydroanthracene exhibit primary isotope effects: The rates of oxidation of substituted toluenes show only small substituent effects. In the reactions of dihydroanthracene and fluorene, the MnO_2 product is consumed in a subsequent reaction that appears to form a charge-transfer complex. The rate-limiting step in all of the reactions is hydrogen atom transfer from the substrate to a permanganate oxo group. The enthalpies of activation for the different substrates are directly proportional to the ΔH for the hydrogen atom transfer step, as is typical of organic radical reactions. The ability of permanganate to abstract a hydrogen atom is explained on the basis of its ability to form an 80 ± 3 kcalmol^{-1} bond to H–, as calculated from a thermochemical cycle.

Polymerization

The kinetics of oxidation of a macromolecule, poly(ethylene glycol) [PEG] by ceric sulfate in sulfuric acid medium has been studied by means of UV-visible spectrophotometry. The observed difference in rates of oxidation has been explained in terms of cage formation. The oxidation of PEG proceeded without the formation of a stable intermediate complex. The order with respect to the concentrations of PEG and ceric sulfate has been found to be one and the overall order is two. The effects of acid concentration, sulfate ion concentration, ionic strength and temperature on the rate of the oxidation reaction have been studied. The thermodynamic parameters for the oxidation reaction have also been presented. Based on the experimental results, a suitable kinetic expression and a plausible mechanism have been proposed for the oxidation reaction.

Reaction of *trans*-1,3-diphenyl-1-butene (D), the *trans* ethylenic dimer of styrene, with trifluoromethanesulfonic acid in dichloromethane has been performed at temperatures lower than room temperature using a stopped-flow technique with real time UV-visible spectroscopic detection. The main product of the reaction was the dimer of D. A transient absorption at 340 nm has been assigned to 1,3-diphenylbutylium, a model for the polystyryl cation. Other absorptions at 349 nm and 505 nm have also been observed and were assigned to an allylic cation, 1,3-diphenyl-1-buten-3-ylium, resulting from hydride abstraction from D. This species was very stable at temperatures lower than −30°C. A general mechanism was proposed based on a kinetic study of the reactions involved.

Determination of metals and molecules at low concentration

A new and simple method for measuring peroxides in a single living cell has been developed, and the generation of peroxides upon ultraviolet (UV) irradiation was measured in human and pig pidermal keratinocytes. The method was based on the fact that the nonfluorescent dye dihydrorhodamine 123 reacts in the presence of peroxides, such as H_2O_2, and changes into fluorescent rhodamine 123, and hence the fluorescence intensity is proportional to the amount of reacted peroxide. The epidermal kerationcytes were loaded with the dihydrorhodamine under a fluorescence microscope and exposed to UV radiation. Taking C as the content of peroxides generated within the cell and I as the increase influence (radiation intensity × time = photons cm^{-2}), the following empirical relationship was established: $C = C_s(1 - \exp^{-kI})$, where C_s is the content of peroxides at the saturation state, and k is a kinetic parameter. The dependence of the two parameters on wavelength in the range 280–400 nm was studied. In human keratinocytes C_s had a peak at 310 nm and a small peak (shoulder) at 380 nm, while k increased gradually toward shorter wavelengths. In pig keratinocytes, on the other hand, k had a peak around 380 nm and a shoulder at 330 nm, while C_s remained unchanged. Aminotriazole, an inhibitor of catalase, and low temperatures increased the stationary levels of peroxide generation in pig keratinocytes upon UV irradiation, indicating that the reaction used for measuring intracellular peroxides is competitive with the intrinsic reactions in scavenging peroxides.

A first simultaneous EPR and visible spectrophotometric study is reported on the interaction of the stable free radical 1,1-diphenyl-2-picrylhydrazyl (DPPH) dissolved in ethanol with thioglycolic acid (HSCH$_2$CO$_2$H, TGA). The results of the kinetic studies at room temperature allow us to assume 1:1 stoichiometry of the reaction between DPPH and TGA giving 1,1-diphenyl-2-picrylhydrazine (DPPH) and thioglycolic disulfide. The linear plots of EPR and UV-visible responses versus the quantity of added TGA are used to find the DPPH molar absorptivity at 520 nm to be 12350 ±3% 1mol^{-1}cm^{-1} which may be used as a criterion for the purity of the material itself. It was also found that the paramagnetic and optical properties of a 30-year old sample gave results suggesting that in the solid state DPPH is a fairly stable material.

Reaction kinetics, substitution effects, structure–reactivity relationships

The kinetics of the addition of arenesulfinic acids to 4-substituted 2-nitroethenylarenes was studied by means of UV spectrophotometry. The effect of 4-substituting groups in benzenesulfinic acids, and the change in reactivity of the nitroethylene system due to typical electron-donating and electron-withdrawing groups were investigated. The substituent effect on benzenesulfinic acid fits Hammett's equation, *p*-value at 298 K being −1.12. Kinetic studies were carried out at 288–308 K, and the activating energy and the enthalpy of activation were determined.

The kinetics of the addition of unsubstituted and substituted benzenesulfinic acids to various 2-halogenonitroethenylarenes was studied by means of UV spectroscopy. The effect of 4-substituents in benzenesulfinic acids and the change in reactivity of the nitroethylene system in 2-halogeno-2-nitroethenyl-arenes in the presence of various substituents in the benzene ring and various halogens at the double bond were investigated. Kinetic studies were made at 288–308 K, and the activating energy and enthalpy of activation were determined.

Effect of pH, pressure, temperature and ionic strength on reaction kinetics

The degradation kinetics of xanthate in homogeneous solution as a function of pH at 5°C, 20°C and 40°C was systematically studied by UV-visible spectrophotometric measurements. The results indicate that the degradation of ethyl xanthate is rapidly increased with decreasing pH at pH < 7. At pH 7–8, the maximum half-life of the xanthate appears. The degradation was faster at pH 9–10, but at pH > 10 the half-lives of xanthate once again increase. The investigations were also extended to different media other than pure water, such as, 0.1 M $NaClO_4$, 0.1 M $NaNO_3$, 0.1 M NaCl as well as in the supernatants of flotation tailings of sulfide minerals. The rate constants of xanthate degradation were calculated and presented together with half-lives and activation energies of xanthate degradation. The degradation products and reaction mechanisms are discussed based on experimental results.

The kinetics of the association reaction of the phenoxy radical with NO were investigated using a flash photolysis technique coupled to UV absorption spectrometry. Experiments were performed at atmospheric pressure, and theoretical calculations showed that the rate constant is at the high-pressure limit above 50 torr (6.7×10^3 Pa) for temperatures below 400 K. Upon increasing the temperature, the reaction was found to be reversible, and the equilibrium kinetics have been studied at seven temperatures between 310 and 423 K.

Validation of molecular orbital calculations

The suitability of Gaussian distribution functions to describe the shape and temperature dependence of the UV absorption continua of peroxy radicals has been investigated. The ethylperoxy radical was used as a test case. Its 298 K absorption continuum was found to be best described by a semilogarithmic Gaussian distribution function. A linear Gaussian distribution function performed less well but still adequately described the continuous absorption. The temperature dependence of the ethylperoxy radical UV absorption continuum was also well predicted. Analogous results obtained for the methylperoxy radical support these conclusions. A theoretical comparison of the semilogarithmic and linear Gaussian distribution functions is given and a potential energy diagram of the ethylperoxy radical derived. The experimentally determined absorption cross-sections of HO_2 have been reanalysed. It is shown that either the measurements at short wavelengths are in error or an unidentified electronic transition of HO_2 exists.

Enzymatic reactions

A one-step spectrophotometric method for monitoring nucleic acid cleavage by ribonuclease. H from *E. coli* and type II restriction endonucleases has been proposed. It is based on recording the increase in the UV absorbance at 260 nm during the course of enzymatic reaction. Duplexes stable under the reaction conditions were chosen as substrates for the enzymes being studied. In order to obtain duplex dissociation following their cleavage by the enzyme appropriate temperature conditions were selected. The spectrophotometric method may be applied for rapid testing of the nuclease activity in protein preparations as well as for precise quantitative analysis of nucleic acid degradation by enzymes. This method may be successfully employed in kinetic studies of nucleic acid–protein interactions.

See also: **Biomacromolecular Applications of UV-Visible Absorption Spectroscopy; Colorimetry, Theory; Dyes and Indicators, Use of UV-Visible Absorption Spectroscopy.**

Further reading

Atkins PW (1998) *Physical Chemistry*, 6th edn. Oxford: Oxford University Press.

Barrow GM (1986) *Introduction to Molecular Spectroscopy*, 17th edn. London: McGraw-Hill Book Company.

Denney RC and Sinclair R (1991) *Visible and Ultra Violet Spectroscopy*. New York: Wiley.

Kempt W (1989) *Organic Spectroscopy*, 2nd edn. London: Macmillan Education.

Ladd M (1998) *Introduction to Physical Chemistry*, 3rd edn. London: Cambridge University Press.

Pilling MJ and Seakins PW (1996) *Reaction Kinetics*, 1st edn. London: Oxford Science Publication.

Pérez-Bendito D and Silva M (1988) *Kinetics Methods in Analytical Chemistry*, 1st edn. Chichester: Ellis Horwood.

Vemulapalli K (1993) *Physical Chemistry*. New York: Prentice-Hall International.

Chemical Shift and Relaxation Reagents in NMR

Silvio Aime, **Mauro Botta**, **Mauro Fasano** and
Enzo Terreno, University of Torino, Italy

> MAGNETIC RESONANCE
> Applications

Introduction

Chemical shift and relaxation reagents are substances that, when added to the systems under study, enable the spectroscopist to tackle a specific problem of spectral assignment, stereochemical determination or quantitative measurement. They are usually represented by paramagnetic compounds, and their effects may exploited by the solute and the solvent, manifested as large shifts and/or pronounced relaxation enhancement of the lines. Both effects reflect the response of the nuclear dipole to the local magnetic field of an unpaired electron in the reagent, whose magnetic moment is 658.21 times higher than that of the proton.

A knowledge of the basic theory of the paramagnetic interaction is useful to pursue a better exploitation of the effects induced by the paramagnetic perturbation.

In principle, the interaction of a nucleus with an unpaired electron (hyperfine interaction) has the same nature as a nucleus–nucleus interaction, i.e. it is determined by a direct, through-space dipole–dipole coupling (H_D) and an indirect, through-bonds contact (scalar) coupling (H_S):

$$H = H_D + H_S$$

The hyperfine shift arises from components present in both terms which are invariant with time in the external magnetic field; these contributions are indicated as 'dipolar' or 'pseudocontact' shifts when resulting from dipolar interactions and 'Fermi' or 'contact' shifts when resulting from scalar interactions, respectively. The time-dependent fluctuations in either term will result in nuclear relaxation ('dipolar' and 'scalar' relaxation, respectively). Although, in theory, correct expressions for H_D and H_S can be derived, the elusive nature of the orbiting electrons and the number of their interactions make it very difficult to develop an accurate theory for shift and relaxation promoted by paramagnetic substances. In general one may conclude that, while measurements of the effects promoted by paramagnetic species are easy, interpretation of data is difficult.

According to their dominant effect on the NMR spectra of surrounding nuclei, paramagnetic reagents can be classified as chemical shift or relaxation reagents. Organic free radicals and metal ions with isotropic electron density, such as Mn^{2+} (five electrons in five d orbitals) or Gd^{3+} (seven electrons in seven f orbitals) do not give rise to pseudocontact shifts and are commonly regarded as relaxation probes. On the other hand, anisotropic ions with short electron relaxation times, such as all other lanthanide ions, can be referred to as chemical shift probes. Species such as Ni^{2+}, Fe^{3+} and Cr^{3+} give rise to both effects and defy classification.

Chemical shift reagents

In 1969, Hinckley suggested that the use of paramagnetic lanthanide complexes might be very useful to simplify unresolved ^1H, NMR resonances, thus allowing otherwise intractable spectra to be easily interpreted (**Figure 1**). Although the introduction of high-field magnets and two-dimensional methods have largely reduced the need for such qualitative use of 'lanthanide induced shift' (LIS) measurements, the exploitation of the large local fields generated by some lanthanide ions is still an item of strong interest in several areas of NMR applications.

When a substrate interacts with the lanthanide complex, it magnetically active nuclei *feel* the presence of the unpaired f electrons; their LIS is basically determined by two properties of the lanthanide complex, namely (i) its Lewis acid behaviour and (ii) the number of unpaired electrons. The commercially available lanthanide shift reagents (LSRs) are mainly tris(β-diketonate) complexes, having the 2,4-pentanedione chelate structure (**Table 1**). These neutral complexes are soluble in organic solvents and they promptly form acid–base adducts with substrates endowed with Lewis base behaviour. In fact alcohols, ethers, amines and nitriles reversibly interact with the LSR to form paramagnetic adducts whose ^1H NMR shifts may be far away from the diamagnetic values. As the substrate B in fast exchange between the LSR-bound and free forms, the effect will be averaged on all the molecules of the substrate. The

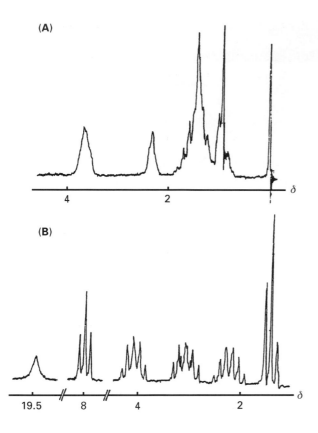

(A)

(B)

Figure 1 Simplification of the low-field ^{1}H NMR spectrum of 1-pentanol by Eu(thd)$_3$. The clustered signals of the methylene chain are totally separated into clearly defined multiplets. (A) normal spectrum; (B) after addition of Eu(thd)$_3$

Table 1 Lanthanide diketonate chelates used as NMR shift reagents

R	R'	Ln	Acronym
t-C$_4$H$_9$	t-C$_4$H$_9$	Pr	Ln(thd)$_3$ or
		Eu	Ln(dpm)$_3$
		Dy	
		Ho	
		Yb	
t-C$_4$H$_9$	CF$_3$	Eu	Eu(pta)$_3$
t-C$_4$H$_9$	n-C$_3$F$_7$	Eu	Eu(fod)$_3$
		Pr	
n-C$_3$F$_7$	n-C$_3$F$_7$	Eu	Eu(tfn)$_3$
C$_2$F$_5$	C$_2$F$_5$	Eu	Eu(fhd)$_3$
CF$_3$	n-C$_3$F$_7$	Eu	Eu(dfhd)$_3$

thd = tris(2,2,6,6-tetramethyl-3,5-heptanedionato);
dpm = tris(dipivaloylmethanato);
pta = tris(1,1,1-trifluoro-5,5-dimethyl-2,4-hexanedionato);
tfn = tris(1,1,1,2,2,3,3,7,7,8,8,9,9,9-tetradecafluoro-4,6-nonanedionato);
fhd =tris(1,1,1,2,2,6,6,7,7,7-decafluoro-3,5-heptanedionato);
dfhd = tris(1,1,1,2,2,3,3,7,7,7-decafluoro-4,6-heptanedionato);
fod = tris(1,1,1,2,2,3,3-heptafluoro-7,7-dimethyl-4,6-octanedionato).

resulting shifts will depend upon the strength of the interaction, the geometry of the adduct and the LSR to substrate ratio.

The acid character of the lanthanide chelate and the donor and steric properties of the substrate define the strength of the acid–base adduct. Usually more stable complexes are found as the ionic radius of the metal decreases. Fluorinated diketonates increase the Lewis acidity of the lanthanide ion with respect to the protonated derivatives. This explains why Eu(fod)$_3$ (see **Table 1**) is the mostly commonly used LSR.

Contributions to the lanthanide-induced shift

The shift induced by the lanthanide ion on the resonances of the organic substrate results from three contributions.

- A diamagnetic term corresponding to the coordination shift. On the ^{1}H resonances it is usually upfield and rather small.
- The contact or Fermi term, which depends on the delocalization of unpaired electron density on the substrate nuclei, and has then been related to the value of the hyperfine coupling constant.
- The dipolar or pseudocontact term, which depends on the spatial proximity of a given nucleus to the paramagnetic centre and on the anisotropy of the magnetic susceptibility tensor. In order to extract structural information from LIS data there has to be an accurate determination of the latter term.

The first step in the procedure consists of the determination of the all-bound shifts $\Delta_i^{\text{LSR–S}}$. In the presence of the equilibrium between the substrate (S) and the lanthanide shift reagent (LSR)

$$\text{LSR} + \text{S} \overset{K_i}{\rightleftharpoons} \text{LSR–S}$$

the observed $\Delta\delta_i$ are proportional to the term $([\text{LSR–S}]/[\text{S}])\Delta_i^{\text{LSR–S}}$. Thus $\Delta\delta_i$ are measured for various LSR to S ratios (ρ) and graphs are obtained by plotting the observed shifts $\Delta\delta_i = \delta_i^{\text{obs}} - \delta_i^{\text{o}}$ (where δ_i^{o} is the chemical shift measured in the absence of the LSR or, better, in the presence of diamagnetic La or Lu analogue) as a function of ρ.

If an axial symmetry of the dipolar magnetic field in the LSR–S adduct is assumed, the dipolar shifts are proportional to the term $(3 \cos^2\phi - 1)/r_i^3$. In this case, the principal magnetic axis of the lanthanide is taken as collinear with the lanthanide–ligand bond. r_i is the distance between a given nucleus and the paramagnetic ion and ϕ is the angle between the r_i vector and the Ln–ligand bond. Many systems have been successfully investigated on the basis of this simplifying assumption. The occurrence of axial symmetry is justified as a result of time averaging effects. This means that the fast rotation of the substrate along the coordination axis introduces an effective axial symmetry, where ϕ now refers to the angles between r_i and the rotation axis rather than the actual symmetry axis.

In the more general case of a rhombic, nonaxial symmetry, the dipolar contribution to the (all-bound) LIS is described as follows:

$$\Delta_i^{\text{LSR}-\text{S}} = K_1 \frac{(3 \cos^2 \theta_i - 1)}{r_i^3} + K_2 \frac{\sin^2 \theta_i \cos 2\Omega_i}{r_i^3}$$

where r_i, θ_i and Ω_i are the usual spherical coordinates of the i nucleus i the reference frame of the principal magnetic susceptibility tensor. K_1 and K_2 are coefficients related to the anisotropy of the magnetic susceptibility tensor. Several computer programs have been written to find the molecular geometry that best fits the set of experimental Δ_i values.

A decrease in temperature causes an increase in the lanthanide-induced chemical shifts, which usually provides an enhanced resolving power of the LSR. This behaviour results (i) from an increase of the concentration of the adduct and (ii) from the temperature dependence of the paramagnetic shift (the dipolar shift shows a $1/T^2$ dependence, whereas the contact term displays a $1/T$ dependence).

Diketonates are insoluble in water. Diglycolates have been proposed as substitutes for studies in aqueous solutions. Another possibility deals with the use of lanthanide aquo-ions, and perchlorates and nitrates have been used in several cases: they are soluble at pH < 6 and provide good shifts for a number of anionic functions. Eventually, good results in aqueous solutions have been obtained by using EDTA chelates which work at pH values as high as 10.

Chiral shift reagents

It is well established that a method to distinguish enantiomers by NMR is to convert them into diastereoisomers by means of appropriate optically active reactants. Thus it was suggested that the use of chiral LSR (LSR*) can induce shift differences in the spectra of enantiomers:

$$\text{LSR}^* + \text{R} \rightleftharpoons \text{LSR}^*\text{–R}$$
$$\text{LSR}^* + \text{S} \rightleftharpoons \text{LSR}^*\text{–S}$$

R and S give the same NMR spectrum as they are mirror images, whereas LSR*–R and LSR*–S are diastereoisomers and then are characterized by different NMR properties. In the presence of fast exchange between free and bound forms, R and S will give rise to different signals as they represent average values between the free and the bound forms. The enantiomeric shift difference between the $\Delta\delta$ LIS for the R and S configuration is usually indicated as $\Delta\Delta\delta$. The observed $\Delta\Delta\delta$ values depend on the equilibrium constants for the formation of the labile complexes between the enantiomers and the chiral reagent, and on the shift differences between the diastereomeric complexes.

The first chiral LSR was reported by Whitesides and consisted of a camphor-based Eu^{3+} complex. Since then, several chiral LSR* have been proposed and many of them are commercially available (**Figure 2** and **Table 2**).

In some cases, it has been found that enantiomeric resolution has been obtained also with the use of an achiral reagent such as $Eu(fod)_3$. This behaviour has been explained in terms of the formation of ternary 1:2 adducts which are no longer mirror images (such as LSR–R–R and LSR–R–S). In principle, another explanation may be possible on the grounds that $Eu(fod)_3$ may be considered as a racemic mixture of enantiomers.

Lanthanide shift reagents for alkali metal ions

An interesting extension of the use of LSR deals with the separation of the resonances of $^7Li^+$, $^{23}Na^+$ and $^{39}K^+$ present in different biological compartments. Such NMR-active metal cations are of clinical importance, but from their routine NMR spectra it is not possible to distinguish whether they are in the intra- or the extracellular compartments. The addition of a paramagnetic complex that selectively distributes in one compartment only, can remove such signal degeneracy. In fact, the interaction with the paramagnetic species induces a shift in other resonances of the alkaline ions present in the same compartment as the LSR (**Figure 3**). Up to now in laboratory practice, four complexes have been

Table 2 Chiral shift reagents of the β-diketonato type

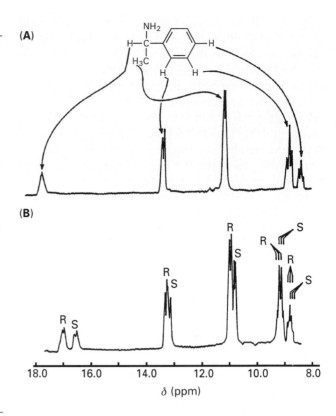

(1) R = –C(CH₃)₃
(2) R = –CF₃
(3) R = R¹
(4) R = 77%R² + 23%R¹

(5) R′ = R″ = R¹
(6) R′ = R¹, R″ = R³
(7) R′ = 77%R² + 23%R¹; R″ = R³
(8) R′ = R¹; R″ = 77%R² + 23%R¹

Figure 2 ¹H NMR spectra of α-phenylethylamine in CCl₄ after addition of tris[(3-*tert*-butylhydroxymethylene)-*D*-camphorato]-europiumᴵᴵᴵ. Molar ratio LSR* to substrate 0.5: (A) S-form; (B) racemic mixture.

studied in detail and applied to answer several questions of biomedical interest: the Dy(III) complexes of PPP⁵⁻ (triphosphate) and TTHA⁶⁻ (triethylenetetraaminehexaacetate) and the Dy(III) and Tm(III) complexes of DOTP⁵⁻ (tetraazacyclododecane-*N, N′, N″, N‴* tetrakis (methylenephosphonate)).

The most effective reagent so far reported is the chelate Dy(PPP)₂⁷⁻, first introduced by Gupta and Gupta in 1982. However, this metal complex has been proved to present several disadvantages that prevent its use in living animals or humans. First of all it is rather toxic, probably due to irreversible ligand dissociation promoted *in vivo*, with subsequent release of the metal ion. Furthermore, the complex is involved in several protonation equilibria and shows competition with endogenous divalent cations (Mg²⁺, Ca²⁺) which reduces its resolution ability. The Dy(TTHA)³⁻ complex is much less toxic, as a consequence, probably, of the high stability constant of Dy³⁺ with this multidentate ligand.

However, the reduced value of the negative charge on the complex and the fact that it is mainly localized on unbound carboxylic groups, away from the paramagnetic centre, makes this metal chelate less effective in removing the signal degeneracy. This implies that higher doses of the LSR are needed for an optimum resolution. The Dy(III) and Tm(III) complexes of DOTP represent an important

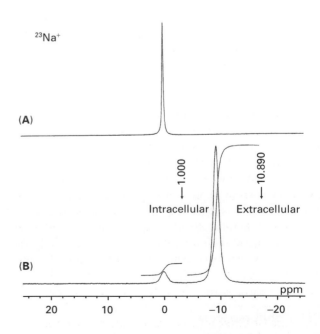

Figure 3 ²³Na NMR spectra at 2.1 T and 39°C of whole human blood before (A) and after (B) the addition of Dy(DOTP)⁵⁻ (5 mM). Intra- and extracellular sodium resonances are resolved.

improvement in the search for safe and effective LSRs for metal cations of biological relevance. In fact, these metal chelates are extremely resistant to dissociation processes over a wide range of pH and present in axial symmetry and interaction sites for the cations very close to their 4-fold axis of symmetry, which ensures a maximum shifting efficiency. Indeed $Dy(DOTP)^{5-}$ produces a magnitude of shift analogous to that observed for $Dy(PPP)_2^{7-}$ at the same concentration. However, the efficiency of the Dy(III) chelate has been shown to be markedly quenched by stoichiometric amounts of Ca^{2+}, and thus the Tm(III) complex, less sensitive to these effects, has found wider applications.

In conclusion, it is worth commenting that, although several studies have been possible by the use of the available LSRs, much remains to be done in terms of synthetic strategy and ligand design in the search for safer and more effective compounds.

Relaxation reagents

An alternative exploitation of paramagnetic complexes as auxiliary NMR reagents deals with their effect on the relaxation of substrate nuclei. Indeed, in the presence of a relaxation reagent the electron–nucleus relaxation may easily become so efficient that it predominates over the intrinsic relaxation processes of a given nucleus. Such effects may be exploited either on the solute or on the solvent resonances. A number of applications based on the use of relaxation reagents have been reported, ranging from structural assignments to nuclear Overhauser effect (NOE) quenchers for correct intensity measurements, from the assessment of the binding to proteins to the separation of resonances from different compartments. A particularly important class of relaxation reagents is represented by the contrast agents for magnetic resonance imaging (MRI).

The first example dealing with the use of paramagnetic species to shorten the relaxation time of the solvent nuclei deals with the very first detection of the NMR phenomenon, as Bloch used iron(III) nitrate to avoid saturation of the water proton signal. Then in the 1970s it was rather common, in organic and organometallic chemistry, to add to the substrate solution a reagent such as $Cr(acetylacetonate)_3$ ($Cr(acac)_3$) or $Fe(acac)_3$ in order to shorten the T_1 of ^{13}C and ^{15}N resonances without introducing broadening of the signal. In **Figure 4** the ^{13}C intensities of the $Fe(CO)_5$ signal, measured at 2.1 T, in the presence of variable amounts of $Cr(acac)_3$, are reported. At this field the T_1 value of the ^{13}CO groups is very long as it is determined by the chemical shift anisotropy term; by

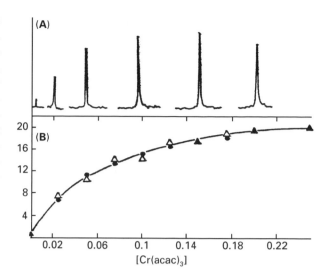

Figure 4 ^{13}C NMR spectra (A) and integrated intensities (B) of a $Fe(CO)_5$ sample with different added concentrations of $Cr(acac)_3$. Different symbols refer to independent measurements.

increasing the magnetic field strength it becomes markedly shorter. Thus, the availability of high magnetic fields has limited to some extent the use of relaxation reagents for the purpose of an overall shortening of the relaxation times.

The presence of a relaxation reagent induces a dominant relaxation path for all the resonances of a given substrate present in solution and this may be exploited for quantitative NMR determinations. In fact, this method eliminates the problems associated with a correct intensity measurement when the various resonances are endowed with different relaxation times and different NOE enhancements.

Another application of relaxation reagents deals with the selective removal of a given spin-coupling pattern, i.e. the interaction with a paramagnetic reagent produces a so-called chemical decoupling mechanism. In fact, in an AX coupled system the multiplicity of each resonance may be removed when the mean lifetime of the coupled nucleus in a given state (α or β) becomes shorter than the inverse of the coupling constant J_{AX}. Therefore, the occurrence of efficient relaxation induced by the paramagnetic reagent causes the removal of the effective coupling.

Theory of paramagnetic relaxation

The quantitative description of relaxation effects induced by paramagnetic species presents theoretical problems that are far from trivial. However, it is customary to resort to the simplified approach developed earlier by Solomon and Blombergen to account for the relaxation enhancement of solvent nuclei of aqueous solutions of paramagnetic metal ions. On the basis of this approach the observed solvent relaxation rate is given by the sum of three contributions:

$$R_1^{obs} = R_{1p}^{is} + R_{1p}^{os} + R_1^{o}$$

where R_1^{o} is the relaxation rate in the absence of the paramagnetic compound, R_{1p}^{is} represents the paramagnetic contribution due to the exchange of water molecules in the inner coordination sphere with water molecules in the bulk, and R_{1p}^{os} is the diffusion-controlled contribution from water molecules in the outer coordination sphere of the paramagnetic centre. In the absence of a coordinative interaction between the paramagnetic metal ion and the substrate–solvent molecules, only the latter term contributes to the observed paramagnetic relaxation enhancement. In this case the relaxation enhancement is exclusively due to the electron–nucleus dipolar interaction between the metal ion and the solvent molecules diffusing in the outer coordination sphere of the complex. This interaction is modulated by the translational diffusive motion of solute and solvent and by the electronic relaxation time. An analytical expression for this contribution has been derived by Freed and co-workers in a study of the relaxation enhancement of solvent molecules in solutions containing stable nitroxide radicals.

$$R_{1p}^{os} = C^{os} \left(\frac{1}{aD} \right) [7J(\omega_S) + 3J(\omega_H)]$$

In the expression above, C^{os} is constant for a given observed nucleus, a is the distance of closest approach between the paramagnetic centre and the diffusing solvent molecules and D is the relative solute–solvent diffusion coefficient. The dependence on the electronic relaxation time is expressed by the nonLorentzian spectral density functions $J(\omega)$.

As far as the inner sphere contribution is concerned, R_{1p}^{is} is determined by the relaxation time T_{1M} of the nuclei of the substrate–solvent complex and by its lifetime τ_M, weighted by the molar ratio of the bound substrate–solvent:

$$R_{1p} = \frac{[C]q}{[S]} \frac{1}{T_{1M} + \tau_M}$$

where [C] is the molar concentration of the paramagnetic reagent, q is the number of interacting

sites, and [S] is the concentration of the substrate–solvent complex. In the case of dilute water solutions, when the relaxation enhancement is observed on the solvent nuclei, [S] = 55.54 M.

The T_{1M} value in the paramagnetic reagent–substrate adduct is determined by (i) the dipolar interaction between the observed nucleus and the unpaired electron(s) and (ii) the contact interaction between the magnetically active nucleus and the unpaired electron density localized in the position of the nucleus itself. An analytical form is given by the Solomon–Bloembergen equation (here for T_1; an analogous expression is derived for T_2):

$$\frac{1}{T_{1M}} = \frac{2}{15} \left(\frac{\mu_0}{4\pi} \right) \frac{\hbar^2 \gamma_S^2 \gamma_H^2}{r^6} S(S+1)$$

$$\times \left[\frac{3\tau_{c1}}{1 + \omega_H^2 \tau_{c1}^2} + \frac{7\tau_{c2}}{1 + \omega_S^2 \tau_{c2}^2} \right]$$

$$+ \frac{2}{3} S(S+1) \left(\frac{A}{\hbar} \right)^2 \left(\frac{\tau_e}{1 + \omega_S^2 \tau_e^2} \right)$$

where S is the electron spin quantum number, γ_H and γ_S are proton and electron magnetogyric ratios, respectively, r is the distance between the metal ion and the given nucleus of the coordinated substrate, ω_H and ω_S are the proton and electron Larmor frequencies, respectively. $\tau_{c1,2}$ and τ_e are the relevant correlation times for the time-dependent dipolar and contact interactions, respectively:

$$\frac{1}{\tau_{c1,2}} = \frac{1}{\tau_R} + \frac{1}{\tau_M} + \frac{1}{\tau_{S1,2}}$$

$$\frac{1}{\tau_e} = \frac{1}{\tau_{S1}} + \frac{1}{\tau_M}$$

where τ_R is the reorientational correlation time and τ_{S1} and τ_{S2} are the longitudinal and transverse electron spin relaxation times.

The contact term is often negligible, especially when dealing with a paramagnetic ion bound to a macromolecule because the hyperfine coupling constant $A/\hbar$ is in the MHz range, whereas the coefficient of the dipolar term is one order of magnitude larger.

The protons relaxation enhancement (PRE) method

Upon interacting with a macromolecule the relaxation induced by paramagnetic species usually displays remarkable changes, primarily related to the increase of the molecular reorientational time τ_R on going from the free to the bound form. This may result in a strong increase of the R_{1p}^{is} term, from which it is possible to assess the affinity (and the number of binding sites) between the interacting partners. Quantitatively, in the presence of a reversible interaction between the paramagnetic species and the macromolecule, the observed enhancement depends on both the molar fraction of the macromolecular adduct χ^b and the solvent–water relaxation in the all-bound limit R_{1p}^b. The increase in the relaxation rate is expressed by the enhancement factor ε^*:

$$\varepsilon^* = \frac{R_{1p}^*}{R_{1p}} = \frac{\left(R_1^{obs} - R_1^o\right)^*}{\left(R_1^{obs} - R_1^o\right)}$$

The asterisk indicates the presence of the macromolecule in solution. The enhancement factor ranges from 1 (no interaction) to $\varepsilon^b = R_{1p}^b/R_{1p}$ in the all-bound limit (strong excess of macromolecule).

$$\varepsilon^* - 1 = \chi^b\left(\varepsilon^b - 1\right)$$

The determination of the binding parameter K^A (association constant) and n (number of independent sites characterized by a given K_A value) from the mass action law is straightforward. Given the association equilibrium between the metal complex M and the protein (enzyme) E,

$$M + E \rightleftharpoons ME$$

$$K_A = \frac{[ME]}{[M]_f[E]_f} = \frac{[ME]}{([M]_t - [ME])([nE]_t - [ME])}$$

$$\varepsilon^* = \chi^b\varepsilon^b + (1 - \chi^b) = \frac{[ME]}{[M]_t}\varepsilon^b + \frac{[M]_t - [ME]}{[M]_t}$$

where n time [E] gives the concentration of each class of sites on the macromolecule. Superscript b and subscripts f and t indicate bound, free and total, respectively.

The experimental procedure consists of the determination of the enhancement factor ε^* through two distinct titrations. In the first titration, a rectangular hyperbola describing the change of ε^* as the $[E]_t$ to $[M]_t$ ratio increases is reported (**Figure 5**). The treatment of the obtained binding isotherm yields the value of ε^b and the product nK_A. In the second titration, the behaviour of $\varepsilon^*$is monitored at fixed macromolecule concentration by changing the metal complex concentration. Results are conveniently expressed under the form of a Scatchard's plot:

$$r/[M]_f = n\,K_A - rK_A$$

The value of r may easily be calculated once the all-bound enhancement ε^b is known. In the same way, the free $[M]_f$ concentration is obtained from the total $[M]_t$ complex concentration.

$$r = \frac{(\varepsilon^* - 1)[M]_t}{(\varepsilon^b - 1)[E]_t}$$

$$[M]_f = \left(1 - \frac{\varepsilon^* - 1}{\varepsilon^b - 1}\right)[M]_t$$

By plotting $r/[M]_f$ vs. r a straight line is obtained whose x-axis intercept gives the n value and whose slope is equal to $-K_A$.

Contrast agents for magnetic resonance imaging

Magnetic resonance imaging (MRI), one of the most powerful tools in modern clinical diagnosis, is based on the topological representation of NMR parameters such as proton density and transverse and

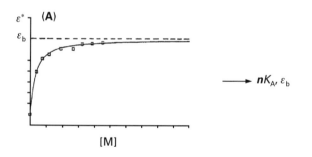

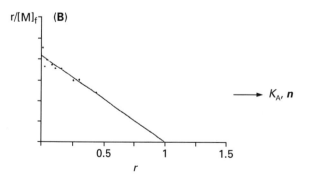

Figure 5 Direct (A) and inverse (B) PRE titrations: by plotting ε^* vs. macromolecule concentration a rectangular hyperbola is obtained, whose analysis yields the ε^b value together with the nK_A product. The Scatchard's plot obtained from the inverse titration allows the determination of n and K_A separately.

longitudinal relaxation times. Differences in these parameters allow impressive anatomical discrimination to be made, and make it possible to distinguish pathological from healthy tissues. The potential of MRI is further strengthened by the use of suitable contrast agents (CAS), compounds that are able to alter markedly the magnetic properties of the region where they are distributed. Among them, paramagnetic Gd^{3+} complexes (seven unpaired electrons) are under intense scrutiny because of their ability to enhance the proton relaxation rates of solvent water molecules (**Figure 6**). Their use provides the physicians with further physiological information to be added to the impressive anatomical resolution commonly obtained in the uncontrasted images. Thus, administration of Gd-based contrast agents has entered into the pool of diagnostic protocols and is particularly useful to assess organ perfusion and any abnormalities in the blood–brain barrier or in kidney clearance. Several other applications, primarily in the field of angiography and tumour targeting, will soon be available in clinical practice. Nowadays about 35% of MRI examinations make use of contrast agents, but this percentage is predicted to increase further following the development of more effective and specific contrast media than those currently available.

In order to be used as a CA for MRI a Gd^{3+} complex must fulfil two basic requirements: (i) it must have at least one coordinated water molecule in fast exchange (on the NMR relaxation timescale) with the solvent bulk, and (ii) the Gd^{3+} ion must be tightly chelated to avoid the release of the metal ion and the ligand which are both potentially harmful. **Figure 7** reports the schematic structures of four ligands whose Gd^{3+} chelates are currently used as CAs for MRI. The overall PRE of water protons ($R_{1p}^{is} + R_{1p}^{os}$, see 'Theory of paramagnetic relaxation' section above) referred to a 1 mM concentration of Gd^{3+} chelate is called *millimolar relaxivity* (hereafter *relaxivity*). At the magnetic field strength currently employed in MRI applications, the relaxivity of a Gd^{3+} complex is roughly proportional to its molecular weight, i.e. it is determined by the value of the molecular reorientational time τ_R. It follows that the search for high relaxivities has been addressed to systems endowed with slow tumbling motion in solution. This condition may be met either (i) by linking a Gd^{3+} chelate to high molecular weight substrates like carboxymethylcellulose, polylysine or dendrimers, etc. or (ii) by forming noncovalent adducts with serum albumin. In the latter case the Gd^{3+} chelates are designed in order to introduce on the surface of the ligand suitable functionalities able to strongly interact with the serum protein. In addition to providing high relaxivities (which, in turn, allows

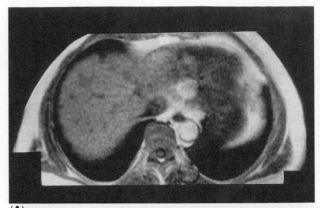

(A)

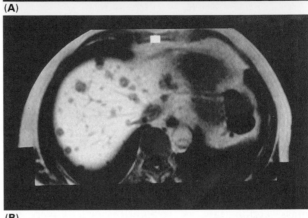

(B)

Figure 6 Male patient, 54 year old, 67 kg with multiple hepatic metastases from a neuroendocrine tumour of the pancreas: images are taken before (A) and 90 minutes after (B) intravenous administration of 0.1 mmol kg^{-1} of a liver-specific Gd^{3+} chelate.

the administered doses of CA to be reduced) the formation of adducts between Gd^{3+} complexes and albumin are of interest for the design of novel angiographic experiments for which a reduced clearance time and a better compartmentalization in the circulating blood is required.

List of symbols

a = distance of closest approach between the paramagnetic centre and the diffusing solvent molecules; D = relative solute–solvent diffusion coefficient; **H** = hyperfine shift; $\mathbf{H}_S$ = hyperfine shift scalar coupling; $\mathbf{H}_D$ = hyperfine shift dipole coupling; K_i = interaction constant between LSR and S; K_1 and K_2 = coefficients related to anisotropy of magnetic susceptibility tensor; K_A = association constant; n = number of independent sites characterized by a given K_A value; q = number of interacting sites; r_i = distance between nucleus and paramagnetic ion; r_i, θ_i and Ω_i = spherical coordinates of i nucleus in the reference frame of the principal magnetic

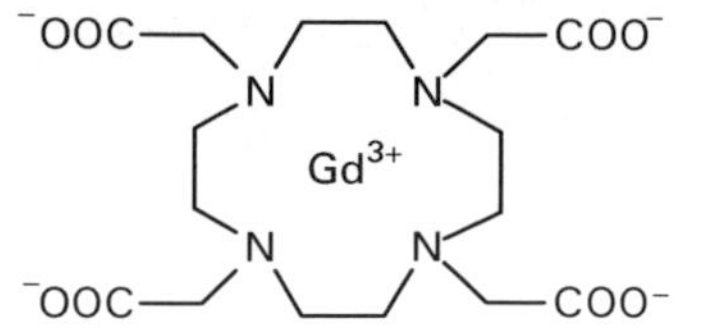

[Gd–DOTA]
DOTAREM® (Guerbet)

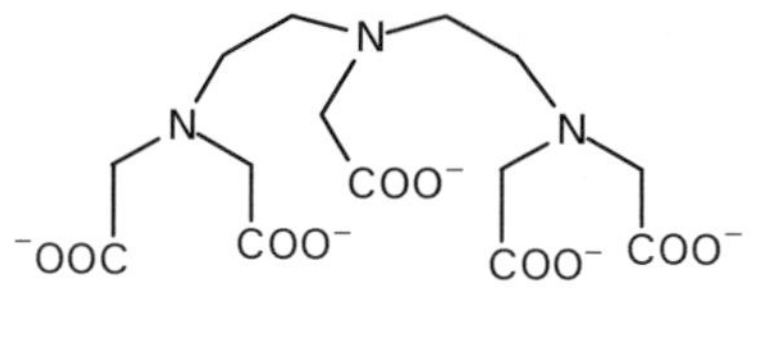

Gd^{3+}
$[Gd\text{–}DOTA]^{2-}$
MAGNEVIST® (Schering)

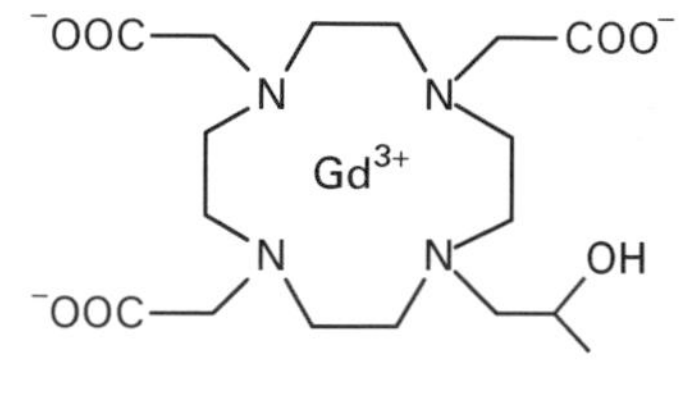

[Gd–HPDO3A]
PROHANCE® (Bracco)

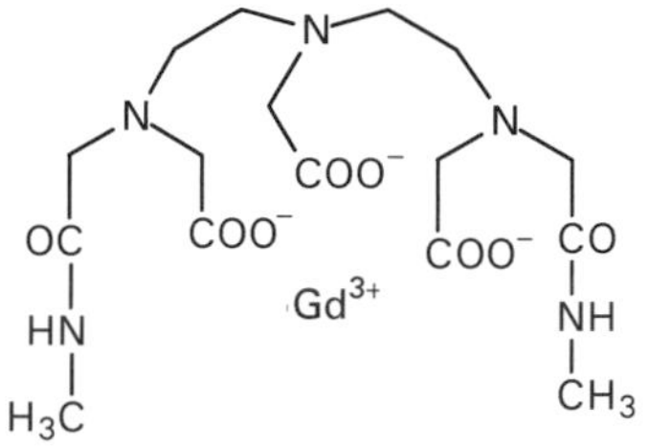

[Gd–DTPABMA]
OMNISCAN® (Nycomed)

Figure 7 Schematic structure of the Gd^{3+} chelates used in clinical practice. DOTA = 1,4,7,10-tetraazacyclododecane-N,N′,N″,N‴-tetraacetic acid; DTPA = diethylenetriaminepentaacetic acid; HPDO3A = N‴-(2-(1-hydroxypropyl))-1,4,7,10-tetraazacyclododecane-N,N′N″-triacetic acid; DTPABMA = N,N″-bis(N-methylcarbamoylmethyl)-diethylenetriamine-N,N′,N″-triacetic acid.

susceptibility tensor; R_1° = relaxation rate in absence of paramagnetic compound; R_{1p}^{os} = diffusion-controlled contribution to relaxation rate from water molecules in the outer coordination sphere of the paramagnetic centre; R_{1p}^{is} = paramagnetic contribution to relaxation rate due to exchange of water molecules in the inner coordination sphere with those in the bulk; R_{1p}^{b} = solvent–water relaxation in the all-bound limit; S = substrate; T_{1M} = relaxation time of the protons of the water molecule in the first coordination sphere of the complex; T_1 = relaxation time; χ^b = molar fraction of macromolecular adduct; $\Delta_i^{LSR\text{–}S}$ = all-bound shift of the LSR-S complex; $\Delta\delta_i$ = observed LSR-induced shifts; $\Delta\Delta\delta$ = enantiomeric shift difference between the $\Delta\delta$ LIS for R and S configuration; ε^* = enhancement factor; ϕ = angle between r_i vector and Ln–ligand bond or rotation axis; γ_H and γ_S = electron and proton magnetogyric ratios; ρ = LSR to S ratio; $\tau_{c1,2}$ and τ_e = relevant correlation times for time-dependent dipolar and contact interactions; τ_M = residence lifetime of the water molecule in the first coordination sphere of the complex; τ_R = reorientational correlation time; τ_{S1} and τ_{S2} = longitudinal and transverse electron spin relaxation times; ω_H and ω_S = proton and electron Larmor frequencies.

See also: **Contrast Mechanisms in MRI; EPR Spectroscopy, Theory; MRI Applications, Biological; MRI Applications, Clinical; MRI Applications, Clinical Flow Studies; MRI Theory; NMR Relaxation Rates; NMR Spectroscopy of Alkali Metal Nuclei in Solution; Nuclear Overhauser Effect.**

Further reading

Aime S, Botta M, Fasano M and Terreno E (1998) Lanthanide (III) chelates for NMR biomedical applications. *Chemical Society Reviews* 27: 19–29.

Bertini I and Luchiant C (1986) *NMR of Paramagnetic Molecules in Biological Systems*. Menlo Park: Benjamin/Cummings.

Cockerill AF, Davies GLO, Harden RC and Rackham DM (1973) Lanthanide shift reagents for nuclear magnetic resonance spectroscopy. *Chemical Reviews* 73: 553–588.

Martin ML, Delpuech J-J and Martin GJ (1980) *Practical NMR Spectroscopy*, pp 377–409. London: Heyden.

Peters JA, Huskens J and Raber DJ (1996) Lanthanide induced shifts and relaxation rate enhancements. *Progress in NMR Spectroscopy* 28: 283–350.

Sherry AD and Geraldes CFGC (1989) Shift reagents in NMR spectroscopy. *In: Bünzli J-CG and Choppin GR (eds) Lanthanide Probes in Life, Chemical and Earth Science*. Amsterdam: Elsevier.

Von Ammon R and Fisher RD (1972) Shift reagents in NMR spectroscopy. *Angewandte Chemie, International Edition in English* 11: 675–692.

Chemical Structure Information from Mass Spectrometry

Kurt Varmuza, Laboratory of Chemometrics, Vienna University of Technology, Austria

MASS SPECTROMETRY
Methods & Instrumentation

Mass spectra of chemical compounds have a high information content. This article describes computer-assisted methods for extracting information about chemical structures from low-resolution mass spectra. Comparison of the measured spectrum with the spectra of a database (library search) is the most used approach for the identification of unknowns. Different similarity criteria of mass spectra as well as strategies for the evaluation of hitlists are discussed. Mass spectra interpretation based on characteristic peaks (key ions) is critically reported. The method of mass spectra classification (recognition of substructures) has interesting capabilities for a systematic structure elucidation. This article is restricted to electron impact mass spectra of organic compounds and focuses on methods rather than on currently available software products or databases.

Introduction

Mass spectrometry (MS) is the most widely used method for the identification of organic compounds in complex mixtures at the nanogram level and below. NMR and, sometimes, also infrared spectroscopy provide better structural information than MS data but these spectroscopic methods cannot (or only with a substantial loss of sensitivity) be coupled with chromatographic separation techniques. Despite the great amount of information contained in mass spectra it is difficult to extract structural data because of the complicated relationships between MS data and chemical structures. The fragmentation processes which finally result in the measured data characterize MS as a chemical method, in contrast with NMR and IR. Chemical effects are, in general, more difficult to describe and to predict than physical ones. The occurrence of rearrangement reactions means that structure identification from mass spectra is often difficult or impossible; furthermore functional groups do not always produce the same peaks. This background explains why structure elucidation in organic chemistry is mainly based on ^{13}C NMR data, while H

NMR, IR, UV–VIS and MS data are considered only to a much lesser extent.

The aim of spectra evaluation can be either the identification of a compound (assuming the spectrum is already known and available) or interpretation of spectral data in terms of the unknown chemical structure (when the spectrum of the unknown is not available).

Identification is performed best by library search methods based on spectral similarities; a number of MS databases and software products are offered for this purpose and are routinely used.

The more challenging problem is the interpretation of a mass spectrum, which is still a topic of research projects in chemometrics and computer chemistry. No comprehensive solutions are yet available and these methods are only rarely used in routine work. Four groups of different strategies have been applied to the complex problem of substructure recognition (or the recognition of more general structural properties) from spectral data. (1) *Knowledge-based methods* try to implement spectroscopic knowledge about spectra–structure relationships into computer programs. Because of the lack of generally applicable rules this approach was not successful in MS. However, spectroscopic knowledge has been extensively applied in other methods to guide the construction of mathematical models, and for optimizing model parameters. (2) Appropriate *interpretive library search* techniques can be used to obtain structural information if the unknown is not contained in the library. (3) *Correlation tables* containing characteristic spectral data (key ions) together with corresponding substructures have met with only limited success because a specific structural property does not always give the same spectral signals. (4) *Spectral classifiers* are algorithms based on multivariate classification methods or neural networks; they are constructed for an automatic recognition of structural properties from spectral data.

In principle, mass spectral data are fully determined by the chemical structure of the investigated compound and the experimental spectroscopic conditions:

$$\text{mass spectral data} = F(\text{chemical structure,} \\ \text{experimental spectroscopic conditions})$$

For problems of practical interest, however, the relationship F can neither be formulated as a mathematical equation nor as an algorithm. Consequently, methods for the prediction of the mass spectrum from the chemical structure have had only very limited success. A similar difficult situation arises for the inverse problem, spectra interpretation, **Figure 1**. Only for selected cases can the relationship

$$\text{structural property} = f(\text{selected mass spectral data})$$

be established and applied to mass spectra of unknowns. Some relationships f for different structural properties have been developed by the use of statistical methods or neural networks rather than by the influence of spectroscopic knowledge.

Because of the above-mentioned difficulties the intervention of the human expert is still required in mass spectra interpretation. However, this situation causes a number of drawbacks. Human experts are not sufficiently available and are expensive; furthermore their success rate can hardly be quantified. Automated instruments with high throughput are able to produce huge amounts of data, and analytical interest in complex mixtures (environmental chemistry, food chemistry, combinatorial synthesis) is rapidly increasing. Thus the extensive use of computer-assisted methods for data interpretation is highly desired.

Characteristic mass spectra signals

A number of masses in the low mass region (*key ions*) are considered to be characteristic for certain substructures or classes of compounds. In addition, mass differences (*key differences*) between the molecular ion and abundant fragment ions or between abundant fragment ions are often related to functional groups. Correlation tables containing such spectral data and the corresponding chemical structures are contained in several textbooks on MS. The use of these tables is widespread and often helpful in MS; however, the capability of this approach must not be overestimated.

The following example demonstrates the potentials and drawbacks. From the NIST Mass Spectra Library a random sample containing 5000 compounds with a phenyl group (class 1) and 5000 compounds without a phenyl group (class 2) have been selected; from these data it was estimated whether the signal at m/z 77 (corresponding to the ion $C_6H_5^+$) can be used to predict the presence or absence of a phenyl group in the molecule. Let n_1 be the number of compounds from class 1 exhibiting a peak at m/z 77 in a given intensity interval, and n_2 be the corresponding number for compounds from class 2. Assuming for an unknown compound that the a priori probability for belonging to class 1 is 0.5 the probability $p(\text{phenyl} \mid I_{77})$ can be calculated from the signal at m/z 77 as

$$p(\text{phenyl} \mid I_{77}) = n_1/(n_1 + n_2) \qquad [1]$$

Table 1 shows results for this example using four intensity intervals. Depending on the intensity of the peak at m/z 77, the probability for the presence of a phenyl group decreases or increases. If a peak at m/z

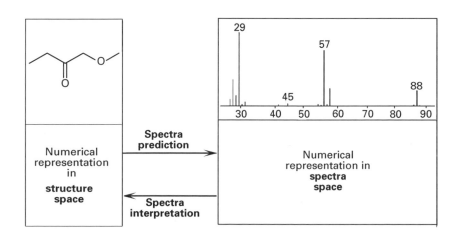

Figure 1   Relationships between mass spectrum and chemical structure.

Table 1 Significance of a peak at m/z 77 for the presence of a phenyl group in a molecule

Intensity interval (%B)	n_1	n_2	$p(\text{phenyl} \mid I_{77})$
$I \leq 1$	497	2486	0.17
$1 < I \leq 5$	738	983	0.43
$5 < I \leq 30$	2283	1288	0.64
$30 < I \leq 60$	898	182	0.83
$60 < I$	584	61	0.91
Sum	5000	5000	

The terms n_1 and n_2 are the number of spectra (compounds) with a peak at m/z 77 in the given intensity interval for compounds that contain a phenyl group and those that do not, respectively; $p(\text{phenyl} \mid I_{77})$ is the a posteriori probability for the presence of a phenyl group (assuming an a priori probability of 0.5)

77 is absent or has an intensity not higher than 1% of the base peak (%B) the probability for the presence of a phenyl group is still 17%. Intensities above 60%B strongly indicate a phenyl group (91% correct assignments to class 1). However, most spectra have peak intensities between these limits and therefore do not allow a reliable answer about the presence or absence of a phenyl group. A systematic examination of about 40 key ions between m/z 30 and 105 can be summarized as follows: (1) only a small number of key ions provide significant structural information, and (2) the absence of a key ion in general does not indicate the absence of the corresponding compound class.

Mass spectra similarity search

Overview

A basic assumption in library search systems is the hypothesis that 'similar spectra often correspond to similar structures'. A search for spectra in a library can be guided by two types of criteria: (1) a search profile is defined by some logical restrictions, for example "peak height at m/z 77 > 90%B and peak height at m/z 105 > 30%B". The result is a not-ordered hitlist containing all spectra that exactly match the search profile. (2) A numerical similarity between two spectra is used to find the reference spectra most similar to the spectrum of the unknown (**Figure 2**). In this case the resulting hitlist is ordered according to the value of the similarity criterion. Of course any type of search can be combined with additional search criteria (relative molecular mass range, restrictions to the molecular formula, name fragments, etc.). Unfortunately, the capabilities for chemical structure searches are still very limited in commercially available MS databases.

The performance and effectiveness of a spectral library search depends on the quality of the given spectrum, the content of the spectral library, the

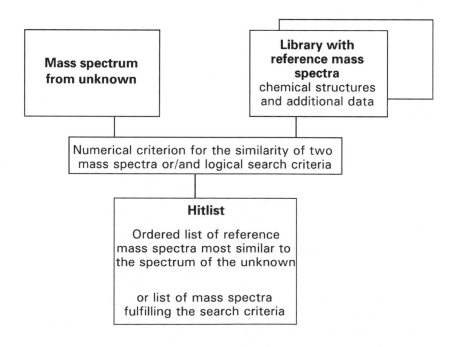

Figure 2 Spectral library search.

Table 2 Common quality problems with mass spectra in databases

Spectrum reduced to a small number of (large) peaks
Limited mass range (start mass too high)
Peaks present far above the relative molecular mass
Poor dynamics of peak intensities
Wrong isotope peak intensities
Tilted peak intensities
Additional peaks from impurities or instrument background
Small peaks missing
Wrong mass assignment
Illogical mass differences
Spectrum does not match with given compound
Missing experimental data
Missing structure data
Too many replicate spectra

used similarity criterion and the implemented software technique. A fundamental restriction is given by the limited structural information present in mass spectra, for instance isomers can often not be distinguished by their mass spectra.

Libraries

Any spectroscopic retrieval system in chemistry depends on an appropriate structural diversity of the reference data. Commercially available MS databases today contain about 40 000 up to almost 300 000 spectra. Some libraries contain a number of replicate spectra (collected from different sources), especially for common compounds. The compositions of existing libraries (except a few dedicated small spectra collections) are not systematic. Some compound classes are very well represented (for instance hydrocarbons) while others are not but may be of major interest to a particular user. Good software support for building user-libraries is therefore essential. **Table 2** lists common problems of quality deficiencies in MS reference data.

Peak search

A simple and evident method for library search of mass spectra applies user-selected restrictions for the presence of peaks. Initially, a characteristic peak in the spectrum of the unknown is selected (typically one at a high or unusual mass number and with a not too small intensity) and an intensity interval for reference spectra is defined. The peak search software answers with the number of reference spectra containing a peak at the given mass and in the given intensity interval. This procedure is continued until a reasonable number of hits is reached (in cases where

no reference spectrum fulfills all restrictions it is possible to go back one step). The resulting list of spectra can be ordered by applying an additional similarity criterion. The method allows (and requires) the interaction of the user; consequently spectroscopic knowledge (for instance about the background peaks) can be easily considered; however, automated processing of spectra is not possible. The use of appropriate sorted index files avoids time-consuming sequential searches and makes the method very fast. An example for this method is presented in **Figure 3**.

Spectra similarity criteria

An infinite number of different similarity measures for matching two mass spectra is possible. In reality, about a dozen similarity criteria have been described and tested; however, full details of the algorithms that are actually implemented in commercial software are seldom described. Some similarity criteria are purely mathematically oriented, others are influenced by spectroscopic ideas. A single criterion cannot fit all requirements: on the one hand, a compound which is contained in the library should be retrieved even if the spectrum deviates in some parts from the reference spectrum; on the other hand, the user expects hits with chemical structures similar to the unknown even though they exhibit different spectra. In the ideal case the similarity criterion has a practicable sensitivity to chemical structures: identical compounds measured under different spectroscopic conditions appear more

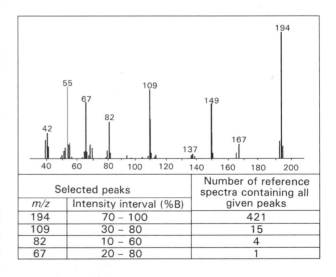

Selected peaks		Number of reference spectra containing all given peaks
m/z	Intensity interval (%B)	
194	70 – 100	421
109	30 – 80	15
82	10 – 60	4
67	20 – 80	1

Figure 3 Peak search example using the NIST mass spectral database. The mass spectrum of a hypothetical unknown is from caffeine contaminated with a phthalate. Manual selection of relevant peaks easily allows the spectroscopist to consider probable contaminations (peaks at m/z 149, 167). The correct solution is found after the input of four peaks by excluding the typical phthalate peaks. Note the wide intensity intervals applied. Should a peak at m/z 149 be required the correct compound is not found.

similar than two compounds that are slightly different in their structure.

The two main types of mass spectral data, peak heights and masses, have different reliability. While the presence of a peak at a certain mass number is well reproducible the peak height often varies greatly under slightly different experimental conditions. Spectra similarity criteria have to consider this aspect.

For similarity searches an efficient ordering of the spectra in the library is not possible and, therefore, a rather time-consuming sequential search is necessary. Some speed-up is possible by a pre-search in which, for instance, reference spectra are quickly eliminated that do not have a minimum number of peaks in common with the spectrum of the unknown. Before calculating the spectra similarity the spectrum of the unknown is treated by cleaning procedures, for instance peaks at m/z 28, 32 and 40 (probably from N_2, O_2 and Ar respectively) and peaks from column bleeding are removed, and the peak heights are normalized to percents of the base peak. The selection of relevant peaks and weighting of peak heights is described below. Let $_UA_m$ and $_RA_m$ be the (weighted) abundances at mass m in the spectrum of the unknown (U) and the reference spectrum (R), respectively. Summation in Equations [2] to [4] is over all selected masses m.

The most used similarity criterion for mass spectra is based on the *correlation coefficient*:

$$S_1 = \frac{[\Sigma_U A_m {}_R A_m]^2}{\sqrt{\left[\Sigma^1 \left(_U A_m\right)^2 \Sigma \left(_R A_m\right)^2\right]}} \qquad [2]$$

Figure 4 explains this similarity criterion: each mass is considered as a point in a coordinate system with one axis for the abundances of the unknown and the other for the reference. For identical spectra all points are located on a straight line that passes through the origin ($S_1 = 1$). If the two spectra are different the correlation coefficient for the regression line – as given in Equation [2] – is a measure of the similarity of the spectra. The value of S_1 ranges from 0 to 1 because the regression line is forced to pass through the origin and all abundances are 0 or positive. The important range of S_1 between 0.9 and 1 can be widened by calculating S_1^2; for practical reasons S_1 or S_1^2 are often multiplied by 100.

A peak list with nominal (integer) masses can be considered as a point in a q-dimensional space, with q equal to the maximum relevant mass number. The coordinates of the point are given by the abundances (weighted peak heights). Because the value of q is high the q-dimensional space cannot be visualized

directly but can be handled by rather simple mathematics. A two-dimensional simplification is used in **Figure 5** to demonstrate this view of MS data. Masses 43 and 58 have been selected for the coordinates; each spectrum (from the example in **Figure 4**) can be represented by a point or by a vector (starting at the origin). Similarity criterion S_1 is equal to the cosine of the angle between two spectral vectors.

Alternative similarity measures are the *Euclidean distance* S_2:

$$S_2 = \sqrt{\left[\Sigma \left(_U A_m - {}_R A_m\right)^2\right]} \qquad [3]$$

and the *city block distance* S_3:

$$S_3 = \Sigma \mid {}_U A_m - {}_R A_m \mid \qquad [4]$$

The similarity criterion used in the PBM (Probability Based Matching System) software is based on so-called uniqueness values for masses and abundances. The uniqueness of a signal is equivalent to the information content (as defined in information theory) and is given by the negative logarithm of the probability of the occurrence of this signal; this means that a low probability corresponds to a high uniqueness. Probabilities for masses and intensity intervals have been estimated from the database used. Matching of spectra is performed by using selected peaks that exhibit highest uniqueness values for mass and intensity.

The MS database system MassLib uses a composite similarity criterion containing the similarity criterion S_1 together with measures for the number and intensity sums of peaks that are common or not in both spectra. The algorithm has been optimized to give good results for identification as well as for interpretation.

In the STIRS (Self-training Interpretive and Retrieval System) system for 26 data classes, specific similarity criteria are defined using characteristic masses or mass differences. For each of these criteria the most similar reference spectra are searched with the aim of obtaining information about the presence of substructures.

Weighting

Intensity scaling and mass weighting is used to consider the different significances of mass numbers and peak heights. In general, peaks in the higher mass range are more important than peaks in the lower mass range; because large peaks dominate the values of most similarity criteria, peak heights are often scaled to enhance the influence of small peaks. A

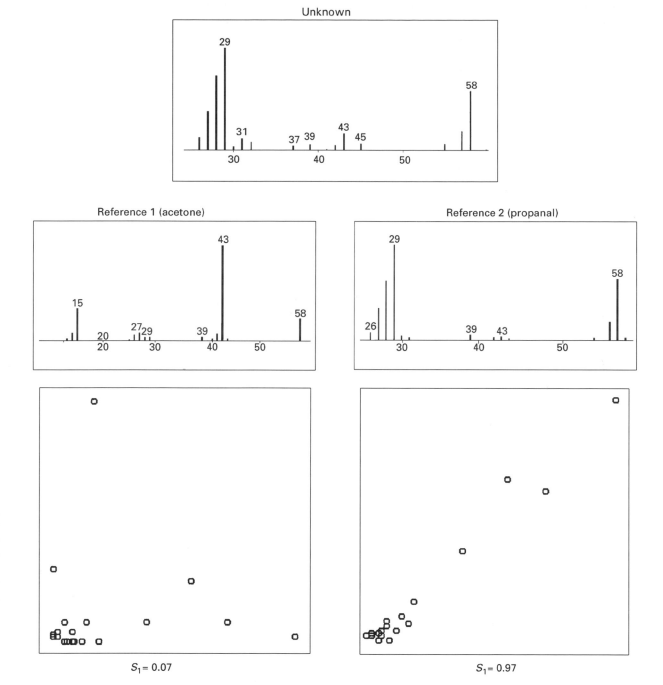

Figure 4 Comparison of the mass spectrum from an unknown with two reference spectra using the similarity measure S_1 (based on the correlation coefficient). Axes in the two plots at the bottom: horizontal, peak height %B of unknown; vertical, peak height %B of reference. Each point corresponds to a mass number with a peak in at least one of the two compared spectra.

general transformation of intensities I_m at mass m is given by Equation [5]

$$A_m = (m)^v (I_m)^w \qquad [5]$$

Optimum values for v and w depend on the used similarity citerion; for S_1 (Eqn [1]) good results have been obtained with $v = 2$ and $w = 0.5$ [see Stein (1994) in the Further reading section].

Selection of revelant masses

The selection of masses used to calculate a spectral similarity may greatly influence the result. The user

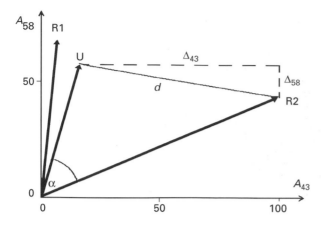

Figure 5 Mass spectra can be considered as points or vectors in a multidimensional spectral space. For simplicity only two mass numbers (43, 58) have been selected in this example. A_m, abundance (peak height in %B) at mass m; U, spectrum from unknown; R1, reference spectrum of propanal; R2, reference spectrum of acetone. Measures for spectral similarity are the Euclidean distance (d), the city block distance ($\Delta_{43} + \Delta_{58}$) or the cosine of angle α (equivalent to S_1 in Eqn [2]).

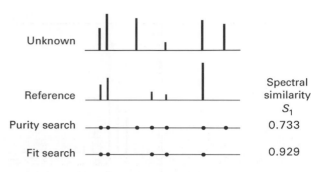

Figure 6 Different selection of masses for the calculation of spectral similarity (schematic). The fit search considers only masses with peaks in the reference spectrum and therefore is insensitive to impurities in the unknown.

peak at mass m is considered to be significant if its intensity is higher than the average intensities at masses $m - 14$ and $m + 14$.

Figure 7 presents an example of an MS library search in which the unknown was contained in the library.

often can decide between three modes for mass selection (**Figure 6**).

(a) All masses for which at least one of the compared spectra has a peak with an intensity above a defined threshold are considered. High similarity values are obtained only if the spectra are almost identical; this approach therefore is suited for *identity searches* (sometimes also called *purity searches*).

(b) Only masses with a peak in the reference spectrum are considered. High similarity values are obtained if the reference spectrum is almost completely contained in the spectrum of the unknown. Non-matching peaks in the unknown do not influence the result; therefore this approach is routinely applied if the unknown may contain impurities or is a mixture of compounds (*fit search, reverse search*).

(c) Only masses with a peak in the spectrum of the unknown are considered. This method is not tolerant of additional peaks in the reference spectrum. From a high similarity value one may conclude that parts of the chemical structure of the reference compound are present in the unknown (*interpretive search, forward search*).

Simple selection methods of peaks such as 'k highest peaks in the spectrum (or in given mass intervals)' have been applied to save data storage and computation time but are no longer important. A special selection of masses is used in the MassLib system: a

Mass spectra classification

Overview

When the unknown compound is not contained in the spectral library then pure identification methods are less useful. For 'unknown unknowns' interpretive search systems or classification methods are required to obtain structural information that can be used for constructing molecular structure candidates. Such candidates are usually created manually by applying spectroscopic–chemical knowledge and intuition. A serious drawback of this strategy is that the solution is rarely complete and the procedure hardly can be documented or verified.

The most important systematic approach for structure elucidation of organic compounds is still based on the DENDRAL project, in which, for the first time, artificial intelligence principles have been applied to complex chemistry problems (**Table 3**). The central tool is an isomer generator software capable of generating an exhaustive set of isomers from a given molecular formula. The generator also has to consider structural restrictions, which are usually obtained from spectral data. Substructures which have to be present in the unknown molecular structure are collected in the so-called goodlist while forbidden substructures are put into the badlist. Mass spectrometry can contribute to this approach in various aspects: the molecular formula can be determined from high resolution data, and structural information can be derived even from low resolution data.

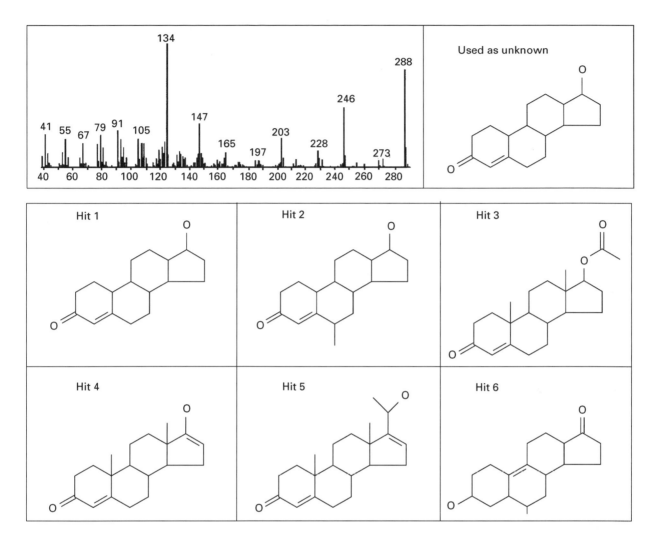

Figure 7 Example of a library search by MassLib. A mass spectrum from testosterone has been considered as unknown and searched in a library consisting of 130 000 spectra (including duplicates). A mass spectrum from testosterone is contained in this library and has been found as the first hit (most similar spectrum). The other hits are from the same compound class and demonstrate the interpretive capabilities of this system.

The two most important computer-assisted strategies for the recognition of structural information from mass spectral data are:

(a) the structures in the hitlist from a spectra similarity search are used to estimate the probability of substructures in the unknown.
(b) Random samples of mass spectra are first characterized by a set of variables and then methods from multivariate statistics or neural networks are applied to develop spectral classifiers.

Substructure recognition by library search

Library search systems have been developed or adjusted for the purpose of obtaining structurally similar compounds in the hitlist. In such methods an evaluation of the molecular structures of the hitlist compounds can provide useful information about the presence or absence of certain substructures in the unknown.

It is not trivial to define or select substructures that should be considered for this purpose. For the STIRS system, considerable effort has gone into the search for substructures that can be successfully classified by the implemented spectral similarity search. The MassLib system uses a predefined set of 180 binary molecular descriptors to characterize the similarity of structures. In most investigations a more or less arbitrary set of substructures, functional groups or more general structural properties (compound classes) has been considered. Self-adapting methods that automatically analyse the molecular structures in the hitlist (for instance by searching for frequent and large substructures) have not been used up to now in MS.

Table 3 Scheme for systematic structure elucidation

(1) Determine the molecular formula of the unknown

(2) Derive structural restrictions as substructures from spectral and other experimental data and from pre-knowledge about the unknown

(3) Make consistency checks of the structural restrictions

(4) Add all substructures that are considered to be present in the unknown to the goodlist. Add all substructures that are considered to be absent in the unknown to the badlist

(5) Generate all isomers for the given molecular formula that contain all substructures from the goodlist but none from the badlist

(6) Test the generated molecular candidate structures:
   (a) by a comparison of predicted spectra with measured spectra
   (b) by a comparison of predicted properties with measured properties
   (c) by considering more complicated structural restrictions that could not be handled by the isomer generator

(7) If the number of survived molecular candidates is too large try to create additional structural restrictions and continue at step 3

The number of occurrences of a certain substructure in the hitlist is compared with the corresponding number for the library and a probability is derived for the presence of that substructure in the unknown. This classification method is a variant of the well-known 'k-nearest neighbour classification'. Each mass spectrum is considered as a point in a multidimensional space; the neighbours nearest to the spectrum of the unknown correspond to the most similar reference spectra in library search. If the majority of k neighbours (k is typically between 1 and 10) contain a certain substructure then this substructure is predicted to be present in the unknown. A drawback of this approach is the high computational effort necessary for classifying an unknown because a full library search is required. The performance has been described by Stein (1995, see Further reading section) as 'sufficient to recommend it for routine use as a first step in structure elucidation'.

Multivariate classification

A *mass spectral classifier* is a part of a computer program that uses the peak list of a low resolution mass spectrum as input and produces information about the chemical structure as output. For such a classification procedure a number of methods are available in multivariate statistics. Many of them have already been applied to various problems in chemistry; classification of mass spectra (with the aim of recognizing chemical compound classes) was one of the pioneering works in chemometrics.

Application of most multivariate classification methods first requires a transformation of the data to be classified (the mass spectrum) into a fixed number of variables (so-called features). Besides the simple use of the peak intensities for these variables more sophisticated transformations have been applied. Problem-relevant *numerical spectral features* x_j are defined as linear or nonlinear functions of the peak intensities; the definitions are based on spectroscopic and/or mathematical concepts. **Table 4** contains a list of frequently used spectral features for mass spectra. It has been shown that such features are more closely related to chemical structures than the original peak data; an example will illustrate this. A characteristic fragment in a homologous series of compounds may not appear at the same mass but may be shifted by a multiple of 14 mass units, corresponding to $(CH_2)_n$. The often used modulo-14 summation features (also called mass periodicity spectra) consider this fact by an intensity summation in mass intervals of 14. Other spectral features are based on intensity ratios and reflect competing fragmentation pathways; autocorrelation features contain information about mass differences between abundant peaks. The stability of molecular ions and some functional groups can be characterized by features that describe the distribution of peaks across the mass range. The appropriate generation and selection of spectral features is considered to be the most essential part in the development of spectral classifiers.

The development of a classifier (**Figure 8**) for a particular substructure requires a random sample of spectra which is selected from a spectral library. Typically some 100 spectra from compounds containing the substructure (class 1) and some 100 spectra from compounds not containing the substructure (class 2) are necessary. One part of the

Table 4 Types of numerical spectral features; a set of spectral features can be used to classify a mass spectrum as to whether a certain substructure or a more general structural property is present or absent in the molecule

Modulo-14 intensity sums	$x_j = \sum I_{m+14n}$	$n : 0, 1, 2, \ldots, j,$ $m := 1, 2, \ldots, 14$
Logarithmic intensity ratios	$x_j = \ln I_m / I_{m+\Delta m}$	$I_m = \max (I_m, 1)$
Autocorrelation	$x_j = \sum I_m I_{m+\Delta m} / \sum I_m I_m$	$j, \Delta m : 1, 2, \ldots, 50$ m : a defined mass range
Distribution of peaks	$x_j = \sum I_{even} / \sum I_m$	even : all even masses m : all masses
Relative base peak intensity	$x_j = 100 / \sum I_m$	m : all masses

x_j, spectral feature; I_m, intensity in %base peak at mass m; Δm, mass difference

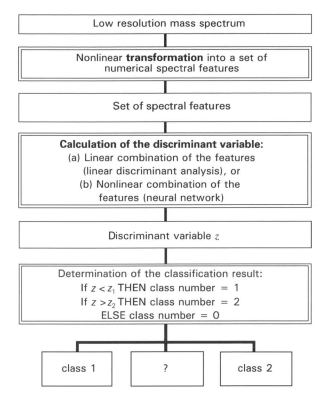

Figure 8 Classification scheme for mass spectra. The mass spectrum is transformed into a set of typically 5 to 20 variables (spectral features). A linear or nonlinear mathematical combination of the features results in a discriminant variable. The value of the discriminant variable together with the required maximum statistical risk for wrong answers determines which of the three possible answers is given: 'class 1' (substructure present), 'class 2' (substructure absent) or 'answer not possible because error risk is too high'.

random sample is used for a training of the classifier, the other part for testing it. Classification is based on the value of a discriminant variable z which is defined either as a linear function of the q selected spectral features x_j:

$$z = b_1 x_1 + b_2 x_2 + \cdots + b_q x_q \qquad [6]$$

or as a nonlinear function which is usually implemented as a neural network. During the training process the parameters of the classifier can be directly calculated (as for instance by multiple linear regression or partial least-squares regression) or they have to be adjusted iteratively (if a neural network is used). The aim of the training is to obtain values for z as defined by target values (for instance +1 for class 1 and −1 for class 2). The test set (which has not been used for the training) serves to estimate optimum classification thresholds z_1 and z_2 for

applications of the classifier to unknowns by the scheme:

IF $z < z_1$ THEN assign compound to class 1

IF $z > z_2$ THEN assign compound to class 2

ELSE reject classification answer [7]

Simple yes/no classifiers without the possibility to refuse the classification answer are not adequate for the recognition of substructures from mass spectra.

Today's performance of mass spectra classification by multivariate methods can be summarized as follows: (1) only a rather small number of substructures can be recognized with a low error rate. (2) Predictions of the absence of a substructure are usually more accurate than predictions of its presence. (3) Erroneous classifications cannot be avoided completely; therefore the intervention of a human expert and the parallel use of other spectra interpretation methods are advisable. (4) For small molecules a systematic and almost complete structure elucidation is sometimes possible by mass spectra classification and by application of the obtained structural restrictions in automatic isomer generation.

An example for this approach of structure elucidation is presented in **Figure 9**, where ethyl 2-(2-hydroxyphenyl)acetate has been considered as unknown. The molecular formula and the mass spectrum are given. Application of software MSclass resulted in substructures that are assumed to be present (goodlist) and others that are assumed to be absent (badlist); only substructures relevant to the molecular formula are shown. Considering these structural restriction, six isomers are possible, including the correct solution. The isomer generator used was MOLGEN. The computation time on a Pentium 233 MHz is more than 20 hours for the generation of all isomers, but only 4 s when the structural restrictions of the goodlist and badlist are considered.

Conclusions

The best worldwide performance has been claimed for more than one commercial MS database system. However, more neutral observers state that automated spectra interpretation systems have a rather limited scope. Spectra library search systems are now widely used in MS laboratories and do a good job with routine problems. They are not so useful with complex problems or if the unknown is not contained in the library; however, current research promises considerable improvements of these methods in the future. The routine application of library

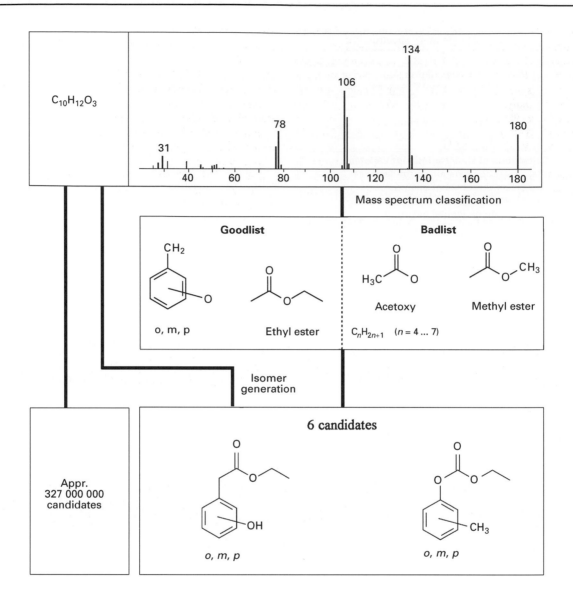

Figure 9 Systematic structure elucidation using the molecular formula of the unknown, structural restrictions from automatic mass spectra classification and exhaustive isomer generation. Ethyl 2-(2-hydroxyphenyl)acetate is the 'unknown'.

search methods and spectra classification liberate the human spectroscopist from time-consuming, simple work or at least provides valuable preliminary suggestions for the expert.

List of symbols

A_m = abundance at mass m (weighted peak intensity); b = parameter of discriminant function; d = distance; I_m = intensity (peak height) at mass m in %base peak; m = mass number; n = number of compounds or spectra; p = probability; q = number of variables (masses, spectral features); R = reference spectrum; S = similarity between two mass spectra; U = spectrum from unknown; x = spectral feature; z = discriminant variable.

See also: **Chromatography-MS, Methods; Forensic Science, Applications of Mass Spectrometry; Fragmentation in Mass Spectrometry; Laboratory Information Management Systems (LIMS); Medical Applications of Mass Spectrometry; Pyrolysis Mass Spectrometry, Methods.**

Further reading

Adams MJ (1995) *Chemometrics in Analytical Spectroscopy.* Cambridge: The Royal Society of Chemistry.

Clerc JT (1987) Automated spectra interpretation and library search systems. In Meuzelaar HLC and Isenhour TL (eds) *Computer Enhanced Analytical Spectroscopy,* Vol 1, pp 145–162. New York: Plenum Press.

Davis R and Frearson M (1987) *Mass Spectrometry (Analytical Chemistry by Open Learning).* Chichester: Wiley.

Drablos F (1987) Symmetric distance measures for mass spectra. *Analytica Chimica Acta* **201**: 225–239.

Gasteiger J, Hanebeck W, Schulz KP, Bauerschmidt S and Höllering R (1993) Automatic analysis and simulation of mass spectra. In Wilkins CL (ed) *Computer-Enhanced Analytical Spectroscopy*, Vol 4 pp 97–133. New York: Plenum Press.

Gray NAB (1986) *Computer-Assisted Structure Elucidation*. NewYork: John Wiley.

Massart DL, Vandeginste BGM, Buydens LMC, DeJong S, Lewi PJ and Smeyers-Verbeke J (1997) *Handbook of Chemometrics and Qualimetrics: Part A*. Amsterdam: Elsevier.

McLafferty FW, Loh SY, and Stauffer DB (1990) Computer identification of mass spectra. In Meuzelaar HLC (ed) *Computer-Enhanced Analytical Spectroscopy*, Vol 2 pp 163–181. New York: Plenum Press.

McLafferty FW and Turecek F (1990) *Interpretation of Mass Spectra*, 4th edn. Mill Valley: University Science Books.

Owens KG (1992) Application of correlation analysis techniques to mass spectral data. *Applied Spectroscopy Reviews* **27**: 1–49.

Stein SE and Scott DR (1994) Optimization and testing of mass spectral library search algorithms for compound identification. *Journal of the American Society of Mass Spectrometry* **5**: 859–866.

Varmuza K and Werther W (1996) Mass spectral classifiers for supporting systematic structure elucidation. *Journal of Chemical Information and Computer Science* **36**: 323–333.

Chemically Induced Dynamic Nuclear Polarization

See **CIDNP, Applications.**

Chemometrics Applied to Near-IR Spectra

See **Computational Methods and Chemometrics in Near-IR Spectroscopy.**

Chiroptical Spectroscopy, Emission Theory

James P Riehl, Michigan Technological University, Houghton, MI, USA

ELECTRONIC SPECTROSCOPY
Theory

All interactions involving light with chiral molecules discriminate between the two possible circular polarizations (left = L and right = R). In the absence of a perturbing static electric or magnetic field, the light emitted by a molecular chromophore will be partially circularly polarized if the emitting species is *chiral* or *optically active*. In this context these two terms have identical meaning, and describe a molecular structure in which the mirror image isomers (enantiomers) are not superimposable. The experimental technique of analysing the extent of circular polarization in the light emitted from chiral molecules has been variously referred to as circularly polarized emission (CPE), emission circular intensity differentials (ECID), or circularly polarized luminescence (CPL). In this article we will use the acronym CPL to describe this spectroscopic technique. In addition, it will not be necessary in the discussion presented here, to use the more specific terms of circularly polarized fluorescence or circularly polarized phosphorescence. Although CPL and circular dichroism (CD) spectroscopy, i.e. the difference in absorption of left versus right circular polarization, share many common characteristics, they differ from one another in two main ways. First, due to the Franck–Condon principle, CPL probes the chiral geometry of the excited state in the same way that CD probes the ground state structure, and secondly, CPL measurements reflect molecular motions and energetics that take place between the excitation (absorption) and emission. Both of these effects have been exploited in CPL measurements, and will be a major focus of what is presented here.

Circularly polarized luminescence transition probabilities

In CPL spectroscopy one is interested in measuring the difference in the emission intensity (ΔI) of left circularly polarized light (I_L) versus right circularly polarized light (I_R). By convention this difference is defined as follows

$$\Delta I \equiv I_L - I_R \qquad [1]$$

Just as in ordinary luminescence measurements, the determination of absolute emission intensities is quite difficult, so it is customary to report CPL measurements in terms of the ratio of the difference in intensity, divided by the average total luminescence intensity.

$$g_{lum} = \frac{\Delta I}{\frac{1}{2}I} = \frac{I_L - I_R}{\frac{1}{2}(I_L + I_R)} \qquad [2]$$

g_{lum} is referred to as the luminescence dissymmetry ratio (or factor). The extra factor of $\frac{1}{2}$ in Equation [2] is included to make the definition of g_{lum} consistent with the previous definition of the related quantity in CD, namely, g_{abs}.

$$g_{abs} = \frac{\Delta \varepsilon}{\varepsilon} = \frac{\varepsilon_L - \varepsilon_R}{\frac{1}{2}(\varepsilon_L + \varepsilon_R)} \qquad [3]$$

where in this equation ε_L and ε_R denote, respectively, the extinction coefficients for left and right circularly polarized light.

The time dependence of the intensity of emitted light of polarization σ, at a particular wavelength, λ, for a specific transition $n \rightarrow g$ of a molecule with orientation Ω (in the laboratory frame) may be expressed

in terms of the transition probability, W_{gn}^σ as follows

$$I_\sigma(\lambda, t) = (hc/\lambda)N_n(\Omega, t)W_{gn}^\sigma(\Omega)f_\sigma(\lambda) \qquad [4]$$

where $N_n(\Omega, t)$ denotes the number of molecules in the emitting state $|n\rangle$ with orientation Ω at time t, and $f_\sigma(\lambda)$ is a normalized lineshape function. The number of molecules in the emitting state with orientation Ω depends on the orientation at the time of excitation ($t = 0$) with respect to the polarization π and direction of the exciting beam.

$$N_n(\Omega, t) = N_e^\pi(\Omega : \Omega_0, t)\eta_n \qquad [5]$$

η_n denotes the fraction of molecules that end up in state $|n\rangle$ that were initially prepared in the intermediate state $|e\rangle$ by the excitation beam. This quantity is assumed to be independent of orientation. In the most general case we would also need to describe the time dependence of any conformational changes that take place between the time of excitation and emission, but for simplicity this will not be considered in the treatment presented here. The differential intensity of left minus right circularly polarized light may now be expressed as follows

$$\Delta I(\lambda, t) = (hc/\lambda)N_n(\Omega, t)\Delta W_{gn}(\Omega)f(\lambda) \qquad [6]$$

where we have introduced the differential transition probability ΔW_{gn}

$$\Delta W_{gn}(\Omega) \equiv W_{gn}^L(\Omega) - W_{gn}^R(\Omega) \qquad [7]$$

The probability of emitting a right or left circularly polarized photon may be related in the usual way to molecular transition matrix elements through Fermi's Golden Rule. Under the assumption that the emitted light is being detected in the laboratory 3 direction (see **Figure 1**), and allowing for electric dipole and magnetic dipoles in the expansion of the molecule–radiation interaction Hamiltonian we obtain the following expressions

$$W_{gn}^L(\Omega) = K(\lambda^3)[|\mu_1^{gn}|^2 + |\mu_2^{gn}|^2 + |m_1^{gn}|^2 + |m_2^{gn}|^2 \\ - 2i(\mu_1^{gn}m_1^{gn} + \mu_2^{gn}m_2^{gn})] \qquad [8]$$

$$W_{gn}^R(\Omega) = K(\lambda^3)[|\mu_1^{gn}|^2 + |\mu_2^{gn}|^2 + |m_1^{gn}|^2 + |m_2^{gn}|^2 \\ + 2i(\mu_1^{gn}m_1^{gn} + \mu_2^{gn}m_2^{gn})] \qquad [9]$$

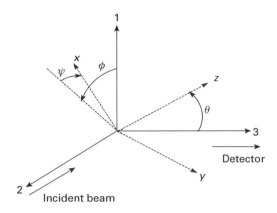

$$\hat{R} \equiv \begin{pmatrix} \cos\psi\cos\phi - \cos\theta\sin\phi\sin\psi & -\sin\psi\cos\phi - \cos\theta\sin\phi\cos\psi & \sin\theta\sin\phi \\ \cos\psi\sin\phi + \cos\theta\cos\phi\sin\psi & -\sin\psi\sin\phi + \cos\theta\cos\phi\cos\psi & -\sin\theta\cos\phi \\ \sin\theta\sin\psi & \sin\theta\cos\psi & \cos\theta \end{pmatrix}$$

Figure 1   Laboratory (1, 2, 3) and molecular (x, y, z) coordinate systems, and the Euler rotation matrix, (R).

where $K(\lambda^3)$ is a proportionality constant, 1 and 2 refer to laboratory axes and the electric dipole transition moment μ^{gn} and the imaginary magnetic dipole transition moment m^{gn} are defined as follows.

$$\mu_1^{gn} \equiv \langle g|\mu_1|n \rangle$$
$$m_1^{gn} \equiv \langle g|m_1|n \rangle \qquad [10]$$

The differential transition rate is, therefore,

$$\Delta W_{gn}(\Omega) = K(\lambda^3)[2i(\mu_1^{gn}m_1^{gn} + \mu_2^{gn}m_2^{gn})] \qquad [11]$$

The final connection between molecular properties and experimental observables requires knowledge of the orientational distribution of the emitting molecules with respect to the direction and polarization of the excitation light and the direction of detection, and the time dependence of this distribution.

Circularly polarized luminescence from solutions

There have been very few attempts at measuring CPL from oriented samples due to the inherent problems associated with measurement of circular polarization in the presence of linear polarization. For this reason we restrict the discussion of the orientation dependence of CPL to randomly oriented molecular samples such as occur in liquid solutions. Of course, even if the ground state distribution is randomly oriented

(isotropic), the emitting state may not be, due to the photoselection of the emitting sample by the excitation beam. We will denote an orientational distribution by brackets $\langle ... \rangle$. Substitution of Equations [5] and [11] into Equation [6], and allowing for an ensemble of orientations we obtain the following

$$\Delta I(\lambda, t) = \left(\frac{2ihc\eta_n}{\lambda}\right) K(\lambda^3)\langle N_e^\pi(\Omega : \Omega_0, t) \\ \times [\mu_1^{gn}m_1^{gn} + \mu_2^{gn}m_2^{gn}]\rangle f(\lambda) \qquad [12]$$

We may formally separate the time dependence of the orientational distribution, $\eta(\Omega,t)$, from the number of molecules excited at time $t = 0$ by an excitation pulse as follows

$$N_e^\pi(\Omega : \Omega_0, t) = N_e^\pi(\Omega_0)\eta(\Omega, t) \qquad [13]$$

$N_e^\pi(\Omega_0)$ may be calculated from knowledge of the initial distribution, and direction and polarization of the excitation beam.

$$N_e^\pi(\Omega_0) = \kappa^2 \left| \mu^{eg} \cdot \pi \right|^2 \qquad [14]$$

where μ^{eg} is the absorption transition vector. Final expressions for the formal relationship between the time-dependent circularly polarized luminescence intensity or g_{lum} require knowledge of experimental geometry, and $\eta(\Omega,t)$.

For purposes of illustration, we examine here the most common situation of random (isotropic) ground state distributions, and 90° excitation/emission geometry with unpolarized excitation. If the emission is detected in the laboratory 3 direction, then the excitation is polarized in the 13 plane. Equation [13] may then be written as

$$N_e^\pi(\Omega_0) = \kappa^2 \left[\left| \mu_1^{eg} \right|^2 + \left| \mu_3^{eg} \right|^2 \right] \qquad [15]$$

and the final formal expression (in the laboratory coordinate system) is obtained by substituting this result into Equation [12].

$$\Delta I(\lambda, t) = \left(\frac{2ihc\eta_n \kappa^2}{\lambda} \right) K(\lambda^3) \left\langle [\left| \mu_1^{eg} \right|^2 \right.$$
$$\left. + \left| \mu_3^{eg} \right|^2]_0 [\mu_1^{gn} m_1^{gn} + \mu_2^{gn} m_2^{gn}]_t \, \eta(\Omega, t) \right\rangle f(\lambda) \qquad [16]$$

where we have explicitly labelled the square brackets to indicate that the absorption takes place at time 0 and the emission at time t.

The ensemble average over molecular orientations implied in Equation [16] may be performed if $\eta(\Omega, t)$ is known. We first examine the case in which the emitting sample is 'frozen', i.e. the orientational distribution of emitting molecules is identical to that prepared by the excitation beam. In this case $\eta(\Omega, t) = \exp(-t/\tau)$, where τ denotes the emission lifetime, and, since the orientations are fixed, the subscripts on the square brackets may be eliminated. The orientational average may be performed by relating the transition moments in the molecular coordinate system to the laboratory system using the appropriate elements of the Euler matrix (see **Figure 1**). If one assumes, for example, that the absorption transition moment defines the molecular z-axis, then one must evaluate terms such as the following

$$\left\langle \left| \mu_1^{eg} \right|^2 \mu_1^{gn} m_1^{gn} \right\rangle =$$

$$\int_0^{2\pi} \int_0^{2\pi} \int_0^\pi |R_{1z}|^2 R_{1z} R_{1z} \, d\theta d\psi d\phi \left| \mu_z^{eg} \right|^2 \mu_z^{gn} m_z^{gn} \qquad [17]$$

This photoselection of emitting distributions also affects the measurement of total luminescence, and the

development of the formal expression for $I(t)$ parallels that presented above. The final result for $g_{lum}(t)$ is obtained by evaluating all of the integrals implied in Equation [16], and the related expression for $I(t)$. In this particular simple model, the time-dependence is only in the decay of the emitting state, and is identical for ΔI and I, so that g_{lum} is time independent. This resulting general expression for the 'frozen' limit is

$$g_{lum}(\lambda) = 4i \frac{f_{CPL}(\lambda)}{f_{TL}(\lambda)} \left[\frac{7\mu_x^{gn} m_x^{gn} + 7\mu_y^{gn} m_y^{gn} + 6\mu_z^{gn} m_z^{gn}}{7|\mu_x^{gn}|^2 + 7|\mu_y^{gn}|^2 + 6|\mu_z^{gn}|^2} \right] \qquad [18]$$

where we have allowed for different lineshapes for CPL and TL, and an arbitrary absorption transition direction. This result simplifies under conditions in which the absorption and emission transition directions are parallel. Furthermore, the reader is referred to the 'Further reading' section concerning the results expected for other excitation/emission geometries. It should be noted that if the luminescence is partially linearly polarized, the measurement of CPL is problematic due to experimental artifacts. For this reason it is usually the case that the excitation/emission geometry and incident polarization are chosen so as to eliminate the possibility of linear polarization in the luminescence.

The other useful limiting case is the situation in which the orientation of emitting molecules is isotropic (random). This limiting case is appropriate for small molecular systems with relatively long emission lifetimes. If the distribution prepared by the excitation beam has been completely randomized by the time of emission, then Equation [16] reduces to the following

$$\Delta I(\lambda, t) =$$
$$\left(\frac{2ihc\eta_n \kappa^2}{\lambda} \right) K(\lambda^3) \left\langle \mu_1^{gn} m_1^{gn} + \mu_2^{gn} m_2^{gn} \eta(\Omega, t) \right\rangle f(\lambda) \qquad [19]$$

where we have assimilated the absorption strength and other parameters associated with the excitation into the constants. The orientational averaging in this equation involves the product of only two elements of the Euler matrix. Just as in the previous discussion, the time dependence of the total luminescence and the CPL for this model are identical. The final formal relationship for this 'isotropic' limit is

$$g_{\text{lum}}(\lambda) = 4i \frac{f_{\text{CPL}}(\lambda)}{f_{\text{TL}}(\lambda)} \left[\frac{\mu^{gn} \cdot m^{gn}}{|\mu^{gn}|^2} \right] \qquad [20]$$

The major difference between this result and the frozen result is the fact that, in Equation [18], the different components of the emission transition vectors contribute unequally to the observable. In principle, this would allow one to investigate the chirality of molecular transitions along specific molecular directions, but to date this has not been exploited.

Equation [20] illustrates one of the operational principles of CPL (and CD) spectroscopy. In most situations one is interested in studying transitions that are formally forbidden. If the transition is allowed, the dipole strength, which is proportional to the denominator of Equation [20], is large. Since the rotatory strength represented by the numerator of Equation [20] is generally quite small, because of the magnitude of the magnetic dipole transition moment, g_{lum} (or g_{abs}) for allowed transitions is usually very small. In CPL spectroscopy this leads to the practical result that one does not generally study highly luminescent dyes, or other strongly allowed transitions, but rather, the most interesting molecular structural information comes from the study of weakly or mildly luminescent transitions in which the inherent molecular chirality is closely connected with the emissive chromophore.

Circularly polarized luminescence from racemic mixtures

One of the most useful applications of CPL spectroscopy has been in the study of racemic mixtures. This experiment is possible, because, even though the ground state is racemic, the emitting state can sometimes be photoprepared in an enantiomerically enriched state by use of circularly polarized excitation. If the racemization rate of the excited state is less than the emission rate, then the emission will be circularly polarized. We consider a racemic sample containing equal concentrations of R and S enantiomers. The preferential absorption of circularly polarized light is related to the CD through the absorption dissymmetry ratio g_{abs} which was defined in Equation [3]. Rewriting this equation for the CD of the R enantiomer we obtain the following

$$g_{\text{abs}}^{\text{R}} = \frac{\Delta\varepsilon}{\varepsilon} = \frac{\varepsilon_{\text{L}}(R) - \varepsilon_{\text{R}}(R)}{\frac{1}{2}(\varepsilon_{\text{L}}(R) + \varepsilon_{\text{R}}(R))} \qquad [21]$$

The number of R enantiomers that absorb left circularly polarized light, $N_{\text{L}}(R)$ is proportional to $\varepsilon_{\text{L}}(R)$, and similarly for $\varepsilon_{\text{R}}(R)$. Thus, we may make the following substitution

$$g_{\text{abs}}^{\text{R}} = \frac{N_{\text{L}}(R) - N_{\text{R}}(R)}{\frac{1}{2}(N_{\text{L}}(R) + N_{\text{R}}(R))} \qquad [22]$$

since all of the proportionality constants cancel. The number of R enantiomers that would absorb right circularly polarized light is exactly equal to the number of S enantiomers that would absorb left circularly polarized light, i.e. $N_{\text{R}}(R) = N_{\text{L}}(S)$. This substitution yields the following

$$g_{\text{abs}}^{\text{R}} = \frac{N_{\text{L}}(R) - N_{\text{L}}(S)}{\frac{1}{2}(N_{\text{L}}(R) + N_{\text{L}}(S))} = \frac{2\Delta N_{\text{L}}}{N_{\text{L}}} \qquad [23]$$

ΔN_{L} is the differential population $(R - S)$ of excited enantiomers that would result from left circularly polarized excitation.

In general, an exact description of CPL from racemic mixtures requires consideration of the competition between racemization and emission. Under the assumption that racemization is much slower than emission, we may write the following expression for CPL from a racemic mixture under left circularly polarized excitation as follows

$$g_{\text{lum}}^{\text{L}}(\lambda) = 4i \frac{f_{\text{CPL}}(\lambda)\Delta N_{\text{L}}}{f_{\text{TL}}(\lambda)N_{\text{L}}} \left[\frac{\mu^{gn}(R) \cdot m^{gn}(R)}{|\mu^{gn}(R)|^2} \right] \qquad [24]$$

Substituting from above we obtain the following expression

$$g_{\text{lum}}^{\text{L}}(\lambda) = \frac{1}{2} g_{\text{abs}}^{\text{R}}(\lambda') g_{\text{lum}}^{\text{R}}(\lambda) \qquad [25]$$

where we have explicitly labelled the excitation wavelength as λ'. Examination of Equation [25] shows that the measurement of CPL from racemic mixtures depends on the product of the CD and CPL. Since the measurement of g_{lum} is generally limited to magnitudes greater than 10^{-4}, the magnitude of the intrinsic dissymmetry ratios must usually be greater than 10^{-2} for this unique measurement to be feasible. This restriction has limited applications of this technique to racemic lanthanide complexes, since intraconfigurational f $\leftrightarrow$ f transitions that obey magnetic dipole transition rules (i.e. $\Delta J = 0, \pm 1$) often are associated with dissymmetry ratios greater than 0.1.

Time-resolved circularly polarized luminescence

For the simple model enantiopure systems described above, it was concluded that the time dependence of the CPL and total luminescence were identical, and, therefore, the dissymmetry ratio contained no dynamic molecular information. This, of course, would not be the case if intramolecular geometry changes, that would effect the chirality of the molecular transitions, were occurring on the same time scale as emission. However, no such examples of this type of study have yet appeared. Time-resolved CPL measurements have been useful in the study of racemic mixtures of lanthanide complexes in which racemization or excited state quenching is occurring on the same time scale as emission.

It was assumed in the previous section, that the enantioenriched excited state population, which was prepared from a racemic ground state distribution by a circularly polarized excitation beam, was maintained until the time of emission. If this is not the case, then the differential excited state population is time-dependent. Neglecting orientational effects, we may derive the following expression for the time-dependence of the excited state population following an excitation pulse of left circularly polarized light

$$\Delta N^{\mathrm{L}}(t) = \Delta N^{\mathrm{L}}(0)\exp(-2k_{\mathrm{rac}}t - k_0 t) \quad [26]$$

where k_{rac} is the racemization rate and k_0 is the excited state decay rate. Since the total number of emitting molecules decays as

$$N^{\mathrm{L}}(t) = N^{\mathrm{L}}(0)\exp(-k_0 t) \quad [27]$$

one can express the time-dependence of the luminescence dissymmetry as follows

$$g^{\mathrm{L}}_{\mathrm{lum}}(t) = g^{\mathrm{L}}_{\mathrm{lum}}(0)\exp(-2k_{\mathrm{rac}}t) \quad [28]$$

This technique has been used in a number of studies in which the decay of g_{lum} has been analysed to extract the racemization rate constant.

It is also possible to extract dynamic information concerning racemization rates through measurement of steady-state CPL. Again neglecting time-dependent orientational effects, one may integrate Equations [26] and [27] over long times ($0 \to \infty$) and obtain the following result

$$g^{\mathrm{L}}_{\mathrm{lum}}(\lambda) = \tfrac{1}{2} g^{\mathrm{R}}_{\mathrm{abs}}(\lambda') g^{\mathrm{R}}_{\mathrm{lum}}(\lambda)\left(\frac{k_0}{2k_{\mathrm{rac}} + k_0}\right) \quad [29]$$

This equation has the expected limiting behaviour. If $k_{\mathrm{rac}} \ll k_0$ then the result of Equation [25] is obtained, and if $k_{\mathrm{rac}} \gg k_0$ then the magnitude of g_{lum} becomes negligibly small.

The second application of time-resolved CPL is in the study of enantioselective (or more properly, diastereomer-selective) quenching. In these experiments an enantiopure quencher molecule is added to a racemic solution. The interaction of the chiral quencher with the individual enantiomers is such that the excited state of one of the enantiomers is quenched more rapidly than the other. This process is depicted schematically in **Figure 2**, where we have assumed the quencher is an R enantiomer, and k_{RS} and k_{RR} denote the quenching rate constants. This process may be described using the following simple kinetic model

$$R^* + R_{\mathrm{Q}} \underset{k^+_{-\mathrm{diff}}}{\overset{k^+_{\mathrm{diff}}}{\rightleftarrows}} [R^* : R_{\mathrm{Q}}] \overset{k^+_{\mathrm{ET}}}{\longrightarrow} [R : R_{\mathrm{Q}}{}^*]$$
$$\overset{k^+_{\mathrm{P}}}{\longrightarrow} R + R_{\mathrm{Q}} \quad [30]$$

$$S^* + R_{\mathrm{Q}} \underset{k^-_{-\mathrm{diff}}}{\overset{k^-_{\mathrm{diff}}}{\rightleftarrows}} [S^* : R_{\mathrm{Q}}] \overset{k^-_{\mathrm{ET}}}{\longrightarrow} [S : R_{\mathrm{Q}}{}^*]$$
$$\overset{k^+_{\mathrm{P}}}{\longrightarrow} S + R_{\mathrm{Q}} \quad [31]$$

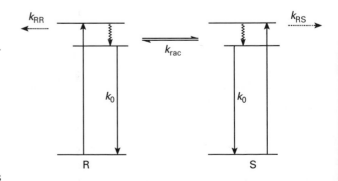

Figure 2 Schematic energy level diagram for measurement of CPL from a racemic mixture.

These reactions are modelled in terms of a diffusional step (k_{diff}^- or k_{diff}^+), resulting in the formation of a bimolecular encounter complex. This is followed by competing pathways for dissociation of the complex ($k_{-\text{diff}}^-$ or $k_{-\text{diff}}^+$) or energy transfer (k_{ET}^- or k_{ET}^+). Deactivation of the excited quencher is described by k_{P}. The observed CPL is directly proportional to the difference in excited state concentrations of the two enantiomers, and this may be related to the specific rate constants introduced above. If, for example, we make the reasonable assumption that the diffusion and deactivation are independent of chirality, one can derive the following expression for $\Delta N(t)$

$$\Delta N(t) = [S^*](t) - [R^*](t) = [S]_0 e^{-k_{\text{obs}}^{\text{RS}} t} - [R]_0 e^{-k_{\text{obs}}^{\text{RR}} t}$$
[32]

where

$$
\begin{aligned}
k_{\text{obs}}^{\text{RS}} &= k_{\text{RS}}[S] + k_0 \\
k_{\text{obs}}^{\text{RR}} &= k_{\text{RR}}[R] + k_0
\end{aligned}
$$
[33]

and

$$
\begin{aligned}
k_{\text{RR}} &= \frac{k_{\text{ET}}^+ k_{\text{diff}}}{k_{\text{ET}}^+ + k_{-\text{diff}}^+} \\
k_{\text{RS}} &= \frac{k_{\text{ET}}^- k_{\text{diff}}}{k_{\text{ET}}^- + k_{-\text{diff}}^-}
\end{aligned}
$$
[34]

If the quenching reactions are far from diffusion control (i.e. $k_{\text{ET}} \ll k_{-\text{diff}}$), then one can express the diastereomeric quenching rate constants in terms of equilibrium constants for formation of the diastereomeric ion pairs.

$$
\begin{aligned}
k_{\text{RR}} &= K^+ k_{\text{ET}}^+ \\
k_{\text{RS}} &= K^- k_{\text{ET}}^-
\end{aligned}
$$
[35]

Thus, the time-dependence of $\Delta I(t)$ or $g_{\text{lum}}(t)$ may be used as a probe of the fundamental differences in the interaction of chiral molecules.

Spectra–structure correlations

The theoretical discussion presented above has focused on developing formal relationships between the molecular transition integrals and the experimental observables. These are, of course, of primary importance in interpreting CPL measurements in terms of inter- and intramolecular processes, including molecular reorientation, racemization and chemical reactions which affect the net molecular chirality. These processes are all well understood, and specific information is most often obtained within an appropriate model system. The more fundamental aspects of molecular chirality associated with the molecular transition matrix elements is less well developed. Although there have been a few successes in interpreting the sign and magnitude of a CPL spectrum in terms of the identity and structure of a specific enantiomer, in this regard, it is certainly not the case that the technique is as well developed as CD spectroscopy. Even in the case of fairly highly symmetric chiral lanthanide complexes, the level of computation and development of f-f intensity theory is not yet at the stage where one can associate *a priori* the sign of an individual transition with the identity of the enantiomer.

List of symbols

f = lineshape; g_{lum} = luminescence dissymmetry ratio; I = emission intensity; k = rate constant; n→g = transition; N = number of molecules; t = time; W = transition probability; ΔI = difference in emission intensity; ε = extinction coefficient; η = fraction of molecules; λ = wavelength; μ = absorption transition vector; π = polarization of excitation; σ = polarization of emission; Ω = orientation.

See also: **Chiroptical Spectroscopy, General Theory; Chiroptical Spectroscopy, Oriented Molecules and Anisotropic Systems; Circularly Polarized Luminescence and Fluorescence Detected Circular Dichroism; Luminescence, Theory.**

Further reading

Richardson FS and Riehl JP (1977) Circularly polarized luminescence spectroscopy. *Chemical Reviews* **77**: 773–792.

Riehl JP and Richardson FS (1986) Circularly polarized luminescence spectroscopy. *Chemical Reviews* **86**: 1–15.

Riehl JP and Richardson FS (1993) Circularly polarized luminescence. *Methods in Enzymology* **226**: 539–553.

Riehl JP (1994) Excited state optical activity. *Analytical Applications of Circular Dichroism*. Amsterdam: Elsevier.

Steinberg I (1975) Circular polarization of fluorescence. In: Chen RF and Edelhoch H (eds) *Biochemical Fluorescence: Concepts*. New York: Marcel Dekker.

Chiroptical Spectroscopy, General Theory

Hans-Georg Kuball, **Tatiana Höfer** and **Stefan Kiesewalter**, Universität Kaiserslautern, Germany

Copyright © 1999 Academic Press

Introduction and molecular description of absorption and circular dichroism

The interaction of light with an oriented molecule or an ensemble of oriented molecules can be treated as a scattering process with photons or semiclassically as an induction process of electric or magnetic moments by an electromagnetic field. As a result of this interaction the incident light beam undergoes changes in its state of polarization and its intensity. Using Gō's description, the complex optical rotation $\hat{\phi}$ can be given as a sum of the contributions of individual molecules:

$$\hat{\phi} = \phi + \mathrm{i}\theta = \frac{Nd}{8\pi^2 V}\int f(\alpha, \beta, \gamma) H'_{21}(\alpha, \beta, \gamma) \sin\beta \, \mathrm{d}\alpha\mathrm{d}\beta\mathrm{d}\gamma$$

[1]

with

$$H'_{21} = -\frac{2\pi\mathrm{i}e^2}{\hbar m^2 c\omega}\sum_n \left\{ \frac{\langle p'_2 \exp(-\mathrm{i}kx'_3)\rangle_{0n}\langle p'_1 \exp(\mathrm{i}kx'_3)\rangle_{n0}}{\omega_{n0} - \omega - \mathrm{i}\eta} \right.$$
$$\left. + \frac{\langle p'_1 \exp(\mathrm{i}kx'_3)\rangle_{0n}\langle p'_2 \exp(-\mathrm{i}kx'_3)\rangle_{n0}}{\omega_{n0} + \omega + \mathrm{i}\eta} \right\}$$

[2]

and

$$\langle p'_j \exp(\pm\mathrm{i}kx'_3)\rangle_{nm} = \langle n \,|\, \sum_\mu p'_{j\mu} \exp(\pm\mathrm{i}kx'_{3\mu})|\, m\rangle$$

$$j = 1, 2$$

[3]

Here, $f(\alpha, \beta, \gamma)$ is the orientational distribution function of an ensemble of molecules which is normalized by $1/8\pi^2 \int f(\alpha, \beta, \gamma)\sin\beta\mathrm{d}\alpha\mathrm{d}\beta\mathrm{d}\gamma = 1$, α, β, γ are the Eulerian angles, N denotes the number of molecules in the volume V, d is the optical path length, $\sum_\mu$ represents the sum over all electrons of a molecule, m and $-e$ are the mass and the charge of the electron, p'_j is the operator of the linear momentum, $\omega = 2\pi c\bar{\nu}$, c is the velocity of light in a vacuum, $k = \omega/c$, η is a positive damping parameter representing the natural

line shape, $\omega_{n0} = 2\pi c\bar{\nu}_{n0} = \hbar^{-1}(E_n - E_0)$, and E_n and E_0 are the energies of the states $|n\rangle$ and $|0\rangle$. Primed symbols refer to the space-fixed and unprimed letters to the molecule-fixed coordinate system. ϕ is the optical rotation and θ represents the ellipticity of the light beam emerging from the sample. H'_{21} is an element of the interaction matrix H'_{ij} of a photon and an atom or an oriented molecule. The incident light is propagating in the x'_3 direction and is polarized in the $x'_1 x'_2$ plane. Using the relation

$$\langle p'_2 \exp(-\mathrm{i}kx'_3)\rangle_{0n}\langle p'_1 \exp(\mathrm{i}kx'_3)\rangle_{n0}$$
$$= \langle p'_1 \exp(\mathrm{i}kx'_3)\rangle_{0n}\langle p'_2 \exp(-\mathrm{i}kx'_3)\rangle_{n0}$$

[4]

where the wave functions $|j\rangle$ are chosen to be real. Expanding the exponentials of Equation [2] into a series which converges very rapidly if the extension of the molecule is small compared to the wavelength of the incident light, it follows that the complex optical rotation is given by

$$\hat{\phi} = -\frac{\mathrm{i}Nd}{2\pi\hbar cV}\int f(\alpha, \beta, \gamma)\sin\beta \, \mathrm{d}\alpha\mathrm{d}\beta\mathrm{d}\gamma$$
$$\times \left\{ \sum_n \frac{\omega}{\omega_{n0}^2 - \omega^2 - 2\mathrm{i}\eta\omega}\left[\omega_{n0}D'^{0n}_{12}(\alpha, \beta, \gamma) \right.\right.$$
$$\left.\left. + 2\mathrm{i}\omega R'^{0n}_{123}(\alpha, \beta, \gamma) \right]\right\}\Bigg|$$

[5]

where D'^{0n}_{12} is a coordinate of the transition moment tensor. Loxsom has discussed the effect of an increasing extension-to-wavelength ratio. The third rank tensor R'^{0n}_{123}, antisymmetric with respect to the first two indices, can be replaced by a symmetric pseudo-tensor R'^{0n}_{ij} of rank two which is called the tensor of rotation. This tensor of rotation can be decomposed into a magnetic dipole and an electric quadrupole contribution

$$R'^{0n}_{ij} = \frac{1}{4}\sum_{r,s}\langle \mu_r\rangle_{0n}(\varepsilon_{rsi}\langle C_{sj}\rangle_{n0} + \varepsilon_{rsj}\langle C_{si}\rangle_{n0})$$

[6]

with

$$\langle C_{sj}\rangle_{n0} = -i\sum_r \varepsilon_{sjr}\langle m_r\rangle_{n0} - \frac{\omega_{n0}}{c}\langle Q_{sj}\rangle_{n0} \quad [7]$$

and

$$\mu_i = -e\sum_\nu x_{i\nu} \quad Q_{ij} = -\frac{1}{2}e\sum_\nu x_{i\nu}x_{j\nu}$$

$$m_r = -\frac{e}{2mc}\sum_\nu\sum_{i,j}\varepsilon_{rij}\,x_{i\nu}\,p_{j\nu} \quad [8]$$

Here, $\langle u_i\rangle_{0n}$ and $\langle m_i\rangle_{n0}$ represent the coordinates of the electric and magnetic dipole transition moment vectors, respectively, $\langle Q_{ij}\rangle_{n0}$ is a coordinate of the electric quadrupole transition moment tensor, and ε_{ijk} is the Levi–Civita tensor. Operators depending on the dynamic variables of the nuclei are omitted because they do not contribute to the effect in the approximation used.

The complex optical rotation $\hat{\phi}$ contains the elliptical birefringence and dichroism which can be given for small effects as a superposition of a linear and a circular birefringence and dichroism:

$$\hat{\phi} = \phi_{LD} + \phi_{CB} + i[\theta_{LB} + \theta_{CD}] \quad [9]$$

with the linear birefringence (LB) and dichroism (LD) effects

$$\theta_{LB} = -\pi\bar{\nu}d(n_1 - n_2) \qquad \phi_{LD} = \frac{1}{4}(\ln 10)(\varepsilon_1 - \varepsilon_2)cd$$
$$[10]$$

and the circular birefringence (CB) and dichroism (CD) effects

$$\phi_{CB} = \pi\bar{\nu}d(n_L - n_R), \qquad \theta_{CD} = \frac{1}{4}(\ln 10)(\varepsilon_L - \varepsilon_R)cd$$
$$[11]$$

Here, ε_1, ε_2 and n_1, n_2 are the molar decadic absorption coefficients and the refractive indices for a linearly polarized light beam with a polarization parallel and perpendicular to the optical axis of a uniaxial sample, and ε_L, ε_R and n_L, n_R are the absorption coefficients and the refractive indices for the left and right circularly polarized light beams, respectively. In the

approximation used the CD/CB and LD/LB effects are additive. From the theory of the Cauchy–Rieman differential equations it can be shown that the Kramers–Kronig transforms derived with microcausality as the only presupposition are also valid for a system of oriented molecules:

$$\theta_{LB}(\omega) = -\frac{2}{\pi}\oint\frac{\bar{\omega}\phi_{LD}(\bar{\omega})}{\bar{\omega}^2 - \omega^2}\,d\bar{\omega} \quad [12]$$

$$\phi_{CB}(\omega) = \frac{2}{\pi}\oint\frac{\omega\theta_{CD}(\bar{\omega})}{\bar{\omega}^2 - \omega^2}\,d\bar{\omega} \quad [13]$$

The frequency dependence given in Equations [2] and [5] is based on a situation where a natural line width is found. In all cases where optical activity has been analysed, e.g. in a gas or in a liquid or solid phase, the bandwidth is determined by the broadening of rotational states – for which the theory of optical activity is not well developed and no experimental analysis exists – or by the existence of librational states etc. Therefore, broadened rotational states or librational states can be introduced for which the spectrum can be considered as a quasi-continuum. With the assumption of the Born–Oppenheimer approximation the states $|0\rangle$ and $|n\rangle$ can then be given as a product of vibronic states and librational states λ_{Nn} and λ_{Kk} as follows

$$|0\rangle = |Nn\rangle|\,\lambda_{Nn}(\omega)\rangle \text{ and } |n\rangle = |Kk\rangle|\lambda_{Kk}(\omega)\rangle \quad [14]$$

where $|Nn\rangle$ and $|Kk\rangle$ represent the vibronic electronic ground and excited states, respectively. n and k are quantum numbers for the vibrational states in the electronic states N and K, respectively. An integration over the quasi-continuum of the librational states ω and the neglect of the natural bandwidth η yields the spectral functions $F^{NnKk}(\bar{\nu})$ and $G^{NnKk}(\bar{\nu})$ for the absorption and the circular dichroism effect which are equal for most of the experimental conditions given. With the new line shape of the absorption band $F^{NnKk}(\bar{\nu})$ and the circular dichroism $G^{NnKk}(\bar{\nu})$, the coordinates of the absorption ε_{ij} (Eqn [15]) and circular dichroism tensor $\Delta\varepsilon_{ij}$ (Eqn [16]) are given by

$$\varepsilon_{ij}(\bar{\nu}) = \frac{B\bar{\nu}}{4}\sum_n\sum_{Kk}D_{ij}^{NnKk}F^{NnKk}(\bar{\nu}) \quad [15]$$

and

$$\Delta\varepsilon_{ij}(\bar{\nu}) = B\bar{\nu}\sum_n\sum_{Kk}R_{ij}^{NnKk}G^{NnKk}(\bar{\nu}) \quad [16]$$

where for the vibronic transition $|Nn\rangle \to |Kk\rangle$ the dipole transition moment tensor is defined by

$$D_{ij}^{NnKk} = \langle \mu_i \rangle_{NnKk} \langle \mu_j \rangle_{KkNn} \qquad [17]$$

and the rotational strength tensor R_{ij}^{NnKk} is given analogously to the rotational strength tensor in Equations [6] and [7] by replacing $|0\rangle \to |n\rangle$ with $|Nn\rangle \to |Kk\rangle$ and

$$B = \frac{32\pi^3 N_A}{10^3 hc \ln 10} = 7.653 \times 10^{40} \text{ cgs} \qquad [18]$$

N_A is Avogadro's number, c the velocity of light and h is Planck's constant. Together with the Kramers–Kronig transforms (Eqns [12] and [13]) the frequency dependence of the optical rotation tensor can be obtained as follows

$$M_{ii}(\bar{\nu}) = \frac{288\pi N_A \bar{\nu}}{hc} \sum_n \sum_{Kk} R_{ii}^{NnKk} J_2^{NnKk}(\bar{\nu}) \qquad [19]$$

$$J_2^{NnKk}(\bar{\nu}) = \oint \frac{\bar{\nu}' G^{NnKk}(\bar{\nu}') d\bar{\nu}'}{\bar{\nu}'^2 - \bar{\nu}^2} \qquad [20]$$

$J_2^{NnKk}(\bar{\nu})$ in Equation [19] is the principal value of the integral over the singularity in the wavenumber region of an absorption process.

Absorption and circular dichroism in isotropic media

For isotropic solutions the orientational distribution function $f(\alpha, \beta, \gamma)$ in Equation [5] is equal to one and the absorption coefficient $\varepsilon(\bar{\nu})$, the circular dichroism $\Delta\varepsilon(\bar{\nu})$, and the molar optical rotation $[M(\bar{\nu})]$ (optical rotatory dispersion – ORD) are one third of the trace of the corresponding tensor or pseudotensor of the second rank, respectively,

$$\varepsilon(\bar{\nu}) = \frac{1}{3}\sum_i \varepsilon_{ii} = \frac{B\bar{\nu}}{12} \sum_n \sum_{Kk} D^{NnKk} F^{NnKk}(\bar{\nu})$$
$$= \sum_K \varepsilon^{NK}(\bar{\nu}) \qquad [21]$$

with the dipole strength of the transition $|Nn\rangle \to |Kk\rangle$

$$D^{NnKk} = \sum_i D_{ii}^{NnKk} = \sum_i \langle \mu_i \rangle_{NnKk} \langle \mu_i \rangle_{NnKk} = \langle \vec{\mu} \rangle^2_{NnKk} \qquad [22]$$

and

$$\Delta\varepsilon(\bar{\nu}) = \frac{1}{3}\sum_i \Delta\varepsilon_{ii} = \frac{B\bar{\nu}}{3} \sum_n \sum_{Kk} R^{NnKk} F^{NnKk}(\bar{\nu})$$
$$= \sum_K \Delta\varepsilon^{NK}(\bar{\nu}) \qquad [23]$$

with the rotational strength R^{NnKk} of the transition $|Nn\rangle \to |Kk\rangle$ derived from the rotational strength tensor

$$R_{ij}^{NnKk} = -\frac{1}{2}i \sum_{r,l} \langle \mu_r \rangle_{NnKk} \cdot \langle m_l \rangle_{KkNn}$$
$$\times \left[\delta_{rl}\delta_{ij} - \frac{1}{2}\delta_{ri} \cdot \delta_{lj} - \frac{1}{2}\delta_{rj} \cdot \delta_{li} \right]$$
$$- \frac{1}{4}\frac{\omega_{KkNn}}{c} \sum_{r,s} \langle \mu_r \rangle \left[\varepsilon_{rsi}\langle Q_{sj}\rangle_{KkNn} + \varepsilon_{rsj}\langle Q_{si}\rangle_{KkNn} \right] \qquad [24]$$

δ_{ij} is the Kronecker symbol. The contribution of the electric quadrupole transition moments cancels in the isotropic solution. Then a scalar product of the electric and magnetic dipole transition moments follows for the rotational strength

$$R^{NnKk} = \sum_i R_{ii}^{NnKk} = \sum_i \text{Im}\{ \langle \mu_i \rangle_{NnKk} \langle m_i \rangle_{KkNn} \}$$
$$= \text{Im}\{ \langle \vec{\mu} \rangle_{NnKk} \cdot \langle \vec{m} \rangle_{KkNn} \} \qquad [25]$$

The ORD is now given by

$$[M(\bar{\nu})] = \frac{1}{3}\sum_i M_{ii} = \frac{288\pi N_A \bar{\nu}}{3hc} \sum_n \sum_{Kk} R^{NnKk} J_2^{NnKk}(\bar{\nu})$$
$$= \sum_K [M^{NK}(\bar{\nu})] \qquad [26]$$

If vibronic states n, k do not contribute to the ORD, there follows from Equation [26] with the dispersion

$$J_2^{NK}(\bar{\nu}) = \frac{\bar{\nu}^2}{\bar{\nu}_{NK}^2 - \bar{\nu}^2} \qquad [27]$$

the well-known Rosenfeld equation.

The right sides of Equations [23] and [26] lead to an interpretation of the CD and ORD spectra as a sum of contributions of the electronic transitions. To every absorption band $\varepsilon^{NK}(\bar{\nu})$ belongs a $\Delta\varepsilon^{NK}(\bar{\nu})$ and a corresponding ORD curve $[M^{NK}(\bar{\nu})]$ as presented in **Figure 1**. It is evident from **Figure 1** that the optical rotation $[M^{NK}(\bar{\nu})]$ possesses a larger half bandwidth than $\Delta\varepsilon^{NK}(\bar{\nu})$. Thus $[M(\bar{\nu})]$ can only be measured as a sum of contributions of many transitions independent of the chosen wavelength even in the wavelength region far away from absorption bands. In contrast to this it is often possible to measure the CD curve of only one or an overlap of CD curves of a few transitions. The smaller bandwidth of the CD bands is the basis for the answer to the often-discussed question of why a CD measurement has advantages over an ORD measurement. In principle, one can measure either the circular dichroism or the optical rotatory dispersion because one measurable quantity can be evaluted from the other with the Kramers–Kronig relations (Eqns [12] and [13]) if the frequency region measured is sufficiently large. From the numerical point of view the calculation of an ORD curve from a CD curve is easier than vice versa because of the smaller half bandwidth of the CD curves.

The different absorption bands, to which the CD and ORD belong, can be classified by the symmetry of the molecular states involved especially by their polarization, i.e. their transition moment direction which is defined by Equations [6], [8], and [17] and visualized in **Figure 2**. Experimentally the polarization can be determined by the linear dichroism (Eqn [10]) measurement usually described in polarized spectroscopy in terms of the degree of anisotropy R:

$$R = \frac{\varepsilon_1 - \varepsilon_2}{\varepsilon_1 + 2\varepsilon_2} = \frac{\varepsilon_1 - \varepsilon_2}{3\varepsilon} \qquad [28]$$

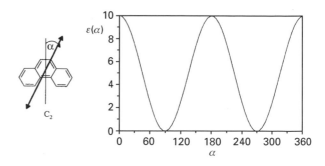

Figure 2 Dependence of the polarized absorption on the angle α between the transition moment direction along the C_2 axis of phenanthrene and the plane of polarization of the light. The transition moment direction is defined by a vector $\langle\vec{\mu}\rangle_{n0}$ parallel to the direction of maximum absorption. The light beam propagates perpendicular to the plane of the molecule.

Furthermore, the dissymmetry factor

$$g(\bar{\nu}) = \frac{\Delta\varepsilon(\bar{\nu})}{\varepsilon(\bar{\nu})} \qquad [29]$$

contains information about the kind of the Cotton effect (CE) of a compound and is a hint whether an effect is measurable or not. Nowadays, the lowest limit for the sensitivity with commercial instruments is about 10^{-7}.

From the Kramers–Kronig transformation and from the quantum-theoretical description, sum rules for the rotational strength analogous to the Kuhn–Reichert sum rule for the absorption can be derived. This means that, for the sum over the rotational strength of all transitions of the CD spectrum of a molecule, one obtains

$$\sum_{N,n,K,k} R^{NnKk} = 0 \qquad [30]$$

From this result it follows that the CD and ORD values decrease when going to wavelengths longer than those of the microwave spectral region or to wavelengths shorter than those of the vaccum UV. This can be understood from the fact that the ratio of the extension of the molecules and the wavelengths is either very much smaller or larger than for the UV-visible spectral region.

Vibrational progressions in the CD and UV-visible spectra of an electronic transition

From Equations [21] and [23] it also follows that the UV ($\varepsilon^{NK}(\bar{\nu})$) and the CD ($\Delta\varepsilon^{NK}(\bar{\nu})$) band are built up

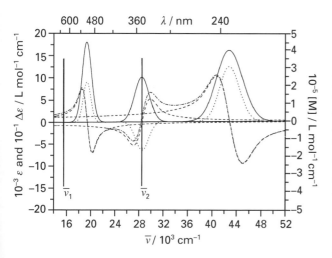

Figure 1 UV ($\varepsilon(\bar{\nu})$, ——), CD ($\Delta\varepsilon(\bar{\nu})$, ·····) and ORD ($[M(\bar{\nu})]$, -·-·-·) spectra of a compound as a sum of contributions $\varepsilon^{NK}(\bar{\nu})$(——), $\Delta\varepsilon^{NK}(\bar{\nu})$ (······), and $[M^{NK}(\bar{\nu})]$(----) of different transitions (chromophores) $|N\rangle \rightarrow |K\rangle$.

by a sum of different vibrational progressions

$$\varepsilon^{NK}(\bar{\nu}) = \frac{B\bar{\nu}}{12} \sum_n \sum_k D^{NnKk} F^{NnKk}(\bar{\nu}) \qquad [31]$$

and

$$\Delta\varepsilon^{NK}(\bar{\nu}) = \frac{B\bar{\nu}}{3} \sum_n \sum_k R^{NnKk} F^{NnKk}(\bar{\nu}) \qquad [32]$$

In **Figure** 3 a term scheme and the corresponding CD and UV spectra for an example where two progressions contribute to the UV-visible and the CD spectrum are constructed. The vibrational progressions can contribute according to Equations [31] and [32] to the UV and CD spectra of an electronic transition with vibrations of different symmetry and thus with differently polarized bands and CD contributions of different sign. This is analogous to the electronic spectra of a molecule where transitions of different polarization, i.e. different transition moment directions, contribute to the spectra (Eqn [21]). This can be proven experimentally by a frequency-dependent degree of anisotropy (Eqn [28]). In order to eliminate the effect of the vibrational fine structure one has to integrate over the whole band of the electronic transition $|N\rangle \rightarrow |K\rangle$. The dipole strength D^{NK} and the rotational strength R^{NK} include information about the electronic transition from the electronic ground state N to the electronic excited state K and are of interest for the characterization of the transition:

$$D^{NK} = \frac{12}{B} \int \frac{\varepsilon^{NK}(\bar{\nu})}{\bar{\nu}} \mathrm{d}\bar{\nu} \qquad [33]$$

$$R^{NK} = \frac{3}{B} \int \frac{\Delta\varepsilon^{NK}(\bar{\nu})}{\bar{\nu}} \mathrm{d}\bar{\nu} \qquad [34]$$

The dimension is given, as still used even today, in cgs units. A frequency-independent dissymmetry factor g can also be defined with the help of the dipole strength and rotational strength of the electronic transition

$$g = \frac{4R^{NK}}{D^{NK}} \qquad [35]$$

For an allowed electric and allowed magnetic transition $|N\rangle \rightarrow |K\rangle$ the Franck–Condon factor $\langle n|k\rangle$ can

be introduced and the rotational strength (Eqn [34]) is then given by

$$R^{NK} = \sum_k R^{NnKk} = \mathrm{Im}\big\{\langle\vec{\mu}\rangle_{NK} \cdot \langle\vec{m}\rangle_{KN}\big\} \sum_k |\langle n \mid k\rangle|^2$$
$$= \mathrm{Im}\big\{\langle\vec{\mu}\rangle_{NK} \cdot \langle\vec{m}\rangle_{KN}\big\} \qquad [36]$$

$\langle\vec{\mu}\rangle_{NK}$ and $\langle\vec{m}\rangle_{KN}$ are the electric and magnetic transition moments for the 0-0 transition, respectively (a 0-0 transition means a transition where neither excited vibrational nor excited rotational states of the ground and excited electronic states are involved). Analogously, the dipole strength D^{NK} can be evaluated by

$$D^{NK} = \sum_k D^{NnKk} = \langle\vec{\mu}\rangle_{NK} \cdot \langle\vec{\mu}\rangle_{KN} \sum_k |\langle n \mid k\rangle|^2$$
$$= \langle\vec{\mu}\rangle_{NK} \cdot \langle\vec{\mu}\rangle_{KN} \qquad [37]$$

For forbidden electronic transitions, intensity is stolen from other allowed electronic transitions and thus further terms have to be added in Equations [36] and [37] (Herzberg–Teller theory of vibronic coupling).

Classification of absorption and circular dichroism spectra and of chromophores

Classifications of absorption and CD bands

Electronic transitions $|N\rangle \rightarrow |K\rangle$ can be classified by the irreducible representations of the wave functions involved or more conveniently for a chemist by the orbitals which are involved, i.e. the non-bonding n, bonding σ, π, and antibonding n*, σ*, and π* orbitals, e.g. n $\rightarrow \pi$*, $\pi \rightarrow \pi$*, and $\sigma \rightarrow \sigma$*, etc. Additionally it is very convenient to classify the transitions as electrically or magnetically allowed or forbidden, i.e., $|\langle\vec{\mu}\rangle_{NK}|$ and $|\langle\vec{m}\rangle_{KN}|$ is different from or equal to zero.

More important for CD spectroscopy is the classification of chromophores with respect to symmetry operations of the second kind (i, σ, and S$_n$, with $n>2$). An inherently dissymmetric chromophore possesses no symmetry element of the second kind and thus is chiral by itself. In an inherently symmetric chromophore a symmetry element of the second kind exists and thus it is achiral by itself. In **Figure** 4 the enone chromophore is shown as inherently symmetric and inherently dissymmetric. The local symmetry of a

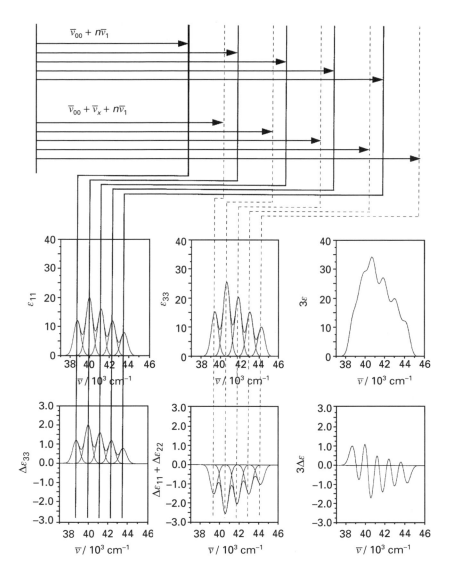

Figure 3 Term scheme for an absorption and the corresponding CD band with contributions of one allowed ($\bar{\nu} = \bar{\nu}_{00} + n\bar{\nu}_1$) and one forbidden ($\bar{\nu} = \bar{\nu}_{00} + \bar{\nu}_x + n\bar{\nu}_1$) vibrational progression. $\bar{\nu}_{00}$ is the 0–0 transition band. $\bar{\nu}_1$ and $\bar{\nu}_x$ are a totally symmetric and a non-totally symmetric vibration, respectively. $n = 1, 2, \ldots$.

chromophore cannot be sharply defined because the chromophore itself is only a qualitative quantity. For CD bands a characterization by the size of the dissymmetry factor g is possible. For an inherently dissymmetric chromophore the dissymmetry factor g is about or larger than 10^{-2}, for an inherently symmetric chromophore $g < 10^{-2}$ (mostly $\approx 10^{-4}$) if the transition is electrically allowed and magnetically forbidden. $g > 10^{-2}$ (mostly 5×10^{-3} to 10^{-1}) if the transition is magnetically allowed and electrically forbidden, also for an inherently symmetric chromophore.

What is a chromophore?

A chromophore is that part of the molecule where the absorption proceeds and where the main change of the geometry or electron density, etc. appears after

the excitation process. The area of the chromophore is not very well demarcated from the residual parts of the molecule which are not involved in the absorption process. The shading in **Figure 5** demonstrates the continuous variation from the centre of the excitation to the undisturbed residue of the molecule. A carbonyl chromophore with an $n\pi^*$-transition and an olefin chromophore with a $\pi\pi^*$-transition are regarded as two distinct chromophores (**Figure 5**). If the two chromophores are located in a molecule next to each other, an influence on the position of the band and on the intensity of the absorption results. If the interaction is strong enough, the two chromophores have to be considered as one new chromophore.

The number of chromophores in a molecule is very large because of the large number of transitions

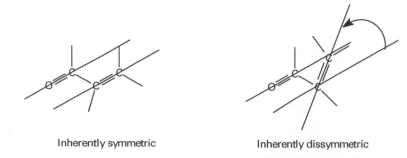

Inherently symmetric   Inherently dissymmetric

Figure 4 The inherently symmetric (left) and inherently dissymmetric (right) enone chromophore.

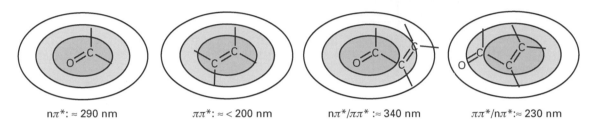

$n\pi^*: \approx 290$ nm $\pi\pi^*: \approx\ < 200$ nm $n\pi^*/\pi\pi^* :\approx 340$ nm $\pi\pi^*/n\pi^*:\approx 230$ nm

Figure 5 The enone chromophore as an example for the interaction of two chromophores, i.e. the 'en' and 'one' chromophore.

localized in different parts of the molecule. Also in a group, like the carbonyl group, there is an infinite number of chromophores. One important chromophore for CD spectroscopy besides the $n\pi^*$-transition is a $\pi\pi^*$-transition, located in the carbonyl area and at wavelength lower than 150 nm.

According to Snatzke a molecule can be divided into several spheres (**Figure 6**). The first sphere is the chromophore itself which can be either chiral or achiral. The second sphere is the neighbourhood to the chromophore, the third sphere the area following the second sphere, and so on. The criteria by which the spheres are differentiated are the distance and the interaction between the parts of the molecules like conjugation etc. Usually changes in the second sphere can affect the sign and size of the CE whereas substitutions in spheres farther away lead to smaller effects.

The quantitative description of the rotational strength

Methods for the description of the optical activity are either based on an estimation of the sum over the rotational strengths of all transitions weighted by a dispersion term as in the case of frequency-dependent optical rotation (ORD; denoted as ϕ or $[M]$) or of the rotational strength of one transition in the case of circular dichroism, denoted as $\Delta\varepsilon$ or θ (Eqn [32]). Numerical quantum-mechanical calculations of the rotational strength R^{NK} for a few transitions $|N\rangle \rightarrow |K\rangle$ of a molecule lead nowadays to results of

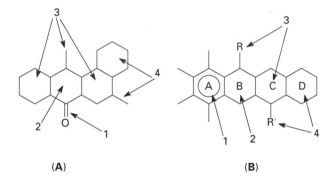

(A)   (B)

Figure 6 Two examples for a decomposition of a molecule into spheres according to Snatzke.

increasing reliability and thus a determination of the absolute configuration becomes possible with a CD measurement over the spectral region to which the analysed transitions belong. But this is only a development in the right direction, and further advances in computational abilities and resources are necessary. The possibility for numerical calculation is not always available in the chemist's daily work today. From this point of view there is no way to renounce the qualitative concepts and semi-qualitative methods used up to now to correlate sign and absolute configuration. Therefore, the following techniques will be discussed in more detail:

1. the quantum-mechanical calculation of the rotational strength,
2. the polarizability theory,
3. the one-electron mechanism,

4. the model of independent groups (MIG) of Tinoco,
5. the exciton model (exciton chirality method).

Quantum mechanical calculation of the rotational strength

In order to calculate the rotational strength, reasonably good wavefunctions for all electronic states should be available. Calculation with a number of semiempirical and *ab initio* techniques have been performed with more or less success. A review of used techniques has been given by Woody in Nakanishi *et al.* As discussed there, for the calculation of the rotational strength, the electric transition moments can be either evaluated with the dipole-velocity method ($\langle N|p_i|K\rangle$, with the linear momentum operator p_i) which leads to an origin-independent rotational strength R^{NK} but overestimates errors from inaccurate wavefunctions far away from the atomic centre, or with the dipole-length method ($\langle N|\mu_i|K\rangle$) which leads to an origin-dependent rotational strength R^{NK} and is not so sensitive to other errors of the wavefunctions. Up to now the dipole-velocity method has been the method of choice for numerical quantum-mechanical calculations. More recent analyses and calculations by Grimme do not support this choice, in contrast to the leading opinion. It seems to us that one may choose the dipole-velocity method or the dipole-length method depending on the kind of molecule and the quality of the calculated wavefunctions. Furthermore, the extent of taking into account singly and doubly excited configurations is a problem when trying to achieve good results for the direction of the electric and magnetic dipole transition moments and thus for the rotational strength. A new development can also be found in the use of the density functional theory for the calculation of R^{NK} by Grimme.

The polarizability theory

Kirkwood expressed the rotatory power in terms of the polarizabilities and anisotropies of the polarizabilities of groups in the molecule. First-order perturbation theory leads in this case to a perturbation potential expressing the interaction of two transition moment dipoles i, j at a distance R_{ij}, where this distance is large compared to the separation of charge in the dipoles. Such terms are important if the molecule contains two or more strong absorption bands near the frequency region where the rotatory dispersion is calculated originating from allowed electric dipole transitions. But even the strongest absorption bands may contribute little to the optical activity by this mechanism if the chromophoric groups are unfavourably disposed relative to each other. In the last few years there has been an improvement in the calculation of the optical rotation with Kirkwood's polarizability theory by using chirality functions as a possibility to find symmetry-adapted functions as shown by Haase and Ruch for asymmetric methane or allene derivatives. The thus obtained electric dipole transition moments are grouped together to polarizabilities or to polarizability tensor coordinates, respectively. The basis for the description is a second-order perturbation theory of the optical rotation.

The one-electron mechanism

Historically, the one-electron mechanism was introduced by Condon and co-workers as an optical activity which originates from the excitation of a single electron in a field of suitable dissymmetry. In the MO concept, the one-electron mechanism is an excitation of one electron to a singly-excited electronic configuration of an inherently symmetric chromophore perturbed by chiral surroundings. Depending on the symmetry of the chromophore the transition is either electrically and/or magnetically dipole allowed and without the perturbation fulfills the condition that the dot product of the electric and magnetic dipole transition moments of this transition is zero. By the perturbation with groups/atoms in chiral surroundings a magnetically and/or electrically allowed dipole is induced by borrowing intensity from other transitions of the same chromophore. Often the language 'intensity is stolen from other transitions' has been used in the literature, e.g. 'contributions to the rotational strength of the $n \to \pi^*$ transitions are stolen by this mechanism from electric-dipole-allowed $\sigma \to \sigma^*$ or $\pi \to \pi^*$ transitions'. There are two possibilities for a perturbation: On the one hand electronic interactions are responsible for the intensity borrowing which is often called the dynamic coupling model. On the other hand vibrational processes change the geometry and, by taking into account the dependence of $\langle \vec{\mu} \rangle_{0n}$ and/or $\langle \vec{m} \rangle_{n0}$ on nuclear coordinates, intensity is borrowed by the transition $N \to K$ via nuclear vibrations from other allowed transitions of the chromophore. Then different vibrational progressions contribute to the UV and CD spectra as shown in **Figure 3**. This mechanism is often called the static coupling or vibronic coupling model. Depending on the type of transitions, a nomenclature system has been introduced by Moscowitz and co-workers and Weigang where the CD of a transition (a) without intensity borrowing is called a case I CD, (b) with intensity borrowing via one forbidden electronic dipole transition is called a case II CD, and (c) with intensity borrowing via electrically and magnetically forbidden transitions is called a case III CD. Whereas in the first case only progressions with totally symmetric vibrations contribute to the CD, in cases II and III progressions with non-totally symmetric vibrations and vibrational CD bands of the same or different sign can also contribute to an electronic CD band as shown by Weigang and in Snatzke by Moscowitz.

The model of independent groups for the calculation of the rotational strength of large molecules or polymers

Tinoco's aim with the model of independent groups (MIG method) was to calculate the total rotational strength R^{NK} for a transition $|N\grave{O} \rightarrow |K\grave{O}$ for a large molecule or a polymer possessing a large number of equal and different groups, i.e. chromophores for which the circular dichroism is given by

$$
\begin{aligned}
\Delta\varepsilon(\bar{\nu}) &= \frac{B\bar{\nu}}{3}\sum_K R^{NK} \sum_n \sum_k |\langle n \mid k\rangle|^2 \, F^{NnKk}(\bar{\nu}) \\
&= \sum_K R^{NK} F^{NK}(\bar{\nu})
\end{aligned}
\tag{38}
$$

The excited state may or may not be degenerate or accidentally degenerate. In the case of an n-fold degeneracy of the state $|A\grave{O}$, the excited states $|L\grave{O}$, linear combinations of the degenerate states $|A\grave{O}$, contribute to the rotational strength (Eqn [39]):

$$
R^{0A} = \sum_{L=1}^{n} R^{NL} = \sum_{L=1}^{n} \mathrm{Im}\big\{\langle\vec{\mu}\rangle_{NL} \cdot \langle\vec{m}\rangle_{LN}\big\}
\tag{39}
$$

For the evaluation of the rotational strength the molecule is decomposed into different groups i which are at a distance $|\vec{R}_{ij}| = |\vec{R}_i - \vec{R}_j|$ and oriented differently with respect to each other. The operator of the interaction between two groups i, j is given by the potential $\hat{V}_{ij}$. Depending on the assumption for the interaction between the groups, different approximations for the wavefunctions of the groups are necessary. In the simplest case of completely isolated groups the wave function of the molecule can then be expressed as a product of the wavefunctions of the groups. It is assumed that the group electronic wavefunctions do not overlap, which means that there is no electron exchange between groups. Only the interaction between two groups at a time is considered. Furthermore, it is assumed that singly and doubly excited states on different groups are possible.

To calculate the total rotational strength for a transition it is necessary to describe the electric and magnetic dipole transition moment operator in an adequate way. Whereas the electric moment operator is the sum of the moments of the different groups, for the magnetic moment operator two contributions have to be taken into account. Analogously to the electric moments, there is a contribution from the magnetic moment of every group. The second contribution comes from the interaction of groups. In a simplified picture a 'moving electron' in one group leads to a magnetic moment in the other group and gives rise to a large contribution to the rotational strength, which resembles the exciton coupling and the coupled oscillator model as discussed in a later section. The rotational strength of a non-degenerate system with the assumptions given above is obtained as follows:

$$
\begin{aligned}
R^{0A} = \sum_i \Bigg[& \mathrm{Im}\langle\vec{\mu}\rangle_{i0a} \cdot \langle\vec{m}\rangle_{ia0} \\
& -2\sum_{j\neq i}\sum_{b\neq a} \frac{\mathrm{Im}\, V_{i0a,j0b}\big\{\langle\vec{\mu}\rangle_{i0a}\cdot\langle\vec{m}\rangle_{jb0}\bar{\nu}_a + \langle\vec{\mu}\rangle_{j0b}\langle\vec{m}\rangle_{ia0}\bar{\nu}_b\big\}}{h(\bar{\nu}_{b0}^2 - \bar{\nu}_{a0}^2)} \\
& -\sum_{j\neq i}\sum_{b\neq a} \frac{\mathrm{Im}\, V_{iab,j00}\big\{\langle\vec{\mu}\rangle_{i0a}\cdot\langle\vec{m}\rangle_{ib0} + \langle\vec{\mu}\rangle_{i0b}\cdot\langle\vec{m}\rangle_{ia0}\big\}}{h(\bar{\nu}_{b0} - \bar{\nu}_{a0})} \\
& -\sum_{j\neq i}\sum_{b\neq a} \frac{\mathrm{Im}\, V_{i0b,j00}\big\{\langle\vec{\mu}\rangle_{i0a}\cdot\langle\vec{m}\rangle_{iba} + \langle\vec{\mu}\rangle_{iab}\cdot\langle\vec{m}\rangle_{ia0}\big\}}{h\bar{\nu}_{b0}} \\
& -\sum_{j\neq i} \frac{\mathrm{Im}\, V_{i0a,j00}\big\{\langle\vec{\mu}\rangle_{iaa} - \langle\vec{\mu}\rangle_{i00}\cdot\langle\vec{m}\rangle_{ia0}\big\}}{h\bar{\nu}_{a0}} \\
& -\frac{2\pi}{c}\sum_{j\neq i}\sum_{b\neq a} \frac{V_{i0a,j0b}\bar{\nu}_{0a}\bar{\nu}_{0b}\big\{(\langle\vec{R}_j\rangle - \langle\vec{R}_i\rangle)\cdot(\langle\vec{\mu}\rangle_{j0b}\times\langle\vec{\mu}\rangle_{i0a})\big\}}{h(\bar{\nu}_{b0}^2 - \bar{\nu}_{a0}^2)} \Bigg]
\end{aligned}
\tag{40}
$$

The coupling $V_{isr,jtv}$ between the groups i, j with their singly- or doubly-excited configurations r, s and t, v is given by the matrix elements

$$V_{irs,jtv} = \int \varphi_{ir}\varphi_{is}\hat{V}_{ij}\varphi_{jt}\varphi_{jv}\,\mathrm{d}\tau \qquad [41]$$

The potential energy operator $\hat{V}_{ij}$ of Equation [41] stands for the electrostatic interaction between the groups i and j. φ_{ir}, φ_{is}, ... etc. are group wavefunctions, where the first index represents the group and the second index the electronic state of the group r, s, ... in i and t, v, ... in j. $\langle\vec{\mu}\rangle_{i0a}$ and $\langle\vec{m}\rangle_{ia0}$ are the electric and the magnetic dipole transition moments of an excitation of a group i for a transition from the ground state $|0\rangle$ to the excited state $|a\rangle$ ($|0\rangle \to |a\rangle$). $\bar{\nu}_{a0}$ and $\bar{\nu}_{b0}$ are the energies given in wavenumbers of the transitions $|0\rangle \to |a\rangle$ and $|0\rangle \to |b\rangle$, respectively.

The first term of Equation [40] contributes to the optical activity of a molecule if the transition $|N\rangle \to |K\rangle$ is electrically and magnetically allowed and the chromophore is inherently dissymmetric. This term represents the optical activity of isolated groups. In the second term the electric and magnetic dipole transition moments on different groups contribute by coupling the groups in their excited states by electrostatic interaction. The interaction of electric and magnetic dipole transition moments of different transitions of one isolated group are described with the third and the fourth term of Equation [40]. The fifth term shows the dependence on the difference of the electric dipole moment of the ground and excited state. This term is – as the fourth term – very small because of the division by $\bar{\nu}_{a0}$ instead of a difference like $\bar{\nu}_{a0}^2 - \bar{\nu}_{b0}^2$. The coupling of the electric dipole transition moments of two different groups in the sixth term represents the exciton model of two groups if their excited states are non-degenerate. This is the term which is the origin for the exciton chirality method of Harada and Nakanishi if the two excited states of the coupled electric transition moments are obtained by absorption in different spectral regions. This term corresponds to the theory of coupled oscillators. The interaction between degenerate groups does not contribute to the rotational strength of a polymer here. This effect is discussed in the last section for a system of two chromophores.

Rules for the determination of the absolute configuration

Chirality functions

For molecules with an inherently symmetric chromophore or molecules which can be formally

decomposed into an achiral skeleton and achiral groups, so-called ligands, chirality functions have been developed which allow a quantitative description of chiral phenomena as a function of phenomenological parameters of the ligand. Additionally these functions allow the correlation of the sign of the chirality phenomenon and the absolute configuration of the molecule.

A systematic analysis of such chirality functions, taking advantage of the symmetry properties of the skeleton, has been given by Ruch and Schönhofer. The ligands, the number of which is determined by the chosen binding sites and the symmetry of the skeleton, are at positions where symmetry operations applied to the skeleton permute the ligands on the different possible binding sites and in this way create new distinguishable molecules. The binding site of a ligand is correlated with an argument for the ligands in the function, i.e. the arguments $\lambda(l_1)$, $\lambda(l_2)$, etc. correspond to the positions 1, 2, etc. in **Figure 7**. In the approximation of the so-called 'qualitative completeness', the chirality function $\chi(l_1,l_2,l_3,l_4)$ for allene derivatives with a skeleton of D_{2d} symmetry and with four binding sites for the ligands which depend on two phenomenological parameters $\lambda(l_i)$, $\mu(l_i)$ for each ligand l_i can be given by

$$\begin{aligned}\chi(l_1,l_2,l_3,l_4) = {} & \eta_1[\lambda(l_1) - \lambda(l_2)][\lambda(l_3) - \lambda(l_4)] \\ & + \eta_2[\mu(l_1) - \mu(l_2)][\mu(l_1) - \mu(l_3)] \\ & \times [\mu(l_1) - \mu(l_4)][\mu(l_2) - \mu(l_3)] \\ & \times [\mu(l_2) - \mu(l_4)][\mu(l_3) - \mu(l_4)] \quad [42]\end{aligned}$$

η_1 and η_2 are two additional phenomenological parameters. For the optical rotation the phenomenological parameters for a large number of ligands have been determined. Ligand parameters of one and the same ligand are different if they are bound to achiral skeletons of different symmetry and different sites, i.e. a determined parameter for one special ligand can only be taken for compounds with identical skeletons. There are further restrictions for the

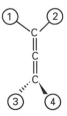

Figure 7 Achiral skeleton with D_{2d} symmetry and the four substitution positions 1 to 4 for an allene.

description of the optical rotation by chirality functions because of some experimental and theoretical reasons. One restriction is the assumption that the ligands at a binding site of the skeleton have to be invariant under the symmetry operation of the site group. This means that a ligand at a binding site must possess sufficient symmetry to make all properties invariant under the symmetry operation of the skeleton. In spite of this restriction, the chirality function allows us in an excellent way to prove whether molecular theories are suitable to describe chirality phenomena or they allow us to write these theories in a way that we can see immediately that they describe a chirality phenomenon. An example of such an application is the rewriting of Kirkwood's polarizability theory of the optical rotation of methane and allene derivatives by Haase and Ruch.

Sector rules

In order to correlate the absolute configuration of a molecule with the sign of the Cotton effect of a transition, empirical or theoretical rules are needed. One form of rules are the so-called sector rules which are applicable to inherently symmetric skeletons/chromophores. In this case the molecules have to be decomposed into a symmetric skeleton/chromophore and perturbing atoms. The potential V of the perturbing atoms is expressed as shown in Schellman's general theory for sector rules by symmetry-adapted functions as

$$V = \sum_{v,i} V_i^v \qquad [43]$$

where the V_i^v belong to the ith row of the vth representation of the symmetry group of the inherently symmetric skeleton. The circular dichroism in the chromophore is then induced by the term V_i^v of Equation [43] which belongs to the pseudoscalar representation of the symmetry point group of the chromophore. With this perturbation within the one-electron theory of Condon, Altar, and Erying the induced CD can be evaluated as a function of the positions of the perturbing atoms or groups. A plausible derivation according to Schellman's description for a chromophore of C_{2v}/D_{2h} symmetry represented by the plane shown in **Figure 8** can be done as follows: A perturbing atom at the position $P(x, y, z)$ (right) makes the entire system chiral with a certain handedness. The molecule is chiral as long as the perturbing atom is not positioned in one of the symmetry planes of the achiral chromophore, e.g. in $P(0, y, z)$ (middle). When the perturbing atom is shifted to the position $P(-x, y, z)$ (left), the mirror image of

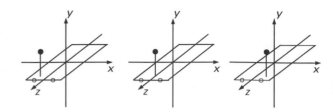

Figure 8 Scheme for developing a sector rule.

the system in the right part of the figure is obtained. For the resulting molecule a sign change in the Cotton effect has to be measured because this molecule is the enantiomer of the initial molecule. If the existence of chiral zero points can be excluded the sign of the CE can only be changed when the perturbing atom crosses one of the symmetry planes of the model molecule (**Figure 8**). All atoms – except fluorine – induce the same sign for the Cotton effect. This means, however, that a sector rule for this model exists which is a quadrant rule for C_{2v} and an octant rule for D_{2h} symmetry. The quadrant rule can change to an octant rule if it is taken into account that, e.g. a new apparent symmetry plane can be introduced into the molecule by the nodal plane of the π^* orbital of the carbonyl group. The question whether experimentally a quadrant or an octant rule has to be used for the carbonyl chromophore has been discussed in the literature for a very long time.

Figure 9 shows six examples of inherently symmetric chromophores in which the region where atoms induce a Cotton effect of different sign are indicated as black or white. The higher the symmetry the larger is the number of sign changes when moving a perturbing atom in space. As a consequence, rules for chromophores with high symmetry are not very reliable.

For the most popular rule, the octant rule for the carbonyl chromophore, an application is

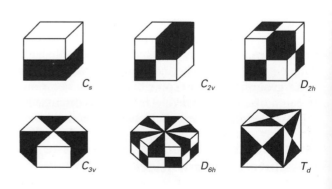

Figure 9 Schematic representation of the sector rules for the skeletons or chromophores of C_s, C_{2v}, D_{2h}, C_{3v}, D_{6h}, and T_d symmetry according to Schellman.

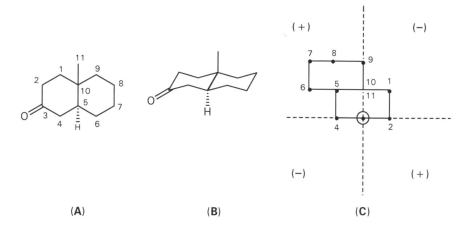

Figure 10 Octant projection of a compound with a carbonyl group showing a CE induced by the second sphere. Each atom outside the symmetry planes contributes to the CD. The contributions of atoms which are in mirror image positions compensate.

demonstrated in **Figure 10** for a cyclic ketone. To perform a sector rule the region around the chromophore is divided into sectors by symmetry or nodal planes: Atoms in a symmetry or a nodal plane do not contribute to the CD, the contributions of atoms in neighbouring sectors are always opposite in sign. Spectra of an aromatic system where a sector rule can be applied to the third sphere are given as further examples in **Figure 11**.

Helicity rules

Sector rules fail if a chromophore is inherently dissymmetric and the rotational strength is only determined by its internal structure. From the definition of a chromophore given above it is easy to imagine that the delimitation of an inherently dissymmetric from an inherently symmetric chromophore is not straightforward. Furthermore, it is difficult to choose atoms in a molecule with which a helical arrangement can be approximately defined. If there is at least one full turn of a helix constituted by a dissymmetric arrangement of groups or atoms that are positioned exactly on a helical line, then the sense of helicity can be assigned unequivocally and a correlation of the sign of the CE can be experimentally or theoretically determined to derive a general rule. Otherwise only a helicity with atoms which are approximately positioned on a helical line can be determined. It has been pointed out in Janoschek by Snatzke that such an approximately determined helicity of a molecule – for which a helicity rule can be applied – depends on the viewing angle. In general this means, in order to find a helical order in a molecule, one has to select a direction in a chromophore and then has to choose atoms within the chromophore which lay approximately on a helical line with respect to the chosen viewing angle. But the question is which viewing

angle has to be chosen, because there are many directions for the selection of an approximate helical order for which one can try to develop a helicity rule. In most cases chemical intuition from experimental experience can help to solve this problem. In some cases the CD of an oriented chromophore in an anisotropic sample (ACD) can help to decide whether there is a helical structure along a special direction or not as discussed by Kuball and Höfer.

Helicity rules have been developed which can be applied to inherently dissymmetric chromophores, such as, e.g. the enone chromophore shown in **Figures 12** and **13**, where the chosen atoms are only approximately on a helical line. Many papers have been published concerning the enone chromophore as summarized in Patai and Rappoport by Gawronski.

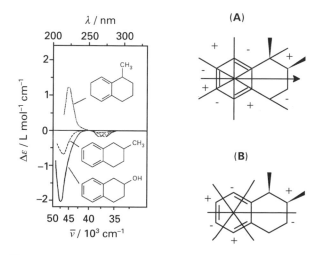

Figure 11 The benzene chromophore as an example for the application of sector rules for the third sphere. On the right the schematic representation of the sector rules for the long wavelength ^1L_b transitions (A) and for the short wavelength ^1L_a transitions (B) are given. The plane of the aromatic system is a nodal plane.

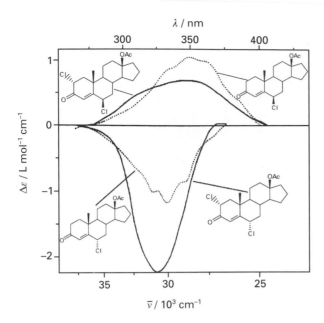

λ / nm

Figure 12 The enone chromophore as an example for the application of helicity rules.

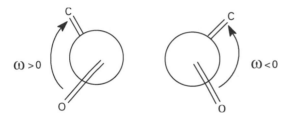

Figure 13 Helicity of the enone chromophore.

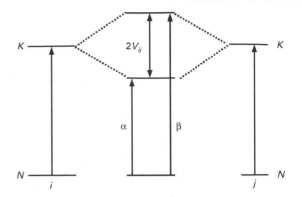

Figure 14 Davidov splitting $2V_{ij}$ of the two degenerate excited states of two interacting identical chromophores in a compound of C_2 symmetry for $\vartheta < \vartheta_0$.

In the simple cases the torsional angle of the enone chromophore determines the sign of the Cotton effect where for a P helix ($\omega > 0$) and for an M helix ($\omega < 0$) a positive and a negative CE for the $n\pi^*$ transition result, respectively. Substitution at the chromophore may have dramatic effects and can be responsible for the requirement of a modification of the rule.

The exciton chirality method

Two or more chromophores located nearby in space can constitute a chiral system. An electric dipole–electric dipole interaction between the chromophores shifts and splits the energy levels of their excited states with respect to the interaction-free states. In the case of two identical chromophores, e.g. in a molecule with at least, or even approximately, the point symmetry group C_2, the degeneracy is removed (**Figure 14**).

As a consequence the UV spectrum is built up from two transitions with different transition moment directions as shown for βR, $\beta' R$-dimethyl-mesobilirubin-XIIIα (**Figure 15**) which gives rise to a frequency-dependent degree of anisotropy R (**Figure 16**). In the case of a molecule of C_2 symmetry the transitions are polarized perpendicular to each other, one transition parallel to the C_2 symmetry axis and the other one in the plane perpendicular to this symmetry axis. For both of these transitions, using the shape of the UV band of the transition of the dipyrrinone fragment, the absorptions $\varepsilon^{\alpha}(\bar{\nu})$ and $\varepsilon^{\beta}(\bar{\nu})$ (**Figure 17**) for βR, $\beta' R$-dimethylmesobili-rubin-XIIIα are given by

$$\varepsilon(\bar{\nu}) = F^{NK}\left(\varepsilon^{\alpha}(\bar{\nu}) + \varepsilon^{\beta}(\bar{\nu}) - \left(\varepsilon^{\alpha}(\bar{\nu}) - \varepsilon^{\beta}(\bar{\nu})\right)\cos\vartheta\right)$$

[44]

where ϑ is the angle between the dipole transition moments of the fragments and F^{NK} is a factor which takes into account a change of intensity of the monomer band by the interaction in the molecule (**Figure 17** left) built up of the two fragments. In the spectral regions of the absorption bands belonging to the exciton transitions, a positive and a negative CE appears as shown in **Figure 17** (right). Both CEs superpose to a couplet which can be described by bands possessing the shape of the UV spectra as follows

$$\Delta\varepsilon(\bar{\nu}) = \frac{4R^{NK\alpha}}{D^{NK\alpha}}\left(\varepsilon^{\alpha}(\bar{\nu}) - \varepsilon^{\beta}(\bar{\nu}) - (\varepsilon^{\alpha}(\bar{\nu}) + \varepsilon^{\beta}(\bar{\nu}))\cos\vartheta\right)$$ [45]

where $R^{NK\alpha}$ is the rotational strength of the transition α (**Figure 14**) and $D^{NK\alpha}$ the corresponding

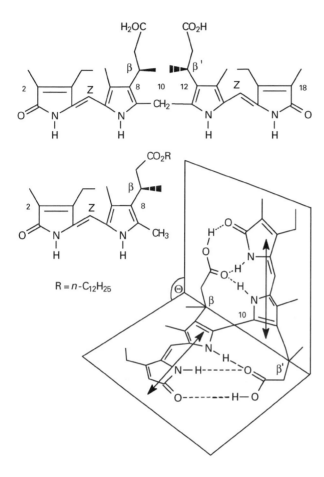

Figure 15 Linear and folded structure of $\beta R, \beta' R$-dimethylmes-obilirubin-XIIIα of M helicity and the dipyrrinone ester as the building unit (fragment). Θ is the angle between the mean molecular planes of the dipyrrinone units. The arrows indicate the transition moment directions within the dipyrrinone units which enclose the angle ϑ. The x_2^* axis is chosen to be parallel to the C_2 symmetry axis which is the angle bisector through C10 between the mean molecular planes of the dipyrrinone units. In the solution there is an excess of the P conformer in $P \rightleftarrows M$.

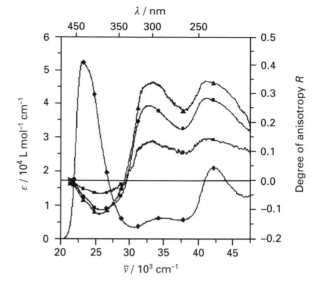

Figure 16 The UV spectrum ($\blacklozenge$) ($T = 80°C$) and the temperature dependent degree of anisotropy of $\beta R, \beta' R$-dimethylmesobi-lirubin-XIIIα in ZLI-1695 ($T = 28°C$ ($\blacktriangle$), 43°C($\bullet$), and 65°C ($\blacksquare$)).

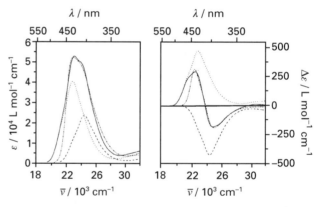

Figure 17 Exciton coupling of the two chromophores of $\beta R, \beta' R$-dimethylmesobilirubin-XIIIα dissolved in ZLI-1695 in the CD and UV spectrum. The dotted and the dashed lines are the calculated contributions of the two exciton bands. The dashed–dotted lines are the calculated CD and UV spectra.

dipole strength. The fitting of the CD and UV bands of $\beta R, \beta' R$-dimethylmesobilirubin-XIIIα (**Figure 15**) according to Equations [44] and [45] allows to determine the angle ϑ between the dipole transition moments of the fragments.

Using a simple version of the exciton theory, the exciton couplet built up by the CD curves of two identical chromophores i and j of different signs can be described assuming spectral functions of Gaussian form by

$$\Delta \varepsilon(\bar{\nu}) = \frac{2\sqrt{\pi}\bar{\nu}_0^2}{2.296 \times 10^{-39}\Delta\bar{\nu}^2}\left[\frac{\bar{\nu}_0 - \bar{\nu}}{\Delta\bar{\nu}}\right]$$
$$\times \exp\left\{-\left[\frac{\bar{\nu}_0 - \bar{\nu}}{\Delta\bar{\nu}}\right]^2\right\}\vec{R}_{ij} \cdot \left(\langle\vec{\mu}\rangle_{iNK} \times \langle\vec{\mu}\rangle_{jNK}\right)V_{ij} \quad [46]$$

Here, ν_0 is the centre of the exciton bands and $\Delta\nu$ the half band width of the single exciton bands α and β. The transition moment of the transition α or β can be expressed by

$$\langle\vec{\mu}\rangle^{\alpha,\beta} = \frac{1}{\sqrt{2}}\left(\langle\vec{\mu}\rangle_{iNK} \mp \langle\vec{\mu}\rangle_{jNK}\right) \quad [47]$$

where the upper sign belongs to the α and the lower sign to the β band. The pseudoscalar product $\vec{R}_{ij}\cdot(\langle\vec{\mu}\rangle_{iNK} \times \langle\vec{\mu}\rangle_{jNK})$ in Equation [46] determines the sign of the couplet. The Davidov splitting $2V_{ij}$

(**Figure 14**) between the energy states N, K of the groups i and j is given by

$$V_{ij} = \frac{(\langle \vec{\mu} \rangle_{iNK} \cdot \langle \vec{\mu} \rangle_{jNK}) - 3R_{ij}^{-2}(\langle \vec{\mu} \rangle_{iNK} \cdot \vec{R}_{ij})(\langle \vec{\mu} \rangle_{jNK} \cdot \vec{R}_{ij})}{R_{ij}^3} \qquad [48]$$

where $\vec{R}_{ij}$ is a vector pointing from group i to group j and the Davidov splitting is smaller than the half-band width. For derivation of the exciton theory it is assumed that the interacting groups are far away from each other and thus the distance of the chromophores $R_{ij} = |\vec{R}_{ij}|$ can be chosen independently from the extension of the chromophores. This is a presupposition not fulfilled in most cases of application of the theory. Furthermore, one has to take into account an effect given above but often not realized in the literature: A change of sign of the couplet appears without a change of the helicity if the sequence of the states resulting from the Davidov splitting $2V_{ij}$ (Eqn [48]) is interchanged (**Figure 14**; sign change of V_{ij}). The sign change depends on the angle ϑ between the transition moment directions. For $V_{ij} = 0$ there is a chiral zero of the 'chirality function $\Delta\varepsilon(\bar{\nu})$' (Eqn [46]) i.e. $\Delta\varepsilon = 0$, in spite of the fact that the system is chiral. V_{ij} is zero when $\vartheta = \vartheta_0$ and $\cos \vartheta_0 = R_{ij}^{-2} |\langle \vec{\mu} \rangle_{iNK}|^{-1} \cdot |\langle \vec{\mu} \rangle_{jNK}|^{-1} (\langle \vec{\mu} \rangle_{iNK} \cdot \vec{R}_{ij})(\langle \vec{\mu} \rangle_{jNK} \cdot \vec{R}_{ij})$. If ϑ is about 90° and thus near a chiral-zero position only an experimental assignment of the α and β transitions can help to avoid an incorrect conclusion. This is also of importance because the vector $\vec{R}_{ij}$ is strongly dependent on the origin not only in its length but also in its orientation with respect to the molecular skeleton. This fact makes it difficult to estimate the sign of the couplet because of the chiral zero ($V_{ij} = 0$) given by the uncertainty of the chosen ϑ. This problem has to be considered from the point of view that in most applications of the exciton chirality method the theoretical condition that the distance between the chromophores is large compared to their extension is not well fulfilled.

Taking these discussed problems into account, the following rule holds: Positive (P) and negative (N) helicities lead always to positive (positive CE at the longer wavelength side and negative CE at the shorter wavelength side) and negative couplets of the CD as long as the angle ϑ between the transition moment direction is smaller than ϑ_0 with $0 \leq \vartheta \leq \vartheta_0$. Here the helicity of the structure is defined as positive, when the geometrical arrangement gives rise to a clockwise rotation of the electric dipole transition moment in the foreground if looking along the axis about which the rotation is performed and vice versa. Choosing suitable values for $|\vec{R}_{ij}|$ and taking care that the dipole strength of the absorption band of the exciton couplet is approximately twice the dipole strength of the monomer's absorption which build up the inherently dissymmetric molecule, reliable conclusions for the absolute configuration can be obtained. So far, the exciton chirality method is a useful method for establishing absolute configurations and conformations of organic compounds in solution.

In order to improve the reliability of the conclusion from the CD and UV measurements of isotropic solutions, the assignment of the exciton transitions should be determined experimentally. For the evaluation of the correct dihedral angle, polarized spectroscopy can be used as a very suitable additional information for the assignment. Since, in most cases, this information has not been taken into account, the value of such an analysis is demonstrated in the following with βR, $\beta' R$-dimethylmesobilirubin-XIIIα and its constituting unit dipyrrinone (**Figure 15**) according to Bauman and co-workers. Due to the assumed C_2 symmetry of βR, $\beta' R$-dimethylmesobilirubin-XIIIα the transition moments either lie parallel to the C_2 symmetry axis or in a plane perpendicular to it.

From the degree of anisotropy R in **Figure 14** it follows that the β and the α transitions are of $A \rightarrow A$ and $A \rightarrow B$ symmetry, respectively, for $\vartheta < \vartheta_0$. From the temperature dependence of the degree of anisotropy (**Figure 16**) with the help of a computational simulation the tensor coordinates of the absorption tensor and the dipole strength tensor D_{ii}^* with respect to the principal axes of the order tensor can be obtained. Furthermore, information about the orientational order of the molecule in the used anisotropic system as stretched polymers or ordered liquid crystal phases can be obtained too.

To determine the transition moment directions in the dipyrrinone units and of the α-transition of βR, $\beta' R$-dimethylmesobilirubin-XIIIα which are not determined by symmetry, the ratio V (Eqn [49], derived from a simple version of the exciton theory) was used as further information to select possible mathematical solutions for the transition moment direction of the dipyrrinone unit.

$$V = \frac{D_{22}^{*\beta}}{D_{11}^{*\alpha} + D^{*\alpha}} = \frac{1 + \cos \vartheta}{1 - \cos \vartheta} \approx \frac{\varepsilon^\beta(\bar{\nu}_{\max})}{\varepsilon^\alpha(\bar{\nu}_{\max})} \qquad [49]$$

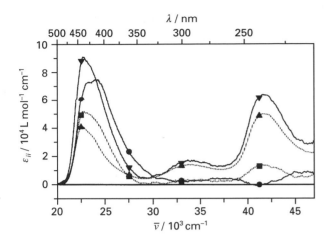

Figure 18 The tensor coordinates ε_{11}^*(······■······), ε_{22}^* (——●——), ε_{33}^*(······▲······) of the absorption tensor and the sum $\varepsilon_{11}^* + \varepsilon_{33}^*$ (——▼——) of $\beta R, \beta'$ R-dimethylmesobilirubin-XIIIα in ZLI-1695. $\varepsilon_{11}^* + \varepsilon_{33}^*$ is the long wavelength (α band: A → B) and ε_{22}^* is the short wavelength (β band: A → A) exciton band.

This ratio determined from the $\varepsilon^\alpha(\bar{\nu})$ and $\varepsilon^\beta(\bar{\nu})$ curves of **Figure 17** and the $\varepsilon_{22}(\bar{\nu})$ and $\varepsilon_{11}(\bar{\nu}) + \varepsilon_{33}(\bar{\nu})$ curves from **Figure 18** allows to check the reliability of the experimental results obtained from UV and CD spectroscopy in isotropic media and from polarized UV spectra. In the example given here, the ratio V from both results is equal within 30%, which is in fair agreement with the expected accuracy of the determined reduced spectra. In **Figure 18** the reduced UV spectra for the exciton coupling are given. The analysis for this special example demonstrates how useful it can be to determine and analyse polarized spectroscopy for the use of the exciton chirality method. In particular, the angle of the chiral zero can be estimated to avoid errors resulting from interchanging the position of the α and β states in the term scheme (**Figure 14**).

List of symbols

c = velocity of light; c = concentration in mol L⁻¹; d = optical path length; D^{NnKk} = dipole strength for the transition $|Nn\rangle \rightarrow |Kk\rangle$; D_{ij}^{NnKk} = transition moment tensor for the transition $|Nn\rangle \rightarrow |Kk\rangle$; $-e$ = charge of the electron; E_n = energy of the state $|n\rangle$; f = orientational distribution function; F^{NK} = factor taking into account a change of intensity of the monomer bands; F^{NnKk} = spectral function for the absorption band of the transition $|Nn\rangle \rightarrow |Kk\rangle$; g = dissymmetry factor; G^{NnKk} = spectral function of the vibronic transition $|Nn\rangle \rightarrow |Kk\rangle$; h, $\hbar = h/2\pi$ = Planck's constant; H'_{21} = matrix element; l_i = ligand; $\langle m_r \rangle_{n0}$ = coordinate of the magnetic dipole transition

moment for the transition $|0\rangle \rightarrow |n\rangle$; m = mass of the electron; $\langle \vec{m}_{ia0} \rangle$ = magnetic dipole transition moment of group i for the transition $|0\rangle \rightarrow |a\rangle$; $\langle \vec{m} \rangle_{NnKk}$ = magnetic dipole transition moment for the transition $|Nn\rangle \rightarrow |Kk\rangle$; M_{ii} = optical rotation tensor; $[M]$ = molar rotation; n, k = vibrational states; $n_{1,2, L,R}$ = refractive indices for linearly or circularly polarized light; N = number of molecules in a volume V; N_A = Avogadro's number; N,K = electronic states; p_i = operator of the linear momentum; $\langle Q_{sj} \rangle_{n0}$ = electric quadrupole transition moment for the transition $|0\rangle \rightarrow |n\rangle$; R = degree of anisotropy; R^{0A} = rotational strength for the transition $|0\rangle \rightarrow |A\rangle$; R^{NnKk} = rotational strength for the electronic transition $|Nn\rangle \rightarrow |Kk\rangle$; $\langle \vec{R}_i \rangle$ = distance vector of group i from the origin of the coordinate system; $\vec{R}_{ij}$ = distance vector between two groups i, j; R_{ij}^{NnKk} = rotational strength tensor for the transition $|Nn\rangle \rightarrow |Kk\rangle$; $V_{irs, jtv}$ = interaction potential between groups i, j; V = potential of the perturbing atoms; $\hat{V}_{ij}$ = potential energy operator; $2V_{ij}$ = Davidov splitting; x_i, x'_i = axes of the molecule fixed or space fixed coordinate system; α, β, γ = Eulerian angles; δ_{ij} = Kronecker symbol; $\Delta \varepsilon$ = CD of an isotropic system; $\Delta \varepsilon_{ii}(i = 1, 2, 3)$ = coordinates of the CD tensor; $\Delta \varepsilon_{ij}$ = circular dichroism tensor; ε_{ij} = absorption tensor; ε_{ijk} = Levi Civita tensor; ε = molar decadic absorption coefficient; $\varepsilon_{1,2, L,R}$ = absorption coefficients for linearly or circularly polarized light; η_1, η_2 = phenomenological parameters; η = positive damping parameter representing the natural line shape; θ = ellipticity; $\theta_{LB,CD}$ = ellipticity related to linear birefringence or circular dichroism; ϑ = angle between the electric dipole transition moments of two fragments; λ = wavelength; $\lambda_{Nn}, \lambda_{Kk}$ = librational states; $\lambda(l_i), \mu(l_i)$ = phenomenological parameters for ligand l_i; $\langle \mu_r \rangle_{0n}$ = coordinate of the electric dipole transition moment for the transition $|0\rangle \rightarrow |n\rangle$; $\langle \vec{\mu} \rangle_{NnKk}$ = electric dipole transition moment for the transition $|Nn\rangle \rightarrow |Kk\rangle$; $\langle \vec{\mu} \rangle_{i0a}$ = electric dipole transition moment of group i for the transition $|0\rangle \rightarrow |a\rangle$; $\bar{\nu}$ = wavenumber; $\bar{\nu}_x$ = wavenumber for the (k^{th}+1) band of a progression; $\bar{\nu}_{00}$ = wavenumber for the origin of a progression; $\bar{\nu}_0$ = centre of the exciton bands; ϕ = optical rotation; $\phi_{LD,CB}$ = rotation related to linear dichroism or circular birefringence; $\hat{\phi}$ = complex optical rotation; φ_{ij} = group wave functions; $\chi(l_1, l_2, l_3, l_4)$ = chirality function; $\omega = 2\pi c\bar{\nu}$ = frequency.

See also: **Biochemical Applications of Raman Spectroscopy; Biomacromolecular Applications of Circular Dichroism and ORD; Chiroptical Spectroscopy, Emission Theory; Chiroptical Spectroscopy, Oriented Molecules and Anisotropic Systems;**

Circularly Polarized Luminescence and Fluorescence Detected Circular Dichroism; Induced Circular Dichroism; Magnetic Circular Dichroism, Theory; ORD and Polarimetry Instruments; Rotational Spectroscopy, Theory; Vibrational CD Spectrometers; Vibrational CD, Applications; Vibrational CD, Theory.

Further reading

Barron LD (1982) *Molecular Light Scattering and Optical Activity*. Cambridge: Cambridge University Press.

Bauman D, Killet C, Boiadjiev SE, Lightner DA, Schönhofer A and Kuball H-G (1996) Linear and circular dichroism spectroscopic study of β, β'-dimethylmesobilirubin-XIIIα oriented in a nematic liquid crystal. *Journal of Physical Chemistry* 100: 11546–11558.

Charney E (1979) *The Molecular Basis of Optical Activity – Optical Rotatory Dispersion and Circular Dichroism*. New York: Wiley.

Condon EU, Altar W and Eyring H (1937) One-electron rotatory power. *The Journal of Chemical Physics* 5: 753–775.

Grimme S (1996) Density functional calculations with configuration interaction for the excited states of molecules. *Chemical Physics Letters* 259: 128–137.

Gō N (1965) Optical activity of anisotropic solutions I. *Journal of Chemical Physics* 43: 1275–1289.

Haase D and Ruch E (1973) Quantenmechanische Theorie der optischen Aktivität der Methanderivate im Transparenzgebiet. *Theoretica Chimica Acta* 29: 189–234.

Harada N and Nakanishi K (1983) *Circular Dichroic Spectroscopy – Exciton Coupling in Organic Stereochemistry*. Oxford: Oxford University Press.

Janoschek R (ed) (1991) *Chirality – From Weak Bosons to the α-Helix*. Berlin: Springer Verlag.

Kuball H-G and Höfer T (1999) Chiroptical spectroscopy, oriented molecules and anisotropic systems. In Lindon J, Tranter G and Holmes JL (eds) *Encyclopedia of Spectroscopy and Spectrometry*. London: Academic Press.

Kuball H-G, Karstens T and Schönhofer A (1976) Optical activity of oriented molecules II. Theoretical description of the optical activity. *Chemical Physics* 12: 1–13.

Kuball H-G, Altschuh J and Schönhofer A (1979) Optical Activity of Oriented Molecules III. The absorption process and the vibrational coupling effects for the tensor of rotation. *Chemical Physics* 43: 67–80.

Kuball H-G, Neubrech S and Schönhofer A (1992) Optical activity of oriented molecules. α, β-unsaturated steroid ketones and their sector rules. *Chemical Physics* 163: 115–132.

Loxsom FM (1969) Optical rotation of helical polymers: periodic boundary conditions. *Journal of Chemical Physics* 51: 4899–4905.

Moffit W and Moscowitz A (1959) Optical activity in absorbing media. *The Journal of Chemical Physics* 30: 648–660.

Nakanishi K, Berova N and Woody RW (eds) (1994) *Circular Dichroism–Principles and Applications*. New York: VCH Publishers.

Patai S and Rappoport Z (eds) (1989) *The Chemistry of Enones*. New York: John Wiley & Sons.

Rodger A and Norden B (1997) *Circular Dichroism & Linear Dichroism*. Oxford: Oxford University Press.

Ruch E and Schönhofer A (1970) Theorie der Chiralitätsfunktionen. *Theoretica Chimica Acta* 19: 225–287.

Schellman JA (1966) Symmetry rules for optical rotation. *Journal of Chemical Physics* 44: 55–63.

Snatzke G (ed) (1967) *Optical Rotatory Dispersion and Circular Dichroism in Organic Chemistry*. London: Heyden and Son.

Snatzke G (1968) Circulardichroismus und optische Rotationsdispersion – Grundlagen und Anwendung auf die Untersuchung der Stereochemie von Naturstoffen. *Angewandte Chemie* 80: 15–26.

Snatzke G (1979) Circulardichroismus und absolute Konformation: Anwendung der qualitativen MO-Theorie auf die chiroptischen Phänomene. *Angewandte Chemie* 91: 380–393.

Tinoco I (1962) Theoretical aspects of optical activity part two: polymers. *Advances in Chemical Physics* 4: 113–160.

Weigang OE (1965) Vibrational structuring in optical activity II. "Forbidden" character in circular dichroism. *Journal of Chemical Physics* 43: 3609–3618.

Chiroptical Spectroscopy, Oriented Molecules and Anisotropic Systems

Hans-Georg Kuball and **Tatiana Höfer**, Fachbereich Chemie der Universität Kaiserslautern, Germany

ELECTRONIC SPECTROSCOPY
Theory

Introduction

Homochirality, such as the dominance of the left-handed α-amino acids owing to symmetry breaking arising from the nuclear weak interaction and/or other mechanisms, is the basis of biomolecular enantioselection. The 'wrong' chiral form can, for example as a compound in a drug, lead to unwanted and even dangerous effects. Thus, the phenomenon of 'chirality' itself as well as new ideas for exploiting it are of current interest.

Nowadays, chirality is often discussed from the point of view of such basic phenomena as charge conjugation, parity and time reversal (CPT symmetry). But for a 'way in' to chirality in chemistry and physics, the different occupation of space by two enantiomers, inherent in the definition of Kelvin, and its extension to statistical systems by Avnir covers most of the problems. Kelvin's definition provides a background for a hierarchy of chiral objects or structures through four levels: atoms, molecules, phases with long-range orientational and positional order, and the shape of objects. Long-range orientational and positional order in a phase can lead to a microscopic chiral structure of particles in the phase. The shape of a phase is its macroscopic appearance, e.g. the crystal habit. Chirality that is caused by long-range orientational and positional order will be called 'suprastructural phase chirality' in the following.

Methods are required to detect chirality. A so-called 'chirality observation' answers the question whether the system is chiral or not with yes or no. Then, a 'chirality measurement' yields a value and a sign for a quantity that gives information about the chirality of the molecules or the phases but, in general, gives no measure for chirality itself.

One of the most popular methods for detecting chirality is the measurement of the optical activity. A chiral object rotates the plane of polarization of linearly polarized light and generates ellipticity in the region of absorption bands because of the different refractive indices and absorption coefficients for left- and right-circularly polarized light – circular birefringence and dichroism. The main interest in circular dichroism (CD) spectroscopy is based on three applications. First, CD spectroscopy can be applied as a very specific and sensitive fingerprint method in analytical chemistry; CD spectrometer units have even been developed as detection devices in chromatography. Secondly, the determination of the absolute configuration of a molecule is of great interest in spite of the fact that the CD method is not an absolute method in contrast to Bijvoet's method. Thus, many rules, so-called sector and helicity rules, for different classes of compounds have been developed for electronic transitions in the visible and UV spectral region to correlate the CD with the absolute configuration. For vibrational transitions in the IR spectral region, the existence of such correlation is the exception, not the rule.

The circular dichroism, $\Delta\varepsilon$, of an isotropic system is one third of the trace of a pseudotensor of second rank:

$$\Delta\varepsilon = \tfrac{1}{3}(\Delta\varepsilon_{11} + \Delta\varepsilon_{22} + \Delta\varepsilon_{33}) \qquad [1]$$

To increase the quantity of information available for a molecule, the three coordinates $\Delta\varepsilon_{ii}$ ($i = 1,2,3$) of $\Delta\varepsilon$ in Equation [1], i.e. the CD of oriented molecules in anisotropic samples (ACD), can be measured.

There is a third problem for which chirality information is of current interest: anisotropic phases are often stabilized by chiral structures. Apart from chiral structures with enantiomorphic crystals of chiral compounds, suprastructural chirality exists in liquid crystal phases built up by chiral molecules as in the cholesteric phases and the smectic C* phases. Even liquid crystalline phases with suprastructural chirality originating in achiral, so-called banana-shaped molecules, seem to be possible. Anisotropic polymer films with chiral structures have been found. It can be anticipated that chiroptical spectroscopy with anisotropic chiral systems will lead to new questions and answers.

Birefringence and dichroism of chiral isotropic and chiral anisotropic phases

Optical rotatory dispersion and circular dichroism in isotropic dissymmetric phases

The interaction of light with an isotropic dissymmetric solution leads to a left- and a right-circularly polarized light beam for which distinct refractive indices n_L and n_R as well as absorption coefficients ε_L and ε_R exist owing to the diastereomeric interaction between light and the chiral system. The plane of polarization of the incident linearly polarized light is rotated by an angle ϕ (optical rotation in rad) that is proportional to the circular birefringence $n_L - n_R$,

$$\phi = \pi \bar{\nu} d (n_L - n_R) \quad (\text{rad}) \qquad [2]$$

where d is the path length of the sample in cm and $\bar{\nu}$ is the wavenumber of the incident light in cm^{-1}. The specific rotation $[\alpha]$ for a fluid or solid compound is defined with respect to the density ρ (g cm^{-3}) or for a solution with respect to a concentration q in g/(100 cm^3 solution) and the angle α in degree:

$$[\alpha] = \frac{10\alpha}{\rho d} = \frac{10^3 \alpha}{qd} \quad \left(\frac{\text{deg cm}^3}{\text{g dm}}\right) \qquad [3]$$

The molar rotation $[M]$ can be expressed by

$$[M] = \frac{[\alpha]M}{100} = \frac{100\alpha}{cd} \quad \left(\frac{\text{deg cm}^3}{\text{mol dm}}\right) \qquad [4]$$

where M is the molecular mass in g mol^{-1} and c (mol L^{-1}) the concentration. In the region of an absorption band, both circularly polarized waves superpose and the light becomes elliptically polarized with an ellipticity $\tan\theta$ given by

$$\tan\theta = \frac{10^{-\varepsilon_L cd/2} - 10^{-\varepsilon_R cd/2}}{10^{-\varepsilon_L cd/2} + 10^{-\varepsilon_R cd/2}} \cong \theta$$

$$= \frac{\ln 10}{4}(\varepsilon_L - \varepsilon_R)cd = 0.5757 \, \Delta\varepsilon \, cd \qquad [5]$$

Instead of the circular dichroism $\Delta\varepsilon = \varepsilon_L - \varepsilon_R$, often the molar ellipticity $[\theta]$ is used:

$$[\theta] = \frac{100\theta}{cd} \approx 3300 \, \Delta\varepsilon \qquad [6]$$

Because there are two independent waves propagating through the sample (**Figure 2**), the Lambert–Beer law does not hold,

$$I = I_0 \, 10^{-\bar{\varepsilon}cd} \cosh\left(\frac{2\,303}{2}(\varepsilon_L - \varepsilon_R)\,cd\right) \qquad [7]$$

but the deviation is in most cases negligible. $\bar{\varepsilon}$ is the average absorption of the sample:

$$\bar{\varepsilon} = \tfrac{1}{2}(\varepsilon_L + \varepsilon_R) \qquad [8]$$

The chirality measurements (**Figure 1**) circular dichroism (CD) and optical rotatory dispersion (ORD), the wavenumber dependence of the molar rotation, are designated together as the Cotton effect (CE). CD and ORD are quantitatively correlated by a Kramers–Kronig transformation. The Kramers–Kronig transformation can be used to check CD measurements by ORD measurements and vice versa by comparing calculated and experimental results.

Anisotropic samples

The anisotropy of systems is a widespread phenomenon in nature. The description as anisotropic means that a material has different properties along different directions in space. This anisotropy, caused by long-range orientational and positional order, can be homogeneous, statistically homogeneous or periodically homogeneous. These microscopic structures, e.g. the screw axis of a NaClO$_3$ crystal or the pitch of the helical structure in a cholesteric or smectic C* phase, have to be taken into consideration when, for

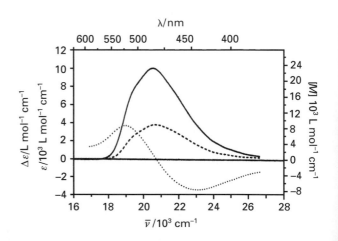

Figure 1 A positive Cotton effect (CE), i.e. the CD (– – –) and ORD (•••), and a UV absorption band (——).

example, an internal periodicity yields internal reflection of circularly polarized light in the visible or infrared spectral region.

The concept of eigenstates of light

The interaction of light with a material, described by the Jones or Mueller calculus, leads to the concept of eigenstates. These states are determined by the symmetry of the material that is penetrated by the light (**Table 1**). An eigenstate of light refers to light that propagates through a material without changing its state of polarization. For isotropic materials, all states of polarization, and hence also unpolarized light, are eigenstates. Inherently dissymmetric isotropic solutions possess left- and right-circularly polarized light as the two orthogonal eigenstates. For anisotropic uniaxial and biaxial materials, the eigenstates are two orthogonal linearly polarized light beams, whereas in complex systems, e.g. in chiral anisotropic systems, the eigenstates are two orthogonal elliptically polarized light beams (**Figure 2**).

The linearly, elliptically or circularly polarized eigenstates propagate with different velocities because of the different refractive indices. Different absorption of the two independent light beams leads to dichroism. Thus, each of the eigenstates has its own and distinct refractive index and molar decadic absorption coefficient (**Table 1**). The plane light waves of the two eigenstates interfere after passing the sample to form an elliptically polarized light beam with an ellipticity and an azimuth different from that of the incident light (**Figure 2**), in general. Its intensity may be less than that of the incident light beam. Linear and circular eigenstates and thus linear and circular birefringence and dichroism of a sample can be seen as two limiting cases of elliptical birefringence and elliptical dichroism.

Table 1 Material constants for birefringence and dichroism

Two orthogonal eigenstates of light	Refractive indices (n) and absorption coefficients (ε)	Birefringence	Dichroism	Mean absorption
Linearly polarized light	n_1, ε_1; n_2, ε_2	$n_1 - n_2$	$\varepsilon_1 - \varepsilon_2$	$\frac{1}{2}(\varepsilon_1 - \varepsilon_2)$
Elliptically polarized light	n_{EL}, ε_{EL}; n_{ER}, ε_{ER}	$n_{EL} - n_{ER}$	$\varepsilon_{EL} - \varepsilon_{ER}$	$\frac{1}{2}(\varepsilon_{EL} + \varepsilon_{ER})$
Circularly polarized light	n_L, ε_L; n_R, ε_R	$n_L - n_R$	$\varepsilon_L - \varepsilon_R$	$\frac{1}{2}(\varepsilon_L + \varepsilon_R)$

Chiral anisotropic systems

A chiral system is a system that is not superposable on its mirror image. The number of independent measurements of different kinds needed to identify the sample as chiral depends on the knowledge of the system. If it is known, for example, that the system is an isotropic dissymmetric phase, one optical measurement is sufficient, whereas for an anisotropic uniaxial system more than one measurement is demanded. For the latter case the wavenumber dependence of the ellipticity and the rotation of the plane of polarization can be measured. Here the dispersion of the circular birefringence and dichroism can be, but needs not be, different from that of the linear birefringence and dichroism (**Figure 3**). Another possibility is the measurement of the rotation of the plane of polarization and the ellipticity for different azimuth orientations of the sample about the axis of propagation of the light beam.

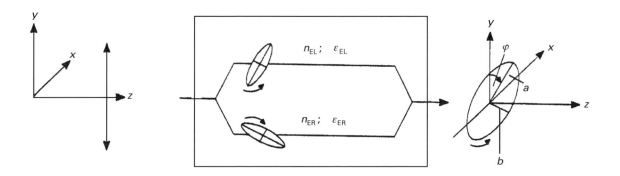

Figure 2 Schematic representation of the two eigenstates of light (see **Table 1**) propagating through an anisotropic phase of chiral molecules with or without suprastructural chirality.

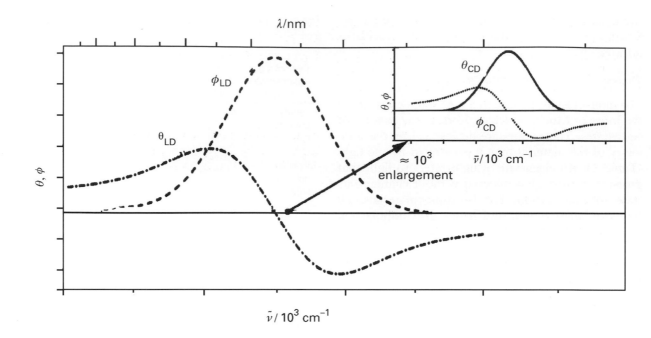

Figure 3 Dispersion of linear and circular birefringence and dichroism, ϕ_{LD}, θ_{LD}, ϕ_{CD}, θ_{CD} ($\Delta\varepsilon$) within an absorption band of a chiral molecule.

The circular dichroism of oriented molecules: ACD spectroscopy

Phenomenology of the CD of anisotropic phases (ACD) without suprastructural chirality

The two eigenstates of light emerging from a homogeneous sample superpose in accordance with their different optical path lengths. The azimuth and the ellipticity are then given by the Stokes vector $\underline{s}^d$,

$$\underline{s}^d = \mathbf{F} \cdot \underline{s}^0 \qquad [9]$$

where $\mathbf{F}$ is the Mueller matrix and $\underline{s}^0$ the Stokes vector of the incident light. In $\underline{s} = \{s_0, s_1, s_2, s_3\}$, s_0 is the intensity whereas $\vec{s} = \{s_1, s_2, s_3\}$ describes the state of polarization (the notation { } indicates a column vector). For a nondepolarizing medium,

$$\mathbf{F} = \exp(-2nd\mathbf{A}) \quad \text{with}$$

$$\mathbf{A} = \begin{pmatrix} c' & b_1 & b_2 & b_3 \\ b_1 & c' & a_3 & -a_2 \\ b_2 & -a_3 & c' & a_1 \\ b_3 & a_2 & -a_1 & c' \end{pmatrix} = \mathbf{A}' + c'\mathbf{I} \quad [10]$$

is obtained. d is the pathlength and n the number of particles in a unit volume. I is the identity matrix. It is convenient to rewrite the Mueller matrix in the form

$$\mathbf{F} = \exp(-2ndc')\mathbf{\Lambda}, \quad \text{with} \quad \mathbf{\Lambda} = \exp(-2nd\,\mathbf{A}') \quad [11]$$

As shown by Schönhofer and co-workers $\mathbf{\Lambda}$ can be rewritten as a matrix polynomial of third degree:

$$\mathbf{\Lambda} = c_0(\mathbf{A}')^3 + c_1(\mathbf{A}')^2 + [(\lambda^2 - \kappa^2)c_0 + c_2]\mathbf{A}' + [(\lambda^2 - \kappa^2)c_1 + c_3]\mathbf{I} \quad [12]$$

$$c_0 = -\frac{1}{\kappa^2 + \lambda^2}\left[\frac{1}{\kappa}\sinh(2nd\kappa) - \frac{1}{\lambda}\sin(2nd\lambda)\right]$$

$$c_1 = \frac{1}{\kappa^2 + \lambda^2}[\cosh(2nd\kappa) - \cos(2nd\lambda)]$$

$$c_2 = -\frac{1}{\kappa^2 + \lambda^2}[\kappa\sinh(2nd\kappa) + \lambda\sin(2nd\lambda)]$$

$$c_3 = \frac{1}{\kappa^2 + \lambda^2}[\kappa^2\cosh(2nd\kappa) + \lambda^2\cos(2nd\lambda)] \quad [13]$$

with $\genfrac{}{}{0pt}{}{\kappa}{\lambda} = \{\frac{1}{2}\{((\vec{b}^2 - \vec{a}^2 + 4(\vec{a} \cdot \vec{b}^2)^{1/2} \pm (\vec{b}^2 - \vec{a}^2)\}\}^{1/2} \geq 0$. For a homogeneous medium with large linear and circular birefringence and dichroism effects, all elements of the differential Mueller matrix $\mathbf{A}'$ (for the birefringence $\vec{a} = \{a_1, a_2, a_3\}$ and for the dichroism $\vec{b} = \{b_1, b_2, b_3\}$ in Equation [12]) have to be taken to evaluate the azimuth and ellipticity of the emerging light. The elements a_2 and b_2 are responsible for the optical rotation and for the circular dichroism, respectively, and a_1 and a_3 represent the anisotropy of the refractive index of linear polarized light whereas b_1 and b_3 belong to the corresponding linear dichroism. Because there are six unknown parameters for the anisotropy (Eqn [12]), six different measurements made by rotating the sample and/or measuring with a light beam propagating in different directions within the sample are needed to determine the optical constants. Measuring the dispersion of the ACD or the optical rotatory power of an oriented ensemble (AORD) does not yield new and independent information (**Figure 3**).

The ACD measurement $\Delta\varepsilon^A$ with light propagating parallel to the optical axis of a uniaxial phase is a chirality measurement, but one has to take care to avoid artefacts. A check on the reliability of such a measurement can be obtained by measuring the CD with different sample azimuths rotated around the direction of propagation of light. The CD signal has to be independent of the orientation of the azimuth of the sample. Measurements of the optical effects with a light beam propagating perpendicularly to the optical axis are very difficult to perform because of the artefacts that may be caused by interference with the linear birefringence and dichroism. At present it does not seem possible to overcome the experimental problems with instruments working on the basis of polarization modulation. These experimental problems can be solved using other techniques, e.g. by measuring the state of polarization of the light, but solving the experimental problem does not solve the whole problem. A measurement with a light beam propagating perpendicularly to the optical axis is not a chirality measurement; that is, measuring two anisotropic phases of enantiomers with a light beam propagating perpendicularly to the optical axis does not yield values that are equal in their absolute value and different in sign, because of the interference with linear effects. The question is whether the measured quantity can be decomposed into an achiral and a chiral part ($\Delta\varepsilon_2^A$), i.e. whether chirality information can be extracted from the data obtained. For a uniaxial system, the following relation can be derived from b_2:

$$\Delta\varepsilon = \tfrac{1}{3}(\Delta\varepsilon^A + 2\Delta\varepsilon_2^A) \qquad [14]$$

For light polarized parallel and perpendicular to the optical axis and propagating perpendicularly to this

axis, with the molar decadic absorption coefficients ε_1 and ε_2 for a uniaxial system, an equation analogous to Equation [14] is obtained:

$$\varepsilon = \tfrac{1}{3}(\varepsilon_1 + 2\varepsilon_2) \qquad [15]$$

A measurable macroscopic property of the anisotropic phase is then given by the tensor coordinates Y_{kl} as given by Kuball and colleagues

$$Y_{kl} = \sum_{i,j} g_{ijkl}\, X_{ij} \qquad [16]$$

(Three tensors are of particular importance for ACD spectroscopy: the transition moment or the absorption tensor (D_{ij}^{NK} or ε_{ij}), the rotation or the circular dichroism tensor (R_{ij}^{NK} or $\Delta\varepsilon_{ij}$), and the order tensor (g_{ij33}). For molecules without any symmetry, the principal axes of the three tensors do not coincide: x_i^+ for D_{ij}^{NK} or ε_{ij}, x_i^o for R_{ij}^{NK} or $\Delta\varepsilon_{ij}$, and x_i^* for g_{ij33}.)

In Equation [16] X_{ij} is a tensor of rank 2 that describes a molecular property. With the orientational distribution function of the molecules in the anisotropic sample $f(\alpha, \beta, \gamma)$, where α, β, γ are the Eulerian angles, the orientational distribution coefficients g_{ijkl} (Eqn [17]) define a set of two different tensors of rank 2 with respect to the transformation of the molecule-fixed (g_{ijkl}: indices ij for each pair of kl) as well as the space-fixed (g_{ijkl}: indices kl for each pair of ij) coordinate system:

$$g_{ijkl} = \frac{1}{8\pi^2}\int f(\alpha, \beta, \gamma) a_{ik}\, a_{jl} \sin\beta\, d\alpha\, d\beta\, d\gamma \qquad [17]$$

a_{ij} is an element of the transformation matrix from the space-fixed $x_i{}'$ to the molecule-fixed x_i coordinate system. For a sample possessing at least a C_3 symmetry axis (uniaxial system) and a light beam propagating along this direction, the circular dichroism of the anisotropic sample $Y_{33} = \Delta\varepsilon^A$ is obtained:

$$\Delta\varepsilon^A = \sum_{i,j} g_{ij33}\, \Delta\varepsilon_{ij} \qquad [18]$$

$X_{ij} = \Delta\varepsilon_{ij}$ is the circular dichroism tensor of the molecules. According to Equation [14], the CD value $\Delta\varepsilon_2^A$ for light propagating perpendicularly to this axis, the optical axis, can be evaluated. Because of the interference with linear birefringence and dichroism, the present authors have not yet been able to measure $\Delta\varepsilon_2^A$.

For light polarized parallel to the optical axis and propagating perpendicularly to it follows,

$$\varepsilon_1 = \sum_{i,j} g_{ij11}\, \varepsilon_{ij} \qquad [19]$$

$X_{ij} = \varepsilon_{ij}$ is the absorption tensor of the molecules. The correlation between the second pair of indices of the orientational distribution coefficient and the numbering of the tensor coordinates of the measurable quantity gives rise to an interesting phenomenon. For the ACD ($\Delta\varepsilon^A$, Eqn [18]) the indices of the coordinates g_{ijkl} are determined by the direction of propagation of light with respect to the molecular frame $kl = 33$. This is due to the fact that the CD effect for molecules is given by the projection of the molecular properties on the propagation direction of the light. For polarized absorption spectroscopy, the projection of the molecular properties onto the direction of polarization of the light beam is decisive, i.e. $kl = 11$, 22. This means that the numbering of the absorption tensor is obtained from the polarization direction of the light with respect to the molecular frame, whereas for the ACD the direction of propagation of the light wave is decisive.

The experimental determination of the molecular tensors $\Delta\varepsilon_{ij}$ and ε_{ij}

The equations for the ACD   The result of a CD measurement of an anisotropic sample is a weighted sum of the three quantities $\Delta\varepsilon_{ii}$ (Eqns [18] and [20]). The weighting factors are the orientational distribution coefficients. For a light beam propagating along a direction of at least a C_3 symmetry axis of a sample, the circular dichroism $\Delta\varepsilon^A$ (Eqn [18]) is a chirality measurement; this condition is fulfilled for a chiral nematic phase unwound by an electric field when the light beam propagates along the optical axis (x'_3), the nematic director. If the orientation of the principal axes of the order tensor g_{ij33} relative to the molecular frame is temperature-dependent, one may use for the ACD:

$$\Delta\varepsilon^A = \sum_{i,j} a_{ij}^2 g_{jj33}^* \,\Delta\varepsilon_{ii}^o \qquad [20]$$

The eigenvalues $\Delta\varepsilon_{ii}^o$ describe the molecular property $\Delta\varepsilon_{ij}$ in its system of principal axes. The terms a_{ij} are the elements of the matrix that transforms the x_i^* coordinates into the x_i^o coordinates.

For uniaxial systems the order parameters S^* and D^* may be introduced instead of g_{ij33}

$$S^* = (3g_{3333}^* - 1)/2 \qquad [21]$$

$$D^* = \sqrt{3}(g_{2233}^* - g_{1133}^*)/2 \qquad [22]$$

They refer to the optical axis of the sample, which is chosen as the space-fixed x'_3 axis, and to the molecule-fixed x_i^* coordinate system. Whereas S^* characterizes the order of the orientation of the molecule-fixed x_3^* axis – the orientation axis – with respect to the optical axis of the uniaxial phase, the parameter D^* is a measure of the deviation from a rotational symmetric distribution function for the molecules about their x_3^* axis.

The principal axes of g_{ij33} are fixed by symmetry for molecules with a point symmetry group different from C_1, C_2, C_i, C_s and C_{2h}. For the latter point groups, the orientations of the x_i^* axes with respect to the molecular skeleton have to be determined experimentally; the orientations of the x_i^* axes vary, in general, with temperature. The two order parameters S^* and D^* can be measured, albeit with the use of some approximations, for example by IR, VIS, UV, or NMR spectroscopy.

If the temperature dependence of the orientation of the x_i^* axes can be neglected, or if these axes are fixed by symmetry, $\Delta\varepsilon^A$ can be conveniently described as

$$\Delta\varepsilon^A - \Delta\varepsilon = (\Delta\varepsilon_{33}^* - \Delta\varepsilon)S^* + \frac{1}{\sqrt{3}}(\Delta\varepsilon_{22}^* - \Delta\varepsilon_{11}^*)D^* \qquad [23]$$

From $\Delta\varepsilon^A - \Delta\varepsilon$ and S^* and D^* measured as functions of temperature, the quantities $3(\Delta\varepsilon_{33}^* - \Delta\varepsilon) = 2\Delta\varepsilon_{33}^* - \Delta\varepsilon_{22}^* - \Delta\varepsilon_{11}^*$ and $\Delta\varepsilon_{22}^* - \Delta\varepsilon_{11}^*$ can be obtained. Because $\Delta\varepsilon_2^A$ is not available experimentally at present $\Delta\varepsilon$ (Eqn [1]) can be used to calculate the tensor coordinates $\Delta\varepsilon_{ii}^*$.

To account for the contributions of the different vibrational transitions, the anisotropic rotational strength R^A for the electronic transition $|N\rangle \rightarrow |K\rangle$ is obtained by integrating over the spectral region of an absorption band:

$$R^A = \frac{3}{B}\int \frac{\Delta\varepsilon^A(\bar{\nu})}{\bar{\nu}}\, d\bar{\nu} = 3\sum_{i,j} g_{ij33}R_{ij}^{NK};$$

$$\frac{3}{B} = 22{\cdot}96 \times 10^{-40} \quad \text{(cgs)} \qquad [24]$$

For an isotropic solution, $g_{ij33} = \frac{1}{3}\delta_{ij}$ and R^A becomes the rotational strength for the transition $|N\rangle \rightarrow |K\rangle$:

$$R^{NK} = \frac{3}{B}\int \frac{\Delta\varepsilon(\bar{\nu})}{\bar{\nu}}\, d\bar{\nu} = \sum_i R_{ii}^{NK} \qquad [25]$$

δ_{ij} is the Kronecker symbol. The tensor coordinates for the rotational strength tensor (tensor of rotation) can be evaluated experimentally as

$$R_{ij}^{NK} = \frac{1}{B} \int \frac{\Delta \varepsilon_{ij}(\bar{\nu})}{\bar{\nu}} \, d\bar{\nu} \qquad [26]$$

For allowed electric dipole transitions, the contributions from vibrational coupling can be neglected. In the case of a forbidden electric or a magnetic dipole transition there should be contributions from the vibronic coupling and intensity-borrowing from other transitions via vibrations of the molecules.

The magnitude of the dissymmetry factor $g = 4R^{NK}/D^{NK}$ can be 10^{-1}, 10^{-3}, 10^{-4} for an inherently dissymmetric chromophore, an $n\pi^*$ transition and a $\pi\pi^*$ transition, where D^{NK} is the dipole strength (see Eqn [31]).

The equations for the polarized absorption For an anisotropic sample, the molar decadic absorption coefficient for light polarized parallel (ε_1) and perpendicular (ε_2) to the optical axis of a uniaxial sample follows from Equations [19] and [15]:

$$\varepsilon_k = \sum_{i,j} g_{ijkk} \, \varepsilon_{ij} = \sum_{i,j} a_{ij}^2 g_{jjkk}^* \, \varepsilon_{ii}^+, \quad k = 1, 2 \quad [27]$$

ε_{ii}^+ are the absorption coefficients of a completely oriented system of molecules for light beams linearly polarized parallel to the x_i^+ axes (principal axes of ε_{ij}). The a_{ij} in Equation [27] are the elements of the matrix that transforms the x_i^* coordinates into the x_i^+ coordinates. The orientational distribution coefficients g_{ij33} have to be used in the following instead of g_{ij11} because, compared to the experimental situation of the CD measurement, the sample is rotated about an axis perpendicular to the propagation direction of the light beam by 90°. The ε_{ii}^+ terms are related to the absorption coefficient of the isotropic solution, analogously to $\Delta\varepsilon$, by Equation [28]:

$$\varepsilon = \tfrac{1}{3}(\varepsilon_{11} + \varepsilon_{22} + \varepsilon_{33}) \qquad [28]$$

If there is no temperature dependence of the x_i^* axes, an appropriate description of the anisotropic absorption is given by

$$\varepsilon_1 - \varepsilon = (\varepsilon_{33}^* - \varepsilon)S^* + \frac{1}{\sqrt{3}}(\varepsilon_{22}^* - \varepsilon_{11}^*)D^* \qquad [29]$$

Furthermore, for the degree of anisotropy follows,

$$R = \frac{\varepsilon_1 - \varepsilon_2}{3\varepsilon} = \tfrac{1}{2} \sum_{i,j}(3g_{ij33} - \delta_{ij})q_{ij}; \quad q_{ij} = \frac{\varepsilon_{ij}}{3\varepsilon} \quad [30]$$

where the q_{ij} are the squares of the direction cosines of the transition moment direction with respect to the x_i axes in the case of a pure polarized transition. For mixed absorption bands the q_{ij} are a measure of mixing. If the molecule-fixed coordinate system is arbitrarily chosen, nondiagonal elements occur in Equation [30]. The dipole strength is given by

$$D_k^A = \frac{12}{B} \int \frac{\varepsilon_k(\bar{\nu})}{\bar{\nu}} \, d\bar{\nu} = 3 \sum_{i,j} g_{ij33} D_{ij}^{NK} \qquad [31]$$

$k = 1, 2$ and with $g_{ij33} = \tfrac{1}{3}\delta_{ij}$ follows $D^{NK} = \Sigma_i \, D_{ii}^{NK}$ for an isotropic solution. The transition moment tensor

$$D_{ij}^{NK} = \frac{4}{B} \int \frac{\varepsilon_{ij}(\bar{\nu})}{\bar{\nu}} \, d\bar{\nu} \qquad [32]$$

can be calculated from the coordinates of the absorption tensor (Eqn [32]).

The tensor coordinates of the circular dichroism tensor $\Delta\varepsilon_{ij}^*$ The meaning of the diagonal elements of the tensor (Eqns [18] and [36]) can be visualized according to **Figure 4**. $\Delta\varepsilon_{33}^*$ can be obtained either by calculating a mean value of three CD measurements for completely oriented molecules with light beams polarized parallel to the three planes, schematically represented in **Figure 4**, or one CD measurement with an ensemble of molecules in a uniaxial system with $S^* = 1$ or a system with at least a C_3 symmetry axis with a light beam propagating parallel to the optical axis or the C_3 symmetry axis. The coordinates $\Delta\varepsilon_{11}^*$ and $\Delta\varepsilon_{22}^*$ have to be evaluated from the temperature-dependent $\Delta\varepsilon^A(T)$ and $\Delta\varepsilon$. For a direct measurement analogous to that of $\Delta\varepsilon_{33}^*$ a new orientational distribution is needed in which there is a symmetric rotational distribution about either the x_1^* or x_2^* axis. However, this procedure is nothing more than changing the numbering of the axes, because only that coordinate $\Delta\varepsilon_{ii}^*$ can be measured directly about which the orientational distribution function possesses a rotationally symmetric distribution.

The $\Delta\varepsilon_{ii}^*$ can also be different in sign and thus one may say that by 'looking' at the molecule from different directions one 'sees' $\Delta\varepsilon_{ii}^*$ with different sign. Changing the line of sight, the now observed $\Delta\varepsilon_{ii}^*$ varies strongly and can also change its sign. One might

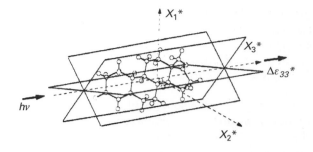

Figure 4 Schematic representation of the measurement of $\Delta\varepsilon_{33}^*$.

loosely interpret this as 'seeing' the chirality of the molecule from different directions. This measurement must be done under conditions given in **Figure 4**.

The molecular description of the circular dichroism and the absorption tensor

The equations for molecules in an orientated state From semiclassical or quantum field theory, the rotational strength R_{ij}^{NnKk} for a transition from the electronic ground state N with the vibrational state n to the electronic excited state K with the vibrational state k, i.e. $|Nn\rangle \rightarrow |Kk\rangle$, follows from the development of Kuball and co-workers:

$$R_{ij}^{NnKk} = \tfrac{1}{4} \sum_{r,s} \langle \mu_r \rangle_{NnKk} (\varepsilon_{rsi} \langle C_{sj} \rangle_{KkNn}$$
$$+ \varepsilon_{rsj} \langle C_{si} \rangle_{KkNn}) \qquad [33]$$

The tensor $\langle C_{sj} \rangle_{KkNn}$ can be decomposed into a symmetric and antisymmetric part, the magnetic dipole and the electric quadrupole moment:

$$\langle C_{sj} \rangle_{KkNn} = -\mathrm{i} \sum_r \varepsilon_{sjr} \langle m_r \rangle_{KkNn} - \frac{\omega_{KkNn}}{c} \langle Q_{sj} \rangle_{KkNn}$$
$$[34]$$

$$\mu_i = -e \sum_\nu x_{i\nu}; \quad Q_{ij} = -\tfrac{1}{2} e \sum_\nu x_{i\nu} x_{j\nu};$$
$$m_r = -\frac{e}{2mc} \sum_\nu \sum_{i,j} \varepsilon_{rij} x_{i\nu} p_{j\nu} \qquad [35]$$

$\langle \mu_r \rangle_{NnKk}$ and $\langle m_r \rangle_{KkNn}$ represent the electric and the magnetic dipole transition moment, respectively, $\langle Q_{sj} \rangle_{KkNn}$ is the electric quadrupole transition moment. $-e$ is the charge, m the mass of the electron. Σ_ν means

the sum over all electrons. Operators depending on the dynamic variables of the nuclei are omitted because they do not contribute to the effect in the approximation used. ε_{ijk} is the Levi–Civita tensor. $\varepsilon_{123} = \varepsilon_{312} = \varepsilon_{231} = 1$, $\varepsilon_{132} = \varepsilon_{213} = \varepsilon_{321} = -1$; all other tensor coordinates ε_{ijk} are equal to zero. With the rotational strength tensor R_{ij}^{NnKk} (Eqn [33]) and the spectral function $G^{NnKk}(\bar{\nu})$ of the vibronic transition $|Nn\rangle \rightarrow |Kk\rangle$, the circular dichroism tensor $\Delta\varepsilon_{ij}$ for an oriented molecule can be expressed as

$$\Delta\varepsilon_{ij} = B\bar{\nu} \sum_n \sum_{Kk} R_{ij}^{NnKk} G^{NnKk}(\bar{\nu}) \qquad [36]$$

$$B = \frac{32\pi^3 N_A}{10^3 hc \ln 10}$$

N^A is the Avogadro constant and c is the velocity of light. The absorption of light by an oriented molecule requires knowledge of the electric dipole transition moment tensor (dipole strength)

$$D_{ij}^{NnKk} = \langle \mu_i \rangle_{NnKk} \langle \mu_j \rangle_{KkNn} \qquad [37]$$

and the corresponding absorption tensor ε_{ij}

$$\varepsilon_{ij} = \tfrac{1}{4} B\bar{\nu} \sum_n \sum_{Kk} D_{ij}^{NnKk} F^{NnKk}(\bar{\nu}) \qquad [38]$$

$F^{NnKk}(\bar{\nu})$ is the spectral function for the absorption band of the vibronic transition $|Nn\rangle \rightarrow |Kk\rangle$ which, in the approximation that is used here, is equal to $G^{NnKk}(\bar{\nu})$.

The $\Delta\varepsilon_{ij}$ term can be decomposed into an electric dipole/magnetic dipole and an electric dipole/electric quadrupole contribution:

$$\Delta\varepsilon_{ij}(\bar{\nu}) = \Delta\varepsilon_{ij}^{\mu m}(\bar{\nu}) + \Delta\varepsilon_{ij}^{\mu Q}(\bar{\nu}) \qquad [39]$$

The decomposition is, except for special circumstances, origin dependent. From the diagonal elements $\Delta\varepsilon_{ii}$ the quantities $\Delta_i(\bar{\nu})$ defined by

$$\Delta_1(\bar{\nu}) = \tfrac{1}{2} [\Delta\varepsilon_{11}(\bar{\nu}) - \Delta\varepsilon_{22}(\bar{\nu}) - \Delta\varepsilon_{33}(\bar{\nu})] \qquad [40]$$

and cyclic permutation of the indices 1, 2, 3 can be calculated. The $\Delta_i(\bar{\nu})$ terms yield information about the electric dipole ($\langle \mu_i \rangle_{NnKk}$), magnetic dipole ($\langle m_i \rangle_{KkNn}$), and electric quadrupole ($\langle Q_{ij} \rangle_{KkNn}$)

transition moments:

$$\Delta_i(\bar\nu) = -\tfrac{1}{2} B\bar\nu \sum_n \sum_{Kk} \mathrm{Im}\{\langle\mu_i\rangle_{NnKk}\langle m_i\rangle_{KkNn}\}$$
$$\times\, G^{NnKk}(\bar\nu) + \Delta\varepsilon_{ii}^{\mu Q}(\bar\nu) \qquad [41]$$

The decomposition of R_{ii}^{NK} corresponding to Equation [39] yields

$$R_{11}^{\mu m} = \mathrm{Im}\{\langle\mu_2\rangle_{NK}\langle m_2\rangle_{KN} + \langle\mu_3\rangle_{NK}\langle m_3\rangle_{KN}\} \quad [42]$$

$$R_{11}^{\mu Q} = -\langle\mu_2\rangle_{NK}\langle Q_{31}\rangle_{KN} + \langle\mu_3\rangle_{NK}\langle Q_{21}\rangle_{KN} \quad [43]$$

by integration over the absorption band. R_{22}^{NK} and R_{33}^{NK} can be obtained by cyclic permutation of the indices. The terms $\langle\mu_i\rangle_{NK}$, $\langle m_i\rangle_{KN}$, and $\langle Q_{ij}\rangle_{KN}$ are the electric dipole magnetic dipole and electric quadrupole transition moments for the 0–0 electronic transition $|N\rangle \to |K\rangle$. By integration of Equations [40 and 41], wavenumber-independent quantities Δ_i can be obtained with the tensors of rotation given by Equations [44] and [45] and cyclic permutation of the indices 1, 2, 3:

$$\Delta_1 = \tfrac{1}{2}\left[R_{11}^{NK} - R_{22}^{NK} - R_{33}^{NK}\right] \qquad [44]$$

$$\Delta_i = -\tfrac{1}{2}\mathrm{Im}\{\langle\mu_i\rangle_{NK}\langle m_i\rangle_{KN}\} + R_{ii}^{\mu Q} \qquad [45]$$

Vibronic coupling in ACD spectroscopy   Forbidden electric dipole transitions or allowed magnetic dipole transitions possess no or nearly no intensity of absorption in a spectrum. There are a number of different mechanisms by which a transition can gain intensity. One mechanism is the coupling of an electronic transition to the nuclear motions. By the weak dependence of the state functions $|Nn\rangle$, $|Kk\rangle$ on the nuclear coordinates, electronic states within the molecules are coupled. The transition moment of the forbidden transition $|N\rangle \to |K\rangle$ increases, as normally described, by borrowing intensity from other allowed electronic transitions via normal vibrations, i.e. the variation of the normal coordinates. Within the adiabatic approximation, the coordinates of the absorption tensor ε_{ii}^* are intensified by the Herzberg–Teller mechanism via different normal vibrations which – depending on the irreducible representation to which the normal coordinate belongs – couple transitions with different transition moment

directions. This can be demonstrated experimentally by measuring a frequency-dependent degree of anisotropy (Eqn [30]). With the Herzberg–Teller method the intensification of $\Delta\varepsilon^A$ (Eqn [46]) follows from Kuball and co-workers

$$\Delta\varepsilon^A(\bar\nu) = B\bar\nu \sum_{i,j} g_{ij33}\Bigg(R_{ij}^{NK} G^{NK}(\bar\nu)$$
$$+ \sum_\mu R_{ij}^{NK,\mu} G^{NK,\mu}(\bar\nu) + \sum_{\mu,\nu}\Big[R(1)_{ij}^{NK,\mu\nu} G_1^{NK,\mu\nu}(\bar\nu)$$
$$+ R(2)_{ij}^{NK,\mu\nu} G_2^{NK,\mu\nu}(\bar\nu)\Big]\Bigg) \qquad [46]$$

R_{ij}^{NK} gives the contribution of the 0–0 transition to the CD band. The coefficients $R_{ij}^{NK,\mu}$, $R(1)_{ij}^{NK,\mu\nu}$ and $R(2)_{ij}^{NK,\mu\nu}$ determine the size of the intensity borrowing via the μth and ν^{th} vibrations from other electric dipole-allowed transitions with a corresponding transition moment direction. The frequency dependence of the progressions $G^{NK}(\bar\nu)$, $G^{NK,\mu}(\bar\nu)$, $G_1^{NK,\mu\nu}(\bar\nu)$, and $G_2^{NK,\mu\nu}(\bar\nu)$ are described by

$$G^{NK}(\bar\nu) = \sum_{n,k} |\langle n|k\rangle|^2 G^{NnKk}(\bar\nu) \qquad [47]$$

$$G^{NK,\mu}(\bar\nu) = \sum_{n,k} \langle n|k\rangle\langle k|Q_\mu|n\rangle G^{NnKk}(\bar\nu) \qquad [48]$$

$$G_1^{NK,\mu\nu}(\bar\nu) = \sum_{n,k} \langle n|k\rangle\langle k|Q_\mu Q_\nu|n\rangle G^{NnKk}(\bar\nu) \quad [49]$$

$$G_2^{NK,\mu\nu}(\bar\nu) = \sum_{n,k} \langle n|Q_\mu|k\rangle\langle k|Q_\nu|n\rangle G^{NnKk}(\bar\nu) \quad [50]$$

This means analogously to the UV absorption, each CD tensor coordinate $\Delta\varepsilon_{ii}^*$ is intensified by normal modes of different irreducible representations by a different amount, i.e. each $\Delta\varepsilon_{ii}^*$ ($i = 1,2,3$) borrows intensity from other electronic transitions of suitable symmetry (Eqns [47] to [50]).

Because of the importance of intensity borrowing, a forbidden electric and magnetic dipole transition of a compound of symmetry D_2 will be analysed as an example. For the transition $|N\rangle \to |K\rangle$ between an electronic and vibrational ground and an electronic and vibrational excited state A $\to$ A, only the coefficient $R(2)_{ij}^{NK,\mu\nu}$ in Equation [46] is different from zero. With Equation [50], the frequency dependence of the CD and ACD is determined by a progression starting on false origins of one vibration $\mu = \nu = 1,2\ldots$ of an irreducible representation b_i ($i = 1,2,3$):

$$\bar{\nu}_k = \bar{\nu}_{00} + \bar{\nu}_{0\mu}(b_i) + n_k \bar{\nu}_0(a) \qquad [51]$$

$n_k = 0,1,2\ldots$ are the quantum numbers for the kth excited state of the involved totally symmetric vibration with the wavenumber $\bar{\nu}_0(a)$. $\bar{\nu}_k$ is the wavenumber for the $(k + 1)^{th}$ band of the progression. The false origin is $\bar{\nu}_{00} + \bar{\nu}_{0\mu}(b_i)$ and every vibration of the irreducible representation b_i can cause intensity borrowing from an allowed electric dipole transition A $\rightarrow$ B$_i$. The coupling scheme (**Figure 5**) for an A $\rightarrow$ B$_i$ transition contributes with four terms. The intensity borrowing for the corresponding UV band is achieved by the same progressions (Eqn [51]). The transition moment direction of the progressions is determined by the transition moment direction of the coupled electronic transitions A $\rightarrow$ B$_i$. For the A $\rightarrow$ A transition $\Delta\varepsilon_{11}$, ε_{33} and ε_{22} borrow intensity via vibrations of irreducible representation b_1 and b_2. $\Delta\varepsilon_{22}$, ε_{11} and ε_{33} borrow the intensity via vibrations of symmetry b_3 and b_1 whereas $\Delta\varepsilon_{33}$, ε_{11} and ε_{22} gain the intensity from the coupling with b_3 and b_2.

For each chromophore, the reduced CD spectra, the tensor coordinates $\Delta\varepsilon_{ii}$, and the reduced UV spectra, the tensor coordinates ε_{ii} gain intensity (**Figure 6A and 6B**) by a special amount of coupling. By substituting the molecule in the neighbourhood of this chromophore, another coupling is achieved and the reduced spectra $\Delta\varepsilon_{11}$, $\Delta\varepsilon_{22}$, $\Delta\varepsilon_{33}$ as well as ε_{11}, ε_{22}, ε_{33} change. Thus, in both cases, $\Delta\varepsilon$, $\Delta\varepsilon^A$ ε, and R are different. With this simple model a sign change in $\Delta\varepsilon$ and possibly also in $\Delta\varepsilon^A$ can be obtained without changing the absolute configuration, thus the sector or helicity rules for this chromophore break down, because the sector rule is now applied to the CD of vibrational progression for which the rule was not developed. Applying the rule to measurements in isotropic solution, one has to presuppose that the same progression determines $\Delta\varepsilon$ for both compounds.

From ACD spectroscopy each tensor coordinate $\Delta\varepsilon_{ii}$ is obtained directly. This means that the sector rules or helicity rules can always be applied to that $\Delta\varepsilon_{ii}$ for which the rule has been developed.

Application of ACD and polarized UV spectroscopy

In addition to the analytical application of the ACD spectroscopy as a fingerprint technique, structural differences far from the chromophore can also be detected. A complete analysis of the ACD, the CD, the UV and the polarized UV spectra requires the order tensor or at least an estimation of the order parameters. The variation of $\Delta\varepsilon^A$ with temperature for an $n\pi^*$ transition of androst-4,17-dien-3-one ([1]; **Figure 7**) and the knowledge of g_{ii33}^* from ^{2}H NMR allows the calculation of $\Delta\varepsilon_{ii}^*$ with respect to the principal axis of the order tensor. The tensor coordinates $\Delta\varepsilon_{ii}^*$ have different signs and are shifted with respect to each other, which indicates the vibronic effect of intensity borrowing. The UV spectrum of [1] as well as the coordinates $\Delta\varepsilon_{ii}^*$ have a low spectral resolution in comparison to the CD spectrum $\Delta\varepsilon$. The different signs of $\Delta\varepsilon_{ii}^*(\bar{\nu})$ and their shifts against each other 'burn holes' into the spectrum calculated from Eqn [1]. Thus as often found in CD spectroscopy, an apparent high spectral resolution of the CD spectrum is obtained in contrast to the UV spectrum.

As a second example, $\Delta\varepsilon^A$ for a uniformly polarized charge transfer transition of 1,4-bis(R)-1-phenyl-ethylamino)-9,10-anthraquinone [2] is shown in **Figure 8**. There is no vibronic coupling and therefore no difference in the frequency dependence of the tensor coordinates. Furthermore, the tensor coordinate $\Delta\varepsilon_{33}^* \approx 0$ because the transition is polarized parallel to the x_3^* direction (long axis of the anthraquinone skeleton). This indicates that for [2] the x_3^* axis is approximately a principal axis of the circular dichroism tensor. Of interest, but not shown for [2], is the large

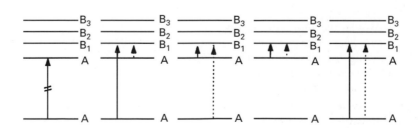

Figure 5 Coupling scheme for intensity borrowing of the circular dichroism and absorption tensor coordinates of a forbidden A $\rightarrow$ A transition via the coefficient $R(2)_{ij}^{NK,\mu\nu}$ with one vibration b_1 from an allowed electric (——) and magnetic (•••) dipole transition of A $\rightarrow$ B$_1$. Quadrupole contributions are neglected for this discussion.

contribution of the electric dipole/electric quadrupole term (Eqn [39]) to $\Delta\varepsilon^*_{ii}$ in the wavenumber region below $25\,000$ cm^{-1}.

With the steroid 3-methyl-3α-5α-3,3′,4′,5′-tetrahydrobenzo[2,3]cholest-2-en-1-one [3] and its diastereomer [4] the breakdown of a helicity rule was demonstrated by Frelek and co-workers (**Figure 9**). From X-ray analysis it is known that the helicity of the enone chromophore is opposite in sign in [3] and

[4] and thus the fact that $\Delta\varepsilon < 0$ in both cases contradicts the helicity rule for cisoid enones either for [3] or [4]. From ACD spectroscopy follows that the enhancement of the $\Delta\varepsilon^*_{ii}$ by vibronic coupling for [4] yields a sign change of $\Delta\varepsilon$ without change of the absolute configuration of [4]. This is the first example where this effect has been proven.

At the end of the section on applications one must point out that it is impossible, at present, to obtain

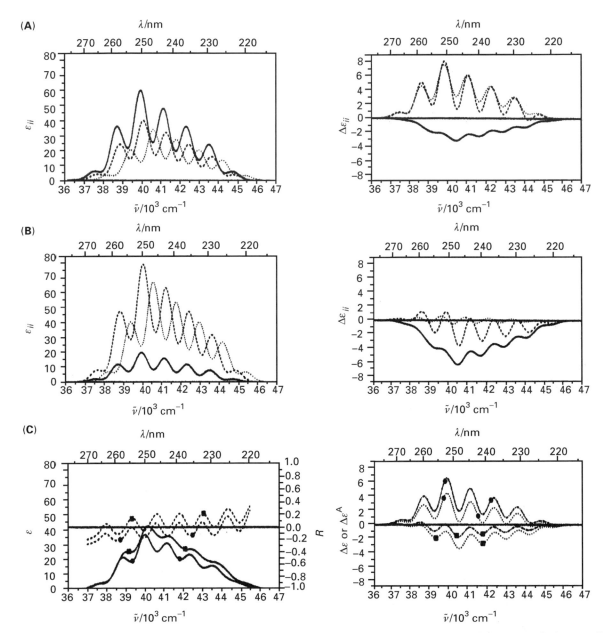

Figure 6 Calculated UV (left) and CD (right) spectra of the A → A transition for a molecule of symmetry D_2 for two different strength of vibronic coupling. The intensity of the vibrational progressions is changed from $\varepsilon_{11} : \varepsilon_{22} : \varepsilon_{33} = 3 : 1 : 1$ (A left) to $\varepsilon_{11} : \varepsilon_{22} : \varepsilon_{33} = 1 : 2 : 2$ (B): ($\Delta\varepsilon_{11}(\bar{\nu})$, $\varepsilon_{11}(\bar{\nu})$ (——), $\Delta\varepsilon_{22}(\bar{\nu})$, $\varepsilon_{22}(\bar{\nu})$ (----), and $\Delta\varepsilon_{33}(\bar{\nu})$, $\varepsilon_{33}(\bar{\nu})$ (······). The UV spectra ε (——) and the degree of anisotropy R (----), (C, left), the CD spectra $\Delta\varepsilon$ (······) and ACD spectra $\Delta\varepsilon^A$ (---) (C, right) are given for the cases shown in (A) and (B). ((●) intensity ratio $\varepsilon_{11} : \varepsilon_{22} : \varepsilon_{33} = 3 : 1 : 1$; (■) intensity ratio $\varepsilon_{11} : \varepsilon_{22} : \varepsilon_{33} = 1 : 2 : 2$.) R and $\Delta\varepsilon^A$ have been calculated for the order parameter $S^* = 0.460$ and $D^* = 0.121$ where D^* is calculated from Luckhurst's relation $D^* = F(S^*, \delta)$ with $\delta = 0.5$. According to Eqn [51] the assumed 0–0 transition and false origins are $\bar{\nu}_{00} = 37100$ cm^{-1}; $\bar{\nu}_{01}(b_1) = 1100$ cm^{-1} $\bar{\nu}_{02}$ $(b_2) = 560$ cm^{-1}; $\bar{\nu}_{03}(b_3) = 430$ cm^{-1}; $\bar{\nu}_0(a) = 1200$ cm^{-1}; the intensity ratio within the progressions is chosen equally for b_1, b_2, b_3: 1 : 6 : 10 : 8 : 6 : 4 : 1.

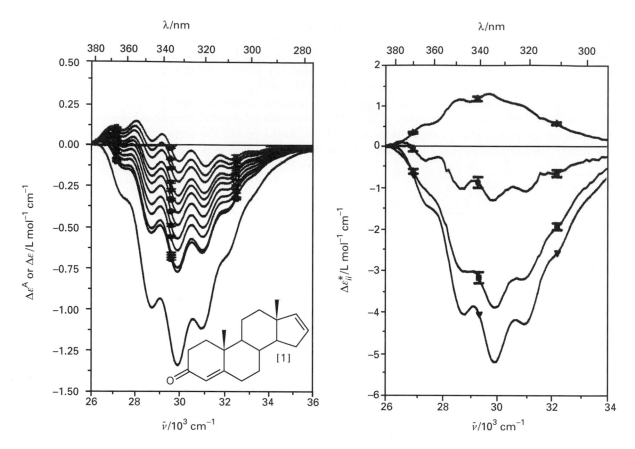

Figure 7 $\Delta\varepsilon$ (——) ($T = 75°C$) and temperature-dependent $\Delta\varepsilon^A$ (from above $T/°C = 31.2$ to 64.0) (left) and the tensor coordinates $\Delta\varepsilon_{11}^*$ (■), $\Delta\varepsilon_{22}^*$ (●), $\Delta\varepsilon_{33}^*$ (▲) and $\Delta\varepsilon_{11}^* + \Delta\varepsilon_{22}^*$ (▼) (right) of androst-4,17-dien-3-one [1] in the 'nematic' liquid crystal phase ZLI-1695 (Merck).

experimentally the circular dichroism $\Delta\varepsilon_{ij}$ or the rotational strength tensor R_{ij}^{NnKk} with respect to their own principal axes. For a molecule of D_2 symmetry, the orientation of the principal axes of R_{ij}^{NnKk} is symmetry determined.

Isotropic phases with suprastructural phase chirality

With ACD and CD spectroscopy, information can be obtained about suprastructural chirality of isotropic phases, e.g. cubic phases with suprastructural chirality. There is no interference with linear effects and thus $\Delta\varepsilon$ and $\Delta\varepsilon^A$ can easily be measured. A well-known example is crystalline $NaClO_3$. Another known example is that of liquid crystalline blue phases (O^2, O^8). Both with light in the region where λ is of the order of the periodicity within the phase and with wavelengths very different from the periodicity, a CD spectrum is obtained from which a detailed analysis of the phase structure has been derived, e.g. by Hornreich and colleagues.

Anisotropic phases with suprastructural phase chirality

Chiral anisotropic phases of chiral molecules

Anisotropic phases are often stabilized by chiral structures. Before discussing phenomena presented by such structures, the question must be answered of how chirality information about molecular chirality and suprastructural phase chirality is to be obtained from chirality measurements of anisotropic phases. Ruch and Schönhofer defined the notion of chirality measurement for an isotropic system: 'a chirality measurement of a property is a measurement with an ensemble of non-oriented molecules and this measurement leads for enantiomers to a result of the same value but with an opposite sign'. Extending this definition to anisotropic systems one might say that 'a chirality measurement of a property with an ensemble of oriented molecules, i.e. the anisotropic phase, is a measurement with an ensemble of non-oriented supermolecules – i.e. an ensemble of "anisotropic phases" – and this measurement leads for an enantiomer of such an ensemble of supermolecules to a

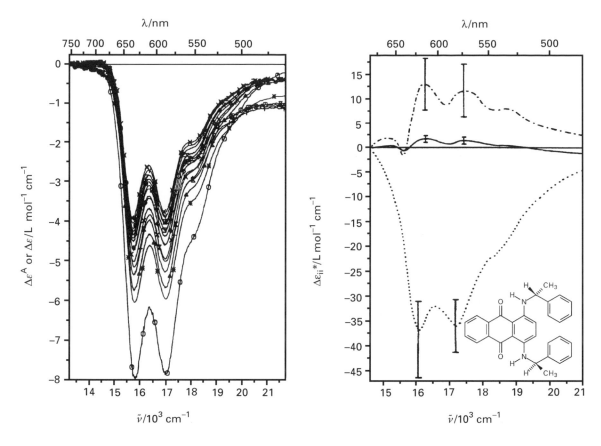

Figure 8 $\Delta\varepsilon$ (O)(T = 75°C) and temperature-dependent $\Delta\varepsilon^A$ (from above T/°C = 28 to 68) of 1,4-bis(R)-1-phenylethylamino)-9,10-anthraquinone [2] (left) and the tensor coordinates $\Delta\varepsilon^*_{11}$ (–•–) $\Delta\varepsilon^*_{22}$ (•••), $\Delta\varepsilon^*_{33}$(——) (right) of the charge transfer transition in the 'nematic' liquid crystal phase ZLI-1695 (Merck).

result of the same value but with an opposite sign.' Furthermore, for oriented molecules the anisotropy of chirality measurements has been demonstrated with ACD spectroscopy. In analogy to molecular properties, one must expect to find an anisotropy for any chirality measurement of a chiral anisotropic phase. Considering the chiral anisotropic phase as a supermolecule, as done before, a chirality measurement with CD spectroscopy is a measurement with the phase rotated about the direction of the propagating light beam analogously to **Figure 4**. Do such definitions hold in general? One question arises immediately from the knowledge that, for oriented achiral molecules of symmetry C_s, C_{2v}, D_{2d} and S_4 with suitable achiral positional and orientational order, nondiagonal elements of the rotational strength tensor R^{Nn}_{ij} can lead to a CD effect. This effect has been shown by Hobden using optical rotation measurements with crystals ($\overline{4}2m$ or $\overline{4}$, for example) without enantiomorphism. Is this an effect of chirality? Not really, because the CD is zero if the above definition of a chirality measurement/observation is applied.

For a homogeneous sample, the phenomenological constants a_2 and b_2 are responsible for the circular effects (Eqns [9] to [13]). But what does 'homogeneous' mean in a sample possessing positional and orientational order? Whether a phase represents a homogeneous or an inhomogeneous sample depends on the method used for the analysis. For an optical measurement, as for example with a cholesteric phase or a smectic C* phase, this depends on the chosen wavelength of the light, i.e. on whether the wavelength is close to or far from the dimension of the periodicity of the phase. In the first case Equations [9] to [13] are not valid because no 'internal reflection', originating from a periodicity of the inhomogeneity, has been taken into account. Berreman's theory and its extension for samples with very small pitch are discussed by Oldano and colleagues. The optical constants a_1, a_3, b_1 and b_3 that are responsible for local linear birefringence and dichroism are then according to the de Vries model the main origin for optical effects caused by the chirality of the sample.

Equations [9] to [13] integrated over the periodicity of a helical phase can also be applied to describe a

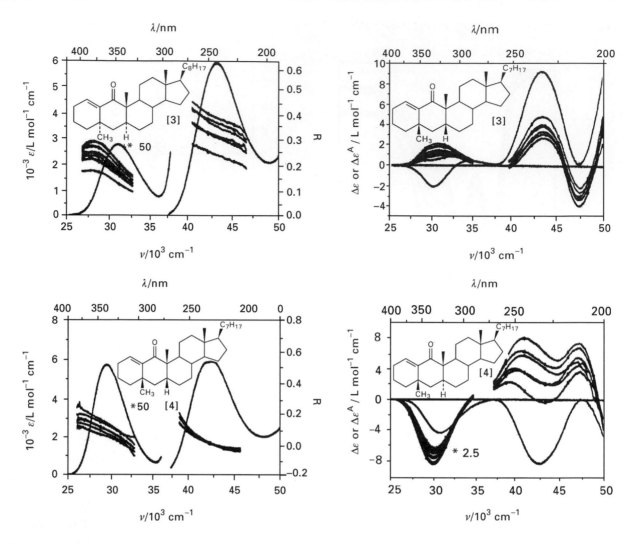

Figure 9 UV spectra ($T = 75°C$), temperature dependent degree of anisotropy ($T/°C = 28$ to 67.5) (left), $\Delta\varepsilon$ ($T = 75°C$) and temperature dependent $\Delta\varepsilon^A$ ($T/°C = 28$ to 67.5) (right) of the cisoid enones 3-methyl-3α,5α,3,3',4',5'-tetrahydrobenzo[2,3]cholest-2-en-1-one [3] (above) and its 3β-diastereomer [4] (below) in the 'nematic' liquid crystal phase ZLI-1695 (Merck).

circular dichroism often termed 'liquid crystal induced circular dichroism' (LCICD) in cases where the pitch $p \gg \lambda$ or $p \ll \lambda$. The CD originates from the helical arrangement of the absorbing molecules owing to their orientation in an anisotropic sample. The molecules and thus the chromophores dissolved in the anisotropic helical phase that are responsible for the CD can be chiral or even achiral. The LCICD is always magnitudes larger than the molecular CD in isotropic or anisotropic systems without suprastructural chirality. LCICD is found, for example, in cholesteric and chiral smectic C* liquid crystalline phases as shown by Gray. The sign of the LCICD provides information about the handedness of the phase.

Comparatively few optical investigations of chiral structures in anisotropic phases are known for two reasons. First with CD spectrometers based on polarization modulation, an artificial CD is often measured because of incorrect superposition of linear and circular effects in the signal processing. Secondly, the method for the decomposition of a measured quantity into the chiral and achiral part is not well developed. In the case of small linear and circular effects it is sufficient to expand the exponential (Eqn [11]) up to the first order. The linear approximation for $\mathbf{F}(\mathbf{A}')$ (Eqn [9]) can be applied and the contributions of linear and circular effects are additive, as for example for polymer or liquid crystalline polymer films. By measuring the state of polarization as a function of the azimuth of the sample, the CD originating from a chiral structure in the films can be evaluated. The expansion is, in general, not appropriate, because the linear effects are three to five orders of magnitude larger than the circular effects.

Chiral anisotropic phases of achiral molecules

In inorganic salts such as $NaClO_3$ discussed above, or quartz, chiral structures are built up from achiral species. An interesting field has been opened by the analysis by Takezoe and colleagues of the so-called 'banana shaped' molecules, which build up chiral domains together with phase separation, although, as is generally believed, they consist of achiral molecules. The analysis of such chiral phases by CD or ACD spectroscopy opens up a new field for an analysis of suprastructural chirality.

List of symbols

A = differential Mueller matrix; c = concentration (mol L^{-1}), velocity of light; d = path length (cm); D^* = order parameter $\sqrt{3}(g^*_{2233} - g^*_{1133})/2$; D^{NK}_{ij}, D^{NnKk}_{ij} = dipole transition moment tensor; $f(\alpha,\beta,\gamma)$ = orientational distribution function; F = Mueller matrix; F^{NnKk}, G^{NnKk} = spectral functions for $|Nn\rangle \rightarrow |Kk\rangle$; g_{ijkl} = orientational distribution coefficients; h = Planck constant; I = identity matrix; I = light intensity; $[M]$ = molar rotation; M = molecular mass (g mol^{-1}); n = refractive index, number density; N_A = Avogadro constant; q = concentration (g/100 cm^3); R^A = anisotropic rotational strength of a transition; R^{NK}_{ij}, R^{NnKk}_{ij} = rotational strength tensor; $\underline{s}$ = Stokes vector; S^* = order parameter = $(3g^*_{3333} - 1)/2$; X_{ij}, Y_{kl} = tensors of second rank for the molecules and the phase, respectively; α, β, γ = Euler angles; $[\alpha]$ = specific rotation; $\Delta\varepsilon$ = circular dichroism; $\Delta\varepsilon_{ij}$ = circular dichroism tensor; ε = absorption coefficient; ε_{ijk} = Levi–Civita tensor; $[\theta]$ = molar ellipticity; $\mu = \kappa$, λ = roots of the characteristic polynomial of Det $(A' - \mu I) = 0$; $\bar{\nu}$ = wavenumber; ρ = density (g cm^{-3}); ϕ, α = optical rotation in rad, degree.

See also: **Chiroptical Spectroscopy, Emission Theory; Chiroptical Spectroscopy, General Theory; NMR in Anisotropic Systems, Theory; Raman Optical Activity, Theory; Vibrational CD, Theory.**

Further reading

Azzam RMA and Bashara NM (1977) *Ellipsometry and Polarized Light*. Amsterdam: North-Holland.

Berreman DW (1972) Optics in stratified and anisotropic media: 4×4-matrix formulation. *Journal of the Optical Society of America* 62: 502–510.

Demus D, Goodby J, Gray GW, Spiess H-W and Vill V (eds) (1998) *Handbook of Liquid Crystals*. Weinheim: Wiley VCH.

De Rossi U, Dähne S, Meskers SCJ and Dekkers HPJM (1996) Spontane Bildung von optischer Aktivität in J-Aggregaten mit Davydov-Aufspaltung. *Angewandte Chemie* 108: 827–830.

Frelek J, Szczepek WJ, Weiss HP, Reiss GJ, Frank W, Brechtel J, Schultheis B and Kuball HG (1998) Chiroptical properties of the cisoid enone chromophore. *Journal of the American Chemical Society* 120: 7010–7019.

Guo J-X and Gray DG (1995) Induced circular dichroism as a probe of handedness in chiral nematic polymer solutions. *Liquid Crystals* 18: 571–580.

Hobden MV (1967) Optical activity in a non-enantiomorphous crystal silver gallium sulphide. *Nature* 216: 678.

Hornreich RM and Shtrikman S (1983) Theory of light scattering in cholesteric blue phases. *Physical Review A* 28: 1791–1807.

Katzenelson O, Zabrodsky Hel-Or H and Avnir D (1996) Chirality of large random supramolecular structures. *Chemistry – a European Journal* 2: 174–181.

Kuball H-G, Schultheis B, Klasen M, Frelek J and Schönhofer A (1993) Circular dichroism of oriented molecules. Electric dipole/magnetic dipole and electric dipole/electric quadrupole contribution for cholest-4-en-3-one. *Tetrahedron: Asymmetry* 4: 517–528.

Luckhurst GR and Gray GW (eds) (1979) *The Molecular Physics of Liquid Crystals*. London: Academic Press.

Mezey PG (ed) (1991) *New Developments in Molecular Chirality*. Dodrecht: Kluwer.

Michel J and Thulstrup EW (1995) *Spectroscopy with Polarized Light*. New York: VCH.

Nakanishi K, Berova N and Woody RW (eds) (1994) *Circular Dichroism – Principles and Application*. New York: VCH.

Oldano C and Rajteri M (1996) Optical activity of small-pitch helical-shaped dielectric media. *Physical Review B* 54: 10 273–10 276.

Quack M (1989) Structure and dynamics of chiral molecules. *Angewandte Chemie International Edition in English* 28: 571–586.

Samori B and Thulstrup EW (eds) (1988) *Polarized Spectroscopy of Ordered Systems*. NATO ASI Series, Series C: Mathematical and Physical Sciences 242. Dordrecht: Kluwer.

Schönhofer A and Kuball H-G (1987) Symmetry properties of the Mueller-matrix. *Chemical Physics* 115: 159–167.

Schönhofer A, Kuball H-G and Puebla C (1983) Optical activity of oriented molecules IX. Phenomenological Mueller matrix description of thick samples and of optical elements. *Chemical Physics* 76: 453–467.

Sekine T, Niori T, Watanabe J, Furukawa T, Choi SW and Takezoe H (1997) Spontaneous helix formation in smectic liquid crystals comprising achiral molecules. *Journal of Materials Chemistry* 7: 1307–1309.

Snatzke G (ed) (1967) *Optical Rotatory Dispersion and Circular Dichroism in Organic Chemistry*. London: Heyden.

Zabrodsky H and Avnir D (1995) Continuous symmetry measures. 4. Chirality. *Journal of the American Chemical Society* 117: 462–473.

Chromatography-IR, Applications

George Jalsovszky, Hungarian Academy of Sciences, Budapest, Hungary

Introduction

A chromatography–IR spectroscopy combination can be approached from two angles. The method can be seen as (i) a chromatograph with a specific IR-spectroscopic detector, to ensure selective detection or (ii) as an IR spectrometer preceded by a chromatographic device to provide samples of appropriate purity. In any case, the essence is a separation and subsequent identification of pure samples. Thus, a chromatography–IR spectroscopy combination is applicable to any analytical problem in which the components of a composite sample (mixture) are to be identified, and it can be expected that a chromatographic method can be applied successfully for separation and IR spectroscopy can give at least pieces of information for identification.

Before the introduction of FT-IR instruments, the spectroscopic measurement was always carried out off-line, separately from the chromatographic process, with the insertion of an often-tedious sample-preparation step, which limited the applicability. The multiplex and throughput advantages of FT-IR spectroscopy have significantly improved the sensitivity and reduced the time demand of IR-spectroscopic measurement, making it possible to perform chromatographic separation with simultaneous, on-line IR-spectroscopic detection and identification, opening up thereby an extremely wide field of application. The chromatographic methods that offer the possibility of on-line spectroscopic detection are gas chromatography (GC), high-performance liquid chromatography (HPLC) and supercritical-fluid chromatography (SFC). Thin-layer chromatography (TLC) can also be combined 'on-line' with IR spectroscopy insofar as spectra can be taken directly from the chromatographic plate.

Chromatographs and FT-IR spectrometers can be interfaced by two distinct approaches. In the first, the effluent from a chromatographic column is passed through a flowcell (often called 'light pipe' in GC/FT-IR) together with the chromatographic carrier, where infrared spectra are measured continuously. In the second approach the carrier is removed, the eluates are deposited on a suitable substrate (as such or isolated in a matrix), and the infrared measurements are made some time after deposition. Thus, when discussing the applications of GC/FT-IR, HPLC/FT-IR and SFC/FT-IR techniques, flowcell techniques and deposition techniques will be treated separately. TLC/FT-IR can be accomplished *in situ* or by transferring the eluates from the plate to an IR-transparent material, but these techniques will be discussed only briefly in a single section.

GC/FT-IR – flow cell (light pipe) approach

This method is applicable to any sample that can be evaporated in the injector of the GC, separated into its components on the column, and which, mixed with the carrier gas, remains in the gas phase and suffers no decomposition in the heated transfer line and the flowcell (light pipe) of the GC/FT-IR interface. If this is not the case, HPLC or SFC should be used for separation. However, this restriction still leaves an extremely wide choice of possible applications, of which only some examples are discussed below.

Industrial applications

In industrial processes starting materials, intermediates and end products are regularly analysed, in which GC/FT-IR may have its due share. For example, solvents are recycled after distillation or other purification steps. The purity of solvents is checked before recycling by gas chromatography, where eventual impurities, once known, are identified on the basis of retention times. In the case of unknown impurities GC/FT-IR or GC/MS methods must be used for identification, of which the former is particularly useful if aromatic isomers are to be distinguished. Another important area is monitoring reactions like catalytic oxidation or polymerization.

Petrochemistry

The distinction of aliphatic isomers and homologues in crude oil and gasoline fractions by IR spectroscopy is a hard task. However, the method is very powerful if the identification of aromatic compounds in oil fractions is required. The discrimination ability of

IR spectroscopy is also high when the substitutional isomers of polycyclic aromatic hydrocarbons (PAH) are to be identified. The method is also applicable in the analysis of antioxidants used in oils and fuels.

Pollutants in the environment

One of the first applications of GC/FT-IR was, in 1979, the identification of alkyl-9-fluorenones in diesel exhaust particulate extracts. Some years later in the EPA laboratory of Athens, GA, USA and then in the EPA laboratory of Las Vegas, GC/FT-IR was applied to identify toxic substances in environmental-sample extracts. In the latter laboratory, the minimum identifiable quantities of 30 priority pollutants and 25 other environmental contaminants were determined. The method was shown to be applicable for the analysis of hazardous waste and industrial wastewater extracts. The gas-phase spectra of 47 compounds of environmental importance were also taken, and their characteristic absorption frequencies and specific absorbances were published.

Important groups of pollutants are pesticides, fungicides and insecticides. Their toxicity has made it necessary to monitor their presence in the food chain, soil or water. Analytical and monitoring techniques first relied on the GC analysis of a limited number of target compounds. However, in the identification of other residues that were suspect but not among the preset target compounds, IR-spectroscopic information often proved to be vital. There are three problems making this analysis difficult: the large number of components, the presence of a number of conformers or isomers with widely varying degree of toxicity or biochemical activity, and the possibility of the formation of new degradation products. GC/FT-IR is suitable for tackling all the three, being particularly useful in the differentiation between isomers (often superior to MS) and in the identification of degradation products.

Of the other pollutants, polychloro-dibenzodioxins have attracted special attention, where a crucial point is the differentiation between isomers, since some of them are extremely poisonous. Of these compounds the vapour-phase IR spectra of 22 tetrachloro-dibenzodioxins (TCDDs) were taken to prove that the TCDD isomers have unique IR spectra. Indeed, spectra–structure correlations showed that each TCDD isomer has characteristic, individual asymmetric and symmetric COC stretching frequencies, on the basis of which they can be safely distinguished at low-microgram concentrations from other, less-toxic isomers.

GC/FT-IR' can also be used to identify pesticide degradation products, including low-volatility pesticides and their residues in biological samples. For example, phenols and 2,4-dichlorophenoxyacetic acid derivatives were analysed in body fluids of humans exposed to pesticides. More recent topics of analysis are complex marine mixtures, nitrated polycyclic aromatic hydrocarbons in environmental samples, aromatic isomers in marine samples and fatty acid methyl esters in the marine environment.

Natural products of vegetable origin (flavours, fragrances and essential oils)

Essential oils are complex mixtures for which conventional GC techniques (retention index) are insufficient for complete identification. The need for the information supplied by the infrared spectrum justifies here the use of GC/FT-IR. In a similar manner to pesticides, particular flavours are often associated with one of the possible several isomers or conformers.

For example, the essential oil of Anthemis nobilis L. is a mixture of terpenoids and saturated and unsaturated C_4 and C_5 acids esterified by saturated or unsaturated C_3 to C_6 alcohols. GC/MS with electron-impact ionization fails to distinguish the angelates and tiglates of unsaturated alcohols, whereas these geometric isomers have widely different characteristic frequencies in the infrared. The same holds for the distinction between crotonates and methacrylates or between linear and branched ethers. A further interesting example is the case of α-pinene and γ-terpinene with very similar MS but sufficiently specific IR spectra. GC/FT-IR has been used to analyse a number of other essential oils, like the flavour components in kiwi, peppermint oil, lemon oil or the essential oil extracted from celery seed. An analogous task was the identification of 43 components in the headspace of coffee samples.

Biomedical applications

GC/FT-IR can be used in biomedicine to detect and identify medicaments, toxic compounds and narcotics as well as their metabolites in biological fluids (blood, urine, bile, etc.). Of course, samples must be pre-treated prior to GC/FT-IR analysis to have relatively clean extracts. The metabolic pathways of drugs or potential drugs can also be established by analysing biological fluids for the decomposition products. An interesting example is the metabolism of trifluoromethylethyl benzhydrol, an effective hepatic enzyme inducer. Many of its metabolites are polyhydroxy compounds, the isomers of which could be distinguished only by GC/FT-IR.

In toxicology it is very important to identify the product that is responsible for fatal poisoning. A

GC/FT-IR method has been elaborated, e.g. for the identification of paraquat, a herbicide, and for the diagnosis of fatal poisoning due to the vapours of volatile solvents (acetone, diethyl ether, trichloroethylene, etc.) used by young drug addicts. In the analysis of narcotics and drugs of abuse (heroin, cannabis, LSD, cocaine, amphetamines, etc.) GC/FT-IR is essential in identifying the numerous isomers, and also applicable in the analysis of barbiturates or stimulants abused as drugs. By identifying impurities, the 'fingerprint' of drugs can be determined, which can serve as a basis of checking the geographical origin of drugs of abuse.

Polymer analysis

GC/FT-IR is highly applicable in the study of polymers. On heating the sample the polymer is degassed or (partly or completely) decomposed, and thus gases are produced. Pyrolysis can be used to determine the structure of polymers on the basis of the resulting fragments. An example is the pyrolytic study of polybutadiene in which a number of monosubstituted vinyl compounds can be distinguished on the basis of their characteristic C–H out-of-plane frequencies. A similar approach was used in the analysis of elastomers applied in the automobile industry and of epoxy–phenolic resins. Pyrolysis-GC/FT-IR is applicable in the study of the thermal degradation of thermoresistant amide, imide and diimide polymers. Although not directly related to polymer chemistry, it is noted that substructure components of various aquatic humic substances can also be identified by this method. Thermogravimetric analysis (TGA) can also be combined with GC/FT-IR: an example is a study of the decomposition of ethylene–vinyl acetate copolymers by TGA/GC/FT-IR analysis.

Miscellaneous applications

GC/FT-IR can be used for amino acid profiling with *tert*-butyldimethylsilyl derivatization, for the investigation of food-packaging materials, the analysis of biologically active wheat straw extracts with allopathic activity, that of *cis*- and *trans*-octadecanoic acids in margarines and of sugar units in disaccharides, for the identification of small molecular migrants extracted from γ-irradiated plastic laminates or the radiolysis products of γ-irradiated medazepam solution.

GC/FT-IR – subambient solute trapping (direct-deposition) approach

In this direct-deposition (DD) technique, the eluates are deposited on a moving ZnSe window cooled by liquid nitrogen, and the spectra are measured continuously, a short time after deposition. The resulting spectra are very different from gas-phase spectra, being more similar to spectra of samples prepared as a KBr disk or Nujol mull; thus they may be searched against standard condensed-phase reference libraries, as was shown for the example of barbiturates (barbital, apobarbital, butabarbital, phenobarbital). This is an advantage over matrix isolation (MI) detection (see below). Detection limits for on-line measurements are almost two orders of magnitude lower than in the case of a flowcell interface, and the limits can be pushed down by post-run spectrum recording, which is easy to perform with DD interfaces.

An interesting application of this technique was the analysis of β-agonists in cattle. These compounds are a group of N-phenylethanolamines illegally used in the treatment of veal and cattle. In this study tissue samples were analysed for the presence of clenbuterol, mabuterol and salbutamol, all of which were trimethylsilylated before analysis. The limit of detection was ~1 to 2.5 ng μL^{-1}. Another important application is in forensic toxicology, where the compound must be identified by at least two independent methods. Here, GC/DD/FT-IR is a good second technique in addition to GC/MS. A characteristic application is the distinction between amphetamines and metamphetamines in post-mortem analyses.

GC/FT-IR – matrix-isolation (MI) approach

The isolation and structure elucidation of complex mixtures require high-resolution chromatographic separation coupled with advanced spectroscopic techniques. This is offered by matrix isolation (MI), based on mixing the eluate with a noble gas (e.g. argon) and depositing the mixture on a metal surface cooled to a temperature of 10 K, where the eluate molecules are trapped in a matrix of the noble gas. The analysis of essential oils may require such an approach. For instance, peppermint oil contains (–)-menthol as the main component causing a cooling sensation, whereas other isomers (neomenthol, isomenthol and neoisomenthol), though structurally similar, have no such effect. Mass spectrometry is unable to differentiate between these isomers; however, the MI/FT-IR spectra show characteristic differences (see **Figure 1**). Soya bean oil is an important vegetable oil, in which the presence of unsaturated fatty acids causes instability. GC/MI/FT-IR can be applied successfully to determine the presence and structure of any unsaturated species. In essential oils used for aroma, flavour and fragrance purposes, stereochemical

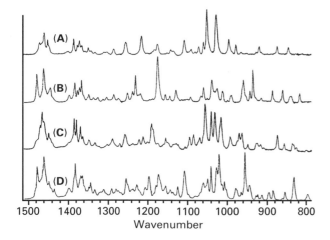

Figure 1 MI/FT-IR spectra (fingerprint region) of (A) menthol, (B) neomenthol, (C) isomenthol and (D) neoisomenthol, 25 ng each. Reproduced with permission of the Society for Applied Spectroscopy from Coleman III WM and Gordon BM (1989) Examinations of the matrix isolation infrared spectra of organic compounds: Part XIII. *Applied Spectroscopy* **43**: 303.

conformations and configurations play an important role in sensory responses. For instance, geraniol and nerol in coriander oil differ only in the *trans* vs. *cis* conformation of the double bond, which is readily distinguished on the basis of the MI/FT-IR spectrum.

Due to the low-nanogram sensitivity of the GC/MI/FT-IR technique, minor components of natural essential oils can be detected and safely identified on the basis of spectral differences due to positional isomers. Similarly to the light pipe approach, or even more so, this technique is fitted to the identification of the isomers of polycyclic aromatic hydrocarbons (PAH), e.g. in soil samples. Picogram levels of polychlorinated biphenyls and chlorinated dibenzo-*p*-dioxins can be determined as well as nitrocresols in air-sample extracts.

HPLC/FT-IR – flowcell approach

The first flowcell experiment was reported in 1975, and some years later in 1979 Vidrine reviewed the most important points of this technique, limited at that time to size-exclusion chromatography (SEC). The use of flowcell methods with normal- and reversed-phase chromatographic separations has proven more difficult, since for a successful coupling the spectroscopic properties of the solvent system should meet some minimum requirements. In this respect deuterated solvents seemed to be promising, and numerous applications have been reported on normal-phase, flowcell HPLC/FT-IR, in which deuterated or halogenated solvents were used. Hydrocarbons, esters, ketones, phenols, amines and

azarenes were easily separated and detected. To increase chromatographic performance, small amounts of polar modifiers can be added to the mobile phase without seriously affecting IR-detection ability. However, gradient elution, or traces of water, buffer or alcohol proved to be incompatible with the flowcell method. To overcome this problem, the aqueous effluent can be extracted with an IR-compatible solvent and the resulting solution measured in the flowcell.

Using microbore LC columns, a separation method was developed for a series of weakly basic cyclic and noncyclic secondary amines, which were identified as components of coal-derived solvents. As shown in another study, using normal HPLC on silica gel with Freon-113 elution, model mixtures of aliphatic and aromatic hydrocarbons and nonpolar constituents can be separated in coal-liquefaction process solvents. Encouraging results were obtained on both semipreparative (4.6 mm i.d.) and microbore (1 mm i.d.) columns. According to further reports, carbamate pesticides, polymer additives and solvent-refined coal were analysed, and the components of bergamot oil can be identified. GPC/FT-IR can be used to detect components of cold-rolling oil and to analyse polymers, whereas SEC/FT-IR can be applied to the analysis of coal liquids and to improve detection and identification of proteins.

HPLC/FT-IR – solvent-elimination approach

With the removal of the mobile phase, IR detection no longer restricts the selectivity of the chromatographic process through solvent-transparency requirements. Solvent elimination is more difficult with reversed-phase chromatography because of the presence of water. This problem can be overcome by extraction with organic solvent, and deposition of the organic solution in a known manner, or by a postcolumn reaction with 2,2-dimethoxypropane, converting water into methanol and acetone, which have sufficient volatility for the deposition technique.

In the first realization of solvent-elimination HPLC/FT-IR the solution eluting from the column is concentrated in a short heated tube, then passed dropwise onto a cup filled with KCl powder, held in a carousel. After the evaporation of solvent, the IR spectrum can be taken in diffuse reflection (DR). In another approach (called 'buffer memory') the effluent is deposited on a KBr plate as a narrow, continuous band, and IR spectra are measured in transmission. This method was illustrated for the example of polystyrene and Carbowax 6000. According to a third technique, using microbore columns, eluates are again deposited on KCl, and IR spectra are measured in DR. The performance was investigated on a test dye solution (Butter Yellow and Indophenol Blue), and on a mixture of nitrobenzene and 4-chloronitrobenzene.

It is worth noting that over the 15 years of the history of HPLC/FT-IR only a limited number of real applications have been reported in scientific papers, most of them dealing with improvements on the technique or investigations on its performance. Of the few applications, the analysis of chloropyrene isomers and congeners, organics in water and polymer-composition distribution can be mentioned.

SFC/FT-IR – flowcell approach

The development of SFC/FT-IR has closely parallelled that of HPLC/FT-IR. In both cases the first demonstration of the technique was the flowcell approach, and the solvent-elimination techniques (using transmission and diffuse-reflection arrangements) appeared slightly later. The flowcell approach in SFC appeared to be superior to the same in HPLC for two basic reasons: (i) the preferred SFC mobile phases transmit IR radiation over a larger frequency range than many HPLC solvents, and (ii) density programming in SFC poses fewer difficulties in solvent subtraction than the solvent-gradient technique used in HPLC analysis.

The choice of a suitable mobile phase is subject to the following conditions: (i) minimum absorption in the mid IR, (ii) adequate solvation of the molecules of interest in SFC, (iii) high solute diffusivities to permit high column efficiencies and (iv) a high degree of selectivity as a function of pressure variation. Whereas the most common candidate, supercritical carbon dioxide, fulfils many of these conditions, its polarity and solvent strength are only moderate. However, adding small amounts of more polar compounds to carbon dioxide may result in increase of solvent strength without substantially compromising IR measurement (although some authors are of different opinion). Still, even without modifiers, the absorption of carbon dioxide may cause problems, for instance in the identification of hydroxy compounds (the OH stretching region coincides with a strong carbon dioxide absorption band). Thus, some of the authors prefer using xenon as the mobile phase, which has no IR absorption.

The first demonstration of SFC/FT-IR coupling was reported in 1983 with the use of a wide-bore fused-silica capillary column, supercritical carbon dioxide and a flowcell of 1 cm pathlength. With this setup, easily identifiable spectra can be obtained for 3 µg each of anisole, acetophenone and nitrobenzene.

An interesting application is the analysis of ultraviolet-curing inks and coatings. These formulations contain oligomer, reactive diluent and photoinitiator, where the oligomers are low- to medium-molecular weight multifunctional, unsaturated materials. SFC/FT-IR is an ideal method for their analysis, since it allows high-resolution separation of mixtures at low temperatures, while the components are readily identified on the basis of their characteristic IR spectra. This is shown, e.g., by the analysis of the volatile components of citrus oil and that of model steroid mixtures, pyrethrins (naturally occurring esters with insecticidal activity). Typical identification limits, as studied with caffeine as model, are in the 2 ng range. In a commercial preparation of pyrethrin extract, the six insecticidally active components were separated by SFC, since thermal degradation was observed when attempting GC separation. IR spectra of the components taken in the flowcell clearly reveal the structural differences between them. Practical applications of SFC/FT-IR include the determination of non-volatile compounds from microwave susceptor packaging that may migrate into the heated food. Another application is the analysis of fibre finishes on fibre/textile matrices.

SFC/FT-IR – solvent-elimination approach

In contrast to flowcell interfaces, solvent-elimination approaches lead to spectra free of solvent interferences. Various sampling techniques are possible: the sample can be deposited on a flat ZnSe plate, on a smooth metallic substrate or a thin layer of powdered alkali halide salt, whereas the spectrum can be taken in conventional transmission, external-reflection or diffuse-reflection arrangement. In one of the first applications a synthetic mixture of three quinones was separated on a microbore column packed with silica, using a mobile phase of 5% methanol in supercritical carbon dioxide. The peaks were collected on a plate on which a layer of KCl powder was deposited, and then spectra were measured by a diffuse-reflection accessory. Test measurements on acenaphthenequinone (AQ) showed later that conventional transmission spectra of samples on flat infrared-transmitting windows give the best compromise between high sensitivity, correct relative band intensities and adherence to the Lambert–Beer law.

The most up-to-date device is a real-time (on-line) SFC/FT-IR interface based on a commercial direct-deposition GC/FT-IR system, described by Norton and Griffiths, which allows spectra of subnanogram quantities of analytes to be measured over the complete mid-IR region. It is possible to get identifiable spectra from injections as low as 600 pg for strong IR absorbers and between 1 to 10 ng for weaker IR absorbers. The interface also allows for usage of polar modifiers in the mobile phase.

TLC/FT-IR

This combination can be approached in two ways: either the FT-IR measurement is performed *in situ*, directly on the TLC plate, or the analytes are transferred from the plate to an IR-transparent substrate prior to the FT-IR measurement. Interfaces for both methods are now commercially available. NIR and photoacoustic FT-IR detection have also been applied, although to a lesser extent, in the coupling of TLC with IR spectroscopy.

Conclusion

It is important to note that scientific papers or reviews are not the most appropriate sources of information on the application of chromatography-IR spectroscopy combinations. These techniques, like most of the techniques of instrumental analysis, are widely applied just as an analytical tool in research laboratories and in industry. Typical accounts of these applications are progress reports, technological prescriptions or patent applications, rather than scientific papers.

When considering the application of chromatography-IR spectroscopy combination techniques, three questions should be answered: (i) is the problem the analysis of a composite sample? (ii) can the sample be separated by a chromatographic method? (iii) is there a chance that the IR spectrum of the components will help identification? If the answer is 'yes' to these questions, a chromatography-IR spectroscopy combination is a promising approach to the analytical problem.

See also: **ATR and Reflectance IR Spectroscopy, Applications; Chromatography–IR, Methods and Instrumentation; Fourier Transformation and Sampling Theory; Industrial Applications of IR and Raman Spectroscopy; IR Spectral Group Frequencies of Organic Compounds; IR Spectrometers; IR Spectroscopy Sample Preparation Methods; Medical Science Applications of IR; Polymer Applications of IR and Raman Spectroscopy; Quantitative Analysis.**

Further reading

Coleman III WM and Gordon BM (1994) Analysis of natural products by gas chromatography/matrix isolation/infrared spectrometry. In: Brown PR and Grushka E (eds) *Advances in Chromatography* 34: 57–108.

Fuoco R, Pentoney L and Griffiths PR (1989) Comparison of sampling techniques for combined supercritical fluid chromatography and Fourier transform infrared spectrometry with mobile phase elimination. *Analytical Chemistry* 61: 2212–2218.

Griffiths PR, Pentoney SL, Giorgetti A and Shafer KH (1986) Hyphenation of chromatography and FT-IR spectrometry. *Analytical Chemistry* 58: 1349A–1366A.

Herres W (1987) *HRGC-FTIR: Capillary Gas Chromatography–Fourier Transform Infrared Spectroscopy: Theory and Applications.* Heidelberg: Hüthig Verlag.

Lacroix B, Huvenne JP and Deveaux M (1989) Gas chromatography with Fourier transform infrared spectrometry for biomedical applications. *Journal of Chromatography* 492: 109–136.

Norton KL and Griffiths PR (1995) Performance characteristics of a real-time direct deposition supercritical fluid chromatography–Fourier transform infrared spectrophotometry system. *Journal of Chromatography A* 703: 503–522.

Ragunathan N, Krock KA, Klavun C, Sasaki TA and Wilkins CL (1995) Review: Multispectral detection for gas chromatography. *Journal of Chromatography A* 703: 335–382.

Somsen GW, Morden W and Wilson ID (1995) Review: Planar chromatography coupled with spectroscopic techniques. *Journal of Chromatography A* 703: 613–665.

Taylor LT and Calvey EM (1989) Supercritical fluid chromatography with infrared detection. *Chemical Reviews* 89: 321–330.

Vidrine DW (1979) Liquid chromatography detection using FT-IR. In: *Fourier Transform Infrared Spectroscopy – Applications to Chemical Systems*, Vol. 2, pp 129–164. New York: Academic Press.

Chromatography-IR, Methods and Instrumentation

Robert L White, University of Oklahoma, OK, USA

VIBRATIONAL, ROTATIONAL & RAMAN SPECTROSCOPIES
Methods & Instrumentation

Linking chromatographic separation methods with IR spectroscopic detection of separated components results in powerful chemical analysis tools. As with any hyphenated analysis method, combining mixture component separation with infrared spectroscopic detection provides capabilities that surpass those of the constituent techniques. Unlike conventional chromatography detectors, IR analysis provides detailed structural information regarding complex mixture components. In favourable cases, mixture components can be unequivocally identified based on their chromatographic retention times and IR spectra. The performance characteristics of chromatography-IR combinations are largely determined by the interface employed. The ideal chromatography-IR interface would link the two techniques without sacrificing the performance of either. In practice, instrument performance tradeoffs are unavoidable. A variety of interfaces have been developed in attempts to minimize these tradeoffs when gas chromatography, liquid chromatography, supercritical fluid chromatography, and thin layer chromatography are combined with IR spectroscopic detection.

Inherent in all chromatography-IR combinations is the need for rapid IR spectrum measurement. Fourier transform IR spectrophotometry (FT-IR) provides rapid scanning and is the preferred IR spectroscopic technique for detection of species separated by chromatographic methods. Instrumental requirements for chromatography-IR interfaces depend on many factors. In particular, the physical state of matter that must be analysed determines which spectroscopic sampling techniques can be employed. For example, gas chromatography detectors must provide a means of analysing flowing vapours. Liquid and supercritical fluid chromatography detection requires analysis of flowing liquids. IR detection of thin layer chromatography solutes presents the particularly difficult problem of analysing species adsorbed on infrared opaque substrates.

Chromatogram generation

Regardless of the type of chromatography implemented or the specific chromatography-IR interface employed, separated mixture components are ultimately represented by a plot of detector response as a function of separation time or migration distance. There are two common procedures for converting structure-specific IR spectroscopic information into chromatogram plots. The Gram–Schmidt vector orthogonalization method uses subtle variations in interferogram data acquired during FT-IR scans to detect solute elutions. The functional group chromatogram method is more computationally intensive and requires interferogram Fourier transformation and calculation of absorbance spectra, but can be used to elucidate mixture component structural features.

For Gram–Schmidt chromatogram generation, a portion of each acquired FT-IR interferogram is selected to represent a multidimensional vector. The number of interferogram data points used to form vectors determines the dimensionality of the vector space and thus the number of orthogonal axes needed to define the space. Vectors derived from interferograms obtained when no solute is present in the IR beam are similar and are confined to the same region of the vector space. By using the Gram–Schmidt orthogonalization method, a few of these background vectors are used to calculate orthogonal axes that are representative of the portion of the vector space in which background vectors are found. Chromatogram generation is accomplished by comparing the orientation of measured interferogram vectors with the background subspace. Interferograms acquired when solute is in the IR beam have vector components outside of the background subspace. The sum of vector residuals remaining after removing those components that lie in the background space are plotted as a function of elution time or migration distance to yield a chromatogram. Gram–Schmidt chromatogram intensity is therefore a measure of how much interferogram vectors differ from the background subspace, a quantity not necessarily related to IR absorbance. The signal-to-noise ratio of Gram–Schmidt generated chromatograms depends on the number of interferogram points used to define vectors, the number of orthogonal axes used to specify the background subspace, and the interferogram segment used to create vectors.

Functional group specific chromatograms are generated from chromatography-IR absorbance spectra by plotting the integrated absorbance within selected wavelength regions as a function of elution time or migration distance. Significantly more calculations are required to create functional group specific chromatograms than to generate Gram–Schmidt chromatograms. However, functional group specific chromatograms provide much more information regarding separated species structures than Gram–Schmidt chromatograms. Typically, multiple wavelength regions are integrated to produce several functional group specific chromatograms. When separated species contain functionalities that absorb within a selected wavelength region, a peak representing the elution of this substance will appear in the chromatogram. Total IR absorbance chromatograms are generated by integrating spectra over the entire mid-infrared range (e.g. 4000–400 cm^{-1}). The total IR absorbance chromatogram contains peaks representing all infrared absorbing solutes.

Gas chromatography-IR

The most commonly employed chromatography-IR combination is between gas chromatography (GC) and FT-IR. Interfaces for GC-IR can be categorized as either flow cell (light pipe) or solute trapping. **Figure 1** illustrates the three GC-IR interfaces described in the following sections.

Light pipe GC-IR

The first on-line GC-IR interface consisted of a heated flow-through gas cell or 'light pipe'. Simple and

inexpensive light pipe interfaces constitute the most common method employed for combining gas chromatography with infrared spectrophotometry. In these combinations, IR radiation from an FT-IR source is directed through a light pipe, which is a glass tube with a gold-coated inner surface through which the gas chromatographic effluent flows. The light pipe contains no chromatographic stationary phase. Therefore, it does not contribute to mixture component separation. In fact, the light pipe dead volume results in degradation of chromatographic resolution. To minimize chromatographic resolution loss, small light pipe volumes are employed. Typical GC-IR light pipes used with capillary gas chromatography columns have volumes of about 100 μL. To maximize IR radiation flux and optical path length, light pipes with diameters of about 1 mm and lengths of 12–15 cm are commonly employed. Light pipe ends must be sealed in order to conduct GC effluent without loss. This is accomplished by attaching IR transparent alkali halide windows to the light pipe ends. IR transparent gases such as He and N$_2$ are employed for the gas chromatographic mobile phase so that acquired spectra represent separated mixture components only. Vapour phase IR spectra obtained by light pipe GC-IR contain sharp spectral features because solute–solute interactions are minimal. When small molecules are detected at sufficiently high spectral resolution, vibration bands may exhibit rotational fine structure.

Subambient solute trapping GC-IR

Mixture components can be removed from GC effluent by trapping them on a cold surface. Subambient solute trapping GC-IR interfaces take advantage of the large differences between the volatilities of separated mixture components and GC carrier gas. When GC effluent is directed onto a surface maintained at a temperature at which mixture components condense but carrier gas does not, mixture components can be isolated from the GC carrier gas stream. In practice, mixture components are deposited on a cooled IR transparent window that is continuously moved during the chromatographic separation so that eluting solutes are deposited at different locations. IR analysis is achieved by measuring transmittance spectra of the frozen solutes. In contrast to light pipe GC-IR, trapped solute infrared measurements are not restricted to the time required for species to elute from the chromatographic column. Extended signal-averaging can be used to improve spectrum quality because IR analysis of frozen solutes is conducted off-line. Unlike light pipe GC-IR, subambient solute trapping GC-IR yields condensed phase IR spectra.

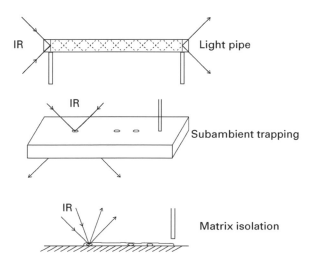

Figure 1 Gas chromatography-IR interfaces.

Condensed phase spectra contain broader features than vapour phase spectra and reflect significant solute–solute interactions.

Matrix isolation GC-IR

Matrix isolation GC-IR is a variation of the subambient solute trapping technique that yields significantly different spectral features. Rather than freezing mixture components, molecules are trapped within an inert argon matrix. Interactions between solute molecules are virtually eliminated by the argon matrix. As a result, spectra contain much sharper features than those obtained by subambient solute trapping without matrix isolation. Instead of depositing solutes on a cooled infrared transparent window, matrix isolated species are formed on a reflective gold surface under vacuum and maintained near 10 K. Argon matrix gas is provided by using a pre-mixed GC carrier gas composed of helium and about 1% argon. Helium is not condensed at 10 K and is removed by the vacuum system. The argon and separated mixture components condense on the cryogenic surface. Because argon is present in the carrier gas at much higher concentration than mixture components, molecules deposited on the surface are surrounded by condensed argon. The argon matrix isolates individual solute molecules and prevents solute–solute interactions. IR spectra for separated mixture components are measured in a similar manner to that described for subambient solute trapping. However, reflectance rather than transmittance techniques are employed for IR analysis. Matrix isolation IR spectra exhibit sharp features because solute–solute interactions and molecular rotations are absent.

Liquid chromatography-IR

Compared to GC-IR, liquid chromatography (LC)-IR measurements are often more challenging because mobile phases are not necessarily infrared transparent. Due to the fact that mobile phase concentration greatly exceeds that for mixture components in LC effluent, even mobile phases that have weak IR absorptivities can present significant problems. LC-IR interfaces can be categorized as flow cell or mobile phase elimination. **Figure 2** illustrates the three LC-IR interfaces described in the following sections.

Flow Cell LC-IR

LC-IR flow cells consist of two IR transparent windows separated by a spacer with a thickness that determines the cell path length. Alkali halide windows are commonly used with non-polar LC mobile

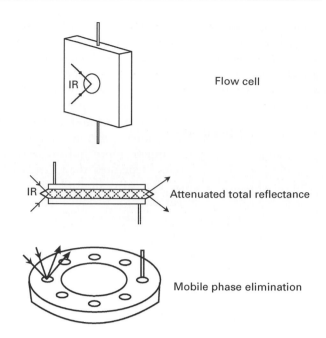

Figure 2 Liquid chromatography-IR interfaces.

phases whereas zinc selenide windows are used with polar mobile phases. Unlike GC-IR, optimum flow cell path lengths for LC-IR are dictated by the absorptivity of the mobile phase. Path lengths of 0.1–0.2 mm are commonly employed with moderately absorbing mobile phases. For highly absorbing mobile phases (e.g. water), path lengths of about 10 μm are often required. Mixture component IR spectra are obtained from flow cell measurements after subtracting contributions from the mobile phase.

Aqueous solution IR analysis by flow cell LC-IR is particularly difficult because of the need for extremely short flow cell path lengths. Short path length flow cells are easily clogged and internal pressure fluctuations caused by mobile phase movement through the cell can appreciably change the IR beam path length. Fortunately, attenuated total reflection (ATR) provides an alternative method for attaining short path length IR spectrum measurements. ATR flow cells consist of a high refractive index infrared transparent material through which the IR beam is passed and around which LC mobile phase flows. When the incident IR beam angle at the ATR material/mobile phase boundary exceeds the critical angle, it is totally reflected. At each internal reflection, the electric field of the IR radiation penetrates slightly into the LC mobile phase. This 'evanescent wave' can be absorbed by species dissolved in the mobile phase. The penetration depth of IR radiation at each reflection varies with wavelength and is typically on the order of μm. By using appropriate length ATR materials, multiple reflections can be achieved to provide reproducible IR beam path lengths of 10–20 μm. Thus, by using ATR, short flow cell path lengths can be attained without the disadvantages of clogging and pressure fluctuations suffered by conventional flow cell designs.

Mobile Phase Elimination LC-IR

One method of avoiding the problems associated with the fact that most LC mobile phases absorb IR radiation is to remove these solvents prior to IR analysis. Like subambient solute trapping GC-IR, this approach exploits the differences between solute and mobile phase volatilities. By heating the end of the LC column, the mobile phase can be vaporized and removed by aspiration. Mixture components can then be deposited onto an IR transparent substrate. In early interface designs, solutes were deposited on IR transparent potassium chloride powder contained in sample cups. In this system, multiple sample cups are contained in a rotating carousel. The carousel is rotated during LC separation to collect individual mixture components as they elute from the LC column. IR analysis is performed off-line by using diffuse reflectance optics. Like solute trapping GC-IR interfaces, there are no limits to the extent of signal averaging that can be applied to measure mobile phase elimination LC-IR mixture component spectra.

Column end heating can be an effective method for removing organic mobile phases such as *n*-hexane, but is not very effective for reverse phase separations that employ aqueous mobile phases. One method used to remove solutes from aqueous mobile phases is by extracting them into an organic solvent followed by solute deposition on an IR transparent substrate and solvent evaporation. However, on-line continuous solvent extraction greatly complicates LC-IR interfaces.

Microbore LC-IR

Although most LC separations are conducted with analytical scale columns, microbore LC columns provide better resolution and are becoming increasingly popular. The quantity of mobile phase needed for a separation by microbore LC is much less than that required for analytical scale LC separations. In addition, optimal flow rates for microbore LC are significantly lower than for analytical scale LC. These features of microbore LC simplify both flow cell and mobile phase elimination approaches to LC-IR. In addition, because small quantities of mobile phase are employed for microbore LC separations, expensive mobile phases that may not otherwise be practical can be employed. For example, the C–H stretching region of solute IR spectra can be accessed in reverse phase flow cell LC-IR measurements by employing deuterated solvents such as D_2O and CD_3CN as mobile phases. Because small amounts of mobile phase are employed for microbore LC-IR, less solvent must be removed to achieve mobile phase elimination. Rather than using a carousel of sample cups, a single IR transparent window (or reflective surface) that is moved beneath the column exit can be employed to collect mixture components.

Supercritical fluid chromatography-IR

Supercritical fluid chromatography (SFC) is often employed for separations involving non-volatile or thermally labile species that cannot be separated by GC or LC. Supercritical mobile phases (typically CO_2) have viscosities and solute diffusivities that are intermediate between those for gases and liquids. Supercritical CO_2 is created by subjecting the gas to high pressures. SFC-IR flow cells are made from high pressure UV flow cells by replacing the quartz windows in these cells with infrared transparent windows. Alternatively, light pipe flow cells similar to those employed for GC-IR can be used for SFC-IR. In either case, regions in which CO_2 strongly absorbs are opaque in spectra measured by using flow cell SFC-IR. Mobile phase elimination SFC-IR techniques are particularly effective because supercritical CO_2 mobile phase readily vaporizes when exposed to ambient conditions. The same methods used for mobile phase elimination LC-IR are applicable to mobile phase elimination SFC-IR. Matrix isolation SFC-IR can be achieved by using the same apparatus employed for GC-IR but with CCl_4 instead of argon as the matrix substance. To prevent CO_2 condensation, the matrix isolation deposition surface temperature must be maintained at 150 K for matrix isolation SFC-IR.

Thin Layer Chromatography-IR

Two FT-IR sampling methods are commonly employed to measure IR spectra of mixture components separated by thin layer chromatography (TLC). Diffuse reflectance and photoacoustic IR spectroscopy are techniques that can be employed when sample materials are opaque.

Diffuse reflectance TLC-IR

Diffuse reflectance (DR) IR analysis can be used to obtain IR spectra of species separated by TLC without removing them from the TLC plate. TLC plates containing spatially separated mixture components are passed through an optical system that focuses the incident IR beam onto a small spot on the plate and collects the resulting diffusely scattered IR radiation. **Figure 3** illustrates the basic principle. When the focused beam strikes a region of the TLC plate that contains solute, the resulting diffuse reflectance spectrum is characteristic of the solute and the TLC stationary and mobile phases. Solute spectra are obtained after subtracting the IR spectral features of the stationary and mobile phases. Little solute spectral information can be obtained in wavelength regions in which the stationary phase is highly absorbing. In addition, interactions between solutes and the stationary phase often result in absorbance band wavelength and intensity differences relative to spectra measured for the same substance pressed into a KBr pellet.

Photoacoustic TLC-IR

In contrast to most other IR analysis techniques, the photoacoustic (PA) IR spectroscopic signal is a direct measure of the amount of radiation absorbed by the sample. Consequently, samples need not be transparent or reflective to be analysed by this method. However, TLC separated solutes must be physically removed from the TLC plate prior to analysis because PA-IR requires samples to be sealed in an air-tight chamber. As with diffuse reflectance measurements of TLC separated solutes, PA-IR spectra contain features representing stationary phase and residual mobile phase in addition to the solute. Spectral subtraction is employed to isolate solute spectra from PA-IR spectra. The main disadvantage of PA-IR compared to diffuse reflectance is that the locations of mixture components on the TLC plate must be known before infrared analysis can be performed. When solutes cannot be detected visually, an

additional detector must be used to track solute migrations.

Solute transfer TLC-IR

IR spectra of TLC-IR solutes obtained by DR-IR and PA-IR exhibit band intensity distortions in spectral regions in which stationary and mobile phases absorb and may not resemble those obtained for the same substances by KBr pellet transmittance measurements. To avoid stationary and mobile phase interferences, solute transfer methods can be used to physically move separated mixture components from the TLC plate to an infrared transparent substrate. Solute transfer is accomplished by subjecting the developed TLC plate to a second solvent that is applied orthogonally to the direction of the chromatographic mobile phase flow. This causes separated mixture components to move to the edge of the TLC plate where they are collected on IR transparent KCl powder. **Figure 3** illustrates the basic principle. Solute spectra subsequently obtained by diffuse reflectance IR analysis are similar to those obtained by KBr pellet transmittance measurements.

Chromatography-IR interface characteristics

Compared to mobile phase elimination techniques, flow cell chromatography-IR interfaces are simpler and less expensive. However, flow cell chromatography-IR yields optimal results only when the mobile phase is IR transparent. Detection limits in the low nanogram range can be attained by light pipe GC-IR. Due to mobile phase absorbances, detection limits for flow cell LC-IR and SFC-IR combinations are higher than for light pipe GC-IR. In fact, flow cell LC-IR detection limits vary considerably and are primarily determined by the mobile phase absorptivity.

In general, mobile phase elimination chromatography-IR methods provide lower detection limits than flow cell approaches. The primary goal of mobile phase elimination is to remove interfering absorbances from acquired IR spectra. In addition to removing chromatographic mobile phase, these interfaces concentrate solutes prior to IR analysis, which enhances solute infrared absorbances. Furthermore, off-line IR analysis by mobile phase elimination chromatography-IR facilitates increased signal averaging when compared to flow cell interfaces. When spectral noise sources are random, FT-IR spectrum signal-to-noise ratios increase by the square root of the increase in the number of scans averaged. The advantages of increased solute concentration and extended signal averaging are

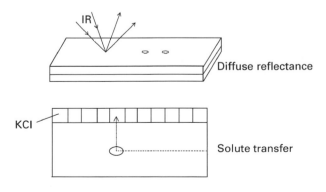

Figure 3 Thin layer chromatography-IR interfaces.

evidenced by the fact that typical solute trapping GC-IR detection limits are about ten times lower than those provided by light pipe GC-IR despite the fact that GC-IR mobile phases are infrared transparent.

See also: **ATR and Reflectance IR Spectroscopy, Applications; Fourier Transformation and Sampling Theory; IR Spectrometers; IR Spectroscopy, Theory; Matrix Isolation Studied By IR and Raman Spectroscopies; Photoacoustic Spectroscopy, Applications; Photoacoustic Spectroscopy, Theory; Quantitative Analysis; Surface Studies By IR Spectroscopy.**

Further reading

Erickson MD (1979) Gas chromatography/Fourier transform infrared spectroscopy applications. *Applied Spectroscopy Reviews* 15: 261–325.

Fujimoto C and Jinno K (1992) Chromatography/FT-IR spectrometry approaches to analysis. *Analytical Chemistry* 64: 476A–481A.

Griffiths PR and deHaseth JA (1986) *Fourier Transform Infrared Spectrometry*. New York: John Wiley & Sons.

Herres W (1987) *HRGC-FTIR: Capillary Gas Chromatography–Fourier Transform Infrared Spectroscopy: Theory and Applications*. Heidelberg: Huthig.

Jinno K and Fujimoto C (1996) Microcolumn liquid chromatography in hyphenated techniques (microcolumn LC/FTIR). *Methods for Chromatography* 1: 199–211.

Kalasinsky VF and Kalasinsky KS (1992) HPLC Detection using Fourier transform infrared spectrometry. In: Patonay G (ed) *HPLC Detection*, pp 127–161. New York: VCH.

McClure GL (1993) General methods of sample preparation for infrared hyphenated techniques. In: Coleman PB (ed) *Practical Sampling Techniques for Infrared Analysis*, pp 165–215. Boca Raton: CRC.

White RL (1987) Gas chromatography–Fourier transform infrared spectrometry. *Applied Spectroscopy Reviews* 23: 165–245.

White RL (1990) *Chromatography/Fourier Transform Infrared Spectroscopy and its Applications*. New York: Dekker.

Wurrey CJ and Gurka DF (1990) Environmental applications of gas chromatography/Fourier transform infrared spectroscopy (GC/FT-IR). *Vibrational Spectra and Structure* 18: 1–80.

Chromatography-MS, Methods

WWA Niessen, hyphen MassSpec Consultancy, Leiden, The Netherlands

MASS SPECTROMETRY
Methods & Instrumentation

Introduction

Chromatography is an important tool in purification and isolation of compounds from (complex) mixtures prior to mass spectrometric analysis. Mass spectrometry can be used to detect the compounds separated by chromatography. This can be performed off-line, i.e. after fraction collection, but an even more powerful and versatile technique is achieved when the separation is coupled on-line to mass spectrometric detection. The on-line combination of a chromatographic separation method and mass spectrometry (MS) is an important and versatile tool in many areas of analytical chemistry. In principle, it enables the mass spectrometric characterization of components in complex mixtures after separation, with minimal or no sample losses. Both major and minor components in a mixture are amenable to such a technique. Furthermore, MS offers advantages in quantitative analysis, i.e. enhanced selectivity by the use of high-resolution MS or tandem MS (MS/MS) and/or the use of stable-isotopically labelled internal standards, which optimally mimic the properties of the analytes during sample pretreatment, separation and detection.

Initially, in the mid 1960s the on-line combination of gas chromatography (GC) and MS was developed, first for packed columns and soon after their introduction, also for capillary columns. The introduction of quadrupole mass analysers greatly facilitated GC–MS, because of their faster scanning and less stringent vacuum restrictions. They are now the most widely applied mass analysers for on-line chromatography-MS. The success of GC–MS initiated in the mid 1970s, research into the on-line combination of liquid chromatography (LC) and MS. Subsequently, other separation techniques were successfully coupled to MS, e.g. supercritical fluid chromatography (SFC), thin-layer chromatography, and various electromigration techniques like capillary zone electrophoresis (CZE) and electrochromatography. The actual breakthrough of LC–MS as a routinely applicable analytical technique is taking place now in the mid 1990s.

General instrumentation

The general instrumentation for on-line chromatography–MS consists of a conventional chromatograph, the outlet of which is connected to an interface and a mass spectrometer (see **Figure 1A**). The main function of the interface is to protect the high-vacuum system of the mass analyser from too high gas loads from the chromatograph. Simultaneously, the interface should transfer as much of the analytes eluting from the column as possible to the mass analyser. While initially interfaces were developed to keep as much of the carrier gas or mobile phase out of the high vacuum, the development of LC–MS especially resulted in changes in this function of the interface. In the current LC–MS interfaces, the mobile phase plays an essential role in analyte transfer and ionization.

Gas chromatography–mass spectrometry

The on-line capillary GC–MS combination can be achieved in two ways, i.e. via direct coupling or via an open-split coupling.

The gas flow through a typical 0.25-mm-ID capillary GC column matches the allowable gas load of a simple single-stage MS vacuum system. Since the components eluting from the GC column are already in the vapour state, immediate introduction of the capillary column effluent into the MS ion source in electron ionization mode is possible. Although frequently applied, this direct coupling has some disadvantages. The column outlet is at high vacuum, which results in changing retention times relative to other GC detectors, such as the flame ionization detector. Furthermore, the gas load to the ion source changes during the temperature programme of the GC analysis. This may influence the tuning of the ion source parameters. The problems related to the effect of the solvent load on the lifetime of the heated filament are easily overcome by switching off the filament during the first few minutes of the analysis.

An alternative to direct coupling is the open-split coupling; a schematic diagram of such a device is shown in **Figure 2**. The column is connected with

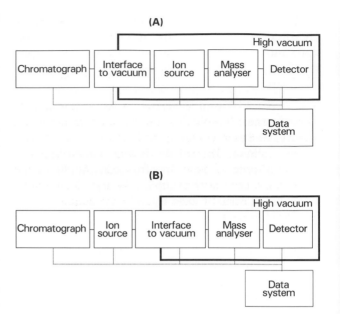

Figure 1 Schematic diagrams of chromatography–mass spectrometry systems: (A) GC–MS system, (B) atmospheric-pressure ionization LC–MS system.

the ion source via a fixed restriction. A make-up gas flow is provided either to raise the column flow to match the restrictor throughput or to rapidly remove the excess of carrier gas. As a result, the column outlet is at atmospheric pressure, just as in a conventional GC detector. By increasing the makeup flow it is possible to efficiently divert high vapour loads from the solvent away from the MS.

Liquid chromatography–mass spectrometry

For a number of reasons, the on-line coupling of LC–MS appeared far more difficult to accomplish than the GC–MS coupling.

1. The gas load to the vacuum system resulting from evaporating the mobile phase in LC is generally much higher than in GC.
2. The mobile phase composition (containing nonvolatile additives such as phosphate buffers) is often incompatible with MS.

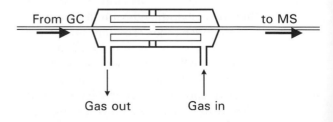

Figure 2 Open-split coupling for GC–MS.

3. The transfer of polar and ionic analytes from the liquid to the gas phase is difficult.

In order to solve the first problem a number of LC–MS interfaces have been developed, some of which are discussed in more detail below. The second problem must be solved at the chromatography side, e.g. by changing to volatile buffers or the use of column switching. The third problem is largely solved by the advent of new liquid-based soft ionization techniques, especially thermospray and electrospray, which were actually developed in the course of LC–MS development.

A variety of interfaces was developed for on-line LC–MS coupling. A number of these found wide application when they became commercially available, e.g. the moving-belt interface, direct-liquid introduction, thermospray, particle-beam and continuous-flow fast-atom bombardment (Cf-FAB), but are hardly used any longer due to the introduction of interfaces based on atmospheric-pressure ionization (API), i.e. electrospray and atmospheric-pressure chemical ionization (APCI). Before discussing the API-based interfaces in detail, brief attention is given to three of the older LC–MS interfaces.

In a particle-beam interface (**Figure 3A**), the column effluent is pneumatically nebulized into an atmospheric-pressure desolvation chamber. This is connected to a momentum separator where the analytes are transferred to the MS ion source while the low molecular mass solvent molecules are efficiently pumped away. The analyte particles hit the heated source surface, evaporate and can be ionized by electron or chemical ionization. The evaporation step obviously limits the application range of the interface to not-too-polar analytes.

In a thermospray interface (**Figure 3B**), the column effluent is rapidly heated in a narrow bore capillary such that partial (*ca.* 90%) evaporation of the solvent is achieved inside the capillary. As a result, a mist of vapour and small droplets is formed in which the heated droplets further evaporate and ions are generated, either by the thermospray ionization process based on ion evaporation or by solvent-mediated chemical ionization initiated by electrons from a heated filament or a discharge electrode. The excess vapour is pumped away directly from the ion source.

In a Cf-FAB interface (**Figure 3C**), a small mobile-phase stream, containing glycerol as a FAB matrix, is flowing towards a FAB target where the effluent is bombarded by fast atoms or ions to achieve analyte ionization. Non-evaporated mobile phase constituents are collected in a wick, which is a compressed paper pad which has to be regularly renewed.

(A)

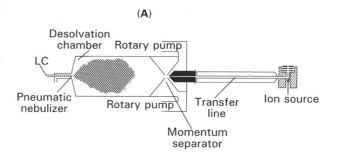

(B)

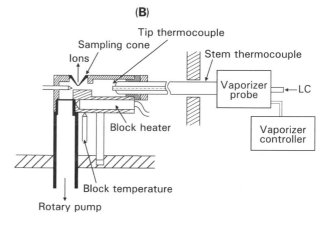

(C)

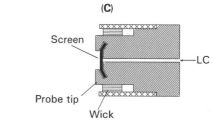

Figure 3 Older types of LC–MS interfaces: (A) particle-beam interface, (B) thermospray interface, and (C) target area of continuous-flow fast-atom bombardment interface.

Interfaces based on atmospheric-pressure ionization

After initial attempts in the mid 1970s, LC–MS interfaces based on API were developed in the mid 1980s. An API mass spectrometer differs from a conventional instrument because ions are generated in an atmospheric-pressure region and subsequently introduced into the high vacuum mass analyser. The difference is clearly depicted in **Figure 1B**: the ion source and the interface to vacuum are interchanged. In an API interface, the sample can be introduced in two different ways and ions can be produced in two different ways, i.e. by electrospray or by APCI. After ionization, the ions generated are transferred to the mass analyser via an API interface (**Figure 4**), which is independent of the mode of sample introduction

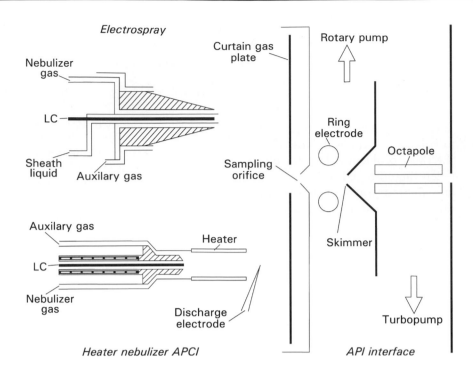

Figure 4 Atmospheric-pressure interface for LC–MS with electrospray and heated nebulizer APCI spray probes.

or ionization. LC–MS interfaces based on API nowadays are by far the most widely used.

API interfacing thus comprises a number of steps:

1. Nebulization of the column effluent into small droplets followed by their rapid evaporuntion. Nebulization is achieved either by pneumatic nebulization in combination with a heated zone (APCI), or by (pneumatically assisted) electrospray nebulization due to the action of a high electric field on the liquid in a narrow bore needle. Schematic diagrams of the spray probes are shown in **Figure 4**.
2. Generation of ions, either by gas-phase ion–molecule reaction initiated by electrons from a corona discharge needle, where the solvent acts as the reagent gas (APCI), or by ion evaporation of preformed ions in solution (electrospray ionization).
3. Sampling of ions from the atmospheric-pressure region into a two-stage differentially pumped interface, which enables sufficient pressure reduction and transfers the ions to the mass analyser in a high-vacuum chamber.
4. Mass analysis of the ions and subsequent detection.
5. Data acquisition and handling.

Electrospray interfacing and ionization

The electrospray spray probe typically consists of a 100 μm i.d. stainless-steel or fused-silica capillary surrounded by a coaxial tube for the nebulization gas. The nitrogen nebulization gas allows the introduction of flow rates in excess of 10 μL min^{-1} (pneumatically assisted electrospray or ionspray). The spray probe is positioned in the atmospheric-pressure ion source at a short distance from a counter electrode. A 2–3 kV electric field is applied between the needle and the counter electrode. This results in the formation of a fine aerosol of charged droplets from the spray probe. These droplets are rapidly evaporating, assisted by a concurrent, countercurrent or perpendicular heated nitrogen gas stream, resulting in an increase in droplet charge at the surface. This leads to two competing processes: field-induced Rayleigh or electrohydrodynamic droplet disintegration, leading to an explosion of the initial droplet into a series of highly charged smaller offspring droplets, and field-induced ion evaporation, leading to desorption of preformed ions from the bulk solution into the gas phase. The ions generated are sampled by the mass analyser.

Although current electrospray interfaces can be used with flow rates of up to 1 mL min^{-1}, lower flow rates, e.g. 40–200 μL min^{-1}, are preferred in most LC–MS applications, thus requiring the use of either a postcolumn solvent splitter or microbore LC column.

Electrospray ionization is especially suited for the analysis of compounds that preform ions in solution, e.g. by acid–base chemistry, i.e. organic acids and bases, peptide and proteins, oligonucleotides, etc. For analytes with a molecular mass exceeding 500 Da and multiple protonation or deprotonation sites, multiply charged ions may be generated by electrospray ionization. This enables the mass analysis of biomacromolecules by means of a quadrupole mass analyser with a limited mass range. As such, electrospray plays an important role in the characterization of peptides and proteins by MS. Given the limited sample availability in many of these applications, nano-electrospray devices have been developed which allow the introduction of 0.01–1 μL min^{-1} of protein solutions into the MS.

APCI interfacing and ionization

The APCI probe consists of a liquid introduction capillary surrounded by a coaxial tube for pneumatic nebulization. After pneumatic nebulization of the column effluent, the droplets are swept through a heated zone in order to achieve evaporation of the droplets. The solvent vapours are subsequently used as reagent gas in chemical ionization, initiated by electrons from a corona discharge electrode kept at a few kV. In positive-ion mode, a series of gas-phase ion–molecule reactions leads to the formation of protonated solvent molecules which may react with the analyte molecules having a higher proton affinity than the solvent. The protonated analytes are sampled and subsequently mass analysed.

The APCI spray probe is readily applicable to a wide flow-rate range, e.g. between 0.2 and 2 mL min^{-1}. It allows the ionization of analytes in a wide polarity range. It can be used with both typical reversed-phase LC mobile phases (mixtures of water and methanol or acetonitrile) and pure organic solvents. APCI is a highly efficient ionization technique.

API interfacing in more detail

After the generation of ions in the atmospheric-pressure ion source by either electrospray or APCI, the ions are sampled into the multistage differentially pumped vacuum system. Different ion sampling devices are applied by different instrument manufacturers: a glass capillary with metallized ends, a heated stainless-steel capillary, or a sampling orifice or cone. The low-pressure side of the ion sampling device acts as a nozzle in a molecular-beam source: expansion of the gas–vapour–ion mixture takes place. The core of the expanding jet, primarily containing the constituents with the higher mass, is sampled by a skimmer. In order to preferentially sample the ions in the jet, a ring electrode is positioned between the nozzle and the skimmer. The region between the nozzle and skimmer is kept at a pressure of ca. 100 Pa. After sampling by the skimmer, a second expansion stage takes place in the ion optics region, kept at a pressure of 0.1–1 Pa by means of a turbomolecular pump. The ion optics region contains a quadrupole, hexapole or octapole ion guide, which efficiently transports the ions towards the mass analyser, housed in the third pumping stage, kept at a pressure below 10^{-3} Pa by means of a second turbomolecular pump.

The transfer of ions from an atmospheric-pressure region through a series of differentially pumped vacuum chambers can be used to collisionally induce fragmentation of these ions in the transition region, e.g. by increasing the potential difference between the nozzle and the skimmer or between the skimmer and the multipole ion guide. This in-source CID approach is successfully applied in structure elucidation of unknown compounds. In order to obtain well-interpretable mass spectra after in-source CID, either pure compounds or well-separated compounds should be used, because (in contrast to CID in an MS/MS instrument) this CID is done without prior precursor-ion selection.

Mass analysis API interfacing

Although the mass analysis after API interfacing is in most cases performed using a (triple) quadrupole instrument, API devices have also been developed (and are commercially available) for other types of mass analysers, i.e. ion-trap, time-of-flight (TOF), magnetic sector and Fourier-transform ion-cyclotron resonance (FT-ICR) instruments as well as a number of hybrid instruments, such as quadrupole–TOF and magnetic sector–ion-trap instruments. In particular, the combination of API with ion-trap and TOF instruments is receiving considerable attention because of the favourable price to performance ratio of these instruments in a number of applications, compared to triple-quadrupole instruments. The extremely high resolution achievable with FT-ICR can especially be useful in determining charge states of multiply charged protein ions after fragmentation.

Data acquisition and handling

The mass spectrometer has two general modes of data acquisition:

1. The full-scan acquisition mode, where a series of complete mass spectra is acquired as a function of time.

2. The selected-ion monitoring mode, where the intensity of selected ions with particular m/z is acquired as a function of time.

Choosing between these two modes is a matter of striking a compromise in the particular application between maximizing the measurement time of a particular m/z (in SIM mode), which is favourable for sensitivity, and the information content (in full-scan mode), which is favourable in identification and/or structure confirmation.

The data acquisition results in a three-dimensional data-array with ion intensity, m/z, and time or scan number as the three dimensions. This data-array can be handled in various ways. For the total-ion chromatogram, the ion intensities in each scan are summed irrespective of the m/z of the ions and the summed intensity, i.e. the total-ion current, is plotted as a function of time. In this mode the mass spectrometer acts as a universal sector. In GC–MS, the total-ion chromatogram largely resembles the chromatogram obtained via flame-ionization detection. In a base-peak chromatogram, the intensity of the base peak irrespective of the m/z, i.e. the most intense peak in each mass spectrum, is plotted as a function of time. This mode can be useful for peak finding in applications with high background levels, e.g. in LC–MS with API interfacing. In a mass chromatogram, the ion intensity for an ion or a series of ions with particular m/z is plotted as a function of time. In this mode the mass spectrometer acts as a specific or selective detector. And finally, in the mass spectrum, for each scan number, the ion intensity can be plotted as a function of the m/z. Summed, averaged, and/or background-corrected mass spectra over a series of scans can be made as well. The mass spectrum can be computer-searched against a mass spectral library.

Additional and advanced techniques

The performance of on-line chromatography–MS can for some applications be greatly enhanced by the use of additional techniques. On-line chromatography–MS should be considered as an integrated and hyphenated approach, where all system parts should be optimally tuned for the complete system. The combination of a potentially high-efficiency separation and a potentially high-selectivity detection enables a number of problem-oriented analytical strategies, e.g. directed at identification of minor impurities in complex samples, requiring highly efficient separation combined with MS in full-scan mode, offering the highest information content or

directed at high-throughput target compound analysis, where the separation efficiency is deliberately reduced to save analysis time and the high selectivity afforded by either high-resolution MS or tandem MS is applied selectively to detect the compounds of interest within the matrix.

Analyte derivatization

Capillary GC–MS is an extremely powerful approach, combining the high separation efficiency of the capillary GC column with the identification power of the MS in electron-ionization mode. However the applicability range of GC is limited to relatively volatile compounds. In order to widen the applicability range, precolumn analyte derivatization strategies are often applied to enhance the volatility of the analytes. Methylation, silylation and acetylation reactions are most often applied for the analysis of compounds with amine, (poly-) hydroxy and/or carboxylic acid functional groups. Furthermore, derivatization to pentafluorobenzyl derivatives is applied to enhance the sensitivity of analytes in electron-capture negative-ion chemical ionization.

Tandem mass spectrometry

A powerful tool to enhance the analytical potential of combined chromatography–MS especially in LC–MS, is tandem MS (MS/MS). MS/MS was first introduced as a way to induce fragmentation of ions generated in the ion source, e.g. by soft ionization methods, and after their mass selection. In such an experiment, the first mass spectrometer selects a particular precursor ion, which is (collisionally) dissociated, and the product ions are analysed with the second mass spectrometer. This is the product-ion scan mode which is primarily used to elucidate the structures of ions.

For the analytical use of MS/MS, other scan modes can be useful as well as the product-ion scan mode. The various modes are explained and summarized in **Table 1**. The precursor-ion and neutral-loss scan modes are particularly useful in screening. They are part of a powerful analytical strategy applicable in impurity, degradation product and/or metabolite profiling, as well as other applications where searching for structurally related compounds is important. After an appropriate ionization mode is found, a product-ion mass spectrum is obtained for the parent compound. In this spectrum, possible common precursor ions and/or common neutral losses have to be identified.

As an example, the screening for unknown metabolites of heptabarbital can be considered (see **Figure 5**). The fragmentation of heptabarbital under

Table 1

Mode	MS1	MS2	Application
Product-ion	Selecting	Scanning	To obtain structural information by CID of ions produced in the source
Precursor-ion	Scanning	Selecting	To monitor compounds which in CID give an identical fragment
Neutral-loss	Scanning	Scanning	MS1 and MS2 are scanning at a fixed m/z difference; to monitor compounds that lose a common neutral species
Selective reaction monitoring (SRM)	Selecting	Selecting	To monitor a specific CID reaction

CID is also shown in **Figure 5**. Knowing that most of the metabolism of this drug primarily involves the cycloheptene ring, the barbituric acid related product ion at m/z 157 can be used as a common precursor ion to screen for compounds characterized by metabolism of the cycloheptene ring. Alternatively, the neutral loss of 156 Da can be applied in screening.

Subsequently, these precursor-ion and/or neutral-loss scans are performed, preferentially with on-line chromatographic separation. In this way, the m/z for related compounds can be found. Product-ion mass spectra are obtained based on these m/z. Interpretation of the mass spectra leads to identification of the related compounds, impurities or metabolites. Neutral-loss scans of 176 Da and 80 Da can be applied to screen for O-glucuronide and aryl-O-sulfate conjugates of drugs, respectively.

SRM is extremely useful in quantitative environmental and bioanalyses, where interfering sample matrix components may adversely influence the achievable detection limits. The enhanced confidence level in SRM, based on the monitoring of a specific chemical reaction in the gas phase, made LC–MS/MS in SRM mode the analytical method of choice in quantitative bioanalysis during drug development and clinical trials by pharmaceutical industries. In addition, the monitoring of a specific collision-induced reaction by SRM significantly improves the selectivity, resulting in greatly improved signal-to-noise ratios and enabling a limited sample pretreatment prior to LC–MS/MS analysis, thereby increasing the sample throughput.

Coupled column chromatography

One of the major difficulties in LC–MS coupling is that the prolonged use of nonvolatile mobile-phase additives such as phosphate buffers and sodium dodecyl sulfate, is prohibited. In most applications, the nonvolatile additives can successfully be replaced by a volatile alternative, such as ammonium acetate and heptafluorobutyric acid. When this is not possible without significant influence on the chromatographic resolution, the use of column-switching approaches may be useful, especially in target-compound analysis. The chromatography is performed on a first column under optimized chromatographic conditions. Compounds of interest are trapped onto short trapping columns, which are switched in-line during appropriate retention time windows. These loaded trapping columns are subsequently eluted with a mobile phase and/or flow rate readily amenable to the LC–MS interface, either directly to the LC–MS/MS system (phase-system switching) or to a second column (coupled-column chromatography). These strategies may also provide (significant) improvements in selectivity. In this respect, the use of immunoaffinity chromatography as a first step can be extremely useful.

Capillary zone electrophoresis–MS

LC–MS interfaces have successfully been applied for the coupling to MS of other liquid-based separation techniques, such as SFC, CZE and electrochromatography. The coupling of the highly efficient CZE to MS has achieved most attention because of the high promise of this combination in the analysis of ionic compounds, peptides and proteins. On-line identification of minor impurities made observable by the high separation efficiency achievable in CZE would be extremely useful. On-line CZE–MS is relatively easy to accomplish via an electrospray interface, either via the use of a coaxial sheath liquid in order to make up the sub-μL min⁻¹ flow rate from the CZE to the 5–10 μL min⁻¹ required in

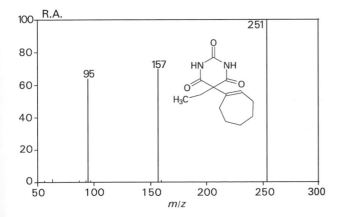

Figure 5 Product-ion mass spectrum of heptabarbital.

conventional electrospray interfaces, or via the use of a nano-electrospray interface that can accommodate sub-µL min^{-1} flow rates. However, as a result of the small injection volumes, the concentration detection limits reported are generally in the low ppm range. Various on-line strategies have been proposed to improve these detection limits, e.g. on-line preconcentration via isotachophoresis, solid-phase extraction and/or liquid–liquid electroextraction. In general, these strategies are target-compound directed and therefore appear not to be particularly useful in general impurity profiling.

Application areas of chromatography-MS

Both GC–MS and LC–MS are widely applied in many fields. As indicated, GC–MS is limited to relatively volatile compounds. Within this restriction, GC–MS is widely applied to the identification and quantitation of compounds of environmental and pharmaceutical interest. In areas such as the characterization of oil fractions, environmental monitoring, e.g. of chlorinated pesticides like DDT and also of chlorinated dibenzodioxins, dibenzofurans and related compounds, and in a number of routine analytical procedures, e.g. in a number of clinical assays, in doping analysis in sports and for forensic purposes, GC–MS shows unsurpassed analytical possibilities, in terms of both sensitivity and identification power.

LC–MS finds wide application in the analysis of compounds that are not amenable to GC–MS, i.e. compounds that are highly polar, ionic and thermolabile, as well as (bio)macromolecules. In environmental applications, LC–MS is applied, often in combination with off-line or on-line solid-phase extraction, to identify pesticides, herbicides, surfactants and other environmental contaminants. LC–MS plays a role in the confirmation of the presence of antibiotic residues in meat, milk and other food products. Furthermore, there is a substantial role for LC–MS in the detection and identification of new compounds in extracts from natural products and the process control of fermentation broths for industrial production of such compounds, e.g. for medicinal use. LC–MS technology is also widely applied in the characterization of peptides and proteins, e.g. rapid molecular-mass determination, peptide mapping, peptide sequencing and the study of protein conformation and noncovalent interactions of drugs, peptides and other compounds with proteins and DNA. However, the most important application area

of LC–MS is in the pharmaceutical drug development area, where LC-MS is involved in almost every stage of development, e.g. rapid characterization based on the molecular mass of medicinal combinatorial libraries, the detection and identification of drug metabolites, reaction by-products, and drug degradation products. LC-MS/MS in SRM mode is currently the method of choice in high-throughput quantitative bioanalyses for bioavailability studies, pharmacokinetic and pharmacodynamic studies, and during clinal trials.

See also: **Atmospheric Pressure Ionization in Mass Spectrometry; Biochemical Applications of Mass Spectrometry; Chemical Ionization in Mass Spectrometry; Chemical Structure Information from Mass Spectrometry; Fast Atom Bombardment Ionization in Mass Spectrometry; Forensic Science, Applications of Mass Spectrometry; Hyphenated Techniques, Applications of in Mass Spectrometry; Ion Molecule Reactions in Mass Spectrometry; Ion Trap Mass Spectrometers; Isotopic Labelling in Mass Spectrometry; Medical Applications of Mass Spectrometry; MS–MS and MS^n; Peptides and Proteins Studied Using Mass Spectrometry; Quadrupoles, Use of in Mass Spectrometry; Sector Mass Spectrometers; Thermospray Ionization in Mass Spectrometry; Time of Flight Mass Spectrometers.**

Further reading

Barceló D (ed) (1996) *Environmental Applications of LC–MS*. Amsterdam: Elsevier Science.

Cai J and Henion JD (1995) Capillary electrophoresis–mass spectrometry, *Journal of Chromatography A* 703: 667.

Cole RB (ed) (1997) *Electrospray Ionization Mass Spectrometry*. New York: Wiley.

Karasek FW and Clement RE (1988) *Basic Gas Chromatography–Mass Spectrometry: Principles and Techniques*. Amsterdam: Elsevier Science.

Kitson FG, Larsen BS and McEwen CN (1996) *Gas Chromatography and Mass Spectrometry, a Practical Guide*. London: Academic Press.

Niessen WMA and Van der Greff J (1992) *Liquid Chromatography–Mass Spectrometry*, New York: Marcel Dekker (second edition, authored by WMA Niessen, will appear by the end of 1998).

Niessen WMA and Voyksner RD (ed) (1998) *Current Practice of LC–MS*, Special Issue of *Journal of Chromatography A*, Vol. 794.

Pinkston JD and Chester TL (1996) Putting opposites together: Guidelines for successful SFC-MS, *Analytical Chemistry* 67: 650A.

Chromatography-NMR, Applications

JP Shockcor, Du Pont Pharmaceuticals Co. Newark, DE, USA

MAGNETIC RESONANCE
Applications

Introduction

High-performance liquid chromatography (HPLC) has become a routine tool for the separation of complex mixtures. However, structural information on substances separated using HPLC is limited by the detector system employed. The most common HPLC detectors, refractive index, UV, radiochemical, fluorescence, and electrochemical, provide little or no structural information. As a result, structure elucidation required the isolation of the analyte from the matrix, followed by off-line spectroscopic characterization. With the advent of HPLC-MS the ability to detect and identify substances at low concentrations without the need for an isolation step became possible. Although this has simplified structure elucidation to a great extent, there are often circumstances where HPLC-MS alone is insufficient for complete characterization of a compound and further studies by NMR are required. Logically the next step in instrument development would be directly coupling HPLC and NMR yielding the hyphenated technique HPLC-NMR.

History

In the late 1970s and early 1980s, a number of attempts to couple these techniques were carried out. However, these studies suffered from the low sensitivity of the NMR spectrometer systems then available. Also, because of dynamic-range problems, there was a need to use expensive deuterated solvents for the HPLC because the solvent suppression methods in use at the time could not cope with fully protonated solvents. The reduction of HPLC-NMR to routine use was slow in developing and not practically achieved until technical improvements in electronics, higher magnetic fields strengths, advanced solvent suppression sequences, and improved instrumental sensitivity, made it feasible to interface an HPLC directly to an NMR spectrometer.

In addition, investigations into coupling other types of chromatography with NMR have been carried out. These include capillary electrophoresis, capillary electrochromatography, supercritical fluid chromatography and capillary HPLC. However there are no commercially available systems based on these separations (late 1999).

Technical improvements

There have been many major technical improvements in NMR sensitivity in recent years. Foremost has been the development of ultra-high-field-strength magnets (by late 1999, 900 MHz). High dynamic range analog–digital converters, which give benefits in situations where large solvent peaks are present along with small analyte signals, and digital filtering techniques that allow spectral windows to be limited to include only the region of interest without the problem of signals and noise from outside this spectral region folding in have also contributed to increases in sensitivity. Another important improvement is oversampling. That is, the collection of digital data points at a rate faster than that required to satisfy the Nyquist criterion of twice the highest desired spectral frequency. In theory, for an oversampling factor of n, a gain in dynamic range of $\log_2(n)$ is obtained, i.e. for an oversampling of 8 times, an effective gain of 3 bits in ADC resolution of the signals results. In practice, the oversampled signal is simply averaged over the n measurements to restore the same number of data points corresponding to the Nyquist criterion. This prevents folding of noise or artifacts which would have been in the extended spectral region resulting from a consequence of the oversampling being equivalent to a spectral region n times wider than required to satisfy the Nyquist criterion. Thus a second consequence of oversampling is an improved signal–noise ratio from removal of folded noise when the spectral region is truncated.

The HPLC-NMR system

A block diagram of a typical instrumental set-up for HPLC-NMR is shown in **Figure 1**. This comprises a high-resolution NMR spectrometer with its superconducting magnet into which is placed a dedicated NMR flow probe, a standard HPLC system controlled by PC-resident software and a flow control unit which enables the system to be operated in four main modes as listed below.

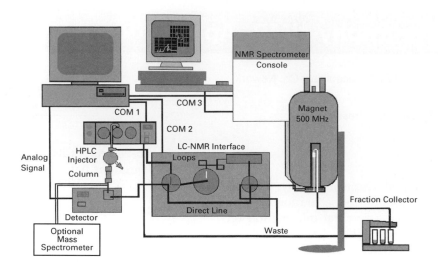

Figure 1 Block diagram of a typical HPLC-NMR system.

- On-flow
- Stop-flow
- Time-sliced stop-flow
- Peak collection into capillary loops for post-chromatographic analysis.

The simplest method of operation is continuous-flow detection. This mode of operation is generally only feasible when using ^{1}H or ^{19}F NMR for detection unless enriched compounds are used. If on-flow detection is required during a solvent gradient elution, the NMR resonance positions of the solvent peaks will shift as the solvent proportions change. For effective solvent suppression, it is therefore necessary to determine these solvent resonance frequencies as the chromatographic run proceeds. This is accomplished by measuring a single exploratory scan as soon as a chromatographic peak is detected in real time during the chromatographic run and then applying solvent suppression irradiation at these frequencies as the peak elutes. The data from an on-flow experiment performed on a sample of human urine after dosing with paracetamol [1] is shown in **Figure 2**. The data are plotted in a pseudo-2D format with the axes being the chemical shift in ppm and the retention time of the chromatographic run. Slices extracted from the 2D plot show the 1D spectra of the glucuronide [2] and sulfate [3] conjugates of paracetamol (**Figure 3**).

If the retention times of the compounds to be separated are known, or if they can be detected using refractive index, UV (including diode arrays), radiochemical or fluorescence detectors, stop-flow HPLC-NMR becomes an option. Upon detection the PC controlling the liquid chromatograph allows the pumps to continue running, moving the peak of the

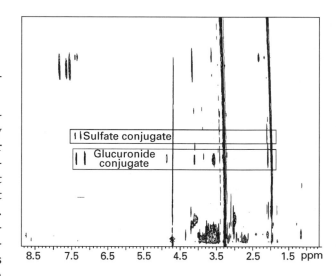

Figure 2 Pseudo-2D plot of on-flow ^{1}H HPLC-NMR data obtained on human urine after dosing with paracetamol [1]. The resonance from the glucuronide [2] and sulfate [3] conjugate metabolites of paracetamol are indicated.

interest into the NMR probe. Once the pumps have stopped, normal high-resolution NMR is possible. It could be argued that the long length of capillary tubing connecting the HPLC system to the NMR probe will cause a significant loss of resolution in the separation. In fact this is not a problem. The NMR detection cell volume is typically 60–120 μL, and this represents the limiting factor in the chromatographic resolution. The practicality of the stop-flow approach has been amply demonstrated and although several separate stops are often made in each chromatographic run, the quality of the resulting NMR spectra are such that good structural information can be obtained. Even long 2D experiments, which provide

[1] Paracetamol

[2] Glucuronide conjugate
of paracetamol

[3] Sulfate conjugate
of paracetamol

correlation between NMR resonances, based on mutual spin–spin coupling such as COSY or TOCSY and heteronuclear correlation studies such as HMQC can be performed. It is in this stop-flow mode that HPLC-NMR is most commonly used.

There are two further special categories of stop-flow experiment. These comprise the ability to store eluting fractions in capillary loops for later off-line NMR study ('peak picking') and the very useful ability to stop the flow at short intervals over a chromatographic peak to 'time slice' different parts of a chromatographic run. This is analogous to the use of diode-array UV spectroscopy to obtain spectra from various positions within an eluting peak to determine peak purity. The time-slicing methods may be useful if there is poor chromatographic separation, if the compounds under study have weak or no UV chromophores or if the exact chromatographic

retention time is unknown. It is also possible to time slice through an entire chromatographic run producing the equivalent of an on-flow experiment with higher signal to noise. The data from such a time slicing experiment has been referred to as a total NMR chromatogram (tNMRc).

In all the modes described above programmed gradient elution profiles can be performed and no compromise needs to be made in the chromatographic conditions, with the exception of those outlined in the following section.

HPLC-NMR solvents and chromatographic considerations

Another major problem in development of HPLC-NMR involved the use of deuterated solvents. Conventional high-resolution NMR spectroscopy routinely uses deuterated solvents. However, these solvents are thought to be prohibitively expensive if they have to be used for elution in HPLC-NMR

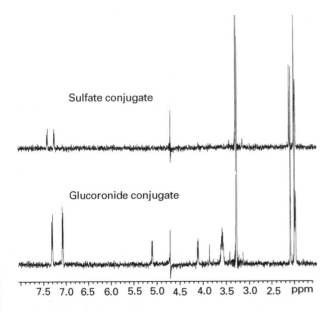

Sulfate conjugate

Glucoronide conjugate

7.5 7.0 6.5 6.0 5.5 5.0 4.5 4.0 3.5 3.0 2.5 ppm

Figure 3 1D slices extracted from the on-flow experiment shown in **Figure 2**. The top trace is the sulfate conjugate [2] and the bottom trace is the glucuronide conjugate [3].

because of the volumes involved. It is always possible to use solvents such as CCl_4, which contain no protons, for normal-phase chromatography but this then requires the use of HPLC-NMR probes which have a separate external deuterium lock channel. Since most chromatographic methods are reverse-phase, they involve the use of H_2O and an organic solvent such as acetonitrile or methanol. These give rise to large signals in the NMR spectrum which can obscure the spectrum of the analyte. The solution to this problem is solvent suppression which can be performed quite efficiently on modern NMR spectrometers. So efficiently, in fact, that the use of deuterated solvents is no longer a necessity. However, D_2O is used rather than H_2O to make the multiple solvent suppression easier and to serve as the field-lock substance. The cost of D_2O is relatively modest at ~$300 per litre.

One of the most common organic solvents used in reverse-phase HPLC separations is acetonitrile, which gives rise to a singlet resonance in the ^1H NMR spectrum at about δ 2.0. This singlet and that arising from residual H_2O in D_2O can be easily suppressed by presaturation. However, these suppression methods based on presaturation leave the ^{13}C satellite peaks from the 1.1% of molecules with the naturally abundant ^{13}C isotope at the methyl carbon. These satellite peaks are often much larger than the analyte peaks of interest in an HPLC-NMR study and so it is often necessary to suppress these also. This has been achieved in two ways. It is possible to set the suppression irradiation frequency over the central peak and the two satellite peaks in a cyclic fashion or, if an inverse-geometry probe is used, which includes a ^{13}C coil, then ^{13}C decoupling is possible. This will collapse the satellite peaks under the central peak, enabling conventional single-frequency suppression. The WET solvent suppression method incorporates selective RF pulses, pulse-field-gradients and optional ^{13}C decoupling to provide a robust solvent suppression method well suited to HPLC-NMR. **Figure 4** shows an example of WET solvent suppression on the ^1H HPLC-NMR spectrum of the desmethyl–glucuronide conjugate of naproxen [5] from human urine. The top trace shows the unsuppressed spectrum, while the bottom trace shows the same sample with WET solvent suppression.

Use of a fully deuterated solvent system, acetonitrile-d_3 and D_2O, is also an option when looking at very small quantities of material with resonances in the region near δ 2.0. Although the cost might seem to be prohibitive, at ~$1500 per litre for acetonitrile-d_3, the solvents for a typical HPLC run using both D_2O and acetonitrile-d_3 cost only $12 to $15.

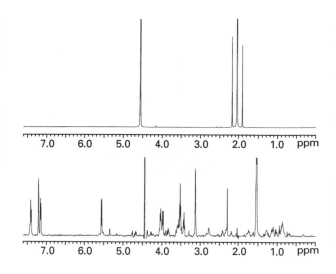

Figure 4 An example of solvent suppression using WET. The top trace shows the data with no solvent suppression and the bottom trace shows the data with WET solvent suppression applied.

Other chromatographic considerations, outlined below, should be considered when designing an HPLC separation method.

- Use buffers that have as few ^1H NMR resonances as possible.
  - Trifluroacetic acid.
  - Ammonium phosphate.
  - Ammonium acetate.
  - Deuterated formic acid ($130 per 5 g).

- Use ionpair reagents that have as few ^1H NMR resonances as possible.
  - Ionpairs with *n*-butyl groups create 4 additional resonances.
  - Ionpairs with *t*-butyl groups create 1 additional resonance.

- Whenever possible use HPLC-gradient methods.
  - Run at a 5% change per min maximum.

- Use columns which will be suitable for the above criteria.
  - Reverse-phase C_n.
  - 2.0 to 4.6 mm diameter.

Sensitivity of HPLC-NMR

The sensitivity of HPLC-NMR and the expected detection limits have been estimated for both 1D ^1H NMR and 2D NMR. All NMR experiments were carried out using a Bruker Avance 500 NMR spectrometer operating at an ^1H observation frequency

[4]   AZT

of 500.13 MHz. The spectral data were measured using a dedicated ^{1}H–^{13}C inverse-geometry HPLC-NMR probe with a 120 μL active volume. All NMR spectra were measured at 25°C. The total probe volume including the inlet line and the outlet to waste was measured to be 800 μL. Samples were injected directly into the NMR probe and hence the tests were of probe sensitivity rather than the overall sensitivity of HPLC-NMR. The S/N ratio is defined as the peak height of a given signal multiplied by 2.5 and divided by the peak-to-peak height of the noise measured over a given spectral range of 200 Hz. Spectra were measured on a standard sample of 0.1% ethylbenzene in $CDCl_3$. The ^{1}H S/N calculation, based on the signal for the methylene quartet at 2.7 ppm, gave a value of 237:1.

A typical organic molecule 3′-deoxy-3′-azidothymidine (AZT) [4] was used to test the absolute detection limit of the same HPLC-NMR probe. Having determined the total volume of the NMR probe system to be 800 μL, a solution containing 1μg of AZT in 120 μL was injected directly into the probe. The solvent was 80% D_2O–20% acetonitrile. The S/N ratio was measured on the ribose H1′ proton in the ^{1}H NMR spectrum at δ 6.2. The detection limit for this ^{1}H signal was defined as a signal height to peak-to-peak noise ratio of 3.0 (i.e. a S/N of 7.5 by the definition given earlier). Thus a detection limit of 500 ng in 64 scans was calculated.

AZT [4] was also used to test the ^{1}H sensitivity limit for a ^{1}H–^{1}H 2D TOCSY spectrum. A spectrum was obtained using a sample of 1 μg of AZT in 120 μL of 80% D_2O–20% acetonitrile and acquired over 4.0 hr. Examination of peak volumes shows that the detection limit of this experiment for 16 hr of acquisition (i.e. overnight) is about 500 ng.

The data acquisition parameters used were designed to ensure full T_1 relaxation so that any relative integral values would be quantitative.

However, these are not the parameters, which give maximum S/N values. If the ^{1}H NMR relaxation times are assumed to be about 1.5 s, and if the data had been collected in a manner designed to yield maximum S/N (i.e. with a pulse repetition rate of about $1.3T_1$), then this would result in a S/N enhancement by a factor of about 1.4. In addition, because of the faster pulsing rate, more scans could be accumulated in the same experimental time leading to a further S/N enhancement factor of about 1.7. Thus the overall S/N might be expected to increase by a factor of about 2.4. In addition, for the 2-dimensional experiments, it is possible to halve the number of T_1 increments and extrapolate the data using linear prediction. This allows the acquisition of twice as many scans per increment for the same overall experiment time resulting in a 1.4 increase in S/N.

These detection limits will be lowered as new technical advances in NMR become available. These include higher magnetic field strengths and the future availability of HPLC-NMR cyro-probes, probes cooled to cryogenic temperatures, which reduce the noise figure and can provide as much as a 4.0 increase in S/N.

HPLC-NMR probe design

The flow probe is the heart of the HPLC-NMR system. The main design criteria for such probes have been defined as follows:

- The geometry of the NMR cell should allow flow characteristics that give spectral resolution sufficient to allow spin-coupled multiplets to be resolved. This enables detailed structural elucidation to be carried out.
- NMR line broadening from magnetic susceptibility and other cell-wall effects should be avoided.
- The coil design and position should provide the highest possible NMR sensitivity.

The achievable signal–noise (S/N) ratio of a NMR detector is a function of a number of parameters expressed in Equation [1].

$$S/N = \gamma . N . I(I+1) . (B_0/T)^{3/2} . f . (QV_S/b)^{1/2} . n^{-1} \quad [1]$$

Therefore S/N can be increased by doing the following:

- Increase the sample volume and hence the number of detected nuclei, N.
- Increase the magnetic-field strength B_0.

- Increase the filling factor f of the NMR coil (this is V_s/V_c where V_s is the sample volume and V_c is the volume inside the detector coil).
- Increase the quality factor, Q, of the RF coil.
- Reduce the receiver bandwidth b.
- Operate at reduced temperature T.
- Improve preamplifiers by reducing their noise figure n.

All these are for a given nucleus with gyromagnetic ratio γ and spin quantum number I.

For HPLC-NMR probes there is a compromise which has to be made in that the detection volume should be as small as possible to get optimum chromatographic resolution and hence this can only be compensated for by increasing the filling factor. This is achieved by fixing the RF coil directly onto the outside of the NMR cell. If this is done, it is of course not possible to spin the sample to improve magnetic-field inhomogeneities. However, in practice this is not a problem because the smaller the sample volume, the better the shimming. Additionally the use of a computer-controlled orthogonal shim sets has reduced the necessity to spin the sample by greatly improving non-spinning line shapes. One major factor in determining the sensitivity or peak heights is the observed line shape. If the peaks have wide bases, then a significant part of the signal intensity is found in this part of the peak and poor signal-to-noise (S/N) results. Thus good shimming, especially with respect to the 'hump test' (the linewidth of the chloroform ^{1}H NMR peak at the height of the ^{13}C satellite signals relative to that at 20% of the height of the ^{13}C satellite signals) is a prerequisite for good S/N and essential to successful HPLC-NMR operation.

In superconducting magnets, the main magnetic field is parallel to the long axis of the NMR flow cell and the flow direction and the RF is applied to a Helmholtz coil on the side of the probe insert. This coil design has a lower Q (quality factor) than a horizontal solenoid coil. A horizontal coil would be preferred on sensitivity grounds but shim systems are designed for vertical sample tubes, as normally samples are inserted vertically from the top of the magnet. The NMR flow cell fixed in a vertical position has the advantage of inducing laminar flow which helps remove air bubbles from the cell.

NMR flow cells have so far had much larger volumes than, for example, UV detection cells because of the lower sensitivity of NMR. The effects that large-volume NMR cells have on chromatographic peak dispersion have been investigated using a modified fluorescence detector. The selection of the detection volume of an NMR flow cell is a compromise

between the needs of the NMR sensitivity and the HPLC resolution. Typically the optimum detection volume is 1–5 μL when an HPLC column of 250×4 mm is used. In contrast the usual detected volume in a conventional NMR tube is rarely less than 60 μL. It has been shown that the NMR resonances are broadened at different flow rates using detectors with a range of volumes. At flow rates of 1 mL min^{-1} an increase of 0.14 Hz in linewidth is obtained only with flow cell volumes <120 μL. The broadening is due to an effective shortening of the spin–spin relaxation time, T_2, due to flow. Thus,

$$1/T_2\,(\text{obs}) = 1/T_2 + 1/t \qquad [2]$$

where t is the flow rate. The selection of an HPLC-NMR probe is thus a compromise between sensitivity and resolution. The two most common flow cell volumes are 60 μL and 120 μL. In general the 60 μL flow cells are best for on-flow experiments and stop-flow using small-diameter columns with low flow rates. The 120 μL flow cell is more suited to stop-flow experiments using conventional HPLC columns and flow rates.

Applications

HPLC-NMR has been applied to a wide range of analytical problems. These include:

- The characterization of endogenous and xenobiotic metabolites directly from a biological matrix.
- Characterization of metabolites from *in vitro* studies.
- The dynamic study of reactive metabolites.
- Characterization of natural products from complex mixtures.
- Polymer characterization.
- Characterization of reaction impurities from small- and large-scale systems.

In order to illustrate the utility of HPLC-NMR and provide examples of typical data from HPLC-NMR experiments, the characterization of the human urinary metabolites of the anti-HIV drug efavirenz (Sustiva®) is described below.

Efavirenz [6] is a non-nucleoside inhibitor of HIV-1 reverse transcriptase. The metabolism of efavirenz in rats was studied using directly coupled HPLC-NMR-MS. This hyphenated technique utilizes an eluent splitter to also incorporate a mass spectrometer into the HPLC-NMR system in order to obtain additional information on molecular weight and to act as a detector for selection of peaks on which to perform

[5] Desmethyl - glucuronide
conjugate of naproxen

[6] Efavirenz

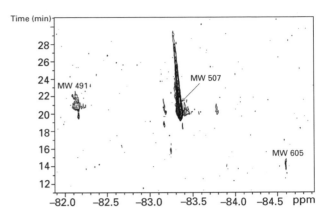

Figure 5 Pseudo-2D plot of on-flow ^{19}H HPLC-NMR-MS data obtained on rat urine after dosing with efavirenz. The data supply the ^{19}F chemical shift, retention time, and mass of the metabolites.

[7]

NMR experiments. The sample for analysis was prepared by solid-phase extraction of 5 mL of urine obtained from rats after dosing with 800 mg kg^{-1} of efavirenz, with 0–24 hr collection. The extract was dried and reconstituted with 100 μL of 80% D$_2$O and 20% acetonitrile-d_3. A 40 μL injection was made onto a 3.9 × 150 Waters Symmetry C$_{18}$ column. A gradient elution from 80% D$_2$O and 20% acetonitrile-d_3 to 50% D$_2$O and 50% acetonitrile-d_3 over 20 min at a flow rate of 0.8 mL min^{-1} was employed for separation. Using a splitter immediately after the column, 95% of the sample went to the UV detector and onto the NMR spectrometer while 5% went to a Finnigan LCQ ion-trap mass spectrometer equipped with an ESI probe operating in the positive-ion mode. The system was plumbed to allow the peak to reach the UV detector and the mass spectrometer at the same time.

The initial experiment, an on-flow ^{19}F detected HPLC-NMR-MS run, takes advantage of the CF$_3$ group present in the parent drug. Since there are no endogenous fluorinated compounds in rat urine, responses in the spectrum must arise from metabolites of efavirenz. This experiment provided the retention times of the metabolites that were of sufficient concentration to be detected. The ^{19}F chemical shift and the mass of the metabolites were also obtained from this single experiment. **Figure 5** shows the data as a pseudo-2D plot. The peaks of interest are labelled with their molecular mass.

The next phase of the experimental procedure is to obtain stop-flow ^1H HPLC-NMR spectra at the retention times determined in the on-flow ^{19}F HPLC-NMR experiment. The 1D and TOCSY spectra obtained on the minor metabolite with retention time 14.2 min, mass 605 Da and ^{19}F chemical shift − 84.6 ppm are shown in **Figure 6**. The mass-spectral data showed a loss of 80 Da from the parent mass indicative of SO$_3$ loss. The loss of the methine proton from the cyclopropyl ring, seen clearly from the TOCSY spectrum, and the downfield shift of the remaining methylene protons indicates addition of OH to the cyclopropyl ring. The singlet at δ 6.47 is consistent with reduction of the alkyne. The spin system observed between δ 3 and δ 4 ppm is consistent with cysteinylglycine conjugation. Based on these data and the observed molecular mass of 605 Da, this peak is assigned as [7].

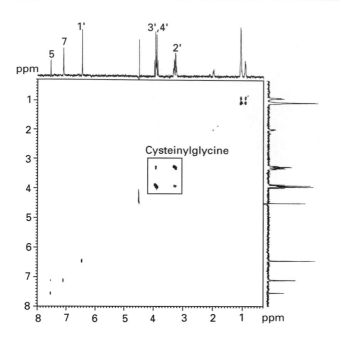

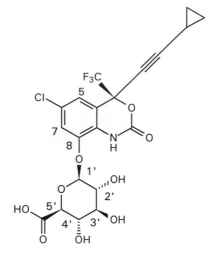

[8] 8-*O*-glucuronide
conjugate of efavirenz

Figure 6 1D and TOCSY HPLC-NMR spectra of the 8-O-sulfate, cysteinylglycine di-conjugate of efavirenz.

The 1D and TOCSY data obtained on the major metabolite with retention time 20.5 min, mass 507 Da and ^{19}F chemical shift −83.3 ppm is shown in **Figure 7**. These data clearly show the presence of a glucuronide conjugate. The observed changes in the aromatic region of the spectrum are consistent with the 8-O-glucuronide conjugate of efavirenz [8].

Co-eluting with the major metabolite is a component with mass 491 Da and ^{19}F chemical shift −82.1 ppm. Using the mass spectrometer as a detector the stop flow was executed when the desired mass, 491 Da, was observed. While the spectrum (**Figure 8**) is dominated by the major metabolite, the 8-O-glucuronide conjugate of efavirenz, it is possible to assign several resonances to a glucuronide conjugate. The assignment of this metabolite as the N-glucuronide conjugate [9] is consistent with the chemical shift of H1′, δ 4.75, and with the observed molecular mass.

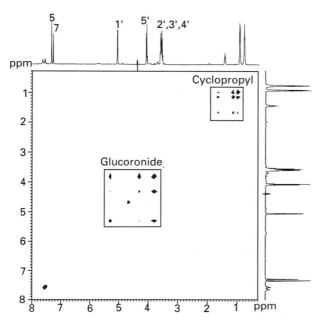

Figure 7 1D and TOCSY HPLC-NMR spectra of the 8-O-glucuronide conjugate of efavirenz.

[9] *N*-glucuronide
conjugate of efavirenz

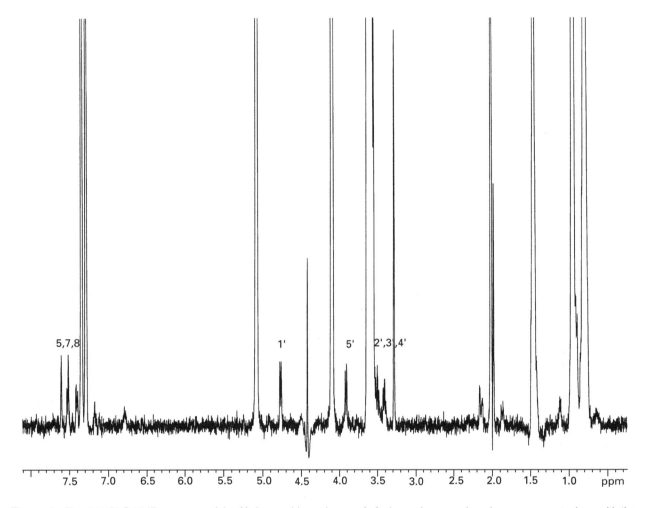

Figure 8 The ^{1}H HPLC-NMR spectrum of the *N*-glucuronide conjugate of efavirenz shown as the minor component, along with the major component the 8-O-glucuronide conjugate.

Summary

The experiments described serve to illustrate how HPLC-NMR can rapidly provide information on the structure of drug metabolites. These same methods have been employed in the study of endogenous compounds in biofluids, natural products, polymers and many other complex mixtures.

The high cost of the HPLC-NMR system is of course a factor which must be taken into consideration. Since most laboratories have seen the value of NMR in its traditional form, it is only necessary to add an HPLC system, an appropriate flow probe and a flow-control unit to an existing spectrometer to enable HPLC-NMR experiments to be performed. The cost of these accessories, or even the cost of a complete system dedicated to HPLC-NMR, is offset by the efficiency of the method. The laborious extraction of minor components from complex mixtures followed by off-line analysis or in the case of synthetic drugs the synthesis of radiochemically labelled materials with the numerous problems associated with handling and disposing of radiolabelled samples make the cost associated with HPLC-NMR less of a factor.

There are a number of limitations to the utility of HPLC-NMR. The greatest of these is the limitation on the amount of material that can be loaded on the column and consequently moved into the NMR probe. HPLC-NMR is not well suited to the less-sensitive NMR techniques like HMBC, which provides critical structural information via long-range interactions. Thus it is still necessary to isolate minor components from complex mixtures on occasion and HPLC-NMR should not be viewed as a replacement for conventional NMR. It is a tool which enhances the utility of NMR and like any tool it can be used properly and effectively or misused. Although it has experimental limitations when compared to conventional NMR, many structural problems can be solved using HPLC-NMR without resorting to tedious isolation methods.

The advantages of HPLC-NMR can be summarized as:

- Speed of analysis by elimination of the isolation step.
- Ability to obtain structural information and high-quality spectral data from complex matrices.
- The lack of need for perfect chromatographic separation.
- Synergy when directly coupled with mass spectrometry.

List of symbols

b = receiver bandwidth; B_0 = magnetic-field strength; f = coil filling factor; I = spin quantum number; n = oversampling factor; n = preamplifier noise figure; N = number of detected nuclei; Q = coil quality factor; T = temperature; T_1 = relaxation time; T_2 = spin–spin relaxation time; V_c = detector-coil volume; V_s = sample volume; γ = gyromagnetic ratio; τ = flow rate.

See also: **Biofluids Studied By NMR; Chromatography-IR, Applications; Chromatography-MS, Methods; Drug Metabolism Studied Using NMR Spectroscopy; Fourier Transformation and Sampling Theory; NMR Data Processing; NMR Pulse Sequences; Solvent Suppression Methods in NMR Spectroscopy.**

Further reading

Albert K (1995) On-line use of NMR detection in separation chemistry. *Journal of Chromatography A* 703: 123–147.

Albert K (1995) Direct on-line coupling of capillary electrophoresis and ¹H NMR spectroscopy. *Angewandte Chemie International Edition in English* 34: 641–642.

Albert K, Kunst M, Bayer E, Spraul M and Bermel W (1989) Reverse-phase high-performance liquid chromatography-nuclear magnetic resonance on-line coupling with solvent non-excitation. *Journal of Chromatography* 463: 355.

Albert K, Braumann, U, Tseng L-H, *et al.* (1994) On-line coupling of supercritical fluid chromatography and proton high-field nuclear magnetic resonance spectroscopy. *Analytical Chemistry* 66: 3042–3046.

Behnke B, Schlotterbeck G, Tallarek U, *et al.* (1996) Capillary HPLC-NMR coupling: high resolution NMR spectroscopy in the nanoliter scale. *Analytical Chemistry* 68: 1110–1115.

Griffiths L (1995) Optimization of NMR and HPLC conditions for LC-NMR. *Analytical Chemistry* 67: 4091–4095.

Laude DA Jr and Wilkins CL (1984) Direct-linked analytical scale high-performance liquid chromatography/nuclear magnetic resonance spectroscopy. *Analytical Chemistry* 56: 2471–2475.

Lindon JC, Farrant RD, Sanderson PN, *et al.* (1995) Separation and characterization of components of peptide libraries using on-flow coupled HPLC-NMR spectroscopy. *Magnetic Resonance in Chemistry* 33: 857–863.

Lindon JC, Nicholson JK and Wilson ID (1996) Direct coupling of chromatographic separations to NMR spectroscopy. *Progress in Nuclear Magnetic Resonance Spectroscopy* 29: 1–49.

Lindon JC, Nicholson JK, Sidelmann UG and Wilson ID (1997) Directly coupled HPLC-NMR and its application to drug metabolism. *Drug Metabolism Reviews* 29: 705–746.

Pullen FS, Swanson AG, Newman MJ and Richards DS (1995) "On-line" liquid chromatography/nuclear magnetic resonance mass spectrometry – a powerful spectroscopic tool for the analysis of mixtures of pharmaceutical interest. *Rapid Communications in Mass Spectrometry* 9: 1003–1006.

Shockcor JP, Unger SE, Wilson ID, Foxall PJD, Nicholson JK and Lindon JC (1996) Combined HPLC, NMR spectroscopy, and ion-trap mass spectrometry with application to the detection and characterization of xenobiotic and endogenous metabolites in human urine. *Analytical Chemistry* 68: 4431–4435.

Shockcor JP, Wurm RM, Frick LW, *et al.* (1996) HPLC-NMR identification of the human urinary metabolites of (−)-cis-5-fluoro-1-[2-(hydroxymethyl)-1,3-oxathiolan-5-yl]-cytosine, a nucleoside analogue active against human immunodeficiency virus (HIV). *Xenobiotica* 26: 189–199.

Smallcombe SH, Patt SL and Keifer PA (1995) WET solvent suppression and its application to LC-NMR and high resolution NMR spectroscopy. *Journal of Magnetic Resonance Series A* 117: 295–303.

Sweatman BC, Farrant RD, Sanderson PN, *et al.* (1995) Evaluation of the detection limits of directly coupled 600 MHz ¹H AND ¹H-¹³C HPLC-NMR spectroscopy. *Journal of Magnetic Resonance Analysis* 1: 9–12.

Wolfender J-L, Rodriguez S and Hostettmann K (1998) Liquid chromatography coupled to mass spectrometry and nuclear magnetic resonance spectroscopy for the screening of plant constituents. *Journal of Chromatography A* 794: 299–316.

Chromium NMR, Applications

See **Heteronuclear NMR Applications (Sc–Zn).**

CIDNP Applications

Tatyana V Leshina, **Alexander I Kruppa** and
Marc B Taraban, Institute of Chemical Kinetics and
Combustion, Novosibirsk-90, Russia

MAGNETIC RESONANCE
Applications

The phenomenon of chemically induced dynamic nuclear polarization (CIDNP) consists of the manifestation of unusual line intensities and/or phases of signals of radical reaction products in the NMR spectrum when reaction takes place directly in the probe of the spectrometer. These anomalous lines (enhanced absorption or emission of NMR signals), which reflect the populations of nuclear spin states deviating from the Boltzmann condition, are observed within the time range of nuclear relaxation times of the diamagnetic molecules (T_{1N}), which are as a rule, several seconds to tens of seconds. Subsequently, the NMR spectrum re-acquires its usual form. In 1967, two research groups in Europe (J. Bargon, H. Fischer, and U. Johnson) and the USA (H. Ward and R. Lawler) discovered independently that this phenomenon is directly associated with the free radicals involved in the process. Later on, it was shown that this also pertains to radical ions and triplet excited states of molecules.

According to currently existing theory, the nonequilibrium intensities in NMR spectra are generated as a result of electron–nuclear interactions in the so-called radical pair. These are the pairs of paramagnetic particles that may originate during the homolysis of a molecule under the action of heating, light or ionizing radiation, as well as resulting from single-electron transfer between the donor and acceptor molecules and radical encounters in the bulk preceding the recombination.

It follows from the above list that, in principle, the CIDNP phenomenon can be observed for virtually all types of radical reactions. To date, various CIDNP techniques have been employed to clarify the role of radical species in: redox reactions of organic compounds, nucleophilic and electrophilic substitution, addition and elimination reactions, isomerization processes, thermolysis and photolysis of organic peroxides and azo- and diazocompounds, a number of photosensitized processes, including the reactions of organoelement compounds of silicon, germanium, tin, mercury, aluminium, lithium, sodium, potassium, magnesium and organoxenon compounds, polymerization processes and a number of rearrangement reactions.

Note that prior to CIDNP studies, in certain cases, the involvement of paramagnetic species in the reactions had not even been anticipated. A good example is the process of sensitized *cis–trans* photoisomerization of substituted olefins. Moreover, CIDNP has brought unambiguous evidence of earlier-suggested single-electron transfer as the first stage of aliphatic nucleophilic substitution in the reaction of alkyl- and benzylhalides with alkyl derivatives of lithium.

The analysis of CIDNP effects allows information to be obtained on the structure and reactivity of active short-lived (from nanoseconds to microseconds) paramagnetic species (free radicals and radical ions, triplet excited molecules), and on molecular dynamics in the radical pair. The analysis makes it possible to distinguish between the bulk and 'in-cage' stages of complex chemical reactions. CIDNP data also provide information on the multiplicities of reacting states, which is of utmost importance for understanding photochemical processes.

CIDNP is not a direct method of detection of the structure of radical species: the information of hyperfine interaction (hfi) constants, *g*-factors, and lifetimes of paramagnetic intermediates, is obtained on the basis of the analysis of the phases and intensities of polarized lines in NMR spectra, as well as from CIDNP kinetics, and the dependence of polarization efficiency on the magnetic field strength applied to the sample during reaction. However, the simplicity and reliability of the identification of polarized lines in the NMR spectra ensure high plausibility of CIDNP-based structural and kinetic information.

Thus nowadays CIDNP is one of the powerful techniques for the investigation of elementary stages of complex radical chemical reactions.

Nuclei demonstrating the CIDNP phenomenon

Chemical polarization is observed only for nuclei possessing a magnetic moment (spin). **Table 1** lists the nuclei for which CIDNP effects have been detected to date. At present, the majority of CIDNP experiments have been carried out for hydrogen (^1H) and fluorine (^{19}F) nuclei. Next in incidence are ^{13}C and ^{31}P. Unless

Table 1 Nuclei which have been used in CIDNP studies

Nucleus	Spin
^{1}H	1/2
^{2}H	1
^{13}C	1/2
^{15}N	1/2
^{19}F	1/2
^{31}P	1/2
^{119}Sn	1/2

otherwise specified, all further consideration will employ the parameters characteristic for ^{1}H CIDNP.

CIDNP observation techniques

The chemical reactions are most commonly run directly in the probe of an NMR spectrometer. Such chemical polarization formed in the magnetic fields of NMR spectrometers (with a strength of several tesla) is usually called CIDNP in high magnetic fields.

There exist other techniques of CIDNP observation, as well. In these techniques the reaction under study is carried out in a separate magnet, as a rule, with variable field strength, and afterwards the reaction mixture is transferred rapidly (with the characteristic time lower than T_{1N}) to the probe of the NMR spectrometer. The transfer is often performed by a flow system; however, other possibilities also exist.

It is necessary to note that CIDNP formed in high and low magnetic fields, such as the geomagnetic field, are two essentially different physical phenomena.

Types of CIDNP effects

According to their manifestation, CIDNP effects can be subdivided into two types: net effects (enhanced absorption and emission of NMR signals), and multiplet effects, consisting of the redistribution of the line intensities between specific components of multiplet signals in the NMR spectrum (**Figure 1**).

CIDNP effects in high magnetic fields

For the chemical application of CIDNP it is important to be aware of the fact that in high magnetic fields the nonequilibrium population of nuclear spin sublevels is formed as a result of transitions solely between singlet (S) triplet zero (T_0) states or a radical pair (**Figure 2**). These transitions occur due to interaction of an electron spin with spins of magnetic nuclei in the radical pair and they are also dependent on the difference in g-factors of radicals in the radical pair.

Theory shows that in this case the probabilities of the formation of the singlet (recombining) state of a radical pair are different for α and β projections of nuclear spins. As a result, spins with a certain orientation predominantly appear in the 'in-cage' recombination product, while spins with an opposite orientation occur in products resulting from the escape of radicals into the bulk (**Figure 3**). According to the existing Kaptein rules modified by Closs, the sign of the net CIDNP effect (N) in high magnetic fields is defined by the following set of parameters:

$$N = \mu \cdot \varepsilon \cdot \Delta g \cdot \mathrm{hfi} \cdot \lambda$$

where μ is the multiplicity of the precursor of the radical pair ('+' for triplet and '−' for singlet precursor), ε is '+' for 'in-cage' and '−' for escape recombination products, Δg is the sign of the difference in the g-factors of the radical with polarized nucleus and radical partner in the radical pair, hfi is the sign of the hyperfine constant of a given nucleus in the radical under study, and λ is '+' for the radical pair recombining from an S-state and '−' for the radical pair recombining from a T-state (usually employed for radical ion reactions). The sign of N reflects the phase of the NMR signal of a given nucleus: '+' for enhanced absorption (A) and '−' for the emission (E).

There are also qualitative rules employed for the analysis of multiplet effects:

$$M = \mu \cdot \varepsilon \cdot \mathrm{hfi}_1 \cdot \mathrm{hfi}_2 \cdot J \cdot \gamma,$$

where μ, ε, are defined as in the previous case, hfi_1 and hfi_2 are the signs of hyperfine constants of nuclei 1 and 2 in the radical forms, J is the sign of the spin–spin coupling constant of these nuclei in the molecule, γ is '+' if these nuclei belong to the same radical, and '−' if they belong to different radicals in the radical pair. The sign of M reflects two types of multiplet effect: '+' for E/A and '−' for A/E type effects.

Analysis of the detected net and multiplet effects is also performed by means of a comparison of the experimentally observed effects with those that are calculated.

For example, the above-mentioned qualitative rules are applicable for the analysis of CIDNP effects for all nuclei from **Table 1**, except for ^{119}Sn, which has a number of peculiarities stipulated by its anomalous magnetic resonance parameters (hyperfine constant more than 0.16 T, g-factor 2.0160). To analyse CIDNP effects of this nucleus one has to employ a special theoretical approach.

Thus, analysis of the experimental sign of the CIDNP effects basically allows the above enumerated parameters of the radical pair to be singled out

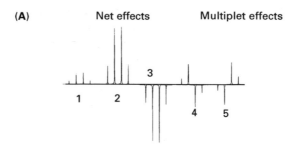

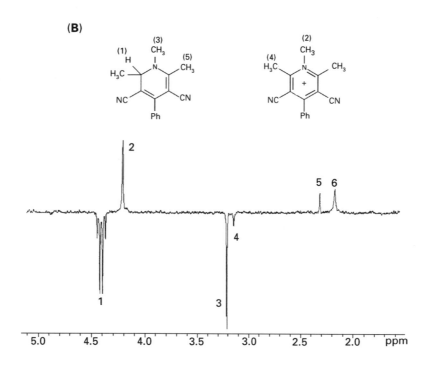

Figure 1 (A) Schematic presentation of the types of CIDNP effects: (1) equilibrium NMR spectrum, (2) enhanced absorption (A), (3) emission (E_2), (4) multiplet effect (A/E), (5) multiplet effect (E/A). (B) Experimental example of time-resolved (100 μs after laser pulse) 250 MHz ^{1}H CIDNP spectrum, detected during the photolysis of substituted 1,2-dihydropyridine in the presence of 2,5-dichlorobenzoquinone: line assignment is shown on the structures; signal (6) is attributed to HDO admixtures in CD_3CN. The equilibrium signals were removed using the presaturation pulse sequence preceding the laser pulse.

provided that all the others are known in advance. For example, the chemical shift of the polarized signal in an NMR spectrum and its multiplet structure often reveal the structure of the radical precursor of the polarized compound. In this case, this assumption is easily verified by substituting the magnetic resonance

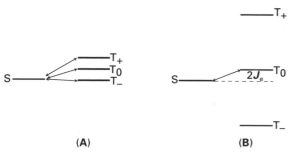

Figure 2 Schematic presentation of the radical pair energy levels in (A) low ($H_0 \approx$ hyperfine interaction constant) and (B) high ($H_0 \gg$ hyperfine interaction constant) magnetic fields.

Figure 3 Distribution of chemical polarization between geminate ('in-cage') and homogeneous products of radical reaction.

parameters of the suggested radical in the above expressions for the rules. Another possibility is the definition of the multiplicity of the reacting state for known compositions of the radical pair.

The enhancement coefficient serves as a basic quantitative feature of CIDNP. One usually distinguishes between the observed enhancement coefficient and the absolute one. The observed enhancement coefficient is defined as the ratio of the intensities of the polarized NMR signal to the equilibrium one recorded after the completion of the reaction. The time dependence of observed polarization on time (dI^*/dt) has two components, $(dI^*/dt)_{react}$ and $(dI^*/dt)_{relax}$, reflecting the contributions from chemical reaction and relaxation of CIDNP. As a first approximation, the relaxation of CIDNP occurs through spin–lattice relaxation, which suggests the following expression for the observed polarization:

$$\frac{dI^*}{dt} = \left(\frac{dI^*}{dt}\right)_{react} - \frac{I^*(t) - I_0(t)}{T_{1N}}$$

where I^* and I_0 are the intensities of polarized and equilibrium signal, respectively; $(dI^*/dt)_{react} = dI_0/dt \cdot m$, where m is the absolute enhancement factor of the product polarization. This expression implies that the observed polarization depends on the magnetic resonance parameters of the radical pair (hyperfine coupling, g-factors) through m, radical relaxation times in the diamagnetic state, and the kinetics of the buildup of polarized products.

The absolute enhancement coefficient of CIDNP differs from the observed one in that the former is dependent only on the radical pair parameters. At present, the absolute enhancement coefficient can be reliably measured only for the case of photochemical reactions with the employment of specially elaborated time-resolved CIDNP techniques.

From the viewpoint of experimental techniques, the major potentialities are now realized for research into CIDNP in photochemical reactions. In this case, time-resolved techniques (so-called flash-CIDNP) are widely used. These techniques utilize laser-pulse photoexcitation with pulse detection of CIDNP effects. The solution of the sensitivity versus time-resolution dilemma stipulates the following restrictions of the attainable time resolution: the duration of the detecting radiofrequency pulse should not be shorter than 100 ns, otherwise the sensitivity of the NMR method is lost. Shorter pulses can be employed only to study the processes demonstrating an extreme absolute enhancement coefficient of CIDNP. The latter are characteristic only for a limited number of model

reactions, for example the photolysis of cyclic ketones.

The improvement of time resolution is possible when using both a standard NMR spectrometer and modified equipment. These techniques differ by the methods of mathematical processing of the experimental data when estimating the kinetic parameters.

Even in the simplest case of a microsecond time range, the time-resolved techniques allow one to observe homogeneous and geminate processes separately. The homogeneous processes include bimolecular recombination in the bulk, degenerate electron exchange reactions, proton exchange, and various rearrangement reactions which also occur in the bulk.

CIDNP effects in low magnetic fields

A low magnetic field is defined as one in which the energy of the Zeeman interaction is of the same order of magnitude as the hyperfine interaction (see **Figure 2**). Therefore, a nonequilibrium population of nuclear states may arise not only via S–T$_0$ transitions, but also through S–T$_-$ and S–T$_+$ transitions between energy levels of a radical pair. The observed sign of CIDNP will predominantly depend on the difference in efficiencies of S–T$_{+,-}$ transitions.

As **Figure 2** suggests, the efficiency of S–T$_-$ and S–T$_+$ transitions also depends essentially on the magnitude and sign of the electron exchange interaction (J_e) in the radical pair. However, experimental studies on CIDNP in low magnetic fields show that for a radical pair comprising neutral radicals in nonviscous solutions, J_e does not contribute significantly to the resulting polarization. The exceptions are radical ion pairs and biradicals.

To unravel the mechanisms of the elementary stages of radical reactions by means of the CIDNP method in low magnetic fields, one usually analyses the dependencies of CIDNP efficiencies on the magnetic field strength (field dependence of CIDNP). In this case, the data on radical pair composition are obtained on the basis of comparison of experimentally measured and theoretically calculated CIDNP field dependencies.

Comparison of CIDNP techniques in high and low magnetic fields for investigation of elementary mechanisms of radical reactions

The basic distinction in the potential of application of these two techniques of CIDNP observation is stipulated by the difference in the mechanisms of their formation. Thus, formation of CIDNP effects in high magnetic field is not accompanied by flips of either electron or nuclear spin (for S and T_0 states the projection onto the magnetic field direction is equal to zero). Here, the radical reaction actually plays the role of dispatcher, directing radicals with α and β projection of nuclear spin to different products (see **Figure 3**). Thus, the formation of at least two products is a key element for CIDNP manifestation. This condition essentially narrows the range of processes open for investigation. Thus, so-called crypto-radical reactions, which are rather widespread in chemical practice, fall outside of the consideration range. To study these processes it is recommended CIDNP in a low magnetic field is used.

In the case of CIDNP in low magnetic field, the true nonequilibrium population of nuclear sublevels results from simultaneous flips of both electron and nuclear spins (see **Figure 2**). Therefore, in this case, the formation of the single product does not preclude the observation of CIDNP. A number of other reasons also stipulate the greater versatility of chemical polarization in low magnetic fields, namely, the greater observed enhancement coefficients as compared to those in a high magnetic field and the potential and to detect net polarization formed in the radical pair with identical radicals.

Examples of CIDNP applications to the structure determination and properties of paramagnetic species (free radicals, radical ions, biradicals, carbenes, macromolecules)

CIDNP in structural chemistry

By now the analysis of CIDNP effects has produced a large set of useful structural data on the signs of

paraffin constants, g-factor values of the radicals, and the energies of electron exchange interaction in the radical pairs.

Of special value is the information on the so-called magnetic resonance parameters (paraffin interactions and g-factors) that cannot be estimated easily by other magnetic resonance techniques in the case of short-lived radicals. The data obtained allow inferences to be made on the electronic structure of paramagnetic species (σ- or π-radical) and hence about their reactivity.

Among the main sources of such data are photolysis and thermoses reactions of various organic peroxides, demonstrating ^1H, ^{19}F and ^{13}C CIDNP effects. As a rule, the mechanisms of these processes have been determined previously, and the multiplicity of the reacting state is also known. Thus, employing the rules for the analysis of multiple and net CIDNP effects in high magnetic fields (see above) one can sequential determine either the signs of the paraffin constants for the radical pairs or the sign of Δg.

To define the values of the paraffin constants and the g-factors of the radical pairs the entire polarized spectrum is stimulated and compared with the experimental one. This approach has been employed to obtain the values of the paraffin constants and g-factors listed in **Table 2**.

For example, it follows from **Table 2** that the bonds in the cyclopropyl free radical are flexed, and the proton in position 1 protrudes out of the plane of the three-membered ring, since the hyperfine constant of this proton is much lower than that observed in a methyl radical (2.3 mT).

Aside from hyperfine constants and g-factors, the analysis of the multiplet effect in the photolysis products of cyclopropyl peroxide and, cyclopropene has defined the sign of the vicinal proton–proton

Table 2 Magnetic properties of free radicals defined by means of CIDNP

Radical	Hyperfine interaction constant		Spin density, ρ_π	g-factor
(fluorobenzoyloxyl radical, positions 1,2,3,4)	^{19}F: ^{13}C:	$a_4 > 0$ $a_2 > 0$	$\rho_1 > 0$ $\rho_2 < 0$ $\rho_3 < 0$ $\rho_4 > 0$	
(fluorophenyl radical, positions 1,2,3,4)	^{19}F:	$a_2 > 0$ $a_3 > 0$ $a_4 < 0$		
(acetyloxyl radical CH_3)	^1H:	$a < 0$ (−2.3 mT)		2.0058
(cyclopropyl radical, positions 1,2)	^1H:	$a_1 < 0$ (−0.65 mT) $a_2 > 0$ (2.34 mT)		

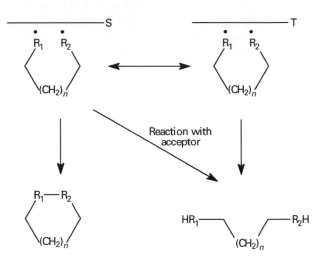

$a_\alpha > 0$

$a_\beta \approx 0$

$a_\alpha < 0$

$a_\alpha > 0$

Figure 4 Alternative structures of the 1,2-diphenylcyclopropane radical cation with the expected signs of the hyperfine interaction constants, a.

spin–spin coupling constant ($J_{H\text{-}H} > 0$) which was unknown earlier.

The analysis of CIDNP effects has also made possible the determination of the structures of a number of radical ions of cyclic compounds (benzonorbornadiene, trialkylcyclopropanes, *syn*- and *anti*-paracyclophanes, etc.).

For example, the structure of the radical cation of 1,2-diphenylcyclopropane has been assigned on the basis of the analysis of ^1H CIDNP data formed in the act of photoinduced reversible electron transfer from cyclopropane to chloranil. The choice has been made between 'closed' and 'open' structures of the radical cation of 1,2-diphenylcyclopropane (**Figure 4**). The observed CIDNP effects of 1,2-diphenylcyclopropane (absorption of aromatic *ortho*-, *para*-, and H_α, and the emission of H_β) comply only with the 'open' structure.

CIDNP is also a method of determining nuclear spin–lattice relaxation times of free radicals and radical ions (T_{1R}). This approach was used to determine the T_{1R} values of the CH_2-protons of benzyl (3.5×10^{-4} s), dichloromethyl (4.5×10^{-4} s), and *t*-butyl (2.4×10^{-4} s) free radicals.

CIDNP effects in reactions involving carbenes and biradicals

Paramagnetic species possessing two unpaired electrons (biradicals, carbenes, and their analogues) have been intensively studied using CIDNP techniques in high and low magnetic fields. The involvement of biradicals in organic and biochemical reactions is postulated in a number of cases, but the physical methods (EPR, laser spectroscopy) have allowed the detection of their formation predominantly in the photolysis of cyclic ketones. These species have also served as models for detailed studies of CIDNP peculiarities for biradicals, as well as the distinction of these effects from the case of free radicals.

A major peculiarity is the contribution of $S–T_{+,-}$ transitions to CIDNP, which is due to the electron

exchange interaction between the spins of unpaired electrons (see the CIDNP effects in low magnetic fields section above).

If the processes taking place in a biradical allow the possibility of the selection from nuclear spin states, the $S–T_0$ transitions can also manifest themselves.

Reference data on biradicals of cyclic ketones point to the appearance of CIDNP effects in a wide range of magnetic field strengths. The manifestations of various mechanisms of CIDNP formation depend on the structure of the biradical.

As a rule, the field dependence of CIDNP shows up as curves with extremum, which are located in the magnetic field with the strength of about J_e.

Comparatively low values of J_e (< 0.1 T) are characteristic of flexible biradicals with a long carbon chain (C_8 and more) separating the radical centres. Therefore, one might expect the manifestation of an $S–T_-$ mechanism (for $J_e > 0$, $S–T_+$ mechanism) in low magnetic fields, while in a high magnetic field the $S–T_0$ mechanism will manifest itself.

Large J_e values are typical of short biradicals (C_7 and shorter). In this case the extremum will appear in a magnetic field similar to that of conventional NMR spectrometers, $J_e > 1$ T. Therefore, in this case only the $S–T_{+,-}$ mechanism of CIDNP formation will appear.

There are also so-called rigid biradicals whose structure does not possess a flexible carbon chain. In this case, J_e may be either very large or negligibly small, the latter case reflecting the orthogonality of the orbitals of unpaired electrons. In both cases, it is reasonable to expect the manifestation of the $S–T_0$ mechanism of CIDNP formation.

In the former case, the $S–T_0$ mixing results from the existence of well-known nonradiative transitions.

Here, the energy of J_e is counterbalanced by a pool of vibrational degrees of freedom.

The manifestation of an $S-T_0$ mechanism of CIDNP formation in high magnetic fields has been observed recently for biradicals containing Group 14 elements (Si, Ge).

At present, the investigation of CIDNP field dependencies for products of transformation of biradicals is the most reliable indirect method of demonstrating the involvement of biradicals in chemical reactions. The structure of the biradical can be inferred by comparison of observed with calculated field dependences. For this purpose, one might employ various approximations of the radical pair described in a number of specialized monographs.

Reference data show the examples of CIDNP manifestations of triplet (ground) and singlet (excited) states of carbenes. The triplet state is observed in the reactions of diphenylcarbene and methylene, while the singlet state has also been detected in the case of methylene. The analysis of CIDNP effects has led to the following conclusions. In contrast to the generally accepted viewpoint expressed in the chemical literature, not only the triplet but also the singlet state of carbenes reacts via hydrogen or halogen atom abstraction from the partner rather than insertion into the corresponding bond.

Recently, ^1H CIDNP effects have also been observed in the reactions of carbenes containing silicon and germanium with various scavengers. In this case, the radical stages have been detected in the reaction of singlet (ground) and triplet (excited) states of dimethylsilylene and -germylene.

CIDNP as a method of investigation of spin and molecular dynamics in radical pairs

Research into spin dynamics is presented in numerous papers devoted to the definition of the range of values of J_e in biradicals. This approach also allows an estimate to be made of the range of J_e values in radical ion pairs.

A prominent example of the potential of the method for investigating molecular dynamics is the recent research into the process of protein folding by means of photo-CIDNP observed in stopped flow experiments. It has been shown that the use of the CIDNP technique in stopped flow experiments has certain advantages over the conventional NMR studies using flow systems. The sensitivity gain due to the CIDNP enhancement coefficient allows the folding processes to be studied on a time range two orders of magnitude shorter than those observable by mean of conventional NMR methods. In the process under study, CIDNP results from photoinduced electron transfer between a flavin additive and tyrosine and other amino acids of the protein (hen egg lysozyme). After the refolding, the CIDNP disappears, since after the conformational changes in protein structure amino acids become inaccessible to the flavin species.

At present, CIDNP is virtually a unique technique for the detection of so-called weak interactions (with the energy lower than several kT) between the reagents in chemical reactions. This includes charge transfer complexes, exciplexes and donor–acceptor complexes. As a rule, the involvement of these complexes is confirmed by the data on multiplicity of the reacting state obtained on the basis of CIDNP analysis (see corresponding sections in this article on the analysis of CIDNP effects). For example, the inconsistency between the multiplicity estimated on the basis of CIDNP data and that expected for the known set of photochemical parameters has justified the suggestion of a triplet exciplex in the photooxidation of Hantzsch ester by quinones.

The alteration of the multiplicity in the reaction of silylene wit CCl_4 observed in the 7,7-dimethyl-7-silanorbornadiene derivative in the presence of triphenylphosphine has served as proof of the complexation between dimethylsilylene and Ph_3P.

CIDNP in reactions involving biologically relevant molecules

The CIDNP technique has made a decisive contribution to the widely discussed question of the role of single electron transfer stages in biochemical processes. These are primarily the oxidation of organic substrates which occur under the action of UV irradiation and electron acceptors, as well as the enzymatic processes involving oxygenases.

The CIDNP technique has been used to unravel the elementary mechanisms of the oxidation of NADH coenzymes and their synthetic analogues, and substituted 1,4-dihydropyridines, in the presence of electron acceptors. The underlying mechanism of NADH transformation is the transition from the dihydropyridine ring to pyridine.

In this case, the basic question is whether this process occurs via the single act of hydride transfer, or the redox reaction includes the sequence of several elementary stages involving paramagnetic species.

Application of ^{1}H and ^{15}N CIDNP techniques has allowed the demonstration that the process under study indeed includes the sequence of radical stages. For unsubstituted dihydropyridines, this is a chain of 'e–H$^+$–H$^•$' transformations, while for substituted species and NADH, the sequence 'e–H$^•$' has been observed.

The CIDNP technique has also been used to reveal the role of radical species in the processes of phototransformations of coenzyme B_{12} model compounds and *all-trans*-retinal.

List of symbols

a = hyperfine interaction constant; A = absorption; E = emission; I^* and I_0 = intensities of the polarized and equilibrium signal; J = spin–spin coupling constant; J_e = electron exchange interaction; m = absolute enhancement factor of product polarization; M = multiplet effect; N = net CIDNP effect; S = singlet state; t = time; T_0 = triplet zero state; T_{1N} = nuclear relaxation time; T_{1R} = nuclear spin–lattice relaxation time in a radical; γ = parameter for location of the magnetic nuclei in the radicals; Δg = difference in the g-factors of the radical with polarized nucleus and the radical partner; ε = parameter for 'in-cage' or escape recombination; λ = parameter for recombination from S- or T-state; μ = multiplicity.

See also: **Chemical Applications of EPR; Chemical Exchange Effects in NMR; EPR, Methods; Heteronuclear NMR Applications (Ge, Sn, Pb); NMR Pulse Sequences.**

Further reading

Closs GL and Miller RJ (1978) Photoreduction and photodecarboxylation of pyruvic acid. Application of CIDNP to mechanistic photochemistry. *Journal of the American Chemical Society* 100: 3483–3494.

Egorov MP, Ezhova MB, Kolesnikov SP, Nefedov OM, Taraban MB, Kruppa AI and Leshina TV (1991) ^{1}H CIDNP study of the addition reaction of Me$_2$E (E = Si, Ge) to a carbon–carbon triple bond. *Mendeleev Communications* 4: 143–145.

Goez M (1996) Optimization of flash CIDNP experiments. *Journal of Magnetic Resonance, Series A* 123: 161–176.

Hore PJ, Winder SL, Roberts CH and Dobson CM (1997) Stopped-flow photo CIDNP observation of protein folding. *Journal of the American Chemical Society* 119: 5049–5050.

Kruppa AI, Taraban MB, Leshina TV, Natarajan E and Grissom CB (1997) CIDNP in the photolysis of coenzyme B$_{12}$ model compounds suggesting that C–Co bond homolysis occurs from the singlet state. *Inorganic Chemistry* 36: 758–759.

Kruppa AI, Taraban MB, Polyakov NE, Leshina TV, Lusis V, Muceniece D and Duburs G (1993) The mechanism of the oxidation of NADH analogues. 2. N-Methyl substituted 1,4-dihydropyridines. *Journal of Photochemistry and Photobiology A: Chemistry* 73: 159–63.

Kruppa AI, Taraban MB, Shokhirev NV, Svarovsky SA and Leshina TV (1996) ^{119}Sn CIDNP: Calculation and experiment. *Chemical Physics Letters* 258: 316–322.

Kruppa AI, Taraban MB, Svarovsky SA and Leshina TV (1996) Paramagnetic intermediates in the photolysis of 2-methylpropanoyltripropylstannane studied by means of multinuclear CIDNP. *Journal of the Chemical Society, Perkin Transactions* 2 10: 2151–2156.

Lawler RG and Halfon MA (1974) A flow reactor for use with an unmodified high resolution NMR spectrometer. Application to CIDNP. *Review of Scientific Instruments* 45: 84–86.

Lepley AR and Closs GL (1973) In: Closs GL (ed) *Chemically Induced Magnetic Polarization*. New York: Wiley.

Muus LT, Atkins PW, McLauchlan KA and Pedersen JB (1977) In: Pedersen JB (ed) *Chemically Induced Magnetic Polarization*. Dordrecht: Reidel.

Polyakov NE and Leshina TV (1990) Study of CIDNP field dependencies in the photoreduction of quinones by amines. *Journal of Photochemistry and Photobiology A: Chemistry* 55: 43–51.

Richard C and Granger P (1974) *Chemically Induced Dynamic Nuclear and Electron Polarization—CIDNP and CIDEP*. Berlin: Springer Verlag.

Roth HD and Herbertz T (1993) The structure of simple vinylcyclopropane radical cations. Evidence for conjugation between alkene and cyclopropane group. *Journal of the American Chemical Society* 115: 9804–9805.

Salikhov KM, Molin YuN, Sagdeev RZ and Buchachenko AL (1984) In: Molin YuN (ed) *Spin Polarization and Magnetic Effects in Radical Reactions*. Amsterdam: Elsevier.

Circularly Polarized Luminescence and Fluorescence Detected Circular Dichroism

Christine L Maupin and **James P. Riehl**, Michigan Technological University, Houghton, MI, USA

ELECTRONIC SPECTROSCOPY
Applications

The measurement of the usually small net circular polarization in the luminescence from chiral molecules began with the pioneering studies of Professor L.J. Oosterhof at the University of Leiden in the late 1960s. In recent years this technique has developed into a reliable and useful spectroscopic tool for the study of a wide variety of chemical systems. The phenomenon has been referred to as circularly polarized luminescence (CPL), circularly polarized emission (CPE), and by other names. Here the most common acronym, CPL, is used to describe this experimental technique. The measurement of the differential emission intensity due to differences in the absorption of left versus right circularly polarized light is also discussed. This is commonly referred to as fluorescence-detected circular dichroism (FDCD). Whereas FDCD spectroscopy (like conventional circular dichroism (CD)) probes the chiral structure of the ground state, CPL is a probe of the chirality of the excited state. These types of experiments are not redundant, since changes of geometry do occur on electronic excitation. Furthermore, both of these experimental techniques provide information concerning molecular dynamics and energetics that occur between the time of excitation and emission.

Instrumentation for measurement of circularly polarized luminescence

Circularly polarized luminescence spectroscopy (CPL) has been used for more than 30 years as a probe of the molecular stereochemistry of chiral luminescent molecules and complexes. Unlike CD, CPL measurements have not, as yet, become routine. This is primarily due to the lack of a commercially available instrument. The measurement of the generally small net circularly polarized luminescence, relative to the total luminescence, is difficult owing mainly to the presence of optical artefacts from various sources, and to electronic problems associated with the measurement of weak differential signals. The first CPL measurements were made using analogue phase-sensitive detection systems. Measuring the generally small CPL signal in this way is particularly difficult because of electronic problems associated with lock-in amplifiers. All of the current instruments use photomultipliers operating in photon-counting mode, and some type of gated photon counting system to determine the circular polarization. This is the type of system that will be described here.

Steady-state CPL

Since no commercial instrumentation is available, all CPL measurements that have been published have been measured on instruments that were designed and constructed in individual laboratories. The basic configuration of a system designed to measure CPL is given in **Figure 1**. A laser system or a UV lamp used in combination with an excitation monochromator or a combination of filters can be used as the exciting source. Care should be taken in ensuring that the polarization of the excitation light is such that there will be no linear polarization in the emission since this is known to cause artefacts in the detection of circular polarization. In most luminescence experiments the emission is collected at an angle of 90.0° from the direction of the excitation beam to avoid detecting light from the excitation source. In some applications of CPL spectroscopy, it is necessary to control the excitation polarization. **Figure 1**, for example, shows the optics necessary to produce circularly polarized excitation.

The luminescence from the chiral sample is first directed through a photoelastic polarization modulator (PEM) and then a linear polarizer before it passes through any focusing lenses or filters because these optical components may be birefringent. The linear polarizer ensures that the detected light entering the monochromator will always be polarized in the same direction. This prevents artefacts associated with the emission monochromator being polarization sensitive. The PEM is composed of an isotropic, clear optical material that produces an oscillating birefringence at a fixed frequency. In CPL measurements, the PEM is operated as an electrically controlled advancing-retarding quarter-wave plate at a predetermined frequency (f). This introduces a periodic phase shift at frequency f, such that either right or

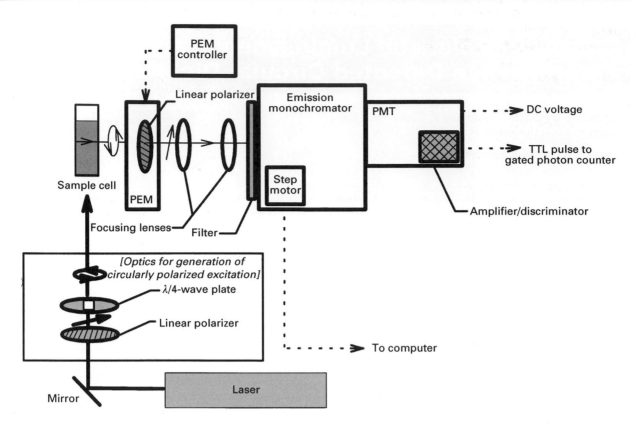

Figure 1 Schematic diagram of an instrument designed to measure circularly polarized luminescence.

left circularly polarized emitted light is converted to linearly polarized light oriented at 45° to the PEM crystal axis. Therefore, the PEM and linear polarizer allow the passage of alternately left then right circularly polarized light. The emitted linearly polarized light is then focused onto the entrance slits of an emission monochromator. Background signal from stray excitation and room light can be reduced by placement of a long pass filter in front of the monochromator. The light exiting the monochromator is detected by a photomultiplier tube operating in photon-counting mode. The signal from the photomultiplier tube is sent into an amplifier-discriminator. These devices are capable of providing simultaneous dual output for TTL pulses that are generated for photon spikes, as well as a voltage proportional to the total emission intensity.

The TTL pulses are analysed by a photon counter using the reference signal at frequency f to direct the pulses into two separate counters. The time dependence of the phase shift imparted by the PEM is such that it is exactly quarter-wave retarding or advancing only at the maximum or minimum of the periodic oscillation. For this reason, it is common to select finite time-windows centred at the peak of the modulation cycle for the counting of photon pulses, and not to count the pulses in the 'dark' interval. Even

when these windows are set to collect 50% of the incoming TTL pulses, only a slight error (< 5%) in the accuracy of this measurement is introduced. On alternate half-cycles of the modulation cycle, the TTL pulses are directed into two separate counters corresponding to the transmission of left (I_L) or right (I_R) circularly polarized light through the PEM. In CPL spectroscopy one commonly reports the ratio of the differential emission intensity at wavelength, λ.

$$\Delta I(\lambda) = I_L(\lambda) - I_R(\lambda) \qquad [1]$$

to the total emission intensity

$$I(\lambda) = \frac{1}{2}(I_L(\lambda) + I_R(\lambda)) \qquad [2]$$

We thus define the luminescence dissymmetry ratio, g_{lum}, as follows:

$$g_{lum}(\lambda) = \frac{\Delta I(\lambda)}{\frac{1}{2}I(\lambda)} = \frac{I_L(\lambda) - I_R(\lambda)}{\frac{1}{2}(I_L(\lambda) + I_R(\lambda))} \qquad [3]$$

I, ΔI, and g_{lum} may all be calculated from the values of the two counters. Because of the difficulty in determining absolute emission intensities, it is most common to report CPL results in terms of g_{lum}.

Calibration and standards

The determination of the precise state of polarization of the circularly polarized emission, and of the accuracy of a gated photon-counting system, may be accomplished by a number of methods. The simplest and most convenient approach for calibration is to use a standard to measure g_{lum}. For example, several groups have used the chiral and commercially available NMR chiral shift reagent tris(3-[trifluoromethyl-hydroxymethylene]-d-camphorato)europium-(III) ($Eu(facam)_3$; Aldrich) in dry dimethyl sulfoxide (DMSO). This calibration can be performed by exciting the solution of $Eu(facam)_3$ in the visible or ultraviolet, and measuring g_{lum} at several different wavelengths. This complex has known g_{lum} values of +0.072 at 612.8 nm for the strong emissive transition $^5D_0 \rightarrow {}^7F_2$, and −0.78 for the $^5D_0 \rightarrow {}^7F_1$ transition at a wavelength maximum of 595 nm. Calibration of the gated photon counting system could also be accomplished by passing completely left and right circularly polarized light through the PEM. This should in theory yield g_{lum} values of +2 and −2, respectively. However, the value of exactly 2 (or −2) is never obtained owing to imperfect circular polarization and the inaccuracies associated with the finite time-window for the gated photon counter, as described above.

One of the main advantages of the photon-counting system over lock-in detection is the fact that, if the instrument is working properly, i.e. obeying Poisson photon statistics, then it can be shown that the distribution of measured g_{lum} values should obey a Gaussian distribution. It can also be shown that the standard deviation in g_{lum} is equal to the inverse square root of the total number of photon counts:

$$\sigma = \sqrt{\frac{1}{N}} \qquad [4]$$

Measuring a large number of g_{lum} values and seeing whether the distribution is Gaussian is another convenient way of checking whether the instrument is working properly.

The intensity of luminescence will obviously vary greatly from one sample to the next. Some samples will be highly luminescent (i.e. give high count rate) while others will be weakly luminescent (i.e. give low count rate). For samples that are highly luminescent, accurate g_{lum} measurements can be achieved in a short time. For example, if a sample is emitting 10^6 photons s^{-1}, then using Equation [4], g_{lum} with a standard deviation of 10^{-4} requires 100 s to be measured. Weakly luminescent samples require much longer collection times to achieve the same standard deviation. If the count rate is only 10^4 photons s^{-1}, then more than 5 h of data collection are required.

Time-resolved CPL

Advances in digital counting technology and high-speed computers have allowed several research groups to expand the measurement of CPL into the time domain. These advances have been led by the research groups of Professor H.P.J.M. Dekkers at the University of Leiden and Professor F.S. Richardson at the University of Virginia. The instruments used are based on modifications of steady-state instruments such as the system described above, with the major instrumental change being that the excitation source must be pulsed. The principal problem in time-resolved CPL is how to measure the decay of the polarization with an instrument based on a polarization modulator that may be oscillating much more rapidly than the lifetime of the decay. Two approaches to this problem have been successful. If the precise phase of the polarization modulator is known, then an excitation pulse can be coordinated with the periodic oscillation in such a way that the flashes occur at precise times in both halves of the modulation cycle. To ensure that the device is analysing for circular polarization at each decay point, the time between the excitation and the decay measurements is varied. Just as described above, for each finite time interval centred at time t in the decay, photon pulses corresponding to $I_L(t)$ and $I_R(t)$ are accumulated, and by simple calculation, $g_{lum}(t)$ is determined. Alternatively, it is possible to completely decouple the excitation flashes from the modulation cycle and accumulate enough flash events that the complete modulation cycle is sampled for every finite time channel. This latter method has the advantage that the complete time decay for each excitation pulse contributes to the measurement.

Applications for circularly polarized luminescence

Since, as described, there are no commercial sources of CPL instrumentation, the applications of this spectroscopic technique have been limited to the study of specific research goals of those groups who have built the instrumentation. Nevertheless, the applications have covered a fairly wide area of chemistry, including organic, inorganic, polymer and biochemical systems. It should be mentioned that most applications of CPL have involved forbidden spectroscopic transitions. This is because, in these experiments, one is always measuring small differences in luminescence intensity. These differences are obviously easier to detect if the individual intensities themselves are not very large, i.e. if they are not associated with allowed transitions.

This section highlights examples of different kinds of applications, and the reader is referred to one of the review articles listed at the end of this article for a more detailed discussion of specific chemical systems in which CPL has been used.

Applications to organic chromophores

Many of the early applications of CPL involved analysis of the circular polarization from $^1n\pi^*$ states of chiral ketones. Pyramidal distortion of the >C=O group upon going from the ground state to the excited/emitting state in CD spectroscopy is well known, and comparative CD and CPL studies have been used to probe these excited state geometry changes. As an example, **Figure 2** shows the CPL spectrum (upper curve) and total luminescence spectrum (lower curve) for an α-diketone, d-camphorquinone dissolved in ethanol. Note that the form of presentation of CPL results parallels that usually seen for CD studies. In this example, the luminescence dissymmetry ratio, g_{lum}, at the peak maximum for this $n \leftarrow \pi^*$ transition is approximately −0.01. g_{abs} measured from the CD spectrum of this compound for the $n \rightarrow \pi^*$ transition is approximately −0.009, and the differences between these measurements reflects geometry changes within the dicarbonyl moiety upon electronic excitation. In bichromophoric molecules one may use the well-known exciton coupling model to interpret CPL results in exactly the same way that this model is applied to CD measurements. In a very few cases, for example, β,γ-enones, chirality rules have been developed that relate the sign and magnitude of the CPL signal for closely related compounds to specific aspects of chiral structure. These attempts at spectra–structure correlations are clearly not as well developed as they are for CD spectroscopy.

Inorganic applications

As described above, CPL spectroscopy is particularly suited for the study of forbidden transitions, so the potential for applying this technique to intraconfigurational transitions in metal complexes is very high. In the case of transition metals, however, applications of CPL have been quite limited. The reasons for this are many. Certainly, the preparation of resolved chiral transition metal complexes is quite difficult,

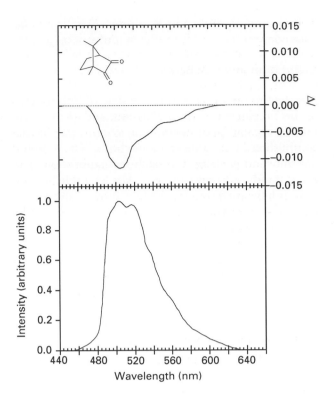

Figure 2 CPL spectrum (upper curve) and total luminescence spectrum (lower curve) for d-camphorquinone in ethanol.

and often the racemization under UV or visible illumination may not be neglected. In addition, not many transition metal complexes are luminescent in liquid solution. The luminescence from 'frozen' samples can be quite high, but measurement of CPL under conditions where the sample is highly scattering and nonisotropic are fraught with experimental difficulties. There have been a number of interesting applications of CPL to chiral transition metal complexes. For example, Ru(1,10-phenanthroline)$_3^{2+}$ may easily be prepared and resolved. This compound possesses D_3 symmetry and, as such, exists as Δ and Λ enantiomers. The broad luminescence from this complex centred at 600 nm under UV excitation is fairly intense with a measured g_{lum} value of approximately 0.0007. g_{lum} varies significantly in the low-wavelength shoulder of the emissive transition, indicating that the precise nature of the metal–ligand states involved are quite complicated. There have also been a few examples of CPL measurements from crystalline samples; however, the requirements here are that the sample be chiral and luminescent, and occur in a space group that is uniaxial. This combination is very restrictive.

By far the most common class of inorganic compounds that have been studied by CPL spectroscopy are luminescent lanthanide complexes. There are several reasons for this special interest. Complexes of

Tb(III), Eu(III), and to lesser extent, Dy(III), are luminescent in the visible region of the spectrum, particularly if water molecules can be excluded from the first coordination sphere. The transitions involved are isolated f $\leftrightarrow$ f transitions, and even though the absorptions are very weak, all three of these ions have absorptions in the visible, which overlap convenient strong laser lines. In many cases it is also possible to excite allowed transitions in the ligand orbitals, and populate the lanthanide ion energy levels through efficient energy transfer. Lastly, and most importantly, the states derived from the various f-electron configurations are well characterized by the total angular momentum quantum number J, and transitions that obey magnetic dipole selection rules, that is $\Delta J = 0, \pm 1$, are often associated with very large dissymmetry ratios.

An example of CPL from a chiral lanthanide complex is given in **Figure 3**. This spectrum was obtained by exciting the sample of the macrocyclic Tb(III) complex with 488 nm excitation from an argon-ion laser. The spectral region displayed corresponds to a transition from the $^5D_4$ excited state to the $^7F_5$ state in the ground-state manifold. This transition satisfies the magnetic dipole selection rules introduced above. The ground state for Tb(III) is the $^7F_6$ state. The various peaks shown in the spectrum correspond to different crystal field transitions within the term-to-term transition. The chiral pendant amines in the macrocyclic ligand are chiral, and this results in only one of the possible C_4 orientations around the central metal ion being formed. Confirmation of this latter conclusion may be obtained by varying the incident polarization from left to right circularly polarized, and the result for this system is that no effect is seen in the CPL spectrum. As can be seen in **Figure 3** $|g_{lum}|$ for magnetic dipole-allowed transitions may be very large. The values seen in **Figure 3** approach 0.4; this should be compared to the results presented in **Figure 2** (0.01), which represents approximately the maximum value that has been measured for organic chromophores.

Unlike transition metal compounds such as Ru(phen)$_3^{2+}$, which are composed of achiral ligands but can be resolved into chiral enantiomers, no such examples of the resolution of chiral lanthanide compounds with achiral ligands have appeared in the literature. This is because lanthanide complexes are generally more labile than transition metal complexes. Just like transition metal complexes, however, it is possible in some cases to perturb the ground-state equilibrium concentration through the addition of a chiral environment compound. This is sometimes referred to as the Pfeiffer effect. An example of such an experiment is presented in **Figure 4**. In this

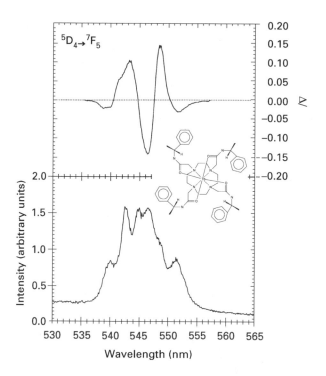

Figure 3 CPL spectrum (upper curve) and total luminescence spectrum (lower curve) for an aqueous solution of a chiral macrocyclic Tb(III) complex.

experiment, D- and L-enantiomers of glucose have been added to aqueous solutions of Tb(2,6-pyridinedicarboxylate) (Tb(DPA)$_3^{3-}$). This complex is known to possess D_3 symmetry, so it occurs in solution as a mixture of Δ and Λ enantiomers.

Racemic mixtures

Even though the ground-state distribution of a mixture of chiral molecules is racemic, in some cases it is still possible to measure the CPL if the excited state can be enriched in one of the enantiomers. If, for example, the excitation beam is circularly polarized, one of the enantiomers with be excited preferentially over the other, and the excited state will not be racemic. This experiment, which has been referred to as CPE/CPL, relies on discrimination in absorption and emission. Obviously, it is also important that the nonracemic excited-state distribution created by the excitation beam is maintained during the excited state's lifetime. An example of such an experiment is given in **Figure 5**. The CPL spectrum given in **Figure 5** was collected while exciting an aqueous solution of Eu(DPA)$_3^{3-}$ with circularly polarized 472.8 nm light from an argon-ion laser. These complexes do not racemize on the emission timescale, so one may write down the following equation relating the measured luminescence dissymmetry ratio for

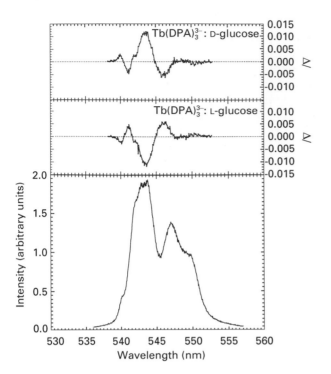

Figure 4 CPL spectra and total luminescence (lower curve) for an aqueous solution of Tb(DPA)$_3^{3-}$ containing 0.01 M D-glucose (upper curve) and L-glucose (middle curve).

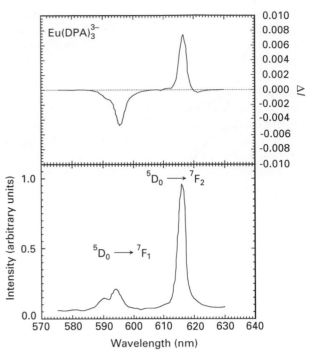

Figure 5 CPL spectrum (upper curve) and total luminescence (lower curve) for an aqueous solution of Eu(DPA)$_3^{3-}$ using circularly polarized 472.8 nm excitation.

left circularly polarized excitation to the intrinsic absorption and emission dissymmetry ratio for one of the enantiomers as follows:

$$g_{lum}^{L}(\lambda) = \frac{1}{2} g_{abs}^{\Delta}(\lambda') g_{lum}^{\Delta}(\lambda) \qquad [5]$$

In this equation we have been careful to denote that the wavelength of excitation (λ') is different from the wavelength of emission detection.

Measurement of CPL from racemic mixtures is not a technique that can be applied to all racemic solutions. For moderately luminescence systems g_{lum} values of approximately 10^{-4} are measurable. This means that the intrinsic absorption and emission dissymmetry ratios need to be on the order of 10^{-2} for this experiment to be successful. Although there have been a couple of examples using this technique in organic systems, by far the most widely studied systems are racemic lanthanide complexes, such as given in **Figure 5**, because of the large g_{lum} and g_{abs} values that may exist for certain f $\leftrightarrow$ f transitions. It is also useful to perform these experiments in a time-resolved mode and as a function of temperature to determine racemization rate constants.

There are other ways to generate enriched excited states from racemic ground states. If, for example, a

chiral excited state quencher molecule is added to a racemic solution, the rate of quenching of the two enantiomers may be different. This kind of experiment has been performed both in the steady-state and time-resolved modes. At time $t = 0$ following an unpolarized excitation pulse directed at a racemic mixture, the concentration of excited enantiomers will be equal. The diastereomeric selective quenching results in a time-dependent population that becomes increasingly richer in one of the enantiomers. Results from a typical experiment of this type are given in **Figure 6**. Here g_{lum} is plotted as a function of time for an aqueous solution of Tb(DPA)$_3^{3-}$ into which a small amount of Δ-Ru(phen)$_3^{2+}$ has been added. The different quenching rates for the diastereomeric pairs depend in a complicated way on mutual diffusion, reorientation, electronic overlap and diastereomer structure. These types of experiments may be performed with varying donor–quencher pairs, in different solvents, and as a function of temperature and pressure to probe intimate details of this excited-state chiral discrimination.

Biological and polymer systems

Almost all biological systems are chiral, so it is natural to expect that CPL studies involving luminescent biological material will be important. Unlike CD, where all of the absorbing chromophores contribute

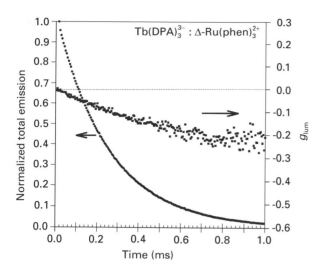

Figure 6 Time dependence of g_{lum} at 544 nm following excitation at 330 nm for an aqueous solution of $Tb(DPA)_3^{3-}$ into which a small amount of Δ-$Ru(phen)_3^{2+}$ has been added.

to the resultant CD spectrum, luminescence from protein and other biological systems is generally only seen from a very small number of emitting chromophores. Thus, CPL studies involving biological systems generally only provide specific structural information localized around the emitting chromophore, and not the longer-range information on secondary and tertiary structure that is routinely obtained from CD studies. This is true both for intrinsic protein fluorescence and in studies in which metal ion or organic fluorescence probes are used.

Some of the very early applications of CPL spectroscopy involved chiral polymeric systems. In particular, the CPL from achiral dye molecules dissolved in cholesteric liquid crystals has been used to probe chirality changes. CPL from chromophores attached to a chiral poly-amino acid may also be used to study exciton coupling between aromatic chromophores. In these polymeric systems, it is often observed that g_{lum} is quite large. In some cases it is so large, in fact, that detection using static quarter-wave plates is possible.

Instrumentation for measurement of fluorescence-detected circular dichroism

In fluorescence-detected circular dichroism (FDCD) the difference in total emission intensity is used to monitor differences in circularly polarized absorption. This spectroscopic technique was developed in the early 1970s in the laboratory of Professor I. Tinoco at the University of California,

Berkeley. In the simplest application of FDCD a photomultiplier tube is placed at 90° to the direction of emission detection in a modified commercial or custom-built CD instrument and the luminescence intensity in phase with the oscillating left and right circularly polarized absorption beam is monitored. For many applications, including recent uses of FDCD as detectors in HPLC, this type of experimental setup is sufficient. The major technical challenge in the construction of an FDCD instrument is associated with the fact that it is very difficult to produce pure circularly polarized excitation without any residual polarization. Furthermore, if this linear polarization varies in phase with the incident circular polarization, serious artefacts may result. To reduce this problem, a somewhat more complicated experimental setup is required. A schematic diagram of an FDCD instrument designed to deal with the linear polarization problem is shown in **Figure 7**.

The incident (absorption) beam is wavelength-selected by an excitation monochromator and then converted to alternately left and right circular polarization by use of a linear polarizer and quarter-wave modulator. The quarter-wave modulator may be a PEM as described above for CPL measurement, or a Pockels cell. This luminescence in phase with the incident polarization modulation is collected by two photomultiplier tubes (PMT1 and PMT2) oriented at right angles to each other and the incident beam. If the excitation is linearly polarized, then the sum of

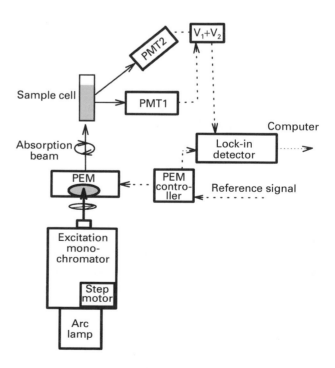

Figure 7 Schematic diagram of FDCD instrument.

the fluorescence intensities from the two detectors due to this artefact gives a signal $(V_1 + V_2)$ that is independent of the linear polarization. So even if this polarization changes as the incident polarization varies from left to right, the resultant artefact signal subtracts out of the difference in fluorescence intensity, $F_L - F_R$. The artefact signal does not subtract out of the total luminescence intensity, $F_L + F_R$, but, if present, it may be reduced by rotation of the PEM (or adjustment of the Pockels cell). The signals from the photomultiplier tubes are added and either these are input to a lock-in detector (as shown) or, if digital counting techniques are employed, the TTL pulses are analysed by a gated photon-counting system as described previously.

To relate the measurement of FDCD to conventional CD, it is necessary to take into account that, if differential absorption of the circularly polarized excitation beam is taking place, then the intensity of the absorption beam as a function of distance into the sample may be different for the two circular polarizations. The relationship between CD and FDCD measurements for isotropic systems containing only one optically active fluorophore may be expressed as

$$\Delta\varepsilon = \varepsilon_L - \varepsilon_R = \frac{g_F(1 - 10^{-A})}{cd\,10^{-A}\ln 10} \qquad [6]$$

where A denotes the absorbance of the sample, c is the concentration, and d is the path length. Processing of FDCD signals under conditions where there are more than one chiral chromophore is more complicated. Calibration of an FDCD instrument may be accomplished by measuring the signal for an achiral fluorescent molecule to ensure that a value of zero is obtained, and then by measuring a standard compound with a known CD such as d-camphorsulfonic acid.

Applications of fluorescence-detected circular dichroism

FDCD measurements have the potential for providing valuable spectroscopic and structural information for systems in which conventional CD measurements are problematic. Of particular interest are complex systems containing multiple absorption chromophores but only one fluorescent chromophore. It is also possible to use the intrinsically higher sensitivity of luminescence detection to determine

the CD of samples for which only very small quantities are available. FDCD has also shown promise in high-pressure liquid chromatographic detection of luminescent chiral molecules, although no commercial applications of this technology are yet available.

Although interesting and important applications for the FDCD technique have been undertaken, the technical problems associated with the generation of exactly equal intensities of circularly polarized exciting light, and the artefacts associated with residual linear polarization, have severely limited the widespread use of this technique. These problems are not insurmountable, and it is anticipated that modification by the manufacturers of commercial CD instruments to include FDCD attachments should result in considerably more applications of this technique to interesting and important chemical problems.

List of symbols

A = absorbance; c = concentration; d = path length; f = frequency; g_{abs} = absorption dissymmetry ratio; g_{lum} = luminescence dissymmetry ratio; I = light intensity; J = total angular quantum number; t = time; ε = extinction coefficient; λ = wavelength; σ = standard deviation.

See also: **Biochemical Applications of Fluorescence Spectroscopy; Biomacromolecular Applications of Circular Dichroism and ORD; Chiroptical Spectroscopy, Emission Theory; Induced Circular Dichroism; Luminescence, Theory; ORD and Polarimetry Instruments; Vibrational CD Spectrometers; Vibrational CD, Applications; Vibrational CD, Theory.**

Further reading

Richardson FS and Riehl JP (1977) Circularly polarized luminescence spectroscopy. *Chemical Reviews* 77: 773–792.

Riehl JP (1994) Excited state optical activity. In: Purdy N and Brittain HG (eds) *Analytical Applications of Circular Dichroism.* Amsterdam: Elsevier.

Riehl JP and Richardson FS (1986) Circularly polarized luminescence spectroscopy. *Chemical Reviews* 86: 1–15.

Riehl JP and Richardson FS (1993) Circularly polarized luminescence. *Methods in Enzymology* 226: 539–553.

Steinberg I (1975) Circular polarization of fluorescence. In: Chen RF and Edelhoch H (eds) *Biochemical Fluorescence: Concepts.* New York: Marcel Dekker.

Cluster Ions Measured Using Mass Spectrometry

O Echt, University of New Hampshire, Durham, NH, USA

TD Märk, Leopold Franzens Universität, Innsbruck, Austria

MASS SPECTROMETRY
Applications

Introduction

Clusters are defined as homogeneous aggregates of atoms (A_n) or molecules (M_n), although mixed (heterogeneous) aggregates (X_nY_m or X_nY^+) are also considered as clusters. The size range of interest starts at the dimer ($n = 2$) and reaches up to 10^6 to 10^8 constituents per cluster. Depending on the nature of the building blocks, the binding energy between the constituents can be rather weak, as in the case of van der Waals bound rare gas clusters (10–100 meV), or rather large, as in the case of covalently bonded carbon clusters (several eV). Gas-phase clusters can be produced either by fragmentation of solid material (laser desorption or sputtering into vacuum) or by aggregation of single particles (supersonic jet expansions, gas aggregation); the clusters may be either neutral or ionized. Owing to the statistical nature of these methods, the cluster sample (beam) usually comprises a broad distribution of different cluster sizes (a notable exception being the formation of C_{60}). Thus virtually all experiments – in particular those concerned with the properties of a single cluster size, or with the evolution of cluster properties as a function of size – rely on mass spectrometry as a major tool for investigating this unique form of matter.

Besides high mass resolution for the proper identification of a specific cluster ion (**Figure 1**), one of the most important requirements for cluster studies is a large mass range (see **Figure 2** which shows a photoionization time-of-flight mass spectrum of caesium oxide clusters). This has led to the sophisticated and improved use of established techniques and to the innovative development of new concepts in mass spectrometry (electrospray ion sources, reflection type mass spectrometers, etc.), which nowadays are also of great use in other fields such as medicine and biology for the investigation of large biomolecules. Besides identification of cluster ions and cluster ion distributions, the other major use of mass spectrometry in cluster science is the selection of mass-analysed monodisperse cluster ions in order to allow the

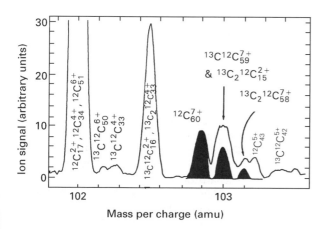

Figure 1 Section of high resolution C_{60} mass spectrum obtained by multistep electron impact ionization. The black peaks are calculated isotopomers of the septuply charged C_{60}^{7+} ion normalized to the first isotopomer of the experimental run. After Scheier P and Märk TD (1994) *Physical Review Letters* **73**: 54.

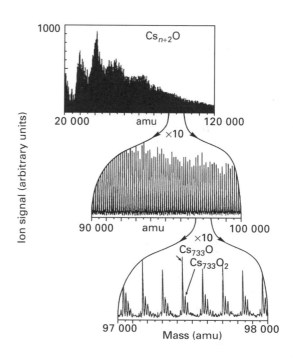

Figure 2 Section of mass spectra for Cs/O_2 clusters using photoionization time-of-flight mass spectrometry shown on three different mass scales. After Martin TP, Bergmann T, Näher U, Schaber H and Zimmermann U (1994) *Nuclear Instruments and Methods* **B88**: 1.

investigation of cluster ion properties such as the structure, energetics, stability and reactivity. While size-selected neutral cluster targets are usually unavailable, mass spectrometry plays a crucial role in their investigation as well. For example, the information gained from photoelectron spectroscopy of mass-selected metal cluster anions pertains to the neutral species. Likewise, the sizes of neutral clusters that are interrogated by charge transfer reactions or resonance-enhanced ionization can be determined in a mass spectrometer, provided fragmentation does not occur.

Atomic structure

The geometric arrangement of atoms is of paramount interest in cluster science. Mass spectrometry has provided essential clues to cluster structure and morphology, even though the approach is indirect. In short, one attempts to relate anomalies in the size distribution of clusters to anomalies in their properties, such as stability, reactivity or ionization energy.

Figure 3 shows a mass spectrum of krypton clusters. Neutral clusters are grown by expansion of the gas through a nozzle into vacuum. Several abundance anomalies, often called 'magic numbers', are observed. Two questions arise: (1) Assuming the numbers indicate enhanced stability, is there a growth sequence that can consistently explain them? (2) Do the anomalies exist before ionization, or are they caused by dissociation following ionization?

Atoms in rare gas solids arrange in a close-packed structure, each atom being surrounded by 12 nearest neighbours. Clusters of rare gases are expected to

assume compact shapes. Hence, it is reasonable to assume that clusters grow layer by layer around a core comprising 13 atoms, and that closed shells are particularly stable. The abundance of these clusters will be enhanced if clusters are warm enough to evaporate atoms.

However, the 13-atom core may be either a cuboctahedron cut out of the fcc crystalline solid, or an icosahedron (**Figure 4**). Either way, the total number of atoms in complete-shell clusters would be 13, 55, 147, 309, 561; these cluster sizes are, indeed, particularly abundant in the mass spectra. **Figure** 3 reveals several additional magic numbers. They indicate closure of subshells that occurs whenever a facet of the cluster is covered by a layer of atoms. A cuboctahedral cluster has 14 facets, the icosahedral one has 20, leading to distinct subshell closures. For argon through xenon in this size range, only the

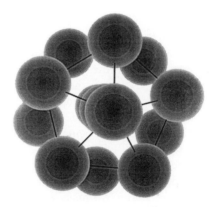

Icosahedron

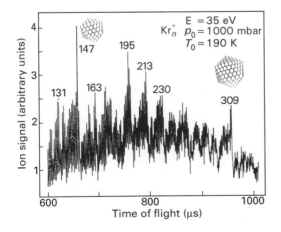

Figure 3 Mass spectrum of krypton clusters, grown in a supersonic expansion and ionized by electron impact. The cluster size and assumed cluster structure of particularly prominent mass peaks is indicated.

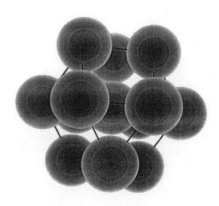

Cuboctahedron

Figure 4 Schematic view of a 13-atom icosahedron and a 13-atom cuboctahedron.

icosahedral growth sequence can explain the observed anomalies.

Concerning ionization, theory and a large number of experiments indicate that it is essentially impossible to 'softly' ionize a rare gas cluster or, in fact, any cluster with van der Waals bonding. Dissociation will lead to enhanced intensities for particularly stable charged clusters; metastable decay will cause the relative intensities to become time dependent (**Figure 5**). Likewise, mass spectra of small clusters formed in the charged state, e.g. by sputtering, may deviate considerably from spectra of clusters formed in the neutral state. For larger clusters, however, neither the system (Ar, Kr, Xe) nor the cluster formation/ionization process affects the anomalies.

Magic numbers in spectra of calcium and strontium clusters also indicate icosahedral growth, while aluminium clusters appear to adopt an fcc structure with octahedral morphology. Magic numbers observed when metal atoms are grown around a cluster of known geometry, C_{60}, nicely confirm the power of the closed-shell or closed-subshell concept.

Mass spectra of transition metal clusters do not exhibit any abundance anomalies. However, if they are mixed with reactive gases prior to expansion into vacuum, the number of molecules being absorbed exhibits pronounced size effects. Mass spectra may directly reveal the number of reactive sites on the cluster surface; Co_{19}, for example, offers six highly reactive sites towards ammonia. This number tends to increase with cluster size, but local minima are observed at certain cluster sizes, indicating closed shells or subshells in agreement with an icosahedral growth sequence.

The stoichiometry of NaCl cluster cations is $(NaCl)_n Na^+$. A pronounced sequence of magic numbers is observed if $n = 13, 22, 37, 62, 87, \ldots$. Assuming the ionic arrangement in the cluster is that of the crystal (rock salt, fcc lattice), compact shapes can be obtained by cutting cuboids out of the crystal. The number of ions in such a cuboid is odd only if the number of ions along each edge is odd. The numbers n listed above correspond to cuboids with $3\times3\times3$, $3\times3\times5$, $3\times5\times5$, $5\times5\times5$ and $5\times5\times7$ ions. Several other alkali halide clusters show the same pattern.

A magic number in mass spectra of carbon clusters has ultimately led to the discovery of a new form of crystalline carbon. **Figure 6** displays spectra of carbon clusters formed by laser vaporization into helium gas with subsequent expansion into vacuum. Only even-sized clusters are observed above $n = 30$. Smaller clusters, which are known to form chains and rings, occur for odd and even sizes. A structural transition takes place near $n = 30$. Ion chromatography has revealed that various structural isomers may co-exist. The abundance anomalies at $n = 60$ and 70 become more pronounced if the helium gas density is raised (**Figure 6**, upper spectrum). The structure of carbon clusters in this size range can be described as that of a graphitic sheet which is curved and closed upon itself by replacing 12 hexagons with pentagons. C_{60} is the smallest of these so-called fullerenes in which all pentagons can be separated by hexagons; it has icosahedral symmetry. C_{70} would be the next larger fullerene with no adjacent pentagons. The unusually large enhancements of C_{60} and C_{70} in **Figure 6** reflect their inertness towards growth. Later research has led to the synthesis of macroscopic amounts of C_{60} and larger fullerenes. They are stable in air and, unlike other clusters, do not react with each other.

Fullerenes are hollow; C_{60} can cage atoms as large as xenon. The 'endohedral' nature of a complex may be demonstrated by its stability. $(HeC_{60})^+$, formed by collisions of C_{60}^+ with helium gas, survives

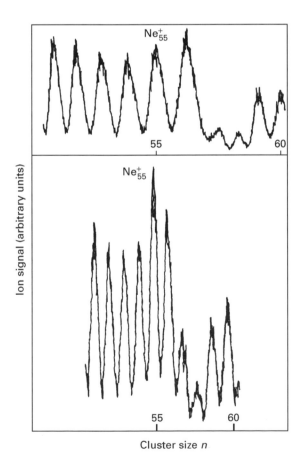

Figure 5 Section of a Ne_n^+ cluster ion mass spectrum in the vicinity of the magic number size $n = 55$ measured approximately 68 μs (upper trace) and 118 μs (lower trace) after the primary ionization event. After Märk TD and Scheier P (1987) *Chemical Physics Letters*, **137**: 245.

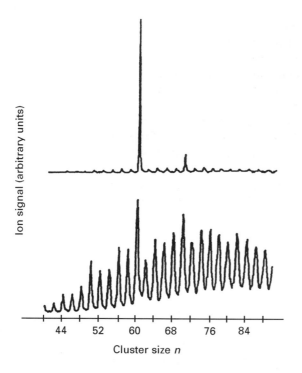

Figure 6 Mass spectrum of carbon clusters formed by laser vaporization into helium gas of high pressure (upper spectrum) and low pressure (lower spectrum). Reproduced with permission of Macmillan Magazines Ltd from Kroto HW, Heath JR, O'Brien SC, Curl RF and Smalley RE (1985) *Nature* **318**: 162.

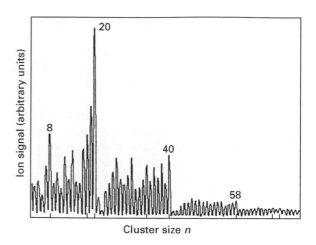

Figure 7 Mass spectrum of sodium clusters, 'softly' ionized by low-intensity light from a UV lamp. After Knight WD, Clemenger K, deHeer WA, Saunders WA, Chou MY and Cohen ML (1984) *Physical Review Letters* **52**: 2141.

neutralization and re-ionization. The number of carbon atoms required to cage an atom depends on the size of the latter: photoexcited $(C_nK)^+$ will lose C_2 units until, below $n = 44$, potassium escapes. $(C_nCs)^+$, however, requires at least 48 carbon atoms.

Abundance anomalies as discussed above relate to unique atomic structures. Molten clusters will not exhibit magic numbers. This has been utilized to determine melting temperatures of clusters in the gas phase. If sodium clusters are thermalized in warm helium and then allowed to effuse into vacuum, no magic numbers related to atomic structure are observed. If the gas is cooled to 307 K (83% of the bulk melting temperature), magic numbers appear for clusters of size $n \geq 6000$. They extend to smaller clusters if the temperature is lowered further. Smaller clusters melt at lower temperatures. Molten sodium clusters show a different sequence of magic numbers as explained in the following section.

Electronic structure

Key features of the electronic structure of some non-transition metal clusters were first revealed through mass spectrometry. **Figure 7** shows a mass spectrum of sodium clusters, grown from vapour in cold helium gas and ionized by light from a UV lamp. Abundance anomalies are observed at $n = 8$, 20, 40, 58. If, instead, alkali clusters are ionized by an intense pulsed laser in the visible or near UV, abundance anomalies are observed at $n = 9$, 21, 41, 59. It appears that cluster stability is enhanced whenever the number of valence electrons, n_e, is 8, 20, 40, 58, ... The mass spectrum in **Figure 7** is believed to reflect the abundance and, indirectly, the stability distribution of neutral clusters. On the other hand, multiphoton excitation leads to excessive dissociation of charged cluster ions, thus revealing the stability of cluster cations. Copper and silver clusters directly formed in the charged state by sputtering do indeed show the same magic numbers 9, 21, 41,.... Experiments involving other mono-, di- and trivalent metals provide further support for this hypothesis. For example, Cu_{13}^- and Ag_{43}^{3+} and Al_{13}^- (also see **Figure** 8) form local abundance maxima in their respective distributions.

The observed 'magic numbers' find an explanation in the 'jellium model', which assumes the valence electrons to move freely throughout the cluster, only confined by an effective potential that is, thanks to the shielding by the other free electrons, approximately constant throughout the cluster. This quasi-free electron approximation may seem naive, but it successfully explains basic properties of many metallic solids. Making the approximation that the effective potential is spherically symmetric, one immediately predicts electronic shell structure: spherical symmetry gives rise to a degeneracy of the single-electron quantum states, the energy does not depend on the magnetic quantum number. Whenever a 'shell' of electrons is filled, addition of a monovalent atom

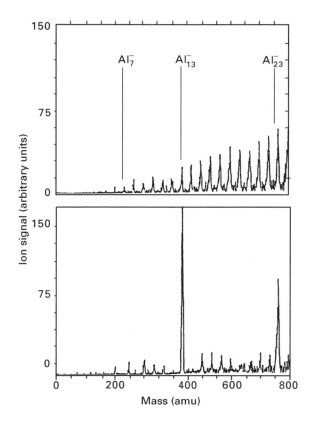

Figure 8 Mass spectrum of aluminium cluster anions before (upper) and after (lower) reaction with oxygen. Al_{13}^- is least reactive. After Leuchtner RE, Harms AC and Castleman Jr. AW (1989) *Journal of Chemical Physics* **91**: 2753.

implies that its valence electron has to occupy a higher state, making this extra atom less tightly bound. The energetic order of the first few shell closures is consistently predicted to occur at $n_e = 2$ (1s state), 8 (1p, in the notation used in nuclear physics), 18 (1d), 20 (2s), 34 (1f), 40 (2p), 58 (1g), etc. for the simple square-well potential as well as for more sophisticated potentials calculated self-consistently. Some of these gaps, like that between the 1d and 2s states, may be quite small, explaining the absence of an abundance anomaly at $n_e = 18$.

Many related effects have been studied in detail. Dissociation energies have been derived from metastable decay rates of alkali cluster ions. The gaps between electronic shells have been probed by measuring the size dependence of the ionization energy. Reduced ionization energies are, indeed, found for systems with 9, 21, 41, etc. valence electrons. The anomalies are much less than predicted, but refined effective (spherically nonsymmetric) potentials can explain this trend as well as the observation of some additional fine structure in the size distribution and in the ionization energy. Additional information

has been obtained from photoelectron spectroscopy of neutral and negatively charged clusters. Again, the basic features of these spectra and their size dependence can be rationalized within the jellium model.

Classically, the ionization energy $E_{ion}(R)$ of a metallic sphere of radius R is expected to converge to the bulk value W (the work function) as $E_{ion}(R) = W + \text{const } R^{-1}$ (for a compact cluster of size n, the radius will scale as $n^{1/3}$). This behaviour is indeed observed for most metal clusters, except for the local anomalies mentioned above. Mercury, however, behaves differently (**Figure 9**). For small sizes, the ionization energies are significantly larger than expected, but a transition to the expected asymptotic behaviour occurs between $n \cong 20$ and 70 ($R \cong 5–8 \text{Å}$). This phenomenon has been assigned to a transition from an insulator to a metallic system.

The jellium model also provided a motivation to develop high-resolution mass spectrometry of very large clusters. Although the gaps between electronic states will gradually disappear with increasing level density, i.e. with increasing cluster size, they were predicted to decrease in a non-monotonic fashion owing to interference effects between different types of semiclassical orbits of the quasi-free electrons in the clusters. These 'supershells', originally predicted for nucleonic systems much larger than naturally occurring nuclei, have been confirmed for clusters of sodium, lithium and gallium containing thousands of valence electrons.

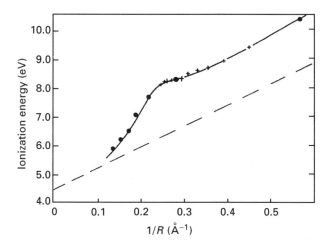

Figure 9 Ionization energy of mercury clusters (solid circles, crosses) versus (estimated) reciprocal cluster radius. The slow convergence to the behaviour expected classically for a metallic sphere (dashed line) is attributed to the occurrence of an insulator-to-metal transition in the size range between 20 and 70 atoms. After Rademann K, Kaiser B, Even U, and Hensel F (1987) *Physical Review Letters* **59**: 2319.

Cluster ion reactivity

Two categories of reactions have been studied that require the temporal evolution of the cluster ion composition to be known. The first comprises reactions of (neutral or ionized) clusters with neutral or ionized atoms, molecules, clusters or surfaces (not to mention reactions of photons or electrons with the clusters which, however, are usually discussed in the context of ionization and dissociation reactions); the second deals with reactions inside (excited) cluster ions. In the first case the mass spectrometer in various combinations serves as the chemical reactor, whereas in the second the cluster itself may be considered to be the chemical laboratory.

Cluster ion reactions

One of the interesting questions addressed in this field is the influence that solvation, the degree of aggregation and the proximity of solute complexes have on the reactions and properties of reactive species. Huge changes (by many orders of magnitude) of the corresponding reaction rate constants may occur when going from the gas phase to the cluster environment; or very specific selectivity may be observed in reaction channels for metal cluster ions reacting with ligands (see **Figure 8**). Moreover, research on the reactivity of naked metal cluster ions is of special interest as these systems are considered to be potential catalysts since, in general, they are highly unsaturated. Fe_4^+ ions, produced by sputtering, thermalized and trapped in the storage cell of a FT-ICR mass spectrometer, were shown to react sequentially with ethylene, forming benzene precursors. The existence of these precursors and the occurrence of a complete catalytic cycle has been shown in a MS^5 experiment, where the various intermediates are successively identified.

Reactions between clusters may lead to fusion-like reactions, such as $C_{60}^+ + C_{60} \rightarrow C_{120}^+$, where fusion occurs only for collision energies exceeding a barrier of about 70 eV. On the other hand, they may cause fragmentation and even multifragmentation reactions, the latter showing an activation energy of about 70 eV for C_{60}. Studies involving various types of tandem mass spectrometer combinations have, in addition, dealt with the rapidly growing topic of cluster ion/surface reactions. Besides reflection, surface-induced dissociation, charge exchange and surface-induced reactions have been observed, thus allowing study of the reactivity and, in addition, the structure of these ions. For instance, collisions of size-selected stoichiometric and nonstoichiometric acetone ions with a stainless steel surface covered by hydrocarbon always give, besides typical fragmentation patterns expected for weakly interacting van der Waals cluster ions, the protonated acetone ion as the major secondary ion. Using deuterated acetone it is possible to distinguish two completely different reaction routes leading to this major product ion, one involving an H abstraction (pick-up) reaction with the surface adsorbates and one involving an ion molecule reaction within the cluster ion (intracluster reactions). The branching ratio for these two reactions depends strongly on the cluster size. Again, the ultimate goal is to study new reactions in the novel non-equilibrium environment present under the 'nano-shock' conditions shortly after cluster impact on the surface.

Reactions in cluster ions

One of the most fascinating aspects of cluster reactivity is given by those unimolecular reactions that occur inside of an energized cluster (ion) after excitation/ionization by, for example, electrons or photons. Whereas inelastic interactions of free electrons with gas-phase atoms or molecules result in changes of the electronic configuration and/or nuclear motion, interactions of electrons with clusters may, in addition, (i) comprise multiple collisions of the incoming (scattered) electron(s) and (ii) involve subsequent intra- and intermolecular reactions within the 'cluster reactor'. This sequence of events leads to the formation of excited cluster ions (which may differ in composition and stoichiometry from the initial neutral precursor), involving the deposition of excess energy into various degrees of freedom. Energy and charge exchange between different sites in the cluster may, in turn, cause additional spontaneous reactions not present in ordinary gas-phase molecules. These spontaneous reactions, occurring on time scales ranging from a few vibrational oscillations up to the millisecond time regime, will contribute strongly to the final fragmentation pattern observed in mass spectrometry. They are also responsible for the fact that measured cluster mass spectral distributions are often quite different from the distribution of clusters in the probed neutral beam and are usually only snapshots of the changing ion compositions after a primary ionization or excitation event (see **Figure 5**). Here, from the large number of different types of reactions, only four examples will be discussed, namely two internal ion molecule reactions and two unimolecular cluster decay reactions.

Reactions in the cluster may lead to the production of ions not produced in the case of the monomer, and vice versa. For instance, electron impact ionization of CO_2 does not give O_2^+ fragment ions, whereas electron impact ionization of CO_2 clusters yields $(CO_2)_m O_2^+$ ions by the following reaction sequence: via dissociative ionization of a single CO_2 molecule within the cluster, the incoming electron produces an O^+ fragment ion. In a second step, this fragment ion, via an ion molecule reaction with a neighbouring CO_2 molecule, forms O_2^+ plus CO.

In contrast, for polyatomic molecules, fragment ion abundances are usually much smaller if the ionization and subsequent fragmentation occur within a cluster. For instance, for a propane cluster, the relative abundance of the parent molecular ion $C_3H_8^+$ and the abundances of the first-generation fragment ions $C_3H_n^+$, $C_3H_6^+$, $C_2H_5^+$, $C_2H_4^+$, are significantly enhanced; those of the second-generation fragment ions $C_3H_5^+$, $C_3H_4^+$, $C_2H_3^+$, $C_2H_2^+$ are diminished in comparison to the situation with the monomer. This is due to energy transfer from the originally excited propane molecular ion, via quenching reactions, to the cluster constituents.

The common unimolecular (spontaneous) decay in the metastable time regime for (excited) cluster ions is monomer evaporation, although in certain cases dimer and trimer ejection have also been observed. Whereas electronic predissociation and barrier penetration are likely mechanisms for the decay of small cluster ions, vibrational predissociation is the dominant metastable dissociation mechanism for larger cluster ions, leading in some cases — owing to the large amount of energy stored — to sequential decay series. Moreover, because cluster ions of a specific size produced in the ion source of a mass spectrometer by electron impact of a neutral cluster beam will comprise a broad range of internal energies (owing to the statistical nature of the complicated ionization process and the possibility of cascading from different neutral precursors), the decay of such an ion ensemble will not be governed by a single decay constant k as predicted by RRKM theory for a polyatomic ion excited to a specific energy. The experimental results show, in accordance with predictions using the concept of an evaporative ensemble (EEM), a nonexponential function for the time dependence of the apparent metastable decay rate (**Figure 10**). Moreover, this EEM model can be also used to describe the experimentally observed increase of the apparent decay rate with cluster size. Deviations observed from the prediction are due to the special nature of 'magic number' cluster sizes. It is thus not surprising that these anomalously small or large decay rates agree in a mirror-like fashion with enhanced or depleted ion abundances in the mass spectrum.

In addition to the single monomer evaporation based on the statistical predissociation mechanism,

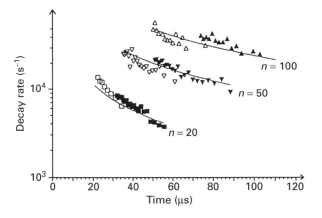

Figure 10 Measured and predicted (solid line) apparent metastable decay rates (metastable fraction divided by the experimental time window) for three different cluster ion sizes and two different rare gases (open symbols, argon; filled symbols, krypton) versus time since electron ionization. After Ji Y, Foltin M, Liao CH and Märk TD (1992) *Journal of Chemical Physics* **96**: 3624.

several unusual metastable fragmentation reactions involving the ejection of more than one monomer have been observed. It has been demonstrated by mass spectrometric techniques that these decay reactions are due to the decay of isolated excited states within the cluster ion, including vibrational states, excimers and isomers.

Hot surfaces may 'boil off' electrons if the corresponding activation energy is not much higher than that for the competing reaction, evaporation of atoms. For energetic reasons, this 'thermionic emission' process is not observed for ordinary molecules unless they are negatively charged (autodetachment). The cohesive energy per building block tends to increase with increasing cluster size, while the ionization energy decreases. Also, a large system can store excitation energy more effectively than a small molecule; the cluster can serve as its own heat bath. Hence, delayed electron emission from neutral clusters excited by photons or by collisions is a viable reaction. This process produces asymmetric peaks in conventional time-of-flight (TOF) mass spectra, because the arrival time of an ion is usually measured with respect to the time at which the parent molecule was excited. **Figure 11** (top) shows a TOF mass spectrum of a C_{60}/C_{70} mixture, using multiphoton excitation at 266 nm. To select the time of ionization separately from the time of flight through the instrument, the extraction field is modulated in time such that only ions formed within a certain time interval (4–5 μs in this case) reach the detector, thus giving rise to narrow mass peaks that reflect the rate of formation within the chosen time interval (**Figure 11**, lower). It

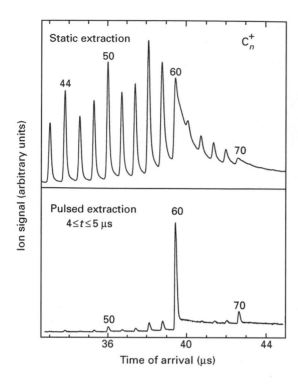

Figure 11 Time-of-flight mass spectrum of multiphoton-ionized fullerenes. Upper spectrum: static extraction, delayed ions give rise to asymmetric peaks. Lower spectrum: pulsed extraction, delayed ions formed within a chosen time window are bunched into narrow peaks.

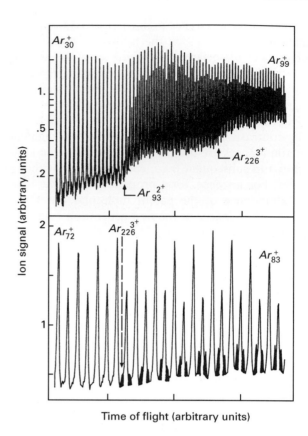

Figure 12 Mass spectrum of argon clusters ionized by electron impact. Doubly and triply charged clusters are observed above characteristic appearance sizes.

becomes apparent that C_{58}^+ and other fragment ions also have delayed components. C_{60}^+ may form hundreds of microseconds after excitation. Delayed ionization has been observed for neutral refractory metal clusters containing as few as five atoms, for some metal oxides and for metallocarbohedrenes such as Ti_8C_{12}. The phenomenon is ubiquitous for negatively charged clusters owing to their reduced activation energy for electron emission.

Multiply charged clusters

Systems with a high charge density become unstable with respect to dissociation into charged fragments (fission). This effect limits the net charge of molecules, small clusters, and droplets of a given size, as well as the size of natural and man-made atomic nuclei. Multiply charged cluster ions are easily identified by their non-integer size-to-charge ratios (**Figures 1** and **12**). However, for each charge state z there is a characteristic appearance size n_z below which the species becomes undetectable, as shown in **Figure 12** for argon. For more tightly bound metal clusters, $n_{z=2}$ may be as small as 2, but the concept of an appearance size is still meaningful for higher charge states.

The effect has been observed for various atomic and molecular van der Waals clusters, hydrogen-bonded clusters and metal clusters. Seven-fold sodium clusters, for example, are not observed below $n_7 \cong 445$. Under typical experimental conditions, multiply charged clusters are, like their singly charged counterparts, thermally excited; they are found to evaporate atoms or molecules in a statistical, unimolecular process if their size is well above the appearance size. In the vicinity of the latter, however, fission competes with evaporation. Below the appearance size, the rate of fission greatly exceeds the time needed for mass spectrometric detection. The appearance size can be somewhat reduced if colder multiply charged clusters are generated. Ultimately, however, the clusters become unstable and the fission barrier vanishes.

The size distribution of fission fragments can be determined mass spectrometrically. Distributions tend to be asymmetric, but there are considerable variations in the details. Several studies have been devoted to fullerenes. C_{60}^{z+} is observed with charge states up to $z = 9$. Spontaneous (unimolecular) dissociation of C_{60}^{3+}, as well as producing C_{58}^{3+} and

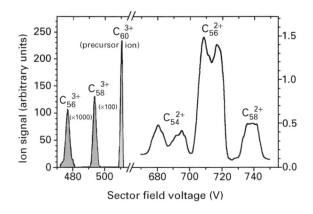

Figure 13 MIKE scan of C_{60}^{3+} in a double focusing sector field mass spectrometer, showing in addition to the parent ion spontaneous fission into different fragment ions. The broadening and splitting for the charge separation reactions arises from the large kinetic energy release in these reactions. Reproduced with permission of Taylor & Francis from Scheier P, Dünser B, Wörgötter R *et al* (1996) *International Review of Physical Chemistry* **15**: 93.

smaller fragments, produces C_{54}^{2+}, C_{56}^{2+} and C_{58}^{2+} via a 'charge separation' reaction (**Figure 13**). These MIKE spectra (scan of the electric sector field voltage) reveal the energy released upon fissioning. C_{54}^{2+} shows the greatest broadening because it is related to the heaviest fission partner, C_6^+. Detection of the light fission fragments is difficult because of their large recoil velocity, but C_6^+ has been directly identified.

Cluster anions

Often a single molecule cannot bind an excess electron, although it is quite possible to trap

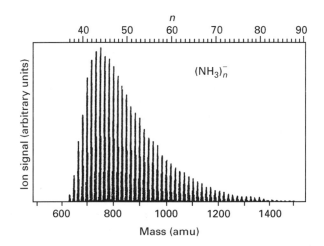

Figure 14 Mass spectrum of negatively charged $(NH_3)_n$ clusters showing a sharp onset around the minimum appearance size $n_{min} = 35$. After Haberland H, Schindler HG and Worsnop PR (1984) *Berichte der Bunsengesellschaft für Physikalische Chemie* **88**: 270.

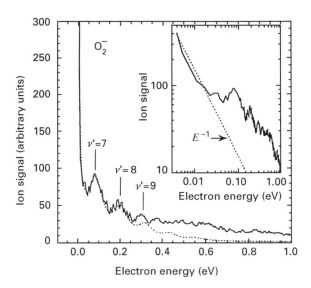

Figure 15 Measured (full line) and calculated (broken line) O_2^+ anion signal produced by electron attachment to an oxygen cluster beam as a function of electron energy. Inset: Same data on a log–log scale demonstrating the E^{-1} dependence predicted by s-wave scattering theory (dashed line). After Matejcik S, Kiendler A, Stampfli P, Stamatovic A and Märk TD (1996) *Physical Review Letters* **77**: 3771.

electrons in liquids and solids of these molecules. Negatively charged clusters of these substances are of particular interest since it is possible to study details of this trapping as a function of aggregation. It has been shown by mass spectrometry that three mechanisms are responsible for the occurrence of negatively charged stoichiometric cluster anions in these cases: (i) stabilization by collision if accommodation of the electron affinity leads to autodetachment for a single isolated molecule (e.g. O_2); (ii) stabilization by solvation if Franck–Condon problems lead to dissociation (e.g. CO_2); and (iii) generation of a solvated electron (polaron) if the electron affinity of the monomer is zero (e.g. H_2O). Whereas in the first two cases attachment of an electron to the cluster leads to the production of monomer anions and higher analogues, in the last case only cluster anions above a certain minimum size n_{min} can be observed (see **Figure 14** showing $n_{min} = 35$ for ammonia clusters). It turns out that this minimum size in the observed mass spectrum may depend on the method of anion production. Cluster anions of volatile compounds can be formed in the gas phase either by injection of free electrons into the expansion zone of a supersonic jet or by attachment of electrons to a beam of pre-existing neutral clusters. For water clusters one obtains $n_{min} = 2$ in the former and $n_{min} \cong 11$ in the latter mode. Calculations show that the electron is localized in a 'surface state' for the very small water cluster anions ($n = 2-8$); vertical detachment energies measured by photodetachment methods are, in fact, very low. Thus long-lived anions below $n \cong 11$ can only be produced in the special (cold) environment of the expanding jet.

As well as production of stoichiometric cluster anions, nonstoichiometric fragment cluster anions have also been observed (e.g. $(O_2)_mO^-$ for oxygen). Whereas these fragment anions can usually be produced by attaching electrons to pre-existing clusters at energies similar to the known resonances for the monomer (e.g. at around 6 eV and higher for (O^-/O_2), the stoichiometric anions are, in addition, often produced with very high probability at energies close to zero. **Figure 15** shows the region of the 'zero-energy' resonance of oxygen clusters as measured with a high-resolution spectrometer. In addition to the zero-energy peak ascribed to s-wave capture by the entire cluster, further peaks are observed; these are attributed to vertical attachment of the incident electron to a single molecule within the cluster, thus allowing probing of the vibrational levels of a single oxygen anion within the cluster.

See also: **Ionization Theory; Thermospray Ionization in Mass Spectrometry.**

Further reading

Andersen HH (1997) *Small Particles and Inorganic Clusters*. Berlin: Springer-Verlag.

Bernstein ER (1990) *Atomic and Molecular Clusters*. Amsterdam: Elsevier.

Bréchignac C and Cahuzac Ph (eds) (1995) *Comments on Atomic and Molecular Physics* 31: 137–477.

Castleman Jr. AW and Bowen KH (1996) Clusters: structure, energetics, and dynamics of intermediate states of matter. *Journal of Physical Chemistry* 100: 12911–12944.

de Heer WA (1993) The physics of simple metal clusters: experimental aspects and simple models. *Review of Modern Physics* 65: 611–676.

Deng R and Echt O (1998) Efficiency of thermionic emission from C_{60}. *Journal of Physical Chemistry A* 102: 2533–2539.

Dresselhaus MS, Dresselhaus G and Eklund PC (1996) *Science of Fullerenes and Carbon Nanotubes*. San Diego: Academic Press.

Haberland H (1994) *Clusters of Atoms and Molecules: Theory, Experiments, and Clusters of Atoms*. Springer Series in Chemical Physics 52. Berlin: Springer-Verlag.

Haberland H (1994) *Clusters of Atoms and Molecules II: Solvation and Chemistry of Free Clusters, and Embedded, Supported and Compressed Clusters*. Springer Series in Chemical Physics 56. Berlin: Springer-Verlag.

Ingólfsson O, Weik F and Illenberger E (1996) The reactivity of slow electrons with molecules at different degrees of aggregation: gas phase, clusters and condensed phase. *International Journal of Mass Spectrometry and Ion Processes* 155: 1–68.

Jortner J (1992) Cluster size effects. *Zeitschrift für Physik D* 24: 247–275.

Märk TD (1995) Mass spectrometry of clusters. In: Cornides I, Horváth G and Vékey K (eds) *Advances in Mass Spectrometry*, Vol 13, pp 71–94. Chichester: Wiley.

Martin TP (1996) Shells of atoms. *Physics Reports* 273: 199–241.

Näher U, Bjørnholm S, Frauendorf S, Garcias F and Guet C (1997) Fission of metal clusters. *Physics Reports* 285: 245–320.

Ng CY, Baer T and Powis I (1993) *Cluster Ions*. Chichester: Wiley.

Cobalt NMR, Applications

See **Heteronuclear NMR Applications (Sc–Zn).**

Colorimetry, Theory

Alison Gilchrist and **Jim Nobbs**, University of Leeds, UK

ELECTRONIC SPECTROSCOPY
Theory

Colorimetry is the science of the measurement of colour. It involves the replacement of subjective responses, such as 'light blue', 'rich dark purple', 'bright gold', with an objective numerical system. This study began with the work of Young, Helmholtz and Maxwell in the early nineteenth century, who recognized the principles of additive and subtractive colour mixing, and proposed the trichromatic nature of human colour vision. The science began to be formalized in 1931, when the Commission International de l'Eclairage (CIE) recommended a system of colour specification based on the three tristimulus values X, Y and Z. The current CIE reference document relating to colorimetry is CIE Publication 15.2-1986 (Edition 3 is in preparation); for more practical information on colorimetric measurement, ASTM document E308-96 should be consulted.

Illumination, object, observer

Three things contribute to our perception of the colour of an object: the nature of the illumination, the optical properties of the object itself and the response of the human eye (**Figure 1**). The nature of the illumination can be characterized by the spectral power distribution $S(\lambda)$ of the light source, the relative intensity of the illumination at each wavelength in the visible spectrum. The object reflects a certain fraction of the incident light and this can be characterized by the reflectance spectrum $R(\lambda)$. The intensity of light entering the eye, $I(\lambda)$, is the product of these terms. Thus in order to measure colour, and to specify colour by numbers, it is necessary to specify each of these three components of the colour trio. In 1931, the CIE recommended standard illuminants for use in colorimetry, published data representing the standard observer and recommended standard optical geometries for use in colour-measuring instruments.

Standard illuminants (A, C, D65, F)

In 1931 the CIE recommended three standard sources and Illuminants (A, B and C) where a source is a physically realizable light source and an illuminant is a table of data (spectral power distribution) representing a source. A is intended to represent an incandescent tungsten filament lamp (colour temperature 2856 K); B and C may be produced by filtering the light from source A to produce a representation of noon sunlight (B) or average daylight (C). More recently, the CIE has recommended the D series as illuminants only, representing daylight in various phases. D75 (colour temperature 7500 K) represents north sky daylight, D65 (colour temperature 6500 K) represents average daylight and D55 (colour temperature 5500 K) represents sunlight plus skylight. D65 is now the reference illuminant for use in colorimetry (although the graphic arts industry tends to prefer D55) and so should be included in any colour specification. Unfortunately, illuminant D65 is not realizable in practice, and so colour-matching cabinets generally contain a daylight simulator (a filtered fluorescent lamp). A further series of florescent lamps, the F series, were also defined, although none is specified as a standard illuminant. There are three types: normal, broad-band and three-band. The most important are F2 (normal, known as 'cool white'), F7 (broad-band, known as 'artificial daylight' and commonly used for colour matching) and F12 (three-band, with high energy efficiency, and so often used in shops). **Figure 2** shows the relative spectral power distribution of the more commonly used illuminants.

Standard observers (2-degree, 10-degree)

The experiments that established the basis of colorimetry were carried out separately by J. Guild at the

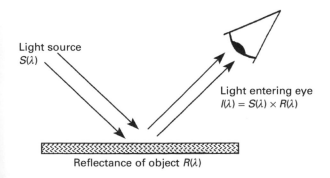

Light source
$S(\lambda)$

Light entering eye
$I(\lambda) = S(\lambda) \times R(\lambda)$

Reflectance of object $R(\lambda)$

Figure 1   How colour reaches the eye.

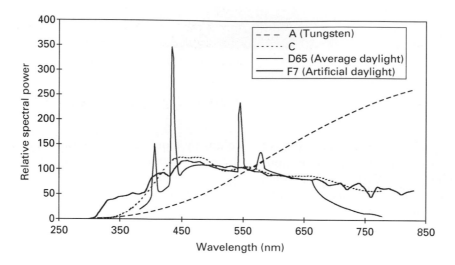

Figure 2 Relative spectral power distributions of common CIE illuminants.

National Physical Laboratory (UK) and W.D. Wright at Imperial College London in a famous series of experiments in the late 1920s. These experiments were dedicated to characterizing the response of the human eye to different wavelengths of light, and were carried out by colour matching using red, green and blue lights. A set of red, green and blue light sources are often called primary sources since the appearance of any one of the sources cannot be matched by blending together light from the other two. A possible experimental arrangement is illustrated in **Figure 3**. Each worker used several observers (Guild 7 and Wright 10) who adjusted the mixture of light from three primaries until it matched the appearance of the monochromatic light.

These early experiments were carried out using the central area of the retina, the fovea. This region contains the highest concentration of colour-sensitive cells, and occupies the central 2° field (based on the angle to the line of sight, drawn from the centre of the lens to the foveal pit). The standard observer derived from Guild's and Wright's experiments is therefore known as the 2° standard observer (1931). The colour matching functions $\bar{r}(\lambda)$, $\bar{g}(\lambda)$ and $\bar{b}(\lambda)$ represent the amounts of light from the red, green and blue primaries, in tristimulus units, needed to match unit intensity of light with a narrow band of wavelengths centred on λ. Tristimulus units are defined such that a mixture of equal units of the red, green and blue primaries will match an equal-energy white. The values of the colour-matching functions are normalized so that there is an equal area under each of the spectral curves. The values are plotted against wavelength in **Figure 4**.

It was found that not all the colours produced by monochromatic light could be matched by a simple additive mixture of light from three primaries. For some wavelengths it was necessary to add light from one of the primaries to the monochromatic light, reducing the colour saturation of the monochromatic light. The appearance of the desaturated mixture could then be matched by a combination of the other two primaries. When this occurs, the amount of the primary added to the monochromatic light appears as a negative number in the colour specification.

Later workers observed that although the fovea is the region most sensitive to colour, most everyday objects occupy a larger area of the retina. The 10° standard observer (1964) was thus derived in a similar manner and is now the preferred observer function. The true cone sensitivities were not measured directly until 1979, but correspond very well to those expected from the colour-matching functions found by Guild and Wright.

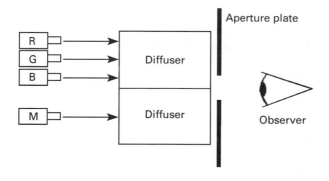

Figure 3 Possible arrangements of Guild's and Wright's experiments.

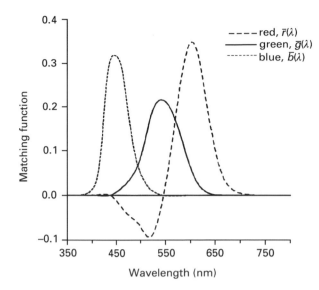

Figure 4 Guild's and Wright's colour-matching functions.

RGB specification

The RGB system of colour specification was based on the trichromatic theory of colour vision and the colour-matching experiments carried out by Guild and Wright. It had been suggested by Maxwell in 1860 that if the specifications of the red, green and blue lamps used in such experiments were known, then the amounts of each light required to match a sample colour would provide a numerical specification of that colour. These amounts may be found either by passing the light from the sample through filters matching the RGB primaries, or by calculation from the reflectance values of the sample at a series of wavelengths through the visible spectrum:

$$R = k\left[I(\lambda_1)\,\bar{r}(\lambda_1) + I(\lambda_2)\,\bar{r}(\lambda_2) + \cdots\right]$$

$$\text{or} \quad R = k\sum_{\lambda=380}^{\lambda=780} I(\lambda)\,\bar{r}(\lambda)$$

$$G = k\left[I(\lambda_1)\,\bar{g}(\lambda_1) + I(\lambda_2)\,\bar{g}(\lambda_2) + \cdots\right]$$

$$\text{or} \quad G = k\sum_{\lambda=380}^{\lambda=780} I(\lambda)\,\bar{g}(\lambda)$$

$$B = k\left[I(\lambda_1)\,\bar{b}(\lambda_1) + I(\lambda_2)\,\bar{b}(\lambda_2) + \cdots\right]$$

$$\text{or} \quad B = k\sum_{\lambda=380}^{\lambda=780} I(\lambda)\,\bar{b}(\lambda) \quad [1]$$

Although RGB values are a valid representation of the colour of an object viewed under specified conditions, for practical reasons these values are rarely used. For a significant proportion of colours one of the three values is negative, and this was considered a distinct disadvantage.

XYZ specification (1931)

The XYZ system of colour specification was recommended by the CIE in 1931 and was closely based on the RGB system. The difference lies in the fact that R, G and B were real light sources of known specifications. X, Y and Z are theoretical sources which are more saturated than any real light source, and allow matching of any real colour using positive amounts of the three primary sources. The X, Y and Z stimuli are defined so that:

- The XYZ tristimulus values of all real colours are positive.
- The Y tristimulus value is proportional to the luminance.
- A mixture of equal amounts of X, Y and Z has the same colour appearance as an equal-energy white.
- The X, Y, Z values of the visible colours have the widest possible range of values.
- The X stimulus represents a red more saturated than any spectral red.
- The Y stimulus represents a green more saturated than any spectral green.
- The Z stimulus represents a blue more saturated than any spectral blue.

X, Y, Z values for the 2° observer may be calculated from RGB values using the equations

$$X = 0.49R + 0.31G + 0.20B$$
$$Y = 0.17697R + 0.81240G + 0.01063B$$
$$Z = 0.00R + 0.01G + 0.99B \quad [2]$$

Alternatively, XYZ values may be calculated from reflectance values using similar equations to those for RGB, but using weighting functions $\bar{x}(\lambda)$, $\bar{y}(\lambda)$ and $\bar{z}(\lambda)$ in place of $\bar{r}(\lambda)$, $\bar{g}(\lambda)$ and $\bar{b}(\lambda)$. The CIE nomenclature is to denote the 10° observer function with a subscripted '10' (Eqn (3)). The XYZ system is an improvement over the RGB system in that it eliminates negative numbers in colour specification. However, it is still based on the mixing of coloured lights and not on the visual appreciation of colour. In particular, the visual colour difference between two samples is not linearly related to the difference in XYZ values:

$$X_{10} = k\big[I(\lambda_1)\,\bar{x}_{10}(\lambda_1) + I(\lambda_2)\,\bar{x}_{10}(\lambda_2) + \cdots\big]$$

$$\text{or}\quad X_{10} = k\sum_{\lambda=380}^{\lambda=780} I(\lambda)\,\bar{x}_{10}(\lambda)$$

$$Y_{10} = k\big[I(\lambda_1)\,\bar{y}_{10}(\lambda_1) + I(\lambda_2)\,\bar{y}_{10}(\lambda_2) + \cdots\big]$$

$$\text{or}\quad Y_{10} = k\sum_{\lambda=380}^{\lambda=780} I(\lambda)\,\bar{y}_{10}(\lambda)$$

$$Z_{10} = k\big[I(\lambda_1)\,\bar{z}_{10}(\lambda_1) + I(\lambda_2)\,\bar{z}_{10}(\lambda_2) + \cdots\big]$$

$$\text{or}\quad Z_{10} = k\sum_{\lambda=380}^{\lambda=780} I(\lambda)\,\bar{z}_{10}(\lambda) \qquad [3]$$

L*a*b* colour space (1976)

CIE $L^*a^*b^*$ colour space is based on the opponent theory of colour vision, which arose from the ideas of Hering. Hering pointed out that of the thousands of words used to describe hue, only four are unique; red, green, yellow and blue. They are unique because they cannot be described without using their own colour name, yet any other hue can be described using one or more words from this set. When taken together with white and black, they form a group of six basic colour properties that can be grouped into three opponent pairs, white/black, red/green and yellow/blue. The concept of opponency arises from the observation that no colour could be described as having attributes of both redness and greenness, or of both yellowness and blueness. A red shade of green just does not exist.

It seems sensible to associate these three opponent pairs with the three axes of a three dimensional colour space. On a colour chart, the amount of redness or greenness of a colour sensation could be represented by the distance along a single axis, with pure red lying at one extremity and pure green lying at the other. In a similar way yellow and blue are opposite extremities of a second axis that could be placed perpendicular to the red/green direction. The third axis would go from white to black and lie in the plane normal to the other two. CIE $L^*a^*b^*$ (1976) space uses three terms L^*, a^* and b^* to represent colour, as shown in **Figure 5** and described below.

L^* The vertical axis represents lightness; 100 represents a perfect white sample and 0 a perfect black.

a^* The axis in the plane normal to L^* represents the redness–greenness quality of the colour. Positive values denote redness and negative values denote greenness.

b^* The axis normal to both L^* and a^* represents the yellowness–blueness quality of the colour.

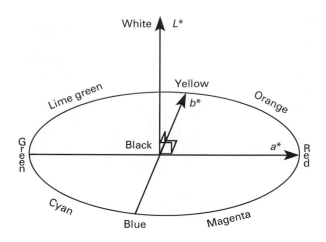

Figure 5 CIE $L^*a^*b^*$ colour space.

Positive values denote yellowness and negative values denote blueness.

$L^*a^*b^*$ values are calculated from XYZ values using the following equations:

$$F_X = \left(\frac{X}{X_0}\right)^{1/3} \qquad [4a]$$

unless

$$\left(\frac{X}{X_0}\right) \leq 0.008\,856$$

when

$$F_X = 7.787\left(\frac{X}{X_0}\right) + \frac{16}{116} \qquad [4b]$$

where X_0 is the X tristimulus value of a perfect white sample under the chosen illuminant. Values of F_Y and F_Z are calculated in a similar manner. Then:

$$L^* = 116F_Y - 16$$
$$a^* = 500[F_X - F_Y]$$
$$b^* = 200[F_Y - F_Z] \qquad [5]$$

Since $L^*a^*b^*$ values for a sample may change for a different illuminant and observer combination, any colour specification must always include the illuminant and observer for which the values were calculated.

CIE $L*C*h°$ coordinates

The coordinates of the points representing the colour of a sample can also be expressed in cylindrical coordinates of lightness L^*, chroma C^* and hue $h°$, as illustrated in **Figure 6**. This system corresponds more closely to the natural description of colour.

L^* Lightness (see above)
C^* Chroma:

$$C^* = \sqrt{\left(a^{*2} + b^{*2}\right)}$$

$h°$ Hue angle:

$$h° = \tan^{-1}\left(\frac{b^*}{a^*}\right) \qquad [6]$$

Neutral grey samples would have C^* values between 0 and about 5 units. The hue angle is always measured anticlockwise from the positive a^* axis. It is traditional to associate one of the psychological primary hues with each axis direction, as shown in the diagram, although in reality the angles of the pure hues do not lie exactly along the axis directions.

Colour difference

As the CIE $L^*a^*b^*$ system was designed to be visually uniform, it is fairly simple to devise an equation for the total colour difference between a trial and standard sample. The total colour difference is the distance between the two points representing those colours in the colour space, as shown in **Figure 7**. The distance, expressed as ΔE^*_{ab}, is determined using the laws of right-angle triangles:

$$\Delta E^*_{ab} = \sqrt{\Delta L^{*2} + \Delta a^{*2} + \Delta b^{*2}} \qquad [7]$$

where

$$\Delta L^* = L^*_{trl} - L^*_{std}$$
$$\Delta a^* = a^*_{trl} - a^*_{std}$$
$$\Delta b^* = b^*_{trl} - b^*_{std} \qquad [8]$$

The overall colour difference may also be split into lightness, chroma and hue terms, which again correspond more closely to the natural description of colour:

$$\Delta C^* = C^*_{trl} - C^*_{std} \qquad [9]$$

$$\Delta H^* = \sqrt{\Delta E^{*}_{ab}{}^2 - \Delta L^{*2} - \Delta C^{*2}} \qquad [10]$$

It is worth remembering that the scaling of the colour space was set so that a value of $\Delta E^*_{ab} = 1$ between the colour of two samples should be just visually perceptible. This means that if a large

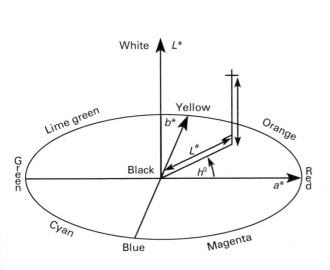

Figure 6 CIE $L^* C^* h°$ colour coordinates.

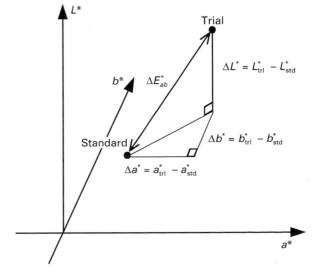

Figure 7 The CIE $L^*a^*b^*$ colour difference calculation.

number of people were asked to judge the colour difference between the two samples, about 50% of observers would say that there was a difference in colour, the other 50% would say the two colours matched. This is the point of maximum argument.

Comparisons between the majority visual decision and a numerical assessment, based on the condition that ΔE_{ab}^* must be less that 1 for a match, show that the instrument decision will disagree with the majority visual decision about 19% of the time. On average, an individual colourist will disagree with the majority decision about 17% of the time when judging pairs of panels with fairly small colour differences. Pass/fail decisions based on the CIE $L^*a^*b^*$ (1976) equation with a limit of $\Delta E_{ab}^* = 1$ are, on average, slightly less reliable than visual judgements made by a single colourist. However, a colour-measuring instrument has the great advantage of giving much more repeatable decisions.

A number of scientists have devised more reliable or optimized colour difference equations. The simplest method of improvement is to assign individual tolerances to ΔL^*, Δa^* and Δb^*. Alternatively, individual tolerances can be assigned to ΔL^*, ΔC^* and ΔH^* (differences in lightness, chroma and hue). The most successful equations are based on the ΔL^*, ΔC^*, ΔH^* system of colour splitting and the CMC (l, c) and CIE 94 equations are of this type. The CMC equation was developed by the Colour Measurement Committee of the Society of Dyers and Colourists (Bradford, UK) and has been adopted as the ISO standard for small colour difference assessment of textile materials. In 1994 the CIE suggested the simpler CIE 94 equation for evaluation. The published evidence collected so far by the CIE suggests that pass/fail decisions made using either of the optimized colour difference equations are considerably more reliable than those based on ΔE_{ab}^* or on the judgement of a single human colourist. Comparison with decisions made by panels of observers show that the equations disagree with the majority decision about 13% of the time. Neither equation has yet been adopted as a CIE standard as it has been shown that further improvements can still be made.

Colour and appearance

The systems of colorimetry described above can only specify the colour of an isolated sample viewed against a neutral grey background. However, almost all colours are viewed against different backgrounds, with different fields of view, viewing environments, and states of adaptation to that environment. The surrounding colours or patterns will affect the perceived colour of the sample in ways which cannot

be predicted using the CIE $L^*a^*b^*$ system. More complicated models have been developed that take these factors into account; the CIE have recently recommended the CIECAM97s model for further testing. Colour appearance models are becoming increasingly important in these days of digital images and global communication. The faithful reproduction of an image from computer screen to printed hard copy requires an effective colour appearance model at the heart of the colour management system.

Summary

The use of colour-measuring instruments allows specification and communication of colour by means of international standard terms such as CIE $L^*a^*b^*C^*h^\circ$. Colour difference values can also be determined instrumentally and, if optimized colour difference equations are used, can out-perform a single trained colourist. However, the visual appreciation of colour and colour difference is still a subjective response, affected by many factors. Care must be taken to include measurement conditions in the interpretation of colour measurement data. The gold standard answer must always be that which agrees with the majority of a group of human observers. Whether this can be achieved by optimized equations based on CIE $L^*a^*b^*$ colour space or by colour appearance models remains to be established.

List of symbols

a^* = redness/greenness; b^* = yellowness/blueness; c^* = chroma; $F_XF_YF_Z$ = functions of XYZ values used in deriving $L^*a^*b^*$ values; h° = line angle; $I(\lambda)$ = intensity of light entering the eye; k = normalization constant; L^* = lightness; L^*, a^*, b^* = values in the CIE $L^*a^*b^*$ system; L^*, C^*, h° = values in the CIE $L^*C^*h^\circ$ system; $r(\lambda)$, $g(\lambda)$, $b(\lambda)$ = Guild/Wright colour-matching functions; $R(\lambda)$ = reflectance; R,G,B = values in the RGB system; $S(\lambda)$ = spectral power distribution of light source; $\bar{x}(\lambda)$, $\bar{y}(\lambda)$, $\bar{z}(\lambda)$ = weighting functions replacing $\bar{r}(\lambda)$, $\bar{g}(\lambda)$, $\bar{b}(\lambda)$ in the XYZ system; X,Y,Z = tristimulus values in the XYZ system; ΔE_{ab}^* = distance function in the CIE $L^*a^*b^*$ system.

See also: **Art Works Studied Using IR and Raman Spectroscopy; Colorimetry Theory; Dyes and Indicators, Use of UV-Visible Absorption Spectroscopy.**

Further reading

Billmeyer FW Jr and Saltzman M (1981) *Principles of Colour Technology*, 2nd edn. New York: Wiley.

CIE Publication 15.2 (1986) *Colorimetry*. Vienna: CIE.

Hunt RWG (1987) *Measuring Colour*. Chichester: Ellis Horwood.

Hunter RS (1975) *The Measurement of Appearance*. New York: Wiley-Interscience.

Judd DB and Wyszecki G (1975) *Colour in Business, Science and Industry*, 3rd edn. New York: Wiley.

Kuehni RG (1997) Color: *An Introduction to Practice and Principles*. New York: Wiley-Interscience.

McDonald R (ed) (1997) *Colour Physics for Industry*, 2nd edn. Bradford: Society of Dyers and Colourists.

Nassau K (ed) (1998) *Color for Science, Art and Technology*. Amsterdam: Elsevier.

Wright WD (1969) *The Measurement of Colour*, 4th edn. London: Hilger.

Wyszecki G and Stiles WS (1982) *Color Science Concepts and Methods, Quantitative Data and Formulae*, 2nd edn. New York: Wiley.

Computational Methods and Chemometrics in Near-IR Spectroscopy

Paul Geladi, Umeå University, Sweden
Eigil Dåbakk, Foss Sverige, Sollentuna, Sweden

VIBRATIONAL, ROTATIONAL & RAMAN SPECTROSCOPIES
Methods & Instrumentation

Introduction

In near-infrared and related spectroscopic methods, spectra consisting of spectroscopic measures (absorbance, transmittance, reflectance) at a large number of wavelengths, frequencies, energies or wavenumbers (also called by the general term 'variables') are produced. A few hundred to a few thousand variables are not uncommon, even with fairly simple, medium-priced instruments. For most uses it is not the study of the spectra itself (spectroscopy) but extracting information about the samples (spectrometry) that is important. This requires a multivariate treatment. Some of the multivariate techniques used for this are presented here. Much of the explanation in this text will be based on figures showing spectra or plots visualizing analysis parameters and results. Since space is limited, equations are avoided in most instances and the reader is referred to the quoted literature. Some lesser used multivariate techniques will be given just as literature references. Literature on the near-infrared technique and the related infrared, Raman and Fourier transform techniques is easily found in this Encyclopedia.

Raw spectra

Near-infrared (NIR) spectroscopy/spectrometry is the study of spectra recorded on instruments in the region 780–2500 nm. Many instruments add the visual region of 400–780 nm to this, so that colour information in the spectra can also be studied. Other instruments are more restrictive, e.g. between 850 and 1050 nm. Technical details about instrumentation can be found elsewhere.

It is important to obtain a visual impression of the measured spectra and this is done in line plots. In these plots the horizontal axis is wavelength or wavenumber and the vertical axis can be transmission, reflectance or some type of calculated absorbance. Sometimes it may be useful to study a scatter plot of two raw spectra. Many researchers and industries prefer to work with transformed spectra and smoothing and derivation are used for that purpose (see **Figure 1**). Although raw spectra can show a great deal about classes of samples and outliers, some of the finer details cannot be found in them. Therefore, a more thorough, often multivariate analysis is needed. **Figure 2** shows a typical data matrix obtained from NIR measurement. Some instruments only use 19 well selected wavelength bands and filters to give only 19 (or even fewer) variables. At the other extreme, some instruments give several thousands of wavelength variables.

Exploration and classification

The first multivariate operation to be done on a data matrix (see **Figure 2**) is exploration of the data. Exploration of multivariate space is done by factor analysis methods, of which principal component analysis (PCA) is the easiest choice. The data are projected to a matrix with less and more meaningful variables called scores (latent variables) (see

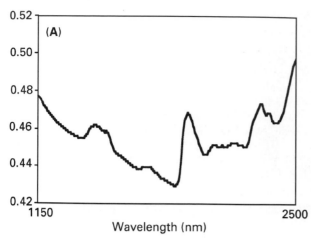

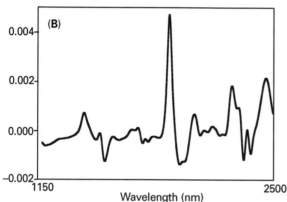

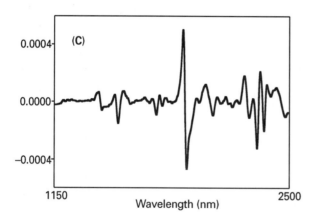

Figure 1 (A) A typical NIR spectrum and its (B) first and (C) second derivatives. The raw spectrum is in absorbance units.

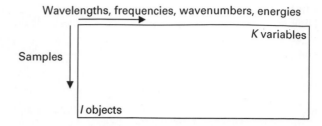

Figure 2 A typical data matrix from an NIR study. The number of objects *I* may be a few tens to a few hundreds; *K* may be a few tens to several thousand.

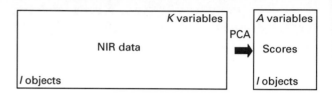

Figure 3 Principal component analysis (PCA) projects the *K* variables into *A* principal component scores containing all meaningful information about the samples.

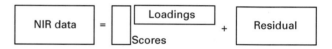

Figure 4 PCA produces a score matrix and a loading matrix for *A* components and a residual matrix containing mainly noise. This is a data reduction.

plots of one loading vector or as scatter plots of two loading vectors.

Any analysis of NIR data should start with the detection of outliers in score plots. The outliers should be identified and removed. After this, grouping of objects may be observed. Distinct groups, overlapping groups and gradients between groups may be found. Grouping may be supervised, meaning known in advance, or it may be unsupervised i.e. found only after the analysis. With supervised grouping, one uses the score plot to check whether the expected (*a priori* known) groups really occur in the NIR spectra. Two different types of coffee are expected to have different spectra and if no difference can be found, the spectroscopic technique chosen may not be appropriate. Unsupervised grouping may show storage or sample treatment differences. A single type of coffee may give grouped spectra because different storage conditions have resulted in different humidity or because the grinding was performed in different machines.

Studies of score plots may be used for studying grouping, ageing, deterioration, erroneous processing

Figure 3). In **Figure 4,** a more complete view of PCA is given. Scatter plots of the scores are called score plots. Score plots are 2D windows in multivariate space. The points in the plots are the objects. In this way, outliers may be detected and groupings of objects may be found. The loadings explain how the original variables contribute to the scores. They can be used to study spectral properties, or to select wavelengths. Loadings are usually shown as line or counterfeiting of all kinds of organic products, e.g. food and feed products, cellulose, paper and textiles, wood products, fossil fuels and biofuels, seeds and growing crops, pharmaceuticals, packaging materials and environmental samples. An industrial application is to keep the spectra of satisfactory batches or products in a file and to compare new spectra to see if they still fit the 'satisfactory' group. A large collection of classification methods (sometimes called 'pattern recognition') is found in the literature, but many of these methods are less useful for NIR data because of the huge number of variables compared with objects. Also in this respect, a data reduction to scores is useful, allowing the use of the classical clustering and pattern recognition techniques on a limited number of score vectors. Statistical tests can be applied to the residual matrix.

Calibration

One of the main purposes of measuring NIR data is the determination of chemical composition or physical properties in a quantitative way. It was found very early that NIR spectra contain information on the water, protein, carbohydrate and fat content of food products. Measuring a NIR spectrum is quick and cheap (30–60 samples per hour for solids) while the wet chemical methods for measuring water, fat, protein, and carbohydrates are slow and expensive. Hence if a function $y = f(x)$ can be found where x is the NIR spectrum and y is the desired chemical concentration, then time and money are saved. As an extra advantage, the NIR spectral measurement can be automated to function continuously on a process stream. Constructing and testing the functions $y = f(x)$ is calibration. Using the function $f(\ )$ to find a value of y with only x known is prediction. Many methods are available for finding $f(\ )$ and many diagnostics are available for checking its quality. Generally, for the calibration step, accurate and precise measures of y and x are required. However, noise can never be avoided. Therefore, it is impossible to find an exact function $f(\ )$ and approximations have to be found. Linear approximations are easy to calculate, more easily interpreted and robust. Therefore, a linear model $y = xb + f$ is often adequate. One important aspect to point out is that in the equation $y = xb + f$ both x (the spectra) and y (the response) have systematic and random noise.

There are many different techniques for building transfer functions between spectra and quantities (responses) measured by a reference method. Several techniques are found in the literature, but not all of them are frequently applied. The equation given above can be extended to $y = Xb + f$ or even to

$Y = XB + F$ (see **Figure** 5), with y ($I \times 1$) a vector of response values for many samples, Y ($I \times M$) a matrix of response values for many responses, X ($I \times K$) the spectral data for the samples, b ($K \times 1$) a vector of regression coefficients, B ($K \times M$) a matrix of regression coefficients for the M responses, f ($I \times 1$) a vector of residuals and F ($I \times M$) a vector of residuals for many responses. One may consider percentages of fat, water and protein as three different responses ($M = 3$). The traditional least-squares solution for b (B) is called multiple linear regression or ordinary least squares. The method was used in the pioneering days of NIR, when there were only few wavelength bands available.

The selected linear regression methods differ in the way the regression coefficients are estimated. They also have different properties for outlier detection and other diagnostics. In the early days it was found that wavelengths produced by spectrophotometers caused problems in a regression situation because of collinearity between the variables. There is also the problem of finding a least-squares solution to the equation above when the number of samples is less than the number of variables. Some selected methods are outlined below.

Stepwise multiple linear regressions (SMLR)

SMLR is a way of cleverly reducing the number of wavelengths in order to find a least-squares solution, based on the meaningful assumption that repeated and useless information may be present in some wavelengths. In stepwise multiple linear regression, one starts with selecting the wavelength amongst all available ones, that correlates the best to the measured response. When the first wavelength is found, another wavelength is selected that increases the degree of explanation maximally and so on, until a stop criterion is fulfilled.

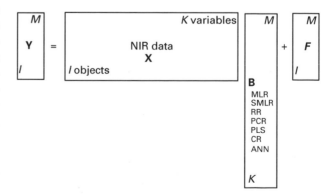

Figure 5 A regression model is built between the matrix **X** with NIR spectra and the responses **Y**. The regression coefficients **B** have to be calculated by MLR, SMLR, PCR, PLS, RR, ANN, etc. **F** is the residual containing the noise.

Principal component regression (PCR)

PCR uses the scores from a principal component analysis to avoid the problems of many variables and collinear variables described earlier. The PCA part also gets rid of some of the noise in the spectral data as an extra advantage. The regression solution becomes easier to calculate and more stable, and one obtains predictions that are more reliable. In addition, the PCA scores give the possibility of outlier detection, both for samples in the calibration data and for later samples for prediction.

Partial least-squares regression (PLS)

PLS is similar to PCR, in that regression is done on scores. The difference is that the responses y (Y) are used to find scores that have a large covariance between X and y (Y). One advantage of PLS over PCR is that the number of required components is reduced. Since PLS is used on scores, these can be used for the detection of outliers and groupings, as was explained for PCA. In this way, PLS automatically gives access to a number of diagnostic tools for outlier detection (in X and y). PCR and PLS add an important parameter to the regression model: the number of components used in the model, A. This is an extra choice to be made. Too few components give a bad model and too many components give a model that is sensitive to noise. PLS, PCR and MLR variants are the most commonly used regression methods for quantification of multivariate spectral data. Other methods are ridge regression (RR) and continuum regression (CR).

Artificial neural networks (ANNs)

The use of ANNs is increasing. ANNs are especially used for pattern recognition and nonlinear calibration. There are many types of ANN. However, it seems that the back-propagation ANNs are the most useful for calibration purposes. In linear calibrations, the traditional linear regression techniques will more often give stable and robust results as compared with ANNs. However, in nonlinear situations and with large heterogeneous data sets, ANNs work very well.

Locally weighted regression (LWR)

LWR is based on the idea that even in a nonlinear situation there exist regions where a local model is close to linear. In LWR the neighbours of a spectrum to be used for prediction are used to build a local regression model. This model can be determined by PCR or PLS. There are different techniques for selecting appropriate neighbours, using different distance criteria and weighting (see **Figure 6**).

In addition to the building of regression models, testing them is also important. Therefore, the data are split into a training set (size I) on which the model is built and a test set (size J) that is used for testing the model in a simulated 'real' situation (see **Figure 7**). Checking how well this works is called

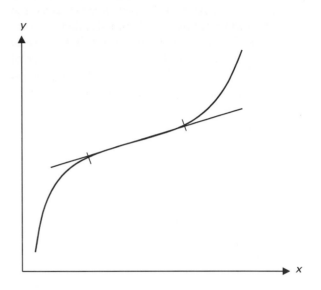

Figure 6 Non-linear relationship between the variables x and y. However, when the range is narrowed, local linear models can be used.

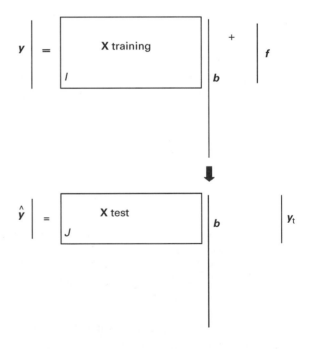

Figure 7 The training set (I objects is used to find the regression coefficients vector b. With a test set (J objects), the predicted responses can be calculated using the same b. Usually the responses for the test set are kept in the background for diagnostic checking.

validation and is explained in its own section below. Calibration is very important for NIR data since the classical interpretation of the spectra (as is done for IR and UV spectra) is nearly impossible. Because of this, one may state that there is a very close relationship between an NIR instrument and the calibration model, making an instrument useless if no calibration model is available. A complete flow chart for NIR data analysis is given in **Figure 8**.

Validation

It is very important to test the quality of a regression model constructed by PCR, PLS or any other regression method. The model for the training set (see **Figure 7**) can be improved by using more PCR or PLS components, but some of these will only describe noise. With an appropriate test set, a check can be made whether the model also has predictive properties. A revealing test is to plot the predicted values for the test set against the measured values in a scatter plot. In this plot, all points should ideally fall on the diagonal. Outliers, systematic errors, groupings and noisy behaviour are all detected easily. Sometimes numbers are needed. The predicted responses; $\hat{y}$ for the test set can be compared with the measured values y_t. This is done by constructing a sum of squares PRESS $= (y_t - \hat{y})(y_t - \hat{y})$ and the corresponding standard deviation RMSEP $= [\text{PRESS}/J]^{1/2}$. A useful plot is that of PRESS (prediction residual sum of squares) or RMSEP (root mean square error of prediction) against the number of components in the model (see **Figure 9**). When I and J are very small because of lack of samples, cross validation may be used advantageously. **Figure 10** shows a typical predicted versus observed plot.

Calibration transfer

The building of a good, robust calibration model with excellent prediction properties is extremely important and sometimes very expensive and time consuming. In many uses of NIR instruments and calibration models, long series of measurements are required and one has to be sure that the calibration model works well over long periods of time. This is not easy since instrument ageing, parts replacement and repairs may create the need for a new calibration model. Another situation arises when different instruments are used to do the same task in parallel. In order to avoid repeating the calibration, the techniques of calibration transfer have been developed. Some calibration transfer techniques are based on keeping individual instrument differences small or

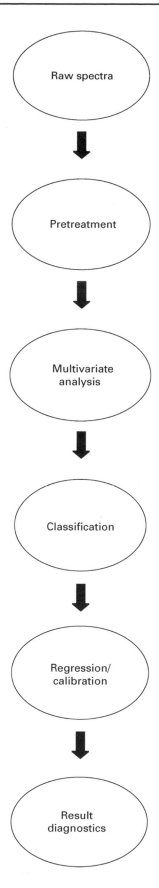

Figure 8 A flow chart for a complete NIR data analysis.

compensating in the instrument hardware itself. Others work by calculations for making the spectra from one instrument (the field instrument) look like those of another one (the master instrument). In this way, the calibration model from the master instrument can still be used successfully.

Pretreatment of the spectra

Several pretreatments are used in the processing of spectral data. The usual mean centring, scaling and nonlinear transformations are part of normal procedure related to classification and regression. Some other techniques were developed specifically with the nature of the NIR measurement in mind. The main purpose is always to improve the performance of a classification or calibration. Pretreatments can also be used to compress spectral information and ease computation on large sets of data, as is the case with Fourier and wavelet transforms. Multiplicative signal correction (MSC) was originally developed for NIR reflectance data in order to correct for different pathlengths and scatter arising from different particle sizes in the samples. This is done by adjusting baselines and slopes. The basis for MSC is simple. Each spectrum is compared with a reference spectrum, usually the mean spectrum of all good calibration samples. The sample spectrum is assigned the same offset and average slope as the reference spectrum. Orthogonal signal correction (OSC) works in a similar way to MSC but uses the responses for orthogonalization, thus ensuring that all information related to the responses stays in the transformed spectra. Derivatives (see **Figure 1**) encompass basically the same thing as MSC. A second derivative of a spectrum removes baseline offsets and linear slopes. Derivation (or differentiation) is a widely used, simple and intuitive data pretreatment. For digitized spectra, different smoothing algorithms can also be applied, e.g. using different widths of derivation.

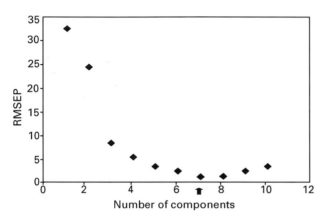

Figure 9 A plot of RMSEP against the number of components used in the model gives a minimum (arrow) indicating how many components should be in the model and how low the prediction error can be made.

List of symbols

b = vector of regression coefficients; $\mathbf{B}$ = matrix of regression coefficients for many responses; f = vector of residuals; $\mathbf{F}$ = vector of residuals for many responses; PRESS = prediction residual sum of squares; RMSEP = root mean square error of prediction; X = spectral data for samples; y = vector of response values for many samples; $\mathbf{Y}$ = matrix of response values for many responses; y_t = measured value; $\hat{y}$ = predicted response.

See also: **Calibration and Reference Systems (Regulatory Authorities); FT-Raman Spectroscopy, Applications; IR Spectrometers; IR Spectroscopy, Theory; Near-IR Spectrometers; Raman Spectrometers.**

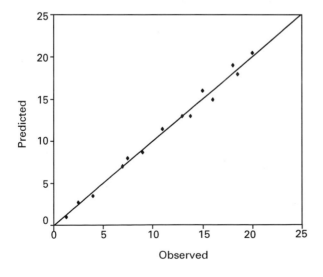

Figure 10 Plotting predicted values of the response against observed values give an idea of prediction quality. All points should fall close to the diagonal. Individual outliers or deviating groups can easily be detected.

Further reading

Beale R and Jackson T (1990) *Neural Computing: an Introduction.* Bristol: Institute of Physics Publishing.

Beebe K, Pell R and Seasholtz M-B (1998) *Chemometrics. A Practical Guide.* New York: Wiley.

Brown P (1993) *Measurement, Regression and Calibration.* Oxford: Clarendon Press.

Davies A and Williams P (eds) (1996) *Near Infrared Spectroscopy: The Future Waves.* Chichester: NIR Publications.

Geladi P and Dåbakk E (1995) An overview of chemometrics applications in NIR spectrometry. *Journal of Near-Infrared Spectroscopy* 3: 119–132.

Johnson R and Wichern D (1982) *Applied Multivariate Statistical Analysis.* Englewood Cliffs, NJ: Prentice Hall.

Jackson J (1991) *A User's Guide to Principal Components.* New York: Wiley.

Joliffe J (1986) *Principal Component Analysis.* Berlin: Springer.

Malinowski E (1991) *Factor Analysis in Chemistry*, 2nd edn. New York: Wiley.

Mardia K, Kent J and Bibby J (1979) *Multivariate Analysis.* London: Academic Press.

Martens H and Naes T (1989) *Multivariate Calibration.* Chichester: Wiley.

Wold S, Esbensen K and Geladi P (1987) Principal component analysis. *Chemometrics and Intelligent Laboratory Systems* 2: 37–52.

Contrast Mechanisms in MRI

IR Young, Imperial College of Science, Technology and Medicine, London, UK

> **MAGNETIC RESONANCES**
> **Theory**

Introduction

The key factor which has led to the huge growth and clinical acceptance of magnetic resonance imaging (MRI) is really quite fortuitous. This is that tissues have quite different values of T_1 and T_2, and that, more importantly, the relaxation time constants of a great many diseased tissues are quite markedly different from those of the normal structures surrounding them. Clinical MRI seeks to highlight these differences at the expense of all other considerations. As a result, all the methods used in traditional NMR have been investigated to find out if they can deliver maximal differentiation between normal and abnormal tissue. The manipulations are performed through complex sets of instructions issued to the various components of a machine at the correct times which are known as 'sequences'. In the following sections we discuss the most important of these, and their variants.

Thereafter a number of phenomena are described which are both artefacts of images if care is not taken, and sources of desirable contrast under other circumstances. Included among these are such things as the effect of flow, a matter of great concern in medical imaging where data about the cardiovascular system is of substantial importance in many clinical conditions. Some of the implications of apparently recherché observations such as that of diffusion are very significant in the diagnosis of unexpected disease situations.

An important, though unfortunately brief, note discusses the single most significant artefact in MRI, arising from the physiological measurements of the patient. The problem of controlling the effect of respiratory motion in particular has delayed the application of MRI to the imaging of much of the thorax and abdomen very substantially.

Standard sequences

Most clinical MRI revolves around the contrast generated by three very standard sequences, partial saturation, spin echo and inversion recovery, which are familiar to those working on traditional high-resolution spectroscopy, but having a resonance and significance in clinical MRI not found elsewhere. The notation used in this article for times and other sequence features is that proposed by the American College of Radiology, and generally used in human MR studies.

(Partial) saturation recovery sequence

This is the classic α–TR–α sequence of NMR in imaging form. The sequence form, showing how the various gradients and RF pulses are manipulated for a spin-warp data acquisition, is sketched in **Figure 1A**. The dotted lines indicate that the phase encoding gradient is varied from one acquisition to the next. The theoretical relationship for the sequence is:

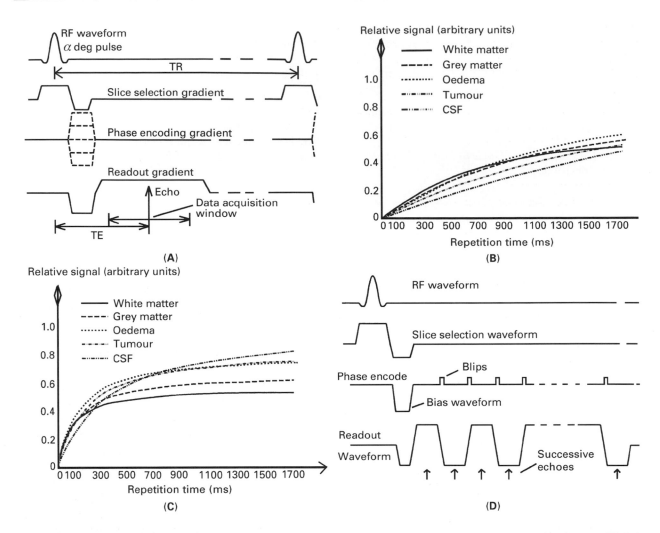

Figure 1 (Partial) saturation recovery sequence. (A) Sequence configuration. A flip angle of α is assumed in this diagram. (B) Calculated signal characterization for the tissues, the parameters for which are given in **Table 1**. α in this case is 90°. (TE = 20 ms for the purposes of this plot.) (C) The same characteristics as in **Figure 1B**, but with α = 30°. (TE = 20 ms for the purposes of this plot.) (D) Blipped echo planar sequence. An echo is formed during each excursion of the readout gradient, with a line of k-space being acquired at the same time. The bias pulse in the phase encoded waveform is used to control the starting point in k-space of the image acquisition.

$$S = \frac{S_0[1 - \exp(-\mathrm{TR}/T_1)]\exp(-\mathrm{TE}/T_2^*)\sin\alpha}{1 - \exp(-\mathrm{TR}/T_1)\cos\alpha} \quad [1]$$

where S is the observed signal, and S_0 is that which is available after full relaxation and immediately prior to excitation. In this sequence TR is the time between successive excitations, while TE is the time from the centre of the excitation to the centre of the echo formed by the refocusing effects of the gradients. α is the actual flip angle. Note that this relationship is much simplified from that derived by solution (see below) of the Bloch equations, by assuming $T_1 \gg T_2$ and that off-resonance effects and spatial selection effects are ignored.

What this means in practice in terms of signal differences and behaviour is shown in **Figure 1B**. This

Table 1 Tissues, and their parameters, used in the description of the main sequences

Tissue	T_1 (ms)	T_2 (ms)	Relative available proton density
White matter	670	85	0.7
Grey matter	920	95	0.8
Oedema	1060	150	0.9
Tumour	1410	200	0.9
CSF	2500	500	1.00
Fat	265	50	0.6

The relative available proton density column indicates the relative peak signal observable from the tissue. Cerebrospinal fluid (CSF) which has a signal near to that of pure water in normal subjects is taken as unity.

(and other similar figures to follow) uses the values in **Table 1** for typical normal and abnormal brain tissue. It is the differences between these components that decide the efficacy of MRI. The values in **Table 1** are typical of those at 1.0 T. The coefficient, ν, relating

echo time (TE) can result in very little contrast indeed in normal organs (as **Figure 2B** suggests).

Figure 2A shows a typical sequence structure for a spin-echo acquisition. The simplified relationship governing the sequence is:

$$S = \frac{S \exp(-\text{TE}/T_2)[1 - \exp(-\text{TR}/T_1)]\{\sin\alpha\cos\beta\exp(-\text{TE}/2T_1) - \sin\beta[1 - \exp(-\text{TE}/2T_1)(1 - \delta\cos\alpha)\}}{1 + \exp(-\text{TR}/T_1)(\cos\alpha\cos\beta + \sin\alpha\sin\beta)} \quad [2]$$

field and relaxation time constant variation for grey matter and white matter can be taken to be ~0.3, for blood and oedema as ~0.2, for tumour as ~0.1 and for cerebrospinal fluid (CSF) as being ~0.0. Note that this implies a tendency for relative contrast levels to fall as the main field (B_o) increases. (Note: ν is the coefficient in the term ($T_1\alpha B_\text{o}^\nu$).)

Figure 1C shows how the contrasts obtained are radically affected by reducing α from 90° (the value in **Figure 1B**).

Echo planar imaging is an extreme example of this sequence. In this procedure, after a single excitation, one of the gradients (the readout gradient) is switched to form a series of echoes during each of which one line of k-space is acquired, while the equivalent of the phase encoding gradient is incremental in some way. Though the original form of the sequence was rather different, the sequence form in current use ('blipped' echo planar) is shown in **Figure 1D**.

Spin-echo sequence

The partial-saturation sequence is typically used with short values of TE (except in special circumstances) and is thus primarily dependent for its contrast on differences in T_1 (i.e. it has 'T_1-weighting'). The spin-echo sequence, though it is poor from the point-of view of discriminating between normal anatomical structures, has proved to be extraordinarily effective in identifying a very wide range of pathological conditions, once it had been successfully implemented using relatively long TR and TE to give 'T_2-weighting'. Initially applied to normal anatomy by the current author more than 2 years before its successful application clinically, it had seemed more or less completely uninteresting for MRI. Unlike inversion recovery, which generated spectacular contrast between normal tissues such as grey and white matter (see below), an inappropriate choice of the

and

$$\delta = \frac{1 - \exp(-\text{TR}/T_1)}{1 + \exp(-\text{TR}/T_1)(\cos\alpha\cos\beta + \sin\alpha\sin\beta)}$$

where α and β are respectively the excitation and refocusing pulse flip angles. When these angles are respectively 90° and 180° (and α in Equation [1] is 90°) the relationships for the partial saturation and spin-echo sequences look misleadingly similar, with the difference between T_2^* and T_2 being the only apparent change. Spin echo was very popular in the early days of clinical MRI because it was robust, and tolerant of many machine defects. In practice, while still regarded as an essential screening tool, it is now seen as being a relatively blunt instrument in terms of the subtlety of contrast that can be obtained with it.

Figures 2C and **2D** show contrast characteristics for a value of TR less than that of typical normal tissue, and a value of TR substantially longer, respectively. The characteristics suggest how choice of sequence parameters can emphasize, or eliminate, contrast between any two tissues.

The fast spin-echo (rapid acquisition relaxation enhanced (RARE)) sequence is the spin-echo equivalent of echo planar imaging, and its structure is shown in **Figure 2E**. Contrast is manipulated by choosing at which point in the echo train the central point of k-space is acquired.

Inversion recovery sequence

Although two of the very early developers of MRI placed substantial weight on it, the inversion recovery sequence was regarded through much of the development of MRI in the 1980s as being a difficult sequence to operate, and saddled with long acquisition times. The sequence is shown schematically in **Figure 3A**, and is governed by the relationship:

$$S = S_\text{o}\frac{\{[1 - \exp(-(\text{TR} - \text{TI})/T_1)]\exp(-\text{TI}/T_1)\cos\beta\sin\alpha - [-\exp(-\text{TI}/T_1)][\sin\alpha + \exp(-(\text{TR} - \text{TI})/T_1)]\}\exp(-\text{TE}/T_2)}{1 - \cos\beta\cos\alpha\exp(-\text{TR}/T_1) + \sin\alpha\sin\beta\exp(-(\text{TR} - \text{TI})/T_1)\exp(-\text{TI}/T_2)}$$

$$[3]$$

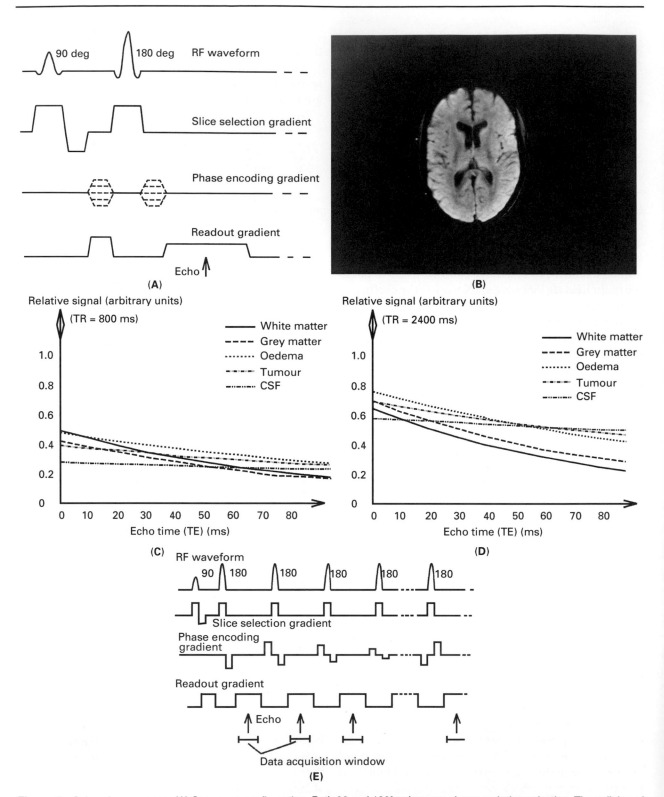

Figure 2 Spin echo sequence. (A) Sequence configuration. Both 90 and 180° pulses are shown as being selective. The splitting of the phase encoding gradient (one half being in the opposite sense to the other) is quite normal, as is the placing of the warping gradient before the 180° pulse. These help a little with artefact generation. (B) Spin-echo image taken with TR 860 ms and TE 20 ms, showing how small the contrast between normal tissues can become if an inappropriate choice of parameters is made. (Slice 4 mm; matrix 256 × 192; field of view 22 cm.) (C) Signal characteristics for the spin-echo sequence, for the tissues given in **Table 1**. α = 90° and β = 180°; in this case TR (800 ms) was about half that of the mean of the T_1 values of white and grey matter. (D) Signal characteristics as in (C), but with TR being three times that of the mean of the T_1 values of grey and white matter at 2400 ms. (E) Schematic drawing of the RARE sequence (shown in this case for an echo train of 8: the letters 'ETL' in a sequence description refer to this value). Actual sequences are much more complex and subtle.

(where TI is inversion time) which simplifies to the familiar form:

$$S = S_\mathrm{o} \left[1 - 2\exp\left(-\frac{TI}{T_1}\right) + \exp\left(-\frac{TR}{T_1}\right) \right] \exp\left(-\frac{TE}{T_2}\right) \quad [4]$$

when $\alpha = 90°$ and $\beta = 180°$. A gradient-recalled echo data acquisition is assumed for simplicity (as the equation, in its spin-echo form, becomes very cumbersome).

Tissue characteristics for the sequence, with a TR assumed to be long compared with the T_1 of both grey and white matter, are shown in **Figure 3B** for the same tissues as before, but with the addition of fat.

This sequence is unique in having both positive and negative signals and exploitation of this leads to its versatility. **Figure 3C** shows an image acquired with a typical TI in the brain using real reconstruction (i.e. retaining the sense of the signal signs). **Figure 3D** shows the same data reconstructed as a magnitude map. The TI was such that grey and white matter on the one hand, and CSF on the other, have signals of alternate sign. Regions where two tissues with alternate magnetization overlap show reduced intensities as their signals oppose each other. The black rims round the ventricles are very obvious, but the unwary investigator can readily mistake more subtle differences, due to this effect, as having clinical significance.

The special property of inversion recovery, that a value of TI can be chosen which results in no signal from anything with the appropriate T_1, has turned out to be a surprisingly powerful tool. From Equation [4], when $TR > 3T_1$ (or thereabouts), in effect the null point is obtained when $TI / T_1 \approx \ln 0.5$. The use of this form of cancellation was first applied to fat (in the short tau inversion recovery (STIR) sequence). Because, generally speaking, tissues with long T_1 times also have substantial T_2 times (practically all of which are longer than that of fat), in practice this means that tissue contrasts dependent on T_1 and T_2 are cumulative for practically all components, giving the approach additional diagnostic power. At the other extreme, cancellation of CSF might not seem to be as obviously beneficial. However, CSF behaves not unlike pure water, and its T_1 and T_2 are very long. Artefacts from its very large signals (due to tissue pulsatility, in the main) contaminate very many of the spin-echo acquisitions, particularly those with longer TEs, and elimination

of the CSF signals has turned out to have substantial diagnostic benefits, particularly in the cervical spine.

Two inversions (with the cancellation of signals from two tissues) have also been studied using the sequence shown in **Figure 3E**, which results in the relationship:

$$S = S_\mathrm{o} \left[1 - 2\exp\left(-\frac{T_\mathrm{B}}{T_1}\right) + 2\exp\left(-\frac{T_\mathrm{A} + T_\mathrm{B}}{T_1}\right) \right. \\ \left. - \exp\left(-\frac{TR}{T_1}\right) \right] \exp(-TE_2) \quad [5]$$

For simplicity all flip angles are assumed to be either 90° or 180° as appropriate. The sequence has had limited clinical application. An example of its use, to cancel the signals for white matter and CSF while using a long TE data acquisition to attenuate the fat signal, is shown in **Figure 3F**. The residual tissue is predominantly grey matter. Attempts to extend the method to use more inversions still are unproductive as the signal-to-noise ratio of the images becomes unacceptable.

Artefacts

Radiologists are trained to be suspicious of even slightly unusual results, and to examine them very carefully for the possibility that they are artefacts. Spectroscopists (and other more traditional investigators using NMR) are less aware of the possibility, though more conscious that there may be problems in multidimensional Fourier transform NMR. Nevertheless, such terms as 'F1 noise' used to describe something which is actually artefact is indicative of users' perceptions.

The utility of the majority of clinical MRI scans is, however, not limited at all by signal-to-noise ratio concerns, even in quite low fields, but by artefact. Since this is a function of signal, arising, in most instances, from any instability at all in the data set, in effect the value of MRI images obtained from a study is quite substantially independent of field level, but dictated by the performance of the machine in use, and the skill of the operators in positioning the patient, choosing the sequences to use, and in persuading the patient to remain still.

Artefact analysis is a huge topic in its own right, and no more than a very cursory glance at it is possible here. Two different forms of artefact will be considered in very broad outline, to illustrate two quite different errors that can occur. These are those arising from slice selection problems, and those due to

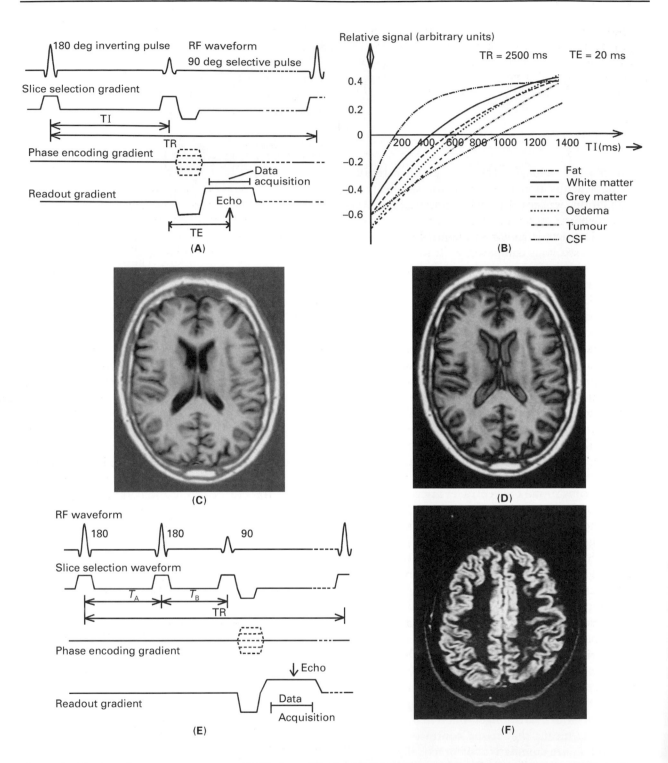

Figure 3   The inversion recovery sequence. (A) The form of the inversion recovery sequence. The initial inverting 180° pulse is normally selective (as shown) but in some situations this is omitted. The data acquisition shown is a gradient recalled echo, but, very frequently, a spin echo is used. (B) Characteristics of the inversion recovery for different values of TI at fixed, relatively long, TR (greater than twice the mean value of the T_1 values of grey and white matter). Because of its importance in some inversion recovery sequences the fat characteristic is also included. (C) Inversion recovery image using real reconstruction (TR 2720 ms, TI 600 ms, TE 30 ms) (128 × 256 matrix, 25 cm field of view; 2 acquisitions; 6 mm slice). (D) Magnitude reconstruction of the same inversion recovery image data as in (C). The sequence was designed so that grey and white matter and CSF have opposite signs. (E) Sequence for a double inversion recovery acquisition. The terms on the figure are those used in Equation [5]. A gradient-recalled echo data recovery is shown, though spin echoes are very frequently used. (F) Double inversion recovery image of the brain of a volunteer with long TE (80 ms), leaving the cortex of the brain as the sole remaining tissue with substantial signal. (TR 6000 ms; T_A 2300 ms; T_B 245 ms; 3 mm slice; 128 × 256 matrix.)

motion artefact (which, ranging from the gross to the microscopic, is the most pervasive of all).

To illustrate slice selection problems, it is useful to begin with the fuller expression for selective excitation [2] for a partial saturation sequence:

$$S = S_0 \frac{\exp(-\text{TE}/T_2)[1 - \exp(-\text{TR}/T_1)]\sin\alpha}{1 - \exp(-\text{TR}/T_1)\cos\alpha - [\exp(-\text{TR}/T_1) - \cos\alpha]\exp(-\text{TR}/T_2)} \qquad [6]$$

Note that this ignores effects due to the spatial localization process. Assuming the rather benign shape for the RF spectrum shown in **Figure 4A** (which has relatively sharp edges and little overshoot) **Figures 4B** and **4C** show, respectively, the predicted slice shapes for two tissues, one with $\text{TR} = 2T_1$, the other with $\text{TR} = 0.2T_1$. In a typical head image in which the range of T_1 values is ~12 (from fat at 250 ms to CSF at 3000 ms), the slice shape changes quite markedly from place to place across the image. Regions in which partial volume effects mean that a single voxel can include two tissues with dramatically different T_1 values result in very different shapes at either side of a single slice. It should be borne in mind that in slice selective MRI, the slice thickness dimension is usually the largest one in a voxel by a substantial margin (at least a factor of 2, and more usually 5–10). Measurements of tissue parameters made from such data must be treated with great care, and the appreciation that the signals may well be badly contaminated by unwanted material.

Motion artefact has proved to be one of MRI's greatest challenges, and the most intractable of problems. It is generally associated with abdominal motion, but can occur anywhere due to either voluntary or involuntary motion of the patient, or regular physiological effects such as blood pulsatility. Motion can be very large, or very small.

The periodicity of the effects is given by (for motion in the phase encoding direction; other directions have similar forms of relationship):

$$\frac{\Delta y}{y_f} = FN\text{TR} \qquad [7]$$

where Δy is the spacing in the y (phase encoding) direction, y_f is the size of the field of view in that direction, F is the frequency of the motion, N the number of averages of each phase encoding step which is acquired and TR is the repetition time.

Their amplitude is given by:

$$S_m = \left| F\left[J_m\left(\frac{\pi A \Delta G_y}{y_v G_{y\max}} \right) \right] \right| \qquad [8]$$

for the mth ghost. F is the Fourier transform function, J_m the mth-order Bessel function of the first kind, ΔG_y, is the gradient movement, $G_{y\max}$ the maximum value of the gradient, A the amplitude of the motion and y_v the length of a voxel in the y-direction. This assumes regular motion, which holds only to a limited extent. Irregular behaviour is largely unmodellable and probably not productively so in general.

A variety of methods are used to attempt to control the effects, ranging from rapid imaging with breathholding, to the relatively sophisticated probability assessments used in the respiratory ordered phase encoding method, or attempts to use internal markers to detect motion (as in the Navigator methods). These techniques are aimed primarily at controlling effects in the phase encoding direction(s). In the read-out or slice selection directions, methods such as motion artefact suppression techniques, which rely on the use of multiple gradient pulses to

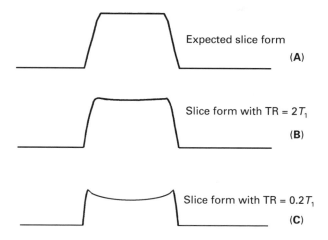

Figure 4 Slice selection artefact. (A) Target slice shape in a partial saturation data recovery (equivalent to the spectral distribution of the RF pulse). This is rather more rounded than is achieved in the best formulations, but has less in the way of undershoot and oscillation at the edges, than is typically achieved. (B) Predicted slice shape for a tissue with $\text{TR} = 2T_1$. Note that T_2 effects are ignored. The pulses in (A) and (B) are scaled in the same way. (C) Predicted slice shape for a tissue with $\text{TR} = 0.2T_1$. This pulse is scaled at twice that of those above.

refocus tissues moving with first, second, third and potentially more motion orders, are quite effective.

Susceptibility effects

Susceptibility effects have a significance for MRI that goes far beyond the concerns high-resolution spectrosocpists have about them as a potential source of line broadening and loss of resolution in their spectra. This is because, as with so many things in MRI, they are also a means of generating contrast, and of recognizing disease patterns. Apart from effects in tissue adjacent to air (as in the sinuses) and, to some extent, trabecular bone, both components having significantly different susceptibilities to that of tissue, the important applications of susceptibility relate to the presence of blood. While oxygenated blood is diamagnetic (possibly more so than normal tissue), deoxygenated blood is mildly paramagnetic, while the degradation products of blood (found in haemorrhagic deposits) become increasingly strongly paramagnetic as they change from methaemoglobin, through haemosiderin to ferritin.

Susceptibility effects cause changes to both expected signal intensities and their phases, both of which can be observed in clinical MRI. The phase changes given by the mean field deviations in voxels with sides x_m, y_m, z_m and field deviation ΔB_{0xyz} at a point x, y, z inside them are:

$$\varphi = \frac{\gamma TE \int_{x=0}^{x_m} \int_{y=0}^{y_m} \int_{z=0}^{z_m} \Delta B_{0xyz}\, dx\, dy\, dz}{x_m y_m z_m} \quad [9]$$

while the amplitude differences are:

$$S = \left[1 - \exp\left(-\frac{TR}{T_1}\right)\right] \exp\left(-\frac{TE}{T_2}\right)$$
$$\times \int_{x=0}^{x_m} \int_{y=0}^{y_m} \int_{z=0}^{z_m} S_{xyz} \exp(\gamma \Delta B_{0xyz} TE) dx\, dy\, dz \quad [10]$$

where S_{xyz} is the available signal from location dz, dy, dx and ΔB_{0xyz} is the local field deviation. TE is the echo time as usual. Relaxation effects have been assumed to be constant over the whole voxel.

For a linear gradient in the slice selection direction, for example, which is, as noted before, the largest dimension of typical voxels, the result is:

$$S = [S_o \, \text{sinc} \, \theta]_{-\theta_0}^{+\theta_0} \quad [11]$$

where $\theta = \gamma G_s s TE$, with G_s being the erroneous gradient in the slice selection direction and θ_o is the value of θ at the voxel boundary.

Susceptibility effects have recently taken on a great significance in MRI, in the form of the suggested basis of the blood oxygen level-dependent approach to the observation of brain function. The experiment is modelled as arising from the diffusion of water molecules in regions with changing concentrations of paramagnetic deoxyhaemoglobin.

Flow effects

Apart from its tendency to generate artefacts, flow is very well visualized by MRI. The approaches used to measure flow are variants of the gradient method and have been developed as a quantitative method by Bryant et al. who used the flow sequence shown in **Figure 5**. Using the parameters shown in the figure the velocity of flow (v) is measured as a phase difference (φ) and given by

$$\varphi = \gamma G \delta \Delta v \quad [12]$$

Flow velocity is then linearly related to phase, though care has to been taken with the choice of sequence parameters to ensure that the velocities being measured do not exceed the available range (as the unrestrained signal pattern is a repetitive sawtooth, with consequent 'wrap round', and potential ambiguity).

Flow is used in a wide range of applications in MRI of which the most important is angiography. Basically there are two strategies used for this. One is 'phase contrast' based on a strategy derivative from Equation [12] but using very short echo partial saturation (or gradient recalled echo) sequences. The alternative method is the 'time of flight' approach, in which spins outside a region to be studied are tagged in some way, and their future progress monitored.

Diffusion effects

Diffusion effects have long been studied in NMR. The pulse sequence used is identical to that of **Figure 5**, though the gradient amplitudes are much larger. The relationship for the signal appropriate for an experiment involving tissue diffusion is:

$$S = S_0 \left[1 - \exp\left(-\frac{TR}{T_1}\right)\right] \exp\left[-\frac{TE}{T_2} - D\gamma^2 G^2 \delta^2\left(\delta - \frac{\Delta}{3}\right)\right] \quad [13]$$

which can be reformulated as:

$$S = S' \exp\left(-\frac{TE}{T_2} - bD^*\right) \qquad [14]$$

where

$$b = \gamma^2 G^2 \delta^2 \left(\delta - \frac{\Delta}{3}\right)$$

D^* (the apparent diffusion coefficient) can be calculated from the results of two experiments which are identical except for the values of b. (Note: the qualification by which this experiment is assumed to measure D^* and not D, the true diffusion coefficient, is a reflection of the uncertainty about the real diffusion process in tissue, as is further suggested in the next paragraph.)

However, the model for diffusion on which this theory is based assumes a uniform isotropic medium. In actuality, as evaluated in human subjects, diffusion in tissue is anything but isotropic. The coefficient can appear to be quite different depending on the direction of sensitization. This topic is much too large to be covered adequately here, but the anisotropy is of surprising clinical importance, particularly in the study of stroke and the development of myelination in the infant brain.

Magnetization transfer contrast (MTC)

Another method which MRI has adopted from traditional NMR is MTC. This seeks to investigate chemical and magnetization exchange between what are typically NMR visible and invisible spin pools. Normally the system is modelled as having two pools (A, visible, and B, invisible in the experiment being performed), though three or more might be necessary to explain what really happens. This concept was applied to MRI, and it is possible to map the different behaviours of various tissues. The model is shown in **Figure 6A**, and the form of sequence used to implement the technique in **Figure 6B**. The prolonged irradiation that is used is applied off-resonance, and in the traditional experiment the amplitudes of the residual signal are plotted against the offset frequencies (at constant RF amplitude).

In MRI, useful contrast improvements are achieved by MTC in angiography and in haemorrhage studies. The time needed to achieve saturation of the pool with the broad-line has to be minimized in MRI (for thermal deposition reasons), and the effective time constant which applies is that appropriate to the

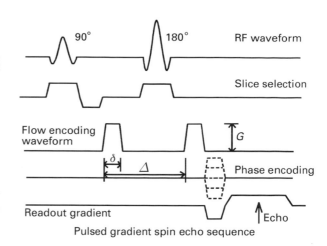

Figure 5 Sequence for flow measurement. The parameters used in Equation [12] are marked on the figure. Note that the sensitizing pair of gradient pulses can be included in any of the gradients (or any combination of them).

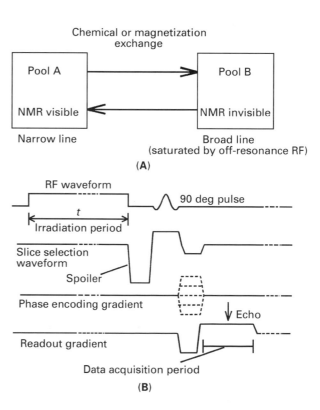

Figure 6 Magnetization transfer technique. (A) Two-pool model used for the evaluation. This is certainly a simplification of reality, although imaging data quality does not justify further elaboration. (B) MTC imaging sequence. A gradient-recalled echo data acquisition is used once more. The spoiler shown is used to dephase residual transverse magnetization after the end of the off-resonance irradiation period (of length t). Practical sequences of all types use pulses like this to eliminate the unwanted signals from spins pursuing complex patterns of excitation and recovery.

signal relationship:

$$S_t = S_{0ASAT} + (S_o - S_{0ASAT}) \exp\left(-\frac{t}{T_{1ASAT}}\right) \quad [15]$$

S_t is the signal after a period of irradiation t, S_{0ASAT} is the residual signal after prolonged irradiation, and T_{1ASAT} the time constant which defines the development of saturation. T_{1A} the time constant of the free pool of spins (in the absence of any exchange) is given by:

$$\frac{S_o}{S_{0ASAT}} = \frac{T_{1A}}{T_{1OBS}} \quad [16]$$

where T_{1OBS} is the value of T_1 measured in the absence of any RF irradiation. The experiment needed to derive the data is one in which the signals are measured after various durations of irradiation and the effects in tissue can be substantial.

A common factor of both diffusion weighting and MTC can on the one hand be a highly desirable effect but can on the other hand be an artefact of other measurements, affecting, for example, the accuracy of measured time constants if not carefully monitored.

Conclusion

This has been a very brief review of a large, complex and still quite fast growing area. Contrast is a matter of obsessive importance to radiologists and as a result will continue to be investigated as long as new possibilities for its generation can be found. For anything outside the scope of this article, readers seeking more information are referred to the burgeoning literature. Useful major sources are given in the Further reading section, and there are a great many more books discussing clinical contrast for particular tissue systems and structures.

List of symbols

A = amplitude of motion; B_o = main electromagnetic field; D = time diffusion coefficient; D^* = apparent diffusion coefficient; F = frequency of motion or Fourier transform function; G_s = gradient in the slice selection direction; G_{ymax} = maximum gradient value; J_m = mth-order Bessel function of the first kind; S = observed signal; S_m = signal of the mth order artefact; S_o = signal after full relaxation; S_{0ASAT} = residual T_{1A} signal after prolonged irradiation; S_t = signal after irradiation period t; T_1 = longitudinal relaxation time; T_{21} = spin–spin relaxation time; T_A = first inversion period in a double inversion recovery sequence; T_{1A} = time constant of free pool of spins; T_{1ASAT} = time constant that defines development of saturation; T_{1OBS} = T_1 in the absence of any RF irradiation; T_B = second inversion period in a double inversion recovery sequence; y_f = field-of-view size in f-direction; y_v = voxel length in y-direction; α = excitation pulse flip angle; β = refocussing pulse flip angle; γ = gyromagnetic ratio; δ = pulse duration; ΔG_y = gradient movement; Δy = spacing in y-direction; θ = spin phase angle; θ_o = value of θ at the voxel boundary; ν = velocity of flow; φ = phase change.

See also: **Diffusion Studied Using NMR Spectroscopy; MRI Applications, Biological; MRI Applications, Clinical; MRI Applications, Clinical Flow Studies; MRI Instrumentation; MRI of Oil/Water in Rocks; MRI Theory; MRI Using Stray Fields.**

Further reading

American College of Radiology (ACR) (1983) *American College of Radiology Glossary of NMR Terms.* Chicago: ACR.

Bryant DJ, Payne JA, Firmin DN and Longmore DB (1984) Measurement of flow with NMR imaging using a gradient pulse and phase difference technique. *Journal of Computer Assisted Tomography* 8: 588–593.

Collins AG, Bryant DJ, Young IR, Gill SS, Thomas DGT and Bydder GM (1988) Analysis of the magnitude of susceptibility effects in diseases of the brain. *Journal of Computer Assisted Tomography* 12: 775–777.

Foster MA and Hutchinson JMS (eds) (1987) *Practical NMR Imaging.* Oxford: IRL Press.

Grant DM and Harris RK (eds) (1996) *Encyclopedia of NMR.* Chichester, UK: Wiley.

Hajnal JV, Doran M, Hall AS, Collins AG, Oatridge A, Pennock JM, Young IR and Bydder GM (1991) Magnetic resonance imaging of anisotropically restricted diffusion of water in the nervous system: technical, anatomic and pathologic considerations. *Journal of Computer Assisted Tomography* 15: 1–18.

Le Bihan D (1995) *Diffusion and Perfusion Magnetic Resonance Imaging.* New York: Raven Press.

Mansfield M (1977) Multiplanar image formation using NMR spin echoes. *Journal of Physics C: Solid State Physics* 10: L55–58.

Ogawa S, Lee T-M, Nayak AS and Glynn P (1990) Oxygenation-sensitive contrast in magnetic resonance image of rodent brain at high magnetic fields. *Magnetic Resonance in Medicine* 14: 68–78.

Partain CL, Price RR, Patten JA, Kulkarni MV and James AE (eds) (1988) *Magnetic Resonance Imaging,* 2nd edn. Philadelphia: W.B. Saunders.

Pattany PM, Philips JJ, Chiu LC, Lipcamon JD, Duerk JL, McNally JM and Mohapatra SN (1987) Motion artifact suppression technique (MAST) for MR imaging. *Journal of Computer Assisted Tomography* 11: 369–377.

Stark DD and Bradley W (1992) *Magnetic Resonance Imaging,* 2nd edn. St. Louis MO: Mosby Year Book (two volumes).

Copper NMR, Applications

See **Heteronuclear NMR Applications (Sc–Zn).**

Cosmochemical Applications Using Mass Spectrometry

JR De Laeter, Curtin University of Technology, Perth, Australia

MASS SPECTROMETRY
Applications

Cosmochemistry is the field of science which investigates the composition and evolution of material in the universe, and in particular in the Solar System. Cosmochemistry is intimately related to nuclear astrophysics, because almost all the chemical elements were synthesized by nuclear reactions in the interior of stars. The distribution of the abundances of the elements provides the basis for an understanding of the chemical and nuclear processes which determined their formation and evolution. The discipline of cosmochemistry therefore embraces the universe both in space and time. Cosmochemistry requires a diverse range of observations and measurements of the elements, and in particular their isotopes, in order to understand the processes and events that shaped their past. Variations in isotope abundances are a vast information source, and their retrieval from mass spectra are of fundamental importance to cosmochemistry because of their unique isotopic signatures. Mass spectrometric isotopic abundance measurements are therefore crucial to unravelling our cosmochemical history. The role of mass spectrometry in providing isotopic information in areas such as 'cosmic' abundances, isotopic anomalies, cosmochronology and planetary science, which are essential to our understanding of cosmochemistry, is the subject of this article.

Cosmic abundances

Cosmic abundance tables of the chemical elements provide the basic data for theories of cosmochemistry and nuclear astrophysics. The abundances are based on a variety of terrestrial, meteoritic, solar, stellar and cosmic ray data. **Figure 1** shows a schematic curve of the elemental abundances (normalized to Si equal to 10^6 atoms) as a function of mass number A. This curve is often referred to as the 'cosmic' abundance curve, but in reality the data is primarily of Solar System origin, and the curve is therefore representative of main sequence stars of similar age and mass as the Sun. However, since the overwhelming majority of stars have essentially solar compositions, the term 'cosmic' is of some relevance. Thus, first we need to determine the distribution of the elements and then proceed to an understanding of the processes which produced the relative abundances of these elements in the Solar System – and by inference in other parts of the universe. So the task of cosmochemistry, and its associated discipline nuclear astrophysics, is one of immense scope and importance.

Since the 1950s, when the elemental abundance distribution depicted in **Figure 1** became available, significant improvements have occurred in the accuracy with which the Solar System abundances are known. In part this is owing to the realization that Type I carbonaceous chondrites – made up of low-temperature condensates that escaped the many alteration processes which affected other classes of meteorites – closely approximate the condensable fraction of primordial Solar System material, and so represent our best opportunity of obtaining an abundance distribution which approximates reality. Isotope dilution mass spectrometry (IDMS) is an effective analytical method for determining trace element concentrations, because of the excellent sensitivity, accuracy and precision of the technique, and because a quantitative recovery of the element concerned is not required. The abundance of the

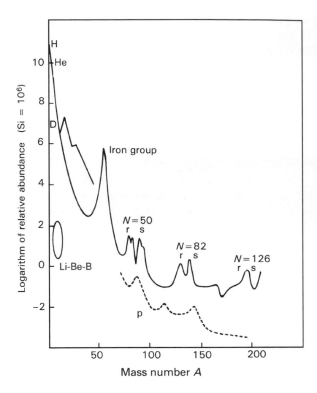

Figure 1 A schematic curve of atomic aboundance (relative to Si = 10⁶) versus mass number A for the Sun and similar sequence stars. The symbols s, r and p stand for the slow and rapid neutron capture processes and the proton capture process, respectively.

element is determined from the change produced in its isotopic composition by the addition of a known quantity of a tracer of that element whose isotopic composition is different from that of the naturally occurring element. IDMS has been used extensively in refining the Solar System abundance curve because it is ideally suited to analysing small samples – such as the Type C-I carbonaceous chondrites. Meteoritical elemental abundances can now be determined by IDMS at the 3% uncertainty level, although most of the Solar System abundances have an uncertainty of 10% or greater.

The canonical model of nucleosynthesis, which was developed to explain the elemental abundance distribution, is based on the assumption that presently constituted matter is derived from a mixture of numerous components synthesized by a number of processes with a variety of nuclear histories. The chemical elements are synthesized through a continuous process in which a fraction of the material is cycled through successive generations of stars. Apart from hydrogen, most of the helium and some lithium and boron, which were produced during the high temperature, high density stage of the universe which is referred to as the 'big bang',

the remaining elements were synthesized in the interior of stars by a combination of thermonuclear reactions (for the 'light' elements with $Z < 20$), nuclear statistical equilibrium processes (for the iron peak group of elements), and the slow (s) and rapid (r) neutron capture processes (for the 'heavy' elements with $Z > 31$).

The products of nucleosynthesis are returned to the interstellar medium by supernovae and by mass loss from red giants and novae. The processed and expelled material, isotopically enriched to varying degrees, is then incorporated in later generations of stars, where further nuclear processing takes place. Thus the Solar System was formed from material that has been processed by a number of nuclear processes in a variety of stellar environments over the lifetime of the galaxy.

Isotopic anomalies

The conventional wisdom of the late 1960s, based on the apparent isotopic homogeneity of Solar System materials, was that the memory of interstellar chemistry had been erased by a hot gaseous homogenization of pre-existing material from which the Solar System condensed. Technological advances in mass spectrometry, particularly in ion microprobe analysis, have enabled the isotopic composition of extremely small meteoritic samples to be measured with high precision in relation to 'standard' samples, and these analyses have revealed the presence of isotopic anomalies, some of which can be correlated with nucleosynthetic mechanisms occurring in stellar sites. Isotopically anomalous material provides compelling evidence that the Solar System is derived from compositionally different and imperfectly mixed nucleosynthetic reservoirs. Whilst considerable isotopic homogenization undoubtedly took place during the formation of the solar nebula and in planetary formation, some remnants of primordial material have been preserved in refractory inclusions and prestellar grains found in primitive meteorites.

Isotopic abundances are said to be anomalous if the measured isotopic ratios cannot be related to the terrestrial (and by inference solar) isotopic composition of elements through a mass fractionation relationship resulting from a natural physiochemical process, and/or the mass spectrometer measurement itself. The only variations in the isotope abundances in terrestrial materials occur by radioactive decay and by the physiochemical processes mentioned above. Thus terrestrial samples serve as a good example of homogenized Solar System material against which anomalous isotopic material can be compared.

The first isotopic anomaly was discovered in 1960 in the noble gas Xe. The anomalies were of a dual nature – the so-called 'special' anomaly in ^{129}Xe which was shown to be caused by the presence of the extinct radionuclide ^{129}I, and the 'general' anomalies which occurred in most of the other isotopes (**Figure 2**). The pattern of enrichment in the heavy isotopes ^{136,134,132,131,130}Xe, which is particularly pronounced in carbonaceous chondrites, became known as carbonaceous chondrite fission xenon, because it was hypothesized that these anomalies were produced by the spontaneous fission of a super-heavy element. Isotopic anomalies in Xe, and also Ne, made little impression on the cosmochemical community because the systematics of the noble gases were poorly understood and other explanations for the anomalies were forthcoming. It was not until isotopic variations were found in the abundant element oxygen in 1973 that the reality of isotopically anomalous material in meteorites was fully accepted.

Refractory inclusions

The fall of the carbonaceous meteorite Allende in 1969 was fortuitous in that it contained refractory inclusions – a logical location for the survival of presolar material – in sufficient quantity to enable extensive mass spectrometric investigations to be carried out on a wide range of elements. These inclusions condensed from the cooling gas of solar composition and comprise refractory oxides such as spinel, perovskite, pyroxene and melilite and may contain isotopically anomalous material since they probably condensed before isotopic homogenization was complete. Oxygen consists of three isotopes, and thus in principle it is possible to distinguish between physiochemical and nuclear effects. The former produces fractionation that is linearly proportional to the relative mass difference $\Delta m/m$ and thus the measured δ^{17}O values would be approximately half the δ^{18}O values. Terrestrial samples fall on this fractionation line of slope 0.5 (**Figure 3**), whereas meteoritic materials yield values that define a line of slope unity. This unexpected result was explained by arguing that a component of almost pure ^{16}O, produced by nucleosynthesis, was incorporated in meteorites. The δ notation is extensively used in mass spectrometric measurements to distinguish isotopic variations in a sample from a terrestrial standard. In the case of oxygen the terrestrial standard is called 'standard mean ocean water' or SMOW. The δ values are customarily given as per mil deviations from the standard.

The elements of the iron peak group provide information on the late evolution of massive stars by nuclear statistical equilibrium processes. Any isotopic anomalies found in these elements would reflect variations in the quasi-equilibrium conditions of stellar interiors, provided that the isotopic

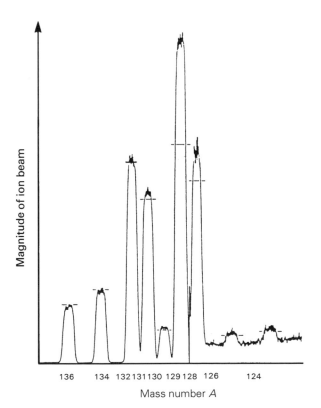

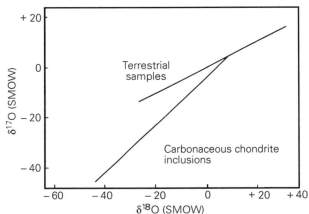

Figure 2 The mass spectrum of Xe extracted from the Richardton meteorite. The horizontal lines show the comparison spectrum of terrestrial Xe. The isotopes 124, 126 and 128 have been measured at 10 times the sensitivity of the remaining Xe isotopes.

Figure 3 The isotopic composition of O in inclusions from carbonaceous chondrities which define a line of slope approximately unity in contrast to terrestrial samples which fall on a fractionation line of slope 0.5. SMOW is a terrestrial standard against which the per mil deviation δ of the oxygen isotopes is measured.

abundances were not affected during their injection into the solar nebula. High-precision thermal ionization mass spectrometry (TIMS) has revealed correlated anomalies in the neutron-rich isotopes of Ca, Ti, Cr, Ni and Zn from refractory element-rich inclusions in carbonaceous chondrites. The magnitudes of these anomalies in Ca–Al-rich inclusions from the Allende meteorite are of the order of ≥ 1‰. No single set of nuclear statistical equilibrium conditions can be found to explain the relative excesses in all the elements in these inclusions. However, a multi-zone mixing model has been proposed in which material from a number of stellar zones, with a variety of neutron enrichments, is mixed during the ejection of that material from a star. The magnitudes of the anomalies are affected by dilution with normal Solar System material, and by chemical fractionation processes that occur after nucleosynthesis has terminated.

Figure 4 shows the isotopic anomalies of the iron peak elements predicted by the multi-zone mixing model as compared with the average excesses as observed in Ca–Al-rich inclusions. The match between the two data sets is impressive, except for Fe and Zn. In the case of Fe no significant anomalies have been measured, but the multi-zone mixing model only predicts a ^{58}Fe excess of approximately 1 part in 10^4, which is at the limit of present mass spectrometric capability. In the case of Zn, the excess in ^{66}Zn is approximately an order of magnitude less than that expected. This can be explained in terms of the volatility of Zn with respect to the other iron peak elements, as it would be the last of these elements to condense from the expelled stellar material. The correlation between anomalies in neutron-rich isotopes in the iron peak elements can therefore be explained in terms of the nuclear statistical equilibrium processes, which took place at a late stage in the evolution of massive stars.

The isotopic anomalies detected in the iron peak elements are extremely small (**Figure 4**), since the anomalous material is diluted with material with a terrestrial isotopic composition. This contamination results from the fact that the meteorite inclusion is taken into solution before chemically extracting the relevant element in a form suitable for conventional TIMS analysis. However, ion microprobe mass spectrometry can be used to analyse small meteoritic inclusions in situ without the need of chemical processing. This enables single inclusions to be analysed for a variety of elements, whilst maintaining the petrographic context of the sample. The carbonaceous chondrites Murchison and Murray also contain refractory inclusions such as corundum and hibonite, but they are invariably small and difficult

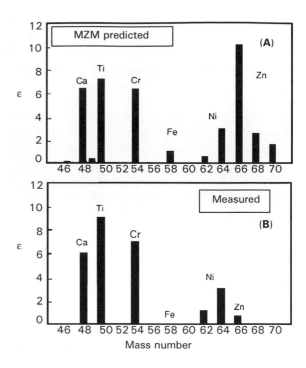

Figure 4 (A) Isotopic excesses of several iron peak elements as predicted by the multi-zone mixing model. They are given as ε units (parts in 10^4) relative to normal terrestrial isotopic composition. (B) Average excesses observed in normal Allende inclusions are displayed.

to analyse by TIMS. Ion microprobe mass spectrometry revealed that Ti anomalies in these hibonite inclusions were an order of magnitude greater than those found in Allende, presumably because hibonite has a higher condensation temperature than the assemblages of the Allende inclusions. Both excesses and deficits were measured in ^{50}Ti and these anomalies were correlated with anomalies in ^{48}Ca (**Figure 5**). A number of morphological, chemical and isotopic features show marked correlations in Murchison refractory inclusions, which provides strong evidence that nucleosynthetic anomalies have survived through discrete inclusion-forming events in the early Solar System.

Presolar grains

The discovery of presolar grains was made possible by the development of chemical procedures in which carbonaceous meteorites were subjected to a stringent acid digestion regime. Carbon compounds such as diamonds, SiC and graphite were isolated in this manner and identified through their distinctive δ^{13}C pattern and anomalous noble gas component. These carbonaceous phases are samples of interstellar matter which provide a window into the prehistory of the Solar System.

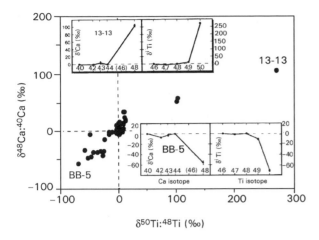

Figure 5 Ca and Ti isotopic compositions in hibonite $(Ca[AlMgTi]_{12}O_{19})$ inclusions displayed a large range from excesses in the most neutron-rich isotopes to deficits, while the other isotopes are close to solar proportions (defined as deviations, δ^i-Ca or δ^i-Ti, in ‰). The most anomalous grains are both from the Murchison meteorite, 13-13 and BB-5 which have large excesses and large depletions respectively in ^{48}Ca and ^{50}Ti (insets). The ^{48}Ca and ^{50}Ti anomalies are clearly correlated and are inferred to come from neutron-rich supernova ejecta.

Although diamonds have the highest abundance of prestellar grains (400 ppm by weight of the bulk meteorite), their average size is only 2 nm, which is currently too small for microanalysis. The diamonds exhibit a characteristic enrichment in the heavy and light isotopes of Xe (Xe-HL) as shown in **Figure 6**. The diamonds may have been produced by supernova shock waves passing through molecular clouds or from chemical vapour deposition. Individual grains of SiC may be up to 20 nm in size, which enables ion microprobe analysis of individual grains. These presolar grains contain measurable concentrations of a number of trace elements which has enabled isotopic anomalies in Si, C, N, Mg, Ti, Ba, Nd and Sm to be identified. Most of the heavy elements show evidence of s-process nucleosynthesis (**Figure 6**), whilst ^{12}C and ^{15}N enrichments show evidence of a mixture of CNO-cycle material with isotopically normal C and N. A minor fraction of the SiC grains have excesses in the decay products of short-lived radionuclides, indicating that they may represent C-rich zone material from the outer layers of a supernova. The third carbonaceous presolar grain material, graphite, represents less than 2 ppm by weight of these primitive meteorites, and only a fraction is isotopically anomalous. Graphite spherules show a large range in anomalous C, and also contain small inclusions of refractory carbides of Si, Ti, Zr, Mo and W. Some s-process anomalies occur in Zr, suggesting that some of the graphite grains were produced in Asymptotic Giant Branch stars.

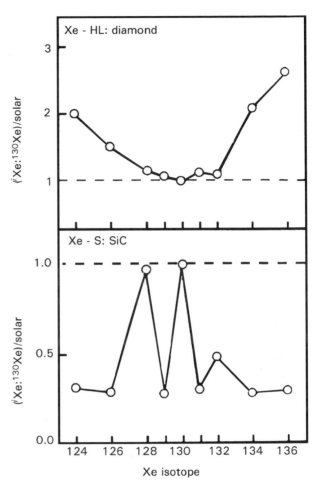

Figure 6 Progressive enrichment of isotopically exotic Xe components led to the discovery of interstellar diamond and SiC. The light and heavy components of Xe-HL cannot be produced in the same nucleosynthetic event and are probably the result of mixing. The Xe-S component from SiC reflects a mixture between the composition produced in s-process nucleosynthesis and a near-normal component of Xe.

It was fortuitous that the resistance to acid digestion of presolar carbonaceous material led to their discovery. In the digestive procedures adopted, silicates and oxides are destroyed. It is hypothesized that most interstellar grains in the solar nebula are oxides, but apart from corundum (Al_2O_3) grains it has not been possible to isolate interstellar oxides. Furthermore the abundance of corundum grains is very small, although they do show isotopic anomalies in ^{26}Mg, ^{17}O and s-process Ti.

Cosmochronology

One of the most challenging tasks in cosmochemistry is to place a time scale on the events that have occurred in the formation and evolution of the Solar

System and for the nucleosynthesis of the chemical elements. **Figure** 7 is a schematic diagram of the major epochs in the history of the cosmos, as described in the figure caption.

The dating of terrestrial materials (geochronology) does not provide any significant information on cosmochronology, but as early as 1953 an estimate for the age of formation of the Solar System was obtained from U–Pb isotopic analyses on meteoritic lead. Subsequently meteorites were dated by the U–Pb technique to give an age of $4.551 \pm 0.004 \times 10^9$ y, and the fine-scale separation of events that took place during planetary formation can be resolved down to a few million years.

Numerous chronological investigations of lunar materials were carried out in the 1970s resulting in the chronology depicted in **Figure** 8. Crystallization ages of $\sim 4.6 \times 10^9$ y have been obtained from lunar rocks, so that there is a good correspondence between the meteoritic and lunar data for the time of planetary formation. The melt rocks derived from impact metamorphism gave ages of 3.85–4.05×10^9 y. This is interpreted as the result of a major bombardment of the Moon by small planetary bodies, which created the lunar basins and implies widespread bombardment of the Solar System during this time. After the termination of this bombardment phase there was decreasing activity until approximately 3×10^9 y ago, after which the lunar surface appears quiescent.

To obtain accurate dating it is necessary to select a decay scheme whose half-life is of similar magnitude as the age of the event to be measured. Fortunately nature has provided a number of radionuclides with half-lives from 10^5 to 10^8 y which have given us the

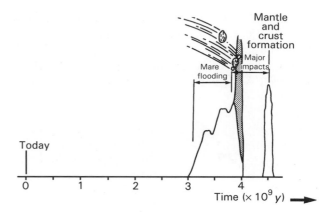

Figure 8 Schematic diagram showing the chronology of major lunar events.

ability to place tight radiometric constraints on the chronology of nebular and planetary events in the early Solar System, and even on nucleosynthetic events that influenced the presolar molecular cloud before its collapse (**Table 1**). The radionuclides listed in **Table 1** are extinct in that they cannot be directly observed at the present time. Their prior presence must be inferred from an excess in their daughter product linked to the magnitude of the parent:daughter ratio.

The first mass spectrometric evidence of the presence of an extinct radionuclide in meteorites was ^{129}I, which, with a half-life of 16×10^6 y, produced

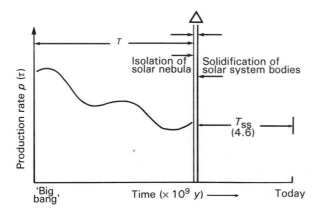

Figure 7 Schematic diagram of the history of the universe. The first epoch T represents the time period from the 'big bang' to the isolation of the solar nebula from galactic nucleosynthesis and the solidification of Solar System bodies, while the third epoch T_{ss} represents the age of the Solar System. The shape of the production rate for nucleosynthesis is for illustrative purposes only.

Table 1 Short-lived radionuclides

Parent isotope	Daughter isotope	10^{-6} Half-life (y)	Early solar system abundance
^{26}Al	^{26}Mg	0.74	^{26}Al:$^{27}Al \approx 5 \times 10^{-5}$
^{53}Mn	^{53}Cr	3.7	^{53}Mn:^{55}Mn ≈ 0.1–6.7×10^{-5}
^{60}Fe	^{60}Ni	1.5	^{60}Fe:$^{56}Fe \approx 4 \times 10^{-9}$
^{107}Pd	^{107}Ag	6.5	^{107}Pd:$^{108}Pd \approx 2 \times 10^{-5}$
^{129}I	^{129}Xe	16	^{129}I:$^{127}I \approx 1 \times 10^{-4}$
^{146}Sm	^{142}Nd	103	^{146}Sm:^{144}Sm ≈ 0.005–0.015
^{244}Pu	Xe fission	81	^{244}Pu:^{238}U ≈ 0.004–0.007
^{41}Ca	^{41}K	0.10	^{41}Ca:$^{40}Ca \approx 1.5 \times 10^{-8}$
^{182}Hf	^{182}W	9	^{182}Hf:$^{180}Hf \approx 2 \times 10^{-4}$
^{36}Cl	^{36}Ar	0.30	^{36}Cl:$^{35}Cl \approx 1.4 \times 10^{-6}$
^{92}Nb	^{92}Zr	35	^{92}Nb:$^{93}Nb \approx 2 \times 10^{-5}$
^{99}Tc	^{99}Ru	0.21	^{99}Tc:$^{99}Ru \approx 1 \times 10^{-4}$
^{205}Pb	^{205}Tl	15	^{205}Pb:$^{204}Pb \approx 3 \times 10^{-4}$

an excess of ^{129}Xe (**Figure 2**). In the 1970s a study of refractory inclusions from Allende gave an excess in ^{26}Mg, the magnitude of which was correlated with the abundance of Al in the samples (as shown in **Figure 9**). A plausible explanation is that ^{26}Al, with a half-life of 0.74×10^6 y, was present in these minerals and that the decay was in situ. **Figure 9** is a typical isochron diagram, so useful in geochronology, and the slope of the linear array enables a formation age Δ of $\sim 10^6$ y to be calculated. Another important discovery was that of excess ^{107}Pd in iron meteorites, resulting from the decay of ^{107}Pd (with a half-life of 6.5×10^6 y), which gave a value for Δ of $\sim 10^7$ y.

Table 1 lists the extensive isotopic evidence for these extinct radionuclides, which demonstrates that the time between the cessation of nucleosynthesis and the formation of meteorites was so short that it supports the hypothesis that a 'last minute' injection of nucleosynthetic products may have accompanied a supernova shock wave which triggered the collapse of the solar nebula. In this case it is likely that some of the nucleosynthetic products may not have been well mixed in the molecular cloud from which the Solar System evolved.

Unlike the situation with extinct radionuclides, there are few long-lived radionuclides which are appropriate to measure the period of nucleosynthesis in that part of the interstellar medium from which our Solar System formed about 4.6×10^9 y ago. Various models have been developed to measure the mean age of nucleosynthesis T, and ^{232}Th, ^{235}U, ^{238}U and ^{187}Re have been used to calculate T, but the estimation of nuclear production rates is a difficult undertaking for each of these r-process nuclides. If one adds the age of the Solar System T_{ss} to estimates of the mean age of nucleosynthesis T, taking into account big bang nucleosynthesis and stellar and galactic evolution, an age of the universe of $\sim 14 \pm 4 \times 10^9$ y can be obtained.

Planetary science

Mass spectrometry is ideally suited to the investigation of planetary atmospheres and cometary material in terms of both elemental and isotopic abundances. Mass spectrometers have been an integral component of space probe instrumentation because they are mechanically robust, have a low power consumption, yet are sensitive and versatile.

Dust particle impact mass spectrometers were carried on space probes that approached close to Halley's comet in 1986. These are time-of-flight mass spectrometers of the design shown in **Figure 10**. The data confirm that the dust from the comet is essentially solar in both elemental and isotopic composition. A double focussing mass spectrometer was also used to determine the composition, density and velocity of the neutral gases and of low-energy cometary ions. This mass spectrometer showed that the neutral gas composition of the coma of Halley's comet was dominated by water vapour and other light element constituents. The third mass spectrometer carried on the Halley space probes was designed to determine the composition, density, energy and angular distribution of ions in

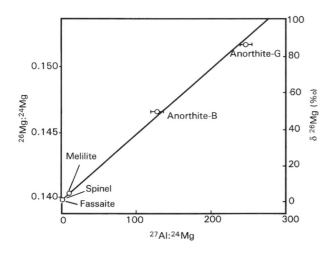

Figure 9 An 'isochron' diagram of ^{26}Mg:^{24}Mg versus ^{27}Al:^{24}Mg for minerals from an Allende inclusion. The linear dependence of the magnitude of the anomalous ^{26}Mg on the Al:Mg ratio can be interpreted as resulting from the decay of the radioactive nuclide ^{26}Al.

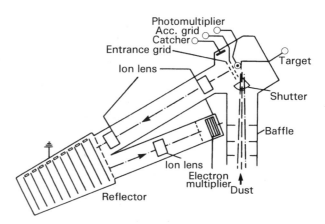

Figure 10 Schematic diagram of the time-of-flight dust-particle impact mass spectrometer which was mounted in a space probe to examine the mass spectra of dust particles from Halley's comet.

the solar wind and cometary plasma. This ion mass spectrometer consists of two independent sensors – a high-energy range spectrometer and a high-intensity spectrometer. As the space probe Giotto approached comet Halley, H^+, C^+, H_2O^+, CO^+ and S^+ were found in a diffuse shell-like distribution out to 3×10^5 km from the comet. As Giotto penetrated the atmosphere to within 1300 km to the nucleus of the comet, the main ion species were H^+, H_2^+, C^+, OH^+, H_2O^+, H_3O^+, CO^+ and S^+. The gas and dust stream emitted from comet Halley contains a higher proportion of volatiles than is found in meteorites, which is consistent with the comet's original location in the cooler regions of the Solar System. Isotopic measurements show that O, S, C, Mg, Si and Fe in cometary and stratospheric dust particles are typical of those found in other Solar System material, implying that the dust has the same nucleosynthetic reservoir as the rest of the Solar System. On the other hand, isotopic anomalies in H and Mg confirm meteoritic evidence that some solid particles survived the homogenization processes that occurred in the early history of the Solar System.

The particle flux that is part of the expanding corona of the Sun is known as the solar wind. The elemental and isotopic composition of the solar wind was studied by various space craft in the period 1965–1975, all of which carried electrostatic positive ion analysers. An ideal opportunity for further study was afforded by the Apollo mission to the Moon, in that aluminium foil collectors were exposed on the lunar surface in the Apollo 11–16 missions which were then analysed for the noble gases by gas source mass spectrometry on the returned Al foil.

The lunar soil is an ideal repository for implanted solar wind elements, as are certain gas-rich meteorites. Deuterium is depleted relative to the terrestrial standard in these materials, the D:H ratio of $<3 \times 10^{-6}$ being consistent with the hypothesis that D is converted into ^3He in the proto-Sun. Ion probe mass spectrometry has been used to study Mg, P, Ti, Cr and Fe which are present to enhanced levels in lunar minerals, indicating an exposure age of approximately 6×10^4 y. The isotopic data indicate that the light isotopes of a number of elements have been preferentially lost from lunar material because of volatilization by micrometeorites or solar wind bombardment. There is some indication, from a study of Ne in gas-rich meteorites, of a large solar flare irradiation during the early history of the Solar System, perhaps related to the T-Tauri phase of the Sun.

Isotopic studies of Gd, Sm and Cd from lunar samples show depletions in those isotopes which possess large thermal neutron capture cross sections.

This has not only enabled the integrated neutron flux and neutron energy spectrum in the lunar samples to be determined, but has provided information on the stability of the lunar surface by analysing these elements and the noble gases in lunar core material.

Interplanetary dust particles (IDPs) can be collected on adhesive surfaces by U-2 aircraft at altitudes of 20 km, and a fraction of these have been identified as of non-terrestrial origin. Some of these IDPs have been analysed by gas source mass spectrometry using a double focussing instrument employing the Mattauch–Herzog geometry (**Figure 11**). The measured $^3He{:}^4He$ and $^{20}Ne{:}^{22}Ne$ ratios are lower than those observed in the solar wind. The sensitivity of static gas source mass spectrometry is such that stepwise heating of these IDPs has the potential to elucidate the thermal history of the stratospheric particles, which enables cometary and asteroidal origins to be differentiated.

Mass spectrometers were an integral component in the scientific payload of Viking Landers 1 and 2 and were designed to measure the composition and structure of Mars' upper atmosphere. Carbon dioxide is the major constituent of the atmosphere, whilst the isotopic composition of C and O in the Martian atmosphere is similar to that of the terrestrial atmosphere. However, ^{15}N is enriched in Mars' atmosphere by approximately a factor of 1.6. Certain meteorites have been identified as samples from the Martian crust. These SNC meteorites, including the orthopyroxenite ALH84001, on which claims for the evidence of life on Mars have been based, have produced a large quantity of isotopic evidence which has been used in conjunction with the Viking Lander data. Since meteorites are delivered to our doorstep,

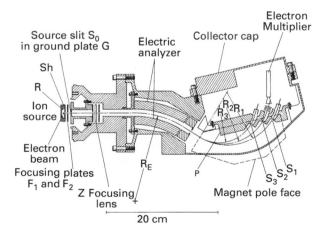

Figure 11 Schematic diagram of the high-performance, double focussing mass spectrometer used to measure the isotopic composition of the noble gases on IDPs.

free of charge as it were, it is important to acquire and classify new samples because of their potential value to planetary science.

Meteorites have provided the bulk of the data on which the 'cosmic' abundance distribution has been established, which led directly to the canonical theory of nucleosynthesis. A range of nucleosynthetic sites is now available for laboratory study and nucleosynthetic models can be compared with measured isotopic compositions. Isotopic data on refractory inclusions and prestellar grains from primitive meteorites are a new source of astrophysical data, giving fresh insights into nuclear processes in stars. Isotope abundance measurements have also been invaluable in giving a cosmochronological time scale to the major events in the history of the universe. Mass spectrometers have been an essential component of the payload of space probes to the planets and Halley's comet, and technological advances have even enabled the noble gases in IDPs to be measured isotopically.

List of symbols

H = high; L = low; m = mass; p = proton capture process; r = rapid neutron capture process; s = slow neutron capture process; T = time period from big bang to isolation of the solar nebula; T_{ss} = the age of the Solar System; δ = deviation from terrestrial composition(‰); Δ = time period from isolation of solar nebula to solidification of Solar System bodies; Δm = difference in mass; ε = 1 part of 10^4.

See also: **Fluorescent Molecular Probes; Geology and Mineralogy, Applications of Atomic Spectroscopy; Inorganic Compounds and Minerals Studied Using X-Ray Diffraction; Interstellar Molecules, Spectroscopy of; Isotope Ratio Studies Using Mass Spectrometry; Isotopic Labelling in Mass Spectrometry; Labelling Studies in Biochemistry Using NMR; MRI of Oil/Water in Rocks; Time of Flight Mass Spectrometers.**

Further reading

Anders E and Grevesse N (1989) Abundances of the elements: meteoritic and solar. *Geochimica et Cosmochimica Acta* 53: 197–214.

Anders E and Zinner E (1993) Interstellar grains in primitive meteorites: diamond, silicon carbide and graphite. *Meteoritics* 28: 490–514.

Bernatowicz TJ and Zinner E (eds) (1996) *Astrophysical Implications of the Laboratory Study of Presolar Materials*, Proceedings no. 402. New York: American Institute of Physics.

Cowley CR (1995) *An Introduction to Cosmochemistry*. Cambridge: Cambridge University Press.

De Laeter JR (1990) Mass spectrometry in cosmochemistry. *Mass Spectrometry Reviews* 9: 453–497.

De Laeter JR (1994) Role of isotope mass spectrometry in cosmic abundance studies. *Mass Spectrometry Reviews* 13: 3–22.

Ireland TR (1995) Ion microprobe mass spectrometry. In Hyman M and Rowe M (eds) *Techniques and Applications in Cosmochemistry, Geochemistry and Geochronology in Advances in Analytical Chemistry*, Volume II, pp 51–118. Greenwich (UK): JAI Press.

Krankowsky D and Eberhardt P (1990). In: Mason JW and Horwood E (eds) *Comet Halley Investigations, Results and Interpretations*, Volume 1, pp. 273–296. New York: Simon and Schuster.

Nier AO (1996) Planetary atmospheres with mass spectrometers carried on high speed probes or satellites. *Rarefied gas dynamics. Progress in Astronautics and Aeronautics* 51: 1255–1275.

Dairy Products, Applications of Atomic Spectroscopy

See **Food and Dairy Products, Applications of Atomic Spectroscopy.**

Data Sampling Theory

See **Fourier Transformation and Sampling Theory.**

Diffusion Studied Using NMR Spectroscopy

Peter Stilbs, Royal Institute of Technology, Stockholm, Sweden

MAGNETIC RESONANCE
Applications

The magnetic field gradient spin-echo method

Under the influence of a non-uniform magnetic field, otherwise equivalent nuclei at different locations give rise to different NMR frequencies. A linear magnetic field gradient along one sample axis will thus quantitatively map NMR frequency with a location along this axis. This is the basic tool used for NMR imaging and microscopy. However, it can also be used for the detection of transport phenomena such as diffusion and flow. The traditional and most widespread NMR method for measuring diffusion is based on the Hahn spin-echo experiment in such a field gradient. Under its influence a phase dispersion results in the rotating frame after application of the initial 90° RF pulse. However, if the NMR frequency remains the same during the whole experiment sequence a second RF pulse refocuses the spin dispersion and leads to the formation of a spin-echo. Normally a 180° pulse is used (applied at time t after the first RF pulse), and the spin-echo will thus appear at a time $2t$ (**Figure 1**). The effect of diffusion on this experiment is easy to work out – the random molecular movement from self-diffusion will partially change the location and NMR frequency of all nuclei in the sample in a random fashion, leading to an incomplete refocusing of the echo. Originally these concepts and experiments were developed and performed in static magnetic field gradients (hence the notation SGSE – static gradient spin-echo). For SGSE experiments the echo attenuation is given by

$$A(2\tau) = A(0) \exp(-2\tau/T_2) \exp(-\gamma^2 g^2 \tau^3 D/3) \quad [1]$$

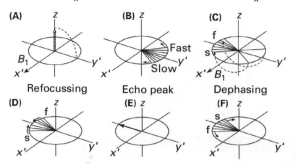

Excitation (90°$_x$-pulse) Dephasing Inversion (180°$_x$-pulse)

(A)   **(B)**   **(C)**

Refocussing   Echo peak   Dephasing

(D)   **(E)**   **(F)**

Figure 1 The basic 90°$_x$–180°$_x$ spin-echo experiment. In the case of diffusion measurements the cause of 'fast' and 'slow' spin vectors is that nuclei in different locations precess at different frequencies under the influence of the imposed magnetic field gradient. A useful supplementary picture of the phenomenon is to visualize a formation of a helical magnetization pattern in the sample along the gradient direction during the first gradient pulse, and the unwinding of this magnetization helix during the second gradient pulse. At the time of the echo peak, the net transverse magnetization is sampled and becomes the NMR signal.

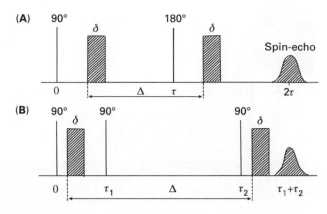

Figure 2 The pulsed-gradient spin-echo (PGSE) experiment (A), and its stimulated echo analogue (B). Radiofrequency pulse phase cycling is required to isolate the stimulated echo from four other echo peaks that occur as a result of three RF pulses (at times $2\tau_1$, $2\tau_2 - 2\tau_1$, $2\tau_2 - 2\tau_1$ and $2\tau_2$).

where T_2 represents the irreversible spin–spin relaxation time, g the strength of the applied gradient, D the self-diffusion coefficient of the molecule in question and γ the magnetogyric ratio of the monitored nucleus. It is evident that the echo attenuation effect is greatest for protons, but experiments on other nuclei are feasible.

Numerous developments of these procedures have been made since their initial discovery by Hahn in the early 1950s. Around 1965 the pulsed-gradient spin-echo (PGSE) experiment was suggested and tested. In its basic form it relies on the use of two linear magnetic field gradient pulses (of amplitude 'g', duration 'δ' and separation 'Δ'; **Figure 2A**. Under these conditions the amplitude of the spin-echo attenuates from its full value according to the so-called Stejskal–Tanner relation:

$$A(2\tau) = A(0) \exp(-2\tau/T_2)$$
$$\times \exp[-D(\gamma g \delta)^2 (\Delta - \delta/3)] \quad [2]$$

The experiment is always made at a fixed value of τ, so as to keep the T_2-attenuation effects constant. The main reasons for using pulsed gradients at the time were primarily to allow a variation of the observation timescale of the experiment [here = $(D - \delta/3)$], to provide more experimental flexibility (in that Δ, δ and g can readily be varied over a large range of

values), and to detect the echo in a relatively homogeneous magnetic field, thereby avoiding the necessity of using the (noisy) broadband spectrometer electronics needed for the detection of very sharp echoes. Fourier transform techniques in NMR that were developed some years later made it possible to make a frequency separation of the components of the echo (FT-PGSE), so as to provide a direct pathway to multicomponent self-diffusion measurements (**Figure 3A**). Of course, the homogeneous magnetic field during the PGSE detection period is an essential component in this context.

The T_2 relaxation is the main limiting factor of the magnetic field gradient experiment, especially when studying slow diffusion. Unfortunately there is an intrinsic correlation between short T_2 values and low D values in a given system. For this reason, the PGSE technique is not normally applicable to solids, and it is often also inapplicable to most heteronuclei. The overwhelming majority of studies are made on protons in solution.

In macromolecular systems in solution (where T_1 is often much longer than T_2) one can often lessen the effects of T_2 relaxation by instead applying the three RF-pulse stimulated echo (STE) experiment (normally performed with three 90° pulses, at times 0, τ_1 and τ_2 – the stimulated echo appears at time $\tau_1 + \tau_2$; **Figure 2B**. Here the echo attenuation is given by

$$A(\tau_1 + \tau_2) = \tfrac{1}{2} A(0) \exp[-(\tau_2 - \tau_1)/T_1]$$
$$\times \exp(-2\tau_1/T_2) \exp[-D(\gamma g \delta)^2 (\Delta - \delta/3)] \quad [3]$$

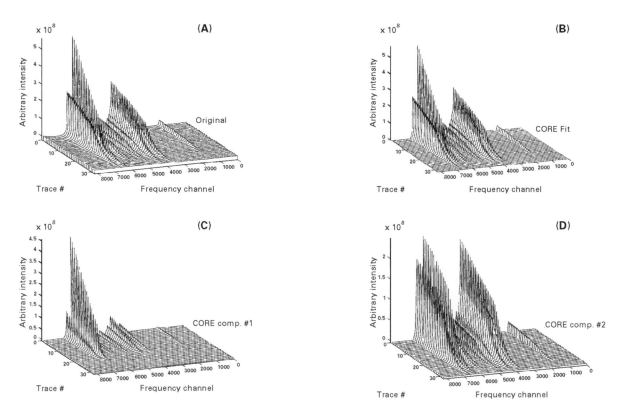

Figure 3 (A) A typical proton FT-PGSE data set from a complex solution sample (gelatin and two surfactants in D_2O, at 300 MHz). The experiment was based on a stimulated-echo PGSE sequence and made at fixed values of τ, Δ and δ, while varying the gradient current. The residual proton water peak is suppressed already at the lowest gradient setting. Figures (B)–(D) illustrate the application of a CORE analysis of this data set. CORE stands for COmponent REsolved spectroscopy. It is a computational approach that applies a global least-squares analysis to the whole data set, using the prior knowledge that component band shapes remain constant (see text). Band shape regions that are below 2% of the maximum intensity were omitted from the analysis. Two band shape components (gelatin and a common surfactant one) are found, which suggests that both surfactants diffuse at the same rate in this system and indicates the formation of a common aggregate.

One should note that T_2 relaxation can be effectively suppressed, since it only occurs between the two first RF pulses (the spacing of which only needs to be slightly longer than the duration of the gradient pulse) and after the third. However, there are further advantages in using the STE experiment, especially in homonuclearly coupled spin systems. By using a short first RF pulse interval one will also suppress effects of J-modulation. This is most often very signal-enhancing since for complex spin systems J-modulation has effectively the same attenuation effect as very efficient T_2 relaxation. In addition, short-τ_1 FT-PGSTE spectra resemble normal absorption spectra, since the period during which J-modulation evolves is made small compared with $1/J$, where J represents the magnitude of the homonuclear spin coupling.

PGSE methods are the overwhelmingly dominant modes for measuring self-diffusion by NMR. It should be mentioned, however, that an RF field gradient analogue of the magnetic field gradient experiment has also been developed, but its use has not yet become widespread.

Other NMR techniques may be useful in certain situations, but are normally less powerful, or suffer from problems with regard to interpretation. Relaxation by paramagnetic species and intermolecular spin relaxation of protons, for example, are significantly dependent on translational diffusion rates. Separating these effects from other relaxation paths and translating them into self-diffusion data of the components is never straightforward.

Direct imaging of relaxation of the diffusional profile of a concentration gradient by NMR microscopy methods is sometimes a useful alternative approach to PGSE. This is especially true in heterogeneous systems, where basic PGSE techniques may fail owing to rapid T_2 relaxation. One should note, however, that only one component can normally be imaged in a single experiment and that the quantity measured is mutual diffusion, rather than self-diffusion. The mutual diffusion coefficient is not directly translatable into a molecular displacement in space – rather it relates to the relative motions of the different components in the system.

Recent methodological developments

In just a decade or so, the performance level of PGSE instrumentation has increased by roughly two orders of magnitude with regard to most aspects, and the actual hardware has become commercially available almost to the level of a standard spectrometer option. Consequently, PGSE methodological development and experimental application have greatly increased. To a large extent this is coupled to parallel developments with regard to field gradient equipment in NMR microscopy and imaging, and to the development of pulsed field gradient methods for coherence selection and solvent suppression in high-resolution NMR.

Required equipment for high-performance PGSE on modern pulsed NMR spectrometers is a probe with self-shielded field gradient coils (i.e. having an auxiliary coil of opposite polarity – wound in such a way so as to cancel out the pulsed field gradients outside the actual sample volume) and a high-performance current-regulated gradient driver that can produce subsequent gradient pulses that match at the ppm level with regard to area/shape. Present-day PGSE instrumentation is often capable of producing high-resolution FT-PGSE spectra at maximum gradient settings of 100–1000 Gauss cm^{-1} (1–10 T m^{-1}). This is typically made at maximum gradient current levels of 10–40 A, applied during 1–10 ms.

A general problem with PGSE measurements is that rapid switching of high gradient currents create eddy currents in surrounding metal (probe housing, wires, supercon magnet bore) near the gradient coils. These effects, which last of the order of milliseconds, create disruptive time-dependent magnetic field gradients shortly after gradient switching. They can greatly disturb the experiment, but are dramatically suppressed by the above mentioned use of self-shielded gradient coils. Further suppression can be achieved by selecting a larger-bore supercon magnet type, or probe housing metals other than aluminium (e.g. nonmagnetic stainless steel). Notably, eddy current problems are almost completely absent in PGSE measurements made in resistive iron magnet geometries.

Eddy current suppression can also be achieved by using shaped (ramped) gradient pulse shapes (i.e. with much lower rise/fall times than for 'rectangular' pulses). One should then note that the functional form of the Equations [2] and [3] assumes the use of rectangular gradient pulses and that the appropriate actual analytical expressions may be considerably more complicated if this is not the case. However, when the separation of the gradient pulses (Δ) is very large compared with their duration (δ) the expressions converge to a situation where only the pulse area matters. It can then merely be substituted as a 'corrected' quantity into, for example, Equations [2] and [3]. Two other remedies for the eddy current problem have also been suggested and used: (i) the so-called LED (longitudinal eddy delay) pulse sequence, where a further 90° pulse is appended to the experiment, so as to 'store away' the magnetization in the z direction (where it is unaffected by magnetic field gradients) until the magnetic field gradients from eddy currents have decayed. A further 90° pulse then reads back the longitudinal magnetization to create the echo in the transverse plane (ii) the 'bipolar pulse pair', where two gradient pulses of equal strength, but of opposite polarity, are sandwiched closely around a 180° degree RF pulse. For large gradient pulse separations this pair (for which eddy current effects cancel out to first approximation) effectively corresponds to a single (monopolar) gradient pulse of twice the length. Both approaches require appropriate RF phase cycling to work. For the LED pulse appendix to the stimulated echo sequence, no less than 32 cycling steps are needed. In the author's experience neither of these two latter methods are normally even needed nor recommendable if one uses a properly designed shielded gradient coil probe in a wide-bore magnet.

Considerable progress has also recently been made with regard to adequate data evaluation in the FT-PGSE experiment. The key point is that the entire band shape of a given component attenuates exactly the same amount with regard to the pertinent field gradient parameters in Equations [2] and [3], provided the experiment is made under conditions of a constant RF pulse interval setting. Consequently, a global minimization approach using this prior knowledge will result in direct extraction of the respective band shapes of components in the composite FT-PGSE data set [**Figure 4A–4D**]. The main advantages of this approach is that all the information in the data set is accounted for. Therefore the effective signal-to-noise ratio of the experiment increases an order of magnitude as compared with an evaluation based on peak heights alone. It also allows confident application of the FT-PGSE method to multicomponent systems with extensive NMR signal overlap and closely similar component self-diffusion rates. Within the powerful computing environment of present-day NMR spectrometers this appears to be the natural way to proceed, even for routine work.

It has recently become popular to present results in a 2D manner, with spectral information on the x and z axes (frequency and intensity, respectively) and diffusion information on the y axis. This display mode for FT-PGSE results has been named DOSY (diffusion-ordered spectroscopy). Its introduction

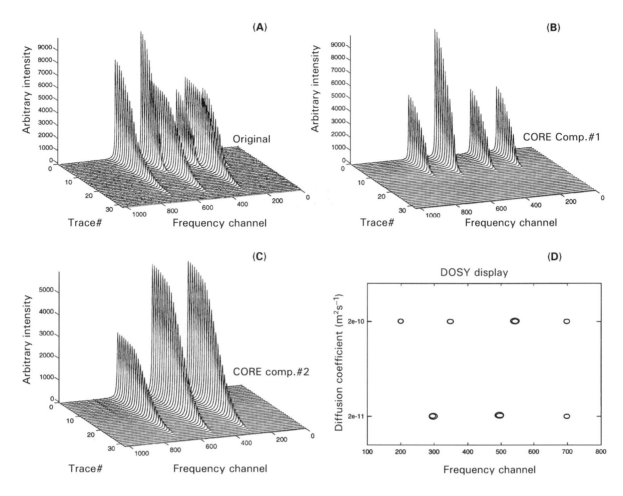

Figure 4 (A) A synthetic FT-PGSE data set from a sample with two components that diffuse at rates of 2.0×10^{-10} and 2.0×10^{-11} m² s⁻¹ and (B,C) its CORE analysis. (D) The corresponding DOSY display, as based on the CORE analysis. In the diffusion direction, the width of the DOSY peak will reflect either or both of the error limits of the D value and the width of the D value distribution, as a result of, for example, the polydispersity of the component in question.

has, unfortunately, caused considerable confusion in the literature and the unjustified implication of new types of experiments. One should note that a 'DOSY-display' is most often just based on interpolated individual single-frequency FT-PGSE data (or variants thereof) that are evaluated without taking account of prior knowledge about constant component band shapes with echo attenuation level. Nonetheless, although the DOSY display mode does not add any new information, it is appealing to the eye and has therefore undoubtedly stimulated the use and development of FT-PGSE-based techniques.

Finally one should note that two new variants of the original (static gradient) spin-echo experiment have recently been found to be quite useful, especially for systems where very slow diffusion and very short spin–spin relaxation rates prevail. One is based on the fact that commonly available superconducting magnets produce a dipolar-like stray field that has a very strong, stable and linear field gradient (at a lower than nominal, but still quite strong, magnetic

field) a few 10 cm above or below the normal sample position. Thus, by pulling the probe out to the appropriate location one can monitor diffusion through the echo attenuation. In principle, this is possible on most modern supercon spectrometers. The notation STRAFI is often used for this type of stray field setup. The second variant is still quite unusual, and is based on sandwiching two supercon magnets in an 'anti-Helmholz' fashion. A preferred notation for both experimental setups is supercon-fringe field steady-gradient spin-echo (SFF-SGSE). One should note that this technique is not limited to protons–multinuclear measurements (e.g. protons and ⁷Li) are straightforward in the same multinuclear probe. The intrinsic problems with SGSE experiments are (a) under the influence of a strong magnetic field gradient the RF pulses only excite a thin slice of the sample – the echo will therefore be quite weak and (b) actual echo detection is made in an inhomogeneous magnetic field, therefore the signal is very sharp and Fourier transformation to separate its different components is not possible. In addition, the spectrometer must have broadband detection electronics, data acquisition filters disabled and a fast ADC to properly detect and sample the echo spike.

Applications of NMR self-diffusion measurements

Self-diffusion data is intrinsically a very direct source of information. For unrestricted diffusion during a selected time span, the characteristic self-diffusion coefficient (D) simply translates into a mean square displacement in space ($\langle \Delta r^2 \rangle$) during the observation time (Δt) through the Einstein relation: $\langle \Delta r^2 \rangle = 6D\Delta t$. One should note that under normal measurement conditions of PGSE techniques the mean square displacement of even large macromolecules during the characteristic time span of the experiment is generally very much larger than the average macromolecular diameter. Therefore the quantity ($\langle \Delta r^2 \rangle$) normally requires no further modelling and D values thus become very easy to interpret.

Diffusion in simple liquids

The NMR FT-PGSE approach is the only method that can provide quantitative information on multicomponent self-diffusion. Such data are central as a reference for, for example, theories of simple liquids.

Binding studies

In binding studies the self-diffusion approach typically relies on a relative comparison of time-averaged self-diffusion rates between a 'bound' and 'free' state, meaning that in the (simplest two-state situation the effective self-diffusion coefficient $D(\text{obs})$ will be given by $D(\text{obs}) = p\, D(\text{bound}) + (1-p)\, D(\text{free})$ where the degree of binding (p) may assume values in the range $0 < p < 1$. This equation can be rewritten into the form $p = [D(\text{free}) - D(\text{obs})]/[D(\text{free}) - D(\text{bound})]$. Hence, such experiments do provide the information needed to obtain binding isotherm information, for example, by a simple comparison of experimentally available self-diffusion coefficients. It is evident that the method is most suited for binding/aggregation of small molecules to larger entities and is most accurate at intermediate p values. Numerous such studies have been presented.

Non-Gaussian diffusion in macromolecular systems

One should note that the Einstein relation $\langle \Delta r^2 \rangle = 6D\Delta t$ only holds in the limit of large Δt.

'Large', in this context, has a different meaning for small molecules and macromolecules. Uhlenbeck and Ornstein derived a generalized relation, in their classical treatment of the subject back in the 1930s:

$$\langle \Delta r^2 \rangle = 6D\{\Delta t - \tau_c(1 - \exp[1 - \Delta t/\tau_c])\}$$

where the correlation time $\tau_c = m/f$, f being the coefficient of friction and m the mass of the Brownian particle. For small molecules in liquids the correlation time is of the order of a nanosecond, while the shortest imaginable NMR spin-echo experiments will exceed a fraction of a millisecond, but are more typically tens or hundreds of milliseconds. For macromolecules, however, this correlation time may be such that it affects (and thus becomes measurable by) the typical NMR PGSE experiment, when using appropriately selected measurement parameters (strength, duration and separation of gradient pulses). Only a few studies along these lines have appeared to date.

Non-Gaussian diffusion in heterogeneous systems

In porous and heterogeneous systems, penetrant diffusion is not necessarily isotropic or Gaussian on a given timescale. PGSE techniques have been used since the early 1960s for characterizing such aspects of system morphology through the effects of non-Gaussian diffusion on the echo attenuation in the PGSE experiment. Restricted diffusion manifests itself in a variation of the apparent diffusion coefficient with the observation time (i.e. $\Delta - \delta/3$ in the PGSE experiment). A large number of system geometries and experimental conditions have been considered and analysed. Several complex pulse sequences have also been suggested for eliminating the effect of local magnetic field gradients that arise from susceptibility variations in the sample.

Another important recent development of experimental PGSE techniques applied to heterogeneous systems has been the discovery and utilization of so-called diffraction effects with regard to the attenuation of the echo, i.e. weak echo level maxima and minima that occur at high attenuation levels. In a system of spherical pores, for example, the system geometry manifests itself in a first echo maximum when the inverse of the pore diameter matches the gradient setting $\gamma g\delta/2\pi$. Mainly methodological work has been presented so far. Notably, the technique has also found use for the investigation of cell suspensions such as, for example, erythrocytes.

List of symbols

D = self-diffusion coefficient; g = applied gradient strength, amplitude; J = magnitude of homonuclear spin coupling; T_1 = longitudinal relaxation time; T_2 = irreversible spin–spin relaxation time, transverse relaxation time; δ = duration of magnetic field gradient pulse; Δ = separation of magnetic field gradient pulses; γ = magnetogyric ratio; τ = time delay between RF pulses.

See also: **Magnetic Field Gradients in High Resolution NMR; MRI Applications, Biological; MRI Applications, Clinical; MRI Theory; MRI Using Stray Fields; NMR Microscopy; NMR Pulse Sequences.**

Further reading

Callaghan PT (1991) *Principles of Nuclear Magnetic Resonance Microscopy.* Oxford: Clarendon Press.

Callaghan PT and Coy A (1994) PGSE and molecular translational motion in porous media. In: Tycko P (ed) NMR *Probes of Molecular Dynamics*, pp 489–523. Dordrecht: Kluwer Academic.

Canet D (1997) Radiofrequency field gradient experiments. *Progress in Nuclear Magnetic Resonance Spectroscopy* 30: 101–135.

Hahn EL (1950) Spin echoes. *Physics Review* 80: 580–594.

Johnson CS (Jr) (1994) Transport ordered 2D-NMR spectroscopy. In: Tycko R (ed) NMR *Probes of Molecular Dynamics*, pp 455–488. Dordrecht: Kluwer Academic.

Kärger J, Pfeifer H and Heink W (1988) Principles and application of self-diffusion measurements by nuclear magnetic. In: Waugh JS (ed) *Advances in Magnetic Resonance*, pp 2–89. San Diego: Academic Press.

Kimmich R (1997) NMR *Tomography Diffusometry Relaxometry*. Berlin, Heidelberg, New York: Springer.

Lindblom G and Orädd G (1994) NMR studies of translational diffusion in lyotropic liquid crystals and lipid membranes. *Progress in Nuclear Magnetic Resonance Spectroscopy* 26: 483–515.

Norwood NJ (1994) Magnetic field gradients in NMR – friend or foe. *Chemical Society Reviews* 23: 59–66.

Price WS (1996) Gradient NMR. In: *Annual Reports on NMR Spectroscopy*, pp 51–142. New York: Academic Press.

Price WS (1997) Pulsed-field gradient nuclear magnetic resonance as a tool for studying translational diffusion. Part 1. Basic theory. *Concepts in Magnetic Resonance* 9: 299–336, and Part 2 Experimental aspects. *Concepts in Magnetic Resonance* 10: 197–237.

Söderman O and Stilbs P (1994) NMR studies of complex surfactant systems. *Progress in Nuclear Magnetic Resonance Spectroscopy* 26: 445–482.

Stilbs P (1987) Fourier transform pulsed-gradient spin-echo studies of molecular diffusion. *Progress in Nuclear Magnetic Resonance Spectroscopy* 19: 1–45.

Waldeck AR, Kuchel PW, Lennon AJ and Chapman BE (1997) NMR diffusion measurements to characterize membrane transport and solute binding. *Progress in Nuclear Magnetic Resonance Spectroscopy* 30: 39–68.

Drug Metabolism Studied Using NMR Spectroscopy

Myriam Malet-Martino and **Robert Martino**,
Université Paul Sabatier, Toulouse, France

MAGNETIC RESONANCE
Applications

The purpose of this article is to briefly review the scope and potential of NMR as a non-invasive technique for studying drug metabolism. NMR has been used for the study of drug metabolism *in vitro* (biofluids, tissue extracts, biopsies, intact cells, perfused organs) and *in vivo* (animal models and human subjects). NMR spectroscopy is unique in its ability to permit both these kinds of studies. We shall see that the advantages and limitations of NMR as a tool for drug metabolism and disposition studies are different depending on the approach chosen (*in vitro* or *in vivo* studies). The first part of this article deals with the various isotopes used in such studies. The second part is a general discussion of the value and difficulties encountered with NMR in investigations on drug metabolism. In the third part, we have chosen some specific examples to emphasize the interest of NMR.

Isotopes used in the study of the metabolism of drugs by NMR

There are several NMR active nuclei (**Table 1**) that can be routinely used in drug metabolism and disposition. The merits and disadvantages of some of these nuclei are discussed below.

Fluorine-19 (^{19}F)

The majority of studies in the literature concern ^{19}F NMR mainly because of the favourable NMR characteristics of this nucleus, including a nuclear spin of $\frac{1}{2}$, relatively narrow lines, 100% natural abundance, high sensitivity (83% that of proton), large chemical shift range and short longitudinal relaxation times (T_1) which permit rapid pulsing with a corresponding improvement in the signal-to-noise ratio per unit time. Moreover, the negligible level of endogenous fluorine eliminates interfering background signals and dynamic range problems. As a number of fluorinated drugs are currently in clinical use, ^{19}F NMR offers a powerful method of monitoring their pharmacokinetics and metabolism either *in vitro* or *in vivo*. The main requirement is that drugs and metabolites should be present in minimal concentrations, near 10^{-3} mM, for studies in biofluids or 5×10^{-2} mM for *in vivo* studies.

Proton (^{1}H)

The ^{1}H nucleus is present in all drugs, and has the highest NMR sensitivity of any stable nucleus. Moreover, it is the most abundant (99.98% natural abundance) of the two natural isotopes (^{1}H and ^{2}H or deuterium) of the hydrogen atom. However, the small chemical shift range and the extensive multiplicity due to homonuclear *J*-coupling sometimes make the observation and quantification of drug metabolites difficult. Moreover, ^{1}H NMR spectra of biofluids or tissues contain many intense resonances from water, proteins and lipids that must be reduced or eliminated. A simple method for water suppression (which nevertheless implies a treatment of the biological sample) consists in freeze-drying the sample and redissolving it in D_2O. Several NMR techniques are used to achieve the suppression of water and macromolecule resonances. The form of water suppression most commonly used is presaturation of the H_2O resonance using a secondary irradiation field at the solvent resonance frequency. A second approach takes advantage of the short T_2 (transverse relaxation time) of tissue water, proteins and some fats, and consists in applying a spin-echo sequence that will suppress the broad resonances of water and macromolecules.

Presaturation and spin-echo sequences can also be combined. Despite these drawbacks, several studies have shown that this nucleus can be exploited for the detection and measurement of drugs and metabolites in biofluids, i.e. for *in vitro* studies. The main requirements are that metabolites should have suitable protons and be present at concentrations 0.01–0.1 mM. As ^{1}H is a poor nucleus for the *in vivo* monitoring of drugs and metabolites, *in vivo* ^{1}H NMR spectroscopy has so far had little application. The reasons for this are manifold. Because of factors such as tissue heterogeneity, restricted molecular mobility, magnetic field inhomogeneity over the relatively large sample volume and the use of lower magnetic field strength spectrometers, *in vivo* signals linewidths are substantially broader than those obtained *in vitro*, in biofluids for example. There is thus excessive signal overlap due to the very large number of signals that occupy a small chemical shift range and the signals from endogenous metabolites may obscure the detection and quantification of the drug and/or its metabolites. Moreover, the replacement of H_2O by D_2O becomes impractical for studies carried out on whole animals or humans since high concentrations of D_2O are toxic and D_2O is very expensive. More elaborate NMR techniques than those reported above are thus required for solvent suppression and spectral editing.

Carbon-13 (^{13}C)

One advantage of ^{13}C NMR is that its large chemical shift range enables reliable structural identification. However, although the carbon nucleus is found in all drugs, the magnetically active isotope (^{13}C) is neither abundant nor particularly sensitive. These drawbacks combined with the low concentrations of drugs and their metabolites in biological systems make this nucleus in its natural abundance rather hard to detect. Improved sensitivity can be attained with polarization transfer experiments which partly transfer the favourable properties of the proton to ^{13}C. Furthermore, ^{13}C-enriched drugs can be synthesized to increase the NMR sensitivity. However, to obtain a full picture of the chemical changes taking place, several key positions usually need to be labelled. This may be chemically difficult, presupposes an *a priori* knowledge of the metabolism of the particular drug and is rather expensive in practice. It should be stressed that the label still needs to be present at millimolar concentrations, rather than the trace quantities that are typically required for radiolabelling studies. Because of these difficulties, the use of ^{13}C for *in vitro* or *in vivo* NMR studies of drug metabolism and disposition has been very limited.

Table 1 NMR isotopes for drug metabolism studies

Isotope	Spin	Natural abundance (%)	NMR receptivity[a]	Chemical shift range (ppm)	Minimum concentration (mM) for in vitro studies[b]	Minimum concentration (μmol g⁻¹) for in vivo studies
^1H	1/2	99.98	100	15	0.01	0.1
^2H	1	0.016	0.00015	15	1.2	1.5
^7Li	3/2	92.58	27	12		0.1
^{10}B	3	19.58	0.389	200		2[c]
^{11}B	3/2	80.42	13.3	200		5–10
^{13}C	1/2	1.11	0.018	250	0.1–1[d]	50[d]
^{14}N	1	99.63	0.1	350	1[e]	
^{15}N	1/2	0.37	0.00038	350	1[d]	
^{17}O	5/2	0.037	0.0011	2500		
^{19}F	1/2	100	83.3	500[f]	0.001–0.005	0.05–5
^{31}P	1/2	100	6.6	800 (30)[g]	0.01	0.2–1
^{195}Pt	1/2	33.8	0.34	10 000	10	

[a] Receptivity (D) relative to proton (H) is a relative sensitivity figure used to compare signal areas theoretically obtainable at a given magnetic field strength: $D_X = [\gamma_X{}^3 N_X I_X (I_X+1)] / [\gamma_H{}^3 N_H I_H (I_H+1)] \times 100$ where γ is the magnetogyric ratio, N the natural abundance and I the spin quantum number.

[b] In biofluids.

[c] With an enriched drug and if an indirect method of observing protons coupled to the ^{10}B nuclei is employed.

[d] With an enriched drug.

[e] Only nitrogen centres with relatively symmetrical electron distributions can be studied.

[f] For organofluorine compounds.

[g] Chemical shift range of endogenous phosphorus compounds.

Phosphorous-31 (^{31}P)

The ^{31}P nucleus is readily detected as it has a 100% natural abundance and is relatively sensitive. The presence of endogenous phosphates and derivatives (phosphomonoesters, phosphodiesters, etc.) may interfere with signals from phosphorated drugs and their metabolites. In practice, this is not a large obstacle as there are relatively few endogenous compounds that produce detectable signals. In the studies carried out to date, no interference in NMR signals between phosphorus-containing drugs or metabolites with endogenous compounds, apart from inorganic phosphate, have been described. As phosphorus-containing drugs are fairly rare, only a few ^{31}P NMR drug studies have so far been carried out.

Lithium-7 (^7Li)

The major lithium isotope, ^7Li, has a high NMR sensitivity. Since ^7Li is not a spin $I = \frac{1}{2}$ but a spin $I = \frac{3}{2}$ nucleus, it possesses a quadrupole moment, which gives rise to broad spectral lines. ^7Li NMR is not hampered by interference with endogenous signals. The only studies to date involve the *in vivo* determination of the pharmacokinetic profile of lithium salts.

Boron-10 and boron-11 (^{10}B, ^{11}B)

NMR is a potentially valuable technique for evaluation of boron neutron capture therapy (BCNT)

agents since both boron isotopes are magnetically active. The ^{11}B isotope has better sensitivity (16.5 versus 2% relative to ^1H) and higher natural abundance (80.4 versus 19.6%) than ^{10}B. Even though ^{10}B is the active nucleus for neutron capture in BCNT, ^{11}B is more appropriate for NMR studies except if an indirect method of observing protons coupled to the ^{10}B nuclei is employed.

Other nuclei

Deuterium (^2H), tritium (^3H), ^{17}O, ^{15}N, ^{14}N and ^{195}Pt suffer from several problems which cannot be overcome easily, including low sensitivity (^{17}O, ^{15}N, ^{14}N, ^2H, ^{195}Pt), poor resolution and broad signals (^{17}O, ^{14}N, ^2H), and the radioactivity (^3H) associated with the relatively high concentrations required for NMR studies. The sensitivity of detection can be improved by isotopic enrichment combined with enhancement by polarization transfer from proton. Nevertheless, all these problems are reflected in the limited use of these nuclei in NMR studies both *in vitro* or *in vivo*.

Advantages and limitations of NMR for drug metabolism studies

In vitro studies

Many analytical techniques, especially chromatographic methods, are in routine use for the analysis of biofluids. The reasons for the additional use of NMR spectroscopy are manifold and lie in the unique properties of NMR spectroscopy (**Table 2**).

NMR is non-destructive and enables the direct study of any intact biofluid without resort to treatment. Problems of extraction recovery and chemical derivatization, and those that may be encountered with pH-sensitive metabolites, are consequently avoided. NMR spectroscopy is thus particularly suited to the study of delicate samples.

The method is non-specific and unexpected substances are not overlooked during the investigation, since all metabolites (provided they bear the nucleus under observation and they are present at sufficient concentrations) are detected simultaneously in a single analysis. This contrasts with chromatography which usually requires some prior knowledge of the structures of metabolites to optimize sample preparation and/or detection. Novel metabolites may therefore be missed. NMR also avoids the use of a number of different chromatographic techniques, which is sometimes necessary when metabolites have different chemical structures. This is an important attribute of NMR in the search for novel metabolites when, often, the analyst will have no idea of the type of molecule to look for.

The NMR spectrum contains much structural information, and the observed chemical shifts and spin–spin coupling patterns can provide valuable clues about the structure of novel metabolites. Even though metabolites must still be extracted and purified for unequivocal elucidation of the structure, the NMR 'behaviour' nevertheless can give a good estimate of the structures of unknown compounds.

Although optimization of the quantification of drugs and metabolites in biofluids by NMR is somewhat tedious [choice of a suitable standard, determination of the T_1, checking that excitation is uniform over the whole frequency range when a large spectral width is observed (often the case in ^{19}F and ^{13}C

NMR), taking into account the heteronuclear nuclear Overhauser effect when proton decoupling is applied, validation of the assay, etc.] it is nevertheless fairly easy to obtain quantitative data, and NMR can be used routinely in the same way as HPLC. Moreover, given a good signal-to-noise ratio, NMR can quantify substances accurately and reproducibly. Quantification in plasma and bile may nevertheless be affected by the binding of drugs or metabolites to macromolecules. This results in substantial signal broadening and gives a reduced signal-to-noise ratio or even leads to NMR-invisibility. Some sample pretreatment may be required in such cases.

Set against these advantages, however, are a number of disadvantages which need to be taken into account.

NMR spectrometers are expensive instruments, with the cost rising steeply with the field strength of the magnet used; higher field strengths provide greater spectral dispersion and yield better sensitivity. However, the magnets have long useful lifetimes (~ 15 years) and the electronic components may be upgraded continuously at reasonable cost.

Although there is little or no spectral interference from endogenous molecules using ^{19}F, ^{31}P, ^7Li and ^{13}C (since ^{13}C-enriched drugs are required) NMR, the problem of peak overlap is critical in ^1H NMR since the signals of a large number of endogenous compounds are present in a relatively narrow chemical shift range. Moreover, suppression of the intense signal from water is a prerequisite for ^1H NMR spectroscopy of biofluids. Resonances of the drug and metabolites must occur in spectral regions which are free of signals from endogenous metabolites. With a spectrometer operating at a field $\geq$ 9.4 T, and taking the necessary precautions, drug metabolite resonances may be measured on the entire spectrum, except between 3.5 and 4.1 ppm as this range contains many overlapped resonances from endogenous molecules. In most biofluids, the region of the

Table 2 Advantages and drawbacks of *in vitro* NMR for drug metabolism studies

Advantages	Drawbacks
No destruction of the sample	NMR spectrometers are expensive instruments
No sample pretreatment necessary	Critical problem of peak overlap in ^1H NMR. Resonances of the drug and metabolites must occur in spectral regions which are free of signals from endogeneous metabolites
No degradation of unstable metabolites due to sample treatment	
All metabolites are detected simultaneously in a single analysis	
Does not require preselection of conditions based on a knowledge of the compound to be analysed	The volume of biofluid needed for NMR analysis may be large
An analytical technique does not have to be designed for each newapplication	NMR is relatively insensitive; this limits its use to the observation of drugs that are present at relatively high concentrations
NMR gives both structural and quantitative information	
Little or no spectral interference from endogenous molecules in ^{19}F, ^{31}P, ^7Li and ^{13}C NMR	

spectrum deshielded from the water signal is relatively low in interfering resonances, so that aromatic or heterocyclic drugs can often be studied with relative ease. The positive aspect of this limitation is that information can be obtained on alterations in composition of the biofluid after drug administration since the ^{1}H NMR spectrum of a biofluid gives a characteristic pattern of resonances for a range of important endogenous metabolites.

The volume of biofluid needed for NMR analysis ranges between 0.3 to 0.4 or 1.5 to 2.5 mL with 5 or 10 mm tubes respectively. This can be a hindrance for pharmacokinetic studies which require numerous plasma samples or for difficult-to-obtain biofluids such as bile, cerebrospinal fluid (CSF) or those from neonates. Recent progress in the use of flow-injection NMR probes will alleviate this disadvantage.

Compared with most chromatographic and other spectroscopic techniques, NMR is relatively insensitive, which represents the principal drawback of the technique and limits its use to the observation of those drugs that are present at relatively high concentrations. It is difficult to give a precise figure for the minimum concentration requirement because there are many factors that influence the signal-to-noise ratio. The theoretical detection limit for ^{1}H NMR is around 10^{-3} mM for 600 MHz spectrometers using a 5 mm probe, but in practice it is often difficult to accurately quantify substances found at concentrations below 0.1 mM. With one currently available spectrometer (7 T, corresponding to a proton resonance frequency of 300 MHz) the detection threshold in a 10 mm diameter tube is 5×10^{-3} mM and 10^{-2} mM for 10 to 15 h of ^{19}F NMR or ^{31}P NMR recording. With a 11.7 T spectrometer (corresponding to a proton resonance frequency of 500 MHz) a detection threshold of 1 to 2×10^{-3} mM has been achieved in 12 h of ^{19}F NMR data acquisition in 10 mm tubes. However, the accuracy and reproducibility of the NMR concentration determinations are generally ≤10% for concentrations ≥5×10^{-2} mM for ^{19}F and ^{31}P NMR. For lower concentrations, the accuracy and reproducibility decrease and depend on the measurement time and the resulting signal-to-noise ratio in the spectra.

However, in spite of these difficulties, NMR spectroscopy has now become a technique of demonstrated utility and of unique ability in the analysis of biofluids. Moreover, the recent development of using separation methods in conjunction with NMR spectroscopy has given rise to a radical new approach in biofluid metabolite analysis. The coupling of HPLC to NMR in 1992 in the form of a commercially available instrument has been shown to be of great use in separating and determining the structure of drug metabolites in biofluids. The sensitivity of conventional HPLC-NMR provides a detection limit of about 150 ng of analyte for an overnight data accumulation. This technique has been extended to other chromatographic techniques such as supercritical fluid chromatography and to the use of nuclei other than ^{1}H or ^{19}F which form the basis of most studies so far because of their high NMR sensitivity. Another major recent development has been the combination of both NMR and mass spectrometry to an HPLC system, which will be of great help to elucidate the structures of drug metabolites in biofluids.

The advantages and limitations of NMR for biofluids also apply to the analysis of excised tissue samples or intact cells packed in NMR tubes. For quantitative studies, the metabolites of interest should not be degraded by enzymes even at 4 °C. For the analysis of cells, one must make sure that the density of the cellular suspension does not change with time, i.e. that there is no sedimentation in the NMR tube. This can be checked by acquisition of successive data blocks. The spectra are then examined individually, and if there is little or no change in metabolite levels, spectra can be added together to improve the signal-to-noise ratio. In such samples, there is often considerable signal broadening owing to inhomogeneities in the medium analysed and signals from compounds with similar chemical shifts may merge into a single peak. The ^{19}F NMR detection threshold at 470 MHz was found to be about 5 nmol g^{-1} of tissue or cellular suspension for 1–2 h measurement on 0.5–1 g samples. Recently, it has been shown that it is possible to obtain high-resolution ^{1}H NMR spectra of biopsy tissues simply by spinning at the magic angle ($\theta = 54.7°$) at modest speeds (around 3 kHz). This could open up new routes of monitoring drug metabolism.

In vivo studies (Table 3)

The main interest of *in vivo* NMR is obvious. Measuring drug levels in plasma does not always reflect the drug concentration at the receptor sites, which are generally located in the tissue cells of the target organ. Consequently, a method that allows one to measure the concentration of drugs and their metabolites in situ may be extremely useful.

In vivo NMR spectroscopy can be used repetitively on the same patient since it is noninvasive. The ability to monitor the pharmacokinetic of a drug and to evaluate the potential modifications in drug metabolism could prove extremely valuable in patient management.

However, there are three main obstacles to the use of NMR *in vivo*. The first and major limitation is

Table 3 Advantages and drawbacks of *in vivo* NMR for drug metabolism studies

Advantages	Drawbacks
Measurement of drug and /or metabolites levels in target organ	Low sensitivity. Tissue concentration necessary for *in vivo* detection of a drug is 1–10 times that necessary *in vitro*
Can be used repetitively on the same patient	Only molecules with unrestricted mobility are visible
	Prolonged data acquisition may be required
	The observed signals may have various anatomical origins
	Poor resolution. Chemically similar compounds may not be readily distinguished
	Difficult in making accurate quantitative measurements

the lack of sensitivity. The minimum detectable concentration (**Table 1**) depends on many factors, including magnetic field strength, intrinsic sensitivity, presence of background signal, data collection time and sample size. Tissue concentration necessary for *in vivo* detection of a drug is typically 1–10 times that necessary *in vitro*. Order-of-magnitude detection limits for endogeneous metabolites are 0.2 μmol g^{-1} for ^{31}P and 0.1 μmol g^{-1} for ^{1}H. The *in vivo* ^{19}F NMR detection limit of 5-fluorouracil (FU) and its metabolites has been estimated at around 0.05 μmol g^{-1} with a 7 T vertical bore magnet for a 1 h measurement and 0.1 μmol g^{-1} with a 1.5 T whole-body spectrometer for total measurement time that is tolerable (30 min) for patients. It should not be forgotten that these detection limits are for molecules with unrestricted mobility, which is not always the case since drug and/or metabolites are often strongly bound to macromolecules or cell membranes.

The insensitivity of NMR has two consequences. First, prolonged data acquisition may be required to obtain satisfactory spectra. This is because the confident detection of resonances requires a good signal-to-noise ratio, which is proportional to the square root of the number of scans performed. The kinetics of drug metabolism or elimination must therefore not be too rapid. Second, the surface coils used for the detection of drugs and metabolites *in situ* are relatively large, particularly in humans. The observed signals may thus have various anatomical origins (e.g tumour tissue and normal tissue). To get a specific pharmacokinetic profile from one tissue, spatially localized spectroscopy using field gradients to define the volume of interest is useful. However, the relatively low concentrations of metabolites encountered have so far limited the use of this method for drug metabolism studies.

The second limitation is the lack of resolution. NMR signals observed *in vivo* are broad, and chemically similar compounds may not be readily distinguished.

The third limitation of *in vivo* NMR is the difficulty in making accurate quantitative measurements. The only quantitative studies of drug metabolism carried out so far have concerned fluorine- or lithium-containing compounds with ^{19}F or ^{7}Li NMR. For quantitative determinations, signal areas are normalized relative to that of an external reference – a capillary or a sphere filled with a standard of known concentration and volume. Within a particular experiment this method can estimate the relative concentrations of metabolites with an accuracy of about 10%, rising to around 40% for the estimation of absolute concentrations. The absolute concentration in human brain of fluorinated neuroleptics or lithium salts has been determined by performing an additional separate calibration experiment (immediately before and after each experiment) with a phantom of known concentration and volume placed in the position of the tissue. However, since the tissue volume is not readily measured and since the loading on the coil and hence the nutation angle are assumed to be the same for the phantom as for the tissue, the method tends to be rather inaccurate. Another method consists in using a calibrated internal standard such as tissue water.

Some recent applications of NMR in drug metabolism studies

Most of the ^{19}F NMR studies are related to anaesthetic, psychoactive and antineoplastic drugs. The distribution of anaesthetics in the brain and the pharmacokinetics of their elimination are still a subject of controversy. The feasibility of *in vivo* ^{19}F NMR studies of halothane in humans has been recently demonstrated. Resonances attributable to this compound were observed up to 90 min after withdrawal of the anaesthetic agent in eight patients.

Fluoxetine (Prozac®) is a widely used antidepressant drug. In patients receiving daily doses of 20 or 40 mg, brain concentration continued to increase well after the clinical effects were evident (2 weeks), seemed to level off after about 6 to 10 months of treatment and was about 20 times the concentration in plasma. It thus appears that there is no correlation between brain concentration and clinical response. A combined *in vivo* and *in vitro* study on several patients showed that the single ^{19}F resonance observed *in vivo* contains, in roughly equal proportion, both the drug and its active metabolite (norfluoxetine) (**Figure 1**). The comparison of *in vivo* and *in vitro* quantifications (1–11 µg of fluoxetine + norfluoxetine per mL of tissue *in vivo* versus 12–19 µg mL^{-1} of tissue in extracts) suggested that a substantial fraction of the drug and its metabolite is not visible *in vivo*.

Most of the ^{19}F NMR studies on fluorinated drugs have concerned FU. This drug is an ideal case for ^{19}F NMR studies for three reasons. High doses are administered to patients. The fluorine atom remains intact during biotransformation. Its metabolism, and more especially its catabolic pathway, leads to compounds that are structurally different from the parent compound. The ^{19}F NMR signals are therefore displayed over a large spectral width with no peak overlap (except for fluoronucleotides *in vivo*). The number of *in vitro* and *in vivo* ^{19}F NMR studies published reflects the contribution of the technique to the knowledge of FU metabolism. The following two examples emphasize the interests in ^{19}F NMR.

FU injected at high doses by continuous intravenous perfusion during 4 to 5 days provoked a cardiac toxicity. For commercial purposes, FU is solubilized in a solution of sodium hydroxide at two different pH (8.6 or 9.2). The ^{19}F NMR analysis of these solutions revealed the presence of about a hundred signals of fluorinated 'impurities', among which fluoroacetaldehyde and fluoromalonic acid semi-aldehyde were identified (**Figure 2**). These products come from the degradation of FU and are formed with time in the basic medium which is indispensable to the solubilization of this antitumour agent. These two compounds are highly cardiotoxic on the isolated perfused rabbit heart model since fluoro-acetaldehyde is metabolized into fluoroacetate which is a violent cardiotoxic and neurotoxic poison. The intensity of the cardiotoxicity of FU commercial solutions, tested on the isolated perfused rabbit heart model, is a function of the pH of the solution, and the preparations at pH 9.2 are less cardiotoxic than the preparations at pH 8.6. Nevertheless, the non-negligible frequency of cardiotoxic accidents observed after injection of the commercial solutions at pH 9.2 was too significant to be explained by the levels of fluoroacetaldehyde and fluoromalonic acid semi-aldehyde found in these preparations. The possibility of an eventual metabolism of FU itself into fluoroacetate has thus been explored. It has long been postulated, without ever being demonstrated, that the main catabolite of FU, α-fluoro-β-alanine, could be transformed into fluoroacetate. Using *in vitro* ^{19}F NMR and the isolated perfused rat liver model, it was found that the last catabolite of FU is not α-fluoro-β-alanine. The metabolism continues and leads to two new catabolites, fluoroacetate and

fluoromalonic acid semi-alcohol (**Figure 3**). The conclusion of this study was twofold. To limit the cardiotoxicity of FU, it would be better to use in clinical treatments a new dosing form of FU which is devoid of degradation products (lyophilized material for example). However, to suppress this toxicity it would be necessary to act on the degradation pathway of FU by blocking it before the formation of α-fluoro-β-alanine.

In vivo ^{19}F NMR has been used to monitor FU metabolism in the liver metastases of colorectal cancer patients. The patients were treated with a continuous low dose intravenous infusion of FU until the point of refractory disease, at which time interferon-α was added with the objective of modulating FU activity. **Figure 4** shows the spectra obtained in a 1.5 T whole body spectrometer with a 16 cm diameter surface coil positioned over hepatic metastases of a patient treated with 300 mg of FU m^{-2} day^{-1}. The first NMR scan (**Figure 4A**) was performed after 1 week of FU infusion. A large signal of catabolite (α-fluoro-β-alanine) and a small resonance from FU were observed. The patient went on to have an objective partial response but later developed progressive disease. At this time (**Figure 4B**), the FU peak had disappeared and the catabolite signal had reduced. The next scan (**Figure 4C**), after a week on interferon-α, showed a new anabolite peak. A week later, the last scan (**Figure 4D**) showed FU and catabolite signals. At this stage the patient had a symptom and tumour marker response and stabilization of tumour size.

The ^{1}H NMR spectrum of a human urine sample from a volunteer who had ingested the anti-inflammatory drug flurbiprofen is shown in **Figure 5A**. The complexity of the spectrum precluded any detailed structural or quantitative analysis. For example, the ability to discern resonances from the aromatic protons of flurbiprofen metabolites is severely limited. However, a stopped-flow HPLC-NMR experiment at 30.5 min afforded the ^{1}H spectrum shown in **Figure 5B** which corresponds to the two diasteromeric forms of the β-D-glucuronic acid conjugate of 4'-hydroxyflurbiprofen. The resonances of the *para* substituted phenyl ring appear at 6.91 and 7.42 ppm and the other three aromatic resonances as a complex multiplet at 7.19 ppm. The β-D-glucuronic acid H1 proton appears as two resonances at 5.49 and 5.52 ppm. The remaining β-D-glucuronic acid resonances are located between 3.4 and 3.6 ppm. The methyl resonances of the propionic acid sidechain of flurbiprofen are two doublets at 1.49 and 1.51 ppm. The two methine signals are observed at about 3.94 ppm as overlapped multiplets. This example emphasizes the interest in HPLC coupled to NMR spectroscopy for the identification of drug metabolites.

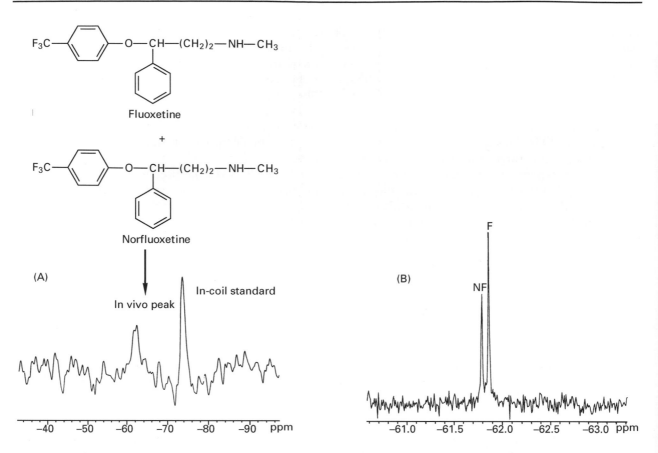

Figure 1 The ^{19}F NMR spectra (A) from the head of a patient on 20 mg per day of fluoxetine for 1 month, (B) of the extract of a small section of temporal cortex from a deceased patient on 40 mg per day of fluoxetine for 19 weeks. F = fluoxetine, NF = norfluoxetine. Adapted with permission of Williams and Wilkins from Komoroski *et al* (1994) *In vivo* ^{19}F spin relaxation and localized spectroscopy of fluoxetine in human brain. *Magnetic Resonance in Medicine* **31**: 204–211.

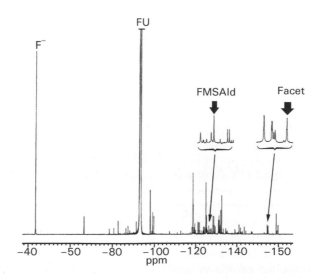

Figure 2 The ^{19}F NMR spectrum of a commercial solution of 5-fluorouracil (pH 9.2). F⁻, fluoride ion; FU, 5-fluorouracil; FMSAld, fluoromalonic acid semi-aldehyde; Facet, fluoroacetaldehyde.

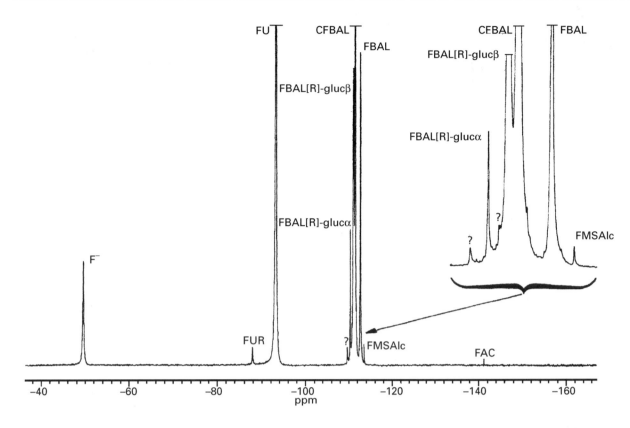

Figure 3   The ^{19}F NMR spectrum of a perfusate (concentrated 20-fold by lyophilization) from an isolated perfused rat liver treated with 5-fluorouracil at a dose of 45 mg per kg b.w. for 3 h. F$^-$, fluoride ion; FUR, 5-fluorouridine; FU, 5-fluorouracil; FBAL, α-fluoro-β-alanine; CFBAL, N-carboxy-α-fluoro-β-alanine; FBAL[R]-glucα, FBAL[R]-glucβ, adducts of α-fluoro-β-alanine with α-glucose or β-glucose; FMSAlc, fluoromalonic acid semi-alcohol; FAC, fluoroacetate.

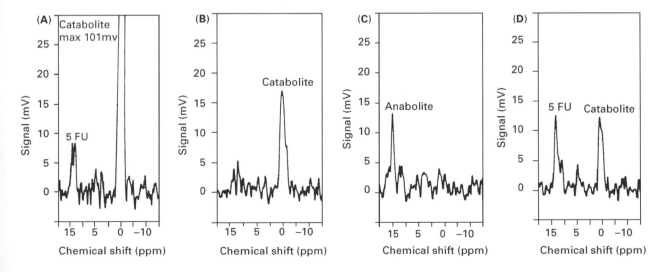

Figure 4   A series of ^{19}F NMR spectra of liver metastases from a colorectal cancer patient treated with 5-fluorouracil at a dose of 300 mg m^{-2} day^{-1} (A, B) and 5-fluorouracil + interferon-α (C, D). 5FU, 5-fluorouracil. mV: millivolts (arbitrary values). Reproduced with permission of Kluwer Academic from Findlay *et al* (1993) The non-invasive monitoring of low dose, infusional 5-fluorouracil and its modulation by interferon α using *in vivo* ^{19}F MRS in patients with colorectal cancer. *Annals of Oncology,* **4**: 597–602.

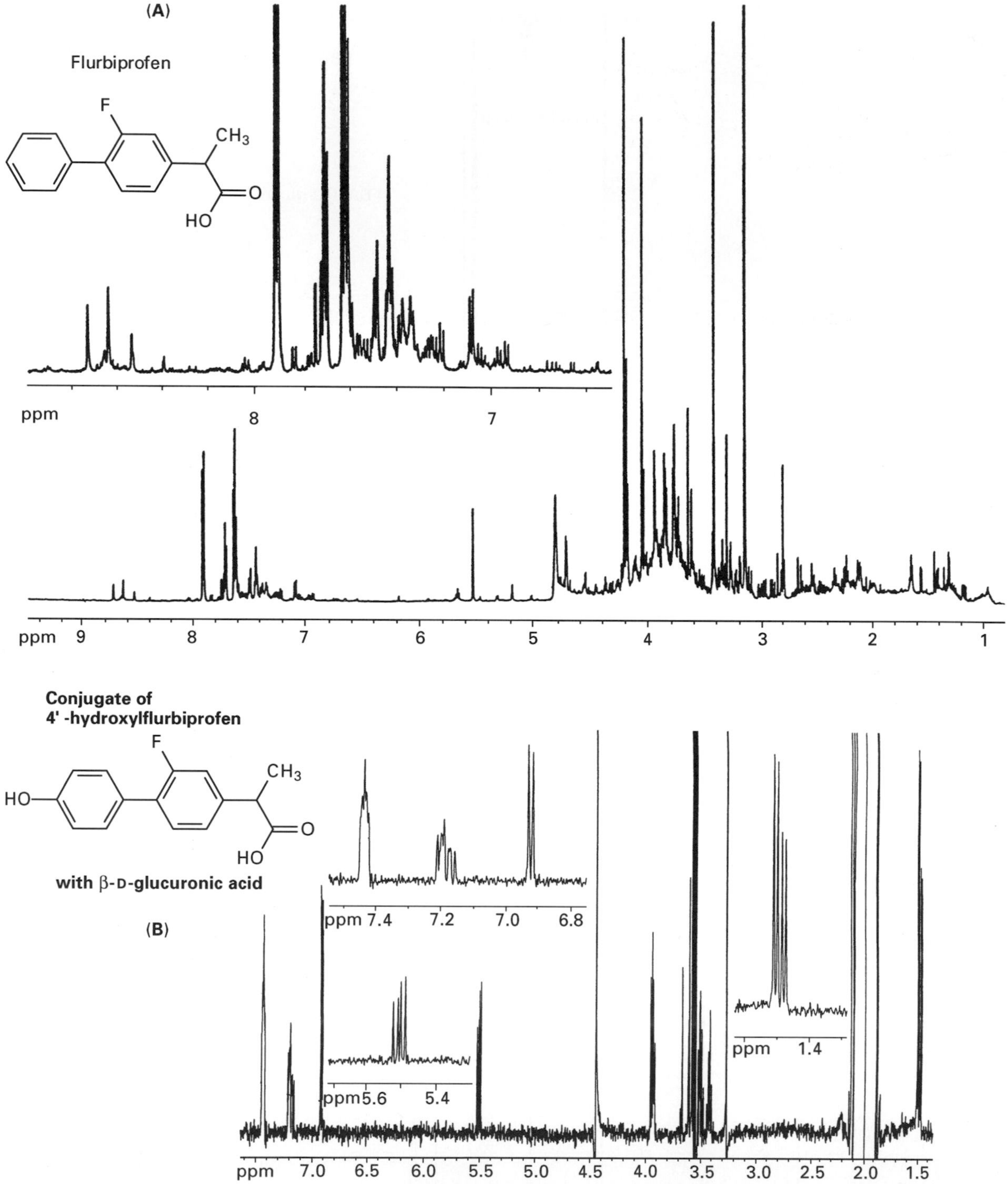

Figure 5 The 600 MHz ^{1}H NMR spectra of (A) a sample of human urine collected after the oral ingestion of 200 mg flurbiprofen, (B) the 30.5 min retention time species corresponding to the β-D-glucuronic acid conjugate of 4'-hydroxyflurbiprofen. Reprinted from (A) Spraul *et al* (1993). Coupling of HPLC with ^{19}F and ^{1}H NMR spectroscopy to investigate the human urinary extraction of flurbiprofen metabolites. *Journal of Pharmaceutical and Biomedical Analysis*, **11** (10), 1009–1015; (B) Lindon *et al* (1996) Direct coupling of chromatographic separations to NMR spectroscopy. *Progress in NMR Spectroscopy*, **29**, 1–49, with kind permission from Elsevier Science-NL, Sara Burgerhartstraat 25, 1055 KV Amsterdam, The Netherlands.

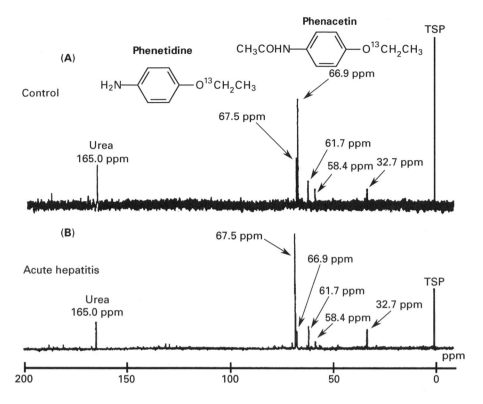

Figure 6 The ^{13}C NMR spectra of human urine taken 1 h after oral administration of ([1-^{13}C]ethoxy)phenacetin to a healthy subject (A) and to a patient with acute hepatitis (B). TSP, 3-(trimethylsilyl)propionate (internal standard). Adapted with permission of The Pharmaceutical Society of Japan from Kajiwara *et al* (1996) Studies on ^{13}C phenacetin metabolism II. A combination of breath test and urine test of *in vivo* metabolites in the diagnosis of liver disease. *Chemical Pharmaceutical Bulletin* **44**: 1258–1260.

^{13}C-labelled phenacetin has been used for the diagnosis of liver disease. In the urine of healthy subjects, the –OCH$_2$– signal of phenacetin was higher than that of its metabolite, phenetidine, whereas in patients with acute hepatitis the situation was the reverse (**Figure 6**). A return to the normal profile was observed during convalescence. The anticancer agent temozolomide labelled with ^{13}C was noninvasively detected in mice tumours by a selective cross-polarization transfer ^{13}C NMR method which has the advantage of higher sensitivity with lower power deposition. The resonance of temozolomide was observed but not the signal of its active metabolite, 5-(3-methyltriazen-1-yl)imidazole-4-carboxamide (**Figure 7**).

^{31}P NMR spectroscopy has been used to analyse biofluids from patients treated with the antineoplastic drug ifosfamide, widely used in the treatment of soft-tissue sarcomas and paediatric malignancies. Its metabolism is complex (**Figure 8**). The spectrum in **Figure 9** emphasizes the wealth of information on drug metabolism that can be gained from *in vitro* studies. About 20 phosphorus-containing compounds were detected. In addition to the already known metabolites (unchanged ifosfamide, carboxyifosfamide, 2-dechloroethylifosfamide, 3-dechloroethylifosfamide, isophosphoramide mustard,

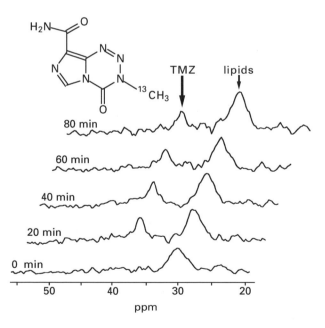

Figure 7 *In vivo* ^{13}C NMR spectra of a RIF-1 tumour before and after an intraperitoneal injection of ^{13}C-temozolomide (TMZ). Adapted with permission of Williams and Wilkins from Artemov *et al* (1995) Pharmacokinetics of the ^{13}C-labelled anticancer agent temozolomide detected *in vivo* by selective cross polarization transfer. *Magnetic Resonance in Medicine,* **34**: 338–342.

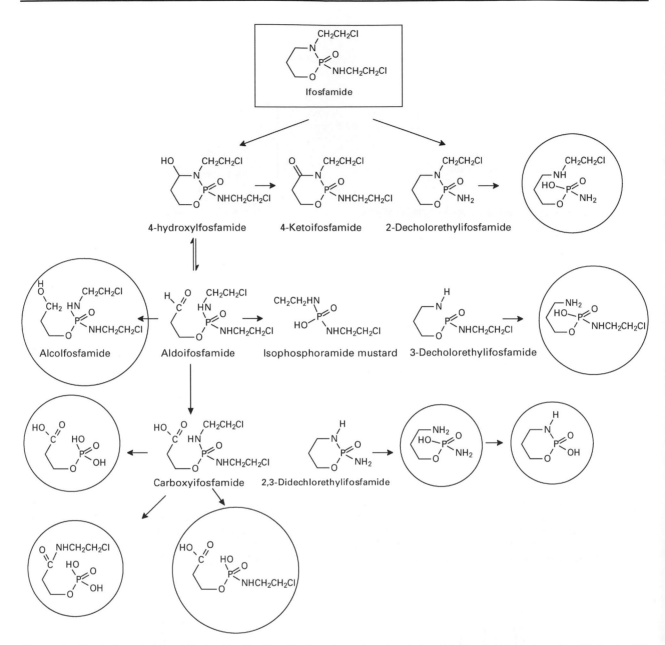

Figure 8 Metabolic pathways of ifosfamide. Compounds ringed are phosphorated metabolites that have been identified using ^{31}P NMR.

ketoifosfamide) eight other compounds were identified. Furthermore ^{31}P NMR spectroscopy has been used to detect and quantify ifosfamide *in vivo* in rat tumours. In addition to the resonances of the phosphorus-containing endogenous metabolites, only one broad signal could be detected for ifosfamide and, probably, also some of its metabolites. Carbogen breathing increased the amount of ifosfamide taken up by the tumour by 50% (**Figure 10**).

The concentration of Li in serum is not an adequate measure of Li efficacy. Localized spectroscopy that permits the measurement of Li in the brain may be a better approach of efficacy and/or neurotoxicity. Lithium was detected noninvasively *in vivo* in the brain and in the calf muscle of normal volunteers and of psychiatric patients. The lithium levels in both muscle and the brain rose more slowly than in serum. The concentration of Li in the brain is typically 0.1–0.6 mM. Therapeutic serum levels (ranging between 0.4–1.0 mM) are typically reached after 1 or 2 days of lithium treatment whereas clinical efficacy is not observed for several more days. The ^{7}Li NMR data strongly suggested that this delay may be due to the relatively slow accumulation of lithium in the brain (after 3–7 days of treatment). After the lithium treatment was stopped, the elimination half-time was slower in the brain (32–48 h) than in the serum or muscle (~24 h).

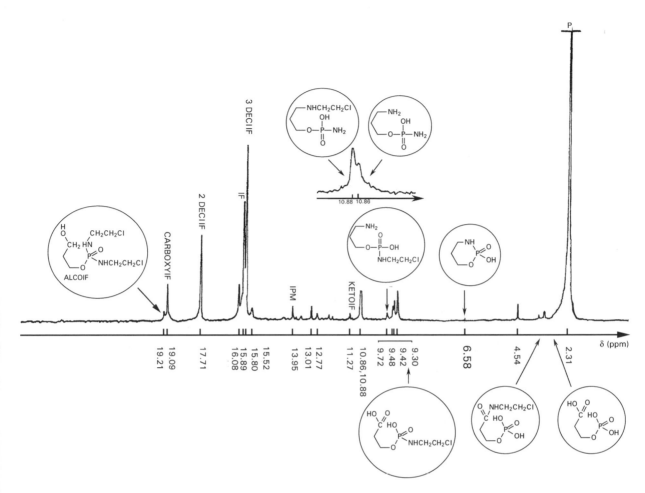

Figure 9   The ^{31}P NMR spectrum of a urine sample from a patient treated with ifosfamide (3 g m^{-2} injected intravenously over 3 h). The urine sample was collected 8–16 h after the start of the infusion, frozen immediately after collection and concentrated 17-fold by lyophilization. Chemical shifts are related to external 85% H$_3$PO$_4$. ALCOIF, alcoifosfamide; CARBOXYIF, carboxyifosfamide; 2DEC1IF, 2-dechloroethylifosfamide; IF, ifosfamide; 3DEC1IF, 3-dechloroethylifosfamide; IPM, isophosphoramide mustard; KETOIF, ketoifosfamide. Compounds ringed are phosphorated metabolites that have been identified using ^{31}P NMR.

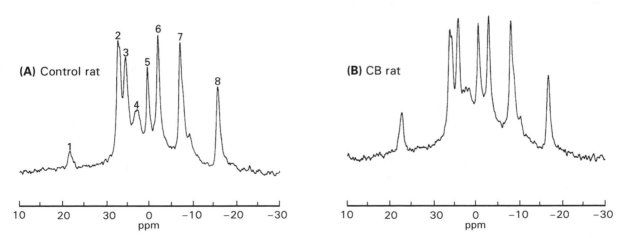

Figure 10   The ^{31}P NMR spectra of a GH3 prolactinoma. (A) Ifosfamide administered after 10 min breathing air. (B) Ifosfamide administered during tenth minute of breathing carbogen (CB). The spectra were acquired 15 min after intravenous ifosfamide. Peak assignments: 1, ifosfamide; 2, phosphomonoesters; 3, P$_i$; 4, phosphodiesters; 5, phosphocreatine; 6, γ-phosphate of NTP; 7, α-phosphate of NTP; 8, β-phosphate of NTP. Reproduced with permission of Churchill Livingstone from Rodrigues *et al* (1997) *In vivo* detection of ifosfamide by ^{31}P MRS in rat tumours: increased uptake and cytoxicity induced by carbogen breathing in GH3 prolactinoma. *British Journal of Cancer* **75**: 62–68.

Conclusion

The increasing number of NMR studies of drug metabolism and pharmacokinetics testifies to the growing interest of the method. For *in vitro* studies (biofluids in particular), NMR can now be considered as complementary to, and even in some cases replacing, other analytical techniques such as chromatography. *In vivo* experimental and clinical applications of NMR are still at an early stage of development, and the techniques will need to be improved before localized spectra and accurate absolute quantification are routinely available.

See also: **Biofluids Studied by NMR; Cells Studied by NMR; Chromatography-NMR, Applications; *In vivo* NMR, Applications, ^{31}P; *In vivo* NMR, Applications, Other Nuclei; ^{31}P NMR; Perfused Organs Studied Using NMR Spectroscopy; Solvent Suppression Methods in NMR Spectroscopy.**

Further reading

Bell JD, Gadian DG and Preece NE (1990) NMR studies of drug metabolism and disposition. *European Journal of Drug Metabolism and Pharmacokinetics* 15: 127–133.

Everett JR (1991) High-resolution nuclear magnetic resonance spectroscopy of biofluids and applications in drug metabolism. *Analytical Proceedings* 28: 181–183.

Findlay MPN and Leach MO (1994) *In vivo* monitoring of fluoropyrimidine metabolites: magnetic resonance spectroscopy in the evaluation of 5-fluorouracil. *Anti-Cancer Drugs* 5: 260–280.

Heimberg C, Komoroski RA, Newton JEO and Karson C (1995) ^{19}F-MRS: a new tool for psychopharmacology. *Progress in Psychiatry* 47: 213–234.

Komoroski RA (1994) *In vivo* NMR of drugs. *Analytical Chemistry* 66: 1024A–1033A.

Kuesel AC, Kroft T and Smith ICP (1991) Nuclear magnetic resonance spectroscopy. *Analytical Chemistry* 63: 237R–246R.

Lindon JC and Nicholson JK (1997) Recent advances in high-resolution NMR spectroscopic methods in bioanalytical chemistry. *Trends in Analytical Chemistry* 16: 190–200.

Lindon JC, Nicholson JK and Wilson ID (1996) Direct coupling of chromatographic separations to NMR spectroscopy. *Progress in NMR Spectroscopy* 29: 1–49.

Malet-Martino MC and Martino R (1992) Magnetic resonance spectroscopy: a powerful tool for drug metabolism studies. *Biochimie* 74: 785–800.

Preece NE and Timbrell JA (1990) Use of NMR spectroscopy in drug metabolism studies: recent advances. *Progress in Drug Metabolism* 12: 147–203.

Dyes and Indicators, Use of UV-Visible Absorption Spectroscopy

Volker Buss, University of Duisberg, Duisberg, Germany
Lutz Eggers, Gerhard-Mercator-Universität, Duisburg, Germany

ELECTRONIC SPECTROSCOPY
Applications

It seems appropriate to start an article on UV-visible absorption spectroscopy and dyes with a consideration of the fundamental processes underlying vision, for reasons which will presently become evident.

When electromagnetic radiation with wavelengths between 800 and 400 nm hits the human retina, a chain of events is initiated which eventually results in the excitation of the visual nerve and the perception of light or colour in the brain. The molecular basis for this process is the photoreceptor protein rhodopsin, which forms part of the membrane layers inside the cone and rod cells of the retina. The chromophore of rhodopsin, i.e. the coloured part of an otherwise colourless protein, is a highly unsaturated entity, 11-cis-retinal protonated Schiff base (**Scheme 1**). Through the action of light the Schiff base is transformed into an electronically excited state from which it returns, in a very fast reaction, to the ground state with the chromophore in the all-*trans* configuration (**Scheme 2**). Though there are many different visual pigments found in vertebrate eyes, with absorption maxima ranging from 415 nm in chicken to 575 in humans, nature uses the same chromophore, protonated retinal Schiff base,

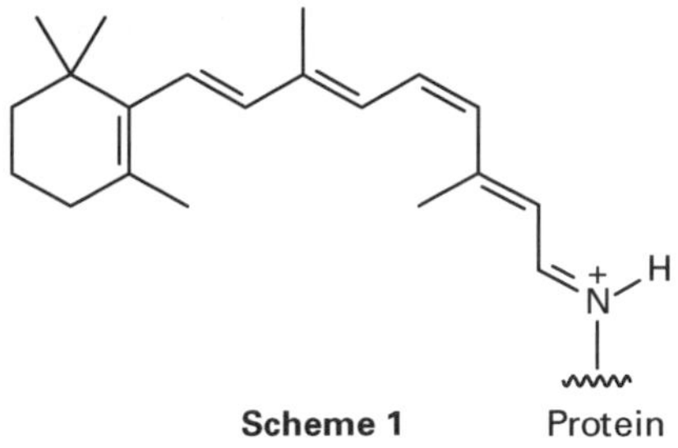

Scheme 1   Protein

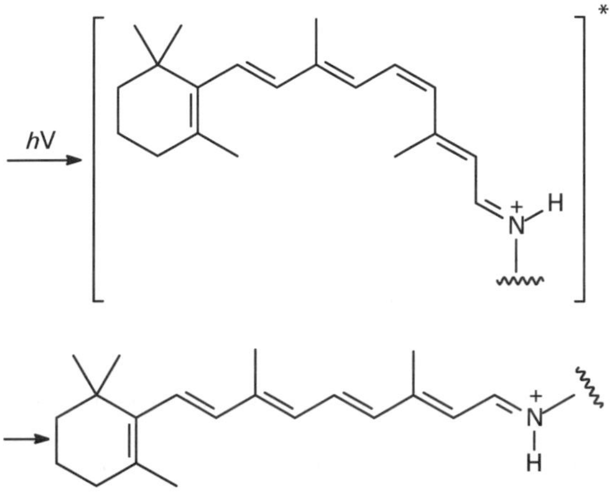

Scheme 2

throughout. In addition, different pigments are found in the same species, a structural prerequisite for colour vision much like the light-sensitive layers of a colour film coated with different dyes.

How these spectral changes are brought about is not completely understood. The chromophore without the attached protein absorbs at 440 nm in ethanol. Embedding the chromophore in the protein changes the position of the absorption maximum, a consequence of the different environment. This environment consists of the amino acid residues of the protein, differing from species to species, which may be charged or uncharged. There are other counterions and water molecules which might induce solvatochromic effects. An amino acid placed in a particular position might induce deprotonation. For nature, this presents a perfect example of tuning the spectral characteristics of the visual pigment according to its needs. Conversely, one might consider the chromophore an extremely sensitive indicator of its molecular environment.

All molecular properties are a function of the environment. Often these changes are subtle and hidden to the human eye. Sometimes, especially when dyes are involved, they are striking. Thus, dyes may be used as indicators of solvent polarity, of the pH or of the redox potential. Dyes will change their colours when they are adsorbed on surfaces or when they

form complexes with other species or with themselves, i.e. when they aggregate. Spectral change, which is what we perceive as the colour change of a dye, always has a molecular basis. It provides yet another window for studying molecular structure and molecular interactions.

Spectral characteristics of dye molecules

When light travels through a homogenous isotropic solution of an absorbing species, the fraction of light that is absorbed is proportional to the number of molecules in the light path according to the Beer–Lambert equation

$$\log_{10}(I_0/I) = \varepsilon \cdot c \cdot l$$

in which I_0 is the intensity of the incident and I of the emergent radiation. With the concentration c in moles per litre and the path length l in cm, the proportionality constant ε is the molar absorptivity or molar extinction coefficient. While ε is a measure of the *intensity* of the absorption, the wavelength λ at which the absorption occurs, or better its inverse, the wavenumber $\bar{v}$, is a measure of the *energy* which is absorbed, according to the well-known relation

$$E = hc/\lambda = hc\bar{v}$$

in which h is Planck's constant and c the speed of light. The range of UV-visible absorption extends from 800 to 200 nm (12 500 to 50 000 cm^{-1}) which corresponds to energies from ca. 35 to 145 kcal mol^{-1} (1 kcal = 4.184 kJ). In molecules absorbing in this wavelength range these energies are sufficient to create short-lived electronically excited states. In such a state the electronic wavefunction does not correspond to the most stable arrangement of the electrons in the field of the nuclei; as a consequence the molecule returns, within 10^{-7} to 10^{-9} seconds, to the electronic ground state. Factors that affect the two states differently, e.g. chemical substitution, protonation or change of the solvent, will induce spectral shifts either to longer ('bathochromic shift') or shorter ('hypsochromic shift') wavelengths.

Empirical rules are widely employed to correlate chemical structure changes with spectral shifts. The spectra of short conjugated double bond systems can be predicted based on the topology and the presence of substituents (Woodward rules); linear free-energy relationships have been developed to correlate substituent and spectral changes in closely related dye molecules. Other rules originate from basic quantum-mechanical principles, e.g. the linear relation between the absorption maxima of cyanine dyes and the length of the conjugated chain; the perturbation of a conjugated carbon chain by substituents of different electronegativity (perturbational molecular orbital theory); or the effects of steric crowding and, as a consequence, non-planar distortion of the molecule.

For a more quantitative description, theories may be applied which range from the purely π–electron, but highly successful PPP-theory and the semi-empirical all-valence electron theories such as CNDO/S and AM1 to *ab initio* methods. Only the latter allow the calculation of the energies without requiring recourse to experimental data. While it is possible today to calculate molecular ground state energies with an accuracy of a few kilocalories, depending on the size of the molecule and the computational effort expended, the uncertainty is much larger in the excited state. The *ab initio* calculation of absorption wavelengths which come within 50 nm of the experimental value must still be considered successful.

Calculated energies refer to the isolated molecule in the (dilute) gas phase. Every solvent will change these data to a degree. Solvent shifts can be striking when there is significant charge rearrangement between the ground and excited state. For computing these effects, one may allow specific interactions, e.g. through hydrogen bonds, of the dye with the solvent, though this approach is limited in the number of the solvent molecules that can be considered. In other approaches the dye molecule is specifically solvated and then embedded into a solvent continuum which in its response to the solute is treated as a polarizable dielectric.

A measure of the absorption intensity is the oscillator strength f, a dimensionless quantity which is obtained by integrating over the absorption band:

$$f = 4.315 \times 10^{-9} \int \varepsilon \mathrm{d}\bar{v}$$

f can also be calculated from a theoretical quantity, the transition dipole moment M,

$$f = (8\pi m_e \, \bar{v}/3he^2)|M|^2$$

where m_e and e are, respectively, the mass and charge of the electron and $\bar{v}$ is the absorption wavenumber. M, a vector quantity, is a measure of the charge displacement caused by the electronic excitation. When the absorption can be attributed to the excitation of an electron from one molecular orbital to another with an energy sufficiently different from all other excitations of the molecule the magnitude and direction of M can be estimated from multiplying the two orbitals which are involved in the excitation (**Figure 1**). This procedure gives atom-centered transition charge densities which are used to calculate the transition dipole moment. These 'back-of-an-envelope' calculations readily explain why the oscillator strength increases with the size of the chromophore, or why it is larger in the stretched conformation of a molecule than when it is coiled. The fixed orientation of M with respect to the molecular frame is also the physical basis for dichroism, i.e. the orientation dependence of the extinction coefficient of an ordered ensemble of dye molecules.

Electronic excitation in the UV-visible range is accompanied by changes in the vibrational and rotational state of the molecule as well. Since the spacing between rotational and vibrational levels is small relative to the spacing between electronic

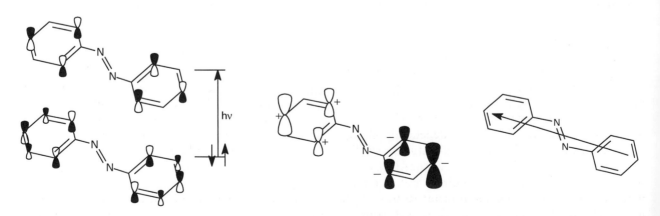

Figure 1 Highest occupied and lowest unoccupied molecular orbitals of azobenzene (schematic, left), their product, giving the transition density (middle) and the resulting transition moment vector (right).

states, a typical UV-visual spectrum will appear as a band, or an envelope curve, of close-lying 'rovibronic states', the band shape being determined by the relative intensities of those states. Strong vibrations of the excited state may give rise to 'vibrational fine structure' of an absorption band, which may change as a function of the solvent and temperature.

Dyes as indicators of solvent polarity

Any solvent will change the wavelength, intensity and shape of an absorption band compared to the spectrum in the gas phase to a degree, a consequence of the different perturbation of the electronic ground and excited states of the solute by the solvent. The term 'Solvatochromism' has been coined to describe the change in position of an UV-visible absorption band due to a change in the polarity of the medium. A hypsochromic or blue shift with increasing solvent polarity is called negative solvatochromism, whereas a bathochromic or red shift is called positive solvatochromism. It is unreasonable to expect any macroscopic parameter, such as the dipole moment of a solvent or its refractive index, to correlate quantitatively with solvent effects which depend in addition on interactions at the molecular level. Therefore, empirical scales have been devised to correlate solvent polarity and spectral shifts.

One of the most popular and successful scales has been developed by Dimroth and Reichardt. It is based on the pyridinium-N-phenoxide betaine [3], which exhibits one of the largest solvatochromic effects ever observed. The solvatochromism of this dye is negative since its ground state is considerably more polar than the excited state and is stabilized by polar solvents. Thus, in diphenylether the dye absorbs at 810 nm and appears blue-green, whereas in water the absorption is centred at 453 nm, giving an orange impression. The transition energy, expressed in kcal mol^{-1}, is the so-called $E_T(30)$ value of the solvent. $E_T(30)$ values have been determined and tabulated for more than 270 pure solvents and many different solvent mixtures. Protonation converts the dye (**Scheme 3**) into a phenol; as a consequence, $E_T(30)$ values cannot be measured for acidic solvents, such as carboxylic acids.

Another solvent polarity scale, Kosower's Z-scale, is based on the highly solvent-dependent charge transfer absorption of the *para*-substituted pyridinium salt (**Scheme 4**). The ground state of this dye is best described as a highly polar tight ion-pair, whereas the less polar charge-transfer form is more important in the excited state. Thus, the solvatochromism of the pyridinium salt is negative, like Reichardt's dye. In fact, the $E_T(30)$- and the Z-scale

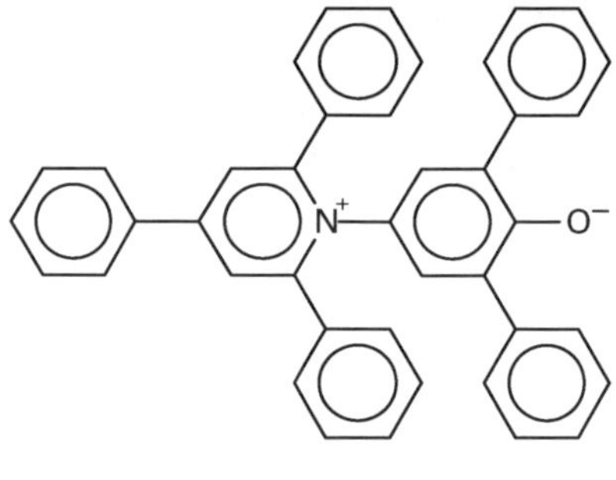

Scheme 3

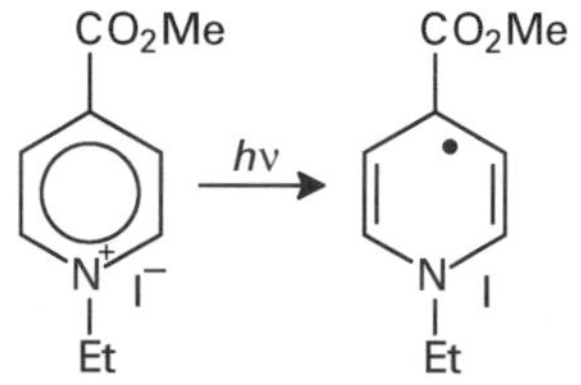

Scheme 4

show a very good correlation, and similar anomalies with respect to theoretical polarity functions, which is not surprising because both use similar molecular models to 'measure' solvent polarity.

For converting experimental transition energies $\bar{v}$ (in cm^{-1}) into $E_T(30)$ and Z-values, the following equation may be used:

$$E_T(30) = Z = hcN_A\bar{v}$$
$$= 2.8591 \times 10^{-3}\,\bar{v}$$

in which N_A is Avogadro's constant and both $E_T(30)$ and Z have the dimensions kcal mol^{-1}. The use of normalized, dimensionless values 'E_T^N' has been recommended in order to avoid the non-SI unit. In this scale, water and trimethylsilane (TMS) are taken as reference solvents, with assigned values of E_T^N of 1.00 and 0.00, respectively.

$E_T(30)$ values show a good, often linear, correlation with a large number of other solvent sensitive processes, such as reaction rates and shifts of chemical equilibria. The betaine dye (**Scheme 3**) and specially designed derivatives are useful molecular probes in the study of micellar interfaces, microemulsions and phospholipid bilayers, of rigid rod-like isocyanide polymers, and the retention behaviour in reversed-phase chromatography. In addition to its solvatochromic behaviour, the dye is sensitive to temperature ('thermosolvatochromism') and pressure changes ('piezosolvatochromism') and also to the presence of electrolytes ('halosolvatochromism').

Dyes as indicators of chemical equilibria

For a dye to act as indicator of a chemical equilibrium, it has to meet two requirements; first, it must be able to participate in the reaction. That is, in an acid/base equilibrium the dye itself must have acidic or basic properties; in a redox reaction, the dye must exist in different oxidation states; in a complexation reaction involving e.g. metal ions, the dye must have complexing properties. Second, the different forms of the dye involved must absorb at different wavelengths, i.e. they must have different colours. Applications of dyes as indicators are of enormous importance in analytical chemistry.

One of the oldest applications of UV-visible spectroscopy is found in acid–base titrations and the determination of pK values. Several types of organic dyes are very sensitive to changes in the concentration of H_3O^+ (or OH^-) ions, e.g. azo dyes, phthaleines or triphenylmethane dyes. As a typical example, consider the protonation of methyl orange (Scheme 5), which occurs at a position on the aza bridge which allows the delocalization of the positive charge into the p-dimethylamino group. As a consequence, one of the aromatic rings becomes part of a cyanine type chromophore (eight π-electrons delocalized over seven atomic centres), with a concomitant bathochromic shift of the absorption spectrum and a colour change from yellow to red.

The intense red, dianionic form of phenolphthalein (Scheme 6), which is stable in weakly basic solution, can be formulated as a resonance hybrid but is really another example of a cyanine-type chromophore, with oxygen instead of nitrogen. In acidic or strongly basic solutions, the negative charges are localized and the compound is colourless.

During titration, the sharp change of the pH near the end point is indicated by the visible colour change as the added indicator dye is converted from the basic into the acidic form (or vice versa). Since the end point of the titration does not necessarily coincide with neutrality of the solution, an indicator should be employed with a pK value that corresponds most closely to the pH of the end point. The concentrations (or better activities) of both the acidic and the basic form of the indicator are then the same and true mixed colour is observed. For a determination of the indicator pK, the extinction coefficient ε of the dye is measured in buffered solutions at different pH values. According to the relationship

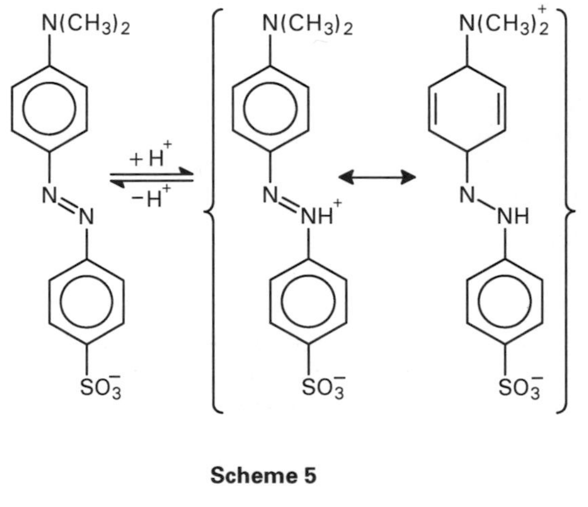

$$pK = pH + \frac{\varepsilon - \varepsilon_{base}}{\varepsilon_{acid} - \varepsilon}$$

the pK can be calculated from ε and the extinction coefficients of the indicator in strongly alkaline (ε_{base}) and strongly acidic solutions (ε_{acid}). All measurements should be performed in dilute solutions, and at wavelengths sufficiently distant from the isosbestic point where $\varepsilon_{base} = \varepsilon_{acid}$ and the extinction does not change with the pH.

Similar considerations apply to redox titrations, where the indicator dye should possess differently coloured oxidation states and a redox potential which is appropriate for the reaction to be studied. A large number of dyes are known which are suitable for this purpose. Ferroin (Scheme 7), a complex between iron(II) and phenanthrolin with a deeply red colour, upon oxidation forms a pale blue iron(III) complex. Another redox indicator is diphenylamine (Scheme 8), which is colourless in the reduced state. It is oxidized in acidic solution irreversibly to diphenylbenzidine (Scheme 9) which can be further oxidized reversibly to the intensely coloured diphenylbenzidine violet (Scheme 10).

Comparable to an acid/base titration the end point of a redox titration is reached when the concentrations of the reduced and the oxidized form of the dye are the same. The latter situation is somewhat more complex, however, because most redox reactions involve the exchange of protons, i.e. the equilibria are pH dependent.

In the determination of metal ions by ligand binding with ethylenediaminetetraacetic acid, EDTA, which has become known as complexometric or chelatometric titration, the end point may be indicated by so-called metal indicators. For this method to work, the metal to be titrated must form a less stable complex with the dye than with the complexing agent. Titration with the complexing agent binds all free metal in solution, until finally the dye is demetallated, too, which is manifest in the colour change.

Scheme 5

A very popular indicator is Eriochrome Black T (**Scheme 11**), which forms complexes with many different metals (the chromium complex is also used to dye protein and polyamide fibres). Between pH values of 7 and 11, the dye changes its colour from blue in the uncomplexed to red in the complexed form.

The end point of a titration may be indicated, besides the chemical transformation of a dye indicator, by physical processes, such as the adsorption of eosin from solution on the surface of a silver halide precipitate.

Dyes as indicators of molecular environment

Any medium other than the vacuum will affect the absorption spectrum of a dye, through its bulk properties as a dielectric, which does not imply a fixed geometry or stochiometry, and through specific interactions, such as electrostatic interaction, π–π stacking, or hydrogen bonding. The combination of any of these between a dye and an organic substrate

may lead to the formation of an entity with a more or less well-defined supramolecular geometry, which exhibits characteristic spectral changes. As an example, consider the complex formed between *p*-nitrophenol and α-cyclodextrin (**Figure 2**). Cyclodextrins are water-soluble cyclic oligomers of amylose with a torus-like structure and a strongly hydrophobic inner lining. While they form complexes with many different guests, the complexation of dyes can be quantitatively followed by absorption spectroscopy. Usually, the low resolution of the spectra is not sufficient to develop a detailed structural model of these complexes, and other methods have to be used to implement the UV-visual spectroscopic data. The most accurate information is obtained from an analysis of X-ray data (**Figure 2**) which displays the orientation of the dye inside the cyclodextrin host. In contrast to the solid state the condition in solution is dynamic and more complex, since more species are involved, including the water molecules which are equilibrating between the solvent, the solute and the pore. However, the formation of a

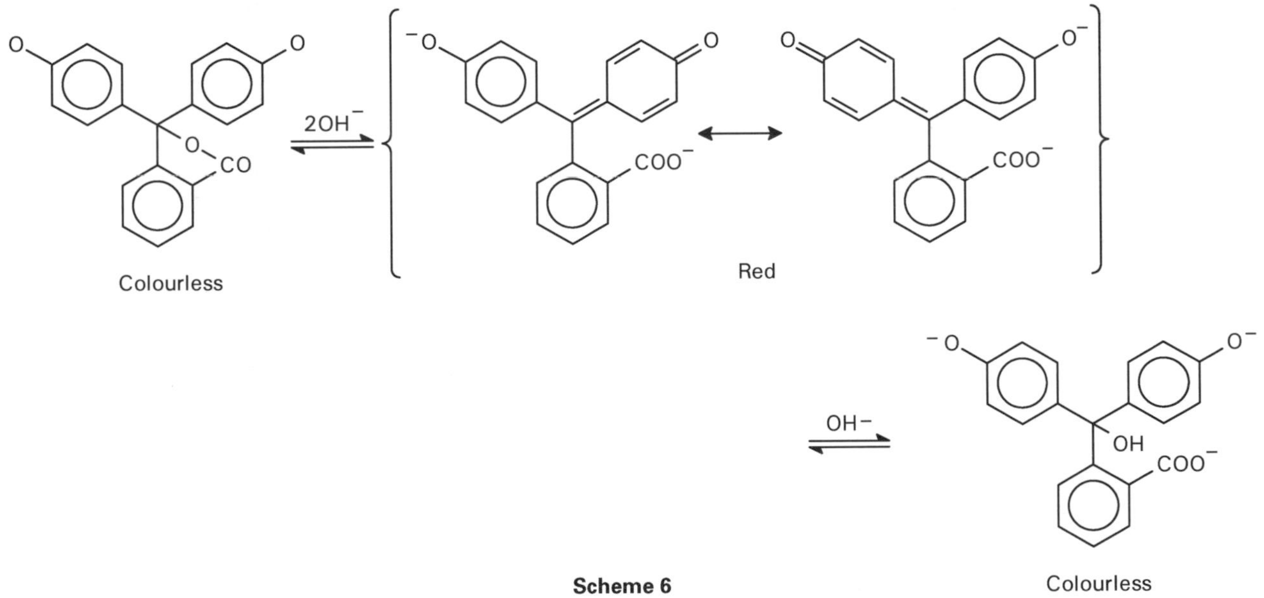

Scheme 6

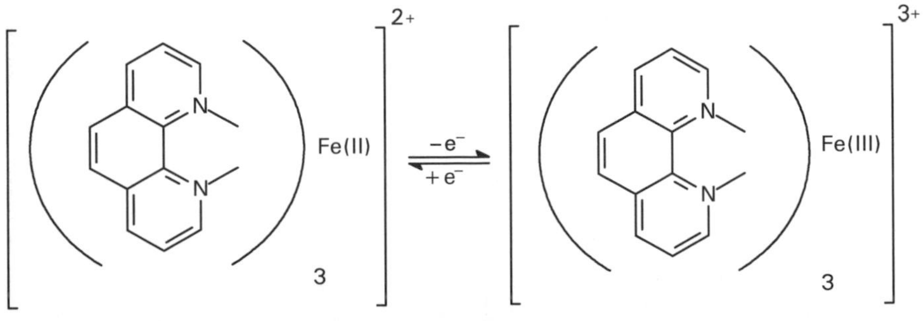

Scheme 7

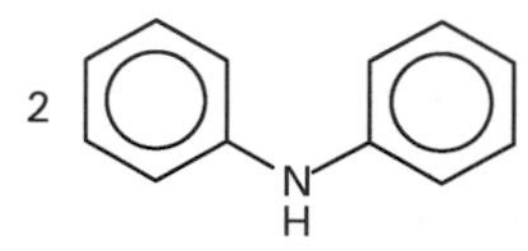

Scheme 8

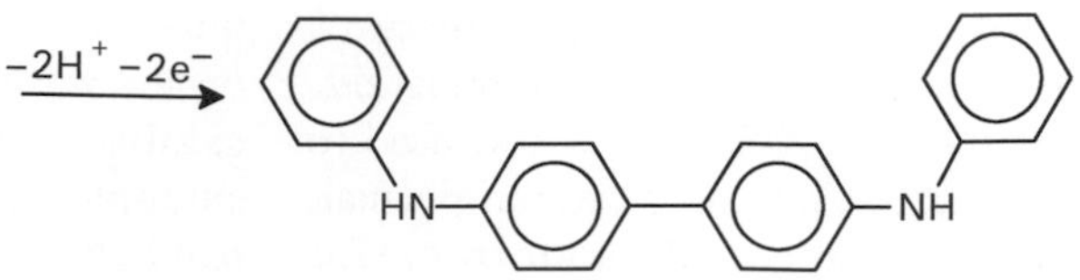

$$-2H^+ -2e^-$$

Scheme 9

$$\frac{-2H^+ -2e^-}{+2H^+ +2e^-}$$

violet

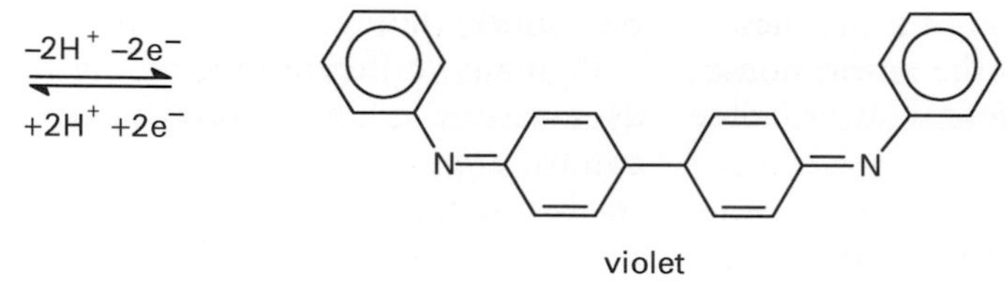

Scheme 10

structured host–guest complex is supported by other spectroscopic methods, such as 2-dimensional NMR spectroscopy.

The situation depicted above is typical for many cases where the complex formation between dyes and organic substrates is studied as a model for intermolecular interaction, the dye functioning at the same time as the complexed ligand and the indicator of this interaction. The literature dealing with this topic is vast and ranges from histochemistry, where tissue is stained for diagnostic purposes, and biochemistry, where the complexation with dyes is used to model enzyme–ligand interactions and study possible binding sites, to organic polymers where dyes act as indicators of macromolecular conformations or of phase transitions.

Dyes, and their absorption changes, can be used to sense chirality and differentiate between two enantiomers 'with the naked eye', i.e. without the use of a polarimeter or any other device that measures optical activity. In principle every diastereomeric interaction between a chiral medium and the two enantiomers of a chiral dye must lead to different spectral characteristics. In practice, however, the effects are

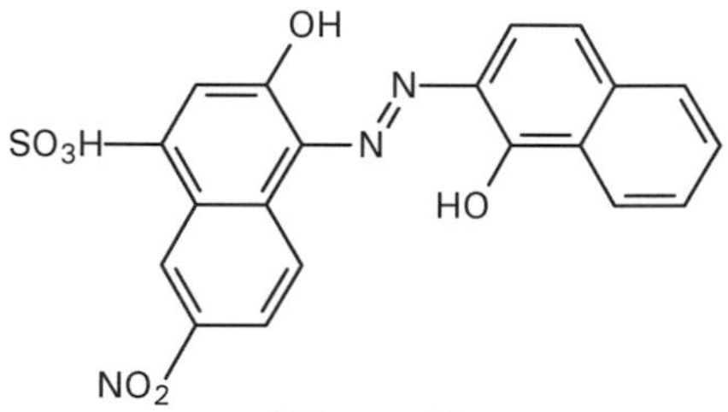

Scheme 11

usually too small to be observed. The situation is different when diastereomeric supramolecular structures are involved. Thus, the colour of a phenol based bowl shaped agent (a 'calixarene') with two indophenol chromophores and a chiral binaphthyl subunit (**Figure 3**) turns blue-violet when it forms a 1:1 complex with (*R*)-phenyl-glycinol, while the corresponding complex with the (*S*)-isomer has a much smaller

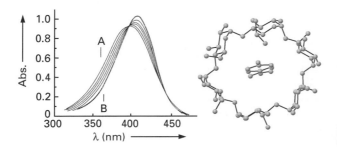

Figure 2 Spectral change, A to B, upon addition of α-cyclodextrin to an aquous solution of *p*-nitrophenolate (left; after Saenger W (1980). *Angewandte Chemie* **92**: 343). X-ray structure of the α-cyclodextrin-*p*-nitrophenol complex (right: after Harata K (1977) *Bulletin of the Chemical Society of Japan* **50**: 1416).

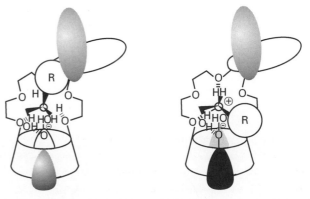

Figure 3 Possible structures of diastereomeric complexes formed between the bowl-shaped chiral calixarene and an (*S*)-(left) and an (*R*)-configurated aminoalcohol. The shaded entities at the bottom represent the indophenol chromophores, the ones on top the binaphthyl unit. R is a phenyl group of the chiral amine (Reproduced with permission from Kubo Y, Hirota N, Maeda S, Tokita S (1998) *Analytical Sciences: the International Journal of the Japan Society for Analytical Chemistry* **14**: 183).

association constant and exhibits mainly the spectrum of the uncomplexed chiral sensor, which is red. As a consequence, it is possible to detect chirality by simple visible inspection. Similarly, supramolecular structures involving different kinds of non-bonded interactions have been invoked to account for the ability of chiral phases to discriminate between enantiomers in chromatographic separation techniques.

For a study of the molecular microenvironment, but also as promising new optical materials in solar energy conversion, for dye lasers, as optical switches or in non-linear optics, silicate glasses doped with dye molecules have been prepared by the sol–gel method. The formation of the porous solid from the initially liquid organic solution has been followed by UV-visible spectroscopy, and the acid/base reaction of the dyes with the inorganic matrix has been studied. Solvatochromic dyes have been used to determine the $E_T(30)$ values of different sol–gel materials. One advantage of these so-called xerogels is the immobilization of the embedded dye in a rigid cage, which prevents aggregation processes which are typical for concentrated dye solutions (see next section). Another aspect is the increased photochemical stability and fluorescence efficiency of the dyes. For the development of biochemical sensors, whole enzymes have been encapsulated in xerogels and their activity shown by the formation of dyes inside the glass from coupled enzyme reactions.

Dyes as indicators of aggregation and molecular order

Dyes have a natural tendency to aggregate, a consequence of their typically flat geometries which allows for maximum contact when the aggregation takes place side by side, in a sandwich manner. The aggregation is effected by non-bonding interactions, including π–π stacking, and is supported by hydrophobic effects when the solvent-accessible molecular surface is reduced as a consequence of aggregation. Aggregation is promoted by increasing the dye concentration in a solution or lowering the temperature. It is one of the most common causes for deviation of a solution from the Beer–Lambert law. Often a change of the solvent from polar to less polar or vice versa will be sufficient to effect dye aggregation or deaggregation. Aggregates may form spontaneously in solution or in the presence of polyelectrolytes, such as polyphosphates, or biological macromolecules, which may influence or even control the formation of the aggregate. Aggregates may form on solid surfaces, such as mica, or they may be assembled as monomolecular layers on glassy surfaces. Dye aggregates grown on the surface of silver halide grains absorb light of suitable wavelength leading to latent images in the halide; this so-called spectral sensitization is the basis for photography. Dyes tend to concentrate at liquid interfaces and interact with surfactants leading to distinct spectral changes. Finally, dyes may adjust to the highly ordered environment present in liquid crystals, an area of huge technical importance.

Conspicuous colour changes take place when dyes aggregate. The spectral shifts caused by aggregation can be to longer or to shorter wavelengths relative to the monomeric dye absorption. When dyes dimerize, a splitting of the absorption band may be observed. Formation of trimers, tetramers or larger oligomers is accompanied by increasing spectral shifts. The limiting value corresponding to an aggregate of 'infinite' size is called a 'J-band' characteristic of a J-aggregate when lying on the long wavelength side of the monometer; if it appears on the high-energy side, it is called an 'H-band' (**Figure 4**).

These spectral changes are the result of the individual dye molecules interacting with each other in the close contact of the aggregate. Theoretical models based on the Coulomb interaction of the transition moments have been put forward to correlate the spectral shifts with certain geometrical parameters of the aggregate. Thus the hypsochromic shift observed in H-aggregates results when the dye contacts are mainly side by side. A lateral shift of the dye molecules against each other or a stacking into a brickstone work arrangement will lead to the bathochromic shift of a J-aggregate. While the H-band is usually sharp and intense due to the high order of the aggregate, the J-band may appear broad and featureless due to the loose and more irregular structure, but exceptions are well documented.

For studying dyes adsorbed on solids by UV-visible spectroscopy transparent solutions are necessary.

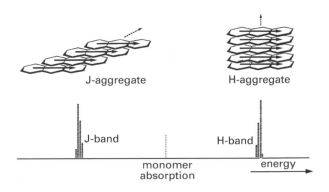

Figure 4 Stacking of dye molecules (with transition moment vectors) which leads to a J-aggregate with red-shifted absorption relative to the monomer (left) and to an H-aggregate with blue-shifted absorption (right).

There are many inorganic colloids suitable for this purpose, such as silicas, charcoal, clays, zeolites or metal oxides, but also inorganic and organic polymers and even aggregates, such as micelles. Flocculation or aggregation of smaller particles may occur due to the adsorption of a charged organic dye on a polyelectrolyte which reduces the elecrostatic repulsion by charge compensation.

The number of publications dealing with this topic is increasing rapidly. For studying the interaction of dyes with an enzyme the adsorption on solid surfaces with different binding characteristics may serve as a model. In this case the dyes are used in their classical role as indicators. Dyes are well suited for this purpose because of their strong tendency towards chemi- and physisorption. When the UV-visible properties of the adsorbed species change sufficiently the immediate comparison with dye molecules dissolved in a homogeneous solution is possible. Also, dynamic aspects of the adsorption of molecules on these surfaces can be examined by electronic absorption techniques.

The colour change accompanying the self-aggregation of dyes on solid surfaces is called metachromasy. Other properties of adsorbed dyes can also change allowing new or improved applications, such as enhanced stability against environmental effects such as oxygen or water or higher quantum yields for a photochemical conversion. New technologies are developing based on such processes. Solar energy conversion in photovoltaic cells, where charge carriers are generated by photoexcitation of dyes deposited on titanium dioxide surfaces, is a good example.

Complex equilibria between the dye and its surroundings may take place in heterogenous liquid mixtures, involving self-association, binding with ionic surfactants, micellar aggregation and micellar solubilization. These equilibria are concentration and temperature dependent, and involve entities with sometimes unspecific spectroscopic properties which leave plenty of room for discussion. Still, much can be learned from studying how dyes distribute in heterogenous systems since this can serve as models for the biodistribution of drugs in tissues and membranes and their behaviour at the lipophilic barrier. Also, the photosensitizing properties of a dye may increase when brought into a lipophilic environment.

H-aggregation of methyl orange has been postulated to account for the hypsochromic shift of the dye in a water-in-oil microemulsion, with the dye molecules in a parallel orientation at the interface beween the water and the oil phase. The observation of salt-induced strongly red-shifted J-bands in aqueous solutions of Rose Bengal, an anionic xanthene dye, in the presence of zwitterionic surfactants was explained on the basis of head-to-tail aggregates of the dye molecules in novel premicellar aggregates.

Highly ordered phases, so-called liquid crystals, are formed by a variety of rod-like, rigid dipolar organic molecules. Liquid crystals are also called the fourth state of matter since they combine the long-range order of a crystalline solid with the fluidity of a liquid. When a dye is dissolved in a liquid crystal the molecules will form ordered aggregates, with their long axes preferentially aligned with the liquid crystal molecules. These crystals are dichroic, i.e. they have different absorptivites for light polarized parallel and perpendicular to that axis, from which the orientation of the transition moment vector relative to the molecular framework can be determined.

The dichroism is the physical basis for the use of dyes in LCDs or liquid crystal displays: a small amount (0.1–1%) of the dichroic dye (the 'guest') is dissolved in a nematic or cholesteric liquid crystal (the 'host'). Different orientations of the dye aggregates are achieved by application of an electric field through transparent electrodes. The dye will for example align with its long axis parallel to the long axis of a nematic liquid crystal. If this solution is aligned in a parallel mode of the electrodes and has a positive dielectric anisotropy, the cell will be coloured in the off position, when the dye absorbs strongly, and become colourless upon application of an electric field, when the alignment is changed.

List of symbols

c = molar concentration; e_e = charge of electron; $E_T(30)$ = solvent transition energy; f = oscillator strength; I = intensity; l = path length; m_e = mass of electron; M = transition dipole moment; ε = molar extinction coefficient.

See also: **Biomacromolecular Applications of UV-Visible Absorption Spectroscopy; Biomacromolecular Applications of Circular Dichroism and ORD; Colorimetry, Theory; ORD and Polarimetry Instruments**.

Further reading

Fabian J and Hartmann H (1980) *Light Absorption of Organic Colorants*. Berlin: Springer.

Rao CNR (1975) *Ultraviolet and Visible Spectroscopy*. London: Butterworth.

Reichardt C (1995) Solvatochromic dyes as solvent polarity indicators. *Chemical Reviews* 94: 2319–2358.

Sommer L (1989) *Analytical Absorption Spectrophotometry in the Visible and Ultraviolet*. Oxford: Elsevier.

Suppan P and Ghoneim N (1997) *Solvatochromism*. Cambridge: The Royal Society of Chemistry.

Zollinger H (1991) *Color Chemistry*. Weinheim: VCH.

EELS, Applications

See **High Resolution Electron Energy Loss Spectroscopy, Applications.**

Electromagnetic Radiation

David L Andrews, University of East Anglia, Norwich, UK

FUNDAMENTALS OF SPECTROSCOPY
Theory

Electromagnetic radiation is the technical term for light; not just visible light, but any light, ranging right through from radio frequencies to gamma rays. As the name suggests, light of all kinds is radiated through conjoined electric and magnetic fields, as shown by Maxwell in the mid-nineteenth century. To fully appreciate the nature of electromagnetic radiation, however, we first have to consider both its wave-like and photonic aspects.

Waves and photons

In the case of monochromatic (single-frequency) radiation, light propagates as a wave with a well-defined repeat or wavelength λ (**Figure 1**). Travelling at the speed of light, c, these waves oscillate at a characteristic frequency ν given by c/λ. The electric field, the magnetic field and the direction of propagation are mutually perpendicular, and for convenience may be chosen to define a set of Cartesian axes (x, y, z), respectively. With propagation in the z direction, the electric field vector points in the x direction and oscillates in the xz plane:

$$E(z, t) = E_0 \sin(kz - \omega t) \qquad [1]$$

where E_0 is the field amplitude and where also for conciseness we introduce $k = 2\pi/\lambda$ and $\omega = ck = 2\pi\nu$.

The magnetic field vector accordingly points in the y direction and oscillates in the yz plane:

$$B(z, t) = B_0 \sin(kz - \omega t) \qquad [2]$$

with amplitude B_0. Given that the electric and magnetic fields oscillate in phase and in mutually perpendicular planes, equations [1] and [2] together satisfy Maxwell's equations provided the amplitudes are related by $E_0 = cB_0$. In fact most spectroscopy involves the interaction of matter with the electric rather than the magnetic field of the radiation, because of its stronger coupling – usually by a factor of a thousand or more – with atomic and molecular charge distributions.

The speed of travel of an electromagnetic wave as described above can be understood in terms of the motion of any part of the waveform, such as the crest. The interval Δt between the arrival of two successive crests at any given point is given by $\omega \Delta t = 2\pi$, while at any time their spatial separation Δz also satisfies $k \Delta z = 2\pi$. So the speed of propagation $\Delta z/\Delta t = \omega/k = c$, which *in vacuo* takes the value $c_0 = 2.9979 \times 10^8$ m s^{-1}.

On the radiation travelling through any substance, the electronic influence of the atoms or molecules traversed by the light reduces the propagation speed to a value less than c_0; then we have $c = c_0/n_\lambda$ where

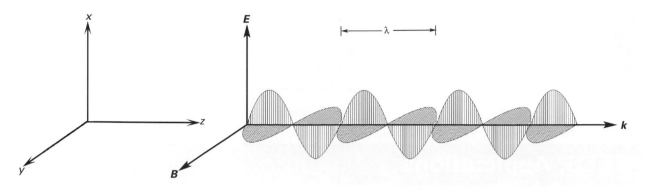

Figure 1 Oscillating electric field E and magnetic field B associated with mononchromatic radiation.

n_λ is the refractive index of the medium for wavelength λ; accordingly, $k = 2\pi n_\lambda/\lambda$. The significance of the refractive correction is greatest in the solid or liquid phase, especially at wavelengths close to an optical absorption band of the medium.

A fundamental paradox in the nature of electromagnetic radiation, to some extent apparent even in the earliest scientific studies by Newton and others, is that it exhibits not just wave-like but also particle-like (corpuscular) properties, and both prove to be of key importance in spectroscopy. In particular, it is only through the association of discrete units of energy with electromagnetic radiation of any given frequency that we can properly understand atomic and molecular transitions and the appearance of spectra.

In the modern quantum representation of light developed by Planck, Einstein, Dirac and others, we now understand the twin wave and particle attributes through a description in terms of *photons* (a term in fact first introduced by the thermodynamicist Lewis). As such, electromagnetic radiation of a given frequency ν is seen to propagate as discrete units of energy $E = h\nu$, where h is the Planck constant.

With the key concepts in place, we can now take a look at some of the more detailed aspects of electromagnetic radiation in two stages — first by more fully enumerating the properties of photons in general terms, and then by examining those more specific features that relate to particular wavelength or frequency regions of the electromagnetic spectrum.

Photon properties

Mass Photons are elementary particles with zero rest mass – necessarily so, since, from special relativity theory, no particle with a finite mass can move at the speed of light.

Velocity The speed of light is normally quoted as speed *in vacuo*, c_0, with refractive corrections applied as appropriate; the free propagation of any photon also has a well-defined direction, usually denoted by the unit vector $\hat{k}$.

Energy Photon energy is linked to optical frequency ν through the relation $E = h\nu$ (where $h = 6.6261 \times 10^{-34}$ J s). Each photon essentially conveys an energy E from one piece of matter to another, for example from a television screen to a human retina.

Frequency The optical frequency ν expresses the number of wave cycles per unit time. Also commonly used in quantum mechanics is the circular frequency $\omega = 2\pi\nu$ (radians per unit time), in terms of which the photon energy is $E = \hbar\omega$ where $\hbar = h/2\pi$. The lower the optical frequency, the more photons we have for a given amount of energy; and the larger the number of photons, the more their behaviour approaches that of a classical wave (this is one instance of the 'large numbers' hypothesis of quantum mechanics). It is for this reason that electromagnetic radiation becomes increasingly wave-like at low frequencies, and why we tend to think of radiofrequency and microwave radiation primarily in terms of waves rather than particles.

Wavelength The wavelength λ of the electric and magnetic waves is given by $\lambda = c/\nu$. In spectroscopy, common reference is made to its inverse, the *wavenumber* $\bar{\nu} = 1/\lambda$, usually expressed in cm^{-1}.

Momentum Each photon carries a linear momentum p, a vector quantity of magnitude $h/\lambda = \hbar k$ pointing in the direction of propagation. It is then convenient to define a *wave vector or propagation vector* $\mathbf{k} = k\hat{k}$ such that $p = \hbar k$. Since the photon momentum is proportional to frequency, photons of high frequency have high momenta and so exhibit the most particle-like behaviour. X-rays and gamma rays, for example, have many clearly ballistic properties not evident in electromagnetic radiation of lower frequencies.

Electromagnetic fields The electric and magnetic fields, *E* and *B* respectively, associated with a photon are vector quantities oriented such that the unit vectors ($\hat{E}$, $\hat{B}$, $\hat{k}$) form a right-handed orthogonal set.

Polarization For plane-polarized (also called linearly polarized) photons, the plane within which the electric field vector oscillates can sit at any angle to a reference plane containing the wave vector, as shown in **Figures 2A** and **2B**. Other polarization states are also possible: in the right- and left-handed circular polarizations depicted in **Figures 2C** and **2D**, the electric field vector sweeps out a helix about the direction of propagation. Elliptical polarization states are of an intermediate nature, between linear and circular. Together, the wave vector and polarization of a photon determine its *mode*.

Spin Many of the key properties of photons as elementary particles relate to the fact that they have an intrinsic spin $S = 1$, and so are classified as *bosons*

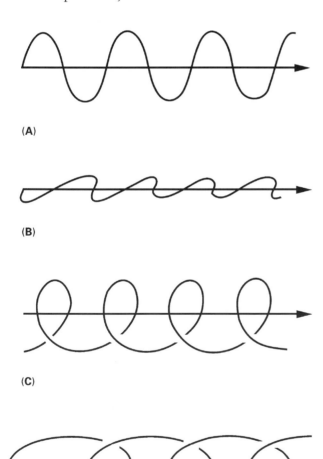

(A)

(B)

(C)

(D)

Figure 2 Polarization states: (A) and (B) plane; (C) and (D) circular.

(particles with integer spin as opposed to half-integer spin particles of matter such as electrons). As such, photons collectively display a behaviour properly described by a Bose–Einstein distribution. At simplest, this means that it is possible for their oscillating electromagnetic fields to keep in step as they propagate. Through this, coherent beams of highly monochromatic and unidirectional light can be produced; this is of course the basis for laser action.

Angular momentum The intrinsic spin of each photon is associated with an angular momentum, a feature that plays an important role in the selection rules for many spectroscopic processes. Circularly polarized photons have the special property of quantum angular momentum: the two circular polarization states, left- and right-handed, respectively carry +1 or −1 unit of angular momentum, $\hbar$.

Regions of the spectrum

We can now delineate the properties of the various spectral regions shown in **Figure 3**. Looking across the electromagnetic spectrum, it is immediately apparent that there is enormous difference in scale between the extremes of wavelength (or frequency), and it is not surprising to find that they encompass an enormous range of characteristics.

Gamma-rays This is a region of highly penetrating radiation with the highest optical frequencies (exceeding 3×10^{19} Hz), and an upper bound of 10 pm on the wavelength. Spectroscopy in this region primarily relates to nuclear decay process, as in the Mössbauer effect.

X-rays This region encompasses both 'hard' X-rays (wavelengths down to 10 pm) and less penetrating 'soft' X-rays (wavelengths up to 10 nm) with optical frequencies lying between 3×10^{19} and 3×10^{16} Hz. With the capacity to produce ionization by electron detachment, photon energies in this region are often reported in electronvolts (eV), and run from around 10^5 eV down to 100 eV (1 eV = 1.602×10^{-19} J). X-ray absorption and fluorescence spectra as such mostly relate to atomic core electronic transitions.

Ultraviolet With UV wavelengths running from 10 nm up to the violet end of the visible range at around 380 nm, this region is commonly divided at a wavelength of 200 nm into the 'far-UV' region (wavelengths below 200 nm, often referred to as 'vacuum UV' because oxygen absorbs here) and the near-UV (200–380 nm). With photon energies from 100 eV down to less than 10 eV – the latter being typical of the lowest atomic or molecular ionization

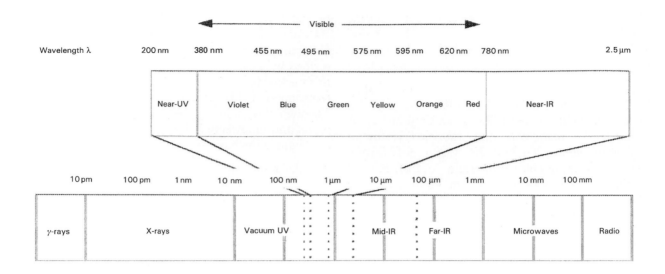

Figure 3   The electromagnetic spectrum.

energy – photoionization processes are associated with many of the techniques of ultraviolet spectroscopy. At the lower end of this region, photon energies are comparable with valence bond energies. As such they are commonly scaled by Avogadro's number and reported in units of kJ mol^{-1}; for example, the wavelength of 240 nm corresponds to 500 kJ mol^{-1}, a typical bond energy. One other form of division often applied to the near-UV region relates to its photobiological effects and applies principally to solar radiation; UV-A (320–380 nm) is the relatively safe region closest to the visible range; UV-B (280–320 nm) signifies radiation that can produce extensive tissue damage; UV-C radiation with wavelengths below 280 nm is potentially more damaging but is mostly filtered out by atmospheric gases that absorb here, the most important being ozone.

Visible   The visible range extends from approximately 380 nm (violet) to 780 nm (red) (**Table 1**). The precise divisions are a little arbitrary and depend on individual perceptions, but the wavelengths given in **Table 1** are a reasonable guide. Note that the link with perceived colour is not 1:1 – for example, light containing an equal mixture of red and green wavelengths appears yellow, although no yellow wavelengths are present. The visible range spans near enough an octave of frequencies and is a region in which photon energies are comparable with the bond energies of some of the weaker chemical bonds, running up from around 150 kJ mol^{-1} at the red end to over 300 kJ mol^{-1} at the other. Most of the spectroscopy in this range of wavelengths relates to electronic transitions unaccompanied by chemical change, and of course all absorption or fluorescence processes responsible for colour.

Table 1   The visible spectrum

Colour	Wavelengths (nm)
Red	780–620
Orange	620–595
Yellow	595–575
Green	575–495
Blue	495–455
Violet	455–380

Infrared   This is another region commonly subdivided, in this case into the near IR (wavelengths running from the red end of the visible spectrum at 780 nm out to 2.5 μm), the mid-IR (from 2.5 to 50 μm) and the far IR (50 μm to 1 mm). The absorption of near-IR radiation is commonly associated with the excitation of low-lying electronic excited states and overtones or combinations of molecular vibrations. In spectroscopic connections, the unqualified term 'infrared' generally refers to the mid-range, where spectral positions are usually cited by reference to wave-numbers $\bar{\nu} = 1/\lambda$. In the mid-IR range between 200 and 4000 cm^{-1}, it is principally vibrational transitions that accompany the absorption of radiation, with the corresponding nuclear motions less localized at the lower-wavenumber end of the scale. The far-IR region beyond relates mostly to low-frequency molecular vibrations and inversions, or rotations in small molecules.

Microwave   With wavelengths in the 1–100 mm range and frequencies on the GHz (10^9 Hz) scale, spectroscopy in the microwave region relates primarily to transitions involving molecular rotation, or others involving states of different electron spin orientation (electron spin resonance).

Radio With radiation of wavelengths exceeding 100 mm we finally run into the radio wave region, where frequencies are commonly reported in MHz (10^6 Hz). Photon energies here are too small to lead to transitions associated with electronic or nuclear movement, but they can produce transitions between spin states (nuclear magnetic resonance).

As a caveat by way of conclusion, it should be pointed out that the nature of transitions studied by a particular spectroscopic technique is not always so obviously linked with a particular optical frequency or wavelength region if the elementary interaction involved in the spectroscopy entails more than one photon. For example, Raman spectroscopy allows molecular vibrational transitions to be studied with visible or ultraviolet light, while multiphoton absorption of IR radiation can lead to electronic excitations.

List of symbols

B = magnetic field vector; B_0 = magnetic field amplitude; c = speed of light; c_0 = speed of light *in vacuo*; E = photon energy; E = electric field vector; E_0 = electric field amplitude; h = Planck constant; $\hbar$ = Planck constant/2π; k = wave vector; n_λ = refractive index at wavelength λ; p = photon momentum; S = spin; t = time; λ = wavelength; ν = frequency; $\overline{\nu}$ = wavenumber; ω = circular frequency.

See also: **Light Sources and Optics.**

Further reading

Ditchburn RW (1976) *Light*, pp 631–640. London: Academic Press.

Fishbane PM, Gasiorowicz S and Thornton ST (1993) *Physics for Scientists and Engineers*, extended version, pp 1009–1047. New Jersey: Prentice-Hall.

Goldin E (1982) *Waves and Photons*, pp 117–134. New York: Wiley.

Hakfoort C (1988) Newton's optics: the changing spectrum of science. In: Fauvel P, Flood R, Shortland M and Wilson R (eds) *Let Newton Be!*, pp 81–99. Oxford: Oxford University Press.

Sheppard N (1990) Chemical applications of molecular spectroscopy – a developing perspective. In: Andrews DL (ed) *Perspectives in Modern Chemical Spectroscopy*, pp 1–41. Berlin: Springer-Verlag.

Electron Magnetic Resonance Imaging

See **EPR Imaging.**

Electronic Components, Applications of Atomic Spectroscopy

John C Lindon, Imperial College of Science, Technology and Medicine, London, UK

ATOMIC SPECTROSCOPY
Applications

Atomic absorption spectroscopy and atomic emission spectroscopy have found application in many areas of materials science. The electronics industry requires materials of high purity and hence there is a need to monitor trace impurity levels in materials used for electronic components. As a consequence, various techniques of atomic spectroscopy have been applied to the analysis of materials used in the manufacture of electronic components. A number of recent studies are listed in the Further reading section and these therefore provide an entry point to the scientific literature. The reader should also consult other articles in this encyclopedia for related topics.

For example, the analysis of electronic grade silicon has been studied using atomic emission spectroscopy from inductively coupled plasma and the level of aluminium and phosphorus in such material has also been determined using an indirect atomic absorption method. The growth of europium on palladium surfaces has also been investigated using atomic emission spectroscopy.

The use of inductively coupled plasma–atomic emission spectroscopy (ICP-AES) has been applied to the analysis of traces of certain rare earth metals in highly pure rare earth matrices and to the levels of approximately 25 impurities in sintered electronic ceramics.

The impurity levels and profiles from the surface into the bulk sample have been probed using a combination of secondary-ion mass spectrometry and electrothermal atomic absorption spectroscopy in composite materials such as CdZnTe.

See also: **Atomic Absorption, Methods and Instrumentation; Atomic Absorption, Theory; Atomic Emission, Methods and Instrumentation; Atomic Spectroscopy, Historical Perspective; Inductively Coupled Plasma Mass Spectrometry, Methods; Materials Science Applications of X-Ray Diffraction; X-Ray Fluorescence Spectroscopy, Applications.**

Further reading

Bertran F, Gourieux T, Krill G, Alnot M, Ehrhardt JJ and Felsch W (1991) Growth of Eu on Pd(111) – AES, photoemission and RHEED studies. *Surface Science* 245: L163–L169.

Daskalova N, Velichkov S, Krasnobaeva N and Slavova P (1992) Spectral interferences in the determination of traces of scandium, yttrium and rare earth elements in pure rare earth matrices by inductively coupled plasma atomic emission spectrometry. *Spectrochimica Acta* B47: E1595–E1620.

Gerardi C, Milella E, Campanella F and Bernadi S (1996) SIMS-ETAAS characterization of background impurities in CdZnTe bulk samples. *Material Science Forum* 203: 273–278.

Gupta PK and Ramchandran R (1991) Indirect atomic absorption spectrometric determination of phosphorus in high purity electronic grade silicon using bismuth phosphomolybdate complex. *Microchemical Journal* 44: 34–38.

Morvan D, Amouroux J and Claraz P (1984) Analysis of electronic and solar grade silicon by atomic emission spectroscopy from inductively coupled plasma. *Progress in Crystal Growth and Characterization* 8: 175–180.

Uwamino Y, Morikawa H, Tsuge A, Nakane K, Iida Y and Ishizuka T (1994) Determination of impurities in sintered electronic ceramics by inductively coupled plasma atomic emission spectrometry. *Microchemical Journal* 49: 173–182.

Ellipsometry

GE Jellison Jr, Oak Ridge National Laboratory, Oak Ridge, TN, USA

ELECTRONIC SPECTROSCOPY
Applications

Ellipsometry is a technique often used to measure the thickness of a thin film. Generally speaking, the measurement is performed by polarizing the incident light beam, reflecting it off a smooth sample surface at a large oblique angle and then re-polarizing the light beam before the intensity of the light beam is measured. Since the process of reflecting light off a smooth sample surface generally changes linearly polarized light into elliptically polarized light, the technique has been called 'ellipsometry'.

The earliest ellipsometry measurements (ca 1890) were used to determine the optical functions (refractive index n and extinction coefficient k, or equivalently, absorption coefficient α) for several materials. In the 1940s it was discovered that a single-wavelength nulling ellipsometry measurement could be used to determine the thickness of certain thin films very accurately. Since that time, single-wavelength ellipsometry has evolved to be the standard of thickness measurement for several industries, including the semiconductor industry.

Ellipsometry experiments produce values that are not useful by themselves: computers must be used to obtain useful quantities such as thin-film thickness or the optical functions of materials. The advent of modern computers has resulted in the invention of several spectroscopic ellipsometers and the creation of more realistic analysis programs required to understand spectroscopic ellipsometry data.

Any ellipsometer (see **Figure 1**) consists of five elements: (1) a light source, (2) a polarization state generator (PSG), (3) a sample, (4) a polarization state detector (PSD), and (5) a light detector. The light source can be a monochromatic source, such as from a laser, or a white light source, such as from a xenon or mercury arc lamp. The PSG and

PSD are optical instruments that change the polarization state of a light beam passing through them and they contain optical elements such as polarizers, retarders and photoelastic modulators. In most ellipsometry experiments, light from the PSG is reflected from the sample surface at a large angle of incidence ϕ. Spectroscopic ellipsometers use a white light source and a monochromator (either before the PSG or after the PSD) to select out specific wavelengths. Some spectroscopic ellipsometers image the white light from the PSD onto a detector array, thereby allowing the whole spectrum to be collected simultaneously.

Any ellipsometer will only measure characteristics of the light reflected from or light transmitted through the sample. Ellipsometers do not measure film thicknesses or optical functions of materials, although these parameters can often be inferred very accurately from the ellipsometry measurements. Data analysis is an essential part of any ellipsometry experiment.

Polarization optics

Several optical elements will change the polarization state of a light beam interacting with them. Linear polarizers (see **Figure 2A**) transmit only one polarization of the light beam; the azimuthal angle of the polarizer will determine the azimuthal angle of the light polarization. The best linear polarizers in the visible part of the spectrum are prism polarizers made from birefringent crystals such as calcite, magnesium fluoride or quartz, which have refractive indices that depend upon the direction in the material.

Retarders or compensators (see **Figure 2B**) are another common type of optical element used in ellipsometers. In combination with a polarizer, a retarder can produce circularly or elliptically polarized light if the direction of the linear polarization is not parallel to a major axis in the retarder. Static retarders, such as quarter-wave or half-wave plates produces a fixed amount of retardation at a fixed wavelength. Circularly polarized light can be produced from a polarizer–quarter-wave retarder pair if the light polarization is ±45° with respect to the fast axis of a retarder.

Another common retarding element used in spectroscopic ellipsometers is the photoelastic modulator (PEM), shown schematically in **Figure 2C**. The drive element is a crystal quartz bar, which is cut so that an a.c. voltage V applied to the front and back faces causes the bar to vibrate along its long axis at the frequency of V. The optical element is a bar made of fused quartz that is placed in intimate contact with the crystal quartz bar so that it vibrates at the same frequency. Both bars are sized so that the vibrations are resonant, making the frequency of the PEM extremely stable. A polarizer–PEM pair produces dynamically elliptically polarized light if the polarizer

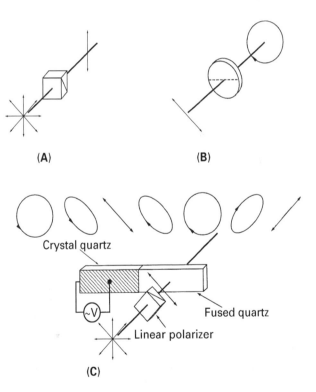

Figure 2 Several optical elements used in ellipsometers that alter the polarization state of a light beam passing through them. (A) A linear polarizer, which transforms any light beam (polarized or unpolarized) to linearly polarized light. (B) A retarder, which transforms linearly polarized light to elliptically polarized light. In certain cases, the light can be circularly polarized. (C) A photoelastic modulator, which changes linearly polarized light into dynamically elliptically polarized light.

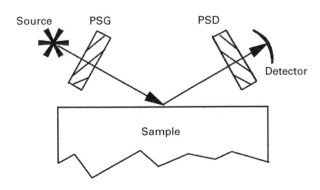

Figure 1 Schematic diagram of an ellipsometer. The PSG is the polarization state generator and the PSD is the polarization state detector.

is not lined up with the vibration axis of the PEM; generally, the polarization state of the emergent light beam will cycle through linearly polarized, circularly polarized and elliptically polarized states.

The Stokes vector representation of light polarization has often been used for ellipsometry measurements. (Another representation is the Jones vector representation, of which details can be found in the book by Azzam and Bashara listed under Further reading). In the Stokes representation, the polarization state of a light beam is given by its four-element Stokes vector,

$$\boldsymbol{S} = \begin{bmatrix} I_0 \\ Q \\ U \\ V \end{bmatrix} = \begin{bmatrix} I_0 \\ I_0 - I_{90} \\ I_{45} - I_{-45} \\ I_{rc} - I_{lc} \end{bmatrix} \quad [1]$$

All the elements of the Stokes vector are intensities and therefore are real. The total intensity is I_0, while I_0, I_{90}, I_{45}, and I_{-45}, are the intensities of linearly polarized light at 0°, 90°, 45°, and −45°, respectively. The fourth element of the Stokes vector V is the difference between the intensities of right circularly polarized light I_{rc}, and left circularly polarized light I_{lc}. A linear polarizer can only transform a light beam into a linear polarization state; therefore the fourth element of the Stokes vector for linearly polarized light will be $V = 0$. Generally, a retarding optical element is needed to transform the polarization state of a light beam to one where $V \neq 0$. The Stokes representation can also represent partially polarized light. In general,

$$I_0 \geq \sqrt{Q^2 + U^2 + V^2} \quad [2]$$

where the equality holds when the light beam is totally polarized.

Each optical element must transform the polarization state from one four-element vector to another four-element vector; therefore, optical elements are represented by 4 × 4 Mueller matrices, where all the elements are real. The intensity of the light beam passing through an ellipsometer can therefore be written using matrix notation as

$$I = \boldsymbol{S}_{PSD}^{T} \, \mathbf{M} \, \boldsymbol{S}_{PSG} \quad [3]$$

where S_{PSG} is the Stokes vector representing the polarization state for light coming from the PSG, **M**

is the Mueller matrix for the light interaction with the sample, and S_{PSD}^{T} is the transpose of the Stokes vector representing the effect of the PSD. In the most general case, 16 elements are required to describe the light interaction with a sample. Fortunately, many of these elements are normally 0 or equal to another element in **M**. Since detectors can only measure the light intensity coming from an instrument, Equation [3] emphasizes the fact that ellipsometers can only measure elements of **M**.

Ellipsometry data analysis

Ellipsometry measurements are not useful by themselves but can be extremely useful if the measurements are interpreted with an appropriate model. Ultimately, ellipsometry results are always model dependent. Fortunately, the physics of light reflection from surfaces is well understood, and very detailed and accurate models can be made using classical electromagnetic theory based on the Maxwell equations.

Figure 3 shows a schematic of light reflection from a sample surface. The light beam is incident upon the sample surface at an angle of incidence ϕ. The specularly reflected beam comes from the sample surface, also at angle ϕ. The incident and reflected beams define a plane, called the plane of incidence. This in turn defines two polarization directions: p for the light polarization parallel to the plane of incidence (in the plane of the paper), and s for the light polarization perpendicular to the plane of incidence (perpendicular to the plane of the paper). All azimuthal angles are defined with respect to the plane of incidence, where positive rotations are defined as clockwise rotations looking from the light source to the detector.

If the sample surface is isotropic and has no film or other overlayer ($d = 0$ in **Figure 3**), then one can use the Maxwell equations to calculate the complex reflection coefficients:

$$r_p = \frac{N_s \cos(\phi_0) - N_0 \cos(\phi_s)}{N_s \cos(\phi_0) + N_0 \cos(\phi_s)} \quad [4a]$$

$$r_s = \frac{N_0 \cos(\phi_0) - N_s \cos(\phi_s)}{N_0 \cos(\phi_0) + N_s \cos(\phi_s)} \quad [4b]$$

In Equations [4], N_0 and N_s ($= n_s + ik_s$, where n_s is the refractive index and k_s is the extinction coefficient) are the complex indices of refraction for the ambient (usually air, where $N_0 = 1$) and the substrate, respectively. The quantities ϕ_0 and ϕ_s are complex angles, determined from the Snell law [$N_0 \sin(\phi_0) = N_s \sin(\phi_s)$]. The reflection ratios r_p and r_s are complex, indicating that light reflected from a surface will generally undergo a phase shift.

If the sample near-surface region consists of a single film (see **Figure 3**), the composite reflection coefficients can be calculated from the Airy formula:

$$r_{s,p} = \frac{r_{1s,p} + r_{2s,p}e^{-2ib}}{1 + r_{1s,p}r_{2s,p}e^{-2ib}} \qquad b = \frac{2\pi d N_f \cos(\phi_f)}{\lambda} \quad [5]$$

where d is the thickness of the film, N_f is the complex refractive index of the film, and ϕ_f is the complex angle within the film defined by the Snell law. The quantities $r_{1s,p}$ and $r_{2s,p}$ are the complex reflection coefficients calculated using Equations [4a] and [4b] for the air–film interface and the film–substrate interface, respectively.

Reflections from more complicated layer structures can be calculated using matrix methods. If all the media in the calculation are isotropic, the matrix formulation of Abelés (using 2×2 complex matrices) can be used to calculate the composite r_s and r_p.

If any of the media are birefringent, many of the implicit assumptions made above are no longer valid. For example, for isotropic media the s and p polarization states represent eigenmodes of the reflection; that is, if the incoming light is pure s or p polarized, then the reflected light will be pure s or p polarized. If any of the media in the sample are birefringent, this assumption is no longer generally valid. As a result, one must also calculate the cross-polarization coefficients r_{sp} and r_{ps} as well as $r_{ss} \equiv r_s$ and $r_{pp} \equiv r_p$. If $r_{sp} \neq 0$ or $r_{ps} \neq 0$, the 2×2 matrix methods must be replaced by more complicated 4×4 matrix methods.

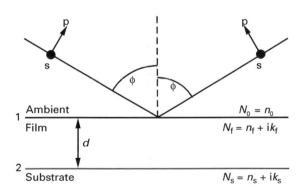

Figure 3 A schematic diagram of light reflecting from a sample surface. The s and p vectors indicate the direction of s and p polarized light, as defined by the plane of incidence. The angle of incidence is ϕ, and N_0, N_f and N_s are the complex refractive indices of the ambient, film and substrate, respectively.

Up to this point, a clear distinction has been drawn between the parameters measured by an ellipsometer (the sample Mueller matrix **M**) and the calculated parameters obtained from classical electromagnetic theory, which can be expressed as elements of the sample Jones matrix:

$$\mathbf{J} = \begin{bmatrix} r_{pp} & r_{ps} \\ r_{sp} & t_{ss} \end{bmatrix} \quad [6]$$

Fortunately, the two representations are related, so long as it can be assumed that the sample does not depolarize the incident light beam. In this case, a Mueller–Jones matrix $\mathbf{M}_{MJ}$ can be calculated from the elements of the Jones matrix (Eqn [6]),

$$\mathbf{M}_{MJ} = \mathbf{A}\,(\mathbf{J} \otimes \mathbf{J}^*)\,\mathbf{A}^{-1} \quad [7a]$$

where

$$\mathbf{A} = \begin{bmatrix} 1 & 0 & 0 & 1 \\ 1 & 0 & 0 & -1 \\ 0 & 1 & 1 & 0 \\ 0 & -i & i & 0 \end{bmatrix} \quad [7b]$$

For isotropic samples ($r_{sp} = r_{ps} = 0$), the normalized sample Mueller–Jones matrix is given by

$$\mathbf{M} = \mathbf{M}_{MJ} = \begin{bmatrix} 1 & -N & 0 & 0 \\ -N & 1 & 0 & 0 \\ 0 & 0 & C & S \\ 0 & 0 & -S & C \end{bmatrix} \quad [8a]$$

where

$$N = \cos(2\psi) \quad [8b]$$

$$S = \sin(2\psi)\sin(\Delta) \quad [8c]$$

$$C = \sin(2\psi)\cos(\Delta) \quad [8d]$$

$$\rho = \frac{r_p}{r_s} = \tan(\psi)e^{i\Delta} = \frac{C + iS}{1 + N} = \rho_r + i\rho_i \quad [8e]$$

The angles ψ and Δ are the traditional ellipsometry angles, and are the naturally measured parameters for nulling ellipsometers (see below). The matrix **M** simplifies considerably when the sample is isotropic: **M** is block-diagonal, with eight elements equal to 0, and only two parameters (such as $\rho = \rho_r + i\rho_i$ or ψ and Δ) are needed to specify $\mathbf{M}$, since $N^2 + S^2 + C^2 = 1$.

For birefringent samples, $r_{sp} \neq r_{ps} \neq 0$, and the off-block diagonal elements of $\mathbf{M}$ are no longer 0, nor are the on-block diagonal elements of $\mathbf{M}$ so simply defined. However, Equation [7a] is still valid as long as the sample is nondepolarizing. In this case, six parameters are required to specify all 16 elements of the normalized sample Mueller matrix, given by Equation [8e] and

$$\rho_{ps} = \frac{r_{ps}}{r_{ss}} = \tan(\psi_{ps})e^{i\Delta_{ps}} = \frac{C_{ps} + iS_{ps}}{1 + N} \quad [8f]$$

$$\rho_{sp} = \frac{r_{sp}}{r_{ss}} = \tan(\psi_{sp})e^{i\Delta_{sp}} = \frac{C_{sp} + iS_{sp}}{1 + N} \quad [8g]$$

Up to now, we have assumed that no optical element depolarizes the light. Usually, this is true, but there are some cases where depolarization must be considered. (1) If the input light beam illuminates an area of the sample where the film thickness(es) is (are) not uniform, quasi-depolarization can occur. (2) If the sample substrate is transparent, then light reflecting from the back surface will contribute an intensity component to the light beam entering the PSD that is not phase-related to the light reflected from the front face and the light beam will be quasi-depolarized. (3) If the sample is very rough, then some of the light reaching the PSD will not have an identifiable polarization state or cross-polarization can occur in nominally isotropic systems. All of these effects invalidate the Mueller–Jones matrix representation of the sample surface shown in Equation [7a].

Types of ellipsometers

There are many different kinds of ellipsometers, some of which are shown schematically in **Figure 4**. All ellipsometers measure one or more quantities that can be related to the complex reflection coefficient ratio ρ shown in Equations [8].

Nulling ellipsometer

The oldest and most common type of ellipsometer is the nulling ellipsometer, shown schematically in **Figure 4A**. The light source for a modern nulling ellipsometer is usually a small laser, but other monochromatic sources can also be used. The PSG is a polarizer–retarder pair, and the PSD is a linear polarizer, historically called the analyser. If the quarter-wave plate is oriented at 45° with respect to the plane of incidence, the intensity of the light beam incident upon the detector is given by

$$I = I_0[1 - N\cos(2\theta_a) + \sin(2\theta_a)(C\sin(2\theta_p) + S\cos(2\theta_p))] \quad [9]$$

where θ_p and θ_a are the azimuthal angles of the polarizer and analyser, respectively, and N, S and C are the quantities given in Equations [8]. The intensity given in Equation [9] has a null at two sets of angles: $(2\theta_p, \theta_a) = (270° - \Delta, \psi)$, $(90° - \Delta, 180° - \psi)$. Therefore, nulling ellipsometry measurements are made by rotating the PSG polarizer and the PSD polarizer (analyser) until the light intensity reaching the detector is a minimum. The angles ψ and Δ are determined from the azimuthal angles of the PSG and PSD polarizers.

The nulling ellipsometer is one of the simplest ellipsometers and it is capable of very accurate measurements with proper calibration. However, it normally uses a single-wavelength source and it is not easily made spectroscopic. Moreover, measurement times are slow, so this type of instrument is not useful for spectroscopic ellipsometry or for fast time-resolved measurements.

Rotating analyser ellipsometer (RAE)

The most common type of spectroscopic ellipsometer is the rotating analyser (or rotating polarizer) ellipsometer, shown schematically in **Figure 4B**. The PSG is a linear polarizer, as is the PSD, and one of the optical elements (the analyser in **Figure 4B**) is physically rotated, making the light intensity at the detector a periodic function of time. For an analyser (PSD) rotating at an angular frequency of ω and with the polarizer (PSG) set at an azimuthal angle of θ_p, the intensity is given by

$$I(t) = I_0[1 + \alpha\cos(2\omega t + \delta_c) + \beta\sin(2\omega t + \delta_c)] \quad [10a]$$

where

$$\alpha = \frac{\cos(2\theta_p) - N}{1 - N\cos(2\theta_p)} \quad [10b]$$

$$\beta = \frac{C\sin(2\theta_p)}{1 - N\cos(2\theta_p)} \quad [10c]$$

The quantity δ_c is a constant phase shift. The coefficients α and β can be determined using demodulation techniques or Fourier analysis. Normally, $\theta_p = 45°$, where $\alpha = -N$, and $\beta = C$.

In its normal form, the RAE includes only polarizers, although a retarder can be added to either the PSG or the PSD. If only polarizers are used, this instrument is very easy to make spectroscopic, since no optical component other than the sample is wavelength dependent. Furthermore, the measurement time depends principally on the rotation speed of the rotating element. For a rotation speed of 100 Hz, measurements can be completed in 40 ms. In some configurations of the rotating polarizer ellipsometer, a spectrograph is placed after the PSD and the light is detected by a photodiode array. This ellipsometer can collect an entire spectrum with a single series of measurements, typically taking about 1s to collect and analyse each spectrum.

The RAE measures two quantities, N and a linear combination of C and S. If there is no retarder, the RAE measures N and C, and will therefore be subject to large errors in Δ if Δ is near 0° or 180° and an uncertainty of the sign of Δ. The use of a quarter-wave retarder shifts the inaccuracy in Δ to the regions near ±90°.

Photoelastic modulator ellipsometer (PME)

Another common spectroscopic ellipsometer is based on the photoelastic modulator (PEM), shown in **Figure 4C**. Normally, the PSG consists of a polarizer–PEM pair, where the polarizer is oriented at 45° with respect to the vibration axis of the PEM (see **Figure 2C**). The intensity of the light arriving at the detector is a function of time given by

$$I(t) = I_0[I_{dc} + I_X \sin(A\sin(\omega t)) + I_Y \cos(A\sin(\omega t))]$$
$$[11a]$$

where

$$I_{dc} = 1 - N\cos(2\theta_a) \qquad [11b]$$

$$I_X = S\sin(2\theta_a) \qquad [11c]$$

$$I_Y = \sin(2\theta_m)\left[\cos(2\theta_m) - N\right] - C\cos(2\theta_m)\sin(2\theta_a)$$
$$[11d]$$

The quantity A is the Bessel angle of the PEM modulation, and is proportional to the amplitude of modulation; normally, this is set at 2.4048 radians. The frequency of the PEM is denoted by $2\pi\omega$, and is normally ~50 kHz. The quantities θ_m and θ_a are the azimuthal angles of the modulator and the PSD polarizer, respectively.

The time dependence of the light intensity from the PME is considerably more complicated than that from the RAE, since the basis functions are $\sin(A\sin(\omega t))$ and $\cos(A\sin(\omega t))$, rather than $\sin(\omega t)$ and $\cos(\omega t)$ as for the RAE. However, these functions can be expressed in terms of a Fourier–Bessel infinite series:

$$\sin(A\sin(\omega t)) = 2\sum_{k=1}^{\infty} J_{2k-1}(A)\sin((2k-1)\omega t)$$
$$[12a]$$

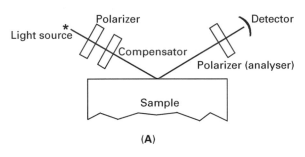

(A)

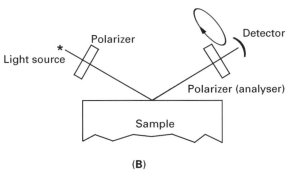

(B)

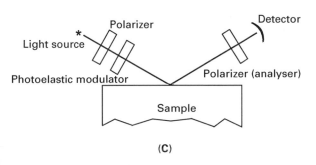

(C)

Figure 4 Schematic diagrams of three ellipsometers in common use. (A) The nulling ellipsometer, which normally uses monochromatic light. (B) The rotating analyser ellipsometer. (C) The polarization modulation ellipsometer.

$$\cos(A(\sin(\omega t))) = J_0(A) + 2 \sum_{k=1}^{\infty} J_{2k}(A) \sin(2k\omega t)$$

[12b]

Therefore, a demodulation or Fourier analysis of the waveform of Equation [11a] can also be used to get the coefficients I_X and I_Y.

The PME is more complicated than the RAE, since the modulation amplitude of the PEM must be calibrated and its small but significant static retardation measured at all wavelengths used. Furthermore, the frequency of the PEM is about 500 times faster than the rotation speed of the RAE; this improves the time resolution to ~1 ms, but requires faster electronics for data collection. In its normal configuration discussed above, the PME can measure S and either N or C. However, if the PSD polarizer is replaced with a Wollaston prism polarizer and both channels are detected (measuring a total of four quantities), then it is possible to measure N, S, and C simultaneously. This implementation is called the two-channel spectroscopic polarization modulation ellipsometer (2C-SPME).

Other configurations

The ellipsometer configurations discussed above are particularly useful if the sample is isotropic. If the sample is anisotropic, then many of the simplifications used above are not valid, and many more measurements must usually be made, since the normalized sample Mueller matrix contains six independent quantities. One such implementation is the two-modulator generalized ellipsometer (2-MGE) which uses two PEMs operating at different frequencies. The 2-MGE is spectroscopic and is capable of measuring all six independent elements of the sample Mueller matrix simultaneously. Other configurations are discussed in the review article by Hauge.

Application: Silicon dioxide films on silicon

One of the most useful applications of ellipsometry has been the routine measurement of silicon dioxide (SiO_2) film thicknesses grown on silicon. Since the refractive index of SiO_2 is well known, and does not depend significantly on film deposition technique, nulling ellipsometry measurements are usually sufficient to determine film thickness.

Figure 5 shows a plot of ψ versus Δ for thin-film SiO_2 grown on silicon for a wavelength of 633 nm (the wavelength of a HeNe laser). The angle of incidence used for the calculation was 70°, the film

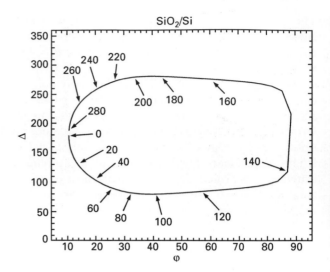

Figure 5 The ψ–Δ trajectory for a thin film of SiO_2 on silicon. The angle of incidence $\phi = 70°$, $n_f = 1.46$, $N_s = 3.86 + i0.018$, and the wavelength $\lambda = 633$ nm (HeNe laser).

refractive index $n_f = 1.46$, and the complex refractive index of silicon $N_s = n_s + ik_s = 3.86 + i\,0.018$. This plot shows the trajectory that ψ–Δ follows as the film thickness increases from zero to ~280 nm, where the position along the trajectory for several thicknesses is noted in the figure. For zero film thickness, the value of Δ is very close to 0° or 180°, owing to the very small value of k_s. If other film refractive indices were included in this plot, it would be seen that they would converge to this point. It is therefore very difficult to determine the refractive index and the thickness of a very thin film from ellipsometry measurements. Note that the curve comes around on itself near 283 nm. Therefore, any ellipsometry measurement of SiO_2/Si can only be used to determine the film thickness modulo the repeat thickness (283 nm). For film thicknesses from 140 to 160 nm, the measurements are very sensitive to film thickness, crossing the $\Delta = 180°$ point for a thickness of 141.5 nm (half the repeat thickness).

The same data are plotted in a different manner in **Figure 6**. Here, the complex ρ ($= \tan(\psi)\, e^{i\Delta}$, see Equation [8e]) is plotted versus film thickness. Near zero thickness, $\mathrm{Im}(\rho)$ (which is proportional to $\sin(\Delta)$) is very small. As the thickness approaches 140 nm, $-\mathrm{Re}(\rho)$ gets very large and $\mathrm{Im}(\rho)$ changes sign ($\mathrm{Im}(\rho) = 0$ at half the repeat thickness); this corresponds to the region of large ψ in **Figure 5**, with the change in the sign of $\mathrm{Im}(\rho)$ corresponding to Δ going through 180°. Again, the value of ρ repeats itself after 283 nm.

Figures 7A–C show the measured values of of ρ for three thicknesses of SiO_2 films grown on silicon

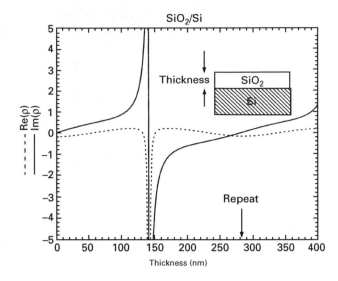

Figure 6 The complex reflection ratio ρ for the system described in **Figure 5**.

Table 1 The results of the fitting procedure on the spectroscopic ellipsometry data taken on several thin-film SiO$_2$/Si samples, shown in **Figures 7A–C**. The refractive index is calculated from Equation [13] assuming $\lambda_0 = 92.3$ nm

Sample	Film thickness (nm)	A	Refractive index (600 nm)	Interface thickness (nm)	χ^2
a	66.4 ± 0.1	1.149 ± 0.002	1.475	0.0 ± 0.2	0.18
b	188.9 ± 0.3	1.115 ± 0.003	1.464	1.4 ± 0.2	1.21
c	321.5 ± 0.4	1.139 ± 0.003	1.472	1.0 ± 0.2	0.84
Fused silica	–	1.099	1.458	–	–

(66 nm, 190 nm and 321 nm) at an angle of incidence of 65°. There are several points where Im(ρ) = 0 and where –Re(ρ) is a maximum: ~335 nm for the 66 nm film; ~305 nm for the 190 nm film; and ~315 nm and ~505 nm for the 321 nm film. Using spectroscopic ellipsometry for this system is similar to sampling several points along the ψ–Δ trajectory shown in **Figure 5**, although the precise values of the null points of Im(ρ) will not scale, since N_f and N_s are both functions of wavelength.

Spectroscopic ellipsometry data analysis

It is easy to see that there are more than 100 data points in each of **Figures 7A–C**, but only a few parameters to be determined from the data; the problem is overdetermined. Therefore, fitting procedures must be used to determine film thicknesses and other properties of thin films from spectroscopic ellipsometry. This process has three steps: (1) model the near-surface region of the sample; (2) parametrize or determine the optical functions of each layer; (3) fit the data using a realistic figure of merit to measure the 'goodness of fit'. For the case of SiO$_2$ grown on silicon, a four-media model is used, consisting of air/SiO$_2$/interface/silicon. The optical functions of the SiO$_2$ layer are assumed to follow the Sellmeier approximation:

$$n^2 = 1 + \frac{A\lambda^2}{\lambda^2 - \lambda_0^2} \qquad [13]$$

where the two possible fitting parameters are the amplitude A and the resonance wavelength λ_0. The interface is assumed to be a thin layer whose optical functions are those of a Bruggeman effective medium, consisting of 50% SiO$_2$ and 50% Si, and the optical functions of silicon are taken from the literature. Using this model, three parameters are determined in the fitting procedure (listed in **Table 1**): the film thickness d_f, A from Equation [12] and the thickness of the interface $d_{\text{interface}}$; the resonant wavelength $\lambda_0 = 92.3$ nm. The third step of the analysis involves the actual fitting of the calculated values of ρ with the measured values of ρ. This can be done using a Levenberg–Marquardt algorithm to solve for the fitted parameters $d_{\text{interface}}$, d_f and A using the reduced χ^2 as the figure of merit:

$$\chi^2 = \frac{1}{N - m - 1} \sum_{i=1}^{N} \frac{(\rho_{\exp}(\lambda_i) - \rho_{\text{calc}}(\lambda_i, \mathbf{z}))^2}{\delta\rho_{\exp}^2(\lambda_i)} \qquad [14]$$

The sum in Equation [13] is taken over N wavelength points (λ_i), $\mathbf{z}$ is a vector of the fitted parameters (in this case, $d_{\text{interface}}$, d_f and A), and m is the dimensionality of $\mathbf{z}$ (i.e. 3). The quantities in the denominator ($\delta\rho_{\exp}^2(\lambda_i)$) are the errors in the experimental data. The χ^2 is a good metric for the 'goodness of fit' to ellipsometry data. If χ^2 is near 1 the fit is a good fit, but if it is much greater than 1 the model does not fit the data. After the parameters are determined, it is also necessary to calculate the error limits as well as the correlation coefficients of all the fitted parameters.

The results of the fits to the data presented in **Figures 7A–C** are shown in **Table 1**, as well as the errors of each of the fitted parameters and the final χ^2 of the fit. All χ^2 values are near 1, indicating that the model fits the data. The errors listed for the thickness are all small, showing that the film thickness is a very well determined parameter for these data sets. The parameter A is a measure of the SiO$_2$ refractive index,

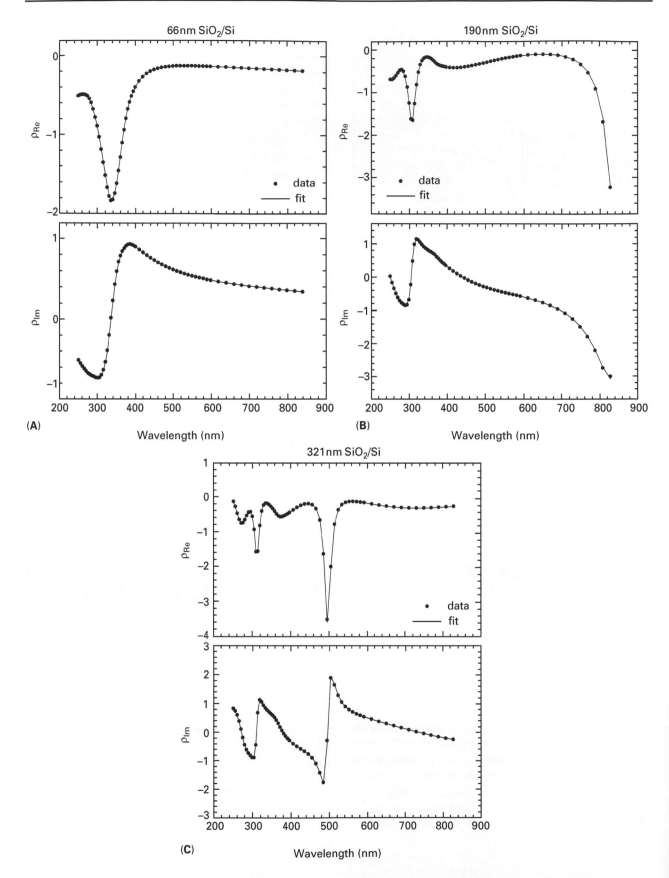

Figure 7 The spectroscopic values of the complex reflection ratio ρ for three samples of SiO$_2$ films grown on silicon as measured by spectroscopic ellipsometry. The three different thicknesses are (A) 66nm, (B) 190nm, and (C) 321nm.

showing that thin-film SiO_2 can have a slightly larger refractive index than bulk fused silica.

List of symbols

A = Bessel angle of PEM modulation, amplitude; $C = \sin(2\psi)\cos(\Delta)$; d = film thickness; I_0 = total intensity element of Stokes vector; I_n = element of Stokes vector, intensity of light linearly polarized at n degrees; J_n = Bessel functions; $\mathbf{J}$ = Jones matrix; k = extinction coefficient; m = dimensionality of z; $\mathbf{M}$ = Mueller matrix, normalized Mueller–Jones matrix; $\mathbf{M}_{MJ}$ = Mueller–Jones matrix; n = refractive index; N_0, N_f, N_s = complex refractive indices; r_p, r_s, etc. = complex reflection coefficients; $S = \sin(2\psi)\sin(\Delta)$; $\mathbf{S}$ = Stokes vector; V = voltage, I_{rc}–I_{lc} element of Stokes vector; z = vector of fitted parameters (Eqn [14]); α = absorption coefficient; δ_c = constant phase shift; Δ = ellipsometry angle; θ = azimuthal angle; λ = wavelength; ρ = complex reflection ratio; ϕ = angle of incidence; ϕ_f, ϕ_s = complex angles; χ^2 = figure of merit; ψ = ellipsometry angle; ω = angular frequency.

See also: **Chiroptical Spectroscopy, Oriented Molecules and Anisotropic Systems; Fibres and Films Studied Using X-Ray Diffraction; Fluorescence Polarization and Anisotropy; Light Sources and Optics; Linear Dichroism Theory.**

Further reading

Azzam RMA (1995) *Ellipsometry*, In: Bass M (ed) *Handbook of Optics II*, Chapter 27. New York: McGraw-Hill.

Azzam RMA and Bashara NM (1987) *Ellipsometry and Polarized Light*. Amsterdam: Elsevier Science.

Chipman RA (1995) Polarimetry. In: Bass M (ed) *Handbook of Optics II*, Chapter 22. New York: McGraw-Hill.

Collins RW (1990) Automatic rotating element ellipsometers: calibration, operation, and real-time applications. *Review of Scientific Instruments* **61**: 2029–2062.

Drévillon B (1993) Phase modulated ellipsometry from the ultraviolet to the infrared: *in-situ* application to the growth of semiconductors. *Progress in Crystal Growth and Characterization of Materials* **27**: 1–87.

Hauge PS (1980) Recent developments in instrumentation in ellipsometry. *Surface Science* **96**: 108–140.

Jellison GE Jr (1993) Characterization of the near-surface region using polarization-sensitive optical techniques. In: Exharos GJ (ed) *Characterization of Optical Materials*, pp 27–47. Stoneham, MA: Butterworth-Heinemann.

Jellison GE (1998) Spectroscopic ellipsometry data analysis: measured vs. calculated quantities. *Thin Solid Films* **313–314**: 33–39.

Kliger DS, Lewis JW and Randall CE (1990) *Polarized Light in Optics and Spectroscopy*. New York: Academic Press.

Tompkins HG (1993) *A User's Guide to Ellipsometry*. New York: Academic Press.

Enantiomeric Purity Studied Using NMR

Thomas J Wenzel, Bates College, Lewiston, ME, USA

MAGNETIC RESONANCE
Applications

Introduction

Living organisms have a remarkable ability to distinguish enantiomers. Enzymes and cell surface receptors are handed, such that enantiomers often exhibit different properties in living systems. In extreme cases one enantiomer might be beneficial to a living organism, whereas the other might be harmful. Therefore, it is critical that methods exist to determine the enantiomeric purity and assign absolute configurations of chemical compounds. Various types of chiroptical spectroscopy are used for this. NMR spectroscopy is one of the most common and powerful methods for the determination of enantiomeric excess (typically to 1% of the minor enantiomer) and absolute configurations. There are generally considered to be three broad categories of reagents suitable for chiral resolution in NMR spectroscopy: chiral derivatizing agents, chiral solvating agents, and lanthanide shift reagents.

Chiral derivatizing agents are optically pure compounds that undergo reactions with the enantiomers being analysed. Most derivatization reactions involve the formation of covalent bonds; however, some examples, most notably with carboxylic acids and amines, rely on the formation of soluble salts.

The resulting complexes are diastereoisomers, which may then exhibit different chemical shifts in their NMR spectra. Derivatization agents must react qu antitatively with both enantiomers, with complete retention of configuration of both the substrate and resolving agent or with complete stereospecificity of any perturbations that occur. Useful chiral derivatizing agents usually have a discrete and specific NMR signal (e.g. a methyl group, fluorine, or phosphorus singlet in the ^{1}H, ^{19}F, or ^{31}P NMR spectrum) that is conveniently monitored for enantiomeric resolution. Most chiral derivatizing agents function with compounds that are readily modified, such as amines, alcohols, and carboxylic acids. In many instances it is possible to assign absolute configurations either by observing trends in the data on similar families of compounds, or through a detailed analysis of the geometric differences that lead to distinctions in the chemical shift values of the diastereoisomers.

Chiral solvating agents are additives that react *in situ* with the compound being analysed. Rather than forming covalent bonds or ion pairs, chiral solvating agents associate with the substrate through combinations of van der Waals forces. Hydrogen bonding, charge-transfer complexation of electron-rich and electron-deficient aromatic rings, and steric effects are important interactions. The change in chemical shifts that occur in the spectrum of a compound in the presence of a chiral solvating agent are denoted as $\Delta\delta$. The extent of enantiomeric resolution that might then be observed for a shifted resonance is denoted as $\Delta\Delta\delta$.

Chiral solvating agents function by two possible mechanisms: Equations [1] and [2] represent the association of a pair of enantiomers with a chiral solvating agent (CSA), in which K_R and K_S denote the association constants for the R and S enantiomers

$$R + CSA \rightarrow R\text{-}CSA \qquad K_R \qquad [1]$$

$$S + CSA \rightarrow S\text{-}CSA \qquad K_S \qquad [2]$$

respectively. These reactions should occur rapidly on the NMR time scale such that the spectrum of the substrate is a time average of the bound and unbound form. A typical observation is that K_R and K_S are not equal, so that one enantiomer spends more time free in the bulk solvent than the other. The different time-averaged solvation environments may result in differences in the NMR spectrum. In addition, the two complexes (R-CSA and S-CSA) are diastereoisomers and therefore can have different chemical shifts irrespective of the relative magnitude of the association constants. In many cases, it is reasonable to assume

that both mechanisms partially contribute to the enantiomeric resolution. It is sometimes possible to determine which of the two mechanisms predominates. With certain chiral solvating agents, the mechanism that leads to the enantiodiscrimination is well understood and absolute configurations can be assigned on the basis of the relative shifts.

Enantiomeric resolution with chiral solvating agents is dependent on the relative concentrations of the reagents. Usually, enantiomeric resolution is best when the concentration of solvating agent is greater than or equal to that of the enantiomers. Temperature is also an important variable. Lowering the temperature usually, but not always, increases the association of solvating agent and substrate, thereby increasing the enantiomeric resolution.

Lanthanide shift reagents are in actuality chiral solvating agents; however, their behaviour is in some ways unique and they are typically considered a separate category of chiral resolving agents. Chirality is achieved by complexing the lanthanide ion with a chiral ligand. Electron-rich hard Lewis bases (typically oxygen- and nitrogen-containing compounds) interact through a donor–acceptor process with the positive lanthanide ion. Binuclear lanthanide–silver complexes have been developed for use with soft Lewis bases such as olefins, aromatics, phosphines, and halogenated compounds. With the binuclear reagents, the soft Lewis base associates with the silver ion.

Paramagnetic lanthanide ions induce substantially larger shifts, and often larger enantiomeric resolution, than is observed with other chiral derivatizing or solvating agents. Enantiodiscrimination can occur because of differences in association constants and/or because the resulting complexes are diastereoisomers. An advantage of lanthanide shift reagents is the broad range of compounds for which they are potentially effective. A disadvantage is that the chiral ligands in the lanthanide complexes are not geometrically rigid but instead are conformationally mobile. As a substrate binds, the chiral ligands alter their spatial position to accommodate the additional ligand. This effectively eliminates the ability to understand the basis of the enantiodiscrimination at the molecular level, so that assigning absolute configurations is difficult and only accomplished by observing trends on similar compounds with known configurations.

Chiral derivatizing agents

Perhaps no reagent for enantiomeric resolution is more widely known than α-methoxy-α-trifluoromethylphenylacetic acid (MTPA), otherwise known as Mosher's reagent [1]. Either the acid or acid chloride form is suitable for derivation of alcohols into esters and amines into amides. MTPA possesses all the classical features of an effective chiral derivatizing agent. The methoxy singlet in the ^{1}H NMR spectrum and the trifluoromethyl singlet in the ^{19}F NMR spectrum often show enantiomeric resolution. Enantiomeric distinction is usually observed only when the site of chirality is near the hydroxy or amine group at which derivatization occurs. In addition, by analysing the geometry of the resulting derivatives, it is often possible to explain the order of shifts in the ^{1}H NMR spectrum. An example is shown in **Figure 1** for the MTPA derivatives of secondary carbinols and primary amines.

A variety of chiral resolving reagents that are structurally similar to Mosher's reagent have been reported. These include examples in which the phenyl group of [1] has been replaced with naphthyl or anthryl groups; the methoxy group with acetoxy, methyl, ethyl, or hydroxy groups or a fluorine atom; or the trifluoromethyl group with a cyano group or a hydrogen atom. Collectively, these reagents are effective for amines, alcohols, β-amino alcohols, and thiols. Several have been used to resolve the spectra of amines through the formation of diastereoisomeric salts. The compounds α-methoxy-α-(1-naphthyl) acetic acid and α-methoxy-α-(9-anthryl) acetic acid, which possess a more sterically hindered group than [1], often provide a larger degree of enantiomeric resolution in the NMR spectra and can be used to analyse compounds with more remotely disposed chiral centres. The ^{19}F NMR spectrum of derivatives of α-cyano-α-fluoro-phenylacetic acid, in which the fluorine atom is closer to the chiral centre than in MTPA, show enantiodiscrimination in compounds with remotely disposed chiral centres. Enantiomeric excesses of primary alcohols that are highly hindered or have the asymmetric centre at the β carbon have been determined using a sterically hindered compound such as α-methoxy-α-(9-anthryl)-acetic acid.

The choice of solvent is critical when diastereoisomeric salts are formed. Polar solvents solvate the ions, thereby inhibiting ion pair formation and reducing enantiomeric resolution. Non-polar solvents such as chloroform or benzene are preferable. A similar observation occurs as well with many chiral solvating agents and lanthanide shift reagents, and non-polar solvents are preferred.

Axially chiral carboxylic acids such as [2] and [3] can be used to prepare diastereoisomers of secondary alcohols. The NMR spectra of the diastereoisomers show enantiomeric resolution and it is possible to assign absolute configurations. Monocyclic, chiral carboxylic acid compounds such as tetrahydro-5-oxo-2-furancarboxylic acid and 2,2-diphenylcyclo-propane-carboxylic acid have been used to derivatize secondary alcohols for the purpose of enantiomeric resolution. Similarly, bicyclic derivatizing agents such as camphanic acid, camphanic acid chloride, or camphor sulfonyl chloride have been used to derivatize amines or alcohols to their respective amides or esters for the purpose of enantiomeric resolution.

By using optically pure alcohols such as methylmandelate, α-phenylethanol, 2-(trifluoromethyl)benzhydrol, 1,1′-binaphthalene-8,8′-diol, or ethyl 2-(9-anthryl)-2-hydroxyacetate, or optically pure amines such as methylbenzylamine, (1-naphthyl)ethylamine, 9-(1-aminoethyl)anthracene,

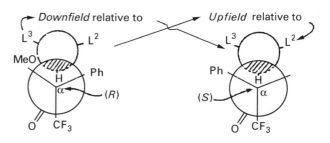

Figure 1 Configurational correlation model for (R)-MTPA and (S)-MTPA derivatives of secondary carbinols and amines. Reprinted with permission from Dale JA and Mosher HS (1973) Nuclear magnetic resonance enantiomer reagents. Configuration correlations via nuclear magnetic resonance chemical shifts of diastereomeric mandelate, O-methylmandelate, and α-methyl-α-trifluoromethylphenylacetate (MTPA) esters. *Journal of the American Chemical Society* **95**: 512–519. Copyright (1973) American Chemical Society.

[1]

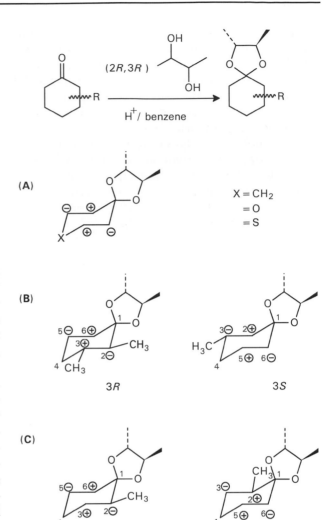

2-(diphenylmethyl)pyrrolidine, 1,2-diphenyl-1,2-diaminoethane, or 9-(1-amino-2,2-dimethylpropyl)-9,10-dihydroanthracene as the chiral derivatizing agent, it is possible to assess the enantiomeric purity of chiral carboxylic acids. Amine derivatives of the naturally occurring monoterpene (+)-3-carene have been used for this purpose as well. In certain cases it is possible to assign absolute configurations with these reagents.

Whereas there are a wide variety of effective chiral derivitizing agents for amines, alcohols, and carboxylic acids, there are considerably fewer options for ketones. The absolute configuration of six-membered cyclic ketones can be determined using butane-2,3-diol or the related 1,4-dimethoxybutane-2,3-diol. Enantiomeric resolution is observed in the ^{13}C or ^{1}H NMR spectra of the resulting butane-2,3-diol acetal. Absolute configurations can be assigned according to the model in **Figure 2**, provided the cyclohexane ring of the acetal has a preference for the chair form. The model does not work for acyclic ketones, cyclopentanones, or cycloheptanones.

The ^{31}P nucleus is especially well suited for observing enantiodiscrimination in diastereoisomers and a variety of phosphorus-containing chiral reagents (**Figure 3**) have been developed [4]–[11]. Certain of these reagents [5, 6, 8] function with chiral primary alcohols. A particularly interesting example is the use of [8] to resolve the ^{2}H and ^{31}P spectra of derivatives of [1-^{2}H] ethanol (CH$_3$CHDOH). Other of the phosphorus-containing reagents function with amines, secondary or tertiary alcohols, and thiols. The phosphoric acid derivative [7] can be used to resolve amines through the formation of diastereomeric salts.

The ^{77}Se nucleus has a spin of $\frac{1}{2}$, reasonable natural abundance (7.5%), adequate sensitivity (6.98 × 10^{-3} compared with ^{1}H), a large chemical shift range, and is extremely sensitive to its electronic environment. Compound [12] is an effective chiral derivatizing agent for chiral carboxylic acids. Compounds such as 5-methylheptanoic acid and lipoic

Figure 2 (A) The model, (B) both enantiomers of 3-methylcyclohexanone acetalysed with (2R, 3R)-butane-2,3-diol and fitted in the model, and (C) both enantiomers of 2-methylcyclohexanone acetalysed with (2R, 3R)-butane-2,3-diol and fitted in the model. Reprinted with permission from Lemiere GL, Dommisse RA, Lepoivre JA *et al* (1987) Determination of the absolute configuration of six-membered-ring ketones by ^{13}C NMR. *Journal of the American Chemical Society* **109**: 1363–1370. Copyright (1987) American Chemical Society.

acid, which have remotely disposed chiral centres, and α-deuterophenylacetic acid, which is enantiomeric by virtue of replacement of a hydrogen atom with deuterium, show enantiomeric resolution in the ^{77}Se NMR spectrum upon derivatization with [12].

Metal complexes other than the lanthanides can sometimes be used to effect chiral resolution in the NMR spectra of enantiomers. Zinc porphyrins with chiral ligands have been used to distinguish N-substituted, anionic amino acids and amino acid esters. Techniques such as NOESY and ROESY can be used to determine relative interatomic distances. In metal complexes containing a chiral ligand of fixed geometry, such techniques can be used to definitively establish the absolute configuration of the ligand.

Chiral metal cations such as $[Ru(bipy)_3]^{2+}$ and $[Ru(phen)_3]^{2+}$ can be enantiomerically resolved through the formation of diastereoisomeric salts with [13]. Both the metal complex and [13] possess D_3 symmetry, which presumably accounts for the discriminating interactions.

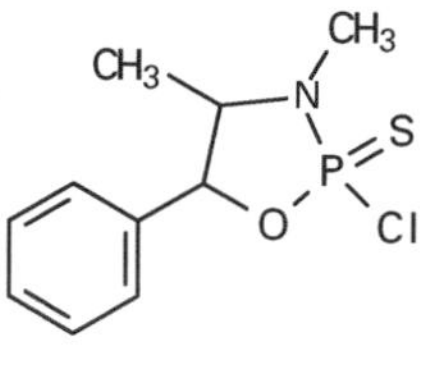

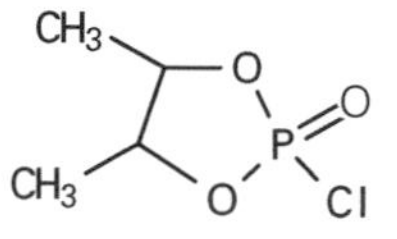

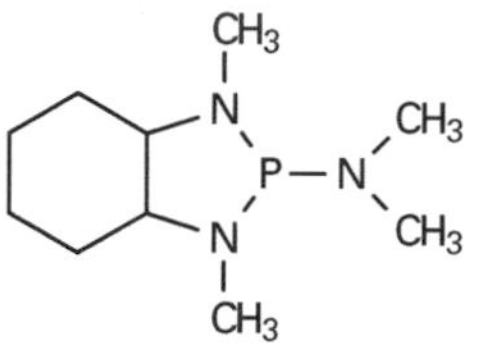

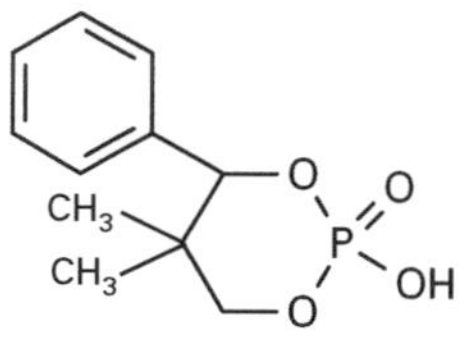

Chiral solvating agents

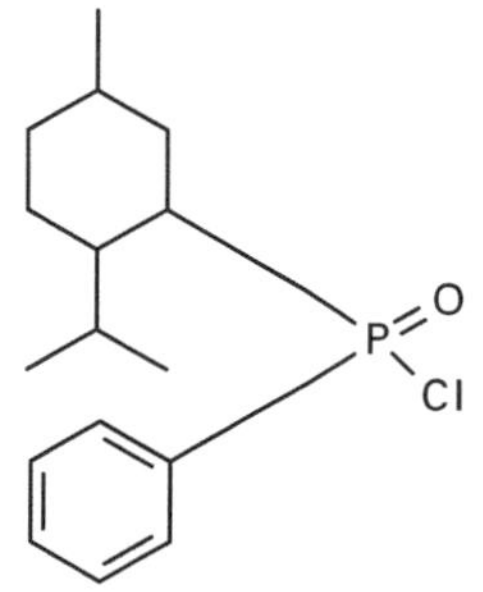

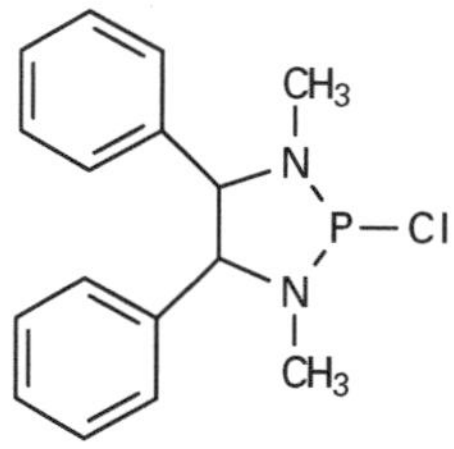

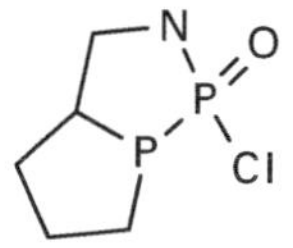

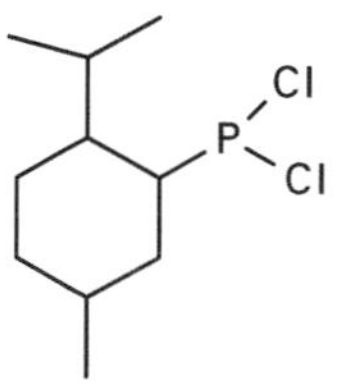

The first application of a chiral solvating agent involved the observation of distinct ^{19}F NMR resonances for 2,2,2-trifluoro-1-phenyl ethanol [14] dissolved in (R)-α-phenylethylamine. Optically pure [14] has since been applied more widely as a chiral solvating agent for a range of compounds, including oxaziridines, N,N-dialkyl arylamine oxides, and sulfoxides. (R)-(1)-(9-Anthryl)-2,2,2-trifluoroethanol [15], which is similar to [14], is often a better chiral solvating agent than [14]. Compound [15] is also effective with γ-lactones. **Figure 4** shows the association of racemic sulfoxides and lactones with [14] and [15]. Shielding of the aryl group causes the group *cis* to it to occur at higher field than the enantiomer with the same group in the *trans* position. Similar structural considerations explain the relative shift orders in the spectra of oxiridines and N,N-dialkylarylamine oxides in the presence of [14] and [15].

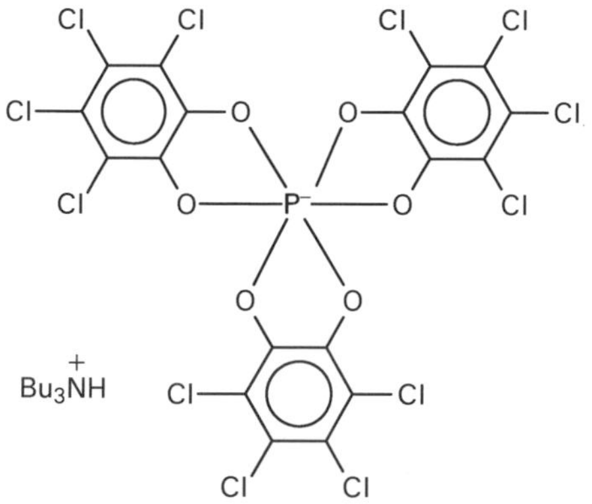

3,5-Dinitrobenzoyl derivatives of leucine [16] and phenylglycine, and 1-(1-naphthyl)ethylurea

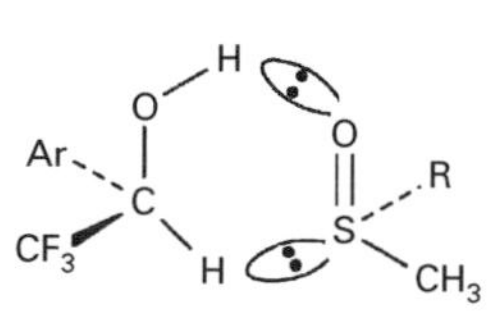

Figure 3 Phosphorus-containing chiral derivatizing agents. [4] 2-chloro-3,4-dimethyl-5-phenyl-1,3,2-oxazophospholidine-2-sulfide; [5] 2-chloro-4,5-dimethyl-1,3,2-dioxaphospholane-2-oxide; [6] 2-dimethylamino-1,3-dimethyloctahydro-1H-1,3,2-benzodiazaphosphole; [7] 5,5-dimethyl-2-hydroxy-4-phenyl-1,3,2-dioxaphosphonan-2-oxide; [8] L-menthyl(phenyl)phosphoryl-chloride; [9] 2-chloro-1,3-dimenthyl-4,5-diphenyl-1,3,2-diazaphosphole; [10] dichloro(menthyl)phosphine.

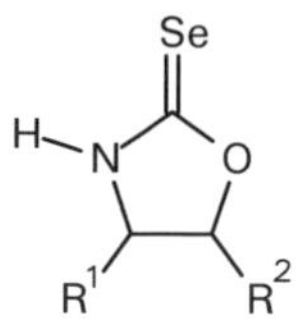

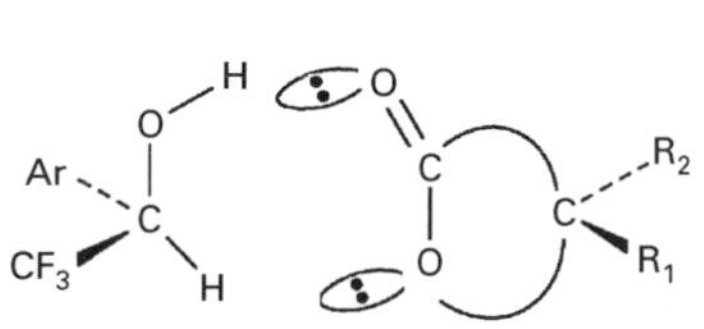

Figure 4 Association of racemic sulfoxides and lactones with 2,2,2-trifluoro-1-phenyl ethanol [14] and 1-(9-anthryl)-2,2,2-trifluoroethanol [15].

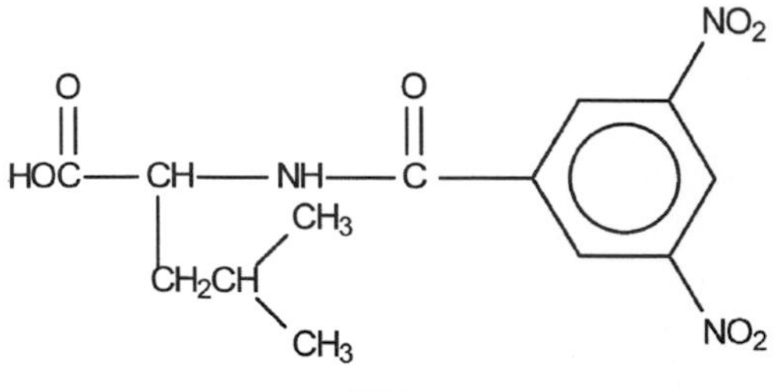

[16]

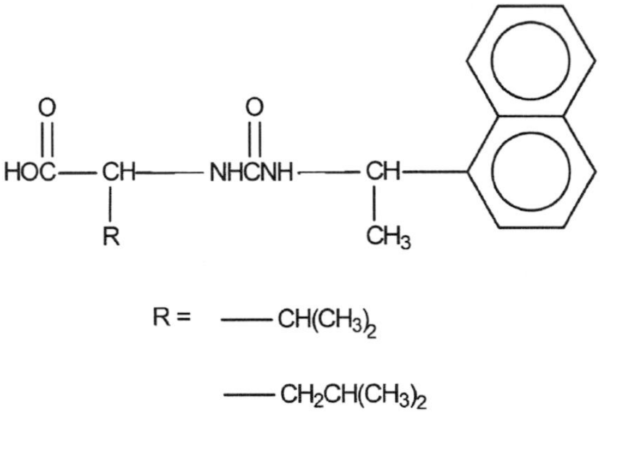

R = —CH(CH_3)_2

—CH_2CH(CH_3)_2

—C(CH_3)_3

[17]

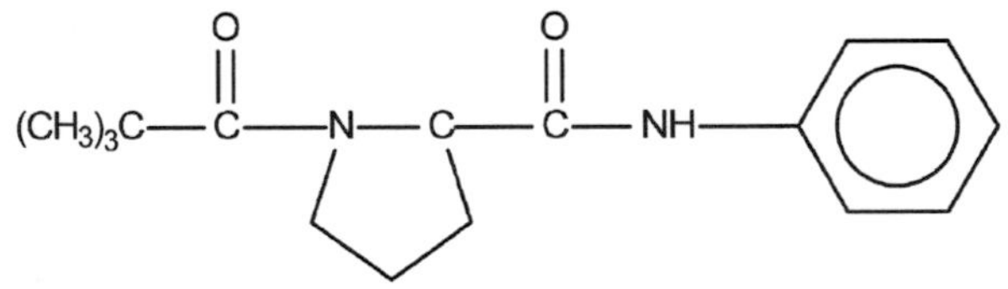

[18]

derivatives of amino acids such as valine, leucine, and t-leucine [17] were originally developed and exploited in the liquid chromatographic separation of enantiomers. The chromatographic phases are potentially suitable for separating thousands of different enantiomers. Organic-soluble analogues, achieved by the conversion of the acid functionality into an ester, are useful NMR chiral solvating agents for a wide range of substrates. The related 3,5-dinitrobenzoyl derivative of phenylethylamine is also useful as a chiral solvating agent. The dinitrobenzoyl ring is especially suited to charge-transfer complexation with aromatic residues of the substrate, the amide functionalities provide sites of partial negative and positive charges, and the R group of the amino acid provides a site for steric hindrance. The naphthyl ring of [17] and the aromatic moiety of the proline-containing reagent [18] will form charge-transfer associations with compounds that have been converted into their dinitrobenzoyl derivatives.

Figure 5 shows the interactions of aromatic rings and hydrogen bonding groups that occur between [16] and [18].

Aromatic diols such as [19] and [20] are axially chiral. Compound [20] is an especially effective chiral solvating agent for sulfoxides, phosphine oxides, alcohols, lactams, and amines. With sulfoxides, it is possible to assign absolute configurations. Quinine has been used as a chiral solvating agent for hemiacetals and β-hydroxyesters. The compound t-butylphenylphosphinothioic acid [21] is a useful chiral solvating agent for a broad range of compounds, including tertiary amine oxides, alcohols, thiols, and amines.

Other classes of chiral solvating agents function through the formation of host–guest complexes. Most notable among these are chiral crown ethers and cyclodextrins, although host–guest chemistry is an active area of research and reports describing the synthesis of specialized hosts designed to effect specific molecular recognition are increasingly common. Compounds [22]–[25] are examples of chiral crown ethers that induce enantiodiscrimination in the NMR spectra of chiral substrates. Substituents on the crown are responsible for the chirality. The 18-crown-6 unit in [22] and [23] is especially well suited for inclusion complexation of protonated primary amine functionalities.

The chirality in [22] is imparted by the axial chirality of the binaphthyl group; however, steric hindrance from the substituent phenyl groups promotes a greater degree of enantiomeric selectivity. The

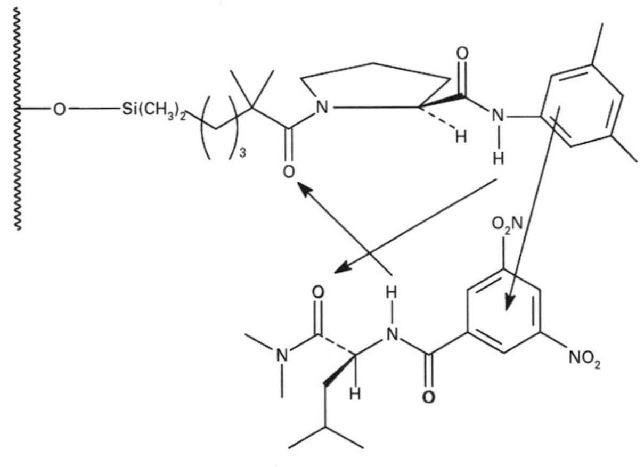

Figure 5 Interactions proposed to account for the retention of the more highly associating enantiomer in [16] and [18]. Reprinted with permission from Pirkle WH, Murray PG, Rausch DJ and McKenna ST (1996) Intermolecular 1H–1H two-dimensional nuclear Overhauser enhancements in the characterization of a rationally designed chiral recognition system. *Journal of Organic Chemistry* 61: 4769–4774. Copyright (1996) American Chemical Society.

[19]

[20]

$(CH_3)_3C-\overset{\overset{\displaystyle O}{\|}}{P}-SH$

[21]

ethers is often enhanced in more polar solvents. The phenolic crown ethers [24] and [25], which each have two sites of chirality from *cis*-cyclohexane-1,2-diol or 1-phenylethane-1,2-diol units, are effective chiral hosts for neutral amines or aminoethanols.

Cyclodextrins are cyclic oligomers, the most common of which contain six (α), seven (β), and eight (γ) D-glucose units (**Figure 6**). The molecules have a tapered cavity with the secondary hydroxyl groups positioned at the larger opening and the primary hydroxyl groups at the smaller. Chiral discrimination most often occurs through differences in hydrogen bonding of the substrate at the secondary opening. Underivatized cyclodextrins are water soluble. Organic-soluble cyclodextrins can be obtained by alkylation of the hydroxyl groups.

The underivatized cyclodextrins are rather unusual in their ability to function as water soluble chiral solvating agents. Enantiomeric resolution is observed in the NMR spectra of a wide range of water-soluble cationic and anionic substrates. Organic-soluble cyclodextrins are one of only a few reagents that can be used to enantiomerically resolve the spectra of hydrocarbons such as trisubstituted allenes, α-pinene, and aromatic hydrocarbons. Amines, alcohols, and carboxylic acids can also be resolved with organic-soluble cyclodextrins.

A set of host compounds of considerable current interest are calix(*n*)arenes [26] and the related resorcarenes [27]. Both sets of compounds form cone-shaped molecules. Calixarenes contain *para* substituted phenol units connected by methylene bridges. The most common have four, six, or eight phenolic units, and by varying the nature of the substituent group or derivatizing the hydroxyl groups

mannitol unit is responsible for the chirality of [23]; however, enantiomeric discrimination is promoted through steric effects of the *t*-butyl group. Crown ethers can be used as chiral solvating agents in a variety of solvents, including chloroform, acetonitrile, and methanol. Unlike many other chiral solvating agents, enantiodiscrimination with crown

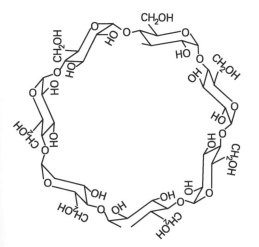

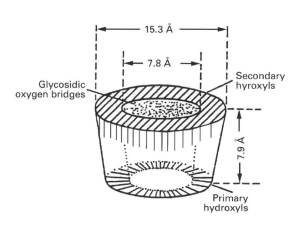

Figure 6 Structure of β-cyclodextrin. Reprinted with permission from Li S and Purdy WC (1992). Cyclodextrins and their applications in analytical chemistry. *Chemical Reviews* **92**: 1457–1470. Copyright (1992) American Chemical Society.

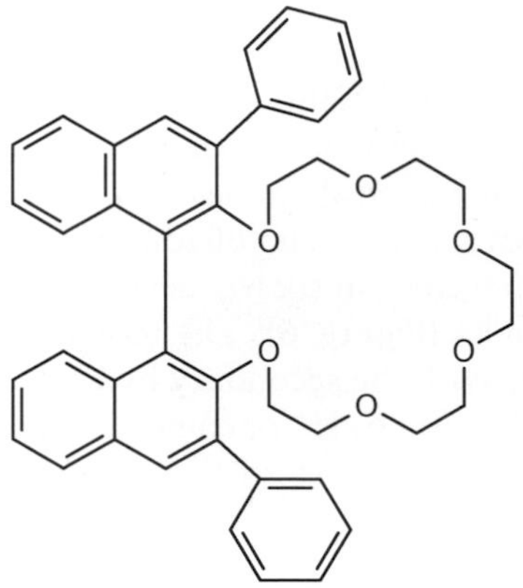

[22]

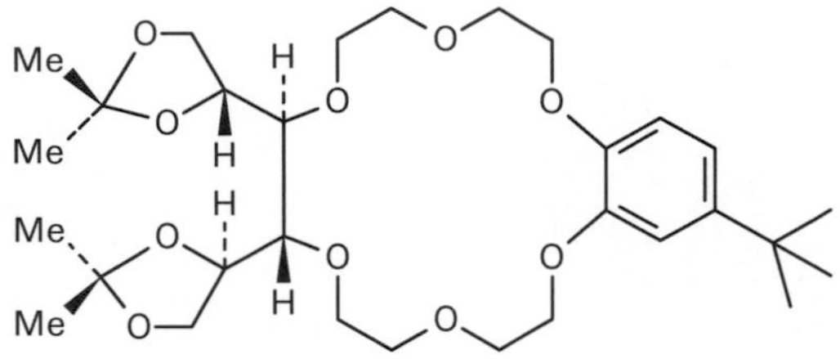

[23]

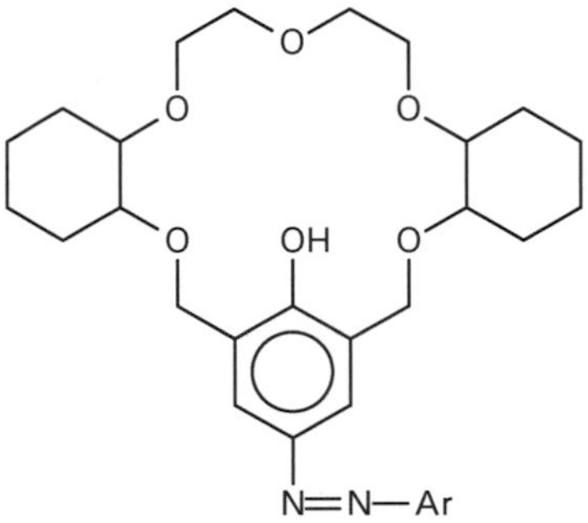

[24]

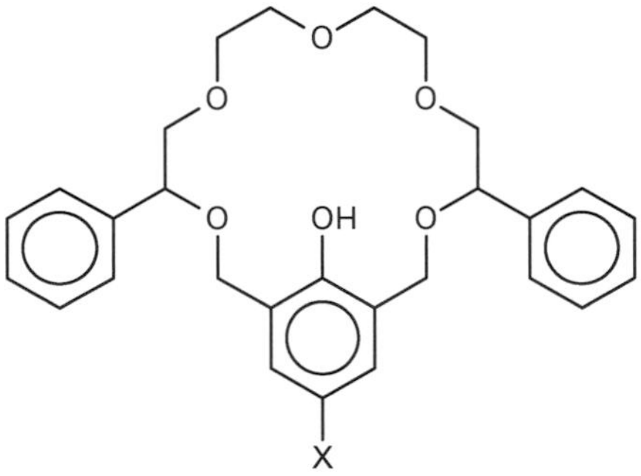

[25]

an endless variety of calixarenes can be prepared. Chirality can be achieved by one of two means. Inherently chiral calixarenes can be obtained by replacing one of the ring hydrogens on one of the phenol units with a substituent group or by synthesizing a compound in which each of the phenol rings has a different R group. Alternatively, one can attach optically pure R groups or derivatize the hydroxyl groups with an optically pure reagent. The calix resorcarene [27], in which chirality is achieved by attaching optically pure proline units to the aromatic ring, is suitable for enantiomeric resolution of certain water-soluble aromatic alcohols, amino acid esters, and carboxylic acids in water. Given the current interest in this family of compounds, further developments in this area can be expected.

Chiral liquid crystals consisting of poly-γ-benzyl-L-glutamate have been used in various organic solvents to enantiomerically resolve the NMR spectra of alcohols, amines, carboxylic acids, esters, ethers, epoxides, tosylates, halides, and hydrocarbons. Enantiomeric pairs exhibit different ordering properties in the liquid crystal, leading to different chemical shifts in the ^{1}H NMR spectra. A more sensitive way to probe distinctions among enantiomers in these systems is to examine deuterium quadrupolar splittings. The disadvantage is that deuterium labelling of the compound under study is then required, although procedures are available for conveniently incorporating deuterium-containing substituents into amines, amino acids, and alcohols.

Lanthanide shift reagents

Chiral lanthanide shift reagents are noteworthy because they are effective at inducing enantiomeric resolution in the NMR spectra of such a wide range of compounds. Organic soluble lanthanide shift reagents for hard Lewis bases (primarily oxygen- and nitrogen-containing compounds) consist of tris-β-diketonate complexes in which the ligands are chiral. The most common metals used in such studies are Eu(III), Pr(III), and Yb(III). The lanthanide complex with 3-(t-butylhydroxymethylene)-D-camphor was the first one to be studied. A large number of chiral lanthanide shift reagents have since been evaluated,

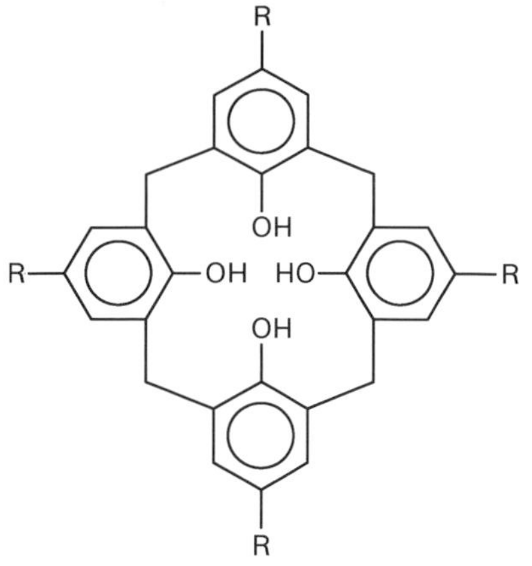

[26]

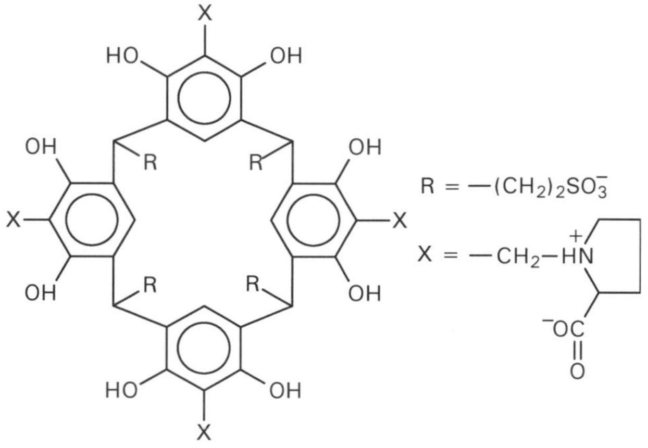

$R = -(CH_2)_2SO_3^-$

$X = -CH_2-\overset{+}{H}N$

[27]

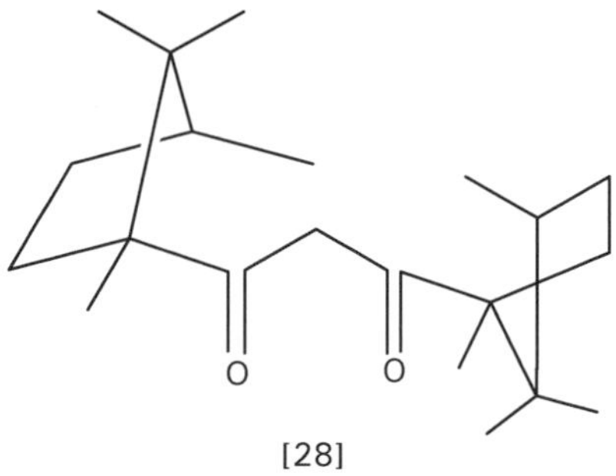

[28]

alcohols, ketones, esters, lactones, aldehydes, carboxylic acids, ethers, amines, nitrogen heterocycles, amides, lactams, sulfoxides, sulfines, sulfones, sulfoximes, sulfilimines, sulfenamides, thiocarbonyls, phosphorus oxides, and chiral metal complexes that contain ligands with functional groups suitable for direct association with the lanthanide ions.

Chiral lanthanide shift reagents for soft Lewis bases have been prepared by mixing a chiral lanthanide tris-β-diketonate such as Ln(facam)$_3$ or Ln(hfbc)$_3$ with a silver β-diketonate such as Ag(fod) (fod = 2,2-dimethyl-6,6,7,7,8,8,8-heptafluorooctane-3,5 dione). Addition of the two reagents leads to the formation of a binuclear reagent as illustrated in Equation [3]. The binuclear reagent consists of an ion pair between a lanthanide tetrakis chelate anion and a silver cation.

$$Ln(\beta\text{-}dik)_3 + Ag(\beta\text{-}dik)$$
$$\rightarrow Ag[Ln(\beta\text{-}dik)_4] \qquad [3]$$

The silver ion coordinates with the soft Lewis base and the lanthanide induces the shifts. Chiral binuclear reagents have been applied to chiral olefins and aromatics. Binuclear reagents formed using the dcm ligand are not effective for chiral resolution, presumably because this ligand is too sterically hindered to permit formation of the tetrakis chelate anion. Experimentation is often needed to find the correct combination of lanthanide and silver β-diketonate to cause enantiomeric resolution.

Tetrakis chelate anions formed when either a silver or potassium β-diketonate is mixed with a lanthanide tris β-diketonate are effective shift reagents for organic-soluble salts. Adding an organo-halide or fluoroborate salt leads to precipitation of silver halide or potassium fluoroborate, resulting in an organic-soluble ion pair. These reagents have been used to enantiomerically resolve the spectra of chiral isothiouronium and sulfonium ions.

and those that are generally most effective and commercially available include complexes with 3-trifluoroacetyl)-D-camphor (facam or tfc), 3-heptafluorobutyryl-D-camphor (hfbc or hfc), and D,D-dicampholylmethane (dcm) [28]. The shifts in the spectra of substrates are usually larger with chelates of hfbc than with facam; however, the degree of enantiomeric resolution exhibits no consistent pattern. Chelates with dcm are particularly effective at causing enantiomeric resolution. This may be owing, in part, to the steric bulk of the ligand which results in enhanced enantiodistinction of binding substrates.

Lanthanide tris-β-diketonates rely on association of a lone pair of electrons with the tripositive lanthanide ion, and substantially larger enantiomeric resolution is observed in non-coordinating solvents. In addition, the reagents are hygroscopic and so the presence of water will significantly reduce the binding of substrates. Examples of the classes of compounds for which lanthanide tris-β-diketonates have been utilized for enantiomeric resolution include

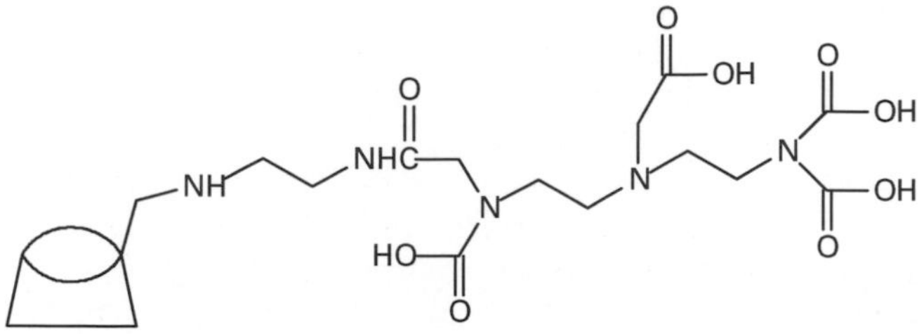

[29]

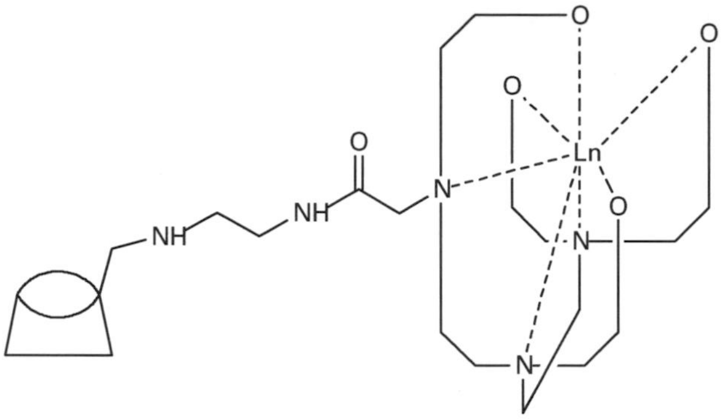

Complex with [29]

Water-soluble, chiral lanthanide shift reagents with ligands such as (S)-[(carboxymethyl)oxy]succinic acid, (R)-propylene-1,2-diaminetetraacetic acid, (S,S)-ethylenediamine-N,N'-disuccinic acid, N, N, N'',N'-tetrakis-(2-pyridylmethyl)-(R)-propylenediamine, and (S,S)-((1S,2S)-diaminophenylethylene)-N,N'-disuccinic acid have been developed. These are especially effective for resolving underivatized α-amino or carboxylic acids. In some cases, consistent trends in relative shift magnitudes are observed which allows the assignment of absolute configurations.

Achiral lanthanide shift reagents such as Eu(fod)$_3$ have been used in a variety of instances, including with MTPA esters, to enhance the resolution in the spectra of diastereoisomers formed from derivitization with an optically pure reagent. The relative shifts in the spectra of the diastereoisomers often show characteristic trends, enabling the assignment of absolute configurations. Achiral lanthanide shift reagents can also be added to solutions of chiral solvating agents and substrates to enhance enantiomeric resolution. Two mechanisms can explain the enhancement.

The first is observed when the chiral solvating agent has a high association with the achiral lanthanide complex. Bonding of the CSA to the lanthanide (Ln) effectively creates a chiral lanthanide shift reagent (Ln–CSA) (Eqn [4]). This species then interacts with the chiral substrate (S) to cause enantiomeric resolution (Eqns [5] and [6]). The chiral substrate can directly associate with either the lanthanide ion (Eqn [5]), or, as observed with certain optically pure chiral carboxylate species, the lanthanide-bonded chiral solvating agent (Eqn [6]).

$$Ln + CSA = Ln \cdot CSA \qquad [4]$$

$$Ln \cdot CSA + S = S \cdot Ln \cdot CSA \qquad [5]$$

$$Ln \cdot CSA + S = Ln \cdot CSA \cdot S \qquad [6]$$

The second mechanism is favoured when the substrate has a much stronger association with the lanthanide complex than the chiral solvating agent. If the enantiomers have different association constants with the chiral solvating agent, the enantiomer less favourably associated with the solvating agent is more available for complexation with the lanthanide and exhibits larger lanthanide-induced shifts (Eqns [7] and [8]).

$$S + CSA = CSA \cdot S \qquad [7]$$

$$S + Ln = S \cdot Ln \qquad [8]$$

This mechanism has been observed with perfluoro-alkyl carbinols [14] and [15], dinitrobenzoyl derivatives of amino acids [16], 1-(1-naphthyl)-ethylurea derivatives of amino acids [17], and chiral crown ethers [22] and [23]. The lanthanide shift reagent is usually $Ln(fod)_3$. With crown ether–ammonium salt combinations, lanthanide tetrakis chelate anions of the formula $Ln(fod)_4^-$ are used.

Lanthanide ions have been covalently attached to either the primary or secondary side of cyclodextrins through a diethylenetriaminepentaacetic acid (DTPA) ligand [29]. Adding Dy(III) to the cyclodextrin-DTPA derivatives can enhance the enantiomeric resolution in the spectra of substrates that form host–guest complexes with the cyclodextrin. Enhancements in enantiomeric resolution are larger with the secondary derivative, which is consistent with the observation that enantiodiscrimination with cyclodextrins usually involves interactions with the hydroxyl groups on the secondary opening of the cavity.

Conclusions

The number of available chiral derivatizing agents, solvating agents, and lanthanide shift reagents provides the possibility of enantiomerically resolving the NMR spectra of a wide variety of compound classes. Furthermore, in many instances the mechanism that leads to the enantiomeric resolution is well enough understood that it is possible to assign absolute configurations based on the relative shift order. Since none of the reagents is ever 100% effective, experimentation with several resolving agents may be necessary. Given the increasing interest in knowing and controlling the optical purity of chemicals, a better understanding of existing systems and the development of additional reagents for enantiomeric discrimination in NMR spectroscopy seems to be assured.

See also: **Chemical Shift and Relaxation Reagents in NMR; Chiroptical Spectroscopy, General Theory; Heteronuclear NMR Applications (O, S, Se, Te); Liquid Crystals and Liquid Crystal Solutions Studied By NMR; NMR Relaxation Rates; Nuclear Overhauser Effect; ^{31}P NMR; Parameters in NMR Spectroscopy, Theory of; Vibrational CD, Applications.**

Further reading

Lehn JM (1995) *Supramolecular Chemistry*. Weinheim: VCH Verlagsgesellschaft.

Morrison JD (ed) (1983) *Asymmetric Synthesis*, Volume 1. New York: Academic Press.

Parker D (1991) NMR determination of enantiomeric purity. *Chemical Reviews* 91: 1441–1457.

Pirkle WH and Hoover DJ (1982). In Eliel EL and Wilen SH (eds) *NMR Chiral Solvating Agents*, Topics in Stereochemistry, Volume 1, pp 263–331. New York: Wiley-Interscience.

Stoddart JF (1988). In Allinger NL, Eliel EL and Wilen SH (eds), *Chiral Crown Ethers*, Topics in Stereochemistry, Volume 17, pp 207–288. New York: Wiley-Interscience.

Sullivan GR (1978) In Elien EL and Wilen SH (eds) *Chiral Lanthanide Shift Reagents*. Topics in Stereochemistry, Volume 10, pp 287–329. New York: Wiley-Interscience.

Webb TH and Wilcox CS (1993) Enantioselective and diastereoselective molecular recognition of neutral molecules. *Chemical Society Reviews* 22: 383–395.

Wenzel TJ (1987) *NMR Shift Reagents*. Boca Raton (FL): CRC Press.

Environmental and Agricultural Applications of Atomic Spectroscopy

Michael Thompson, Birkbeck College, London, UK
Michael H Ramsey, Imperial College of Science
Technology and Medicine, London, UK

ATOMIC SPECTROSCOPY
Applications

We consider the determination of the concentration of elements in various materials studied in agricultural and environmental applications, by the use of the following methods: atomic absorption spectroscopy (AAS) using a flame (FAAS) or a graphite furnace (GFAAS) as an atom cell; inductively coupled plasma atomic emission spectroscopy (ICPAES); inductively coupled plasma mass spectrometry (ICPMS) and X-ray fluorescence (XRF). The analytical characteristics of the methods as normally practised are compared with the requirements of fitness for purpose in the examination of soils and sediments, waters, dusts and air particulates, and animal and plant tissue. However, there are numerous specialized techniques that cannot be included here.

Introduction

Environmental and agricultural studies involve a number of different materials that need to be analysed. The principal test materials encountered by the analyst in this sector are soils, sediments, air, dusts, water, plant material and animal tissue. For completely practical reasons, we need to know the composition of these media over wide concentration ranges, from major constituents to elements at ultratrace levels. One is especially interested in the way elements are transferred between media, their biological role and metabolic pathways, and their ultimate fate.

Almost invariably in these applications we can be satisfied if the analytical result obtained is within a range of 90–110% of the true result. This modest fitness-for-purpose requirement is an outcome of the high uncertainties in sampling of environmental materials. In these circumstances, the total uncertainty (sampling plus analytical) is not improved by the use of high-accuracy analytical methods. Overall, therefore, it is a more effective use of resources to analyse a larger number of samples with only moderate accuracy.

Atomic spectroscopy is almost exclusively the method of choice in the determination of concentrations of elements in agricultural and environmental studies. There are several reasons for this: (a) for the most part no separations are needed before the measurement, only a straightforward dissolution procedure; (b) test materials can be analysed in rapid succession; (c) elements can be determined at suitably low concentrations; (d) apart from atomic absorption, atomic spectroscopy methods are suitable for determining many elements simultaneously; (e) modern atomic spectroscopy methods can easily fulfil the accuracy requirement specified above. No other methods can match this range of performance. Where speciation studies require a preliminary separation of the analytes, atomic spectroscopy is often the method used in 'hyphenated' methods to provide a very sensitive and virtually specific detector for almost any element.

In the following sections we consider the key features of the atomic spectroscopy methods, how they relate to user requirements, and how they are applied to specific types of test materials. Although only laboratory-based methods are considered here, it must be recalled that field methods sometimes find worthwhile applications: XRF is one atomic spectroscopy method that is available as a truly portable instrument.

Detection limits

The detection limit is (informally) the lowest concentration of the analyte that can be reliably detected, and is a reflection of the precision of the instrumental response obtained by the method when the concentration of the analyte is zero. Obviously, if the uncertainty range of the measurement (at some specified level of probability) includes zero, we are not sure that the analyte has been detected. However, detection limits as supplied in the literature and manufacturers' brochures can be misleading to the unwary. Their magnitude depends critically on the conditions under which the precision was estimated. Manufacturers and method developers usually quote 'instrumental detection limits', where the precision is estimated on the pure solvent (or other matrix) in the shortest possible time, with no adjustments at all to the instrument.

Such detection limits usually need to be adjusted before they can be applied to the analysis of real materials, *inter alia* to allow for the 'dilution' of the analytes. For instance, in the analysis of a soil, a 0.1 g test portion may be treated with reagents and the analytes liberated into solution may be made up to a volume of 10.00 mL for presentation to the instrument. The analytes have been diluted by a factor of 100, in that an element at a concentration of 100 ppm in the soil will be presented to the instrument at a concentration of 1 ppm in the test solution. In addition, in 'real analysis', many features other than short-term instrumental variation play a part in the variability of results. Thus, detection limits estimated under repeatability or reproducibility conditions may be considerably higher than simple dilution-adjusted instrumental detection limits, by as much as an order of magnitude. We here consider instrumental detection limits, adjusted for dilution and with an extra factor of 5 to convert them roughly to repeatability detection limits.

The detection limit also depends critically on details of the instrumentation. For example, magnetic sector ICPMS provides detection limits lower than those of quadrupole instruments by between one and two orders of magnitude.

Among the atomic spectroscopy methods, flame and furnace methods have very variable detection limits, because they are affected by variations among the elements in terms of efficiency of atomization and efficiency of ionization. ICPAES and ICPMS show a smaller range of detection limits, because atomization is uniformly close to 100% efficient, and the efficiency of ionization accounts for much of the remaining differences. Thus, in ICPMS, elements more difficult to ionize (e.g. nonmetals) have higher detection limits. The outcome is that FAAS, being fast, cheap and robust, is used for a limited range of more abundant elements. GFAAS, being slow and requiring much care, but with much lower detection limits for some elements, is used on a much more restricted scale, typically for low trace elements such as lead and cadmium in food and drink. ICPAES is fast and robust and moderately priced, and is widely used as the workhorse method for major and trace elements, with instrumental detection limits typically 1–50 ppb (parts per 10^9). ICPMS is unique in the very low detection limits obtainable in simultaneous analysis, typically 0.1–10 ppt (parts per 10^{12}), but it requires a high dilution factor (typically 500–1000), is more expensive, and is more prone to downtime. In XRF methods, the detection limit is poor for light elements but roughly constant at a few ppm for elements heavier than iron, and the method usually requires only small dilution factors.

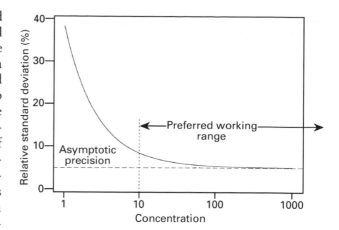

Figure 1 Relationship between precision of measurement (as relative standard deviation) and concentration of analyte (expressed in units of multiples of detection limit). The graph shows the sharp fall in relative standard deviation just above the detection limit, and the comparatively steady values in the preferred working range.

Asymptotic precision

The other aspect of precision that affects the judgement of instrumental analysis can be called 'asymptotic precision', which is precision estimated at a concentration orders of magnitude above the detection limit. Under that condition we usually find that relative precision is concentration invariant, i.e. the relative standard deviation (RSD) is constant. Thus, in general, with increasing concentration the asymptotic precision always increases, but the relative standard deviation falls sharply, from about 30% at the detection limit, to close to the asymptotic level at 20–100 times the detection limit (**Figure 1**). Usually we try to arrange matters so that we make measurements at these near-asymptotic precisions.

Asymptotic precisions for atomic spectrometry methods depend on the type of instrumentation, but also on other factors such as the protocols used for calibration and measurement, and whether internal standardization has been used. Typical between-run values as RSDs are: FAAS, 1.5%; GFAAS, 5%; ICPAES, 1%; ICPMS, 3%; XRF, 0.5–1%.

Abundance–detection limit (ADL) diagrams

A convenient way of summarizing the applicability of the various methods to various types of test materials, over the range of relevant elements, is the use of abundance–detection limit (ADL) diagrams. For each element in these diagrams, the detection limit of the method (d), is plotted against the median concentration found in the sampling medium in question (the abundance, a). (The detection limit is adjusted

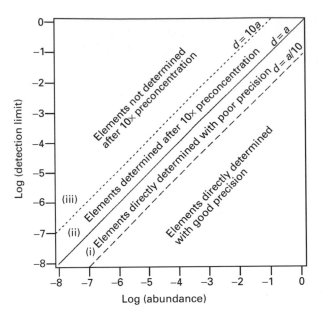

Figure 2 The layout of an ADL (abundance–detection limit) diagram, showing the for interpretation for elements that plot in the various regions of the diagram. The concentration units used for both axes are mass fractions.

for a likely dilution factor and with an extra allowance to convert instrumental detection limits into between-run detection limits.) Both variables are expressed as mass fractions (e.g., 1 ppm $\equiv 10^{-6}$) and plotted on $\log_{10}$ scales. (Frequency distributions of concentrations of trace elements in randomly selected agricultural and environmental media are usually strongly skewed to the positive tail, so lower concentrations are well represented by the median. In any event, information on percentiles other than the median is difficult to compile.)

ADL diagrams are divided into four zones (**Figure 2**) by three lines: (i) $d = a/10$; (ii) $d = a$; and (iii) $d = 10a$. Elements plotting to the right of and below line (ii) can be determined with at least moderately good precision at normal concentrations. Line (i) represents concentrations at 10× the detection limit, at which the elements can be determined with a relative precision of about 5%. Elements plotting to the right of and below this line can therefore be determined with completely acceptable precision, and those plotting between lines (i) and (ii) with moderate precision. Elements between lines (ii) and (iii) can be determined with moderate precision only after a 10× preconcentration, while elements above or to the left of line (iii) can be determined only if preconcentrated to a degree greater than tenfold. These diagrams will be used to compare the capabilities of the various methods.

Interference effects

Another figure of merit that describes the quality of the data produced in atomic spectroscopy is 'trueness', which, in the absence of analytical blunders, is synonymous with lack of interference. Interference is defined as the influence on the analytical signal relating to an analyte caused by constituents of the test materials other than the analyte (the concomitants). Broadly, atomic spectroscopy is affected by interference to a remarkably low extent. However, interference is not absent and usually has to be taken seriously. Fortunately, there are well-established analytical techniques for obviating the effects of interference that are eminently applicable to atomic spectroscopy.

Interference effects are often spectral in origin. In atomic emission and absorption they result when a concomitant gives rise either to photons close to the same wavelength as the analyte (spectral overlap) or to photons of another wavelength that inadvertently arrive at the detector (stray light). Methods for overcoming such spectral problems usually involve either a separate estimation of the background to a spectral line, or estimation of the concomitant concentration at another region of the spectrum and the calculation therefrom of an appropriate correction term. These features are built into modern instruments and their software, but the correction procedure may degrade other aspects of performance such as the detection limit. Comparable problems occur in mass spectrometry.

Nonspectral interferences are called matrix effects. In free-atom methods, matrix effects often reflect changes in the efficiency of atomization of the analyte or of excitation of the separate atoms produced in the atom cell. In FAAS and ICPAES, changes in the physical characteristics of the test solutions (e.g., surface tension, viscosity, density) may additionally affect the efficiency of the nebulizer. However, matrix effects seldom cause errors of greater than ±5% and can often be ignored in agricultural and environmental studies. In GFAAS, by contrast, the composition of the matrix is crucial in determining the atomization efficiency, and needs to be carefully controlled. Likewise, in XRF the gross composition of the pellet or bead may affect the intensity of fluorescent X-rays.

There are a number of convenient and effective ways of handling matrix effects. Where the gross composition of the test materials is effectively constant, it is sufficient to calibrate the instrument with matrix-matched calibrators. Even when test materials are also variable, they may sometimes be matched with calibrators by swamping the native matrix with a much larger quantity of a matrix modifier. This technique is commonly used in FAAS, GFAAS and XRF methods. If matrix matching is not possible, the method of analyte additions is always applicable, if somewhat laborious, to methods with linear calibration functions. In XRF a set of complex correction equations based on the major composition is used to correct for matrix effects.

Performance enhancement by chemical pretreatment

The detecting power of all of these methods can be improved by preconcentration after the initial decomposition of the test material. This procedure, effected by methods such as ion-exchange, solvent extraction or solid-phase extraction, is very effective. It often has the additional advantage of removing the analytes from their original matrix and thereby considerably reducing matrix effects. Concentration by a factor of 10 is straightforward, and by 100 is usually feasible. For the atomization methods it is convenient, after the initial separation, to return the analytes to a small volume of aqueous medium. Trace elements can also be analysed directly in organic solvents. This is not a popular option, although there are no inordinate difficulties about the procedure. Different options for presenting the separated analytes exist in XRF, for instance trace analytes can be concentrated by passing the test solution through an ion-exchange membrane and analysing the membrane in a specially designed holder.

Another widely used method of improving detection limits is the injection of certain elements in the gas phase. The elements Ge, Sn, Pb, As, Sb, Bi, Se and Te all readily form hydrides in acidic aqueous media on reduction with sodium tetrahydroborate. As the hydrides are not very soluble in water, they can be readily transferred into the gas phase and thus completely separated from the matrix. The efficiency of hydride formation is high (often approaching 100%), so that detection limits for these elements can be improved by a factor of 100 without problems.

Speed of analysis

Speed of analysis is one of the attractive features of the atomic spectroscopy methods described here. An important discrimination can be made, however, between methods that determine elements sequentially and those, clearly much faster overall, that can determine a whole suite of elements simultaneously. In methods involving nebulization, the critical determinant is the time taken for the system to 'wash out', so that material from one test solution is completely replaced by its successor in the analytical system and a new stable signal established. It usually takes only a few seconds subsequently to make a measurement of the signal intensity. Such sample changeovers can be as short as 15 s, but are usually longer (1–2 min) because of poor attention to design in this aspect of commercial instrumentation. In GFAAS the speed of analysis is largely controlled by the time needed for the furnace to cool sufficiently before introduction of the next sample. XRF exhibits different properties. It may require a substantial time for the signal to be collected (minutes), whereas the exchange of one test material for the next takes only seconds (because only solids in special holders are moved).

The question of speed is related to the reliability of the instrumentation. Reliable methods with low operating costs can readily be automated and can left unattended, e.g. overnight. In those circumstances, speed may be a secondary consideration. XRF methods come into this category. For methods that are more prone to problems during running (and so cannot be left unattended) or that use expensive consumables (e.g. argon for ICP methods), speed is usually of the essence.

Instrumental reliability and stability

Reliability is also a key feature in the comparison of methods. Some methods are inherently reliable and stable, e.g. XRF. The signal obtained from an analyte in a particular material will vary only slightly from run to run, so that calibrations have to be updated only infrequently. The situation is quite different in methods requiring nebulization and/or atomization, because there the stability of the signal depends on such variables as power input, gas flows, etc. In ICP methods particularly, the efficiency of atomization can occasionally vary suddenly due to partial blockages in the nebulizer or in the injector tip of the plasma torch. Vigilance is required for early detection of such problems. In addition, in ICPMS, drift may be caused by gradual occlusion of the orifice in the sampling cone (part of the plasma–MS interface) by material passing through the plasma. This problem must be minimized by the use of high dilutions (i.e. at least 1:500) or by the use of flow injection to deliver the test solutions in short slugs, or its effects may be alleviated by the employment of an internal standard.

Ease of calibration

Two features contribute to ease of calibration in analytical methods, (a) linearity over a wide concentration range, and (b) use of solutions (rather than solids) to present the analyte to the instrument.

Linearity makes calibration simple by limiting to two the number of calibrators necessary to estimate the calibration function for a single analyte. A nearly (but not necessarily completely) linear calibration over a considerable concentration range (e.g. more than four orders of magnitude) is also essential in multielement calibration, where major and trace constituents are determined simultaneously.

Preparation of calibrators in solution can be effected by simple mixing of the correct amounts of the constituents in an appropriate matrix. There is little problem with either homogeneity or speciation of the analytes in that instance. This consideration applies directly to conventional ways of operating AAS, ICPAES and ICPMS. It also applies to XRF methods that employ fused beads, where the test material is homogenized in a flux. (The end-product is, of course, a solid solution). Where solid calibrators are used, more problems are usually encountered. The most common of such are problems of creating a sufficiently homogeneous mixture of the analytes in a realistic matrix, and problems of speciation of the analyte. Fortunately, neither of these is a serious problem in routine XRF, although they may be in specialist applications of ICPAES and ICPMS, such as laser ablation work.

There are special problems associated with multielement calibration when more than a few elements are required. First, it is necessary to provide a set of calibrators that span the range of the test materials adequately. This is simple if there are no constraints on the design of the calibrator set. Often there are such constraints. For instance, for major constituents, it is not realistic (and sometimes not even possible) to have all of the major constituents at their maximum concentration in a single calibrator. In consequence, methods such as serial dilution of a single concentrated solution cannot be undertaken, and more complex designs are required. Constraints due to minor interferences apply to methods such as ICPAES. In some styles of calibration for major and trace elements simultaneously, there are minor deviations from linearity and small interference effects from major constituents that have to be taken into account. Finally, the calibration set must be designed to avoid major matrix mismatches with the test materials.

In XRF it is common for practitioners to calibrate with reference materials. As these are natural materials, designs cannot be made optimal. It would in fact be easier and more accurate to use pure stoichiometric compounds to synthesize calibrators, just as in liquid solution work, and this strategy would allow optimal designs.

Soils and sediments

Soil is the key medium in environmental and agricultural studies of the distribution of the elements. Soils and sediments are the main repositories of trace metals. For conventional analysis by AAS and ICP methods, it is necessary to bring the analytes into solution. This is usually effected by treatment of the test material with strong mineral acid mixtures. In many instances a complete dissolution of the soil is not required. Most elements of agricultural or environmental import can be sufficiently solubilized by evaporation with nitric and perchloric acids, followed by a dissolution with hydrochloric acid. Any elements remaining undissolved after this type of treatment will hardly be of significance to living organisms. An attempt at studying the bioavailability of significant metals can be made by the use of 'selective extractants'. These are designed to enable experimenters to distinguish between categories such as 'readily exchangeable', 'organic bound', etc.

By making some reasonable assumptions about dilution factors and the relationship between instrumental and repeatability detection limits, it is possible to construct ADL diagrams for soils analysed by the various atomic spectroscopy methods. These are shown in **Figures 3–6** for a selection of key elements. A surprisingly uniform picture emerges, with only ICPMS being suitable for the direct determination of elements such as As, Cd, Hg, Mo, Se and U, a range of important essential or toxic elements of low abundance. However, preconcentration or other sensitivity-enhancing methods can usually render these elements capable of determination by AES or AAS methods. Of the methods considered here, perhaps the least applicable is FAAS, which is also probably the most laborious, being a single-element method.

Water

Fresh water is in many ways an ideal medium for analysis by the atomization methods. It is ready for direct analysis with virtually no pretreatment, and no dilution is required. Matrix effects are usually negligible or easily handled. The method *par excellence* for water analysis is ICPMS (**Figure 7**), which can be used to determine most of the elements directly. Another important method here is GFAAS which, although undoubtedly slow, can be used very effectively for a range of useful trace elements. The main problem with the remaining methods (**Figure 8**) is that many important elements are present in fresh waters at very low concentrations, below the

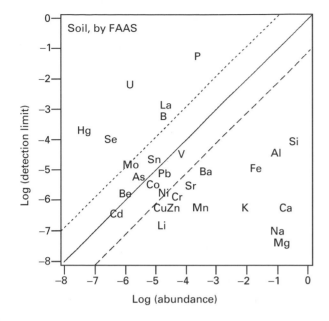

Figure 3 ADL diagram for soil analysed by flame atomic absorption spectroscopy. (For explanation see **Figure 2**.)

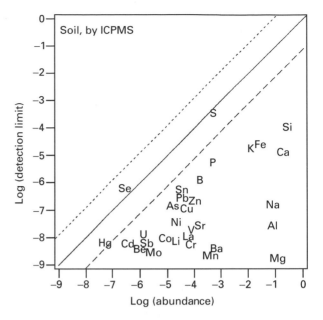

Figure 5 ADL diagram for soil analysed by inductively coupled plasma mass spectrometry. (For explanation see **Figure 2**.)

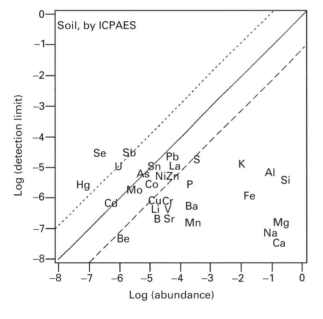

Figure 4 ADL diagram for soil analysed by inductively coupled plasma atomic emission spectroscopy. (For explanation see **Figure 2**.)

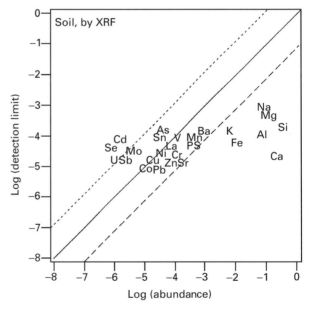

Figure 6 ADL diagram for soil analysed by X-ray fluorescence spectroscopy. (For explanation see **Figure 2**.)

detection limits for direct determination by most AAS and AES methods. For most elements this drawback can readily be overcome by preconcentration. Simple evaporation is effective for 10–20-fold preconcentration, without causing serious matrix effects. Matrix separations are also simple and useful. For example, elements such as Cd, Cu, Fe, Mn, Pb and Zn can be determined by FAAS or ICPAES after solvent extraction or ion exchange.

Sea water and other saline waters pose some problems for the atomization methods. Nebulizers for the ICP tend to become blocked with salt encrustations with a disastrous effect on sensitivity. This can be overcome in direct analysis by flow injection techniques, or by the use of 'high-solids' nebulizers. Dilution also helps here but, of course, degrades the detection limits. The sampling cone in ICPMS equipment is also prone to be gradually occluded when

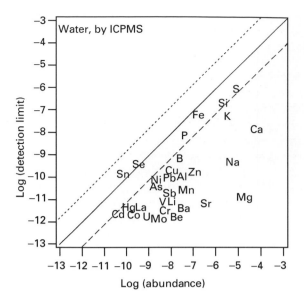

Figure 7 ADL diagram for water analysed by inductively coupled plasma mass spectrometry. (For explanation see **Figure 2**.)

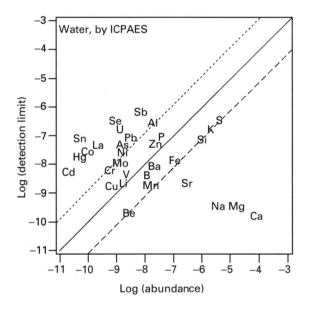

Figure 8 ADL diagram for water analysed by inductively coupled plasma atomic emission spectroscopy. (For explanation see **Figure 2**.)

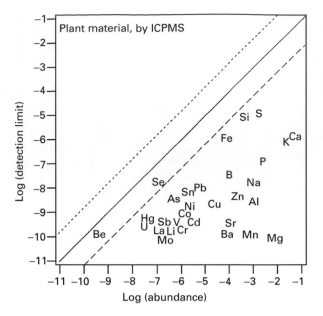

Figure 9 ADL diagram for plant material analysed by inductively coupled plasma mass spectrometry. (For explanation see **Figure 2**.)

preliminary process such as chemical separation or chromatography.

Air and dust

Metallic elements are present in air mostly as particulate material, which can be removed *inter alia* by filtration. Dusts are simply air particulates that have collected naturally by gravitation. Atomic spectrometry methods are well adapted to the analysis of used air filters. For instance, filters of the cellulose acetate type can conveniently be destroyed by oxidation with nitric/perchloric acid while the metallic elements are thereby solubilized. The resulting aqueous solution, of small volume, can be analysed simultaneously for many elements by ICPAES or ICPMS. The former is quite capable of determining most elements at their threshold concentration after the filtration of (say) 10 litres of air. Alternatively, air filters have a suitable form for direct analysis by XRF in special filter holders.

Biological tissues

Animal and plant tissues are usually prepared for elemental analysis by destruction of the organic matrix by oxidation, either in a furnace (dry ashing) or by the use of oxidizing acids (wet ashing). The two methods have similar outcomes in most respects, except that there is a tendency for volatile elements

high-solids solutions are sprayed. Again, preconcentration techniques that provide a separation of the analytes from the matrix are very useful in this context.

The speciation of certain elements in water is important, as it throws light on the transport of the elements between other media, e.g. the soil–plant interface. In this context, atomic spectrometry has the role of providing a sensitive and virtually specific detector for the individual elements, after some

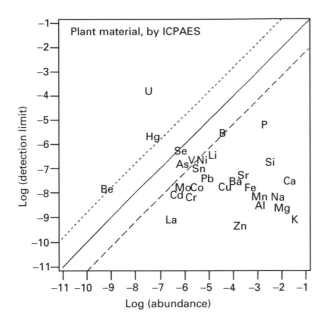

Figure 10 ADL diagram for plant material analysed by inductively coupled plasma atomic emission spectroscopy. (For explanation see **Figure 2**.)

rial. Concentrations of elements in animal tissues are broadly similar, although there are contrasts in certain cases.

See also: **Atomic Absorption, Methods and Instrumentation; Atomic Absorption, Theory; Atomic Emission, Methods and Instrumentation; Atomic Spectroscopy, Historical Perspective; Calibration and Reference Systems (Regulatory Authorities); Environmental Applications of Electronic Spectroscopy; Food & Dairy Products, Applications of Atomic Spectroscopy; Food Science, Applications of Mass Spectrometry; Food Science, Applications of NMR Spectroscopy; Inductively Coupled Plasma Mass Spectrometry, Methods; X-Ray Fluorescence Spectrometers; X-Ray Fluorescence Spectroscopy, Applications.**

such as As and Se and even Cd to be lost during dry ashing, especially if the temperature is allowed to rise above 500°C. The normal outcome is for the elements to be nebulized from a dilute solution of hydrochloric acid. Matrix effects are not usually troublesome in this sector, although some materials high in calcium, potassium or phosphorus may require special attention. **Figures 9** and **10** show the applicabilities of ICPMS and ICPAES for plant mate-

Further reading

Alloway BJ and Ayres DC (1997) *Chemical Principles of Environmental Pollution*, 2nd edn. London: Blackie.

Boumans PWJM (ed) (1987) *Inductively Coupled Plasma Emission Spectrometry*. New York: Wiley.

Fifield FW and Haines PJ (1995) *Environmental Analytical Chemistry*. London: Blackie.

Gill R (ed) (1997) *Modern Analytical Geochemistry*. Harlow: Longman.

Haswell SJ (ed) (1991) *Atomic Absorption Spectrometry: Theory, Design and Applications*. Amsterdam: Elsevier.

Thompson M and Walsh JN (1989) *Handbook of Inductively Coupled Plasma Spectrometry*, 2nd edn. Glasgow: Blackie.

Environmental Applications of Electronic Spectroscopy

John W Farley, University of Nevada, Las Vegas, NV, USA

William C Brumley, National Exposure Research Laboratory, US EPA, Las Vegas, NV, USA

DeLyle Eastwood, Air Force Institute of Technology, Wright-Patterson AFB, OH, USA

ELECTRONIC SPECTROSCOPY
Applications

Introduction

Electronic spectroscopy is widely used to detect environmental contamination. Environmental applications of electronic spectroscopy involve challenging analytical problems. Both qualitative identification and quantitative determination are performed for analytes which may occur at concentrations ranging from parts per hundred to parts per quadrillion. The analytes occur in a very wide range of matrices. Electronic spectroscopy consists of monitoring the absorption of light by the sample or monitoring the emission of light, often after excitation of the sample by an appropriate light source or laser beam. Both inorganic and organic analytes can be detected. Real-world samples are often complex mixtures, and therefore environmental applications may involve coupling a separation technique with electronic spectroscopy. Separation techniques include liquid chromatography (LC), gas chromatography (GC), high-performance liquid chromatography (HPLC), capillary electrophoresis (CE), and immu-nochemical analysis. Emerging techniques include the increasing use of hybrid techniques and the development of highly parallel detection instruments.

The analytical problem

The environmental analytical problem generally presents two questions: what substances are present in an environmental sample and how much of each of the substances is there? Sometimes these questions can be answered by spectroscopic techniques in a direct manner with little sample handling. However, the range of needs encompasses a vast array of matrices and levels of determination. To indicate some scope to the problem consider a list of potential matrices: soil, sediment, water, plants, animals, fly ash, sludge, waste water, leachates, food, blood, urine, hair (fur), drinking water, commercial pesticide formulations, air, dust, automobile and truck exhaust, and rain. The various matrices require extraction techniques, cleanup techniques, and often a final separation or detection technique. Contaminant levels range from percent levels (part per hundred) to parts per quadrillion (pg kg^{-1}). To give some realism to these numbers, for example, pesticides in formulations may be present at percentage levels. In another context, contaminants at Superfund sites are often found in the parts per million (µg g^{-1}) level. (A Superfund site is an identified site under the authorities of the Comprehensive Environmental Response, Compensation and Liability Act where releases of hazardous substances had already occurred or might occur and posed a serious threat to human, health, welfare or the environment.) Near the other extreme, the European Union regulates phenoxyacid herbicides at the 100 parts per trillion (pg g^{-1}) level in drinking water.

The applications, therefore, fall into two broad categories: qualitative identification and quantitative determination of substances or classes of compounds in environmental matrices. The qualitative identification is usually accomplished by acquiring a full UV-visible absorption or emission spectrum of the compound of interest. This spectrum is then compared with libraries of spectra for potential matches. Usually, the specificity of this match does not produce the same level of certainty as could be obtained from a mass spectrum that has been properly matched to a library spectrum. Nevertheless, within the context of a limited number of compounds and a particular analysis, some degree of certainty is obtained. In some favourable cases (e.g. fluorescence of oils), UV-visible identification has prevailed in legal proceedings and shown similar degrees of certainty to GC-MS. In UV-visible identifications, comparison is often made to site-specific standards, rather than against a library. The certainty of identification by UV-visible can be enhanced by the association of a retention time or migration time achieved by employing one or more separation techniques in combination with the spectroscopic data obtained on-the-fly.

The identification problem or the specificity of the determination can be a complicated matter because of the large number of analytes, matrix components, and the variability of environmental matrices depending upon location. There are now over 11 million organic compounds recognized by the *Chemical Abstracts Registry*, with over 30 000 compounds considered as chemicals of commerce, and perhaps 500 to 1000 routinely measured as members of industrial solvents, disinfection by-products, insecticides, PNAs, herbicides, organochlorine pesticides, PCBs, phenols, anilines, benzidines, and potential endocrine-disrupting compounds.

The quantitative determination is usually based on the absorption of light at some fixed wavelength or wavelengths, the selection of a wavelength for quantitation from the diode array detection (DAD) data, or, in fluorescence detection, the emission of light at some fixed wavelength (or band) while exciting at a fixed wavelength. For even better quantitation, one monitors the area under a peak or area of the whole spectrum. Thus one may monitor absorption at 214 nm for the sensitive detection of many compounds. Alternatively, one may monitor emission at 520 nm with excitation at 488 nm for the sensitive fluorescence detection of fluorescein and fluorescein-related compounds.

Inorganic compounds

Electronic fluorescence techniques may be used to detect inorganic compounds, which exist as ions in solution, using one of several methods. Photoluminescence (or simply luminescence) refers to the photoexcitation of an analyte in solution, caused by absorption of visible or ultraviolet radiation, followed by emission at a longer wavelength. Luminescence is classified as either fluorescence or phosphorescence. Fluorescence is an allowed radiative transition from the first excited singlet state, while phosphorescence is a spin-forbidden transition from the first excited triplet state. Typical fluorescence lifetimes are nanoseconds to sub-microseconds, while typical phosphorescence lifetimes are microseconds to seconds. For many inorganic substances, fluorimetric methods are the best analytical methods.

(1) If the inorganic ion luminesces in solution it may be detected simply by placing the ion in an appropriate solution. This technique works primarily for rare earths and uranyl (UO_2^+)

compounds. When combined with an inorganic reagent (e.g. HCl, HBr, etc.), the technique works for some group 3–5 elements.

Since the rare earths and uranyl compounds luminesce in solution, these ions may be detected by 'native fluorescence'. Even though the bare ion fluoresces in these cases, the luminescence intensities can be greatly enhanced by appropriate complexation. The spectrum may be broad or narrow, depending upon the element and the nature of the electronic transition involved: the fluorescence spectrum is broad for Ce(III), Pr(III), and Nd(III), because they result from electron transitions from the 5d to the 4f shell. For example, cerium salts fluoresce in a broad band from 330 to 402 nm. In contrast, the luminescence spectra of Sm(III), Eu(III), Td(III), and Dy(III) are narrow, consisting of a few characteristic bands because they result from transitions of the 4f electron, and the f electron is little perturbed by its environment. For example, Gd has a single bright, narrow band at 311 nm. Sensitive fluorimetric methods have been developed for Pr, Nd, Sm, Eu, Gd, Tb, Dy, Er, and Tm. Detection limits range from 50 picograms to a microgram depending on the rare earth. Uranyl (UO_2^+) ions phosphoresce in several bands in the 460–620 nm range. The uranyl ion is typically detected in acid solution. The fluorescence efficiency depends on pH and on impurities. Many organic compounds can quench the luminescence from uranyl. In phosphoric acid or phosphate solutions uranyl can be detected at 5 parts per trillion.

(2) If the inorganic ion does not luminesce in solution, the ion may be combined with an organic ligand to yield a metal chelate. In favourable cases, the isolated ligand does not fluoresce, but the metal chelate does fluoresce, allowing for sensitive detection of the ion. In other cases where the complexing agent luminesces, the metal complex is still shifted enough in peak position to be detected even in the presence of the free agent. Over 40 different metals have been determined by this technique. When this method is applicable it is very sensitive and specific. For example, gold may be detected by its formation of a complex with Rhodamine B in 0.4 M HCl. Using excitation and detection wavelengths of 550 and 575 nm, respectively, as little as 20 ppb of Au can be detected, a level 25 times more sensitive than atomic absorption. In another example, cyanide (CN^-) can be detected at levels of 0.2 µg by complexing with *p*-benzoquinone to yield an intense green

fluorescence. Sulfide, thiosulfate, thiocyanate, ferrocyanide, and 26 other ions had no effect.

Several kinds of reaction between inorganic analyte and the organic reagent are possible: binary or ternary complex formation, substitution, redox, catalytic, and enzymatic. This technique can sometimes be applied to both cations and anions. Examples of commonly used reagents are 8-hydroxyquinoline, azo reagents, Schiff's bases, hydrozones, hydroxyflavones and chromone derivatives, and anthraquinones.

(3) If neither of the above approaches works, 'indirect detection' may be employed. The inorganic ion may quench the fluorescence of a second compound, in which case the analyte is detected as a decrease in the fluorescence of the second compound. For example, most assays for fluoride (F^-) are based on the decrease in fluorescence of an Al or Zr chelate due to subsequent fluoride complex formation (AlF_6^- or ZrF_6^-). Quenching is generally less sensitive and less selective than fluorescence. Alternatively, the analyte may cause the release of a ligand which then reacts to form a fluorescent product.

Most fluorimetric determinations of inorganic species are equilibrium methods in which the reaction goes to completion. However, in some cases kinetic methods are used, in which the initial reaction rate is linearly related to the initial analyte concentration. For example, platinum(IV) can be detected at 0.2–0.6 ppm levels using a reaction with di-2-pyridylketone hydrozone, yielding a blue autooxidation product, whose rate of appearance is measured at 435 nm with an excitation wavelength of 359 nm.

Organic compounds

Organic compounds are often of major concern in environmental applications. In this section we discuss the detection of organic compounds by electronic absorption and luminescence spectroscopy.

The technique of UV-visible absorption spectroscopy is a mature technology which is widely used for detection. A beam of light whose wavelength is in the visible or ultraviolet range traverses a sample containing the sample. The signal is detected as a decrease in the intensity of the light. UV-visible absorption spectroscopy has a moderate sensitivity and a low to moderate specificity. The specificity is low because many compounds absorb in the same general wavelength range. Because of the low specificity, this technique can be used for real-world samples of low to moderate complexity. The technique works best for unsaturated compounds (aromatic, polycyclic aromatic or heterocyclic). In these cases, often there is enough vibrational structure for relatively good specificity. It works for dyes and for colorimetric reaction products.

The second technique is UV-visible luminescence, meaning either fluorescence or phosphorescence. When this technique is applicable it can be very sensitive, especially with laser excitation. In aqueous solution, the sensitivity can be parts per billion to parts per trillion. Luminescence is more sensitive than absorption because in the former case one is detecting an increase from a very small background, while in the latter case one is detecting a decrease from a large background. The technique is applicable to aromatics, most polycyclic aromatics and their derivatives, and some heterocycles. For many polycyclic aromatic hydrocarbons, the vibrational structure is sufficient to greatly enhance specificity. It can be applied to many other analytes, using fluorimetric reagents. The selectivity can be enhanced by varying the wavelength of excitation and the wavelength of detection, or by using time or phase resolution.

One of the problems that arises in luminescence studies of large molecules is that the spectrum may be relatively broad and unstructured. Accordingly, in a real-world sample there can be interferences caused by other compounds. One technique to improve the specificity is 'synchronous luminescence', in which both the excitation wavelength and detection are scanned, usually with a fixed difference between the two wavelengths. This tends to sharpen the spectra, reducing the problem of interferences and thereby improving the specificity of the technique. On theoretical grounds, it makes more sense to scan the two wavelengths with a fixed difference in the photon energy, rather than a fixed difference in the wavelength.

Luminescence is applicable to polyaromatic compounds, fluorescent dyes, fluorimetric reaction products, polychlorinated biphenyls (PCBs), phenols, 50% of pesticides, many semivolatiles, many nonvolatiles, and petroleum oils. For luminescence to work, the compound must fluoresce. However, many large molecules absorb but do not fluoresce in the visible or UV. For such non-fluorescing compounds, absorption spectroscopy must be used instead (with much less sensitivity) or, alternatively, the organics can sometimes be complexed.

In practical applications, these spectroscopic techniques are often used for field screening, in which samples are screened in the field rather than being transported to the laboratory for traditional analysis, which can mean a delay of up to a month in obtaining the results of analysis. Recent technical advances include: more compact lasers, miniaturized optical hardware, new types of detectors, increased use of fibre optics, and better computer software for spectral data processing and pattern recognition.

The problem of spectral congestion

Many real-world samples consist of a large number of substances whose spectra are relatively broad and relatively featureless. This limits the ability to discriminate between compounds with similar or substantially overlapping spectra. There are two techniques that can be used to discriminate between compounds that are spectrally overlapping. One technique is discrimination based on fluorescent lifetime. Two compounds that spectrally overlap can be distinguished by their fluorescence lifetimes, provided that their fluorescent lifetimes are sufficiently different. The second technique is chromatographic discrimination; i.e. the use of a separation technique.

Separations coupled with electronic spectroscopy

In some cases, e.g. forensic oil identification, the spectra of mixtures of fluorescent compounds may themselves be of interest for fingerprinting purposes, especially when combined with chemometric or pattern recognition techniques. Many real-world environmental problems involve large numbers of compounds, and the spectra of individual compounds are needed for identification and quantitation. Detection techniques are often preceded by a separation step. Historically, the workhorse separation technique of environmental analysis has been capillary GC combined with MS or with specialized GC detectors such as the flame ionization detector and the electron capture detector, or more recently combined with the liquid chromatography or capillary electrophoresis (LC-MS or CE-MS). In recent decades, the introduction into commerce and the environment of increasingly polar and nonvolatile compounds has resulted in new interest in liquid separations for environmental problems, resulting in the growth of liquid separation methods of analysis, and it is especially useful for polar or thermally labile compounds or for nonvolatile high relative molecular mass compounds. The following discussion is organized by type of separation technique or other analytical technique rather than by compound type.

The most common application of electronic spectroscopy has been as the detection technique for liquid separations where absorption spectroscopy is the workhorse technique. Absorption spectroscopy is well suited for liquid separations because most of the solvents of choice exhibit large UV transparent regions that enable the sensitive detection of compounds of interest. This is in contrast to other types of detectors that might universally detect both the solvent and the analyte of interest. For example, infrared detection is often compromised by either the presence of water or other solvents because of strong absorptions that obscure key portions of the useful wavelength range. Even mass spectrometry requires special interfaces to remove the solvent preferentially before introduction into the ion source even though solvent molecular masses are generally below 100 Da.

The detection limits for absorption spectroscopy are approximately 1×10^{-5} absorption units (AU), varying with the extinction coefficient as a function of wavelength. (The absorption measured in AU is defined as the common logarithm of the initial to the final light intensity.) This results in detection limits in the ppm to low ppb range in favourable cases, providing sensitivity adequate for environmental analysis. In fluorescence detection, the possibility of detecting a single molecule exists. For environmental samples, often fluorescing interferences limit the sensitivity and in these cases pretreatments or liquid separations remove these interferences. In practical terms, fluorescence detection for appropriate molecules can be 100 to 1000 times more sensitive than absorption detection. As a result, detection limits in the low parts per trillion range can be achieved for selected analytes, limited only by excitation source intensities or fluorescence from capillary or solvent background fluorescence.

HPLC

The most widely used detection technique for HPLC involves absorption at either a fixed wavelength or using a DAD that allows for wavelength selection and continuous or selected scans. In the filter photometric format, a wavelength such as 220 nm is selected by means of a filter placed in front of the deuterium lamp. **Figure 1** illustrates the detection of a series of polynuclear aromatic (PNA) compounds using their absorption at 220 nm. Common solvents used in reversed-phase separations (i.e. a hydrophobic adsorbent with water/organic solvent as mobile phase) have UV transparence from ~200 nm on up. A similar situation holds for normal phase separation (i.e. a polar adsorbent with hexane/polar solvent as mobile phase) where UV transparency is counted on. When the solvent choice includes toluene or some other absorber then a different wavelength or different detection technique must be chosen.

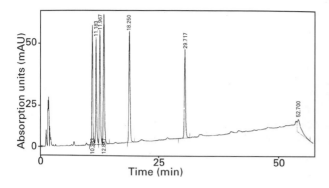

Figure 1 HPLC separation of PNAs with detection by absorption spectroscopy at 220 nm.

When the DAD is used, a selection of wavelengths may be available for individual monitoring as well as the selection of the best wavelength for quantitation in consideration of background noise that may be largely chemical noise or interferences from the matrix. In the case of PNAs, the full absorption spectrum for benzo[*a*]pyrene is given in **Figure 2**.

Molecules that do not possess the appropriate chromophore may be detected when derivatized with a chromophore with high extinction coefficients at peaks in the visible spectrum to facilitate detection. For example, carboxylic acids can be derivatized with 7-methoxy-4-bromomethylcoumarin to give them a chromophore. Specialized books may be consulted for further information concerning derivatization. Other techniques include resorting to wavelengths below 200 nm under special conditions. In some cases, indirect detection can be used based on the presence of a background UV absorber in the mobile phase. The analyte displaces some of this absorber under certain conditions, resulting in a decrease in the absorption level in the detector cell which can be measured in the same manner as an increase in absorption. Further information on this technique can be obtained from specialized references on the subject.

When the analyte of interest possesses fluorescence character then an extremely sensitive detection can be designed based on the absorption/emission properties of the molecule. In **Figure 3**, the fluorescence detection of benzo[*a*]pyrene is illustrated using a excitation of 325 nm with an emission at 425 nm. In this manner, low pg quantities of compound can be detected, making this one of the most sensitive detections in analytical chemistry. Increasingly, charged coupled devices (CCDs) are being utilized to give even higher sensitivity especially with laser excitation, and whole spectra rather than single wavelengths are being recorded. Techniques such as fluorescence lifetime detection with phase or frequency modulation can also be used to separate peaks that would otherwise appear to overlap.

Capillary electrophoresis (CE)

CE has emerged as a pre-eminent separation technique for ionic analytes based on free zone electrophoresis. The separation is based on differing mobilities among the ions. In analogy to HPLC, an absorption detection is a fairly universal detector. The DAD response for a group of phenols is illustrated in **Figure 4**. The influence of ionization on the absorption spectrum of phenols is evident in the shift to longer wavelengths that results from the anionic character. One of the chief drawbacks of CE applications using absorption detection has been the relatively high concentration detection limit (i.e. ppm range rather than ppb range of HPLC). This is a direct result of the capillary format of the separation. In capillary separations, normally tens of nanolitres of sample are injected on-column. This is a 1000-fold smaller volume than in HPLC. In addition, the capillary format usually demands on-column detection to avoid band broadening from post-column detection approaches. The diameter of the capillary is about 75 μm, resulting in a very short optical path. There are specialized techniques for overcoming these limitations and more specialized references should be consulted.

The concentration detection limit problem can be overcome for molecules possessing fluorescence. In the case of phenols, illustrated in **Figure 5**, a frequency-doubled laser operating at 244 nm will enable sensitive detection using laser-induced fluorescence (LIF). In this case, a filter arrangement allows detection of emission light in the 355 to 425 nm range. The use of lasers in this context improves detection limits because of the light intensity as well as the intrinsic narrow beam which allows efficient focussing of the light onto the capillary bore carrying the sample.

Immunochemical analysis

Immunochemical analysis is a convenient approach for environmental analysis provided appropriate antibodies and format have been developed. Often, the assay depends on a colorimetric determination for its quantitative result. The advantage of immunochemical analysis is its specificity as well as its speed. Specific references should be consulted on applications.

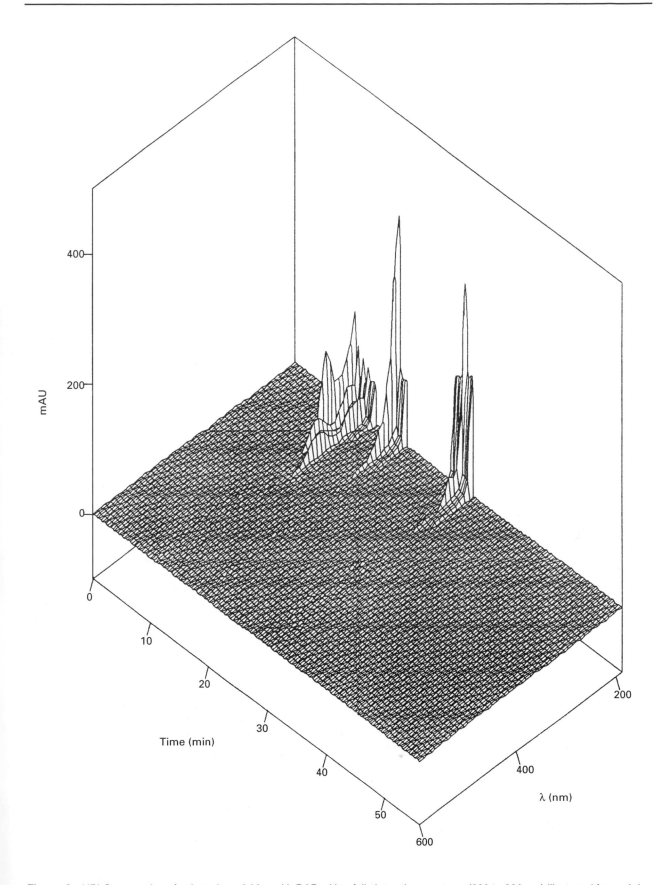

Figure 2 HPLC separation of selected pesticides with DAD with a full absorption spectrum (200 to 600 nm) illustrated for each in a three-dimensional format.

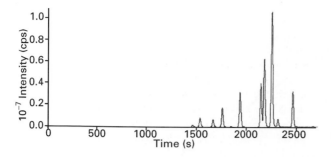

Figure 3 HPLC separation of selected PNAs using fluorescence detection with excitation at 325 nm (xenon lamp) and emission at 425 nm.

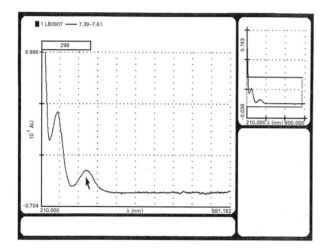

Figure 4 DAD detection of phenols in capillary electrophoresis. The arrow indicates the wavelength position of 298 nm.

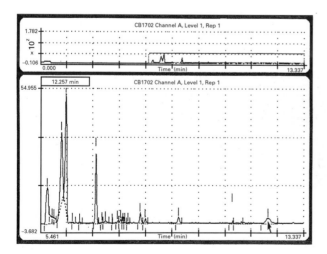

Figure 5 CE-LIF detection of phenols using a frequency-doubled laser. Excitation at 244 nm and emission at 355–425 nm. The arrow indicates the migration time position of 12.257 min.

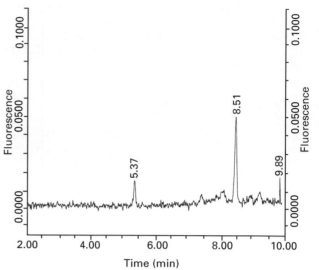

Figure 6 CE-LIF detection of probe peak (fluorescein-labelled 2,4-dichlorophenoxyacetic acid) in an immunoassay format of herbicide analysis. Excitation at 488 nm and emission at 510–530 nm.

An emerging format for the assay depends on CE-LIF determination of a fluorophore-labelled analyte (probe) in either a competitive or noncompetitive assay with the unlabelled analyte as they find binding sites with the antibody. **Figure 6** illustrates the concept for a fluorescein-labelled analogue of 2,4-dichlorophenoxyacetic acid (2,4-D) where the change in the amount of the probe peak indicates whether unlabelled 2,4-D was present in the sample.

Gas chromatography

One detection technique for GC, called atomic emission detector (AED), is of fairly common occurrence and makes use of spectroscopic properties for detection. The elements of a compound that elutes from the column are determined in a He plasma based on the emission spectra of the individual elements (e.g. C, H, O, N, S, Cl, etc.). Reference to more specialized works should be made. The AED technique is mentioned for the sake of completeness, although since AED relies on atomic spectroscopy emission, it is outside the scope of this article, which is mostly restricted to molecular electronic spectroscopy.

Thin-layer chromatography

Thin-layer chromatography (TLC) is usually coupled to either absorption or fluorescent detection of complexed species and is common in many fields of analysis. The applications of TLC to environmental analysis have been less frequent than with other forms of separation. It has recently been demonstrated that UV-visible absorption, near-IR absorption, fluorescence, and phosphorescence can be accomplished for hundreds of spots on a single TLC plate by using CCD detection.

Emerging techniques

The application of molecular electronic spectroscopy as a detector for separations is expected to continue to grow. Optical techniques are a primary detection approach for the hybrid technique of capillary electrochromatography which combines aspects of CE and HPLC, especially capillary LC. To increase throughput of sample handling, arrays of capillaries have been built with optical detection via CCD cameras. Using CE-LIF approaches, hundreds of capillaries can operate simultaneously on samples with simultaneous detection. Further advances are expected in the areas of remote sensing, immunochemical assays, capillary arrays, and detector sensitivity.

See also: **Chromatography-NMR, Applications; Chromatography-MS, Methods; Environmental and Agricultural Applications of Atomic Spectroscopy; Luminescence, Theory.**

Further reading

Acree Jr WE, Tucker SA and Fetzer JC (1991) Fluorescence emission properties of polycyclic aromatic compounds in review. In: *Polycyclic Aromatic Compounds*, Vol 2, 75–105. Lausanne, Switzerland: Gordon and Breach.

Analytical Chemistry. This journal is generally useful as a survey of the field. In odd years, they review applications, while on even years they review fundamentals (i.e. techniques).

Camalleri P (ed) (1993) *Capillary Electrophoresis: Theory and Practice*. Boca Raton, FL: CRC Press.

Fernandez-Gutierrez A and Munoz de la Pena (1985) Determination of inorganic substances by luminescence methods. In: Schulman SG (ed) *Molecular Luminescence Spectroscopy Methods and Applications*, Vol I, Chapter 4. New York: Wiley Interscience. Later volumes in the series include Vol II (1987) and Vol III (1993).

Grosser ZA, Ryan JF and Dong MW (1993) Environmental chromatography methods and regulations in the United States of America. *Journal of Chromatography*. **642**: 75–87.

Guilbault GG (1990) Inorganic substances. In: Guilbault GG (ed) *Practical Fluorescence*, Chapter 5. New York: Marcel Dekker.

Karcher W, Devilliers J, Garrigues Ph and Jacob J *Spectral Atlas of Polycyclic Aromatic Compounds*. Vol I (1983), II (1988), and III (1991). Boston: Kluwer Academic.

Lakowicz JR (1991) *Topics in Fluorescence Spectroscopy*, Vol I. Plenum Press. Later volumes include Vol II Principles (1992), Vol II *Biochemical Applications* (1993), and Vol IV *Probe Design and Chemical Sensing* (1994).

Lawrence JF (ed) (1984) *Liquid Chromatography in Environmental Analysis*. Clifton, NJ: Humana Press.

Sweedler JV, Ratzlaff KL and Bonner Denton M (1994) *Charge Transfer Devices and Spectroscopy*. New York: VCH Press.

Vo-Dinh T and Eastwood D (ed) (1990) *Laser Techniques in Luminescence Spectroscopy*. ASTM STP 1066. Philadelphia: American Society for Testing and Materials.

Wehry EL (ed) (1981) *Modern Fluorescence Spectroscopy*. New York: Plenum Press.

EPR Imaging

LH Sutcliffe, Institute of Food Research, Norwich, UK

> **MAGNETIC RESONANCE**
> **Applications**

Introduction

The basic requirement for an EMRI experiment is that a species having unpaired electrons be present. Although free radicals and some transition metal compounds fulfil this requirement, there are not many circumstances in which these are present naturally in sufficient concentration to give a suitable signal; this is in contrast to MRI where the ubiquitous proton can be used to study most materials. Fortunately, the high sensitivity of EPR compared with NMR (arising from the difference in electron and nuclear magnetogyric ratios) enables paramagnetic material to be introduced in low concentrations into systems under investigation with minimum interference. However, NMR has an advantage over

EPR in that pulse techniques can be used to boost the signal-to-noise ratio: both the irradiating frequency and the applied magnetic field gradients can be pulsed. The main reason why most EPR experiments are carried out in the continuous-wave (CW) mode is because spin–lattice relaxation times for paramagnetic materials are of the order of microseconds (compared with hundreds of milliseconds for proton NMR) and this causes considerable difficulties in achieving the short times required for the pulse experiment. A further difficulty is the spectral bandwidth of an EPR spectrum: this, coupled with a large line width (three orders of magnitude greater than in NMR) means that very large magnetic field gradients (100–1000 times greater than in MRI) have to be applied. However, for the few systems not presenting these difficulties, FT-EMRI will become the method of choice. The ability to employ pulse techniques for both the irradiating frequency and for the field gradients also allows a much wider variety of experiments to be carried out with MRI compared with EMRI.

Spin probe

The paramagnetic materials added to systems to obtain signals in EMRI are referred to as spin probes. These can be transition metal compounds such as those of manganese(II) or vanadium(IV) but they are more likely to be organic free radicals. In addition to having high thermal, chemical and metabolic stability (for biological applications), these radicals should preferably have spectra with very narrow line widths in order to maximize the signal-to-noise ratio. The peak-to-peak line width of a first-derivative spectral line can be represented by the equation:

$$\Delta B_{PP}(m_I) = A + Bm_I + C(m_I)^2$$

where A, B and C are constants and m_I is the magnetic quantum number of a given line. The constant A is dependent on the g factor anisotropy and experimental factors, B upon the g factor and the nuclear hyperfine interaction anisotropies and C upon the nuclear hyperfine interaction anisotropy. EPR spectra are normally recorded in the first-derivative mode but, obviously, the absorption mode is more convenient for EMRI. The most common stable free radicals used as spin probes are nitroxyls and there is both g factor and hyperfine interaction anisotropy of the ^{14}N nucleus. A typical set of values at X-band for a nitroxyl radical having unresolved proton hyperfine structure is $A = 0.116$, $B = -0.010$ and $C = 0.006$ mT. Thus the observed line width is of the order of 0.1 mT and there is little scope for reducing the line width further for imaging at X-band: a nitroxyl radical in a nitrogen-purged low-viscosity solvent can show a resolution of proton hyperfine couplings with observed line widths smaller than 0.02 mT. The three lines of a nitroxyl radical, in addition to having a large bandwidth, also represent a 'waste' of signal since it is more practical to image a single line. Some improvement can be made by substituting ^{14}N with ^{15}N and/or by reducing the proton hyperfine interaction by deuteration. However, there is a soluble commercial single-line free radical based upon the $^{\bullet}C(Ar)_3$ structure which has a line width of only 0.006 mT in a deoxygenated solvent: the sharp line is observed because of the small g factor and hyperfine anisotropies for a carbon-centred radical. One of the main aims in biological studies is to spin-label drugs and to image their organ compartmentation, and hence there are some difficult chemical problems to be solved. For some studies, it is possible to use solid particles as spin probes in which case very narrow single lines are attainable. An example is lithium phthalocyanine which has a line width of 0.002 mT in a deoxygenated system. Other single-line solids are special forms of coal, such as fusinite, and carbohydrate chars.

Theory

Resonance imaging depends on the application of magnetic field gradients to a sample. If two samples are placed at different positions of the spectrometer magnetic field then only one of them will be at resonance, the second one can be brought to resonance by changing the spectrometer field or by applying a magnetic field gradient. Since field gradients can be applied in three dimensions, it is possible to determine the relative positions of both the samples, i.e. to produce an image. Mathematical methods are of central importance in any type of image reconstruction and they were developed some time ago for radiographic tomography. The necessity of using CW techniques in EMRI limits the number of imaging strategies that can be employed.

Spatial imaging

Two-dimensional images of spin density in the xz or yz planes (z axis is parallel to the main magnetic field) can be obtained: for a $n \times n$ image, $n\pi/4$ projections are required. Most spatial images of paramagnetic centres with multiline spectra have been constructed by selecting a single line, using a limited field gradient to avoid overlap with other lines. The resolution can be enhanced by deconvolution of the line shape.

Spectral-spatial imaging

The advantages of spectroscopic imaging compared with spatial imaging methods are:

(i) the problems of resolution and gradient magnitudes owing to the presence of hyperfine lines can be overcome;
(ii) separating spectral and spatial components by assigning them to different imaging axes reduces the resolution degradation arising from the finite width of the spectral lines;
(iii) asymmetric and multiline spectra can be imaged whereas only symmetric line shapes should be used for spatial imaging

Images are obtained as a function of sample position by recording spectra at a series of magnetic field gradients which correspond to projections in the spectral-spatial plane. A pseudo-object of length B is examined in the spectral dimension and length L in the spatial dimension. An angle (α) defines the orientation of a projection relative to the spectral axis. The maximum field gradient, G_{max}, is related to B and L by

$$G_{max} = \tan(\alpha_{max})B/L$$

The maximum magnetic field swept for each projection is given by

$$\text{Sweep width} = 2^{1/2} B \cos \alpha$$

For given values of G_{max} and B the ratio $\tan(\alpha_{max})/L$ is fixed. Most image reconstruction algorithms require projections over 180° with equal angular increments. If a complete set of projections is used for spectral-spatial imaging, the number of projections determines the values of α_{max} and G_{max}. Thus for 64 projections, $L = 0.8$ cm and $B = 3.2$ mT, $\alpha_{max} = 88.59°$, $G_{max} = 162.5$ mT and the sweep width required is 183.9 mT. Spectral-spatial images can be reconstructed from incomplete sets of projections and this has the advantage of reducing both the maximum field and sweep width requirements. Assuming experimental data were obtained for 60 out of 64 projections and estimates were made for the 'missing' projections, then G_{max} and the sweep width requirements are reduced to 32.4 and 37.0 mT, respectively.

Projection reconstruction methods

Important image reconstruction procedures are:

(i) back projection;
(ii) simultaneous, algebraic and iterative least-squares;
(iii) two-dimensional and filtered back projection;
(iv) use of incomplete data sets from which it is also possible to obtain slices from the image.

Modulated magnetic field gradients

These gradients, B_{GM}, provide fields in addition to the main magnetic field that have both time- and space-dependent amplitudes. The time-independent spectrum can only be recorded in the z plane where the amplitude of B_{GM} is zero. Images can be obtained without the use of difficult mathematical deconvolution by scanning the plane of the field gradient across the sample.

X-Band imaging

Imaging at about 9 GHz very convenient since most commercial EPR spectrometers operate at this frequency and the imaging system can be fitted as an accessory. Gradient coils can be fitted to the outside or to the inside of a standard resonant cavity. **Figure 1** shows the arrangement for the former: to generate sufficiently large gradients, ~1 kW of power is needed and hence provision for water cooling has to be made. **Figure 2** shows how small coils, used for microscopy, can be placed inside the cavity. X-Band frequencies can be used for non-polar samples up to 10 mm in diameter but polar samples are restricted to 1.5 mm diameter because of the absorption of the microwave energy. An estimate of the achievable resolution is a pixel of $10 \times 10 \times 10$ μm that contains 1 mM spins. Note that the latter concentration cannot be increased appreciably without causing line broadening from spin–spin interactions.

An example of two-dimensional spectral-spatial imaging is shown in **Figure 3**. This depicts the projection–reconstruction image obtained from four different-sized specks of DPPH (diphenylpicrylhydrazyl), each of which gives an exchange-narrowed single spectral line. **Figure 4** is an example of a two-dimensional spectral-spatial image, showing a phantom that contains solutions of three types of free radicals which each have some hyperfine structure.

X-Band EMRI has been applied to the study of the diffusion and distribution of oxygen in models of biological systems with the aim of studying diffusion

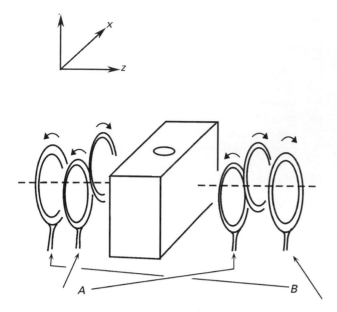

Figure 1 The arrangement of figure-of-eight (A) and anti-Helmholtz (B) coils fitted to the outside of a rectangular resonant cavity, to provide gradients $\delta B_z/\delta_x$ and $\delta B_z/\delta_z$ respectively.

and metabolism *in vivo*. The method is based on the fact that oxygen causes paramagnetic broadening of the spin probe lines. Other applications have been (i) for the determination of depth profiles of E' defects in X-irradiated silica, (ii) for radiation chemistry, where a dosimeter has been designed using crystalline L-α-alanine which can be used to check the spatial variation of the radiation, (iii) a variation of this theme to study the distribution of alkyl and alkoxy radicals in γ-irradiated polypropylene, (iv) to study diffusion and transport of nitroxyl radicals, for example into zeolites, into solid polymers and into the caryopsis of wheat seeds and (v) to study paramagnetic humic substances found in soil. **Figure 5** shows an image obtained from the soaking of these into a wheat grain.

There are several strategies for EMRI microscopy, and an example is shown in **Figure 2**. Arrangements of this type are capable of giving a resolution of 50 μm. EMRI microscopy can be used for (i) small-scale dosimetry of materials exposed to high-energy radiation, (ii) ceramic and semiconductor technology

(A)   **(B)**

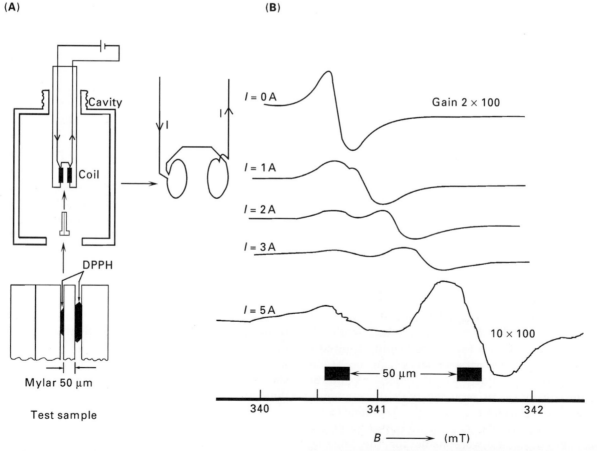

Figure 2 (A) A cylindrical EPR cavity fitted with anti-Helmholtz coils and a PTFE sample holder. The test sample comprises two portions of DPPH powder separated by a 50 μm mylar film. (B) EPR spectra of DPPH for currents ranging from zero to 5 A in the gradient coils. The signal separation of about 1 mT corresponds to a field gradient of 20 T m^{-1}. Reproduced with permission from Ikeya M and Miki T (1987). *Japanese Journal of Applied Physics* **26**: L929.

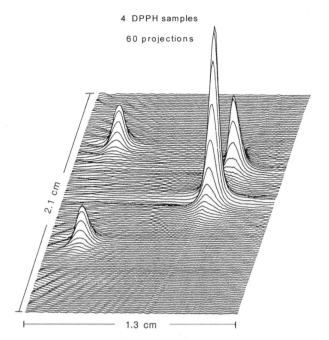

Figure 3 A 2D spatial X-band image of four specks of solid DPPH. Sixty EPR spectra were obtained with maximum field gradients $\delta Bz/\delta z$ = 10 mT cm^{-1} and $\delta Bz/\delta y$ = 6 mT cm^{-1}. The spectra constitute a full set of projections over 180°. Reproduced with permission from Eaton GR and Eaton SS (1995) Introduction to EPR imaging using magnetic-field gradients. *Concepts in Magnetic Resonance* **7**: 49.

and (iii) the distribution of nitrogen centres in diamonds.

L-Band imaging

Most current biological imaging systems operate at L-band frequencies (~1 GHz) which allows larger and more 'lossy' samples to be studied (microwaves of this energy can penetrate about 6 mm into a biological specimen) than is possible at X-band. A fairly standard spectrometer can be adapted for this work but loop-gap resonators, external surface coils and re-entrant cavities are more useful than a standard resonant cavity.

A non-biological application of L-band EMRI has been the study of the oxidation of cylindrical coal samples at 150°C where it was found that there is a homogeneous distribution of radicals across the samples. An important biological application is the study of oxygen concentrations ('oximetry') in a perfused rat heart using implanted solid paramagnetic particles: this rapid (1–5 min) multisite *in vivo* spectroscopy can be applied to a wide range of problems in experimental physiology and medicine. Other biological applications involve the imaging of soluble free-radical distribution: some examples include (i) a three-dimensional image of the head of a living rat,

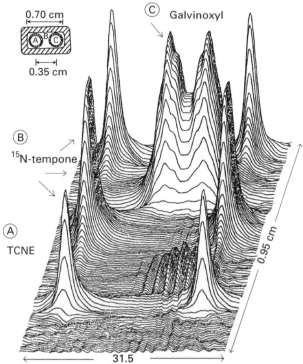

Figure 4 Spectral-spatial X-band image of a composite sample of the tetracyanoethylene (TCNE) radical in tube A, the galvinoxyl radical tube C, and ^{15}N-labelled tempone in region B. The maximum magnetic field gradient was 30 mT cm^{-1}. The image was reconstructed iteratively from 60 experimental projections and four missing projections. Reproduced with permission from Eaton GR and Eaton SS (1995) Introduction to EPR imaging using magnetic-field gradients. *Concepts in Magnetic Resonance* **7**: 49.

showing that the nitroxyl-deficient part corresponds to the brain, and (ii) accurate images of the vessel lumen and wall of a perfused rabbit aorta. When a glucose char (having a single EPR line) is introduced into an isolated biological organ, the resolution obtainable is 0.2 mm (**Figure 6**) whereas a soluble nitroxyl spin probe allows a resolution of only 1–2 mm. The single-line soluble probe $^{\bullet}$C(Ar$_3$) has allowed images of a rat kidney to be visualized with a resolution of 100 µm for samples up to 25 mm. As in MRI, data acquisition can be gated to overcome the problems associated with rapidly-moving organs such as a beating heart.

A biologically important naturally occurring free radical is nitric oxide but, unfortunately, it has very broad EPR lines and thus cannot be imaged. However, its generation in rat hearts subjected to global ischaemia has been mapped by using the spin trap bis(*N*-methyl-D-glucamine dithiocarbamate)iron(II). The latter forms a stable nitroxyl free-radical with nitric oxide that can be imaged.

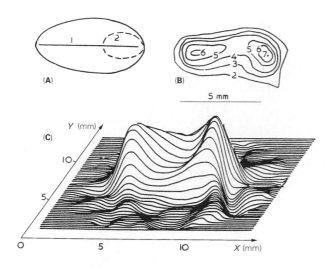

Figure 5 (A) The schematic structure of a wheat grain (1 is the vascular bundle and 2 is the germ). (B) The spatial distribution of the humic substance signal in the wheat grain. (C) The relief surface of the spin density distribution in the wheat grain. Reproduced with permission of Academic Press from Smimov AI, Yakimchenko, Golovina OE, Bekova SK and Lebedev YS (1991) *Journal of Magnetic Resonance* **91**: 386.

Four-dimensional spectral-spatial EMRI has been developed (1000 projections were used to construct $32 \times 32 \times 32 \times 32$ pixel images) and, while considerable data acquisition times are needed, it is feasible to obtain spectral information for a biological structure such as a rat heart.

High-field imaging

A high-field imager using a superconducting magnet operating at 5 T (140 GHz) has been constructed that derives its very large field gradients from ferromagnetic discs and wedges. This type of instrument is aimed at studying processes that have spatial ranges of less than 10 μm such as diffusion, defect formation in irradiated samples and polymer fracture under stretching and loading. High spectral resolution can be achieved because of the high absolute point sensitivity. A commercial EPR spectrometer is available for W-band (95 GHz) that can be adapted for high-field EMRI.

Radiofrequency imaging

It has been stated above that it is difficult to obtain images at L-band frequencies of lossy objects larger than 25 mm. Thus to image large samples, such as live animals, lower frequencies have to be used. There is also an added incentive to work at radiofrequencies because of the low cost of components compared with those used at microwave frequencies.

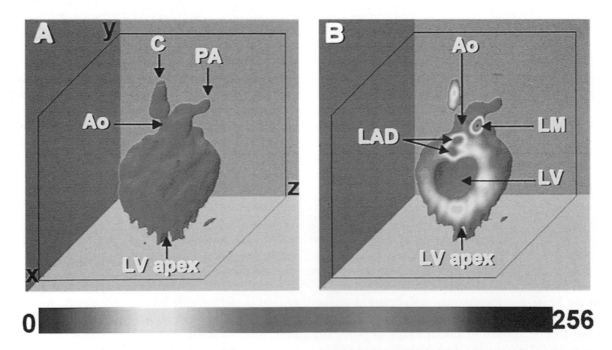

Figure 6 Three-dimensional images ($25 \times 25 \times 25$ mm^3) of an ischaemic rat heart infused with a glucose char suspension. (A) Full view of the heart. (B) A longitudinal cut out showing the internal structure of the heart. C is the cannula; Ao the aorta; PA the pulmonary artery; LM the left main coronary artery; LAD the left anterior descending artery; and LV the left ventricular cavity. Reproduced with permission from Kuppusamy P, Wang P and Zweir JL (1995). *Magnetic Resonance in Medicine* **34**: 99. (See Colour Plate 18)

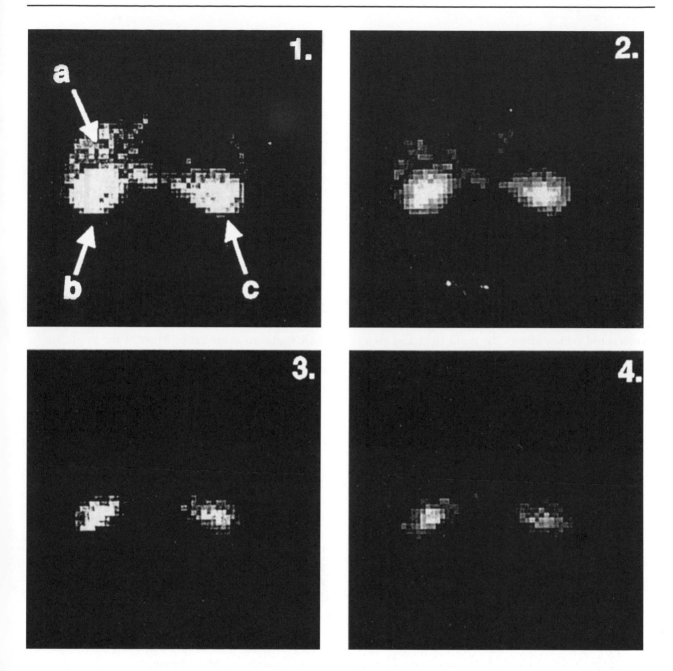

Figure 7 Set of proton Overhauser images from the abdomen of a supine, 350 g, anaesthetized rat, following the administration of a solution of hydrogencarbonate-buffered 3-carboxy-2,2,5,5-tetramethylpyrrolidine-I-oxyl (PCA). Image matrix size 64 × 64; field-of-view, 10 × 10 cm; slice thickness, 2 cm; field strength, 10 mT; EPR Irradiation frequency, 238.7 MHz. Image no. 1 was collected 1 min after administration of PCA and the remaining images were collected at 6, 11 and 16 min post-injection respectively. The labels are (a) liver; (b) right kidney (c) left kidney. Reproduce with permission of Wiley from Lurie DJ (1995). In Grant DM and Harris RK (eds) *The Encyclopedia of Nuclear Magnetic Resonance*.

The consensus of opinion is that the most suitable frequency is about 300 MHz, which has a penetration depth of about 7 cm in systems of high relative permittivity. Theory suggests that the signal-to-noise ratio and resolution of EMRI are very much worse than they are for MRI. However, the high point sensitivity of EMRI compared with MRI provides advantages for probe experiments and thus EMRI

should include, at some level, a spectral dimension. At radiofrequencies the line widths of nitroxyl spin probes are similar to those at X- and L-bands since the dominant factor for these probes is the anisotropic hyperfine interaction which is field independent. It should be noted that, for nitroxyls, account has to be taken of the Breit–Rabi effect. As might be expected, the experimental approach has features in common with MRI; for example, 'bird-cage' resonators can be used. Other useful types of resonators are surface coils, re-entrant cavities and loop-gap resonators. The main magnetic field required for 300 MHz is about 11 mT, which can be obtained from a simple air-cored Helmholtz design. Because a non-ferrous magnet is used, the main field windings can also be used for modulation at a frequency of about 4 kHz. The x, y and z gradient coils can be wound on concentric cylinders and an advantage of the bird-cage resonator is that it can be mounted concentrically inside the cylinder, whereas a loop-gap resonator would have to be mounted across the cylinder making sample access more difficult. A bird-cage resonator has been used to implement longitudinally detected EPR, which has the advantage of giving less noise from animal motion. The latter can also be minimized by automatic frequency control of the radiofrequency. A multipole magnet has been constructed which allows the main field and two of the gradient fields to be generated; the third gradient is obtained by inserting an air-cored coil into the magnet. In another design, two Helmholtz loops are splayed to produce the z direction field gradient.

A variety of studies using RF EMRI have been carried out on living mice. These include (i) oximetry in tumour tissues and the response to perfluorocarbons for their potential use in radiotherapy, (ii) the detection of hydroxyl radicals (by spin trapping) in γ-irradiated tumours, (iii) the distribution of nitroxyl radicals with distinct pharmacological abdominal compartment affinities and (iv) quantitative measurement of the viscosity of a non-vascular compartment.

Overhauser imaging

The broad lines encountered in EPR lead to poor spatial resolution compared with MRI and hence it is unlikely that a whole-body human-size imager is possible using the techniques described above. There is, however, an imaging strategy that could lead to a large-sample imager by taking advantage of the Overhauser Effect – techniques using the effect are also known as PEDRI (proton–electron double resonance imaging) and DNP (dynamic nuclear polarization) imaging. This is done by using a double resonance technique in which low-field proton MRI is combined with EPR. The EPR spectrum of a free radical is irradiated during the collection of a proton NMR image. The Overhauser effect produces an enhancement (by a factor of about 100) of the NMR signal in the parts of the sample that contain free radicals. The big advantage is that MRI resolution is achieved without having to use large field gradients but the disadvantage is that spectral information cannot be

obtained. To overcome this drawback, an instrument has been constructed that combines Overhauser imaging with EMRI. In Overhauser imaging, an EPR frequency lower than 300 MHz has to be used to avoid heating of a biological sample. When nitroxyl spin probes are being used it is possible to irradiate each of the three ^{14}N manifolds simultaneously to achieve a maximum Overhauser enhancement. Resonators are similar to those used in radiofrequency imaging, namely loop-gap or bird-cage structures. Normally, a MRI image is obtained and subtracted from an Overhauser MRI image for the slice under study. This is done by using an interleaved pulse sequence, each alternate NMR excitation being preceded by EPR irradiation. An example of images obtained in this way is shown in **Figure 7**.

By cycling the main magnetic field, large samples can be imaged by reducing the EPR frequency required to about 60 MHz without impairing the NMR signal-to-noise ratio. Most of the applications of Overhauser imaging are similar to those given above for radiofrequency imaging except that it is possible to study much larger samples. Oximetry, however, is carried out differently since it is possible to measure quantitatively the reduced Overhauser enhancement caused by the oxygen. An interesting aspect of Overhauser imaging is that it can be carried out at fields as low as that of the earth, i.e. 0.06 mT. The reason is that the enhancement is still high for magnetic fields lower than the hyperfine coupling of the spin probe, ~1.5 mT (70 MHz) for nitroxyl radicals. Thus it is possible to build an imager having a main field of about 0.3 mT that is regarded as biologically safe, is protected against magnetic perturbations and is of low cost.

See also: **EPR, Methods; Magnetic Field Gradients in High Resolution NMR; MRI Instrumentation; MRI Theory; Nuclear Overhauser Effect; Radio Frequency Field Gradients in NMR, Theory.**

Further reading

Eaton GR, Eaton SS and Ohno K (eds) (1991) *EPR Imaging and In Vivo EPR*. Boca Raton, FL: CRC Press.

Eaton GR and Eaton SS (1995) Introduction to EPR imaging using magnetic-field gradients. *Concepts in Magnetic Resonance* 7: 49.

Fairhurst SA, Gillies DG and Sutcliffe LH (1990) Electron spin resonance imaging. *Spectroscopy World* 2: 14.

Ikeya M and Miki T (1987) *Japanese Journal of Applied Physics* 26: L929.

Kuppusamy P, Wang P and Zweier JL (1995) *Magnetic Resonance in Medicine* 34: 99.

Lurie DJ (1995) Imaging using the electronic Overhauser effect. In: Grant DM and Harris RK (eds) *The*

Encyclopedia of Nuclear Magnetic Resonance. Chichester: Wiley.

Lurie DJ (1996) Commentary: electron spin resonance imaging studies of biological systems. *British Journal of Radiology* 69: 983.

Smirnov AI, Yakimchenko OE, Golovina HA, Bekova SK and Lebedev YS (1991) *Journal of Magnetic Resonance* 91: 386.

Sutcliffe LH (in the press) *Physics and Medicine and Biology.*

EPR Spectroscopy, Theory

Christopher C Rowlands and **Damien M Murphy**,
Cardiff University, UK

MAGNETIC RESONANCE
Theory

This article will discuss the concept and outline the theory of electron paramagnetic resonance (EPR), sometimes known as electron spin resonance (ESR), spectroscopy. Paramagnetism arises as a consequence of unpaired electrons present within an atom or molecule. It can be said that EPR is the most direct and sensitive technique to investigate paramagnetic materials. It will not be possible within this article to fully expound on all the theoretical aspects of EPR and so the reader is encouraged to refer to the excellent texts listed in the Further reading section.

Consequently this work is aimed at the professional scientists who would like to expand their knowledge of spectroscopic techniques and also at undergraduates with the intention of allowing them to fully appreciate the scope, versatility and power of the technique.

Basic principles of the EPR experiment

The electron is a negatively charged particle that moves in an orbital around the nucleus and so consequently has orbital angular momentum. It also spins about its own axis and therefore has a spin angular momentum given by

$$\text{spin angular momentum} = h/2\pi[S(S+1)]^{1/2} \quad [1]$$

where S is called the spin quantum number, with a value of $\frac{1}{2}$, and h = Planck's constant (6.626×10^{-34} Js). If we restrict ourselves to one specified direction, the z direction, then the component of the spin

angular momentum can only assume two values and consequently $S_z = M_S h/2\pi$. The term M_S can have $(2S + 1)$ different values of $+S$, $(S - 1)$, $(S - 2)...(-S)$. If the possible values of M_S differ by one and range from $-S$ to $+S$ then the only two possible values for M_S are $\frac{1}{2}$ and $-\frac{1}{2}$.

Associated with the electron spin angular momentum is a magnetic moment μ and it is related to the spin as follows

$$\mu = g_e h/4\pi m_e [S(S+1)]^{1/2} \quad [2]$$

where m_e = mass of electron, e = electron charge, S = spin quantum number and g_e is the spectroscopic factor known as the g factor (which has a value of 2.0023 for a free electron). The above equation can be rewritten as

$$\mu = -g_e\mu_B \, S/\hbar \quad [3]$$

where $\mu_B = eh/2m_e$ and $\hbar = h/2\pi$. The z axis component μ_z will then be

$$\mu_z = -g_e\mu_B M_S \quad [4]$$

μ_B is the Bohr magneton and has the value 9.2740×10^{-24} J T^{-1}. The negative sign arises because of the negative charge on the electron which results in the magnetic dipole being in the opposite direction to the angular momentum vector. The electron is restricted to certain fixed angular positions that correspond to the z component of the spin

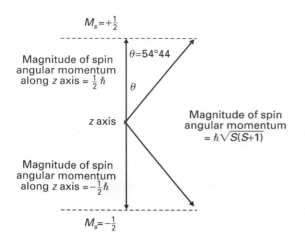

Figure 1 Allowed values of the magnitude of the spin angular momentum along the z axis for $S = \frac{1}{2}$.

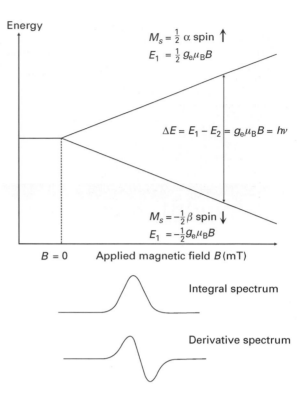

Figure 2 The Zeeman energy levels of an electron ($S = \frac{1}{2}$) in an applied magnetic field for a fixed microwave frequency.

angular moment, S_z is a half integral number (M_S) of $h/2\pi$, units. Since $S = \frac{1}{2}$ then $[S(S + 1)]^{1/2}$ cannot be $\frac{1}{2}$ and consequently the spin angular momentum vector S and the magnetic moment vector $\boldsymbol{\mu}$ can never be aligned exactly in the field direction and they maintain a constant angle θ with the z axis and hence the applied field (**Figure 1**).

Hence, for a single unpaired electron, where $S = \frac{1}{2}$, the magnitude of the spin angular momentum along the z axis will be given by $+\hbar/2$ and $-\hbar/2$. For a free electron the spin angular momentum can have two possible orientations and these give rise to two magnetic moments or spin states of opposite polarity. In the absence of an external magnetic field the two spin states are degenerate. However, if an external magnetic field is applied then the degeneracy is lifted, resulting in two states of different energy. The energy of the interaction between the electron magnetic moment and the external magnetic field is given by

$$E = -\mu_z B \qquad [5]$$

where B = the strength of external magnetic field and μ_z = magnetic dipole along the z axis. In quantum mechanics the $\boldsymbol{\mu}$ vector is replaced by the corresponding operator leading to the following Hamiltonian

$$H = -g_e \mu_B \boldsymbol{B} \boldsymbol{S}_z \qquad [6]$$

so that the energy is given by $E = -g_e\,\mu_B B M_S B$. For a single unpaired electron $M_S = \pm \frac{1}{2}$ and this gives rise to two energy levels known as Zeeman levels. The

difference between these two energy levels is known as the Zeeman splitting (see **Figure 2**):

$$E_1 = \tfrac{1}{2} g_e \mu_B B \qquad M_S = +\tfrac{1}{2} \quad \alpha \,\text{spin} \uparrow \qquad [7]$$

$$E_2 = -\tfrac{1}{2} g_e \mu_B B \qquad M_S = -\tfrac{1}{2} \quad \beta \,\text{spin} \downarrow \qquad [8]$$

Since $\Delta E \propto B$, the difference between the two energy levels is directly proportional to the external applied magnetic field. Transitions between the two Zeeman levels can be induced by the absorption of a photon of energy, $h\nu$, equal to the energy difference between the two levels:

$$\Delta E = g_e \mu_B B = h\nu \qquad [9]$$

where h = Planck's constant and ν = frequency of electromagnetic radiation. The existence of two Zeeman levels, and the possibility of inducing transitions from the lower energy level to the higher energy level, is the very basis of EPR spectroscopy. We can see, therefore, that the position of absorption varies directly with the applied magnetic field. EPR spectrometers may operate at different fields and frequencies (two common frequencies being ~9.5

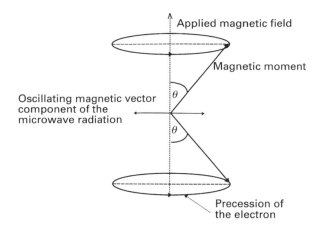

Figure 3 Two allowed electron spin orientations and the corresponding Larmor precession for $S = \frac{1}{2}$.

and ~35 GHz, known as X and Q band respectively). It is therefore far more convenient to refer to the absorption in terms of its g value:

$$g = \Delta E / \mu_B B = h\nu / \mu_B B \qquad [10]$$

Larmor precession, spin populations and relaxation effects

It was noted above that the two spin states were at a constant angle to the z axis. However, in the presence of a magnetic field, the magnetic moment of the electron cannot align with the applied field and a turning couple is set up which results in a precession of the spin states around the z axis (**Figure 3**). This gives a physical mechanism by which spins can interact with electromagnetic radiation. When the magnetic component of the microwave radiation, which is orthogonal to the applied field, is at a frequency equal to that of the *Larmor precession* then the two are said to be in resonance, and energy can be transferred to the electron, inducing $\Delta M_S = \pm\frac{1}{2}$ transitions.

The incident radiation induces transitions not only from the lower to the higher energy states but can also induce emission with equal probability. Consequently the extent of absorption will be proportional to the population difference between the two states. At thermal equilibrium the relative *spin populations* of the two Zeeman levels is given by a Boltzmann equation:

$$N_U / N_L = e^{-\Delta E / kT} \qquad [11]$$

where k = Boltzmann constant, T = absolute temperature (K), N_L = number of electrons in the lower level, N_U = number of electrons in the higher level and

ΔE = difference in energy between the two energy levels. At room temperature (300 K) and in a magnetic field of 300 mT the populations of the two Zeeman levels are almost equal, but a slight excess exists in the lower level and it is this that gives rise to a net absorption. However, this would very quickly lead to the disappearance of the EPR signal as the absorption of energy would equalize these two states. For the experiment to proceed there has to be a process by which energy is lost from the system. Such processes are known as *relaxation processes*.

To maintain a population excess, electrons in the upper level must be able to return to their low energy state. Therefore they must be able to transfer their excess spin energy either to other species or to the surrounding lattice as thermal energy. The time taken for the spin system to lose $1/e$ of its excess energy is called the relaxation time. Two such relaxation processes are:

(a) spin–lattice relaxation (T_{1e}): this process is due to lattice motions such as molecular tumblings in solids, liquids and gases which have a frequency comparable to that of the Larmor precession and this provides a pathway which allows the electron's excess energy to be transferred to the surroundings.

(b) spin–spin relaxation (T_{2e}): the excess spin energy is transferred between paramagnetic centres either through dipolar or exchange coupling, from one molecule to another. This mode of relaxation is important when the concentration of paramagnetic species is high (spins are close together). If the relaxation time is too fast then the electrons will only remain in the upper state for a very short period of time and give rise to a broadening of the spectral line width as a consequence of Heinsenberg's uncertainty principle.

In general $T_{1e} > T_{2e}$ and the line width depends only on the spin–spin interactions (T_{2e}). However, if, in certain circumstances, both spin–spin and spin–lattice relaxations contribute to the EPR line width (ΔH), then the resonance line width can be simply written as

$$\Delta H \propto 1/T_{1e} + 1/T_{2e} \qquad [12]$$

When T_{1e} becomes very short, below ~10^{-7} s, its effects on the lifetime of the species in a given energy level makes an important contribution to the linewidth. In some cases the EPR lines are broadened beyond detection. The term T_{1e} may also be inversely proportional to the absolute temperature ($T_{1e} \propto T^{-n}$) with n depending on the precise relaxation mechanism. In such cases, cooling the sample increases T_{1e} and usually leads to detectable lines. For this reason, EPR experiments are frequently recorded at liquid nitrogen (77 K) or helium (4 K) temperatures.

Basic interaction of the electron with its environment

An expression for the spin state energies of an electron whose only interaction was with an applied magnetic field (a free electron) was derived above (Eqn [6]). This represents an idealized model as in the real situation the electron suffers a variety of electrostatic and magnetic interactions which complicate the resonances. Consequently we must be able to account for these deviations. Magnetic resonance spectroscopists seek to characterize and interpret EPR spectra and these deviations quantitatively using a device known as a *spin Hamiltonian*. The EPR spectrum is essentially interpreted as the allowed transitions between the eigenvalues of this spin Hamiltonian. The Hamiltonian contains terms which reflect the interactions of the spins of electrons and nuclei with the applied magnetic field and with each other, e.g.

$$H = \mu_B B g S + \Sigma_j I A S + \Sigma_j g_n \mu_N B I \quad [13]$$

Analysis of a spectrum amounts to identifying which interactions are involved. The origin of the various terms in this spin Hamiltonian will be examined below. For example, changes in the g factor are shown for a variety of paramagnetic species in **Table 1**.

The g tensor: significance and origin

The main reason for the deviation in g values comes from a spin–orbit coupling that results in an orbital contribution to the magnetic moment. This arises from the effect of the orbital angular momentum L which is non-zero in the case of orbitals exhibiting p, d or f character. In this case the spin is no longer exactly quantized along the direction of the external field and the g value cannot be expressed by a scalar quantity but becomes a tensor. The orbital angular momentum L is associated with a magnetic momentum given by

$$\mu_L = \mu_B L \quad [14]$$

Table 1 Isotropic g value variations for a series $S = \frac{1}{2}$ paramagnetic species. For organic radicals the g value deviations (from g_e) are usually small, less than 1%, but for systems that contain heavier atoms, such as transition metal ions, the variations can be much larger

Species	g value
e⁻	2.0023
$CH_3^\bullet$	2.0026
Anthracene radical cation	2.0028
Anthracene radical anion	2.0029
1,4-Benzoquinone radical anion	2.0047
SO_2^-	2.0056
$HO_2^\bullet$	2.014
$\{(CH_3)_3C\}_2NO$	2.0063
Copper(acetonyl acetate)	2.13
VO(acetonyl acetate)	1.968
Cyclopentadienyl $TiCl_2AlCl_2$	1.975

For a system with a doublet $(S = \frac{1}{2})$ non-degenerate electronic ground state, the interaction with the external magnetic field can be expressed in terms of a perturbation of the general Hamiltonian by the following three terms

$$H = g_e \mu_B B S + \mu_B B L + \lambda L S \quad [15]$$

The first and second terms correspond, respectively, to the electron Zeeman and orbital Zeeman energies. The third one represents the energy of the spin–orbit coupling where λ is the spin–orbit coupling constant which mixes the ground state wavefunctions with the excited states. The extent of the interaction between L and S mainly depends on the nature of the system considered and in many instances this interaction is stronger than that between the magnetic field and the orbital angular momentum. Depending on the strength of the molecular electric field, two limiting cases can be distinguished:

(a) Strong fields: L must align itself along the field so that only S can orient itself with respect to the external magnetic field and contribute to the paramagnetism. In this case $g = g_e$. Many organic radicals or systems in which the unpaired electron is in a molecular orbital delocalized over a large molecule experience this situation.

(b) Weak fields: L is no longer under the constraint of this weak field and the spin–orbit coupling $L + S$ can take place, giving a resultant total angular momentum:

$$J = L + S \qquad [16]$$

associated with the magnetic moment:

$$\mu_J = -g_J \mu_B J \qquad [17]$$

g_J is called the Landé g factor. This situation occurs in the rare earth elements

When intermediate fields are present in the paramagnetic centre, L is only partially blocked by the molecular field (transition metal ions, inorganic radicals). The system must be treated in terms of the perturbation Hamiltonian in Equation [6] which can be written as

$$H = \mu_B B g S \qquad [18]$$

The g_e scalar value reported in Equation [6] is now replaced by g, a second rank tensor which represents the anisotropy of the interaction between the unpaired electron and the external magnetic field and also outlines the fact that the orbital contribution to the electronic magnetic momentum may be different along different molecular axes.

The g tensor may be depicted as an ellipsoid whose characteristic (principal) values (g_{xx}, g_{yy}, g_{zz}) depend upon the orientation of the symmetry axes of the paramagnetic entity with respect to the applied magnetic field (**Figure 4**). The most general consequence of the anisotropy of g, from an experimental point of view, is therefore that the resonance field of a paramagnetic species, for a given frequency, depends on the orientation of the paramagnetic centre in the field itself. The g value for a given orientation depends on ϕ and θ values according to the relation $g^2 = g_{xx}^2\cos^2\phi\sin^2\theta + g_{yy}^2\sin^2\phi\cos^2\theta + g_{zz}^2\cos^2\theta$. Accordingly, the Zeeman resonance will occur at field values given by

$$B_{res} = h\nu/\mu_B(g_{xx}^2\cos^2\phi\ \sin^2\theta$$
$$+ g_{yy}^2\sin^2\phi\ \cos^2\theta + g_{zz}^2\cos^2\theta)^{-1/2} \qquad [19]$$

In the most general case, the resonance observed for a paramagnetic centre in a single crystal is obtained at distinct field values of B_x, B_y and B_z when the magnetic field is parallel to the x, y or z crystal

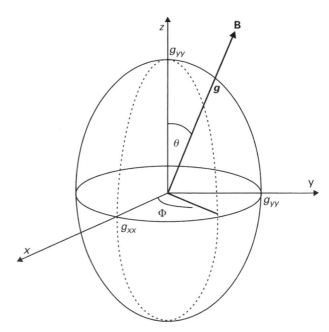

Figure 4 Orientation of the magnetic field **B** with respect to the **g** tensor ellipsoid in the crystallographic frame x, y, z. The characteristic angles θ and ϕ define the orientation of **B** which leads to the **g** (θ, ϕ).

axis respectively. The g values corresponding to these three orientations (g_{xx}, g_{yy} and g_{zz}) are the principal (diagonal) elements of the g tensor. When the principal axes of the crystal are not known, the g tensors can be labelled g_1, g_2 and g_3. Absolute determination of the g values may in principle be carried out by independent and simultaneous measurement of B and ν using a gaussmeter and a frequency meter, respectively, following the equation

$$g = h\nu/\mu_B\ B \qquad [20]$$

The **A** tensor: significance and origin

The source of this splitting is the magnetic interaction between the electron spin and neighbouring nuclear spins, which gives rise to hyperfine structure in the spectra. It arises because a nearby magnetic nucleus gives rise to a local field, B_{local}, which must be compounded with the applied magnetic field **B** to satisfy the EPR condition $h\nu = g_e\mu_B B$. We must now rewrite this equation as

$$h\nu = g_e\ \mu_B(B + B_{local}) \qquad [21]$$

and clearly the value of B required to achieve resonance will depend on B_{local} (i.e. the magnetic moments of the nuclei). Several nuclei possess spin and corresponding magnetic moments. The nuclear spin quantum number (I) of a given nucleus can assume integral or half-integral values in the range 0–6. The magnetic moment μ_n associated to a nucleus is collinear with the spin vector I according to the relation

$$\mu_n = g_n \mu_N I \qquad [22]$$

which is similar to Equation [4]; g_n is the nuclear g factor and μ_N the nuclear magneton which is smaller than the Bohr magneton by a factor of 1838, i.e. the ratio of the mass of a proton to that of an electron.

When the paramagnetic centre contains one or more nuclei with non-zero nuclear spin ($I \neq 0$), the interaction between the unpaired electron and the nucleus with $I \neq 0$ produces further splittings of the Zeeman energies and consequently there are new transitions, which are responsible for the so-called hyperfine structure of the EPR spectrum.

Let us consider the proton as a case in point; it is a spinning positive charge and has its own associated magnetic field and consequently can interact with both the external magnetic field (cf. NMR) and that of the electron, i.e. it has a nuclear spin $I = \frac{1}{2}$ and $M_I = \pm\frac{1}{2}$. This leads to a situation whereby the energy of the system can either be lowered or raised depending on whether the two spins are parallel or antiparallel. The energy of an electron with spin quantum number M_S and a nucleus with spin quantum number M_I is given by

$$E(M_I, M_S) = g_e \mu_B B M_S - g_N \mu_N B M_I + h A_0 M_S M_I \qquad [23]$$

where A_0 is the isotropic coupling constant. The first term in the equation gives the contribution due to the interaction of an electron with an applied field, giving rise to two electron Zeeman levels. The second term is the contribution due to the interaction of the nucleus with the applied magnetic field, the nuclear Zeeman levels (Figure 5). The final term is the energy of interaction between the unpaired electron and the magnetic nucleus. If we now substitute the values for M_S and M_I then the interaction, between an unpaired electron ($S = \frac{1}{2}$) and a single proton ($I = \frac{1}{2}$), gives rise to four energy levels, E_1 to E_4. These are calculated to be

$$E_1 = \tfrac{1}{2} g_e \mu_B B - \tfrac{1}{2} g_N \mu_B B + \tfrac{1}{4} h A_0 \qquad \begin{matrix} M_S & M_I \\ +\tfrac{1}{2} & +\tfrac{1}{2} \end{matrix} \quad [24]$$

$$E_2 = \tfrac{1}{2} g_e \mu_B B + \tfrac{1}{2} g_N \mu_N B - \tfrac{1}{4} h A_0 \qquad \begin{matrix} +\tfrac{1}{2} & -\tfrac{1}{2} \end{matrix} \quad [25]$$

$$E_3 = -\tfrac{1}{2} g_e \mu_B B + \tfrac{1}{2} g_N \mu_N B + \tfrac{1}{4} h A_0 \qquad \begin{matrix} -\tfrac{1}{2} & -\tfrac{1}{2} \end{matrix} \quad [26]$$

$$E_4 = -\tfrac{1}{2} g_e \mu_B B - \tfrac{1}{2} g_N \mu_N B - \tfrac{1}{4} h A_0 \qquad \begin{matrix} -\tfrac{1}{2} & +\tfrac{1}{2} \end{matrix} \quad [27]$$

The selection rules for the allowed transitions between the energy levels are

$$\Delta M_I = 0 \quad \text{and} \quad \Delta M_S = \pm 1 \qquad [28]$$

Thus two resonance transitions can occur at

$$\Delta E_A = E_1 - E_4 = g_e \mu_B B + \tfrac{1}{2} h A_0 \qquad [29]$$

$$\Delta E_B = E_2 - E_3 = g_e \mu_B B - \tfrac{1}{2} h A_0 \qquad [30]$$

These two possible transitions give rise to two absorption peaks which, at constant applied microwave frequency, occur at magnetic fields of values

$$B_1 = \frac{h\nu}{g_e \mu_B} - \frac{h A_0}{2 g_e \mu_B} = \frac{h\nu}{g_e \mu_B} - \frac{a}{2} \qquad [31]$$

$$B_2 = \frac{h\nu}{g_e \mu_B} + \frac{h A_0}{2 g_e \mu_B} = \frac{h\nu}{g_e \mu_B} + \frac{a}{2} \qquad [32]$$

where $a = h A_0 / g_e \mu B$ and is the hyperfine space coupling constant (HFCC) given in mT. The spectrum obtained for such a system would be a doublet centred at the g value and separated by the hyperfine coupling constant (hfc) a. If we extend the number of equivalent nuclei then the complexity of the spectrum increases in a manner described in Table 2. A further example is given in Figure 6 for the radical anion of naphthalene.

It is now necessary to examine the mechanisms by which the hyperfine interactions arise. Two types of electron-spin–nuclear-spin interactions must be considered, of isotropic and anisotropic nature.

The isotropic interaction   This concerns exclusively s-type orbitals or orbitals with partial s character (such as hybrid orbitals constructed from s-type

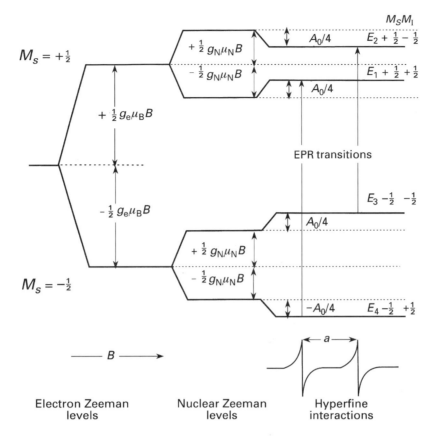

Figure 5 Energy level manifold in a high magnetic field, resulting from the interaction of an unpaired electron ($S = \frac{1}{2}$) with a nucleus of $I = \frac{1}{2}$. Note that a > 0. $|a/2| < V_n$.

Table 2 Line intensities for a series of nuclei with different values of I. For the case of $I = \frac{1}{2}$ the line intensities can be described by the expansion of $(x + 1)^2$ and hence follow Pascal's triangle

Nuclei	Intensity of lines								
$I = \frac{1}{2}$, e.g. ¹H									
Number of equivalent nuclei									
1	1	1							
2	1	2	1						
3	1	3	3	1					
4	1	4	6	4	1				
$I = 1$, e.g. ¹⁴N									
Number of equivalent nuclei									
1	1	1	1						
2	1	2	3	2	1				
3	1	3	6	7	3	1			
4	1	4	10	16	19	16	10	4	1
$I = \frac{3}{2}$, e.g. ⁶³,⁶⁵Cu									
Number of equivalent nuclei									
1	1	1	1	1					
2	1	2	3	4	3	2	1		

orbitals) because only these orbitals have a finite probability density at the nucleus. This mechanism is a quantum interaction related to the finite probability of the unpaired electron at the nucleus and is termed the Fermi contact interaction. The corresponding isotropic coupling constant a_{iso} is given by

$$a_{iso} = (8\pi/3\, g_e\, g_n\, \mu_B\, \mu_N)\, |\, \psi(0)\, |^2 \qquad [33]$$

where g_n and μ_N are the nuclear analogues of g_e and μ_B, respectively, and $|\, \psi(0)\, |^2$ is the square of the absolute value of the wavefunction of the unpaired electron evaluated at the nucleus.

Since s orbitals have a high electron density at the nucleus, the hyperfine coupling constant will be large and since s orbitals are also symmetrical it will be independent of direction. The spherical symmetry of s orbitals accounts for the isotropic nature of the contact interaction.

Thus in the case of the hydrogen atom ($M_I = \frac{1}{2}$) with 100% spin density in the 1s orbital, the EPR spectrum is composed of two resonance lines

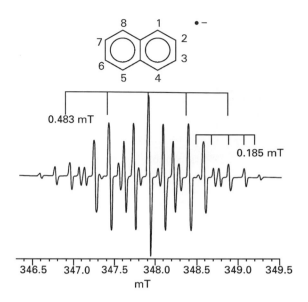

Figure 6 EPR spectrum of the radical anion of naphthalene in tetrahydrofuran as solvent. The radical was formed by dissolving naphthalene in sodium-dried tetrahydrofuran under vacuum and passing the resultant solution over a sodium metal film, again under vacuum conditions. The number of lines is given by $2nI + I$ where I equals the nuclear spin and n the number of nuclei. HFCC for 1,4,5,8 = 0.483 mT and for 2,3,6,7 = 0.185 mT.

separated by approximately 50.8 mT. In the case of naphthalene, the smaller spin density associated with the ^{1}H s orbital results in a smaller isotropic coupling.

The anisotropic interaction   In non-spherically symmetric orbitals (p, d or f) there is no electron density at the nucleus and the electron is to be found some distance away. The interaction between it and the nucleus will be as two magnetic dipoles and consequently the interaction will be small and dependent on the direction of the orbital with respect to the applied magnetic field as well as to their separation. The coupling then arises from the classical dipolar interaction between magnetic momenta whose energy is given by

$$E = \mu_s \mu_n / r^3 - 3(\mu_s r)(\mu_n r)/r^5 \qquad [34]$$

where r is the vector relating the electron and nuclear moments and r is the distance between the two spins. The above equation gives the energy of interaction of two bar magnets if their size is small compared with the distance between them. Moving to quantum mechanics we construct the Hamiltonian by replacing the magnetic moments by the appropriate operators. The quantum mechanics analogue of Equation [34]

is then obtained by replacing μ_s and μ_n by their expressions given in Equations [4] and [22]:

$$H_{aniso} = -g_e \mu_B g_n \mu_B (I\cdot S/r^3 - 3(I\cdot r)(S\cdot r)/r^5) \qquad [35]$$

Equation [35] must be averaged over the entire probability of the spin distribution. H_{aniso} is averaged out to zero when the electron cloud is spherical (s orbital) and comes to a finite value in the case of axially symmetric orbitals (p orbitals, for instance). In addition, in the case of very rapid tumbling of the paramagnetic species (as occurs in a low viscosity solution) the anisotropic term of the hyperfine interaction is averaged to zero and the isotropic term is the only one observed.

Combination of isotropy and anisotropy   Since any orbital may be considered as a hybrid of suitable combinations of s, p or d orbitals, so also may a hyperfine coupling be divided into a contribution arising from p or d orbitals (anisotropic) and s orbitals (isotropic). In general both isotropic and anisotropic hyperfine couplings occur when one or more nuclei with $I \neq 0$ are present in the system. The whole interaction is therefore dependent on orientation and must be expressed by the second rank A tensor as given in Equation [13]. The A tensor may be split into an anisotropic and isotropic part as follows

$$\mathbf{A} = \begin{vmatrix} A_1 & 0 & 0 \\ 0 & A_2 & 0 \\ 0 & 0 & A_3 \end{vmatrix} = a_{iso} + \begin{vmatrix} T_1 & 0 & 0 \\ 0 & T_2 & 0 \\ 0 & 0 & T_3 \end{vmatrix} \qquad [36]$$

with $a_{iso} = (A_1 + A_2 + A_3)/3$. In a number of cases, the second term matrix of Equation [36] is a traceless tensor ($T_1 + T_2 + T_3 = 0$) and has the form ($-T, -T, +2T$). The anisotropic part of the A tensor corresponds to the dipolar interaction as expressed by the Hamiltonian in Equation [13]. The s and p character of the orbital hosting the unpaired electrons are given by the following relations:

$$c_S^2 = a_{iso}/a_0 \qquad \text{and} \qquad c_P^2 = b/b_0 \qquad [37]$$

where a_0 and b_0 are the theoretical hyperfine coupling constants (assuming pure s and p orbitals for the elements under consideration). The terms a_{iso} and b are the experimental isotropic and anisotropic coupling constants respectively.

The *D* tensor: significance and origin

The spin Hamiltonian described in Equation [13] applies to the case where a single electron ($S = \frac{1}{2}$) interacts with the applied magnetic field and with surrounding nuclei. However, if two or more electrons are present in the system ($S > \frac{1}{2}$), a new term must be added to the spin Hamiltonian (Eqn [13]) to account for the interaction between the electrons. At small distances, two unpaired electrons will experience a strong dipole–dipole interaction, analogous to the interaction between electron and magnetic dipoles, which gives rise to anisotropic hyperfine interactions. The electron–electron interaction is described by the spin–spin Hamiltonian given in Equation [38]:

$$H_{ss} = \mathbf{SDS} \qquad [38]$$

where $\mathbf{D}$ is a second rank tensor (the zero-field parameter) with a trace of zero. As with the $\mathbf{g}$ and $\mathbf{A}$ tensors, the $\mathbf{D}$ tensor can also be diagonalized so that $D_{xx} + D_{yy} + D_{zz} = 0$. Equation [38] can be added to Equation [18] to obtain the correct spin Hamiltonian for an $S > \frac{1}{2}$ system (in the absence of any interacting nucleus):

$$H = \mu_B \mathbf{BgS} + \mathbf{SDS} \qquad [39]$$

Since the trace of $\mathbf{D}$ is zero, calculation of the energy state for a system with $S = 1$ requires only two independent parameters which are designated D and E. The spin coupling is direct in the case of organic molecules in the triplet state and biradicals, but occurs through the orbital angular momentum in the case of transition metal ions. In the latter case, the D and E terms depend on the symmetry of the crystal field acting on the ions:

$$H = D\left(S_z^2 - \frac{S^2}{3}\right) + E(S_x^2 - S_y^2) \qquad [40]$$

For axially symmetric molecules, the calculated shape of the $\Delta M_S = 1$ lines are given in **Figure 7A**. The separation of the outer lines is $2D'$ (where $D' = D/g\mu_B$) while that of the inner lines is D (E is zero in this case). The theoretical line shape for a randomly oriented triplet system with $E \neq 0$ is shown in **Figure 7B**. The separation of the outermost lines is

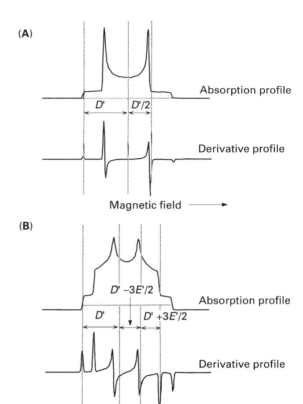

Figure 7 Theoretical absorption and first derivative spectra of the $\Delta M_S = 1$ region of a randomly oriented triplet. (A) $S = 1$ for a given value of D ($E = 0$) and isotropic g. (B) $S = 1$ where D is 6.5 times larger than E and g is isotropic.

again $2D'$ whereas that of the intermediate and inner pairs is $D' + 3E'/2$ and $D' - 3E'/2$ respectively. As the zero-field interactions become comparable to, and larger than, the microwave energy, the line shape exhibits severe distortions from the simulated case in **Figure 7**. Well known examples of $S = 1$ include the excited triplet state of naphthalene with $D/hc = 0.1003$ cm^{-1}, $E/hc = -0.0137$ cm^{-1} and $g = 2.0030$.

Line shapes and symmetry considerations in EPR spectra

An important consideration in EPR is the effect of symmetry on the line shapes. In many cases the samples investigated by EPR are polycrystalline materials, composed of numerous small crystallites that are randomly oriented in space. The resultant *powder* EPR spectrum is the envelope of spectra corresponding to all possible orientations of the paramagnetic species with respect to the magnetic field; provided the resolution is adequate, the magnitude of the $\mathbf{g}$ and $\mathbf{A}$ tensor components can be extracted from the powder spectra whereas no information can be obtained on the orientation of the tensor principal axes.

The profile of the powder spectrum is determined by several parameters, including the symmetry of the **g** tensor, the actual values of its components, and the line shape and the line width of the resonance. Concerning the symmetry of the **g** tensor, three possible cases can be identified.

Isotropy of g

In this case, the **g** tensor is characterized by $g_{xx} = g_{yy} = g_{zz} = g_{iso}$ and a single symmetrical line is observed. This simple situation is not very often encountered in powders except for some solid state defects and transition metal ions in highly symmetric environments such as O_h or T_d.

Axial symmetry of g

Paramagnetic species isolated in single crystals exhibit resonances at typical magnetic fields which depend on their orientation and are given by Equation [19]. In the particular case of axial symmetry of the system, e.g. (C_{4v}, D_{4h}), if z is the principal symmetry axis of the species and θ the angle between z and the magnetic field, the x and y directions are equivalent and the angle ϕ becomes meaningless (**Figure 4**). Equation [19] reduces to

$$B_{res} = h\nu/\mu_B (g_\perp^2 \sin^2\theta + g_\parallel^2 \cos^2\theta)^{-1/2} \quad [41]$$

where $g_\parallel = g_{zz}$ and $g_\perp = g_{yy} = g_{xx}$ are the **g** values measured when the axis of the paramagnetic species is, respectively, parallel and perpendicular to the applied magnetic field.

The powder spectrum is the envelope of the individual lines corresponding to all possible orientations in the whole range of ϕ. Assuming the microcrystals are randomly distributed, simple considerations show, however, that the absorption intensity, which is proportional to the number of microcrystals at resonance for a given θ value, is at a maximum when $\theta = \pi/2$ ($B_\perp$) and a minimum for $\theta = 0$ ($B_\parallel$); this allows the extraction of the $g_\parallel$ and $g_\perp$ values, which correspond to the turning points of the spectrum. **Figure 8** gives a schematic representation of the absorption curve and of its first derivative for a polycrystalline sample containing a paramagnetic centre in axial symmetry. The variation in the spectral profile for the same species as M_I varies is also shown in **Figure 9**.

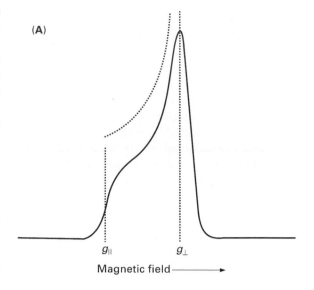

(A)

$g_\parallel$ $g_\perp$

Magnetic field ⟶

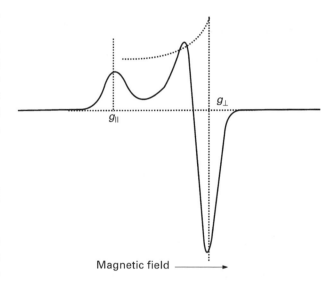

(B)

$g_\parallel$ $g_\perp$

Magnetic field ⟶

Figure 8 Powder EPR spectrum of a paramagnetic species with $I = 0$ in axial symmetry. (A) absorption profile, (B) first derivative profile. The dotted lines have been calculated by assuming a zero line width whereas the solid lines correspond to a finite line width.

Orthorhombic symmetry of g

Three distinct principal components are expected in the case of a system with orthorhombic symmetry, e.g. (D_{2h}, C_{2v}). For polycrystalline samples, the absorption curve and its first derivative exhibit three singular points, corresponding to g_1, g_2 and g_3 (**Figure 10**). For powder spectra the assignments of g_1, g_2 and g_3 to the components g_{xx}, g_{yy} and g_{zz} related to the molecular axes of the paramagnetic centre is not straightforward and must be based on

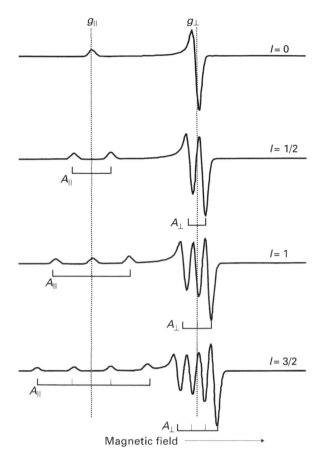

Figure 9 Effects of nuclear spin on the calculated powder EPR spectra for axial symmetry. The value of I varies from 0 to $\frac{3}{2}$. Note that the g values remain constant (illustrated by dotted lines). $\boldsymbol{g}_{\parallel}$ is split by $\boldsymbol{A}_{\parallel}$ and $\boldsymbol{g}_{\perp}$ is split by $\boldsymbol{A}_{\perp}$. The $\boldsymbol{A}_{\parallel}$ and $\boldsymbol{A}_{\perp}$ values remain constant.

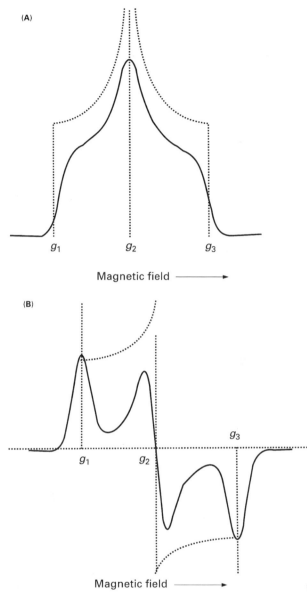

Figure 10 Powder EPR spectrum of a paramagnetic species in orthorhombic symmetry. (A) Absorption profile, (B) first derivative profile. The dotted and solid lines have the same significance as described in **Figure 7**.

theoretical grounds or deduced from measurements of the same paramagnetic species in a single crystal. The situation becomes much more complicated in the presence of nuclei with $I \neq 0$. In several cases an initial interpretation of the experimental data can be achieved by using a simple analysis for the g tensor and neglecting second-order effects since the A tensor has the same type of angular dependence as the g tensor. Provided the principal axes are the same, then each of the three possible lines of the g tensor (g_1, g_2 and g_3) will be split into a number of lines that depend on the nuclear spin with the spacing corresponding to the appropriate component of the A tensor (A_1, A_2 or A_3); g_1 is split by A_1, g_2 is split by A_2 and g_3 is split by A_3 (**Figure 11**).

essential knowledge of the technique and is a useful stepping stone for the articles on methods and instrumentation and applications.

Summary

The basic principles of EPR spectroscopy have been outlined. Although far from being an exhaustive treatment of the subject, it provides the reader with

List of symbols

a = hyperfine coupling constant; A_0 = theoretical isotropic hyperfine coupling constant; A = hyperfine tensor; g = g factor; g_e = g factor of the free electron; g_J = Landé g factor; g_N = nuclear g factor; $g_{\parallel}$ and $g_{\perp}$ = refer to the parallel and perpendicular values

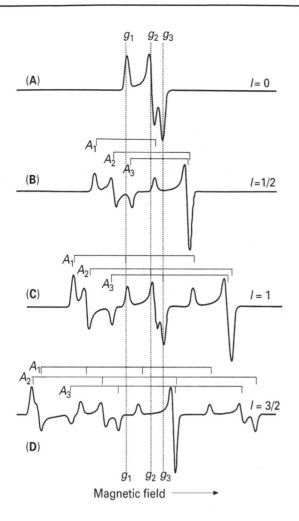

Figure 11 Effects of nuclear spin on the calculated powder EPR spectrum of a species with orthorhombic symmetry. The spectrum is anisotropic both in g and A (g_1 is split by A_1, g_2 is split by A_2 and g_3 by A_3). The value of I varies from 0 to $\frac{3}{2}$. It should be noted that when I assumes an integer value (as in C), the features of the original $I = 0$ spectrum (A) are maintained in the centre of the spectrum. When I has half-integer values the hyperfine values are symmetrically disposed about the centre.

with respect to axis of symmetry; g_{xx}, g_{yy}, g_{zz} = refer to values in the respective crystallographic directions; h = Planck's constant; $\hbar = h/2\pi$; I = nuclear spin quantum number; J = total angular momentum; k = Boltzman constant; L = orbital angular momentum vector; m_e = mass of the electron; N_L and N_U = electron populations in the lower and upper energy levels respectively; r = distance between the spins; r = vector relating electron and nuclear moments; S = electron spin quantum number; S_z = vector of the electron spin quantum number in the z direction; T_{1e} and T_{2e} = spin–lattice and spin–spin relaxation times, respectively; α and β = refer to electron spin; μ = magnetic moment; μ_B = Bohr magneton; μ_N = nuclear magneton; ν = frequency.

See also: **Chemical Applications of EPR; EPR Imaging; EPR, Methods; NMR in Anisotropic Systems, Theory; NMR Principles; NMR Relaxation Rates.**

Further reading

Abragam A and Bleaney B (1970) *Electron Paramagnetic Resonance of Transition Ions* Oxford: Oxford University Press.

Atherton NM (1993) *Principles of Electron Spin Resonance*, PTR Prentice Hall Physical Chemistry Series. Englewood Cliff, NJ: Prentice-Hall.

Bolton JR, Weil JA and Wertz JE (1994) *Electron Paramagnetic Resonance: Elementary Theory and Practical Applications.* New York: Wiley-Interscience.

Che M and Giamello E (1987) In: Fieno JLG (ed) *Spectroscopic Characterization of Heterogeneous Catalysts. Part B: Electron Paramagnetic Resonance*, Vol 57. Amsterdam: Elsevier Science Publishers.

Mabbs FE and Collison D (1992) *Electron Paramagnetic Resonance of d-Transition Metal Compounds*, Vol 16 in Studies in Inorganic Chemistry. Amsterdam: Elsevier.

Pilbrow JR (1993) *Transition Ion Electron Paramagnetic Resonance.* Oxford: Clarendon Press.

Symons MRC (1978) Chemical and biological aspects of electron spin resonance spectroscopy. *Chemical Aspects of Physical Techniques.* New York: Van Nostrand Reinhold.

Plate 1 Roman die, ca. AD 300, from archaeological excavations at Frocester Villa, Gloucester, UK. Raman spectroscopy has suggested the origin of the die as sperm whale ivory.

Plate 2 Selection of ornamental jewellery consisting of three bangles assumed to be ivory but which were shown spectroscopically to be composed of modern resins, and a genuine ivory necklace.

Plate 3 'Ivory' cat, which was identified spectroscopically as a modern imitation composed of poly(methyl methacrylate) and polystyrene resins with added calcite to give the texture and density of ivory.

Plate 3 Reproduced with permission from Edwards HGM and Farwell DW, Ivory and simulated ivory artefacts: Fourier-transform Raman diagnostic study, *Spectrochimica Acta, Part A*, **51**: 2073–2081 © 1995, Elsevier Science B.V.

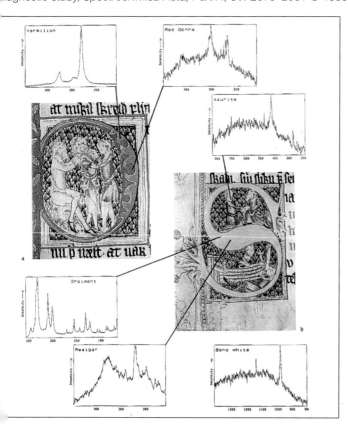

Plate 4 (left) The historiated initials (a) 'P' and (b) 'S' from the Icelandic Jónsbók with the Raman spectra of pigments contained therein. The combination of vermillion and red ochre in the 'P' should be noted. The spectrum of bone white has been obtained from the letter 'H'(not shown). Reproduced with permission from Bent SP, Clark RH, Daniels MAM, Proter CA and Withnay R. Identification by Raman microscopy and visible reflectance spectroscopy of pigments on an Icelandic manuscript, (1995) *Studies in Conservation* **40**: 31–40.

Plate 5a (below) Elaborately historiated initial 'R' in sixteenth-century German choir book, with Raman microscopy spectra of selected pigmented regions.

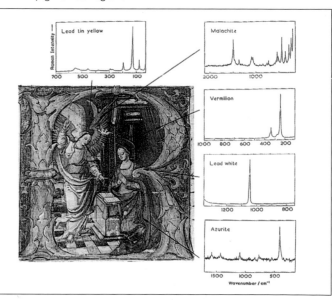

Plate 6 (right) Holy Sepulchre Chapel, Winchester Cathedral. Wall painting of ca. 1175–85 on the east wall depicting the Deposition, Entombment, Maries at the Sepulchre and the Harrowing of Hell. Reproduced with permission from Edwards HGM, Brooke and Tait JKF, An FT-Raman spectroscopic study of pigments of medieval English wall paintings, *Journal of Raman Spectroscopy* 95–98 © 1997 John Wiley and Sons Ltd.

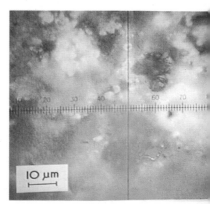

Plate 5b (right) Portion of top of column of historiated initial 'R' shown in 5a the individual pigment-grains can be clearly seen under the x100 magnification and can be separately identified using Raman microscopy. 5a and b, reproduced with permission from Clark RJH (1995) Raman microscopy: application to the identification of pigments on medieval manuscripts. *Chemical Society Reviews* 187–196. © The Royal Society of Chemistry.

Plates 1–6 *See Art Works Studied using IR and Raman Spectroscopy.*

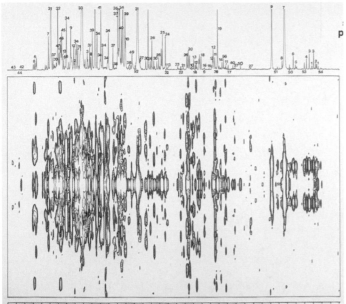

a

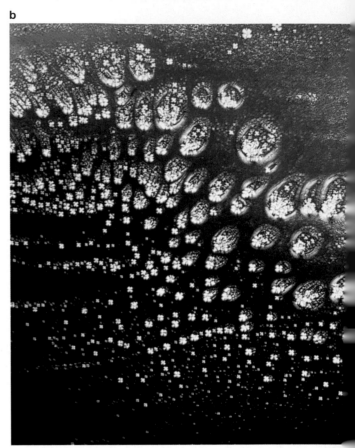

Plate 7 (left) A two-dimensional ^{1}H-NMR spectrum of human urine, demonstrating the complexity of such mixtures. This was measured using the J-resolved pulse sequence and results in a contour plot of spectrum intensity as a function of two frequency axes. The horizontal axis represents the usual ^{1}H NMR chemical shift range and the vertical axis covers the ^{1}H-^{1}H spin coupling range. Each ^{1}H NMR spin-coupled multiplet is rotated such that the overlap between closely spaced signals is minimised, thus aiding interpretation. *See Biofluids Studied by NMR*. Reproduced with permission from John Lindon.

b

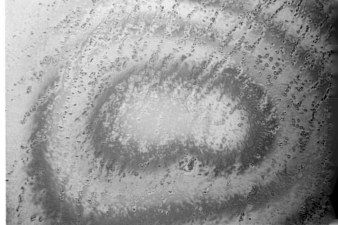

Plates 8 a and b (above left and right) Polarised light micrograph of liquid crystals. *See Chiroptical Spectroscopy, Oriented Molecules and Anisotropic Systems; Liquid Crystals and Liquid Crystal Solutions Studied by NMR*. Reproduced with permission from Science Photo Library.

Plate 9 Light micrograph of a liquid crystal display (LCD) of the type used to represent numerical figures. Although a liquid crystal (LC) can flow like a fluid its molecular arrangement exhibits some order. A LCD has a film of LC sandwiched between crossed polarizers & set on top of a mirror. In the 'off' state light is able to traverse both polarizers to reach the mirror because it gets rotated through 90 degrees by the LC. In the 'on' state an electric field applied across the LC alters its molecular alignment & hence its polarizing properties; light cannot traverse both polarizers to reach the mirror & the display appears black. *See Chiroptical Spectroscopy, Oriented Molecules and Anisotropic Systems; Liquid Crystals and Liquid Crystal Solutions Studied by NMR*. Reproduced with permission from Science Photo Library.

Plate 10 Polarised light micrograph of nematic liquid crystals. *See Chiroptical Spectroscopy, Oriented Molecules and Anisotropic Systems; Liquid Crystals and Liquid Crystal Solutions Studied by NMR*. Reproduced with permission from Science Photo Library.

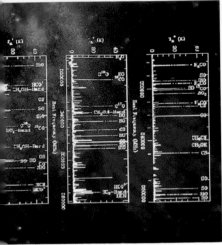

Plate 11 The molecular composition of massive star-forming regions in the galaxy. The false color background shows the heart of a massive star-forming region in the constellation of Orion obtained by the Hubble Space Telescope. The red colour is the emission of molecular hydrogen excited by shocks driven by the powerful wind of a newly formed star in the center of the image. The bluish color is scattered light from this young star reflected in surrounding dust in its stellar nursery. The white spectrum shows the results of a ground-based line survey of this region and reveals the presence of molecules such as methanol, formaldehyde, carbon monoxide, sulfur monoxide, and sulfur dioxide. *See Cosmochemical Applications Using Mass Spectrometry; Interstellar Molecules, Spectroscopy of.*

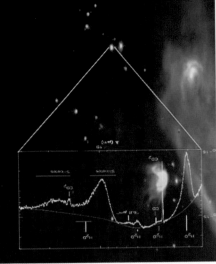

Plate 12 The composition of ices in low-mass star forming regions. This is a false color image of an infrared image obtained by the European ISO satellite (Infrared Space Observatory). The colors indicate emission by warm dust and gas heated by nearby stars in the star-forming molecular cloud Rho Ophiucus. Two bright objects just below the center are newly formed solar-type stars. The spectrum displayed on top was taken by a spectrograph on board ISO towards the indicated source. This spectrum reveals the presence of various simple molecules such as water, carbon monoxide, carbon dioxide in ice-form in these regions. The composition of these ices is very similar to those of comets in the solar system and eventually, these ices are expected to coagulate together forming comets in the budding planetary system around this newly formed star. *See Cosmochemical Applications Using Mass Spectrometry; Interstellar Molecules, Spectroscopy of.*

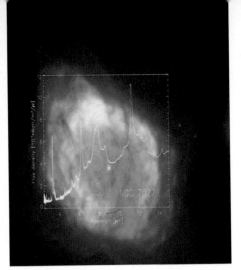

Plate 13 Large molecules in the ejecta from a dying star. When solar-type stars grow old they shed much of their mass in the form of a wind much of that in the form of complex molecules. Eventually, the star becomes very hot and sets this ejected material aglow, forming so-called planetary nebulae. The material ejected during this stage will mix with other material present in space and eventually new stars and planetary systems like our own will form these stellar ashes. The false color image shows the emission of molecular hydrogen (red) and atomic hydrogen (white) in the planetary nebula NGC 7027 obtained by the ISO satellite (Infrared Space Observatory). This spectrum reveals the presence of large complex molecules called polycyclic aromatic hydrocarbons. These molecules are formed in a process akin to that of sooting flames in terrestrial environments such as car engines. *See Cosmochemical Applications Using Mass Spectrometry; Interstellar Molecules, Spectroscopy of.*

Plates 11, 12 and 13 Reproduced with permission from A. G. G. M. Tielens.

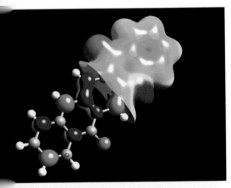

Plate 14 (left) A molecule of the pigment quinacridone. The coloured surfaces represent key molecular properties predicted using computational chemistry. *See Dyes and Indicators, Use of UV-Visible Absorption Spectroscopy.* Reproduced with permission from Molecular Simulations (www.msi.com).

Plate 15 (left) Sir Frederick William Herschel (1738–1822) German-British astronomer, and discoverer of Uranus, 1781. Reproduced with permission from Mary Evans Picture Library.

Plate 16 (Right) Sir Chandrasekhara Venkata Raman. Indian (1888–1970) physicist, Nobel Laureate in physics 1930 and discoverer of the Raman effect, a scattering of light by a sample which results in a small fraction of the light being of a different frequency. This phenomenon has applications in analytical chemistry and molecular structure. *See Electromagnetic Radiation; Vibrational, Rotational and Raman Spectroscopy, Historical Perspective.* Reproduced with permission from Mary Evans Picture Library.

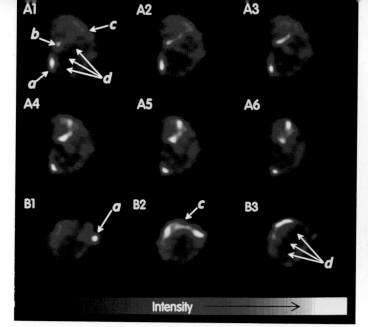

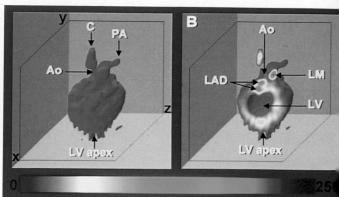

Plate 17 (left) Imaging of a rat kidney perfused with 0.5 mM triaryl-methyl (TAM) radical. A few representative slices (24x24 mm^2), obtained from the 3D spatial image are shown. A1-A6 are vertical slices (0.75 mm apart), B1-B3 are traverse slices (0.75 mm apart). The images show the structure of cannula (a), renal artery (b), cortex (c), and calysis (d). *See EPR Imaging.*

Plate 18 (above) Three-dimensional images (25 x 25 x 25 mm^3) of an ischaemic rat heart infused with a glucose char suspension. (A) Full view of the heart. (B) A longitudinal cut out showing the internal structure of the heart. C is the cannula; Ao the aorta; PA the pulmonary artery; LM the left main coronary artery; LAD the left anterior descending artery; and LV the left ventricular cavity. *See EPR Imaging.*

Plates 17 and 18 Reproduced with permission from Periannan Kuppusamy/EPR Laboratories Johns Hopkins University School of Medicine.

Plate 19 (left) Small angle X-ray fibre diffraction pattern recorded from Plaice fish muscle. *See Fibres and Films Studied Using X-Ray Diffraction.* Reproduced with permission from W. Fuller and A. Mahendrasingam.

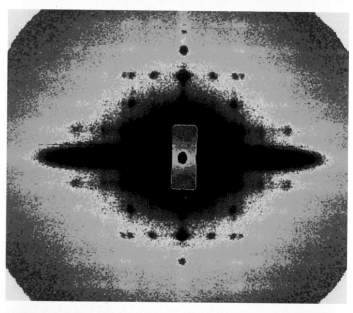

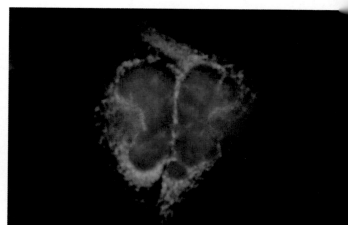

Plate 20 An example of a fluorescent molecular probe. Endogenous alkaline phosphatase activity in the zebrafish brain was localized with an endogenous phosphatase detection kit. Enzymatic cleavage of the phosphatase substrate yields a bright yellow-green fluorescent precipitate at the site of enzyme activity. *See Fluorescent Molecular Probes.* Reproduced with permission from Greg Cox, Molecular Probes, Inc.

Plate 21 A fluorescent molecular probe. Fixed and permeabilized osteosarcoma cells were simultaneously stained with the fluorescent lectins, Alexa 488 concanavalin A (Con A) and Alexa 594 wheat germ agglutinin (WGA). Con A selectively binds a-mannopyranosyl and a-glucopyranosyl residues, whereas WGA selectively binds sialic acid and N-acetylglucosaminyl residues. The nuclei were counterstained with blue-fluorescent Hoechst 33342 nucleic acid stain. *See Fluorescent Molecular Probes.* Reproduced with permission from Greg Cox, Molecular Probes, Inc.

EPR, Methods

Richard Cammack, King's College, London, UK

MAGNETIC RESONANCE
Methods & Instrumentation

Introduction

Electron paramagnetic resonance (EPR) is also known as electron spin resonance (ESR) or electron magnetic resonance (EMR) spectroscopy; no single name has become generally accepted. It is a means of investigating materials that are paramagnetic, i.e. having unpaired electrons. These include organic free radicals, compounds of transition metal ions and defects in solids. Since the development of the X-band continuous-wave spectrometer in the 1950s and 1960s, the basic design of the EPR spectrometer has remained relatively unchanged, while a wide variety of applications in physics, chemistry and biochemistry have been devised. Examples are the observation of radical intermediates in chemical reactions; the determination of electron density distributions on radicals; the observation of rapid motion of spin labels in solution; and the determination of coordination geometry of metal ion sites. Recent advances in solid-state microwave electronics have led to instrument extensions in many different ways, for specialist applications. Examples are ENDOR and TRIPLE and ESEEM, to measure hyperfine interactions, and low-frequency EPR imaging.

The principles of typical applications will be described, with reference to the relevant equipment and sample requirements.

Characteristics of the EPR spectrum

EPR involves the interaction of electromagnetic radiation, usually in the microwave region, with the paramagnetic material in a magnetic field. Because paramagnetic centres are less common than nuclear spins, the method is more selective than the analogous technique of NMR. There are some differences in practice between NMR and EPR that result from the much greater magnetic moment of the electron compared with the nuclear spins. The bandwidth of the spectrum is much greater than the parts-per-million scale of NMR. It is impossible to adjust the frequency over such a wide range, owing to the geometry of microwave waveguides, so it is the usual practice to observe the EPR spectrum at fixed frequency by sweeping the magnetic field. The relaxation rates of electron spins are faster than for nuclear

spins, so that in some cases (particularly transition metals such as iron) it is necessary to cool the sample to cryogenic temperatures to observe the spectrum.

The spectrum is characterized by a number of parameters which give information about the nature of the paramagnetic centres and their surroundings.

Zeeman interaction

The g factor defines the energy of the Zeeman interaction. This splits the energy levels of the paramagnet (**Figure 1**). At the resonance condition the energy ΔE required to reverse the direction of the electron spin in a magnetic field B_0 is equal to the energy of the microwave quantum $h\nu$, given by

$$\Delta E = h\nu = g\mu_B B_0$$

where h is Planck's constant and μ_B the Bohr magneton. In an EPR spectrum the signals are obtained by increasing the B_0 field. The g factor is inversely proportional to B_0.

For a free electron in a vacuum, g is a constant that is very precisely known ($g_e = 2.002\ 319\ 3044...$). In a paramagnetic molecule, g varies from this, under the influence of spin–orbit interactions. These increase considerably with atomic number, and are particularly important for lanthanides. In the spin Hamiltonian formalism the effects of spin–orbit coupling are described by treating g as a spectroscopic variable, the 'spectroscopic splitting factor', which is characteristic of the paramagnetic centre. Therefore the EPR spectrum may be considered as a spectrum of g factors.

Hyperfine interaction

The interaction between the magnetic moment of the electron and nuclear spins is a valuable feature of EPR spectra. It introduces splittings of the electron energy levels (**Figure 1B**). Each nuclear spin I induces a splitting into $[2I + 1]$ levels. The magnitude of the splitting is defined by the hyperfine coupling constant, A, which is expressed in energy units (usually MHz). Where there are several nuclei, for example protons in a radical, each level is further split by each nuclear hyperfine interaction. The

Electron spins in an applied field

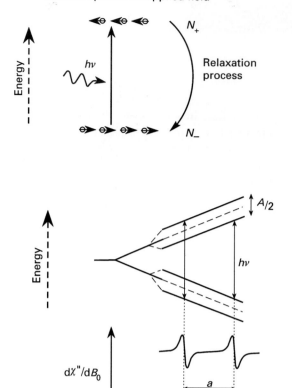

Figure 1 (A) Principle of electron paramagnetic resonance. (B) Effect of hyperfine splitting with a nucleus, $I = \frac{1}{2}$.

Table 1 Some nuclei with magnetic moments

Isotope	% Natural abundance	Nuclear spin, I	μ_N
^{1}H	99.985	$\frac{1}{2}$	+ 2.793
^{2}H	0.015	1	+ 0.857
^{13}C	1.11	$\frac{1}{2}$	+0.702
^{14}N	99.63	1	+0.404
^{15}N	0.37	$\frac{1}{2}$	−0.283
^{17}O	0.037	$\frac{5}{2}$	−1.89
^{19}F	100	$\frac{1}{2}$	+2.629
^{31}P	100	$\frac{1}{2}$	+1.132
^{33}S	0.76	$\frac{3}{2}$	+0.643
^{55}Mn	100	$\frac{5}{2}$	+3.444
^{57}Fe	2.19	$\frac{1}{2}$	+0.090
^{59}Co	100	$\frac{7}{2}$	+4.649
^{63}Cu	69.09	$\frac{3}{2}$	+2.226
^{65}Cu	30.91	$\frac{3}{2}$	+2.385
^{95}Mo	15.72	$\frac{5}{2}$	−0.913
^{97}Mo	9.46	$\frac{5}{2}$	−0.933

hyperfine coupling can be observed as splitting of the EPR spectrum, if the coupling is greater than the line width. The magnitude of the splitting is given the symbol a, expressed in magnetic field units.

The value of A depends on the nuclear moment, and the extent of interaction of the unpaired electron spin density with the nucleus; A has a sign as well as magnitude. If $A > 0$, the state in which electron and nuclear spins align antiparallel is of lower energy. Measurements of the magnitude and sign of hyperfine couplings provide some of the most detailed experimental evidence for electronic structures of molecules. They have been used to verify the results of molecular orbital calculations.

In EPR spectra of transition ions, the hyperfine interaction gives rise to characteristic splittings, for example for copper ($I = \frac{3}{2}$, 4 lines) and manganese ($I = \frac{5}{2}$, 6 lines) (**Figure 2B**). It can be used to identify paramagnetic centres and determine the electron density distribution over radicals. Isotopic substitution can be used (**Table 1**) to identify specific metals, for example ^{57}Fe for natural iron.

Interactions with nuclei of ligands, sometimes called superhyperfine splittings, allow the determination of the electron density distribution in transition metal complexes. Nuclei with spins greater than $S = \frac{1}{2}$ have a quadrupole moment and this gives rise to further splitting of the energy levels.

Interactions between electrons (fine structure interactions)

The magnetic moment of the electron is 658 times that of the proton, so that interactions between electrons are of greater energy than those between nuclear spins, and are detectable over greater distances. This has a number of consequences. For example, when two or more electron spins are present in a paramagnetic centre such as a transition ion with spin $S > \frac{1}{2}$ or an organic triplet state ($S = 1$), coupling of the electron spins gives rise to fine structure splittings (known as zero-field splittings) which strongly influence the apparent g factor. If the spin is an integral value, the splitting between the energy levels may be larger than the X-band microwave quantum, and no EPR signal is detected. However, for an odd number of electron spins there are always energy levels that can only be split by the magnetic field. As a consequence paramagnets with half-integral spin, such as high-spin Fe^{III} and Co^{II}, are

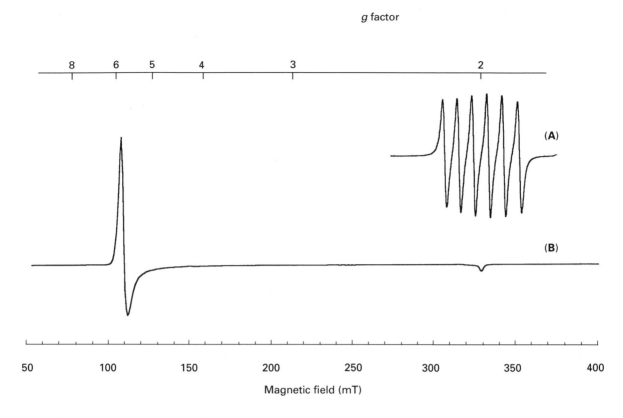

g factor

Magnetic field (mT)

Figure 2 EPR spectra of transition ions (A) manganese ions in aqueous solution; note the six-line hyperfine splitting owing to the ^{55}Mn nucleus ($I = \frac{5}{2}$); (B) high-spin Fe^{III} in metmyoglobin, frozen at 77 K; note the g factors at 6.0 and 2.0.

EPR-detectable. For the latter ions, with $S = \frac{5}{2}$, g factors may extend from 10 to 0. EPR spectra of transition ions are often measured in the solid state, so that the spectrum is broadened by anisotropy of the g factor.

Electron spin–spin interactions occur between paramagnetic centres in solution and in the solid state. This causes splitting or broadening of the EPR signal. For solid paramagnetic materials the broadening may be so great as to render the X-band EPR spectrum undetectable. For EPR spectra in the solid state, the material must usually be *magnetically diluted*, for example doped into a diamagnetic host material. Paramagnets in biological materials such as proteins are usually fixed within the protein matrix and are thus detectable. Spin–spin interactions give rise to characteristic changes in line shape, and enhancement of the electron-spin relaxation rate, from which it is possible to estimate distances between the paramagnetic centres.

Anisotropy

The Zeeman, dipolar hyperfine and fine structure interactions are all anisotropic, meaning that the energy of interaction depends on the angle between the paramagnetic centre and the applied magnetic field. In solution spectra of small molecules, the anisotropy is averaged, and spectral lines are narrow. Note, however, that the isotropic hyperfine interaction does not average to zero and so there is still hyperfine splitting (e.g. **Figure 2A** and **Figure 3A**). For randomly oriented samples in the solid state, (so-called powder spectra) the distribution of molecules in different orientations gives rise to characteristic broadening of the line shape (**Figure 2B** and **Figure 3C**).

Effects of temperature, electron-spin relaxation and spectral linewidths

Temperature has a number of effects on the EPR spectrum. The populations of the electron energy levels, N_- and N_+ (**Figure 1A**) differ by a small amount because of the Boltzmann distribution:

$$\frac{N_+}{N_-} = \exp\left(\frac{\Delta E}{k_B T}\right) = \exp\left(\frac{g\mu_B B_0}{k_B T}\right)$$

where k_B is the Boltzmann constant and T the absolute temperature. At room temperature and a magnetic field of 0.32 T, the difference is about 0.07% of the unpaired electrons. It increases at

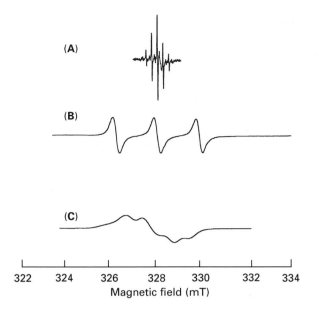

Figure 3 EPR spectra of radicals (A) benzoquinone anion radical in solution; (B) the nitroxide TEMPOL in solution of photosynthetic reaction centres.

higher microwave frequencies (hence higher fields), and at lower temperatures.

The spin–lattice relaxation, with a characteristic time T_1, is responsible for maintaining the population difference between levels, N_- and N_+. The spin–spin relaxation time T_2 reflects the lifetime of the excited state and its effect on the line width. If the electron-spin relaxation rate is too rapid, the lifetime of the excited state is short and the EPR spectrum becomes broadened. At high temperatures the spectrum may become too broad for detection, hence the use of cryogenic temperatures for some transition ions. However, if the spin–lattice relaxation is too slow, the population difference $N_- - N_+$ cannot be maintained, and the amplitude of the signal is attenuated, a situation known as microwave power saturation. Electron-spin relaxation times may be estimated by measuring the amplitude of the signal as a function of applied microwave power.

Operation of the spectrometer

The outline of a conventional continuous-wave EPR spectrometer is shown in **Figure 4**. The term 'continuous-wave (CW)' refers to the fact that microwaves are applied throughout the measurement. It is designed to optimize the weak EPR signals originating from the sample, against a background of noise from the source. The microwave source is a reflex klystron (a type of thermionic valve) or a Gunn diode (a semiconductor device). The detector is a micro-

wave diode. The operating frequency is typically in the X-band (~9.5 GHz). This frequency is a compromise; higher frequencies increase the ratio $N_-:N_+$ (and hence sensitivity) but require smaller cavities and thus smaller sample size (and hence number of spins being measured). A typical sample volume is of the order of 0.1 cm^3.

The microwave circuit is known as the microwave bridge, to emphasize that it is a balanced circuit. The source is tuned to the resonant frequency of the cavity. Microwaves pass through waveguides to the cavity via a circulator, which ensures that signal arising from the sample in the cavity passes to the detector. The reference arm serves to balance the signal with the input microwave power. The circuit is optimally tuned by matching its impedance with the cavity using the iris, to a condition known as critical coupling. In this situation the intensity of the microwave field B_1 at the sample is enhanced by the Q-factor of the cavity, typically several thousand-fold. When electron paramagnetic resonance occurs in the sample, a small amount of power is transmitted to the detector and is amplified and recorded.

The homogeneous B_0 field is provided by an electromagnet. EPR spectra are obtained by modulation of the field by a small amount, at a frequency of 100 kHz. A lock-in amplifier detects the EPR signal in phase with the modulation. Modulation and phase-sensitive detection is a standard method of noise rejection and enhances the signal-to-noise ratio, and hence the sensitivity of the technique, by as much as 10^5. It gives rise to the characteristic first-harmonic (first-derivative) line shape. Although the absorption spectrum could be derived from this by integration, spectra are usually retained as the derivative form as it emphasizes narrow hyperfine splittings and other features.

Temperature control

The form of the paramagnetic material may be solid, liquid or gas, and so appropriate temperature control of the sample is needed. The temperature of the sample may be controlled by flowing gas through a quartz vacuum jacket. For low temperatures, helium is used. The sample may be held in a Dewar of liquid nitrogen or helium. For smaller resonator designs such as loop-gaps, the whole resonator is immersed in a cryostat containing helium or nitrogen.

Resonators

The quality, or Q factor, of the microwave cavity or resonator is a measure of its efficiency at concentrating microwave energy. For a resonant a.c. circuit it is given by

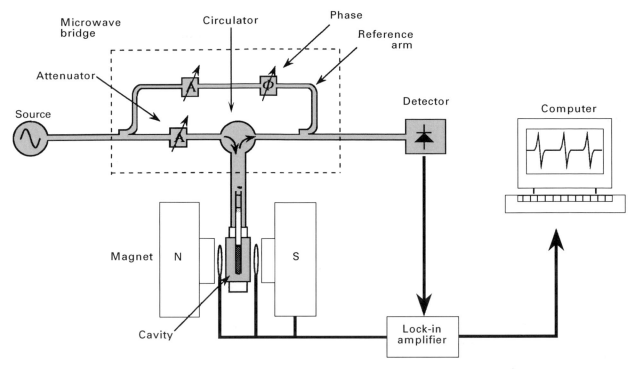

Figure 4 Outline of a typical continuous-wave EPR spectrometer.

$$Q = \frac{2\pi(\text{Energy stored})}{(\text{Energy dissipated per cycle})}$$

The filling factor η is the fraction of the microwave magnetic field B_1 which bisects the sample.

Two examples of resonators used in EPR are illustrated in **Figure 5**. Rectangular cavities [**Figure 5(A)**] are of metal which is coated with silver and gold for high conductivity. The sample is positioned at the maximum of the microwave magnetic field, B_1. It may be irradiated with ultraviolet or visible light, in this example through a waveguide of length λ which prevents microwave leakage. The applied magnetic field B_0 is modulated with Helmholtz coils in the walls of the cavity. The B_0 and B_1 fields are oriented at right angles, which gives a maximum magnetic resonance interactions for $S = \frac{1}{2}$ paramagnets. For paramagnets with an integer spin, the resonant transition is forbidden in this geometry. Parallel mode cavities are used for this situation, in which B_1 is parallel to B_0.

In the loop-gap resonator (**Figure 5B**), the coupling loop acts as an antenna to transmit microwave power to and from the bridge to the split-ring resonator element. The number of gaps is variable, depending on the sample size. The microwave magnetic field B_1 circulates in the ring, while the electric component is mostly confined to the gaps. Magnetic field

modulation is also by external coils (not shown). Loop-gap resonators have better filling factors than rectangular cavities, so that the resonator can be smaller and the sample tube larger. They are particularly effective for lossy samples such as aqueous solutions.

Other designs of cavity are possible, including cylindrical resonators constructed of an inert dielectric material such as sapphire (alumina).

Computer control and data acquisition

Most operations of the instrument are under computer control. The spectrum is recorded as a series of data points, along a linear axis of magnetic field. The signal-to-noise ratio can be improved by repetitive scanning and signal averaging. Fourier transform techniques are normally only used for analysis of pulsed EPR experiments. There is a range of different software for analysis of the spectroscopic data, in particular for simulation.

Sample holders and sample preparation

The holders for EPR samples are constructed of a diamagnetic, nonconductive material such as quartz. Electrically conductive materials such as metals cannot be used as they reflect microwaves and prevent penetration of microwaves into the sample.

It is not possible to measure EPR spectra of large aqueous samples at frequencies above 1 GHz

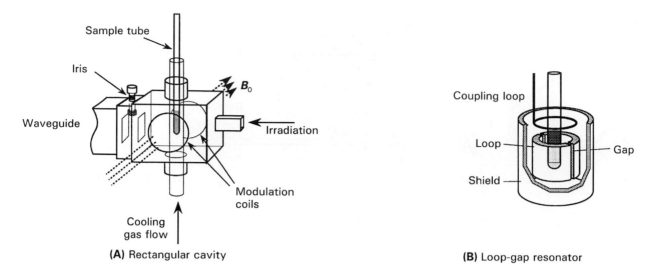

Figure 5 Resonators for EPR spectroscopy. (A) Rectangular cavity; (B) loop-gap resonator.

because of microwave power losses. For measurements in an X-band rectangular cavity, specially constructed flat cells are used, with a thickness of about 0.3 mm (**Figure 6**). They are placed in the centre of the cavity, which is a nodal plane of the electric vector, hence minimizing dielectric losses.

Oxidation–reduction reactions are used to generate organic or inorganic compounds in their paramagnetic oxidation states, for example quinones in their semiquinone states, or transition metal ions in their paramagnetic states (**Table 2**). Electrolytically generated radicals may be obtained by means of electrodes placed in the cell above and below the cavity.

Table 2 Examples of paramagnetic transition ions

Metal ion	Paramagnetic oxidation states	Other states
Vanadium	V^{IV}	V^V, V^{III}
Chromium	Cr^{III}, Cr^{VI}	
Manganese	Mn^{II}, MN^{IV}	Mn^{III}
Iron	Fe^{III}	Fe^{II}
Cobalt	Co^{II}	Co^I, Co^{III}
Nickel	Ni^I, Ni^{III}	Ni^{II}
Copper	Cu^{II}	Cu^I
Molybdenum	Mo^V	Mo^{IV}, Mo^{VI}

Note – other oxidation states and isotopes exist for these elements.

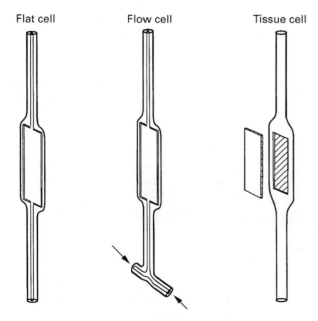

Figure 6 Sample holders for aqueous samples.

For kinetic measurements, paramagnets may be generated by mixing solutions and flowing them into a cell inside the cavity (**Figure 6B**). For short-lived states of transition metals in solution, samples may be obtained by rapid freezing, in which samples are mixed and squirted into a bath of isopentane at 140 K.

Multifrequency EPR

Although the majority of EPR spectra are still taken at X-band frequency, there are advantages for some applications to use lower or higher microwave frequencies, or a range of frequencies. To operate at each frequency it is usually necessary to use a different microwave bridge and cavity. These frequencies are identified by their frequency bands, as listed in **Table 3**.

In high-frequency spectrometers, superconducting magnets are used. Up to 100 GHz, the microwaves

Table 3 Microwave wavebands and frequencies

Band	Frequency (GHz) (app.)	λ (mm)	Field for g = 2 (mT)	Typical applications
Radiofrequency	0.25	1200	8.93	EPR imaging
L	1	300	35.7	Aqueous sample EPR
S	4	75	143	Hyperfine splittings
X	9.5	31.6	339	Routine spectroscopy
K	24	12.5	857	
Q	35	8.6	1250	Electron spin-spin interactions
V	65	4.6	2322	
W	96	3.12	3430	g Factor anisotropy in radicals
D	140	2.14	5000	Large zero-field splittings
G	250	1.2	8931	Even spin system

The definitions of bands follow EPR convention and the frequencies are approximate.

may be obtained by mixing with lower-frequency radiation in a heterodyne arrangement. For higher frequencies, the radiation may be regarded as far infrared, and Fabry–Pérot resonators are used which consist of two parabolic mirrors to concentrate the radiation. In low-frequency spectrometers the magnetic field is produced by Helmholtz coils and the resonator is a coil similar to that used in NMR.

Some features of the EPR spectrum increase with magnetic field, such as the separation between g factors. Others, such as the separation between hyperfine splittings, do not change, to first order. By altering the microwave frequency it is possible to resolve the effects of multiple electron–nuclear and electron–electron interactions. The EPR spectra of organic radicals at X-band (9 GHz) are dominated by hyperfine splittings by protons and other nuclei. At W-band (96 GHz) the spectra are dominated by the anisotropy of the g factor.

Electron–nuclear double resonance (ENDOR) and TRIPLE resonance

The hyperfine interactions with nuclei adjacent to the paramagnetic centre are often too small to be observed because they are smaller than the EPR line width. Electron–nuclear double resonance is a technique which detects these interactions. In addition to the microwave field B_1 a radiofrequency field, B_2, is applied at right angles. A commercial design uses a coil wound around the sample tube in a special cavity.

The energy levels and transitions relevant to the ENDOR measurement are illustrated in **Figure 7**. The microwave field, at the EPR resonant frequency, partially saturates the signal, equalizing the populations of the $m_S = \pm \frac{1}{2}$ levels. The radiofrequency is swept, and when the latter is at the resonant frequency of the interacting nuclear spins this provides further pathways for electron spin relaxation

and the EPR signal is enhanced. The resulting spectrum consists of a pair of lines. The positions of the lines depend on the relative magnitudes of the hyperfine coupling, A, and the nuclear Zeeman frequency, ν_n (the NMR frequency of the nucleus) (**Figure 7**). In the case where $\nu_n > A/2$, as for protons, which have a large nuclear moment, the lines will be centred at ν_n and separated by A. When $A/2 > \nu_n$, as in the case of strong couplings to other nuclei, the ENDOR spectrum consists of two lines centred at $A/2$ and separated by $2\nu_n$.

Hyperfine interactions of the order of a few megahertz can be resolved and provide information about nuclei at distances up to 0.5 nm. An additional advantage of ENDOR is that the number of lines is additive, rather than multiplicative. If there are numerous nuclear spins in the paramagnetic molecule, each type of hyperfine splitting will give just one pair of lines and so the ENDOR spectrum is simpler than the EPR spectrum. This feature is particularly useful in the interpretation of the complex spectra of radicals.

TRIPLE resonance is another method which resolves further a spectrum containing multiple hyperfine couplings. It is analogous to the Overhauser effect in NMR. By continuously applying radiofrequency radiation at the resonant frequency of one line in the spectrum, other lines in the ENDOR spectrum are suppressed or enhanced if they are connected to the same transition. This method makes it possible to determine the sign of A, which is used in molecular orbital calculations.

Pulsed EPR

Pulsed EPR spectrometers are now commercially available and have some of the same electronic features as NMR spectrometers. The layout of the microwave bridge is similar to that of **Figure 4**, but in addition a pulse programmer controls the

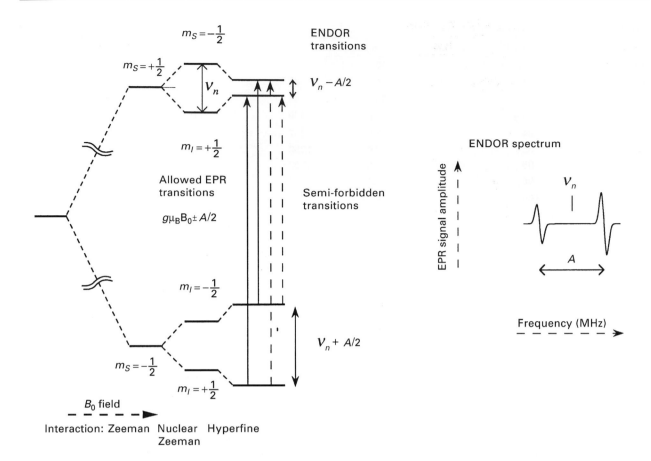

Figure 7 Energy levels for an electron interacting with an $I = \frac{1}{2}$ nucleus, showing the transitions relevant to ENDOR and electron-spin echo envelope modulation (ESEEM).

microwave supply through diode switches. The microwave pulses, of duration 5–50 ns, are amplified by a travelling-wave tube amplifier, to a power up to 1 kW. The cavity has a lower Q-factor than for CW EPR to allow the microwave field to build up more quickly. The main differences between pulsed EPR and pulsed NMR arise from the broad bandwidth of the EPR spectrum.

Flash photolysis

For the spectra of radicals that are dispersed over a narrow range of frequency, it is possible to detect a free-induction decay (FID). This makes it possible to derive a spectrum by Fourier transformation and observe the spectrum of a transient species induced, for example, by a laser flash which is synchronized with the microwave pulse programmer (**Figure 8A**).

Normally, however, the spectrum is too broad, or the FID decays too rapidly. In this case it is possible to detect an electron-spin echo after a two-pulse ($\pi/2$–π–echo) sequence. The spectrum of the flash-induced species is derived by repeating the measurement over a range of magnetic field strength. The

lifetime (which should be a minimum of about 100 ns) is measured by repeating the electron-spin echo sequence, while the interval between the flash and the pulses is systematically incremented.

Electron-spin relaxation rates

The values of T_1 and T_2 are measured in different experiments which observe the decay of the electron-spin echo with time; T_1 is measured with an inversion-recovery experiment as in NMR, while T_2 is observed by a two-pulse echo sequence, as in **Figure 8A**. The amplitude of the echo decreases with increasing delay time τ, owing to loss of 'phase memory'. The spin–spin relaxation is affected by various factors, including molecular motion and spin–spin interactions between paramagnets.

Hyperfine couplings

Electron-spin echo envelope modulation (ESEEM) spectroscopy is a pulsed technique that detects weak hyperfine and quadrupolar couplings in the solid state. It is complementary to ENDOR in that it gives

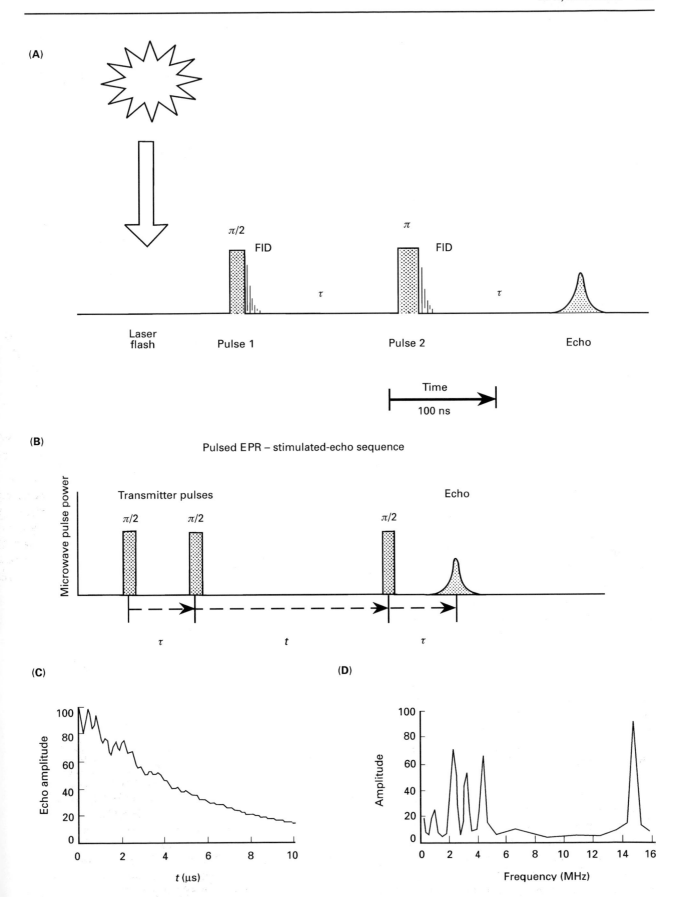

Figure 8 Pulsed EPR experiments. (A) flash photolysis, (B)–(D) ESEEM.

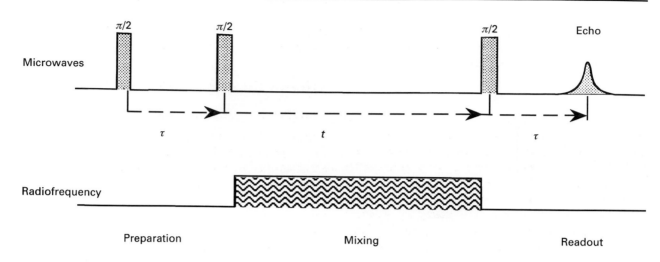

Figure 9 Principle of Mims pulsed ENDOR.

the best results in the intermediate regime of hyperfine couplings where $A \approx \nu_N$; in addition, it is particularly sensitive to couplings with quadrupolar nuclei such as ^2H and ^{14}N. The Mims three-pulse echo sequence is similar to the two-pulse sequence (**Figure 8B**) except that the π pulse is split into two $\pi/2$ pulses separated by time T. The decay of the echo is now determined by the slower spin–lattice relaxation time T_1 allowing echoes to be detected over a longer period of time. During the evolution of the spin echo, hyperfine coupling frequencies cause dephasing of the electron spins, and so the echo amplitude is modulated with time (**Figure 8C**). This can be considered as being due to an interference between the frequencies of the allowed transitions, $\Delta m_S = \pm 1$, with the 'semi-forbidden' transitions, $\Delta m_S = \pm 1$, $\Delta m_I = \pm 1$, which are allowed owing to quadrupolar interactions and anisotropy of the hyperfine interaction. Fourier transformation of the time-domain spectrum yields the spectrum of hyperfine frequencies (**Figure 8D**).

Two-dimensional pulsed EPR experiments are also possible, such as HYSCORE (hyperfine sublevel correlation) spectroscopy. This uses a pulse sequence as in **Figure 8B**, separated by an inversion pulse into two variable time intervals. Fourier transformation in the two time-domains gives a 2D spectrum, in which the effects of multiple hyperfine couplings can be resolved.

Pulsed ENDOR is an alternative technique, in which radiofrequency pulses at NMR frequencies are used to perturb the spin-echo sequence (**Figure 9**). Hyperfine interactions lead to loss of coherence of the electron spins during the mixing period, so the amplitude of the echo is decreased. A plot of echo amplitude against RF frequency yields a spectrum analogous to that of a CW ENDOR spectrometer.

Types of material studied

Free radicals

Organic free radicals can be produced by one-electron oxidation or reduction reactions, by irradiation processes or by homolytic cleavage of a chemical bond. They may occur in the solid, liquid or gaseous states. Organic radicals in solution give complex EPR spectra. The hyperfine structure of their spectra provides information about the type of radical and the distribution of the unpaired electron in it. Free radicals tend to be unstable and to decay by recombination processes. However, some radicals, particularly those in the solid state, are stable. If they have a half-life of at least a few seconds they can readily be examined by conventional EPR spectroscopy, using rapid-scan coils to sweep the B_0 field. If they are more short-lived, the radical may be stabilized by cooling to low temperatures, or chemically trapped with a spin trap which reacts with the unstable species to produce a more stable radical.

Spin traps

Ideally, a spin trap is a diamagnetic compound, which is introduced into a biological system without perturbing it and which reacts rapidly and specifically with a transient free radical to form a stable radical that can be identified from its EPR signal. These requirements are somewhat contradictory. The two most commonly-used spin traps for organic radicals are nitroso compounds and nitrones (**Table 4**). These react with radicals to form nitroxides. The nitroso compounds give an adduct in which the radical added is bonded directly to the nitroxide nitrogen, so that the hyperfine splittings in the EPR signal are more diagnostic of the original radical; however, the adducts are less stable and may decay within minutes. Nitrones tend to give more stable adducts which are more difficult to identify.

Spin labels

Spin labels are stable paramagnetic molecules that are introduced into a biological sample, and extend the use of EPR to situations where no endogenous radicals exist. The commonest type of spin labels are nitroxides (aminoxyl) radicals, though other types of radicals and transition metal ions have been used. The advantage of the specific labelling approach is that it makes it possible, through judicious labelling, to investigate the local environment and motion of defined parts of the system under study.

Transition metals

Transition metals are paramagnetic in certain oxidation states (**Table 2**). Although this paramagnetism is stable, the sample may have to be cooled to low temperatures for the EPR signal to be observed.

Defects in solids

Examples of defects are cation vacancies (V centres), anion vacancies (F centres) and impurities in crystals. Defects can be generated in various ways, such as irradiation with ultraviolet or ionizing radiation, or by imperfect crystallization. An example is a defect (latent image) generated in photographic emulsion by light irradiation. In addition, finely divided solid

materials commonly show radical centres due to homolytic cleavage of chemical bonds.

Information from EPR spectroscopy

Identification of the paramagnet

The type of paramagnetic species may be recognized from the characteristic features of its spectrum such as g factors and hyperfine splittings.

For free radicals in solution, examination of the pattern of hyperfine splittings can provide information about the number of interacting nuclei and hence the structure of the radical. For example, in the benzoquinone radical (**Figure 3A**), the splitting by four equivalent protons gives rise to five lines with amplitudes in the ratio 1:4:6:4:1.

A powerful technique is to substitute atoms in the molecule with nuclei having different nuclear spins, for example deuterium ^2H ($I = 1$) for ^1H or ^{13}C ($I = \frac{1}{2}$) for ^{12}C. Examples of nuclear spins and moments are given in **Table 1**. The observed change in the hyperfine splitting pattern is conclusive evidence of interaction with the nucleus.

Quantification

The concentration of unpaired electrons can be determined from the integrated intensity of the EPR spectrum by double integration or by numerical simulation of the spectrum. For paramagnets with $S = \frac{1}{2}$, measured under non-saturating conditions, the

Table 4 Types of spin traps

Type / spin trap	Structure	Suitable application
2-Methyl-2-nitrosopropane (MNP)		C,S-centred radicals
5,5-Dimethyl-1-pyrroline-N-oxide (DMPO)		C,S,O-centred radicals
N-tert-Butyl-α-phenylnitrone (PBN)		C,S,O-centred radicals
Fe^{II}(diethyldithiocarbamate)$_2$		Nitric oxide

integrated intensity for a given concentration is the same, regardless of the nature of the paramagnetic species. Hence the concentration can be referred to a standard material such as Cu^{II}.

Spin Hamiltonian parameters

The parameters of the EPR spectrum are described formally by a spin Hamiltonian equation, which summarizes the energies of the different types of interaction. For example, the Hamiltonian for a spin S interacting with a nuclear spin I is

$$\mathcal{H} = g\mu_B B_0 S + AIS$$

where the operators g and A are shorthand notation for 3×3 matrices or tensors. The EPR spectrum may be interpreted by computer simulation of the energy of interaction with the magnetic field B_0. This provides information about the electronic state of the paramagnet. For radicals, the information can be used in detailed molecular orbital calculations of electronic structure. For transition metal ions, EPR provides information about the coordination geometry, spin state and types of ligands.

Redox and spin states

Changes in oxidation states of paramagnets, by adding or removing unpaired electrons, cause the EPR signals to appear and disappear. This can be used to monitor oxidation–reduction reactions, and to measure reduction potentials. Generally the EPR spectrum is a useful indication of the oxidation–reduction state of the species observed. Often the presence of an EPR signal is conclusive, though in the case of semiquinones they may be anion radicals (formed by electron addition) or cation radicals (formed by removal of an electron). The EPR spectra of Ni^I and Ni^{III} are similar. In these cases further oxidation–reduction experiments are needed to distinguish them.

For transition ions such as Fe^{III}, the EPR spectrum is a good indicator of spin state, in this case high spin ($S = \frac{5}{2}$) or low spin ($S = \frac{1}{2}$). The spectra of the high-spin states are influenced by zero-field splittings, in this case giving g factors of up to 10.

Distances

The hyperfine coupling between an electron spin and a nucleus or between electron spins consists of two components: an exchange interaction (acting through chemical bonds, and generally isotropic), and a dipolar interaction. The dipolar component is anisotropic and its magnitude is inversely proportional to the (distance)^3, and this can be used to estimate distances. In favourable cases, hyperfine couplings from ENDOR can be used to determine distances of protons up to 0.5 nm. Electron spin–spin interactions can be observed as splittings in the EPR spectrum over distances up to 1.5 nm, and relaxation effects at distances up to 4 nm.

Orientation

If the molecules of a paramagnet are orientated, as in a single crystal, the EPR signal consists of narrow lines, which shift as the crystal is rotated in the magnetic field. The angular dependence can be measured precisely using a goniometer. Some ordering is seen in fibres and layered materials, though less precisely than in crystals. From this orientation dependence it is possible to estimate the spin Hamiltonian parameters, yielding detailed information about the axes of the g and A matrices, and hence the orbital states.

Motion

The line width of the EPR spectrum of a radical such as a nitroxide in solution is sensitive to motion in the range of correlation times 10^{-7}–10^{-11} s. This is due to the anisotropy of the ^{14}N hyperfine splitting. If the motion is rapid enough to average this anisotropy, the spectrum shows narrow lines. For slower motion the spectrum broadens out. The rate of motion is determined by simulation, or by comparison with labelled molecules of known correlation time. By means of an experimental technique known as saturation-transfer EPR the spectra may be made to be sensitive to correlation times up to 10^{-3} s, as found, for example, in membrane-bound proteins.

EPR dosimetry

The build-up of radicals in irradiated solids forms a method of estimating the absorbed dose of ionizing radiation. A standard used in EPR dosimeters is the amino acid alanine, in pellet form, which gives a linear relationship between radical signal amplitude and absorbed dose over several orders of magnitude. Where there has been accidental radiation exposure, the EPR of radicals in hydroxyapatite in tooth enamel has been used to estimate the absorbed dose. Irradiation of foods such as meat may be detected by radicals formed in bone.

EPR dating

Buried archaeological samples that contain crystalline materials show radical EPR signals due to the effects of the radiation in the soil. The time of burial can be estimated by taking measurements of the radioactive isotopes in the surrounding soil and observing the signal amplitude after applying an equivalent dose of radiation to the sample. The decay of the radical signal over time depends on the stability of the defects in the material. Dating works best for crystalline material such as silica; bone is unreliable because of recrystallization. The technique has been used for samples dated between 10^5 and 10^6 years, which is between the ranges possible for radiocarbon and potassium dating.

Medical applications

Oxygen radicals have been widely investigated in biomedical systems and are of great interest because of their cytotoxic effects and involvement in processes such as inflammation. They may be observed by spin traps (**Table 4**). Nitric oxide is also generated in conditions such as inflammation, and can be trapped by iron complexes such as iron diethyldithiocarbamate.

Oximetry exploits another feature of the spectra of radicals in solution: that they are broadened by paramagnetic interactions with oxygen in solution. With narrow-line radicals it is possible to measure oxygen in human tissues at concentrations below 10^{-5} mol dm^{-3}.

List of symbols

A = hyperfine coupling constant; B_0 = static magnetic field; B_1 = microwave magnetic field; B_2 = radiofrequency magnetic field; g = electron g factor; h = Planck's constant; $\mathcal{H}$ = Hamiltonian operator; I = nuclear spin operator; k = Boltzmann constant; m_I = nuclear spin angular momentum; m_S = electron spin angular momentum; N_-, N_+ = populations of electron energy levels separated by a magnetic field; Q = quality factor of a resonator; S = electron spin operator; T = absolute temperature; T_1 = longitudinal (spin–lattice) relaxation time; T_2 = transverse (spin–spin) relaxation time; η = filling factor of a resonator; λ = microwave wavelength; μ_B = Bohr magneton. In an EPR spectrum the signals are obtained by increasing the B_0 field at fixed frequency; ν = microwave frequency; ν_n = nuclear Larmor frequency; τ = delay in pulse sequence.

See also: **Chemical Application of EPR; Chiroptical Spectroscopy, Oriented Molecules and Anisotropic Systems; EPR Imaging; EPR Spectroscopy, Theory; Fluorescence Polarization and Anisotropy; Fourier Transformation and Sampling Theory; Laboratory Information Management Systems (LIMS); NMR in Anisotropic Systems, Theory; NMR Pulse Sequences; NMR Relaxation Rates; Nuclear Overhauser Effect; Spin Trapping and Spin Labelling Studied Using EPR Spectroscopy.**

Further reading

Abragam A and Bleaney B (1970) *Electron Paramagnetic Resonance of Transition Ions*. Oxford: Oxford University Press.

Atherton NM (1993) *Principles of Electron Spin Resonance*. Chichester: Ellis Horwood.

Bencini A and Gatteschi D (1989) *EPR of Exchange Coupled Systems*. Berlin: Springer-Verlag.

Berliner LJ and Reuben J (1993) *EMR of Paramagnetic Molecules: Biological Magnetic Resonance*, Vol 13. New York: Plenum.

Czoch R and Francik A (1989) *Instrumental Effects in Homodyne Electron Paramagnetic Resonance Spectrometers*. Chichester: Ellis Horwood.

Dikanov SA and Tsvetkov YD (1992) *Electron Spin Echo Envelope Modulation (ESEEM) Spectroscopy*. Boca Raton, FL: CRC Press.

Eaton GR, Eaton SS and Salikhov KM (1997) *Foundations of Modern ESR*. Singapore: World Scientific Publishing.

Eaton GR, Eaton SS and Ohno K (1991) *ESR Imaging and In Vivo ESR*. Boca Raton, FL: CRC Press.

Hoff AJ (ed) (1989) *Advanced ESR: Applications in Biology and Biochemistry*. Amsterdam: Elsevier.

Keijzers CP, Reijerse EJ and Schmidt J (1989) *Pulsed ESR – A New Field of Applications*. Amsterdam: North Holland.

Kurreck H, Kirste B and Lubitz W (1988) *Electron Nuclear Double Resonance Spectroscopy of Radicals in Solution*. Weinheim: VCH Publishers.

Lowe DJ (1995) *ENDOR and ESR of Metalloproteins*. Berlin: Springer.

Mabbs FE and Collison D (1992) *Electron Paramagnetic Resonance of d Transition Metal Compounds*. Amsterdam: Elsevier.

Pilbrow JR (1990) *Transition Ion Electron Paramagnetic Resonance*. Oxford: Oxford University Press.

Poole CPJ (1983) *Electron Spin Resonance*, 2nd edn. New York: Wiley.

Specialist Periodical Reports on ESR Spectroscopy. Cambridge: Royal Society of Chemistry.

Weil JA, Bolton JR and Wertz JE (1994) *Electron Paramagnetic Resonance: Elementary Theory and Practical Applications*. New York: Wiley.

ESR Imaging

See **EPR Imaging.**

ESR Methods

See **EPR, Methods.**

ESR Spectroscopy, Applications in Chemistry

See **Chemical Applications of EPR.**

ESR Spectroscopy, Theory

See **EPR Spectroscopy, Theory.**

Exciton Coupling

Nina Berova, Columbia University, New York, NY, USA
Nobuyuki Harada, Tohoku University, Sendai, Japan
Koji Nakanishi, Columbia University, New York, NY, USA

ELECTRONIC SPECTROSCOPY
Theory & Applications

Circular dichroic spectroscopy (CD) is a powerful method for determining the absolute configurations and confirmations of chiral molecules. In particular, the exciton-coupled circular dichroism arising from through-space chromophore–chromophore interactions, called exciton coupling, is one of the most sensitive spectroscopic tools for determining the molecular chirality or molecules in solution. The theoretical analysis of the exciton coupling phenomena is based on the coupled oscillator theory and molecular exciton theory. Since exciton-coupled CD is basically a nonempirical method, the derived conclusions are nonambiguous. It is a microscale technique of great versatility applicable to compounds ranging from small molecules to various types of ligand–receptor complexes. However, the extent of its potential has yet to be explored. Here the theory and applications of exciton-coupled CD are outlined.

When two chromophores with an identical or similar excitation energy are in close spatial proximity to one another and chirally disposed, the individuality of a monomer chromophore transition is lost. The chromophore excited state can delocalize over all chromophores within the system and becomes an exciton. The excitons interact and couple, thus giving rise to a characteristic pair of intense CD bands with opposite signs and comparable band areas, called a CD exciton couplet. The coupling of two identical chromophores is illustrated in **Figure 1** by a steroidal 2,3-bis-*para*-substituted benzoate. The intense intramolecular charge-transfer (CT) bands stemming from the ^1L_a transition of *para*-substituted benzoates give rise to the main electronic absorption band; it is the through-space coupling between these CT bands which is used for determining the absolute configuration between two hydroxyl groups. Although there is some free rotation around the C–O bonds, ester bonds are known to adopt the *s-trans* confirmation.

Furthermore, X-ray crystallographic data, energy calculations, etc., show that the ester C=O is *syn* with respect to the methine hydrogens. Thus, the intramolecular CT transition polarized along the long axis of the benzoate chromophore is almost parallel to the alcoholic C–O bond. Hence the exciton CD data which represent the absolute chirality between the transition moments also represent the chirality between any two hydroxyl groups, provided that their spatial distance permits the two chromophores to couple. When the absolute sense of chirality between the C(2)–O and C(3)–O bonds constitute a clockwise twist as shown by the Newman projection in **Figure 1**, it is defined as positive, and vice versa.

The outline of exciton-coupled CD in a system consisting of two identical chromophores (a degenerate case) is as follows:

(i) Chromophores exhibit intense π–π^* absorptions.

(ii) The excited state of the system is split into two energy levels, α and β, by exciton interaction between the two chromophores, and the energy gap $2V_{ij}$ is called the Davydov splitting.

(iii) In the UV-visible spectra, this coupling gives rise to two-component spectra of the same sign

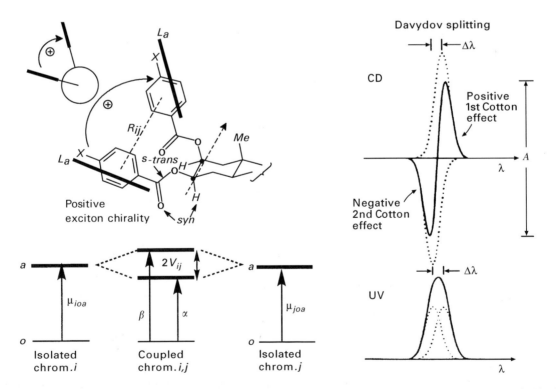

Figure 1 Exciton coupling of two identical chromophores of steroidal 2,3-bisbenzoate with positive chirality; the carbinyl hydrogens are represented *syn* to the ester carbonyl and the ester groups is in an *s-trans* conformation. Also shown are the splitting of the excited states of isolated chromophores *i* and *j* into α and β states with the Davydov splitting $2V_{ij}$, summation CD and UV curves (solid line) of two component curves (dotted line).

Table 1 Definition of exciton chirality for a binary system

	Qualitative definition	Quantitative definition	Cotton effects
Positive exciton chirality		$R_{ij} \cdot (\mu_{ioa} \times \mu_{joa})\, V_{ij} > 0$	**Positive first and negative second Cotton effects**
Negative exciton chirality		$R_{ij} \cdot (\mu_{ioa} \times \mu_{joa})\, V_{ij} < 0$	**Negative first and positive second Cotton effects**

which usually appears as a single absorption maximum with double intensity. In contrast, in CD, this excitation to two split energy levels gives rise to two CEs of opposite signs; summation of the two curves of opposite signs separated by energy gap $\Delta\lambda$ results in a CD with two extrema, a couplet (commonly called a 'split CD'). The extrema at longer and shorter wavelengths are called the first CE (Cotton effect) and second CE, respectively.

(iv) The amplitude of exciton split CE is defined as the A value ($A = \Delta\varepsilon_1 - \Delta\varepsilon_2$), where $\Delta\varepsilon_1$ and $\Delta\varepsilon_2$ are intensities of the first and second CEs respectively.

(v) If the electric transition moments μ of the two chromophores constitute a right-handed screw, a positive couplet with positive first and negative second CEs are observed, and vice versa (**Table 1**).

(vi) The sign and intensity of exciton split CEs are governed by 'exciton chirality', which is theoretically defined as

$$R_{ij} \cdot (\mu_{ioa} \times \mu_{joa}) V_{ij}$$

where R_{ij} is interchromophoric distance vector from i to j, μ_{ioa} and μ_{joa} are electric transition dipole moments of excitation $o \rightarrow a$ of groups i and j, respectively, and V_{ij} is interaction energy between two groups i and j.

(vii) The absolute configuration of two chromophores can be determined, provided that the direction of the transition moment in the chromophore is established.

The cage compound [1] with two anthracene moieties, (6R,15R)-(+)-6,15-dihydro-6,15-ethanonaphtho[2,3-c]pentaphene (**Figure 2B**), provides a typical example of exciton-coupled CD. Of the four

absorption bands of anthracene in the UV-visible region, the band at 252 nm ($\varepsilon = 204\,000$, ^1B_b transition) is the most intense and is polarized along the long axis of the chromophore. The UV of cage molecule [1] (**Figure 2B**) resembles that of anthracene (**Figure 2A**) with the expected bathochromic shift (λ_{max} 267.2 nm, $\varepsilon = 268\,600$). The CD of (+)-[1] shows two intense CEs of opposite signs at the ^1B_b transition: a first positive CE, $\lambda_{ext} = 268.0$ nm, $\Delta\varepsilon = +931.3$, and a second negative CE, $\lambda_{ext} = 249.7$ nm, $\Delta\varepsilon = -720.8$; $A = +1652.1$ (these CEs are one of the strongest ever observed!). The positive A value indicates that the two ^1B_b transition moments along the long axes of the anthracene moieties constitute a clockwise screwsense. This leads to the 6R, 15R absolute configuration shown for (+)-[1].

Both exciton-coupled CD and X-ray techniques are nonempirical methods for determining the absolute configuration of chiral compounds. Although these two methods differ in that CD is for solutions whereas X-ray is for crystals, both methods naturally give the same absolute configuration.

The application of CD exciton coupling formalism to structural analysis started as the 'dibenzoate chirality rule', which became the 'exciton chirality method' when it was expanded to include the coupling between a variety of chromophores for determining the absolute configurations and/or confirmations. As shown schematically in **Figure 1**, 'exciton coupling' is also present in UV-visible spectra, except that the electronic spectrum simply appears as a single maximum with double intensity; only in rare cases when the energy gap $2V_{ij}$ between the α and β levels is substantial will the spectra exhibit double maxima. **Figure 3A** (plotted in cm^{-1}) shows the broad UV-visible band with $\lambda_{max} \approx 390$ nm of Schiff base [1]; upon protonation this is converted into the cyanine dye [2] with delocalized electrons which give rise to a red-shifted band with narrow half-band width ($\lambda_{max} \approx 515$ nm). The coupling of two such cyanine dye chromophores

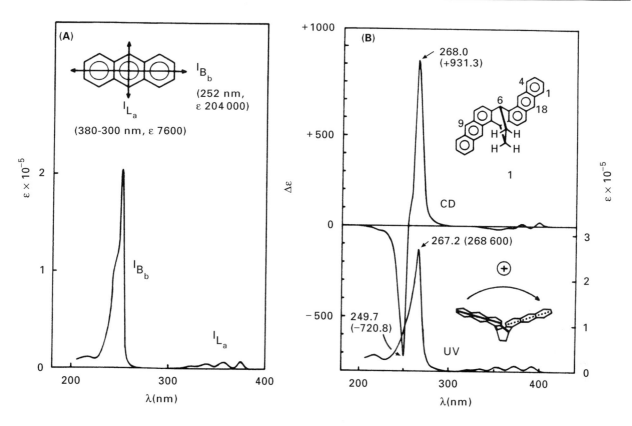

Figure 2 (A) UV spectrum of anthracene in EtOH and polarization of the UV transitions. (B) CD and UV spectra of (6*R*,15*R*) -(+)-6,15-dihydro-6,15-ethanonaphtho[2,3-*c*]pentaphene [1] in 0.18% dioxane–EtOH. Adapted from Harada N *et al. Journal of the American Chemical Society* **98**: 5408.

give rise to an electronic spectrum with a rare double maximum. The CD spectrum exhibits an intense negative couplet with $A = 462$. That the integrated areas or rotational strength of the negative and positive CEs are comparable is a manifestation of the 'sum rule', which states that summation of the positive and negative rotational strengths over the entire CD spectrum should be zero. The 'sum rule' is one of the requirements of the exciton coupling mechanism. The observed regional sum rule, instead of a sum rule covering the entire CD range, is due to the well-separated cyanine dye maxima. In order to emphasize the similarity between electronic and chiroptical spectroscopy, the term 'exciton-coupled CD' rather than 'exciton chirality CD' is used in the following.

Chromophores for exciton-coupled CD

Some aspects of practical importance are summarized as follows:

(1) Sample amount: μg scale.
(2) Solvent: acetonitrile, methylcyclohexane. Alcoholic solvents should be used with caution when dealing with acylated substrates because of the possibility of ester exchange.
(3) Any chromophore with strong intensity and known direction of μ can be used.
(4) The sign of the couplet or 'split CD' is determined by the absolute sense of twist of μ.
(5) Amplitude A is proportional to ε^2, or to a first approximation A and ε have a linear relation.
(6) Amplitude A is inversely proportional to the square of R, the interchromophoric distance.
(7) Additivity relation holds for A values in case of identical chromophores, and for the entire CD curves when the interacting chromophores are different (see below).
(8) There is no exciton coupling when the angles of interacting μs are 0° or 180°; in 1,2-bisacylates, the A value is maximum when the projection angle of interacting μ is ~70°.
(9) The interacting chromophores do not have to be identical. Moreover, they do no have to be in the same molecule, e.g. a ligand and its receptor could also interact and give exciton-coupled CD. Chromophores are introduced into chiral molecules lacking suitable chromophores. When only one chromophore is present in the

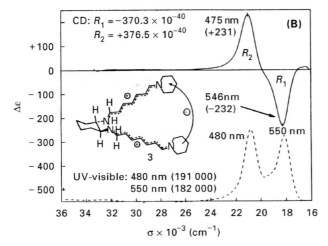

Figure 3 (A) UV-visible spectra of mono-Schiff base [1] and monocynanine dye [2] in CH_2Cl_2; (B) UV-visible and CD spectra and rotational strength R in cgs units of biscyanine dye derivative of (1S, 2S)-*trans*-1,2-cyclohexanediamine [3] in CH_2Cl_2–TFA.

substrate molecule, the newly introduced chromophore is selected so that its absorption maximum will be close to that of the preexisting chromophore.

The chromophores used for determining absolute configuration or conformation by exciton-coupled CD should have intense $\pi \rightarrow \pi^*$ transitions with a known direction of the coupled transition moments; thus chromophores with high symmetry are more suited. Some chromophores referred to in this section are summarized in **Table 2**.

The UV spectra of *para*-substituted benzoates and the direction of the transition moments are shown in **Figure 1**. Among the *para*-substituted benzoates, *p*-dimethylaminobenzoate (**Table 2**, [3]), is highly recommended because of the red-shifted maximum and large ε value. *Ortho*- and *meta*-substituted benzoates are not suited because their transition moments are

not parallel to the alcoholic C–O bond; consequently, the exciton chirality will depend on the conformation around the C–O bond. *para*-Substituted benzoyl chromophores can be used for derivatizing *primary* amino groups, e.g. the intramolecular CT band of the benzamido group, $\lambda_{max} \approx 225$ nm (ε = 11 200), is again polarized along the long axis of the chromophore; however, for *sec*-amino groups the benzamido chromophores cannot be used unless the conformation around the C–N bond is established. The chromophore of 2,3-naphthalenedicarboximide [12] exhibits an intense ^1B_b transition around 260 nm, which is polarized along the long axis of the chromophore. This makes it ideally suited for exciton coupled CD because the long axis-polarized ^1B_b transition moment is exactly parallel to the C–N bond of the amine moiety; the intense ε and its fluorescence are further attributes (see below). Since the maxima of benzamido-2,3-naphthalenedicarboximido [12] and benzoate chromophores [1]–[4] are close to one another, combination of these chromophores is useful for determining the absolute configuration of amino alcohols.

Dienes, enones, ene-esters, ene-lactones, diene-esters, other benzenoids

The transition moments of the $\pi–\pi^*$ band or conjugated dienes, enones, etc., are almost parallel to the long axis of chromophores (**Figure 4**). Phenylacetylene and benzonitrile chromophores with their intramolecular CT bands around 230 nm are also ideally suited because their CT transitions are polarized exactly parallel to the long axis of chromophores.

Polyacenes

As depicted in **Figure 2A**, the ^1B_b transition of polyacene chromophores is ideally suitable for observing exciton coupled CD. See below for a description of the fluorescent 2-anthroate chromophore.

Examples of exciton-coupled CD and natural products

The CD of the plant hormone abscisic acid exhibits a positive couplet arising from the enone–dienoic acid coupling (directions of electric transition moments are shown by thick lines). The positive twist between these two moieties shown in **Figure 4** directly indicates that the absolute configuration of the *tert*-OH is pointing down, or α. Similarly, the negative couplet of dendryphiellin F shows the dienoate side-chain to be β whereas that of manumycin shows the triene amide chain to be α. In the case of lithospermic

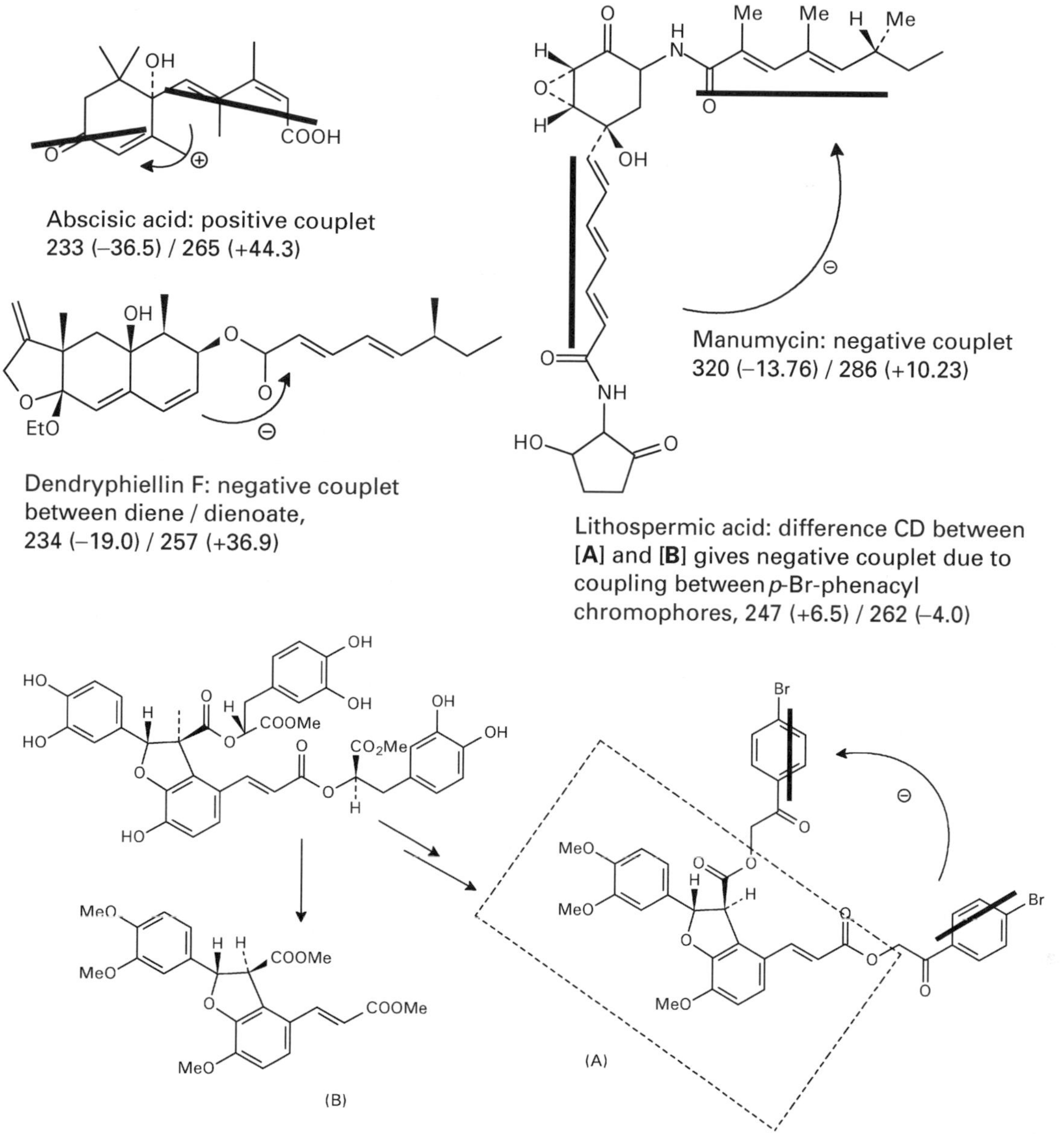

Abscisic acid: positive couplet
233 (−36.5) / 265 (+44.3)

Dendryphiellin F: negative couplet
between diene / dienoate,
234 (−19.0) / 257 (+36.9)

Manumycin: negative couplet
320 (−13.76) / 286 (+10.23)

Lithospermic acid: difference CD between
[A] and [B] gives negative couplet due to
coupling between p-Br-phenacyl
chromophores, 247 (+6.5) / 262 (−4.0)

(A)

(B)

Figure 4 Examples of natural products exhibiting an exciton-coupled CD.

acid, hydrolysis of its heptamethyl ether followed by p-bromophenacylation yielded derivative A, while methanolysis of the same heptamethyl ether gave fragment B. Subtraction of the CD of B from that of A gave a negative couplet; moiety B within the structure of A is indicated by the dotted rectangle. This difference CD shows that the coupling between the two bromophenacyl groups is negative. This defines the helicity between the two ester side-chains in lithospermic acid and hence its absolute configuration. As in any other spectroscopy, difference CD offers a powerful tool for extracting the necessary information hidden in a far more complex structure or a ligand–receptor complex.

Chromophores with intense absorptions

The A values of split CD curves are proportional to the absorption intensity ε of the chromophore. Hence the stronger the absorption the larger is the A value or the more sensitive is the measurement. The 2-anthorate chromophore (**Table 2**, [8]) with an exceptionally intense absorption is a powerful chromophore for exciton-coupled CD measurements. The 2-anthroate and 2-naphthoate [6] chromophores are of lower symmetry than para-substituted benzoates [1]–[4]; however, in both cases the orientation of the strong electric transition moment, ^1B_b, is almost parallel to the long axis of the chromophore and

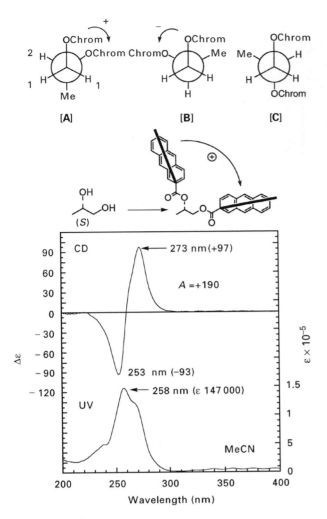

Figure 5 Conformations and UV and CD spectra (in MeCN) of bis-2-anthroyl-(2S)-1,2-propanediol

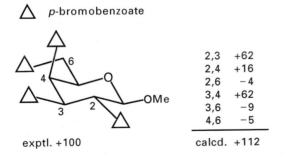

Figure 6 Amplitudes and methyl 2,3,4,6-tetrakis[O-(p-bromobenzoyl)]-β-D-galactopyranoside, experimental (in MeOH) and calculated.

common moiety in many natural products and synthetic intermediates and hence assignment of the absolute configuration of the chiral centre is important. Derivatization of (2S)-1,2-propanediol with anthroylimidazole furnishes the bisanthroate with a positive CD bisignate curve (MeCN) at 263 nm with $A = +190$. Three distinct conformations A, B and C are conceivable for the bisanthroate. A and B lead to opposite CD exciton couplets, whereas exciton coupling in conformer C should be almost zero and therefore this conformer can only influence the amplitude and not the sign of the bisignate curve. Of the remaining two, B is not expected to contribute much because of the extra steric interactions between the methyl and the chromophores. That conformer A is the major form is supported by NMR coupling constants of $J_{1,2} = 6.8$ Hz (trans) and 3.7 Hz (gauche). The very intense UV is reflected in the large A value, which is almost 10-fold that of bis-p-bromobenzoate [2], i.e. +21 (in EtOH).

Pairwise additivity relation in exciton-coupled CD

An aspect of fundamental importance in exciton-coupled CD is the principle of pairwise additivity. Although Brewster and Kauzmann proposed the 'principle of pairwise interactions as a basis for an empirical theory of optical rotatory power' in 1961, it was not possible to secure experimental data to prove this since they were addressed to rotations at 589 nm (Na D-line), where values are not only extremely small but also represent summation of contributions of all stereogenic centres. However, subsequent experimental results have established that the additivity relation holds for both monochromatic systems and bichromophoric systems.

Pairwise additivity in systems containing identical chromophores The amplitude A of the split CD curve for a compound containing two or more identical chromophores can be approximated by the summation of each interacting basis pair. Thus, in a trischromophoric system containing three identical chromophores I, II and III, the observed amplitude can be approximated by the sum of three interacting bischromophoric pairs (basis pairs): $A_{total} = A_{I,II} + A_{II,III} + A_{I,III}$. This is illustrated for a sugar acylate containing four p-bromobenzoates in **Figure 6**. Six interacting pairs of benzoates are present in this derivative, 2,3-, 2,4-, 2,6-, etc. The A values of the six pairs of p-bromobenzoates are shown in the table in **Figure 6**; the summation gives a value of +112, which is in excellent agreement with the experimental value of +100. The A values (i.e. constants) of all 12 conceivable pairs for hexopyranosides have been measured and tested against 42 cases of bis-, tris- and

therefore almost parallel to the C–O bond (**Figure 5**). Moreover, its strong fluorescence not only facilitates microscale manipulations but also should make it attractive for application for fluorescence CD (FDCD). The terminal acyclic 1,2-diol is a

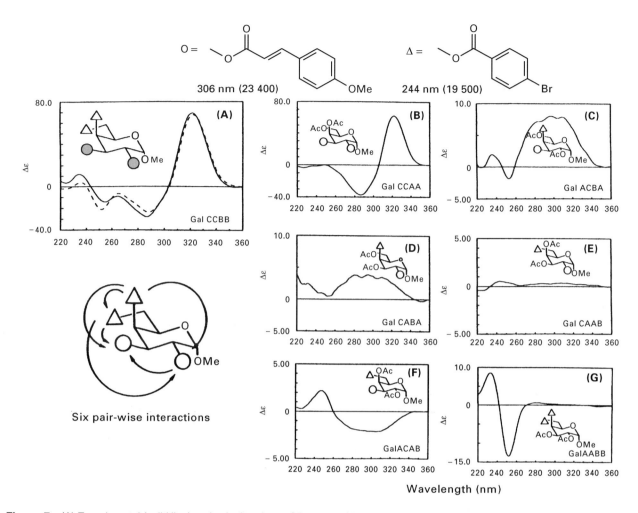

Figure 7 (A) Experimental (solid line) and calculated sum CD spectra (dashed line) of α-methyl galactyopyranoside 2,3-bis-*p*-methoxycinnamate 4,6-bis-*p*-bormobenzoate (GalCCBB). (B)–(G) six experimental pairwise exciton-coupled CDs of four interacting chromophores in GaLCCBB.

tetrakis-acylates and in all cases excellent agreement is seen. This additivity relation is also corroborated by theoretical calculations. The *A* values for *O*-acylates and *N*-acylates are identical, and hence they are applicable to all hexopyranosides (but not to *N*-acetyl sugars since the confirmation of the acylated *N*-acetyl bond is not known). The additivity principle also leads to a simple microscale method for determining the branching point in oligosaccharides, namely the sugars are permethylated, hydrolysed and the liberated hydroxyls which are tags of the linkage point are bromobenzoylated; each sugar benzoate is then identified from the *A* value (microgram level).

The pairwise additivity relation also holds for *p*-phenylbenzylates [11], thus showing that the conformation of the benzylate closely resembles that of the benzoate; the benzylates can readily be oxidized to the corresponding *p*-phenylbenzoates [5], which give rise to another set of data. The advantage of benzylates is that they can derivatize hindered OH groups, including tertiary.

Pairwise additivity in systems containing two different chromophores Two different chromophores can still exhibit exciton coupling when the absorption maxima differ by ~100 nm. Furthermore, the pairwise additivity relation has been shown to hold for systems containing two different chromophores, as exemplified in **Figure 7** for the α-methyl galactopyranoside Gal CCBB substituted with *p*-bromobenzoate (λ_{max} = 245 nm) at C-2,3 and *p*-methoxycinnamate (λ_{max} = 311 nm) at C-4,6. The four-symbol descriptor designates the substituents at positions 2, 3, 4 and 6, their sequence corresponding to the above order of locants: A = *O*-acetyl, C = *O*-(4-methoxycinnamoyl) and B = *O*(4-bromobenzoyl). The CD of the six galactosides containing the interacting basis sets are also shown in **Figure 7**. As shown in **Figure 7A**, there is excellent overlap between the experimental and calculated curves. This agreement holds for all permutations of the galactose, glucose, mannose series, desoxy sugars and amino sugars with primary amino groups, a

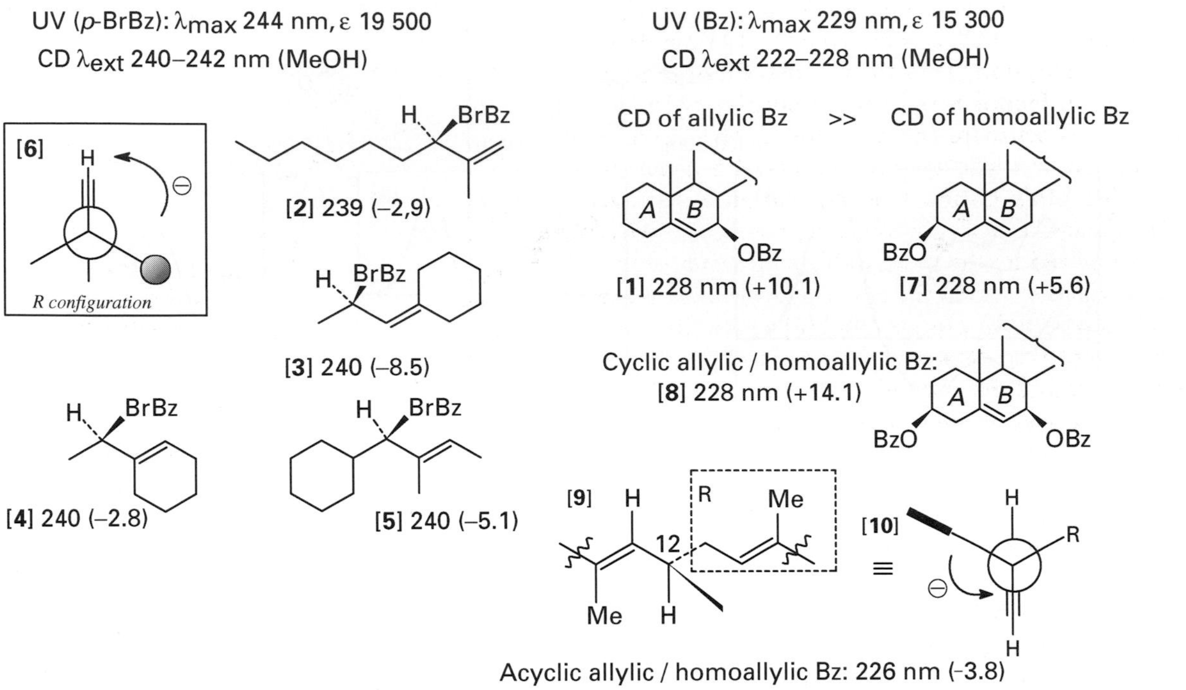

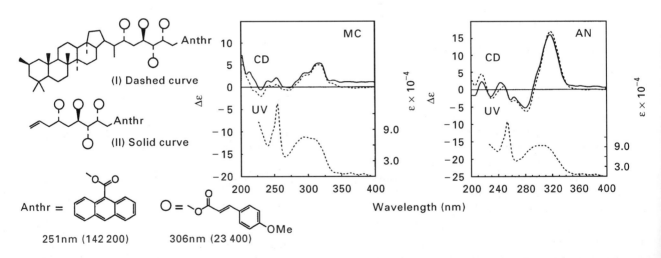

UV (p-BrBz): λmax 244 nm, ε 19 500
CD λext 240–242 nm (MeOH)

UV (Bz): λmax 229 nm, ε 15 300
CD λext 222–228 nm (MeOH)

[6] *R configuration*

[2] 239 (−2,9)

[3] 240 (−8.5)

[4] 240 (−2.8)

[5] 240 (−5.1)

CD of allylic Bz   >>   CD of homoallylic Bz

[1] 228 nm (+10.1)

[7] 228 nm (+5.6)

Cyclic allylic / homoallylic Bz:
[8] 228 nm (+14.1)

[9]

[10]

Acyclic allylic / homoallylic Bz: 226 nm (−3.8)

Figure 8 Absolute configurations of allylic and homoallylic alcohols, cyclic and acyclic.

(I) Dashed curve

(II) Solid curve

Anthr =   251nm (142 200)

O =   306nm (23 400)   OMe

Figure 9 Reference CD spectra for an acyclic 1,3,-polyol and a bacteriohopanepentol in methylcyclohexane (MC) and acetonitrile (AN).

total of 150 cases. These 150 characteristic curves could provide valuable data for the development of an alternative means to conventional methylation analysis of oligosaccharides by tagging free hydroxyls as bromobenzoates or phenylbenzylates, and then derivatizing the hydroxyl groups liberated upon methanolysis. This approach requires no reference sample and, moreover, would also establish the absolute stereochemistry of the individual sugar moieties. As the additivity relation holds for other coupled systems involving various multichromophoric systems, including cage structures, its proper application greatly expands the utility of exciton-coupled CD.

Allylic and homoallylic alcohols, cyclic and acyclic

As chromophores with a difference in λmax of ~100 nm can still couple, this means that the observation of only one branch of the couplet suffices to determine the absolute configuration. Such a case is the allylic benzoates, where the sign of the first CE ascribable to the benzoate, λmax = 229 nm [1] represents the chirality between the benzoate chromophore and the double bond; it is not necessary to observe the CE arising from the 195 nm double bond absorption. Thus the positive CE of steroid [1] shown in **Figure 8** reflects the positive twist between the 7β-benzoate and the 5,6-ene. The CD of a series of acyclic allylic alcohols with established configurations, [2]–[5], show that the relation can be extended to acyclic cases. In this case, the signs of CEs are rationalized by the most stable conformer of the allylic acylates in which the allylic hydrogen adopts a *syn* relation with the double bond [6] so that the sign of the 'couplet' is determined by the twist between the benzoate and the double bond, as in cyclic allylic benzoates. Unsubstituted benzoates are preferred in these studies since the maxima are closest to the ene maxima.

The CE of homoallylic benzoate [7] is also in accord with the chirality, although its intensity is lower than that of allylic benzoate [1]. The additivity again holds, as seen for steroid derivative [8]. Application of this additivity between allylic and homoallylic benzoates was used to determine the configuration of moiety [9] in an acyclic natural product.

Acyclic 1,2-, 1,3- and mixed 1,2-/1,3-acyclic polyols and aminopolyols

Acyclic 1,2-, 1,3- and mixed 1,2-/1,3-polyols and aminopolyols are frequently present in natural products, including polyene macrolides. Because of the lack of general methods for determining their relative and absolute configurations, the stereostructures of most of them remain undetermined. However, a library of such polyols up to pentaols is now available for all permutations of stereorelations; in this library, the primary hydroxyl or amino groups are derivatized by the anthroyl chromophore [10] while the other hydroxyl or amino groups are derivatized by [6]; as in other cases, the CD properties of O and N derivatives are very similar so that the same set of data can be used. **Figure 9** shows that the CD of the mixed polyol moiety present in the hopanoid [I] is superimposable on that of the reference mixed polyol derivative [II] in two solvents, the nonpolar methylcyclohexane (MC) and more polar acetonitrile (AN); this determines the absolute configuration of the side-chain on a microgram scale.

Exciton-coupled CD due to pre-existing chromophores

The well-known bitter principle of the wood of *Quassia amara* L., quassin, has two enone chromophores within its structure shown in **Figure 10**. Accordingly, its CD exhibits a typical couplet at 266 nm (Δε = +10.4) / 242 nm (Δε = −9.5) arising

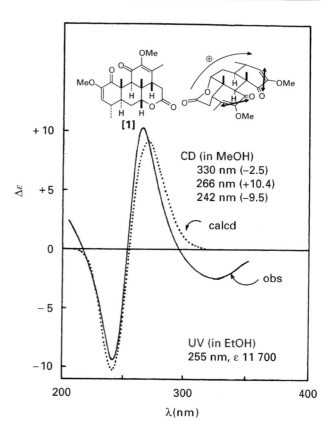

Figure 10 Absolute configuration of quassin.

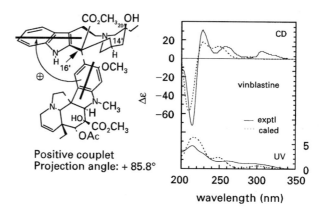

Figure 11 Experimental (in MeCN) and calculated CD spectra of vinblastine.

from the twist between the two chromophores, thus directly establishing its absolute configuration as depicted. The excellent agreement between the CD calculated from the UV and geometry of the molecule and observed CD not only establishes the absolute configuration of quassin but also proves that the observed CD of quassin around 250 nm is due to exciton coupling. This conclusion is consistent with the absolute configurational assignment of quassin

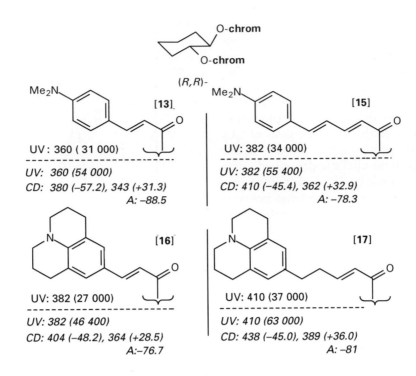

Figure 12 UV-visible spectra of chromophores [13], [15], [16], [17] and CD spectra of diesters formed from 1(*R*), 2(*R*)-cyclohexanediol, in MeCN.

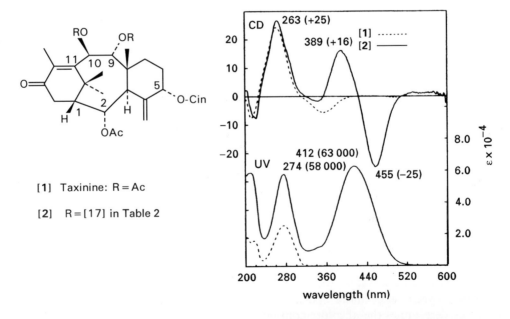

Figure 13 UV-visible and CD spectra of taxinine [1] and its bis-ester formed with chromophore [17] in MeCN.

based on biosynthetic considerations and X-ray analysis of biogenetically related compounds. The negative CE at 330 nm is due to the $\pi \rightarrow \pi^*$ transition of the enone chromophores. It should be noted that despite the negative 330 nm CE, the optical rotation at the sodium D-line (589 nm) is positive: $[\alpha]_D + 34.5°$ (*c* 5.09, CHCl$_3$). Therefore, the D-line rotation, which is simply the reading at 589 nm tak-

en in the optical rotatory dispersion (ORD) mode, is governed by the intense positive Cotton effect at 266 nm. The CD of abscisic acid described in **Figure 4** is very similar to that of quassin.

The clinically important anticancer drug vinblastine also incorporates two chromophores, the indole and indoline moieties, which are coupled through space. The biologically active molecule is

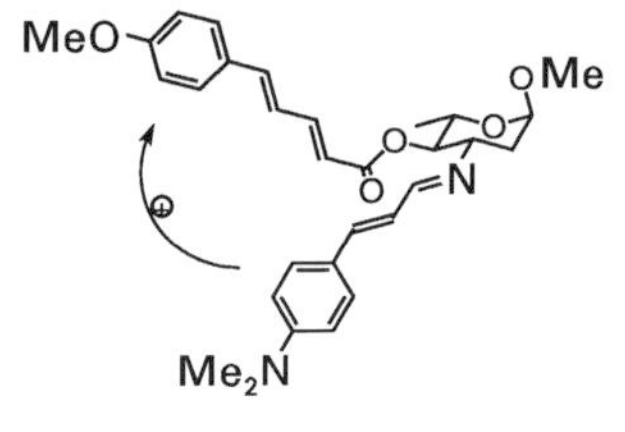

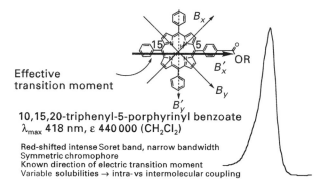

Acosamine derivative

CD neutral: 370 (+62) / 330 (−48), A +110
CD Protonated: 469 (+35) / 350 (−25), A +60

Figure 14 CD of acosamine derivatized with chromophores [14] and [20]/[21], in MeCN.

Effective transition moment

10,15,20-triphenyl-5-porphyrinyl benzoate
λ_{max} 418 nm, ε 440 000 (CH$_2$Cl$_2$)

Red-shifted intense Soret band, narrow bandwidth
Symmetric chromophore
Known direction of electric transition moment
Variable solubilities → intra- vs intermolecular coupling

Figure 15 Tetraarylporphyrins.

that depicted in **Figure 11**, while its biologically inactive diastereomer at C-16′ has a different CD. The UV spectrum has an intense band at 214 nm (ε = 46 200) with a shoulder at 228 nm which arises from the ^1B_a transitions of the indoline (217 nm) and indole (225 nm) chromophores; this is accompanied by the intense positive CD couplet in this region. The weaker 260 nm (ε = 15 000) band is due to the ^1L_a transition of the indoline moiety, while the overlapping bands at 293 nm (ε = 10 500) are from the ^1L_a and ^1L_b transitions of the indole chromophore and the ^1L_b transition of the indoline chromophore. Theoretical calculations of the UV and CD of the strong bands centred around 220 nm using atomic coordinates based on MM2 and SCF-CI-DV MO are in excellent agreement with the observed CD couplet, thus proving that the couplet, which is used empirically to differentiate the bioactive isomer from the inactive isomer, is indeed due to exciton coupling between the two moieties.

Red-shifted chromophores for hydroxyl groups

Since the natural substrate itself frequently contains a chromophore or may be contaminated with some UV-absorbing substance, the use of chromophores with absorbance shifted to the red is advantageous in

order to avoid complications due to unnecessary interactions. Chromophores [13]–[17] and [18]–[21] (**Table 2**) with red-shifted intense maxima can be used for derivatizing hydroxyl and amino groups, respectively. **Figure 12** lists the UV–visible and CD data for derivatives of 1(R),2(R)-cyclohexanediol; in all cases the A values are intense and the spectra appear in regions that usually are transparent.

Figure 13 demonstrates the advantage of using such chromophores. The yew tree constituent taxinine shares the same skeleton as taxol and taxotere, prominent anticancer compounds. The highly strained enone group of the skeleton shows a strong CE at 262 nm arising from a π→π* transition and a weaker CE at 354 nm from the π→π* transition. The 262 nm band overlaps with conventional chromophores. However, [17] with its maximum at 410 nm is completely removed from the enone absorptions. Hence the C-9 and C-10 bis-esters formed from this chromophore show typical exciton couplet in a region remote from enone absorption. The negative couplet agrees with the absolute configuration of the taxane skeleton deduced by other means, including exciton-coupled CD of a derivative using a more conventional chromophore, the interpretation of which was not as clearcut.

Red-shifted chromophores for amino groups

Chromophores [18] and [19] (**Table 2**) are simply imines or Schiff bases formed in high yield by reacting primary amines with p-dimethylaminobenzaldehyde and p-dimethylaminocinnamaldehyde under mild conditions and in the presence of unprotected hydroxyl groups. Addition of a drop of trifluoroacetic acid to the UV or CD cell converts them into protonated Schiff bases, which exhibit intense and very sharp absorptions in the red due to their cyanine dye structure; moreover, addition of a small amount of water will hydrolyse the protonated Schiff base bond, leading to recovery of the starting amine. The derivative of acosamine is prepared by derivatizing the amino group and then the hydroxyl group with the respective chromophores [20] and [14] (**Figure 14**). The exciton couplet is intense with an amplitude of 110. Acidification shifts the couplet to longer wavelength because of the shift of the imine band from 360 to 460 nm; however, the intensity decreases owing to the diminished overlap between the chromophores, from 330 / 360 nm in the neutral form to 330 / 460 nm in the protonated derivative. If the coupling is between two protonated Schiff base moieties, i.e. between two cyanine dyes, then the couplet becomes well-separated and is greatly intensified (see **Figure 3**).

Table 2 Absorption maxima and intensities of some chromophores used for derivatizations of hydroxyl and amino groups

Chromophore	Type		Chromophore	Type

Primary and secondary OH

[1] X = H — 229 nm, 15 300 (EtOH)

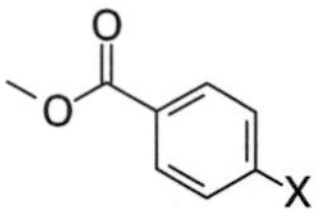

[14] X = OMe — 333 nm, 40 400 (MeOH)

[2] X = Br — 244 nm, 19 500 (EtOH)

[15] X = NMe₂ — 382 nm, 34 000 (MeCN)

[3] X = NMe₂ — 307 nm, 28 200 (MeCN)

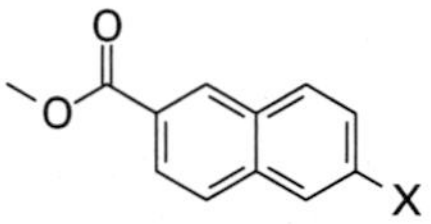

[16] 382 nm, 27 000 (MeCN)

[4] X = OMe — 257 nm, 20 400 (EtOH)

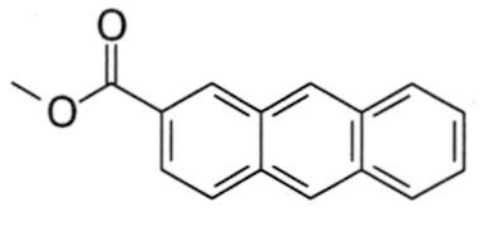

[5] 270 nm, 20 700 (CH₃CN)

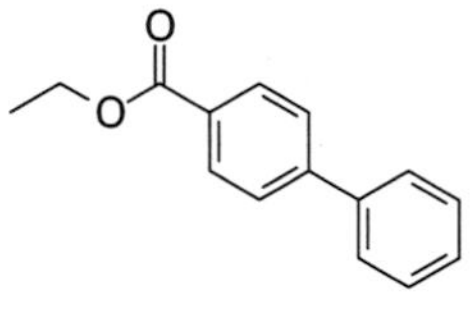

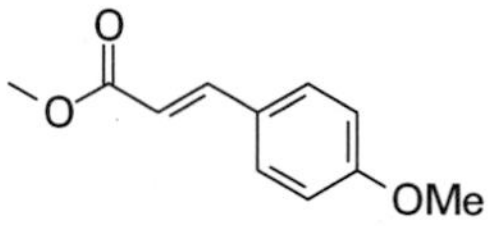

[17] 410 nm, 37 000 (MeCN)

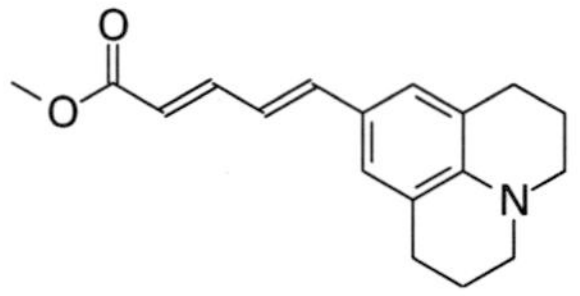

Red-shifted, selective for primary NH₂ (neutral and protonated)

[6] X = H — 234 nm, 58 000 (CH₃CN)

[18] TFA — 305 nm, 24 300 (MeCN)

[7] X = OMe — 237 nm, 48 000 (CH₃CN)

[19] 395 nm, 51 700 (MeCN)

[8] 258 nm, 93 000 (CH₃CN)

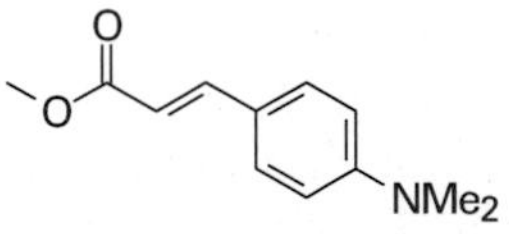

[20] TFA — 361 nm, 37 000 (MeCN)

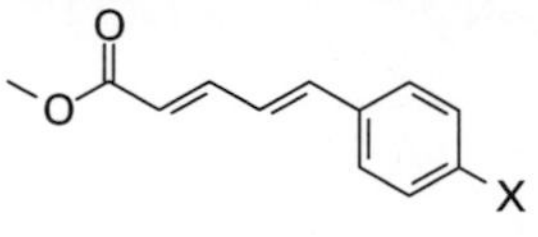

[9] 306 nm, 23 400 (MeCN)

[21] 460 nm, 64 500 (MeCN)

Selective for primary OH

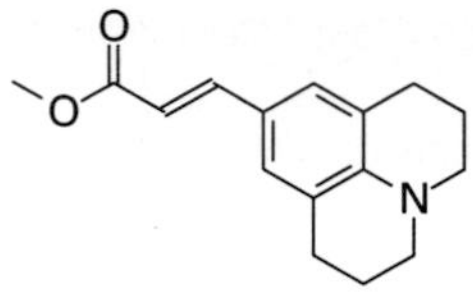

[10] 251 nm, 142 200 (CH₃CN)

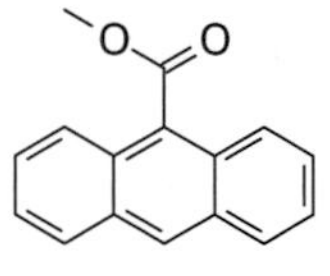

[22] *Porphyrins for OH* — 418 nm, 440,000 (CH₂Cl₂)

OH including tertiary and hindered OH

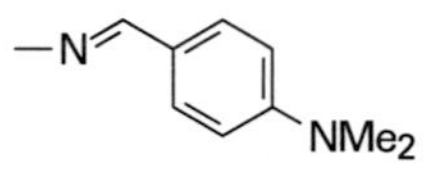

[11] 253 nm, 20 300 (CH₃CN)

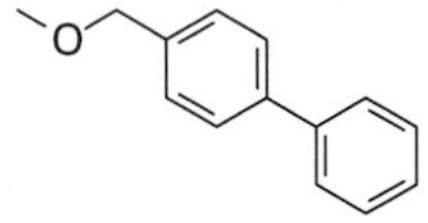

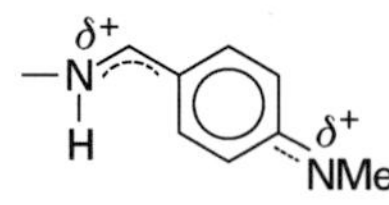

Selective for primary NH₂

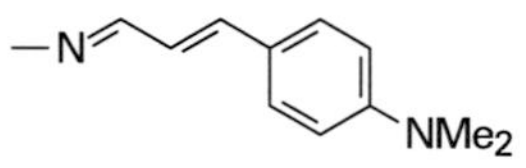

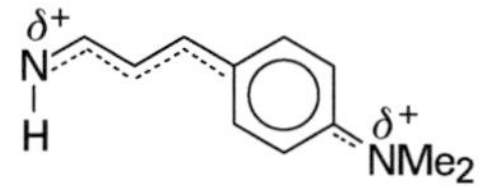

[12] 258 nm, 64 000 (CH₃CN)

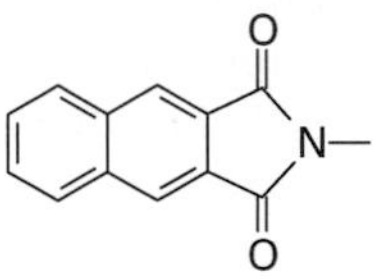

[23] 419 nm, 550 000 (CH₂Cl₂)

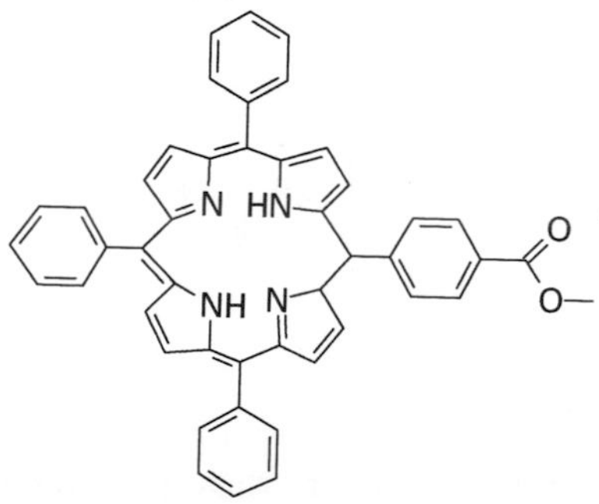

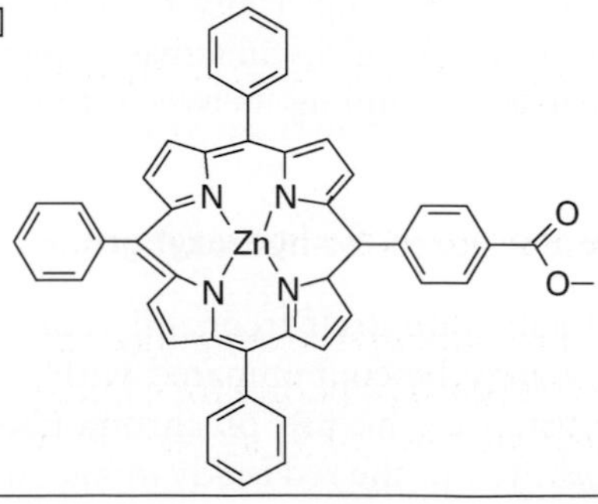

Red-shifted for OH

[13] 360 nm, 31 000 (MeCN)

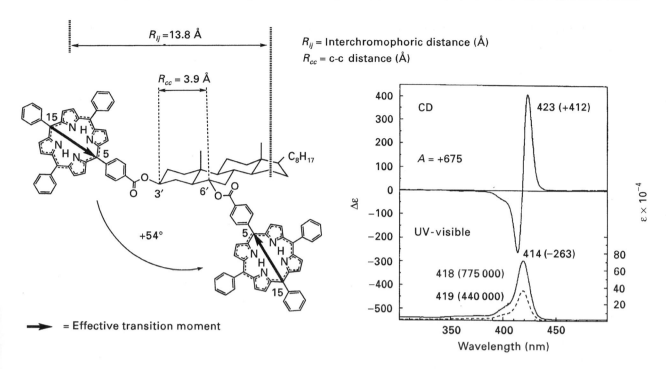

Figure 16 Porphyrin/porphyrin exciton coupling and UV-visible and CD spectra in CH_2Cl_2 ($c = 1.0$ µM) of 5α-androstane -3α,17β-diol bis[p-(10',5',20'-triphenyl-5'-porphyrinyl)benzoate] (solid line) and UV-visible spectrum of 5-(4'-carboxyphenyl)-10,15,20-triphenylporphyn in CH_2Cl_2 (dashed line).

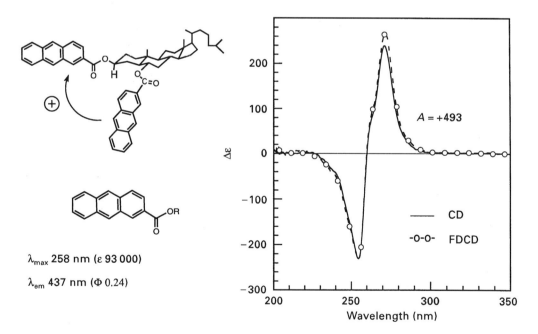

Figure 17 Comparison of conventional CD (solid line) with the CD converted from FDCD (dashed line) for 5α-cholestane-3β,6α-diol bis(2-anthroate) in MeCN.

Porphyrins, a very promising chromophore

The porphyrin chromophores exemplified by [22] and [23] are powerful chromophores for exciton-coupled CD because of the very intense ε of the red-shifted and sharp Soret band at 418 nm (**Figure 15**). The effective direction of the electric transition moment can be regarded as running parallel to the C–O bond of the derivatized alcohol as in any other *para*-substituted benzoates. The solubilities can be modified over a wide range through the appended aromatic rings at C-10, -15 and -20; the triphenyl derivative depicted is hydrophobic whereas a trimethylpyridinium derivative is hydrophilic. Thus the pairing of a hydrophobic or hydrophilic porphyrin derivative with the polarity of the solvent will lead either to intramolecular coupling or to intermolecular stacking. This results in great versatility in the use of porphyrins in exciton-coupled CD. Attachment of trimethylpyridinium porphyrin to a brevetoxin B derivative, a molecule with a rigid skeletal structure, shows intramolecular exciton coupling even over a distance of ~40–50 Å (in methanol). The very efficient intramolecular coupling arising from the intense ε is illustrated in **Figure 16**. The amplitude of the positive couplet of the 3,6-bis-substituted steroid is +675 at an interchromophoric distance of 13.8 Å; the amplitude of the 3,6-bis-p-dimethylaminobenzoate [3], in contrast, is +89. Zinc porphyrins (e.g. [23]) are even more promising because of their more intense absorptions; furthermore, the ability of the central metal atom to coordinate with amino groups adds further versatility to their use in exciton-coupled CD.

Exciton-coupled fluorescence CD

This field is still fairly new. However, exciton-coupled CD detected either by absorption (conventional CD) or by emission both give the same CD curve, as shown in **Figure 17**. The greatest advantage of fluorescence-based CD exciton coupling is that the sensitivity is increased 50–100-fold under favourable conditions involving two identical fluorophores, and that it selectively records the coupling between fluorophores and is not perturbed by nonfluorescent chromophores. However, this new field requires further studies, from the viewpoint of both theory and application.

Quantum mechanical theory of exciton coupled CD

CD spectra and rotational strength of Cotton effect

The rotational strength R, a theoretical parameter representing the sign and intensity of a CD Cotton effect, is expressed in Equation [1] and is obtained from the observed CD spectra.

$$R = 2.296 \times 10^{-39} \int \Delta\varepsilon(\sigma)/\sigma \mathrm{d}\sigma \quad \text{(cgs units)} \quad [1]$$

where σ is wavenumber.

The rotational strength R is also formulated by the Rosenfeld equation:

$$R = \mathrm{Im}\{\langle o|\mu|a\rangle \cdot \langle a|\boldsymbol{M}|o\rangle\} \quad [2]$$

where Im denotes the imaginary part of the terms in brackets, $\langle\ \rangle$ denotes the integration over configuration space and μ and $\boldsymbol{M}$ are operators of electric and magnetic moment vectors, respectively. The dot $\cdot$ stands for scalar product of two vectors and o and a are wavefunctions of ground and excited states, respectively. Thus the rotational strength R is equal to the imaginary part of the scalar product of electric and magnetic transition moments.

Provided a CD Cotton effect is approximated by the Gaussian distribution, Equation [3] is obtained:

$$\Delta\varepsilon(\sigma) = \Delta\varepsilon_{\max} \exp\{-[(\sigma - \sigma_o)/\Delta\sigma]^2\} \quad [3]$$

where $\Delta\varepsilon_{\max}$ is the maximum intensity of the Cotton effect, σ_o is the central wavenumber of the Cotton effect, and $\Delta\sigma$ is the standard deviation of the Gaussian distribution. From Equations [1] and [3]:

$$R = 2.296 \times 10^{-39} \sqrt{\pi} \Delta\varepsilon_{\max} \Delta\sigma/\sigma_o \quad [4]$$

From Equations [3] and [4], the calculated CD curve is formulated as

$$\Delta\varepsilon(\sigma) = [\sigma_o/(2.296 \times 10^{-39}\sqrt{\pi}\Delta\sigma)]$$
$$\times R \exp\{-[(\sigma - \sigma_o)/\Delta\sigma]^2\} \quad [5]$$

where $\Delta\sigma$ can be evaluated from observed UV-visible spectra. Provided the rotational strength R is calculated by Equation [2], the CD spectrum can be reproduced by theoretical calculation.

UV-visible spectra and dipole strength

The dipole strength D representing the intensity of UV-visible bands is estimated from the observed spectra:

$$D = 9.184 \times 10^{-39} \int \varepsilon(\sigma)/\sigma \, \mathrm{d}\sigma \quad [6]$$

The dipole strength is formulated as

$$D = \{\langle o|\mu|a\rangle\}^2 \quad [7]$$

Hence, the dipole strength D is equal to the square of the electric transition moment μ.

Molecular exciton theory of a binary system with two chromophores.

In the exciton coupling system composed of two identical chromophores i and j, exciton wavefunctions are expressed as Equations [8] and [9], where each chromophore undergoes excitation $o \to a$

$$\text{ground state :} \quad \phi_{io}, \quad \phi_{jo} \quad [8]$$

$$\text{excited state :} \quad \phi_{ia}, \quad \phi_{ja} \quad [9]$$

The Hamiltonian operator of the coupling system is formulated as:

$$H = H_i + J_j + H_{ij} \quad [10]$$

where H_i and H_j are the Hamiltonian of groups i and j, respectively, and H_{ij} is the interaction energy term between two groups i and j.

Wavefunction and energy of ground state

The ground-state wavefunction and energy of a binary system are

$$\Psi_o = \phi_{io} \ \phi_{jo} \quad [11]$$

$$E_o = 0 \quad [12]$$

Wavefunction and energy of singly excited state

The singly excited state of the binary system splits into two energy levels, α and β states. For the α-state,

$$E^{\alpha} = E_a - V_{ij}, \Psi_a{}^{\alpha} = (1/\sqrt{2})\{ \phi_{ia} \ \phi_{jo} - \phi_{io} \ \phi_{ja}\} \quad [13]$$

where V_{ij} is the interaction energy between two groups i and j and is approximated by a point dipole approximation method:

$$V_{ij} = \mu_{ioa}\mu_{joa}R_{ij}^{-3}\{e_i \cdot e_j - 3(e_i \cdot e_{ij})(e_j \cdot e_{ij})\} \quad [14]$$

where μ_{ioa}, μ_{joa} and R_{ij} are absolute values of vectors μ_{ioa}, μ_{joa} and R_{ij}, respectively; e_i, e_j and e_{ij} are unit

vectors of μ_{ioa}, μ_{joa} and R_{ij}, respectively. For μ_{ioa}, μ_{joa} and R_{ij}, see Equations [17] and [29].

For the β-state,

$$E^{\beta} = E_a + V_{ij}, \Psi_a\beta = (1/\sqrt{2})\{ \phi_{ia} \ \phi_{jo} + \phi_{io} \ \phi_{ja}\} \quad [15]$$

These equations indicate that the binary system has two electronic transitions $o\to\alpha$ and $o\to\beta$, in the UV-visible spectrum. If $V_{ij} > 0$, the α-state is lower in energy than the β-state, and therefore the transition $o\to\alpha$ locates at longer wavelengths and the transition $o\to\beta$ at shorter wavelengths.

Electric transition moment of a binary system

The electric dipole moment operator μ of a whole system is defined as

$$\mu = \Sigma\mu_i = \Sigma \Sigma er_{is} \quad [16]$$

where μ_i is the electric dipole moment operator of group i, e is the elementary charge and r_{is} is the distance vector of electron s in group i from the origin. The electric transition moment $\langle a|\mu|o\rangle\alpha$ of the transition $o\to\alpha$ is expressed as

$$\langle o|\mu|a\rangle^{\alpha} = \mu_{oa}{}^{\alpha} = \int \Psi_o\mu\Psi_a{}^{\alpha}\mathrm{d}\tau$$
$$= (1/\sqrt{2})(\mu_{ioa} - \mu_{joa}) \quad [17]$$

where $\mu_{ioa} = \int \phi_{io}\mu_i\phi_{ia}\mathrm{d}\tau_i$ and $\mu_{joa} = \int \phi_{jo}\mu_j\phi_{ja}\mathrm{d}\tau_j$. These are electric transition moments of the transition $o\to a$ in groups i and j, respectively. The electric transition moment of the transition $o\to\beta$ is similarly expressed as

$$\langle o|\mu|a\rangle^{\beta} = \mu_{oa}{}^{\beta} = \int \Psi_o\mu\Psi_a\beta\mathrm{d}\tau$$
$$= (1/\sqrt{2})(\mu_{ioa} + \mu_{joa}) \quad [18]$$

Magnetic transition moment of a binary system

The magnetic moment operator M of a whole system is formulated as

$$M = \Sigma M_i = (e/2mc)\Sigma \ \Sigma r_{is} \times p_{is} \quad [19]$$

where m is the mass of an electron, c is the velocity of light, p_{is} is the linear momentum of electron s in group i, and $\times$ represents the vector product of two vectors.

The magnetic moment operator is further changed as

$$M = (e/2mc)\Sigma\, \boldsymbol{R}_i \times \boldsymbol{p}_i + \Sigma\boldsymbol{m}_i \qquad [20]$$

where $\boldsymbol{R}_i$ is distance vector of group i from the origin and $\boldsymbol{p}_i$ and $\boldsymbol{m}_i$ are linear momentum and internal magnetic moment operators of group i, respectively. The magnetic transition moment $\langle a|M|o\rangle^\alpha$ of the excitation $o\to\alpha$ is calculated as

$$\langle a|\boldsymbol{M}|o\rangle^\alpha = \int \Psi_a{}^\alpha \boldsymbol{M}\Psi_o \mathrm{d}\tau$$
$$= (1/\sqrt{2})\{(e/2mc)\boldsymbol{R}_i \times \boldsymbol{p}_{iao} + \boldsymbol{m}_{iao}$$
$$- (e/2mc)\boldsymbol{R}_j \times \boldsymbol{p}_{jao} - \boldsymbol{m}_{jao}\} \qquad [21]$$

where $\boldsymbol{p}_{iao}$ and $\boldsymbol{m}_{iao}$ are linear momentum and internal magnetic moment of group i, respectively.

For the linear momentum of a group, the next equation is very useful:

$$\boldsymbol{p}_{oa} = -(2\pi \mathrm{i}mc/e)\sigma_o\mu_{oa} \qquad [22]$$

where i is the symbol for 'imaginary', σ_o is excitation energy expressed in wavenumber units and μ_{oa} is electric transition moment of transition $o\to a$. Applications of this equation to group i:

$$\boldsymbol{p}_{ioa} = (2\pi \mathrm{i}mc/e)\sigma_o\mu_{ioa} \qquad [23]$$

Accordingly,

$$\langle a|\boldsymbol{M}|o\rangle^\alpha = (1/\sqrt{2})\{\mathrm{i}\pi\sigma_o\boldsymbol{R}_i \times \mu_{ioa} - \mathrm{i}\pi\sigma_o\boldsymbol{R}_j$$
$$\times \mu_{joa} + \boldsymbol{m}_{iao} - \boldsymbol{m}_{jao}\} \qquad [24]$$

In a similar way, the magnetic transition moment of excitation $o\to\beta$ is calculated as

$$\langle a|\boldsymbol{M}|o\rangle^\beta = (1/\sqrt{2})\{\mathrm{i}\pi\sigma_o\boldsymbol{R}_i \times \mu_{ioa} - \mathrm{i}\pi\sigma_o\boldsymbol{R}_j$$
$$\times \mu_{joa} + \boldsymbol{m}_{iao} - \boldsymbol{m}_{jao}\} \qquad [25]$$

Dipole strength a binary system

The dipole strength D^α for the excitation $o\to\alpha$ is expressed as

$$D^\alpha = (1/2)(\mu_{ioa} - \mu_{joa})^2 \qquad [26]$$

Similarly,

$$D^\beta = (1/2)(\mu_{ioa} + \mu_{joa})^2 \qquad [27]$$

Rotational strength a binary system

The rotational strength R^α of the α–state is calculated from Equations [2], [17] and [24]:

$$R^\alpha = \mathrm{Im}\{\langle o|\mu|a\rangle^\alpha \cdot \langle a|\boldsymbol{M}|o\rangle^\alpha\}$$
$$= (1/2)\mathrm{Im}\{(\mu_{ioa} - \mu_{joa}) \cdot (\boldsymbol{m}_{iao} - \boldsymbol{m}_{jao})\}$$
$$+ (1/2)\pi\sigma_o\boldsymbol{R}_{ij} \cdot (\mu_{ioa} \times \mu_{joa}) \qquad [28]$$

where $\boldsymbol{R}_{ij}$ is the interchromophoric distance vector from group i to group j, and is defined as

$$\boldsymbol{R}_{ij} = \boldsymbol{R}_j - \boldsymbol{R}_i \qquad [29]$$

Similarly,

$$R^\beta = \mathrm{Im}\{\langle o|\mu|a\rangle^\beta \cdot \langle a|\boldsymbol{M}|o\rangle^\beta\}$$
$$= (1/2)\mathrm{Im}\{(\mu_{ioa} + \mu_{joa}) \cdot (\boldsymbol{m}_{iao} + \boldsymbol{m}_{jao})\}$$
$$- (1/2)\pi\sigma_o\boldsymbol{R}_{ij} \cdot (\mu_{ioa} \times \mu_{joa}) \qquad [30]$$

In the case of $\pi\to\pi^*$ transitions of common molecules, internal magnetic transition moments $\boldsymbol{m}_{iao}$ and $\boldsymbol{m}_{jao}$ are negligible. Therefore, rotational strengths are approximated as

$$R^\alpha = +(1/2)\pi\sigma_o\boldsymbol{R}_{ij} \cdot (\mu_{ioa} \times \mu_{joa}) \qquad [31]$$

$$R^\beta = -(1/2)\pi\sigma_o\boldsymbol{R}_{ij} \cdot (\mu_{ioa} \times \mu_{joa}) \qquad [32]$$

These equations indicate that the Cotton effects of α- and β–states have equal intensity but are of opposite signs, and therefore the exciton CD satisfies the sum rule:

$$\Sigma R_\mathrm{A} = R^\alpha + R^\beta = 0 \qquad [33]$$

The rotational strength is proportional to the triple product of interchromophoric distance and electric transition moments of groups i and j. Therefore, provided chromophores exhibiting intense π–π^* transitions are employed, intense exciton CD Cotton effects are observable. The rotational strengths of exciton CD satisfy the origin independence as shown in Equation [32].

CD spectra of *N*-mer and dimer: quantitative definition of CD exciton coupling

The exciton theory is also applied to UV-visible and CD spectra of N-mers having N identical chromophores. When N chromophores undergoing intense π–π^* transitions ($o \rightarrow a$) interact with one another, the excited state splits into N energy levels. The wavenumber σ_k of the kth excitation is expressed as

$$\sigma_k - \sigma_o = 2\Sigma\Sigma\, C_{ik}\, C_{jk}\, V_{ij} \qquad [34]$$

where the coefficients C_{ik} and C_{jk} can be obtained by solving the secular equation of Nth order. The rotational strength of R^k is similarly formulated as

$$R^k = -\pi\sigma_o\Sigma\Sigma\, C_{ik}\, C_{jk}\, \boldsymbol{R}_{ij} \cdot (\mu_{ioa} \times \mu_{joa}) \qquad [35]$$

For the N-mer, the CD curve is formulated as

$$\Delta\varepsilon(\sigma) = \{\sigma_o/(2.296 \times 10^{-39}\sqrt{\pi}\Delta\sigma)\}\Sigma\, R^k$$
$$\times \exp\{-((\sigma - \sigma_k)/\Delta\sigma)^2\} \qquad [36]$$

The Taylor expansion of Equation [36] against $\sigma_k/\Delta\sigma$ around $\sigma_o/\Delta\sigma$ yields the second term of the expansion as

$$\Delta\varepsilon(\sigma) = \{2\sigma_o/(2.296 \times 10^{-39}\sqrt{\pi}\Delta\sigma)\}$$
$$\times \exp\{-((\sigma - \sigma_o)/\Delta\sigma)^2\}\{(\sigma - \sigma_o)/\Delta\sigma\}$$
$$\times \Sigma R^k\{(\sigma_k - \sigma_o)/\Delta\sigma\} \qquad [37]$$

From Equations [34], [35] and [37], the CD equation of the N-mer is obtained:

$$\Delta\varepsilon(\sigma) = \{4\sqrt{\pi}\sigma_o{}^2/(2.296 \times 10^{-39}\Delta\sigma^2)\}$$
$$\times \{(\sigma_o - \sigma)/\Delta\sigma\}\exp\{-((\sigma_o - \sigma)/\Delta\sigma)^2$$
$$\times \Sigma\{\Sigma\Sigma\, C_{ik}\, C_{jk}\, \boldsymbol{R}_{ij} \cdot (\mu_{ioa} \times \mu_{joa})\}$$
$$\times \{\Sigma\Sigma\, C_{ik}C_{jk}V_{ij}\} \qquad [38]$$

In the case of a binary system, since the coefficients for the α-state are $1/\sqrt{2}$ and $-1/\sqrt{2}$, and for the β-state, $1/\sqrt{2}$ and $-1/\sqrt{2}$, Equation [38] is simplified to

$$\Delta\varepsilon(\sigma) = \{2\sqrt{\pi}\sigma_o{}^2/(2.296 \times 10^{-39}\Delta\sigma^2)\}$$
$$\times \{(\sigma_o - \sigma)/\Delta\sigma\}\exp\{-((\sigma_o - \sigma)/\Delta\sigma)^2$$
$$\times \boldsymbol{R}_{ij} \cdot (\mu_{ioa} \times \mu_{joa})V_{ij} \qquad [39]$$

This is the CD equation of a binary system. The next term of Equation [39],

$$\{2\sqrt{\pi}\sigma_o{}^2/(2.296 \times 10^{-39}\Delta\sigma^2)\}$$
$$\times \{(\sigma_o - \sigma)/\Delta\sigma\}\exp\{-((\sigma_o - \sigma)/\Delta\sigma)^2$$

represents an anomalous dispersion curve with positive and negative extrema. The sign and intensity of exciton CD depend on the quadruple term, $\boldsymbol{R}_{ij} \cdot (\mu_{ioa} \times \mu_{joa})\, V_{ij}$. Therefore, the term $\boldsymbol{R}_{ij} \cdot (\mu_{ioa} \times \mu_{joa})\, V_{ij}$ is adopted as the quantitative definition of exciton chirality. This term is changed to

$$\boldsymbol{R}_{ij} \cdot (\mu_{ioa} \times \mu_{joa})V_{ij} = D_{ioa}D_{joa}R_{ij}{}^{-2}\boldsymbol{e}_{ij} \cdot (\boldsymbol{e}_i \times \boldsymbol{e}_j)$$
$$\times \{\boldsymbol{e}_i \cdot \boldsymbol{e}_j - 3(\boldsymbol{e}_i \cdot \boldsymbol{e}_{ij})(\boldsymbol{e}_j \cdot \boldsymbol{e}_{ij})\} \qquad [40]$$

where D_{ioa} and D_{joa} are transition dipole strengths of groups i and j, respectively. This equation indicates that the exciton CD amplitude is inversely proportional to the square of the interchromophoric distance R_{ij}.

List of symbols

A = amplitude of splitting; c = speed of light; D = dipole strength; E_0 = ground-state energy; H = Hamiltonian operator; J = NMR coupling constant; m = mass of an electron; $\boldsymbol{m}$ = internal magnetic transition moment. M = operator of magnetic moment vector; p_{is} = linear momentum of electron s in group i; r_{is} = distance vector of electrons in group i; R = rotational strength; R_{ij} = interchromophoric distance vector; V_{ij} = interaction energy; $2V_{ij}$ = Davydov splitting; ε = molar absorptivity; λ = wavelength; μ = electric transition moment; σ = wavenumber; Ψ_0 = ground-state wavefunction.

See also: **Biomacromolecular Applications of Circular Dichroism and ORD; Circularly Polarized Luminescence and Fluorescence Detected Circular Dichroism; Induced Circular Dichroism; Magnetic Circular Dichroism, Theory; Vibrational CD Spectrometers; Vibrational CD, Applications; Vibrational CD, Theory.**

Further reading

Berova N, Gargiulo D, Derguini F, Nakanishi K and Harada N (1993) Unique UV-Vis absorption and circular dichroic exciton-split spectra of a chiral biscyanine dye: origin and nature. *Journal of the American Chemical Society* 115: 4769–4775.

Davydov AS (1962) *Theory of Molecular Exciton.* Kasha M and Oppenheimer M, Jr (translators). New York: McGraw-Hill.

Harada N, Chen SL and Nakanishi K (1974) Quantitative definition of the exciton chirality and the distant effect in the exciton chirality method. *Journal of the American Chemical Society* 97: 5345–5352.

Harada N and Nakanishi K (1972) The exciton chirality method and its application to configurational and conformational studies of natural products. *Accounts of Chemical Research* 5: 257–263.

Harada N and Nakanishi K (1983) *Circular Dichroic Spectroscopy – Exciton Coupling in Organic Stereochemistry.* Mill Valley, CA: University Science Books.

Heyn MP (1975) Dependence of exciton circular dichroism amplitudes on oscillator strength. *Journal of Physical Chemistry* 79: 2424–2426.

Kasha M, Rawls HR and Ashraf EL-Bayoumi M (1965) The exciton model in molecular spectroscopy. *Pure and Applied Chemistry* 11: 371–392.

Kirkwood JG (1937) On the theory of optical rotatory power. *Chemical Physics* 5: 479–491.

Mason SF (1979) General models of optical activity. In Mason SF (ed) *Optical Activity and Chiral Discrimination,* pp 1–22. Dordrecht: Reidel.

Moffitt W, Fitts RD and Kirkwood JG (1957) *Proceedings of the National Academy of Sciences of United States of America* 43: 723–730.

Nakanishi K and Berova N (1994) The exciton chirality method. In Nakanishi K, Berova N and Woody R (Eds) *Circular Dichroism – Principles and Applications,* pp 361–398. New York: VCH.

¹⁹F NMR, Applications, Solution State

Claudio Pettinari and **Giovanni Rafaiani**,
Università di Camerino, Italy

MAGNETIC RESONANCE
Applications

Fluorine-19 is spin 1/2, 100% abundant, and has a sensitivity relative to the proton of 0.83. It is a very easy nucleus to observe, and even prior to 1970 a great deal of information was obtained. ¹⁹F NMR has found widespread application in chemistry and biology: for example, the high-resolution ¹⁹F NMR study of polymers in solution furnishes relevant chemical information concerning polymer structures; furthermore, fluorine NMR can be useful for studying the binding of fluorinated molecules and macromolecules and membranes.

One reason for this wide application is that the wide range and the high sensitivity of the fluorine-19 chemical shift makes the resonance a very sensitive means of monitoring conformational, electronic and solvent effects and that it is possible to study fluorine bonded to nearly every element in the periodic table.

There is a sizeable literature on fluorine, and in particular several reviews with extensive compilations of ¹⁹F chemical shifts (solution state) and coupling constants, both in organic and in inorganic chemistry.

The nuclear properties of ¹⁹F are summarized in **Table 1**: the gyromagnetic ratio, 0.94 that of ¹H, indicates a basic resonance frequency of 188.1 MHz on a spectrometer which operates at a field strength of 4.7 T with ¹H NMR at 200 MHz.

Experimental aspects

In ¹⁹F NMR experiments, a problem exists in the choice of chemical shift reference since the reactivity of some fluorine compounds has often precluded their use as internal standard. However, the standard

Table 1   NMR properties of fluorine-19

Spin	1/2
NA	100
R^H	0.83
R^C	4.73×10^3
γ (10^7 rad T^{-1} s^{-1})	25.1815
Ξ, MHz	94.094003

NA is the natural abundance (%); R^H and R^C are the receptivity relative to that of ¹H and ¹³C, respectively; γ is the gyromagnetic ratio; Ξ is the resonance frequency for the reference compound $CFCl_3$ in a magnetic field for which $(Me)_4Si$ has a proton resonance of 100 MHz.

reference compound most used is internal $CFCl_3$, a very volatile and unreactive derivative, which unfortunately, like C_6F_6, has a sizeable temperature and solvent dependence: medium effects are, in fact, greater in ¹⁹F NMR spectroscopy than in that of ¹H NMR, the signal of several compounds being shifted by 2–10 ppm relative to the gas phase when the spectra are recorded in solvents such as n-heptane, $CHCl_3$ or CH_3I.

Unlike ¹H NMR, in ¹⁹F NMR experiments it is not very important to consider the bulk susceptibility contribution to the shielding, the relative error being negligible due to the larger shift for ¹⁹F than for ¹H with the same bulk susceptibility contribution. On the other hand, there are great difficulties in application of broad band decoupling owing to the large $^2J(^{19}F,^1H)$ coupling and the proximity of ¹H and ¹⁹F chemical shift resonance frequencies.

¹⁹F chemical information (for example, the magnitude and sign of $^nJ(^{19}F,X)$ coupling constants) can be efficiently obtained from two-dimensional NMR

methods such as ¹H–¹⁹F, ¹³C–¹⁹F, and ¹⁹F–¹⁹F correlation spectroscopy (COSY). These methods are helpful in determining small long-range coupling constants and in spectral assignment in polyfluoro derivatives.

¹⁹F NMR chemical shifts cover such a wide frequency range that problems can occur with low digital resolution. For example, to cover a spectral width of approximately 400 ppm using a spectrometer which operates at 564 MHz for ¹⁹F (600 MHz for ¹H NMR) requires a data set comprising 512 k points to give a reasonable digital resolution. With modern computer systems this is not a problem. Alternatively, the use of digital filtering or selective pulse methods can negate the need for such large data sets.

Chemical shift

Fluorine chemical shifts cover a very large range (more than 1300 ppm, about two orders greater than ¹H chemical shift) when organic, inorganic, and organometallic compounds are examined. **Table 2** exhibits selected ¹⁹F data all referenced to CFCl₃, a negative sign indicating a higher shielding or a shift to lower frequency. In several experiments, other compounds have been used as a chemical shift reference, so that to draw straightforward conclusions it is often necessary to convert the chemical shift values to the δ scale with CFCl₃ as reference, taking into account whether an external or internal standard has been used.

Isotope effects

Secondary isotope effects are relevant for ¹⁹F chemical shifts: the main known isotope effects have been observed with the standard CFCl₃ (**Figure 1**), its signal being split into four components of 21:21:7:1 intensity, corresponding to CF³⁵Cl₃, CF³⁵Cl₂³⁷Cl, CF³⁵Cl³⁷Cl₂ and CF³⁷Cl₃ compounds, due to the isotope effect of the ³⁵Cl and ³⁷Cl.

In F(H₂O)ₓ–HF–HF₂⁻ species, which generally are obtained when KF is dissolved in H₂O, the substitution of H with D produces a significant high-field ¹⁹F chemical shift. For this reason it is possible to use the ¹⁹F chemical shift changes of KF to determine the percentage of D₂O in H₂O/D₂O mixtures.

Solvent and temperature dependence of fluorine chemical shifts

Fluorine chemical shifts are very sensitive to temperature and solvent even in the absence of chemical changes (e.g. electrostatic solute–solvent interactions such as hydrogen bond formation, dissociation and change in the structure of the compounds examined).

Table 2 Selected ¹⁹F chemical shift in ppm relative to CFCl₃

Compound	δ (ppm)	Compound	δ (ppm)
HF	40	C₆H₅F	−116
F–C≡N	−156	C₆F₆	−163
F–C≡C–F	−95	1,2-C₆F₄	−139
F₂C=CF₂	−135	1,3-C₆F₄	−104
CH₂=CHF	−114	1,4-C₆F₄	−113
CH₂=CF₂	−81	F₂	429
cyclo-C₆F₁₂	−133	F₂CS	108
cyclo-C₅F₁₀	−133	FSiH₃	−217
cyclo-C₄F₈	−135	F₃SiH	−109
cyclo-C₃F₆	−151	F₂SiH₂	−151
XeF₂	258	CF₄	−63
XeF₄	438	CH₃F	−272
XeF₆	550	CH₂F₂	−144
F₂O₂	865	CHF₃	−79
ClF	−448	F₂O	249
CF₃C(=O)CF₃	−85	FClO₃	287
CFBr₃	7	PhSO₂F	66
F–K⁺	−125	(CF₃)₃N	−59
(CF₃)₄C	−63	SiFBr₃	−77
F₃CCl	−33	SiF₂Br₂	−95
F₃CBr	−21	SiF₃Br	−125
CF₂Br₂	7	SiF₄	−164
NF₃	145	SeF₆	55
CHF=CHF (E)	−183	BF₃	−131
CHF=CHF (Z)	−165	AsF₅	−66
CFBr=CFBr (E)	−113	(AsF₆)⁻	−70
CFBr=CFBr (Z)	−95	IF₇	170
CFCl=CFCl (E)	−120	CF₃COOH	−78
CFCl=CFCl (Z)	−105	1-Fluoronaphthalene	−123

Moreover, variable temperature ¹⁹F NMR spectra can be employed to distinguish between species undergoing internal molecular dynamics between conformations: for example, at −130°C, the two

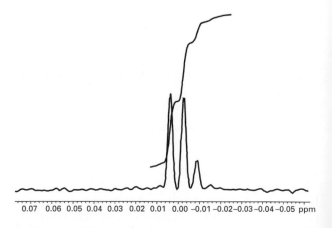

Figure 1   ¹⁹F NMR spectrum of the standard CFCl₃.

rotamers of the $CFBr_2$–CF_2Br molecule exhibit two different ¹⁹F chemical shifts and their ratio can be determined.

As an example of the scale of the temperature dependence of ¹⁹F chemical shifts, **Figure 2** shows the ¹⁹F resonance of 3-trifluoromethylpyrazole at different temperatures.

Effective nuclear charge dependence of fluorine chemical shifts

A good correlation between ¹⁹F chemical shifts and electronegativity has been observed for the shielding in binary fluorides: it was found that for EF_n compounds (E = Se, As, Te, Sb; n = 5 or 6) ¹⁹F resonance frequencies increase with increasing electronegativity. The effect of electronegativity can also be illustrated by the changes of ¹⁹F chemical shifts along the series CH_nF_{4-n} and SiH_nF_{4-n} as the value of n changes from 1 to 3 (**Figure 3**).

An opposite trend (i.e. the shielding increases with increasing electronegativity) has been found in the CF_nCl_{4-n} series: this is probably due to the greater +M (mesomeric) effect of F with respect to Cl.

Analogous simple correlations have also been reported between the ¹⁹F chemical shifts of mixed silicon halides of the type SiF_nY_{4-n} (Y = Cl or Br; n = 1, 2 or 3) and the sum of the electronegativities of the substituents (**Figure 4**).

A similar pattern has been reported for the following pairs of compounds:

$(CF_3)_2O$ $\delta(^{19}F) = -62$ ppm
    $(CF_3)_2S$ $\delta(^{19}F) = -39$ ppm
CF_3CH_2Cl $\delta(^{19}F) = -71$ ppm
    CF_3CH_2Br $\delta(^{19}F) = -68$ ppm
CF_2ClCH_3 $\delta(^{19}F) = -45$ ppm
    CF_2BrCH_3 $\delta(^{19}F) = -37$ ppm

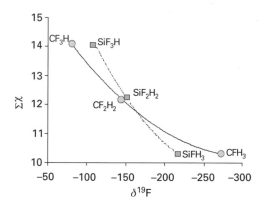

Figure 3   Plot of $\delta(^{19}F)$ against $\Sigma\chi$ (electronegativity of the substituents) for a series of monomeric compounds $CH_{3-n}F_n$ and $SiH_{3-n}F_n$.

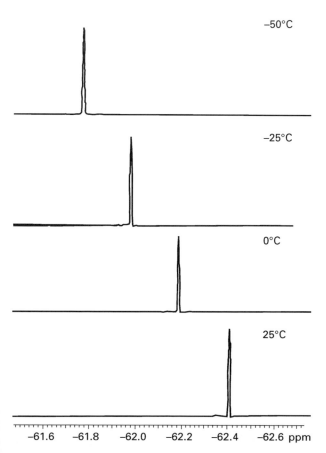

Figure 2   ¹⁹F NMR resonances of the 3-trifluoromethylpyrazole at different temperatures.

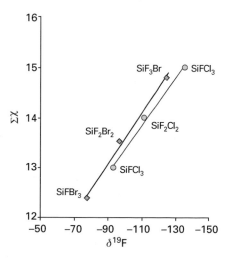

Figure 4   Plot of $\delta(^{19}F)$ against $\Sigma\chi$ (electronegativity of the substituents) for a series of monomeric compounds $SiCl_{3-n}F_n$.

It should always be borne in mind that ¹⁹F shielding is controlled by paramagnetic and anisotropic terms to a greater extent with respect to the -I (inductive) effect of substituents.

The ¹⁹F chemical shift is also dependent on the total charge: ¹⁹F shielding decreases from the neutral to the anionic species; for example, $\sigma(MF_4) > \sigma(MF_6)^{2-}$ (M = Si, Ge, Sn) and also from neutral to cationic species $\sigma(NF_3) > \sigma(NF_4)^+$. In general, the shielding decreases with increasing number of fluorine atoms in the compound.

In poly- and perfluoroalkanes, the following simple relation (with some exceptions) has been found

$$\delta CF_3 > \delta CF_2 > \delta CF$$

which is completely the opposite of that found for ¹H NMR spectra of alkane derivatives.

It is worth noting that the order of ¹⁹F chemical shifts in $FC{\equiv}CCF_3$ corresponds perfectly to the relative electron density of the two groups: the anomaly due to the anisotropy of the triple bond found in the ¹H NMR chemical shift of acetylene is not detected in fluorine resonance.

Substituent and structure dependence of fluorine chemical shifts

The chemical shift of the CF_3 group in a series of 1,1,1,4,4,4-hexafluorobut-2-enes, such as those in Figure 5, clearly indicates that in compounds containing trans-CF_3 groups, $\delta(^{19}F)$ is generally shifted to higher field with respect to isomers containing cis-CF_3 groups.

Geminal, and cis and trans vicinal fluorine in molecules such as $CF_2{=}CFH$ also possess very different chemical shifts, the geminal (−185 ppm) being often the more shielded with respect to cis (−102 ppm) and trans (−129 ppm) vicinal ones.

Aromatic fluorine nuclei are more shielded with respect to perfluoroethylene derivatives, and shielding generally increases with increasing substitution of H with F: for example, $\delta(^{19}F)$ is −116 ppm for C_6H_5F, −139 ppm for 1,2-$C_6H_4F_2$, and −163 ppm for C_6F_6.

The ¹⁹F chemical shift values of fluoroaromatic compounds can be used as a measure of the electronic interaction of ring substituents: for example, it has been shown that the 4-fluorine shift can be related to the π-electron donating or withdrawing properties of the substituent group X in pentafluoroaromatics.

Several studies carried out on monofluorophenylplatinum complexes like that in Figure 6 gave information on the electronic character of the bond between the platinum and the anionic X ligand. By using a simple relationship between the shift of the 4-fluorine atom and the coupling constant between the 2- and 4-fluorine atoms it has been possible to determine empirically the extent of π-interaction in organometallic compounds.

The difference in the chemical shift of the 3- and 5-fluorine atoms and that of 4-fluorine in trans-$[(Et_3P)_2Pt(C_6F_5)X]$ (where X = Me, Cl, Br, I,

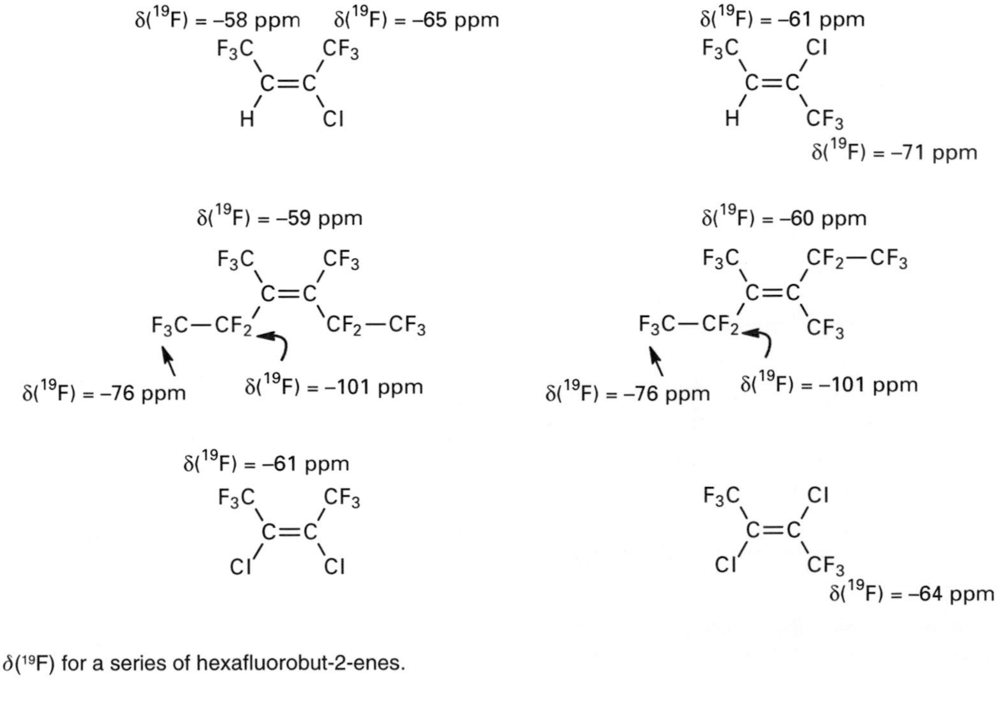

Figure 5 $\delta(^{19}F)$ for a series of hexafluorobut-2-enes.

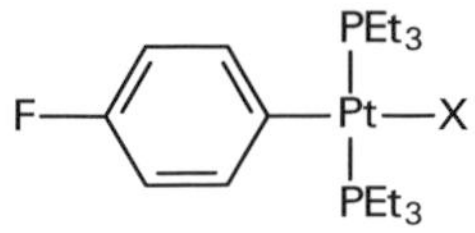

X = Cl, CN, Ph, p–FC₆H₄

Figure 6 Monofluorophenylplatinum(II) complexes.

NO_2, NCS, CN and ONO_2) derivatives has been used as a criterion of π-acceptor interaction. The π-bonding sequence resulted in the following order sequence, in accordance with the *trans* effect:

$$Me < Cl < Br < Br < I < NO_2 < NCS < CN < ONO_2$$

The ¹⁹F chemical shift can be also employed to investigate halide-exchange reactions: from the reaction of Me_3SbF_2 ($\delta = -106.1$ ppm) with Me_3SbCl_2 the mixed halide Me_3SbClF has been obtained ($\delta = -110.9$ ppm). Its ¹⁹F chemical shift value is significantly different from that found for the starting reagent.

Anisotropy of fluorine chemical shifts

In anisotropic media the fluorine chemical shift is strongly dependent on the orientation of the molecule with respect to magnetic field. This fact is very important because many systems widely investigated by ¹⁹F NMR, for example biological matrices, are anisotropic media.

Coupling constants

Coupling constants are very useful for discussing the structure and nature of chemical bonding in compounds containing fluorine. However, to date only a few relative signs of couplings have been reported, and it is not possible in all cases to establish with certainty the effect of the substituents and of the structure on the element–fluorine bond.

Coupling constants involving fluorine atoms are solvent and temperature dependent: for example, for the derivative ClF a difference of ~85 Hz has been found for $^1J(^{35}Cl,^{19}F)$ between the liquid and the gas-phase. The differences are, in general, of the order of 0.1–10%. A dependence on temperature has been observed even when no conformational changes or exchange averaging processes are present. Several investigations, carried out on temperature coefficients of fluorine spin–spin coupling, indicated that these

coefficients can have both signs and can even be as large as 2.3–2.4 Hz per °C.

One-bond couplings

Typical values of one-bond spin–spin coupling constants between ¹⁹F and other nuclei $^1J(X,^{19}F)$ are reported in **Table 3**.

In binary fluorides, it has been shown that the reduced one-bond coupling constant $^1J(X,^{19}F)$ follows a periodic table trend; the value of $J(X,^{19}F)$ generally increases with increasing atomic number of X. The one-bond coupling constants between fluorine and metal atoms vary within a very large range; for example, the coupling between tungsten and fluorine is about 60–70 Hz, whereas that between platinum

Table 3 Magnitudes of some representative one-bond coupling constants, $^1J(X,^{19}F)$

Compound	X	$^1J(X,^{19}F)/Hz$
HF	¹H	530
$Br_2C{=}CFBr$	¹³C	324
$Cl_2C{=}CFCl$	¹³C	303
$Br_2C{=}CF_2$	¹³C	290
$Cl_2C{=}CF_2$	¹³C	289
cis-BrFC=CBrF	¹³C	325
cis-ClFC=CClF	¹³C	299
trans-BrFC=CBrF	¹³C	355
trans-ClFC=CClF	¹³C	290
$CF_2HCF_2SiF_3$	²⁹Si	278
CH_2ClSiF_3	²⁹Si	267
$CHCl_2SiF_3$	²⁹Si	267
CCl_3SiF_3	²⁹Si	264
CH_3SPF_4	³¹P	1032
$C_2H_5SPF_4$	³¹P	1045
$C_6H_5SPF_4$	³¹P	1060
$(CH_3S)CH_3PF_3$	³¹P	925 [$^1J(^{31}P{-}^{19}F_{ax})$]
		1062 [$^1J(^{31}P{-}^{19}F_{eq})$]
$(C_2H_5NH)_2PF_3$	³¹P	694 [$^1J(^{31}P{-}^{19}F_{ax})$]
		875 [$^1J(^{31}P{-}^{19}F_{eq})$]
$(CH_3)_3PF_2$	³¹P	552
$(C_2H_5)_3PF_2$	³¹P	580
$(C_6H_5)_3PF_2$	³¹P	660
$(CH_3)_2P(O)F$	³¹P	980
$(CH_3)_2P(S)F$	³¹P	985
$(CH_3)P(O)F_2$	³¹P	1104
$(CH_3)P(S)F_2$	³¹P	1147
$(C_6H_5)_2PF$	³¹P	905
$(CH_3O)_2PF$	³¹P	1210
WF_6	¹⁸³W	44
WF_5Cl	¹⁸³W	25
trans-WF_4Cl_2	¹⁸³W	20

and fluoride groups ranges from 1100 to 2100 Hz. $^1J(M,^{19}F)$ depends strongly on the structure and co-ordination number of the metal atom M.

$^1J(^{13}C,^{19}F)$, generally negative, has a range centred at about 300 Hz and depends strongly on the nature and electronic properties of substituents. The one-bond coupling constants between phosphorus and fluorine have values between 500 and 1500 Hz. It is interesting to note that the $^1J(^{31}P,^{19}F)$ of RPF₂ groups generally decreases upon coordination to a metal: for example, in CF₃PF₂ the $^1J(^{31}P,^{19}F)$ is 1245 Hz, whereas in the complex (CF₃PF₂)₂Mo(CO)₄ it is 1172 Hz. The $^1J(^{31}P,^{19}F)$ coupling constant can be used to distinguish axial and equatorial fluorinated groups in octahedral derivatives: for example, in the tetrafluoro(pentane-2,4-dionato)phosphorus (V) derivative (**Figure 7**) which contains two sets of magnetically different fluorine atoms, the two coupling constants are 824 (axial) and 741 (equatorial) Hz, respectively.

The $^1J(^{29}Si,^{19}F)$ in silicon fluorides is dependent not only on temperature but also on the time-varying electric field, i.e. dispersion interactions, which alters the net electronic state of derivatives. The values of $^1J(^{29}Si,^{19}F)$ increase with increasing intermolecular dispersion interactions.

$^1J(^{11}B,^{19}F)$ coupling constants are generally very small and dependent on concentration and the nature of the solvent: for example, in acetonitrile the $^1J(^{11}B,^{19}F)$ for the BF₄ ion is 0.39 ± 0.07 Hz, whereas in water it is 1.07 Hz due to ion-pair formation.

Two-bond couplings

Typical values of two-bond spin–spin coupling $^2J(E-X-^{19}F)$ are reported in **Table 4**. *Geminal* coupling constants $^2J(E-X-^{19}F)$ are sensitive not only to the factors reported for the one-bond coupling, i.e. electronegativity, but also to the nature of the intervening atom X, to the F–X–E bond angle, and finally on the stereochemistry of the compound.

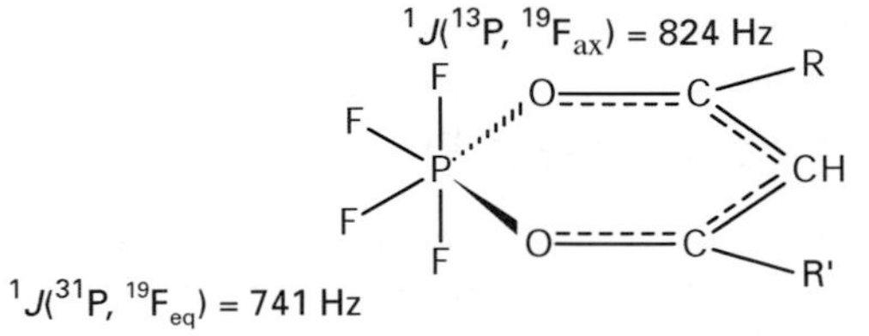

$^1J(^{13}P, ^{19}F_{ax}) = 824$ Hz

$^1J(^{31}P, ^{19}F_{eq}) = 741$ Hz

Figure 7 Fluorine coupling constants in tetrafluoro(pentane-2,4-dionato)phosphorus(V).

Table 4 Magnitudes of some slected geminal coupling constants $^2J(X,^{19}F)$

Compound	X	$^2J(X,^{19}F)$/Hz
Ph₂CF–CH₂F	¹H	48
Ph₂CF–CHF₂	¹H	52
HCF₂–CF₂OF	¹H	56
CH₂F–C(CH₃)OHCO₂Me	¹H	48
CH₃–CHF–CH(OH)–CO₂Et	¹H	47
cis-CHF=CHF	¹H	73
trans-CHF=CHF	¹H	74
CF₃–CH₂–CFHI	¹H	51
CF₃–CFH–CH₂I	¹H	45
CF₃–CH₂–CFHBr	¹H	50
CH₂F–O–CH₂F	¹H	54
CH₂F–CHF–CO₂Et	¹H	47, 48
CH₂F–CH₂–CO₂Et	¹H	55
PhCHF–CCl₃	¹H	42
CHF₂–CF₂–SiMe₃	¹H	54
F₂HPBH₃	¹H	55
(MeS)Ph₂PF₂	¹⁹F	28
WF₅Cl	¹⁹F	73
CF₃PF₂	³¹P	87
CH₂F–CF₂–PCl₂	³¹P	99
CHF₂–CF₂–PCl₂	³¹P	81
CHF₂–CHF–PCl₂	³¹P	83

$^2J(^1H-X-^{19}F)$ and $^2J(^{13}C-X-^{19}F)$

Most of the data reported are for X = C. The $^2J(^1H-X-^{19}F)$ are generally in the range 50–80 Hz. It is very interesting to note that no significant difference has been observed between $^2J(^1H-C-^{19}F)$ and $^2J(^1H-N-^{19}F)$.

In methanes, HCFA′ A″ (A′ and A″ = alkyl, aryl, halides, etc.) a simple relationship (Eqn [1]) between $^2J(^1H-C-^{19}F)$ and the electronegativities $E_{A'}$ and $E_{A''}$ of A′ and A″ substituents has been found on the basis of a large set of experimental data.

$$^2J(^1H-C-^{19}F) = 78.76 + 8.45 E_{A'} E_{A''} - 16.73 (E_{A'} + E_{A''}) \quad [1]$$

It has been possible to distinguish *cis*- and *trans*-1,2-difluoroethylene on the basis of their different *geminal* coupling constant values, the $^2J(^1H-C-^{19}F)$ for the *trans* being greater than that for the *cis* isomer.

The $^2J(^{13}C-X-^{19}F)$ are generally in the range 8–15 Hz and positive.

²J(¹⁹F–X–¹⁹F)

Geminal coupling constants $^2J(^{19}F\text{–}X\text{–}^{19}F)$ are very sensitive to hybridization of the intervening atom X and to the electronegativity of its substituents: the isotropic $^2J(^{19}F\text{–}C\text{–}^{19}F)$ can range from < 5 to 110 Hz for sp^2 carbon, and from ~100 to ~350 Hz for saturated carbon. In $X\text{–}CF_2\text{–}CH_2F$ and $X\text{–}CF_2\text{–}CFYZ$ derivatives (X, Y, Z = halide ≠ F), a relationship [2] has been found:

$$|^2J(^{19}F\text{–}X\text{–}^{19}F)| = 135.3 - 5.95\,\Sigma\chi_i \qquad [2]$$

where χ_i = electronegativity.

Also, in olefinic compounds $XYC=CF_2$, the $^2J(^{19}F\text{–}X\text{–}^{19}F)$ coupling constants are dependent on the nature of substituents. The magnitude of $^2J(^{19}F\text{–}X\text{–}^{19}F)$ is ~50 Hz for X and Y = H or C, whereas values of more than 100 Hz have been observed when X and Y are metals such as Pt or Pd.

In compounds of the type $F_2C=CXY$, a linear correlation between F–F *geminal* coupling constants and fluorine chemical shift has been found. It has been ascribed to a change in the density of the nuclear spin information carrying electrons to the intervening carbon atom due to valence bond resonance structures such as $F_2C^+\text{–}CF=CF\text{–}O^-$.

The dependence of F–F *geminal* coupling constants on the F–C–F angle is very different from that of *geminal* H–H couplings: in fact, the latter increase monotonically with increasing H–C–H angle, whereas the former change markedly with F–C–F angles and have a minimum near 110°. This is due to the fact that the Fermi contact term (FC) is sensitive to F–C–F angle, whereas both the spin dipolar (SD) and orbital (OB) terms seem to be insensitive to the change of F–C–F angle.

Vicinal couplings

Typical values of three-bond spin–spin coupling $^3J(E\text{–}X\text{–}X\text{–}^{19}F)$ are reported in **Table 5**. They are dependent on the electronegativity of substituents, on the bond angle $E\text{–}X\text{–}X\text{–}^{19}F$, and also on the dihedral angle $E\text{–}X\text{–}X\text{–}^{19}F$.

From a comparison of proton–fluorine coupling constants and bond angles in a series of cyclic organic molecules, it has been found that $^3J(^1H,^{19}F)$ generally decreases as the H–C–C–F bond angle increases. In addition, the dihedral angle dependence of $^3J(^1H,^{19}F)$ is very similar to the dependence of $^3J(^1H,^1H)$, i.e. $(^1H,^{19}F)$ has a maximum value when the dihedral angle ϕ between H and F is 0° and 180°,

Table 5 Magnitudes of some slected vicinal coupling constants $^3J(X,^{19}F)$

Compound	X	$^3J\,(X,^{19}F)$/Hz
$Ph_2CF\text{–}CH_2F$	^1H	20
	^{19}F	20
$Ph_2CF\text{–}CHF_2$	^1H	12
	^{19}F	12
$(CH_3)_2CF\text{–}CH(OH)CO_2Me$	^1H	14 (CF–CH), 21(CF–CH₃)
$CH_3\text{–}CF_2\text{–}CO_2Et$	^1H	19
$CF_3\text{–}CH_2\text{–}CO_2Et$	^1H	11
cis-CHF=CHF	^1H	4
trans-CHF=CHF	^1H	20
cis-CHF=CFMn(CO)₅	^1H	10
trans-CHF=CFMn(CO)₅	^1H	25
cis-(C₆H₅)CF=CHCH₃	^1H	22
trans-(C₆H₅)CF=CHCH₃	^1H	36
$CF_3\text{–}CFH\text{–}CH_2I$	^{19}F	11
$CF_3\text{–}CFI\text{–}CH_2\text{–}CF_3$	^{19}F	11
$CF_3\text{–}CFBr\text{–}CO_2Et$	^{19}F	9
$CF_3\text{–}CFCl\text{–}CO_2Et$	^{19}F	7
$CF_2=C=CFCl$	^{19}F	30
$ClCF_2\text{–}CF_2\text{–}C(=O)F$	^{19}F	8(CF₂–C(=O)F) 5(ClCF–CF₂)
$CHF_2\text{–}CH_2\text{–}PH_2$	^{31}P	8
$CHF_2\text{–}CF_2\text{–}PCl_2$	^{31}P	27
$CHF_2\text{–}CH_2\text{–}PCl_2$	^{31}P	13
$CHF_2\text{–}CF_2\text{–}PCl_2$	^{31}P	49

and a minimum value when the dihedral angle is 90°. In olefinic compounds (for example the last six derivatives in **Table 5**) it has been found that $^3J(^1H,^{19}F)_{cis}$ is always smaller than $^3J(^1H,^{19}F)_{trans}$.

The angular dependence of *vicinal* F–F couplings on the $^{19}F\text{–}C\text{–}C\text{–}^{19}F$ dihedral angle is very different from that reported for *vicinal* H–H couplings. Some theoretical studies have suggested that in *vicinal* F–F couplings, although the spin dipolar and orbital terms are a simple function of the dihedral angle ϕ, $^{19}F\text{–}C\text{–}C\text{–}^{19}F$, the Fermi contact term is a function which is strongly dependent both on substituents and ϕ. Therefore, for total *vicinal* F–F couplings (to which the Fermi contact term makes the most important contribution), it is very difficult to find clear relationships involving dihedral angles.

Long-range couplings

Long-range coupling constants have been found to be structurally and stereochemically dependent. $^4J(^{19}F,^{19}F)$ coupling constants can be much larger in magnitude than $^3J(^{19}F,^{19}F)$ and the values range from 0 to 200 Hz. The values of $^4J(^{19}F,^{19}F)$ are diagnostic in several saturated cyclic, unsaturated, and aromatic molecules. For example, from the spectra of *cis*- and *trans*-4H-perfluoromethylcyclohexane, both having an equatorial CF_3 group, the following $^4J(^{19}F,^{19}F)$ coupling constant values have been deduced:

$$^4J(^{19}F_{eq},^{19}F_{eq}) \approx 9 \text{ Hz}; \quad ^4J(^{19}F_{eq},^{19}F_{ax}) \approx 1 \text{ Hz};$$

$$^4J(^{19}F_{ax},^{19}F_{ax}) \approx 27 \text{ Hz}$$

$^4J(^{19}F_{ax},^{19}F_{ax})$ values can be very large due to the through-space interactions of sterically close diaxial fluorine atoms. On the other hand, in aromatic fluorobenzene derivatives the $^4J(^{19}F,^{19}F)$ coupling constant provides only limited structural information: the $^3J(^{19}F,^{19}F)$, $^4J(^{19}F,^{19}F)$ and $^5J(^{19}F,^{19}F)$ coupling constants are often very similar and they can range from 18 to 20, 0 to 20 and 5 to 18 Hz, respectively.

A single large five-bond constant value (~75 Hz) has been observed for the conformationally gauche form of bis(ditrifluoromethyl)tetrachlorethane and it has been ascribed to the close proximity of the fluorines on the two trifluoromethyl groups (**Figure 8**).

In other cases, it has been proposed that some unusually large long range couplings with the nuclei not in close spatial proximity are due to the connection of the fluorine by an intervening π network.

The $^nJ(^{19}F,^{19}F)$ coupling constant between the two fluorine atoms in compound I in **Figure 9** is exceptionally large. It is the highest value reported to date and it also exceeds three- and four-bond coupling constants. This has been attributed to the overlapping nonbonding electron distribution for two F nuclei in close spatial proximity which allows correlation not only of electron-pair bond to electron pair, but also of electron spin.

The $^nJ(^{19}F,^{19}F)$ coupling constant determined for other molecules containing the F–C=C–C=C–F moiety, in which coupling occurs through the s

electrons, are nearly two orders of magnitude smaller; for example, $^5J(^{19}F,^{19}F)$ in $CF_2=CH–CH=CF_2$ is 35.7 Hz, and in compounds II and III in **Figure 9** it is 16.5 and 7.0 Hz, respectively.

In fluorofuranoses (for example compound IV in **Figure 9**), the ¹⁹F chemical shift and the coupling constants often provide a means of assigning the stereochemistry: for example, 1α-F resonates to higher field with respect to 1β-F, the *cis* coupling to H2 is much smaller (4–6 Hz) than the corresponding *trans* coupling (20 Hz), and finally the *trans* coupling to H4 (4–8 Hz) is much larger than the *cis* one (1–2 Hz).

¹⁹F relaxation

Very few reports have been published on fluorine relaxation times: in most cases the results reported for several different molecules indicate that the spin rotation mechanism is dominant at higher temperatures whereas the intermolecular dipole–dipole mechanism is not negligible at lower ones. The chemical shift anisotropy contribution can also be an important factor, whereas the intramolecular dipole–dipole mechanism and scalar coupling contribution seem to be negligible in ¹⁹F relaxation. Some fluorine relaxation times (T_1, T_2 and $T_{1\rho}$) have been used for establishing dynamics in a copolymer of tetrafluoroethene and hexafluoropropene.

Applications

The ever-increasing capabilities of NMR instruments and the high NMR sensitivity allow analytical applications of fluorine NMR to small and big molecules and relevant important information to be gained on chemicals, polymers, membranes, and intracellular environments. Such information is not readily available by other methods. For example, ¹⁹F

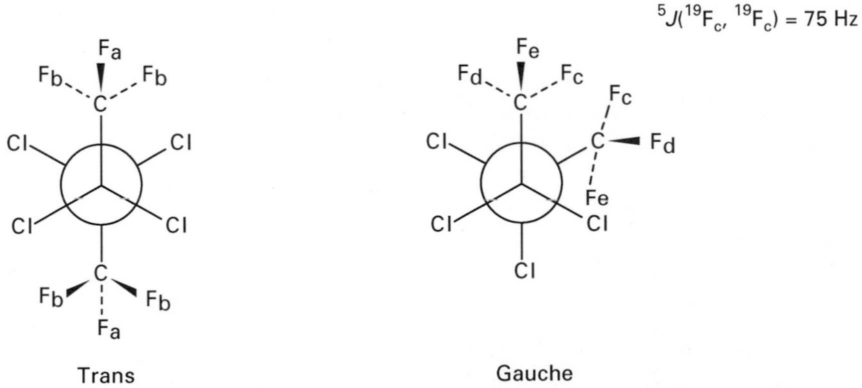

$$^5J(^{19}F_c, ^{19}F_c) = 75 \text{ Hz}$$

Trans   Gauche

Figure 8 The two conformational isomers of bis(ditrifluoromethyl)tetrachloroethane.

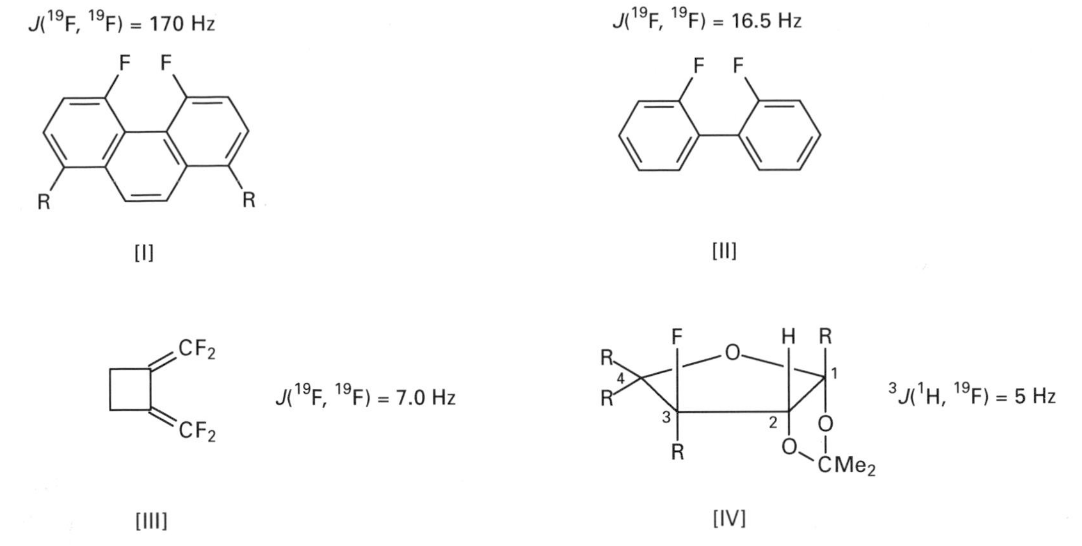

$J(^{19}F, ^{19}F) = 170$ Hz

[I]

$J(^{19}F, ^{19}F) = 16.5$ Hz

[II]

$J(^{19}F, ^{19}F) = 7.0$ Hz

[III]

$^3J(^1H, ^{19}F) = 5$ Hz

[IV]

Figure 9 Long-range coupling constants in different compounds containing fluorine.

spectroscopy has proved to be useful within a large range of applications, from the study of various random and alternating copolymers of fluorosiloxane and hybrid fluorocarbon fluorosiloxane to the study of fluoro complexes of aluminium in aqueous solution and in zeolites. Very recently, ¹⁹F NMR has been applied to study the effect of plasticizers on $LiCF_3SO_3$-containing polymer electrolytes and also to investigate the molecular structure of hydrogen-bonded clusters between F^- and $(HF)_n$.

¹⁹F NMR spectroscopy can be successfully applied to study relevant catalytic reactions: for example, the oxidative addition of $C_6Cl_2F_3I$ to Pd(0) and the *cis*-to-*trans* isomerization (involved in the Stille reaction and other Pd-catalysed syntheses) have been investigated, and it has been shown that the isomerization of *trans*-$[Pd(C_6Cl_2F_3)I(PPh_3)_2]$ reveals a first-order law. A four-pathway mechanism for this isomerization has also been proposed on the basis of ¹⁹F NMR kinetic study.

Several ¹⁹F NMR studies have been reported on ligand exchange reactions of metal complexes (i.e. the $Al(Ttfac)_3$-$Al(bzac)_3$ system) and also on relative polarity and reactivity of N–H, N–Hg and N–Au bonds in intermolecular exchange reactions of fluorine-containing ligands (for example, 2-(4-fluorophenyl)benzimidazole) and their PhHg and Ph_3PAu derivatives.

Another interesting chemical application is the determination of the enantiomeric excesses of chiral acids by ¹⁹F NMR of their esters derived from fluorine compounds.

Molecules containing fluorine that can be transported into cells can be employed as indicators for monitoring intracellular environments. Recently, ¹⁹F

NMR has been used to characterize drug–protein conjugates (immunogens) and to probe drug–protein binding. ¹⁹F NMR spectroscopy has been shown to be a powerful technique for identifying metabolites of fluorine-containing drugs and for evaluating changes in the metabolism of fluoropyrimidines after the use of a biochemical modulator. This technique often allows a correlation between improved therapeutic response and the biochemical effects generated in tissues. For example, in the monitoring of *in vivo* tumour metabolism, after biochemical modulation of 5-fluorouracil by the uridine phosphorylase inhibitor 5-benzylacyclouridine, analysis of the NMR data revealed an increased formation and retention of fluorouracil nucleotides and fluorouridine in colon tumours treated with the regimen containing 5-benzylacyclouridine, and a reduction in 5-fluorouracil catabolites.

See also: **Chemical Exchange Effects in NMR; Macromolecule–Ligand Interactions Studied By NMR; NMR in Anisotropic Systems, Theory; NMR Principles; NMR Relaxation Rates; Nucleic Acids Studied Using NMR; Parameters in NMR Spectroscopy, Theory of; Structural Chemistry using NMR Spectroscopy, Organic molecules.**

Further reading

Dungan CH and Van Wazer JR (1970) *Compilation of Reported 19F Chemical Shifts*. New York: Wiley.

Emsley JW and Phillips L (1971) Fluorine chemical shifts. *Progress in NMR Spectroscopy* 7: 1–523.

Emsley JW, Phillips L and Wray V (1976) Fluorine coupling constants. *Progress in NMR Spectroscopy* 10: 83–756.

Fields R (1972) Fluorine-19 nuclear magnetic resonance spectroscopy. *Annual Reports on NMR Spectroscopy* 5A: 99–304.

Fields R (1977) Fluorine-19 nuclear magnetic resonance spectroscopy. *Annual Reports on NMR Spectroscopy* 7: 1–117.

Jameson CJ (1987) Fluorine. In: Mason J (ed) *Multinuclear NMR*, pp 305–333, New York: Plenum Press.

Mooney EF and Winson PH (1967) Fluorine-19 nuclear magnetic resonance spectroscopy. *Annual Reports on NMR Spectroscopy* 1: 243–311.

Sohár P (1983) *Nuclear Magnetic Resonance Spectroscopy*, vol II, pp 261–268. Boca Raton, Florida: CRC Press, Inc.

Wray V (1980) Fluorine-19 nuclear magnetic resonance spectroscopy (1976–1978). *Annual Reports on NMR Spectroscopy* 10B: 1–507.

Wray V (1983) Fluorine-19 nuclear magnetic resonance spectroscopy (1976–1978). *Annual Reports on NMR Spectroscopy* 14: 1–406.

Far-IR Spectroscopy, Applications

James R Durig, University of Missouri–Kansas City, MO, USA

VIBRATIONAL, ROTATIONAL & RAMAN SPECTROSCOPIES
Applications

Introduction and history

The far-infrared spectral region is the range of wavenumbers where one finds many of the large-amplitude anharmonic vibrations. These include both the symmetric and asymmetric internal torsional modes of many small organic and organometallic molecules, ring puckering vibrations of four- and five-membered rings, and the heavy-atom skeletal bending modes. Additionally, this is the spectral region where one finds lattice vibrations from which information is obtained on intermolecular forces that are of considerable interest to both chemists and the physicists.

Historically, this region has been defined from 200 to 10 cm^{-1}, which is the region below the cutoff of CsI, where it can no longer be used as a dispersive material. Initial investigations in this spectral region began before the turn of the twentieth century in Europe, and it has always been more popular in Western Europe and Japan than in the United States. Even before the use of Fourier transform interferometers became common for this spectral region, there were several laboratories in these other countries where the far-infrared spectral region was studied almost exclusively. The first chemical applications of Fourier transform infrared (FT-IR) spectroscopy were made in the far-infrared spectral region because the interferometric instrumentation required to obtain spectra in this region is much simpler than that needed for the mid- or near-infrared regions. The instrumental advantages in the far-infrared spectral region included a lower tolerance for the mirror drive of a Michelson interferometer, a smaller dynamic range of the interferogram and a longer sampling time, which results in a reduced number of data points.

The far-infrared spectral region initially had its inherent difficulties, such as low-energy sources, poor detectors, lack of suitable optical materials and spectral interference of water vapour. Research workers in this wavenumber area have overcome many of these difficulties and, therefore, spectral investigations in the far-infrared region have become more common. This is particularly true with the commercial availability of Fourier transform spectrometers, which routinely provide spectral data to 30 cm^{-1} with excellent signal-to-noise ratios. However, because of the absorption of water vapour in this spectral region (see **Figure 1**), it is highly desirable to have a vacuum interferometer bench.

Many of the far-infrared spectral studies for chemical information have emphasized the gas phase. This emphasis should not be interpreted as meaning that far-infrared spectral studies of solids are not important. Physicists have put interferometers to use in the study of the solid state to determine optical constants such as the index of refraction, complex indices, phase angle transmission coefficients and the electronic processes in insulation crystals, as well as the intermolecular vibrations of molecular crystals. These studies have been very important in the development of interpretative theories of solids.

With the increased ability to utilize *ab initio* calculations for predicting relative conformational

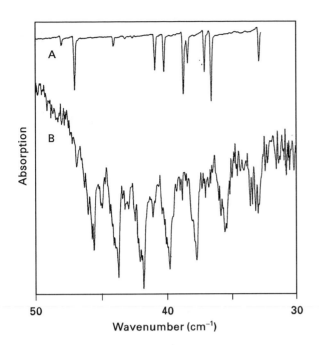

Figure 1 Far-infrared spectra of (A) water vapour and (B) tri-fluoroacetyl fluoride.

stabilities, barriers to internal rotation, small ring inversions and low-wavenumber vibrations, the theoretical predictions are frequently compared with the experimental results in this spectral region. The use of either an incomplete or an incorrect experimental database in comparisons with theoretical results, may lead to erroneous conclusions with regard to the reliability and accuracy of the theoretical results.

Poor experimental databases with regard to the experimental determination of potential barriers and conformational equilibria arise for three basic reasons. First, there is frequently insufficient structural information in many cases. Numerous potential functions governing internal rotation of asymmetric tops (conformational interchange) have been published without the knowledge of the structures of both conformers, particularly of the high-energy conformer. It is extremely important, for the accurate experimental determination of the energy levels and potential functions, that some reasonable structural information be known about all of the exhibited conformations. Furthermore, in the absence of reliable structures determined from experimental techniques, such as those determined from microwave spectroscopy or analysis of the electron diffraction pattern, the reliability of optimized geometries determined from theory cannot be ascertained.

Secondly, there is frequently a lack of complementary experimental data. Owing to symmetry and the resulting selection rules, both infrared and Raman spectroscopic results are often needed to identify the high-energy and low-energy conformers. The Raman spectrum of the gas is almost always neglected, and it is not unusual to find vibrational conformational studies for which spectra are not reported for all three phases.

Thirdly, there is often a lack of high-resolution data. The highest possible resolution available should be used to record the low-frequency spectral data. Many torsional bands are extremely sharp and, therefore, transitions may not be observed at the lower resolution. Additionally, insufficient resolution may result in misleading relative intensities of the Q branches from different energy levels.

Low-wavenumber FT-IR spectroscopy is among the most generally applicable methods used in the study of the conformers for certain types of small molecules with few substituents. Infrared spectra can, and variable-temperature experiments may, be investigated in all phases. The gas-phase band contours observed for the infrared spectra, along with Raman depolarization data, provide considerable information on the molecular symmetry of the conformers. The limitation of the vibrational spectroscopic technique is that it is best applied to relatively simple molecules that contain at least one element of symmetry, and one or perhaps no more than two portions of the molecule capable of producing different conformations upon internal rotation.

One of the earlier uses of the far-infrared spectral region was to determine the barrier to internal rotation of symmetric-top rotors. The most frequently studied molecules were those with one threefold internal rotation, and many of those studies were for methyl tops. The first indication that the rotation around single bonds was not free came from thermodynamic data in the mid-1930s. Barriers were calculated from the thermodynamic data by relating the difference in the observed and statistical entropies by tables involving the barrier height and the reciprocal of the partition function for free rotation. Later, barrier values were obtained from investigations of the microwave spectra of small molecules where the observed perturbations on the pure rotational transitions were correlated with the torsional barrier heights by either the splitting or intensity methods. The splitting method is the most exact method since it depends only on the height of the barrier and usually gives barrier values to a few percent. The microwave intensity method is frequently used for the determination of the fundamental frequencies of asymmetric rotors.

One wishes to obtain the fundamental wavenumber for the torsional mode for the gaseous molecule so that the barrier height in the isolated molecule can be ascertained. However, the dipole moment change associated with the torsional mode may be quite small, and consequently the resulting infrared band intensity may be weak. Therefore, assignments of the torsional modes of molecules in the gas were frequently in error in the initial investigations and the use of isotopic substitution is often necessary to verify the torsional assignments. Also, torsional modes that give rise to B-type infrared band contours may have bands that are very broad and ill-defined because of unresolved excited-state transitions. In favourable cases where the torsional mode gives rise to A-type or C-type band contours with relatively strong Q branches, several excited state transitions may be resolved and not only can the barrier height be obtained but also the detailed shape of the potential well may be ascertained.

When a molecule has a far-infrared spectrum too complicated for quantitative interpretation, estimates of the barrier heights may be obtained by studying the infrared and Raman spectra of the sample in the crystalline state. At liquid-nitrogen temperature, the upper vibrational states are effectively depopulated and only the $1 \leftarrow 0$ torsional transition is observed. The barrier height may be estimated from this single experimental datum, but a detailed analysis of the shape of the potential well is not possible. Additionally, it is sometimes possible to observe the torsional mode in the solid state when it is forbidden for the gaseous molecule. Barriers obtained in the solid state are usually 10–15% higher than those for the gaseous molecules.

Internal rotors

Internal rotors fall into two categories: symmetric and asymmetric tops. For symmetric tops, a rotation about the top-frame bond of $2\pi/n$ (where n is an integer) will bring the top to a position symmetrically equivalent to, or indistinguishable from, the original configuration. Therefore, it is usual to speak of the 'foldness' of the rotor in terms of n. For example, a perfluoromethyl group (CF_3) is a threefold symmetric top (local C_{3v} symmetry), and twofold rotors include $-NO_2$, $-BF_2$ and phenyl groups (local C_{2v} symmetry). When a rotation of 360° (i.e. when $n = 1$) is the only operation that results in a symmetrically equivalent position for the top, the rotor is known as an asymmetric top. For the case of a symmetric frame, the top with the highest degree of symmetry prevails and, when two tops of different foldness are bonded directly to one another, the resultant foldness is the product of the two individual

tops' foldness. For instance, CH_3BH_2 would be classified as a sixfold internal rotor.

The energy minima and maxima for a symmetrical threefold group (CF_3) are 60° apart. The simplest mathematical function that will reproduce such a potential variation upon rotation is a cosine function. If the problem is assumed to be one-dimensional, the quantum-mechanical energy solutions are readily obtainable. The model employed is a rigid symmetric top (CF_3 group) attached to a rigid frame, which may be completely asymmetric. There are four rotational degrees of freedom, three for overall rotation and one for the hindered rotation of the two groups. The axis of internal rotation is usually assumed to coincide with the unique axis of the symmetric top. Since the top has a threefold symmetry axis, the potential energy hindering rotation may be expressed by a Fourier expansion:

$$V(\phi) = \frac{V_3}{2}(1 - \cos 3\phi) + \frac{V_6}{2}(1 - \cos 6\phi) + \cdots$$

where V_3 is the height of the threefold barrier; V_6 the sixfold; and so forth. A positive V_6 makes the minima narrower and the maxima broader, which results in the energy levels near the bottom of the well becoming somewhat more widely separated. A negative V_6 term has the opposite effect. Experimentally it has been found that $0 \leq V_6/V_3 < 0.05$ and $V_6 \gg V_9 \gg$ higher-order terms.

An example of the use of far-infrared spectral data for the determination of the barrier to internal rotation is given by trifluoroacetyl fluoride, CF_3CFO (**Figure 1**). Seven torsional transitions are clearly observed, beginning with the fundamental at 45.65 cm^{-1} and continuing to the $7 \leftarrow 6$ transition at 33.40 cm^{-1}. Utilizing an F number ($h^2/8\pi^2 I_r$ where I_r is the reduced moment of inertia of the top) of 0.5970 cm^{-1} along with the wavenumbers of these seven transitions, the V_3 value is 382 ± 2 cm^{-1} and $V_6 = 8 \pm 1$ cm^{-1}. The intensity of the $4 \leftarrow 3$ transition and higher excited state transitions are significantly higher than expected on the basis of the Boltzmann factors because of the increased anharmonicity in the higher excited states. The statistical uncertainty is very low for both the V_3 and V_6 terms and the V_6 term is very small. This molecule is too heavy for the barrier to be determined by the microwave splitting method. Molecules with two or three equivalent rotors have also been extensively investigated, but the spectral data are much more complex and frequently there is not sufficient data to obtain values for all of the potential constants. Good barrier determination from far-infrared data requires that the internal torsional mode is not mixed (potential energy distribution) with other low-wavenumber bending modes.

The asymmetric rotor is the other type of torsional motion frequently studied. These rotors usually lead to two or more stable conformers. One of the major goals of conformational analysis is the calculation of the energy difference between the two conformers and the energy necessary for interconversion. Four types of information are required to characterize an asymmetric potential function: (1) the approximate dihedral (torsional) angle of each conformer, because the number of torsional energy levels is directly related to the number of potential minima; (2) the approximate relative enthalpy difference between the high- and low-energy conformers, since this is one of the constraints defining the potential functions; (3) the change in molecular kinetic energy as a function of torsional angle; and (4) accurate observation of torsional transition frequencies. A typical spectrum for such a rotor is shown in **Figure 2** where the fundamental torsional mode of the *trans* conformer of FCH_3CFO is observed at 118.9 cm^{-1} with the excited torsional modes falling to lower wavenumbers to the 13 ← 12 transition at 67.0 cm^{-1}. The fundamental torsional mode for the *cis* conformer is assigned at 52.1 cm^{-1} with five successive excited states falling to lower wavenumber with the last one observed at 43.7 cm^{-1}. These data give the following potential constants: $V_1 = 86 \pm 11$, $V_2 = 946 \pm 33$, $V_3 = 407 \pm 4$, $V_4 = 138 \pm 20$, and $V_6 = -14 \pm 7$ cm^{-1} with a *trans* to *cis* barrier of 1297 ± 26 cm^{-1}.

Small ring molecules

Another type of large-amplitude anharmonic vibration is that of the ring puckering modes of small compounds. From the analysis of the frequencies for these ring-bending motions it is possible to obtain the potential surfaces for the interconversion of the different conformers. The equilibrium conformation of four-membered and unsaturated five-membered rings is determined by two major opposing forces. The first is the ring strain, which tends to keep the ring skeleton planar. Puckering the ring decreases the already strained angles, thereby increasing the angle strain. The second is the torsional forces. The torsional repulsions of adjacent groups are at a maximum for a planar ring since the groups are eclipsed; therefore, bending the ring out of the plane will reduce these repulsive forces. It is the delicate balance between these two large forces that determines the ground-state structure of the molecule.

A single substituent on the cyclobutane ring leads to conformational isomers and introduces asymmetry into the potential function for ring inversion. As the ring inverts, the substituent goes from the axial conformation to the equatorial, or vice versa. These two conformations will of course have different energies and the spectra can be interpreted using a potential function of the form

$$V(x) = ax^4 - bx^2 - cx^3$$

where x is the coordinate of the ring inversion. The cubic term is added to the usual quartic-harmonic potential function because of the asymmetry.

The results for chlorocyclobutane will be used as an example of the utility of far-infrared spectra of the gas for the investigation of ring puckering motions. The far-infrared and Raman spectra of gaseous chlorocyclobutane in the region of the ring puckering fundamental are shown in **Figure 3**. From the far-infrared spectrum, this fundamental is observed as well-defined Q branches occurring at 157.55, 149.29, 139.81, 130.12 and 128.08 cm^{-1} with additional, weaker Q branches at 116.46, 115.69, 111.46 and 110.88 cm^{-1}. The first four pronounced Q branches appear to form a reasonable series and are assigned as the first four transitions of the ring puckering mode for the more stable

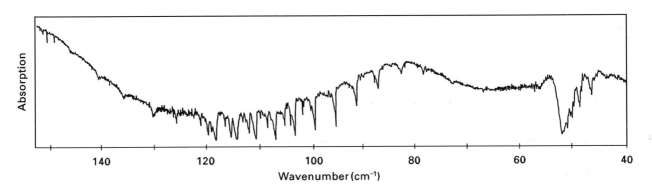

Figure 2 Far-infrared spectrum of fluoroacetyl fluoride, FCH$_2$CFO.

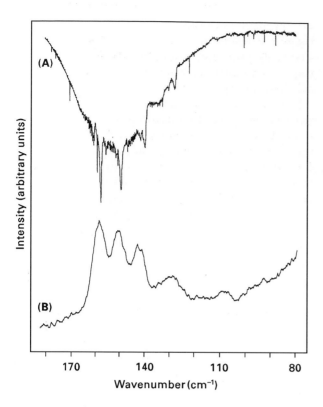

Figure 3 (A) Far-infrared transmission spectrum of chlorocyclopropane; (B) Raman spectrum of gaseous chlorocyclobutane.

equatorial conformer. From the Raman spectrum, the ring puckering fundamental is observed as a weak series of Q branches beginning at ~158 cm^{-1}. Although the fundamental for the axial form can be assigned at 128 cm^{-1} in this Raman spectrum, the 128.08 cm^{-1} Q branch observed in the infrared spectrum is far more definitive.

With these assignments, the potential function governing ring inversion can be calculated with the aforementioned function. The function calculated using this assignment is shown in **Figure 4**. This potential is consistent with an energy difference between the equatorial and axial conformers of 449 cm^{-1} (1.28 kcal mol^{-1}) and a barrier to ring inversion of 827 cm^{-1} (2.36 kcal mol^{-1}). Although the uncertainty in the puckering angles determined at the minima is directly related to the uncertainty in the assumed reduced mass (198 amu), as well as the nature of the molecular motion involved, this potential gives puckering angles of 22° for the equatorial form and 17° for the axial conformer. From the *ab initio* calculations utilizing the 6-31G* basis set, the values for these angles are 25.1° and 20.3°, respectively, indicating that the difference in the two angles should be about 5°, which is the value obtained from the potential function.

In five-membered ring molecules, there are two low-frequency out-of-plane ring motions. These are

usually qualitatively described as the ring-twisting (radial mode) and the ring-puckering (pseudorotational mode) motions. Initially, in order to handle the interpretation of the low-frequency far-infrared data, assumptions were made about the forms of these normal vibrations. If the five-membered ring contains an endocyclic double bond, it has usually been assumed that there is no interaction between the 'high'-frequency ring-twisting mode, which falls around 400 cm^{-1} and the 'low'-frequency ring-puckering mode, which is near 100 cm^{-1}. Therefore, the two out-of-plane ring modes are handled as two one-dimensional problems with the anharmonic, low-frequency ring-puckering transitions interpreted in terms of a one-dimensional potential function of the form $V(x) = ax^4 \pm bx^2$. In recent years it has become increasingly clear that while a one-dimensional model of the ring-puckering motion yields reasonably good barrier height values, it does not allow for the interactions with other vibrational modes, which may often be significant.

Some of the most recent advances in the determination of potential surfaces governing ring inversions in four- and five-membered ring molecules have been in the development of two- and three-dimensional models. Such models then allow for the interaction of the ring puckering mode with other vibrational modes. These interactions have been shown to alter the puckering levels in the excited states of interacting motions and, if neglected, can result in poorly calculated barrier heights and misinterpretation of the far-infrared and Raman spectral features arising from the modes involved.

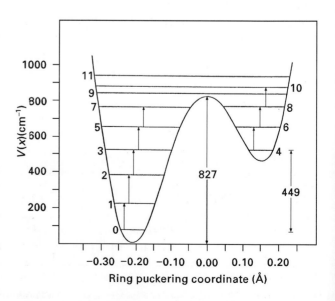

Figure 4 Potential function governing the conformational interchange for axial and equatorial chlorocyclobutane.

Skeletal bending nodes

The third type of low-frequency vibration is heavy-atom skeletal bending modes. Because the vibrations are of very low wavenumbers, there are frequently many excited states populated. Accordingly, the infrared contours in the gas phase can be very complex. In fact, many of these vibrations show a large number of excited states and the band contours are difficult to analyse because of the many vibrational–rotational transitions falling on top of each other. Attention to these types of vibrational modes has been fairly limited.

Gases, liquids and solids

Another type of absorption in the low-wavenumber range is due to the pure rotation of small molecules. The intensity of this type of absorption is frequently 10^4 times that of a low-wavenumber vibration, so very small amounts of HF, HCl, HBr, H_2O (see **Figure 1**) or other similar impurities result in relatively strong absorption from these molecules. This type of far-infrared absorption was studied extensively from the 1930s to the 1960s. The high-resolution infrared data obtained for these molecules made it possible to determine very accurate centrifugal distortional constants.

Only limited studies have been carried out on the far-infrared absorption of molecules in the liquid state, which results in generalized absorption in the region of 30–80 cm^{-1}. However, if the molecule contains hydrogen bonding, a significant amount of information can be obtained for the low-wavenumber vibrations resulting from the absorption of dimers of such molecules. These include the wavenumber for the X⋯H–Y stretching and bending.

Physicists have primarily carried out the spectroscopic study of solids in the far-infrared spectral region. A major exception to this generalization is the study of the lattice modes of molecular crystals. To account for the lattice (phonon) modes of solids, the Bravais space cell is used by molecular spectroscopists to obtain the irreducible representation for the lattice vibrations. The crystallographic unit cell may be identical with the Bravais cell or it may be larger by a multiple of 2, 3 or 4. The relationship between the Bravais cell and crystallographic cell can be obtained from the Hermann–Mauguin X-ray symbol that is used to designate the crystal symmetry. For all crystal structures designated by a symbol P (primitive), the crystallographic unit cell and the Bravais unit cell are identical. Crystal structures designated with capital letters A, B, C or I are double primitive and, therefore, the crystallographic unit cells contain two Bravais cells. Crystal structures designated with the letters R or F are triply and quadruply primitive, respectively, and the crystallographic unit cells contain three and four Bravais cells, respectively.

The potential energy of a crystal can be considered to be made up of the following terms: $V_{\text{Total}} = \Sigma_j V_j + \Sigma_i \Sigma_j V_{ij} + V_L + V_{Lj}$ where V_j is the potential due to the internal coordinates, V_{ij} is the potential due to the correlation field, V_L is the potential for the external degrees of freedom and V_{Lj} represents the potential due to the interaction of the internal modes with the lattice modes. Three symmetries must be considered when studying the vibrational spectrum of a crystal: the molecular symmetry, the site symmetry, and the factor group symmetry. The factor group is not a point group but it is isomorphous with the space group, which means there is a one-to-one correspondence between the two.

Oxamide ($OCNH_2)_2$ will be used as an example of the determination of the frequency of lattice modes. This molecule has molecular symmetry *trans* C_{2h} with space group $P1$. Oxamide has one molecule per unit cell so there are three acoustical translations, which are inactive, and three optical librations, which are only Raman active. The three lattice bands are very prominent features in the Raman spectrum and occur at 106, 134, 157 cm^{-1} (**Figure 5**) with corresponding bands at 100, 137, 156 cm^{-1} for the deuterated compound. An estimate of the range for the force constant of the librations can be calculated associating the highest torsional wavenumber with the largest moment of inertia and the lowest wavenumber with the smallest moment of inertia. The widest possible range for the force constant is found to be 0.14 to 0.79 mdyn Å^{-1}. These values are relatively large for librational force constants and they reflect a high degree of hydrogen bonding for this compound.

For molecular crystals in which the molecules are nearly spherical there will frequently be a high-temperature crystal phase of which the site symmetry is higher than the molecular symmetry. These crystals are referred to as plastic crystals and they have unusual properties such as very high temperatures of melting and low anisotropic properties similar to those of liquids. The theory of lattice vibrations of orientationally disordered solids has been addressed, and the results indicate that such disorder should lead to a broadening of the vibrational bands and that all modes may be active in both the infrared and Raman spectra. Spectra of this type are referred to as density-of-states spectra, since the bands correspond to the flat points in the dispersion curve.

Ionic crystals are the other type of solid that has been studied extensively in the low-wavenumber

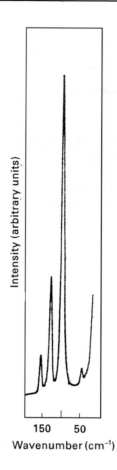

Figure 5 Optical librational modes of oxamide.

spectral range. Of particular interests have been ferroelectrics, where one of the modes can shift from 150 cm^{-1} to nearly zero as the transition temperature is approached.

List of symbols

h = Planck constant; I_r = reduced moment of inertia; $V(x)$ = potential for conformational change; V_n = height of n-fold rotational barrier; $V(\phi)$ = potential for rotation; x = coordinate of ring inversion; ϕ = rotation angle.

See also: **IR Spectrometers**; **IR Spectroscopy, Theory.**

Further reading

Carreira LA, Lord RC and Malloy TB Jr (1979) In: Eliel EL and Allinger NL (eds) *Topics in Current Chemistry*, Vol 82, pp 1–95. Berlin: Springer-Verlag.

Chantry GW (1971) *Submillimetre Spectroscopy; A Guide to the Theoretical and Experimental Physics of the Far Infrared.* London: Academic Press.

Compton DAC (1981) In: Durig JR (ed) *Vibrational Spectra and Structure*, Vol 9, pp 255–404. Amsterdam: Elsevier.

Durig JR and Cox AW (1978) In: Ferraro JR and Basite LJ (eds) *Fourier Transform Infrared Spectroscopy*, Vol 1, pp 215–274. London: Academic Press.

Groner P, Sullivan JF and Durig JR (1981) In: Durig JR (ed) *Vibrational Spectra and Structure*, Vol 9, pp 405–487. Amsterdam: Elsevier.

Hehre WJ, Radom L, Schleyer PVR and Pople JA (1986) *Ab Initio Molecular Orbital Theory.* New York: Wiley.

Killough PM, Irwin RM and Laane JJ (1982) *Journal of Chemical Physics* **76**: 3890.

Martin DH (ed) *Spectroscopic Techniques for Far Infrared, Submillimetre and Millimetre Waves.* Amsterdam: North-Holland.

Mitra SS and Nudelman S (eds), *Far Infrared Properties of Solids.* New York: Plenum Press.

Möller KD and Rothschild WG (1971) *Far Infrared Spectroscopy.* New York: Wiley-Interscience.

Palik ED (1960) A far infrared bibliography. *Journal of the Optical Society of America* **50**: 1329.

Rounds TC and Strauss HL (1978) In: Durig JR (ed) *Vibrational Spectra and Structure*, Vol 7, pp 238–269. Amsterdam: Elsevier.

Wurrey CJ, Durig JR and Carreira LA (1976) In: Durig JR (ed) *Vibrational Spectra and Structure*, Vol 5, pp 176–220. Amsterdam: Elsevier.

Fast Atom Bombardment Ionization in Mass Spectrometry

Magda Claeys and **Jan Claereboudt**,
University of Antwerp (UIA), Belgium

MASS SPECTROMETRY
Methods & Instrumentation

Introduction

The mass spectrometry community uses 1981 as the starting point of fast atom bombardment (FAB) when Barber and co-workers published their first paper. The basis of FAB was laid down in the mid-1970s through pioneering research on static secondary ion mass spectrometry (SIMS) by Benninghoven. At the time of its discovery FAB meant a significant breakthrough in the analysis of polar biomolecules, such as peptides. Until then other ionization techniques such as field desorption, SIMS using ion beams and ^{252}Cf radiation, and laser desorption of ions from surfaces had been applied to the analysis of polar biomolecules with varying degrees of success and experimental difficulty. Since the development of FAB new powerful ionization techniques have emerged, namely matrix-assisted laser desorption ionization (MALDI) and electrospray ionization (ESI), which are routinely used nowadays and have greatly extended the applicability of mass spectrometry to polar biomolecules of high molecular mass ($M_r > 200\,000$). Despite the enormous successes of these last two ionization techniques, it is fair to state that FAB still enjoys an important place in bioanalytical laboratories involved in the analysis of molecules of medium polarity, including, for example, most natural products. A survey of the literature in a representative journal on natural product research, *The Journal of Natural Products*, for the years 1996 and 1997 revealed that of the articles in which mass spectrometric analysis was reported, FAB was used in 44% of the studies. The gas-phase ionization methods, electron impact (EI) and chemical ionization (CI), are still employed in 65% and 15% of the studies respectively, while the use of ESI and MALDI was rather limited (i.e. 5% and 0.7% respectively).

In FAB the sample ions are formed by bombardment of the sample in a liquid matrix with a high-energy beam of atoms (xenon or argon) or ions (caesium). The defining attribute of FAB is the use of a viscous liquid matrix to obtain long-lasting spectra and to 'soften' the sputtering process. The liquid matrix is able to provide continuous surface renewal so that intense primary beams may be used. The overall result is that secondary ion beams with a useful intensity for scanning mass spectrometers may be prolonged to periods of 20 min or more. Interestingly, the importance of the liquid matrix in FAB was not realized by Barber and co-workers when they discovered the technique. In their first article emphasis was laid on the use of fast atoms (argon) instead of ions as energetic bombarding particles and the term 'fast atom bombardment' was introduced to describe the new technique. However, similar spectra to those obtained using a fast neutral atom beam were achieved with better sensitivity using caesium or mercury ions as bombarding particles. The latter technique is known as 'liquid secondary-ion mass spectrometry' (LSIMS), but very often and also in this review the acronym FAB is used to refer to LSIMS.

Instrumentation

The experimental set-up for FAB MS analysis is basically very simple and is illustrated in **Figure 1**. A beam of fast atoms is produced in a saddle field discharge source, called a 'FAB gun', by first ionizing an inert gas (Xe or Ar) to generate ions (Xe$^+$ or Ar$^+$) and accelerating these ions into an appropriate medium where the fast ions can capture an electron, thereby being converted from fast ions (having a kinetic energy due to acceleration) to fast atoms with energies as high as 10 keV. The beam of bombarding particles contains neutral atoms but also ions in various charge states. As illustrated in **Figure 1** the beam of bombarding particles is usually maintained at a large angle relative to the axis of the beam of secondary ions extracted by the ion optics. The intersection of the primary atom–ion beam and the secondary-ion beam is at the focal point of the ion optics. The sample film is mounted on a clean metal tip of the FAB probe which is introduced into the mass spectrometer through a vacuum-lock assembly so that the sample surface rests at the focal point of the instrument.

The experimental set-up for LSIMS analysis is very similar; the only difference is the replacement

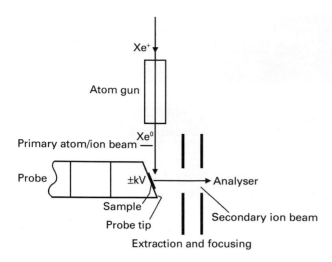

Figure 1 Experimental set-up for FAB.

of the fast atom gun by a caesium ion source which produces a focused beam of fast Cs^+ ions (up to 35 keV). The caesium ions are thermionically emitted from an alkali aluminosilicate solid maintained at high temperature (approximately 1000°C). It is worth mentioning that caesium ion impact has largely replaced FAB in modern instruments. Advantages of the caesium ion source compared to a fast atom gun are that the energy distribution of the beam is narrow, beam focusing can be accomplished and the very low gas load introduced into the ion source.

Mechanism of ion formation

It should be noted that whilst most of the particle-induced desorption techniques are simple to perform, they are only partially understood in theory, i.e. there exists no clear understanding of the mechanisms involved in ion formation from organic molecules in the condensed phase. It is, however, striking that different desorption ionization (DI) techniques [including FAB and LSIMS but also plasma desorption (PD) and laser desorption (LD)] produce reasonably similar mass spectra from nonvolatile organic molecules. Similarities in the spectra produced with incident beams of keV atoms or ions, or MeV particles and photons, can only be explained by similar ion formation and ion dissociation processes occurring after the initially very different physical excitation process. A general mechanism for the DI process can be presented, at least schematically, by the following sequence of events:

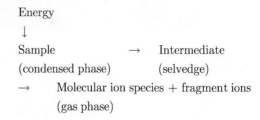

It is evident that the primary processes involved in the energy deposition step are dependent on the type of physical excitation used, and some of the properties of the intermediate phase must depend on the nature of these primary processes. However, the final step, leading to the production of molecular ions and fragment ions, appears to be independent of the primary processes occurring in the condensed phase. It is reasonable to propose that in the processes of ion formation and stabilization, chemical reactions strongly influence the end-products. These reactions are believed to take place in the vibrationally excited, disturbed surface layer, known as the 'selvedge' region. Several models describing the mechanism of FAB are available in the literature. It can be stated that there are almost as many models as there are theoreticians studying the problem.

A unified model for desorption ionization

A comprehensive qualitative model for the DI process, as exemplified in FAB, LSIMS, PD and LD, has been proposed. The main features of this unified model are: (1) irrespective of its origin, the energy deposited at the sample surface is converted from its original form into vibrational energy; (2) desorption of intact molecules and preformed ions which can be described as a vibrational or thermal process (although no equilibrium is implied); (3) ion–molecule reactions (e.g. protonation, cationization and cluster ion formation) and EI occurring in the selvedge region; and (4) dissociation of energetic (metastable) ions well removed from the surface (i.e. in the vacuum region). A schematic illustration of this unified model is given in **Figure 2**.

In the unified model the processes of desorption and ionization are considered separately. Another characteristic of the model is that desorption is followed by chemical reactions of two types occurring in two distinct regions. First, in the selvedge region, fast atom(ion)–molecule reactions and EI can take place. Secondly, in the free vacuum, unimolecular dissociations occur which are governed by the internal energy

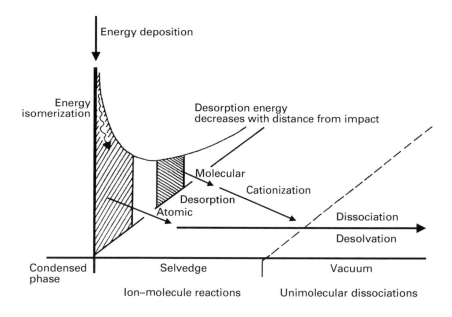

Figure 2 Unified model for desorption ionization. Reprinted with permission from Cooks RG and Busch KL (1983) Matrix effects, internal energies and MS/MS spectra of molecular ions sputtered from surfaces. *International Journal of Mass Spectrometry and Ion Physics* **53**: 111–124. Copyright Elsevier Science.

of the parent ion, the characteristic timescale of the instrument and the structures of the gas phase ions.

Ion formation mechanisms

The four major mechanisms for the formation of ions operating during FAB are: (1) direct desorption of preformed ions; (2) cationization (including protonation) and anionization in which a neutral analyte (M) is observed as adduct with a cation, i.e. [M+Cat]$^+$ or with an anion, i.e. [M+An]$^-$, (3) ion-beam induced processes leading to [M–H]$^+$ and M$^{+\bullet}$ ions; and (4) cluster ion formation. These four mechanisms cover most of the processes observed in FAB and largely describe the majority of ions detected in FAB spectra. Molecular-ion like species which are commonly encountered in FAB spectra are listed in **Table 1**.

Possible mechanisms for the formation of positive ions are given in **Table 2**; analogous mechanisms can be formulated for negative ions. That ion formation is the difficult, energetically demanding step in FAB is indicated by the observation that production of ions in the gas phase is much easier from samples that exist as preformed ions in the condensed phase than it is for neutral molecules. The mechanism by which preformed ions are converted into gaseous ions is conceptually the most straightforward: the molecular agitation generated by the bombarding particle, a process called 'sputtering', releases preformed ions into the gas phase. The types of compound that comprise preformed ions are inorganic salts, organic salts, and strong acids and bases. An

artist's conception of the violent activity at the surface of a matrix solution of the analyte is presented in **Figure 3**.

The next type of ion for which the formation can be easily rationalized is the [M+Cat]$^+$ ions; [M+Na]$^+$ is used as a typical example. However, attachment of other metal ions, especially alkali ions, also readily occurs if these ions are present in the sample. Na$^+$ may be present in the sample as a salt impurity or it may be added deliberately by the analyst; it is a preformed ion which can attach to a neutral molecule (M) either in the gas phase or in the condensed phase. The [M+H]$^+$ ions can be generated by a mechanism analogous to that proposed for [M+Na]$^+$; in this case the preformed ion is the proton. Free

Table 1 Molecular-ion like species in FAB spectra

Positive ions	Negative ions
Cat$^+$ from salt Cat$^+$ An$^-$	An$^-$ from salt Cat$^+$ An$^-$
[M + H]$^+$	[M – H]$^-$
[M + Cat]$^+$, e.g. [M + Na]$^+$, [M + K]$^+$	[M + An]$^-$, e.g. [M + Cl]$^-$
[M + 2Na – H]$^+$ and analogues	[M + Na – 2H]$^-$ and analogues
M$^{+\bullet}$	M$^{-\bullet}$
[M – H]$^+$	[M + H]$^-$
Clusters:	Clusters:
[Cat$_{n+1}$ An$_n$]$^+$	[Cat$_n$ An$_{n+1}$]$^-$
[M(M + H)]$^+$, e.g. sample–sample or sample–matrix clusters	[M(M – H)]$^-$

Table 2   Possible ion formation mechanisms in FAB

Ions formed	Mechanism
Preformed ions e.g. Cat^+	Sputtering $Cat^+An^-(cond.) \rightarrow Cat^+(gas) + An^-(gas)$
$[M + Na]^+$	Attachment of Na^+ in the gas phase $Na^+ (cond.) \rightarrow Na^+(gas)$ $M(cond.) \rightarrow M(gas)$ $M(gas) + Na^+(gas) \rightarrow [M + Na]^+(gas)$
	Attachment of Na^+ in the condensed phase followed by sputtering $M(cond.) + Na^+ (cond.) \rightarrow [M + Na]^+(cond.)$ $[M + Na]^+ (cond.) \rightarrow [M + Na]^+(gas)$
$[M + H]^+$	Attachment of H^+ in the gas phase analogous to Na^+ attachment
	Attachment of H^+ in the condensed phase followed by sputtering, analogous to Na^+ attachment
	Disproportionation $RO\bullet\bullet\bullet HO\ (cond.) \longrightarrow ROH_2^+\ (gas) + RO^-\ (gas)$ HR
$M^{+\bullet}$	Disproportionation $2M(cond.) \rightarrow M^{+\bullet}(cond.) + M^{-\bullet}(cond.)$ $M^{+\bullet}(cond.) \rightarrow M^{+\bullet}(gas)$
	Fast atom–ion beam-induced reactions $M(cond.) - H^\bullet \rightarrow [M-H]^\bullet\ (cond.)$ $[M-H]^\bullet(cond.) + H^+ (cond.) \rightarrow M^{+\bullet}\ (cond.)$ $M^{+\bullet}(cond.) \rightarrow M^{+\bullet}(gas)$
$[M - H]^+$	Fast atom–ion beam-induced reactions $M(cond.) - H_2 \rightarrow [M-H_2]\ (cond.)$ $[M-H_2]\ (cond.) + H^+ (cond.) \rightarrow [M-H]^+\ (cond.)$ $[M-H]^+(cond.) \rightarrow [M-H]^+\ (gas)$

cond. = condensed phase.

protons are probably produced during particle bombardment at the impact site, or can be added deliberately in the sample in the form of a strong acid. Another mechanism for the formation of protonated molecules which has been considered is a dispropor-

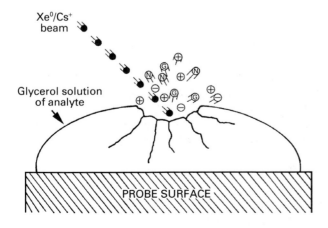

Xe⁰/Cs⁺
beam

Glycerol solution
of analyte

PROBE SURFACE

Figure 3   Artist's conception of the violent activity taking place at the surface of a matrix solution during FAB.

tionation reaction, i.e. perturbation in a hydrogen-bonded system caused by the bombarding particle can result in the sputtering of a $[M+H]^+$ ion and its complementary $[M-H]^-$ anion. True odd-electron molecular ions ($M^{+\bullet}$) are also formed, especially from compounds with low ionization potentials. $M^{+\bullet}$ ions are usually obtained for polyunsaturated molecules with conjugated systems, for example the fullerenes C_{60} and C_{70}. The study of the mechanism of ion formation in FAB is still a research topic of interest. There is no firm evidence that $M^{+\bullet}$ ions are produced by direct EI. Current mechanistic studies focus on molecular hydrogen and hydrogen radical loss from protonated M which yield $[M-H]^+$ and $M^{+\bullet}$ ions respectively. $[M-H]^+$ and odd-electron $M^{+\bullet}$ ion formation in FAB was formerly attributed to gas-phase ionization-like processes but recent studies have demonstrated that fast atom–ion beam induced processes should be considered.

The ion species of the types $[M+H]^+$, $[M-H]^-$ and $[M+Cat]^+$ are very useful for molecular mass determination, which is of key importance in the structure elucidation of natural products. A literature survey in *The Journal of Natural Products* for the years 1996 and 1997 revealed that in 52% of the studies in which FAB was used, additional accurate mass measurement at high resolution was performed on molecular ion species to obtain the precise molecular composition.

Energy deposition

FAB is generally regarded as a 'soft' ionization technique but the question may be asked: what is meant by 'soft'? Molecular-ion like species formed during FAB show little fragmentation, suggesting that they are formed with a low internal energy. Energy deposition during FAB has been addressed by several workers and it has been established that ions are formed with energies varying between 1 and 4 eV, revealing a maximum at 1 eV and a high-energy tail. The classification 'soft' should, therefore, be used with some caution.

Matrix selection and properties

The defining attribute of FAB is the use of a viscous liquid matrix. Consequently, many studies have been devoted to selection of suitable matrices in FAB. General matrix requirements concerning the solvent properties of the matrix are summarized as follows:

(1) the samples must be soluble in the matrix;
(2) only low vapour pressure solvents can be considered in the vacuum of the mass spectrometer;

(3) the matrix must be electrically conductive to avoid charging of the surface layer;

(4) ions from the matrix itself must not interfere with analyte ions in the FAB mass spectrum;

(5) the matrix must be chemically inert.

Lists of useful matrices for FAB are available and their physical and chemical properties have been compiled. In the authors' laboratory where FAB is still routinely applied to the analysis of plant secondary metabolites such as saponins, flavonoid glycosides, fatty acid derivatives and small synthetic peptides ($M_r < 3000$), two matrices are mainly used: glycerol, in the analysis of polar hydrophilic compounds and m-nitrobenzylalcohol (m-NBA) for lipophilic compounds. When glycerol is selected the sample is first dissolved in a cosolvent, methanol or a methanol–water mixture, while in the case of m-NBA dichloromethane is employed as cosolvent to facilitate addition of the sample to the matrix. It is a misconception that FAB can only be applied to the analysis of polar analytes; lipophilic compounds such as fatty acids and their derivatives are well amenable to FAB analysis if a lipophilic matrix is selected. Other matrices that have often been employed in peptide analysis include thioglycerol and a eutectic mixture of dithiothreitol and dithioerythritol (3:1, w/w), known as 'magic bullet'. For negative ion FAB the basic matrices di- and triethanolamine have also been used.

Evaporation of the liquid matrix in the vacuum of the mass spectrometer has to be considered because it can result in significant changes of the physical state of the sample solution. The effect of matrix evaporation on secondary ion formation has been investigated for samples dissolved in glycerol with and without a cosolvent. Depending on the residence time of the sample solution in the vacuum, the bulk and surface concentration of the analyte in the matrix can become supersaturated, analyte molecules can precipitate, and as such the conditions for secondary ion formation can be altered. Fortunately, the solid layers formed by precipitation of analyte molecules at the surface of sample solutions can be removed by sputtering, thereby making the underlying layer amenable to FAB analysis. In other models where evaporation of the liquid matrix was not taken into account, replenishment of the surface layer has been rationalized by diffusion of analyte molecules from the bulk to the surface layer. It is not likely that diffusion, which is not a rapid mechanism, is as important as previously thought to obtaining long-lasting analyte signals in FAB.

Sample preparation for FAB analysis

Impure samples of biological origin cannot be directly submitted to FAB analysis. Polar hydrophilic samples are usually isolated from a biological matrix containing alkali salts by chromatographic procedures employing buffers. When buffers are required for the isolation, preference is given to systems composed of volatile salts, acids and bases. Alkali salts can be eliminated from samples by resorting to simple desalting procedures based on the use of reversed phase cartridges. In the presence of Na^+, for example, acidic compounds containing carboxylic or phenolic groups give rise to multiple sodiated molecular species (i.e. $[M + Na]^+$, $[M − H + 2Na]^+$ and $[M − 2H + 3Na]^+$) leading to decreased detection sensitivity. In order to improve the detection of peptides it is common practice to add a volatile strong acid such as trifluoroacetic acid to the analyte–matrix mixture whereby neutralization of the carboxylate part of the peptide zwitterionic structure results in an enhanced $[M + H]^+$ ion formation. Derivatization strategies to increase the solubility of the analyte in the liquid matrix or to convert the analyte into a salt form with better desorption properties may also be considered. Molecules containing labile groups which are prone to hydrolysis in FAB can be stabilized by using a Li^+-containing matrix.

Continuous-flow FAB

FAB is the basis for an effective coupling technique for liquid chromatography (LC), namely, continuous-flow FAB. A set-up for continuous-flow FAB is given in **Figure 4**. In this technique, a liquid flow of typically $5–10$ μL min^{-1} obtained by splitting the LC flow, is introduced into a heated FAB source via a narrow-bore fused silica capillary. The glycerol matrix is added to the LC effluent to a concentration of approximately 0.5% and subsequently the mixture is directed into the FAB source through a metallic frit. The liquid flow ensures that there is a continuous flow on the probe tip in which previously eluted sample is continuously removed from the area where sputtering occurs. In other designs there is no metal frit but then a cotton wick is used to disperse the solvent and to obtain stable operating conditions. In addition to coupling to LC, continuous-flow FAB has been shown to be a useful technique for the introduction of flow-injected samples and effluents of capillary electrophoresis and microdialysis.

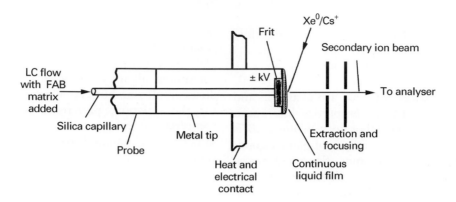

Figure 4   Experimental set-up for continuous-flow FAB.

Combination of FAB with collision-induced dissociation and tandem mass spectrometry

In order to obtain structural information on biomolecules it is common practice to use FAB in combination with collision-induced dissociation (CID) and tandem mass spectrometry. As discussed above FAB is a soft ionization technique producing mainly molecular-ion like species with a low internal energy. These species can be fragmented by CID during which the internal energy of the ions is increased. CID can be performed at low and high collision energies generally leading to different and complementary structural information. The FAB spectrum of compounds of biological origin is usually quite complex for several reasons: peaks due to the FAB matrix are always present but, in addition, impurities and salt forms of the molecular ions are often encountered making the interpretation of a FAB mass spectrum particularly difficult. Using CID and tandem mass spectrometry it is possible to obtain a spectrum of one well-defined molecular ion species allowing the corresponding fragment ions to be determined unambiguously. **Figure 5** illustrates first-order FAB spectra using glycerol as liquid matrix obtained for a flavonoid glycoside, kaempferol-3-O-rutinoside, which was isolated from the leaves of an African medicinal plant *Morinda morindoides*, with and without subsequent desalting. Without desalting several molecular ion species of kaempferol-3-O-rutinoside are observed with low intensity, namely, $[M + H]^+$, $[M + Na]^+$, $[M - H + 2Na]^+$ and $[M - 2H + 3Na]^+$. The presence of Na^+ in the sample is also evident from the matrix peaks including abundant sodiated species. Using desalting a sufficiently intense $[M + H]^+$ signal could be obtained which was amenable to CID and tandem mass

spectrometry. This methodology showed that the ions at m/z 287, 449 and 461 are fragment ions of the protonated molecule.

Selected applications

A number of reviews and overviews are available in the literature. For comprehensive surveys of the literature the biennial reviews of mass spectrometry in the journal *Analytical Chemistry* can be consulted. As it is impossible to review all applications of FAB only selected applications will be mentioned here. FAB has been particularly useful for the analysis of peptides and proteins, providing molecular mass information for large peptides with an upper mass limit of approximately 24 000 Da. It can be stated that in the field of peptides and proteins FAB has now largely been overtaken, but not entirely replaced by, MALDI and ESI. Other biomolecules that have been successfully analysed using FAB are glycoconjugates, including glycopeptides, nucleotides, terpenoid and flavonoid glycosides, and complex lipids such as glycosphingolipids, gangliosides, phospholipids and steroids. The relatively frequent use of FAB compared to other soft ionization techniques (i.e. CI, ESI and MALDI) in natural product research can be attributed to the reliable molecular mass and composition that can be obtained which is of key importance in the structure characterization of unknown compounds. FAB has also been applied in areas that are not of direct biochemical interest, such as the analysis of dyestuffs, organometallics, quaternary ammonium salts, triphenyl phosphonium salts, surfactants and chiral complexes. With regard to this last application FAB has been found to be more suitable than ESI for studying enantioselective intermolecular interactions.

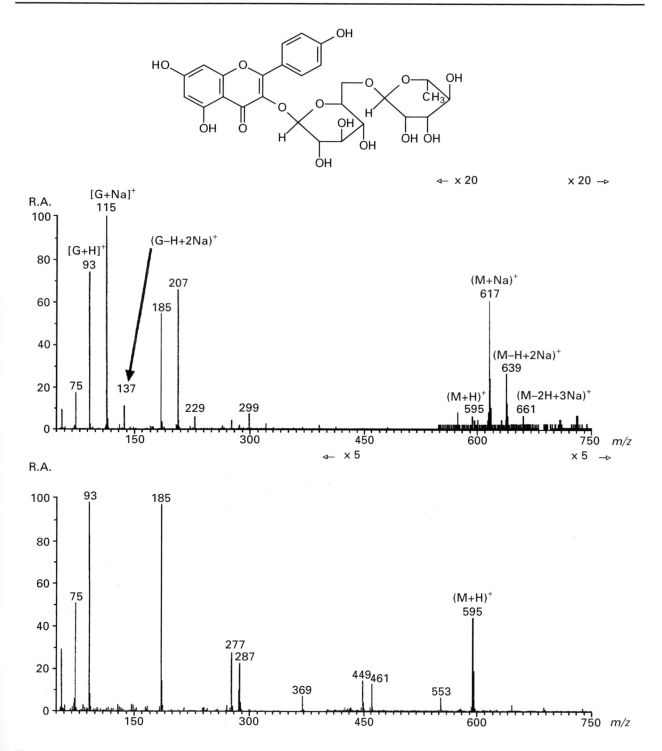

Figure 5 First-order FAB spectra obtained using glycerol (G) as the liquid matrix for kaempferol-3-O-rutinoside, isolated from a medicinal plant, with and without subsequent desalting. Reprinted with permission from Li QM, Dillen L and Claeys M (1992) Positive ion FAB analysis of flavonoid glycosides. Simple procedures for desalting and control of sodium salt contamination. *Biological Mass Spectrometry* **23**: 408–410. Copyright John Wiley and Sons Ltd.

See also: **Biochemical Applications of Mass Spectrometry; Fragmentation in Mass Spectrometry; Ion Energetics in Mass Spectrometry; Ionization Theory; IR Spectroscopy Sample Preparation Methods; MS-MS and MS^n; Organometallics Studied Using Mass Spectrometry; Peptides and Proteins Studied Using Mass Spectrometry; Plasma Desorption Ionization in Mass Spectrometry; Spectroscopy of Ions.**

Further reading

Barber M, Bordoli RS, Sedgwick RD and Tyler AN (1981) Fast atom bombardment of solids (F.A.B.): a new ion source for mass spectrometry. *Journal of the Chemical Society, Chemical Communications* 325–327.

Barber M, Bordoli RS, Elliott GJ, Sedgwick RD and Tyler AN (1982) Fast atom bombardment mass spectrometry. *Analytical Chemistry* 54: 645A–657A.

Caprioli RM and Suter MJF (1992) Continuous-flow fast atom bombardment: recent advances and applications. *International Journal of Mass Spectrometry and Ion Processes* 118/119: 449–476.

Claeys M, Li QM, Van den Heuvel H and Dillen L (1996) Mass spectrometric studies on flavonoid glycosides. In: Newton RP and Walton TJ (eds) *Applications of Modern Mass Spectrometry in Plant Science Research*, pp 182–194. Oxford: Oxford Science Publications.

Cook KD, Todd PJ and Friar DH (1989) Physical properties of matrices used for fast atom bombardment. *Biomedical and Environmental Mass Spectrometry* 18: 492–497.

Cooks RG and Busch KL (1983) Matrix effects, internal energies and MS/MS spectra of molecular ions sputtered from surfaces. *International Journal of Mass Spectrometry and Ion Physics* 53: 111–124.

De Pauw E (1986) Liquid matrices for secondary ion mass spectrometry. *Mass Spectrometry Reviews* 5: 191–212.

De Pauw E, Agnello A and Derwa F (1991) Liquid matrices for liquid secondary ion mass spectrometry-fast atom bombardment: an update. *Mass Spectrometry Reviews* 10: 283–301.

Gower JL (1985) Matrix compounds for fast atom bombardment mass spectrometry. *Biomedical Mass Spectrometry* 12: 191–196.

Hemling M (1987) Fast atom bombardment mass spectrometry and its application to the analysis of some peptides and proteins. *Pharmaceutical Research* 4: 5–5.

Junker E, Wirth KP and Röllgen FW (1992) Effects of matrix evaporation during continuous sputtering in fast atom bombardment. *International Journal of Mass Spectrometry and Ion Processes* 122: 3–23.

Li QM, Dillen L and Claeys M (1992) Positive ion FAB analysis of flavonoid glycosides. Simple procedures for desalting and control of sodium salt contamination. *Biological Mass Spectrometry* 23: 408–410.

Sawada M (1997) Chiral recognition detected by fast atom bombardment mass spectrometry. *Mass Spectrometry Reviews* 16: 73–90.

Fibre Optic Probes in Optical Spectroscopy, Clinical Applications

Urs Utzinger and **Rebecca R Richards-Kortum**,
The University of Texas at Austin, TX, USA

ELECTRONIC SPECTROSCOPY
Applications

Introduction

In the clinical environment, optical techniques have been used for decades (microscope, colposcope, ophthalmoscope, endoscope, laparascope). Integration of spectroscopic devices into existing procedures is an obvious task. Fibre optic cables provide a flexible solution for adequate optical interfacing between the optical and spectroscopic device and the sample to be interrogated *in situ*. Fibre optic probes can be inserted into cavities and tubular structures, put in contact with epithelial surfaces and also inserted into structures that can be punctuated by rigid devices such as needles. Fibre optic devices for optical spectroscopy can be manufactured as flexible catheters with an outer diameter not exceeding 0.5 mm. We will present in this article fibre optic solutions for fluorescence and both elastic and inelastic scattering detection. After describing the fibre optic interface, we will present probes for fluorescence spectroscopy followed by probes for reflectance measurements and side looking. Diffuser tips, refocusing and designs for Raman spectroscopy are discussed towards the end of the article.

The fibre optic interface

A spectroscopic system incorporates a light source, an optical analyser with detector, and a light transport conduit which in many cases is made of fibre optic cables. A separate illumination and collection channel minimizes background signals produced in the illumination fibre (**Figure 1A**). The excitation or illumination light source is typically a laser or a filtered

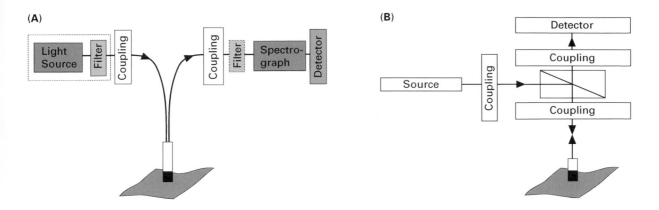

Figure 1 (A) A fibre optic spectroscopy system with separate illumination and collection path is based on an excitation source, which is a laser or a white light source (reflectometry) or a monochromator filtered arc lamp (fluorescence). Optics couple the excitation light into the flexible probe. A probe collects the emitted light. Coupling optics adapt the numerical aperture of the probe to the spectrograph or filter system. An optical detector (charge coupled device (CCD), photodiode array, photomultiplier tube) is read out and digitized. (B) A fibre optic spectroscopy system with a probe that incorporates one optical fibre needs a dichroic beam splitter and well aligned optics to separate excitation and fluorescence light. Reproduced with permission of Optical Society of America Inc. from Greek LS, Schulze HG, Blades MW, Haynes CA, Klein K-F and Turner RFB (1998) Fiber-optic probes with improved excitation and collection efficiency for deep-UV Raman and resonance Raman spectroscopy. *Applied Optics* **37**(1).

white light source such as xenon or mercury lamps. Dielectric bandpass filters (Omega Optical, Inc., Brattleboro, VT, USA), monochromators or double monochromators (ISA Edison, NJ, USA) can be used as filters according to the need for spectral purity. For fluorescence spectroscopy, xenon lamps (ORC, Azusa, CA, USA | Hamamatsu Corp. Bridgewater, NJ, USA) with more than 150 W power consumption need additional cold and hot mirrors to protect the optical parts. The coupling optics adapt the f-number of the light source to the numerical aperture of the fibre and guarantee optimal irradiance into the fibre. The probe transports the collected light to the spectroscopic system. New techniques such as holographic transmission gratings (Holospec, Kaiser Optical Systems Inc., Ann Arbor, MI, USA) and back-illuminated thinned CCDs with high quantum efficiencies (Princeton Instruments, Inc., Trenton, NJ, USA) allow short integration times and sufficient spectral and spatial resolution. Additional filter stages reduce the influence of stray light in the spectrograph. Most fluorescence applications require longpass (Schott Glass Technologies Inc., Duryea, PA, USA) and Raman spectroscopy devices notch filters (Holographic notch filters, Kaiser Optical System Inc).

To achieve smallest probe diameters, single fibre solutions are used in combination with a dichroic beam splitter and with well-aligned coupling optics (**Figure 1B**). Single fibre solutions are limited by the difficulty of reducing back-scattered excitation light and the suppression of parasitic light induced by the illumination path of the probe.

Fibre optic cables

An optical fiber for spectroscopy consists of a silica core, a doped cladding and a protective jacket. Light is transmitted based on the principle of total internal reflection. Half of the maximal angle (α) a fibre can accept light is characterized by the numerical aperture (NA), which is defined by the difference in the refractive indices (n) of core and cladding (**Figure 2**):

$$\mathrm{NA} = n_{\mathrm{media}} \sin \alpha = \sqrt{n_{\mathrm{core}}^2 - n_{\mathrm{cladding}}^2}$$

Improvements in the preform manufacturing (Heraeus Amersil Inc., Duluth, GA, USA) allow

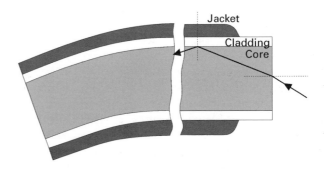

Figure 2 A fibre optic cable for spectroscopy consists of a core material (quartz) and a cladding with a lower refractive index (doped quartz) and a rugged supportive jacket. Light is transported by total internal reflection.

transmittance from 200 nm (solarization-resistant-UV grade fibre, Polymicro Technologies, Inc, Phoenix, AZ, USA) up to 2500 nm (low-OH fibre, Fiberguide Industries, Inc., Stirling, NJ, USA | Ceramoptec, Bonn, Germany). This opens the application of fibre optic probes to UV resonance Raman (UVRR) and IR Raman spectroscopy. Due to bending of the fibre and scattering centres in the fibre, light may exit into the cladding and hit the jacket. Most plastic claddings like Nylon and Tefzel® (DuPont, Wilmington, NJ, USA) produce significant autofluorescence when irradiated with UV light. Polyimide (Thermocoat, Fiberguide Industries Inc.) and metallic coated fibres (gold, aluminium) exhibit minimal fluorescence. Colour centres may be produced during UV radiation in quartz and the fibres fluoresce in a broad band at 450 and 650 nm. This process is reversible but depends on the pulse repetition rate and the pulse energy. It is further known that silica produces Raman intrinsic signals at NIR excitation, which interfere with *in vivo* Raman spectroscopy such that the dynamic range of the detector becomes critical.

Fluorescence probes

A growing number of clinical studies have demonstrated that fluorescence spectroscopy can be used to distinguish normal and abnormal human tissues *in vivo* in the skin, head and neck, genito-urinary tract, gastrointestinal tract, breast and brain. It is well known that fluorescence intensity and lineshape are a function of both the excitation and emission wavelength in samples containing multiple chromophores, such as human tissue. Major fluorescence contributors are structural proteins, NADH, FAD, tryptophane and porphyrins.

Single pixel measurements

The classic fibre optic probe to measure fluorescence consists of at least one excitation and one collection fibre (**Figure 3A**). A quartz shield placed at the distal end of the fibres allows the illuminated and probed areas to overlap. The fraction of overlapping increases with an enlargement of the numerical aperture of the fibres and the thickness of the shield. A larger shield depth requires a larger diameter shield. The collection efficiency β_t for an isotropic fluorescence source can be described as:

$$\beta_t = \frac{\beta_0}{(z + \text{shield}_t)^2}$$

where z is the position along the optical axis of the probe and β_0 a constant which includes the detector efficiency. Previous studies performed in arterial tissue have demonstrated empirically that tissue does not emit isotropically, as a result of high forward scattering. For arterial specimens, tissue fluorescence power decreases with the detector–sample separation distance, R, as $1/R^n$, where $n = 1.1$. A typical thickness of the shield is 1.7–6.4 mm. However, scattered fluorescence could also be detected without a shield. A typical probe consists of excitation fibres, collection fibres, carbon filled epoxy and tubing. A rigid type uses metallic tubing, a flexible type uses shrinking tubing (Zeus Industrial Products Inc., Orangeburg, SC, USA). The shield and sleeve are detachable to allow disinfection of the probe. A probe based on seven 200 μm fibres has a diameter of at least 1.5 mm. The probe with an outer diameter of 4 mm presented in **Figure 3B** has been successfully used by Ramanujam and the authors in studies with more than 100 patients.

InnovaQuartz (InnovaQuartz, Phoenix, AZ, USA) has successfully manufactured fibre optic tips, using CO_2 laser micro machining (**Figure 3C**). Melting and compressing the fibres eliminates the dead space between them. An increase of power density and almost complete overlap between illumination and collection areas has been reported. However, this technique is limited to fibre bundles with a few fibres only.

Multi-pixel measurements

Pitris, Agrawal and co-workers have successfully adopted the principle of single pixel fluorescence measurements into a design of 31 simultaneous measurement locations (**Figure 4A**, **Figure 4B**) without sacrificing spectral resolution. Modern imaging spectrographs for fluorescence applications (Instruments SA, Edison, NJ, USA | Chromex, Albuquerque, NM, USA) allow the simultaneous spectral dispersion of up to 30 input channels with minimal cross talk. Aberration corrected spectrographs project the input slit onto CCD devices with a focal plane size of 30×12 mm (TRIAX, Spectrum One MCR, Instruments SA). Spot scanning and imaging devices increase the spatial resolution further at the cost of spectral resolution. **Figure 4A** illustrates the hexagonal arrangement of the fibre pairs. Adding excitation fibres can increase the illumination intensity. With lamp based excitation sources the focal spot size at the coupling site is determined by the arc size, which is typically in the order of 0.5–3 mm. A 300 W xenon lamp with parabolic reflector (Compact

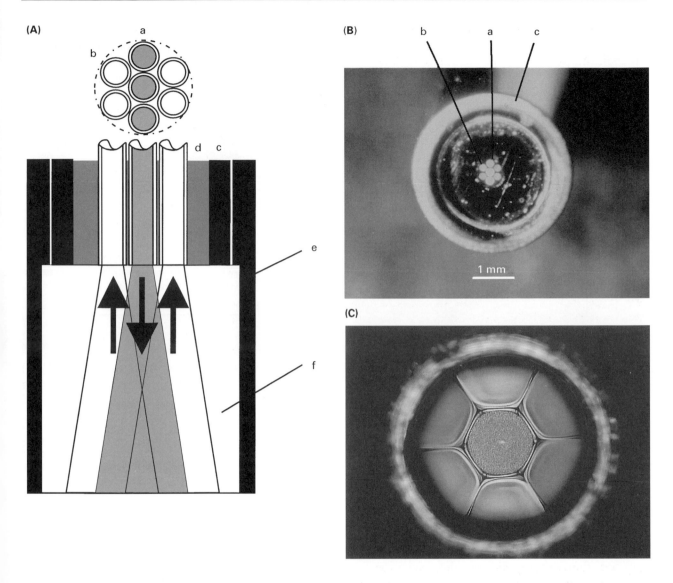

Figure 3 (A) The classic fibre optic probe for fluorescence spectroscopy is based on two or more fibres. Hexagonal packing is a dense arrangement and allows the selection of multiple fibres for different excitation sources and collection channels. A quartz shields permits the overlap of excitation and collection areas. The parts are: (a) excitation fibres; (b) collection fibres; (c) tubing; (d) carbon filled glue; (e) sleeve; (f) shield. (B) A typical fluorescence probe for single point sampling; (a) excitation fibres; (b) collection fibres; (c) sleeve. (C) A hexagonal fibre bundle with a melted and compressed arrangement allows excellent illumination and interrogation overlap. Reproduced by permission of Innova Quartz Inc.

Illuminator 6000CI, ORC) produces a spot diameter of approximately 7 mm after a lens with a focal length of 50 mm. According to the formula for hexagonal packing:

$$n_{\text{fiber}} = 1 + \sum_{k=0}^{m} 6k$$

where m is the total number of rings, 547 fibres with a core diameter of 200 µm can be successfully located in this spot which would lead up to 17 excitation fibres per fluorescence collection fibre. **Figure 4B** shows a multi-pixel probe, which is used by the authors for fluorescence distribution studies. The outer diameter is 25 mm and the shield thickness 6.4 mm.

Separation of illumination and detection spot

Keijzer and co-workers described scattering of fluorescence outside the illumination spot experimentally and theoretically (**Figure 5**) for arterial tissue. β and α absorption bands of oxygenated haemoglobin alter the fluorescence spectra. This effect is evident when the illumination and collection

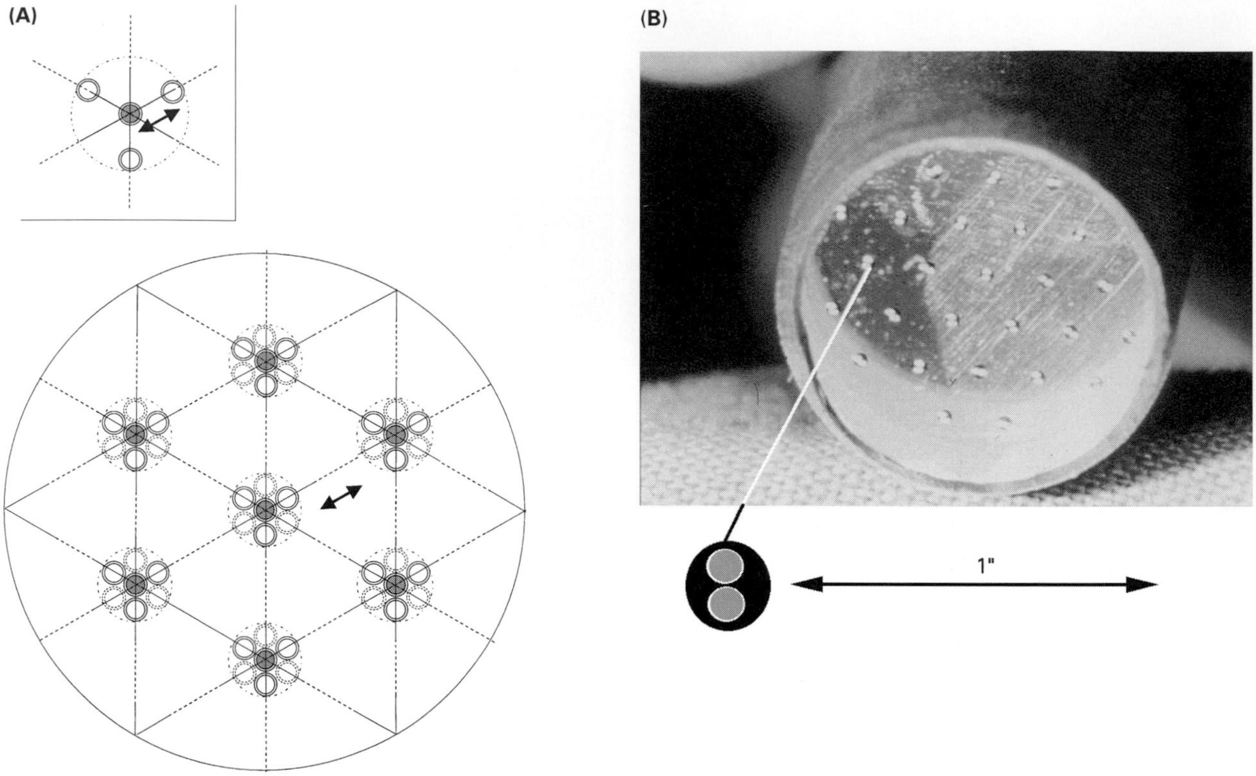

(A)

(B)

Figure 4 (A) A multi-pixel fibre optic design based on the design of **Figure 3A**. At each sample spot one collection fibre is surrounded by one to six illumination fibres. Signal tunneling from in between sample spots can be influenced by the separation of the collection and emission fibres and the shied thickness. (B) A typical fluorescence probe for simultaneous multi-pixel measurements. Black shrinking tubing holds the quartz shield. Emission and collection fibre pairs are arranged hexagonally.

spot do not overlap so the spacing of the collection spots and the shield thickness have to be chosen carefully. In order to enhance the absorption influence on fluorescence spectra, the excitation and collection fibres need to be further separated (**Figure 4A**).

Reflectance probes

The multiply scattered light that escapes the sampling volume is called reflectance. The transport mean free path length of a photon is defined as:

$$\mathrm{mfp}' = \frac{1}{(\mu_a + \mu_s)}$$

where μ_a is the absorption coefficient and μ_s' is the reduced scattering coefficient which is the isotropic equivalent of the anisotropic scattering medium:

$$\mu_s' = \mu_s(1 - g)$$

where g is the average cosine of the scattering angle and μ_s the scattering coefficient. The optical properties μ_s, μ_a and g depend on the chemical and architectural composition of the sample. The chemical composition of tissue depends on the blood supply, the metabolic state and the tissue type. The architecture of the sample is reflected in the shape and size distribution of the scattering particles. All these characteristics vary spatially and are wavelength dependent. Several recent studies have suggested that differences in the optical properties, assessed using diffuse reflectance spectroscopy, can be used to discriminate normal and abnormal human tissues *in vivo* in the urinary bladder, pancreas and the skin.

The goal of fibre optic probes for reflectometry is to measure the reflectance distribution and to derive the optical properties by fitting the data to analytic expressions based on diffusion theory or iterative algorithms based on Monte Carlo modelling. Therefore fibre optic probes consist of a single excitation source and several spatially distributed collection fibres (**Figure 6A** and **Figure 6B**). A probe with a linear alignment constructed by Wang and co-workers

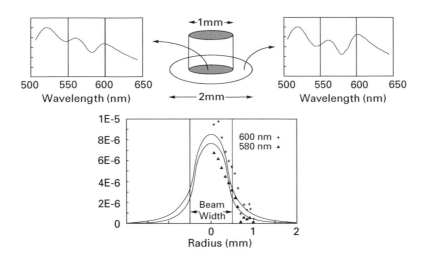

Figure 5 Cross-talk. Fluorescence generated in a 1 mm spot area is detectable from an outside area. It spreads by scattering processes and will be reshaped by absorption processes. Monte Carlo simulations and measured data show that haemoglobin absorption in tissue affects emission intensities at 580 nm more than at 600 nm. Adapted with permission from Keijser M, Richards-Kortum RR, Jacques SL and Feld MS (1989) Fluorescence spectroscopy of turbid media: Autofluorescence of the human aorta. *Applied Optics* **28**(20): 4286–4292.

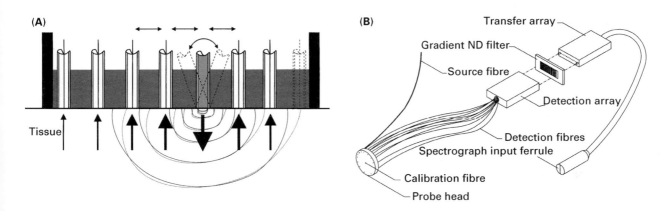

Figure 6 (A) A fibre optic probe for reflectometry. The light scattered from a single excitation fibre is detected by a linear array of collection fibres. Tilting the excitation fibre shifts the profile along the surface by an amount that is determined by the sample's optical properties. (B) A fibre optic probe with a circular fibre arrangement for reflectometry. A 2 cm probe head consist of a central calibration fibre, an excitation fibre and non-equally spaced fibres. Neutral density (ND) filters adapt the light intensity in a transfer array to a smaller dynamic range.

places the detection fibres over a range of ±1 to 10 mfp′ (**Figure 6A**). Nichols and co-workers use a circular arrangement with a similar range (**Figure 6B**). This results in probes with a diameter of approximately 2 cm and an assumption that the optical properties do not vary over this range. However the optical properties on currently investigated target areas (e.g. cervix, ovaries, oral cavity) vary over orders of millimetres (lesion size) and their size is similar to the diameter of the proposed probes. Therefore an accurate measurement of optical properties from small, premalignant lesions is difficult.

Wang and co-workers modified their linear probe design with an oblique incident source fibre and measured the shift of the reflectance profile. This removes the necessity of measuring in absolute units. The circular fibre arrangement of Nichols and others allows a simple calibration of the system by placing a source fibre in the centre of all fibres (**Figure 6B**). Measuring the spectrally resolved reflectance of all fibres simultaneously requires a dynamic range of 4 orders of magnitude. Neutral density filters or other techniques reduce the dynamic range required.

Combined probes for fluorescence and reflectance spectroscopy

It is well known that the absorption and scattering properties of tissues *in vivo* affect both the intensity

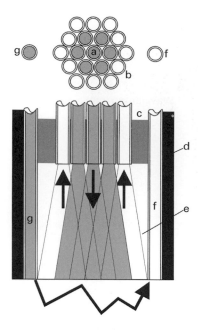

Figure 7   A fibre optic probe that combines fluorescence and reflectance measurements. White light tunnels through the fluoresced sample volume; (a) excitation fibres; (b) fluorescence collection fibres; (c) carbon filled glue; (d) outer tubing; (e) quartz shield; (f) white light illumination; (g) reflectance collection.

and the lineshape of measured fluorescence spectra. Furthermore, measuring both fluorescence and diffuse reflectance spectra may provide additional information of diagnostic value.

A probe that combines the measurement of fluorescence and optical properties was described by Durkin. Scattered light is collected with a white light 'transmission' measurement which travels through the same tissue volume that was excited for fluorescence measurements.

The probe, illustrated in **Figure 7**, consists of a total of 21 optical fibres (200 μm diameter, NA = 0.2) arranged in concentric bundles. The centre bundle contains seven fluorescence excitation fibres and twelve fluorescence collection fibres. At the distal end of the probe the fibres which excite and collect fluorescence are sealed with a quartz shield. This shield is placed in contact with the sample surface, and ensures that the area from which fluorescence is collected is the same as that directly illuminated. The reflectance fibres are flush with the tip of the central shield. Similar attempts have been previously described in the literature.

Side-looking probes and their application

Oblique polishing of individual fibres deflects the output of the fibre with respect to the fibre axis

(**Figure 8A**). If the critical angle for total internal reflection is reached, the light will leave the fibre through the cylindrical side. The shape of the fibre wall focuses the beam in an angular direction resulting in an elliptical focal spot close to the fibre surface. The critical angles for the silica–water and the silica–air interface are 66° and 43.3° respectively. In order to permit the light to leave the fibre sideways at the tip, the jacket needs to be stripped. Depending on the jacket material, it can be dissolved in acid, mechanically abraded or burned away with a lighter.

Illumination analysis (ASAP, Breault Research Organization Inc., Tucson, AZ, USA) shows maximal irradiance increase at a polishing angle of 40° for water and air as surrounding media. The focal position is at a distance of 1.3 times the cladding radius positioned away from the side of the fibre in air and 3.17 times the radius in water. The increase of irradiance is 1.6 in water and 2.36 in air (**Figure 8B**).

Many probe designs are based on the oblique polishing of fibres (InnovaQuartz | Gaser, Visonex Inc., Warner Robins, GA, USA). **Figure 8C** (left) shows a design where the sampling volume of a single excitation fibre overlaps with concentrically arranged fibres. To construct the tip, which will be used in fluids, the fibre bundle is glued together. After its conical polishing, the middle fiber is polished flat. In order to measure on a sample surface the tip can also be enclosed at the distal end with a quartz or sapphire window (**Figure 8C** right).

Enclosing the oblique polished fibre in a glass capillary tube allows the production of minimal diameter deflecting probes. These off the shelf products (MicroQuartz) are used for vaporization and coagulation of tissue, but they could also be used for spectroscopic applications with a single or double fibre system (**Figure 8D**).

If separate illumination and collection fibres are needed, their alignment will be a tedious task. An example of separate illumination and collection fibres is shown in **Figure 8E**. In order to produce a ring probe, fibre pairs are mounted in a cylindrical tube with grooves. A simultaneous investigation of sites along the circumference of the probe is possible (ring probe). The fibre pairs need to be aligned so that the illumination and collection spot overlap. Shrinking tubing or elastic bands could assist in this task.

Diffuser tips

With the medical approval of photosensitive drugs such as Photofrin® (QLT PhotoTherapeutics Inc., Vancouver, BC, Canada) by the US Food and Drug Administration and the plans for clinical trials with

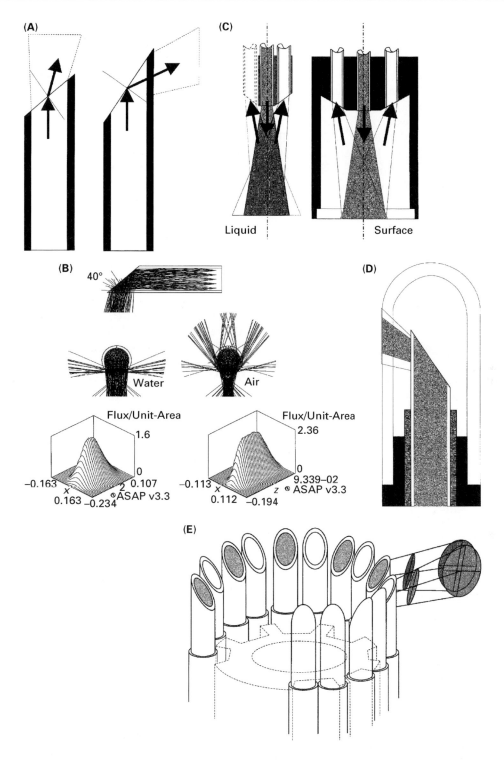

Figure 8 (A) Oblique polished fibres deflect the output beam or focus it at the sidewall of the fibre. (B) Illumination analysis with a uniform spatial and angular light distribution inside the fibre show a focusing in air and aqueous surrounding media. Irradiance increases to a factor of 2 in the focal spot. (C) Two probe designs for the interrogation of liquids (left) and surfaces (right). A central fibre illuminates the sampling volume or the surface. Six surrounding fibre collect the emitted light. Collection and emission are interchangeable. Housing with a thin shield (quartz, sapphire) permits a constant sampling distance for surface measurements. (D) Oblique polished fibres are enclosed in quartz capillary tubing for minimal diameter side-looking probes. A two-fibre arrangement for separate illumination and collection paths can also be manufactured (InnovaQuartz). (E) A circular arrangement of oblique polished fibres allows the fabrication of a ring probe with separate illumination and excitation channels.

Texaphyrin, the market for probes with a homogeneous illumination of large areas in canals and on surfaces is growing (OPTIGUIDE® QLT Photo Therapeutics | LIGHTSTIC® Rare Earth Medical, Yarmouth, MA, USA). Diffusely scattering volumes mounted at the end of fibres or fibre bundles distribute light over large areas. Beside therapeutic applications such as photodynamic therapy (Lambda Plus Photodynamic Laser, Coherent Palo Alto, CA, USA) and coagulation these diffuse light delivery systems can also be used for illumination of spectral measurements.

Scattering particles are titanium (TiO_2) or aluminium oxide (Al_2O_3) which are embedded in a transparent matrix such as optical glue (Epo-tek 305, Epoxy Technology, Inc., Billerica, MA, USA). Flexible containment is produced with micro tubing based on fluoropolymers (PTFE, FEP from Zeus, Orangeburg, SC, USA). Rigid tips are based on ceramic or materials similar to Spectralon (Labsphere, North Sutton, NH, USA). Using UV radiation for fluorescence excitation the autofluorescence and transmission of these materials is of importance. FEP tubing (0.3 mm wall thickness) shows good transmission from 250 nm up to 2 µm (40% at 250 nm, VIS 80–90%) while autofluorescence is at least an order of magnitude lower than tissue fluorescence. Al_2O_3 does not fluoresce in the UV-visible range.

Lightstics® show uniform light distribution along the optical axis: a reflector mounted at the end of the cylindrical scattering volume reflects light that would otherwise leave the probe backwards. The length of the probe determines the concentration of scatterers (**Figure 9A** and **Figure 9B**).

A reflective foil (aluminium, gold) placed around a diffuser tip restricts the angular illumination. The foil is attached with a light wall shrinking tubing (**Figure 9C**) and the result of a 180° illumination is illustrated in **Figure 9D**. Oblique polished fibres probe the light emitted by the sample area. The jacket of the fibres needs to be removed in order to pass the illumination light. A design based on this principle was proposed by the authors for fluorescence measurements in the endocervical canal.

A sapphire tip based ring probe

Sampling along the circumference of the probe is useful to identify the angular position of obstructions in tubes (e.g. arteriosclerosis). A probe initially designed for radial ablation and simultaneous spectroscopic analysis is illustrated in **Figure 10A**. The fibres are glued onto a metallic ring (outer diameter 2.4 mm, inner diameter 2.05 mm). A hollow plug attaches internal flexible tubing for flushing and a guide wire. A custom made sapphire prism (Swiss Jewel, Tenero, Switzerland) is attached to the plug. The output of the fibres (NA = 0.22) is totally deflected at the back surface of the prism. NA is the numerical aperture, whose value is sine of half the angle × index of refraction. An outer flexible silicone tube isolates the tip from the liquid environment. Two different outer diameters were realized: 1.4 and 2.8 mm. **Figure 10B** shows the individual parts partially assembled. In the upper figure the outer tubing is not completely mounted. For spectroscopic measurements, smaller 100 µm fibres were integrated in between the ablation fibres (lower figure). The sealing of such a tip is a tedious task since only small amounts of glue can be used.

Similar tips can be produced with vee-, ring-, orifice-, cup- and chisel jewels (Swiss Jewel, Tenero, Switzerland; Sapphire Engineering Inc., Pocasset, MA, USA; SWIP, Biel, Switzerland).

Refocusing

In order to increase the irradiance or the view of field on the sample area while keeping the distance to the fibres, applications may require converging the excitation beam. Focusing the beam reduces the measurement spot size. If the specimen is put outside the focus, illumination intensities drop while larger areas are illuminated.

Spherical fibre tips

Melting the end of the fibre shapes the exit surface of optical fibres. An almost spherical surface will be created by the surface tension of liquid quartz. Its form is determined by the volume of melted quartz, which is equivalent to the amount of absorbed energy. Several techniques have been used to achieve this deformation; micro furnace, Bunsen microburner, electrical arc and a CO_2 laser beam.

The smallest possible surface curvature is a hemisphere and theoretical results are shown in water and air. The beam exciting a hemispherical fibre tip in air and water is shown in **Figure 11A**. Because the refractive index of silica is 1.46, the focusing power is limited in aqueous media. **Figure 11B** illustrates the focusing of a fibre with a 200 µm core diameter and an NA of 0.22 in air and water. The beam in water is reduced to a quasi-parallel beam. A local concentration of rays can be found at a distance 3 times the distal radius of curvature from the vertex. The maximal irradiance was found to be 3 times the irradiance within the fibre.

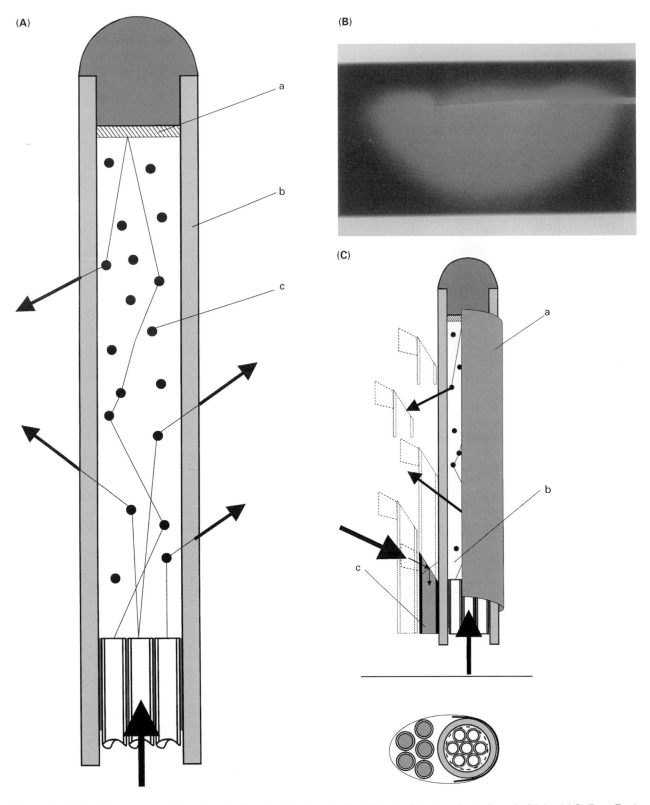

Figure 9 (A) A diffuser mounted as a tip on a fibre bundle allows uniform illumination along the probe axis (Lightstick®, Rare Earth Medical, Inc.). A (metallic) reflector (a) at the diffuser end and the local concentration of scattering particles (c) determine the intensity profile. Tubing (b) made of fluoropolymers has a high transmission and good heat resistance. (B) A Lightstick® with a length of 2 cm and a reflector allows axial illumination with a 180° field of view. (Courtesy of Rare Earth Medical, Inc., West Yarmouth, MA, USA). (C) This line-sampling probe is based on a combination of a diffuser illuminator (b) and oblique polished fibres (c). With the jacket removed from the collection fibre, the light emitted by the diffuser passes and interacts with tubular surfaces. A metallic foil (a) restricts the illumination to a reduced angle.

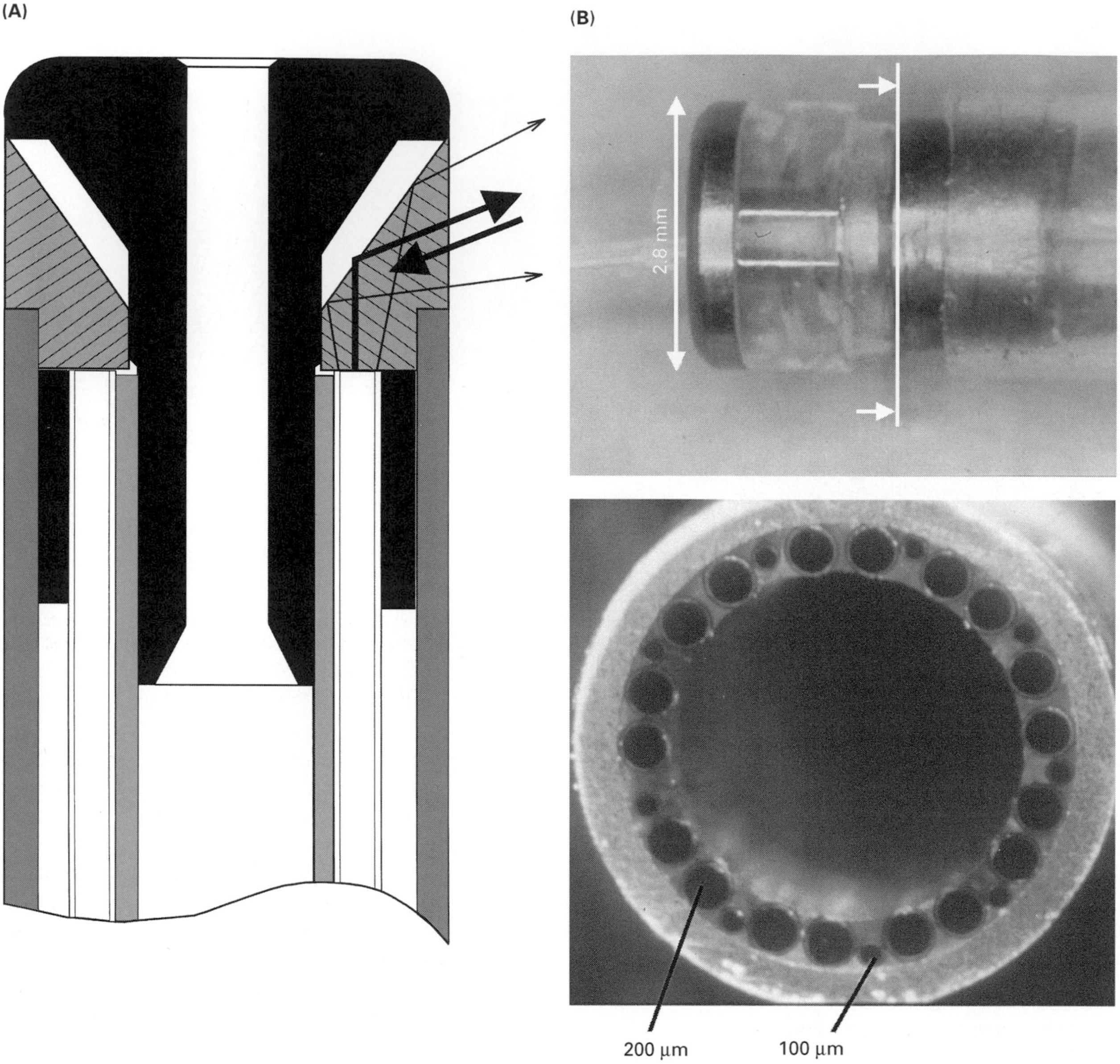

(A)

(B)

Figure 10 (A) A sapphire prism deflects the output of a fibre ring into a radial direction. The probe consists of a plug holding the prism and inner tubing. Fibres are glued into a metallic ring. Outer tubing and glue seal the fibre tip. (B) In the upper picture the sapphire prism and the fibres are partially assembled. The lower figure shows the arrangement of illumination and smaller collection fibres in a metallic ring.

Ball lenses

Spherical ball lenses have been used as collimators in fibre optic connectors. Sapphire spheres with a high refractive power are industrially produced in a wide range of diameters (Sandoz SA, Cugy, Switzerland; Rubis Precis, Charquemont, France). Light emitted from a fibre can be focused with this lens type onto the sample site (**Figure 12A**). Rol found a maximal performance when the ratio of the back surface distance to the fibre and the radius of the sphere is between 1.8 and 2.8 in water and air. Illumination analysis (ASAP) shows a maximal increase of the irradiance by a factor of 4.8 in water and 5.7 in air (**Figure 12B**). If the fibre diameter is increased, spherical aberrations will become important.

A fibre optic probe (**Figure 12A**) with a 200 μm core diameter fibre and a sphere with a radius of 0.4 mm will not exceed an outer diameter of 0.9 mm. The sphere and fibre can be held in a cylinder with a conic inner shape. For side-looking

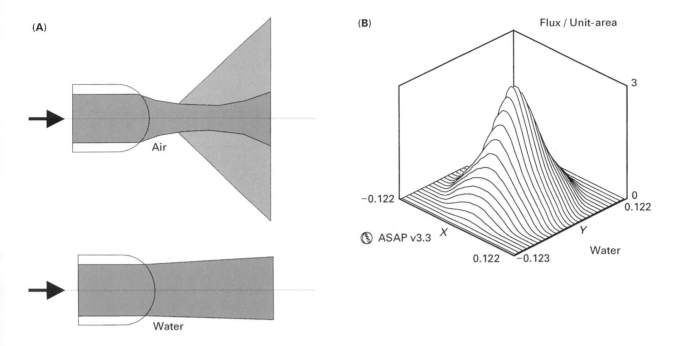

Figure 11 (A) Hemispherically shaped ends of fibres focus the output beam in air and collimate it in water. Adapted with permission from Rol P and Niederer P (1991) High-power laser transmission through optical fibres: Applications to ophthalmology. In: Wolbarsht ML (ed) *Laser Applications in Medicine and Biology*, Vol 5, pp 141–198. New York: Plenum. (B) Maximal irradiance is found at a distance of 3 times the fibre radius in water for quartz fibres with an NA of 0.2. The irradiance increases by a factor of 3.

applications the same principle is used with a half sphere. If the plane is inclined by 45° and the step in the refractive index is equivalent to sapphire–air, the beam will be deflected with total internal reflection at the interface and leave the probe at 90° with respect to the probe axis. A micro quartz tube encloses the half sphere and provides air as the refractive medium.

Spherical and parabolic reflectors

Mirrors can concentrate light. The complex refractive index of dielectrics defines the absorption coefficient and the normal incidence reflectance of a mirror:

$$\hat{n} = n + i\kappa$$

$$\mu_a = \frac{4\pi\kappa}{\lambda}$$

$$R_\perp = \frac{(n-1)^2 + \kappa^2}{(n+1)^2 + \kappa^2}$$

where $\hat{n}$ is the complex index of refraction, n is the index of refraction, κ is the extinction coefficient, λ is the wavelength and $R_\perp$ is the normal incidence reflection.

For side-looking applications, parabolic or spherically polished mirrors focus the output of several fibres onto the same area (**Figure 13A**). Best performance is shown by a parabolic surface. The manufacture of such mirrors requires a high-precision numerically controlled lathe. Surfaces with optical quality can be manufactured with single-point-diamond-cutting machines.

An example of a probe illuminating a ring is shown in **Figure 13B**. A fibre ring bundle is glued into a metallic ring. A plug with the optimized polished surface is centred into this ring and deflects the axial fibre output in a radial direction. A quartz tube shields the reflective surface. Illumination simulations (ASAP) show that the output of 5 fibres overlap (**Figure 13C**). The rotationally symmetric design spreads the light from the fibres in a quasi-elliptical spot around the circumference. The inner and outer concentric rings of fibres overlap at the cylindrical sample surface. This design confines the sampling volume close to the probe surface because outside the focal spot the light diverges rapidly.

Probes for NIR and UVR Raman spectroscopy

Near-infrared Raman scattering can be used as a tool to perform *in situ* histochemical analysis. In biological applications approximately 10^{-10} of the incident

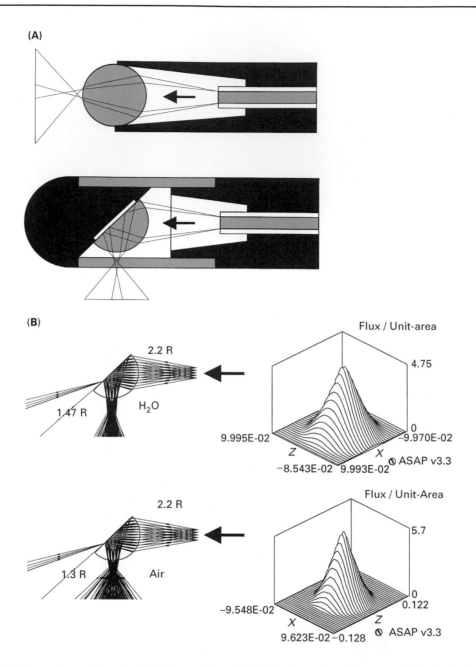

Figure 12 (A) Ball lenses image the output of a fibre and focus it. A deflecting surface is integrated with a half sphere. On this surface light is internally reflected when using a sapphire–air boundary. The deflecting probe is housed in a quartz tubing, while the axial probe will also focus in aqueous media surrounding the exit surface of the sphere. Adapted with permission from Rol P, Utzinger U, Beck D and Niederer PF (1995) Fiber beam shaping and ophthalmic applications. *SPIE Lasers in Ophthalmology II*: **2330**: 56–62. (B) Illumination analysis demonstrates a five-fold increase of irradiance in the focal spot. (B) Illumination analysis demonstrates a five-fold increase of irradiance in the focal spot.

light is Raman scattered and the Raman signal is normally 6 orders of magnitude weaker than typical fluorescence signals. The amount of light that can be delivered to the sample area is limited due to overheating and American National Standard for Safe Use of Lasers (ANSI) safety standards. The design of a fibre optic probe is therefore driven by the need for maximal light collection. Additional background signal originating from the laser source, the fibres and all optical components is crucial in filling the dynamic range of the detector and producing additional Raman signals. These signals must be reduced with filters for sensitive *in vivo* measurements.

A design developed by Myrick and Angel (**Figure 14A**) is based on GRIN lenses (NSG, Tokyo, Japan). Filters are placed in the excitation and

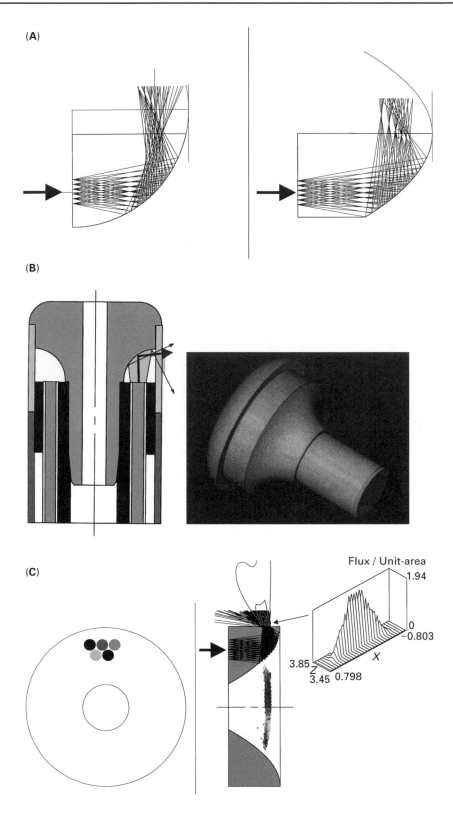

Figure 13 (A) Parabolic or spherical mirrors focus and deflect the fibre output. Rotationally symmetric mirrors with optical surface quality can be manufactured with numerically controlled diamond turning machines. (B) A fibre optic ring probe with a parabolic reflector. As in **Figure 10A** the plug is mounted into the ring of fibres and flexible tubing. A short piece of quartz tube isolates the mirror from the probe environment. (C) Illumination analysis of the output from 5 fibres shows a good overlap of the fibres placed closer and further out from the axis. In the circumferential direction the measurement spot spreads out.

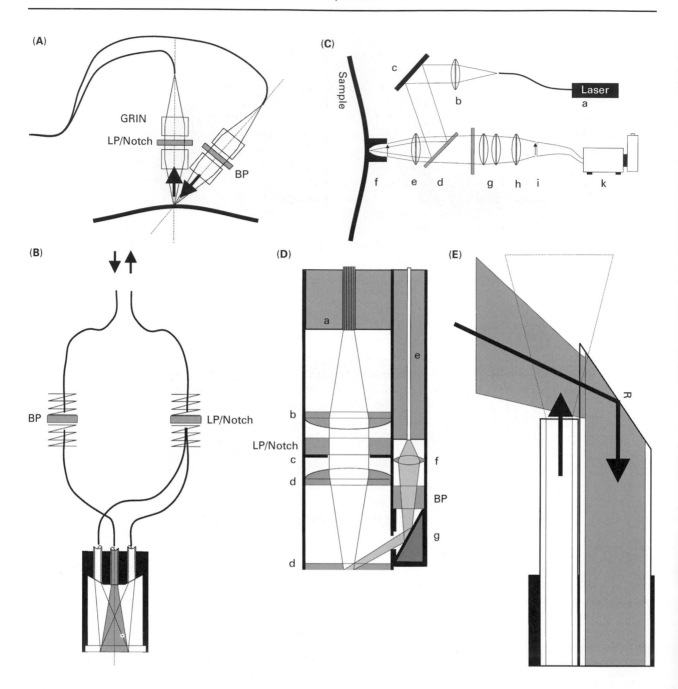

Figure 14 (A) A fibre optic probe for Raman spectroscopy based on separate illumination and collection fibres. The output of the fibres is collimated and filtered and then projected onto the same area on the sample. Bandpass filters are used for the excitation fibres and longpass filters for the collection fibres. GRIN lenses are assembled into the probe in order to maintain small diameters. Adapted with permission from Myrick ML and Angle SM (1990) Elimination of background in fiber-optic Raman measurements. *Applied Spectroscopy* **44(4)**: 565–570. (B) A fibre optic probe based on the design in **Figure 8C**. In-line filters are mounted into spring loaded SMA connectors. The fibre length between the tip and the filters is kept short. Adapted with permission from Nave S, O'Rourke P and Toole W (1995) Sampling probes enhance remote chemical analysis. *Laser Focus World* December: 83–88. (C) A CPC Raman collection system consists of: (a) NIR diode laser; (b) collimating lens; (c) reflective mirror; (d) dichroic mirror; (e) focusing lens; (f) parabolic concentrator; (g) field lenses to prevent vignetting; (h) a focusing lens to couple the Raman signal into the collection fibre bundle (i) and a high efficiency spectrograph with a liquid nitrogen cooled CCD (k). (D) Raman probe. The excitation light is transported through a single fibre (e) and imaged with a biconvex lens (f) through a dielectric bandpass filter via a mirror (g) onto the sample area. The inelastic scattered light is collected behind a quartz window (d) and imaged by lenses (d) and (b) through an aperture stop and a holographic notch filter onto a flexible fibre bundle (a). (E) The UVRR spectroscopy probe incorporates a smaller diameter solarization-resistant-UV-grade quartz fibre for excitation and an oblique polished fibre to collect scattered light close to the excitation fibre: Reproduced with permission from Schulze HG, Bludes MW, Haynes CA, Klein KF and Turner RFB (1998) Fiber-optic probes with improved excitation and collection efficiency for deep UV Raman and resonance spectroscopy. *Applied Optics* **37**(1): 170–180.

emission paths and allow remote analysis with fibre optic cables. The light source is bandpassed to eliminate signals produced in the fibres and a longpass filter reduces Fresnel and elastic scattered light that enters into the collection fibres.

A similar approach is shown in **Figure 14B**. A probe developed by Savannah River Technology Center (Aiken, USA) consists of dielectric filters placed between fibres which are held with spring-loaded SMA connectors. Six collection fibres are arranged around an excitation fibre. They are arranged as shown in **Figure 8C**. A sapphire window isolates the probe from the surrounding environment.

Berger, Tanaka, and other workers have developed a novel design for improved signal collection to allow spectral acquisition in a short time. A compound parabolic concentrator (CPC) is used at the distal end of the probe. They produced a hollow shell, based on a mandrel, electrolytically. The mandrel followed an optimized parabolic form, which was cut with a numerically controlled lathe. Small CPC dimensions with an input aperture of 0.57 mm, exit aperture of 2.1 mm and length of 4.1 mm were achieved (**Figure 14C**). The probe design with the CPC improved the collection efficiency six-fold. The system incorporates a 500 mW tunable 830 nm diode laser, which is coupled with a mirror and a dichroic beam splitter onto the sample site. Inelastic scattered light exciting the CPC is further collimated with the first lens and filtered by the dichroic mirror and additional filters. Field lenses prevent vignetting of the signal. The last lens images the output onto 200 100 μm fibres, which illuminate a f/1.8 spectrograph. The signal is detected with a back illuminated, liquid nitrogen cooled charge coupled device (CCD). Spectral resolution is 13 cm^{-1}.

Mahadevan-Jansen and the authors have successfully measured Raman spectra on the cervix with a fibre optic probe in less than 3 minutes (**Figure 14D**). A diode laser is coupled into a single 200 μm fibre. A small diameter dielectric filter (3–4 mm) rejects out-of-band light (optical density (OD)5) and a gold mirror deflects the focused beam onto the specimen site. The beam has to pass a quartz window, which is a part of the housing. Scattered light from the same spot area is imaged with biconvex lenses on a fibre bundle. Elastic scattered and Fresnel reflected light are rejected with a beam aperture stop and a holographic notch filter (OD6). The optics in the detection arm are 8 mm in diameter and the whole probe diameter is less than 2 cm. A similar spectroscopic detection system is used as described in **Figure 14C**. The spectrograph with the holographic transmission grating (Holospec, Kaiseroptics) and the deep depletion, back illuminated, liquid nitrogen cooled CCD

allow optimal system performance (Princeton Instruments).

All the Raman probes described above include imaging optics to enhance the collection efficiency. This makes it difficult to correct the spectral throughput of the system with standardized radiators such as the tungsten filament lamp. The lamp needs to be imaged onto the location of the measurement site and the NA of the probe has to be filled with the light source. The intensity of the source needs to be reduced by orders of magnitude in order to prevent detector saturation.

The scattering cross-section for Raman events can be increased dramatically if excitation is effected using photons in resonance with the electronic transitions of the chromophore involved in the fibration. For many biologically relevant chromophores these transitions are in the UV. Resonance enhancements are in the order of 10^4–10^5 for UV absorption bands. In order to measure these phenomena with fibre optic probes the problems posed by inherent signals produced in the delivery and collection system had to be overcome. Turner and co-workers have successfully built a fibre optic probe for UV resonance Raman spectroscopy (**Figure 14E**). UV-grade and solarization-resistant-UV-grade optical fibres (Polymicro) are used for the collection and excitation path respectively. Their design incorporates an excitation fibre and an oblique polished fibre with a larger diameter to collect the scattered light (600 μm). This configuration minimizes inner filtering (sample self-absorbance). Because solutions are investigated a reflective surface based on a metallic coating is needed on the collection fibre. Aluminium reflects more than 90% at 250 nm.

Summary

A large variety of optical designs allow optimal illumination and light collection for spectroscopic applications. Specific solutions for fluorescence, reflectance and Raman spectroscopy are available and the possibility to combine them in a single probe is feasible. Future research will lead towards integrated, multifunctional and highly optimized fibre optic probes.

See also: **Light Sources and Optics; Near-IR Spectrometers; Raman Spectrometers.**

Further reading

ANSI (1993) American National Standards for Safe Use of Laser. Orlando FL: American National Safety Institute.

Bigio JJ and Mourant JR (1997) Ultraviolet and visible spectroscopies for tissue diagnostics: fluorescence spectroscopy and elastic-scattering spectroscopy. *Physics in Medicine and Biology* 42(5): 803–814.

Brennan J, Wang Y, Dasari RR and Feld MS (1997) Near-infrared Raman spectrometer system for human tissue studies. *Applied Spectroscopy* 51(2): 201–208.

Durkin AJ (1995) Quantitative fluorescence spectroscopy of turbid samples, PhD Dissertation, University of Texas at Austin.

Farell TJ, Patterson MS and Wilson BC (1992) A diffusion theory model of spatially resolved, steady state diffuse reflectance for the non-invasive determination of optical properties in vivo. *Medical Physics* 19: 879–888.

Keijzer M, Richards-Kortum RR, Jacques SL and Feld MS (1989) Fluorescence spectroscopy of turbid media: Autofluorescence of the human aorta. *Applied Optics* 28(20): 4286–4292.

Mahadevan-Jansen A and Richards-Kortum RR (1996) Raman spectroscopy for the detection of cancers and precancers. *Journal of Biomedical Optics* 1(1): 31–70.

Myrick ML and Angel SM (1990) Elimination of background in fiber-optic Raman measurements. *Applied Spectroscopy* 44(4): 565–570.

Nave S, O'Rourke P and Toole W (1995) Sampling probes enhance remote chemical analysis. *Laser Focus World*, December: 83–88.

Nichols MG, Hull EL and Foster T (1997) Design and testing of a white light, steady state diffuse reflectance spectrometer for determination of optical properties of highly scattering systems. *Applied Optics* 36: 93–104.

Nishioka NS (1994) Laser-induced fluorescence spectroscopy. *Gastrointestinal Endoscopy Clinics of North America* 4(2): 313–326.

Ramanujam N, Mitchell MF, Mahadevan-Jansen A *et al* (1996) Cervical precancer detection using a multivariate statistical algorithm based on laser-induced fluorescence spectra at multiple excitation wavelengths. *Photochemistry and Photobiology* 64(4): 720–735.

Richards-Kortum R (1995) Fluorescence spectroscopy of turbid media. In: Welch AJ and van Gemert MJC (eds) *Optical-Thermal Response of Laser-Irradiated Tissue*. New York: Plenum.

Richards-Kortum R and Sevick-Muraca E (1996) Quantitative optical spectroscopy for tissue diagnosis. *Annual Review of Physical Chemistry* 47: 555–606.

Richards-Kortum RR, Mitchell MF, Ramanujam N,

Mahadevan A and Thomsen S (1994) In vivo fluorescence spectroscopy: potential for non-invasive, automated diagnosis of cervical intraepithelial neoplasia and use as a surrogate endpoint biomarker. *Journal of Cellular Biochemistry* (Suppl) 19: 111–119.

Rol P (1992) Optics for transscleral laser applications. PhD Dissertation Swiss Federal Institute of Technology, Zurich.

Rol P and Niederer P (1991) High-power laser transmission through optical fibers: Applications to Ophthalmology. In: Wolbarsht ML (ed) *Laser Applications in Medicine and Biology* Vol 5, pp 141–198. New York: Plenum.

Rol P, Utzinger U, Beck D and Niederer PF (1995) Fiber beam shaping and ophthalmic applications. *SPIE Lasers in Ophthalmology* II: 2330.

Roy AS (1979) *Concentrating Collectors, Solar Energy Conversion. An Introductory Course*, Dixon AE and Leslie JD (eds) pp 185–252. Oxford: Pergamon.

Shane Greek L, Schulze HG, Blades MW, Haynes CA, Klein KF and Turner RFB (1998) Fiber-optic probes with improved excitation and collection efficiency for deep-UV Raman and resonance Raman spectroscopy. *Applied Optics* 37(1): 170–180.

Tanaka K, Pacheco M, Brennan J, Itzkan I, Berger A, Dasari R and Feld MS (1996) Compound parabolic concentrator probe for efficient light collection in spectroscopy of biological tissue. *Applied Optics* 35(4): 758–763.

Utzinger U (1995) Selective coronary excimer laser angioplasty. Dissertation Swiss Federal Institute of Technology, Zurich.

Wang LH and Jacques SL (1992) Monte Carlo Modeling of Light Transport in Multi-layered Tissues in Standard C, developed at University of Texas M.D. Anderson Cancer Center. Available through lwang@tamu.edu.

Wang LH and Jacques SL (1995) Use of a laser beam with an oblique angle of incidence to measure the reduced scattering coefficient of a turbid medium. *Applied Optics* 34: 2362–2366.

Ward H (1987) Molding of laser energy by shaped optic fiber tips. *Lasers in Surgery and Medicine* 7: 405–413.

Welch AJ, van Gemert MJC, Star WM and Wilson BC (1995) Definition and overview of tissue optics. In: Welch AJ and van Gemert MJC (eds) Optical-Thermal Response of Laser-Irradiated Tissue. New York: Plenum.

Fibres and Films Studied Using X-Ray Diffraction

Watson Fuller and **Arumugam Mahendrasingam**,
Keele University, Stoke-on-Trent, UK

> **HIGH ENERGY
> SPECTROSCOPY**
>
> **Applications**

The fibrous state is a natural habit for those polymer molecules which, because of a high degree of regularity in their monomer sequence, tend to assume helical conformations rather than, for example, the less regular, although nevertheless highly organized, globular structures characteristic of many proteins. Such fibres typically contain regions in which the polymer molecules are arranged with their chain axes parallel to the fibre axis. This orientation is often associated with order in the side-by-side packing of the polymer chains so that the fibre can be regarded as consisting of an array of microcrystallites separated by less ordered regions. Generally there is no preferred orientation of these microcrystallites about the crystal axis parallel to the fibre axis. Consequently an X-ray diffraction pattern from such a crystalline fibre is similar to that which would be obtained from a single crystal if it were rotated through 360° about one of its principal axes during data collection. Although such rotation inevitably results in some loss of 3D information on the crystal structure, this loss is frequently much less significant than might be expected. Indeed the major expansion over the past few years in the use of powder diffraction, where the loss of information because of the complete absence of a preferred crystallite orientation is very much more extensive than that in oriented fibres, provides abundant evidence of the value of the information which can be obtained from studying specimens with a lower degree of order than that of single crystals. In this context it is important to emphasize that the degree of crystallinity in oriented fibres, as indicated by the upper limit in the angle of diffraction for which sharp Bragg diffraction peaks are observed, is often comparable with that observed for single crystals of proteins and other macromolecules. Many polymer materials are manufactured as films with a thickness of a few hundred micrometres. Since crystallite sizes in polymer films are typically a few tens of nanometres or less in lateral dimensions (i.e. ~ ten thousand times less than the film thickness) they can be analysed using similar methods to those used to study fibres.

The relevance of X-ray diffraction studies of fibres and films

Fibre diffraction is one of the most powerful and generally applicable techniques available for the study of macromolecular structure. The validity of this claim may be demonstrated by the following points:

(1) For many biological structures such as muscle and connective tissue, the fibrous state is functionally important and fibre diffraction techniques therefore allow both the structure of the macromolecules themselves and their organization in higher order structures to be investigated with the minimum of disturbance to the *in vivo* structure.

(2) The technological value of many synthetic polymer materials can be related to their molecular conformation and arrangement, typically expressed in terms of the degree of polymer orientation and crystallinity. Fibre diffraction provides the opportunity to investigate these structures in functionally important situations, e.g. in commercially available sheets and fibres.

(3) Because of the high degree of regularity in helical structures and the availability of information on covalent bond geometry from X-ray single crystal studies of model compounds, it is possible to describe the conformation of many fibrous polymers in terms of a relatively small number of parameters, i.e. the torsional angles about the various covalent single bonds. The use of such descriptions can have a dramatic effect on the ratio of observed data to parameters to be determined and hence on the accuracy with which fibrous structures can be determined.

(4) The fibrous state is particularly well suited to the investigation of structural transitions in the conformation of polymer molecules and in their arrangement. Such investigations are of increasing importance in the study of both naturally occurring and synthetic polymers.

(5) While many biological macromolecules can undergo conformational transitions, these are typically between well defined structures. In many cases these structures persist in a wide variety of molecular environments, e.g. the B form of the deoxyribonucleic acid (DNA) double-helix determined from X-ray fibre diffraction is observed with relatively little conformational variation in solution, in association with a number of different proteins and in fibres in a range of crystalline and semicrystalline arrays. Such observations comprise just one aspect of the large body of data which bears testimony to the biological relevance of structural information obtained from diffraction studies of macromolecules in fibres, films and single crystals.

Biological evolution exhibits a wealth of examples of the exploitation of polymer structure, with muscle a particularly impressive and intriguing example. The recognition that these biological structures have been refined over hundreds of millions of years in response to competitive pressures has prompted an increasing interest in using them as a source of ideas for the rational design of man-made materials and with it increasing emphasis on efforts to relate the structure of polymer materials to their physical properties.

The impact of X-ray synchrotron radiation sources

The advent of dedicated X-ray synchrotron radiation sources, beginning with the Daresbury Laboratory Synchrotron Source (SRS) in 1981, dramatically extended the power and scope of X-ray diffraction studies of fibres and films. The high flux per unit area of the beam at the specimen position produced by such sources allowed diffraction patterns to be recorded with exposure times two or three orders of magnitude shorter than those required for a conventional rotating anode X-ray source. As a consequence it became possible to record within minutes complete 2D high angle X-ray fibre diffraction patterns which had previously required exposure times of hours. As third generation sources such as the European Synchrotron Radiation Facility (ESRF), Grenoble, together with corresponding improvements in beam line instrumentation, became available during the 1990s exposure times in the millisecond range became possible with the prospect of recording in real time changes in the X-ray diffraction pattern corresponding to changes in polymer conformation and organization. Such data is of central importance in investigating the relationship between structure and function in biological macromolecules and in relating the structure of industrial polymers to their bulk physical properties. The very high degree of collimation in radiation from sources such as the ESRF together with innovations in beam line optics has made it possible to reduce the diameter of the incident X-ray beam by two orders of magnitude from that typical of rotating anode sources and hence to allow structural variation in polymer materials to be recorded with a spatial resolution of ~1 micrometre. Since the radiation emitted from synchrotron sources comprises a continuous spectrum from X-rays to the far infrared it is possible to select the wavelength used in any particular experiment to match specific features in the absorption spectrum of the material being studied. In particular this 'wavelength tunability' offers the possibility of undertaking anomalous scattering experiments in which the wavelength of the incident X-rays is varied in the region of an atomic absorption edge. This approach has considerable potential but, to date, its application in the study of polymer materials has been limited because of technical difficulties.

Two regimes are distinguished in X-ray diffraction studies of fibres and films according to whether they are in the domain of wide angle X-ray scattering (WAXS) or small angle X-ray scattering (SAXS). In WAXS, X-rays scattered through angles greater than ~1° to the incident X-ray beam are recorded. This typically provides information on structural features with dimensions in the range 5 nanometres to 0.1 nanometres, e.g. the detailed conformation and packing of polymer chains. In SAXS, X-rays scattered through angles less than ~1° to the incident X-ray beam are recorded. This typically provides information on structural features with dimensions in the range 100 nm to 5 nm, e.g. the overall folding and organization of polymer chains and the shape, size and distribution of crystallites. This distinction between the information which can be obtained from WAXS and that which can be obtained from SAXS illustrates the general principle in scattering theory that spacings in a diffraction pattern (i.e. the angle of scatter) are reciprocally related to the dimensions of features in the scattering object to which they relate.

The information which can be obtained in X-ray diffraction studies on fibres and films by the exploitation of synchrotron radiation sources is illustrated in this article by studies of both biological and man-made materials. All the studies were made at either the SRS or the ESRF but the techniques described are typical of those used at the rapidly increasing number of synchrotron sources available throughout the world. Examples of the data which can be obtained from highly ordered fibrous structures are illustrated for a WAXS pattern from DNA in **Figure 1** and a SAXS pattern from muscle in **Figure 2**. The possibility of recording WAXS and SAXS data simultaneously in real-time from the same specimen offers the possibility of relating changes in local polymer structure as revealed by WAXS to larger scale changes in polymer organization as revealed by SAXS. Increasingly, the information obtained from SAXS and WAXS on molecular structure and organization is complemented by the use of other techniques in which the recording of observations is synchronized to SAXS and WAXS measurements, e.g. stress/strain and calorimetric measurements in studies of the response of polymer materials to mechanical and thermal stress.

Experimental facilities and techniques

The radiation emitted from the electron storage rings at synchrotron radiation sources is typically chanelled through a number of beam lines into some 30 or so work stations (**Figure 3**). The work stations vary in having their beam line optics designed to produce a beam with the cross-sectional area and degree of parallelism at the specimen which best matches the demands of the type of experiments to which they are dedicated. For X-ray diffraction studies of fibres and films the most common requirements are a highly monochromatic beam with a wavelength of ~0.1 nm and the highest possible X-ray intensity within a beam diameter ~100 µm. On third genera-

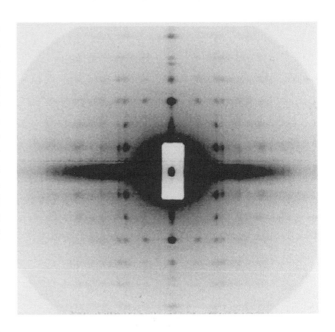

Figure 2 Small angle X-ray fibre diffraction pattern recorded from plaice fish muscle.

tion sources such as the ESRF, microbeams with diameters of ~10 µm or less are produced without unacceptable loss of intensity through collimation or a loss of parallelism because of focusing elements. Loss of parallelism is of particular concern for SAXS studies because the more the incident X-ray beam is focused on the specimen, the more divergent will be the unscattered X-rays passing through the specimen and hence the larger the minimum angle at which scattered X-rays can be recorded and the lower the upper limit in the dimensions of structural features in the specimen on which information can be obtained. Monochromatization and focusing of the main beam are typically achieved using a variety of crystal monochromators and curved mirrors which reflect X-rays at glancing incidence often from surfaces coated with thin metal layers. The increased use of insertion devices such as undulators in straight sections of the electron storage ring in more recent synchrotron sources allows the radiation fed to particular work stations to be matched more closely to specific applications in terms of intensity as a function of wavelength. Such developments are of particular importance for diffraction studies where there is often a need to maximize the intensity in a very narrow wavelength range.

A variety of X-ray cameras have been developed for use on synchrotron radiation beam lines to provide a controlled environment in which the material to be studied can, for example, be subjected to particular conditions of relative humidity, thermal and mechanical stress. The capacity to control relative

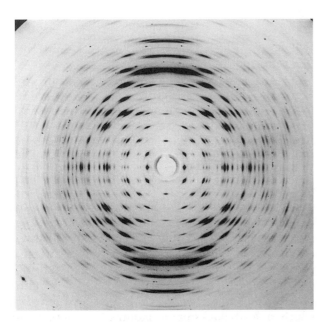

Figure 1 Wide angle X-ray fibre diffraction pattern recorded from the D form of the DNA double-helix.

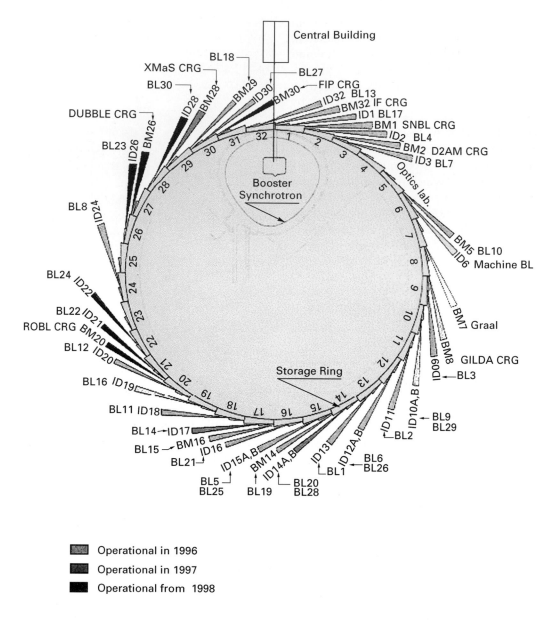

Figure 3   Schematic representation of beam lines and work station at the ESRF.

humidity is of particular importance for biological materials since *in vivo* they are commonly extensively hydrated. The processing of polymer materials typically involves exposing them to thermal and mechanical stress and cameras for studying them under industrial processing conditions typically incorporate drawing rigs able to apply various patterns of mechanical stress at well defined temperatures. The front face of an X-ray camera normally accommodates a final collimating system defining the size of the X-ray beam falling on the specimen mounted close to the final aperture of the collimator. In some cameras the specimen mount incorporates a goniometer driven remotely which allows data to be collected as a function of the tilt of the fibre with respect to the incident X-ray beam. For the collection of WAXS data the distance between the front and rear faces of the camera is ~10 cm. For the collection of SAXS data this distance may be 1 m or even more if data corresponding to periodicities in the specimen of greater than 50 nm are to be collected. The detector on which the diffraction pattern is recorded is placed close to the rear face of the camera which is made from a material with low X-ray absorption such as a 10 μm thick sheet of the plastic Melinex on which a lead backstop is mounted to absorb the component of the incident beam not scattered by the specimen. To reduce degradation of the data due to scattering by air of the X-ray beam not scattered or absorbed by the specimen, the camera is evacuated or the air replaced by helium gas which is a much weaker scatterer of X-rays. Until relatively recently X-ray diffraction data from fibres was commonly recorded on photographic film. However, during the 1990s a variety of electronic area detectors have largely replaced film at synchrotron sources. One of the most important advantages of electronic area detectors of the gas filled multiwire and CCD types is that they allow the diffraction pattern to be displayed in real time so that experiments can be genuinely interactive. In contrast, it may take 10 or 15 min to process film. This time does not of course include scanning of the processed film as a necessary preliminary to accurate determination of the position and intensity of diffraction spots. In contrast, data recorded electronically can be immediately downloaded to an online computer for data reduction and analysis. In deciding between different types of detector it is important to compare parameters such as:

(1) *spatial resolution* – the grain size on X-ray film is typically less than 10 μm and although electronic area detectors generally available have significantly worse resolution than this, they are adequate for recording fibre diffraction data;
(2) *quantum efficiency* – depends critically on the wavelength of the X-rays, therefore while comparisons with film are complex it is fair to say that state of the art electronic area detectors can show efficiency comparable to film;
(3) *dynamic range* – the limited dynamic range of film is one of its major limitations and an area where its performance is generally inferior to electronic area detectors.

Image plates have a spatial resolution comparable to film with a much larger dynamic range. However, they still suffer from the need for scanning which, even with ingenious online carousels and dedicated scanners, still makes it impossible to conduct genuinely real-time interactive experiments. The examples of diffraction data included in this review illustrate the use of the various types of detector.

Examples of the exploitation of X-ray diffraction using synchrotron radiation in the study of fibres and films at high temporal and spatial resolution

These examples are chosen to illustrate the impact of both WAXS and SAXS studies on the characterization of naturally occurring and man-made materials.

These studies exploit both the greatly increased intensity and much reduced divergence in the X-ray beam available from synchrotron sources as compared with that from laboratory rotating anode X-ray sources.

Conformational transitions in the DNA double-helix

The original proposal of a double helical structure for DNA was based on wide angle X-ray fibre diffraction data. Subsequently, a number of distinct conformations of the double-helix have been observed and designated A, B, C, D and Z. The sensitivity of the conformation of the DNA double-helix to its degree of hydration allows conformational transitions to be induced by changing the relative humidity of the fibre environment. The availability of X-ray synchrotron radiation sources offers the possibility of observing the variation in diffraction data during such transitions with the possibility of describing the stereochemical pathway followed during the transition. In an early application of the Daresbury SRS the reversible transition between the B and D conformations was followed through a number of cycles. Selected fibre diffraction patterns from this transition are in **Figure 4**. The transition between the D and B conformations can be followed through the appearance in the original highly crystalline D pattern of reflections corresponding to a semicrystalline intermediate between the D and B forms. As the transition proceeds, these reflections become more intense, corresponding to an increasing fraction of the DNA assuming the semicrystalline phase, and gradually change in position corresponding to a continuous change in helix pitch from the 2.4 nm characteristic of the crystalline D form to the 3.4 nm characteristic of the semicrystalline B form. The diffraction data on which this analysis was based were recorded on photographic film with exposure times of a few minutes. In the more than 10 years since these studies were made major advances in both synchrotron performance and detectors have dramatically decreased the exposure times required for such time-resolved studies.

Changes in molecular organization and interactions during muscle contraction

Muscle is a complex assembly of proteins embracing the metabolic capabilities characteristic of globular proteins and the structural capabilities of fibrous proteins and is thus well suited to perform the dual roles of converting chemical energy into mechanical energy and sustaining tension. Over 40 years ago a combination of electron microscopy and X-ray fibre

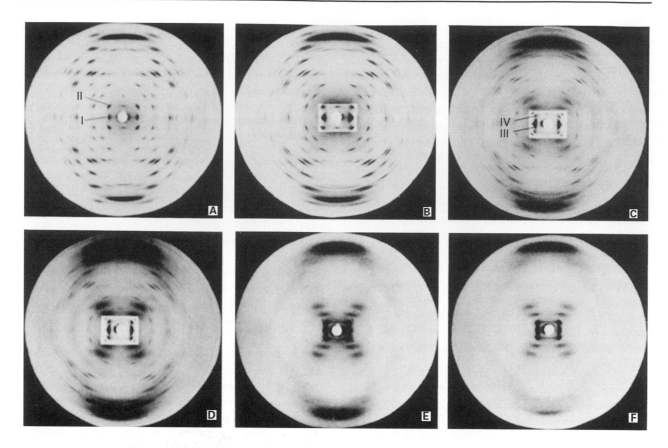

Figure 4 Diffraction patterns illustrating typical stages in the D to B transition in the DNA double-helix. In (A) the reflection marked I is related to the lateral intermolecular separation and that marked II to the helical pitch of molecules in the D form. In (C) the reflection marked III is related to the lateral intermolecular separation and that marked IV to the helical pitch of molecules in the semicrystalline intermediate form.

diffraction showed that muscle consisted of a regular array of actin filaments interpenetrating a regular array of myosin filaments and that during muscular contraction one family of filaments slid over the other. The dimensions of these arrays and the periodicities along the length of the actin and myosin filaments are within the 10 to 100 nm range and X-ray fibre diffraction information on these structures has largely come from the small angle regime. A particularly well defined example of such SAXS data is in **Figure 2**. The high brilliance of synchrotron sources together with recent advances in the efficiency of multiwire area detectors have allowed the variation in diffraction during the contraction cycle, in which a dissected muscle is repeatedly stimulated by an electrical voltage, to be recorded with a time resolution of a few milliseconds. This data is being used to model changes in the interaction between the actin filaments and the heads of the myosin molecules making up the myosin filaments since it is believed that changes in the stereochemistry of these interactions is the origin of the mechanical stress which results in shortening in the length of the

muscle against a load. This modelling is an excellent example of complementarity between an X-ray structure determination of a protein single crystal (in this case the structure of isolated myosin heads) and a small angle X-ray fibre diffraction study of the whole muscle assembly.

Strain-induced crystallization in poly(ethylene terephthalate) (PET)

One of the most challenging issues in the crystallization of polymers is the nature and kinetics of the strain-induced crystallization that takes place when polymers such as PET are mechanically drawn close to the glass transition temperature. From recent experiments at the ESRF and SRS, it has been possible to monitor crystallization in real-time when amorphous PET was drawn at rates comparable to those used in industrial processing. The X-ray camera used in these studies consisted of a 150 mm × 150 mm × 150 mm oven of which three sides were interchangeable so that it was possible to exchange sides depending on the particular application. This, for example, allowed the viewing port in which a video camera was mounted to be varied. The temperature could be controlled to within 1°C and the maximum temperature attainable was 350°C. A strip of polymer material could be clamped between two jaws attached to stepper motors which allowed uniaxial bidirectional drawing at rates up to 72 000% per minute. Diffraction patterns were recorded with exposure times of 40 milliseconds. The variation in WAXS in a typical experiment is shown in **Figure 5**. In this sequence of patterns the sharp crystalline peaks can be seen to progressively emerge above a diffuse background scatter. Simultaneously recorded video camera observations on changes in the position of reference marks on the specimen are also shown in **Figure 5** and allow the precise draw ratio

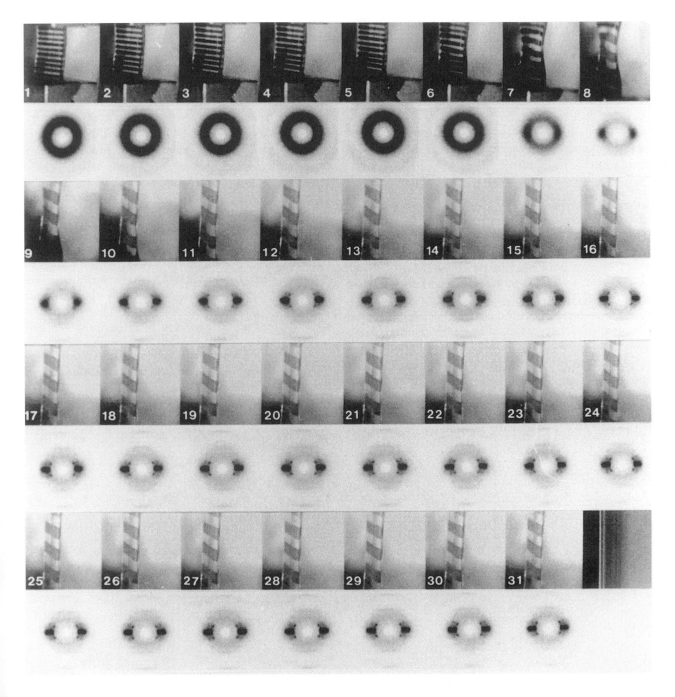

Figure 5 WAXS patterns with corresponding video image of sample of PET recorded during drawing at 90°C at an overall draw rate of 72 000%/min. Each of the 31 frames was recorded with an exposure time of 40 ms. There was essentially no dead time between frames so that the drawing experiment was completed in ~1.2 s.

and draw rate to be calculated at the point in the specimen from which X-ray diffraction data was recorded. Measurements on these patterns allowed the development of polymer orientation and crystallinity to be quantified and hence the kinetics of crystallization to be determined for a number of temperatures and for a variety of draw rates and draw ratios. An important conclusion from these studies, which is at variance with earlier conclusions drawn from studies on samples quenched at various stages during drawing and annealing, is that, at the fast draw rates typical of industrial processing, significant crystallization does not begin until the conclusion of drawing. When the rate constants determined from these experiments were related to theoretical models for the relaxation of deformed chains within entangled networks it became clear that at fast draw rates the onset of crystallization does not occur until the end of draw because the draw rate exceeds the rate of chain relaxation.

Relation between long and short range order during deformation of polyethylene

Instrumentation for simultaneous recording of SAXS and WAXS data is illustrated schematically in **Figure 6**. A hole, ~10 mm diameter, in the centre of the 15 μm thick aluminium foil through which diffraction beams exit the camera is covered with a circular glass cover slip (13 mm in diameter; and 15 μm thick) so that the SAXS and the main beam not scattered by the specimen pass wholly through the glass cover slip. This avoids diffraction of the main beam by the aluminium resulting in spurious sharp diffraction peaks in the WAXS pattern. Both SAXS and WAXS data were recorded using Photonic

Science CCD detectors. An evacuated beam pipe was positioned between the SAXS detector and the glass cover slip mounted on the aluminium foil to minimize scattering by air along the long path length. The WAXS detector was positioned close to the aluminium foil window at the rear of the camera so that SAXS data could still pass unobstructed for collection at the second CCD detector.

A specimen, 10 mm wide and 20 mm long, was cut from a 1 mm thick sample of high density polyethylene and mounted within the camera in jaws attached to stepper motors. The specimen was drawn at a rate of 18 000%/minute at a temperature of 35°C. Strain data, for the region of the specimens from which the SAXS/WAXS data was collected, was calculated from an accurately synchronized continuously recorded video image of the specimen. **Figure 7** shows selected SAXS and WAXS frames recorded from the specimen. In the initial WAXS frame two rings of intensity are clearly visible. These may be indexed as reflections from the orthorhombic form of polyethylene. The initial SAXS frame consists of a largely isotropic ring indicating that the undeformed sample is spherulitic with a lamellar spacing of ~35.0 nm. From the sequence of WAXS patterns it is apparent that increasing strain results in increasing orientation in both of the prominent orthorhombic reflections together with the transient appearance of a monoclinic phase. The SAXS patterns show striking variation in intensity which can be attributed to the formation of microvoids.

Variation in crystallite order and orientation in spherulites of Biopol

X-ray diffraction provides one of the most powerful techniques for the investigation of the spatial

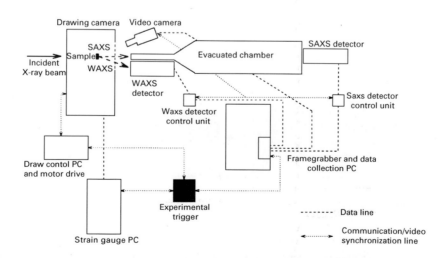

Figure 6 Schematic representation of the experimental arrangement for simultaneous recording of SAXS/WAXS and strain data during mechanical deformation of a sheet of polymer material.

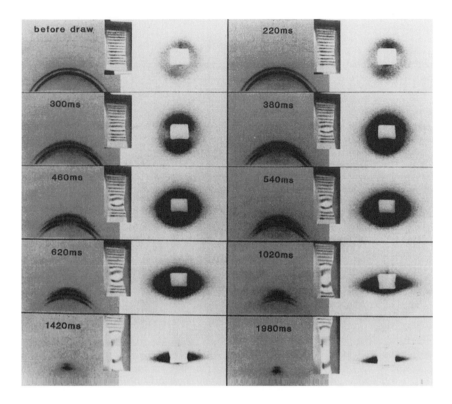

Figure 7 Selected pairs of simultaneous WAXS (left) and SAXS (right) frames with inserts of corresponding synchronously recorded video images which allow strain data to be related to changes in polymer structure and organization of high density polyethylene. Each frame was recorded in 40 ms.

variation of polymer conformation and organization within partially ordered materials. From optical microscopy it is clear that molecular orientation within many polymer materials varies over dimensions of less than 100 μm. Characterization of this variation by X-ray diffraction has been achieved using the microfocus capability of the ESRF which can provide an intense highly monochromatic X-ray beam with a wavelength of 0.092 nm in a beam diameter at the specimen of ~10 μm. Crucial to these experiments was the availability of a computer-controlled X/Y stage which allowed the specimen to be tracked in two dimensions perpendicular to the X-ray beam in steps as small as 1 μm (**Figure 8**). WAXS data were recorded using a Photonics Science CCD detector from spherulites of the optically active thermoplastic polymer Biopol. This material is produced by a variety of microorganisms, which use it for energy and carbon storage. The biological rather than petroleum origins of Biopol, its biodegradability, and its ability to form films with the properties of conventional thermoplastics give it technological importance as well as fundamental interest. The purity of the polymer leads to an absence of heterogeneous nuclei and gives it the capacity to form large spherulites of high crystallinity. X-ray diffraction patterns with exposure times of 1 s were recorded as a spherulite of

Biopol was stepped across a square net of area 150 μm × 150 μm centred on the centre of the spherulite, in horizontal and vertical steps of 10 μm. The 15 × 15 diffraction patterns recorded in this way are assembled as a 2D array in **Figure 9**. This array of patterns vividly demonstrates that all the crystallites have their a-axes oriented along a radius of the spherulite.

Variation in polymer orientation and crystallinity across a plastic bottle wall

Poly(ethylene terephthalate) (PET) is used extensively in the manufacture of fibres, bottles and films. Physical properties, such as strength, transparency and gas-transport properties depend on the degree of polymer orientation and crystallinity and their variation in the artefact.

During blowing of a bottle, the inner surface of a preform is cooled by the expansion of the compressed air used for inflation, the outer surface is cooled by contact with the cold mould whilst the internal thermal energy is raised by strain heating. In addition, the contour shape of the container and the complex blowing process itself produces considerable variation in wall thickness and direction of stretching at different positions in the container. The

Figure 8 Experimental arrangement for recording WAXS data with high spatial resolution. The specimen is mounted on a goniometer on an X/Y stage which allows tracking across the X-ray beam leaving the glass focusing collimator. Data is recorded on the CCD detector. A backstop to absorb unscattered X-rays is suspended immediately before the detector.

Figure 9 X-ray fibre diffraction patterns from the region of the spherulite of Biopol illustrated in the inserted polarizing microscope image. Each pattern was recorded from an area 10 μm in diameter with its position in this figure corresponding to the position in the specimen from which it was recorded. Each crystallite is oriented with its *a*-axis radial; the degree of orientation decreases towards the centre of the spherulite.

polymer therefore experiences a wide variation in draw temperature, draw ratio and draw direction at different positions of the container and at different depths in the wall, resulting in a wide variation of polymer orientation and crystallinity.

The sample for X-ray microbeam examination was prepared by cutting a 1 mm square piece from the container and mounting it on an X/Y stage so that it could be tracked across the X-ray beam. **Figure 10** shows selected frames from a dataset comprising 32 frames. The patterns show a marked variation in crystallinity through the thickness of the wall. At the inside edge there are strong oriented crystalline reflections superimposed on a fairly isotropic amorphous halo. The intensification of the crystal reflections around the equator is consistent with preferential chain alignment along the longitudinal axis of the bottle. The intensity of the crystal reflections diminishes towards the centre of the wall thickness leaving an almost isotropic amorphous diffraction halo. Nearer the outer edge of the wall there is an intensification of the halo on the equator indicating an increasing degree of orientation in the disordered

chains. These observations can be related to variation in the local draw rate, draw ratio and temperature across the wall during fabrication and provide a basis for refining the manufacturing process to produce bottles of improved performance.

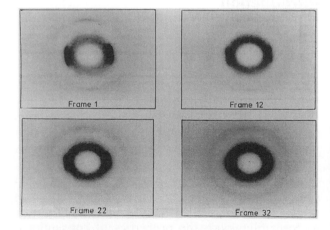

Figure 10 Variation of crystallite size, crystallinity and polymer orientation as a function of position in the wall of a PET bottle. The X-ray diffraction patterns were selected from the 32 frames recorded during the experiment. Frame 1 was collected at the inside face of the bottle wall, whilst frame 32 was collected at the outside face.

See also: **Inorganic Compounds and Minerals Studied Using X-Ray Diffraction; Materials Science Applications of X-Ray Diffraction; Nucleic Acids and Nucleotides Studied Using Mass Spectrometry; Nucleic Acids Studied Using NMR; Polymer Applications of IR and Raman Spectroscopy; Powder X-Ray Diffraction, Applications; Small Molecule Applications of X-Ray Diffraction.**

Further reading

Blundell DJ *et al.* (1996) Characterization of strain-induced crystallization of poly(ethylene terephthalate) at fast draw rates using synchrotron radiation. *Polymer* 37: 3303.

Donald AM and Windle AH (1992) *Liquid Crystalline Polymers*. Cambridge: Cambridge University Press.

Harford J and Squire J (1997) Time-resolved diffraction studies of muscle using synchrotron radiation. *Reports of Progress in Physics* 60: 1723–1787.

Hasnain SS (ed) (1990) *Synchrotron Radiation and Biophysics*. Chichester: Ellis Horwood Limited.

Hughes DJ *et al.* (1997) A simultaneous SAXS/WAXS and stress–strain study of polyethylene deformation at high strain rates. *Polymer* 38: 6427–6430.

Mahendrasingam A *et al.* (1995) Microfocus X-ray diffraction of spherulites of poly-3-hydroxybutyrate. *Journal of Synchrotron Radiation* 2: 308–311.

Mahendrasingam A *et al.* (1986) Time-resolved X-ray diffraction studies of the B to D structural transition in the DNA double helix. *Science* 233: 195–197.

Martin C *et al.* (1997) Investigation of the variation in orientation and crystallinity in poly(ethylene terephthalate) containers using microfocus X-ray diffraction. *Journal of Synchrotron Radiation* 4: 223–227.

Michette A and Pfauntsch S (eds) (1996) *X-rays: The First Hundred Years*. New York: Wiley. See in particular articles by Franks A (The First Hundred Years), Fuller W (X-ray Diffraction), Munro I (Synchrotron Radiation) and Miller A (The Next Hundred Years).

Ryan AJ and Wilkinson N (1997) *Polymer Processing and Structure Development*. Kluwer Academic Publishers.

Field Ionization Kinetics in Mass Spectrometry

Nico MM Nibbering, University of Amsterdam, The Netherlands

MASS SPECTROMETRY
Methods & Instrumentation

Introduction

The mass spectra of molecules, ionized by electron ionization (EI), field ionization (FI) or photoionization (PI) in the gas phase, usually show many peaks with varying intensities. These peaks result from fragment ions generated by a series of competing and consecutive unimolecular dissociation reactions of the molecular ions provided that the latter have obtained during their formation an amount of energy in excess of the ionization energy of the corresponding molecules. The resulting mass spectrum therefore is determined by the relative rates of these unimolecular dissociation reactions, which in their turn are related to the relative abundances of the fragment ions observed. It will be clear that studying the kinetics of ion dissociation can provide an invaluable insight into the foundation of a mass spectrum.

Although trapped-ion mass spectrometry has been used to investigate the extent of fragmentation of ions as a function of time in the time range of μs to ms, another and unique method is that of field ionization kinetics (FIK). The roots of this method date from the work of Beckey at the University of Bonn, Germany, who interpreted (as early as 1961) the broadening of peaks in field ionization (FI) mass spectra, obtained with a single-focusing mass spectrometer, as a consequence of ion lifetimes. Picosecond resolution of these ion lifetimes was achieved in 1971 by Derrick and Robertson at King's College in London, United Kingdom on the basis of a thorough calibration of the time-scale following FI. Since then the development of the experimental techniques to measure such ion lifetimes and of the necessary theory of the corresponding ion decomposition rates have led to the use of the name FIK. With this method the rates of gas phase unimolecular decompositions of ions, occurring at times as short as picoseconds and over a time range of seven orders of magnitude (10^{-12}–10^{-5} s), and corresponding ion lifetimes can be measured.

Following a description of the principle of the FIK method, some illustrative examples of its application in gas-phase ion chemistry studies will be presented.

Principle of the FIK method

FIK experiments are performed on double-focusing mass spectrometers having either the conventional (the electric sector precedes the magnetic sector) or the reversed (the magnetic sector precedes the electric sector) geometry. A schematic diagram of a field ionization double-focusing mass spectrometer with the conventional geometry is given in **Figure 1**.

Ions are generated in a very narrow region close to an anode (in the case of positive ions). This region becomes less narrow as the anode is changed from a tip, to a blade to a 1 μm bare tungsten wire or to a 10 μm activated tungsten wire (**Figure 2**) at a potential V_0 (~8–10 kV) with respect to a slotted cathode at ground potential and at a distance that is increased from 0.25 to 1.5 mm.

If these ions are stable for $\geq 10^{-5}$ s they will acquire a kinetic energy qV_0 during acceleration from the ionization region (where to a good approximation the potential is equal to the anode potential V_0) to the slotted cathode and will pass the electric sector set to transmit ions with a kinetic energy qV_0. Consequently, they will be mass analysed at their correct m/z values if the magnetic field is scanned. In this way, a FI spectrum is obtained which contains peaks that correspond to the ions generated very near the anode, i.e. within approximately 10^{-11} s. Fragment ions m^+ (see Eqn [1]), generated in unimolecular dissociation reactions that occur between the anode and cathode (region I in **Figure 1**) at a potential V_x, will not be transmitted through the electric sector because

they have insufficient kinetic energy $[q(m/M)(V_0 - V_x) + qV_x]$ to be focused by this sector.

$$M^{\bullet +} \rightarrow m^+ + (M - m)^{\bullet} \qquad [1]$$

However, these ions can be transmitted through the electric sector by either decreasing its potential at a constant anode potential V_0 or by increasing the anode potential at a constant electric sector potential set to transmit ions with a kinetic energy qV_0. The most widely used method involves increasing the anode potential.

The determination of the amount, ΔV, with which the anode potential V_0 has to be increased to satisfy Equation [2] permits a calculation of the potential V_x at which reaction [1] has occurred.

$$qV_0 = q(m/M)(V_0 + \Delta V - V_x) + q V_x \qquad [2]$$

It is then possible to calculate the ion lifetime, i.e. the time elapsed between formation of the molecular ion $M^{\bullet +}$ and its subsequent fragmentation at potential V_x, if the potential distribution between the anode and cathode is known. Thus, a scan of the anode potential achieved by increasing ΔV allows one to monitor reaction [1], following FI of M over a continuous time range of $\sim 10^{-11}$–10^{-9} s (region I in **Figure 1**). The required, time-consuming data processing can be speeded up by use of an online computer, which also enables computer controlled data acquisition and multiscan averaging to enhance the signal-to-noise ratio of the usually weak FIK signals.

Decompositions of $M^{\bullet +}$ ions at times longer than 10^{-9} s can be investigated in the focusing region II

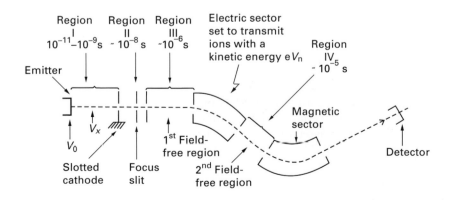

Figure 1 Schematic diagram of a double-focusing field ionization mass spectrometer of conventional geometry (not drawn to scale), showing the four regions in which field ionization kinetic measurements can be made. The stated times refer to an ion source with a blade emitter. Reproduced with permission of Wiley from Nibbering NMM (1984) Mechanistic studies by field ionization kinetics. *Mass Spectrometry Reviews* **3**: 445–477.

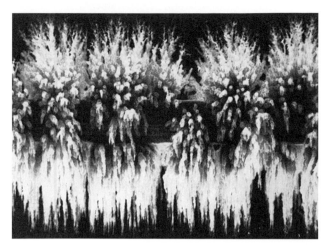

Figure 2 Scanning electron micrograph of a 10 μm tungsten wire covered with carbon microneedles grown by activation.

and in the first and second field-free regions III and IV (**Figure 1**) by using metastable ion methods.

The experimental results take the form of curves of product ion current $I_m(t)$ versus kinetic energy, which are transformed according to the principles outlined above into curves of product ion current $I_m(t)$ versus reaction time t. The product ion current $I_m(t)$ is related to the rate of reaction, $dI_m(t)/dt$, by Equation [3], where Δt is a small interval within which decomposing $M^{•+}$ ions contribute to the product ion current $I_m(t)$ at time t:

$$\frac{dI_m(t)}{dt} = \frac{I_m(t)}{\Delta t} \quad [3]$$

This interval is determined by the energy resolution of the electric sector and depends on the molecular and product ion masses and on the reaction time t, i.e. this time window interval will decrease with increasing mass difference between the molecular and product ions (not unlimited because of the difficulties in increasing the anode potential due to arcing problems) and at shorter decomposition times of the molecular ions because of the increase of the electric field gradient with decreasing distance from the anode.

Division of Equation [3] by the measured mass-resolved molecular ion current I_M at an anode potential V_0, i.e. in the normal FI spectrum, yields the so-called normalized rate $k(t)$:

$$k(t) = I_m(t)/I_M \Delta t \quad [4]$$

Plots of $k(t)$ versus t obtained with different mass spectrometers, different anodes and different ion life-

time calculations have been shown to be reproducible to within 25%.

The normalized rate $k(t)$ is related to the time-independent rate constant $k(E)$ as defined by the quasi-equilibrium theory (QET) in the following way. Suppose that the molecular ions $M^{•+}$ decompose via a single channel to fragment ions m^+. If the $M^{•+}$ ions have acquired an internal energy E upon FI, their abundance $M(t)$ at time t will be given by

$$M(t) = M_0 e^{-k(E)t} \quad [5]$$

where M_0 is the abundance of the molecular ions at the time of their formation. The rate of formation of the fragment ions m^+ is then given by

$$dm^+/dt = k(E)M(t) = k(E)M_0 e^{-k(E)t} \quad [6]$$

The molecular ions $M^{•+}$, however, will have acquired a distribution of internal energies, $P(E)$, upon FI. In this case all individual rates of formation of m^+ given by Equation [6] have to be summed with respect to the internal energy E. This leads to Equation [7] with $P(E)$ as the weighting function and $I_m(t)$ as the product ion current at time t (see above):

$$\frac{dI_m(t)}{dt} = \int_0^\infty P(E)k(E)M_0 e^{-k(E)t} dE \quad [7]$$

Similarly, the molecular ion current I_M is obtained by integration of Equation [5] with respect to the internal energy E, taking $P(E)$ as the weighting function:

$$I_M = \int_0^\infty P(E)M_0 e^{-k(E)t} dE \quad [8]$$

Division of Equation [7] by Equation [8] then gives the relation between the normalized rate $k(t)$ and the time-independent rate constant $k(E)$:

$$k(t) = \frac{\int_0^\infty P(E)k(E)e^{-k(E)t} dE}{\int_0^\infty P(E)e^{-k(E)t} dE} \quad [9]$$

Usually a maximum is observed in the $k(t)$ versus t plots. The maximum has been suggested to represent

$k_{max}(E)$ in the distribution of $k(E)$s and generally occurs at longer times for rearrangement reactions than for simple cleavage reactions. These reactions usually compete with each other continuously over the time range of $10^{-11} - 10^{-5}$ s.

An example of a logarithmic plot of $k(t)$ versus t for the loss of a methyl radical from the molecular ion of pent-3-en-2-ol, $CH_3CH=CHCH(OH)CH_3$, is given in **Figure 3**.

The figure shows two maxima. The first maximum at a molecular ion lifetime of $10^{-10.5}$ s corresponds to the loss of a methyl radical from the unrearranged molecular ion via a direct cleavage reaction, whereas the second maximum at a molecular ion lifetime of $10^{-10.1}$ s is owing to the loss of a methyl radical following a rearrangement reaction of the molecular ion.

The strength of FIK in mechanistic gas-phase ion dissociation studies compared with the conventional EI method is that it gives a time-resolved view of the chemistry and isomerization reactions of ions over a time range of $10^{-12} - 10^{-5}$ s. The details of these processes are largely lost in the EI method which integrates all the events up to 10^{-6} s following ionization, a very broad time range compared with the vibrational periods of chemical bonds ($10^{-14} - 10^{-13}$ s).

The strength of the FIK method in mechanistic ion dissociation studies is shown to full advantage when it is used in combination with stable isotopic labelling. In that case, the fragment ions from competing reactions are nearly equal in mass so that the ratios of their ion currents measured at any time t are practically equal to the ratios of the normalized rates $k(t)$ (see above and Eqn [4]). Most of the mechanistic ion dissociation studies discussed below are therefore based upon changes in relative rates as a function of the ion lifetimes.

Mechanistic ion dissociation studies

Four examples of the application of FIK to mechanistic ion dissociation studies will be presented.

The first deals with the elimination of ethylene from the molecular ion of cyclohexene, the second and third with the dynamics of stereochemically controlled dissociation reactions and the fourth with time-resolved MS-MS to identify the structures of product ions generated at different molecular ion lifetimes.

Elimination of ethylene from the molecular ion of cyclohexene

In both linear and cyclic alkene EI-generated molecular ions, the double bond shifts before or during fragmentation, cause the mass spectra of isomers to be very similar, if not indistinguishable. The same is true for the EI generated molecular ion of cyclohexene, in which specific D-labelling has shown a practically statistical distribution of H and D atoms over the carbon skeleton before or during fragmentation.

Upon FI, four variously deuterated ethylene molecules, C_2H_4, C_2H_3D, $C_2H_2D_2$ and C_2HD_3 are eliminated from cyclohexene-3,3,6,6-d_4.

At 10^{-11} s, more than 50% is lost as C_2H_4, but at longer times a rapid increase of the loss of the deuterated ethylene molecules is observed until at 10^{-9} s a statistical distribution of the H and D atoms has occurred before decomposition. The specific loss of C_2H_4 at 10^{-11} s can readily be explained by a retro-Diels–Alder reaction of the molecular ion to give the ion m/z 58 (Eqn [10]). The interesting observation made, however, is that the maximum rates of the variously labelled ethylene eliminations occur at different times; the order with increasing time is $C_2H_4 < C_2H_2D_2 < C_2HD_3 < C_2H_3D$. This has been rationalized by invoking 1,3-allylic H/D shifts in the molecular ions before the retro-Diels–Alder reaction (Eqn [10]).

As can be seen from Equation [10], the formation of m/z 58, 56, 55 and 57 requires 0, 1, 2 and 3 allylic 1,3-H/D shifts, respectively, which coincides with the shift of the maximum rate of the corresponding ethylene eliminations to longer times.

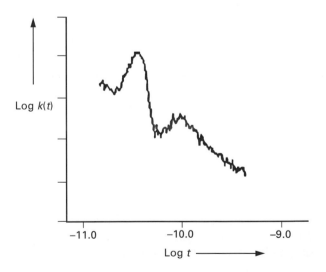

Figure 3 Normalized rate constant $k(t)$ for the loss of a methyl radical from the molecular ions of pent-3-en-2-ol, $CH_3CH=CHCH(OH)CH_3$. The vertical axis covers log $k(t)$ values in the range of 8–12. Reproduced with permission from Zappey HW, Ingemann S and Nibbering NMM (1991) Study of the rate of methyl loss and the structure of the product ion as a function of the lifetime of the molecular ion of penten-3-en-2-ol. *Organic Mass Spectrometry* **26**: 241–246.

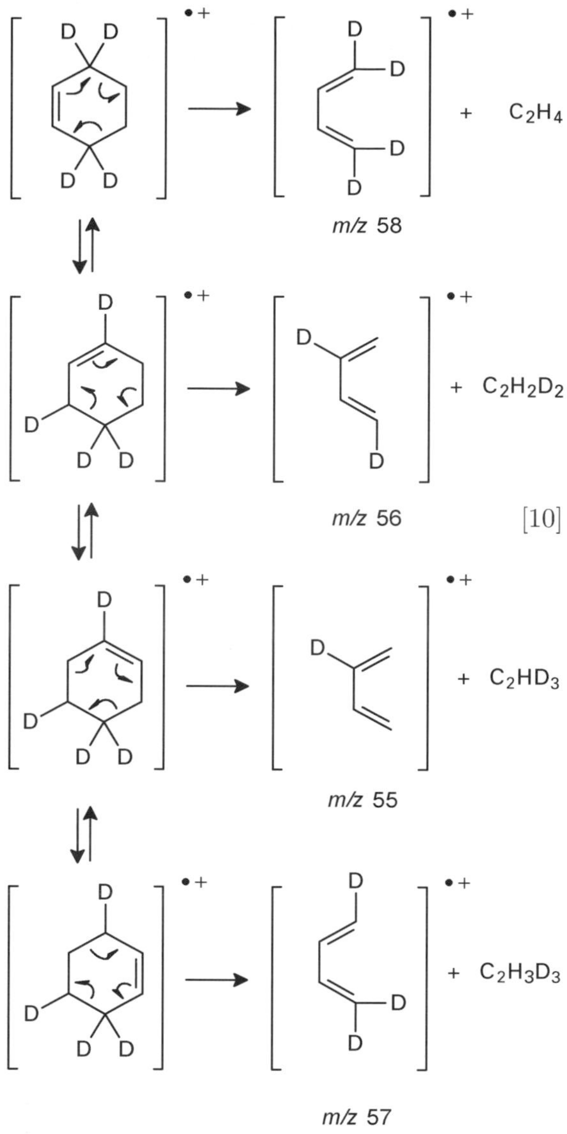

m/z 58

$+ \; C_2H_4$

$+ \; C_2H_2D_2$

m/z 56

[10]

$+ \; C_2HD_3$

m/z 55

$+ \; C_2H_3D_3$

m/z 57

[11]

Loss of a molecule of water from the molecular ions of stereoisomeric cyclohexane diols

An FIK study of the water loss from stereoisomeric cyclohexane diols has shown how elegantly this method can give an insight into the dynamics of unimolecular decompositions of organic ions. This is especially true for the *trans-* and *cis*-1,3- and 1,4-diols, which generate relatively abundant $(M–H_2O)^{•+}$ ions upon FI, in contrast with the *trans-* and *cis*-1,2-diols. The latter compounds yield low-abundant $(M–H_2O)^{•+}$ ions, which is probably owing to a fast cleavage of the C_1–C_2 bond of the molecular ions.

The 1,3-d_2 analogue of *trans*-cyclohexane-1,3-diol and the 1,4-d_2 analogue of the *trans*-1,4-diol

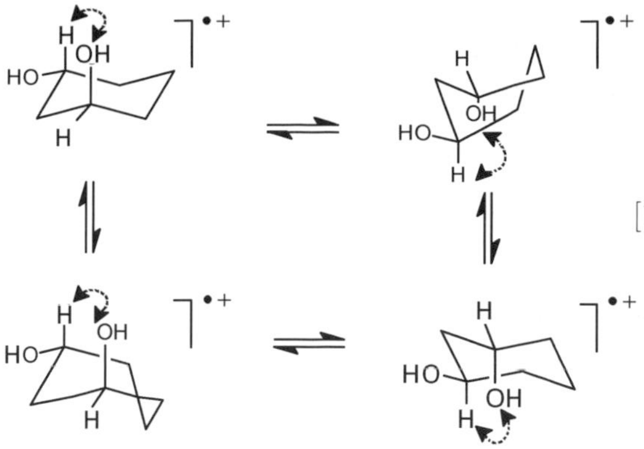

molecular ions predominantly lose water by a transannular 1,3- and 1,4-elimination, respectively, as determined by using FIK. In both cases the normalized rate for this process reaches a maximum value at an ion lifetime of about 110 ps, but the rate is also significantly higher for the *trans*-1,3-diol than for the *trans*-1,4-diol at shorter ion lifetimes. The explanation for these observations is the following: both in the molecular ion of the *trans*-1,3-diol and in that of the *trans*-1,4-diol the hydroxyl group has to come in close proximity of the hydrogen atom to be transferred to effect the water elimination. This is the case for the two identical ground-state chair conformations of the *trans*-1,3-diol, which have a distance of 2.5 Å between the oxygen atom of the hydroxyl group and the hydrogen atom to be transferred, and for the corresponding boat conformation, as shown in Equation [11].

However, for the *trans*-1,4-diol, only the boat conformations have the appropriate geometry for the transannular 1,4-elimination of water, as shown in Equation [12].

Similar arguments apply to the enhanced transannular loss of water from *cis*-cyclohexane-1,3-diol versus that from the *cis*-1,4-diol. The departing water now contains both hydroxyl hydrogen atoms,

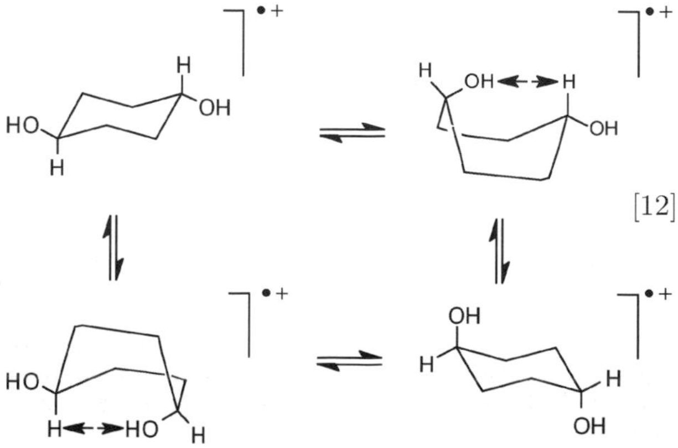

[12]

as shown by D labelling. In both cases the normalized rate for this reaction again reaches a maximum value at ion lifetimes of about 110 ps.

Dyotropic hydrogen rearrangement in the molecular ions of 8,9-disubstituted-tricyclo[5.2.1.0²,⁶] decenes

Upon EI the *exo* and *endo* isomers of the 8,9-disubstituted-tricyclo[5.2.1.0²,⁶]decenes [1] and [2], respectively, fragment essentially differently. The ionized *endo* isomers [2] eliminate C_5H_8 via a retro-Diels–Alder reaction after migration of the hydrogen atoms from positions 8 and 9 to the double bond. This reaction is not observed for the *exo* isomers [1] where the distance between the hydrogen atoms of positions 8 and 9 and the double bond is too large for the hydrogen migration required to effect the retro-Diels–Alder reaction (**Scheme 1**).

From these observations, however, it cannot be concluded whether the double hydrogen migration in ionized [2] occurs in a concerted or stepwise fashion. If the hydrogen atoms would migrate in a concerted fashion, then it is an example of a dyotropic reaction, being defined as an uncatalysed process in which two σ-bonds simultaneously, but not necessarily via a fully symmetrical mode, migrate intramolecularly.

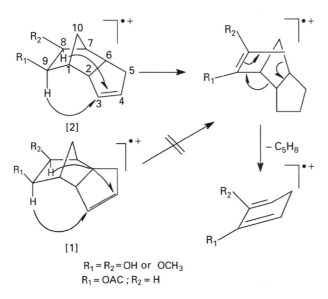

[2]

[1]

R₁ = R₂ = OH or OCH₃
R₁ = OAC ; R₂ = H

Scheme 1 Double hydrogen transfer in the molecular ion of the *endo* isomer [2] of 8,9-disubstituted-tricyclo[5.2.1.0²,⁶]decene (not possible in the molecular ion of the *exo* isomer [1] followed by elimination of C_5H_8 via a retro-Diels–Alder reaction. Reproduced with permission of the American Chemical Society from Kluft E, Nibbering NMM, Kühn H and Herzschuh R (1986) Dyotropic hydrogen rearrangement in the molecular ions of 8,9-disubstituted tricyclo [5.2.1.0²,⁶] decenes as supported by field ionization kinetics. *Journal of the American Chemical Society* **108**: 7201–7203.

Such a suprafacial dyotropic transfer of hydrogen is a thermally 'allowed' $[_\sigma 2_s + _\sigma 2_s + _\pi 2_s]$ pericyclic reaction.

Experimental support for either a concerted or a stepwise double hydrogen migration in ionized [2] before elimination of C_5H_8 via a retro-Diels–Alder reaction (**Scheme 1**) is obtained from application of FIK to both *exo* and *endo* isomers [1] and [2] respectively.

For example, the *exo* isomer [1a] with $R^1 = R^2 = OCH_3$ appears to eliminate, exclusively, C_5H_6 over all molecular ion lifetimes studied, while the corresponding *endo* isomer [2a] eliminates both C_5H_6 and C_5H_8, but not $C_5H_7^\bullet$, see **Figures 4A** and **4B**, respectively.

As expected, the C_5H_8 loss is a delayed process with respect to the C_5H_6 loss because of the required double hydrogen rearrangement, i.e. this channel starts to compete with the C_5H_6 loss at molecular ion lifetimes of $\sim 10^{-10}$ s and becomes dominant beyond molecular ion lifetimes of $\sim 10^{-9.25}$ s. Important, however, is the absence of $C_5H_7^\bullet$ loss, which is expected to occur for a stepwise, but not for a concerted, double hydrogen rearrangement.

Methyl loss from ionized pent-3-en-2-ol, CH₃CH=CHCH(OH)CH₃

A combined FIK and deuterium labelling study has revealed three pathways for the loss of methyl from the molecular ion of pent-3-en-2-ol, $CH_3CH=CHCH(OH)CH_3$. This is shown by the relative rate curves for loss of $CH_3^\bullet$ and $CD_3^\bullet$ from the molecular ion of pent-3-en-2-ol-1,1,1-d_3, $CH_3CH=CHCH(OH)CD_3$ in **Figure 5**.

At short ion lifetimes, $CD_3^\bullet$ is the major loss, whereas at medium ion lifetimes $CH_3^\bullet$ elimination is the major channel. At longer ion lifetimes the $CD_3^\bullet$ loss again becomes increasingly important.

These observations have been rationalized mechanistically in **Scheme 2**.

Route [3] → [4] is operative at ion lifetimes $< 10^{-10.5}$ s and is suggested to lead to oxygen protonated crotonaldehyde, route [3] → [5] → [6] → [7] is dominant at ion lifetimes between $10^{-10.5}$ and 10^{-9} s and is suggested to lead to oxygen protonated methyl vinyl ketone, while route [3] → [5] → [8] → [9] increases at longer ion lifetimes and is suggested to produce the butyryl cation (**Scheme 2**).

The existence of different pathways is also indicated by the normalized rate curve for methyl loss from the molecular ion of unlabelled pent-3-en-2-ol, which contains two distinct maxima at ion lifetimes of $10^{-10.5}$ and $10^{-10.1}$ s after ionization (**Figure 3**).

In the normalized rate curve for CD_3 loss from the molecular ion of $CH_3CH=CHCH(OH)CD_3$ similar

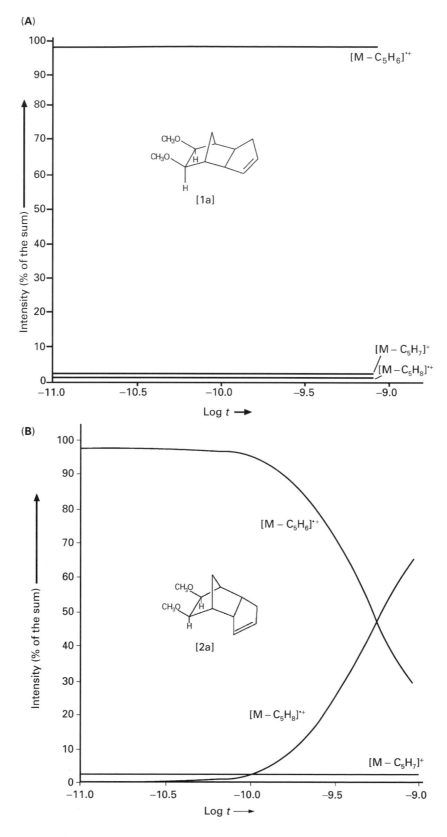

Figure 4 Losses of C_5H_6, $C_5H_7^{\bullet}$ and C_5H_8, expressed in percentages of their sum, as a function of the lifetime of the molecular ions of (A) *exo*- and (B) *endo*-8,9-dimethoxytricyclo[5.2.1.0²·⁶]decenes. Reproduced with permission of the American Chemical Society from Kluft E, Nibbering NMM, Kühn H and Herzschuh R (1986) Dyotropic hydrogen rearrangement in the molecular ions of 8,9-disubstituted tricyclo [5.2.1.0²·⁶] decenes as supported by field ionization kinetics. *Journal of the American Chemical Society* **108**: 7201–7203.

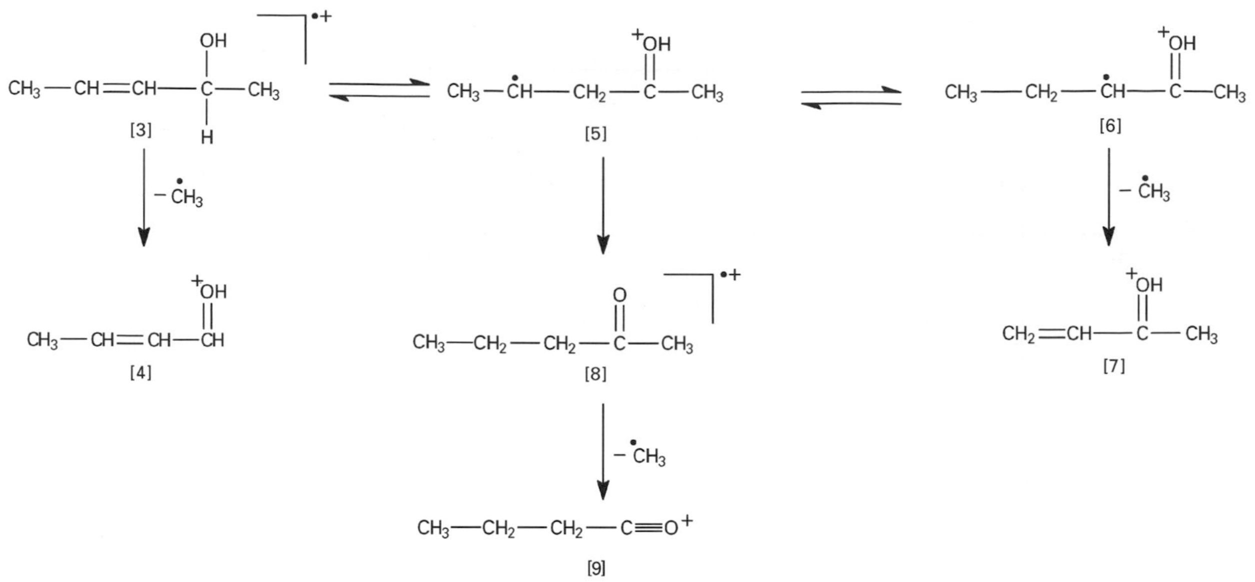

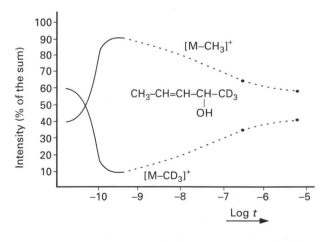

Scheme 2 Routes for methyl loss from ionized pent-3-en-2-ol. Reproduced with permission from Zappey HW, Ingemann S and Nibbering NMM (1991) Study of the rate of methyl loss and the structure of the product ion as a function of the lifetime of the molecular ion of penten-3-en-2-ol. *Organic Mass Spectrometry* **26**: 241–246.

maxima are observed, but in that for CH_3 loss only the maximum at $10^{-10.1}$ s is present. This means that the maximum at $10^{-10.5}$ s corresponds to the route [3] → [4] of **Scheme 2**, as expected for such a direct cleavage reaction, and that the maximum at $10^{-10.1}$ s is owing to the route [3] → [5] → [6] → [7] of **Scheme 2**. The ions [4 and 7] (and also [9]) formed

by the different pathways are known to be non-interconverting species with distinct collision-induced dissociation mass spectra. Collision-induced dissociation experiments on the (M − methyl)$^+$ ions, formed at different molecular ion lifetimes, i.e. at times $< 10^{-10.6}$ and $10^{-10.0}$ s after ionization, have provided experimental support for their structures. The resulting time-resolved MS-MS spectra, in which the peaks are rather noisy owing to their low intensities, are presented in **Figure 6**.

The collision-induced dissociation MS-MS spectra of the independently made reference ions obtained by protonation of crotonaldehyde (ion [4] in **Scheme 2**) and methyl vinyl ketone (ion [7] in **Scheme 2**) are shown in **Figure 7**.

Figure 6A appears to match well **Figure 7A** which provides evidence that most of the (M−methyl)$^+$ ions formed at about $10^{-10.6}$ s have the protonated crotonaldehyde structure, i.e. ion [4] in **Scheme 2**.

Figure 6B is nearly identical to **Figure 7B**, showing that the (M−methyl)$^+$ ions at about $10^{-10.0}$ s have indeed, predominantly, the protonated methyl vinyl ketone structure, i.e. ion [7] in **Scheme 2**.

Figure 5 Losses of $CH_3^\bullet$ and $CD_3^\bullet$ expressed in percentages of their sum, as a function of the lifetime of the molecular ion of pent-3-en-2-ol-1,1,1-d_3, $CH_3CH=CHCH(OH)CD_3$. Reprinted from the *International Journal of Mass Spectrometry and Ion Physics* **38**: Zwinselman JJ, Nibbering NMM, Middlemiss NE, Vajda JH and Harrison AG, A field ionization kinetics and metastable ion study of the fragmentation of some pentenols, 163–179, Copyright 1981, with permission from Elsevier Science.

List of symbols

E = internal energy; I_m = product ion current; I_M = molecular ion current; $k(t)$ = normalized rate constant; $k(E)$ = time-independent rate constant; t = time; V = potential.

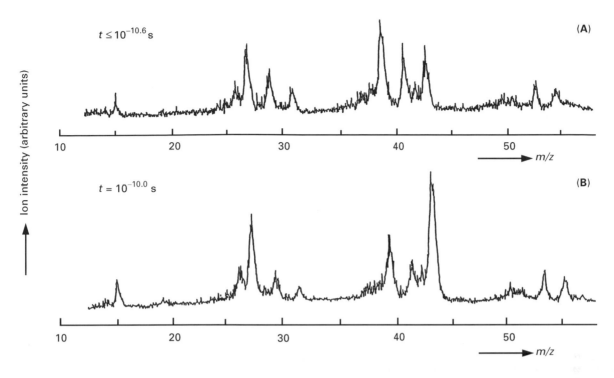

Figure 6 Collision-induced dissociation MS-MS spectra of (M–methyl)⁺ ions, generated upon FI from the molecular ion of pent-3-en-2-ol at molecular ion lifetimes of (A) ≤ $10^{-10.6}$ s and (B) $10^{-10.0}$ s. Reproduced with permission from Zappey HW, Ingemann S and Nibbering NMM (1991) Study of the rate of methyl loss and the structure of the product ion as a function of the lifetime of the molecular ion of penten-3-en-2-ol. *Organic Mass Spectrometry* **26**: 241–246.

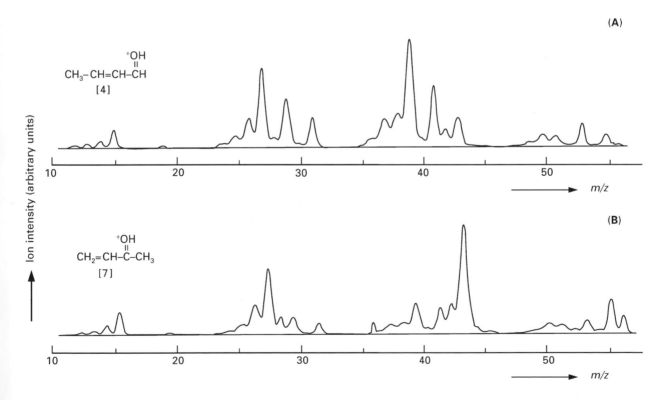

Figure 7 Reference collision-induced dissociation MS-MS spectra of (A) protonated crotonaldehyde [4] and (B) protonated methyl vinyl ketone [7]. Reproduced with permission from Zappey HW, Ingemann S and Nibbering NMM (1991) Study of the rate of methyl loss and the structure of the product ion as a function of the lifetime of the molecular ion of penten-3-en-2-ol. *Organic Mass Spectrometry* **26**: 241–246.

See also: **Field Ionization Kinetics in Mass Spectrometry; Fragmentation in Mass Spectrometry; Ion Dissociation Kinetics, Mass Spectrometry; Ion Structures in Mass Spectrometry; Ionization Theory; Isotopic Labelling in Mass Spectrometry; Metastable Ions; MS–MS and MS^n; Sector Mass Spectrometers; Statistical Theory of Mass Spectra; Stereochemistry Studied Using Mass Spectrometry.**

Further reading

Beckey HD (1961) Measurement of extremely short decay times of organic ions with the field ionization mass spectrometer. *Zeitschrift für Naturforschung* **16a**: 505–510.

Beckey HD (1977) *Principles of Field Ionization and Field Desorption Mass Spectrometry*. Oxford: Pergamon.

Derrick PJ, Falick AM and Burlingame AL (1972) Mechanistic and kinetic studies at 10^{-11}–10^{-5} sec of the unimolecular gas-phase reactions induced by field ionization of cyclohexene and cyclohexene-3,3,6,6-d$_4$. *Journal of the American Chemical Society* **94**: 6794–6802.

Derrick PJ (1977) Field ionization and field desorption. In: Johnstone RAW (ed) *Specialist Periodical Reports on Mass Spectrometry*, Vol 4, pp 132–145. London: The Chemical Society.

Levsen K (1978) *Fundamental Aspects of Organic Mass Spectrometry*. Weinheim: Verlag Chemie.

Nibbering NMM (1984) Mechanistic studies by field ionization kinetics. *Mass Spectrometry Reviews* **3**: 445–477.

Nibbering NMM (1991) Time-resolved studies of unimolecular gas phase ion decompositions by field ionization kinetics. In: Jennings KR (ed) *Fundamentals of Gas Phase Ion Chemistry*, pp 333–349. Dordrecht: Kluwer Academic Publishers.

Flame and Temperature Measurement Using Vibrational Spectroscopy

Kevin L McNesby, Citrus Heights, CA, USA

VIBRATIONAL, ROTATIONAL & RAMAN SPECTROSCOPIES
Applications

Introduction

The application of modern spectroscopic analysis to the study of flames began in the 1920s and 30s with breakthroughs in the understanding of atomic and molecular spectroscopy. The earliest spectroscopic investigations focused on understanding the line and band structure observed in the visible and ultraviolet regions of the spectrum when the light from the flame was dispersed by a grating or prism. One of the great breakthroughs of physics during this period was the understanding that the band structure observed in the emission spectra of flames originated from gas-phase molecular species.

For molecular species, understanding the appearance of flame spectra is simplified by assuming that the total internal energy, E_T, of a gas-phase molecule may be given to first order (Born Oppenheimer approximation) as:

$$E_T = E_{EL} + E_{VIB} + E_{ROT} \qquad [1]$$

Here, E_{EL}, E_{VIB}, and E_{ROT} are respectively the quantized electronic, vibrational, and rotational energy of the molecular species. In the scientific literature, these energies are usually expressed in terms of wavenumbers (the reciprocal wavelength, expressed in centimetres, which is directly proportional to energy, and given the symbol cm^{-1}). All observed spectral features, in emission and absorption, are caused by changes in total energy, ΔE_T, of the individual species present within the flame. Spectral features arise from the emission or absorption of a photon with energy corresponding to the difference between initial and final states of the transition. **Figure 1** illustrates the absorption of a photon corresponding to an electronic transition (**Figure 1A**) and to a vibrational transition (**Figure 1B**) for a diatomic molecule whose interatomic potential energy may be approximated by an anharmonic oscillator.

In general, changes in electronic, vibrational, and rotational energy correspond to emission or absorption of radiation in the visible (and ultraviolet), infrared, and microwave region of the electromagnetic spectrum, respectively. Typically, $\Delta E_{EL} > \Delta E_{VIB} > \Delta E_{ROT}$. For molecular species, changes in E_{EL} generally are accompanied by changes in E_{VIB} and

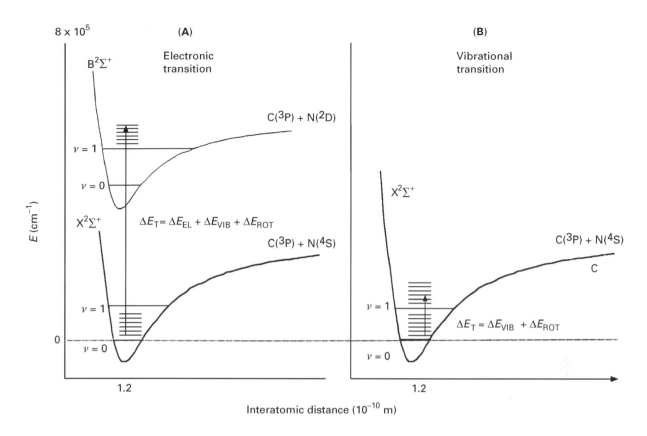

Figure 1 Potential energy diagram for a diatomic anharmonic oscillator. Electronic (A) and vibrational (B) absorption transitions are illustrated. Values for interatomic distance, energy (E), and term symbols are for the radical flame species CN.

E_{ROT}, and changes in E_{VIB} may be (and almost are) accompanied by changes in E_{ROT}. These general trends are illustrated in **Figure 1**. The band structure observed in the visible and ultraviolet spectra of molecular species in flames, in emission and absorption, is therefore understood to be the result of combined changes in electronic, vibrational, and rotational energies.

The visible and ultraviolet radiation from most flames usually accounts for less then 1% of the total emitted energy, with most of the energy emitted by a flame occurring in the infrared region of the spectrum. To see why this is not unexpected, it is useful to compare the radiation emitted by a flame with the radiation emitted by a blackbody. In a system at thermodynamic equilibrium (which, on a macroscopic scale, a flame is not), the distribution of radiation is given by Planck's blackbody radiation law:

$$I_\lambda = [2E_\lambda c_1 \lambda^{-5}/(\exp{[c2/\lambda T]} - 1)]A \, d\lambda \quad [2]$$

Here, I_λ is the wavelength-dependent radiant intensity normal to the surface of the radiator, E_λ is the emissivity at wavelength λ, c_1 and c_2 are the first and second radiation constants and have the values 0.588×10^{-8} Wm^{-2} and 1.438×10^{-2} m K^{-1}, respectively, T is temperature in kelvin (K) and A is the area of the surface in square metres (m^2).

Figure 2 is a plot of Equation [2] at several temperatures. **Figure 2** shows that for a blackbody radiator, at temperatures up to 2300 K, the peak spectral radiancy always occurs in the infrared region of the spectrum. As the temperature increases above 2300 K, the peak of spectral radiancy moves to shorter wavelengths (towards the visible region of the spectrum). It is important to note that for a blackbody, the value of E_λ is equal to 1 at all wavelengths. In flames, the value of E_λ varies (and may approach 1) with wavelength, but is near zero for most wavelengths, indicating that a flame is not a blackbody, and that flame gases may not be in thermodynamic equilibrium. Nevertheless, in some cases where the emissivity of a flame species is known, measurements of spectral radiancy of flames may be used to calculate flame temperatures to within an accuracy of several kelvin.

When the regions of non-zero emissivity ($E_\lambda > 0$) in the flame emission spectrum are expanded along the wavelength scale, these regions exhibit detailed

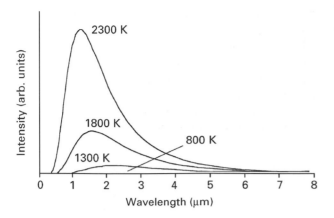

Figure 2 Radiation from an ideal blackbody at several temperatures.

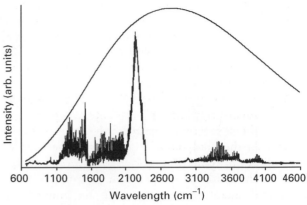

Figure 3 The infrared emission spectrum measured using an FT-IR spectrometer from an 18 torr CH_4/O_2 flame and the blackbody radiation (1173 K) emitted over the same region. The infrared emission spectrum has not been corrected for the responsivity of the detector.

fine structure. In the visible and ultraviolet regions of the flame emission spectrum (in general, radiation with a wavelength between 1 μm and 200 nm), this fine structure represents changes in rotational and vibrational energy which accompany changes in electronic energy. In the more intense infrared region of the flame emission spectrum (in general, radiation with a wavelength between 1 and 30 μm), the observed fine structure is caused by transitions between rotational energy levels which occur with a change in vibrational energy, but with no change in electronic energy. The amount by which the vibrational and rotational energies may change during a transition is governed by selection rules, which are largely dependent upon the symmetry of the species involved in the transition.

Figure 3 shows the raw emission spectrum from a premixed, reduced pressure (18 torr) stoichiometric CH_4/O_2 flame measured using a Fourier transform infrared (FT-IR) spectrometer and the calculated emission from a blackbody at 1173 K over the same spectral region. The peak temperature in this flame was measured (using a fine wire Pt-Pt/10%Rh thermocouple) to be near 2150 K. Comparison of flame emission spectra to calculated blackbody radiance must take into account the emissivity of the different species within the flame, reabsorption of emitted radiation by cold gases outside of the flame region, emission from species at different temperatures along the line of sight of the measurement, chemiluminescence, and the variation in sensitivity of the instrument detector with wavelength. **Figure 3** illustrates that care must be taken when estimating flame temperatures from measured spectra, since results from simple fits of a blackbody radiation curve to a measured spectrum may be inaccurate.

Background

Emission and absorption spectra of flames may be continuous or banded. Continuous emission spectra may be modelled using the Planck blackbody equation (Eqn [2]). Modelling of banded spectra requires an understanding of the statistics that govern the way the population of a species is distributed among available energy levels. It is the dependence of population of molecular energy levels upon temperature, and the influence of this population upon band shape and spectral emissivity and absorption, which makes vibrational spectroscopy a useful tool for flame diagnostics.

The necessary mathematics for species and temperature measurements from infrared spectra of flames are outlined below. The example given is for a diatomic molecule, such as CO. It is assumed that the spectra are measured in absorption at low pressure, where the amount of light absorbed is less than approximately 5% of the incident intensity (often referred to as absorption in an optically thin medium). For extension to more complicated molecules (including more complicated diatomic molecules, such as NO), and to higher pressures, the reader should consult the 'Further reading' section for details.

Population distribution among available energy levels

The classical Maxwell Boltzmann distribution law may be used to approximate the distribution of population among the quantized energy levels of a gas-phase diatomic molecule. For most diatomic molecules (such as CO, but not NO), each separate line shape observed in the infrared spectrum corresponds to a simultaneous change in vibrational and rotational energy. For this reason, the spectral lines that make up the band structure observed in most infrared spectra are often called rovibrational transitions. For purposes here, it is convenient to describe the initial and final energy levels of the transition as rovibrational energy levels.

When describing two energy levels of a given species, the superscript ' denotes the state of higher energy, and the superscript " denotes the state of lower energy. At thermodynamic equilibrium, the ratio of the number of molecules in the rovibrational energy level with vibrational quantum number v' and rotational quantum number J' to the number of molecules in the rovibrational energy level with vibrational quantum number v'' and rotational quantum number J'' is given by:

$$N_{J',v'}/N_{J'',v''} = [(2J' + 1)/(2J'' + 1)]$$
$$\times \exp(-(\Delta E_v + \Delta E_J)hc\,kT) \quad [3]$$

where $N_{J',v'}$ is the population of the higher rovibrational energy level, $N_{J'',v''}$ is the population of the lower rovibrational energy level, ΔE_v is the change in vibrational energy (units of cm^{-1}) and ΔE_J is the change in rotational energy (units of cm^{-1}) which occurs during the transition, h is Planck's constant, k is Boltzmann's constant, and c is the speed of light. The quantity $[(2J'+1)/(2J''+1)]$ accounts for the degeneracy of rotational levels for a given value of J, providing a statistical weight to the levels with a given rotational energy.

For interpretation of measured spectra, it is useful to know the fraction of the total population (N_T) in the energy level from which the transition originates. For absorption (where the transition originates from the level of lower energy), this may be given by:

$$N_{J'',v''}/N_T = (N_{J'',v''}/N_{J=0,v=0})/(\sum N_{J,v}/N_{J=0,v=0})$$
$$[4]$$

where the summation is over all values of J'' and v'' and $N_{J=0,v=0}$ is the population in the rovibrational energy level with rotational quantum number J and vibrational quantum number v both equal to 0. The denominator on the right side of Equation [4] may be written to a first approximation as:

$$\sum N_{J,v}/N_{J=0,v=0} = \sum (2J + 1)$$
$$\times \exp(-(E_v + BJ(J + 1))hc/kT) \quad [5]$$

The first rotational constant, B, is inversely proportional to the moment of inertia of the rotating molecule, and determines the rotational energy level spacing. This means that, in general, the larger the mass of the rotating molecule, the closer is the spacing of the rotational energy levels. The right-hand side of Equation [5] may be factored into rotational and vibrational components:

$$\sum N_{J'',v''}/N_{J-0,v-0} = Q_R Q_V \quad [6]$$

where Q_R and Q_V are, respectively, the rotational partition function and the vibrational partition function:

$$Q_R = \sum (2J'' + 1) \exp(-(B''J''(J'' + 1))hc/kT)$$
$$[7]$$

$$Q_V = \sum \exp(-(E_{v''})hc/kT) \quad [8]$$

The summations in Equations [7] and [8] are over all J and all v, respectively. Substituting Equations [3], [7], and [8] into Equation [4] gives:

$$N_{J'',v''}/N_T = (2J'' + 1)$$
$$\times \exp(-(E_{v''} + B''J''(J'' + 1))hc/kT)/Q_R Q_V$$
$$[9]$$

Equation [9] shows how the fraction of total population in a given rovibrational energy level varies with temperature and rotational and vibrational quantum numbers. For most gas-phase diatomic molecules, if the population in a known rovibrational level is measured, Equation [9] allows the total population (and hence total pressure) of the gas to be calculated.

Absorption of radiation

The Einstein transition probability of absorption, B_{mn} (not to be confused with the first rotational constant B), predicts the energy removed (I_R) from an incident beam of radiation by an optically thin layer of absorbers for a transition from a lower state m to an upper state n as:

$$I_R = I_0 N_m B_{mn} \Delta x h \nu_{mn} \quad [10]$$

where N_m is the number of molecules per unit volume in the energy level from which absorptions occur, I_0 is the energy crossing unit area of absorbers per second, Δx is the absorber thickness, and ν_{mn} is the energy (in cm^{-1}) of the monochromatic radiation exciting the transition. In Equation [10], $I_0 N_m B_{mn}$ is proportional to the number of transitions per second per unit volume produced by the radiation and $h\nu_{mn}$ is proportional to the energy removed from the incident beam per transition.

In an absorption experiment, the intensity of radiation exiting the absorbing medium, I, is described according to the Bouguer–Lambert Law (later restated by Beer for solutions):

$$I = I_0 \exp(-\sigma(\lambda) N_m x) \qquad [11]$$

where x is the path length travelled by the light through the absorbing medium, and $\sigma(\lambda)$ is called the cross section for absorption. This cross section represents the 'effective area' that a molecule presents to the incident photons. When $\sigma(\lambda) N_m x$ is small (optically thin medium), Equation [11] may be rewritten:

$$I_0 - I = I_0(\sigma(\lambda) N_m x) \qquad [12]$$

Integrating Equation [12] over the wavelength range for which the absorption may occur gives:

$$\int (I_0(\lambda) - I(\lambda)) \, \mathrm{d}\lambda = \int I_0(\lambda)(\sigma(\lambda) N_m x) \, \mathrm{d}\lambda \qquad [13]$$

The way in which the cross section for absorption ($\sigma(\lambda)$) varies with wavelength depends mainly upon the total pressure of the gas. For gas pressures above approximately 100 torr, the absorption is observed to occur with a Lorentzian line shape given by:

$$\sigma(\lambda) N_m x = 2.303 A_0 \gamma^2 / ((c/\lambda - c/\lambda_0)^2 + \gamma^2) \qquad [14]$$

where A_0 is the peak spectral absorbance ($-\log(I(\lambda)/I_0(\lambda))$) at the wavelength of maximum light attenuation by the gas, and γ is the half width at half height (HWHH) of the spectral line.

Substituting Equation [14] into Equation [13], assuming that I_0 is invariant with wavelength over the absorption line width, and integrating over the

full line width gives:

$$\int (I_0(\lambda) - I(\lambda)) \, \mathrm{d}\lambda$$
$$= 2.303 I_0(\lambda) \int A_0 \gamma^2 / ((c/\lambda - c/\lambda_0)^2 + \gamma^2) \, \mathrm{d}\lambda$$
$$= 2.303 I_0(\lambda) \pi A_0 \gamma \qquad [15]$$

Substituting Equation [15] into Equation [10], and solving for the peak absorbance, A_0, gives

$$A_0 = N_m \Delta x \times B_{mn} h\nu_{mn} / 2.303 \pi \gamma \qquad [16]$$

The number of molecules in the initial state, N_m, is related to the total number of molecules, N_T, through the vibrational (Q_V) and rotational (Q_R) partition functions:

$$N_m = N_T(2J'' + 1)$$
$$\times \exp(-(E_{\nu''} + B''J''(J'' + 1))hc/kT)/Q_R Q_V \qquad [17]$$

Substituting Equation [17] into Equation [16] gives:

$$A_0 = [\Delta x B_{mn} h\nu_{mn} N_T(2J'' + 1)/2.303\pi\gamma Q_R Q_V]$$
$$\times \exp(-(E_{\nu''} + B''J''(J'' + 1))hc/kT) \qquad [18]$$

Equation [18] is useful for extracting temperature and concentration information from measured values of peak absorbance (A_0) for individual transitions in a rovibrational band. By simultaneously fitting T and N_T to A_0 measured over a rovibrational band, temperature and gas pressure may be obtained. Equation [18] shows that it is necessary to know the half width at half height (HWHH) of each line in the spectrum used in the calculation, and the value of B_{mn}. This is not always trivial, since this value may be temperature and J-value dependent. Additionally, it is important to recognize peak absorbances must be corrected when measured with an instrument of moderate spectral resolution. The method for extracting true peak absorbance from peak absorbance measured at moderate resolution has been treated in detail by Anderson and Griffiths. Substituting Equation [18] into Equation [16], and expressing the result in the form of the Bouguer–Lambert Law gives:

$$I(\lambda)/I_0(\lambda) = \exp[(-\Delta x B_{mn} h \nu_{mn}(\lambda) N_T \gamma /$$
$$\pi Q_R Q_V ((c/\lambda - c/\lambda_0)^2 + \gamma^2))$$
$$\times (2J'' + 1) \exp(-(E_{\nu''} + B'' J''(J'' + 1)) hc/kT)] \quad [19]$$

Equation [19] allows direct comparison between high resolution measurements of transmittance ($I(\lambda)/I_0(\lambda)$) and transmittance calculated from spectral parameters, pressure of the absorbing gas, and the temperature. It should be noted that knowledge of the spectral line HWHH (γ) is required, as well as an instrument capable of measuring the transmittance at high resolution. In practice, values of B_{mn} (usually converted to line strengths) for different rovibrational transitions are taken from the literature or estimated from measurements of total band strength. Also, the Voight line shape profile, which describes the convolution of a Gaussian line shape function (applicable to gases at low pressure) and a Lorentzian line shape function, is usually used to model spectral lines in flames at reduced pressure. As with fits to data based upon Equation [18], when spectra are measured on an instrument of moderate resolution (in general, an instrument resolution $>0.1\gamma$), the instrument line shape function is convoluted with the true line shape (Eqn [19]). When this is the case, this convolution must be included in the model.

Equation [19] describes the fully resolved band structure observed in high resolution infrared spectra of many gas-phase diatomic molecules, such as CO. **Figure 4** shows the transmittance ($I(\lambda)/I_0(\lambda)$) spectrum of CO, calculated using Equation [19], for a constant number of molecules at the temperatures shown in **Figure 2**. **Figure 4** shows that as the temperature increases, and more rotational energy levels becomes populated, the overall shape of the absorption band broadens to cover a wider spectral range. Additionally, because the total number of molecules is divided between a greater number of initial energy levels as temperature is increased, the intensity of individual rovibrational transitions changes with increasing temperature.

Experimental methods narrative

By the 1950s, the mathematical and instrumental methods for determination of temperatures and species concentration from measurements of infrared spectra had been established, in large part spurred by electronics developed during the Second World War, by perfection of the method of commercial

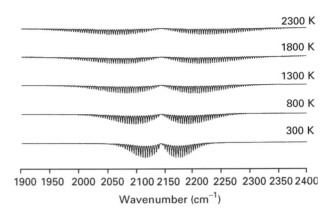

Figure 4 Calculated transmittance of light through a fixed pressure of CO gas at different temperatures. Spectra are offset for clarity (each baseline corresponds to 100% transmission).

replication of diffraction gratings, and by publication in 1945 of Hertberg's book 'Infrared and Raman Spectra of Polyatomic Molecules'. During the 1950s, comparison of infrared emission spectra from high temperature sources in different laboratories was complicated by temperature gradients along the measurement line of sight, and by incomplete spectral parameters for gas absorption at high temperatures. To obtain spectral parameters for gases under study, several efforts were made to study gases in closed cells at controlled temperatures. Most of these efforts at studying gases at high temperature under static conditions employed absorption spectroscopy, in part to minimize self absorption along the line of sight, and also to enable using modulation of the source radiation (allowing discrimination of the high intensity, unmodulated background infrared emission). The success in fundamental studies of band structure and in predicting emissivities and changes in absorption at high temperatures and pressures led to an increase, by the early 1960s, in studies of radiation transfer for systems ranging from jet and rocket motors to high efficiency oil burners. A particular success which resulted from the study of gas-phase emissivities was the determination of thermal stress on NASA rocket motors from exhaust gas radiation.

With the development in the early 1970s of the Fourier transform spectrometer and the computer based fast Fourier transform (FFT), it became possible to achieve high resolution coupled with high energy throughput and phase-sensitive detection. However, the development of laser-based techniques (particularly laser-induced fluorescence (LIF) for measurement of flame radical species, such as OH, HCO, H, O, CH, and C_2) enabling direct measurement of species participating in flame propagation reactions caused a shift in focus of fundamental spectroscopic investigations of flame systems. This shift led to a decrease in the late 1960s and early 1970s in the number of publications describing basic research that applied the techniques of dispersive infrared absorption and emission spectroscopy to flames. This decrease was offset by a considerable body of work on emission studies of hot gas sources, particularly smokestack and waste gas plumes from industrial sources.

By the late 1970s, laser-based techniques employing tunable infrared lasers began to be used for species measurements in flames. The majority of this work used tunable diode lasers (TDLs), semiconductor devices in which the output laser radiation wavelength is tuned by varying the temperature and diode injection current. Initial experiments used TDLs emitting narrow-line width (typically $< 10^{-4}$ cm^{-1})-radiation in the mid-infrared spectral region. The narrow line width usually enables species specificity, even in congested spectral regions. Since the laser line width is typically several orders of magnitude less than the absorption line width, measurements of the fully-resolved absorption transition may be made, enabling the determination of line shape dependence upon pressure and temperature. An additional advantage to using tunable diode lasers, besides very narrows line widths, is the ability to tune the lasers, rapidly (kHz to MHz) over their output wavelength range. This enables phase-sensitive detection that minimizes the effect of the laser output noise, and also enables time-resolved measurements of dynamic systems.

Recently, tunable diode lasers operating in the near-infrared spectral region have begun to see application to the spectroscopy of flames and to flame gases. At this time, commercial availability of TDLs in the mid-infrared spectral range is greater than for the near-infrared spectral range. TDLs operating in the mid-infrared spectral range must be cooled to cryogenic temperatures. TDLs operating in the near-infrared spectral range operate near room temperature, and unlike mid-infrared radiation, the near-infrared radiation may be transmitted over long distances through optical fibres. However, since absorption in the near-infrared spectral region corresponds to a change in vibrational quantum number greater than unity (referred to as an overtone transition), the sensitivity to a given molecule is much less than for mid-infrared spectroscopy, which is usually used to measure rovibrational transitions in the fundamental vibrational band. Typically, mid-infrared TDLs are used when extreme sensitivity is required (ppb range). For many field based techniques which require transportability in rugged environments, near-infrared TDLs are more appropriate (ppm range).

Experimental methods – dispersive and Fourier transform spectroscopy

Emission measurements

Since the mid-1970s, most measurements of emission spectra of steady flames have used Fourier transform techniques. **Figure 5** shows the emission spectrum measured from a premixed, stoichiometric CH_4/O_2 flame (total pressure equal to 18 torr) to which 3% CF_3Br has been added as a flame suppressant. When appropriate, reduced-pressure flames are often studied because at reduced pressure the flame region is expanded, allowing more detailed study. The emission spectrum shown in **Figure 5** was measured using a Fourier transform infrared (FT-IR) spectrometer at a resolution of 1 cm^{-1}.

In **Figure 5**, spectral features from several flame species are identified, but the spectrum is dominated by emission from CO_2 and gaseous H_2O. This is because higher temperatures allow emission over a wide range of rovibrational transitions (see **Figure 4**), from combinations of different vibrational states, and the geometry of the emitting species (specifically in this case H_2O) causes the spectrum to be complicated (relative to that for a diatomic molecule). The measured spectrum in **Figure 5** reports emission from species over a wide range of temperatures, because the line of sight from the flame centre to the detector includes gases at many different temperatures. Also, because many flame species are products of an ongoing chemical reaction (combustion), these species may be produced with vibrational and rotational population distributions which are not in equilibrium with translational

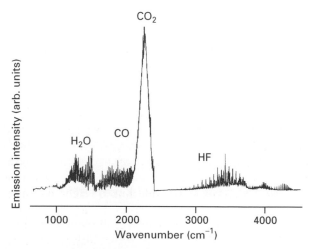

Figure 5 The infrared emission spectrum measured from a premixed, stoichiometric CH_4/O_2 flame (total pressure 17 torr) to which 3% CF_3Br has been added.

temperatures. Because of this, temperatures calculated from emission spectra using expressions similar to Equations [18] or [19] may give misleading results. When these factors are taken into account, temperature and partial pressures of gas species may be extracted from high resolution measurements of infrared emission spectra.

Absorption measurements

Figure 6 shows an absorption spectrum measured through an opposed-flow CH_4/air flame (total pressure 50 torr) using a FT-IR spectrometer at 1 cm^{-1} resolution. It should be noted that this spectrum is less congested and has a slightly more regular appearance than that shown in **Figure 5**. Absorption features from several species are evident in the spectrum. For flame diagnostics using infrared absorption spectroscopy, CO is probably the most widely studied molecule. The fundamental rovibrational absorption band of CO ($v = 0$–1, centred at 2170 cm^{-1}) is well approximately by Equation [19], and has a spacing between rotational lines (2B) of approximately 3.6 cm^{-1}. Modelling of measured spectra is relatively straightforward and fully-resolved rovibrational lines may be measured at moderate instrument resolution (although the reported line shape may be just the instrument line shape function. In addition, since CO is a major product of hydrocarbon combustion, it is present in most flame systems. **Figure 7** shows the spectrum of the first overtone ($v = 0$–2) of CO, measured using a dispersive system, and the fit to the measured spectrum using an equation similar to Equation [19]. Since the resolution of the spectrometer used to measure the spectrum in **Figure 7** was insufficient to fully resolve the line shapes of the individual rovibrational transitions, it was necessary to convolute the instrument function with the true line shape function (Eqn [19]) to obtain an accurate fit of the measured to the calculated spectrum.

A major difficulty in obtaining quantitative absorption measurements of flame species is that absorption of radiation by cold gases in the probe beam line of sight may lead to errors. Currently, several methods of excluding contributions to absorptions by cold gases in the line of sight are used, with most methods employing some type of computer averaged tomography (CAT). Using these method (often referred to as inversion methods), multiple line of sight measurements at different relative orientations are made within a plane, or slice, through the flame. Using these multiple line of sight measurements, numerical methods are used to calculate species concentration at a given point within the plane. Applying this technique to multiple planes, or slices through the flame, can yield a three-dimensional reconstruction of species

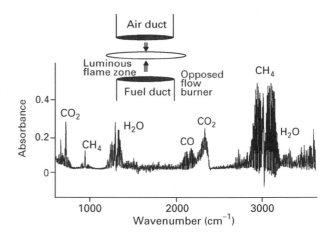

Figure 6 The infrared absorption spectrum measured through a 50 torr opposed-flow CH_4/air flame. The inset shows the burner configuration used to produce the flame.

concentration within the flame. This technique has been used to measure CO in reduced- and atmospheric-pressure flames. A summation of inversion methods used to measure temperatures and species in flames has been given by Limbaugh.

It should be noted that when measuring flame absorption spectra using a FT-IR spectrometer, care must be exercised so that flame emission is not reflected through the interferometer and reported as an absorption signal. To minimize the emitted radiation which reaches the interferometer, the collimated probe infrared beam is brought to a focus at an iris prior to entering the flame region. After passing through the iris, the probe radiation is re-collimated prior to entering the flame. While useful for identification of species within the flame, FT-IR spectroscopy will not yield high spatial resolution unless the infrared probe beam is reduced in cross section (usually by being passed through a set of apertures) prior

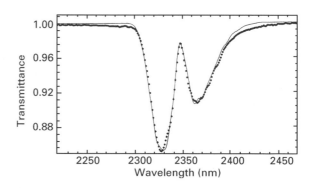

Figure 7 The transmittance spectrum of the first overtone ($v = 0$–2) of CO. The gas was contained in a static cell at 1 atmosphere total pressure and a temperature of 423 K. The dotted line is the measured spectrum, the solid line represents the least-squares fit to the data.

to entering the flame region. This reduction in power of the infrared probe beam energy reduces the signal to noise ratio in the measured spectrum. A typical experimental setup for measuring infrared absorption spectra through a flame using an FT-IR spectrometer is shown in **Figure 8**.

Most commercial spectrometer systems employing broadband sources (typically FT-IR systems) are not capable of resolving the true line shape of an absorbing gas phase species because of limitations to instrument resolution. The instrument yield the convolution of the true absorbing gas line shape and the instrument response function. Computer-based programs used to retrieve species concentration and temperature must either take into account this convolution, or use for calculations the instrument response-corrected peak intensity of the rovibrational transition.

Experimental methods – applications of tunable diode laser absorption spectroscopy (TDLAS)

Tunable diode lasers (TDLs) are semiconductor devices (typically GaAs) which are essentially light emitting diodes constructed with an optical resonator. Lasing is achieved by delivering a small current to the photodiode. Tuning is achieved by changing

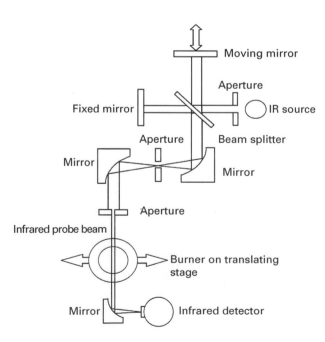

Figure 8 Experimental apparatus for measurement of FT-IR absorption spectra through flames. The system is designed to minimize the amount of flame emission which reaches the interferometer. The burner is mounted on a translating stage (X and Z axes) to allow for tomographic analysis.

the temperature, and hence the Fermi level, of the device. The tuning range varies by type, but is usually between 2 cm^{-1} and 30 cm^{-1}. The useful tuning range (that is, the range over which the device may be rapidly tuned without encountering changes in laser modes, which can mask absorption features) is usually less than 1 cm^{-1}. This restricted tuning range means that the devices are not used in a survey mode. In general, a TDL laser spectrometer system is designed to detect a single gas using a single, specially-fabricated, dedicated, laser diode. Occasionally, absorption transitions from different gases of interest may occur within the useful tuning range of a single laser, but such occurrences are usually fortuitous. Additionally, because of the limited tuning range, calculation of temperature and species concentration may be difficult when only one line is measured.

There are, however, several advantages to using TDLs for measurement of gas-phase flame species. These include high resolution (typically better than 1×10^{-4} cm^{-1}), good spatial resolution (200 to 1 mm), reasonable output power (~1 mW), and the ability to scan over their spectral range on a millisecond or better timescale. Probably the most widely studied molecular flame species by tunable diode laser spectroscopy is CO. In addition to the reasons for study outlined above in the discussion of broadband source methods, CO possesses several fundamental ($v = 0$–1) and hot-band transitions ($v = 1$–2, $v = 2$–3) which occur within several line widths (approximately 0.05 cm^{-1}) of each other. At room temperature, populations of states from which hot-band transitions occur are very low. However, at flame temperatures, populations of vibrational states other than the $v = 0$ state may become appreciable. When temperatures (and also species concentrations) are calculated from simultaneous measurement of a fundamental and a hot-band transition, the technique is referred to as two-line thermometry.

Figure 9 shows several spectra of CO measured in a low-pressure (20 torr), stoichiometric CH$_4$/O$_2$ flame using mid-infrared tunable diode laser spectroscopy. Because the wavelength of the diode laser radiation corresponds to single quantum number changes in CO (in this case a $v = 0$–1 and a $v = 1$–2 transition) simple absorption spectroscopy may be employed. Each spectrum is measured at a different height above the burner surface. For these experiments, the probe laser beam location was fixed, and the burner location was translated vertically. The infrared laser beam diameter through the flame region was 0.5 millimetres. As the diode laser radiation passes through different portions of the flame, the relative areas under the two peaks corresponding

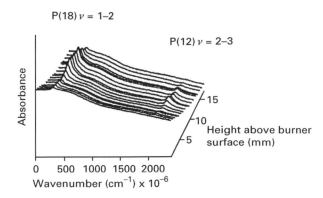

Figure 9 Mid-infrared tunable diode laser spectra of CO, measured through a premixed CH_4/O_2 flame, at a total pressure of 20 torr, as a function of height above the burner surface. The maximum temperature calculated using two-line thermometry was 2150 K.

to absorption from different rovibrational levels changes. **Figure 10** shows the results of fitting an equation similar to Equation [19] to an absorption spectrum from **Figure 9**. Using this method, the CO vibrational temperature was calculated as a function of height above the burner surface ($T_{max} = 2150$ K ±50 K). Additionally, tomographic analysis of the data showed that the error from cold gas absorption, for these experiments, was always less than 10% of calculated temperatures and partial pressures.

Tunable diode laser spectroscopy is also used for detection of many trace gas species in flames and in flame environments. The method used for measurements of small amounts of gases (often to the ppb range) relies on modulation spectroscopy using

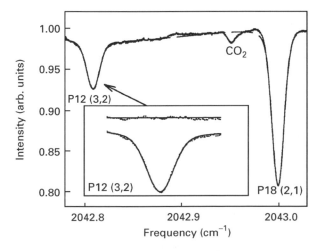

Figure 10 A transmission spectrum measured through a low-pressure, premixed, CH_4/O_2 flame at 20 torr. The main absorption features are the two CO lines used to calculate temperature using the method of two-line thermometry. Calculated temperature was 2150 K.

phase-sensitive detection. The principal advantage to using modulation spectroscopy results from minimization of laser output noise by shifting detection to high frequencies. When phase-sensitive detection is employed using tunable diode lasers, the second derivative of the diode laser probe beam intensity with respect to wavelength is usually measured, since the wavelength at which the second derivative is a maximum coincides with the wavelength of maximum light absorption. The second-derivative signal peak height may be shown to be proportional to absorbance, A:

$$X''/V = kA \qquad [20]$$

Here, X'' is the peak height of the second-derivative signal (volts), V is the DC voltage measured by the detector in the absence of any molecular absorbance, and k is a constant which includes the measuring instrument and optics function. Letting S, denoted as the 2f signal peak to trough height (the distance between the maximum and minimum points on the 2f signal – see **Figure 11C**, equal X''/V, yields:

$$S = (k\alpha)\,LP \qquad [21]$$

The slope of a plot of LP versus S provides the value of $k\alpha$. A calibration of the system using known concentrations of the absorbing gas must be performed to determine the value of $k\alpha$. Once this value is known, gas pressure (P) may be obtained directly from Equation [21]. Care must be exercised so that changes in optical surfaces during measurement (as when measuring corrosive gases such as HF) do not affect the value of $k\alpha$, since this value is instrument-function dependent. For this reason, calibration should be performed at the beginning and end of each measurement set. Additionally, for gas concentrations which attenuate more than 5% of the incident light, the linear relationship between gas pressure and 2f signal peak height may no longer hold.

Modulation spectroscopy using tunable diode lasers (often referred to as 2f spectroscopy) has been extensively discussed in the literature. **Figure 11** gives a generic description of the signal processing employed in these experiments. Briefly, the laser output wavelength is scanned (10 Hz–10 kHz) through a spectral region where the gas being measured absorbs. The output at the detector during the laser scan (with the high-frequency modulation turned off) may be seen in the upper trace of

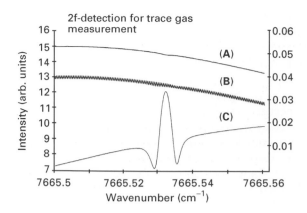

Figure 11 Simulated signal versus wavelength of laser radiation after passing through a gas with an absorption feature near 7665.5 cm⁻¹. Graphs for (A) unmodulated laser radiation; (B) laser radiation with a small-amplitude, high frequency modulation; (C) demodulation at twice the high-frequency modulation in (B).

Figure 11. A high-frequency (10 KHz to MHz) wavelength modulation with an amplitude approximately equal to line width under investigation is superimposed on the laser drive current (middle trace in **Figure 11**). Demodulation at twice the frequency of the high-frequency laser drive modulation yields the second-derivative signal shown in the lower trace in **Figure 11**. It should be noted that the upper trace in **Figure 11** shows that the laser diode output power has a non-linear dependence on laser drive current. While this is non-ideal behaviour, this is a common trait of commercially available laser diodes. The non-linear power output dependence on laser drive current (exaggerated here for illustration) causes the sloping baseline for the second-derivative signal in **Figure 11**. For measurements at extremely low concentrations or for gases with small absorption cross sections, the non-linearity of laser diode output power with laser drive current may affect limits of detection.

Modulation spectroscopy enables discrimination against contributions to light attenuation by scattering from particulate matter using a single laser beam. Attenuation of laser radiation by a ro-vibrational transition of a small gas molecule is detected because the wavelength range of the scan is several times the width of the spectral absorption feature (typically on the order of 0.1 cm⁻¹ at atmospheric pressure). Because light scattering by particulate matter is nearly constant over the very small wavelength range of the laser scan, the change in detector signal intensity with the change in wavelength is effectively zero in the absence of any absorbing gas. However, because the 2f signal is also proportional to the background

signal intensity, the second-derivative signal is divided by the DC signal to account for light scattering by particles.

While modulation techniques enable trace gas detection, special care must be exercised when making measurements in flames, especially when using near-infrared laser radiation. Significant errors in measured concentrations may arise because of beam steering as the laser radiation passes through the flame. In some cases, the beam steering may be severe enough to cause loss of signal at the detector. Secondly, great care must be exercised when measuring 2f spectra through regions of differing temperature and pressure, since the 2f signal is extremely sensitive to changes in line width of the absorbing species.

Conclusion

Infrared spectroscopy continues to be a valuable tool for measuring species concentrations and temperatures in flames and combustion gases, especially in field-based studies. The advent of compact, fibre-coupled, tunable diode lasers operating at room temperatures is expanding the use of vibrational spectroscopy beyond the laboratory, and providing a useful complement to broadband methods currently in place.

List of symbols

A = area of radiator surface; A_0 = peak spectral absorbance; B = first rotational constant; B_{mn} = Einstein transition probability of absorption for states m, n; c = speed of light; c_1, c_2 = radiation constants; E = internal energy of gas-phase molecule; E_λ = emissivity; h = Planck's constant; HWHH = half width at half height; I_0 = energy crossing unit area of absorbers/second; I_R = energy removed from incident beam by absorbers; I_λ = radiant intensity at wavelength λ; J = rotational quantum number; k = constant; k = Boltzmann's constant; N = energy-level population; Q_R = rotational partition function; Q_V = vibrational partition function; S = signal peak-trough height; T = temperature (K); V = DC voltage from detector for no absorbance; x = path length through absorbing medium; X'' = peak height of second-derivative signal; Δx = absorber thickness; γ = half width at half height of spectral line; v = vibrational quantum number; v_{mn} = energy of exciting radiation; $\sigma(\lambda)$ = absorption cross-section.

See also: **Laser Spectroscopy Theory; Rotational, Spectroscopy Theory; Solid State NMR, Rotational Resonance; Vibrational, Rotational and Raman Spectroscopy, Historical Perspective.**

Further reading

Anderson RJ and Griffiths PR (1975) Errors in absorbance measurements in infrared Fourier transform spectrometry because of limited instrument resolution. *Analytical Chemistry* **47**: 2339–2347.

Anderson RJ and Griffiths PR (1977) Determination of rotational temperatures of diatomic molecules from absorption spectra measured at moderate resolution. *J. Quant. Spectrosc. Radiat. Trans.* **17**: 393–401.

Barnes AJ and Orville-Thomas WJ (eds) (1977) *Vibrational Spectroscopy – Modern Trends*. Amsterdam: Elsevier Scientific Publishing Company.

Bomse DS, Hovde DC, Oh DB, Silver JA and Stanton AC (1992) Diode laser spectroscopy for on-line chemical analysis. In: Bomse DS, Brittain H, Farquharson S, Lerner JM, Rein AJ, Sohl C, Todd TR and Weyer L (eds) *Optically Based Methods for Process Analysis*. Proc. SPIE **1681**: 138–148.

Daily JW (1997) Laser induced fluorescence spectroscopy in flames. *Prog. Energy Combust. Sci.* **23**: 133–199.

Daniel RG, McNesby KL and Miziolek AW (1996) The application of tunable diode laser diagnostics for temperature and species concentration profiles of inhibited low pressure flames. *Applied Optics* **35**.

Dasch (1994) *Applied Optics* **33**.

Fried A, Henry B, Wert B, Sewell S and Drummond JR (1998) Laboratory, ground-based and airborne tunable diode laser systems: performance characteristics and applications in atmospheric studies. *Applied Phys. B* **67**: 317–330.

Gaydon AG (1948) *Spectroscopy and Combustion Theory*, p 129. London: Chapman & Hall Ltd.

Gaydon AG (1974) *The Spectroscopy of Flames*, 2nd edn. London: Chapman & Hall Ltd.

Gaydon AG and Wolfhard HG (1953), *Flames: Their Structure, Radiation and Temperature*. London: Chapman & Hall Ltd.

Griffiths PR and de Haseth JA (1986) *Fourier Transform Infrared Spectroscopy*. New York: Wiley.

Hanson RK, Varghese PL, Schoenung SM, and Falcone PK (1980) Absorption spectroscopy of combustion gases using a tunable IR diode laser. *Laser Probes for Combustion Chemistry*. ACS Symposium Series 134. Washington DC: American Chemical Society.

Hertzberg G (1950) *Spectra of Diatomic Molecules*. New York: Van Nostrand & Co.

Hertzberg G (1950) *Electronic Spectra of polyatomic molecules*. New York: Van Nostrand & Co.

Limbaugh CC (1985) The infrared emission-absorption method for temperature and species partial pressure determination in flames and plumes. In: Wormhoudt Joda (ed) *Infrared Methods for Gaseous Measurements*. New York: Marcel Dekker Inc.

McNesby KL (1998) Informal survey of *Proceedings of the International Symposia (International) on Combustion* from 1948 to 1996. New York: Academic Press.

McNesby KL, Daniel RG and Miziolek AW (1995) Tomographic analysis of CO absorption in a low-pressure flame. *Applied Optics* **34**: 3318–3324.

McNesby KL, Daniel RG, Widder JM and Miziolek AW (1996) Spectroscopic investigation of atmospheric diffusion flames inhibited by halons and their alternatives. *Applied Spectroscopy* **50**: 126.

Miller JH, Elreedy S, Ahvazi B, Woldu F and Hassanzadeh P (1993) Tunable diode-laser measurement of carbon monoxide concentration and temperature in a laminar methane-air diffusion flame. *Applied Optics* **32**: 6082–6089.

Moore Walter J (1972) *Physical Chemistry*, 4th edn. New Jersey: Prentice-Hall Inc.

Oppenheim UP (1963) Spectral emissivity of the 4.3μ CO_2 band at 1200 K. *Ninth Symposium (International) on Combustion*, pp 96–101. New York: Academic Press.

Penner SS (1959) *Quantitative Molecular Spectroscopy and Gas Emissivities*. Reading, MA: Addison-Wesley Pub. Co.

Reid (1981) Second harmonic detection with tunable diode lasers – comparison of experiment and theory. *Applied Physics B* **26**: 203–210.

Silver JA (1992) Frequency modulation spectroscopy for trace species detection: theory and comparison among experimental methods. *Applied Optics* **31**: 707–717.

Steinfeld JI (1974) *Molecules and Radiation*. New York: Harper & Row.

Townes CH and Schawlow AL (1955) *Microwave Spectroscopy*. New York: McGraw-Hill.

Vanderhoff JA, Modiano SH, Homan BE and Teague MW (1997) Overtone absorption spectroscopy of solid propellant flames: CO and N_2O concentrations. In Kuo KK (ed) *Challenges in Propellants and Combustion 100 Years after Nobel*, pp 876–884. New York: Begell House Inc.

Varghese PL and Hanson RK (1980) Tunable infrared diode laser measurements of line strengths and collision widths of $^{12}C^{16}O$ at room temperature. *Journal of Quantitative Spectroscopy & Radiative Transfer* **24**: 279.

Wilson EB, Jr., Decius JC and Cross PC (1955) *Molecular Vibrations*. New York: McGraw-Hill Co.

Wormhoudt JA and Conant JA (1985) High resolution infrared emission studies from gaseous sources. In: Wormhoudt Joda (ed) *Infrared Methods for Gaseous Measurements*, pp 1–45. New York: Marcel Dekker Inc.

Fluorescence and Emission Spectroscopy, Theory

James A Holcombe, University of Texas at Austin, TX, USA

ATOMIC SPECTROSCOPY
Theory

Like all atomic spectroscopic techniques, emission and fluorescence spectrometry require the production of free atoms (or ions) in the gaseous phase for detection. These two techniques require the population of a particular excited electronic state and determine concentration by monitoring the radiative relaxation process, i.e. by detection of the photon produced. In general, the intensity of the light emitted should be proportional to the gas-phase atom density and, ultimately, to the concentration of the analyte atoms in the sample being introduced to the atomization source. Likewise, it follows that increasing the excited-state population will enhance the sensitivity of the method.

The difference between the two spectroscopic approaches stems from the means by which the excited state is populated. In fluorescence, absorption of radiation at the proper frequency is used for excitation; in emission spectroscopy, collisional energy transfer with other atoms, molecules, ions or electrons produces the excited-state population.

Atomic emission

Excited-state population

Since excited-state populations are achieved by collisions with other charged or neutral particles within the source, the population of any given excited state can be calculated if local thermodynamic equilibrium (LTE) can be assumed to exist within the source. Under LTE, the fractional population of any given excited electronic state, N_j, is given by

$$\frac{N_j}{N_{\text{total}}} = \frac{g_j \exp\left(-E_j/k_{\text{B}}T\right)}{Z(T)} \qquad [1]$$

where g_j and E_j are the statistical weight and excitation energy of the jth state, respectively; k_{B} is the Boltzmann constant and T is the temperature (in kelvin). The partition function $Z(T)$ is the 'sum-over-states', or simply represents the relative population of all other states in the atom:

$$Z(T) = \sum_{i=0}^{\infty} g_i \, \exp\left(\frac{-E_i}{kT}\right) \qquad [2]$$

For relatively cool sources (e.g. $T < 5000$ K), $Z(T)$ can be approximated by the value for the statistical weight of the ground state, g_0, which implies that a majority of the atoms are present in the ground state ($E_0 = 0$). While an elevated population of the excited state is a *necessary condition* to realize a strong emission signal and good analytical sensitivity, it is not a *sufficient condition*. It is also necessary that there exists a high probability that, once populated, the excited state will radiatively relax to a lower energy state. In many instances the transition used terminates in the ground state of the atom or ion, and the line emitted is termed the resonance line. The probability that radiative relaxation will occur is indicated by the magnitude of the oscillator strength (f) or Einstein A coefficient for spontaneous emission. Both of these parameters are indicators of the strength of the emission, with smaller values indicative of low probability or even 'forbidden' transitions. Examples where such transitions might have a low probability include transitions in which an apparent change in electron spin state is needed to accomplish the relaxation, i.e. a 'spin forbidden' transition. As a result of selection rules that dictate transitions with a high probability, not all possible transitions from an excited state to a lower-lying state occur with equal probabilities. In spite of this, the emission spectrum of elements with d or f electrons can exhibit hundreds to thousands of lines in the UV and visible region of the spectrum.

Line widths

While radiative relaxation can produce relatively narrow spectral lines (e.g. 0.0001 nm or less, depending on the source and line), the lines are not truly monochromatic. At the very least each spectral line exhibits a 'natural line width' that is governed by the Heisenberg uncertainty principle

$$\Delta E \, \tau_{\text{r}} = \frac{h}{2\pi} \qquad [3]$$

where ΔE is the uncertainty in the position of the state, τ_r is the excited-state lifetime (typically 1–10 ns) and h is the Plank constant. With rearrangement and substitution of the Plank relationship ($E = h\nu$), the natural line width in hertz can be expressed as

$$\Delta\nu_N = \frac{1}{2\pi\tau_r} = \frac{A_{ji}}{2\pi} \quad [4]$$

where A_{ji} is the Einstein A coefficient for spontaneous emission from the jth to the ith state. With typical excited-state lifetimes of approximately 10 ns, the natural line widths are of the order of 0.00001 nm.

Line widths in most sources are generally broader than the neutral line widths as a result of collisional and Doppler broadening. Collisional broadening is a result of collisions or near collisions between the excited-state atom or ion and other particles in the system that result in a perturbation of the relative location of the energy states of the atom at the time of emission. As a result, elevated pressures and temperatures in the source enhance the impact of collisional broadening on the line width. Doppler broadening is a consequence of phenomena similar to the familiar Doppler effect for sound, which results in change in pitch or frequency as a sound emitting source approaches and then passes an observer, for example for a passing train that is blowing its whistle. In a similar fashion, an emitting atom that is moving towards or away from the observer exhibits a frequency shift in the emitted radiation that is related to the relative direction and velocity of the emitter. Taken over an ensemble of atoms moving in random directions, this produces a broadened spectral line. Like collisional broadening, this is accentuated with increasing temperatures since the average molecular velocities increase with temperature. For sources at atmospheric pressure and temperatures of a few thousand degrees, these broadening events are similar in magnitude and produce spectral line widths in the range of 0.001–0.01 nm.

Other effects that can produce broadening or splitting of the spectral lines include Stark broadening and Zeeman splitting. Stark broadening results from local electric fields from neighbouring ions and electrons or from externally applied, strong electric fields. Zeeman splitting results from the interaction of a magnetic field with the particles or ions.

From Equation [1] it is apparent that elevated temperatures promote population of electronic excited states. However, competing with this enhanced excited state population is ionization:

$$M \rightleftharpoons M^+ + e^- \quad [5]$$

This is an equilibrium condition and can be expressed by the Saha equation:

$$K_i = \frac{n_{M^+} n_e}{n_M} \quad [6]$$

where n represents the number density of the respective species. This equilibrium is quantitatively described by

$$\log K_i = 15.68 + \log \frac{Z_{M^+}}{Z_M} + 1.5 \log T - \frac{540 E_{ion}}{T} \quad [7]$$

where Z represents the partition functions for the respective species and E_{ion} is the ionization potential in eV. In brief, the equation predicts that reduced ionization potentials, low partial pressures of the neutrals in the source and elevated temperatures will enhance the fraction of the neutrals that are ionized. It is for this reason that in some high-temperature sources, such as inductively coupled plasmas, the resonance transitions of the *ions* produce more intense emission lines (and improved analytical detection limits) than neutral atom lines for the same element.

While the above discussion applies to many analytical atomic spectroscopic sources, much of the discussion and quantitative treatment requires the existence of local thermodynamic equilibrium, LTE. Some sources do not exist at LTE. Typically, sources that are at reduced pressure, which limits the rate of collisional distribution of energy, or those where energy is being input at a rapidly changing rate, may not exhibit LTE. As an example, glow discharge sources such as hollow-cathode lamps used in atomic absorption are operated in a low-pressure noble gas environment and exhibit extremely high ionization temperatures but very low kinetic gas temperatures; that is, the temperatures are not equal and, as a result, the system is not considered at LTE. This source, in spite of its kinetic gas temperature of only a few hundred degrees, exhibits an extremely high degree of ionization and population of some very high transition energy excited states.

Calibration curves

In general, the intensity of light emitted from the excitation source should be proportional to the concentration of analyte atoms or ions within the source. This linear relationship should be valid as long as equilibrium processes within the source govern the fraction of the total analyte population. An example where this may not be true can be seen in Equations [5]–[7], where ionization can be dependent on the partial pressure of the free atomic species. In the case of a flame source, low concentrations of the easily ionized elements (e.g. alkali metals) show measurably higher degrees of ionization. As a result, there can be curvature at the low concentration end of the calibration curve. Interestingly, the much hotter inductively coupled plasma does not exhibit the same nonlinearity at low concentrations. This is a result of the nearly constant electron density (n_e), which keeps the degree of ionization nearly independent of concentration.

Since resonance lines are often used for spectrochemical analysis, radiation emitted from the centre of an excitation source may pass through a substantial concentration of atoms in the ground state in the outer regions of the source. As a consequence, some of this light may be absorbed by the analyte atoms in this region. This is referred to as self-absorption. In the extreme, the absorbing atoms can reduce the intensity at the central wavelength of the emitting line to near zero (self-reversal). **Figure 1** shows intensity profiles with increasing analyte concentrations for an atomic emission line subject to self-absorption.

Consider light emitted from the hot centre region of a source (e.g. a flame) at an intensity of $I_{c,\lambda}^0$. The intensity is a function of both the analyte concentration and the wavelength, with maximum intensity emitted at the line centre. Since the intensity is proportional to the concentration c of the analyte element in the sample,

$$I_{c,\lambda}^0 = k_\lambda' c \qquad [8]$$

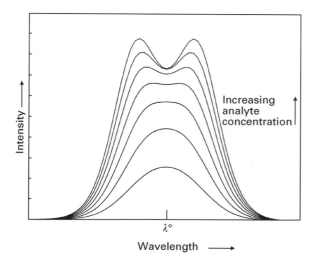

Figure 1 Intensity profiles with increasing analyte concentration (bottom to top) for an emission line subject to self-absorption.

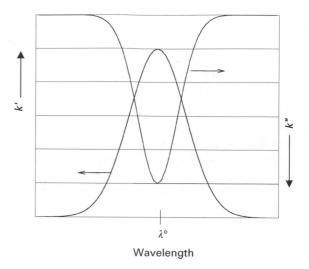

Figure 2 Wavelength dependences of k' and k''. Note that the vertical axis for k_λ'' has been inverted to aid presentation.

where k_λ' is a constant that is wavelength dependent. The resonance emission also passes through the cooler outer regions of the source that also contain analyte atoms. However, they are emitting a negligible amount of light because of the cooler temperatures and the very strong temperature dependence of emission (Eqn [1]). Thus, these cooler atoms can absorb the resonance radiation without contributing significantly to the emission intensity. The absorption of this resonance radiation exhibits a Beer's law type of behaviour, where the intensity passing through this layer is related to the concentration c and the incident intensity $I_{c,\lambda}^0$.

$$I_{c,\lambda} = I_{c,\lambda}^0 \exp(k_\lambda'' c) \qquad [9]$$

where k_λ'' is indicative of the strength of the absorption at any given λ. A graphical representation of the wavelength dependence of k' and k'' is shown in **Figure 2**. (Note: The scale for k'' has been reversed for clearer presentation.) Since $I_{c,\lambda}^0$ is dependent on c (Eqn [8]), the intensity emitted from the source after passing through this absorbing layer can be described as

$$I_c = \int k_\lambda' c \exp(k_\lambda'' c) \, \partial\lambda \qquad [10]$$

From this expression two things become apparent: (a) the intensity detected (I_c) is not linearly dependent on c unless $0 \le k_\lambda'' \ll 1$ (i.e. there is a low probability of absorbing a photon) or c is very small; and (b) as c increases, the deviation from the ideal calibration curve slope of k' increases. This behaviour generates nonlinear calibration curves with greatest curvature at higher concentrations. Other instrumental factors (such as stray light in the spectrometer) can also affect the shape of the calibration curve but will not be dealt with here.

Atomic fluorescence

Excited-state population

Fluorescence, like emission, relies on radiative relaxation of an atom or ion from an excited electronic state to a lower-lying state. Unlike emission spectroscopy, population of the excited state occurs via absorption of a photon. Two of the possible fluorescence schemes are shown in **Figure 3**. If the excitation beam that is absorbed is the same as the detected fluorescent wavelength, it is termed resonance fluorescence. Nonresonance fluorescence denotes the case where the excitation and fluorescence wavelengths are different. The latter excitation/detection approach offers the advantage of minimizing scattered radiation from the excitation beam from reaching the detector. Additionally, by using a nonresonance line for analysis, self-absorption by the source of the fluorescent signal is minimized. (See also the discussion below on calibration curves.)

If the relaxation is a spontaneous emission process occurring after photon absorption has populated the excited state, then it follows that the intensity of the detected radiation will increase for transitions with larger Einstein A coefficients, as in emission spectroscopy. However, population of the excited state is not governed by thermal or collisional processes, but instead by absorption of a photon of the proper frequency. Thus, it is logical that improved absorption should also yield improved sensitivity.

Absorption is a *stimulated* event, that is the number of atoms that are promoted to a higher energy level is dependent not only on the total number of atoms in the low-lying originating state that are capable of absorbing a photon, but also on the photon flux. For example, if one were to double the incident excitation beam intensity, one would expect twice as many atoms to absorb photons. (Note: This implies that the fraction of the total photons absorbed by the analyte is constant, which is consistent with the precepts of absorption where the ratio of I^0/I is independent of I^0.) This dependence on the incident photon flux makes the absorption a stimulated process and the probability of absorbing a photon is governed by the magnitude of the Einstein B coefficient for stimulated absorption. This B coefficient is proportional to the Einstein A coefficient and to f,

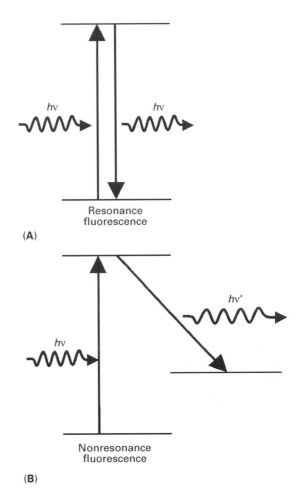

Figure 3 Two fluorescence schemes. (A) Resonance fluorescence: excitation and detection at the same wavelength. (B) Nonresonance fluorescence: excitation and detection at different wavelengths, $\lambda_{det} > \lambda_{exc}$.

the oscillator strength. Hence there is justification for the axiom that 'a strong emitter is a strong absorber' often cited in atomic spectroscopy.

Since sensitivity is dependent on the excitation source intensity, it would seem that one should be able to increase the source intensity and thereby improve the sensitivity of the fluorescence technique for analysis. This is valid with low radiative fluxes. However, there is one more interaction between radiation and matter that must be introduced into this discussion to explain the limitation of this assumption: stimulated emission.

Stimulated emission is the process that occurs when an atom or ion (or molecule, for that matter) is in an excited state and a photon interacts with this atom to stimulate the relaxation to a lower energy state (where $\Delta E = h\nu$). The stimulated photon is emitted at the same wavelength as the incident photon, travels in the same direction as the incident photon and is in phase with the incident photon. As a result of this process, the excited state has been diminished by this interaction with light. Since this is a stimulated process (like absorption), the number of particles that will participate in this interaction will depend on the incident flux and the number of atoms or ions located in the proper excited state. In short, as the excitation source intensity continues to increase, the population of the excited state increases from stimulated absorption, and the number of stimulated emission events also increases as the excited state population increases. Eventually a point is reached where the rate of the stimulated absorption process equals the rate of the stimulated emission and no net increase (or decrease) in population of the excited state occurs. The system is said to be 'saturated'. Additional increases in the incident radiation source intensity will have little impact on either the fraction of atoms or ions in the excited state or the spontaneous emission, i.e. fluorescence signal. As a result, small fluctuations in the source light intensity will not alter the detected signal for a given concentration of atoms in the atomization cell. Usually, source intensities comparable to that produced by lasers are needed to reach saturation.

Since the excitation process is dependent not only on the photon flux of the source but also on the number of atoms or ions in a state capable of absorbing the photon, most absorption/excitation steps involve transitions that originate in the ground electronic state where a majority of the analyte atoms or ions reside. Two-colour (or multiphoton) experiments can also be conducted in which multiple-step excitation is the objective. Similarly to nonresonance fluorescence discussed previously, multiphoton excitation also makes the fluorescent transition energy distinctively different from the excitation energy, and minimizes self-absorption if the terminating levels of the fluorescent wavelength is not the ground state. (See calibration curve discussion below.) Multicolour experiments are almost always restricted to high-intensity sources such as lasers.

Once a photon is absorbed, there is a finite possibility that it will relax radiatively. The fraction of the atoms or ions that fluoresce after they absorb radiation is referred to as the quantum yield, Φ.

Line widths

Parameters discussed previously for line broadening in emission spectrometry are also applicable for fluorescence spectrometry, especially since many of the same atom producers (flames, inductively coupled plasmas, etc.) are used for both techniques. In fluorescence, the width of the absorption profile can be

as critical as that of the fluorescent line profile. This is particularly true when narrow-band excitation sources (e.g. lasers or low-pressure discharge lamps) are employed. At the atomic cloud becomes less transparent, the intensity of the excitation beam that reaches the 'analytical volume' in the source centre also decreases, since it is being attenuated by passing though absorbing atoms in the outer regions of the atomization source. The intensity distribution of this exciting beam is, of course, moderated by k'_λ. Narrow absorption line profiles and narrow excitation line widths (e.g. from laser and low-pressure discharge lamps) accentuate the impact that this absorption will have on the fluorescent intensity.

Calibration curves

As in emission spectrometry, the intensity of the detected radiation should be proportional to concentration in the original sample, and this is generally true at low concentrations. As with emission, curvature at low concentrations can occur as a result of ionization interferences in some sources for the more easily ionized elements.

Also similarly to emission spectrometry, self-absorption (and self-reversal) of the fluorescent spectral lines can take place within the sample 'cell' if a resonance line is the line being monitored and detected and the atom density is sufficiently large. This alone can lead to a flattening of the calibration curve at elevated concentrations similar to that observed with emission spectroscopy. In addition, if the excitation beam is also a narrow line source used at a resonance wavelength of the analyte, the excitation beam intensity can be attenuated. This leads to lower populations of the excited state and decreased fluorescence signals when the beam attenuation becomes significant. In fact, with a line source (in contrast to use of a continuum excitation source) at elevated analyte concentrations, the calibration curve can actually roll over, i.e. exhibit a negative slope at higher concentrations.

List of symbols

A_{ij} = Einstein A coefficient for spontaneous emission from jth to ith state; c = analyte concentration; E_j = excitation energy of the jth state; f = oscillator strength; g_j = statistical weight of the jth state; h = Planck constant; I = intensity of emitted/absorbed radiation; k_B = Boltzmann constant; k'_λ = proportionality constant $(I^0_{c,\lambda} = k'_\lambda c)$; k''_λ = constant expressing strength of absorption at λ; K_i = equilibrium constant in Saha equation; n_X = number density of species X; T = temperature;

$Z(T)$ = partition function; ΔE = uncertainty in energy of a state; Δv_N = natural line width; λ = wavelength; v = line/transition frequency; τ_r = lifetime of excited state; Φ = quantum yield.

See also: **Atomic Absorption, Theory; Atomic Absorption, Methods and Instrumentation; Atomic Emission, Methods and Instrumentation; Atomic Fluorescence, Methods and Instrumentation; UV-Visible Absorption and Fluorescence Spectrometers.**

Further reading

Barnes RM (1976) *Emission Spectroscopy.* Stroudsburg PA: Dowden, Hutchinson and Ross.

Boumans PWJM (1966) *Theory of Spectrochemical Excitation.* Bristol: Adam Hilger.

Bower NW and Ingle JD Jr (1981) Experimental and theoretical comparison of the precision of atomic absorption, fluorescence, and emission measurements. *Applied Spectroscopy,* 35: 317.

Ingle JD Jr and Crouch SR (1988) *Spectrochemical Analysis.* Upper Saddle River: Prentice Hall.

Kuhn HG (1962) *Atomic Spectra.* New York: Academic Press.

Mitchell ACG and Zemansky MW (1961) *Resonance Radiation and Excited Atoms.* New York: Cambridge University Press.

Vickers TJ and Winefordner JD (1972) Flame spectrometry. In: Grove EL (ed) *Analytical Emission Spectroscopy,* Vol 1, Part II. New York: Marcel Dekker.

Winefordner JD, Schulman SG and O'Haver TC (1972) *Luminescence Spectrometry in Analytical Chemistry.* New York: Wiley-Interscience.

Fluorescence Microscopy, Applications

Fred Rost, Ashfield, NSW, Australia

ELECTRONIC SPECTROSCOPY
Applications

This article gives an overview of the current applications of the technique of fluorescence microscopy. The four sections describe, respectively, (1) the basic principles of fluorescence microscopy, (2) the types of information which can be obtained by fluorescence microscopy, (3) the technical ways in which fluorescence microscopy can be adapted to study various chemical species and (4) some examples of the range of biological, mineralogical and artificial specimens that can be studied.

Principles of fluorescence microscopy

Fluorescence microscopy is a technique whereby fluorescent substances are examined in a microscope. It has a number of advantages over other forms of microscopy, offering high sensitivity and specificity.

In fluorescence microscopy, the specimen is illuminated (excited) with light of a relatively short wavelength, usually blue or ultraviolet (UV). The specimen is examined through a barrier filter that absorbs the short-wavelength light used for illumination and transmits the fluorescence, which is therefore seen as bright against a dark background

(**Figure 1**). Because fluorescence is observed as luminosity on a dark background, fluorescent constituents of the specimen can be seen even in extremely small amounts. There are several different modes of fluorescence microscopy, of which the most important is confocal fluorescence microscopy.

In most modern fluorescence microscopes, epi-illumination is employed. This means that the light used for excitation is reflected onto the specimen through the objective, which acts as a condenser. Opaque or very thick objects can be examined using epi-illumination, even the skin of living people.

The position of fluorescence microscopy in relation to other techniques is summarized in **Table 1**. Conventional, transmitted-light, absorption microscopy is appropriate for coloured objects of resolvable size, and instrumentally is the simplest form of microscopy. Colourless, transparent objects can be studied only by retardation techniques (polarization, phase-contrast, interference); these techniques depend upon conversion of phase retardation into changes in intensity that can be seen by the eye. An exception is darkground illumination, which may reveal colourless transparent objects by reflection or refraction at interfaces of differing refractive indices. Darkground microscopy is otherwise suitable mainly for

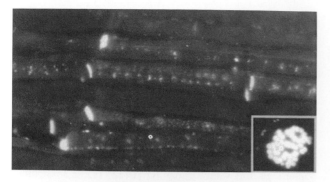

Figure 1 Fluorescence photomicrographs of a section of a plant stem, cut longitudinally (main picture) and transversely (inset, at higher magnification). The tissue was stained with a fluorochrome, Aniline blue, to show the sugar-conducting tissue (phloem). Aniline blue stains specialized regions of the cell walls. Intense fluorescence is most obvious on the transverse (end) walls of elongated cells. The end walls have pores so that there is continuity from cell to cell, forming a continuous tube in which sugars may be moved down the plant from the leaves. A face view of one of the end walls is shown in the inset. There is a ring of fluorescence around each pore. There is also fluorescent staining, to a much lesser degree, on the longitudinal walls. This surrounds pores that allow sugar transport between adjacent tubes. Aniline blue stains 1→3β–glucans. Here it is staining callose, a glucan which is deposited to occlude the pores should the tubes become damaged, as when tissue is cut. This presumably blocks the cut part of the tubes, minimizing loss of sugars from the cut ends of the tubes, and may well also minimize entry of micro-organisms that would be attracted by leaking sugars. The tubes shown in the main picture are 70 μm in diameter. Field size of main picture approximately 500 × 1000 μm (0·5 × 1 mm), of inset approximately 100 × 100 μm. Pictures courtesy of Professor AE Ashford, University of New South Wales.

particulate matter, and (like fluorescence microscopy) can reveal the positions of particles too small to be resolved. Fluorescence microscopy is closely allied to transmission (absorption) microscopy in its range of application, but possesses particular advantages: great sensitivity for detection and quantification of small amounts of fluorescent substances or small particles, and the possibility of application to opaque objects. Since fluorescence involves two wavelength bands (excitation and emission), optical specificity can be substantially increased by careful selection of filter combinations to favour the excitation and emission of some particular fluorophore, and modern developments also permit time-resolution of the fluorescence lifetime.

Fluorescence microscopy, because of its complexity, gives more difficulty than usual in interpretation of the image. Factors which may affect the appearance of the image in a fluorescence microscope are related to the specimen, to the microscope optical system (particularly the filter combination) and to the observer's own optical and neurological characteristics. In particular, the use of a narrow-band barrier filter can be misleading, since it makes everything appear its specific colour, whereas a wide-band or long(wavelength)-pass filter allows differentiation of different colours. Even photography, apparently objective, may be misleading if not interpreted correctly.

The current definitive texts on fluorescence microscopy are those of Rost (see Further reading). There also exists several introductory works, such as that of Abramowicz, and a vast specialized literature. The major texts on confocal fluorescence microscopy are those of Pawley and of Wilson.

Types of information to be obtained

Because many substances are fluorescent, or can be made so, fluorescence microscopy is widely applicable. It has a number of advantages over other forms of microscopy, offering high sensitivity and specificity. Because fluorescence is observed as luminosity on a dark background, specific constituents of the specimen can be seen even in minute amounts. Therefore, fluorescence microscopy is used mainly to detect and localize very small amounts of substances in specimens. The chemical substance being studied is either fluorescent of itself, made so by a chemical process or attached to a fluorescent label.

Table 1 Applicability of fluorescence microscopy, compared with other techniques

| Specimen | Fluorescence | Type of microscopy | | |
		Absorption (transmission)	Retardation (DIC, Pol, etc.)	Reflection (incl. darkground)
Coloured	Suitable	Suitable	Unsuitable	Suitable
Transparent	Impossible	Impossible	Suitable	Impossible
Opaque	Suitable	Impossible	Suitable	Suitable
Dynamic	Suitable	Impossible	Impossible	Impossible
Particles below limit of resolution	Suitable	Impossible	Unsuitable	Suitable

Absorption microscopy is the conventional transmitted-light type. Retardation microscopy includes Nomarski interference-contrast (DIC), phase-contrast and polarization. Reflection microscopy includes darkground.

Fluorescence microscopy offers three significant capabilities. First, its high sensitivity allows very low concentrations of specific substances to be localized. Secondly, specific substances can be localized in structures smaller than the resolution of the microscope. Thirdly, fluorescence is particularly suitable for confocal microscopy, which offers optical sectioning, giving very clear imaging and the possibility of building up 3D reconstructions.

As in other forms of microscopy, there are three basic kinds of fluorescence microscopy: qualitative, quantitative and analytical. Qualitative fluorescence microscopy is concerned with morphology, or with whether something (e.g. an immunological reaction) is present. Quantitative fluorescence microscopy is concerned with finding out how much of a specific substance is present in a specified region of the specimen. Analytical fluorescence microscopy is the characterization of a fluorophore by measurement of excitation and emission spectra or other characteristics such as polarization or decay time. Kinetic studies essentially involve studying the fluorescence parameters described above over a period of time; examples include the study of fading rates, enzyme kinetics, time-resolved fluorescence and phosphorescence.

Photomicrography

Images may be captured by photographic or digital means. Photography is best achieved by fast colour-positive (transparency) film in a small (35 mm) format. Film is certain to be largely replaced by solid-state electronic devices (CCDs) in the near future because these have a higher quantum efficiency, enabling the recording of fluorescence which is weaker and/or the use of a shorter exposure time. Confocal microscopy produces a digital image directly.

Video intensification microscopy (analogue and digitized)

Attachment of a video camera to a fluorescence microscope facilitates viewing of the image, which is normally rather dim. A video camera, particularly one using a CCD, can readily be interfaced to a computer for image analysis. For details of video microscopy, see the book by Inoué and Spring and for an introduction to technique see the book by Herman, both listed under Further reading.

Confocal fluorescence microscopy and 3D imaging

If both the illumination and detection systems are simultaneously focused onto a spot which is scanned over the specimen, this is called confocal fluorescence microscopy, and has several advantages. Most importantly, the confocal microscope has sharp depth discrimination, and therefore produces optical sections in the plane of focus. This is because the detector does not accept all the light from out-of-focus planes and so these are imaged less strongly than the in-focus plane. Background fluorescence is thereby sharply reduced. Three-dimensional data can be obtained by computer reconstruction of a series of optical sections. Confocal fluorescence microscopy offers greatly increased discrimination against background and some increase in resolution, and gives the possibility of optical sectioning and three-dimensional reconstruction. For detailed information, see the books by Gu, Matsumoto, Pawley, and Wilson listed under Further reading.

Other modes

Several other modes of fluorescence microscopy have been developed, including automated fluorescence image cytometry (AFIC), fluorescence resonance energy transfer microscopy (FRETM), two-photon fluorescence microscopy, total internal reflection fluorescence microscopy, and standing-wave fluorescence microscopy.

Combination with other techniques

Because the image seen in the microscope may consist of only a few small fluorescent areas in an otherwise black field, fluorescence microscopy is sometimes supplemented with other forms of microscopy to enable the specimen as a whole to be visualized and to show the position of fluorescent areas in relation to the rest of the specimen. Fluorescence microscopy is easily combined simultaneously or alternately with other methods for increasing object contrast, such as darkground illumination, interference-contrast (DIC) or phase-contrast microscopy. Combined electron microscopy and fluorescence microscopy of the same stain is possible with a small number of stains which contain heavy atoms.

Quantification by microfluorometry

For quantification, microfluorometry has four advantages as compared with microdensitometric methods: distributional error does not occur, so that scanning or two-wavelength techniques are not required; optical specificity can be obtained by selection of appropriate wavelengths both for excitation and for measurement of emission; great sensitivity can be obtained; and thick or opaque objects can be examined. Disadvantages of microfluorometry include the necessity for artificial standards, and (usually) relatively rapid fading of the specimen. It is important to ensure that the background fluorescence is kept to a minimum and that an adequate standard is available.

Microspectrofluorometry

Excitation and/or emission spectra are measured from a specified area of the specimen using a microspectrofluorometer. Such instruments usually have to be assembled by addition of an appropriate light source and monochromators to a microfluorometer. This can be applied either for the identification of unknown fluorophores or to the investigation of the conditions of fluorescence of known fluorophores. The main application to date has been the identification of neurotransmitter amines in situ using formaldehyde-induced fluorescence.

Kinetic studies: time-related and video fluorescence microscopy

Kinetic studies essentially involve studying the fluorescence parameters described above over a period of time; examples include the study of fading rates, enzyme kinetics, time-resolved fluorescence and phosphorescence.

Scanning of the image with a television camera, with the televised image shown on a video monitor, enables an otherwise dim image to be seen more conveniently. Time-lapse video recording can be used to study movements of fluorescent probes in living cells.

The specimen often photobleaches (fades) under irradiation. Photobleaching is put to good use in modern studies of molecular kinetics using video recording and subsequent measurements of fluorescence recovery after photobleaching (FRAP), which gives information about rates of diffusion of fluorescent species back into a bleached area.

Fluorescence lifetime imaging microscopy (FLIM)

In this technique, the instrumentation provides time resolution of the fluorescence process, enabling spatial resolution of changes in fluorescence lifetime caused by environmental changes. This is useful for monitoring intracellular processes which cannot be studied by other means.

Types of fluorescence

Autofluorescence

Autofluorescence (primary fluorescence) is the fluorescence of naturally occurring substances, such as chlorophyll, collagen and fluorite. Most plant and animal tissues show some autofluorescence when excited with ultraviolet light (e.g. light of wavelength around 365 nm). Sometimes this autofluorescence is a nuisance as it may conceal or be confused with a specific fluorescence. The term autofluorescence is sometimes extended to cover the fluorescence of drugs and other exogenous substances which may be present in tissues, e.g. tetracycline antibiotics which accumulate in growing bone. In biology, autofluorescence is often regarded as a nuisance, masking the specific fluorescence of introduced substances, but it usually provides a guide to the morphology of the tissue, and is also worthy of study in its own right. In geology, fluorescence microscopy is widely used to study coal. In minerals, autofluorescence is most commonly due to the presence of trace amounts of activator substances, generally regarded as 'impurities'.

Induced fluorescence

This is observed following a chemical reaction to convert some non-fluorescent substance, already present in tissue, into a fluorescent substance. The most important example is the demonstration of neurotransmitter amines such as noradrenaline, which can be converted into fluorescent substances by treatment with formaldehyde or glutaraldehyde.

Fluorochromy

This is the process in which specimens are stained with fluorescent dyes, as compared with diachromy in which tissues are stained with coloured dyes. An example is shown in **Figure 1**, in which structures in a botanical specimen are stained with a fluorescent dye, Aniline blue. Many dyes which are used as coloured stains are also fluorescent, and can be used as either diachromes or fluorochromes. Hence, many conventionally stained preparations can be examined with either a conventional microscope or a fluorescence microscope. However, the use of fluorescence microscopy extends the range of dyes which can be used, because of the additional possibility of using dyes which are colourless or nearly so but fluoresce in the visible region after excitation with ultraviolet light. Fluorescence also gives increased sensitivity. Staining can also be carried out after chemical treatment, as in the Schiff reaction for carbohydrates and in the Feulgen reaction for nucleic acids. Some staining procedures commonly applied to biological material are listed in **Table 2**. For data and references on fluorochromes and fluorescent probes, see the Molecular Probes catalogue and the books by Kasten, Mason and Rost.

Table 2 Some examples of fluorochromes (fluorescent stains)

Field	Application	Stain
Bacteriology	Acid-fast bacteria	Auramine O
Botany	Plant cell walls	Calcofluor White
Cell biology	Chromosome banding	Quinacrine, Quinacrine mustard
Cell biology	Labels for affinity reactions	FITC, Texas Red isothiocyanate
Cell biology	Membrane lipids	ANS, SITS
Cell biology	Nucleic acids	Acridine Orange, bisbenzimide
Neurology	Nerve cell tracing	DiI
Pathology	Amyloid	Thioflavine T

Immunofluorescence This is a special form of fluorochromy, in which the site of an antigen–antibody reaction is revealed by labelling one of the components of the reaction (the antigen, the antibody or complement) with a fluorescent dye. Such reactions allow specific proteins to be localized rapidly, reasonably accurately and with great sensitivity.

Similar techniques are used for the demonstration of other cytochemical affinity reactions, such as in nucleic acid hybridization and in the binding of lectins, hormones and other substances. I refer to these techniques as indirect fluorochromy, since the fluorochrome is bound indirectly (through at least one intermediate molecule) to the substrate (the substance intended to be stained). Immunofluorescence, nucleic acid hybridization and similar affinity reactions have made an enormous contribution to cell biology.

Double- and triple-staining techniques enable visualization of two or more affinity reactions in the same preparation, using differently coloured fluorochromes for each reaction. Special filter combinations are available to facilitate simultaneous visualization of the different colours, and wedge-free filters allow sequential photography of images in each colour without image shift.

Enzymatically induced fluorescence

In general, enzyme activity is demonstrated by fluorescence microscopy as follows. A substrate is offered to the enzyme, which is allowed to act on the substrate to obtain a reaction product which is localized at the site of enzyme activity and is either fluorescent or easily rendered so. The technique is usually used for qualitative purposes only, to demonstrate the location of enzyme activity, but quantification of enzymatic activity and estimation of Michaelis–Menten constants (K_m) are possible. It is usually

easier to demonstrate the presence of the enzyme protein using immunofluorescence, but of course this does not guarantee that the enzyme system is active.

Applications

Fluorescence microscopy is widely used in biology and medicine, as well as in other fields. The literature is vast, and only a very brief outline of actual or potential applications can be presented here.

Fluorescence techniques can be applied to all kinds of material. Because fluorescence microscopy requires more complex and expensive instrumentation than conventional transmitted-light microscopy, fluorescence microscopy is usually reserved for those applications in which its high sensitivity is of importance: i.e., to examine substances present in low concentrations. Fluorescence microscopy can also be applied to detect particles below the resolution of a light microscope, and in histochemistry to visualize substances which cannot be seen by conventional microscopy – e.g. neurotransmitter amines. Biological material is commonly stained in some manner with a fluorescent stain.

Biological applications

Probably the most important applications of fluorescence microscopy are to the study of living cells and tissues, to protein tracing by the Coons fluorescent antibody technique and to the study of nucleic acids by in situ hybridization.

Fluorescence microscopy has long been used in the study of living tissues (**Table 3**). This is possible because fluorescence microscopy reveals the position of very tiny amounts of fluorescent substances, which

Table 3 Some examples of fluorescent probes used in living tissues

Application	Probes
Bone growth	Tetracycline, chlortetracycline
Calcium ions	Fura-2, Indo-1, Calcium Green
Magnesium ions	Mag-Fura-2
Membrane lipids	ANS, SITS, Nile Red
Mitochondria	Rhodamine 123, DASPI, DASPMI
Neoplasia	Haematoporphryin derivative
Neuronal tracing	Fast Blue, Lucifer Yellow
Nuclei	Acridine Orange, bisbenzimide, ethidium bromide
pH Indication	BCECF, carboxyfluorescein, dicarbocyanines
Streaming (plant cells)	Ethidium bromide
Viability	6-Carboxyfluorescein diacetate

can be introduced into living tissues or cells. Information about the environment of fluorescent probes in the tissues can be obtained by measuring changes in fluorescence spectrum or other characteristics. This technique is now widely applied to the intravital study of the concentration of various inorganic ions. The measurement of pH and of calcium ion concentration are particularly important.

The viability of cells is testable by using fluorescent esters, such as 6-carboxyfluorescein diacetate, which are non-fluorescent and which (being non-polar) can pass through the cell membrane. Inside the cell, the ester is hydrolysed to release the free fluorescent anion (6-carboxyfluorescein), which, if the membrane is intact, accumulates inside the cell. Fluorescence is also widely used to demonstrate the ramifications of nerve cells, by injection or takeup of fluorochromes into or by living cells.

Mineralogical applications

Fluorescence microscopy is widely and routinely used for the study of coal.

Technological applications

Fluorescence microscopy is ideal for studies of porosity in ceramics, using a fluorescent dye. It is also applicable to studies of semiconductors.

List of symbols

K_m = Michaelis–Menten constant.

See also: **NMR Microscopy; Rotational Spectroscopy, Theory; Scanning Probe Microscopy, Applications; Scanning Probe Microscopy, Theory.**

Further reading

Abramowitz M (1993) *Fluorescence Microscopy – the Essentials*. New York: Olympus-America.

Bollinger A and Fagrell B (1990) *Clinical Capillaroscopy*. Toronto: Hogrefe & Huber.

Conn PM (ed) (1990) *Methods in Neurosciences,* Vol III, *Quantitative and Qualitative Microscopy*. San Diego: Academic Press.

Coulton GR and de Belleroche J (eds) (1992) *In situ Hybridization: Medical Applications*. Lancaster: Kluwer Academic.

Gu M (1996) *Principles of Three-dimensional Imaging in Confocal Microscopes*. Singapore: World Scientific.

Haugland RP (1996) In: Spence MTZ (ed) *Handbook of Fluorescent Probes and Research Chemicals*, 6th edn. Eugene, OR: Molecular Probes.

Herman B (1998) *Fluorescence Microscopy*, 2nd edn. Oxford: Bios Scientifica Publishers / The Royal Microscopical Society.

Inoué S and Spring KR (1997) *Video Microscopy*, 2nd edn. New York and London: Plenum Press.

Kasten F (1981) In: Clark, G. (ed) *Staining Procedures*, 4th edn, Chapter 3. Baltimore: Williams & Wilkins.

Kawamura A and Aoyama Y (eds) (1983) *Immunofluorescence in Medical Science*. Tokyo: University of Tokyo Press; Berlin/Heidelberg/New York: Springer Verlag.

Kotyk A and Slavick J (1989) *Intracellular pH and its Measurement*. Boca Raton, FL: CRC Press.

Mason WT (1993) *Fluorescent and Luminescent Probes for Biological Activity*. London: Academic Press.

Matsumoto B (ed) (1993) *Cell Biological Applications of Confocal Microscopy*. New York: Academic Press.

Pawley JB (ed) (1995) *Handbook of Biological Confocal Microscopy*, 2nd edn. New York: Plenum Press.

Polak JM and Van Noorden S (eds) (1983) *Immunocytochemistry*. Bristol: Wright.

Rost FWD (1991) *Quantitative Fluorescence Microscopy*. Cambridge: Cambridge University Press.

Rost FWD (1992) *Fluorescence Microscopy*, Vol I. Cambridge: Cambridge University Press. (Reprinted with update, 1996.)

Rost FWD (1995) *Fluorescence Microscopy*, Vol II. Cambridge: Cambridge University Press.

Rost FWD and Oldfield RJ (1999) *Photography with a Microscope*. Cambridge: Cambridge University Press.

Wilkinson DG (1994) *In situ Hybridization. A Practical Approach*. Oxford: IRL Press.

Wilson T (ed) (1990) *Confocal Microscopy*. San Diego: Academic Press.

Fluorescence Polarization and Anisotropy

GE Tranter, Glaxo Wellcome Medicines Research
Centre, Stevenage, Herts, UK

ELECTRONIC SPECTROSCOPY
Theory

Fluorescence polarization techniques have proved to be a very useful method of investigating the environment and motion of fluorescent molecules and fluorophores within larger macromolecules.

When molecules are illuminated by plane-polarized light, those with their transition electric dipole moment parallel with the plane of polarization will preferentially absorb the light ('photoselection'). Subsequent emission of light from this excited electronic state, as the molecule returns to the original electronic ground state, will be similarly polarized parallel to the transition moment of each molecule. If the molecules are unable to rotate during the lifetime of the excited state, then the plane of polarization of the emitted light will be consistent and directly related to that of the exciting light (their electric vectors will be parallel). However, if the molecules are able to rotate within this time period, then the degree of polarization will be diminished. A similar description can be envisaged for fluorophoric groups within larger molecules; in this case the rotational relaxation involves the tumbling of the complete molecule together with the localized motion of the fluorophore with respect to the molecular environment.

Although rotational relaxation is the main mechanism for loss of polarization, other phenomena can also contribute to depolarization, not least radiationless transfer among fluorophores. Furthermore, if the emission occurs between a different pair of electronic states from that of the original absorption, with differently oriented transition moments with respect to the molecular axes, then a change in polarization may be observed.

Steady-state anisotropy

In a typical fluorescence spectrometer, the excitation beam illuminating the sample and the optical path employed to detect the emission are orthogonal to each other and define a 'horizontal' plane. By employing linear polarizers before and after the sample, the plane of polarization of the excitation beam can be controlled and the polarization components of the emitted light determined. Specifically, by orientating the polarizers as either parallel or perpendicular to the horizontal plane, the plane of polarization of the excitation or detected emission can be described as either parallel to the plane (H, for horizontal) or perpendicular to it (V, for vertical). Four emission spectra can then be acquired:

V_V = intensity of vertical emission from vertical excitation

H_V = intensity of horizontal emission from vertical excitation

V_H = intensity of vertical emission from horizontal excitation

H_H = intensity of horizontal emission from horizontal excitation

In these circumstances, for an *ideal* instrument, the fluorescence polarization, $p(\lambda)$ for a sample at a given wavelength λ is

$$p(\lambda) = \frac{(V_V - H_V)}{(V_V + H_V)} \qquad [1]$$

and the corresponding fluorescence anisotropy, $r(\lambda)$, as

$$r(\lambda) = \frac{(V_V - H_V)}{(V_V + 2H_V)} \qquad [2]$$

In principle, this would allow the fluorescence polarization or anisotropy to be determined with just two spectral acquisitions: V_V and H_V. However, in practice, it is vital to correct for the differing efficiencies of the emission monochromator and detector towards vertically and horizontally polarized light, which may be substantial. Consequently, for isotropic samples, Equation [1] can be amended to

$$p(\lambda) = \frac{(V_V - \kappa(\lambda)H_V)}{(V_V + \kappa(\lambda)H_V)} \qquad [3]$$

where

$$\kappa(\lambda) = V_H/H_H \qquad [4]$$

and similarly for the fluorescence anisotropy,

$$r(\lambda) = \frac{(V_V - \kappa(\lambda)H_V)}{(V_V + 2\kappa(\lambda)H_V)} \qquad [5]$$

The basis for the correction ratio κ is that, for horizontally polarized excitation of isotropic samples, the resulting vertically and horizontally polarized fluorescences detected should be of equal intensity; any difference may be attributed to the polarization dependence of the emission monochromator and detector.

The fluorescence polarization $p(\lambda)$ and anisotropy $r(\lambda)$ both take the value of unity for the limit of complete preservation of polarization during the excited state lifetime (e.g. if the molecules are immobile). Likewise, they both take the value of zero if there is a complete loss of polarization (i.e. a randomization of orientation). However, their values differ for states between these extremes. In principle $-1 < p < 1$ and $-0.5 < r < 1$, although the limits are rarely encountered; negative values may be indicative of processes beyond depolarization through rotational relaxation, such as transitions to other excited states.

If the only significant process for depolarization is rotational relaxation, then the fluorescence anisotropy, for a single molecule or fluorophore, may be given by a form of the Perrin equation:

$$r = \frac{r_0}{(1 + \tau/\phi)} \qquad [6]$$

where τ is the fluorescence lifetime, ϕ is the rotational correlation or relaxation time for the fluorophore's tumbling and r_0 is the intrinsic anisotropy in the absence of rotational relaxation. Small molecules (less than 1000 Da molecular mass) with fluorescence lifetimes of the order of $\tau = 10$ ns are able to tumble with relative ease in low-viscosity solvents (e.g. water), typically $\phi = 0.1$ ns and thus $\tau \gg \phi$ and there is almost complete depolarization during the fluorescence lifetime ($r \to 0$). In contrast, a fixed fluorophore in a large protein may have the same fluorescence lifetime, but a rotational relaxation time of $\phi = 10$ ns, so that $\tau \sim \phi$ and there is significant preservation of the polarization.

As a convenient rule of thumb, the rotational relaxation time (in nanoseconds) for a spherical molecule of mass M (g mol^{-1}) in a medium of viscosity η (g cm^{-1} s^{-1}) is

$$\phi \approx 0.03\,\eta M \quad \text{ns} \qquad [7]$$

with, for example, $\eta = 0.01\,\text{g cm}^{-1}\,\text{s}^{-1}$ for water.

In the case of a small fluorescent molecule binding to a large macromolecule, the bound form should be distinguishable from the unbound form by having a substantially greater fluorescence anisotropy. Likewise, fluorophores rigidly held within a macromolecule would show significantly greater anisotropies than those exposed to the solvent and possessing some degree of motional freedom. Thus, one can employ fluorescence anisotropy to locate groups within a structure. This can prove particularly powerful if an environmental change (including perhaps the binding of a compound) induces a change in rigidity about observable fluorophores. Finally, if the variation of fluorescence anisotropy is plotted against wavelength, spectroscopic bands originating from distinct transitions or fluorophores may be distinguished by their differing anisotropies.

Fluorescence lifetime anisotropy

By the use of fluorescence lifetime instrumentation, one can further determine the evolution of the fluorescence anisotropy with time during and beyond the lifetime of the excited state. In the simplest case of a fluorophore in solution, with a single fluorescence lifetime and depolarization through rotational relaxation alone, the fluorescence anisotropy will decay according to

$$r(t) = r_0 \exp(-t/\phi) \qquad [8]$$

whereas the total emission intensity, $I_T(t)$ will decay as

$$I_T(t) = I_T(t_0)\exp(-t/\tau) \qquad [9]$$

Thus the rotational relaxation and fluorescence lifetime may be directly distinguished.

In more complex cases, perhaps involving binding kinetics or multiple environments, it is possible to deduce characteristic half-lives for the various processes. For example, one may identify subpopulations of molecules that are bound in different modes and even characterize the residual motion of a molecule while it is bound to a site that is itself tumbling. Nonetheless, these complex applications all require a sufficiently robust hypothesis regarding the phenomena being modelled if the interpretations are to be valid.

List of symbols

H_H = intensity of horizontal emission from horizontal excitation; H_V = intensity of horizontal emission from vertical excitation; I_T = total emission intensity; M = molar mass; $p(\lambda)$ = fluorescence polarization at wavelength λ; r, $r(\lambda)$ = fluorescence anisotropy (at wavelength λ); r_0 = intrinsic anisotropy; t = time; V_H = intensity of vertical emission from horizontal excitation; V_V = intensity of vertical emission from

vertical excitation; η = viscosity; $\kappa(\lambda) = V_H/H_H$; τ = fluorescence lifetime; ϕ = rotational correlation time.

See also: **Chiroptical Spectroscopy, Oriented Molecules and Anisotropic Systems; NMR in Anisotropic Systems, Theory.**

Further reading

Clark BJ, Frost T and Russell MA (eds) (1993) *Techniques in Visible and Ultraviolet Spectrometry*, Vol 4, *UV Spectroscopy*. London: Chapman and Hall.

Harris DA and Bashford CL (eds) (1987) *Spectrophotometry and Spectrofluorimetry, a Practical Approach*. Oxford: IRL Press.

Ingle JD and Crouch SR (1988) *Spectrochemical Analysis*. Prentice Hall International.

Miller JN (ed) (1981) *Techniques in Visible and Ultraviolet Spectrometry*, Vol 2, *Standards in Fluorescence Spectrometry*. London: Chapman and Hall.

Fluorescence Spectroscopy in Biochemistry

See **Biochemical Applications of Fluorescence Spectroscopy.**

Fluorescent Molecular Probes

F Braut-Boucher and **M Aubery**, Université, Paris V, France

ELECTRONIC
SPECTROSCOPY
Methods & Instrumentation

Introduction

Fluorescent methods have been extensively developed for analytical purposes. Due to their specificity and sensitivity, fluorescent molecular probes (FMP) are presently largely used for the labelling of living cells. Numerous applications in biomedical research, including pharmacology and toxicology, are in progress. Initially, only flow cytometry was capable of quantifying fluorescent signals in cell biology and obtaining information such as cell type characterization, antigen expression, cell receptor localization, etc. More recently the most important advances have resulted from the combination of probe development for exploring specific cellular functions *in situ* and the use of performing instrumentation. Spectrofluorimetry associates the high sensitivity of fluorescence with the

reproducibility of microplate reading. Moreover, new probes are expected in the near future and attempts are being made to develop automatic procedures for monitoring analysis and producing a variety of tests used in cell biology, microbiology, toxicology and oncology. Several reviews and monographs on molecular fluorescent probes are available in the bibliography and are listed in the Further reading section.

Micromethods are one of the most recent applications of fluorescence and efforts are devoted to the adaptation of apparatus capable of direct measurement on cell culture microplates that can be automatically scanned. In basic research, as well as in pharmaceutical drug development, this advanced technology represents a significant gain in productivity and cost control. This article deals first with probe labelling strategy, choice of suitable devices for microplate reading and data analysis. Secondly, selected applications are discussed.

Fluorescent molecular probes and practical considerations

Fluorescence theory and applications have been documented by Guilbault. According to this author, a fluorescent probe could be defined as 'a chromophore that undergoes changes in absorption characteristics as a function of environmental disturbance'. The molecule absorbs energy, rises to an excited state, and then returns to its normal state, concomitantly emitting energy as photons. This light emission is named fluorescence. The optimum difference between excitation and emission wavelengths (the Stokes shift) lies between 20 and 50 nm.

Several cellular functions are revealed by probes which react stoichiometrically and thus lead to a quantitative relationship between probe and target. In fact, fluorescent probes are versatile tools that can be used to reveal specific cell properties, such as information about cell structure, function, viability, proliferation or differentiation. Probes include fluorescent dyes, antibodies coupled with fluorochromes or fluorogenic substrates, and oligonucleotides. New fluorescent probes are being constantly developed, and additional fluorescent indicators can be found in catalogues or Web pages of a few specialized companies; among them are Molecular Probes and Fluka Biochemicals. A probe is generally selected because of its specific spectroscopic properties: fluorescence lifetime, fluorescence intensity, or functional characteristics: cellular incorporation, behaviour under pH variation. To minimize cellular probe disturbance, i.e. possible chemical interactions between probe and the tested xenobiotic, low probe concentrations (micro- to nanomolar) must be used. Moreover, to avoid photochemically induced reactive oxygen species production, all experiments have to be carried out with as little light present as possible. It is also necessary to consider quenching which results in fluorescence reduction by absorption of emitted light by a fluorophore or other substances present in the medium, such as phenol red.

Currently, there are a wide variety of highly fluorescent extrinsic dyes that may be grouped in two classes:

- fluorescent exclusion probes, which do not penetrate into the cell and so maintain its integrity;
- vital fluorescent probes, which are incorporated into living cells. The latter are more frequently used and can be divided into four sub-classes:

- probes related to membrane integrity;
- probes related to membrane potential;
- probes related to DNA content;
- probes related to ion content.

Fluorescent probes and their derivatives commonly used in microspectrofluorimetry are listed in **Table 1** with their specific spectra in UV, visible and near IR areas. The structures of some of the probes mentioned later are given in **Figure 1**.

Concerning cells

Adherent or non-adherent cells could be obtained from either primary cultures or immortalized cell lines. While better reproducibility is generally obtained with cell lines, with careful monitoring very good results have also been obtained with primary cell cultures (see below). Concerning adherent cells, for better reproducibility, the cell monolayer should be used only when subconfluent. In such controlled condition, the cell outline is clearly visible (**Figure 2A**). Non-adherent cells in suspension must be carefully isolated before distribution into each well (**Figure 2B**). In both cases, monolayer variability or cell clusters should be avoided. Moreover, cellular autofluorescence has to be precisely measured before experiment, although there is a generally low fluorescence background when the labelling procedure is well adapted. The morphology and integrity of the

Table 1 Fluorescent probes commonly used in microspectrofluorimetry

Fluorochrome	Excitation wavelength (nm)	Emission wavelength (nm)
Acridine orange / RNA	460	620
Acridine orange / DNA	485	530
DAPI / RNA	360	510
DAPI / DNA	360	460
DIOC (3) / RNA	482	610
Ethidium bromide / DNA	530	620
Ethidium bromide / RNA	370	620
Hoechst 33258	360	460
Hoechst 33342	365	502
Propidium iodide	530	620
Alamar blue	530	590
Rhodamine 123	485	530
BCECF	485	530
Calcein	485	530
Neutral red	540	635
Resorufin	530	585
Methylumbelliferone(4-MU)	360	460

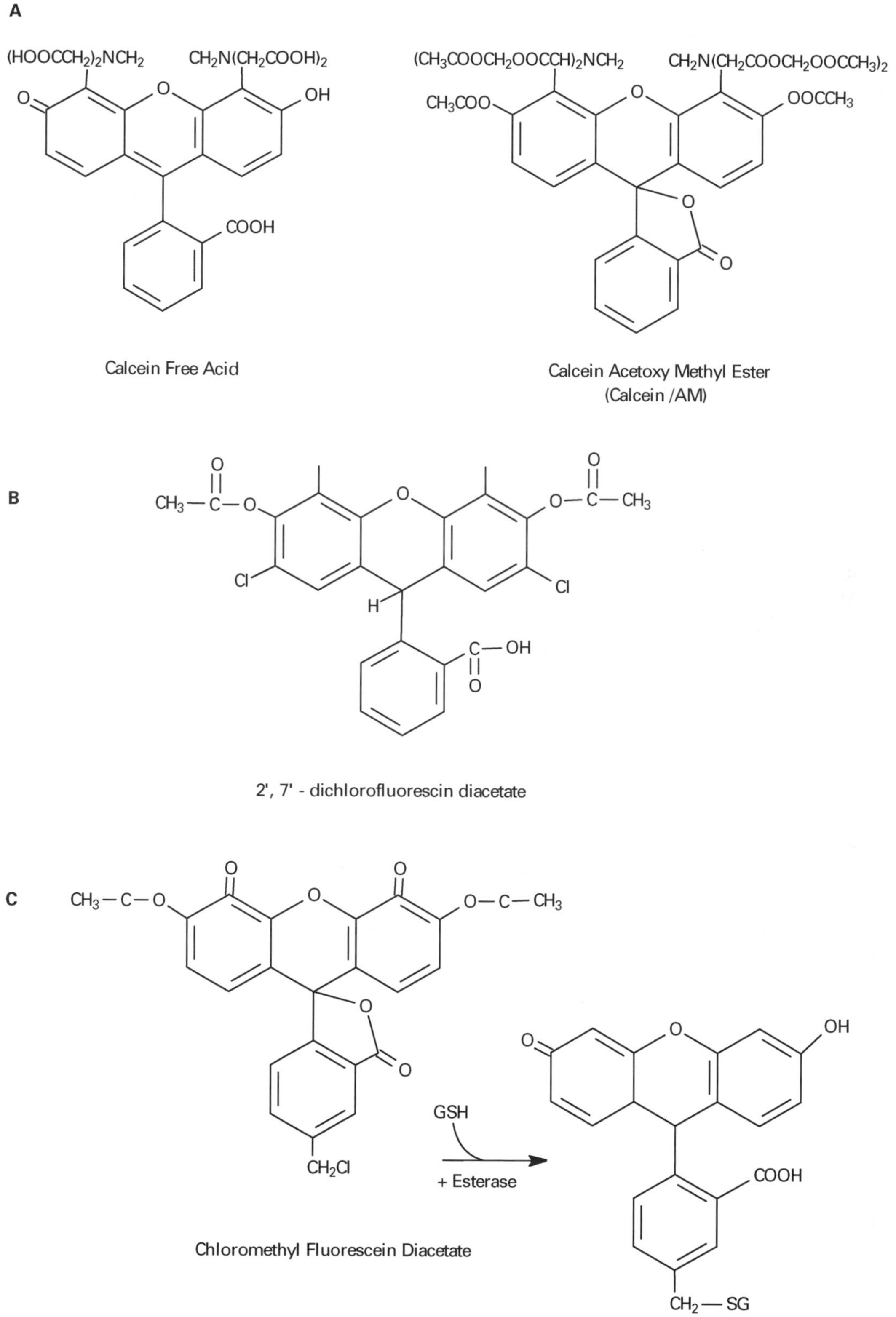

A

Calcein Free Acid

Calcein Acetoxy Methyl Ester
(Calcein /AM)

B

2', 7' - dichlorofluorescin diacetate

C

GSH
——————→
+ Esterase

Chloromethyl Fluorescein Diacetate

Figure 1 Structure of some fluorescent probes used in microspectrofluorimetry.

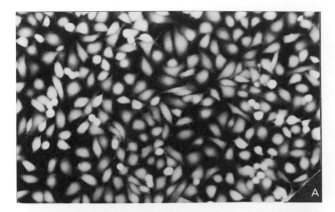

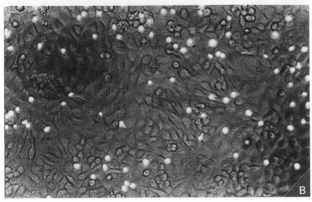

Figure 2 (A) Endothelial cells (magnification × 200) in a portion of monolayer labelled with calcein-AM (5 μM for 30 min); (B) HL 60 (magnification × 200) in suspension labelled with calcein-AM (10 μM for 30 min) adhering onto an endothelial cell monolayer.

cells and their distribution on the bottom of the microplate have to be monitored to see if they are correctly reproduced from one set to another, paying particular attention to optimization of the culture support. The best optical design is represented by plates with a transparent flat bottom and black edges, but accepting some cross contamination between wells, and standard culture supports can be used in most apparatus. An optimal distribution of microplated cells is shown on **Figure 3**. The cell population could vary in terms of staining levels due, in part, to the state of the type of cells. In any case, the methodology has to be adapted to each cell type and controlled by fluorescence microscopy. Besides validating a new procedure, a comparison could also be made by using fluorescence-activated cell sorting (FACS).

Choosing a fluorescent probe

The choice of a probe depends on the function or the major target to be explored. The first step of the labelling procedure is the determination of the concentration–response relationship between the probe and

the fluorescent signal. Toxic effects of probes related to their concentration are easily controlled, as is non-release of the probe, when experiments are carried out in dilute samples. Higher concentration could induce leakage or quenching. The probe stability in the storage solvent and the culture medium has to be controlled. As deleterious effects and structural damage have been observed in loaded cells exposed to high light energy, potential phototoxic effects, i.e. peroxidative damages or necrosis, must be considered. Moreover, probes may be photolabile, and thus it is necessary to expose the sample in the wells for a short time, at the lowest possible light energy. In these experimental conditions, phototoxic interferences are minimized. In most apparatus, the microplate is scanned at a defined wavelength for 0.3 s, which results in limited fading.

General labelling procedure

In all cases, whatever the probe used, concentration–response and kinetics curves should be established. These studies will indicate the optimal concentration and incubation time for the probe, depending on the cell type. As an example, calcein-AM is used at 20 μM for lymphocytes and 5 μM for keratinocytes, at 37°C, for a 30 min incubation period. Before loading, the cells are washed three times with PBS and then incubated with the appropriate concentration of the probe in PBS-BSA (0.2%) for 30 min. After elimination of the non-incorporated probe by centrifugation, labelled cells are distributed over the plate, in accordance with the scheme presented in **Figure 3**. The fluorescence

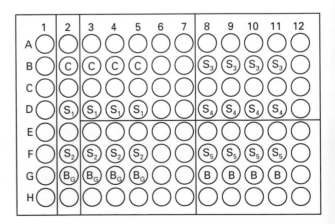

Figure 3 Experimental setup with configuration of the 96-well microtitre plate. B_G, background fluorescence; B, blank (solvent and xenobiotic); C, control (cells loaded with probe without any treatment); S, samples (cells loaded with probe, treatment applied in quadruplicate; five concentrations of xenobiotic could be tested S1 to S5).

is then rapidly monitored using a cytofluorimeter. All manipulations concerning the loading of cells with fluorophores are carried out avoiding exposure to direct light. Generally, probes are aliquoted in a concentrated stock solution (0.1 to 1 mM), and diluted before use of the required final concentration. At low concentrations (below 50 μM), probes have little deleterious effect on cells. The efficiency of probe incorporation is temperature-dependent, with an optimum at 37°C. The excess probe is eliminated by gentle centrifugation and the cell pellet is resuspended in the medium and examined under an inverted microscope. Finally, the fluorescence is measured using a microplate reader.

Selecting a microplate reader

The instrumentation for microspectrofluorimetry consists of two essential units: a source of radiation and a system for measuring the intensity of the fluorescence emission. The basic components of this apparatus are shown on **Figure 4**. The different

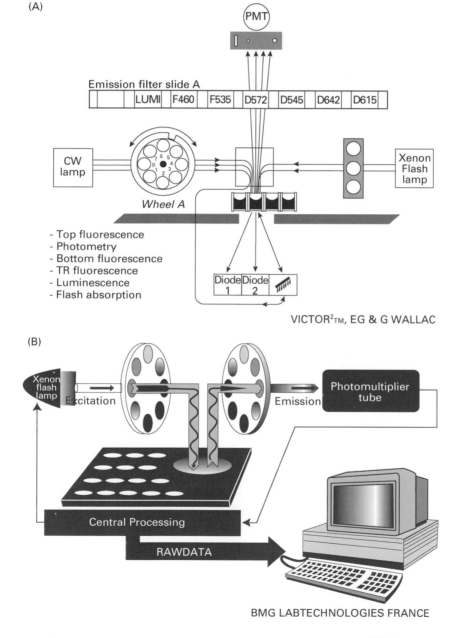

Figure 4 (A) Diagram of an epiillumination pathway in a fluorimetric plate reader. (Reproduced by permission of EG & G WALLAC.) (B) Schematic overview of the main components of the Fluostar. (Reproduced by permission of BMG Labtechnologies, France.)

appliances available from various companies are compared in **Table 2**. As an energy source, the xenon lamp has a better continuum than that of the tungsten lamp in the ultraviolet spectrum. For this reason the xenon lamp is the most widely used source of radiation in spectrofluorimetry.

For quantitative measurements, excitation and emission monochromator can be set at the optimum wavelength for the probe. Difficulties are encountered with apparatus according to the chosen plate reading method above or below the plastic support of the microplate; to counter this, most manufacturers equip their instruments with two types of plate reader. To diminish interference with other biological molecules, longer wavelengths are preferred to shorter ones.

Environmental factors also have to be considered. The main inconvenience using fluorimetry comes from parameters such as (i) temperature: the decrease of fluorescence with temperature is estimated to be 1% per °C, consequently automated temperature control is of the upmost importance, (ii) light bleaching: it is necessary to operate out of direct light, (iii) pH. All appliances should serve as routine instruments and provide a wide range of excitation wavelengths. Most of them are provided with many accessories, such as automatic sample distributor, temperature regulator, shaking system, etc., as listed in **Table 2**.

Collecting data (interpretation)

Each test must be repeated at least three times and each microplate has it own control with quadruple wells for each concentration. The intra-assay variability among quadruple concentrations must be less than 5%. The mean and standard deviation should be used in standard tests. Data could be exported into other programs and dealt with statistically. Plate-reader software, however, does possess graphical functions and data analysis capabilities. A wide range of graphical functions, such as linear regression and statistical tests, are included in most apparatus.

Microspectrofluorimetry: biological applications

In the following section, selected applications of fluorescence are briefly discussed, and experimental results given. Our aim will be restricted to the presentation of some specific probes which target different mechanisms of cytotoxicity, oxidative stress, detoxification potential or cellular adhesive capacities. We demonstrate here that the selection of the most appropriate method for early detection of the cytotoxicity of xenobiotics is directly related to the mechanism of the toxic agent.

Fluorescent probes for studying cell viability

To study cell viability, several probes are proposed, as indicated in **Table 3**. Two examples will be given in this section. They correspond to either membrane damage (calcein-AM) or disturbance of mitochondrial potential (Alamar blue). Cell viability is often estimated by the Neutral red assay which is widely used in classical spectrometry and, more recently, in fluorimetry because of its increased sensitivity. More often than not, the fluorogenic dye calcein

Table 2   Some manufacturers of fluorescent microtitreplate readers: apparatus and specifications

Manufacturers and instruments	Fluorescence reading	Light source and power	Temperature control / plate shaking	Sensitivity (lowest detectable fluorescein concentration)
BMG Fluostar II	340–700 nm	Xenon 75 W	yes / yes	10 fmol/well
EG & G Instruments	340–850 nm Xenon flash	Xenon flash	yes / yes	10 fmol/well
FISHER	340–750 nm	Quartz–halogen 75 W	yes / yes	10 fmol/well
LABSYSTEM Fluoroskan II	320–670 nm	Quartz–halogen 30 W	yes / yes	75 fmol/mL (4–MU)
TECAN	320–670 nm	Xenon 75 W	yes / yes	10 fmol/mL
PACKARD Fluorocount	340–670 nm	Quartz–halogen 150 W	yes / yes	5 fmol/well
MILLIPORE Cytofluor	340–650 nm	Tungsten	yes / yes	10 fmol/mL
DYNEX (Fluorolite 1000)	300–880 nm Cold light fluorimetry	Tungsten/ halogen/xenon	no / yes	10 fmol/mL (Hoechst 33258)

Table 3 Biological applications of fluorescent probes in microspectrofluorimetry

Tests	Fluorescent probes	Targets	References
Cell proliferation	Hoechst 33258 Propidium iodide	DNA / apoptosis	Papadimitriou E and Lelkes P (1993) *Journal of Immunological Methods* **162**: 41–45
Cellular viability	Calcein-AM	Cytoplasmic membrane	Braut-Boucher F *et al.* (1995) *Journal of Immunological Methods* **178**:41–51
Cellular viability	Neutral red	Lysosomes	Noel-Hudson MS *et al.* (1997) *Toxicology in Vitro* **11**: 645–651
Energetic metabolism	Alamar blue Rhodamine 123	Mitochondrial system	Shanan TA *et al.* (1994) *Journal of Immunological Methods* **175**: 181–187
			Lilius H *et al.* (1996) *Toxicology in Vitro* **10**: 341–348.
Oxidative stress	H2DCFDA	Reactive oxygen species	Le Bel CP *et al.* (1992) *Chemical Research and Toxicology* **5**: 227 Braut-Boucher F *et al.* (1998) *Veterinary and Human Toxicology* **40**: 178
Detoxification	Resorufin	Cytochrome P450 activity	Rat P *et al.* (1997) In van Zutphen LFM and Balls M, (eds), *Animal Alternatives, Welfare and Ethics*, pp 813–825. Amsterdam: Elsevier.
Enzymatic systems	CMFDA	Glutathione trans-ferases	Lilius H *et al.* (1996) *Toxicology in Vitro* **10**: 341–348
	Bromobimane		Rat P *et al.* (1995) Cold light fluorimetry: a microtitration technology for cell culture to evaluate anethole dithiolethione and other thiols. *Methods in Enzymology* **252**: 331–341
Adhesion	Calcein-AM	Cytosolic enzymes	Braut-Boucher F *et al.* (1995) A non-isotopic, highly sensitive, fluorimetric, cell-cell adhesion assay using calcein AM-labeled lymphocytes. *Journal of Immunological Methods* **178**: 41–51

acetoxymethylester (calcein-AM) can be used to provide a rapid and sensitive method for measuring cell viability as well as cell–cell and cell–substratum adhesion interactions. Calcein-AM is cleaved by intracellular esterases to produce highly fluorescent and well retained calcein, independently of pH.

The mitochondrial membrane potential can be monitored using several fluorescent probes. These probes are often cationic and accumulate in active mitochondria with a high membrane potential. A review on the use of rhodamine 123 has been published by Slavik. It is stated that the fluorescence intensity of rhodamine 123 in mitochondria of living cells could be quantified by various techniques such as fluorescence microscopy or FACS, and can also be used in microspectrofluorimetry. A new fluorimetric metabolic indicator for detecting mitochondrial activity (which reflects the viability of target cells) is Alamar blue. Alamar blue is a redox indicator that gives fluorescence and visible colour changes in response to cellular metabolic activity. It acts as an intermediate electron acceptor in the electron transport system between the final reduction of O_2 and cytochrome oxidase. Alamar blue spectra are shown in **Figure 5**.

Performing calcein-AM assay in practice To perform the assay, the probe is added at the concentration determined in preliminary experiments to each well before or after cell treatment with the xenobiotic, be it sodium dodecyl sulfate (SDS) (**Figure 6A**) dinitrophenol (DNP) (**Figure 6B**) or doxorubicin (**Figure 6C**). After incubation for 30 min, all wells are gently rinsed before scanning the microplate using an excitation wavelength of 490 nm and an emission wavelength of 530 nm. The calcein fluorescence in treated samples is compared to that of control cells and the relative fluorescence intensity is used to calculate the viability percentage.

Performing Alamar blue assay in practice After treatment with the studied xenobiotic, cells are carefully washed to eliminate proteins which decrease the rate of dye reduction. Alamar blue solution (10% final concentration in PBS) is added to each well and after two hours, supernatants are transferred to a separate 96-well culture plate and the fluorescence is monitored using an excitation wavelength of 560 nm and an emission wavelength of 590 nm.

The cytotoxicities of both DNP and SDS are presented in **Figure 6A** and **B**. For each compound, one can determine the cytotoxic concentration exerting 50% cell damage, i.e. 50% decrease of the fluorescent signal of control cells. Depending on the target, membrane or mitochondrial damage, the best test for SDS is the calcein-AM assay, whereas the Alamar blue assay is the most sensitive one for DNP.

(A)

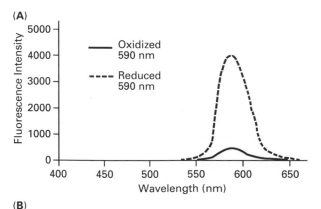

(B)

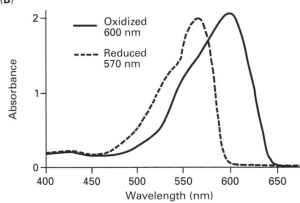

Figure 5 Fluorescence and absorbance spectra of Alamar blue (oxidized or reduced) reprinted from Interchim, France.

Probes for assessing oxidative stress

Intracellular peroxide levels were assessed using 2′,7′-dihydrodichlorofluorescein diacetate (H2DCFDA), an oxidation sensitive fluorescent probe. Once inside the cell, the probe is deacetylated by intracellular esterases forming 2′,7′-dichlorofluorescin (H2DCF), which is oxidized in the presence of a variety of peroxides, to a highly fluorescent compound, namely 2′,7′-dichlorofluorescein (DCF).

Performing H2DCFDA assay in practice　Cells are loaded with 10 μM, H2DCFDA for microplate fluorescent analysis for 30 min. The fluorescence is quantified with the microtitreplate reader using an excitation wavelength of 480 nm and an emission wavelength of 538 nm, with three separate assays per treatment. This protocol was validated using well-known peroxide inducers, such as: H2O2, cumen hydroperoxide and the t-butylhydroperoxide (t-BHP). For each compound, a concentration–response relationship is observed.

The sensitivity of the assay allows the early detection of the formation of reactive oxygen species (ROS) induced by doxorubicin, when applied to a monolayer of endothelial cells. Cells treated with 20 μg/mL of doxorubicin for three hours revealed a 50% increase in ROS measured by the DCFDA assay (**Figure 6D**) which correlates to the cell viability estimated by the calcein-AM assay (**Figure 6C**), as aforementioned.

The study of the cytotoxicity of doxorubicin, a red coloured molecule for which cytotoxic effects involve at least three mechanisms (necrosis/apoptosis, mitochondrial activity and oxidative stress) connected in some way or another, is a prime example to demonstrate the relevance of fluorimetric methods.

Probes for assessing intracellular sulfhydryl levels

An easy and sensitive method to evaluate the function of the cellular detoxification mechanisms, which could possibly be linked to stress conditions, is valuable. Glutathione (GSH), a tripeptide (Glu-Cys-Gly) which plays a key role in numerous metabolic processes, is a major intracellular reducing agent involved in detoxification of reactive species. A number of flow cytometric methods for measuring cellular GSH have been described using probes that form adducts with GSH. Cellular thiol levels could be analysed using either the monobromobimane (mBBr), which forms a fluorescent adduct with sulfhydryl groups, or 5-chloromethylfluorescein diacetate (CMFDA). CMFDA, a non-fluorescent molecule capable of permeating cells, becomes fluorescent after cleavage by cellular esterases. The glutathione transferase catalyses the reaction between CMF and the intracellular thiols retained inside the cell, whereas the unconjugated probe is released in the medium and eliminated. A decrease of fluorescence reveals a decrease of GSH content.

Performing CMFDA assay in practice　In order to validate the CMFDA assay, intracellular sulfhydryl groups of the cells were depleted either by using N-ethylmaleimide (NEM), a thiol depleting agent, or by inhibiting GSH synthesis with buthionine sulfoximine (BSO). An overnight incubation with 2 mM BSO reduces the fluorescent signal intensity by 30%, compared to control cultures, whereas a 2 min incubation with 100 μM NEM induces a more acute depletion of 50%.

Cell–cell adhesion assay

We have previously described a non-isotopic, highly sensitive, fluorimetric, cell–cell adhesion microplate assay using calcein-AM-labelled lymphocytes. To successfully perform the lymphocyte/keratinocyte adhesion assay, the following conditions are recommended:

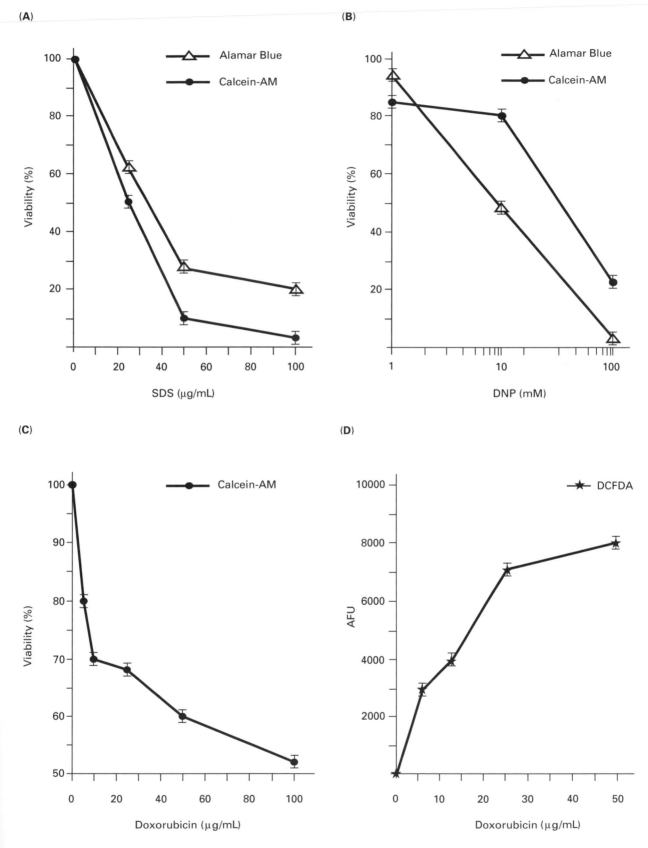

Figure 6 (A) Cytotoxicity of SDS, Calcein-AM and Alamar blue assays; (B) Cytotoxicity of DNP, Calcein-AM and Alamar blue assays; (C) Cytotoxicity of doxorubicin, Calcein-AM assay; (D) Cytotoxicity of doxorubicin, H2DCFDA assay.

- the confluence of the keratinocyte monolayer should be checked in each well;
- the lymphocytes should be labelled with 20 µM calcein AM and incubated for 30 min;
- a defined lymphocyte density per well should be plated onto the monolayer;
- fluorescence should be monitored immediately using the fluorimetric microtitreplate reader to avoid probe release.

The percentage of lymphocytes adhering to the keratinocyte monolayer was calculated as follows:

$$\% \text{ adhesion} = \frac{F_A}{F_T - F_R} \times 100$$

where F_A is the fluorescence of adherent lymphocytes, F_T the fluorescence of total lymphocytes deposited, and F_R the released fluorescence. This equation takes into account the background fluorescence emitted by contol wells in the spectrum considered.

This method can be easily applied to any cells whose adhesive capacities have to be investigated and can also be used to manage cell–extracellular matrix adhesion.

Concluding remarks

The aim of this article was to introduce microspectrofluorimetry, using fluorescence probes, as a new technology in analytical fluorescent applications and provide practical guidelines to the use of this powerful technique. When compared with conventional methods, its main advantages are:

- high sensitivity and specificity;
- simultaneous monitoring of different cellular functions;
- combining two or more probes, if the probes emit at different wavelengths;
- use of coloured molecules without interference;
- easy handling, low cost and possibility of batch studies;
- increasing number of available fluorescent probes and kits;
- avoiding the use of radiolabelled molecules.

Successful applications require the appropriate choice of model system and probes with the aim of defining the target cell functions. Under these conditions, microspectrofluorimetry offers innovative

analytical tools for both research and development projects, as well as rapid and accurate assays for quality control. Recently, cytofluorimetric microtitration was recommended by the European Centre for Validation of Alternative Methods (see Further reading). These methods, based on the selection of appropriate test systems, could give strong evidence for possible molecular mechanisms involved in cellular functions.

See also: **Cells Studied By NMR; Colorimetry, Theory; Dyes and Indicators, Use of UV-Visible Absorption Spectroscopy; Fluorescence Microscopy, Applications.**

Further reading

Brand L and Johnson M (1997) *Methods in Enzymology.* London: Academic Press.

Guibault GG (1996) *Practical Fluorescence*, 2nd edn. New York: Marcel Dekker.

Haugland RP (1996) In: Larrison KD (ed) *Handbook of Fluorescent Probes and Research Chemicals*, 6th edn. Molecular Probes, Eugene, Oregon.

Metezeau P, Ratinaud MH and Carayon P (1994) *Cytometrie par fluorescence. Apports comparatifs de techniques flux, image et confocale.* Paris: Editions INSERM.

Rat P, Chisten MO, Thevenin M, Warnet JM and Adolphe M (1995) Cold light fluorimetry: a microtitration technology for cell culture to evaluate anethole dithiolethione and other biothiols. *Methods in Enzymology,* 40: 252–331.

Rat P, Osseni R, Christen MO, Thevenin M, Warnet JM and Adolphe M (1997) Microtitration fluorimetric assays on living cells: new tools for screening in cell pharmacotoxicology. In: van Zutphen LFM and Balls M (eds) *Animal Alternatives, Welfare and Ethics*, pp 813–825. Amsterdam: Elsevier.

Ronot X, Paillasson S and Muirhead, KA (1996) Assessment of cell viability in mammalian cell lines. In: Al-Rubeai M and Emery AN (eds) *Flow Cytometry Applications in Cell Culture*, pp 177–192. New York: Marcel Dekker.

Slavik J (1994) *Fluorescent Probes in Cellular and Molecular Biology*. Boca Raton, FL: CRC Press.

Wang YL and Taylor LD (1989) Methods in Cell Biology part A: Fluorescent analogs, labeling cells and basic microscopy. *29* and part B: quantitative fluorescence microscopy-imaging and spectroscopy *30* Fluorescence microscopy of living cells in culture, Part A and B, (Eds), Academic Press, Inc., San Diego, New York, Berkely, 498 pp.

White CE and Argauer RJ (1970) *Fluorescence Analysis. A Practical Approach*. New York: Marcel Dekker.

Food and Dairy Products, Applications of Atomic Spectroscopy

NJ Miller-Ihli and **SA Baker**, U.S. Department of Agriculture, Beltsville, MD, USA

ATOMIC SPECTROSCOPY
Applications

Food composition data are critical for consumers and health care professionals to make choices and recommendations based on the nutrient content of foods. Trace elements contribute significantly to human health with 25 elements identified as being of interest. A review of the nutrients for which recommended daily allowances (RDAs) exist shows that 7 of 19 are minerals (Ca, Fe, I, Mg, P, Se, and Zn). In addition, the National Research Council has established estimated safe and adequate daily dietary intakes (ESAD-DIs) rather than RDAs for an additional five elements (Cr, Cu, F, Mn, and Mo) and three electrolytes (Cl, K, and Na) have been identified as essential. Ten additional trace elements are considered essential but no human requirement level has been set for them: As, Ni, Si, B, Cd, Pb, Li, Sn, V, and Co. The status of fluorine as an essential nutrient has been debated and conflicting data exist but its valuable effects on dental health certainly make it of potential interest.

Recent reviews of the state of knowledge of food composition data shows that there is a significant lack of data for specific food groups and for several of the elements of interest, including the seven for which RDAs have been identified. This is owing, in large part, to the lack of suitable analytical methodology, which is why rugged, accurate methods are needed that use modern analytical methodology and instrumentation. Health professionals and the public are increasingly interested in the maintenance or promotion of good health and disease prevention, rather than the diagnosis and cure of diseases. Clearly good nutrition is an important part of good health so nutritionists need high quality food composition data so that they can assess the nutrient content of foods consumed. Recent reviews of food composition data still suggest the mineral composition of foods is not well characterized despite the extensive data found in U.S. Department of Agriculture's Nutrient Data Base for Standard Reference SR-12 (the former USDA Handbook 8). In addition, there is now interest in knowing the exact chemical form of the trace elements in foods since not all forms are biologically active.

Another issue which highlights the importance of continued research for methods for trace elements in foods and beverages is related to the Nutrition Labelling and Education Act of 1990 (NLEA); this is now law in the United States and provides nutritional information to the consumer. Fifteen mandatory nutrients must be listed on the labels, including three trace elements (Ca, Fe, and Na). Thirty-four additional 'voluntary' nutrients which affect human health may be required in the future and 12 of those are trace elements (K, P, Mg, Zn, I, Se, Cu, Mn, F, Cr, Mo and Cl).

Typically, essential elements have been at sufficiently high concentrations that, with the introduction of graphite furnace atomic absorption spectroscopy (GFAAS) more than two decades ago, detectability is not an issue. At present, researchers are interested in 'new' low-level trace elements with known or potential nutritional impact, which are present at lower concentrations and, because of potential interferences in GFAAS, it has been found that inductively coupled plasma mass spectrometry (ICP-MS) coupled with graphite furnace/electrothermal vaporization (ETV-ICP-MS) provides superior detectability. The most often utilized instrumentation for generation of food composition data includes: atomic absorption spectroscopy [flame (FAAS) and graphite furnace (GFAAS)]; flame atomic emission spectroscopy (Na, K); inductively coupled plasma atomic emission spectroscopy (ICP-AES) and ICP-MS.

In this article, the description of applications will highlight the selection of the instrumentation used as well as the sample preparation methods (often the most critical part of a complete analytical method). Calibration strategies will be discussed and any specialized interference effects will be highlighted. Strategies for validating and facilitating technology transfer of the methods will also be discussed and the need for commercial as well as in-house reference materials will be featured.

Sampling (sample collection and processing)

One of the most overlooked aspects of analytical determinations relates to the suitability of the sample provided for analyses. If food composition data are being generated it is very important to know if interest lies with the specific sample at hand or if that sample is supposed to be representative of a larger bulk sample (e.g. 1 jar from a lot totalling 100 000 jars). Work completed in the Food Composition Laboratory (USDA, Beltsville Human Nutrition Research Center, Beltsville, MD) is often in support of the Nutrient Data Laboratory which is responsible for the publication of the U.S. Department of Agriculture's Nutrient Data Base for Standard Reference SR-12 (formerly known as Handbook 8) which is the primary US database for food composition data. As such, the goal is to include data representative of the national US food supply. Because large-scale national sampling is too expensive, samples are often obtained nationwide (if possible) and brands are selected on the basis of popularity (consumption) to try to gain representative samples for analysis. Samples are typically brought into the laboratory and, depending on the form of the sample, a decision is made to either analyse multiple samples or to composite and homogenize the whole of the material (if possible) and to remove separate sample aliquots. Most often, a preliminary evaluation of the homogeneity of a selected number of analytes in the material, serves as a basis to select the sample size and number of replicate samples to be analysed.

Tools are available which facilitate sample homogenization without trace element contamination and equipment that is not available commercially can often be custom made. To dissect meat, titanium blade knives or high purity glass knives may be appropriate. Large-scale mixers (e.g. Robot-Coupe, Jackson, MS) equipped with a plastic-lined bowl and titanium blades to avoid possible contamination from the stainless steel apparatus have been used successfully to blend both common samples and a diet reference material (RM8431) sold by the National Institute of Standards and Technology (NIST, Gaithersburg, MD). Polythene trays and polypropylene storage containers are most often used without risk of contamination. An assessment of contamination risk is summarized in **Tables 1** and **2**. **Table 1** lists those elements prone to contamination while highlighting common sources of contamination. **Table 2** contains a list of materials (from most preferable to least) for use during sample preparation for trace element determinations. To further minimize the likelihood of contamination [particularly for low temperature ($\leq$ 85–95 °C) wet ash preparations] glassware can be silanized; the coating will be retained on the surface of the glassware up to temperatures of approxim-

Table 1   Trace element determinations – risk assessment

Contamination potential	
High potential	Al, Ca, Cr, Cu, Fe, Pb, Sn, Zn
Medium potential	Co, K, Mn, Mo, Na, Ni, V
Low potential	Mg, Se
Sources of contamination	
Glass	Al, Ca, Co, Cr, Cu, Fe, K, Mg, Na
Porcelain	Al, Cu, K, Na, V
Rubber	Co, Cu, Fe, Zn, Sn
Building materials	Al, Ca, Mg
Stainless steel	Co, Cr, Cu, Fe, Mn, Ni, Zn
Cosmetics	Fe, Pb, Zn, Cu
Skin	K, Na, Ni, Cu, Fe
Smoke	Fe, Pb, K, Ni
Haemolysis	Ca, Co, Cu, Fe, K, Mg, Mn, Zn

Table 2   Sample container materials

Material	
Teflon	Most desirable (least trace element contamination)
PFA	
FEP	
Tefzel	
Polyethylene	
Polypropylene	
Quartz	
Platinum	
Borosilicate glass	Least desirable

ately 350°C, providing excellent protection from contaminants leached from the glassware owing to matrix interaction with the glass. In addition, silanization prevents loss of analyte due to adsorption on the container walls.

Contamination control, in a general sense, is not difficult and common sense can often provide protection from sample contamination, especially during the early sample handling stage. In an ideal world, samples are processed in clean rooms (e.g. class 100 clean room) designed to have minimal contamination because the air is filtered through hepa filters. Clean hoods and laminar air flow units can also prove useful, as well as wearing clothing that minimizes particulate generation, selecting powder-free gloves, taking care in cleaning glassware and plasticware with dilute, high purity acid, and covering gas cylinders which may be delivered coated with rust and peeling paint.

Sample preparation (reagents and ashing methods)

One of the most important means of avoiding unnecessary risk for trace element contamination is the judicious selection of reagents for sample preparation. Sub-boiling distilled or double-distilled acid is a good choice for sample treatment. As an example, Pb contaminant concentrations found in nitric acid (65–70% v/v) ranged from < 1 part per trillion (ppt) for sub-boiling distilled to 0.3 ppb for commercial 'high' purity acid. Although ppb levels may be suitable in many instances, sub-boiling distilled nitric acid (Seastar Chemicals, Sidney, British Columbia) provides even lower blanks with levels of < 50 ppt for Ca; < 20 ppt Fe and K; < 10 ppt Cr, Cu, Na, Mg, Ni, Sn, Zn; and < 1 ppt for Cd, Mn, and V, and is therefore highly recommended for ultratrace analyses.

Of course all reagents used throughout the course of any analyses provide potential sources of contamination and must be evaluated. Hydrogen peroxide comes in different levels of purity and some types prove to be particularly problematic for only a select number of elements. In the past, unacceptably high Cr levels have often been found. A peroxide called Perone (DuPont, Wilmington, DE), made for the semiconductor industry, has been evaluated in several laboratories and proved suitable for any number of elements (including Mn, Cu, Zn, Fe, Cr, and Co). Unfortunately this is not sold in small enough quantities to be suitable for the typical analytical laboratory. For analysts using matrix modifiers for GFAAS, the modifier can provide a significant level of contamination, indicating that very high purity materials should be used.

In all instances, environmental considerations are equally as important as reagent purity, when selecting a sample preparation method. In an ideal situation, samples can be handled under clean air conditions under laminar flow to prevent contamination. Potential sources of environmental contamination include the analyst (particles from skin or clothing), sample container material, air, gloves, heating block, muffle furnace, etc.

The two most common ashing procedures for food analyses are wet ashing and dry ashing procedures. Wet ashing may be done using a typical hot plate or heating blocks or it may be done using microwaves under increased pressure. Many modern microwave systems are now under temperature and pressure control and various methods are available in the literature. In a laboratory which processes many samples with one matrix type, microwave digestion can be quite useful. If, however, the matrices are variable, experiment design requires that samples of similar type be put in the microwave together. Otherwise, depending on the carbohydrate content, pressure will build up in the samples at different rates, requiring venting which can be problematic if the microwave system does not have temperature and pressure monitoring control for each vessel. Another great benefit is the use of microwave digestion as a reference to compare with wet ashing results. The microwave system provides the benefit of increased pressure, ensuring almost complete digestion (depending on the sample matrix). When used under optimum conditions, microwave digests provide a reliable dissolution of samples and minimize the risk of sample contamination while saving significant time. Microwave sample digestion is very much a growing field and the number of analysts using this sample preparation tool probably increased 10-fold between 1993 and 1998.

Method selection and method validation

Several factors should be considered in selecting the appropriate sample preparation and instrumental detection method for a particular set of analyses. Factors include detection capability, ease of use, availability of instrumentation, time, cost, etc. One must consider the relative importance of the factors for their application but, most often, detection capability is the top priority. A good general rule is to use the simplest method possible that meets your analysis requirements. For example, if asked to determine Mn, Cu, and Zn in soft drinks, analysts should check to see if it could be carried out with a simple dilution directly using FAAS or simultaneous multi-element ICP-AES (for greater speed). Once a method is selected it should be validated using one or more of the following strategies: (1) analysis of one or more standard reference materials; (2) comparison of results using an alternative method; (3) comparison of results with those from another laboratory; and (4) thorough evaluation of possible matrix interference effects (compare direct calibration with the method of additions). **Table 3** contains a list of commercial reference materials with matrices suitable for food and beverages analyses. One unfortunate aspect of solid commercial materials is that they are almost all freeze-dried and this is not a realistic representation of typical samples (e.g. meats, vegetables, fruits, etc.); analysts prefer to analyse 'as received'. Because these primary controls are so expensive, analysts should consider developing secondary, in-house controls that represent the food type/matrices most often analysed in their laboratory. These materials can be characterized using all the validation strategies listed above, with the assistance of colleagues and by using alternative methods.

Table 3 Commercial reference materials

FCL-ID	Reference material (μg/g dry weight)	Number	Manufacturer	Ca	Cd	Co
R - 0010	Rice Flour SRM 1568	1568	NIST	140 ± 20	0.029 ± 0.004	0.02 ± 0.01
R - 0011	Rice Flour SRM 1568a	1568a	NIST	118 ± 6	0.022 ± 0.002	(0.018)
R - 0020	Wheat Flour SRM 1567	1567	NIST	190 ± 10	0.032 ± 0.007	
R - 0021	Wheat Flour SRM 1567a	1567a	NIST	191 ± 4	0.026 ± 0.002	(0.006)
R - 0030	Total Diet SRM 1548	1548	NIST	1740 ± 70	0.028 ± 0.004	
R - 0040	Carrot Powder		ARC/CL	1700 ± 45	0.0686 ± 0.0031	
R - 0050	Skim Milk Powder		ARC/CL	13040 ± 260	(<0.005)	
R - 0060	Pork		ARC/CL	150 ± 6	0.022 ± 0.0042	
R - 0070	Wheat Flour		ARC/CL	208 ± 4	0.039 ± 0.004	
R - 0080	Potato		ARC/CL	91 ± 6	0.035 ± 0.0016	
R - 0090	Total Diet		HDP/CL	2860 ± 124	0.021 ± 0.003	
R - 0100	Skim Milk (recommended values)		ARC/CL	13000 ± 291	(<0.005)	
R - 0110	Mixed Diet SRM 8431 (recommended values)	8431	NIST	1940 ± 140	0.042 ± 0.011	0.038 ± 0.008
R - 0120	Non-Fat Milk Powder SRM 1549	1549	NIST	13000 ± 500	0.0005 ± 0.0002	(0.004)
R - 0130	Milk Powder IAEA A-11	A-11	IAEA	12900 ± 800	(0.526)	0.005 ± 0.001
R - 0140	Oyster Tissue SRM 1566a	1566a	NIST	1960 ± 190	4.15 ± 0.38	0.57 ± 0.11
R - 0150	Bovine Liver SRM 1577a	1577a	NIST	120 ± 7	0.44 ± 0.06	0.21 ± 0.05
R - 0151	Bovine Liver SRM 1577b	1577b	NIST	116 ± 4	0.50 ± 0.03	(0.25)
R - 0160	Dogfish Muscle and Liver DORM1	DORM1	NRCC		0.086 ± 0.012	0.049 ± 0.014
R - 0170	Dogfish Muscle and Liver DOLT1	DOLT1	NRCC		4.18 ± 0.28	0.157 ± 0.037
R - 0180	Dogfish Muscle and Liver DORM2	DORM2	NRCC		0.043 ± 0.008	0.182 ± 0.031
R - 0190	Dogfish Muscle and Liver DOLT2	DOLT2	NRCC		20.8 ± 0.5	0.24 ± 0.05
R - 0210	Bovine Muscle Powder SRM 8414	8414	NIST	145 ± 20	0.013 ± 0.011	0.007 ± 0.003
R - 0220	Non-defatted Lobster Hepatopancreas LUTS-1 (as bottled)	LUTS-1	NRCC	203 ± 33	2.12 ± 0.15	0.051 ± 0.006
R - 0230	Non-defatted Lobster Hepatopancreas LUTS-1 (dry weight)	LUTS-1	NRCC	1360 ± 220	14.2 ± 1.0	0.34 ± 0.04
R - 0240	Lobster Hepatopancreas TORT-1	TORT-1	NRCC	8950 ± 580	26.3 ± 2.1	0.42 ± 0.05
R - 0250	Spinach Leaves SRM 1570	1570	NIST	13500 ± 300	(1.5)	(1.5)
R - 0251	Spinach Leaves SRM 1570a	1570a	NIST	15270 ± 410	2.89 ± 0.07	0.39 ± 0.05
R - 0260	Apple Leaves SRM 1515	1515	NIST	15260 ± 150	(0.014)	(0.09)
R - 0270	Citrus Leaves SRM 1572	1572	NIST	31500 ± 1000	0.03 ± 0.01	(0.02)
R - 0280	Peach Leaves SRM 1547	1547	NIST	15600 ± 200	0.026 ± 0.003	(0.07)

Table 3 Continued

FCL-ID	Cr	Cu	Fe	I	K	Mg	Mn	Na
R-0010		2.2±0.3	8.7±0.6		1120±20		20.1±0.4	6.0±1.5
R-0011		2.4±0.3	7.4±0.9	(0.01)	1280±8	560±20	20.0±1.6	6.6±0.8
R-0020		2.0±0.3	18.3±1.0		1360±40		8.5±0.5	8.0±1.5
R-0021		2.1±0.2	14.1±0.5	(0.001)	1330±30	400±20	9.4±0.9	6.1±0.8
R-0030		2.6±0.3	32.6±3.6		6060±280	556±27	5.2±0.4	6250±260
R-0040			12.6±0.79		10200±350	350±9	4.90±0.17	485±24
R-0050	0.0079±0.0032	0.58±0.06	4.53±0.61	1.74±0.07	17700±590	1230±24	0.45±0.07	4870±350
R-0060		2.68±0.283	52.84±5.456			(1010)	0.30±0.024	
R-0070	0.028±0.005	2.35±0.18	51±3.79		2200±190	562±10	12.87±0.34	895±42
R-0080	0.098±0.018	3.87±0.18	22.0±2.0		14100±100	747±16	8.1±0.32	
R-0090		3.18±0.19	30.4±0.9		9420±300	785±25	12.9±0.58	7870±570
R-0100		0.59±0.148	4.54±0.855			1230±27	0.44±0.104	
R-0110	0.102±0.006	3.36±0.33	37.0±2.6		7900±420	650±40	8.12±0.31	3120±160
R-0120	0.0026±0.0007	0.7±0.1	1.78±0.10	3.38±0.02	16900±300	1200±30	0.26±0.06	4970±100
R-0130	(0.257)	0.838±0.165	3.65±0.76	(1.5)	17200±1000	1100±80	0.373±0.081	4420±330
R-0140	1.43±0.46	66.3±4.3	539±15	4.46±0.42	7900±470	1180±170	12.3±1.5	4170±130
R-0150		158±7	194±20		9960±70	600±15	9.9±0.8	2430±130
R-0151		160±8	184±15		9940±20	601±28	10.5±1.7	2420±60
R-0160	3.60±0.40	5.22±0.33	63.6±5.3		15900±1000	1210±130	1.32±0.26	8000±600
R-0170	0.40±0.07	20.8±1.2	712±48		10100±1000	1100±150	8.72±0.53	7260±730
R-0180	34.7±5.5	2.34±0.16	142±10				3.66±0.34	
R-0190	0.37±0.08	25.8±1.1	1103±47				6.88±0.56	
R-0210	0.071±0.038	2.84±0.45	71.2±9.2	0.035±0.012	15170±370	960±95	0.37±0.09	2100±80
R-0220	0.079±0.012	15.9±1.2	11.6±0.9		948±72	89.5±4.1	1.20±0.13	
R-0230	0.53±0.08	107±8	77.8±6.0		6360±480	601±28	8.02±0.86	
R-0240	2.4±0.6	439±22	186±11		10410±400	2550±250	23.4±1.0	36700±2000
R-0250	4.6±0.3	12±2	550±20		35600±300		165±6	
R-0251		12.2±0.6			29030±520	(8900)	75.9±1.9	18180±430
R-0260	−0.3	5.6±0.24	(80)	(0.3)	16100±200	2710±80	54±3	24.4±1.2
R-0270	0.8±0.2	16.5±1.0	90±10	1.84±0.03	18200±600	5800±300	23±2	160±20
R-0280	(1)	3.7±0.4	218±14	(0.3)	24300±300	4320±80	98±3	24±2

Table 3 *Continued*

FCL-ID	Ni	P	Pb	Se	Zn
R-0010	(0.16)			0.4 ± 0.1	19.4 ± 1.0
R-0011		1530 ± 80	(<0.010)	0.38 ± 0.04	19.4 ± 0.5
R-0020	(0.18)			1.1 ± 0.2	10.6 ± 1.0
R-0021		1340 ± 60	(<0.020)	1.1 ± 0.2	11.6 ± 0.4
R-0030	(0.41)	3240 ± 40	(0.05)	0.245 ± 0.005	30.8 ± 1.1
R-0040					6.9 ± 0.15
R-0050	0.059 ± 0.022		0.016 ± 0.003	0.082 ± 0.0072	41.7 ± 1.00
R-0060			0.089 ± 0.013	0.0394 ± 0.031	103.80 ± 3.005
R-0070	0.153 ± 0.009	2090 ± 140	0.018 ± 0.007	0.057 ± 0.00545	14.6 ± 0.71
R-0080	0.193 ± 0.043	2370 ± 100	0.026 ± 0.0028		9.0 ± 0.32
R-0090	0.271 ± 0.038		0.043 ± 0.008	0.181 ± 0.017	28.9 ± 1.3
R-0100			0.016 ± 0.003	0.082 ± 0.0077	41.68 ± 1.056
R-0110	0.644 ± 0.151	3320 ± 310		0.242 ± 0.030	17.0 ± 0.6
R-0120		10600 ± 200	0.019 ± 0.003	0.11 ± 0.01	46.1 ± 2.2
R-0130	(0.93)	9100 ± 1020	(0.27)	0.0339 ± 0.0072	38.9 ± 2.3
R-0140	2.25 ± 0.44	6230 ± 180	0.371 ± 0.014	2.21 ± 0.24	830 ± 57
R-0150		11100 ± 400	0.135 ± 0.015	0.71 ± 0.07	123 ± 8
R-0151		11000 ± 300	0.129 ± 0.004	0.73 ± 0.06	
R-0160	1.20 ± 0.30		0.40 ± 0.12	1.62 ± 0.12	21.3 ± 1.0
R-0170	0.26 ± 0.06		1.36 ± 0.29	7.34 ± 0.42	92.5 ± 2.3
R-0180	19.4 ± 3.1		0.065 ± 0.007	1.40 ± 0.09	25.6 ± 2.3
R-0190	0.20 ± 0.02		0.22 ± 0.02	6.06 ± 0.49	85.8 ± 2.5
R-0210	0.05 ± 0.04	8360 ± 450	0.38 ± 0.24	0.076 ± 0.010	142 ± 14
R-0220	0.200 ± 0.034		0.010 ± 0.002	0.641 ± 0.054	12.4 ± 0.8
R-0230	1.34 ± 0.23		0.069 ± 0.011	4.30 ± 0.36	82.9 ± 5.4
R-0240	2.3 ± 0.3	8790 ± 210	10.4 ± 2.0	6.88 ± 0.47	177 ± 10
R-0250	(6)	5500 ± 200	1.2 ± 0.2		50 ± 2
R-0251	2.14 ± 0.10	5180 ± 110	(200)	0.117 ± 0.0009	82 ± 3
R-0260	0.91 ± 0.12		0.470 ± 0.024	0.050 ± 0.009	12.5 ± 0.3
R-0270	0.6 ± 0.3	1300 ± 200	13.3 ± 2.4	(0.025)	29 ± 2
R-0280	0.69 ± 0.09	1370 ± 70	0.87 ± 0.03	0.120 ± 0.009	17.9 ± 0.4

Instrumentation and applications

The literature is the most informative source of insights on how to select the most appropriate sample preparation methods and instrumentation to make accurate and precise trace element determinations. Analysts worldwide are continuously publishing their successes (and often their failures), allowing their colleagues to benefit from their experiences. Since the bulk of trace element determinations for foods and biological samples are done by a handful of techniques, four methods represent 98% of all determinations: (1) flame atomic absorption spectroscopy (FAAS); (2) graphite furnace atomic absorption spectroscopy (GFAAS); (3) inductively coupled plasma atomic emission spectroscopy (ICP-AES); and (4) ICP-mass spectrometry (ICP-MS). Comments on both the sample preparation method and the instrumentation are available to the reader. It is often a good idea to combine the best portions of several different methods when developing a method, and it is always good to avoid mistakes documented by others.

FAAS and GFAAS

An estimated 80% of all currently available trace element food composition data are the result of FAAS analyses after either wet ashing or dry ashing sample pretreatment. FAAS is a simple, robust, and easy to implement tool for the analysis of digests, and calibration can typically be accomplished using aqueous standards. Detection limits are in the sub-ppm range, making this method suitable for a wide range of elements (including Ca, Cu, Fe, Mg, Mn, Zn) in various sample matrices. Please note that Na and K are most often determined using flame emission spectroscopy rather than absorption on an AAS system.

GFAAS provides sub-ppb detection capability with µL-sized sample injections into a platform-containing graphite tube which is resistively heated to high (e.g. 2700°C) temperatures for sample atomization. Over the last decade, so-called STPF (stabilized temperature platform furnace) conditions established by Slavin have been almost universally adopted. STPF conditions call for the use of (1) platform atomization, (2) matrix modification, (3) rapid heating (1500°C s^{-1} or more), (4) pyrolytically coated tubes; (5) fast digital electronics, (6) integrated absorbance measurements (peak area), (7) argon (stop-flow during atomization), and (8) Zeeman (or Smith–Hieftje) background correction. Methods developed with these criteria in mind will facilitate straightforward quantitation using aqueous standards to make external calibration curves, in most cases minimizing

matrix interference effects and reducing the need for using the method of additions. Analytical figures of importance include sensitivity/characteristic mass, detection limit, accuracy, and precision.

The field of trace element analyses in nutrition is one of the most interesting areas. Trace elements serve as structural components of enzymes, vitamins, hormones, and protein-containing tissues. The trace element selenium helps in the defense mechanism against diseases and environmental risk. Selenium is the most promising trace element potentially involved in immune response. Chromium is a cofactor for several enzyme systems, and is required for insulin receptor interaction. Consequently, research was conducted using FAAS to assess the Se and Cr content of eight food categories (cereals, beans, vegetables, greens, fruit, condiments and spices, dried fruits, and edible flowers). Samples were wet ashed using a combination of three acids and samples were analysed using FAAS at 196 nm (Se) and 425 nm (Cr) using an air–acetylene flame. A total of 190 samples were analysed and from the study it was concluded that dried fruits have the highest Cr content (15–43.5 µg per 100 g) and that beans have the highest Se content (48.7–02.5 µg per 100 g). FAAS also recently proved useful in the monitoring of Pb in dinnerware where excess levels can provide increased risk to fetuses, children, and adults. A total of 0.9% of imported ceramic dinnerware and 2.5% of domestic ceramic dinnerware, evaluated over the course of 2 months in 1992, had levels in excess of the 3 ppm allowed limit for plates, saucers, and flatware. This work involved the use of Association of Official Analytical Chemists Official Method 973.32. FAAS was used by scientists at Behrend College (Erie, PA) to evaluate stainless steel cookware as a significant source of Ni, Cr, and Fe for ingestion. Nickel ingestion is potentially dangerous since Ni is implicated with health problems related to allergic dermatitis. Conversely, Cr and Fe are essential nutrients and stainless steel (typically 18% Cr, 8% Ni and 70–73% Fe) may provide a significant source. Sample preparation involved the addition of 5% acetic acid (Fisher), both cold and boiled, in each vessel for 5 min. Next the acetic acid was analysed using FAAS and the manufacturers' standard conditions. Measurable levels of all three elements were determined with only Ni being high enough (0.0–0.1 mg Ni per day) to pose a health threat, leading to a recommendation that Ni-sensitive patients avoid stainless steel cookware and that the industry switch to a non-Ni formulation.

As evidenced by the previous examples, FAAS is a powerful technique but it may not always provide the necessary sensitivity for the determination of trace elements present at extremely low concentrations. This became apparent when the Swedish National Food Administration set out to develop a method for the determination of Pb, Cd, Zn, Cu, Fe, Cr, and Ni in dry foodstuffs after dry ashing at 450°C. Realising that contamination with elements such as Pb, Cr and Ni is probable, care was taken to acid-wash all plasticware associated with the analyses. FAAS was selected for Zn (213.9 nm air–acetylene, oxidizing), Cu (324.7 nm air–acetylene, oxidizing), and Fe (Fe 248.3 nm nitrous oxide–acetylene, oxidizing). Owing to the lower concentrations, GFAAS was used for Pb, Cd, Cr and Ni. Conditions were optimized based on the use of the appropriate resonance line but no one set of instrumental conditions proved acceptable for all four elements. Lead was analysed using platform atomization while Cd was measured in an uncoated tube and Cr and Ni were measured in a standard pyrolitic tube. Unfortunately, not using STPF conditions limited the capabilities of the method and the method of additions was required by most collaborating laboratories to get reasonably accurate results.

GFAAS serves as an excellent method for the direct determination of Pb in degassed cola beverages. Recent reports indicate, however, that GFAAS analyses of cola diluted with a solution of lanthanum to reduce chlorides and other matrix interferences required the use of the method of additions to obtain accurate results. If the authors had done in situ oxygen ashing, and used Pd or Pd combined with magnesium nitrate as a matrix modifier, all matrix interference effects could have been removed. It is likely that the biggest problem with the background was owing to the sugar (carbon) which could have been removed during the thermal pretreatment step using oxygen ashing – permitting the use of aqueous calibration standards rather than requiring the method of additions. Sugars and syrups have recently been analysed directly after diluting ~1 g of sugar per 10 mL 5% nitric acid and using oxygen ashing in the thermal pretreatment step. Instrumental GFAAS detection limits (DL) were 10 pg, corresponding to a method DL of 0.9 ng g^{-1}. Researchers often use ingenious approaches to improve GFAAS performance. This was the case when developing a strategy for the determination of La in food and water samples. The problem is that La has a strong affinity for the graphite, leading to carbide formation and memory effects. Although the graphite has been improved to reduce this, lining the tube with tantalum or tungsten foil can eliminate physical contact with the graphite and lead to increased sensitivity. The tungsten-lined tube provided a detection limit of 7.8 ng

and a characteristic mass of 8.1 ng for La. Precisions were better than 10% RSD and the average accuracy was 90 ± 10%. GFAAS has been used by researchers at the US Food and Drug Administration to successfully determine Se in infant and enteral formulas. The method utilizes sample digestion on a hotplate after addition of magnesium nitrate–nitric acid. Following heating, digests were evaporated to dryness and placed in a 500°C muffle for 30 min to complete ashing. All Se is converted into Se^{4+} by dissolving the ash in HCl (5:1) and holding the solution at 60°C for 20 min. The Se^{4+} was subsequently reduced to Se^0 with ascorbic acid and collected on a membrane filter which was digested in nitric acid using microwave digestion. Following digestion and dilution, Se was determined using GFAAS. The recovery range for Se was 85–127% and analysed reference materials fell within the certified range for Se. The work was performed on a commercial system equipped with only a deuterium background correction. Peak height and peak area measurements both provided accurate results when using nickel nitrate for matrix modification. Finally, GFAAS has proved useful for the determination of Cr and Mo in medical foods. Both wet ashing and dry ashing proved acceptable and the detection limits were 0.24 ng mL^{-1} for Cr and 0.67 ng mL^{-1} for Mo. Both elements could be determined directly off the shelf and neither required the use of a matrix modifier. Optimum ashing temperatures of 1650 and 1600°C were found for Mo and Cr, respectively. The ideal atomization temperatures were 2400 and 2650°C, respectively, and all standard reference materials analysed provided results within the certified concentration range.

ICP-AES and ICP-MS

To assist consumers in maintaining healthy dietary practices and to assure product safety, the Nutrition Labelling Education Act (NLEA) was published in 1990 and the compliance deadline was May, 1994. More than 17 000 food companies were affected by NLEA and labels for more than 250 000 products required modification. As such, Perkin-Elmer researchers developed a multielement ICP-AES method using an axially viewed plasma which they coupled with microwave sample digestion. ICP-AES measurements allowed both trace and macro elements to be determined simultaneously owing to the wide linear dynamic range. In contrast to many official methods, which are analyte and/or matrix specific, and which often require multiple dilutions, this ICP-AES method gives multielement data (seven elements; nine wavelengths), providing an elemental fingerprint which proves useful in adulteration and/ or contamination studies. Measurements were made on a Perkin-Elmer Optima [echelle polychromater with a segmented array charge coupled detector (SCD) equipped with a standard torch, a Scott spray chamber, and a cross-flow nebulizer]. Although only a single 10 ppm multielement standard was successfully used for calibration, the use of three or more standards is recommended to cover such a large range. The use of duplicate wavelengths for Ca and Mg, combined with spike recovery checks, assisted with method validation. Literature reports highlighted the benefit of the Multiwave (Perkin-Elmer, Norwalk, CT) microwave system which provides pressure monitoring in each vessel, allowing different sample types to be digested together. Simultaneous multielement ICP-AES provides significant time-savings and the establishment of compromise conditions does not typically lead to significant deterioration of performance as compared with single element determinations.

In the 1990s, ICP-MS determinations have started to play a significant role in generation of food composition data. This technique provides multielement isotopic data with sub-ppb detection limits with the same convenience as flame AAS or ICP-AES and it is much faster than GFAAS. As interest in low level trace element concentrations continues to grow, more data are generated by ICP-MS and often alternative methods are employed to validate the ICP-MS method and resultant data. Such was the case in the recent study of Pb in green vegetables where GFAAS (dry ashing) and isotope dilution–ICP-MS (ID-ICP-MS) (wet ashing) were both used. The end result was comparable detectability and accuracy for in-house and commercial control materials. The limit of detection for Pb was 1–3 µg kg^{-1} and precisions ranged from 6% RSD for kale containing 500 µg kg^{-1} Pb to 20% RSD for cabbage containing 3 µg kg^{-1} Pb. These methods were used for a four-year study of Pb levels in green vegetables with good success. Another advantage of ICP-MS is that it provides excellent detection capability for non-metals of nutritional interest, including P and I. Methods for I have typically been dependent on radio-iodide measurements and the specialized equipment required is not widely available. In contrast, ICP-MS allows the direct determination of I in solution using highly sensitive, specific, and interference-free detection of monoisotopic iodine (^{127}I). Samples were wet ashed in closed steel bombs using a mixture of nitric and perchloric acids which converted any volatile I species into non-volatile species. The concentrations measured were in the range 0.15–

4.59 µg g^{-1} (dry mass) and the detection limit was 30 ng g^{-1} (dry mass) for a 0.5 g (dry) sample diluted to a final volume of 20 mL. The importance of ICP-MS in food analysis is growing but probably won't be fully realized until the list of the mandatory nutrients for labelling has been expanded.

Acknowledgements

Mention of trademark or proprietary products does not constitute a guarantee of warranty of the product by the U.S. Department of Agriculture and does not imply their approval to the exclusion of other products that may also be suitable. The author would like to acknowledge that this work is based on literature reports provided by many colleagues and regrets that the specified presentation format precluded referencing the many interesting applications individually. The references listed are of general interest and many of them are overviews which include a large number of pertinent literature citations.

See also: **Atomic Absorption, Methods and Instrumentation; Environmental and Agricultural Applications of Atomic Spectroscopy; Food Science, Applications of Mass Spectrometry; Food Science, Applications of NMR Spectroscopy; Inductively Coupled Plasma Mass Spectrometry, Methods; Microwave and Radiowave Spectroscopy, Applications; Microwave Spectrometers.**

Further reading

Brown (1990) *Present Knowledge in Nutrition.* Washington, DC: International Life Sciences Institute, Nutrition Foundation.

Chang S, Rayas-Duarte P, Holm E and McDonald C (1993) Food. *Analytical Chemistry* **65**: 334R–363R.

Greenfield H and Southgate DAT (1992) *Food Composition Data, Production, Management and Use.* Barking: Elsevier Science Publishers.

Howard AG and Statham PJ (1997) *Inorganic Trace Analysis: Philosophy and Practice.* New York: Wiley.

Kingston HM and Jassie LB (1988) *Introduction to Microwave Sample Preparation.* Washington, DC: American Chemical Society Press.

National Research Council (1988) *Designing Foods, Animal Product Options in the Marketplace.* Washington, DC: National Academy Press.

Pauwel J, Stoeppler M and Wolf WR (1998) Proceedings 7th International symposium on biological and environmental reference materials, Antwerp, Belgium, April 21–25, 1997. *Fresenius Journal of Analytical Chemistry* **360**: 275–504.

Stewart K and Whitaker JR (1984) *Modern Methods of Food Analysis*. Westport CT: AVI Publishing Company.

Taylor A, Branch S, Crews H, Halls D, Owen L and White M (1997) Atomic spectrometry update – clinical and biological materials, food and beverages. *Journal of Analytical Atomic Spectrometry* 12: 119R–221R.

Wolf W (1984) *Biological Reference Materials, Availability, Uses and Need for Validation of Nutrient Measurement*. New York: Wiley.

Food Science, Applications of Mass Spectrometry

John PG Wilkins, Unilever Research, Sharnbrook, Bedfordshire, UK

MASS SPECTROMETRY
Applications

The exquisite analytical sensitivity and specificity of MS have found numerous and diverse applications in the field of food science.

Trace analysis for undesirable chemical contaminants in foodstuffs is conducted in many laboratories around the world. This may be for quality control, regulatory or surveillance purposes. MS is frequently the analytical method of choice because of its ability to produce unequivocal data.

Organic compounds sought include naturally derived materials, such as mycotoxins and off-flavours (produced by rancidification or spoilage), and man-made/industrial chemicals, e.g. pesticides, veterinary drugs, environmental contaminants (such as polychlorodibenzo-*p*-dioxins, polychlorinated biphenyls, polynuclear aromatic hydrocarbons, etc.) and food tainting compounds (e.g. 2,4,6-trichloro-anisole, the compound responsible for musty cork taint in wine, arising from the inappropriate use of wood preservatives). GC-MS and HPLC-API-MS are widely used for these types of analyses. Desirable food components present at trace levels, such as nutrients, are also determined using these techniques.

Trace levels of inorganic chemical species, e.g. lead, arsenic, cadmium, are also monitored in foodstuffs, often using ICP-MS. The advantage of MS over AAS is that several elements may be measured simultaneously and the concentrations of individual isotopes may be measured, facilitating metabolism/nutrient studies with stable isotope materials. Precise determination of isotope ratios (e.g. C, N and O) by IRMS is also important in agricultural and food authenticity studies. Accelerator mass spectrometry is used in tracer studies, for the determination of extremely low levels of carbon-14 (and other) isotopes.

The table below summarizes the ranges of analytical sensitivities required for different classes of food analysis.

Much work has been done on the characterization of food flavours and aromas by GC-MS. A significant recent advance has been the use of APCI and drift tube MS techniques for sensitive, on-line ('real time') monitoring of trace volatiles, e.g. in human breath during flavour release studies.

Characterization of unfractionated foodstuffs and related materials for screening or taxonomic purposes may be performed by pyrolysis MS and direct headspace MS. These techniques generate rather simple mass spectra that may be classified using pattern recognition/chemometric software.

MS is also used for the analysis of the more abundant (non-trace) components of food, e.g. oils and fats (triglycerides), proteins and carbohydrates. Analysis of these materials is often challenging, as they may comprise complex mixtures of isomeric compounds. Modern MS techniques (MALDI TOF and electrospray ionization) have become extremely important in protein and peptide studies.

Table 1 Typical analytical sensitivities required for various classes of food contaminant/component

Class	Typical concentraion sought (by weight)
Flavours	% - ppb
Food taints	% - ppq
Pesticide residues	ppm - ppb
Trace elements	ppm - ppb
Mycotoxins	ppb
Dioxins	ppt - ppq

See also: **Biochemical Applications of Mass Spectrometry; Food and Dairy Products, Applications of Atomic Spectroscopy; Food Science, Applications of NMR Spectroscopy; Inorganic Chemistry, Applications of Mass Spectrometry; Peptides and Proteins Studied Using Mass Spectrometry; Time of Flight Mass Spectrometers.**

Further reading

Belton PS, Mellon FA and Wilson RH (1993) Spectroscopy in Food Science, *Spectroscopy in Europe* 5: 8–14.

Crews H (1993) A decade of ICP MS analysis. *International Laboratory* 23: 38–42.

Gilbert J (ed.) (1987) *Applications of Mass Spectrometry in Food Science*. Amsterdam: Elsevier Applied Science.

Horman I (1979) Mass spectrometry in food science. *Mass Spectrometry* 5: 211–233.

Self R, Mellon FA, McGaw BA, Calder AG, Lobley GE and Milne E (1996) The application of mass spectrometry to food and nutrition research. *NATO ASI Series, Series C* 475 (*Mass Spectrometry in Biomolecular Sciences*) 483–515.

Food Science, Applications of NMR Spectroscopy

Brian Hills, Institute of Food Research, Norwich, UK

MAGNETIC RESONANCE
Applications

The applications of NMR in food science have been the subject of several international conferences, reviews and books which are cited in the Further reading section. This intense interest in the NMR spectroscopy of food is driven not only by the commercial importance of food, but also by the intellectual challenge of unravelling the physicochemical properties of this exceedingly diverse and complex group of materials. In order to focus on food aspects, it will be assumed that the reader is familiar with the principles of NMR and magnetic resonance imaging (MRI), which are lucidly explained in other articles of this encyclopedia.

Spatially resolved NMR applications

MRI is beginning to have a major impact on our understanding and control of the growth of food crops, on our assessment of fruit and vegetable quality and in developing the best methods of fruit and vegetable preservation. Moreover, by providing noninvasive, real-time images of moisture and lipid distributions, as well as spatial maps of temperature and food quality factors, MRI has the potential for making a major impact on food processing science. We therefore begin with a brief review of whole-plant functional imaging, then progress on to MRI applications in food processing and storage.

Functional imaging of whole plants

The growth of agricultural crops is affected by many factors, including drought (osmotic stress), luminescence, disease and soil pollutants, such as heavy metals. By permitting noninvasive monitoring of a whole plant under realistic environmental conditions, MRI has become a powerful tool in the plant physiologists' armoury. The development of small seedlings can be observed directly inside adapted NMR tubes. Larger plants can be grown in controlled-environment boxes, which bathe the root system in nutrients, and control humidity, temperature and luminescence. The root system, stem or leaf areas can then be imaged noninvasively. Root imaging is now an established technique, though it is not applicable to all soil types because paramagnetic impurities in the soil can severely shorten the water relaxation time. Nevertheless, three-dimensional (3D) T_1-weighted imaging has been used to distinguish roots and soil in developing pine seedlings, to follow water depletion zones around the tap root, lateral roots and fine roots, and to follow the effects of symbiotic relationships, such as infection with mycorrhizal fungus, on water uptake. The increase in root network volume and surface area can also be measured from the 3D images.

A combination of flow imaging and chemical shift imaging has been used to monitor the effect of environmental stresses on the flux of nutrients through the plant stem. For example, chemical shift imaging has been used to map the spatial distribution of sucrose, glucose, glutamate, lysine, arginine and valine in castor bean seedlings. If, at the same time, the velocity distribution through the vascular bundles in the stem is observed with phase-encoding MRI, the product of the concentration and flow rate gives the nutrient fluxes. The effects of varying osmotic stress and luminescence on the nutrient fluxes can then be monitored nondestructively. **Table 1** lists examples of this type of application.

Imaging quality factors in intact fruit and vegetables

Understanding the development of the whole plant under realistic environmental conditions is only one aspect of food production. The quality of unprocessed, intact fruit, grains or vegetables as they are stored or transported is just as important. Here again MRI provides a useful noninvasive monitor of the effect of storage conditions on the ripening of fruit and vegetables, on the progress of detrimental changes such as disease and bruising and on the effects of preservation processes such as chilling and surface heat sterilization.

The ripening changes in tomato are of commercial importance because they are picked when they are green and transported, internationally, in boxes gassed with ethylene. Unfortunately, on arrival the tomatoes display a wide range of ripeness from green to the red-ripe fruit and require costly (and damaging) hand-sorting. If, when first picked and boxed, the degree of ripeness could be determined noninvasively by MRI, then ripening could be made more uniform and the second sorting stage rendered unnecessary. Motivated by these considerations, Saltveit and co-workers have shown that ripening in the early green stages is associated with increased water

Table 1 Applications of functional imaging to the development of fruit and vegetables[a]

Plant	MRI technique	Phenomenon
Castor bean seedlings	Flow imaging, chemical shift imaging	Effect of environmental stresses on nutrient fluxes
Maize and millet	Proton density mapping, T_2 mapping	Osmotic stress
Wheat grains	Water proton density, velocity and diffusion maps	Relationship between flow and grain development
Red raspberry	Gradient-echo signal intensity	Fungal infection
Strawberry	Proton density T_1, T_2	Fungal infection

Table 2 Representative MRI studies of quality factors in fruit and vegetables[a]

Plant	Quality factor/process	NMR parameter
Red raspberry	Ripening	Spin-echo signal intensity
Strawberry	Ripening, bruising	Spin density, T_2
Grape berries	Ripening, drying	Chemical shift
Red raspberry	Fungal infection	Gradient-echo signal intensity
Barley seeds	Ripening	Diffusion
Wheat grains	Ripening	Diffusion and flow
Tomato	Ripening	Spin density, T_1
Mango	Heat sterilization	Relaxation-weighted signal intensity
Blueberry	Freezing–thawing	T_1 discrimination of sugar and water
Cherry	Ripening	T_1, D and volume selective spectroscopy
Papaya	Heat injury	Spin density, T_1, T_2

[a] For references see Hills (1998) in Further reading section.

content, and hence increased MRI signal intensity, as the locular tissue 'liquefies'. Later stages of ripening are associated with increased image graininess as small air pockets develop in the pericarp tissue.

Strawberries have also been the subject of several MRI investigations. Bruising is associated with a lengthening of the apparent transverse relaxation time (T_2^*) and an increase in apparent water proton spin density, consistent with the reduction of local magnetic susceptibility inhomogeneities as a result of the filling of the intercellular gas spaces with intracellular fluid released through membrane and cell wall damage. A similar increase in T_2^* was found on infection with the fungus *Botrytis cinerea* and permitted the diseased and healthy tissue to be T_2^*-contrasted and the rate of progress of the disease quantified as the rate of increase in the volume of infected tissue. Ripening of strawberries is associated with an increase in T_2 from about 16–25 ms in the immature green fruit to 25–100 ms in the red, ripe fruit. The ability to discriminate diseased, bruised and unripe regions from healthy ripe tissue is essential for the development of on-line MRI quality monitoring. **Table 2** lists a number of other MRI studies on the quality of fruit and vegetables, references to which can be found in the Further reading section.

MRI studies of food processing

Many foods are processed before being packaged, distributed and consumed. The major processing operations include drying, freezing, baking and extrusion, but novel processing methods, such as high-pressure treatment, are under development and can be used either alone or in combination with more conventional processing operations. By providing spatial maps of moisture, temperature and quality factors during processing, in either the off-line or the on-line mode, MRI has the potential of revolutionizing the food processing industry. The scope for this type of study is very wide and will be illustrated by a few key operations, beginning with drying.

Drying   The output of MRI drying studies is a set of moisture profiles at known drying times. The usual strategy is then to fit the profiles with appropriate mass transport models, which are used subsequently to optimize the drying conditions for maximum product quality with minimum energy expenditure. In practice, complexities, such as shrinkage, case hardening and spatially varying diffusivities have meant that the MRI data do not always conform to simple drying models, and this has highlighted the need for more refined theoretical models. **Table 3** lists a number of MRI drying studies on foods such as pasta, potato and apple.

Rehydration   The quality of extruded cereal products, such as pasta, depends on extrusion temperature and the variety of wheat used in the extrusion process. The effect of varying the ratio of hard and soft wheat on the rehydration kinetics of extruded spaghetti has been investigated by MRI. By measuring the moisture profiles at various times as the spaghetti is cooked in hot water it was shown that increasing the amount of soft wheat shifts the rehydration kinetics from a case II situation, where there is a fairly sharp water front ingressing into the pasta, towards rehydration by conventional Fickian diffusion. Other MRI studies have been reported for the rehydration of starch films and air-dried apple tissue. Besides simple Fickian diffusion, the Thomas Windle model provides a useful theoretical tool for analysing the rehydration kinetics, especially near the case II regime. **Table 3** lists a number of MRI rehydration studies.

Freezing   Ice has a proton transverse relaxation time of just a few microseconds, so it does not contribute signal in conventional liquid-state MRI. Freezing is therefore associated with a progressive loss of signal and this provides a powerful tool for investigating freezing kinetics. The freezing of raw potatoz has been imaged at various resolutions. At low spatial resolution an ice front is observed to ingress into the food. But with smaller samples and higher spatial resolutions (~ 40 μm) there appears to be a gradation in signal intensity throughout the sample, indicating that different regions are associated with different amounts of unfrozen water. This observation is, perhaps, a consequence of the different freezing behaviour of starch granules, cell walls, vacuolar and cytoplasmic regions in the potato cell, though the possibility of ice 'channelling' in the extracellular spaces cannot be discounted. Like drying and rehydration, finding suitable theoretical models to quantify this type of behaviour presents a considerable theoretical challenge. **Table 3** includes a number of MRI food freezing studies. The potential of using MRI as an on-line monitor of food freezing has yet to be exploited but clearly has considerable potential.

Freeze-drying   Despite the earlier statement that ice does not contribute signal in conventional MRI, it has, nevertheless, been possible to image the shrinkage of the frozen core in potato during freeze-drying. In this example there is sufficient unfrozen water in the 'frozen' potato to permit the core to be imaged. The absence of liquid water, which would saturate the signal, is another important factor, permitting high receiver gains during image acquisition. The time course of the sublimation of the frozen potato core was shown to follow a simplified freeze-drying model, but no other MRI freeze-drying studies have yet appeared.

Temperature mapping and thermal transport   Most food processing operations involve not only mass

Table 3   Representative MRI processing studies[a]

Food	Process	MRI technique
Model food gel	Drying	2D spin density
Extruded pasta	Drying	Radial imaging
Apple, potato	Drying	1D profiling
Pasta	Rehydration	T_2-weighted radial imaging
Starch gels	Rehydration	STRAFI[b]
Wheat grains	Boiling/steaming	2D spin density
Dry beans	Rehydration	Spin warp imaging
Potato	Rehydration	1D profiling
Pasta, potato	Freezing	Radial imaging
Meat	Freezing–thawing	Magnetization transfer imaging
Potato	Freeze-drying	Spin warp imaging
Cookies	Baking	Spin warp imaging
Potato	Frying	Chemical shift imaging
Model fluid	Extrusion	Flow imaging
Soya bean	Rehydration	Chemical shift imaging

[a] For references see Hills (1998) in Further reading section.
[b] STAFI, stray field imaging.

Table 4 MRI temperature mapping in food systems[a]

Food	Temperature mapping technique
Food gel	T_1 inversion recovery, snapshot FLASH [b]
Agar gel	Diffusion-weighting
Potato	Diffusion-weighting
Carrot	T_1 weighting
Potato pieces in 6% starch	Aseptic processing by chemical shift weighting

[a] For references see Hills (1998) in Further reading section.
[b] FLASH, fast low-angle shot.

transport but also simultaneous heat transport. By providing real-time temperature maps, MRI can also be used to monitor heat transport. All that is required is an NMR parameter, such as a relaxation time, chemical shift or water diffusivity that depends uniquely on temperature over the desired range, so that, after calibration, a map of the parameter, or the parameter-weighted signal intensity, can be converted into a temperature map. **Table 4** lists a number of examples of MRI temperature mapping. Most recently the technique has been applied to the problem of aseptic food processing in which a fluid suspension of food particles is thermally processed in a continuous process. The problem here is knowing whether the central regions of the particles have attained sufficiently high temperature to ensure complete cooking and sterilization.

Imaging quality factors in processed food Processing involves not only mass and heat transport but also changes in quality factors associated with microstructure, phase changes, gelatinization, aggregation and chemical or enzymatic reactions, Some of these changing quality factors are not amenable to NMR measurements, colour and aroma being obvious examples. However, in many cases NMR parameters are sensitive to quality factors so that MRI can be used to image changes in food quality during processing and storage. Chocolate confectionery is a typical example of this type of study. The long-term storage stability of fat-containing foods, such as chocolate, can be affected by fat crystal growth and polymorphic transitions in the solid fat phase. The MRI signal from cocoa butter arises from regions of high lipid chain mobility and so depends both on the solid–liquid ratio and, because each polymorph is associated with different chain mobility, on the polymorphic state of the solid lipid components, For this reason transverse-relaxation time-weighted MRI has been used to distinguish regions of chocolate that have melted and resolidified from those that have remained solid throughout. The migration of liquid lipid components between two phases in multilayer

confectionery products greatly affects the long-term quality of the product. For example, the transport of liquid triglycerides from a praline filling into the surrounding solid chocolate layer in a filled chocolate sweet can alter thermal stability and accelerate blooming on the chocolate surface. Accordingly, MRI has been used to image the transport of liquid lipids into chocolate in a model lamellar, 2-layer chocolate product, consisting of a layer of hazelnut oil and sugar layered over with dark chocolate. The lamellar structure permitted a 100 μm spatial resolution to be achieved and the time course of the lipid migration was followed over an 84 day period for two samples stored at 19 and 28°C. Intensity changes in the profiles were converted to liquid lipid concentrations by calibrating with homogeneous samples of known liquid-fat content. At 19°C, the liquid lipids were observed to accumulate at the interface between the hazelnut oil and the chocolate layers. In contrast, at 28°C, the lipid accumulated at the chocolate surface. Other examples where MRI has been used to image lipid transport and its effect on quality include the migration of fat into bread from peanut butter and the separation of cream from milk over a 9 h period.

The spatial distribution of fat in meat affects quality factors such as texture and has been imaged with chemical shift and T_1-weighted imaging. A number of feasibility studies have shown that MRI can detect bruises in apples, pears, onions and peaches, and detect the deterioration of apples in storage, watercore distribution in apples and core breakdown in pears. The quality of cereal products is also amenable to MRI analysis. The staling of baked biscuit 'sweetrolls' over several days storage has been imaged and shown to be associated with a migration of moisture from the inside to the outside of the sweetroll. **Table 5** lists a number of examples where MRI has been used to image quality factors.

Flow imaging and food rheology

Knowledge of the rheological properties of food pastes, slurries and sauces, such as ketchup, mayonnaise and salad creams, is important both for quality assurance and for optimizing industrial flow and mixing processes. Unfortunately, many food slurries and pastes are opaque and do not lend themselves to flow studies with conventional techniques such as laser Doppler anemometry. Moreover, conventional rheological measurements are model-dependent in that it is necessary to fit the data by assuming a function relationship between the stress and strain (or strain rate) and to assume a set of boundary conditions (such as slip or stick) at the fluid–container

Table 5 NMR studies of food quality factors[a]

Food	Quality factor	NMR technique
Chocolate	Lipid migration	MRI
Model food emulsion	Crystallization kinetics	Localized spectroscopy
Creaming	Phase separation	T_1-weighted MRI
Egg-white and beer foam	Foam stability	MRI profiling
Meat	Curing/brining	^{23}Na NMR
Various fruits	Bruising	T_2- and T_2^*-weighted MRI
Cookies (sweet rolls)	Retrogradation	M_0 maps

[a] For references see Hills (1998) in Further reading section.

interface. This is satisfactory when it is known that the fluid has a simple rheological behaviour, such as that of a Newtonian or power-law fluid. However, many liquid foods do not conform to simple rheological models so their stress–strain relationship, usually called the flow curve, must be deduced in a model-independent way. MRI can do this by imaging the velocity profile at every point in a rheometer, including the boundary layer (at least within one voxel dimension). The rheometer can be of any of the standard types, including a cylindrical tube and rotational viscometers (or couette cell). **Table 6** lists a number of representative MRI rheological studies.

Shear-induced nonrheological changes Shear-induced aggregation in concentrated food emulsions such as dairy creams and shear-induced particle migration are just two examples of microstructural or compositional changes caused by shear. These changes can be studied with MRI methods by combining flow imaging with MRI spatial maps of NMR parameters, such as relaxation times or spin densities, which are sensitive to the microstructural changes. **Table 6** includes a number of examples of this approach.

Flow and quality assurance On-line quality control of food slurries and pastes often requires rapid, empirical, rheological tests based on flow times or flow lengths rather than quantitative measurements of well-defined rheological parameters such as shear viscosities. The Bostwick consistometer (which is used to determine the quality of puréed fruit and vegetable products), the Brabender Visco-amylograph (which is used to test the consistency of starches) and the Torque plasticorder (used for dough consistency), are examples of such empirical rheological testing. MRI studies of the flow within these various test rigs, such as the Bostwick consistometer, have helped provide a rheological rationale for these empirical measurements.

Flow in extruders Pioneering MRI studies on flow in an extruder have been reported by McCarthy and co-workers (see Further reading section) who designed a special nonmagnetic single-screw extruder for operation inside the magnet of an NMR spectrometer, and used carboxymethyl cellulose solution as a model extrudate. The flow patterns were compared with theoretical models of extruder flow for an incompressible Newtonian fluid. This type of study has considerable research potential because extrusion is used widely throughout the food industry for the manufacture of pasta, snacks, confectionery, pet foods and animal feeds. Despite this, extrusion is, perhaps, the least understood of all food processing operations, involving complex series of physical and chemical changes such as mixing, gelatinization, denaturation, evaporation, flavour production (and loss) and rheological changes. To date only the flow aspects of extrusion have been studied by MRI, but, by exploiting relaxation and diffusion weighted imaging techniques, the potential exists for exploring other associated physical and chemical changes.

NMR diffusion studies of food microstructure

The use of pulsed gradients and fringe field gradients in NMR diffusion measurements has been widely documented in the research literature and is reviewed in other articles of this encyclopedia. Provided certain conditions are fulfilled, such as factorization of the effects of relaxation and diffusion and translational invariance, the pulsed-gradient techniques permit measurement of the average displacement

Table 6 Flow imaging studies of food–related materials[a]

Fluid	Apparatus	Phenomenon
Nutrient fluid	Hollow fibre bio-reactor	Starling flow
1.5% Carboxymethyl cellulose solution	Tube flow	Rheology
Tomato juice	Tube flow	Rheology
Fibrous suspensions	Tube flow	Rheology
Carboxymethyl cellulose solution	Extruder	Velocity field in an extruder
Particle suspensions	Tube/couette flow	Shear-induced migration
Xanthan gum	Tube flow	Shear thinning and apparent wall slip
Egg-white, cornflour/water, tomato sauce	Couette, flow/cone and plate flow	Shear-induced phase transitions

[a] For references see Hills (1998) in Further reading section.

propagator, $P_{av}(R,\Delta)$, where R is the distance diffused by a spin in the diffusion time, Δ. Various theoretical models have been developed relating $P_{av}(R,\Delta)$ to microstructure in simple geometries, so that measurements of the average propagator can be used to gain information about food microstructure. In favourable circumstances, microstructural information on the submicrometre distance scale can be accessed by this approach, which greatly exceeds the spatial resolution in NMR microscopy. In the following we consider the application of diffusometry to various food systems, including gels, cellular tissue and emulsions.

Diffusion studies of gels and starch-based systems

The microstructure of starch and gellan gum gels has been investigated by the stimulated-echo pulsed-gradient method as a function of increasing diffusion times, Δ (up to 1010 ms), and at various gradient amplitudes, G, at fixed gradient pulse duration δ and pulse separation, t_1. An effective water diffusion coefficient, $D(\Delta)$ was extracted by assuming the echo attenuation for unrestricted diffusion,

$$\ln[S(q,\Delta)/S(q=0)] = -D(\Delta)(4\pi q)^2(\Delta - \delta/3)$$

The gel microstructure caused the coefficient $D(\Delta)$ to decrease with increasing Δ, and this decrease was fitted with the model of von Meerwall and Ferguson. This model represents the gel micropores as a series of parallel compartments of width, a, separated by barriers of permeability, P. Within each compartment the intrinsic water diffusivity was assigned the value, D_0. Pore sizes, a, of the order of 10–15 μm for starch gels and 5–15 μm for gellan gums were deduced and these varied slightly with gel concentration, and, for gellan gums, with ionic strength. These pore spacings appear to be rather large for these concentrated gel systems and, in the case of starch gels, may reflect the presence of residual granule structure resulting from incomplete gelatinization. The effects of retrogradation on starch gel microstructure were also followed in these studies, and were associated with changes in pore size and water diffusivity. Fringe field diffusion methods have been used to monitor water diffusion in gelatine gels of increasing concentration and in protein aerogels obtained by freeze-drying. The diffusion behaviour in the aerogel system was shown to be anomalous, and akin to diffusion in a fractal space. Echo attenuation of the pulsed-gradient stimulated echo of water in packed beds of potato starch granules appears to conform to diffusion in a lower dimensional space, lying between one and two dimensions, depending on the water content. This could be a consequence of the crystalline A- and B-type starch regions in the granule, channelling the water translational motion, but further work is needed to test this hypothesis.

Diffusion studies of cellular plant tissue

The pulsed-gradient stimulated-echo sequence has been used to investigate the effects of electrical and conventional heating on intercell permeabilities in carrot tissue. The echo attenuation was interpreted with a lamellar cell model in which carrot cells were represented as lamellar regions of width, W, containing water of diffusivity, D, and separated by a barrier of effective permeability, P. Presumably this barrier arises from the combined effect of diffusion through the tonoplast, cytoplasm, plasmalemma and cell wall subcellular compartments. It was found that cooking increases the apparent diffusion coefficient and barrier permeability, which is consistent with the breakdown of cell structure, but there were also small, but statistically significant, differences between electrical and conventional cooking.

Diffusion studies of food emulsions

The determination of droplet-size distributions in water-in-oil or oil-in-water emulsions by pulsed-gradient diffusion measurements is now a routine industrial method for monitoring the quality of food emulsions such as mayonnaise and salad creams. The method was first described in a classic paper by Packer and Rees (see Further reading section) and subsequently developed into a rapid analytical method. The theoretical model assumes spherical droplets and a log-normal size distribution, but other distributions have been considered. The MRI generalization of the method to spatially dependent droplet-size distributions in emulsions undergoing coalescence and/or phase separation has yet to be implemented, although a spatially dependent study of emulsion crystallization has been reported.

Cheese is, in essence, a distribution of fat droplets in a protein–water matrix. The fat droplet size distribution has been studied by the NMR diffusion method and showed a very broad Gaussian size distribution with significant differences between Swiss and cheddar cheeses. In this sample the water signal could be removed by using a long echo time, because the water transverse relaxation time was much shorter than that of the fat. The water diffusion was also shown to be consistent with 1D diffusion along the protein chains in the cross-linked protein matrix of the cheese. In this case the water signal was separated by Fourier transforming the echo obtained at shorter echo times. This resulted in separate peaks for the fat and water.

NMR relaxation studies of food

Relaxation studies of food microstructure

NMR relaxation time measurements provide a unique window to view both food microstructure and the dynamic behaviour of molecules comprising the food matrix. The microstructural aspect arises because different aqueous microscopic compartments are usually associated with different intrinsic water-proton relaxation times. Provided the exchange of water magnetization between compartments is slow on the NMR measurement timescale, the distribution of water-proton relaxation times therefore reflects the number and relative volumes of the microscopic compartments. The proviso that the exchange is slow is important, because if the exchange is fast, then only a single average relaxation time will be observed and all compartmental information is lost. Like diffusion, theoretical models are required to relate the observed relaxation-time distribution to microstructure, and these models are usually based on numerical solution of the Bloch–Torrey equations. The Monte Carlo method is perhaps one of the most powerful solution tools for doing this. A cube is divided into N^3 volume elements and the morphology of the system is defined by assigning compartments (or regions) within the cube characterized by intrinsic relaxation times and diffusion coefficients. Magnetization density vectors are then randomly distributed throughout the cube, which then diffuse between voxels and relax with a rate corresponding to their location. In this way echo-decay amplitudes for the CPMG (Carr–Purcell pulse sequence, Meiboom–Gill modification) and pulsed-gradient diffusion sequences can be calculated and analysed with one of the standard deconvolution software packages.

Numerical solutions of the Bloch–Torrey equations have been used to interpret relaxation data for a variety of microscopically heterogeneous model systems, including randomly packed, water-saturated beds of Sephadex, silica and glass microspheres. Their application to plant cell tissue is perhaps more directly relevant to food. The three major compartments in a plant cell are the vacuole, cytoplasm and cell-wall region, though starch granules comprise an additional compartment in some foods, such as potato. The Bloch–Torrey equations describing relaxation and diffusive exchange between these intracellular compartments have been solved for an idealized three-compartment spherical cell morphology and used to predict the dependence of the relaxation-time distribution and pulsed gradient spin–echo (PGSE) diffusion behaviour of apple tissue, both fresh and during air drying. The latter showed that water is removed mainly from the vacuolar compartment during air-drying, and that this is associated with volume shrinkage of the tissue. The subcellular changes induced by the freezing and drying of potato and carrot have also been investigated in this way.

Relaxation studies of aqueous solutions, gels and glasses

There are at least two proton pools in aqueous solutions of dissolved sugars, proteins and polysaccharides, namely the nonexchanging CH proton pool and the exchangeable pool of protons on water and biopolymer/solute hydroxyl and amino groups. Fast proton exchange between the water and exchangeable biopolymer protons means that any process, such as denaturation or gelation, that changes the rigidity of the biopolymer chains will result in an altered transverse relaxation time of both the exchangeable 'water' and the nonexchanging proton pools. The latter can be observed by replacing water with D_2O, or, at sufficiently high concentrations, as a faster relaxing component in the free induction decay (FID). The transverse relaxation rate of the exchangeable 'water' proton pool can be monitored as a slowly relaxing component in the FID or, more usually, with the CPMG pulse sequence, and can be used to follow the kinetics of gelation and denaturation. This approach has been used to follow the thermal denaturation of whey proteins and the gelation of pectins and starch.

Longitudinal and rotating-frame relaxation in these systems is complicated not only by proton exchange but also by the possibility of secular dipolar cross relaxation between all proton pools. This makes theoretical modelling difficult, although it does not prevent longitudinal relaxation being used empirically to monitor food processing changes.

At lower water contents, below ~ 50% by weight water, the signal from the nonexchanging solute proton pool appears clearly as a faster relaxing component in the FID and can be characterized with an effective T_2 by analysing the decay as either Gaussian or exponential. For example, in concentrated gelatine gels, the fast relaxing component has a T_2 of the order of a few tens of microseconds, and this increases with increasing water content as the gelatine chains are plasticized. Plasticization, which permits high-energy 'strained' interchain conformations to be relaxed, can result in increased local order and changes in texture. For this reason it has been investigated in a number of cereal proteins that are important in controlling the functional behaviour of dough and bread, such as glutenin in wheat and chordein in barley. Retrogradation in starch-based foods is one of the most important examples of plasticization, and permits the partial recrystallization of amylopectin chains into A or B crystallites, and results in the staling of bread and cake products. The retrogradation kinetics can be followed from the concomitant decrease in the water-proton relaxation time, and the lowering of the glass transition temperature by plasticization can be modelled with the polymer science approach.

NMR and food compositional analysis

High-resolution NMR spectral studies

The application of high-resolution NMR methods to molecular structure elucidation of purified proteins, polysaccharides and lipids is too extensive to be treated in depth in this chapter. The use of ^{1}H high-resolution solution-state NMR spectroscopy, in combination with enzymatic debranching and chromatographic purification, for elucidating the molecular structure of complex food polysaccharides, such as amylopectins, carrageenans, pectins and alginates, has been reviewed (see Gil and co-workers in the Further reading section). High-resolution liquid-state ^{1}H and ^{13}C NMR has also been widely used in the compositional analysis of lipids in vegetable and fish oils and **Table 7** indicates the extent of this type of study. Cross-polarization magic angle spinning (CP MAS) spectroscopy and other high-resolution solid-state NMR methods have been used to study structure–function relationships in the solid and gelled states of starches and other polysaccharides. Multinuclear NMR spectroscopy has been especially valuable for understanding cation–polysaccharide interactions in ionic food polysaccharides, and some examples are listed in **Table 8**.

The low-speed MAS technique is worthy of special mention because it overcomes spectral-line broadening artefacts from local field gradients created by differences in magnetic susceptibility across phase boundaries in microscopically compartmentalized foods. This technique has been used to obtain well-resolved spectra of sugars and metabolites in intact fruits, and to resolve water and oil peaks in multiple water-in-oil-in-water emulsions, found, for example, in low fat spreads. The detection of food adulteration is of increasing commercial importance and several high-resolution NMR methods, such as site-specific natural isotope fractionation (SNIF) NMR are being used routinely for food authentication, and have been extensively reviewed (see the Further reading section).

Table 7   Recent NMR studies of food lipids[a]

Lipid	Nucleus	Analytical purpose
Vegetable epoxidized oils	^{13}C	Epoxy acid content
Castor oil	^{13}C	Quantification of castor oil in edible oils
Minuta oil	^{13}C, ^{1}H	Chemical changes during plant flowering and fruiting
Olive oil	^{13}C	Saturated fatty acid content. Grading of olive oils
Pomace oils	^{13}C	Composition of unsaponifiable matter in pomace oil
Fish oils	^{13}C, ^{1}H	Quantification of oils in tuna and Atlantic salmon
Oxidative deterioration of oils	^{1}H	Ratio of olefinic to aliphatic protons as an index of deterioration
Canula (rapeseed) oils	^{13}C, ^{1}H	Colour pigment formation during bleaching and deodorization
Liguloxide	^{13}C, ^{1}H	Characterization of odour in tomato ketchup
Butter, baking and spreading fats	^{13}C	Compositional analysis
Milkfat	^{13}C, ^{31}P	Solid fat content of milkfat fractions and blends
Tripalmitin	^{2}H	Lipid polymorphism and crystallization
Animal and vegetable fats	^{1}H	Solid fat content

[a] For reference see Gil *et al* (1996) in Further reading section.

Table 8   Multinuclear NMR studies of food biopolymers[a]

Nucleus	Biopolymer	Phenomenon
^{23}Na, ^{1}H, ^{13}C	Carrageenans	Role of cation in gelation
^{23}Na, ^{1}H	Pectins	Role of ^{23}Na and degree of esterification on gelation
^{31}P	Starch	Number and location of phosphate ester groups and relationship to viscosity. Location of amylose in granule.
^{87}Rb, ^{133}Cs, ^{127}I	Carrageenans	Role of cation in gelation
^{23}Na, ^{1}H	Xanthan gum	Sequestration of Fe^{2+}. Effect of Na^+ content on texture and stability
^{25}Mg, ^{43}Ca, ^{31}P	β-Casein	Ion binding sites
^{111}Cd, ^{1}H	β-Lactoglobulin	Aggregation–gelation kinetics

[a] For references see Gil *et al* (1996) in Further reading section.

Low-resolution NMR analysis of food lipids

The NMR analysis of the oil content of seeds relies on the differences in proton transverse relaxation times of the constituents. Typical T_2 values of solid macromolecules in seeds are just a few microseconds, and so they appear as the initial, rapidly decaying part of the FID. Water in seeds is strongly adsorbed to macromolecules and so has a short T_2 (of the order of a few milliseconds). In contrast, oil exists as liquid droplets in the seeds and therefore has a relatively long T_2 (of the order of hundreds of milliseconds). This means that the oil content can be determined directly from the FID as the ratio,

$$\text{Seed oil content } (\%, \text{ w/w}) = 100L/(L + fS)$$

where L is the FID signal intensity measured after the solid signal has decayed and near the beginning of the liquid signal (typically after ~75 μs) and S is the FID intensity of the slightly decayed solid signal. The correction factor, f, takes account of the decay during the instrument deadtime (typically the first 5–10 μs). The Hahn-echo pulse sequence is an alternative means of determining oil content. The water signal can be eliminated by choosing an echo time much longer than $5T_2$ of the seed water, the remaining oil signal then being calibrated with standard oil samples.

Alternatively, the echo amplitudes of the full CPMG echo-decay envelope acquired at short interpulse spacing can be analysed as a multiple exponential decay, whose relative weightings are proportional to the water and oil content of the seed. If this measurement is combined with the relative solid to liquid ratio determined from the FID, the complete solid–oil–moisture content of the seeds can be determined. Solid to liquid fat ratios in chocolate can be determined in a similar way. **Table** 7 includes a number of such low-resolution studies of food lipids.

List of symbols

a = pore width; $D(\Delta)$ = water diffusion coefficient; f = correction factor; G = gradient amplitude; L = FID liquid signal intensity; M_0 = initial magnetization; $P_{av}(R, \Delta)$ = average displacement propagator; P = permeability; q = diffusion wavevector defined as $(2\pi)^{-1} \gamma G\delta$; R = distance diffused by a spin; S = amplitude of solid component in FID; S = chemical shift; S = FID solid signal intensity; $S(q, \Delta)$ = stimulated echo amplitude; t_1 = pulse separation time; T_2 = transverse relaxation time; W = width of lamellar regions; Δ = diffusion time; γ = gyromagnetic ratio.

See also: **Contrast Mechanisms in MRI; Diffusion Studied Using NMR Spectroscopy; Food and Dairy Products, Applications of Atomic Spectroscopy; Food Science, Applications of Mass Spectrometry; High Resolution Solid State NMR, ¹³C; Industrial Applications of IR and Raman Spectroscopy; Labelling Studies in Biochemistry Using NMR; MRI Applications, Biological; MRI Instrumentation; MRI Theory; MRI Using Stray Fields; NMR Data Processing; NMR Relaxation Rates; NMR of Solids.**

Further reading

Gil AM, Belton PS, and Hills BP (1996) Applications of NMR to food science. *Annual Reports on NMR Spectroscopy* 32: 1–49.

Hills BP (1998) *Magnetic Resonance Imaging in Food Science.* New York: Wiley.

Martin GJ (ed) (1996) Proceedings of the Third International Conference on applications of magnetic resonance in food science. *Journal of Magnetic Resonance Analysis* 2: 1–264.

McCarthy KL, Kauten RJ and Agemura CK (1992) Application of NMR imaging to the study of velocity profiles during extrusion processing. *Trends in Food Science and Technology* 3: 215–219.

McCarthy MJ (1994) *Magnetic Resonance Imaging in Foods.* New York: Chapman and Hall.

Packer KJ and Rees C (1972) Pulsed NMR studies of restricted diffusion: Droplet size distributions in emulsions. *Journal of Colloid Interface Science* 10: 206–218.

Forensic Science, Applications of Atomic Spectroscopy

John C Lindon, Imperial College of Science, Technology and Medicine, London, UK

Copyright © 1999 Academic Press

Introduction

A number of studies have been published of the use of atomic spectroscopy in the furtherance of forensic science. This short article is intended to provide an outline of recent work that has been published in the open scientific literature. The representative Further reading at the end of this article should allow the reader an entry point into the scientific literature for details of methods and other studies.

Reviews

The *Journal of Analytical Atomic Spectrometry* publishes periodical reviews of atomic spectroscopy, including its application to problems in metallurgy and industrial materials. This review also includes forensic studies and a recent example is given in the Further reading section.

Analysis of glass

Fragments of glass can be transferred to a person during many criminal activities and recent work on the elemental analysis of glass has been reported using atomic emission spectroscopy. Results have been compared with those obtained using X-ray fluorescence.

Gunshot residues

A review of many types of instrumental analysis used for the characterization of gunshot residues has appeared recently. This includes descriptions of the use of atomic absorption spectroscopy, neutron activation analysis and X-ray fluorescence as well as various mass spectrometry techniques after preconcentration of analytes using separation methods such as HPLC and capillary electrophoresis. In addition, atomic absorption spectroscopy has been used to characterize the brass used in different production lots of cartridge cases.

Metal residues

Inductively coupled plasma atomic emission spectroscopy (ICP-AES) has been used to characterize the elemental composition of steel fragments found at crime scenes. In general the levels of chromium, nickel, cadmium, zinc and copper could be used as fingerprints to identify the source of the steel. The trace elemental composition of household aluminium foils can also, in some cases, be used to identify the brand name and production details.

See also: **Atomic Absorption, Methods and Instrumentation; Atomic Absorption, Theory; Atomic Emission, Methods and Instrumentation; Atomic Spectroscopy, Historical Perspective; Forensic Science, Applications of IR Spectroscopy; Forensic Science, Applications of Mass Spectrometry; X-Ray Fluorescence Spectrometers; X-Ray Fluorescence Spectroscopy, Applications.**

Further reading

Crighton JS, Fairman B, Haines J, Hinds MW, Nelms SM and Penny DM (1997) Industrial analysis: metals, chemicals and advanced materials. *Journal of Analytical Atomic Spectrometry* **12**: R509–R542.

Curran JM, Triggs CM, Almirall JR, Buckleton JS and Walsh KAJ (1997) The interpretation of elemental composition measurements from forensic glass evidence. *Science and Justice* **37**: 241–244.

Heye CL and Thornton JI (1994) Firearm cartridge case comparison by graphite furnace atomic absorption spectrochemical determination of nickel, iron and lead. *Analytica Chimica Acta* **288**: 83–96.

Koons RD, Peters CA and Merrill RA (1993) Forensic comparison of household aluminium foils using elemental composition by inductively coupled plasma atomic emission spectrometry. *Journal of Forensic Sciences* **38**: 302–315.

Koons RD, Peters CA and Rebbert PS (1991) Comparison of refractive index, energy dispersive X-ray fluorescence and inductively coupled plasma atomic emission spectrometry for forensic characterization of sheet glass fragments. *Journal of Analytical Atomic Spectrometry* **6**: 451–456.

Meng HH and Caddy B (1997) Gunshot residue analysis – a review. *Journal of Forensic Sciences* **42**: 553–570.

Poolman DG and Pistorius PC (1996) The possibility of using elemental analysis to identify debris from the cutting of mild steel. *Journal of Forensic Sciences* **41**: 998–1004.

Thornton JI (1994) The chemistry of death by gunshot. *Analytica Chimica Acta* **288**: 71–81.

Forensic Science, Applications of IR Spectroscopy

Núria Ferrer, Universitat of Barcelona, Spain

VIBRATIONAL, ROTATIONAL & RAMAN SPECTROSCOPIES
Applications

Introduction

Infrared spectroscopy has been applied to the characterization of molecules. Conventional dispersive infrared spectrometers have been replaced by Fourier-transform infrared equipment, which incorporates a Michelson interferometer and presents a series of advantages over dispersive systems, such as an improvement of energy and the simultaneous measurement of the whole spectral range.

Owing to these improvements, infrared spectroscopy has undergone a marked development in the last few years because of the possibility of adapting new accessories to the spectrometers and, therefore, of analysing samples whose infrared spectrum was impossible to obtain some years ago. Among these accessories are those which can measure specular and diffuse reflection, attenuated total reflectance and microscopes. One area in which the technique has improved dramatically is forensic analysis.

Many laboratories perform forensic analysis nowadays. Some of them seek to determine what is wrong with a process, while others are trying to copy a process from a competitor. The possibility of analysing and identifying microscopic particles, the accurate comparison of samples from different origins and the fact that the sample is not destroyed, have also converted the FT-IR technique into a valuable tool where criminal and social implications are present. Some of the forensic materials that can be identified by infrared spectroscopy are: paints, fibres, textiles, polymers, explosives, varnishes, adhesives, papers, gemstones, drugs, food additives, fur or textile treatments and physiological samples. An important feature is the possibility of analysing solid, liquid or gaseous samples. In criminal investigations it is important to attach the infrared spectrum to each sample, which is unique to every substance, like a fingerprint. In addition, in lawsuits where several parties are involved, infrared spectroscopy can contribute relevant information.

This article shows a number of applications of infrared spectroscopy to the forensic analysis of a range of samples using different accessories and methods. A number of examples are given, although applications are unlimited and any analysis might be of value in forensic research.

Main accessories and methods used in forensic analysis

The two main restrictions of analysis by infrared spectroscopy are size and difficulties regarding handling. Three kinds of sample compartments are available depending on the size of the sample to be placed in the spectrometer:

- macro compartment: this employs a beam between 2 to 10 mm and any type of accessory might be fitted.
- Semimicro compartment: this requires the use of a beam condenser and the spot of the beam in the sample is about 1 or 2 mm.
- micro compartment: the size of the beam is less than 1 mm and a microscope needs to be used.

A beam condenser can be installed in the sample compartment of the conventional equipment, but the microscope needs the infrared beam to go out of the bench to an externally attached accessory. The microscope uses the source and beamsplitter of the infrared equipment to which it is coupled, but it has its own more sensitive detector. Obviously, large samples can also be located on the microscope, but in such cases the measurement area is reduced to less than 100 μm. Therefore, if an analysis of the predominant material in the sample is required, its homogeneity must be assured and defects or pollutants must be avoided.

Depending on the manipulation that the sample requires, a range of accessories might be used. Many samples for forensic analysis must not be destroyed or manipulated either because they need to be kept for further analysis or because of their economic or historic value. KBr pellet measurement requires mixing the sample with the salt and, therefore, the possibility exists that it might be lost. Other methods such as diamond cell or reflection do not require any alteration or mixing of the sample with other substances. In the main, liquid and gaseous samples suffer no risk of destruction and it is only necessary to fit them in cells which are transparent in the desired range of measurement and which are also compatible with the material to be analysed. Solid samples can be mixed with salts in order to make an infrared transparent pellet, or can be placed on a suitable surface for further analysis.

Diamond cell

Diamond cell is one of the most frequently used methods in forensic analysis. It is normally used when samples are very small, but can be separated from the matrix. The particle or fibre, which can be as small as 5 μm, is placed on a diamond window and put under pressure with the help of another diamond window. The pressure of the cell allows the sample to spread over the diamond increasing its surface area while decreasing in thickness. In this way, the infrared radiation is able to go through the sample and, therefore, its infrared spectrum can be obtained by transmission.

Diamond cells are about 4 mm² in aperture and depending on the area of the spot of the spread sample it is possible to utilize a beam condenser or a microscope. Despite the absorption of diamond in the mid infrared, the wavenumber reproducibility that FT-IR is able to achieve makes it possible to subtract the diamond spectrum and to obtain clean spectra of the sample. In forensic analysis this technique is applied to paints, varnishes, inks, fibres, crystals and particles in general, which can be put under pressure and suffer a deformation. To avoid excess of absorption, one of the cells can be removed. In the case of samples such as cork, tobacco and some polymers, which recover their original size after decompression of the cell, the alternative is to use two diamond windows with the sample in between, instead of removing one of them.

Specular reflection

When a sample cannot be manipulated at all, is opaque and has a specular or polished surface, specular reflection is the technique to be applied in either the microscope or the macro compartment.

Diffuse reflection

When the surface of the sample is not specular and there is a certain diffuse component of reflected light, or when analysing a cut gemstone, the analysis is appropriately conducted by using the diffuse reflection technique where all the light emitted by the sample is measured.

Attenuated total reflectance (ATR)

Attenuated total reflectance is applied to samples where the composition of the surface needs to be measured. It is applied to soft samples such as paper, fur, polymers and textiles, which can achieve good contact with the crystal used. It is also used for aqueous samples and viscous or sticky samples such as pastes, soaps and detergents. The only consideration to be borne in mind is the need to obtain good optical contact between the surface of the sample and the crystal of the accessory, which can be made of diamond, ZnSe, Ge, Si, KRS-5, etc.

When the surfaces to be analysed are very small it is possible to use the attenuated total reflectance objective of the microscope. A unique internal reflection over the sample in touch with the crystal is enough to obtain the spectrum. This is the alternative when even a single particle cannot be removed from the matrix, which is the case for very ancient or valuable samples such as manuscripts, books or paintings.

Separation techniques

Determination of complex mixtures is one of the most common problems in infrared analysis. Hyphenated techniques such as chromatography provides a useful way of separating mixtures into individual compounds. Gas chromatography (GC) has been by far the best application of combining infrared spectroscopy and a separation technique. One of the principal reasons has been the use of carrier gases which do not absorb in the infrared region.

Specific analyses

Analysis of paints

Infrared spectroscopy has been widely applied in the forensic analysis of paints. It is particularly useful in car accidents where it is necessary to determine the origin of the paint adhering to a car. In some countries there are car paint libraries, and it is possible to find the model of the car and the year of manufacture by analysing the paint. In the absence of witnesses, or if a vehicle has left the scene of an accident, this analysis can be a crucial element in obtaining a verdict.

Another major application of the technique is in the analysis of ancient paints used in a particular period of history or a particular place. This is essential not only for the study of the pigments and materials used by different civilisations, but also for authenticating works of art, restoration work and in lawsuits involving damaged works of art.

Figure 1 shows two spectra from two sections of a modern painting. The damage to the picture might have been caused by a heterogeneous application of oil or the poor conservation of the picture. A non-homogeneous distribution of oil over the surface was made evident when two sections of the picture were analysed. As it was possible to use part of the picture from under the frame, the amount of sample was sufficient to allow use of the attenuated total reflectance horizontal accessory.

A further application of this technique is in seeking to obtain high quality paints. Changes in pigment composition as well as in solvent composition can have a considerable impact in industry. Therefore, it is important to be able to determine quickly the composition of both solid and volatile fractions. In addition, increasingly stringent controls make it necessary to check for the presence of forbidden solvents in commercial paints.

Figure 2 shows the spectrum of the volatile fraction of a commercial paint. The spectrum is obtained by placing a long strip of aluminium foil on the bottom of the gas cell with some paint on the foil. After a few minutes the number of molecules within the cell corresponding to the volatile fraction is sufficiently great to obtain the spectrum of the gas. In this particular case it was possible to identify xylol as a component of the gaseous fraction. Pigment analysis can be performed by covering a KBr pellet with a very thin layer of paint.

When paint samples are very small, for example in the case of a valuable painting, or an archaeological sample or the paint which remains attached to a lead bullet once it has passed through the bodywork of a car, the most widely used technique is to put small particles on a diamond cell and to obtain the infrared spectrum by transmission. If the sample cannot be removed from the matrix, microscopy with an attenuated total reflectance objective can be used.

Fibre analysis

Forensic analysis of fibres constitutes another major application of infrared spectroscopy. The spherical structure of fibres makes it necessary to flatten the sample before measuring its spectrum, for which it is possible to use either a roller or a diamond cell. Once the fibre is prepared, the variable apertures of the microscope allow the masks to be adapted to the size of the fibre.

Figure 3 shows the spectrum of a fibre found in a pharmaceutical product. Here, it was necessary to know whether the sample came from the filters in the industrial process or from the white coats of the workers. Although it was clear that the three fibres analysed were polyester, the technique was able to identify the kind of polyester and to determine its relation to fibres coming from the filters of the industrial process. In this case the sample consisted of a single fibre and, therefore, the use of a microscope was the only option. If the sample consists of a piece of textile or several fibres, other methods such as attenuated total reflectance, KBr pelleting following shattering after treatment in liquid nitrogen or diamond cell in the beam condenser can be used.

Other applications have been reported, in the field of medicine where suture threads have had harmful reactions in patients, and for analysing single fibres recovered from the scenes of crime.

The appearance of surface defects in animal furs is another reported application of attenuated total reflectance techniques. Figure 4 shows an example of a valuable fur damaged by white spots. The manufacturer claimed that the client was responsible for damage to the fur because it had been stored in an inappropriate environment or because of treatments it had been exposed to. The client, on the other hand, claimed its original treatment during manufacture was to blame. Infrared analysis demonstrated that the white spots came from the migration of animal fat to the surface and bands were comparable to the fat extracted from the natural fur. Other substances, such as silicone, have been found in this kind of sample.

Paper and ink analysis

Application of the infrared technique with microscope to inks in paper has been a useful tool. Examples include comparison of inks used on the same cheque, for example an additional zero to the total sum or in the signature, and documents which are signed and dated, where it is possible to analyse the inks from printers or hand typewriters to determine if they actually existed when the document was signed. Particles of toner from a photocopy can also

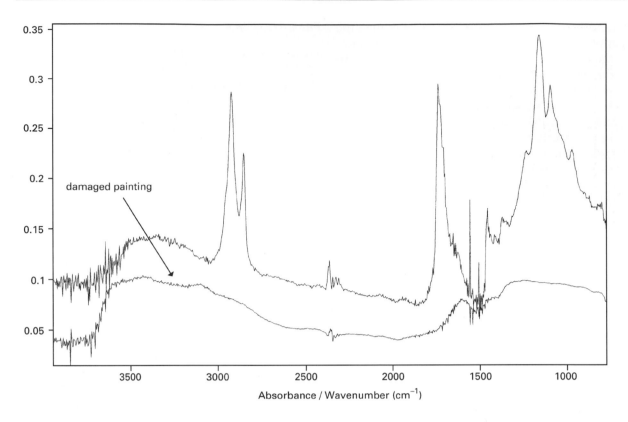

Figure 1 Spectra of two pieces of painting: damaged painting does not show the bands corresponding to oil (ATR horizontal).

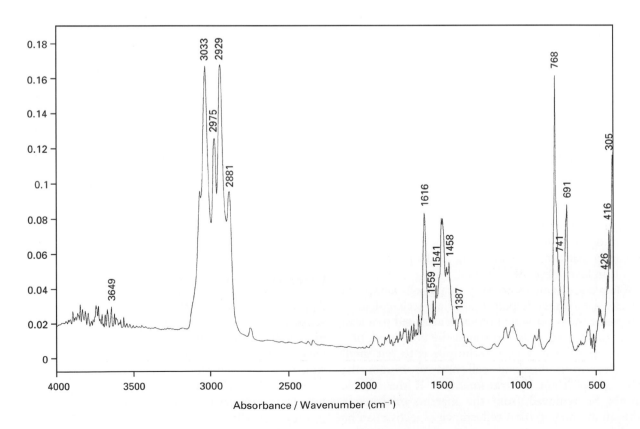

Figure 2 Spectrum of volatile fraction (xylol) of an industrial painting (gas cell).

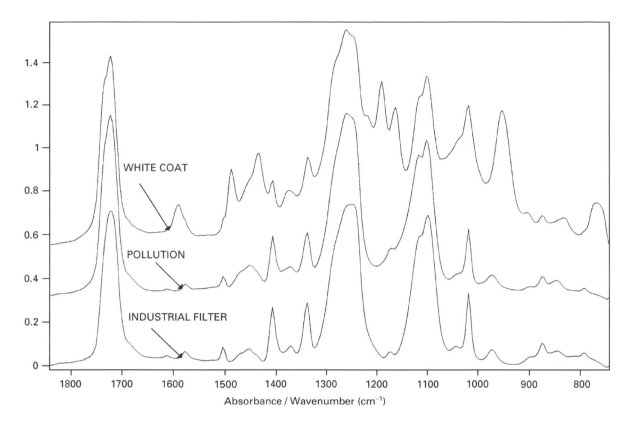

Figure 3 Comparison of fibres from a polluted sample, white coats and an industrial filter (diamond cell in the microscope).

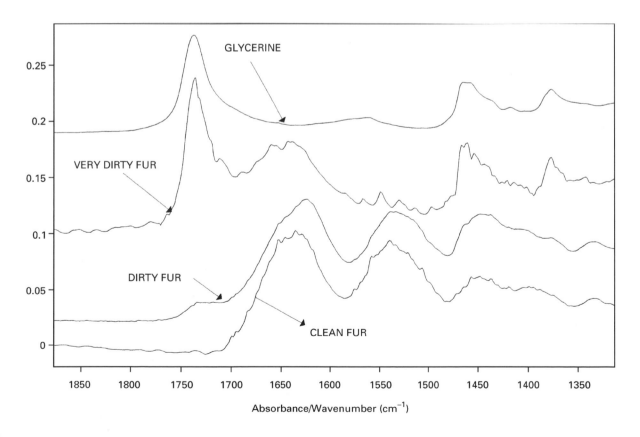

Figure 4 Comparison of spectra of clean and dirty furs with glycerine spectrum (ATR horizontal).

help to solve cases by comparing the particles of the document with the toner from the photocopying machine used.

Figure 5 shows the spectra obtained by transmission microscopy of two cotton fibres, one of which contains a blue ink spot. Comparison of the spectral differences between the clean fibre and that spotted with ink, shows that this difference can indeed be imputed to the ink.

Cellulose fibres also have a highly characteristic infrared spectrum. Acidity of paper due to the presence of acidic inks or biological degradation can be measured by analysing the infrared spectrum. **Figure 6** shows the spectral differences in certain ranges of the spectrum due to corrosion related to the acidity of ancient papers. Carbonate bands at 875 and 1797 cm^{-1} decrease with decreasing pH, carbonyl groups increase with decreasing pH and hydroxyl groups from cellulose decrease with increasing pH. This kind of change arises as a result of poor conservation or the type of paper.

Spots on papers can also be measured. **Figure 7** shows the spectra of two samples from the same piece of paper, one of which is spotted. The attenuated total reflectance objective of the microscope allows us to identify the presence of paraffins. In such cases, a microscope is essential to obtain the infrared spectrum while the attenuated total reflectance objective allows the analysis to be carried out without destroying the document.

Analysis of gemstones

There is an increasing number of gemstones, treated or synthetic, that need analysing and the technique that allows them to be distinguished visually is not particularly reliable. Infrared spectroscopy can provide objective evidence for the characterization of such samples and this is very important for both identification and certification purposes in commercial and legal practices.

Diamonds are one of the more interesting groups of gems to study. Only when diamonds present two parallel faces is it possible to use transmission to determine the infrared spectrum. Normally diamonds, especially those destined for jewellery, have a brilliant cut. This cut affects the light and makes it go out in all directions. In this case the only technique that allows us to obtain the infrared spectrum is diffuse refle.ction. Here the gem is placed horizontally over an aluminium mirror.

Natural diamond without impurities has a characteristic spectrum with bands that are common to both natural and synthetic diamonds, but most diamonds also present bands caused by defects in the crystalline lattice. We can distinguish three different ranges in the infrared spectrum. The first corresponds to the interval between 5000 and 2700 cm^{-1} and is the range where the bands corresponding to hydrogen impurities appear (4499, 4169, 3237, 3107 and 2785 cm^{-1}) and is characteristic of the natural diamond. The second range corresponds to the common bands of the diamond structure but they do not reveal the difference between natural and synthetic diamond (2505, 2443, 2159, 2033 and 1970 cm^{-1}). Finally, the range between 1400 and 1000 cm^{-1} identifies the impurities due to nitrogen as a specific defect in the crystalline lattice. According to the type of nitrogen substitution, there are different kinds of diamonds and it is in this range that it is possible to distinguish between the natural and the synthetic stones. **Figure 8** shows the spectra of a natural diamond with hydrogen and nitrogen impurities, and of a synthetic diamond.

The spectral differences of each diamond are in general great enough to distinguish between them. It is of particular value in identifying a gem with a high commercial value. If an infrared spectrum is attached to a gem, it can be used as proof in case of legal certification. The ratio of the hydrogen or nitrogen bands to the common bands of the second region of the diamond provides a value that is probably unique to the stone.

Another group of gems for which it has been possible to measure the differences between natural and synthetic specimens is emeralds. The water and carbon dioxide content in natural gems means they can be distinguished by measuring both mid and near infrared bands. **Figure 9** shows the infrared spectra of a natural and a fused phase synthetic emerald, where carbon dioxide and water are not present. Nowadays, techniques of manufacture of synthetic emeralds allow water to be incorporated in the lattice. In spite of this, infrared spectroscopy can distinguish between both kinds of water with the help of polarized radiation. **Figure 10** shows the near infrared spectra of a natural and a synthetic emerald using polarized light at 0° and 90°. In general, infrared spectroscopy enables us not only to see the differences between certain synthetic and natural specimens, but also to obtain the fingerprint of each gem with its impurities and, therefore, to use it in judicial and commercial certification.

Polymer analysis

Polymers are a group of substances that can easily be investigated by infrared spectroscopy. Apart from analyses aimed at identifying small particles or the composition of packagings used for food or drinks,

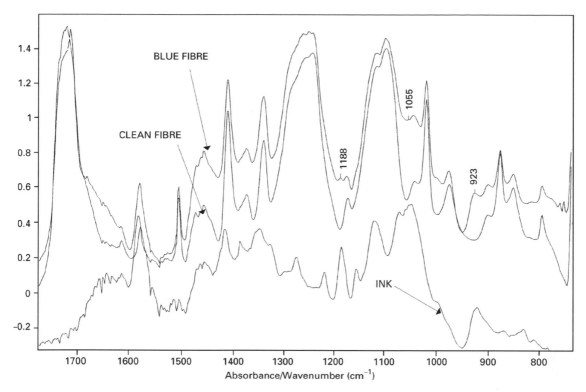

Figure 5 Comparison of the difference between a clean and a blue cotton fibre with the suspected ink (diamond cell in the microscope).

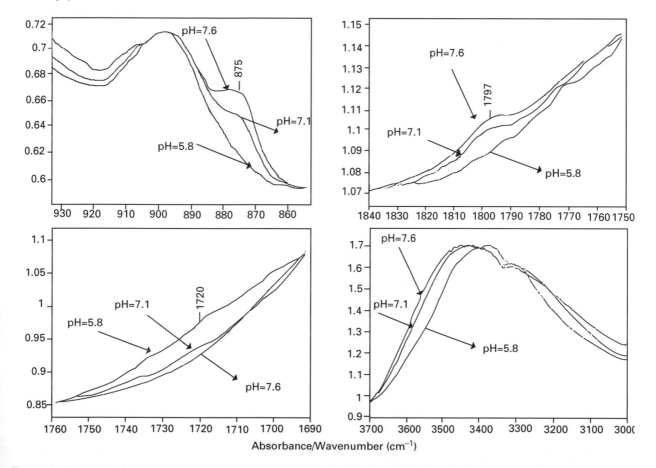

Figure 6 Changes of four different regions of infrared spectrum with pH of paper (diffuse reflection accessory).

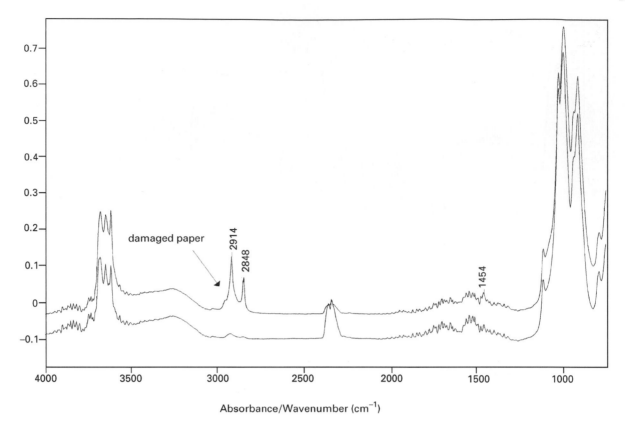

Figure 7 Spectrum of a spot on the surface of a paper identified as paraffin (ATR microscope).

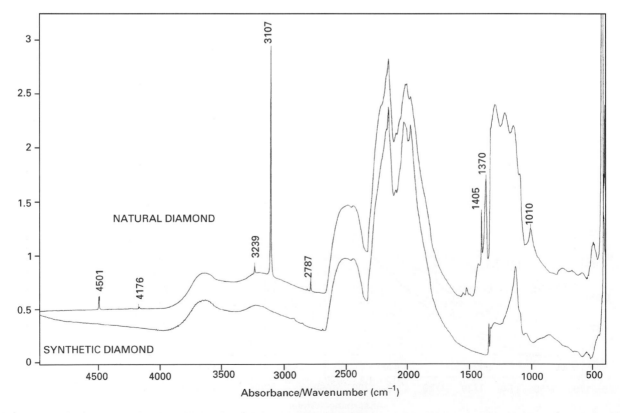

Figure 8 Comparison of a natural and a synthetic diamond (diffuse reflection).

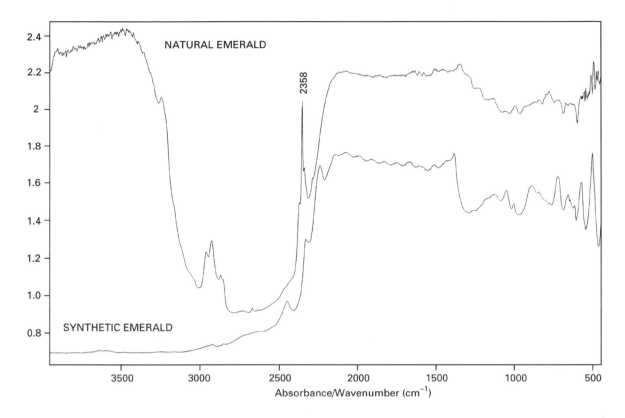

Figure 9 Comparison of a natural and a synthetic emerald (diffuse reflection).

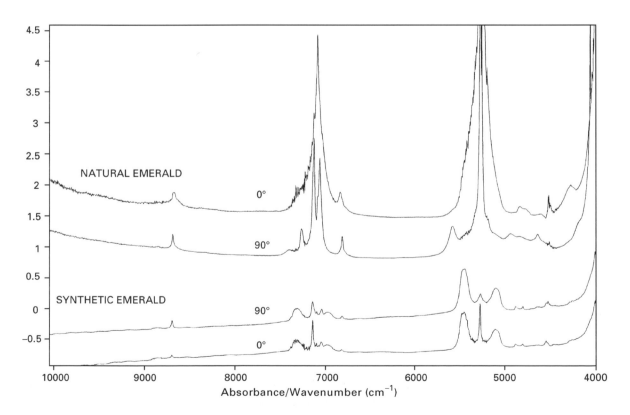

Figure 10 Comparison of bands of water in the near infrared between a natural and a synthetic emerald using polarized light (reflection microscope).

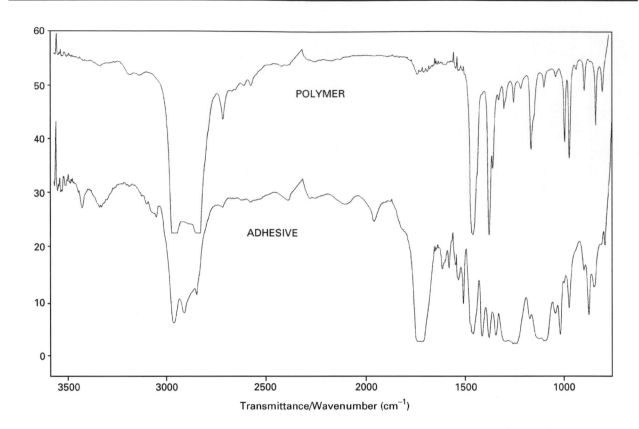

Figure 11 Spectra of the polymer and the adhesive of a multilayer film (transmission microscope).

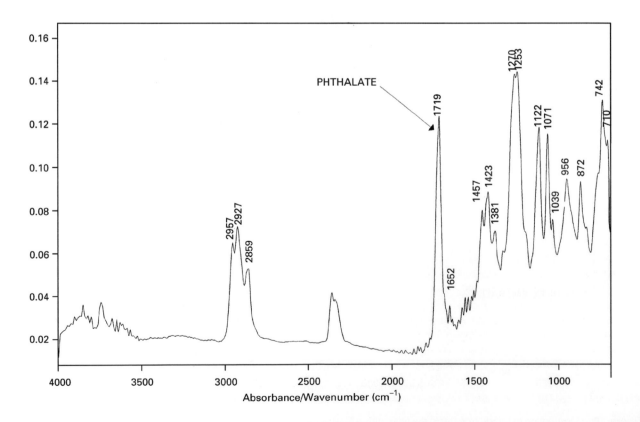

Figure 12 Spectrum of a PVC polymer used in contact with drinking water, showing phthalate bands at 1719 cm⁻¹ (ATR horizontal).

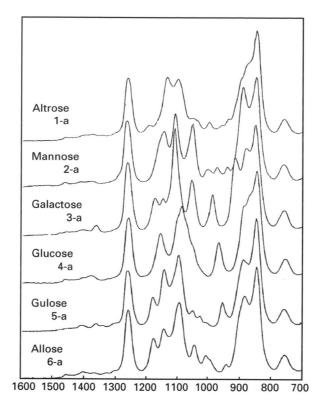

Figure 13 GC-IR spectra of six aldohexoses.

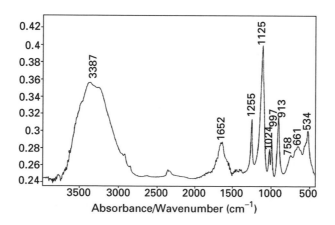

Figure 14 Spectrum of particles on the surface of a frozen fish (diamond cell in beam condenser).

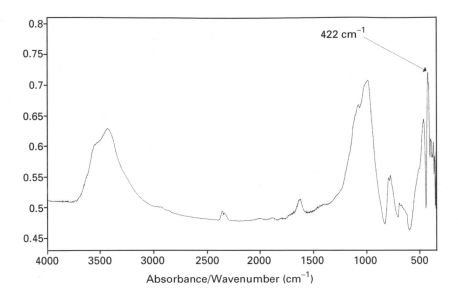

Figure 15 Spectrum of pyrite from the spill of waste in Doñana, showing pyrite band at 422 cm⁻¹ (KBr pellet).

the use of infrared spectroscopy allows multilayer polymers and adhesives used with them to be determined. **Figure 11** shows the infrared spectra of a 50 μm thick packaging where the content was altered because of migration of components from the adhesive between the layers.

The use of phthalates as plasticizers in the polymers of food packagings can lead to the migration of these molecules to the food itself. Therefore, it is essential to check for their presence in such packagings. **Figure 12** shows the attenuated total reflectance spectra of a polymer in contact with drinking water where phthalates are detected.

Analysis of physiological samples

Infrared spectroscopy has also been applied to the identification of particles found in organs such as the lung, kidney or liver. Some studies have been reported in which this technique has been applied to the analysis of malignant cells, by measuring the region of phosphodiesters of lymphocytes. This technique has been applied to the determination of samples of normal and malignant cervical cells. The presence of biomaterials such as polyesters, polyurethanes and silicones originating from implants can also be measured.

In the field of drugs, the analysis of human and animal hairs has been reported. Application of infrared microscopy in determining the presence of drugs in inner cuts of hair samples is a useful method for avoiding false positives from external contamination.

Some minerals, including carbonates and phosphates, from both inner and external parts of bones, have also been measured.

Food analysis

A number of analyses show the possibility of measuring additives used as preservatives in foods. Adulteration of juices, vegetable oils and milk, as well as caffeine content, protein and lipid content or sugar composition are other examples that have been reported.

Figure 13 shows the GC-IR spectra of six aldohexoses. None of the monosaccharides showed identical spectral patterns, so the technique was effective for their identification following separation by GC.

Figure 14 shows the spectrum of particles removed from the surface of frozen fish. The particles were identified as pyrophosphate.

Environmental samples

Analysis of solid samples which might contain particles from a contamination is another area for which infrared spectroscopy might be needed. In cases where it is necessary to prove that a particular waste has been introduced in a certain place, the technique allows the nature of the particles found to be analysed. Compensation is often available for those who have lost crops or cattle, but evidence of pollution must be proved.

Analysis of vapours from polluted soils or from stacks can also be carried out.

Figure 15 shows the spectrum of particles from the spill near the natural park of Doñana (Spain) which caused an ecological catastrophe in 1998. The pool from which the spill originated was full of pyrite waste and a black wave covered crops over an area extending many kilometres. The band at 422 cm⁻¹

corresponds to pyrite (FeS_2). Infrared spectroscopy was able to identify this compound at a distance of many kilometres from the spill.

The identification of compounds that should not occur in a place, makes it possible to impute them to spills or pollution from other places.

Conclusions

To sum up, infrared spectroscopy allows a rapid, initial determination of a substance sent for forensic analysis and can provide evidence that is of great, perhaps decisive, use.

It is often difficult to identify a substance by its infrared spectrum, especially one containing organic molecules, when there is no indication of what it might be. Inorganic molecules are easier to identify, at least those bands corresponding to the anionic part of the molecules: carbonates, silicates, nitrates, sulfates, phosphates, etc. Yet, frequently it is only necessary to compare samples or to follow up a suspicion of the presence of a substance.

Libraries play an important role in forensic analysis, but the best libraries are the ones built up by analysts themselves using suitable accessories and developing their own methods.

See also: **ATR and Reflectance IR Spectroscopy, Applications; Chromatography-IR Methods and Instrumentation; Chromatography-IR Applications; Fibres and Films Studied Using X-Ray Diffraction; Forensic Science, Applications of Atomic Spectroscopy; Forensic Science, Applications of Mass Spectrometry; Geology and Mineralogy, Applications of Atomic Spectroscopy; Industrial Applications of IR and Raman Spectroscopy; Inorganic Compounds and Minerals Studied Using X-Ray Diffraction; Medical Science Applications of IR; MRI of Oil/Water in Rocks; Polymer Applications of IR and Raman Spectroscopy.**

Further reading

Brettel TA and Saferstein R (1997) Forensic science. *Analytical Chemistry 97 Application Reviews* 123R.

Coleman PB (ed) (1993) *Practical Sampling Techniques for Infrared Analysis*. Boca Raton, FL: CRC Press.

Ferraro JR and Krishnan K (eds) (1990) *Practical Fourier Transform Infrared Spectroscopy. Industrial and Laboratory Chemical Analysis*. San Diego: Academic Press.

McKelvy ML, Britt TR, Davis BL, GLillie JK, Lentz LA, Leugers A, Nyquist RA and Putzig CL (1996) Infrared spectroscopy. *Analytical Chemistry* 68: 93R–160R.

Messerschmidt RR and Harthcock MA (eds) (1988) *Infrared Microspectroscopy Theory and Applications*. New York: Marcel Dekker.

Roush PB (ed) (1987) *The Design, Sample Handling and Applications of Infrared Microscopes*. Ann Arbor: ASTM.

Forensic Science, Applications of Mass Spectrometry

Rodger L Foltz, **David M Andrenyak** and **Dennis J Crouch**, University of Utah, Salt Lake City, UT, USA

MASS SPECTROMETRY
Applications

An important component of forensic toxicology is the determination of drugs in biological specimens when the analytical results have potential legal consequences. Mass spectrometry is generally the primary analytical tool for these investigations. This is because mass spectrometry, in combination with chromatographic methods of sample introduction, offers the best combination of high sensitivity and reliable identification. The areas of forensic toxicology where mass spectrometry has played a particularly important role include workplace drug testing, testing of motor vehicle operators for drug impairment, and postmortem toxicology.

Workplace drug testing typically involves the analysis of urine samples from job applicants and/or employees to determine whether they have recently used drugs of abuse such as heroin, marijuana, amphetamines or cocaine. Test samples are normally screened by a relatively inexpensive immunoassay, and all presumptively positive samples are submitted for mass spectral analysis to confirm the presence of the drug or drugs. Similarly, for determination of impairment in motor vehicle operators, blood samples are often analysed by mass spectrometric methods to determine whether the impaired performance was associated with recent use of a psychoactive drug. In postmortem toxicology, additional sample matrices are often analysed by mass spectrometric methods in order to gain the greatest insight into whether drugs contributed to the cause of death. In each of these applications the biological specimens must be collected, transported, stored and analysed by procedures that are scientifically and legally supportable.

The application of mass spectrometry in forensic toxicology typically consists of extraction of the drug or drug metabolites from selected body fluids or tissues, chromatographic separation of the drug from other co-extractants, and mass spectrometric detection of the analyte. The chromatographic separation is most often performed by capillary column gas chromatography (GC), following chemical conversion of the analyte to a volatile derivative that has more favourable chromatographic and mass spectral characteristics. The combination of liquid chromatography and mass spectrometry (LC-MS) can also be used, but its adoption by forensic toxicology laboratories has been slow, primarily because of the high cost of LC-MS instrumentation. Nevertheless, LC-MS offers the important advantages of not requiring derivatization and of being amenable to analysis of high-molecular-mass compounds (such as peptides and proteins) and very polar compounds (such as glucuronide-conjugated metabolites).

GC-MS analyses can be performed using electron ionization (EI) or chemical ionization (CI), followed by either full-scan mass analysis or selected-ion monitoring. Electron ionization is favoured in forensic toxicology because it generates detailed, compound-specific mass spectra, and because more than 100 000 electron ionization reference mass spectra have been published. However, chemical ionization can often provide better sensitivity, depending on the chemical characteristics of the analyte. Most drugs contain basic functional groups that are easily protonated in the gas phase within a chemical ionization source, resulting in a simple mass spectrum dominated by a large protonated molecule (MH⁺) ion peak. Drugs that contain functional groups with high electron affinity, such as halogens, can be ionized selectively and with very high efficiency by electron-capture negative-ion chemical ionization.

Figure 1 compares the mass spectra resulting from analysis of a cocaine-containing sample using electron ionization and positive-ion chemical ionization, and illustrates the features typical of these two types of mass spectra. The electron ionization

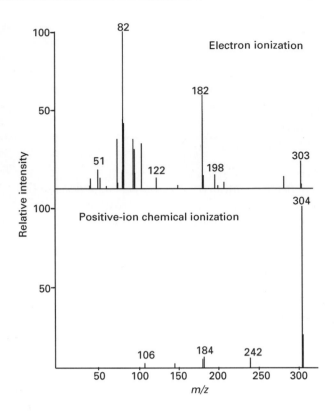

Figure 1 Comparison of the electron ionization and positive-ion chemical ionization mass spectra of cocaine.

mass spectrum has a relatively low-intensity molecular ion (M⁺) peak at m/z 303 and contains many fragment ion peaks. In contrast, the positive-ion chemical ionization mass spectrum shows a single intense peak (m/z 304) that corresponds to the protonated molecule ion (MH⁺). The high sensitivity often afforded by chemical ionization is due to the way the analyte ions are concentrated at a single m/z value, rather than being distributed among many different m/z values.

Both full-scan recording and selected-ion monitoring are widely used in forensic toxicology. Each has advantages. A full-scan spectrum that closely matches a reference spectrum generally constitutes the most reliable identification of an analyte. However, monitoring just a few well-selected ions characteristic of an analyte is more sensitive than full-scan recording and can be a reliable identification if it is combined with a selective chromatographic separation.

Workplace drug testing

The analysis of urine specimens from employees to determine whether they are inappropriately using drugs that can affect job performance falls under the

Table 1 Initial test cutoffs

Drug class	Cutoff (ng mL^{-1})
Marijuana metabolites	50
Cocaine metabolites	300
Opiate metabolites	300
Phencyclidine	25
Amphetamine	1000

Table 2 Confirmation test cutoffs

Drug	Cutoff (ng mL^{-1})
9-Carboxy-Δ^9-THC[a]	15
Benzoylecgonine[b]	150
Morphine	300
Codeine	300
Phencyclidine	25
Amphetamine	500
Methamphetamine	500

[a] The major urinary metabolite of Δ^9-tetrahydrocannabinol, the psychoactive component of marijuana.

[b] The major urinary metabolite of cocaine.

broad category of forensic toxicology because of the legal consequences that can arise. In the United States, workplace drug testing is widely practised, and can be divided into 'regulated' and 'nonregulated' testing. 'Regulated drug testing' refers to programmes that fall under federal regulations, such as the 'Mandatory Guidelines for Federal Workplace Drug Testing Programs' issued in 1988. These 'mandatory guidelines' define how workplace drug testing must be performed, and include the requirement that the testing be performed by laboratories certified by the National Laboratory Certification Program. Another key component of the guidelines requires a two-tier testing process consisting of an initial series of immunoassays followed by confirmatory tests based on gas chromatography–mass spectrometry. For a sample to be reported as positive for a drug, the immunoassay screen must show the presence of a drug class at a concentration above the cutoff values shown in **Table 1** and the GC-MS confirmation assay must show a drug, or metabolite within the drug class, to have a concentration above the cutoff values listed in **Table 2**.

'Nonregulated' workplace drug testing refers to programmes that do not fall under federal regulations. These programmes may include testing for other psychoactive drugs such as benzodiazepines, barbiturates and lysergic acid diethylamide (LSD). Also, nonregulated drug testing programmes sometimes include analysis of other types of specimens, such as hair or sweat.

Federal regulations do not identify specific methods for performing GC-MS confirmations. However, certain mass spectrometric procedures have become widely adopted by laboratories that perform workplace drug testing. A laboratory must first demonstrate and document that each confirmation assay is specific for the target drug and that it is sufficiently sensitive to permit accurate measurement of the drug at its established confirmation cutoff.

Typically, an analytical batch of urine samples will include each of the following: (1) calibration standards with at least one standard at the analyte's cutoff concentration; (2) quality-control samples containing known concentrations of the analyte both above and below the analyte's cutoff; (3) a sample blank containing none of the analyte; and (4) the test samples submitted for confirmation. A specific amount of internal standard is added to each of these samples. Deuterium-labelled analogues, now commercially available for all common drugs of abuse, are generally the first choice as internal standards. After addition of the internal standard, each sample is extracted to separate the analyte from other components of the urine. This step may involve a simple liquid–liquid extraction with a water-immiscible organic solvent, extraction onto a solid adsorbent and elution with an organic solvent, or a combination of these procedures.

Most drugs and metabolites are chemically converted to more volatile and stable derivatives before GC-MS analysis. **Table 3** gives examples of derivatizations commonly used for GC-MS confirmation of drugs of abuse.

Before any test samples are analysed, the mass spectrometer's mass calibration is checked by

Table 3 Examples of derivatization procedures for GC-MS analysis of common drugs of abuse

Drug group	Derivatization procedure
Amphetamines	Acylation with acid anhydrides such as trifluoroacetic anhydride (TFAA), pentafluoropropionic anhydride (PFPA) or heptafluorobutyric anhydride (HFBA)
Opiates	Trimethylsilylation with bis(trimethylsilyl)trifluoroacetamide (BSTFA) or acylation with a fluorinated anhydride
9-Carboxyl-Δ^9-THC	Trimethylsilylation with BSTFA or alkylation with methyl iodide in dimethyl sulfoxide
Benzoylecgonine	Trimethylsilylation with BSTFA
Benzodiazepines	Trimethylsilylation with BSTFA
Barbiturates	Alkylation with trimethylanilinium hydroxide (TMAH)

analysis of a reference material such as perfluorotributylamine (PFTBA), and the measured mass-to-charge (m/z) values for the major PFTBA ions are compared with the corresponding theoretical m/z values. A reference solution containing a known concentration of the analyte or analytes of interest is then injected into the GC-MS system to evaluate the instrument's performance. If the analyte's peak intensity, retention time and chromatographic peak shape are acceptable, analysis of a batch of samples can be initiated.

Fused-silica capillary gas chromatographic columns are universally used for analysis of drugs in forensic laboratories. Typically the columns are coated to a thickness of approximately 0.3 µm with relatively nonpolar methylsilicone or phenylmethylsilicone adsorptive phases and have internal diameters of 0.2–0.3 mm and are 10–30 meters in length. Extracted and derivatized samples injected into the GC-MS are most often detected by electron ionization and selected-ion monitoring (SIM) of three prominent analyte ions and two prominent internal standard ions. The analyte concentration is determined by reference to a calibration curve based on a plot of the ratio of the area of the analyte's quantitating ion to the area of the internal standard's quantitating ion versus analyte concentration. For a urine sample to be confirmed as positive for a drug, the analytical results must meet all of the following criteria:

1. The drug's measured concentration must be equal to, or greater than, the established cutoff concentration for the drug (see **Table 2**).
2. The ratios of peak areas for the monitored ions of the analyte and internal standard must fall within 20% of the corresponding peak area ratios for the calibration standard at the cutoff concentration.
3. The retention times for the analyte and the internal standard must be within 2% of those for the cutoff calibrator.
4. The chromatographic peaks for the analyte and the internal standard must be clearly separate from co-eluting compounds.

The analytical methods most widely practised in laboratories for workplace drug testing have been well validated and are highly reliable. However, some problems continue to be of concern. For example, ingestion of certain legitimate food products can result in a urine sample testing positive for a controlled drug. Poppy seeds, widely used in bakery products, contain low concentrations of morphine, and hemp oil contains low concentrations of Δ^9-tetrahydrocannabinol, the psychoactive component

of marijuana. Efforts to develop analytical methods for testing urine that would permit distinction between ingestion of these food products and nonmedical use of regulated drugs have so far been unsuccessful. However, methods are now available for determining whether urine found to be positive for methamphetamine reflects nonmedical use of methamphetamine or legitimate use of 'Vick's Inhaler', an over-the-counter medication which in the United States contains the *l*-enantiomer of methamphetamine. To make this distinction, the urine must be analysed by a method that can separate and identify each of the optical isomers of methamphetamine. This can be accomplished by derivatizing the extracted methamphetamine with a chiral derivatizing reagent such as N-trifluoroacetyl-L-prolyl chloride. This reaction forms separate diastereoisomers of *d*- and *l*-methamphetamine that are easily distinguished by GC-MS analysis.

Testing of motor vehicle operators for drug impairment

Drug impairment of operators of motor vehicles (cars, buses, trains, airplanes, etc.) is a continuing threat to public safety. To counteract this threat there has been increased effort devoted to developing analytical methods for determining when an operator's impaired performance is due to use of psychoactive drugs. These analytical methods have also played an important role in investigations into how drugs of abuse, as well as some legitimate prescription drugs, can affect a person's ability to operate a motor vehicle safely. To better understand how drugs can affect human performance, controlled clinical laboratory experiments are required. These experiments are designed to record self-reported impairment and to detect physiological, perceptual and cognitive drug-induced changes that may impact the subject's performance. The most useful clinical-study designs incorporate collection of biological samples at predetermined times for measurement of the drug concentration. Here mass spectrometry is used extensively to quantify the studied drug and its metabolites. Drug concentrations can then be compared to the physiological and performance measures to determine whether a dose–response relationship exists for the drug being evaluated. These clinical studies provide invaluable data for assisting in interpreting cases with legal implications.

Mass spectrometry is also used to analyse biological samples collected from subjects suspected of operating a motor vehicle under the influence of a drug. In contrast to workplace drug testing, where urine is the body fluid normally analysed, blood is the preferred body fluid for determining impaired performance. This is because drug concentrations in blood have been found to correlate better with the level of impairment at the time the blood was collected than drug concentrations in urine, which can vary widely depending on the subject's degree of hydration and other variables. For some drugs, collection and analysis of saliva can also be an effective method for determining impairment.

Alcohol is the drug most frequently associated with driving impairment, but its presence in the body is measured by a breathalyser or by analysis of blood using procedures that do not require the use of mass spectrometry. Analysis of blood or saliva for other drugs normally includes an initial screening procedure based on a series of immunoanalyses or by a chromatographic screening method, followed by confirmation and quantitative measurement of the drug by mass spectrometry in combination with gas or liquid chromatography.

The frequency with which specific drugs are detected in impaired motor vehicle operators varies from one country or region to another. In the United States, marijuana is the second most frequently detected drug in impaired drivers, with cocaine probably the third most frequently detected. Benzodiazepines have been shown to affect fundamental driving skills such as coordination and reaction time, and are reported to be frequently detected in drivers suspected of being drug impaired. Amphetamines and chemically related sympathomimetic drugs are sometimes used to reduce fatigue in drivers, but large doses of these drugs may lead to greater risk-taking and aggressive driving. Other drugs that have been detected in motor vehicle operators suspected of driving under the influence (DUI) include opiates, barbiturates, antihistamines and antidepressants.

The confirmation and quantitation of drugs identified in DUI cases is a challenge to the forensic toxicologist because drug concentrations are typically lower in blood than in urine or in tissue samples. The determination of marijuana use by DUI suspects is particularly challenging. Although Δ^9-tetrahydrocannabinol (THC) and its major metabolites may be detected in blood following recent use, the concentrations of these analytes are usually in the low nanogram per milliliter range. THC and its metabolites have active hydroxyl or carboxyl groups that require derivatization prior to GC-MS analysis. Examples of derivatization procedures are shown in **Table 3**. GC-MS with electron ionization may be used for the analysis of THC and its metabolites, but this combination has limited use in forensic cases

because it lacks sufficient sensitivity to routinely measure these analytes in blood samples. Better sensitivity can be achieved using GC-MS with negative-ion chemical ionization following conversion of the cannabinoids to derivatives containing electron-capturing substituents. Sub-nanogram- per -milliliter detection limits have also been achieved by trimethylsilylation of the extracted cannabinoids followed by gas chromatography coupled to a tandem mass spectrometer operated in the positive-ion chemical ionization mode.

Cocaine is rapidly metabolized, primarily to benzoylecgonine and ecgonine methyl ester, and this conversion can continue after the blood is drawn unless an esterase inhibitor such as potassium fluoride is added to the blood immediately. This step is important because the blood concentration of the parent cocaine is a better indicator of the degree of impairment than the concentrations of the inactive metabolites.

Mass spectral identification and quantification of benzodiazepines in blood is complicated by the large number of prescription benzodiazepine drugs available to the public. Treatment of blood extracts with strong acid has been used to convert benzodiazepines to benzophenones, which are more easily identified by GC-MS analysis. However, more recently developed methods using GC-MS with negative-ion chemical ionization or LC-MS with electrospray ionization take less time and offer better sensitivity.

Postmortem toxicology

Postmortem forensic toxicology involves the analysis of drugs and poisons in body fluids and tissues collected from victims that have died under suspicious circumstances. Postmortem forensic toxicology is a key part of a death investigation in that it can provide information addressing the following questions and issues: (1) What drugs or toxins are present in the body? (2) Are the concentrations of the drugs or toxins in the lethal range? (3) Could drugs have caused impairment resulting in a fatal accident?

Currently in the United States there are no federally mandated regulations governing postmortem forensic toxicology. However, several forensic toxicology accreditation programmes have been established, and the Society of Forensic Toxicology and the American Academy of Forensic Sciences have published guidelines for forensic toxicology laboratory procedures and performance.

Unlike workplace drug testing, in which urine samples are analysed for a relatively small number of drugs and drug metabolites, postmortem forensic toxicology often involves analysis for a wide variety of drugs and poisons in many different specimens, including gastric contents, vitreous humor, blood, liver, spleen, muscle and brain tissue. Analysis of blood is particularly important because drug concentrations in blood that are high enough to cause death are known for most drugs. However, drug concentrations in blood are generally lower than in urine or in tissues.

Before extraction and analysis, tissues must be homogenized. Water or buffer solutions such as 0.1 M sodium phosphate, pH 6, are added to the tissue sample prior to homogenization. It is important to record the weight of the tissue and the volume of the fluid in which the tissue is homogenized. This information will be used to express the amount of drug per gram of tissue.

Extraction efficiency and sample cleanliness can be a particular challenge in the analysis of postmortem samples because the samples are often collected from decomposed bodies. Products from decomposition can reduce the extraction efficiency and can produce interfering peaks. Also, tissue homogenates contain lipid material that must be separated from the drug analyte prior to GC-MS analysis. Basic drugs can be efficiently separated from lipid material by back-extraction. In this procedure the drug is extracted from the tissue homogenate into a water-immiscible organic solvent and then back-extracted into a dilute acid solution while the neutral lipid material remains in the organic solvent. The dilute acid solution is then made basic and the drug is re-extracted into an organic solvent.

Another cleanup procedure is to reconstitute the initial extract residue in a mixture of hexane and ethanol–water (80:20 v/v). Most drugs will partition into the ethanol–water layer and lipid compounds will partition into the hexane layer. Precipitation of proteins from the tissue homogenates prior to extraction is also helpful for improving cleanliness of the extract and increasing the extraction efficiency. Reagents that precipitate proteins include acetonitrile, methanol and trichloroacetic acid. Digestion of tissue samples with proteolytic enzymes such as subilisin A can also enhance the recovery of drugs in the extraction procedure.

Following extraction, the analyte is derivatized, if necessary, and analysed by GC-MS. As discussed previously, electron ionization continues to be favoured. However, use of chemical ionization is increasing because it often provides better sensitivity. Also, a chemical ionization mass spectrum will frequently provide valuable complementary information. For example, **Figure 2** shows electron ionization and chemical ionization mass spectra for

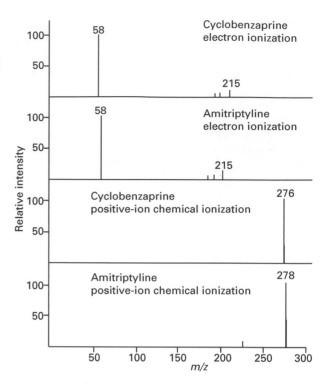

Figure 2 Comparison of the electron ionization and positive-ion chemical ionization mass spectra of cyclobenzaprine and amitriptyline.

two similar drugs, amitriptyline and cyclobenzaprine. The electron ionization mass spectra for the two drugs are virtually identical owing to the prominence of the common fragment ion at m/z 58. The positive-ion chemical ionization mass spectra show protonated molecule ion peaks that differ by two daltons, thereby permitting identification of the two drugs.

Of course, the relevant questions in a death investigation cannot be answered by mass spectrometric analysis and other analytical data alone. Final interpretation should always include careful consideration of all the information relating to the case.

Future developments

The combination of gas chromatography and electron ionization mass spectrometry will continue to be an important analytical tool for identification and quantification of drugs and toxins in most forensic toxicology laboratories. However, the use of liquid chromatography with mass spectrometry is certain to increase, and is likely to become the most widely used method for identification and quantification of drugs and other toxins in forensic toxicology samples. Many variations of liquid chromatography, such as capillary electrophoresis (CE) and capillary electrochromatography (CEC), have been successfully coupled to mass spectrometers and will provide the forensic toxicologist with an array of techniques that will permit many analytical tasks to be performed with improved sensitivity, accuracy and efficiency.

See also: **Biochemical Applications of Mass Spectrometry; Chemical Structure Information from Mass Spectrometry; Chemical Ionization in Mass Spectrometry; Chromatography-MS, Methods; Forensic Science, Applications of Atomic Spectroscopy; Forensic Science, Applications of IR Spectroscopy; Hyphenated Techniques, Applications of in Mass Spectrometry; Ion Trap Mass Spectrometers; Isotopic Labelling Mass Spectrometry; Medical Applications of Mass Spectrometry; MS-MS and MS^n; Negative Ion Mass Spectrometry, Methods; Quadrupoles, Use of in Mass Spectrometry.**

Further reading

Baselt RC and Cravey RH (eds) (1995) *Introduction to Forensic Toxicology*, 4th edn. Foster City, CA: Chemical Toxicology Institute.

Baselt R (ed) (1989) *Advances in Analytical Toxicology*, Vol 2. Chicago, IL: Year Book Medical Publishers.

Curry AS and Thomas CC (1976) *Poison Detection in Human Organs*, 3rd edn. Springfield, IL: Charles C. Thomas.

Karch SB (ed) (1998) *Drugs Abuse Handbook*. Boca Raton, FL: CRC Press.

Moffat AC (ed) (1986) *Clarke's Isolation and Detection of Drugs*. London: The Pharmaceutical Press.

Swotinsky RB (1992) *The Medical Review Officer's Guide to Drug Testing*. New York: Van Nostrand Reinhold.

Yinon J (ed) (1995) *Forensic Applications of Mass Spectrometry*. Boca Raton, FL: CRC Press.

Yinon J (ed) (1987) *Forensic Mass Spectrometry*. Boca Raton, FL: CRC Press.

Forestry and Wood Products, Applications of Atomic Spectroscopy

Cathy Hayes, Trinity College, Dublin, Ireland

Copyright © 1999 Academic Press

ATOMIC SPECTROMETRY
Applications

Introduction

Analytical methods of atomic spectroscopy have been used in forestry and wood product research since their earliest development. Nowadays, almost all of the spectroscopic techniques available are employed in the analysis of metals and trace elements in diverse samples of industrial and environmental origin. The techniques that find most regular application include flame atomic absorption spectroscopy (F-AAS), graphite furnace atomic absorption spectroscopy (GF-AAS), inductively coupled plasma atomic emission spectroscopy (ICP-AES) and, occasionally, also direct current plasma atomic emission spectroscopy (DCP-AES). In many applications F-AAS is a sufficiently sensitive and precise technique; however, in the analysis of some environmental samples for trace elements (forest soils, plant material and water) where concentrations may be very low (of the order of 100 ng mL^{-1}) the greater sensitivity of GF-AAS and ICP/DCP-AES is required. In considering the applications of atomic spectroscopy to forestry and wood products it is worthwhile remembering that the analytical method available for use is often determined by the volume and variety of analyses carried out in individual laboratories.

Atomic spectroscopic techniques in use

F-AAS is probably the most widely used analytical method in forestry and wood product research for two reasons. Firstly, it is a long established technique, machines are relatively cheap and widely available and many of the analyses routinely performed by this method have now been optimized. Secondly, the concentrations of many nutrient elements that are important for forest growth, which are present in soil and water (e.g. Ca, Mg, Al, Fe, K, Si) and in wood treated with metal-containing preservatives (Cu, Cr, As, Zn), are well within the limits of detection (100 ppm) of the technique. Environmental samples of this kind are also usually available in large quantities. F-AAS offers a precise, rapid method of analysis, and automation is possible for large numbers of samples.

GF-AAS, also known as electrothermal atomization atomic absorption spectroscopy (ET-AAS) or heated graphite atomizer atomic absorption spectroscopy (HGA-AAS), is widely available in research analytical laboratories where trace element work is carried out. GF-AAS is possibly the most sensitive method available in the atomic spectrometry range but machines are approximately twice as expensive as F-AAS set-ups. The detection limit for determinations is typically in the order of 10–100 ng mL^{-1}. Elements in soil/water samples most commonly determined by GF-AAS might include Ca and Mg, but also, in particular, toxic Cd, Cr, Cu, Ni and Pb. Details of the furnace program (the temperature, ramp, hold and purge gas flow characteristics of the drying, charring, atomization and cleaning steps) should be reported in published results more often than they are. The heightened sensitivity of this technique requires the careful use of analytical reagents to keep background contamination to a minimum. Matrix matching between samples and calibration solutions is critical in GF-AAS, and matrix modification (through the addition of ammonium nitrate, magnesium nitrate or palladium salts) may be necessary to prevent the loss of analytes before and in the atomization stage.

When DCP/ICP-AES becomes cheaper and therefore more widely available, the advantages offered by simultaneous multielement determinations combined with detection limits comparable to those of F-AAS (100 ng mL^{-1}), the technique will undoubtedly find greater application in the analysis of forestry and wood samples.

Forestry

Forestry research embraces both the monitoring of environmental impacts on forest ecosystems as well as the impact of forestry on surrounding landscapes, in particular on water catchments. Studies carried out have analysed soil to monitor the effects of acid rain on forest nutrition in relation to observed disease susceptibility in trees from different geographical locations. It is postulated that immobilization of essential plant nutrients such as calcium and magnesium occurs in soil where toxic aluminium becomes increasingly mobile owing to the reduced soil pH attributed to acid rain. Routine monitoring of water catchments that surround forestry follows the level of movement of both nutrient elements (N and P) and other forest exudates (Mn, Cu, Zn, Sr and Ba) in run-off. Both F-AAS and GF-AAS are used in these analyses and offer acceptable sensitivity and precision.

Wood and wood products

Timber harvested from commercial forests has several possible end users. Good quality timber (i.e. timber with good growth characteristics) is sawn to produce timber for the construction industry. Non-construction grade timber can end up being processed to make wood pulp and ultimately paper or used as sawn timber, joinery, fencing rails and posts, transmission poles, bulkheads and pilings. Recently, poorer quality timber has been used more and more to make value added commodities such as board materials like medium density fibre board (MDF), oriented strand board (OSB) and particle/chip board. Each of these areas generates the need for further monitoring for various reasons.

The awareness of the need to monitor paper and pulp mill effluent is not novel. The reduction, and even elimination, in the use of chlorine in the bleaching of pulp has focused environmental concern on the other elements present in wood pulp. Pulp and paper mill effluent may contribute to elevated metal concentrations in receiving waters and sediments. Consequently, the wood pulp industry, along with statutory monitoring bodies, make regular use of sensitive and quantitative techniques such as DCP-AES and GF-AAS in attempts to reduce metal levels.

The implication of wood dust and the metals from machined treated wood in the increased instance of carcinoma in people working with wood also requires the use of GF-AAS and F-AAS in the health and safety analysis of dusts from wood related work places (joinery manufacturers, sanding and sawing operations, particle board and plywood industries).

Wood preservatives

The bulk of analytical work performed in the wood sector is associated with wood preservatives. Most temperate timbers are perishable and therefore require treatment with toxic combinations of elements to achieve protection from biodeterioration owing to attack by fungi, insects, termites and, in marine situations, wood-boring crustaceans and molluscs.

The most widely used preservatives are the class of inorganic waterborne salts that contain Cu, Cr, As, B and Zn. A list of commonly used preservatives of this type is presented in **Table 1**.

Preservative penetration and retention

The necessity of monitoring the levels of preservatives (waterborne inorganic salts as well as organic solvents and creosote) in treated timber arises from the fact that such timber is supposed to be treated to specified levels or retentions (kg preservative salt m^{-3} wood), depending on the biological hazard to which it is expected to be exposed in service. There are five classes of biological hazard based on the expected exposure to wetting and these are shown in Table 2.

Technical standards exist which prescribe the levels of preservative salt necessary in timber exposed to a given hazard class. Treatment levels are defined in terms of required penetration depths and must be

Table 1 Common inorganic waterborne salt type wood preservatives; active ingredients and nominal compositions as well as the recommended in-service situation are listed

Preservative	Active ingredient	Nominal % composition	Principal use/country of use
Chromated zinc chloride (CZC)	Zinc as ZnO	80	Out of ground contact
	Chromium as CrO_3	20	USA, Germany
Copper–chromium–arsenic (CCA)	Copper as $CuSO_4 \cdot 5H_2O$ as CuO	33^a, 35^b 18.1^c, 19.6^d, 18.5^e	Ground contact, freshwater and marine situations
	Chromium as $Na_2Cr_2O_7 \cdot 2H_2O$ as CuO_3	41^a, 45^b 65.5^c, 35.3^d, 47.5^e	Worldwide
	Arsenic as $As_2O_5 \cdot 2H_2O$ as As_2O_5	26^a, 20^b 16.4^c, 45.1^d, 34^e	
Copper–chromium–boron (CCB)	Copper as $CuSO_4 \cdot 5H_2O$	36	Out of ground contact
	Chromium as $Na_2Cr_2O_7 \cdot 2H_2O$	40	Continental Europe
	Boron as H_3BO_3	24	
Ammoniacal copper arsenate (ACA)	Copper as CuO	49.8	Ground contact, freshwater and marine situations
	Arsenic as As_2O_5	50.2	USA
Fluor chrome arsenate phenol (FCAP)	Fluoride as F	22	Out of ground contact
	Chromium as CrO_3	37	
	Arsenic as As_2O_5	25	Continental Europe,
	Dinitrophenol	16	USA
Ammoniacal copper, zinc arsenate (ACZA)	Copper as CuO	50	Out of ground contact
	Zinc as ZnO	25	
	Arsenic as As_2O_5	25	Scandinavia, USA

The country of use, is not intended as an exhaustive list.
[a] CCA type 1 (UK).
[b] CCA type 2 (UK).
[c] CCA type A (USA).
[d] CCA type B (USA).
[e] CCA type C (USA).

Table 2 Biological hazard classes for timber exposed in service as defined by the European Standard EN 335-1:1992

Hazard class	General service situations	Description of wetting in service	Occurrence of biological agencies[a]			
			Fungi	Beetles[b]	Termites	Marine borers
1	Above ground, covered (dry)	None	–	U	L	–
2	Above ground, covered (risk of wetting)	Occasionally	U	U	L	–
3	Above ground, not covered	Frequently	U	U	L	–
4	In contact with ground or fresh water	Permanently	U	U	L	–
5	In salt water	Permanently	U	U	L	U

[a] U, Universally present within Europe. L, Locally present within Europe.
[b] The risk of attack can be insignificant, according to specific service situations.

matched to individual wood species' treatability. The more pertinent European and American standards covering these technical requirements are listed in **Table 3**.

Meeting the requirements of these standards generates an enormous amount of routine analysis in the preservative treatment plants that are responsible for correctly treating the timber, and also in research laboratories examining biodeterioration of timber and in industrial development laboratories. Inorganic waterborne elements are readily determined using F-AAS and this is the method recommended in the technical standards governing the analysis of inorganic waterborne preservatives, the European and American examples of which are included in **Table 3**.

Naturally occurring levels of elements

The naturally occurring levels of the elements most frequently analysed in forestry soil and foliage samples and wood/wood product samples are presented in **Table 4**. The normal manner in which samples containing these elements are prepared so as to release the element as an analyte for trace element determination is indicated in the table. Where several analytical techniques are possible this is indicated in the footnotes. The most commonly recommended F-AAS and GF-AAS settings for the same elements are shown in **Table 5** along with potential interferences and their control. From these two tables it is readily appreciated that, for most determinations involving forestry related or wood samples, contamination is not usually problematic and background levels can frequently be ignored.

General instrument settings

Sample instrument settings for F-AAS, GF-AAS and ICP-AES analyses are shown in **Table 6**. However, instrument manufacturers, recommendations should always be followed.

Analytical procedures

General comments

In the use of AAS or AES, quality control can be maintained by the use of standard reference materials in analyses. Although expensive, the practice of including reference samples provides extra control to the data generated. The heterogeneity of elemental distributions in ecological samples requires adequate field sampling to generate a statistically significant number of samples on which accurate determinations can be made. In many instances the pooling of samples after initial processing (grinding, drying, sieving etc.) is possible (and has been reported for

Table 3 National standards referring to preservative treatment of wood and also to the laboratory methods used to analyse the preservative content of wood and treating solutions (including F-AAS)

Standard No	Title
EN 335-1: 1992	*Durability of wood and wood-based products.* Hazard classes of wood and wood-based products against biological attack. Part 1. Classification of hazard classes
EN 335-1: 1992	*Durability of wood and wood-based products.* Hazard classes of wood and wood-based products against biological attack. Part 2. Guide to the application of hazard classes to solid wood
EN 335-3: 1996	*Durability of wood and wood-based products.* Definition of hazard classes of biological attack. Part 3. Application to wood based panels
EN 350-2: 1994	*Durability of wood and wood-based products.* Natural durability of solid wood. Part 2. Guide to natural durability and treatability of selected wood species of importance in Europe
EN 351-1: 1996	*Durability of wood and wood-based products.* Preservative-treated solid wood. Part 1. Classification of preservative penetration and retention
EN 599-1: 1997	*Durability of wood and wood-based products.* Performance of preventive wood preservatives as determined by biological tests. Part 1. Specification according to hazard class
BS 5666-3: 1991	Methods of analysis of wood preservatives and treated timber. Part 3. Quantitative analysis of preservatives and treated timber containing copper/chromium/arsenic formulations
BS 4072: 1978	Specification for wood preservation by means of copper/chrome/arsenic compositions
AWPA Standard P5-98 (1998)	Standards for waterborne preservatives
AWPA Standard C1-98 (1998)	All timber products – preservative treatment by pressure processes
AWPA Standard A11-93 (1993)	Standard method for analysis of treated wood and treating solutions by atomic absorption spectroscopy

EN = European standard; BS = British Standard; AWPA = American Wood Preservers' Association Standard.

soils and herbaceous plant matter) but the decision must be made whether to sacrifice the spatial pattern for a mean concentration. The practice is not recommended for wood. Matrix matching of calibration and test solutions is critical, especially where the more sensitive GF-AAS is concerned. If acid digests have been carried out on samples the same acids must be used in calibration standards. Analytical grade chemicals should be used in all determinations and, where specified, distilled and deionized water may be preferred over distilled water.

Applications of atomic spectroscopy in forestry and wood products

Fine estuarine sediments

Particulate organic matter (POM) is largely associated with fine sediment fractions, i.e. silts (2–20/50 µm) and clays (< 2 µm). Inorganic copper–chromium–arsenic salts (CCA) leached from submerged or floating treated wood in coastal situations also tend to be associated with the organic matter in this fraction. The settlement of fine particulates (POM, silts, clays) where Cu, Cr and As may be present is dependent on the flushing characteristics of the water body as well as the age of the structure from which preservative components are leaching. In slow currents where recently treated timber is present, Cu, Cr and As can be detected several metres from treated wooden emplacements such as pilings, floating jetties and bulkheads. It is important that experimental design takes these facts into account when deciding on a sampling regime.

Fine sediments should be allowed to settle out in volumetric cylinders before subsampling so as to reduce the risk of contamination from metal sieves. Sediment is then dried (150°C) and a subsample wet ashed in a 3:1 mix of nitric acid and perchloric acid (HNO_3–$HClO_4$). *Caution*: perchloric acid is highly explosive when it comes into contact with organic matter. Therefore raw organic samples should always be pre-mixed in warmed nitric acid *before* the addition of perchloric acid. Once perchloric acid has been added, the digest mix should not be allowed to boil dry. Only experienced personnel should handle and store this reagent.

Determinations of Cu, Cr and As can be made on the filtered liquid fraction, using the recommended settings in the manual accompanying the atomic spectrometer used. Elemental levels determined by F-AAS methods from fine coastal sediments have been found in the range 25–2000 µg Cu per g sediment, 40–180 µg Cr per g sediment and 10–100 µg As

per g sediment. The possibility of carrying out hydride generation before the determination of arsenic is an option on many machines.

Algae and animal tissues

Plant and animal material (e.g. algae, crustaceans, molluscs and bivalves) collected to determine the transfer of metals from the environment to biota within habitats can be collected at field sites and frozen until laboratory analysis is possible. Defrosted or fresh material can then be oven dried (150°C) before a subsample is selected and wet ashed in a 3:1 mix of nitric and perchloric acid (HNO_3–$HClO_4$). Note the cautions expressed above for the handling of this mixture.

The filtered leachate is analysed in F-AAS using the settings recommended by the manufacturer of the machine for each element under determination. In cases where the elemental concentrations are likely to be very low (e.g. control samples from 'clean' sites) it may be necessary to carry out analysis using GF-AAS (or ICP-AES if available) with a simpler acid digestion. Typical levels of CCA encountered in coastal biota where preserved wood is submerged are in the range 5–10 µg g^{-1} (As) 5–15 µg g^{-1} (Cr) and 60–150 µg g^{-1} (Cu).

Trees and herbaceous plant material

Foliage samples are collected fresh and then stored cold or frozen until analysis. Dried/defrosted samples are oven-dried at 60–80°C and then ground to a fine powder in a suitable mill. Beware: steel mills may lead to contamination by Fe, Mn, Mo and Co while other mills can lead to contamination by Cu and Zn. Ceramic ball mills may lead to contamination by Al, B, Ca, K or Na.

Ground samples are dissolved in a wet digestion procedure on a heating block using a 3:1:2 mix of HNO_3–H_2SO_4–H_2O_2 (typically 1–5 g in 1.5:0.5:1.0 mL of acid mix). Lanthanum (as lanthanum chloride, 1000 µg mL^{-1}) is added as a releasing agent to prevent phosphate interference in the analysis of Ca and Mg by F-AAS. Aluminium levels may need to be determined by the more sensitive GF-AAS, in which case the chances of contamination need to be minimized by using smaller volumes of acid in the digestion mix (1 mL HNO_3, 0.2 mL H_2SO_4, 0.2 mL H_2O_2).

Forest soils

Soils are collected in the field using stainless steel soil probes which sample to depths of 40–50 cm, maintaining the stratigraphic integrity of the sample.

Table 4 List of common soil minerals and methods for sample preparation and analysis; Naturally occurring levels (in mineral soils, freshwater, etc.) are also included. With the permission of Blackwell Scientific Publications data based on Allen SE (1989) In: Allen SE (ed) *Chemical Analysis of Ecological Materials*, 2nd edn. Oxford: Blackwell Scientific Publications.

Element	Fusion type	Acid digestion[a] (as an alternative to fusion)	Sample size(g) (to give 100 mL solution)	Mineral soil ($\mu g\ g^{-1}$)	Organic soil ($\mu g\ g^{-1}$)	Soil extracts ($\mu g\ g^{-1}$)	Plant materials ($\mu g\ g^{-1}$)	Animal tissues ($\mu g\ g^{-1}$)	Rain water ($\mu g\ L^{-1}$)	Fresh water ($\mu g\ L^{-1}$)
Al	Na_2CO_3 or NaOH		0.1	1–12%	0.05–0.5%	10–200 (mg per 100 g)	0.01–0.1%	0.001–0.02%	2–100	100–2000
B	Na_2CO_3 + 30 mL water, then 20% v/v H_2SO_4 until melt dissolves		0.25	5–50	2–20	0.2–5	10–80	0.4–2	100	10–500
Ca	Any fusion method	HF–$HClO_4$	0.1 (less for calcareous soils)	0.5–2% (excl. calcareous soils)	0.1–0.5%	10–200 (mg per 100 g) (excl. calcareous soils)	0.3–2.5% (high in calcareous species)	0.03–0.3% (excl. bones)	0.1–3 mg L^{-1}	0.1–100 mg L^{-1}
Cl	Na_2CO_3 + $NaNO_3$ to improve flux Dissolve melt in water Adjust pH to 8 with 0.1 M H_2SO_4		1.0	0.004–0.08%	0.03–0.2%	1–40 (excl. saline soils)	0.04–0.4%	0.2–0.8%	1–25 mg L^{-1}	2–100 mg L^{-1}
Co	Any fusion method	HF–$HClO_4$	0.5	1–50	0.2–1	0.05–4	0.1–0.6	0.02–0.1	0.05–0.5	0.5–2.5
Cu	Na_2CO_3 or $KHSO_4$	HF–$HClO_4$	0.25	5–80	6–40	0.1–3	2.5–25	10–100	0.2–2	2–50
Fe	Na_2CO_3	HF–$HClO_4$	0.1	0.5–10%	0.02–0.5%	50–1000	40–500	100–400	5–150	50–1000
Mg	Any fusion method	HF–$HClO_4$	0.1	0.2–2%	0.05–0.3%	4–50 mg per 100 g	0.1–0.5%	0.05–0.2%	0.1–2.0 mg L^{-1}	0.5–20 mg L^{-1}
Mn	Any fusion method If solution is pink–brown add a few drops of H_2SO_3(5%w/v SO_2) and boil before dilution to volume	HF–$HClO_4$	0.1	200–2000	50–500	5–500	50–1000	5–50	0.4–3	1–80
Mo[b]			0.6 or >	0.5–10	0.05–0.5	0.01–0.2	0.1–0.8	0.03–0.3	0.05–0.3	0.1–2
N[c]			0.5–1.0	0.1–0.5%	0.5–1.5%	NH_4^+-N 0.2–3 mg per 100 g NO_3-N 0.1–2 mg per 100 g	1–3%	4–10%	NH_4^+-N 0.1–0.8 mg L^{-1} NO_3-N 0.05–0.4 mg L^{-1}	NH_4^+-N 0.1–2 mg L^{-1} NO_3-N 0.05–3 mg L^{-1}
P	Fusion followed by boiling with H_2SO_4 to convert into ortho-phosphate	HF–$HClO_4$ provided all HF removed or digest with HNO_3 (2 mL) + $HClO_4$(1 mL) + H_2SO_4(0.5 mL) or digest with H_2SO_4–H_2O_2 for 2 h	0.1 0.05 0.2	0.02–0.15%	0.01–0.2%	0.3–8 mg per 100 g	0.05–0.3%	0.3–4% (excl.bone)	2–50 g L^{-1}	5–500 g L^{-1}

Element	Comments	Dissolution								
K	Any fusion method; Fusion reagents may give high blanks	HF–$HClO_4$	0.1	0.3–2%	0.02–0.2%	5–50 mg per 100 g	0.5–5%	0.3–1.5%	0.1–1 mg L^{-1}	0.5–8 mg L^{-1}
Si	NaOH	HF–$HClO_4$	0.1	20–60%	0.2–2%		0.05–0.5%	0.05–0.3%	0.01–0.1	
Na		HF–$HClO_4$	0.1	0.1–2%	0.02–0.1%	2–20 mg per 100 g	0.02–0.3%	0.2–1.0%	0.5–15 mg L^{-1}	2–100 mg L^{-1}
S	Mix soil with 2.5 g Na_2CO_3, preheat to 450°C for 30 min before fusion; Addition of 0.2 g $NaNO_3$ or 0.1 g Na_2O_2 improves the flux		0.5	0.03–0.3%	0.03–0.4%	100–500	0.08–0.5	0.2–0.8%	0.4–4 mg L^{-1} as SO_4^--S	2–150 mg L^{-1} as SO_4^--S
Ti	Na_2CO_3; Not in presence of high soil P; Evaporate solution obtained and add 30 ml 50% v/v H_2SO_4. Evaporate to white fume stage. Add 10 ml 50% v/v H_2SO_4 and filter		0.5	0.2–1.5%	0.01–0.08%	0.1–2	0.4–8	0.01–0.1	0.03–0.2	0.4–4
Zn	Any fusion method	HF–$HClO_4$	0.1	20–300	10–50	1–40	15–100	100–300	1–15	5–50
Cr[d]						10–200	0.05–0.5	0.01–0.3		0.1–0.5
As[d]						0.5–30	0.1–1	0.1–0.5 (accumulates in certain organisms)		0.2–1
Pb[d]						2–20	0.05–3	0.1–3		1–20
Hg[e]						0.1–1	0.005–0.1	0.03–0.3		0.3–3

Note: for the naturally occurring levels all measurements are on a dry weight ($\mu g\ g^{-1}$) basis unless otherwise stated. Dissolution methods (fusion and acid digestion) for the determination of total element concentrations in soil are shown.
a Perchloric acid is explosive if handled incorrectly. Consult safety manuals for information on correct handling.
b This element presents difficulty in determination. Consult specialist texts for methodology.
c Various chemical methods for the determination of NH_3-N, NO_3-N, NO_2-N and tot. N exist. Consult specialist texts.
d Various methods exists. Consult specialist tests for methodology.
e Fusion/dry ashing would cause volatilization of Hg, other methods preferred. Consult specialist texts for methodology.

Table 5 Summary of flame techniques applicable to the analysis of ecological materials. With the permission of Blackwell Scientific Publications, data based on Allen SE (1989) In: Allen SE (ed) *Chemical Analysis of Ecological Materials.* 2nd edn. Oxford: Blackwell Scientific Publications.

Element	Wavelength (nm)	Mode	Flame	Suitable ranges (ppm)	Potential interferences	Control of interference	Comments
Al	309.3	A	N_2O–C_2H_2	0–20	Fe, Ti, Ca, HOAc	Include Fe in standards	Very hot flame essential EA possible
Ca	422.7	E	Air–C_2H_2 O_2–C_2H_2	0–40	Al, P, Si, S	Add releasing agent (Sr or La salt)	A more sensitive than E
		A	Air–C_2H_2	0–10 0–40	As above	As above	
		A	N_2O–C_2H_2	0–40	Na, K	Include ionization buffer	
Co	240.7	A	Air–C_2H_2	0–50	Fe if high	Extract	Prior concn normally reqd.
		EA		0–1			EA preferable
Cr	357.9	EA	–	0–0.1	Ca, Fe	Standard compensation	EA usually essential
Cu	324.8	A	Air–C_2H_2	0–5	None significant		Prior concn desirable EA preferable
Fe	248.3	A	Air–C_2H_2	0–20	Si	Add $CaCl_2$	
Hg	253.7	A	Flameless	0–1	None		Hg vapour is produced which is passed into an abs cell in the light path
Mg	285.2	A	Air–C_2H_2	0–3	Ca, Al, P, Si	Add releasing agent (Sr or La salt)	
Mn	279.5	A	Air–C_2H_2	0–2	None significant		Samples and standards to be in same valency state
				0–20			
Mo	313.3	A	Air–C_2H_2	0–3			Prior concn with organic extraction
K	766.5		Air–C_2H_2 Air–propane	0–10 0–100	Na, Ca, H_2SO_4, HCl	Include in standards Dilute sample	A has no marked advantage over E
Na	589.0		Air–C_2H_2 Air–propane	0–5 0–20	Ca, H_2SO_4, HCl	Include in standards Dilute sample	E preferable
Ni	232.0	EA	–	0–0.2	None serious		EA essential
Pb	217.0	EA	–	0–0.5	None serious, Cl possible		EA desirable
		A	Air–C_2H_2	0–10			Prior concn necessary
Zn	213.9	A	Air–C_2H_2	0–1	None significant		

E=emission, A=atomic absorption, EA=electrothermal atomization.

Before analysis soil may be split into different levels, most metals are associated with the organic fraction in the active root zone 10–15 cm down. Soil should be air-dried for several days (~7) and then ground in a mortar and pestle and passed through stainless steel (2 mm) and 1 mm mesh sieves. After thorough mixing, soils can be stored in polythene bags until analysed.

Available nutrient elements are extracted from ~2 g subsamples which are treated with ammonium acetate (100 mL) and placed on a mechanical shaker for 1 h and then the resulting solution is gravity filftered. Concentrated HCl can be added (2–3 drops) to decrease the bacterial activity in the sample. Lanthanum (as lanthanum chloride at 1000 µg mL^{-1}) is added to the filtrate so as to prevent the interference of phosphate in the determination of Ca and Mg. Soil element concentrations are usually determined by F-AAS. Certain micronutrients such as Mo occur at levels near the limits of detection of F-AAS and are best determined by GF-AAS. Naturally occurring levels of Cr, Pb and Ni are also best determined by the more sensitive GF-AAS. Only chromium absorbance experiences potential

Table 6 Sample instrumental setting for analyses in F-AAS (A) and GF-AAS (B)

(A) F-AAS Settings for the measurement of Cu, Cr and As in CCA treated wood preservatives (BS 4072:1974)

Parameter	Copper	Chromium	Arsenic
Wavelength (nm)	324.8	357.9 or 429.0	193.7 or 197.2
Slit width (mm)	0.08	0.06 to 0.10	0.30
Lamp current (mA)	4	8	7
Scale expansion	up to $\times 10$	up to $\times 10$	up to $\times 10$
Burner	10 cm acetylene	10 cm acetylene	10 cm propane
Burner height (cm)	1.0	0.5	1.4
Acetylene flow rate (P of 0.7 kg cm^{-2})/cm^{-3} min^{-1}	1000	1800	–
Air flow rate (P of 2.1 kg cm^{-2})/L min^{-1}	5	5	–
Hydrogen flow rate (P of 0.7 kg cm^{-2})/cm^{-3} min^{-1}	–	–	1800
Argon flow rate (P of 2.1 kg cm^{-2})/L min^{-1}	–	–	5

(B) GF-AAS Settings for the determination of aluminium in conifer foliage. Reproduced from Lee CE, Cox JM, Foster DM, Humphrey HL, Woosley RS and Butcher D (1997) Determination of aluminium, calcium, and magnesium in Fraser fir (*Abies fraseri*) foliage from five native sites by atomic absorption spectrometry: the effect of elevation upon nutritional status. *Microchemical Journal* **56**: 236–246.

Analytical wavelength: 394.4 nm
Chemical modifier: magnesium nitrate (2 µg as Mg)
Bandpass: 0.40 nm
Background correction: continuum source

Furnace program

	Dry step	Char step	Atomization step	Clean step
Temperature (°C)	110	1100	2300	2500
Ramp (s)	20	20	0^a	1
Hold(s)	15	30	4	4
Purge gas flow	Low	Medium	None	High

These conditions represent improved optimization compared with conditions recommended in manufacturers manual for Al determination

^{a} Maximum power heating.

interference, from Ca and Fe, which is controlled by several means, including the addition of 2% ammonium chloride (NH_4Cl) to samples and standards.

Solid wood

The analysis of solid wood for elements (both naturally occurring and as preservative components) is a relatively straightforward determination using spectroscopic techniques. Some values for naturally occurring elements in hardwoods and softwoods are presented in **Table 7**, the levels of toxic metals reported for spruce are noteworthy. The methodology detailed here for the analysis of wood preserving elements is widely accepted.

Only Ca, Mg and phosphate are reported to occur in appreciable amounts in wood and experiments carried out in the presence of excess Ca, Mg and phosphate showed no interference in typical digest mixes [0.5 M sulfuric acid and 0.3% (w/v) sodium sulfate] in the determination of Cu, Cr, As and Zn. Phosphate

does not interfere with As determination, indicating that an insignificant amount of the phosphate present is ionic. Good extractives are not reported to cause interference in the determination of Cr and Cu and they can enhance the As signal unless the burner height is maintained between 1.3 and 1.5 cm (1.4 cm is frequently used). There is no interelement interference in the determination of Cu, Cr and As solutions. Preservative compounds are extracted from treated wood prepared as thin sections (0.2 mm thick), sawdust (produced by sawing a suitably representative part of the treated wood) or wood flour (prepared by hammer milling wood chips).

The decomposition step is carried out in a mixture of 0.5 M sulfuric acid and 100 volume hydrogen peroxide. The copper(II) sulfate, potassium/sodium dichromate and arsenic pentoxide in CCA preservatives undergo fixation to form potassium/sodium sulfate by-products. Any potential interference in the chromium and arsenic signals from these compounds in the leachate is overcome by an excess of sodium

Table 7 Naturally occurring metal ion concentrations in wood of various kinds; values on a mg per kg dry weight basis

Source	Na	Al	Fe	Mn	Zn	Cu	Pb	Cd	Cr	B
Spruce	76–133	3.2–8.3	1.6–3	32–271	5.7–22.9	0.7–1.3	0.3–1.5	0.2–0.5		
Birch	96	11	21	76						
Pine	35	16	24	76						
Hardwoods						0.2–2.1			0.2–4.6	1.5–7.2
Softwoods						0–3.0			0.2–4.8	0.8–6.8

sulfate (at 0.3% w/v) added to samples and calibration standards. Wood that has been submerged in water for a prolonged period can experience leaching of preservative salts and so the concentration of potassium/sodium salts in treated wood can be very variable.

The analytical sensitivity for arsenic has been improved from 0.60 μg mL^{-1} to 0.10 μg mL^{-1} by the use of an argon–hydrogen flame (instead of air–acetylene) and a 10 cm propane burner (instead of a 10 cm acetylene burner) analysed at a wavelength of 197.2 nm, and this flame is recommended in arsenic analysis. The argon–hydrogen combination is highly explosive and should only be used by trained personnel.

Standard curves can be prepared in the range of 5–35 μg mL^{-1} Cu and 10–70 μg mL^{-1} Cr and As for preservative concentrations in the region of 3% w/v. The acid decomposition described is suitable for samples in the range of 3–8 g.

The comminution of wood probably represents the most time-consuming part of the sample preparation procedure. The production of wood flour requires the tedious reduction of wooden samples to a small enough size (*ca.* matchstick) before feeding into a hammer mill. The hammer mill itself can be a cause of contamination in the form of Cr or Cu from the hammers or the chamber though this is probably insignificant compared with the levels of Cu and Cr in the treated wood. Though wood flour represents the maximum surface area to volume ratio for the action of the acid decomposition, the use of thin wood sections (0.2 mm thick) or sawdust is acceptable providing they are obtained from areas of treated wood deemed representative of the whole, i.e. not from the ends or surfaces, which can be over-treated with respect to the rest of the timber.

Water

The requirements of the analysis of water from various sources is one area of environmental research that is driving efforts to improve the limit of detection of spectroscopic analysis. Elements are frequently in very dilute concentrations and the presence of dissolved salts (chiefly sodium chloride and magnesium salts in marine samples) can offer some interference in sensitive analyses.

The pulping of hardwood and softwood during the manufacture of paper generates waste water that contains elements such as Na, Al, Mn, P, Fe, Mn, Zn, and metals such as Cd, Cu, Cr, Ni, Hg and Pb. These elements and metals are released in mill effluent to receiving waters and elevated concentrations can build up in fine sediments at some distance (up to 12 km) from the effluent source.

Water samples can be handled in several ways but an important precaution is to prevent the sorption of metals to sample bottles. This is achieved by the addition of 0.02 vol.% suprapure nitric acid. Both plasma AES and GF-AAS methods are suitable for the determination of the elements mentioned. Wavelengths (nm) considered more sensitive for elements in organic matrices analysed by DCP-AES include 568.82 (Na), 308.215 (Al), 259.94 (Fe), 403.076 (Mn) and 202.548 (Zn).

The determination of Cr, Ni and Pb require the addition of NH$_4$H$_2$PO$_4$ to samples for matrix modification. Most elements in pulp effluent occur in μg L^{-1} concentrations (As, Ba, Co, Li, Sr, Sn) while some, such as Na, Ca, Mg, Si and K (naturally present in wood or added to pulp during processing in various chemicals), occur in mg L^{-1} concentrations. Mass spectrometry may be employed for certain elements that are present in very low concentrations < 1 μg L^{-1} (Ag, Mo, Sb, Tl) and the combination of elements determined by this method offers a characterization of pulp and paper mill effluent in water. Mercury is determined by cold vapour AAS where the detection limit is ~0.1 μg-Hg L^{-1}.

Air

The concentration of metals in the air of wood-processing factories has recently become a more prominent health and safety issue and the analysis of metal-containing dust samples is carried out by AAS and AES methods.

For the analysis of metal elements in the air of factories that process wood which has been treated with wood preservatives (such as Cu, Cr and B) the finest dust particles are collected by suction onto glass fibre filter paper. The concentration of dust (in mg m^{-3} air) is calculated by dividing the mass of the dust (mass of filter paper after filtering minus the mass of filter paper before filtering) by the volume of air filtered.

Elements are determined in dust by first drying the dust sample over silica gel. Fine dust samples are very mobile and difficulties may be experienced transferring samples. A dust sample is leached in 20 mL HNO$_3$ (1 M) mixed with 1 mL H$_2$O$_2$ (30%) in an Erlenmeyer flask and shaken on a mechanical shaker for 20 h. After filtration, Cu and Cr determinations may be made in F-AAS as described above.

Hexavalent chromium (the more toxic form of Cr) can be determined separately by extraction from dried dust with diphenylcarbazide followed by photometric analysis. Boron is extracted from silica gel dried dust by leaching in the HNO$_3$ (1 M) mix described above. After 20 h in the solution on a mechanical shaker the sample is filtered, extracted with 2-ethylhexa-1,3-diol and curcumin and analysed by photometry.

To date, research has shown that in less than half of the workplaces examined the concentration of dust in workplace air was above the recommended threshold (2 mg dust per m^{-3} of air), and in fewer than a third of cases the levels were five times the recommended threshold value. It has also been shown that the concentration of metals in the dust did not have its origin exclusively in preservative treated wood but that there was a significant load of metals from external air. The concentration of hexavalent chromium in dust was in line with natural levels in air and well below the safe threshold levels of 50 000 ng m^{-3}.

Future developments

Analytical improvements in atomic spectroscopy are most likely to affect the analysis of water and air samples related to environmental impacts of wood-products and the health and safety aspects of working with wood and wood preservatives. Atomic absorption spectrometry is now such a thoroughly tried and tested analytical technique that regular major advances are unlikely.

In spite of the advent of ICP based techniques, F-AAS and GF-AAS continue to be used routinely in many laboratories for the analysis of waters, soils and plant material. As might be expected, most of the significant developments in the AES field have been associated with ICP-AES, and especially with aspects of sample introduction to the ICP, though reports of novel developments in wood related research are not plentiful.

See also: **Atomic Spectroscopy, Historical Perspective; Environmental and Agricultural Applications of Atomic Spectroscopy; Inductively Coupled Plasma Mass Spectrometry, Methods; Quantitative Analysis; X-Ray Fluorescence Spectroscopy, Applications.**

Further reading

Allen SE (1989) In: Allen SE (ed) *Chemical Analysis of Ecological Materials*, 2nd edn. Oxford: Blackwell Scientific Publications.

Holmbom B, Harju L, Lindholm J and Gröning A-L (1994) Effect of a pulp and paper mill on metal concentrations in the receiving lake system. *Aqua Fennica* 24: 93–100.

Lambert M (1969) The determination of copper, chromium and arsenic in preservative treated timber by the method of atomic absorption spectrophotometry. *Journal of the Institute of Wood Science* 4: 27–36.

Lee CE, Cox JM, Foster DM, Humphrey HL, Woosley RS and Butcher D (1997) Determination of aluminium, calcium, and magnesium in Fraser fir (*Abies fraseri*) foliage from five native sites by atomic absorption spectrometry: the effect of elevation upon nutritional status. *Microchemical Journal* 56: 236–246.

Williams AI (1972) The use of atomic absorption spectrophotometry for the determination of copper, chromium and arsenic in preserved wood. *Analyst* 97: 104–110.

Fourier Transformation and Sampling Theory

Raúl Curbelo, Bio-Rad, Cambridge, MA, USA

Copyright © 1999 Academic Press

FUNDAMENTALS OF
SPECTROSCOPY
Theory

In Fourier transform spectroscopy (FTS), the output of the detector can be represented as a value that is a function of the independent variable, usually time. For a given finite measurement time T, it can be considered a transient signal that lasts T seconds. This transient is the interferogram in FT-IR or the free induction decay in FT-NMR. In either case, this transient function contains all the spectral information being measured, but the spectral amplitude for each frequency is encoded in the detector signal and in general is not discernible directly. The Fourier transform is a mathematical operation that can be used to change this function of time into an amplitude that is a function of frequency, which is a representation of the desired spectrum.

Signal processing has always been a critical aspect in spectroscopy and especially in FTS. The generalized use of computers as components in spectrometers to implement the Fourier transform and/or other digital signal processing (DSP) tasks requires, as a first step, that the signals used be discrete amplitude, discrete time-sampled representations of continuous amplitude, continuous time (analogue) signals, such as the output of a detector. The sampling process must meet certain minimum requirements for the resulting sample sequence to contain all the information of the original signal. The output of the sampling process is a continuous amplitude, discrete time representation of the original signal. To obtain a discrete amplitude sequence, the amplitude of each sample is quantized with finite precision by a digital to analogue converter (ADC) and is then represented by a number. The resulting sequence of numbers is the input to the DSP.

The Fourier series

In the artificial case that the spectrometer measures the same sample multiple times, the successive transients will be identical, and the entire sequence will be a periodic signal of period T.

The Fourier series provides a representation of the signal as a sum of sinusoidal components of different frequencies. For a periodic signal of period T it is easy to visualize that the component frequencies will be $f_k = k/T$, where $k = 0, 1, 2, 3 \ldots$, because these sinusoids are the only ones periodic in T. These components are called harmonics of the signal. The sinusoid of frequency f_k is the kth harmonic.

Sinusoids are used to study and represent signals in linear time invariant systems because an input sinusoid generates an output from a linear time invariant system that is a sinusoid of the same frequency. (See Mason and Zimmerman in the Further reading section). The Fourier series representation of a periodic signal $v(t)$ with period T is

$$v(t) = A_0 + \sum_{k=1}^{\infty} \{A_k \cos(2\pi k/T)t + B_k \sin(2\pi k/T)t\} \quad [1]$$

for $k = 1 \ldots \infty$, and $-\infty < t < \infty$; $2\pi k/T$ is the radian or angular frequency of the kth harmonic and $A_k \cos(2\pi k/T)t$ and $B_k \sin(2\pi k/T)t$ are the harmonic components of $v(t)$. A_k is the amplitude of the cosine component of frequency k/T. A_0 is the average of $v(t)$ over a period T, and A_k and B_k are the averages of $2v(t)\cos(2\pi k/T)t$ and $2v(t)\sin(2\pi k/T)t$ respectively, also over a period T.

For an even function of t, that is $v_e(t) = v_e(-t)$ the above simplifies to

$$v_e(t) = A_0 + \sum_{k=1}^{\infty} A_k \cos(2\pi k/T)t \quad [2]$$

For an odd function of t, that is $v_o(t) = -v_o(-t)$ we have

$$v_o(t) = \sum_{k=1}^{\infty} B_k \sin(2\pi k/T)t \quad [3]$$

The Fourier series can be written in a more compact form with the use of the complex exponential function $\exp(i\omega t) = \cos(\omega t) + i \sin(\omega t)$, where ω is the radian frequency $2\pi f$, and $\omega_0 = 2\pi/T$. Then

$$v(t) = \sum_{k=-\infty}^{\infty} V_k \exp(ik\omega_0 t) \quad [4]$$

where V_k is the average of $v(t)\exp(-ik\omega_0 t)$ over the period T:

$$V_k = 1/T \int_{-T/2}^{T/2} v(t) \exp(-ik\omega_0 t)\, dt \qquad [5]$$

V_k has the dimension of voltage, and in the general case will be complex. This expression leads to the Fourier transform, also called the Fourier integral. If we assume $v(t)$ is periodic with period T, and different than zero for $|t| < T/2$, we can define

$$V(k\omega_0) = \int_{-T/2}^{T/2} v(t) \exp(-ik\omega_0 t)\, dt \qquad [6]$$

$V(k\omega_0)$ has dimensions of voltage per unit frequency. It is independent of T, and does not change when T increases, while the Fourier coefficients V_k become smaller as T increases.

$$V_k = (1/T)V(k\omega_0) = (\omega_0/2\pi)V(k\omega_0) \qquad [7]$$

$$v(t) = (1/T) \sum_{k=-\infty}^{\infty} V(k\omega_0) \exp(ik\omega_0 t)$$

$$= \sum_{k=-\infty}^{\infty} V(k\omega_0) \exp(ik\omega_0 t)(\omega_0/2\pi) \qquad [8]$$

For T approaching infinity, ω_0 tends to zero, and if we define $\omega = k\omega_0$, in the limit

$$v(t) = \frac{1}{2\pi} \int_{-\infty}^{\infty} V(\omega) \exp(i\omega t)\, d\omega \qquad [9]$$

and

$$V(\omega) = \int_{-\infty}^{\infty} v(t) \exp(-i\omega t)\, dt \qquad [10]$$

where $V(\omega)$ is the Fourier transform of $v(t)$. The term $V(\omega)$ is also called the frequency spectrum of $v(t)$ or simply the spectrum. This is the Fourier representation of a single pulse, because we defined it by selecting one pulse and moving all others in the sequence to infinity.

In the case of the periodic signal, the coefficients of the Fourier series have the dimension of voltage. The Fourier transform of a pulse $v(t)$, $V(\omega)$, has the dimensions of voltage multiplied by time, and we can think of it as a voltage density spectrum with a dimension of voltage per unit frequency. Therefore, the area under $V(\omega)$ has the dimension of voltage.

Table 1 lists some of the properties of the Fourier transform. The functions $v(t)$ and $u(t)$ are defined for $-\infty < t < \infty$, and their respective Fourier transforms $V(\omega)$ and $U(\omega)$ are defined for $-\infty < \omega < \infty$. In property 6 of **Table 1** the convolution is defined for $-\infty < \xi < \infty$, where ξ is the angular frequency. It is used here to distinguish it from the independent variable ω of the result.

Discrete time signals and sampling

In the development of the Fourier transform above we have assumed that the signal $v(t)$ was a continuous time signal defined for $-\infty < t < \infty$. These signals are also commonly called analogue signals.

Discrete time signals may be generated by a discrete time process or may be the result of a sampling process on a continuous time signal.

The common method to obtain a discrete representation of a continuous time signal is to take periodic samples of the signal, and create a sequence $v[n]$ from the continuous time signal $v_c(t)$ defined by the equality

$$v[n] = v_c(nT_s) \qquad \text{for} \quad -\infty < n < \infty \qquad [11]$$

Table 1 Some properties of the Fourier transform

Function		Fourier transform	
$v(t)$	$-\infty < t < \infty$	$V(\omega)$	$-\infty < \omega < \infty$
$u(t)$		$U(\omega)$	
1 $A\,u(t) + B\,v(t)$		$A\,U(\omega) + B\,V(\omega)$	
2 $v(-t)$		$V(-\omega)$	
3 $v(t/a)$		$a\,V(a\omega)$	for a real positive
4 $v(t-t_0)$		$V(\omega) \exp(-i\omega t_0)$	
5 $dv(t)/dt$		$i\omega\,V(\omega)$	
5 $\int_{-\infty}^{\infty} v(t)u(\tau - t)\, dt$		$V(\omega)U(\omega)$	
6 $v(t)u(t)$		$\frac{1}{2\pi} \int_{-\infty}^{\infty} V(\xi)U(\omega - \xi)\, d\xi = W(\omega)$	
7 $\int_{-\infty}^{\infty} |v(t)|^2\, dt$		$\int_{-\infty}^{\infty} |V(\omega)|^2\, d\omega$	

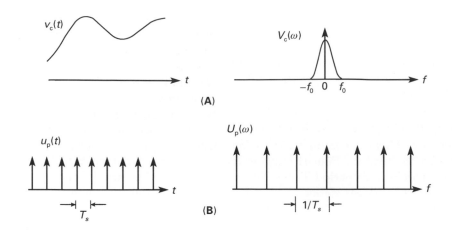

Figure 1   Sampling the continuous signal with a periodic impulse train.

where T_s is the sampling period, and its inverse, $f_s = 1/T_s$, is the sampling frequency, in samples per second.

The sampling operation is not invertible in general, that is given $v[n]$, it is not possible in general to reconstruct $v_c(t)$ because many continuous signals can be represented by $v[n]$. The sampling theorem due to Nyquist removes this uncertainty by imposing a restriction on the signal $v_c(t)$.

Figure 1A shows a portion of the continuous time signal $v_c(t)$ defined for $-\infty < t < \infty$, and the Fourier transform of $v_c(t)$, the spectrum $V_c(\omega)$. We have chosen $v_c(t)$ such that $V_c(\omega) = 0$ for $|f| > f_0$. ($|\omega| > 2\pi f_0$). **Figure 1B** shows a portion of the periodic train of unit impulses $u_p(t)$, defined for $-\infty < t < \infty$, and a portion of the Fourier transform of $u_p(t)$, the spectrum $U_p(\omega)$, $-\infty < f < \infty$.

It is convenient to represent the sampling process in two steps, as shown in **Figure 2**. First the multi-plication of $v_c(t)$ with a periodic train of unit impulses $u_p(t)$, which results in the train of impulses $v_s(t)$ for $-\infty < t < \infty$, and second, the conversion of the train of impulses to a sequence of samples $v[n]$.

The product of the continuous signal $v_c(t)$ with the periodic train of impulses generates the train of impulses $v_s(t)$ that is zero except at integer multiples of T_s.

Figure 2A shows a portion of the train of impulses $v_s(t)$ that has a spectrum $V_s(\omega)$ for $-\infty < \omega < \infty$, where ω is the radian frequency $2\pi f$. This spectrum can be obtained using property 6 of the Fourier transform in **Table 1**, which shows that the transform of the product of two functions is the convolution of the transforms. The convolution of $V_c(\omega)$ with the periodic train of impulses $U_p(\omega)$ of **Figure 1** results in periodically repeated copies of $V_c(\omega)$, separated by $1/T_s$, a portion of which is shown in **Figure 2A**.

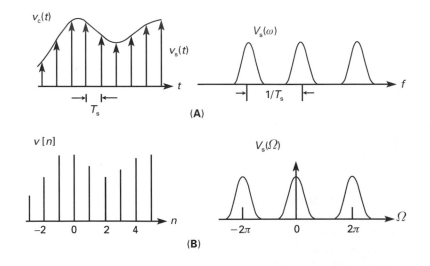

Figure 2   Impulse train converted into a discrete time sequence.

Sampling theorem

The sampling theorem due to Nyquist (see Mason and Zimmermann in the Further reading section) can be visualized from **Figure 2A**. The sampling frequency, $f_s = 1/T_s$, must be at least twice the highest frequency f_0 contained in the signal, otherwise the spectrum of $V_s(\omega)$ will be corrupted by the overlap of adjacent replicas of $V_c(\omega)$ and it will not permit the recovery of the original signal. When the above condition is met, the spectrum of $V_s(\omega)$ contains all the information of the original spectrum $V_c(\omega)$, therefore the original signal is recoverable from $V_s(\omega)$, for example by passing the sampled signal through a low pass filter that rejects frequencies above f_0 but passes all the frequency components below f_0.

Sampling at a rate higher than $1/T_s$ has the undesirable effect of increasing the computational cost of any digital signal process that follows. One exception is the case of the recovery of the original signal with a simple filter. The trade off between sampling rate and filter complexity is more common in the case of narrow band signals and analogue filters.

The discrete sequence $v[n]$ for $-\infty < n < \infty$ in **Figure 2B** is indexed on the integer variable n, which does not have information about the sampling rate and represents the samples of $v_c(t)$ with finite numbers instead of the areas of impulses as in $v_s(t)$. The spectrum of $v[n]$ is $V_s(\Omega)$ for $-\infty < \Omega < \infty$, the discrete Fourier transform of $v[n]$, where Ω is in radians.

Discrete time signals and the Fourier transform

A sequence that represents a sinusoid has the general form for $-\infty < n < \infty$:

$$v[n] = A\exp[i(\Omega_0 n + \phi)] \qquad [12]$$

By analogy with the continuous time case, Ω_0 is called the frequency of the complex sinusoid, and ϕ is called the phase. Here n is a dimensionless integer number. Therefore, the dimension of Ω_0 is radians. One can specify the units of Ω_0 in radians per sample, for a better analogy with the continuous time case.

A more important difference with the continuous time case becomes apparent when we consider the frequency $(\Omega_0 + 2\pi)$. In this case

$$\begin{aligned}
v[n] &= A\exp[i(\Omega_0 + 2\pi)n] \\
&= A\exp(i\Omega_0 n)\exp(i2\pi n) \\
&= A\exp(i\Omega_0 n)
\end{aligned} \qquad [13]$$

It is easy to see that sinusoids and complex exponential sequences that differ in frequency by a multiple of 2π are identical. For these sequences, we only need to consider frequencies in an interval of length 2π, such as $-\pi$ to π, or 0 to 2π.

Discrete time periodic sequences are also different from continuous periodic signals in the behaviour of the frequency of the harmonics. A periodic discrete time sequence can be defined as

$$v[n] = v[n + N] \qquad [14]$$

for $-\infty < n < \infty$, where the period N has to be an integer. A periodic discrete time sinusoid sequence,

$$A\cos(\Omega_0 n + \phi) = A\cos(\Omega_0 n + \Omega_0 N + \phi) \qquad [15]$$

requires that $\Omega_0 N$ must be a multiple of 2π. Therefore

$$\Omega_0 N = 2\pi m \qquad [16]$$

where m is an integer. The same result applies to complex exponential sequences.

A further conclusion from Equations [13] and [16] is that, for a periodic sequence of period N, the only frequencies that can be different are

$$\Omega_m = 2\pi m/N, \text{ for } m = 0, 1, \cdots, N - 1 \qquad [17]$$

Furthermore, low and high frequencies appear differently than in continuous signals. In the discrete time sinusoid $v[n] = A\cos(\Omega_0 n + \phi)$ as Ω_0 increases from 0 to π, $v[n]$ cycles faster, but as Ω_0 increases from π to 2π, $v[n]$ cycles slower and slower, and the sequence for $\Omega_0 = 2\pi$ is the same as the one for $\Omega_0 = 0$.

In general we can define the discrete Fourier transform of the sequence $v[n]$ as

$$V(\Omega) = \sum_{n=-\infty}^{\infty} v[n]\exp(-i\Omega n) \qquad [18]$$

that is valid whenever $v[n]$ meets

$$\sum_{n=-\infty}^{\infty} |v[n]| < \infty \qquad [19]$$

Table 2 Some properties of the discrete Fourier transform

Function		Discrete Fourier transform	
$v(n)$	$-\infty < n < \infty$	$V(\Omega)$	$-\pi < \Omega < \pi$
$u(n)$		$U(\Omega)$	
1 $a\,v[n] + b\,u[n]$		$a\,V(\Omega) + b\,U(\Omega)$	
2 $v[n-m]$ (m an integer)		$\exp(-i\Omega m)\,V(\Omega)$	
3 $\exp(i\Omega_0 n)$		$V(\Omega - \Omega_0)$	
4 $nv[n]$		$i\,dV(\Omega)/d\Omega$	
5 $v[n].u[n]$		$\frac{1}{2\pi}\int_{-\pi}^{\pi} V(\theta)U(\Omega - \theta)\,d\theta$	
6 $\displaystyle\sum_{n=-\infty}^{\infty} |v[n]|^2$		$\frac{1}{2\pi}\int_{-\pi}^{\pi} |V(\Omega)|^2\,d(\Omega)$	

Using similar reasoning to that used for the continuous time case, we can write

$$v[n] = \frac{1}{2\pi} \int_{-\pi}^{\pi} V(\Omega)\,\exp(i\Omega n)\,d\Omega \qquad [20]$$

Equations [18] and [20] form a discrete Fourier transform pair that relate the discrete sample space with the frequency space.

Table 2 lists some of the properties of the discrete Fourier transform. The functions $v[n]$ and $u[n]$ are defined for $-\infty < n < \infty$, and their respective discrete Fourier transforms $V(\Omega)$ and $U(\Omega)$ are defined for $-\pi < \Omega < \pi$. In property 5 of **Table 2** the convolution is defined for $-\pi < \theta < \pi$ where θ is the angular frequency. It is used here to distinguish it from the independent variable Ω of the result.

List of symbols

A_0 = average of $v(t)$ over T; A_k = amplitude of cosine components of frequency k/T; f_k = sinusoid frequency (kth harmonic); T = signal time period; $v_c(t)$ = continuous time signal; $v_c(\pm)$ = continuous time signal; $v[n]$ = discrete sequence; $v(t)$ = periodic signal; $V(\omega)$ = Fourier transform of $v(t)$; θ, ξ = angular frequency; ϕ = phase of complex sinusoid; ω = the radian frequency $2\pi f$; Ω_0 = frequency of a discrete sinusoid.

See also: **FT-Raman Spectroscopy, Applications; Raman and IR Microspectroscopy.**

Further reading

Bracewell RN (1986) *The Fourier Transform and its Applications*, 2nd edn. New York: McGraw-Hill.

Carslow HS (1952) *Introduction to the Theory of Fourier's Series and Integrals*, 3rd edn. New York: Dover Publications.

Dym H and McKean HP (1972) *Fourier Series and Integrals*. New York: Academic Press.

Jerry AJ (1997) The Shannon sampling theorem – its various extensions and applications: a tutorial review. *Proceedings of the Institute of Electrical and Electronic Engineers* 65: 1565–1595.

Mason S and Zimmermann HJ (1960) *Electronic Circuits, Signals and Systems*. New York: John Wiley and Sons.

Nyquist H (1928) Certain topics in telegraph transmission theory. *American Institute of Electrical Engineers Transactions* 617–644.

Oppenenheim AV and Schafer RW (1975) *Digital Signal Processing*. Englewood Cliffs, NJ: Prentice-Hall.

Oppenenheim AV and Schafer RW (1989) *Discrete-Time Signal Processing*. Englewood Cliffs, NJ: Prentice-Hall.

Papoulis A (1962) *The Fourier Integral and its Applications*. New York: McGraw-Hill.

Shannon CE (1949) Communications in the presence of noise. *Proceedings of the Institute of Radio Engineers* 37: 10–21.

Fragmentation in Mass Spectrometry

Hans-Friedrich Grützmacher, Universität Bielefeld, Germany

MASS SPECTROMETRY
Theory

Fragmentation reactions of radical cations

Organic molecules are usually even-electron species with spin paired valence electrons. The one-electron oxidation of these molecules, which is performed by oxidation with certain transition metal salts, by electrochemical oxidation, by photoinduced single-electron transfer (PET), by ionization using high energy radiation or electron impact, results in odd-electron radical cations. Sometimes these organic radical cations form stable salts which can be isolated; a well-known example is 'Wurster's red', the bromide of the N,N-dimethyl-1,4-diaminobenzene radical cation. Mostly, however, organic radical cations are highly reactive, and the study of radical cations as reactive intermediates of organic reactions is an important and active field of research both with respect to reaction mechanism and to synthetic application. In the context of mass spectrometry, the radical cations are generated by electron impact ionization (EI) or by photoionization (PI) in the ion source of the mass spectrometer as energetically excited molecular ions with a rather broad range of excess energy. Because of the low pressure (usually $< 10^{-6}$ mbar) and the short residence time of the ions within the ion source deactivation of the excited molecular ions by collision does not occur and radiative deactivation is too slow as described by a review by Dunbar (see Further reading section). Consequently the excess energy of the radical ions created by EI or PI is used for unimolecular chemical reactions, which eventually results in the formation of fragment ions. The surviving molecular ions and the fragment ions formed constitute the mass spectrum of the compound analysed. This illustrates that 'obtaining a mass spectrum' is principally a kinetic experiment, and kinetics laws postulate that in the case of parallel reactions of the same reaction order, which is inherently first order in the case of mass spectrometric fragmentations, the relative concentrations (intensities I_i) of the individual products match the ratio of the corresponding rate constants k_i (**Scheme 1**).

Therefore, a detailed interpretation of a mass spectrum both for understanding the reactivity of organic radical cations and for a structure analysis of the compound studied requires the knowledge of the rate constant k_i of the individual fragmentation steps and the dependence of k_i on the structure of the reactants and their excess energy.

The unimolecular fragmentation of excited radical cations in a mass spectrometer refers to the general problem of the reaction kinetics of the decomposition of energized species isolated from their environment. The Rice, Ramsperger, Kassel, Marcus (RRKM) theory, the quasi-equilibrium theory (QET) of Rosenstock and Wahrhaftig, and related 'statistical' theories have been developed to deal with this problem. These theories are discussed in detail in the other articles. Using certain prerequisites they result in Equation [1], showing the dependence of the rate constant k_i on the excess energy E, the number s of vibrational degrees of freedom of the reactant ion, the activation energy E_{i0} and the so-called frequency factor ν_i of fragmentation i.

$$k_i(E) = \nu_i \times \left(\frac{E - E_{i0}}{E} \right)^{s-1} \qquad [1]$$

While the oversimplifications used in deriving Equation [1] prevent any genuine calculation of $k_i(E)$ it is useful to demonstrate the factors that influence k_i, and its relation to reaction kinetics in a condensed phase at thermal equilibrium. The activation energy E_{i0} has the same meaning as in condensed phase kinetics, and in the case of cleavage of only one bond of the reactant radical cation during the fragmentation process, E_{i0} corresponds to the enthalpy of reaction ΔH_r. This approximation is valid

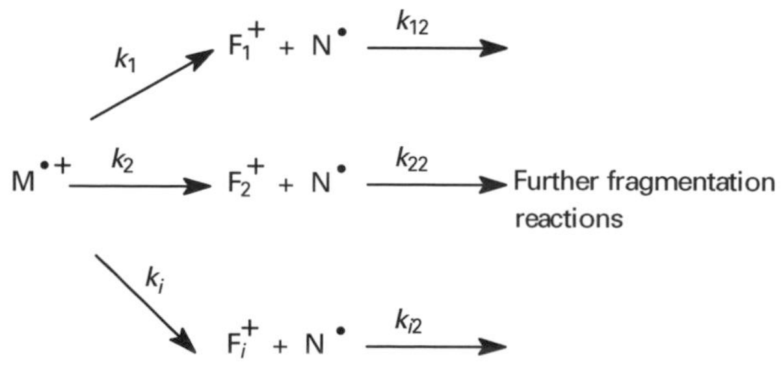

Scheme 1

because the reverse process of the fragmentation, i.e. bond formation between the ionized fragment (a cation) and the neutral fragment (a radical), requires no or only a very small activation energy (the so-called 'reverse' activation energy). If the excess energy E of the reactant ion (the 'parent ion') is very large ($E \gg E_{i0}$) the rate constant k_i approaches the limiting value of ν. The frequency factor ν reflects the frequency of that vibration of the reactant ion which is transformed into translation along the reaction coordinate of the fragmentation. The frequency factor ν is large if the transition corresponds simply to the elongation of the bond to be broken. The situation is more complicated in the case of a fragmentation by a rearrangement of the reactant radical cation which requires the making of new chemical bonds in a particular conformation as well as bond cleavage. As a consequence of the required conformation of the transition state the frequency factors ν of fragmentations by rearrangement are smaller than those of direct bond cleavages, and a rearrangement can compete successfully for the decomposition of the excited reactant ion only if the activation energy E_0 is small. Generally, a small activation energy is expected for an energetically favourable fragmentation that yields stable ionized and neutral fragments, if any extra reverse activation energy can be neglected.

Applying these general considerations to the fragmentation of organic molecules in a mass spectrometer one expects three types of fragmentation reactions. Fragmentations of type I correspond to the direct cleavage of a weak bond of the reactant ion. Since this reaction has a large frequency factor and needs only a small activation energy, the rate constant will be large for any energy regime of the parent ion and the parent ion will always decompose preferably by this pathway. The result is a 'one-peak mass spectrum' that exhibits only the intense signal of the respective fragment ion in addition to a weak peak of the parent ion. Examples of type I fragmentations are found in the mass spectra of many organo-element compounds, which decompose by cleavage of the weak element–carbon bond, and compounds such as N,N-dimethyl-β-phenethyl amine (**Scheme 2**), which fragments into a stable ion $(CH_3)_2N^+=CH_2$ and a stable benzyl radical $C_6H_5CH_2{}^\bullet$ on electron impact ionization, or β-amino ethanol, which fragments into a stable ion $H_2N^+=CH_2$ and a stable ${}^\bullet CH_2OH$ radical.

A type II fragmentation corresponds to decomposition of the parent ion via a rearrangement into a stable fragment ion and a stable neutral fragment. Because of the stable products the reaction enthalpy of this process is usually small, as in the case of the

type I fragmentation. However, the small frequency factor causes a slow rise of the rate constant with the excitation energy of the parent ion. Thus, this rearrangement process is the favoured decomposition pathway for parent ions of low excitation energy. A typical type II fragmentation process is the elimination of H_2O from the molecular ions of n-butanol and its higher homologues. Specific studies have shown that this process is mainly a 1,4-elimination and most likely proceeds as a two step process (**Scheme 3**) with a slow initial H-atom transfer to the hydroxy group.

A type III fragmentation corresponds to the cleavage of covalent bond compounds, e.g. a C–H or a C–C bond. Since these bonds are rather strong the activation energy of this process (comparable to the reaction enthalpy) is relatively large, but the cleavage is entropically favoured by a large frequency factor. Hence, reactant ions with plenty of excess energy follow this fragmentation pathway. In principle there could be a fourth type of fragmentation, an energetically unfavourable rearrangement process that exhibits a large activation barrier and a small frequency factor. Such a fragmentation will be slow for all energy regimes of the reactant ions and cannot compete with the other types of fragmentation. If a process of this type appears in the mass spectrum, there is either no other fragmentation path or a 'hidden' rearrangement of the reactant ion to another isomer precedes the fragmentation. Examples of the first situation are the mass spectra of benzene and polycyclic arenes, which show a distinct signal for the loss of a methyl radical from the parent ion. The only 'simple' fragmentation of these radical ions is the loss of an H atom, which, however, requires about 3–4 eV activation energy because of the strong C_{arene}–H bond. Therefore, the deep-seated rearrangement necessary for the elimination of a methyl radical is possible but most molecular ions of the arene do not fragment at all during their residence time within a

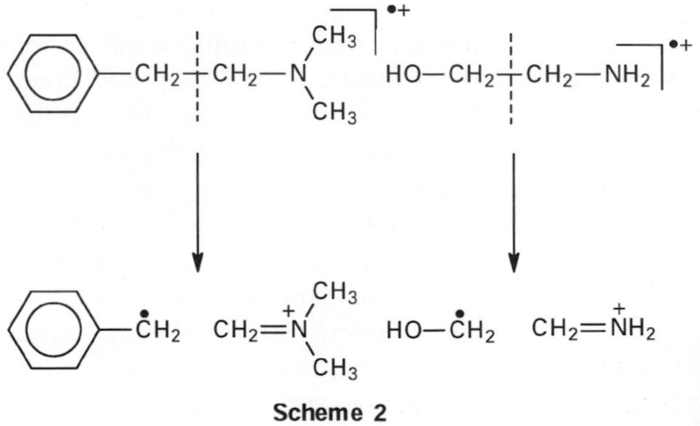

Scheme 2

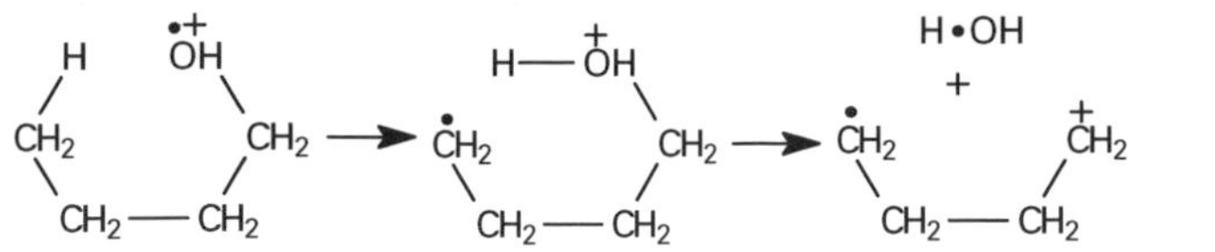

Scheme 3   (Or more stable isomer)

mass spectrometer. An example of the second situation is the loss of a methyl substituent from the molecular ions of xylene which would require cleavage of a strong C_{arene}–C bond and which should not compete effectively with loss of H to generate a stable methyl-benzyl cation (**Scheme 4**). An investigation of the fragmentation mechanism by Grotemeyer and Grützmacher showed (see the Further reading section) that a direct loss of the methyl substituent occurs only for highly excited molecular ions whereas most of the molecular ions rearrange either by H-migration or by a skeletal rearrangement into a methyl cycloheptatriene radical cation before elimination of the methyl group.

The peak pattern of the EI mass spectrum of a standard organic compound arises from competition between type II and type III fragmentations. This will be illustrated below by a few examples to familiarize the reader with the concepts used for the interpretation of the fragmentations of radical ions in a mass spectrometer. However, it has to be remembered that the peak pattern in the mass spectrum of a compound depends very much on the method used for the ionization and on the type of mass spectrometer and scan mode used to acquire the spectrum. The former determines the energy content of the parent ions created in the ion source of the instrument and the latter influences the reaction time allowed for the ions to decompose. If the ionization method chosen supplies only a relatively small amount of excess energy to the molecular ion, the rates of most fragmentations (apart from those of type I) are small and the mass spectrum will consist mainly of the peak for intact parent ions. At long reaction times (> 10 μs) peaks of the product ions of type II fragmentation will appear because of the small activation energies of such processes. 'Soft' ionization

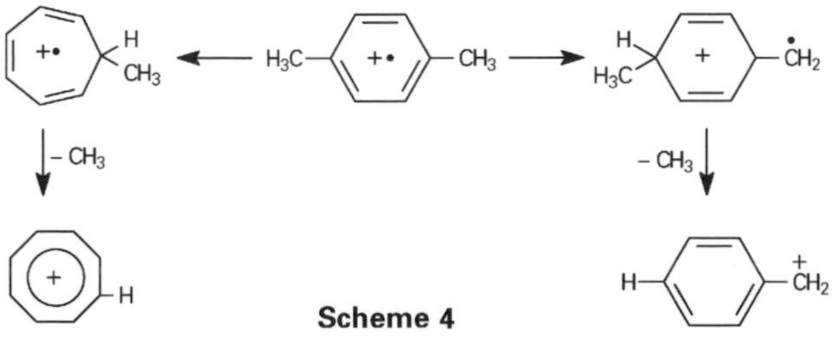

methods that produce this category of spectra are field ionization (FI), charge exchange (CE) ionization with a primary ion of low ionization energy (C_6F_6, CS_2), and multiphoton ionization (MUPI) as described by Grotemeyer and Schlag. In particular, MUPI can be used to steer the ionization into a 'soft' or a 'hard' mode by adjusting the number and energy of the photons absorbed by the molecules of the sample. Ionization of molecules is most often achieved by electron impact (EI) with electrons of 70 eV kinetic energy and is a 'hard' ionization process that produces parent ions having a broad distribution of excess energy up to several eV.

The type of instrument, ion source and scan mode used to obtain the mass spectrum define the reaction time during which the fragment ions that appear in the mass spectrum are generated. A routine sector field mass spectrometer with several kV ion accelerating voltage has a residence time for the ions within the ion source of < 1 μs; this time is extended slightly for a quadrupole mass spectrometer using lower accelerating voltages. All ions appearing in the mass spectrum recorded under standard conditions have to be formed within the ion source, and a short residence time is favourable for the observation of fragmentations with a large frequency factor, i.e. type I and type III fragmentations which originate from highly excited parent ions. However, with *tandem-mass spectrometry* the reaction time can be extended by more than an order of magnitude. At a reaction time of 10 μs or longer the highly excited parent ions have decomposed already and only ions with small amounts of excess energy remain. Since the only fragmentation pathways open to these gently excited ions are those with small activation energies, the decomposition of these so called *metastable ions* occurs predominantly by type I and type II processes. Sometimes this is not very useful for analytical applications of mass spectrometry because the rearrangement and isomerization may obscure the original structure of the molecules analysed, but these low energy processes of the radical cations within a mass spectrometer are more closely related to chemical reactions under thermal conditions and are of eminent importance for an understanding of radical ion chemistry.

Scheme 4

Fragmentation mechanisms and fragmentation/structure relationship

The fragmentation reactions observed in an EI mass spectrometer are usually discussed with respect to their reaction mechanisms and the effect of structural parameters of the compound studied on these mechanisms. The types of arguments used in this discussion follow closely the practice used in mechanistic organic chemistry, considering in particular the stability of individual fragmentation products, intermediates and transition states. Some of the more general fragmentation mechanisms observed in the EI mass spectra of organic compounds are discussed in the following examples.

Example 1: type I fragmentation and charge distribution on the fragments

For a compound of structure X–CH$_2$–CH$_2$–Y where X and Y correspond to functional groups, one expects an easy cleavage of the central C–C bond in the radical cation, generating the fragments X–CH$_2$ and Y–CH$_2$. Only one of these fragments will carry the positive charge while the other one remains as a neutral radical. Stephenson's rule states that in a fragmentation the positive charge appears predominantly with the fragment of lowest ionization energy (IE). This rule is explained easily by the (relative) reaction enthalpy ΔH_r of the competing fragmentations (Scheme 5). Starting from the neutral molecule the ΔH_r for the fragmentation routes a and b are given by Equations [a] and [b], and since the dissociation energy of the central bond is the same for both processes, that with the lowest IE will exhibit the lowest ΔH_r and will be preferred. Examples of this effect are given in Table 1 for neopentyl derivatives (CH$_3$)$_3$ C–CR$_2$–Y and β-phenethyl derivatives for the

Table 1 Ionization energy (IE) and relative intensity (rel. int.) in the 70 eV mass spectra of some neopentyl derivatives (CH$_3$)$_3$C–CR$_2$–Y and β-phenethyl derivatives C$_6$H$_5$–CH$_2$–CH$_2$–Y

Compound	Fragment (CH$_3$)$_3$C		Fragment CH$_2$–Y	
(CH$_3$)$_3$CCH$_2$Y	IE(eV)	rel. int. (%)	IE (eV)	rel. int. (%)
(CH$_3$)$_3$CCH$_2$NH$_2$	6.7	2	6.1	100
(CH$_3$)$_3$CCH$_2$OH	6.7	100	7.56	9
(CH$_3$)$_3$CCH(CH$_3$)OH	6.7	100	6.7	93
(CH$_3$)$_3$CC(CH$_3$)$_2$OH	6.7	15	< 6.7	100
	Fragment C$_6$H$_5$CH$_2$		Fragment CH$_2$–Y	
C$_6$H$_5$CH$_2$CH$_2$NH$_2$	7.2	11	6.1	100
C$_6$H$_5$CH$_2$CH$_2$OH	7.2	100	7.56	3

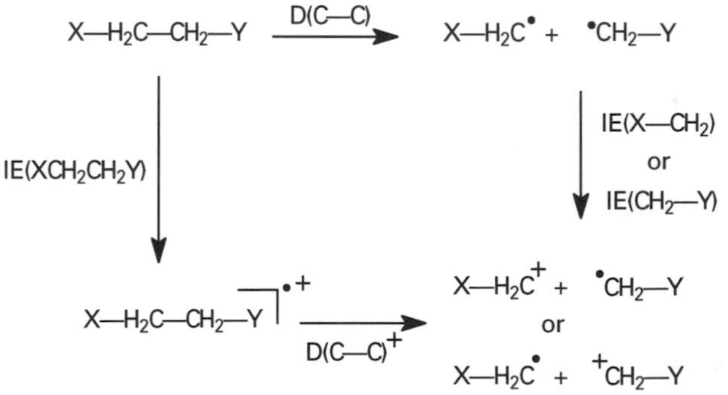

competition between a t-butyl or benzyl group and the fragment CH$_2$–Y.

Example 2: competing bond cleavages and fragmentation by rearrangement

A similar example of the effect of the relative reaction enthalpy ΔH_r on the relative abundance of competing fragmentations is obtained by an estimation of ΔH_r for the (hypothetical) C–C and C–O bond cleavages within the molecular ion of n-butanol and for the loss of a H$_2$O molecule. From the data given in Scheme 6 it is seen that C–C bond cleavage next to the α-C atom carrying the OH group ('α-cleavage') is the energetically most favoured but that a fragmentation of the molecular ion by loss of H$_2$O is even more favoured. The EI mass spectrum of n-butanol indeed exhibits large peaks due to α-cleavage by formation of the $^+$CH$_2$OH (100% rel. int.) and to loss of H$_2$O (~ 50% rel. int.) from the molecular ion as the only abundant primary fragmentation process. However, loss of H$_2$O requires a rearrangement by hydrogen migration and is hindered by a small frequency factor (type II fragmentation). By investigat-

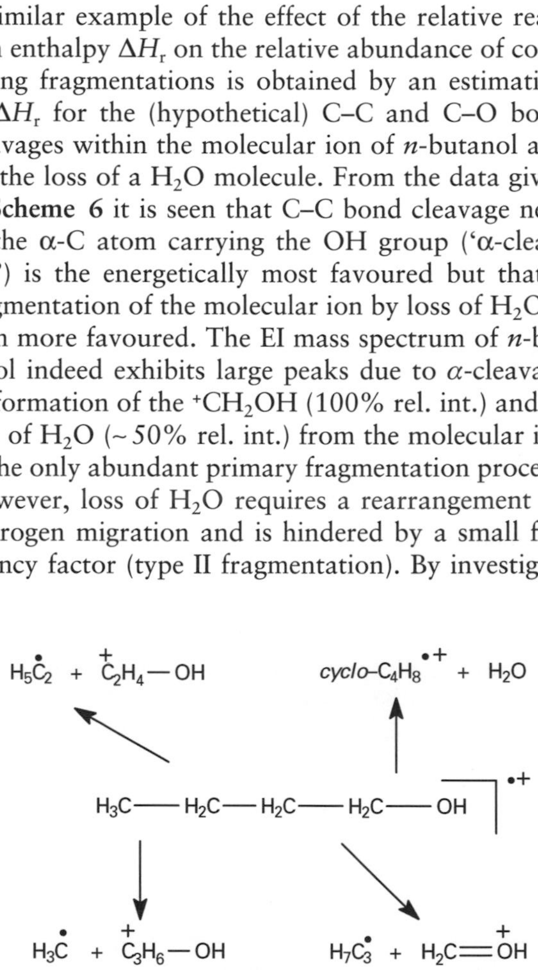

$$\Delta H_r = D(C{-}C) + IE(CH_2{-}X) \text{ or } D(C{-}C) + IE(CH_2{-}Y)$$

$$\Delta\Delta H_r = IE(CH_2{-}X) - IE(CH_2{-}Y)$$

Scheme 5

Scheme 6

ing specifically isotopically labelled *n*-butanols it was shown that loss of H_2O occurs by a 1,4-elimination and is most likely a two-step process, in which the transfer of the hydrogen atom to the OH group in the first step has an additional activation barrier. Hence α-cleavage (a type III fragmentation) is predominant. However, the long-lived metastable molecular ions of *n*-butanol, which have more time for reactions and which are less excited, decompose only by loss of H_2O. If a similar estimation is made for *n*-butylamine it turns out that elimination of NH_3 is not energetically favoured over the α-cleavage of the molecular ion, resulting in the very stable ion $^+CH_2NH_2$ (*m/z* 30). This latter process corresponds to a type I fragmentation, and the 70 eV mass spectrum of *n*-butylamine (as most other *n*-alkyl amines) is a 'one-peak' mass spectrum, exhibiting only one intense signal at *m/z* 30.

Example 3: the McLafferty rearrangement

An example related to Example 2 is the fragmentation of the molecular ions of hexan-2-one by α-cleavage at the carbonyl group and by a McLafferty rearrangement, as shown in **Scheme 7**. α-cleavage at a carbonyl group bond gives rise to a stable acetyl cation CH_3CO^+, *m/z* 43, and a $^{\bullet}C_3H_7$ radical while transfer of a H atom from the γ–CH_2 group to the C=O group accompanied by cleavage of the C(α)–C(β) bond produces the radical cation of the acetone enol and a propene molecule. The latter process exhibits a smaller reaction enthalpy and probably also a smaller activation energy than the α-cleavage, but because of the larger frequency factor v of this simple bond cleavage the acetyl cation is normally the base peak in the mass spectrum at *m/z* 43.

The fragmentation of the radical cation of a carbonyl compound by a rearrangement corresponding to H atom transfer from a γ position to the carbonyl group and cleavage of the bond between the atoms at the α and β positions (**Scheme** 7) is known as the *McLafferty* rearrangement and is a general reaction of excited radical cations of aldehydes, ketones, esters and amides. Because of this the McLafferty rearrangement is one of the most widespread and reliable fragmentations for mass spectrometric structure analysis. Its mechanism is related to the well-known Norrish type II photoreaction of carbonyl compounds and has been studied in much detail. Normally the McLafferty rearrangement is a two-step process composed of a 1,4-hydrogen migration to the carbonyl-O atom followed by bond cleavage in the intermediate. This intermediate corresponds to a unique class of radical cations named *distonic ions*, which are characterized by the positive charge and the radical cation residing in separate molecular orbitals, in contrast to 'normal' radical ions (or *molecular ions*) in which the positive charge (the 'electron hole') and the radical electron are located in the same molecular orbital. Distonic ions and their characteristics have been reviewed by Hammerum and by Kenttämaa (see the Further reading section). The properties of distonic ions will be discussed in more detail in Example 7 below. In the case of the McLafferty rearrangement the distonic intermediate is composed of an oxonium ion (the protonated carbonyl group) and an alkyl radical. The rate of its formation is favoured sterically by the six-membered transition state and depends on the bond energy of the H–X bond (X = C,O,N,S) at the γ position which is cleaved during the process. Any structural feature which lowers the dissociation energy of this bond (for example tertiary C–H bond versus primary C–H bond) enhances the intensity of the product ion of the McLafferty rearrangement in the mass spectrum. The charge may remain with either of the two fragments of the McLafferty rearrangement according to Stevenson's rule; usually this is the oxygen-containing species. Another interesting detail of the mechanism is the possibility that the two fragments stay bound to each other in the gas phase, forming an intermediate *ion–neutral complex*. (The properties of these intermediate ion–neutral complexes will be discussed in Example 7 and have been reviewed by Morton and others; see the Further reading section). During the lifetime of this complex the components may undergo intermolecular reactions, usually H atom transfers that give rise to the 'McLafferty + 1' fragment ion seen in the mass spectra of many esters of long chain alcohols.

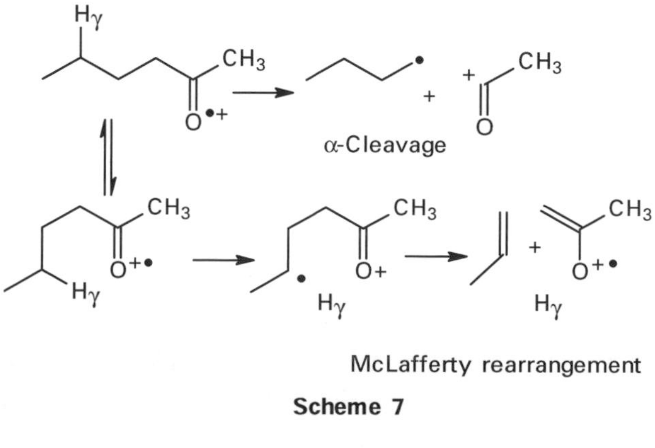

Fragmentations related to the McLafferty rearrangement, and often denoted also as a McLafferty rearrangement, are observed in the mass spectra of other organic compounds with a γ-H atom available to a suitable H acceptor group. An important

α-Cleavage

McLafferty rearrangement

Scheme 7

example of this type of fragmentation is radical cations of long-chain alkyl arenes, which contain the ionized arene systems as the H acceptor group (**Scheme 8**). The 1,4-migration of the H atom from the γ position of the sidechain creates a distonic ion composed of a protonated arene (an *arenium ion*) and an alkyl radical. Subsequent bond cleavage within the sidechain gives rise to an alkene fragment and the tautomer of a methyl arene ('isotoluene'), which usually carries the positive charge. An interesting feature of this fragmentation is proton migration within the arenium ion by successive 1,2-shifts (a 'ring walk') which may initiate further reactions.

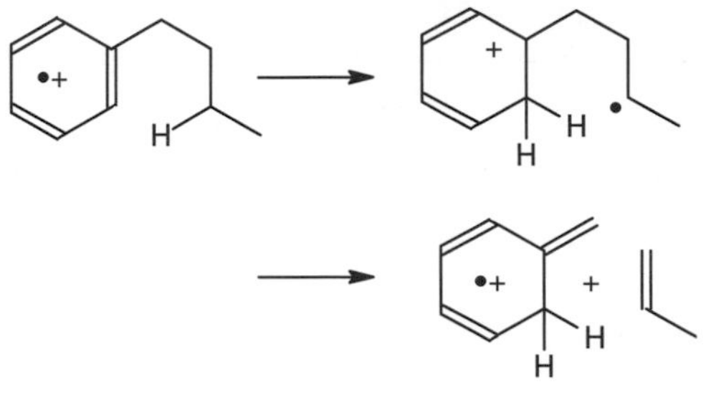

Example 4: the *ortho* effect and steric effects on fragmentations

While the EI mass spectra of the positional isomers of many di- and polysubstituted aromatic compounds are almost identical, the mass spectra of isomers carrying certain substituents in the *ortho* positions are clearly different from the *meta* and *para* isomers. A typical example is the mass spectra of *para-*, *meta-*, and *ortho*-hydroxybenzoic acid. The mass spectrum of the *para* isomer is easily analysed; the main fragment ions arise from a series of simple bond cleavages starting with the loss of an •OH radical by α-cleavage within the carboxy group (**Scheme 9**). Analogous reactions are observed in the mass spectrum of the *meta* isomer. In contrast the mass spectrum of the *ortho* isomer (salicylic acid) is dominated by a signal of the fragment ions $[M-H_2O]^{•+}$. An investigation by McLafferty in 1959, of the mechanism of this *ortho* effect showed that the H_2O molecule lost from the molecular ion is generated by a specific transfer of the H atom of the hydroxy group to the neighbouring carboxy group. The loss of H_2O is energetically very favourable, and the frequency factor for the H rearrangement is large because the H donor (the hydroxy group) and the H acceptor (the carboxy group) are fixed in a sterically favourable arrangement of the *ortho* disubstituted benzene ring. Similar intense

Scheme 9

fragmentations by this *ortho* effect are observed in the mass spectra of many *ortho* substituted arenes, the prerequisites being the presence of a H donor group H–Y– (H–CR_2–, H–O–, H–NR–, H–S–) and the possibility to eliminate a thermodynamically stable molecule H–X (H–Cl, H–OH, H–OR, H–NH_2, etc.) after H transfer to the acceptor group in the *ortho* position. Thus, fragmentation by the *ortho* effect is a very reliable and useful reaction for mass spectrometric structure analysis.

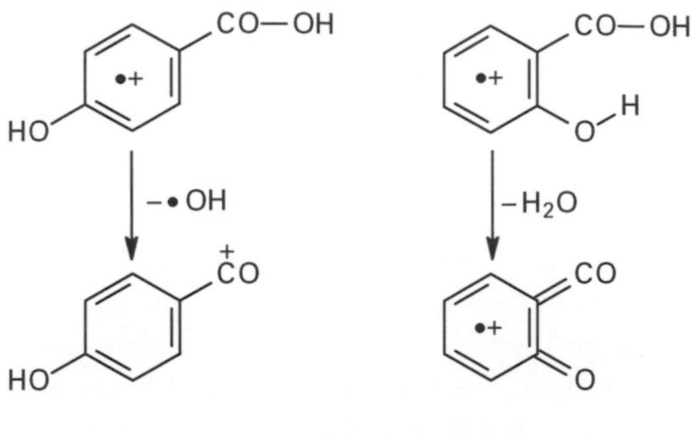

A special aspect of the *ortho* effect, in addition to the 'driving force' of the reaction by elimination of a stable molecule, is the fixed arrangement of the two groups, allowing their intramolecular interaction. This proximity effect or neighbouring group effect, which guarantees a favourable frequency factor v of the fragmentation by rearrangement, is present in many other semi-rigid organic molecules and can be used to differentiate mass spectrometrically between stereoisomers. An early example of this type of effect is the mass spectra of *cis-* and *trans-*cyclohexanediols and their di-O-methyl ethers, which can be easily distinguished by their EI mass spectra because of the stereospecific fragmentations shown in **Scheme 10**. However, it should be noted that this type of effect requires that the molecular ions of the stereoisomers do not interconvert before fragmentation.

Example 5: fragmentations by intramolecular substitution

The generation of the principal fragment ions in the EI mass spectrum of benzalacetone is depicted in **Scheme 11**. These fragment ions can be pictured as arising, at least formally, by a series of simple bond cleavages, with the exception of the formation of the ions $[M-H]^+$. The mass spectra of deuterated derivatives demonstrate that the H atom is lost from the aromatic ring. This is explained by an intramolecular attack of the (nucleophilic) carbonyl group on the ionized benzene ring, eventually giving rise to

Scheme 8

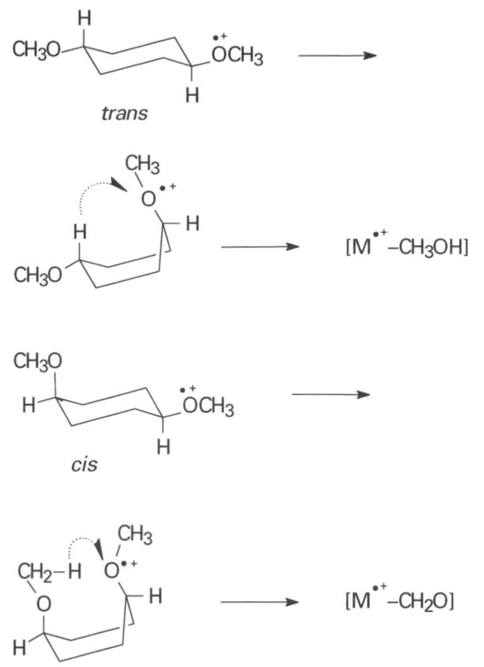

Scheme 10

a very stable benzopyrylium ion via an intermediate related to the σ-complex of aromatic substitution.

This type of fragmentation by an intramolecular substitution has been studied in detail by the author and is a rather common reaction of unsaturated or aromatic radical cations of a suitable structure. The substitution occurs by attack of a nucleophilic group on an ionized C=C double bond or an ionized aromatic ring and is driven by the generation of a stable cyclic fragment ion. By this process a substituent at the attacked position is lost, and the process is particularly abundant in the mass spectra if a weak bond is cleaved, for instance during the loss of a Cl or Br substituent, but methyl or alkyl groups are also easily eliminated.

Interestingly, two definitively different intermediates are generated by the intramolecular substitution of an aromatic ring which exhibit different reactivities depending on whether the positive charge is fixed at the attacking group. In the case of the intermediate created from cinnamic aldehyde, however, both the radical electron and the positive charge are delocalized over the total π-electron system (**Scheme 12**). As a consequence the positive charge may also appear at the (formerly) aromatic ring, giving it the properties of a true σ-complex. In these σ-complexes a proton migrates by successive 1,2-shifts around the ring ('*ring walk*') before further fragmentation. Consequently, substituents which are good 'mass spectrometric leaving groups' because of a small bond energy (such as Cl or Br) are lost not only from the attacked position but also from all other positions after H migration by protolytic bond cleavage (a well-known process of substituted arenes in acidic solutions). The other type of intermediate is represented by the intramolecular substitution within the molecular ion of 2-stilbazole or other related heteroaromatic analogues of stilbene. In this case the positive charge of the intermediate is localized outside of the attacked aromatic ring in a quaternary ammonium ion and only the radical electron is delocalized in the π-electron system. This type of intermediate is therefore a distonic radical cation. The migration of H in radicals by a 1,2-shift is forbidden by the Woodward–Hoffman rules. Hence, from this

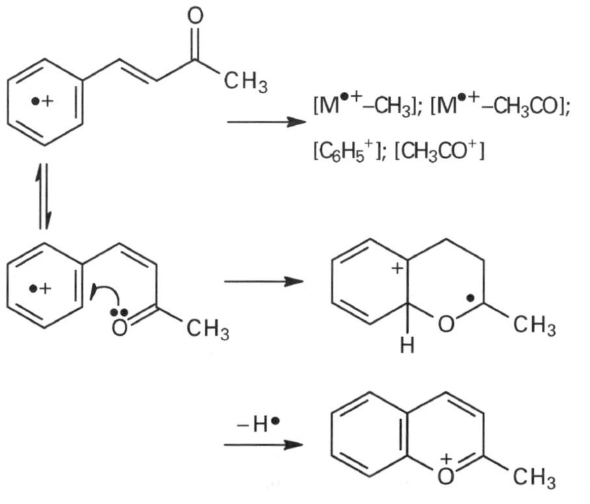

Scheme 11

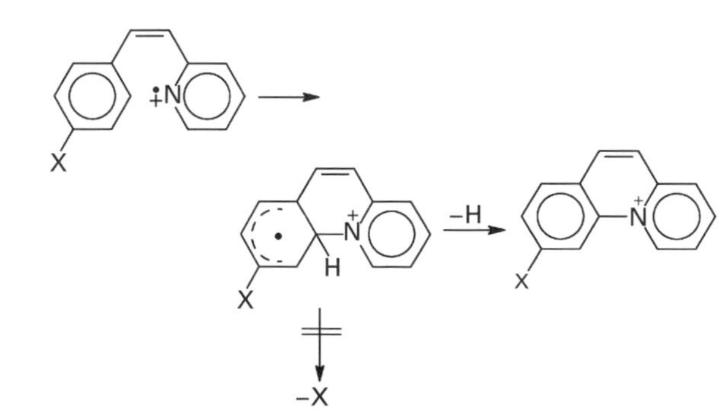

Scheme 12

type of intermediate, substituents (or a H atom) are lost specifically from the attacked position of the aromatic ring.

Example 6: the toluene–cycloheptatriene radical cation rearrangement

The examples of fragmentation reactions of radical cations discussed so far correspond either to direct bond cleavages or to bond cleavages preceded by H migration or intramolecular bond formation. These rearrangements do not alter the connectivity of C and other atoms of the original molecules. However, fragmentations of radical cations by 'skeletal rearrangements' are also known. The most familiar example is the interconversion of the radical cations of toluene and cycloheptatriene, which interchanges C atoms within the C skeleton and which gives rise to stable tropylium ions (**Scheme 13**) after H loss.

The rearrangement of C_7H_8 radical cations has been studied in detail both by experimental methods and by theoretical calculations and the mechanism is well understood. The crucial feature is the initial generation of a distonic ion by a 1,2-H shift from the methyl substituent to the aromatic ring. This is followed by cyclization to a bicyclic norcarane-like structure which eventually ring opens to the cycloheptatriene radical cation. This last step exhibits the highest activation barrier of the rearrangement and makes it a high energy process. However, a fragmentation of the toluene radical cation by a direct C–H bond cleavage also needs a large activation energy. Hence, the rearrangement competes successfully with this fragmentation, and a C_7H_8 molecular ion may undergo several reversible toluene–cycloheptatriene rearrangements before the final fragmentation.

It often is assumed that this type of toluene–cycloheptatriene rearrangement applies to other polysubstituted alkylbenzenes, alkylarenes and even alkyl heteroarenes. Indeed, this rearrangement explains the abundant loss of a whole substituent from an arene molecular ion in spite of the normally strong bond to a $C(sp^2)$ atom of the arene, which has to be cleaved to detach the substituent. Rearrangement into a substituted cycloheptatriene radical cation and proton migrations within the ionized seven-membered ring can place any substituent at the $C(sp^3)$ atom of the cycloheptatriene, facilitating the loss of the substituent X from a >CH–X group. Nonetheless, one must realize that the toluene–cycloheptatriene rearrangement is a high energy process of excited ions, and other, less energy demanding, fragmentation mechanisms of polysubstituted arenes into substituted derivatives of the $C_7H_7^+$ ion may override this rearrangement. Thus, the loss of a methyl radical from t-butylbenzene radical cations occurs by a direct 'benzylic' cleavage, producing the stable cumyl cations, and the loss of a methyl radical from the radical cations of p-xylene is preceded by the toluene–cycloheptatriene rearrangement only for highly excited ions that fragment in the ion source of a mass spectrometer, while the only gently excited metastable p-xylene radical cations decompose by H transfer to the aromatic ring, a proton ring walk to the position of the second methyl group and protolytic bond cleavage (**Scheme 4**).

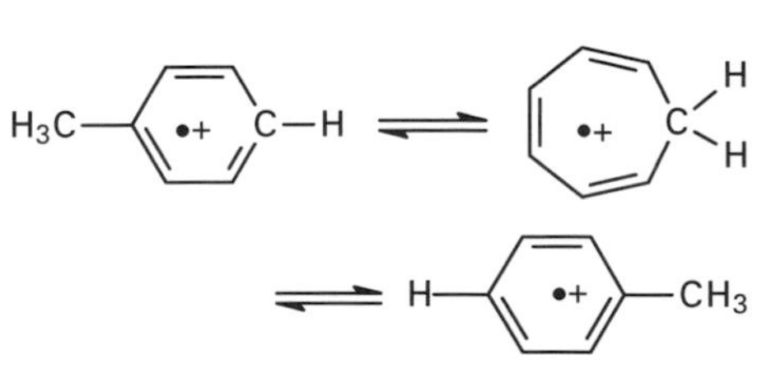

Example 7: distonic radical cations and intermediate ion–neutral complexes

The examples of mass spectrometric fragmentations of radical cations presented above show that in many cases a process which formally appears to be a simple bond cleavage corresponds to a multistep mechanism and includes several intermediates. The reason for this detour in the fragmentation route is usually the lower activation energy of the multistep process, which converts the original radical cation into an isomer more suitable for fragmentation into thermodynamically stable products. This is especially true if decompositions of metastable ions in a mass spectrometer are investigated, since these ions have long lifetimes and hence do not rely only on fragmentations with a large frequency factor. However, it is always essential that the intermediate ions of the multistep mechanism correspond to energetically favourable structures compared with the original radical cation. Two types of such intermediate ions have gained importance in explaining mass spectrometric fragmentations and will be discussed in this last example: distonic ions and intermediate ion–neutral complexes.

As mentioned before, distonic ions are distinguished from normal molecular ions by the occurrence of the positive charge and the radical cation in separate molecular orbitals. Several examples of intermediate distonic ions have been discussed above, and these distonic ions arise usually by hydrogen migration, for example during the initial step of the McLafferty rearrangement. Other modes for the

Scheme 13

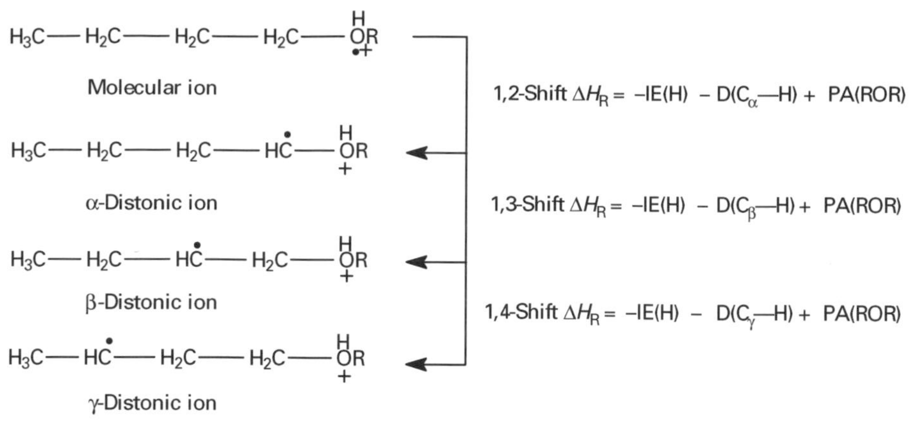

H_3C—H_2C—H_2C—H_2C—$\overset{H}{\underset{\overset{\bullet}{+}}{OR}}$

Molecular ion

1,2-Shift $\Delta H_R = -IE(H) - D(C_\alpha—H) + PA(ROR)$

H_3C—H_2C—H_2C—$\overset{\bullet}{HC}$—$\overset{H}{\underset{+}{OR}}$

α-Distonic ion

1,3-Shift $\Delta H_R = -IE(H) - D(C_\beta—H) + PA(ROR)$

H_3C—H_2C—$\overset{\bullet}{HC}$—H_2C—$\overset{H}{\underset{+}{OR}}$

β-Distonic ion

1,4-Shift $\Delta H_R = -IE(H) - D(C_\gamma—H) + PA(ROR)$

H_3C—$\overset{\bullet}{HC}$—H_2C—H_2C—$\overset{H}{\underset{+}{OR}}$

γ-Distonic ion

Scheme 14

generation of distonic ions are the ring cleavage of cyclic molecular ions or the addition of a radical cation to a C=C double bond. Depending on the minimum number of atoms between the positive charge and the radical site, α-, β-, γ- or even more distant distonic ions are known. A hydrogen shift within an aliphatic molecular ion to a heteroatom generates distonic ions composed of an onium structure and a radical. The stability of these distonic ions relative to the original molecular ion is important for their role as reaction intermediates, and this can be estimated from the thermochemical cycle (**Scheme 14**) generated from the energy needed for homolytic cleavage of the C–H bond, the ionization energy of hydrogen, and of the proton affinity at the heteroatom. For a series of isomeric distonic ions with different distances between the positive charge and the radical site the proton affinity of the heteroatom and ionization energy of the H atom are identical. Hence, the essential feature for the relative stability within this series is the dissociation energy of the C–H bond. It can be shown that this is more or less constant, with exception of the smaller C–H bond energy adjacent to the heteroatom. It follows that in this series of isomeric distonic ions the α-distonic ion should always be the most stable. However, the 1,2-H shift necessary to convert the conventional molecular ion into the α-

distonic isomer is usually associated with a high activation energy and a direct isomerization does not take place. Instead, other distonic ions are formed by less energy demanding 1,4-, 1,5-, 1,6- or even more distant hydrogen shifts, and these distonic ions are eventually transformed into the stable α-distonic isomer by further H shifts.

It is of interest to note that distonic ions also isomerize by shifts of the protonated group that contains the heteroatom. A typical example is the 1,2-NH_3 shift in the α-distonic ion of an *n*-alkyl amine. This shift converts the molecular ion of the *n*-alkyl amine into the molecular ion of an *s*-alkylamine and is responsible for the small but distinct signals at m/z 44 that are found in addition to the expected base peak at m/z 30 in the mass spectra of these amines (**Scheme 15**). Originally it was supposed that the ions were generated by 'β-cleavage' and formation of protonated ethylene imine.

The main factors that determine the relative stability of a distonic ion with respect to molecular ions containing different heteroatoms are, on the one hand, the ionization energy of the molecule and, on the other hand, the proton affinity of the heteroatom at the basic centre of the molecule. Considering these quantities for different heteroatoms it can shown that distonic ions containing an oxonium structure

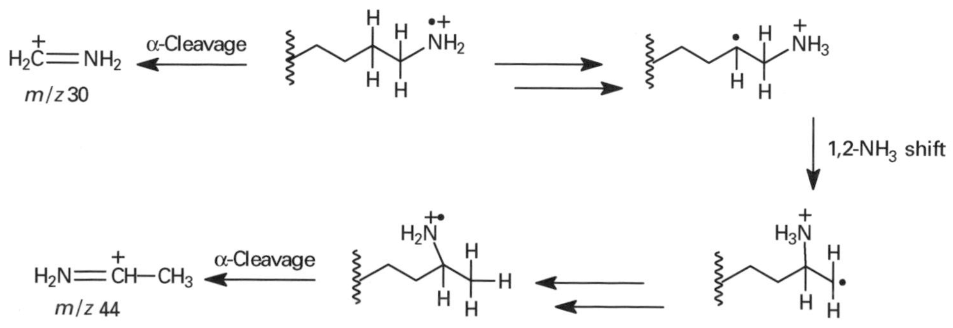

$\overset{+}{H_2C}$=NH_2 ←α-Cleavage— molecular ion → distonic ion

m/z 30

1,2-NH_3 shift

H_2N=$\overset{+}{CH}$—CH_3 ←α-Cleavage— ←— ←— distonic ion

m/z 44

Scheme 15

should be especially stable because alcohols and ethers exhibit rather high ionization energies and also large proton affinities. In *n*-alkylamines the amino group also exhibits a large proton affinity, but the ionization energy of amines is also quite low, favouring the stability of the conventional molecular ion. These expectations are confirmed by high level *ab initio* calculations by Radom (see the Further reading section). Of particular stability is the distonic isomer of the molecular ion of methanol, and it has been shown that the methanol molecular ions may isomerize (catalytically!) into the α-distonic isomer during a biomolecular collision with a neutral methanol molecule (**Scheme 16**). It is possible that a similar isomerization of molecular ions into distonic isomers by collision with a neutral molecule may occur in different systems.

Because parent ions, fragment ions and the neutral fragments travel through the mass spectrometer at the very high speed resulting from the acceleration by the high electric field of the ion source, the relative velocities of a fragment ion and the corresponding neutral fragment after a fragmentation are usually quite small and the fragmentation products drift only slowly apart. Since ions and neutrals attract each other by ion–dipole and ion–induced-dipole forces, the fragment ion and neutral fragment may be held by electrostatic attraction during the first stages of the dissociation in an ion–neutral complex. This gives the components of the intermediate ion–neutral complex of a mass spectrometric fragmentation a chance to undergo an exothermic bimolecular reaction within the complex. After the dissociation, the result of this reaction mimics the fragmentation mechanism of a rearrangement process. However, since the components within the intermediate ion–neutral complex are more or less free to rotate with respect to each other, this fragmentation route by intermediate ion–neutral complexes avoids the small frequency factor of a rearrangement with a sterically strained transition state. Since the bimolecular reaction within the complex has to be fast to compete with the direct dissociation of the ion–neutral complex, only rather simple reactions (without a large additional activation energy) are observed, i.e. exothermic hydrogen or proton transfer or nucleophilic addition of carbenium ions to electron lone pairs or π-electrons.

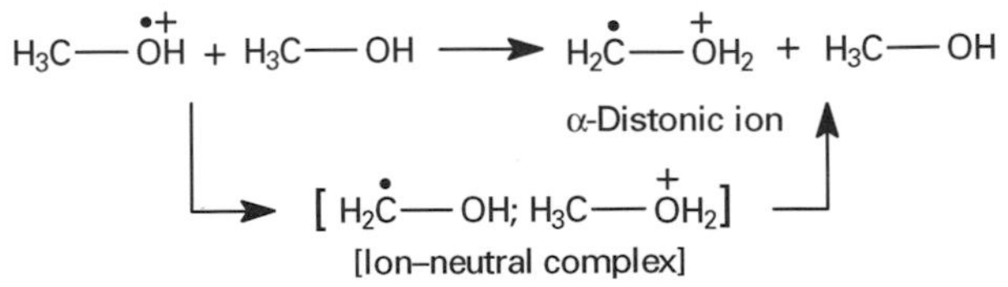

The first fragmentation of this type was described by Longevialle and Botter (see the Further reading section) in the mass spectra of certain amino steroids, in which a fragment ion generated at the right hand side of the steroid molecule transported a proton over a very large intramolecular distance to a functional group at the left hand side of the steroid. Probably the most ubiquitous fragmentation via intermediate ion–neutral complexes is the loss of alkane molecules that originate from H abstractions by an alkyl radical in an ion–neutral complex generated by a simple C–C bond cleavage of an alkyl chain. Traeger and co-workers (see the Further reading section) have shown that this fragmentation is especially abundant in the mass spectra of cycloalkanes carrying alkyl sidechains. Indeed, the EI mass spectra of these compounds are distinguished by a series of peaks at even m/z values which arise by this mechanism. Other interesting examples for this fragmentation mechanism are the migration of onium ions (and other stable carbenium ions) over long molecular distances within ion–neutral complexes before the terminating ion–molecule reaction. This corresponds to a reactive interaction between two functional groups which are far apart in the original molecule. Thus, this type of fragmentation provides an explanation for a formal rearrangement by migration of functional groups within the original molecular ion.

Secondary fragmentation

As discussed in the introduction and shown in **Scheme 1**, the fragmentations of organic molecular ions, which are induced by electron impact in a mass spectrometer and which give rise to the peak pattern of the EI mass spectrum, correspond to a kinetic scheme of parallel and consecutive unimolecular reactions. Any fragment ion that still contains sufficient energy for further fragmentation will do so, and these fragmentations correspond to the reactions of an isolated and energetically excited species described by QET or RRKM theory. However, the radical cations (molecular ions) generated by EI from the neutral molecules may decompose either by elimination of small neutral molecules (loss of H_2O, see Example 2; loss of alkene, see Example 3) or by loss of radicals. In the former case the newly generated secondary fragment ions correspond to radical cations, while in the latter case the fragmentations give rise to secondary even-electron cations. The different electronic nature of the fragment ions has an effect on the preferred pathways for further fragmentations.

The types of fragmentation reactions observed for secondary radical ions are more or less those

Scheme 16

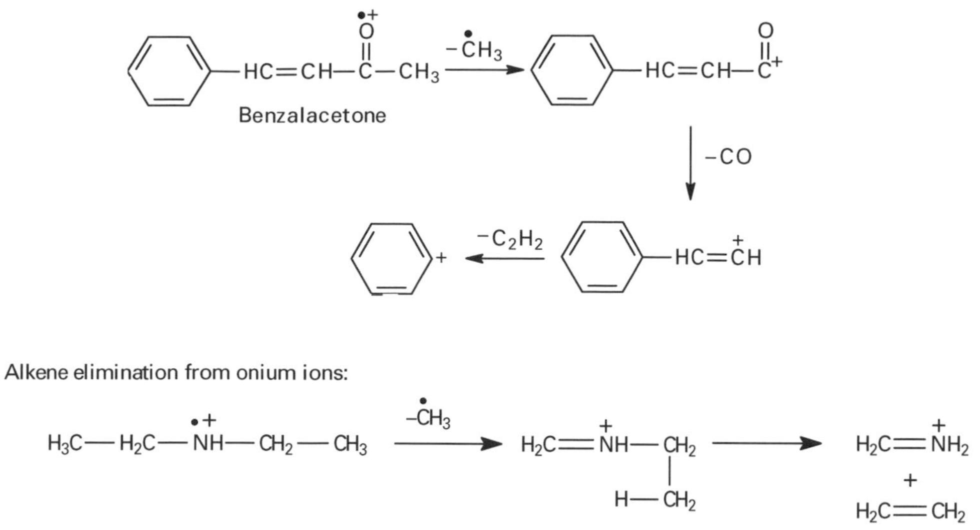

Scheme 17

discussed in the previous section. Thus, the fragmentations of the $[M^{\bullet+}-H_2O]$ fragment ions in the mass spectra of aliphatic alcohols concur with those of the radical cations of the corresponding alkenes (or alkylcyclobutanes), and because of these further fragmentations of secondary fragment ions the mass spectra of long chain alcohols and alkenes are very similar. In comparing the fragmentations of secondary radical cations with radical cations produced directly by EI from stable neutral molecules one has to account for the different excess energy present in radical cations originating from EI and from fragmentation, and, more importantly, for the possibility that fragment ions may correspond to distonic isomers or tautomers of the molecular ions.

Thus, the secondary $C_3H_6O^{\bullet+}$ radical cations created by the McLafferty rearrangement of the molecular ions of hexan-2-one (**Scheme 7**) correspond to the ionized enols of acetone and not acetone radical cations, and further fragmentation may require 're-ketonization'. Similarly, elimination of an alkene fragment from the molecular ions of alkylbenzenes gives rise primarily to an isotoluene radical cation (**Scheme 8**) which differs in reactivity from the toluene molecular ion. In most cases these differences alter the details of the fragmentation mechanisms, but not the principal routes of a further fragmentation.

Secondary fragment ions which are even-electron cations prefer the elimination of neutral molecules in

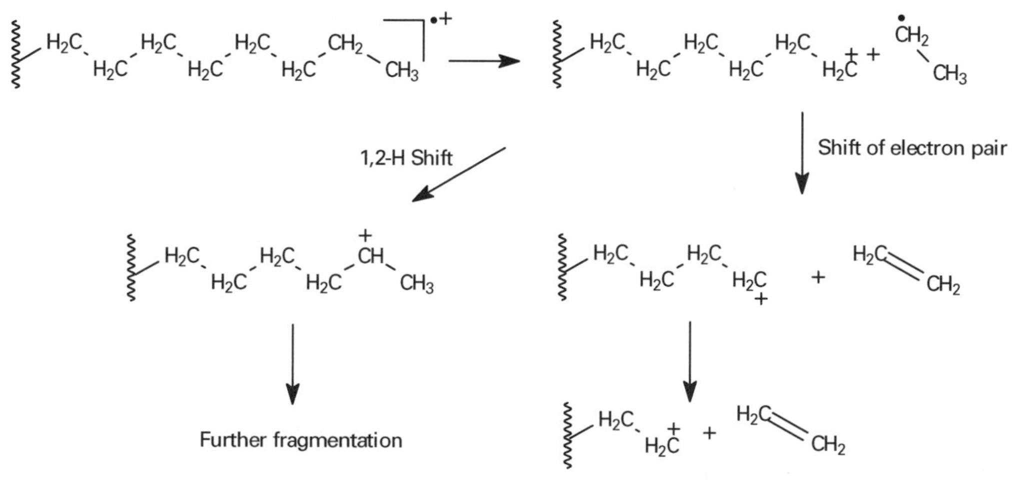

Scheme 18

their further decompositions. This has inspired the formulation of the so-called 'odd–even-electron' rule which states that odd-electron radical cations may yield either odd-electron or even-electron secondary ions on decomposition while even-electron cations produce only smaller even-electron cations as products. However, many exceptions to this rule are known, in particular for the mass spectra of large polyalkylated arenes and of organo-element compounds. In individual cases it is clearly the stability of the product ions which directs further fragmentations. Nevertheless, the rule can still be used as a first guide for the interpretation of mass spectra of organic compounds. The elimination of a small molecule from an even-electron cation is achieved by a shift of an electron pair only or by hydrogen migration. A typical example for the former mechanism is the fragmentation cascade observed for the $[M–CH_3]^+$ ion in the mass spectrum of benzalacetone while the latter fragmentation is typical of alkene eliminations from the onium ions produced by α-cleavage (**Scheme 17**).

Even-electron cations are known to rearrange easily by 'electron-sextet rearrangements' of the Wagner–Meerwein type. In particular, alkyl cations and other hydrocarbon cations rearrange easily by this mechanism. Thus, 1,2-hydrogen shifts and 1,2-alkyl shifts in an alkyl cation compete successfully with a direct fragmentation by alkene loss as depicted in **Scheme 18**. The formation of an unstable primary alkyl cation by direct bond cleavage within a molecular ion is very likely associated with a concerted 1,2-H shift to create a more stable secondary fragment ion. As a consequence of these rearrangements the abundant $C_3H_7^+$ and $C_4H_9^+$ ions observed in the mass spectra of n-alkanes at m/z 43 and m/z 57 correspond to stable isopropyl cations and t-butyl cations.

List of symbols

E = excess energy; E_{i0} = activation energy; k_i = rate constant; v_i = frequency factor; ΔH_r = reaction enthalpy.

See also: **Ion Energetics in Mass Spectrometry; Ion Imaging Using Mass Spectrometry; Ion Structures in Mass Spectrometry; Isotopic Labelling in Mass Spectrometry; Metastable Ions; Photoionization and Photodissociation Methods in Mass Spectrometry; Stereochemistry Studied Using Mass Spectrometry.**

Further reading

Barkow A, Pilotek S and Grützmacher H-F (1995) *European Mass Spectrometry* 1: 525–537.

Bowen RD (1991) *Accounts of Chemical Research* 24: 364–371.

Budzikiewicz H, Djerassi C and Williams DH (1967) *Mass Spectrometry of Organic Compounds*. San Francisco: Holden-Day.

Dunbar RC (1992) *Mass Spectrometry Reviews* 11: 309–339.

Fax MA and Channon M (1988) *Photoinduced Electron Transfer*. Amsterdam: Elsevier.

Gauld JW and Radom L (1997) *Journal of the American Chemical Society* 119: 9831–9839.

Grotemeyer J and Grützmacher H-F (1982) In: Maccoll A (ed) *Current Topics in Mass Spectrometry and Chemical Kinetics*. London: Heyden.

Grotemeyer J and Grützmacher H-F (1982) *Organic Mass Spectrometry* 17: 353–359.

Grotemeyer J and Schlag EW (1988) *Angewandt Chemie, International Edition English* 27: 447–460.

Grützmacher H-F (1993) *Organic Mass Spectrometry* 28: 1375–1387.

Hammerum S (1988) *Mass Spectrometry Reviews* 7: 123–202.

Levsen K (1968) *Fundamental Aspects of Organic Mass Spectrometry*. Weinheim: Verlag Chemie.

Lifshitz C (1997) *International Review of Physical Chemistry* 16: 113–139.

Longevialle P and Botter R (1983) *Organic Mass Spectrometry* 18: 1–8.

Marcus RA (1952) *Journal of Chemical Physics* 20: 359–364.

Mattay J (ed) (1992) In: *Photoinduced Electron Transfer V*, Topics in Current Chemistry, Vol 168.

McAdoo DJ and Morton TH (1993) *Accounts of Chemical Research* 26: 295–302.

McLafferty FW (1966) *Chemical Communications* 78: 80.

McLafferty FW and Gohlke RS (1959) *Analytical Chemistry* 31: 2076–2082.

Morton TH (1983) *Tetrahedron* 38: 3195–3243.

Rosenstock HB and Krauss M (1968) *Adv. Mass Spectrometry* 4: 523–545.

Schwarz H (1981) *Topics in Current Chemistry* 76: 1–31.

Splitter JS and Turecek F (eds) (1994) *Applications of Mass Spectrometry to Organic Stereochemistry*. Weinheim: VCH.

Stirk KM, Kiminkinen LKM and Kenttämaa HI (1992) *Chemical Reviews* 92: 1649–1665.

Traeger JC, McAdoo DJ, Hudson CE and Giam CS (1998) *Journal of American Society Mass Spectrometry* 9: 21–28.

Wurster C and Sendtner R (1879) *Berichte der deutschen chemischen Gesellschaft* 12: 1803–1813.

Yoshida K (1984) *Electrooxidation in Organic Chemistry*. New York: Wiley.

FT-Raman Spectroscopy, Applications

RH Brody, **EA Carter**, **HGM Edwards** and
AM Pollard, University of Bradford, UK

> VIBRATIONAL, ROTATIONAL &
> RAMAN SPECTROSCOPIES
> **Applications**

Introduction

A major advantage proposed for the application of Raman spectroscopy to many problems is the minimal sample preparation required for presentation of the specimen to the spectrometer. Unlike most other analytical techniques, no chemical or mechanical pretreatment is necessary; of particular relevance to biological and biomedical studies is the ability to obtain Raman spectra from specimens which are in their state of natural hydration without further desiccation. The weak Raman scattering of hydroxyl groups and silica means that water and glass will not strongly affect the observation of Raman spectra. In other cases, specimens which have been prepared for optical microscopy and protected using standard glass cuvettes or slide cover-slips can also be studied using a Raman microscope without any changes being effected.

The swamping of weaker Raman spectra by strong fluorescence emission in the visible region of the electromagnetic spectrum is an occupational hazard of Raman spectroscopy using laser excitation, particularly in the blue and green regions (400–520 nm). This arises from the high latent energy of excitation at these wavelengths. In some cases sample fluorescence arises from impurities in the specimen, and in these cases sample treatment may be indicated. This may be a trivial or difficult task, depending on the particular specimen being studied. For example, the swabbing of biological tissue, which is common practice to remove surface contaminants, is neither effective nor admissible when archaeological biomaterials are being studied, which may have involved the absorption and concentration of fluorescent materials over a long period of time.

It has often been reported that prolonged exposure of a specimen to a laser beam will produce a Raman spectrum with enhanced signal-to-noise (the little-understood physical phenomenon of 'fluorescence burnout' which is illustrated in **Figure 1**); by this means, good quality Raman spectra can be recorded in the presence of fluorescent material. However, the removal of the fluorescence emission is only temporary and the method is not always appropriate. Some biomaterials, such as skin and wool, exhibit strong

fluorescence in the blue region of the spectrum and their Raman spectra generally cannot be recorded using laser radiation around 4880 nm.

The choice of exciting line wavelength is therefore of critical importance for the observation of Raman spectra of biomaterials without attendant fluorescence emission. A particular advantage of the move towards red excitation (600–800 nm) and particularly the near-infrared (1064 nm) is the minimization of fluorescence from biological tissues and organs, such as skin, because of lower excitation energies. Other methods of combatting fluorescence have involved the use of pulsed laser excitation and d.c.-chopping or synchronization of the scattered radiation to electronically filter out the background emission.

When coupled to a microscope a Raman instrument with a high spatial resolution (approximately 5 μm) is then available which can produce spectra that are largely fluorescent-free. The Raman microscope exhibits a high signal-to-noise ratio and is ideal for the characterization of weak scattering samples and analysis of specific regions within a sample.

Another attachment that can be effected to a Raman spectrometer is a remote probe. This consists of a fibre optic cable that passes the laser beam to a sample outside the conventional sampling chamber. This can be of use in many different applications, the most obvious of which is where the sample is too large or complex to fit into the instrument. In vivo biological studies utilize a fibre optic probe for the investigation of human tissue. Industrial process monitoring uses a Raman spectroscopic probe for online quality control during manufacture.

The linear dependence between intensity of scattering and species concentration means Raman spectroscopy is not considered to be a sensitive technique for trace quantities of material, unless these are localized, can be identified microscopically and have a large molecular scattering cross-section factor. However, the linearity of dependence between concentration and observed band intensities can be used to good effect in quantitative Raman spectrometry, especially for solution equilibria, for which important physicochemical data can be obtained, such as rate constants, equilibrium constants and enthalpies of chemical bond formation or breakage. Hence,

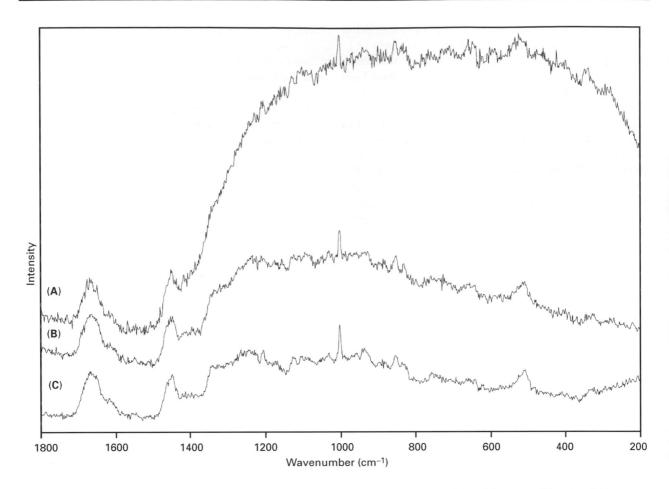

Figure 1 Dissipation of fluorescence in wool on extended exposure to laser irradiation. (A) 0, (B) 30 and (C) 210 s. (633 nm, 5 s acquisition time).

whereas in absorption spectrometry weak bands can be enhanced using larger path lengths, this is not practicable in Raman spectroscopy.

The effect of the input of high energy visible laser radiation into samples with consequent possible increase fluorescence emission may also be marked in local sample heating. This is particularly important for valuable samples that must not suffer damage during analysis; also, coloured materials may absorb the laser radiation particularly effectively and literally burn in the process. For this reason, high laser powers focused onto stable mineral or geological materials are not acceptable for sensitive biomaterials and an experienced Raman spectroscopist would always attempt to use low laser powers initially on samples whose reaction to laser radiation is not known. Even when sample degradation or damage does not occur through laser absorption, for this reason, special devices based on a spinning sample illumination system exist where the species is permitted to have only low residence times in the laser focus, so reducing the localized heating.

With 1064 nm infrared excitation, generally used in FT-Raman studies, the effect of sample heating is noted first in emission at higher wavenumber shifts (> 2000 cm^{-1}), but the problem is exacerbated when the sample itself is being thermally investigated. For this reason, the 'background' emission of infrared radiation at higher wavenumbers with 1064 nm excitation varies with sample temperature until near 200 °C it becomes effectively unworkable, e.g. the Raman spectrum of S_8 at elevated temperatures.

Biological materials

Fluorescence and the inherent weakness of the Raman effect seriously hampered the use of Raman spectroscopy for the routine analysis and investigation of biological samples. Until the commercial development of the FT-Raman technique in 1986 some researchers went to quite extreme lengths to overcome these problems. This is well demonstrated by workers who found it necessary to 'burn out' fluorescence in wool for up to 20 hours before being able to acquire an acceptable Raman spectrum.

The keratins are a specialized group of structural fibrous proteins that are characterized by their high cystine content and can be classified according to the amount of sulfur present in the protein. Structures such as wool, hair, hooves, horns, claws, beaks and feathers are classified as 'hard' keratins because the sulfur concentration in these proteins is greater than 3%. Keratin proteins containing less than 3% sulfur, such as the stratum corneum (the outermost layer of skin), are classified as 'soft' keratins. Although these keratin proteins have a similar molecular composition major spectral differences have been observed in: the intensities of the C–S and S–S stretching vibrations; the conformation of the disulfide bond; and the position, line shape and bandwidth of vibrations associated with the proteins secondary structure.

Figure 2 shows the FT-Raman spectra of wool (**Figure 2A**) and hydrated stratum corneum (**Figure 2B**) over the wavenumber regions 2600–3600 cm^{-1} and 400–1800 cm^{-1}. The presence of water near the ν(CH) stretching region can be clearly

seen; the characteristic amide I, ν(CONH), δ(NH$_2$) and δ(CH$_2$) modes are seen in the region 1200–1700 cm^{-1}. Lipid modes are near 1000–1200 cm^{-1} and the prominent ν(SS) band in wool near 500 cm^{-1} is also noteworthy.

To date, Raman spectroscopy has been used for the characterization of stratum corneum, epidermal membrane and dermal tissue. Of current interest is the administration of therapeutic agents across the skin barrier (transdermal drug delivery); however, there are a number of difficulties which involve the supply, storage and use of biohazardous human material.

Medical/pharmaceutical

The use of Raman spectroscopy in medical diagnostics is an ever-expanding field of application. Again, features such as the nondestructive and noninvasive nature of the technique mean that the technique is well suited to this type of application. Raman spectra have often been masked by fluorescence from some of the naturally occurring constituents of animal tissues. Raman spectra of cells and tissues are

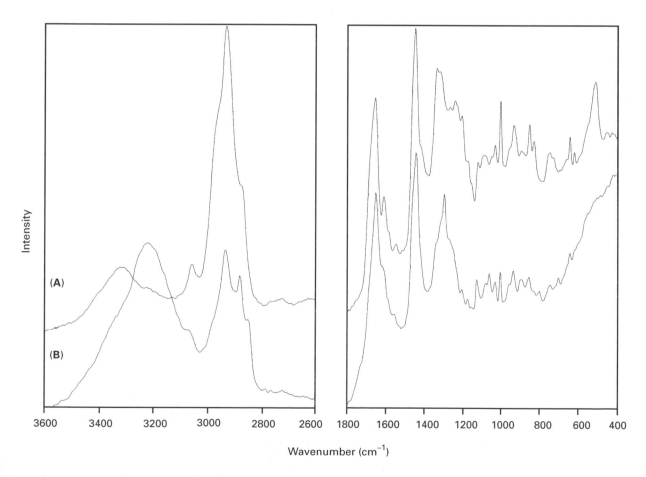

Figure 2 FT-Raman spectra of (A) untreated Merino wool, and (B) hydrated stratum corneum.

usually excited in the near infrared with radiation at 780, 830 or 1064 nm as the probability of exciting fluorescence is reduced at these longer wavelengths.

Raman spectroscopy has been used for the compositional analysis of gallstones and kidney stones; the distribution of the mineral and organic components of a human tooth have been mapped by FT-Raman microspectroscopy; the characterization of normal and pathological tissues including breast, liver, colon and brain. Furthermore, analysis can be performed *in vitro*, on biopsy samples, or *in vivo* whereby spectra are recorded using a fibre optic cable. Examination of complex Raman spectra that contain a large number of bands as in biological tissues can be somewhat subjective, and it is often difficult to detect subtle changes or differentiate between similar spectra. To overcome this problem it is necessary to automate the interpretation of spectra by developing a non-subjective technique for analysis, such as a neural network. To date, neural networking has been successfully used to differentiate between Raman spectra of basal cell carcinoma and healthy skin.

Interactions between hydroxyapatite crystallites, inorganic ions, cellular components and the organic matrix in living tissues form part of the complex process known as biomineralization. An example of biomineralization is the production of calcified tissues such as bone and teeth. The Raman spectrum of bone is dominated by a very intense vibration at 960 cm^{-1}, which is attributed to the phosphate stretch of the hydroxyapatite component.

Bone implants are commonly made of metal coated in a similar material to bone, i.e. synthetic hydroxyapatite, to improve the biocompatility and to aid bonding between the natural bone and the implant. Raman studies of bone implants utilize the spectral difference between bone and that of synthetic hydroxyapatite. The synthetic material lacks many of the characteristic Raman vibrations of bone and those of phosphate have significantly reduced bandwidths. Thus, it is possible to study the interface between the implant and the recipient's bone. Biomineralization of implants coated with materials other than hydroxyapatite can be followed by monitoring the intensity of the phosphate vibration related to those produced by the coating material.

Art and archaeology

The identification of materials that make up an object is important to art historians and archaeologists for a variety of reasons. Information can be provided on trade routes that existed at the time, the development of artistic styles and techniques and the understanding of chemical technology. This information may help to authenticate and date an object, and aid in the process of restoration and conservation. Identification of the products of degradation will help in devising a regime for treatment to arrest or reverse the deterioration, thus aiding the conservator in the preservation of the object. In many cases previous conservation treatments are a problem as the materials used breakdown and damage the object. Often treatment records have not been kept, or are lost, and analysis is required to identify the materials previously employed so they can be successfully removed.

Some problems have been found in analysing archaeological materials from certain burial environments where absorption of coloured and/or fluorescent impurities has taken place. Special care has also to be taken when handling rare or very fragile samples. Often very low laser powers are necessary in order to prevent thermal damage from occurring. These problems are being overcome and Raman spectroscopy is now being applied to the identification of a diverse range of material types including pigments, resins, waxes, gums, ivory and ancient skin.

Pigments

Pigments can be found on a wide range of objects such as paintings, frescoes, manuscripts and china. Often only a very small section of the pigment will remain perhaps only several grains. It is important to be able to identify the pigment (often *in situ*) without harming the object in anyway. Raman microscopy has been found to be one of the best techniques for identifying pigments. It is sufficiently sensitive to analyse pigment grains, generally does not suffer from interference (from surrounding media such as binders) and most importantly it is nondestructive. It is also possible to identify components in a pigment mixture, as the spatial resolution of the technique is about ≤ 2–10 μm. A wide range of pigments have been identified using Raman spectroscopy; for example, it is possible to differentiate between the red pigments: red lead, red ochre and vermilion used in antiquity and to determine the mixture composition quantitatively and nondestructively (see **Figures 3B, 3C and 3E**).

Pigments used in the wall paintings in Winchester Cathedral and Sherborne Abbey have been analysed using FT-Raman spectroscopy. Wall paintings from circa 1175–1185 are found in the Holy Sepulchre Chapel, North Transept of Winchester Cathedral. They illustrate several scenes around the Crucifixion and are amongst the most important paintings of

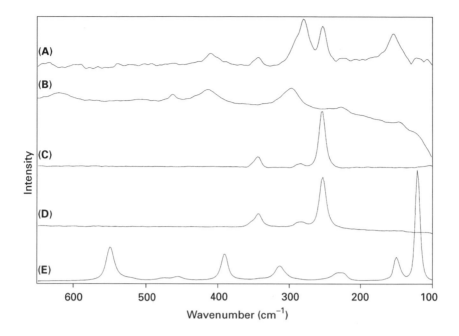

Figure 3 FT-Raman spectra of (A) pigment from wall painting in Winchester Cathedral (baseline corrected) (B) red ochre, (C) vermilion, (D) pigment from wall painting in Sherborne Abbey, and (E) red lead pigment.

their period in the UK. Those found at the Chapter House at Sherborne Abbey are much more fragmentary and date to the late 12th/early 13th Century. The Raman spectra shown in **Figure 3** demonstrate that the pigment from Sherborne Abbey (**Figure 3D**) is pure vermilion (**Figure 3C**) whereas that from Winchester Cathedral (**Figure 3A**) is a mixture of red ochre (**Figure 3B**) and vermilion (**Figure 3C**). Adulteration of pigments was practiced for economic reasons even though it was recognized that the resulting pigment mixture stability could be adversely affected. Also, variation in pigment composition over a painting could indicate to a conservator that some more recent renovation has been undertaken.

Analysis of frescoes at the palazzo Farnese, a mansion at Caprarola, Italy, helped to characterize chemically the biodeterioration that had occurred. A circular courtyard at the mansion has inner walls at ground and first floor levels bearing frescoes painted by Zuccari in the 1560s. In recent years serious biodeterioration through lichen invasion has affected the frescoes. *D. massiliensis* forma *sorediata*, which produces oxalate encrustations which are almost 2 mm thick in some places is responsible for the disfigurement and the chemicals involved in this process have been identified by Raman spectroscopy.

Archaeological resins

Raman spectroscopy can be used to identify different natural resins. By comparing spectra from ancient resins with a database of modern materials it is sometimes possible to identify the source of the ancient sample. An example of this use is the identification of the different forms of the resin 'dragon's blood'. Since ancient times this resin has been used for a diversity of medical and artistic purposes. This rich red resin has been used to make varnishes for a variety of objects including reputedly 16–18th century Italian violins such as those made by Stradivarius. Many cultures have at least one indigenous natural resin which they call 'dragon's blood' but the botanical sources are not necessarily the same. Today, the primary commercial supply of dragon's blood is the resin from the fruit of the Southeast Asian rattan, or cane-palm, *Daemonorops draco* (*Palmae*). In Western antiquity, in the Roman Empire, one of the main sources was *Dracaena cinnabari* L., which was once prevalent around the Mediterranean coast, but now endemically localized to Socotra Island, off the Horn of Africa. **Figure 4** presents the FT-Raman spectra of these two resins. Clearly the spectra are very different and provide a means of identification of a botanical source of 'dragon's blood' resin. Elsewhere, other trees have been tapped for their 'dragon's blood' such as *Eucalyptus resinifera* (actually a gum) in Australia. The FT-Raman technique is capable of distinguishing between fresh resins from these three sources mentioned. Identification of the resin can play a deciding role in the cleaning or stabilization of an object coated in resin and with the provenancing of the artefact.

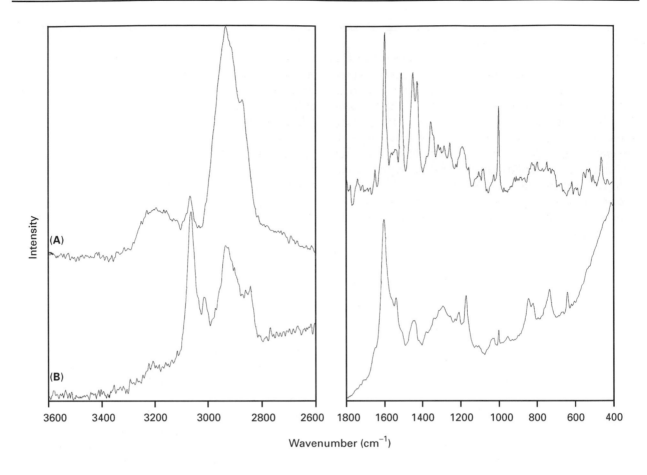

Figure 4 FT-Raman spectra of (A) *Daemonorops draco (Palmae)*, and (B) *Dracaena cinnabari*.

Forensic science

The field of forensic science, scientific analysis used within the legal system, is very wide-ranging; narcotics and explosives are amongst the areas of analysis covered. Many different techniques are used within the field to identify samples found at a crime scene or seized by police. Techniques used must fulfil certain requirements: sensitive enough to deal with often very small or trace samples, provide clear, characteristic results which easily distinguish different materials and minimal sample size and preparation. The latter point is one which Raman spectroscopy can easily fulfil, as the technique is both nondestructive and noninvasive. No sampling is required, which means that further analysis is possible should it be necessary, for example in the case of an appeal. Raman spectroscopy fulfils the other requirements as it is both sensitive and provides characteristic spectra for different substances.

Raman spectroscopy has been applied to a range of different materials of interest to the forensic scientist. A wide range of narcotics can be identified including heroin, cocaine and amphetamines. The wavelength of excitation has to be varied, as some samples are highly fluorescent under certain conditions. For example, heroin fluoresces when excited at 633 nm but a clear, characteristic spectrum can be obtained with resonance Raman using 244 nm excitation. Fluorescence can also be a problem if a narcotic has been 'cut' (mixed) with a fluorescent material. The spectra of these 'street' drugs require baseline correction in order to be successfully compared with a database of 'pure' standards or again a lower wavenumber excitation can be employed. Narcotics often enter the laboratory in a crime scene bag and may also still be in their original packaging – often wrapped in plastic. It has been found that it is possible to take a Raman spectrum of the sample through the packaging (including black plastic), subtracting the spectrum of the packaging to get a spectrum of the sample alone. This is very useful in forensic science as this minimizes any contamination that could occur to the sample during preparation for analysis. Along with the lack of any sample preparation required and with provision of a suitable database, a computerized analytical instrument means that relatively unskilled users could potentially use Raman spectroscopy to analyse any seizure more quickly.

Raman spectroscopy has also been applied to the identification of different explosives. Most explosives, both nitro-containing (i.e. 2,4,6-trinitrotoluene (TNT)) and non-nitro-containing (i.e. TATP), produce high quality, low fluorescence Raman spectra. Some plastic explosives (i.e. Semtex) have some fluorescence originating from the binder materials but this can be overcome by use of anti-Stokes bands and Boltzmann correction of the data.

Raman microscopy has been used for the analysis of gunshot residues. Chemical analysis of the materials that leave the barrel of a gun on firing can give information about the gun, the ammunition, the shooting distance and direction. All of this information is important for investigators of any shooting incident. The residues that leave a gun on firing are deposited on the victim, user, weapon and within the immediate vicinity. Gunshot residues are composed of the metallic components of the bullet, partly burnt propellant particles and particles of elements originating from the primer, the bullet, the propellant and the propellant additives. Raman microspectroscopy has illustrated that different metal anions contained in gunshot residue such as carbon, barium carbonate, lead sulfate, lead oxide and mixtures of these compounds are easily identified.

Real or fake?

An area that combines the interests of historians and forensic scientists is whether an object has been constructed from the material from which it is purported to have been made. Rare or valued materials are often replaced by cheaper, more easily available imitations, some of which are of such good quality that they deceive even expert examination. Examples of these materials are ivory and amber, both having been simulated using a wide range of materials, including modern polymers. In some instances great effort is given by the forger to copy the weight and texture of the original material. Ivory is much heavier than many modern plastics so fillers such as calcite (calcium carbonate) may be added so that the weight then more closely matches that of ivory. On highly carved objects it can be very difficult to identify the material as genuine ivory, for example, or an imitation. The 'hot pin' test has been used to identify modern plastics, over ivory. Plastic materials will tend to melt or burn when a hot pin is applied but it will have no effect on genuine ivory. This rather crude method has one quite obvious drawback in that a plastic object will be marked. Even though the item is not made of ivory it may still be of some

value, which would be affected by any marking. FT-Raman spectroscopy provides a good alternative; a result can be obtained almost as quickly and the object is not damaged. The spectra of ivory and amber are clearly different to polymeric imitations, so a definitive identification can be given where previously it has proved difficult, or impossible, to do so visually. This is illustrated in **Figure 5** which presents the FT-Raman spectra of African elephant tusk, micarta (phenolic thermoset resin), an imitation ivory and calcite. It can be seen that the spectrum of the imitation ivory contains vibrational features which are characteristic of polymer mixtures, in this case polystyrene and polymethylacrylate. The spectrum of micarta indicates that it probably contains either polyethylene or polybutylene terephthalate.

Gemmology

Analytical techniques commonly used in gemmology include X-ray and neutron diffraction, scanning electron microscopy and, more recently, FT-Raman microspectroscopy. Traditional identification is based on the gems' unique physical, chemical and optical properties. These include specific gravity, cleavage, hardness, toughness, fracture, refraction, transparency, lustre and sheen. However, some gemstones have similar properties for example taaffeite ($BeMgAl_4O_8$) and musgravite ($Mg_2Al_6BeO_{12}$) have essentially identical chemical composition and gemmological properties. The advantage of utilizing Raman spectroscopy for analysis lies in the non-destructive nature of the technique, rapid identification and the ability to analyse both carved and rough surfaces.

Other applications include the detection of synthetics and imitations, the detection of composite or assembled stones and the investigation of inclusions to assist in the identification of the origin of the gemstone. In order to hide surface cracks, improve colour or provide protection for soft stones, gemstones may undergo certain enhancement treatments. For example, they may be treated with oil, artificial resins or waxes to fill any fissures or fractures thus improving their clarity. These foreign substances produce distinctive Raman spectra from which their presence may be identified.

Polymers and plastics

Characterization of polymers and their reaction processes can be carried out using Raman spectroscopy. Both qualitative and quantitative information can be provided about stereoregularity, chemical nature,

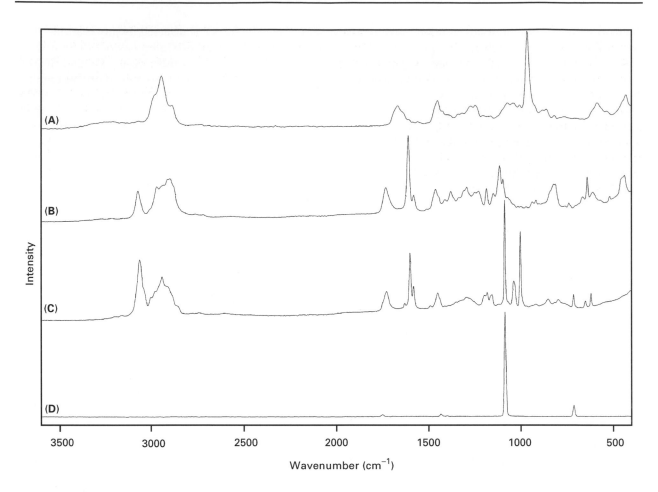

Figure 5   FT-Raman spectra of (A) African elephant ivory, (B) micarta, (C) an imitation ivory and (D) calcite.

orientation, conformation and three-dimensional state of order in a polymer. The polymerization process may also be monitored, allowing the effect of different parameters on the course of the reaction to be analysed. Many different polymers and polymer reactions have been studied using Raman spectroscopy in recent years. For example, epoxysilanes that are used to obtain hybrid polymers via a sol-gel process.

Process control and remote Raman sensing are now possible through the application of fibre optics. Remote measurements can now be taken directly in the reaction vessel. The real-time monitoring of chemical reactions can aid in the streamlining of the process and thus produce cost savings. The minimal fluorescence produced using near-infrared excitation means that FT-Raman spectroscopy is becoming widely accepted as an industrial tool. An example is the analysis of waterborne polymer emulsions, i.e. styrene/butadiene copolymers, which are used as binders in coated paper and carpet manufacture. Raman spectroscopy is useful in this case due to little interference being caused by the aqueous phase or the scattering effects from the emulsion particles,

which can be a problem with other spectroscopic techniques.

Assessment of the mechanical properties and engineering applications of polymers is made by consideration of orientation, conformation and three-dimensional state of order of the polymer during the deformation processes. FT-Raman spectroscopy is also applied to the analysis of transient structural changes induced by polymer deformation. For a clear understanding of a deformation mechanism, characterization of structural changes is required online. Studies have been performed on a wide variety of polymers during elongation and recovery, to monitor phase transitions and alterations in crystallinity and anisotropy. Particularly useful for this purpose has been the simultaneous analysis of mechanical and spectroscopic properties, called rheooptical measurements.

Inorganic/organometallic materials

A typical FT-Raman spectrum covers the wavenumber range 100–3500 cm⁻¹, which encompasses the skeletal vibrational modes of organic molecules normally occurring in the functionality region (1000–2000 cm^{-1}) and the bonds between heavy metals and carbon, oxygen and sulfur which all occur below 500 cm^{-1}. This ability to record the spectra of 'inorganic' and 'organic' components within a sample is a valuable attribute of Raman spectroscopy. For example, the Raman spectrum of mercury (II) sulfide (commonly known as vermillion), shown in **Figure** 3C, exhibits a strong Raman band due to ν(HgS) at 257 cm^{-1} which is not normally seen in infrared spectrometry unless special instrumentation is used.

The ability of Raman spectroscopy to detect coordinate bands in inorganic and organometallic compounds lends itself particularly well to the observation of metal–metal, metal–halogen and metal–phosphorus clusters such as Ag$_4$I$_4$(PMe$_3$)$_4$, metals–oxygen bonds in corrosion studies involving zirconium(IV) oxide and iron(III) oxide in nuclear fuel claddings and steels, and metal–carbon bonds in organometallic compounds used in catalysis.

Studies of metal ion coordination with anion ligands in mixed organic/aqueous solvent systems have provided novel information about the metal ion coordination numbers and mechanism of crystallization processes from Raman spectroscopy. Also, thermodynamic parameters such as enthalpy of band formation or breakage have been evaluated from Raman spectra of solution studied over a range of temperatures.

Geological materials/solid states

Access to the low-wavenumber regions of the vibrational Raman spectra of minerals and solids of geological relevance have yielded information on crystal lattice symmetries and on the effect of ionic impurities. Databases of geological materials now rival those of organic compounds and affiliations to solid-state physics, crystal engineering and to substrate characterization for thin-film devices are numerous.

Some of the most recent affiliations in this area include the analysis of extraterrestrial material in the form of Martian meteorites (Shergotty–Nakhla–Chassigny classification) and in the characterization of simulated soils for planetary exploration.

See also: **Art Works Studied Using IR and Raman Spectroscopy; Astronomy, Applications of Spectroscopy; Biochemical Applications of Raman Spectroscopy; Dyes and Indicators, Use of UV-Visible Absorption Spectroscopy; Forensic Science, Applications of Atomic Spectroscopy; Forensic Science, Applications of IR Spectroscopy; Forensic Science,** **Applications of Mass Spectrometry; Forestry and Wood Products, Applications of Atomic Spectroscopy; Hydrogen Bonding and other Physicochemical Interactions Studied By IR and Raman Spectroscopy; Inorganic Compounds and Minerals Studied Using X-Ray Diffraction; IR and Raman Spectroscopy of Inorganic, Coordination and Organometallic Compounds; IR Spectral Group Frequencies of Organic Compounds; Matrix Isolation Studies By IR and Raman Spectroscopies; MRI of Oil/Water in Rocks; Nonlinear Raman Spectroscopy, Applications; Nonlinear Raman Spectroscopy, Instruments; Nonlinear Raman Spectroscopy, Theory; Polymer Applications of IR and Raman Spectroscopy; Rayleigh Scattering and Raman Spectroscopy, Theory; Surface-Enhanced Raman Scattering (SERS), Applications; Vibrational, Rotational and Raman Spectroscopy, Historical Perspective.**

Further reading

Carmona P, Navarro R and Hernanz A (eds) (1997) *Spectroscopy of Biological Molecules*: Modern Trends. Netherlands: Kluwer Academic Publishers.

Chase DB and Rebelled JF (eds) (1994) *Fourier Transform Spectroscopy from Concept to Experiment*. Cambridge: Academic Press.

Edwards HGM (1996) Raman Spectroscopy Instrumentation. In: *Encyclopedia of Applied Physics*, Vol 16, pp 1–43. New York: American Institute of Physics, VCH Publishers.

Hendra PJ (ed) (1997) Special issue – applications of Fourier transform Raman spectroscopy VI. *Spectrochimica Acta* 53A: 1–128.

Hendra PJ (ed) (1997) Special issue – applications of Fourier transform Raman spectroscopy VII. *Spectrochimica Acta* 53A: 2245–2422.

Hendra PJ and Agbenyega JK (1993) *Raman Spectra of Polymers*. Chichester: Wiley.

Hendra PJ, Jones C and Warnes G (1990) *Fourier Transform Raman Spectroscopy: Instrumentation and Chemical Applications*. New York: Ellis Horwood.

Heyns AM (ed) (1998) *Proceedings of the 16th International Conference on Raman Spectroscopy*. Chichester: John Wiley & Sons.

Lawson EE, Barry BW, Williams AC and Edwards HGM (1997) Biomedical applications of Raman spectroscopy. *Journal of Raman Spectroscopy* 28: 111–117.

Special Issue – Raman Spectroscopy in Art, Medicine and Archaeology. *Journal of Raman Spectroscopy* 28: 77–197.

Special Issue – Raman Spectroscopy (Commemorative Issue). *Journal of Raman Spectroscopy* 26: 587–927.

Turrell G and Corset J (eds) (1996) *Raman Microscopy: Developments and Applications*. London: Academic Press Limited.

Gallium NMR, Applications

See **Heteronuclear NMR Applications (B, Al, Ga, In, Tl).**

Gamma Ray Spectrometers

See **Mössbauer Spectrometers.**

Gamma Ray Spectroscopy, Applications

See **Mössbauer Spectroscopy, Applications.**

Gamma Ray Spectroscopy, Theory

See **Mössbauer Spectroscopy, Theory.**

Gas Phase Applications of NMR Spectroscopy

Nancy S True, University of California, Davis, CA,
USA

MAGNETIC RESONANCE
Applications

NMR spectroscopy has been used to study many aspects of the physics and chemistry of molecules in the gas phase for more than 50 years. The third report of the observation of the magnetic resonance phenomenon, which was published in 1946, described a study of H_2 gas. The signal/noise ratio obtained from a 10 atm H_2 sample (line width 600 ppm at 0.68 T) was ~15/1 in that study. Nuclei in larger molecules have longer relaxation times and in many cases produce NMR spectra with natural line widths of less than 1 Hz. Currently, it is possible to obtain ^1H and ^{19}F spectra with good signal/noise ratios from gases present at partial pressures of less than 1 torr. With this sensitivity, NMR spectroscopy can be used to address many questions about molecular interactions and dynamics in the gas phase. Experimentally measured chemical shifts and spin–spin coupling constants of gas-phase molecules are a benchmark for theoretical calculations of NMR parameters. Temperature- and pressure-dependent relaxation rate constants, T_1, T_2, and $T_{1\rho}$, have been measured for several different nuclei in diatomic and small polyatomic molecules in the gas phase. These data provide information about relaxation mechanisms, rotationally inelastic collision cross sections, coupling constants and intermolecular forces. NMR spectroscopy is an important tool for studying hydrogen bonding in the gas phase. Thermodynamic parameters for keto–enol tautomeric equilibria and conformational equilibria of gases have been determined from measurements of temperature-dependent relative intensity ratios. Temperature- and pressure-dependent rate constants for low-energy intramolecular chemical exchange processes of gases including conformational interconversion, ring inversion and Berry pseudorotation have been measured using NMR spectroscopy. The motion of xenon atoms diffusing in zeolite matrices and the motion of gas-phase molecules in protein cavities can be studied using NMR spectroscopy. Gas-phase NMR spectroscopy can be used to study chemical reactions involving volatile species. Density-dependent chemical shielding and gas-phase spin–lattice relaxation were reviewed a few years ago. The application of gas-phase NMR spectroscopy to chemical exchange processes was reviewed more recently.

Principles

Gas-phase chemical shifts and coupling constants

The general features of high-resolution NMR spectra of gases resemble those of liquids, but the actual resonance frequencies in the gas phase and in solution are different owing to differences in the local environment. A gas sample has a smaller bulk magnetic susceptibility and weaker intermolecular interactions than a liquid. The bulk magnetic susceptibility difference causes the entire gas-phase NMR spectrum to be offset a few ppm from that of the corresponding liquid. The direction and magnitude of the offset depend on the magnetic field strength, the shape of the sample, its orientation in the magnetic field and its chemical composition. The magnetic field experienced by the sample, $B_0(s)$, is $B_0(s) = B_0[1+(4\pi/3 - \alpha)\kappa_s]$, where B_0 is the permanent magnetic field, κ_s is the volume susceptibility, a negative quantity which is much smaller for a gas than for a liquid owing to density differences, and α is a shape factor that is zero for a cylinder aligned along the magnetic field axis and 2π for a cylinder aligned perpendicular to the field axis. For the parallel arrangement of a cylindrical sample in a 7.2 T superconducting magnet, the gas-phase spectrum is shifted a few ppm to higher frequency from the spectrum of the corresponding liquid.

In the gas phase, intramolecular factors primarily determine the relative resonance frequencies of the nuclei in a molecule. In a liquid, intermolecular interactions are also an important factor and the relative resonance frequencies of nuclei located at different sites within a molecule can be phase dependent. The largest phase-dependent shifts in relative resonance frequencies occur for nuclei in positions near the molecule's exterior and for nuclei capable of strong intermolecular interactions such as hydrogen bonding. **Figure 1** shows gas-phase NMR spectra of a series of alcohols obtained at 421 K. The major difference between these spectra and those of the corresponding liquid alcohols is the position of the hydroxyl proton resonance which, in each case, is near 0 ppm relative to gaseous TMS. The hydroxyl proton resonances of the corresponding alcohols in

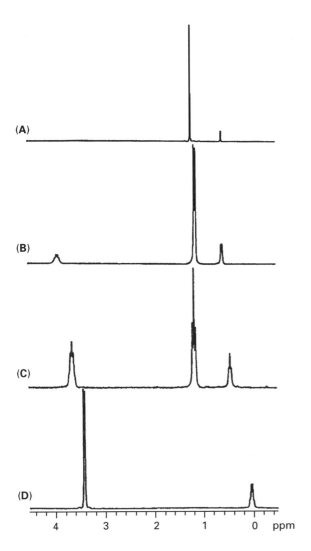

Figure 1 ^{1}H NMR spectra of gaseous *t*-butanol (A), isopropanol (B), ethanol (C), and methanol (D) at their respective vapour pressures at 421.8 K. Reprinted from Chauvel JP Jr and True NS (1985) Gas-phase NMR studies of alcohols. Intrinsic acidities. *Chemical Physics* **95**: 435–441, © 1985, with permission from Elsevier Science.

attention. The chemical shifts of most nuclei at a constant temperature exhibit a linear density dependence for moderate densities up to ~50 atm. Intermolecular effects are deshielding for almost all gas-phase molecules, except for molecules with π^* electronically excited states. The temperature dependence of the chemical shift of nuclei of gas-phase molecules at constant gas density can be used to probe the effects of ro-vibrational shielding. Deshielding occurs with increasing temperature as the population of rotationally and vibrationally excited states increases and the average internuclear distances in the molecule increase.

The phase dependence of spin–spin coupling constants is usually small since spin–spin coupling mechanisms are primarily intramolecular in nature. However, changes in conformational equilibria and molecular geometry can result from intermolecular interactions and can affect the magnitude of spin–spin coupling constants. For example, the magnitude of the spin–spin coupling between two nuclei separated by three bonds is dependent on the effective torsional angle and will be phase dependent if the conformational equilibrium established in the gas phase differs from that in solution. For substituted ethanes such as dimethoxyethane, the average $^3J_{HH}$ coupling constants in the gas phase have been measured as a function of temperature and used to determine conformational equilibria and barriers to internal rotation using models based on the Karplus equation. Solvent effects are determined by comparison with results of similar experiments performed on condensed-phase samples. Small phase-dependent differences in the $^1J_{NH}$ and $^2J_{NH}$ coupling constants of formamide have been observed and can be attributed to changes in bond lengths that occur upon solvation. For molecules that do not exhibit conformational isomerism or form strong intermolecular interactions, gas-phase and solution-phase spin–spin coupling constants are very similar.

Relaxation

Nuclear spin–lattice relaxation of gas-phase molecules occurs primarily via the spin–rotation (SR) mechanism. The magnitude of the magnetic field generated by the rotational motion of the molecule changes at a rate that is dependent on the rotationally inelastic collision frequency. Scalar coupling of the nuclear spin angular momentum to this time-dependent field provides an efficient relaxation pathway that is generally not available in condensed phases where rotational motion is hindered. For medium-sized molecules, ^{1}H and ^{13}C T_1 values are typically on the order of a few hundred milliseconds

the liquid phase are several ppm lower than that of liquid TMS owing to participation in intermolecular hydrogen bonding. In contrast, the position of the resonances of the protons attached to the carbon atoms relative to TMS is similar for both gas-phase and liquid alcohols. For cases where intermolecular interactions are weaker, gas and solution spectra are more similar, but small relative changes are often observed. For example, the limiting ^{1}H methyl chemical shift differences for dimethylacetamide and dimethylformamide are slightly smaller in the gas phase than in solution.

The temperature and density dependences of the chemical shifts of nuclei of gas-phase molecules have received significant theoretical and experimental

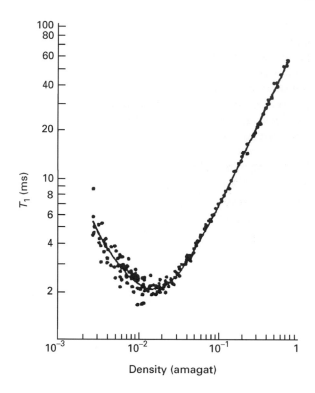

Figure 2 The density dependence of the proton T_1 of ammonia gas at 300 K. The density of an ideal gas at STP is defined as 1 amagat. The solid line is the best fit of a theoretical model that allowed for multiple relaxation times to the data. Reproduced with permission of the American Physical Society from Lemaire C and Armstrong RL (1984) Proton spin longitudinal relaxation time in gaseous ammonia and hydrogen chloride. *Journal of Chemical Physics* **81**: 1626–1631.

and ^{19}F T_1 values are typically on the order of 10 to 100 ms at a gas pressure of 1 atm.

Figure 2 shows the characteristic nonlinear density dependence of T_1 in the gas phase. Most NMR relaxation studies of gases report gas densities in units of amagats where 1 amagat is the density of an ideal gas at 1 atm and 273 K, i.e. 2.687×10^{19} molecules cm^{-3}, or 44.612 moles m^{-3}. At the T_1 minimum, the collision frequency is approximately equal to the Larmor precession frequency of the nucleus. The T_1 minimum can be used to determine the rotationally inelastic cross section, σ^{NMR}, and the average spin–rotation (SR) coupling constant. The density-dependent T_1 data for the ^{1}H resonance of NH$_3$, shown in **Figure 2**, is consistent with a σ^{NMR} of 105(8) Å^2. This value is almost twice as large as the geometric cross section of the molecule, 60 Å^2, indicating that long-range interactions are important in the relaxation process. The effective ^{1}H SR coupling constant of NH$_3$ is 20.3(0.1) kHz, considerably larger than the ^{1}H–^{1}H dipole–dipole coupling constant in this molecule.

Relaxation via dipole–dipole interactions and quadrupolar interactions has also been observed for nuclear spins of molecular gases. Dipole–dipole relaxation is competitive with spin–rotation relaxation of H$_2$ in the gas phase. For H$_2$, the SR coupling constant is ~110 kHz and the dipole–dipole coupling constant is 143 kHz. This mechanism has not been found to contribute significantly to spin–lattice relaxation of larger molecules, where dipole–dipole distances are greater. The lack of observable nuclear Overhauser enhancement (NOE) in gaseous benzene indicated that relaxation via dipole–dipole interactions is not significant at densities between 7 and 81 mol m^{-3} at 381 K. For ^{14}N in nitrogen gas, the dominant relaxation mechanism is through intramolecular quadrupolar interactions. Experiments on ClF established that there is a significant contribution to $T_{1\rho}^{F}$ from modulation through quadrupolar relaxation of ^{35}Cl and ^{37}Cl of $J(^{35}$Cl–F) and $J(^{37}$Cl–F). The dependence of $T_{1\rho}^{F}$ on the spin-lock field strength allowed values of these two spin–spin couplings to be obtained.

Dipole–dipole interactions occurring during collisions provide a relaxation mechanism for atoms in the gas phase. Xenon-129 atoms in the gas phase have spin–lattice relaxation times on the order of minutes which are dependent on gas density, temperature and the amount of other gases in the sample. For xenon-129 adsorbed in cavities inside zeolite matrices relaxation occurs via dipole–dipole interactions with hydrogen atoms that are bound to the sides of the cavities.

Methods

Sample preparation

For many applications, gas-phase samples can be prepared using a routine vacuum line consisting of mechanical and diffusion pumps, a cold trap, thermocouple and capacitance manometer gauges and a manifold with attachments for sample tubes and gas inlets. NMR sample tubes are either sealed with a glass torch or constructed with self-sealing top assemblies. Preparation of samples at elevated pressures requires quantitative transfer operations. With few exceptions, gas-phase NMR samples used to study conformational processes contained a volatile bath gas in addition to the sample molecule, which usually has a low vapour pressure. Molecules with low vapour pressures tend to behave nonideally even at very low pressures and this factor must be taken into consideration when converting measured gas pressures into densities.

Sample diffusion and convection are significant problems in temperature-dependent studies and in relaxation experiments using gases. In most NMR probes the heater is positioned below the bottom of the sample and the top of the sample protrudes above the spinner. With this arrangement a significant temperature gradient occurs along the length of a standard NMR tube when working at elevated temperatures. This causes convection in the sample and makes it impossible to establish, maintain and measure the temperature. Diffusion can cause significant line broadening at low pressures and can result in large systematic errors in relaxation measurements. To minimize these problems, short NMR tubes are frequently used for gas-phase studies. They provide better temperature control and minimize diffusion of magnetized molecules from the active volume region of the receiver during monitoring of the free induction decay. For relaxation studies of gases, the sample must be confined to the active volume of the receiver for accurate measurements of relaxation times and the tube length cannot exceed the receiver coil dimensions. Short tubes can be placed directly inside a standard size coaxial outer tubing to facilitate sample spinning or they can be attached to a spinner via a stem.

For samples containing total gas pressures of 3 atm and below, sample cells constructed from standard thickness 12, 10, or 5 mm o.d. NMR-quality glass tubing have been employed. Sample cells constructed from heavy-wall NMR-quality glass tubing, single-crystal sapphire and vespel can be used for studies of gases at pressures above 1 atm. Special precautions are necessary at all stages of assembling, testing and using any kind of high-pressure NMR apparatus because of the hazards involved. Samples in the 5–50 atm range can be prepared in small-diameter tubes with 2 mm thick walls that can withstand pressures of 50 atm. The maximum pressure is limited by the ratio of the outer diameter to the inner diameter of the tube. Pyrex and borosilicate glasses have been used. Xenon gas has been studied at pressures up to 200 atm in borosilicate glass (3.9 mm o.d./1.2 mm i.d.) Methane and ethane have been studied at pressures up to 300 atm in Pyrex tubes (5 mm o.d./0.5 mm i.d.) Sample cells constructed from single-crystal sapphire epoxied to titanium flanges that attach to vacuum lines have been used to prepare samples containing gases at pressures up to about 130 atm. A newer design allows use with 5 or 10 mm tubes, is light enough to allow spinning in most spectrometers and allows attainment of proton line widths of 0.5 Hz in 300 MHz spectrometers. Heavy-walled Vespel tubes have been used to study reactive systems at elevated pressures. This polyimide material is resistant to chemical attack but heavy walls are required and after several pressurizations the tubes tend to deform.

Specially designed NMR probes capable of withstanding high internal pressures have been described. Improved sensitivity and resolution have been obtained with toroidal probes. They consist of a toroidal-shaped RF detector inside a pressure vessel with RF and gas connectors mounted on the top and bottom, respectively, and are useful for studying pressurized reactor effluent.

Spectral acquisition

Gas-phase NMR spectra can be acquired using procedures similar to those employed for liquid samples. Most gas-phase NMR spectra are obtained without use of a frequency lock. The field drift of most superconducting magnets is only a few hertz per day and does not contribute significantly to observed line widths of spectra acquired in a few hours or less. The absence of a lock necessitates that shimming must be done by observing the magnitude and shape of the free induction decay. Inclusion of several torr of TMS gas ensures a free induction decay of adequate magnitude for shimming. For very high-resolution work or long runs, a deuterated bath gas such as neopentane-d_{12} or isobutane-d_{10} can be used, but the low sensitivity in the lock channel on most spectrometers requires the presence of a partial pressure of at least 0.5 atm of these molecules to achieve a stable lock.

Acquisition of gas-phase NMR spectra is facilitated by rapid spin–lattice relaxation. This is especially true for ^{13}C studies of gases where T_1 values can be hundreds of times shorter than in condensed phases and transients can be acquired much faster. However, relaxation times and mechanisms in the gas phase preclude many experiments involving intermolecular and intramolecular polarization transfer that are routinely performed on solution samples.

Applications

Specific applications of gas-phase NMR spectroscopy to research in heterogeneous catalysis, conformational dynamics and absorption processes in zeolites are described below. The application of gas-phase NMR spectroscopy to many other research areas is described in the Further reading section at the end of this article.

Heterogeneous catalysis

Gas-phase NMR can be used to study the kinetics of reactions that consume or generate volatile species. In a recent study, gas-phase NMR spectroscopy was used to follow the kinetics of the hydrogenation of cis/trans mixtures of perfluoro-2-butenes (R=CF$_3$) and perfluoro-2-pentenes (R=C$_2$F$_5$) over palladium supported on alumina (Eqns [1] and [2]).

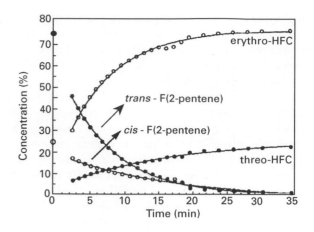

Figure 3 Time dependence of the concentrations of reactants and products of the Pd/Al$_2$O$_3$-catalysed hydrogenation of a 24.8/75.2% *cis/trans* mixture of perfluoro-2-pentene at 65°C obtained by ^{19}F gas-phase NMR. Reprinted with permission from Kating PM, Krusic PJ, Roe DC and Smart BE (1996) Hydrogenation of fluoroolefins studied by gas phase NMR: a new technique for heterogeneous catalysis. *Journal of the American Chemical Society* **118**: 10 000–10 001. © 1996 American Chemical Society.

Samples were prepared in 12 mm NMR tubes and contained a total gas pressure of ~3.1 atm at 333 or 338 K. The time dependence of the composition of the sample was determined from ^{19}F NMR spectra. Since the T_1 values of all the ^{19}F nuclei present are short, it was possible to obtain spectra, the result of 16 transients, with a signal/noise ratio of 1000 in 2.9 s. The data obtained in these experiments would be very labour intensive to obtain by conventional chemical techniques. **Figure 3** shows typical data obtained in this experiment. The *erythro/threo* product ratio was the same as the starting *cis/trans* reactant ratio, indicating that the olefins did not isomerize during the reaction and hydrogenation of the *cis* olefins produces the corresponding *erythro* hydrofluorocarbons while hydrogenation of the *trans* olefins produces the corresponding *threo* hydrofluorocarbons. The rate of hydrogenation was faster for the *trans* isomer of both perfluoro-2-butene and perfluoro-2-pentene, with the rate constant ratio, k_{trans}/k_{cis}, ranging from 2 to 3.4. It was also found that the *cis* isomers are preferentially absorbed on the catalyst, a fact that may account for their lower relative reactivity.

Conformational kinetics

For many gas-phase molecular processes with activation energies in the 5–20 kcal mol^{-1} range, temperature- and pressure-dependent rate constants can be obtained from analysis of exchange-broadened NMR line shapes. Frequently, rate constants can be obtained with accuracy comparable to that obtained in the best liquid-phase studies. This technique can be applied successfully to gases that have at least 1 torr of vapour pressure at temperatures at which slow or intermediate exchange can be observed. Temperature-dependent gas-phase rate constants

and kinetic parameters can be compared with similarly obtained parameters in condensed phases to determine the direction and magnitude of solvent effects on these processes. Gas-phase results also provide a critical test for theoretical calculations. The pressure dependence of rate constants for chemical exchange processes in the gas phase provides information on the microscopic dynamics of these processes in isolated systems.

Rate constants for the chemical exchange processes that occur in alkyl nitrites, cyclohexane, substituted cyclohexanes, sulfur tetrafluoride and formamide are pressure dependent. The mechanism for these thermally initiated, unimolecular gas-phase processes, reported by Lindemann in 1922, involves competition between the reaction and collisional deactivation of the critically energized molecule, A* (Eqn [3]).

$$A + M \xrightarrow{k_a} A^* + M$$

$$A^* + M \xrightarrow{k_d} A \qquad [3]$$

$$A^* \xrightarrow{k(E)} P$$

where k_a, k_d and $k(E)$ are the rate constants for activation, deactivation and reaction. The rate constant for reaction, $k(E)$, is energy dependent. The solution of this mechanism yields the macroscopic pseudounimolecular rate constant, k_{uni} (Eqn [4]).

$$k_{uni} = \frac{1}{[A]}\frac{d[A]}{dt} = \int_{E=0}^{E} \frac{k(E) \cdot k_a/k_d}{1 + k(E)/k_d[M]}\,dE \quad [4]$$

At large [M] the integrand reduces to $k(E)k_a/k_d$ and k_{uni} is a constant. At small [M] the integrand reduces to $k_a[M]$ and k_{uni} is a linear function of the gas density. The zero of energy is taken at the energy at threshold. The details of the fall-off curve, i.e. the rate constant k_{uni} as a function of gas density, provides information about intramolecular energy transfer in the critically energized molecule. If this is ergodic and rapid on a timescale that is short compared to the reaction rate constant, statistical theories such as RRKM theory can be used to calculate $k(E)$. The molecular conditions that favour statistical behaviour are high densities of states in the reactant and efficient coupling mechanisms. Reactions of large molecules at high internal energies can generally be modelled adequately with statistical theories. The processes that can be studied by gas-phase NMR spectroscopy occur at much lower energies where state densities are sparse and coupling mechanisms are inefficient. The shape and location of the experimental fall-off curves of chemical exchange processes provide a test of the applicability of statistical kinetic theories to these low-energy processes.

Figure 4 shows exchange-broadened gas-phase ^{1}H NMR spectra of n-propyl nitrite obtained at 240.6 K as a function of CO_2 bath gas pressures. Each sample contained 2 torr of n-propyl nitrite. Rate constants for the *syn–anti* conformational exchange process, obtained from the spectral simulations, range from 315 s^{-1} at 720 torr to 112 s^{-1} at 11.8 torr. The pressure dependence of these rate constants was compared with predictions using RRKM theory. The agreement obtained indicated that vibrational redistribution in n-propyl nitrite is statistical or nearly so at the average internal energy required for the *syn–anti* conformational interconversion process, which is ~12 kcal mol^{-1}. The observed fall-off curve is dependent on the bath gas used in the study and relative collision efficiencies for activation of the conformational process can be obtained from studies that vary the bath gas. Smaller nitrites, SF_4 and formamide have lower state densities and are currently the subjects of detailed analysis.

Absorption processes in zeolites

Several factors make xenon a very useful probe nucleus for studying absorption processes in

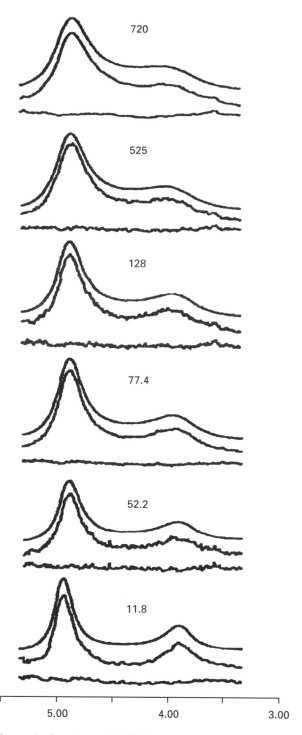

Figure 4 Gas-phase ^{1}H NMR spectra of n-propyl nitrite at 240.6 K. The spectral region shown corresponds to the methylene protons attached to the carbon bonded to the oxygen. Samples contained 2 torr of n-propyl nitrite and various pressures of CO_2. Spectral labels refer to the total sample pressure in torr. In each plot the top trace is the simulated spectrum, the middle trace is the experimental spectrum and the lower trace is the difference between the two. Reprinted with permission from Moreno PO, True NS and LeMaster CB (1990) Pressure-dependent gas-phase ^{1}H NMR studies of conformational kinetics in a homologous series of alkyl nitrites. *Journal of Physical Chemistry* **94**: 8780–8787. © 1990 American Chemical Society.

microporous materials such as zeolites. It is chemically inert, monatomic and of a convenient size, and its chemical shift is very sensitive to environmental factors and is amenable to theoretical modelling. Two of the stable isotopic species of xenon, ^{131}Xe and ^{129}Xe have magnetic moments. ^{129}Xe, which has a spin of $\frac{1}{2}$, a natural abundance of 26.44%, a receptivity 31.8 times greater than ^{13}C and T_1 values on the order of seconds, is the most extensively used. The effects of environmental interactions on the chemical shift of xenon have been modelled using grand canonical Monte Carlo simulations, allowing the determination of the cluster size of xenon in cavities in a zeolite from chemical shift measurements. A recent study of xenon adsorbed in AgA zeolite, which is a system of silicoaluminate cages containing charged Ag clusters separated by 8-membered Si–O–Si rings,

demonstrated that up to 10 different cluster sizes could be distinguished by their differences in chemical shifts. **Figure 5** shows static NMR spectra of ^{129}Xe in AgA zeolite. Spectral labels refer to the zeolite composition, dehydration temperature, and average number of xenon atoms per cage. Chemical shifts are referenced to xenon gas present in spaces between the crystallites. Exchange between cages is slow and the ^{129}Xe NMR spectrum consists of a series of lines corresponding to cages with different numbers of xenon atoms. The ^{129}Xe resonances from high field to low have chemical shifts that range over 200 ppm and are assigned to cages containing from 2 to 8 xenon atoms. Xenon intercage exchange dynamics can be studied using 2D-EXSY experiments. If chemical exchange occurs on the order of the mixing time, off-diagonal cross peaks appear in the 2D spectrum between the

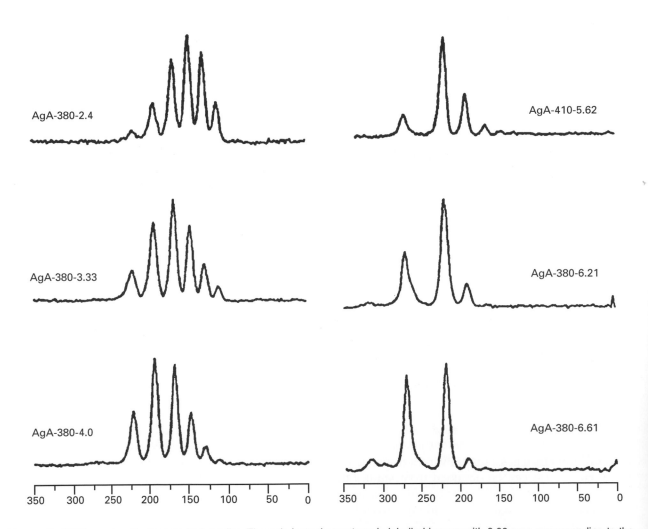

Figure 5 NMR spectra of xenon in AgA zeolite. The axis in each spectrum is labelled in ppm with 0.00 ppm corresponding to the chemical shift of xenon gas in the space between the crystallites in the sample. Spectral labels refer to the zeolite composition, sample preparation temperature, and the average number of xenon atoms per α cage. The progression of lines to lower field corresponds to resonances from 1 xenon atom per α cage to 8 xenon atoms per α cage, respectively. Reprinted with permission from Moudrakovski IL, Ratcliffe CI and Ripmeester JA (1998) ^{129}Xe NMR study of adsorption and dynamics of xenon in AgA zeolite. *Journal of the American Chemical Society* **120**: 3123–3132. © 1998 American Chemical Society.

sites involved in the exchange. The cross peaks provide information about the exchange pathways and their intensities are a function of the exchange rate constant. Rate constants for exchange between cages indicate that sorption energies decrease at higher loading.

List of symbols

$B_0(s)$ = magnetic field experienced by sample; k_a, k_d, $k(E)$ = rate constants for activation, deactivation, reaction; k_{uni} = macroscopic pseudo-unimolecular rate constant; α = shape factor; κ_s = (sample) volume susceptibility; σ^{NMR} = rotationally inelastic cross section.

See also: **High Resolution IR Spectroscopy (Gas Phase) Applications; High Resolution IR Spectroscopy (Gas Phase) Instrumentation; NMR Relaxation Rates; Parameters in NMR Spectroscopy, Theory of; Xenon NMR Spectroscopy.**

Further reading

Armstrong RL (1987) Nuclear magnetic relaxation effects in polyatomic gases. *Magnetic Resonance Review* **12**: 91–135.

Dybowski C and Bansal N (1991) NMR spectroscopy of xenon in confined spaces: clathrates, intercalates, and zeolites. *Annual Reviews of Physical Chemistry* **42**: 433–464.

Jameson CJ (1991) Gas-phase NMR spectroscopy. *Chemical Reviews* **91**: 1375–1395.

True NS (1996) Gas phase studies of chemical exchange processes. In: Grant DM and Harris RK (eds) *Encyclopedia of NMR*, pp 2173–2178. New York: Wiley.

True NS and Suarez C (1995) Gas-phase NMR studies of conformational processes. In: Hargittai M and Hargittai I (eds) *Advances in Molecular Structure Research*, Vol 1, pp 115–155. Stanford, CT: JAI Press.

GC–IR Applications

See **Chromatography-IR, Applications.**

GC–MS Methods

See **Chromatography-MS, Methods; Hyphenated Techniques, Applications of in Mass Spectrometry.**

Geology and Mineralogy, Applications of Atomic Spectroscopy

John C Lindon, Imperial College of Science, Technology and Medicine, London, UK

Copyright © 1999 Academic Press

ATOMIC SPECTROMETRY
Applications

Introduction

With their high level of sensitivity and specificity, atomic absorption and atomic emission spectroscopy methods have been applied to the analysis of elemental metal content in samples of geological, mineral and metal ore relevance. This article highlights some recent studies in the scientific literature and should serve to direct the reader to appropriate articles for further investigation.

Review

A recent review by Balaram gives an account of the various instrumental techniques used in geological analysis. These include atomic absorption spectroscopy and inductively coupled plasma atomic emission spectroscopy as well as X-ray fluorescence, isotope dilution mass spectrometry and neutron activation analysis in addition to recent developments in inductively coupled plasma mass spectrometry especially in rare earth element analysis.

Atomic absorption spectroscopy

A background to this technique is given elsewhere in this Encyclopedia. Tungsten has been determined in ores as its thiocyanate complex after separation using an Amberlite resin. Arsenic has been measured by pre-reduction of As(V) to As(III) using L-cysteine. Much study has been devoted to the quantitation of gold in rocks and ores. This has been achieved using a chelate-forming resin for preconcentration or a two-stage solvent extraction process or by treatment with MnO_2 or $KMnO_4$ and HCl. Additionally, bismuth, indium and lead have been quantified using electrothermal atomic absorption spectroscopy.

Atomic emission spectroscopy

The theory of, and the instrumental methods used in, atomic emission spectroscopy are given elsewhere in this Encyclopedia. The simultaneous determination of hafnium, scandium and yttrium in rare earth element geological materials has been described using separation and concentration stages.

See also: **Atomic Absorption, Methods and Instrumentation; Atomic Absorption, Theory; Atomic Spectroscopy, Historical Perspective; Fluorescence and Emission Spectroscopy, Theory; Inductively Coupled Plasma Mass Spectrometry, Methods; X-Ray Fluorescence Spectrometers; X-Ray Fluorescence Spectroscopy, Applications.**

Further reading

Acar O, Turker AR and Kilic Z (1997) Determination of bismuth and lead in geological samples by electrothermal AAS. *Fresenius Journal of Analytical Chemistry* **357**: 656–660, **360**: 645–649.

Balaram V (1996) Recent trends in the instrumental analysis of rare earth elements in geological and industrial materials. *Trends in Analytical Chemistry* **25**: 475–486.

Celkova A, Kubova J and Stresko V (1996) Determination of arsenic in geological samples by Hg AAS. *Fresenius Journal of Analytical Chemistry* **355**: 150–153.

Lihareva N and Delaloye M (1997) Determination of Hf, Sc and Y in geological samples together with the rare earth elements. *Fresenius Journal of Analytical Chemistry* **357**: 314–316.

Singh AK (1996) Rapid procedure for the determination of gold at sub ppm levels in geological samples by atomic absorption spectrometry. *Talanta* **43**: 1843–1846.

Soylak M, Elci L and Dogan M (1995) Spectrophotometric determination of trace amounts of tungsten in geological samples after preconcentration on an Amberlite XAD-1180. *Talanta* **42**: 1513–1517.

Yokoyama T, Yokota T, Hayashi S and Izawa E (1996) Determination of trace gold in rock samples by a combination of 2-stage solvent extraction and graphite furnace atomic absorption spectrometry—the problem of iron interference and its solution. *Geochemical Journal* **30**: 175–181.

Germanium NMR, Applications

See **Heteronuclear NMR Applications (Ge, Sn, Pb).**

Glow Discharge Mass Spectrometry, Methods

Annemie Bogaerts, University of Antwerp, Belgium

MASS SPECTROMETRY
Methods & Instrumentation

Introduction (synopsis)

This article describes the basic characteristics and methodology of glow discharge mass spectrometry (GDMS). First, the working principles of the glow discharge source will be explained, and its use for mass spectrometry will be clarified, followed by a short historical background of GDMS. Further, the various glow discharge source configurations and mass spectrometers used in GDMS will be outlined. GDMS can be operated in three different electrical operation modes, with either a direct current, radio-frequency powered or pulsed-glow discharge system. The glow discharge can also be 'boosted', by combining it either with a laser, a graphite furnace, a microwave discharge, magnetic fields or external gas jets, in order to improve the analytical results. The three electrical operation modes and the boosting methods will be briefly discussed. Moreover, the quantification methods in GDMS will be described, as well as possible solutions to overcome spectral interferences. Finally, although GDMS is particularly suitable for the analysis of conducting materials, considerable effort has been undertaken also to analyse nonconductors, and three commonly used methods to achieve this will be briefly discussed.

Principle of the glow discharge and its use for mass spectrometry

A glow discharge is a kind of plasma, i.e. a partially ionized gas, consisting of positive ions and electrons, and a large number of neutral atoms. It is formed when a cell, consisting of an anode and a cathode, is filled with a gas at low pressure (e.g. 1 torr; 1 torr = 133.3 Pa). In glow discharges used for mass spectrometry, argon is most frequently used as the filling gas. A potential difference (of the order of 1 kV) is applied between the two electrodes, and

creates gas breakdown (i.e. the splitting of the gas into positive ions and electrons). The ions are accelerated towards the cathode and cause the emission of electrons upon bombardment at the cathode. The electrons arrive in the plasma, and give rise to excitation and ionization collisions with the argon gas atoms. The excitation collisions (and the subsequent decay, with emission of light) are responsible for the characteristic name of the 'glow' discharge. The ionization collisions create new ion–electron pairs. The ions are again accelerated towards the cathode, giving rise to new electrons. The electrons can again produce ionization collisions, creating new electron–ion pairs. Hence, the latter processes make the glow discharge a self-sustaining plasma.

The use of the glow discharge as an ion source for mass spectrometry is based on the phenomenon of sputtering. The material to be analysed serves as the cathode of the glow discharge. The argon ions from the plasma (and also fast argon atoms) that bombard the cathode can also release atoms of the cathode material, which is called sputtering. The sputtered atoms arrive in the plasma where they can be ionized. Thus formed ions of the material to be analysed can be detected with a mass spectrometer, giving rise to GDMS. **Figure 1** illustrates the basic principles of the glow discharge and its coupling to mass spectrometry.

Typical discharge conditions used for GDMS are about 1 kV discharge voltage, an argon gas pressure in the order of 1 torr, and a d.c. discharge current in the mA range. The detection limits of this technique are generally in the ppb range.

Short history of GDMS

The glow discharge has been known as an ion source for mass spectrometry for more than 60 years. Gas discharges were indeed already used in the 1920s and 1930s as ion sources in the first mass spectrographs of

Aston and Thompson. However, the early popularity was followed by a decline into relative obscurity during the next 30 years, due to the development of the simple electron impact ion source. There was indeed more interest at that time for the analysis of organic samples with relatively high vapour pressure; hence simple ionization in the gas phase by an electron beam was largely sufficient. When the interest also shifted to the analysis of inorganic materials with lower vapour pressure, there was again need for other sources, with sufficient energy for atomization and ionization. Since spark and arc discharges were already widely used as excitation sources for atomic emission spectrometry, it was obvious that these sources could also be applied for mass spectrometry. These sources exhibit a high sensitivity, large applicability and only a few spectral interferences, but they do not yield a stable ion population. This led to the reexamination of the glow discharge as a stable, low energy ion source. The first come-back of GDMS was due to Coburn and coworkers; but later, Harrison and co-workers in particular were pioneers in the development of modern GDMS.

Glow discharge source configurations

As mentioned above, the glow discharge is formed in a cell consisting of two electrodes. These two electrodes can be mounted in five different geometries (see **Figure 2**).

The *coaxial cathode* (**Figure 2A**) is the most widely used source configuration in GDMS applications. Samples can be made in pin-form (with a few millimetres exposed to the discharge) or in disk-form (partly shielded so that only the top part is exposed to the discharge). The sample acts as cathode whereas the anode is formed by the cell body itself.

The *planar diode* (**Figure 2B**) is the simplest analytical source. It is used for analysing samples in disk-form. The cathode (sample) and anode are in parallel configuration and are placed inside a tube.

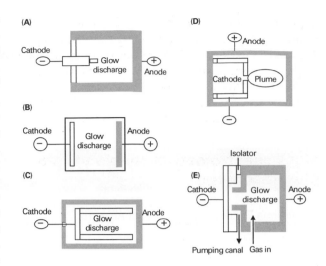

Figure 2 Different glow discharge source configurations: (A) coaxial cathode; (B) planar diode; (C) hollow cathode lamp; (D) hollow cathode plume; (E) Grimm source. Reprinted from Bogaerts A and Gijbels R (1998) Fundamental aspects and applications of glow discharge spectrometric techniques (Review). *Spectrochimica Acta Part B* **53**: 1–42, with permission from Elsevier Science.

In the *hollow cathode lamp* (**Figure 2C**) the cathode forms a cavity rather than a pin or disk. It can be considered as three planar cathodes placed so close to each other that their negative glow regions coalesce into a single negative glow. This results in increased sputtering and ionization/excitation, yielding much better analytical sensitivity. A disadvantage of this source is the extensive machining required to make hollow cathodes from metal samples. Because most of the sputtering occurs at the cathode base, studies have been performed using a disk sample as the base of the cathode.

In the *hollow cathode plume* (**Figure 2D**) the sample is mounted in the base of the hollow cathode, in which an orifice is also made. A highly energetic flamelike plume, where excitation and ionization processes occur, is ejected through this hole. Due to the high atom population, this geometry is also characterized by a high sensitivity. Nevertheless, it is rarely used for practical analyses.

In the *Grimm configuration* (**Figure 2E**) the cell body (anode) approaches the cathode very closely (at a distance smaller than the thickness of the cathode dark space), so that the discharge is constricted to a well-defined part of the sample surface. It is therefore called an 'obstructed discharge'. A similar concept is also used in the standard cells for analysing flat samples in commercial mass spectrometers (e.g. the VG9000 instrument, see below). Moreover, the Grimm source possesses an additional pumping canal close to the cathode, which reduces the pressure near

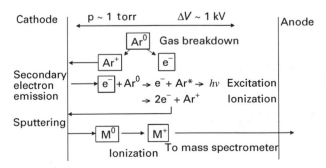

Figure 1 Schematic of the basic processes in a glow discharge.

the cathode, thereby minimizing redeposition. This geometry can, however, only be used for flat samples. Although this source is extensively used in glow discharge optical emission spectrometry (GD-OES; particularly for in-depth analysis), and forms the basic design for all commercial optical emission instruments, it is not so frequently used in GDMS.

Mass spectrometers coupled to the glow discharge

To date, all common mass analysers have been explored for use in GDMS. The first commercial GDMS instruments used a double-focusing mass analyser, permitting the acquisition of high-resolution spectra with high sensitivity. **Figure 3** shows a schematic of the 'VG9000' GDMS instrument, with first a magnetic and then an electrostatic sector (VG Elemental, Thermo Group). This instrument was made commercially available in 1985. About 60 instruments of this type are currently being used worldwide. Beside this VG9000 instrument, two other double-focusing GDMS instruments are used nowadays: the Kratos 'Concept', from which only two instruments have been manufactured, and the 'Element' from Finnigan MAT, which is essentially an inductively coupled plasma mass spectrometer (ICP-MS) that also became recently available with a glow discharge source. **Figure 4** shows a typical

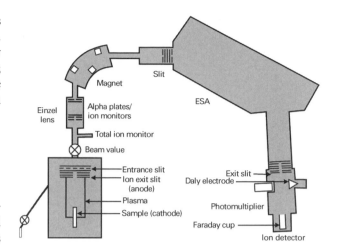

Figure 3 Schematic of the VG9000 glow discharge mass spectrometer (VG Elemental, Thermo Group).

GDMS mass spectrum obtained with the VG9000 glow discharge mass spectrometer.

The wide expansion of modern GDMS began, however, with the quadrupole-based mass analysers, which are mainly being employed for fundamental and development research of GDMS. Nevertheless, this research resulted also in the commercial availability of quadrupole GDMS systems, e.g. the 'VG Gloquad' (VG Elemental, Thermo Group).

Moreover, glow discharges have also been coupled to ion trap mass spectrometers, double and triple

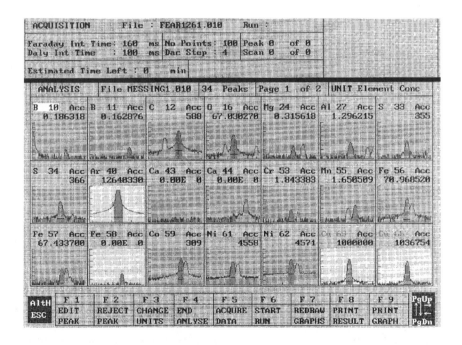

Figure 4 Typical GDMS mass spectrum obtained with the VG9000 glow discharge mass spectrometer (VG Elemental, Thermo), for specific elements in a copper sample, using an argon glow discharge. The VG9000 software provides a separate window for each of the isotopes to be measured, instead of showing the complete mass spectrum. The values shown at the top of each window give the isotopic-corrected concentrations of the elements, in ppm.

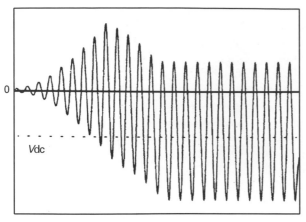

$$V(t) = Vdc + Vrf \sin \omega t$$

Figure 5 Formation of a d.c. bias voltage in an RF-powered glow discharge with a nonconducting sample.

quadrupole instruments, time-of-flight mass spectrometers, and Fourier-transform mass spectrometers, but these investigations can rather be considered as research topics. Indeed, the commercial GDMS systems available at present employ only double-focusing and quadrupole-based mass spectrometers.

Glow discharge electrical operation modes

The simplest and cheapest operation mode is the *direct current* (*d.c.*) mode. Voltages are typically 500–1500 V, yielding electrical currents in the order of mA. This type of discharge mode is the oldest and the most widely used in glow discharge applications. However, it has the serious drawback of not being able to analyse nonconducting samples directly. Indeed, since in a glow discharge the sample to be analysed acts as the cathode, which is sputter bombarded by positive ions, it must be conducting. If not, the surface will be charged, preventing the positive ions from bombarding further. Due to this drawback of the d.c. mode, attention has been drawn during the 1990s to the *radiofrequency* (*RF*) operation mode.

The *radiofrequency* mode is indeed able to analyse nonconductors directly, since the positive charge accumulated during one half-cycle will be neutralized by negative charge accumulation during the next half-cycle, so that no charging occurs. Operation with RF-power of a glow discharge using a nonconducting sample yields a negative d.c. bias voltage on the sample surface (see **Figure 5**). Indeed, during the half-cycles in which the nonconducting electrode is positive, surface charging will occur much faster than in the half-cycles in which the electrode is

negative, due to the much higher mobility of the electrons compared to the positive ions. The self-bias phenomenon permits the establishing of a time-averaged cathode and anode in the glow discharge, so that sputter-bombardment of positive ions on the cathode is still possible. Since the electrons try to follow the RF electric field, they oscillate between the two electrodes and spend more time in the plasma before they are lost, which results in a higher ionization efficiency. This leads to the second advantage of RF discharges, i.e. they can be operated at much lower pressures for the same current than d.c. discharges, which is interesting for reducing redeposition and spectral interferences. The capability of RF-powered GDMS for direct analysis of nonconductors was demonstrated already in the 1970s by Coburn and Kay and by Harrison and co-workers. However, it took until the late 1980s before RF-GDMS was revisited by Marcus and collaborators. Since then, extensive work has been done in this field. RF-discharges used for GDMS have been combined with quadrupole mass spectrometers, a Fourier-transform mass spectrometer, an ion trap mass spectrometer, a time-of-flight system and two sector-based mass spectrometers, but up to now there is no commercial RF-GDMS instrument available.

The third mode of operation of a glow discharge is the *pulsed mode*, which can be employed in combination with a conventional d.c. or with an RF glow discharge. Voltage and current are applied only during short periods of time (generally the millisecond range). Hence, compared to a normal d.c. discharge, higher peak voltages and peak currents can be obtained for the same average power. Therefore, more highly energetic gaseous ions can be produced, yielding more sputtering, a higher concentration of analyte atoms in the plasma and hence better analytical sensitivity. In addition to the better sensitivity, the pulsed mode has a second advantage for mass spectrometry, i.e. the analytically important ions and the interfering ions are formed during a different time in the pulse. By coupling this 'time-resolved' production of ions to a time-resolved detection (time-of-flight mass spectrometer), spectral interferences in the mass spectrum can be reduced. Moreover, the construction of a pulsed dual discharge system allows for simultaneous analyses with two electrodes, rendering possible the *in situ* calibration of an unknown sample against a reference standard. Recently, Harrison and co-workers introduced the microsecond-pulsed glow discharge. Due to the still higher peak currents and voltages that can be obtained during the short pulses, this source exhibits an even better analytical sensitivity. As shown in **Figure 6**, the microsecond-pulsed copper signals are one order of magnitude higher than those in the d.c. glow discharge, and they can still be further enhanced, when the pulse power is increased.

Boosting of the simple GDMS systems

Beside these three electrical operation modes, which make use of the glow discharge in its simple form as a spectroscopic source, the glow discharge can also be employed in a hybrid construction, in combination with lasers, graphite furnaces, microwave discharges or magnetic fields. The common purpose of these constructions is to increase the sputtering (atomization) and/or ionization (or excitation), and hence to improve the analytical sensitivity of the glow discharge.

Laser-based methods

The development of cost-effective laser systems has generated a variety of laser techniques that can be coupled to a glow discharge. The atomization and the ionization/excitation steps occur independently of each other in the glow discharge, and a laser can be employed to enhance either of these two steps.

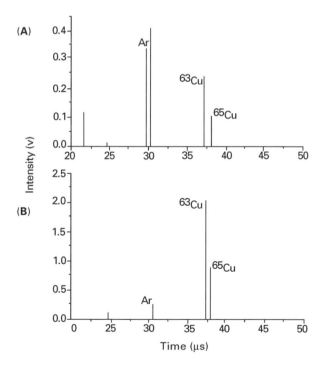

Figure 6 Mass spectra of a d.c. (A) and a microsecond-pulsed (B) glow discharge, at 1 torr gas pressure. Conditions for d.c.: 800 V, 3 mA; conditions for pulsed: 2 kV, 15 µs. Reprinted from Hang W, Baker C, Smith BW, Winefordner JD and Harrison WW (1997) Microsecond pulsed glow discharge time-of-flight mass spectrometry: analytical advantages. *Journal of Analytical Atomic Spectrometry* **12**: 143–149, with permission of the Royal Society of Chemistry.

First, the laser can be utilized to ablate material from the sample cathode, enhancing the atomization step. The laser can also ablate material from a secondary sample (not the cathode) into the glow discharge. In the latter case, the sample must not be conductive, expanding the analytical applications of glow discharges to the analysis of nonconductors without the need to apply RF powers (see above). Beside the enhanced atomization, the possibility of performing spatially resolved measurements is an additional advantage of this method.

Second, the laser can also be used to enhance the excitation/ionization processes in the discharge. The usefulness of this laser enhancement has been demonstrated in a variety of application areas, such as optogalvanic effect spectroscopy (i.e. the laser results in alternations in the ionization rate of the discharge, which are electrically detected by the resulting changes in voltage or current), laser enhanced ionization (i.e. directly measuring the electrons released when the ionization in the discharge is enhanced due to laser photons) and resonance ionization mass spectrometry (i.e. the laser is used for selectively ionization enhancement of sputtered species in the discharge, increasing both the sensitivity and selectivity in GDMS).

Furnace atomization nonthermal excitation spectrometry (FANES)

By analogy to the laser ablation glow discharge technique, FANES also makes use of an external atomization source (i.e. thermal vaporization from a graphite furnace) whereas the atoms created are excited and/or ionized in the glow discharge plasma. This source type is markedly different from a classical glow discharge source. Indeed, sample volatilization is accomplished thermally and the rate of volatilization can be three orders of magnitude higher than typical sputtering rates in a glow discharge. Moreover, this technique is mostly operated at much higher pressures than a conventional glow discharge, which results in a longer residence time for the analyte atoms in the plasma, and hence higher sensitivity.

Microwave boosted glow discharge

Coupling of auxiliary microwave power to a glow discharge, in order to enhance the ionization and excitation, can be performed in two different ways. An antenna can be inserted in the plasma to match the microwave power, or an additional microwave induced plasma can be applied. The latter technique has, in particular, been applied to GD-OES, where considerable enhancement of analytical optical emission lines has been attained. However, this technique is not commonly used for GDMS.

Magnetron discharges

Another method to improve the performance of a conventional glow discharge, is based on magnetic enhancement. One simple device among magnetically enhanced glow discharges is the magnetron glow discharge. In this device, permanent magnets are used to form a magnetic field of a few hundred gauss in the plasma. Electrons in the plasma are forced to move in closed-loop trajectories parallel to the cathode surface. Hence, the electron path length is increased and the ionization of the discharge gas is significantly enhanced. Therefore, the magnetron discharge can operate at much lower pressures than conventional discharges. Lower pressure operation provides higher ion and electron kinetic energies, leading to higher atomization and excitation/ionization efficiencies, and hence a better analytical sensitivity.

Gas-assisted sputtering glow discharges

Another way to increase the analytical sensitivity of a glow discharge is to use a gas-jet discharge. Due to the gas jet impinging on the sample surface, the sputter ablation is improved by both reduced redeposition and increased sample transport. This results in a higher sputtered atom population in the plasma and hence better analytical sensitivity. The gas-jet glow discharge was first developed for glow discharge atomic absorption spectrometry (GD-AAS), but a few applications in GDMS have also been reported.

Quantification in GDMS

To obtain quantitative concentrations in GDMS, two main approaches can be used. The easiest approach is the ion beam ratio method. The assumption is made that the ratio of the ion current for any one isotope with respect to the total ion current (except the signal arising from the discharge gas ions) is representative of the concentration of that isotope in the sample. Since the matrix ion signal is generally large compared to the individual trace species, especially for a high purity metal or semiconductor, the matrix ion current is, to a good approximation, equal to the total ion current and the matrix atoms can be assumed to have a concentration of 100%. Since this method can not correct for the variation in analytical sensitivity among different elements (e.g. due to variations in sputtering and ionization of the elements), it provides only semiquantitative results, i.e. accuracies of a factor of 2–3.

Real quantitative results require the differences in elemental sensitivities to be characterized using standards similar to the material under study. This characterization generates relative sensitivity factors (RSFs) that can be employed to correct the measured ion beam ratios. Since RSFs vary only slightly between matrices of the same general composition, exact matrix matching is not required to yield quantitative results with accuracies of 15–20%. The RSF method of quantification is the most widely employed in GDMS. Generally, the RSFs of different elements in GDMS lie within one order of magnitude, which makes GDMS a technique with rather uniform sensitivity for most elements. Experimental RSFs have been reported in the literature for different kinds of matrices, and some empirical models based on fitting parameters have been developed to predict RSFs. Moreover, it was found that the theoretically calculated RSF values correlated better with the experiment when 1% H_2 was added to the argon discharge gas. This suggests that RSFs could be more accurately predicted theoretically in a gas mixture of Ar + 1% H_2, leading to satisfactory quantitative results without the need to analyse a standard. This might be very interesting, because solid reference materials with known concentrations at the (ultra-) trace level are not commonly commercially available. However, this method is not yet routinely used in practice.

Spectral interferences in GDMS

A common problem in GDMS, as in most mass spectrometric methods, is the occurrence of spectral interferences in the mass spectrum, e.g. by impurity gas ions, various types of cluster ions or multiply charged ions. This problem can partly be overcome by using a high-resolution mass spectrometer, such as a double-focusing instrument or a Fourier-transform mass spectrometer, or by using high purity gases and special purification systems to suppress these interferences. However, some interference problems cannot be overcome in this way. For example, ^{103}Rh (rhodium is monoisotopic) in a copper matrix (^{63}Cu has 69% abundance) is severely limited by the ^{40}Ar ^{63}Cu cluster. Even if one could separate both peaks with a high-resolution mass spectrometer (a resolution of $M/\Delta M = 7620$ is required), the tailing of the huge cluster peak would prevent one from reaching low limits of detection for rhodium. In this case, the problem can be solved using an alternative plasma gas, e.g. neon instead of argon. It is demonstrated that both discharge gases exhibit similar analytical performance if a correspondingly higher pressure is used for neon; and hence, specific interference problems can be overcome. To illustrate this, **Figure 7** presents a part of the mass spectrum (m/z 90–100) of a pure iron sample, both in argon and in neon. It appears that most peak intensities in this mass range are significantly decreased in the neon discharge. This clearly demonstrates that these peaks are interfering peaks due to ArFe$^+$ clusters, which can be avoided by using neon as the discharge gas.

Moreover, another method has recently been proposed in the literature, which tries to suppress the cluster interferences by sampling from a reversed hollow cathode ion source. Indeed, it was found that the analyte ions are characterized by a peak at high energy, whereas argon ions and cluster ions possess a peak at low energy. By sampling only high energy ions, the argon ion and cluster ion interferences can be suppressed. Finally, the clusters can also be used for quantification. Indeed, since argides, dimers and doubly charged analyte ions may be less disturbed by spectral interferences, they can therefore sometimes be better employed for quantification than singly charged analyte ions, as has also been recently demonstrated.

The analysis of nonconductors by GDMS

Since the sample in the glow discharge acts as the cathode bombarded by positive ions, the concept seems to restrict the applications of GDMS to the analysis of electrically conducting materials, because nonconductive materials will be charged. The analysis of conducting materials (e.g. high-purity metals or alloys) forms, indeed, the most important field of application of GDMS. Nevertheless, much effort has also been applied to analyse nonconducting materials with GDMS, to greatly widen the application field. As mentioned above, RF-discharges are often used to analyse nonconductors directly. Nevertheless, there is no commercial RF-GDMS instrument available up to now, which is a serious drawback for routine analysis.

In a d.c. discharge, nonconductors can, however, also be analysed when applying certain modifications. Two methods are reported in the literature. The first consists of mixing the nonconducting sample as a powder with a conductive binder (Cu, Ag, Ga) and pressing it into an electrode. This method is generally well established, as follows from the large number of papers in the literature. However, in addition to the increase in sample preparation time compared to direct analysis of conducting solids, the mixing with the conductive matrix can introduce contamination. Other problems arise from the trapping of water

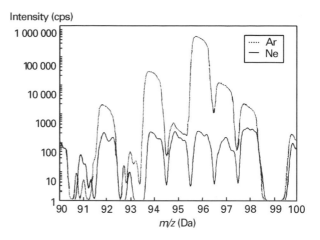

Figure 7 Mass spectra of a pure Fe sample in argon and in neon (m/z 90–100 range). The difference between argon and neon is caused by ArFe$^+$ interferences. Reprinted from Jakubowski N and Stüwer D (1989) Comparison of Ar and Ne as working gases in analytical glow discharge mass spectrometry. *Fresenius' Journal of Analytical Chemistry* **355**: 680–686, with permission of Springer-Verlag, Berlin.

vapour and atmospheric gases in the sample during the compaction process. Also, the time to reach steady state conditions with a composite cathode can be prohibitively long.

The second approach in d.c.-GDMS is the use of a metallic secondary cathode diaphragm in front of the flat nonconducting sample surface. Due to the redeposition of a part of the sputtered metal atoms from the secondary cathode, a very thin conductive layer is formed on the nonconductive material. The sampling depth is large enough (~5 Å, or 0.5 nm) to allow atomization of the nonconducting sample as well. The principle of this method is explained schematically in **Figure 8**. This method is rather new, but has already been successfully applied to the analysis of glass, ceramics, marble, polymers and even atmospheric particulate matter (aerosols).

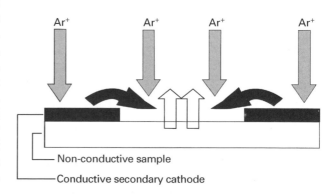

Figure 8 Schematic representation of the sputter-redeposited film formation on the nonconducting sample surface.

See also: **Atomic Absorption, Theory; Inorganic Chemistry, Applications of Mass Spectrometry; Laser Applications in Electronic Spectroscopy; Laser Microprobe Mass Spectrometers; Laser Spectroscopy Theory; Mass Spectrometry, Historical Perspective; Microwave Spectrometers.**

Further reading

Bogaerts A and Gijbels R (1997) Modeling of glow discharges: what can we learn from it? *Analytical Chemistry* 68: 719A–727A.

Bogaerts A and Gijbels R (1998) Fundamental aspects and applications of glow discharge spectrometric techniques. *Spectrochimica Acta Part B* 53: 1–42.

Harrison WW (1988) Glow discharge mass spectrometry. In: Adams F, Gijbels R and Van Grieken R (eds) *Inorganic Mass Spectrometry*, Chapter 3. New York: Wiley.

Harrison WW (1988) Glow discharge mass spectrometry: a current assessment. *Journal of Analytical Atomic Spectrometry* 3: 867–872.

Harrison WW (1992) Glow discharge: considerations of a versatile analytical source. Plenary lecture. *Journal of Analytical Atomic Spectrometry* 7: 75–79.

Harrison WW and Bentz BL (1988) Glow discharge mass spectrometry. *Progress in Analytical Spectroscopy* 11: 53–110.

Harrison WW, Hess KR, Marcus RK and King FL (1986) Glow discharge mass spectrometry. *Analytical Chemistry* 58: 341A–356A.

Harrison WW, Barshick CM, Klingler JA, Ratliff PH and Mei Y (1990) Glow discharge techniques in analytical chemistry. *Analytical Chemistry* 62: 943A–949A.

King FL and Harrison WW (1990) Glow discharge mass spectrometry: an introduction to the technique and its utility. *Mass Spectrometry Reviews* 9: 285–317.

King FL, Teng J and Steiner RE (1995) Glow discharge mass spectrometry: trace element determinations in solid samples. *Journal of Mass Spectrometry* 30: 1061–1075.

Marcus RK (1996) Radio-frequency powered glow discharges: opportunities and challenges. *Journal of Analytical Atomic Spectrometry* 11: 821–828.

Marcus RK, Harville TR, Mei Y, Shick CR (1994) Radio-frequency powered glow discharges: elemental analysis across the solids spectrum. *Analytical Chemistry* 66: 902A–911A.

Gold NMR, Applications

See   **Heteronuclear NMR Applications (La–Hg).**

Hafnium NMR, Applications

See   **Heteronuclear NMR Applications (La–Hg).**

Halogen NMR Spectroscopy (Excluding 19F)

Frank G Riddell, The University of St Andrews, UK

Copyright © 1999 Academic Press

MAGNETIC RESONANCE
Applications

The higher halogens, chlorine, bromine and iodine all possess NMR active nuclei, all of which are quadrupolar. Chlorine has two NMR active isotopes ^{35}Cl (75%) and ^{37}Cl (25%) of which ^{35}Cl is the isotope of choice owing to its higher magnetogyric ratio and natural abundance. Bromine also presents two NMR active isotopes ^{79}Br (50.5%) and ^{81}Br (49.5%). Despite its marginally lower natural abundance, ^{81}Br is the isotope of choice because of its slightly higher magnetogyric ratio. Iodine has only one NMR active nucleus ^{127}I (100%). More NMR studies have been reported on chlorine than on the other two halogens for two reasons: (i) the chloride ion is the most abundant intracellular anion in normal mammalian cells, giving a biological incentive for NMR studies and (ii) both isotopes of chlorine have relatively low quadrupole moments, reducing unfavourable quadrupolar interactions and consequently enhancing visibility relative to the other halogens. This chapter reviews the use of NMR to study halides in biological systems, in solution chemistry and in solids. In particular it concentrates on the use of quadrupolar interactions from the isotopes available. There have been relatively few reviews of halogen NMR (excluding 19F) and as a consequence several leading papers from the primary literature are included in the Further reading section list at the end of this article.

Nuclear properties

The nuclear properties of the five NMR active isotopes of the halogens are presented in **Table 1**.

Quadrupolar relaxation and visibility

The NMR spectra of the halogens are dominated by the fact that all the isotopes are quadrupolar. Indeed, the majority of the reported studies of halogen NMR make use of the resulting quadrupolar interactions. Many of the problems that arise and solutions adopted are similar to those involved with the alkali metals. Quadrupolar nuclei have an asymmetric distribution of charge which gives rise to an electric quadrupole moment. Apart from when the nucleus is in an environment with cubic or higher symmetry, the quadrupole moment interacts with the electric field gradient (EFG) experienced by the nucleus, giving rise to, among other things, quadrupolar relaxation. The strength of the quadrupolar interaction between the quadrupole moment (eQ) and the electric field gradient (eq) is given by the quadrupolar coupling e^2qQ/h. This can take values from very small to hundreds of MHz depending on the magnitudes of Q and q.

In solution, modulation of the EFG at the quadrupolar nucleus by isotropic and sufficiently rapid

Table 1 Nuclear properties of the halogens other than ^{19}F

Isotope	Spin	Natural abundance (%)	10^{-7} Magnetogyric ratio γ (rad T^{-1} s^{-1})	10^{28} Quadrupole moment Q (m^2)	NMR frequency Ξ (MHz)	Relative receptivity D^c
^{35}Cl	3/2	75.53	2.624	-8.2×10^{-2}	9.809	20.2
^{37}Cl	3/2	24.47	2.184	-6.5×10^{-2}	8.165	3.78
^{79}Br	3/2	50.54	6.726	0.33	25.140	2.28×10^2
^{81}Br	3/2	49.46	7.250	0.27	27.100	2.79×10^2
^{127}I	5/2	100	5.390	-0.79	20.146	5.41×10^2

Ξ is the observing frequency in a magnetic field in which ^{1}H is at 100 MHz.
D^c is the receptivity relative to ^{13}C.
Quadrupole moments Q are the least well determined parameters in this table.
Data taken from Mason J (1987). In: Mason J (ed) *Multinuclear NMR*, p 623. New York: Plenum Press.

molecular motions (where $\omega_0\tau_c << 1$) leads to relaxation according to the expression:

$$\frac{1}{T_1} = \left(\frac{3}{40}\right)[(2I+3)/I^2(2I-1)](1+\eta^2/3)e^2qQ/\hbar)^2\tau_c$$

where η is the asymmetry parameter associated with the EFG.

For all of the nuclei in this chapter the quadrupolar couplings for covalently bound halogens are typically many tens for MHz, depending upon the particular case (~70 MHz for chloroalkanes). This can be thought of as arising from the asymmetric distribution of electrons in the covalent bond giving rise to a large EFG at the nucleus. For molecules in solution that tumble with typical correlation times (τ_c) of 10^{-12} s, this gives rise to T_1 and T_2 values that are typically in the submillisecond range. As a consequence, line widths range from the very broad to the invisibly broad with the latter case being the norm. With very broad line widths due to short T_2 values, much if not all of the halide resonance can be lost in the dead time between the pulse and the start of signal acquisition. Thus, quadrupolar relaxation has a devastating effect upon the signals of the covalently bonded halogen nuclei in solution, rendering them invisible in all but the most favourable circumstances. In passing, it is worth noting that the ^{35}Cl line width in a typical small molecule like CCl_4 is about 2000 ppm at an observing frequency of 10 MHz. This has to be compared with a chemical shift range of around 1000 ppm from aqueous Cl$^-$, which is amongst the most shielded cases, to ClO_4^-, which is one of the most deshielded cases.

Quadrupolar couplings are smaller in cases where chlorine is covalently bound to the second and third row elements such as silicon, phosphorus, sulfur and arsenic (Si–Cl ~35 MHz), making observation easier but the lines are still very broad.

The anions themselves in solution present a different picture since they are subject to much lower quadrupolar interactions. In aqueous solution the anions are solvated by hydrogen bonding. At any one instant the pattern of hydrogen bonded water molecules around the anion does not have spherical symmetry but is always close to it. Thus the quadrupolar couplings are much lower than for the covalently bound halogens but are not zero. Typically, in aqueous solution and in the absence of extraneous influences, both isotopes of chlorine show Cl$^-$ line widths of 10–15 Hz, both isotopes of bromine show Br$^-$ line widths of ~400 Hz, ^{127}I$^-$ has a line width of over 1000 Hz.

Covalent compounds with the halogen nucleus at a site of tetrahedral or octahedral symmetry, as in ClO_4^- or IF_6^+, present a low EFG at the halogen nucleus and frequently have relatively sharp resonances. Indeed the spectra of the tetrahedral perhalate ions in acetonitrile solution show exceptionally sharp lines that enable secondary isotope effects on shielding to be observed. The shielding difference between $(Cl^{16}O_4)^-$ and $(Cl^{18}O^{16}O_3)^-$ is 0.090 ppm.

In cases where molecular motion is restricted ($\omega_0\tau_c$ is not $<< 1$) the situation is more complex. The quadrupolar interaction with the nucleus shifts the energies of the Zeeman levels according to the square of the quantum number to a first approximation. Thus, the energy level splittings for a nucleus with $I = \frac{3}{2}$ (e.g. $^{35/37}$Cl and $^{79/81}$Br) become as illustrated in **Figure 1**. With rapid isotropic motion, as described above, the multiple line pattern will collapse into a single line. In the absence of rapid isotropic motion the relaxation rate of the outer transitions, which combine to have 60% of the total intensity, is different from that of the inner transition which has 40% of the total intensity. The frequencies of the outer transitions also shift from the inner transition (dynamic frequency shift). There are three principal consequences of these changes for cases

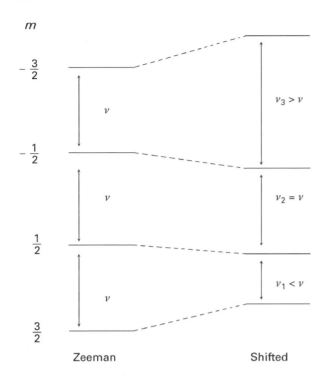

Figure 1 Changes in the energy levels for a spin $\frac{3}{2}$ nucleus subject to a quadrupolar interaction.

where the motion of halogen is restricted. Firstly, line shapes become a double and not a single Lorentzian, secondly two relaxation times are apparent, and finally, where the more rapid relaxation time becomes very short, partial or total invisibility of the signal from the outer transitions may occur. For example, studies on dog erythrocytes showed a relatively sharp ^{35}Cl line (line width ~40 Hz) but only 40% visibility and in renal tubules the ^{35}Cl signal was reported to be invisible. An excellent review of quadrupolar relaxation effects has been given by Springer (listed in the Further reading).

In solids the situation is even more complex with, for example, the possibility of multiple quantum relaxation pathways. A discussion of this is outwith the scope of this article. In general, however, relaxation times for halogens in solids are several orders of magnitude longer than for those in solution and this, in turn, in favourable cases, does allow effects due to the halogen nuclei to be observed.

Biological chloride

Chlorine is the eighth most abundant element in the human body. Although it is present in large amounts as chloride ion in extracellular spaces the intracellular concentration of chloride is high and it is the most abundant intracellular anion in normal mammalian cells. For example, its concentration in the human

erythrocyte is ~150 mM. Chloride is involved in many important biochemical pathways. The most used membrane transport system in mammals is band III protein whose role is to exchange HCO_3^- inside the cell for Cl^- outside, thus facilitating the removal of CO_2. Abnormal chloride transport is implicated in several serious medical conditions such as cystic fibrosis and Duchenne muscular dystrophy. Many good reasons exist, therefore, to develop NMR methods to study biological chloride.

One of the main problems in using NMR to study biological chloride is that the chemical shift of aqueous $^{35}Cl^-$ is essentially independent of the ion's surroundings, making differentiation of intra- and extracellular chloride difficult. Further difficulties arise because of the binding of chloride to proteins, some of which may contain paramagnetic ions, which can greatly increase the line width. If the exchange rate between the free and the bound chloride is slow this leads to a loss of signal.

The first problem can be met by using a contrast reagent (either a shift or a relaxation agent) in one of the compartments, normally extracellular. Aqueous shift reagents that have been employed include Co(II) either as Co^{2+} cations or as the triglycinate complex $[Co(Gly)_3]^-$. Manganese (II) has been shown to be an extremely efficient relaxation agent for $^{35}Cl^-$ in which rapid exchange of chloride for water in the inner ligation sphere brings the chloride ions transiently under the paramagnetic influence of the five unpaired electrons on the manganese.

It has been argued that the relaxation agent Mn^{2+} is superior to Co(II)-based shift reagents. First, the concentrations of Co(II) reagents needed to visualize intracellular or intravesicular chloride are considerably higher than those of Mn^{2+}. Secondly, all contrast reagents induce paramagnetic line broadening of the extracellular ^{35}Cl resonance that is proportional to the square of the magnetic field whilst the induced shift is only linearly dependent on the field. Therefore, on moving to the high fields required to attain satisfactory signal-to-noise, Mn^{2+} has the advantage of requiring even lower amounts of reagent to achieve the same line widths, whilst Co(II)-based shift reagents require even higher concentrations to achieve the same relative separation of resonances.

Cobalt(II) as a shift reagent has been employed to visualize the $^{35}Cl^-$ resonance from chloride ions inside vesicles and erythrocyte ghosts. The latter observation allowed measurements of the kinetics of chloride/sulfate exchange mediated by the band III protein. The same group also used Co(II) to observe Cl^- inside the cells of an alga, *Chlorella pyrenoidosa* and inside the cells of a suspension of cultured plant cells from *Nicotiana tobbaccum*. In both of these

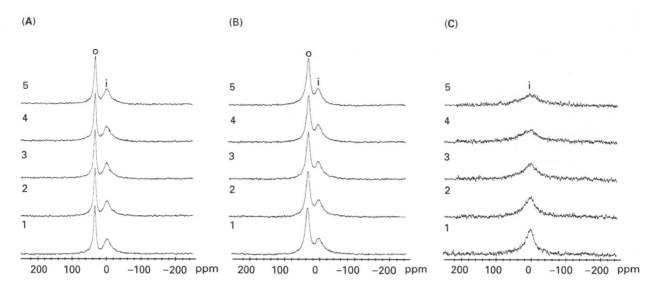

Figure 2 The ^{35}Cl NMR spectra at the mid-points of data accumulation (1) 20, (2) 60, (3) 100, (4) 140 and (5) 180 minutes of human erythrocyte suspensions at 70% haematocrit containing (A) 60 mM [Co(Gly)$_3$]⁻, 40 mM CoCl$_2$ (B) and 20 mM MnCl$_2$ (C)⁻. The symbols i and o denote the resonances originating from the intra- and extracellular halide, respectively. From Lin W and Mota de Freitas D (1996) *Magnetic Resonance in Chemistry* **34**: 768–772. Copyright John Wiley & Sons Limited. Reproduced with permission.

cases relatively sharp intracellular resonances that were suitably separated from the extracellular resonance were observed.

For many years it was believed that the ^{35}Cl⁻ resonance from chloride ions inside human erythrocytes was invisible owing to interaction of the ions with the paramagnetic protein haemoglobin. However, the shift reagent [Co(Gly)$_3$]⁻ and the relaxation agent Mn^{2+} have been used in separate studies that demonstrate the visibility of the intracellular chloride as a line that is ∼700 Hz wide (11.75 T magnet). The intracellular ^{35}Cl⁻ resonance shows a double Lorentzian line shape and two T_2 values of ∼0.09 and 1.2 ms, in keeping with a halide ion undergoing restricted motion. **Figure 2** shows typical spectra using both shift and relaxation agents. In the report using the shift reagent the exchange of intracellular Cl⁻ with extracellular hypophosphite (H$_2$PO$_2$)⁻ as a typical monovalent anion was followed by ^{35}Cl NMR. The exchange followed first order kinetics with a rate constant of $2.4 \pm 0.8 \times 10^{-3}$ min⁻¹. This work shows the potential of ^{35}Cl NMR to study chloride transport kinetics in cellular systems such as erythrocytes.

Manganese(II) has been used in conjunction with model vesicle systems to examine membrane transport induced by natural and synthetic anion carriers.

There are problems, however, with the use of Co^{2+} and Mn^{2+} as contrast reagents because it has been shown that biological membranes are very slowly permeable to these ions. It is, however, claimed that the shift reagent [Co(Gly)$_3$]⁻ does not show

membrane permeability and is, therefore, more suitable for studies on model systems.

The other way that ^{35}Cl NMR can be used in biological systems is in the study of chloride bound to proteins or other macromolecules. Sometimes this can be done by direct visualization of the resonance or from its free induction decay. More commonly it is performed by relaxation time measurements provided that there is rapid exchange of bound and free anions, because bound chloride relaxes very much more rapidly than free chloride. Frequently, double quantum filtered (DQF) spectra are employed which give useful information on chlorine bound to sites on the macromolecule.

The most studied system has undoubtedly been the integral membrane band III protein which mediates the one-for-one exchange of hydrogencarbonate and chloride ions across the membrane, thus playing an integral part in respiration. NMR proved useful in verifying a ping-pong mechanism for the mode of action of this protein. Binding sites for chloride at both surfaces have been identified although the cytoplasmic face has the larger number. Chloride affinities for the binding sites determined by NMR, DQF spectra and ^{35}Cl⁻ and ^{37}Cl⁻ relaxation rates at a variety of field strengths, both in absence and presence of various inhibitors, have also given large amounts of information about the operation of this important protein.

In a typical example, DQF ^{35}Cl NMR was used to study the binding of Cl⁻ to external sites on the surface of red blood cells. A DQF ^{35}Cl signal was observed in cell suspensions containing 150 mM KCl, but the DQF signal was totally eliminated by adding 500 µM 4,4'-dinitrostilbene-2,2'-disulfonate (DNDS), an inhibitor that interferes with Cl⁻ binding to the band III protein transport site. This result shows that only the binding of Cl⁻ to transport sites of band III can give rise to a ^{35}Cl DQF signal from red blood cell suspensions. In accordance with this concept, analysis of the single quantum FID revealed that signals from buffer and DNDS-treated cells could be fitted with a single exponential function, whereas the FID signals of untreated control cells were biexponential. The band III-dependent DQF signal is thus caused, at least in part, by non-isotropic motions of Cl⁻ in the transport site, resulting in incompletely averaged quadrupolar couplings.

Other biological systems that have been investigated by such techniques include sarcoplasmic reticulum membranes, superoxide dismutase, dromedary haemoglobin, the oxygen-evolving complex of spinach photosystem II, the zinc binding sites of human α-2-macroglobulin, alkaline phosphatase and halorhodopsin chromoprotein.

Spin–spin coupling

The rapid quadrupolar relaxation of halogens, in general, effectively masks fine structure arising from scalar coupling between the halogen and other covalently bound magnetically active nuclei. Exceptions occur when the halogen experiences a low EFG as, for example, when it is bound in sites of high symmetry. Thus, for the octahedral IF_6^+ ion, the ^{19}F resonance appears as a sextet due to coupling to the ^{127}I nucleus ($I = 5/2$). Analogously, the ^{19}F spectrum of BrF_6^+ appears as two quartets due to coupling to the two almost equally populated Br isotopes (for both of which $I = 3/2$). In the perhalate ions XO_4^-, when labelled with ^{17}O, it is possible to measure coupling between the halogen and the oxygen. **Figure 3** shows the ^{127}I spectrum of singly ^{17}O-labelled IO_4^-. The six lines arising from coupling between the ^{127}I and the ^{17}O are visible. For non-symmetric sites where direct determination of coupling constants is difficult it is sometimes possible to make use of the scalar contribution to the relaxation of a dipolar nucleus ($I = 1/2$) coupled to the quadrupolar halogen. Typical values of one–bond J couplings to Cl, Br and I are given in **Table 2**.

Residual dipolar coupling in solids

In the solid state the relaxation times of quadrupolar nuclei are typically several orders of magnitude greater than in solution. One important consequence

Table 2 Illustrative values of one-bond coupling constants to Cl, Br and I

Compound	Spin–spin coupling constant (Hz)
HCl	$J(^{35}Cl-^1H) = 41 \pm 2$
PCl_3	$J(^{35}Cl-^{31}P) = 120$
$SnCl_4$	$J(^{35}Cl-^{119}Sn) = 375$
$PbCl_4$	$J(^{35}Cl-^{207}Pb) = 705$
ClF_3	$J(^{35}Cl-^{19}F) = 260$
ClF_6^+	$J(^{35}Cl-^{19}F) = 337$
HBr	$J(^{79}Br-^1H) = 57 \pm 3$
$SnBr_4$	$J(^{81}Br-^{119}Sn) = 920$
BrF_6^+	$J(^{79}Br-^{19}F) = 1575$
SnI_4	$J(^{127}I-^{119}Sn) = 1097$
IF_6^+	$J(^{127}I-^{19}F) = 2730 \pm 15$
ClO_4^-	$J(^{35}Cl-^{17}O) = 83 \pm 3$
BrO_4^-	$J(^{79}Br-^{17}O) = 408 \pm 6$
BrO_4^-	$J(^{81}Br-^{17}O) = 440 \pm 6$
IO_4^-	$J(^{127}I-^{17}O) = 489 \pm 6$

of these longer relaxation times is the ability, sometimes, to determine indirect spin–spin coupling constants, J, involving quadrupolar nuclei. These experiments are performed typically by high resolution cross-polarization magic-angle spinning experiments on spin $\frac{1}{2}$ nuclei that are coupled to the quadrupolar nucleus. The longer quadrupolar relaxation times in solids mean that direct dipolar interactions involving quadrupolar nuclei are not completely averaged by the MAS. This is because the magnetic moments of the quadrupolar nuclei are quantized, not only by the magnetic field, B_0 but also by interaction with the electric field gradient tensor. Consequently the magic angle (described by the familiar $3 \cos^2 \theta - 1 = 0$) is no longer magic and fails to reduce the dipolar interaction to 0. Only since 1993 have examples of residual dipolar coupling of chlorine nuclei to ^{13}C been reported but many examples are now known. The residual dipolar couplings of chlorine to directly bonded ^{13}C are of the order of several hundred Hz and are field dependent, the coupling decreasing with increasing magnetic field. The observed ^{13}C doublets contain overlapping components from coupling of both ^{35}Cl and ^{37}Cl to the ^{13}C nucleus. Analogous residual dipolar coupling between $^{79/81}$Br and ^{13}C has been observed in 1,4-dibromobenzene and 4-bromo-3,5-dimethylpyrazole.

Shielding

From a consideration of the section above on quadrupolar relaxation of covalent compounds it will be realized that, in all but exceptional cases, the large line widths of the resonances give rise to a low precision in the determination of chemical shifts for Cl, whilst shielding studies for Br and I are virtually impossible. Only in molecules in which the halogen is present in a site of high symmetry can chemical shifts (relative to the free halide ion in aqueous solution as standard) be determined with precision. In the case of halogens present in non-ligated anions the chemical shift generally has a low dependence on the accompanying cation. Ligation can, however, have a substantial effect upon the chemical shift (see below). The largest deshieldings, relative to Cl$^-$, are to be found in species such as ClO_3F (978.1 ppm)

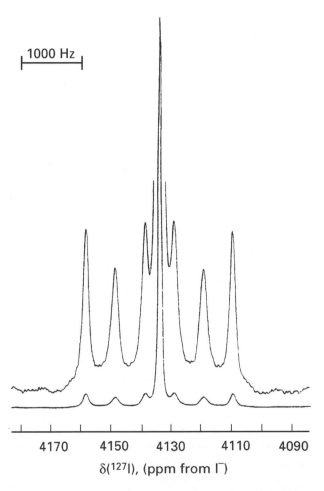

Figure 3 ^{127}I spectrum (50.05 MHz) of 0.1 M ^{17,18}O-enriched Et$_4$NIO$_4^-$ in CH$_3$CN at 24°C. The central line arises from I^{16}O$_4^-$ (46%) and I^{18}O^{16}O$_3^-$ (17%); the sextet arises from I^{17}O^{16}O$_3^-$ (29%) and I^{17}O^{18}O^{16}O$_2^-$ (6%). The vertical expansion reveals the slight variation in line widths of the sextet. From Dove MFA, Sanders JCP and Appelman EH (1995) Comparative multinuclear (^{35}Cl, ^{79}Br, ^{81}Br, ^{127}I and ^{17}O) magnetic resonance study of the perhalate anions XO$_4^-$ (X = Cl, Br or I). *Magnetic Resonance in Chemistry* **33**: 44–58. Copyright John Wiley & Sons Limited. Reproduced with permission.

and ClO_4^- (1002.5 ppm). The corresponding aqueous shifts for perbromate and periodate are 2478 and 4089 ppm, respectively. The normal chemical shift reference for halogen NMR is the frequency of the aqueous halide ion. Differences in the chemical environment of halogens are generally far more readily studied by using relaxation methods than by looking for chemical shift differences.

Halide–ligand interactions

Halide ions may be chelated by ligands that contain suitable arrays of hydrogen-bonding N–H groups. The binding is stronger if the NH is present as part of a positively charged ammonium ion. The ligands (shown as [1] and [2]) form tetra- and hexaammonium ions, respectively, and bind a halide strongly inside the cavity with hydrogen bonds pointing inwards. The chelation of halide ions by ligands of this type is demonstrated by the generation of very substantial chemical shifts for the halide ions either as the ligand is added to an aqueous solution, or as the pH of the solution is lowered to generate the ammonium ion centres. Line widths of the encapsulated ^{35}Cl$^-$ ions range from tens to thousands of Hz, depending upon the particular case. The larger line widths demonstrate that there are appreciable electric field gradients present at the chloride nucleus due to transient deformations of the coordination sphere away from ideal symmetry.

Figure 4 illustrates the use of ^{35}Cl NMR to study the binding of chloride ions to the ligand [1] (as the 4H$^+$ species) as the ratio of free to bound Cl$^-$ is increased. The chloride ion enters the centre of the ligand and is held in position by four inward facing hydrogen bonds from the N$^+$–H groups. The signal near 50 ppm is from the coordinated ^{35}Cl$^-$ and is relatively sharp owing to the tetrahedral site symmetry. The signal at 0 ppm is from free ^{35}Cl$^-$, which at relatively small amounts of free Cl$^-$ is 300 Hz wide due principally to dynamic line broadening that arises from exchange with the bound Cl$^-$. At an equal concentration of bound and free Cl$^-$ the contributions to the line width of the free ^{35}Cl$^-$ of 80 Hz are 36 Hz from exchange and 44 Hz from quadrupolar interactions. At higher relative concentrations of free Cl$^-$ the exchange contribution to the line width diminishes still further but the free ^{35}Cl$^-$ line remains relatively broad probably owing to attraction of the free Cl$^-$ ions to the exterior of the triply positively charged complex.

Halide ions show a weak association with cyclodextrins which can be visualized by the quadrupolar line broadening of the halide resonances. Competition between bromide ions and other ions for the

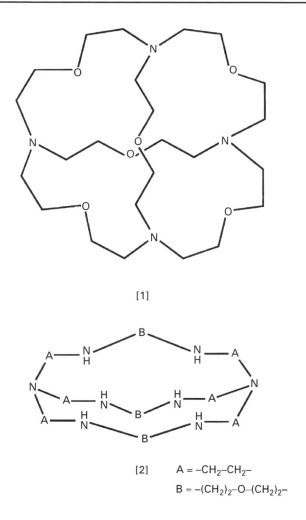

[1]

[2]   A = –CH$_2$–CH$_2$–

        B = –(CH$_2$)$_2$–O–(CH$_2$)$_2$–

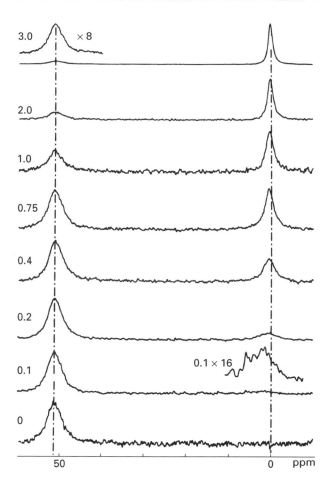

Figure 4 ^{35}Cl NMR spectra (at 39.2 MHz) of aqueous solutions of the Cl$^-$ complex of ligand [1]-4H$^+$ (20 mM) and uncomplexed Cl$^-$ as a function of the ratio of free Cl$^-$ to Cl$^-$ complex. The shifts are given in ppm relative to external aqueous NaCl (0.1 M); pH 1.5 and 20°C. Reprinted with permission from Kintzinger JP, Lehn J-M, Kauffmann E, Dye JL and Popov AI (1983) Anion coordination chemistry – ^{35}Cl NMR studies of chloride anion cryptates. *Journal of the American Chemical Society* **105**: 7549–7553. Copyright 1983 American Chemical Society.

binding site in cyclodextrins can be studied by the reduction in the ^{81}Br$^-$ line width as other electrolytes are added to an aqueous solution containing cyclodextrin and KBr.

Halide–cation interactions

Interaction between ions of opposite charge is an important aspect of the chemistry of electrolyte solutions. The line widths of quadrupolar nuclei are an important probe of such interactions since the closer two ions of opposite charge approach the greater the EFG at their nuclei. This in turn can lead to a stronger quadrupole interaction, more rapid quadrupolar relaxation and larger line widths. Thus studies of the concentration dependence of quadrupolar relaxation are probes for ionic interactions. Such concentration dependence varies markedly with the cation present. Most monovalent ions show a low dependence of quadrupolar relaxation with concentration but this is not the case with alkyl substituted ammonium ions. In one typical such study ^{35}Cl, ^{81}Br and ^{127}I NMR was used to study the interaction of

halide ions with alkylammonium ions. The intrinsic line widths (line width at $\frac{1}{2}$ height corrected for temperature and viscosity) of the halide ions, plotted against molality (m) of the salt, show distinct plateau in the region 1–4 m with the effect being greatest for iodide. These findings suggest the presence of contact ion pairs in these concentration ranges.

Similarly, the T_2 values of ^{35}Cl$^-$ ions show a pH dependence in the presence of polybasic molecules such as polyamines, or amino acids such as histidine or arginine. The relaxation rates increase as the pH is lowered and show steepest changes at pH values equivalent to the pK_a values for the various titratable groups. This is consistent with increasing ionic interactions between the anion as the number of positive charges on the organic substrate is increased.

Halogens in conducting and semiconducting lattices

The ionic mobility of halogen-containing ions in solids, polymeric matrices and solutions may be studied by halogen NMR techniques. The self-diffusion of perchlorate ions in aqueous solutions of $NaClO_4$, $LiClO_4$ and $Mg(ClO_4)_2$ has been studied using pulsed field gradient techniques.

Both ^{81}Br and ^{35}Cl MAS NMR have been used in the investigation of sodium, silver and halo-sodalite semiconductor supralattices. Useful information is available on the environments of the encapsulated clusters within the sodalite lattice. The Na_4X (X = Cl, Br) tetrahedra provide a symmetric environment around the halide and give rise to narrow resonances for specific locations. The spectra are sensitive to the distribution of anion empty cavities or to I^--containing cavities.

^{35}Cl and $^{79\&81}Br$ T_1 measurements between 77 and 700 K shed light on ionic transport in (solid) silver halides. A hopping process due to thermally activated diffusion of defects accounts for the data. The chloride ion conductor $CH_3NH_3GeCl_3$ has a variety of solid state phases and shows electrical conductivity above 398 K. The mobile ion in the high temperature phase was deduced to be Cl^- because a disordered perovskite structure at the Cl site was found by Rietveld analysis of X-ray powder data and confirmed by ^{35}Cl NMR. At temperatures above 364 K a ^{35}Cl resonance becomes visible. This suggests that the rate of isotropic motion at the Cl^- site is becoming comparable with or greater than the quadrupolar coupling, thus reducing the importance of the quadrupolar interactions.

Two-dimensional spectroscopy

Two-dimensional multiple-quantum (2D2Q) ^{35}Cl NMR spectra of systems exhibiting electric quadrupolar effects have been obtained. In such spectra the single quantum spectrum is displayed along one axis against the double quantum spectrum along the other. Such spectra give information about the quadrupolar interactions present in ordered systems.

List of symbols

I = nuclear spin quantum number; Q = quadrupole moment; η = asymmetry parameter; τ_c = correlation time for molecular rotation; ω_0 = Larmor frequency.

See also: Contrast Mechanisms in MRI; Membranes Studied By NMR Spectroscopy; Quadrupoles, Use of in Mass Spectrometry; Relaxometers; Spin Trapping and Spin Labelling Studied Using EPR Spectroscopy.

Further reading

Dove MF, Sanders JCP, and Appelman E (1995) Comparative multinuclear (^{35}Cl, ^{79}Br, ^{81}Br, ^{127}I and ^{17}O) magnetic resonance study of the perhalate anions XO_4^- (X = Cl, Br or I). *Magnetic Resonance in Chemistry* **33**: 44–58.

Drakenberg T, and Forsén S (1983). In: Lambert JB and Riddell FG (eds) *The Multinuclear Approach to NMR Spectroscopy*, NATO ASI Series C, p 405. Boston: Reidel.

Kintzinger JP, Lehn J-M, Kauffmann E, Dye JL and Popov, AI (1983) Anion co-ordination chemistry – ^{35}Cl NMR studies of chloride anion cryptates, *Journal of the American Chemical Society* **105**: 7549–7553.

Lindman B and Forsén S (1976) *Chlorine Bromine and Iodine NMR. Physico-Chemical and Biological Applications.* Heidelberg: Springer Verlag.

Lindman B and Forsén S (1978) The halogens – chlorine, bromine and iodine. In: Harris RK and Mann BE (eds) *NMR and the Periodic Table*. London: Academic Press.

Liu DS, Kennedy SD and Knauf PA (1996) Source of transport site asymmetry in the band-3 anion exchange protein determined by NMR measurements of external Cl^- affinity. *Biochemistry* **35**: 15228–15235.

Liu DS, Knauf PA and Kennedy SD (1996) Detection of Cl^- binding to band-3 by double quantum filtered ^{35}Cl nuclear magnetic resonance. *Biophysics Journal* **70**: 715–722.

Price WS, Kuchel PW and Cornell BA (1991) A ^{35}Cl and ^{37}Cl NMR study of chloride binding to the erythrocyte anion transport protein. *Biophysical Chemistry* **40**: 329–337.

Riddell FG and Zhou Z (1995) NMR visibility of $^{35}Cl^-$ in human erythrocytes. *Magnetic Resonance in Chemistry*, **33**: 66–69.

Roberts JE and Schnitker J (1993) Ionic quadrupolar relaxation in aqueous solution – dynamics of the hydration sphere. *Journal of Physical Chemistry* **97**: 5410–5417.

Shachar-Hill Y and Shulman RG (1992) Co^{2+} as a shift reagent for ^{35}Cl NMR of chloride with vesicles and cells. *Biochemistry* **31**: 6272–6278.

Springer CS (1996) Biological systems, spin-3/2 nuclei. In: Grant DM and Harris RK (eds) *Encyclopedia of Nuclear Magnetic Resonance*, p 940. Chichester: Wiley.

Xu Y, Barbara TM, Rooney WD and Springer CS (1989) Two-dimensional multiple-quantum NMR spectroscopy of isolated half-integer spin systems. 2. ^{35}Cl examples. *Journal of Magnetic Resonance* **83**: 279–298.

Heteronuclear NMR Applications (As, Sb, Bi)

Claudio Pettinari, **Fabio Marchetti** and
Giovanni Rafaiani, Università di Camerino, Italy

> **MAGNETIC RESONANCE**
> Applications

All the elements of group 15 of the periodic table have at least one isotope suitable for study by NMR spectroscopy. The magnetically active isotopes of arsenic, antimony and bismuth (^{75}As, $I = \frac{3}{2}$; ^{121}Sb, $I = \frac{5}{2}$; ^{123}Sb, $I = \frac{7}{2}$; ^{209}Bi, $I = \frac{9}{2}$) have high abundance and relatively high sensitivity; nevertheless they have been little employed in multinuclear NMR spectroscopy owing to their high spin quantum numbers, which produce large quadrupolar moments. In fact, an efficient quadrupolar mechanism dominates their relaxation behaviour and gives rise to extremely broad resonances; this has limited the application of ^{75}As, ^{121,123}Sb and ^{209}Bi NMR spectroscopy to species having a high degree of symmetry. Only a few NMR studies have been reported in the last three decades, and only a small amount of data, often useful in the determination of the molecular structure, is currently available.

Properties of the nuclei

^{75}As is 100% abundant with a receptivity of 2.5×10^{-2} relative to the proton; ^{121}Sb is 57% abundant with a receptivity of 9.2×10^{-2}; and ^{209}Bi is 100% abundant with a receptivity of 0.14×10^{-2}; by contrast, ^{123}Sb is about 43% abundant with a receptivity of 1.9×10^{-2}. In the case of antimony, the nuclide generally chosen for NMR determinations is ^{121}Sb, presumably owing to the lower abundance of ^{123}Sb.

The group 15 nuclides can be studied by NMR spectroscopy only when the E-containing compounds (E = As, Sb, Bi) are highly symmetric, with E in the oxidation state +5. These nuclides are spectroscopically silent in nonsymmetric compounds or in the +3 oxidation state. Narrow line widths arise from quadrupolar nuclei residing at the centre of a highly symmetric ligand environment. In addition, in oxidation state +5, no lone pairs remain in the valence shell and tetrahedral or octahedral symmetry can be achieved through bonding of four or six donor ligands, as in the case of $(SbCl_4)^+$ and $(SbCl_6)^-$. By contrast, when the group 15 elements are in oxidation state +3, there is a lone pair in the valence shell and the structures are generally characterized by substantial electric field gradients at the central atom, which prevents observation of NMR resonance.

The main nuclear properties of group 15 nuclides are shown in **Table 1**.

Experimental aspects

^{75}As, ^{121}Sb, ^{123}Sb and ^{209}Bi NMR spectra can be obtained efficiently using a broad-band probe. The observation frequencies are 85.637 MHz (^{75}As), 119.696 MHz (^{121}Sb), 64.819 MHz (^{123}Sb) and 80.381 MHz (^{209}Bi) when ^{1}H NMR is at 500 MHz. The ideal spectral width setting is 20 kHz, to yield a data point resolution of 4.9 Hz per data point and an acquisition time of 0.205 s. The number of transients that must be accumulated varies with the nuclide under observation and with the concentration of the sample; however, typical values are 14 000 (^{75}As), 7000 (^{121}Sb), 36 000 (^{123}Sb) and 200 000–400 000 (^{209}Bi). A suitable line-broadening

Table 1 Selected nuclear properties of magnetically active isotopes of As, Sb and Bi

Isotope	^{75}As	^{121}Sb	^{123}Sb	^{209}Bi
Nuclear spin quantum number, I	$\frac{3}{2}$	$\frac{5}{2}$	$\frac{7}{2}$	$\frac{9}{2}$
Natural abundance (%)	100	57.25	42.75	100
Magnetogyric ratio (10^{-7} rad T^{-1} s^{-1})	4.5804	6.4016	3.4668	4.2986
Quadrupole moment	0.3	−0.53	−0.68	−0.4
Resonance frequency / MHz referred to ^{1}H TMS resonance at 100 MHz	17.122	23.930	12.958	16.069
Receptivity relative to ^{13}C	143	520	111	777
Receptivity relative to ^{1}H	2.5×10^{-2}	9.2×10^{-2}	1.9×10^{-2}	0.14

parameter can be applied by the exponential multiplication of the free induction decay prior to Fourier transformation. A line broadening factor equal to the original line width will produce the optimum signal-to-noise ratio.

The chemical shift standards generally used (at room temperature) are $0.1\,M$ $[(Et_4N)^+(AsF_6)^-]$ in CH_3CN for ^{75}As, $0.3\,M$ $[(Et_4N)^+(SbCl_6)^-]$ in CH_3CN for $^{121,123}Sb$, and a saturated solution of $[(Me_4N)^+(BiF_6)^-]$ in CH_3CN for ^{209}Bi. In some cases, $0.1\,M$ $[(Et_4N)^+(SbF_6)^-]$ in CH_3CN as standard (0.0 ppm) for antimony has also been used. All the antimony data reported here are referred to $[(Et_4N)^+(SbCl_6)^-]$ at 0 ppm.

Chemical shifts

^{75}As, $^{121,123}Sb$ and ^{209}Bi chemical shift ranges are very large (**Tables 2** and **3**). A positive sign indicates a chemical shift to higher frequency (lower shielding) with respect to the reference compound.

In the case of arsenic(+5) derivatives, the chemical shift range spans approximately 700 ppm, with the tetrahedral oxoanion $(AsO_4)^{3-}$ and the cation $(AsH_4)^+$ occupying the deshielded and the shielded ends of the range, respectively.

Although there is limited chemical shift data in the case of ^{75}As and ^{121}Sb it is possible to observe some general patterns analogous to those found for

Table 2 ^{75}As NMR data for selected arsenic-containing compounds

Compound	Symmetry	Solvent	δ (ppm)	Multiplicity	$^nJ(^{75}As-M)$(Hz)
$[AsF_6]^-$	O_h	CH_3CN	0.0	st	932 $^1J(^{75}As-^{19}F)$
$[AsClF_5]^-$	O_h	CH_3CN	12.2	q of d	ax.897 $^1J(^{75}As-^{19}F)$ eq.1009 $^1J(^{75}As-^{19}F)$
$[cis\text{-}AsF_4Cl_2]^-$	O_h	CH_3CN	−0.3	q	1013 $^1J(^{75}As-^{19}F)$
$[trans\text{-}AsF_4Cl_2]^-$	O_h	CH_3CN	−42.4	qu	1259 $^1J(^{75}As-^{19}F)$
$[fac\text{-}AsF_3Cl_3]^-$					
$[mer\text{-}AsF_3Cl_3]^-$					
$[cis\text{-}AsF_2Cl_4]^-$	O_h	CH_3CN	−102.2	t	985 $^1J(^{75}As-^{19}F)$
$[trans\text{-}AsF_2Cl_4]^-$					925 $^1J(^{75}As-^{19}F)$
$[AsFCl_5]^-$	O_h	CH_3CN	−212.4	d	1005 $^1J(^{75}As-^{19}F)$
$[AsCl_6]^-$	O_h	CH_3CN	−391.8	s	555 $^1J(^{75}As-^1H)$
$[AsF_5Br]^-$	O_h	CH_3CN	102.5	st	
$[AsO_4]^{3-}$	T_d	CH_3CN	369	s	
$[AsH_4]^+$	T_d	HF	−291	q	
$[AsMe_4]^+$	T_d	H_2O	206	s	
$[AsEt_4]^+$	T_d	H_2O	249	s	
$[AsPr^n_4]^+$	T_d	H_2O	230	s	
$[AsBu^n_4]^+$	T_d	H_2O	234	s	
$[AsEt_3Me]^+$	T_d	H_2O	242	s	
$[AsEt_3Pr^n]^+$	T_d	H_2O	243	s	
$[AsEt_3Bu^n]^+$	T_d	H_2O	245	s	
$[AsPr^n_3Me]^+$	T_d	H_2O	225	s	
$[AsPr^n_3Et]^+$	T_d	H_2O	234	s	
$[AsPr^n_3Bu^n]^+$	T_d	H_2O	231	s	
$[AsPr^i_3Et]^+$	T_d	H_2O	258		
$[AsPh_4]^+$	T_d	H_2O	217	s	
$Cs[As(OTeF_5)_6]$	O_h	CH_3CN	−28.9	st	420 $^2J(^{75}As-^{125}Te)$
$NMe_4[As(OTeF_5)_6]$	O_h	CH_3CN	−29.1	st	430 $^2J(^{75}As-^{125}Te)$
$Te(OTeF_5)_3[As(OTeF_5)_6]$	O_h	SO_2	−28.2	st	432 $^2J(^{75}As-^{125}Te)$
$H_3AsW_{12}O_{40}$	T_d	H_2O/dioxane	298	s	
$H_3AsMo_{12}O_{40}$	T_d	H_2O/dioxane	337	s	

Pr^n = n-propyl; Bu^n = n-butyl; Pr^i = isopropyl; st = septet, q = quartet; qu = quintet; t = triplet; d = doublet; s = singlet.

Table 3 ^{121}Sb NMR data for selected antimony-containing compounds

Compound	Symmetry	Solvent	δ (ppm)	Multiplicity	$^nJ(^{121}Sb–M)$ (Hz)
$[SbF_6]^-$	O_h	CH_3CN	86.6	st	1938 $^1J(^{121}Sb–^{19}F)$
$[SbF_5Cl]^-$	O_h	CH_3CN	149.3	q of d	ax. 1859 $^1J(^{121}Sb–^{19}F)$
					eq. 2088 $^1J(^{121}Sb–^{19}F)$
$[cis\text{-}SbF_4Cl_2]^-$	O_h	CH_3CN	183.9	q	2083 $^1J(^{121}Sb–^{19}F)$
$[trans\text{-}SbF_4Cl_2]^-$	O_h	CH_3CN	187.5	qu	1980 $^1J(^{121}Sb–^{19}F)$
$[fac\text{-}SbF_3Cl_3]^-$					
$[mer\text{-}SbF_3Cl_3]^-$					
$[cis\text{-}SbF_2Cl_4]^-$	O_h	CH_3CN	174.4	t	1981 $^1J(^{121}Sb–^{19}F)$
$[trans\text{-}SbF_2Cl_4]^-$					
$[SbCl_5F]^-$	O_h	CH_3CN	114.9	d	2079 $^1J(^{121}Sb–^{19}F)$
$[SbCl_6]^-$	O_h	CH_3CN	0.0	s	
$[SbCl_4]^-$	T_d	SO_2ClF	847.0	s	
$[Sb(OTeF_5)_6]^-$	O_h	SO_2ClF	72.9	s	
$[SbBr_4]^+$	T_d	SO_2ClF	95.3	s	
$[Sb(OTeF_5)_6]^+$	O_h	SO_2ClF	73.1	s	
$NMe_4[Sb(OTeF_5)_6]$	O_h	CH_3CN	72.7	s	
$[SbCl_5Br]^-$	O_h	CH_3CN	−380	s	
$[SbCl_4Br_2]^-$	O_h	CH_3CN	−780	s	
$[SbCl_3Br_3]^-$	O_h	CH_3CN	−1180	s	
$[SbCl_2Br_4]^-$	O_h	CH_3CN	−1590	s	
$[SbClBr_5]^-$	O_h	CH_3CN	−2005	s	
$[SbBr_6]^-$	O_h	CH_3CN	−2430	s	
$[SbS_4]^{3-}$	T_d	H_2O	1032	s	
$[SbMe_4]^+$	T_d	H_2O	780	s	
$[Sb(OH)_6]^-$	O_h	H_2O	296	s	

st = septet, q = quartet; d = doublet; qu = quintet; t = triplet; s = singlet.

other nuclides, such as group 14 elements. For example, even if there seems to be no regularity among the shifts of the AsR_4^+ compounds, the ordering in terms of alkyl substituent is consistent with that found for group 14 and group 15 alkyls: labelling the atoms of the structural unit $[As(alkyl)_4]^+$ as $As_\alpha\text{-}C_\beta\text{-}C_\gamma\text{-}C_\delta$, methyl substitution for one hydrogen at the β position ($\Delta\beta$) deshields by 11 ppm, $\Delta\gamma$ shields by 5 ppm, and $\Delta\delta$ shields by 1 ppm. This trend in shielding effect as one proceeds down the chain is characteristic of methyl substitution.

In $(AsF_{6-n}Cl_n)^-$ ($n = 0–6$) derivatives, the trend observed for F/Cl exchange in ^{75}As chemical shift also suggests an increasing shielding with increasing n, owing to lower electronegativity of chloride with respect to fluoride donor ligands; a further example is the difference in chemical shift for $(AsF_5Cl)^-$ (12.2 ppm) and $(AsF_5Br)^-$ (102.5 ppm).

In the case of the two heteropoly anions $(AsW_{12}O_{40})^{3-}$ and $(AsMo_{12}O_{40})^{3-}$ ('Keggin' anions), the chemical shifts reflect the influence of electron density surrounding the arsenic atom, which is dependent on the nature of peripheral metal atoms: as expected, the As atom will be shielded to a greater degree in the former anion that in the latter, the same trend being observed for phosphate anions. Moreover, both arsenic-containing heteropoly anions display chemical shifts that are upfield from the relatively deshielded arsenate ion $(AsO_4)^{3-}$.

The antimony(+5) chemical shift range spans approximately 3500 ppm. Rather more information is available for ^{121}Sb than for ^{75}As NMR chemical shifts. To date, $(SbS_4)^{3-}$ and $(SbBr_6)^-$ occupy the deshielded and shielded ends of the range, respectively.

In $(SbF_6)^-$ and $(SbCl_6)^-$ the small difference between the antimony shifts (~87 ppm) compared with those for analogous ^{31}P and ^{75}As species (152 and 392 ppm, respectively) is probably due to the electronic excitation energy that in both antimony and, more likely, arsenic is responsible for the anomalous paramagnetic shielding term.

^{121}Sb chemical shifts fall into the normal halogen dependence category and, in the case of the mixed halide compounds $(SbX_nY_{6-n})^-$, it is possible to assign the *cis* and *trans* isomers by the pairwise additivity parameter, calculated using Equation [1]:

$$\delta_{calc} = \frac{n}{6}\,\delta_{SbCl_6^-} + \frac{6-n}{6}\,\delta_{SbBr_6^-} \qquad [1]$$

The dominant contribution to the shielding, σ_p, can be expressed in the form

$$\sigma_p = -\left(\frac{2e^2\hbar^2}{3\,\Delta m\,c^2}\right)\left(\left\langle\frac{1}{r^3}\right\rangle_p P_u + \left\langle\frac{1}{r^3}\right\rangle_d D_u\right) \qquad [2]$$

The large variation in Sb shielding arises from the dominant and negative paramagnetic shielding term, σ_p. The magnitude of σ_p is directly proportional to changes in the inverse mean excitation energy (ΔE^{-1}), the mean inverse cubes of the p and d electron–nucleus distances $(\langle r^{-3}\rangle_{np}$ and $\langle r^{-3}\rangle_{nd})$, and the valence imbalance in the p and d orbitals centred on the Sb atom $(P_u$ and $D_u)$.

The relative chemical shifts of the SbX_4^+ and SbX_6^- ions follow the valence imbalance terms and are dependent upon the electron density and relative electronegativities of the halogens. The electronegativity difference of chlorine and bromine accounts quantitatively for the high-frequency shifts of $(SbCl_4)^+$ (847.0 ppm) and $(SbCl_6)^-$ (0.0 ppm) relative to those of $(SbBr_4)^+$ (95.3 ppm) and $(SbBr_6)^-$ (−2430 ppm), and the formal positive charge, which is expected to reside mainly on the Sb atom, accounts for the high-frequency shift of the $(SbX_4)^+$ cations compared to the $(SbX_6)^-$ anions, which have their formal negative charge located on the halogens.

Only one value is reported for antimony compounds having a symmetry other than octahedral or tetrahedral: $SbCl_5$ resonates at +509 ppm, but the line width is about 8000 Hz. This very broad absorption band illustrates the effect on the rate of quadrupolar relaxation as the geometry around the antimony nucleus is reduced from O_h or T_d to trigonal bipyramidal (C_{3v}).

In the case of the heaviest bismuth nuclide, only a few ^{209}Bi NMR data are available yet. For example, taking $(BiF_6)^-$ (0.0 ppm, 2700 Hz $^1J(^{209}Bi–^{19}F)$) as reference, the anion $[Bi(OTeF_5)_6]^-$ is shifted to +126.7 ppm (2269 Hz $^1J(^{209}Bi–^{125}Te)$) (**Figure 1**), contrary to what is observed in analogous $[As(OTeF_5)_6]^-$ and $[Sb(OTeF_5)_6]^-$ compounds, which exhibit low frequencies with respect to $(AsF_6)^-$ (**Figure 2**) and $(SbF_6)^-$, respectively. This anomalous

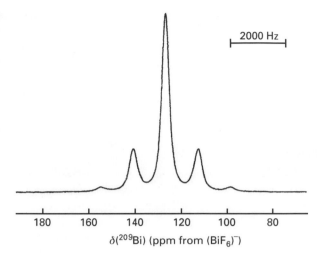

Figure 1 ^{209}Bi NMR spectrum of $[Bi(OTeF_5)_6]^-$. Reproduced with permission from Mercier HPA, Sanders JCP and Schrobilgen GJ (1994) Hexakis(pentafluorooxotellurato)pnictate(V) anions, M(OTeF$_5$)$_6$ (M = As, Sb, Bi): a series of very weakly coordinating anions. *Journal of the American Chemical Society* **116**: 2921–2937. © 1994 American Chemical Society.

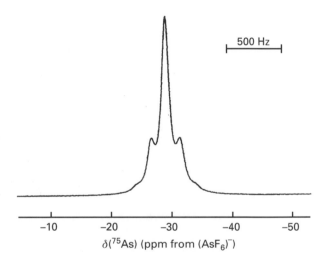

Figure 2 ^{75}As NMR spectrum of $[As(OTeF_5)_6]^-$. Reproduced with permission from Mercier HPA, Sanders JCP and Schrobilgen GJ (1994) Hexakis(pentafluorooxotellurato)pnictate(V) anions, M(OTeF$_5$)$_6$ (M = As, Sb, Bi): a series of very weakly coordinating anions. *Journal of the American Chemical Society* **116**: 2921–2937. © 1994 American Chemical Society.

behaviour could be caused by relativistic effects of the bismuth atom.

Coupling constants

In the case of ^{75}As, only one $J(^{75}As–^1H)$ (555 Hz) value, related to the $(AsH_4)^+$ cation, has been reported, whereas several $^1J(^{75}As–^{19}F)$ values have been obtained from the spectra of $(AsF_nCl_{6-n})^-$ derivatives,

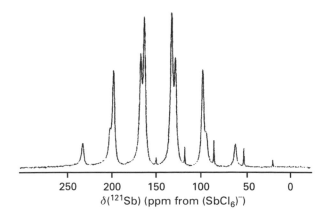

Figure 3 ^{121}Sb NMR spectrum of $[SbF_nCl_{6-n}]^-$. Reproduced with permission from Dove MFA and Sanders JCP (1992) Synthesis and characterization of chlorofluoroantimonates(V). *Journal of the Chemical Society, Dalton Transactions* 3311–3316.

and these range from 897 to 1259 Hz depending on the position of the fluorine atom in the octahedron around arsenic. More recently, several $J(^{75}As–^{125}Te)$ values have been measured (420–430 Hz) in $[As(OTeF_5)_6]^-$ compounds.

Also in $[SbF_nCl_{6-n}]^-$ compounds, some $^1J(^{121}Sb–^{19}F)$ values have been observed (**Figure 3**): the typical range is 1859–2088 Hz and, as with arsenic, the values depend on the fluorine position around the antimony atom.

Relaxation behaviour

For a quadrupolar nucleus, the relaxation under conditions of extreme narrowing can be described by

$$\Delta v_{\frac{1}{2}} = \frac{1}{\pi T_2} = \frac{1}{\pi T_1} = \frac{3\pi}{10}\left(\frac{2I+3}{I^2(2I-1)}\right)$$
$$\times \left(\frac{e^2 qQ}{h}\right)^2\left(1+\frac{\eta^2}{3}\right)\tau_c \qquad [3]$$

where $\Delta v_{1/2}$ is the line width at half-height, T_2 is the spin–spin relaxation time, T_1 is the spin–lattice relaxation time, I is the nuclear spin quantum number, e is the charge on the electron, Q is the nuclear quadrupole moment, q is the electric field gradient (EFG) along the principal z axis, η is the symmetry parameter for the EFG, and finally τ_c is the rotational correlation time.

A careful inspection of several factors in Equation [3] reveals that line widths are dramatically reduced for nuclides having a high value of I or a small value of Q, or both. Moreover, τ_c serves to increase the $\Delta v_{1/2}$ of the quadrupolar anion with increasing ionic/

molecular radius, increased ion pairing and increasing solvent viscosity (decreasing temperature).

The resonance line widths for $(MF_{6-n}X_n)^-$ increase as n increases from 0 to 5 (with the exception of $(SbF)^-Cl_5$), and are also greater for X = Br than for X = Cl. This behaviour reflects the increase in the quadrupolar relaxation rate of the central nucleus as its octahedral environment is distorted by the replacement of F by the larger halogens. An analogous behaviour has been noted in the bromo-chloroantimonates(+5).

Although the larger radius of the $(SbBr_4)^+$ cation should result in a longer τ_c and larger $\Delta v_{1/2}$ values than for $(SbCl_4)^+$, the opposite effect is found; this is attributed to the greater Lewis acidity of the latter cation, which is expected to form with the very weakly basic SO_2ClF solvent a stronger, albeit weak, donor–acceptor bond than $(SbBr_4)^+$. An equilibrium concentration of the adduct leads to an increase in the EFG at the antimony nucleus and increased τ_c, both of which serve to decrease the relaxation time and increase $\Delta v_{1/2}$.

A more complete study of the factors influencing relaxation behaviour has been made on arsonium salts by altering the alkyl group, the temperature, the solvent, the concentration and the counterion. It has been shown that the line width of the resonances increases with increasing size of the alkyl group, i.e. with increasing molecular mass of the species under investigation; in fact, the isotropic motion of a solute in a mobile liquid can be correlated to the correlation time τ_c by

$$\tau_c = \frac{4\pi v a^3}{3k_B T} \qquad [4]$$

where v is the viscosity of the solution and a is the radius of the 'spherical' molecule. Moreover in the case of these small molecules, the hypothesis of extreme narrowing conditions ($\tau_c^2 \omega^2 \ll 1$, i.e. fast molecular motions with respect to the resonance frequency) seems very likely to be valid.

Combining Equations [3] and [4], a linear plot of $\Delta v_{1/2}$ vs the experimentally measurable value v/T is observed. The $\Delta v_{1/2}$ values, unlike ^{75}As chemical shifts, are generally strongly dependent on the type of solvent employed, partly surely owing to ion pairing; in fact, the line widths for $(AsEt_4)^+Br^-$ are reported to be 168, 455, 670 and 700 Hz in water, acetonitrile, dimethylformamide and dimethyl sulfoxide, respectively. Finally, the concentration of the sample influences the arsenic line width, whereas the counterion seems not to play a significant role in temperature variation.

Applications

NMR spectroscopy of ^{75}As, ^{121,123}Sb and ^{209}Bi is arousing increasing interest. ^{75}As, ^{121,123}Sb and ^{209}Bi NMR studies have been found to be very useful for structural characterization in solutions of very weakly coordinating anions such as $[M(OTeF_5)_6]^-$ (M = As, Sb, Bi). These are able to stabilize Meerwein salts (tertiary carbonium salts) and cations such as Cs^+, Rb^+, K^+ and Na^+ can help in the development of the field of stable carbocation chemistry and of protosolvated superelectrophiles.

^{75}As NMR spectroscopy has been applied in solution in elucidating the structures of heteropoly oxometalates $(AsM_{12}O_{40})^{3-}$ (M = W or Mo), currently known as 'Keggin' anions, which are important in both fundamental and applied studies.

In some cases, ^{75}As relaxation of the $(AsF_6)^-$ ion has been employed for probing anion sites in proteins by analysing the band shape of the ^{19}F multiplets in terms of the transition probabilities for ^{75}As relaxation. Outside the motional narrowing limit, the transition probabilities $P_{m,m+1}$ and $P_{m,m+2}$ for ^{75}As relaxation become unequal. However, the actual transition probabilities dominating the line shape are weighted averages of those in the free and bound states, since binding usually occurs under fast exchange limit conditions. In this way, it has been possible to reproduce the experimental ^{19}F line shape for the binding of $(AsF_6)^-$ to human serum albumin.

Finally, a ^{75}As solid-state NMR study has been done on the measurement of isotropic indirect coupling to gain a better understanding of the electronic nature of semiconductor solids of the binary 13–15 type compounds, such as GaAs. ^{75}As NMR has also been used to study the microscopic magnetic behaviour of an incipient heavy-electron compound UAs. It has been reported on the basis of the temperature dependence of the energy scale of the spin fluctuation that the relative importance of the exchange interaction with respect to the single-site dynamics in the moment relaxation generally increases on going from UP to UAs.

List of symbols

a = radius of a spherical molecule; c = speed of light; D_u = d orbital valence imbalance; e = electron charge; $\hbar$ = Planck constant/2π; I = nuclear spin quantum number; J = coupling constant; k_B = Boltzmann constant; m = mass; P_u = p orbital valence imbalance; $P_{m,n}$ = transition probabilities; q = electric field gradient; Q = nuclear quadrupole moment; r = electron–nucleus distance; T = temperature; T_1 = spin–lattice relaxation time; T_2 = spin–spin relaxation time; δ = chemical shift; ΔE = mean excitation energy; $\Delta \nu_{1/2}$ = line width at half-height; η = electric field gradient asymmetry parameter; σ = viscosity; σ = shielding parameter; τ_c = rotational correlation time; ω = angular frequency.

See also: **Nitrogen NMR; NMR Principles; NQR Applications; NQR, Theory; Parameters in NMR Spectroscopy, Theory of; Solid State NMR, Methods; Solid State NMR Using Quadrupolar Nuclei; Structural Chemistry Using NMR Spectroscopy, Inorganic Molecules.**

Further reading

Casteel WJ Jr, Kolb P, Leblond N, Mercier HPA and Schrobilgen GJ (1996) *Inorganic Chemistry* 35: 929–942.

Collins MJ and Schrobilgen GJ (1985) *Inorganic Chemistry* 24: 2608–2614.

Dixon KR (1987) Phosphorus to bismuth. In: Mason J (ed) *Multinuclear NMR*, pp 369–402. New York: Plenum Press.

Dove MFA and Sanders JCP (1992) *Journal of the Chemical Society, Dalton Transactions* 3311–3316.

Drakenberg T (1986) Nuclear magnetic resonance of less common quadrupolar nuclei. In: Webb GA (ed) *Annual Reports on NMR Spectroscopy*, Vol 17, 231–283. London: Academic Press.

Harris RK (1978) In: Harris RK and Mann BE (eds) *NMR and the Periodic Table*, Vol 11, 379–383. London: Academic Press.

Kidd RG (1983) Group V atom NMR spectroscopy other than nitrogen. In: Lambert JB and Riddell FG (eds) *The Multinuclear Approach to NMR Spectroscopy*, NATO ASI Series C, 379–382. Boston: Reidel.

Mercier HPA, Sanders JCP and Schrobilgen GJ (1994) *Journal of the American Chemical Society* 116: 2921–2937.

Morgan K, Sayer BG and Schrobilgen GJ (1983) *Journal of Magnetic Resonance* 52: 139.

Wehrli FW (1979) Nuclear magnetic resonance of the less common quadrupolar nuclei. In: Webb GA (ed) *Annual Reports on NMR Spectroscopy*, Vol 9, p 177. London: Academic Press.

Heteronuclear NMR Applications (B, Al, Ga, In, Tl)

Janusz Lewiński, Warsaw University of Technology, Poland

MAGNETIC RESONANCE
Applications

Introduction

The naturally occurring nuclei of the Group 13 elements, ^{10}B and ^{11}B, ^{27}Al, ^{69}Ga and ^{71}Ga, ^{113}In and ^{115}In, ^{203}Tl and ^{205}Tl, are NMR-accessible magnetically. The receptivities of these nuclei are quite high and, except for thallium, the resonance frequencies are convenient for most multinuclear spectrometers. The relevant NMR properties of these nuclei are summarized in **Table 1**. With the exception of thallium all the nuclei possess a quadrupolar moment. Thallium nuclei with spin-1/2 have a relaxation time much longer than that characteristic of the lighter nuclei. Due to the quadrupolar nature of those nuclei, resonance signals are broadened in all environments, except those where the electric field gradient approaches zero, while the quadrupolar line width effect increases down the group. In the case of ^{115}In, the quadrupole moment is significantly large and only highly symmetric compounds are observable using ^{115}In NMR, even these producing extremely broad lines. From the two naturally occurring boron isotopes, ^{11}B is more advantageous for NMR investigation and only this nucleus is usually studied. This is due to its higher degree of abundance and its lower nuclear spin. For gallium the ^{69}Ga nucleus, due to its significantly lower receptivity and despite its greater abundance, is considerably less favourable than the ^{71}Ga nucleus. ^{115}In and ^{205}Tl have superior receptivities and higher natural abundances than the other natural isotopes of these elements, and are preferred for indium and thallium NMR. Later in this article, the NMR spectroscopy of the preferred nuclei will be discussed.

Nuclei of the Group 13 atoms have been used as NMR probes, especially in view of being able to identify molecular structures, to monitor reaction courses, detect intermediate species, and indicate chemical equilibria in exchange reactions. In general, the relative importance of the heteronuclear NMR technique is dictated by the chemistry of certain elements and because of this ^{11}B NMR and ^{27}Al NMR have been used most extensively. Structural evidence of molecules is mainly deduced from the chemical shift values. Correlations between chemical shifts and chemical parameters, such as coordination environment, different types of π-interaction,

electronegativity, charges, etc. may be observed. Very often no one factor controls the chemical shift, but a combination of several interactions may be balanced to control the chemical shifts. Therefore, conclusions with respect to the structure of compounds in solution can be drawn usually from the chemical shifts when a sufficient amount of experimental data is available for comparison. In some cases, efficient techniques for calculating NMR chemical shifts may provide valuable data, strongly supporting the interpretation of experimental chemical shifts (particularly the individual gauge for the localized orbital method, IGLO, and the gauge-including atomic orbital method, GIAO).

Since a chemical shift is the most informative characteristic of heteronuclear NMR, it is, therefore, of interest to indicate briefly the factors affecting the chemical shift of a nucleus. Generally, the chemical shifts are affected by local diamagnetic (σ_{dia}) and paramagnetic (σ_{para}) contributions to the nuclear screening constant (σ): $\sigma \cong \sigma_{para} + \sigma_{dia}$. The diamagnetic contribution in the Group 13 nuclei is generally assumed to be relatively small compared with the observed chemical shift range, and remains roughly constant for each nucleus despite dramatic changes of the nucleus chemical environment, since in chemical bonding the core electrons are relatively unperturbed. Hence, the paramagnetic shielding term is considered to be responsible for the variation observed in the chemical shifts and it results from the local anisotropic motion of the valence electrons surrounding the nucleus in question. The local paramagnetic term σ_{para} depends to a first approximation on the average electronic excitation energy ΔE, the average radius of the valence orbitals and the symmetry of the distribution of bonding electrons around the nucleus in question. A convenient form of description of the local paramagnetic shielding term is the expression based on Ramsey's nonrelativistic approach:

$$\sigma_{para} \sim \Delta E^{-1} \langle r^{-3} \rangle_{2p} P_u.$$

The value of σ_{para} is larger the lower the energy of the excited states involving a rotation of charge (ΔE is small), the more asymmetric the distribution of p electrons (the P_u term is large) and the closer they are to

692   HETERONUCLEAR NMR APPLICATIONS (B, Al, Ga, In, Tl)

Table 1 NMR properties of nuclei of the Group 13 atoms. Reproduced with permission of Chapman & Hall from *Chemistry of Aluminium, Gallium, Indium and Thallium* (1993) Downs AJ (ed), Chapter 1. London: Chapman & Hall

Property	^{10}B	^{11}B	^{27}Al	^{69}Ga	^{71}Ga	^{113}In	^{115}In	^{203}Tl	^{205}Tl
Natural abundance (%)	19.82	80.18	100	60.108	39.892	4.33	95.67	29.524	70.476
Nuclear spin ($h/2\pi$)	+3	−3/2	+5/2	−3/2	−3/2	+9/2	+9/2	+1/2	+1/2
Magnetic moment (μ/μ_N)	+1.8006	+2.56227	+5.529	+5.541	+1.622258	+1.638215	+2.6886	+3.64151	+2.01659
Relative sensitivity ($^1H = 1.00$)	1.99×10^{-2}	0.17	0.21	6.91×10^{-2}	0.14	0.34	0.34	0.18	0.19
Receptivity ($^{13}C = 1.00$)	22.1	754	1170	237	319	83.8	1890	289	769
Magnetogyric ratio ($\times 10^7$ rad T^{-1} s^{-1})	2.8740	8.5794	6.9704	6.420	8.158	5.8493	5.8618	15.3078	15.4584
Quadrupole moment ($\times 10^{-3}$ m^2)	8.5	4.1	15	17.8	11.2	80	81	–	–
Frequency (MHz) ($^1H = 100$ MHz; 2.3488 T)	10.746	32.084	26.057	24.003	30.495	21.866	21.914	57.149	57.708
Width factor (Al = 1)[a]	0.20	0.31	1.00	5.85	2.32	6.6	6.7	–	–
Normal reference		Et$_2$O.BF$_3$	[Al(OH$_2$)$_6$]$^{3+}$		[Ga(OH$_2$)$_6$]$^{3+}$		[In(OH$_2$)$_6$]$^{3+}$		TlNO$_3$(aq)
Approximate range of chemical shifts (ppm)		250	350		1400		1100		>5500

[a] Width factor = $Q^2(2I + 3) I^2(2I −1)$ and is the nuclear contribution to quadrupole relaxation, normalized to Al = 1.

the origin (the radial term is large). An electronegative neighbouring atom increases the effective nuclear charge on the atom in question, thereby increasing the $\langle r^{-3} \rangle_p$ term and increasing σ_{para}. But even for closely analogous compounds, the electronegativity argument must be used with care since other effects may alter the radius and energy terms in σ_{para} in a complex fashion. Therefore, for Group 13 compounds the substituent effect, for instance, may affect the shielding, often in a way not yet completely understood. According to the above consideration concerning the factors determining the chemical shift, a greater range of chemical shifts may be expected for the three-coordinate than for the four- and higher coordinate Group 13 compounds. In the former there is a p orbital available for π-interaction with the heteroatom lone pair. Due to this interaction a significant variation in orbital angular momentum will be present resulting from the wide variation of strength and number of π-donor ligands. This type of interaction is not available for the four-coordinate complexes since all the valence orbitals are involved in the σ bond formation.

Much less attention has been paid to magnetic relaxation measurement of the Group 13 element nuclei which may provide useful information in the study of molecular motion and interactions, and the symmetry of the nuclear environment. Nuclear magnetic relaxation is usually described both in terms of the methods of coupling of the local magnetic field with the spin system (the spin–lattice and spin–spin relaxation) and the different mechanisms. There are a number of different nuclear magnetic relaxation mechanisms: the quadrupolar mechanism which governs the relaxation behaviour of most quadrupolar nuclei, the spin–spin rotation and the chemical shift anisotropy mechanisms (important for spin-1/2 heavy metal ions), the magnetic dipole–dipole interaction of two nuclei (which is the predominant relaxation mechanism for instance for ^{13}C nuclei with directly bonded hydrogen), the scalar coupling mechanism, and relaxation induced by the presence of unpaired electrons. The contribution of the individual relaxation mechanisms to the overall relaxation rate is additive. Nuclei with spin $I \geq 1$ have unfavourable characteristics arising from the quadrupole moment which interacts with the electric field gradient created by the surrounding electrons. This leads to a decrease of relaxation times and, accordingly, an increase of the line widths.

^{11}B NMR spectroscopy

Boron was among the first elements to attract the attention of NMR spectroscopists in the 1950s. The ^{11}B nucleus has a relatively small quadrupole moment and the NMR line widths are relatively sharp, ranging between 2 and 100 Hz, usually in the 30–60 Hz range. Occasionally, such as for $(R_2N)_2B^+$ borinium cations, the line widths increase greatly to 400–800 Hz owing to the more rapid relaxation associated with an increase in the nuclear field gradient. Boron NMR spectroscopy has found widespread application in boron chemistry. The [11]B chemical shifts, intensity of [11]B resonances and coupling constants between boron and other nuclei are very useful for the elucidation of molecular structures and for control of reactions involving boron compounds. The peak area of a resonance is proportional to the number of boron atoms in the respective chemical environment. The chemical shift is affected by the chemical environment of the boron nucleus under investigation, i.e. the coordination number and the nature of substituents directly attached to the boron atom, and the chemical shift can be readily correlated with structure. The [11]B chemical shifts of boron compounds cover a range of about 250 ppm. The solvent effect on δ [11]B is of small magnitude and can usually be neglected. Noticeable solvent shifts (up to 4 ppm) have been observed only for some polyhedral boranes and related compounds. $BF_3 \cdot OEt_2$ is a reference compound, usually employed as an external standard. Several correlations of δ [11]B with structural parameters in molecules have been established. The number of reported ^{11}B NMR data is enormous. Almost any new boron-containing compound has been subjected to structural analysis by ^{11}B NMR. In general, ^{11}B NMR measurements are applied in two areas: organoboranes and miscellaneous compounds, and boron polyhedral borane derivatives. The first area encompasses research in organic synthesis chemistry and on the structure and dynamics of boron compounds in solution. ^{11}B NMR provides important information on the electronic environment of a boron atom and can be useful in predicting some of the chemical behaviour of boron compounds. The second area is largely a domain of elucidation of the structure and bonding in polyhedral boranes, carboranes, azaboranes, metallaboranes and related compounds.

Organylboranes and miscellaneous compounds

Two-coordinate boron compounds are relatively rare. The [11]B chemical shifts of $(R_2N)_2B^+$ borinium cations cover a narrow range of 35–38 ppm. However, the δ [11]B values of iminoboranes, R–B≡N–R′, fall in the range of 3 to 22 ppm and are sensitive to the nature of the substituents at the site of the imino nitrogen atom. The trend in the [11]B shifts corresponds closely to that observed for the ^{13}C chemical shifts of similarly substituted alkynes.

[11]B chemical shifts have proved to be an extremely valuable tool for distinguishing between three- and four-coordinate boron compounds. Chemical-shift data of representative compounds are given in **Table 2**. Furthermore, a schematic structure representation of compounds [1] – [5] with the corresponding chemical shift values, provides a good illustration of the power of ^{11}B NMR for intimate understanding of the nature of π-interactions in boron compounds. In trialkylboranes of a trigonal planar structure [1] the p orbital remains unoccupied (π-bonding is not an obvious possibility), resulting in a great imbalance of bonding electrons around the boron atom and leading to a large shift of the ^{11}B NMR signal to low field ($\delta \sim 86$ ppm). By contrast, another triorganylborane compound, 7-borabicyclo[2.2.1]heptadiene ([2], δ –5 ppm), exhibits a significant upfield shift, and the magnitude of this shift is rationalized in terms of the interaction of the empty p orbital on boron with the π-clouds of two C=C bonds. ^{11}B NMR spectroscopy has also been commonly employed to provide information about the contribution of π-bonding between boron and carbon centres in many other cyclic and open-chain unsaturated organylboron compounds, i.e. vinylboanes, alkynylboranes, allylboranes, borinenes, boroles and boron derivatives of six-membered rings.

Referring to the R_3B compounds, replacement of the R substituent by a strong electronegative atom or ligand X with a lone pair available for π-interaction results in a high-field shift. For example, the δ [11]B values for three-coordinate $R_2BNR'R''$ and R_2BOR' compounds are found approximately in the range 44–50 ppm and 50–55 ppm, respectively. Hence, the [11]B signal of Et_2B(benzoinato) [4] at 55 ppm is indicative of a three-coordinate environment at boron with the uncoordinated carbonyl group of the bifunctional benzoinato ligand. In this case ^{11}B NMR gives an accurate picture of the π-interaction and is additionally a good indicator of the relative strength of a boron–oxygen π-bond vs. boron–carbonyl dative bond. It is worth noting that the solution structure of compound [4] is consistent with that found in the solid state. From the differences in [11]B chemical shifts of three-coordinate diorganylboranes R_2BX the following order of increasing π-interaction for various X results: $PR_2 < SR < Cl < OR < NR_2 < F$. This is only a qualitative interpretation and a quantitative correlation should not be expected from chemical shift values since the magnitude of the shift may be influenced by other electronic and steric effects, e.g. various functionalities of the ligand X have a remarkable influence on the [11]B nuclear shielding.

Further successive replacement of an alkyl group by the ligand X leads to a progressive upfield shift of ^{11}B resonances and the magnitude of shift depends on the nature of the substituent X. Minor changes in the chemical shift values are observed for sulfide and then for chloride derivatives, and more significant changes are for X = OR, NR_2 and F. A great variety of noncyclic RBX_2 and cyclic $RB(X,X)$ compounds have been studied by ^{11}B NMR spectroscopy. It is interesting to note that for BX_3 compounds the order of increasing substituent effect is somewhat different from that for R_2BX diorganylboranes: SR < Cl < NR_2 < OR < F. ^{11}B NMR is extremely useful in the investigation of the chemistry of three-coordinate diborane compounds, $X'X''B-BX'X''$. The presence of the B–B bond is reflected in the ^{11}B resonance

low-field shift compared with that in the corresponding organylboranes $RBX'X''$.

^{11}B NMR is very useful in the investigation of the rich chemistry of transition metal complexes with boron-containing heterocycles like borole and borabenzene derivatives or various boron–nitrogens and boron–oxygen cyclic compounds. In general, the metal–boron bonding interaction is reflected by a ^{11}B high-field shift compared with that of the corresponding heterocycles. In the triple-decker complexes, the boron of the bridging ligand (to which two metals are bonded) is more shielded than that in the end or capping boron-containing ring to which only one metal atom is bonded.

^{11}B NMR provides valuable information concerning the interaction between three-coordinate boranes

Table 2 ^{11}B chemical shifts of some representative three-coordinate and four-coordinate organylboranes and miscellaneous compounds

Compound	$\delta\ ^{11}B$	Compound	$\delta\ ^{11}B$	Compound	$\delta\ ^{11}B$
CN = 3					
Me_3B	86.6	Ph_3B	68.0	$[MeB(NH)]_3$	34.5
$(CH_2=CH_2)_3B$	56.4	Me_2BPh	77.6	$(MeBO)_3$	33.2
Me—B (ring)	52.8	B—Ph (ring)	84.5	$(^nBuBS)_3$	68.4
Me_2BF	60.1	$MeBF_2$	8.2	BF_3	10.0
Et_2BCl	78.0	$EtBCl_2$	63.4	BCl_3	46.5
Et_2BBr	81.9	$EtBBr_2$	65.6	BBr_3	38.7
Et_2BI	84.4	$EtBI_2$	55.9	BI_3	−7.9
Me_2BNMe_2	44.6	$MeB(NMe_2)_2$	33.5	$B(NMe_2)_3$	27.6
Me_2BOMe	53.0	$MeB(OMe)_2$	29.5	$B(OMe)_3$	18.3
Me_2BSMe	73.6	$MeB(SMe)_2$	66.3	$B(SMe)_3$	61.0
(N,N,N,N diborane ring)	33.7	(O,O,O,O diborane ring)	31.5	(S,S,S,S diborane ring)	68.3
CN = 4					
$Me_3B\cdot NMe_3$	0.1	$H_3B\cdot NMe_3$	−8.3	BMe_4^-	−20.2
$(CH_2=CH_2)_3B\cdot NMe_3$	−3.0	$H_3B\cdot SMe_2$	−20.1	$B(OMe)_4^-$	2.7
$Me_3B\cdot PMe_3$	−12.3	$F_3B\cdot NMe_3$	0.6	BF_4^-	−1.6
$(C_6F_5)_3B\cdot O=C(N^iPr_2)Ph$	−0.1	$Cl_3B\cdot NMe_3$	10.2	BCl_4^-	6.7
$(C_6F_5)_3B\cdot O=C(H)Ph$	5.0	$Br_3B\cdot NMe_3$	−3.3	BBr_4^-	−23.8
$(C_6F_5)_3B\cdot O=C(OEt)Ph$	19.2	$I_3B\cdot NMe_3$	−54.4	BI_4^-	−127.5
(Et,Et / N–N / B / N–N / Et,Et pyrazabole)	2.2	(Me,Me B–O ring)	13.0	(Et,Et B–O ring with OMe)	17.5

[1]
$(\delta \approx 86\ \text{ppm})$

[2]
(R = Me, R' = Ph; δ −5 ppm)

[3]
$(\delta \approx 40\text{–}55\ \text{ppm})$

[4]
(R = Et, R' = Ph; δ 55 ppm)

[5]
(R = Et; δ 22 ppm)

and donor ligands and enables prediction of some of the chemical behaviour. The formation of the four-coordinate Lewis acid–base adducts is evident from the ^{11}B NMR spectrum. When the coordination number of boron increases from three to four, an upfield shift of the ^{11}B resonance is generally observed. The interaction between a borane and a donor ligand often results in a dynamic equilibrium and ^{11}B NMR measurements at variable temperatures are particularly helpful for investigations of this type of equilibrium. The influence of various ratios of donor and acceptor molecules upon the equilibrium can be determined by ^{11}B NMR spectroscopy. However, the utility of ^{11}B NMR to determine the Lewis acid strength of three-coordinate boranes is strongly limited, since the magnitude of chemical shift change upon coordination reflects the π-bonding effect in a parent borane as well as the sensitivity to steric factors. Similarly, an observed order of ligand basicity usually corresponds to certain boron-system affinities and is strongly affected by steric effects. The structures of dialkylboron derivatives of saturated and unsaturated hydroxy carbonyl compounds, Et$_2$B (benzoinato) [4] and Et$_2$B (maltolato) [5], are a good illustration of the importance of π-bonding between boron and oxygen. The ^{11}B signal of [5] at 22 ppm is significantly shifted to high field when compared with that for complex [4] and the chemical shift value falls in the region associated with neutral, four-coordinate boron nuclei. This result is consistent with the chelation of the carbonyl group to the boron atom and the four-coordinate chelate structure [5] in solution. Thus, these data,

based on ^{11}B NMR and X-ray crystallographic studies, provide strong evidence that for organylborane derivatives of unsaturated hydroxy carbonyl compounds the π-interaction of the alkoxide oxygen lone pairs with the chelate-ligand-extended π system may dominate and effectively obstruct the boron–oxygen π-interaction.

Many coupling constants between directly bonded nuclei can be obtained from ^{11}B NMR spectra, especially for four-coordinate boron, where in many cases the quadrupolar relaxation rate is sufficiently slow. The accuracy is somewhat limited because of the line width of the resonance signals. Therefore, the use of a high magnetic field strength is advantageous. One-bond ^{11}B–X coupling constants from 1–5 Hz in BF$_4^-$ up to more than 1000 Hz for X = ^{119}Sn or ^{207}Pb have been observed. For terminal boron–hydrogen bonds, the $^1J(^{11}\text{B–}^1\text{H})$ values are in the range 100–190 Hz, but when a hydrogen bridge is involved $^1J(^{11}\text{B–}^1\text{H})$ is less than 80 Hz. Thus, it is possible to use these coupling constants to determine the coordination at each boron in a boron hydride when resolution permits. Despite the wide use of ^{11}B NMR for the characterization of boron compounds, relatively little attention has been paid to ^{11}B relaxation. Typical values for relaxation times of ^{11}B nuclei are in the range of 10–200 ms, commonly 10–50 ms.

Polyhedral boranes

The development of boron cluster chemistry is intimately coupled with ^{11}B NMR which has been the most efficient method of elucidating the structure and bonding in polyhedral boranes and related clusters. At present, based on a great amount of ^{11}B NMR experimental data, meaningful conclusions with respect to the structure of boron skeleton compounds in solution can be drawn from the chemical shifts. Polyhedral boranes are electron-deficient systems involving multicentre bonding, i.e. B–H–B, B–B–B, B–C–B, and with highly delocalized skeletal electrons. The majority of boron vertices in borane clusters are formally sp^3 hybridized; deviation from the ideal tetrahedral geometry, i.e. the imbalance of bonding electrons, is the most important factor affecting the chemical shift of individual skeletal atoms. In borane cluster compounds the influence of electron density of individual skeletal atoms on chemical shift is significant, but is not the primary factor. [11]B chemical shift values of polyhedral boranes and related clusters span approximately 75 to −60 ppm. Many empirical rules for predicting δ [11]B values have been established which are based on various effects including those related to the coordination number, bridging hydrogens, *endo*- and *exo*-cluster substituents, cluster shape and antipodal atoms. For example, the replacement of a boron nucleus in a polyhedral borane by a heteroatom results in a downfield [11]B chemical shift for the boron across the cage from the heteroatom, usually the so called antipodal effect. The largest effect was observed for the antipodal boron B(10) in the 10-vertex series *closo*-1-EB$_9$H$_9$ heteroboranes, e.g. δ [11]B values of B(10) are −2.0, 28.4 and 74.5 ppm for E = BH^{2-}, CH$^-$ and S, respectively. At present, novel two-dimensional (2D) NMR techniques, i.e. ^{11}B–^{11}B 2D COSY and [1]H–[11]B 2D spectra, usually allow unambiguous ^{1}H and [11]B signal assignments for most polyhedral boranes. In applying the *ab initio*–IGLO–NMR technique, the various competitive structures are subject to *ab initio* structural optimization, following which IGLO calculations are used to predict the sets of [11]B chemical shift values to be expected of each *ab initio* optimized structure.

^{27}Al NMR spectroscopy

The first ^{27}Al NMR measurements were performed in the early 1960s. However, the 1980s and 1990s have brought a considerable increase in investigations of the nature of aluminium compounds using ^{27}Al NMR spectroscopy. The usual reference for chemical shifts is [Al(H$_2$O)$_6$]$^{3+}$ (an acidic solution of Al(III) salt, line width several Hz) or Al(acac)$_3$ in benzene solution (line width ~100 Hz; acac = acetylacetonate). The ^{27}Al NMR chemical shifts of the most often studied organometallic and inorganic aluminium (III) compounds cover a range of approximately 350 ppm, from 290

to −60 ppm. However, there are some rare examples such as the [Cp$_2$Al]$^+$ cation (δ −126.4 ppm) (Cp = C$_5$H$_5$) or the salts with transition metals, i.e. Co(AlCl$_4$)$_2$ (δ −225 ppm). The line width of ^{27}Al resonances is highly sensitive to the symmetry of the aluminium coordination sphere and ranges from a few Hz for highly symmetric species to a few kHz. However, in some cases the resonance may even be undetectably broad and a given spectrum does not display all the response of the aluminium nuclei present. In the standard ^{27}Al NMR spectrum the intensity ratio of resonances cannot usually be interpreted quantitatively with respect to the number of aluminium atoms in the respective chemical environment, due to this strong effect on line widths.

Currently, almost any new organoaluminium compound is routinely characterized by ^{27}Al NMR. Earlier investigations suggested that the observed ^{27}Al chemical shift is diagnostic of the compound coordination state. However, an increasing amount of experimental data has shown that the chemical shift ranges corresponding to the different coordina-

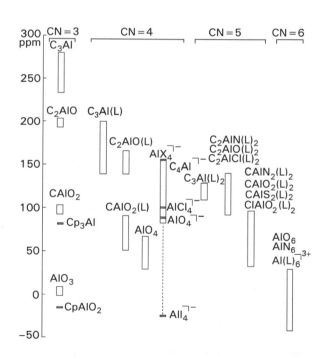

Figure 1 ^{27}Al chemical shift ranges of aluminium compounds, demonstrating the dependence of aluminium shielding on the nature of the coordination environment and the coordination number (CN) of the aluminium atom. L is a monodentate ligand (e.g. NR$_3$, pyridine, OR$_2$, R$_2$C=O, H$_2$O, PR$_3$, SR$_2$) or may denote a neutral chelating group from a monoanionic bidentate ligand (e.g. the carbonyl group in chelate complexes [6] and [9], hence the coordination environment of these complexes is described as C$_2$AlO(L) and C$_2$AlO(L)$_2$, respectively) as well as one of the bridging alkoxide oxygen atoms in associate compounds (e.g. an alkoxide oxygen atom in dimer [8]). X is a ligand carrying one negative charge (e.g. H, R, OR, OH, halogen).

tion states largely overlap and a certain value may not be simply interpreted according to the number of ligands. Shielding data subdivided with respect to the aluminium coordination number, presented graphically in **Figure 1** and for a series of compounds collated in **Table 3**, show that the nature of substituent functions significantly governs the ^{27}Al NMR chemical shifts. Hence, although characterization of organoaluminium compounds by ^{27}Al NMR has recently become routine, the ^{27}Al NMR spectra still pose significant interpretational difficulties. Additionally, the important difficulty in resonance identification is also due to a background signal that in some cases (especially for dilute solutions) may dominate in a measured spectrum (**Figure 2**). This relatively broad 'probe head signal' falls at about 60 ppm and when this region of the spectrum is of interest then it is advisable to use relatively highly concentrated solutions and to identify possible extraneous signals with blank samples.

Because of the relatively fast relaxation of the ^{27}Al nucleus, spin–spin coupling is not generally observed in ^{27}Al NMR spectra. However, a number of coupling constants have been reported. The $^{1}J_{AlH}$ and $^{2}J_{AlH}$ coupling constants were found to be 174 Hz and 6.3 Hz, respectively. The observed $^{1}J_{AlP}$ and $^{2}J_{AlP}$ coupling constants are in the range of 90–290 Hz and 6–30 Hz, respectively. The $^{1}J_{AlF}$ is ~19 Hz, $^{1}J_{AlB}$ is ~9 Hz, $^{1}J_{AlC}$ is ~73 Hz and $^{1}J_{AlN}$ is ~22 Hz. Only a few investigations have been made of the spin–lattice and spin–spin relaxation times of the ^{27}Al nucleus. Relaxation of the ^{27}Al nucleus is dominated by the quadrupolar interaction, as expected from its relatively high quadrupolar moment. For highly symmetrical species the relaxation rates are a few Hz and other relaxation mechanisms may be significant. For example, the relaxation of the ^{27}Al nucleus in the solvate complex $[Al(CH_3CN)_6]^{3+}$ and in anions $[AlCl_nBr_{4-n}]^-$ proceeds by means of the quadrupole and scalar mechanisms.

Organoaluminium (III) and miscellaneous compounds

Various factors that influence the aluminium nuclear shielding may be employed for specifying different

Table 3 ^{27}Al chemical shifts of some representative three-, four-, five- and six-coordinate aluminium compounds

Compound	δ ^{27}Al	Compound	δ ^{27}Al	Compound	δ ^{27}Al
CN = 3					
^{i}Bu$_3$Al	276	^{t}Bu$_3$Al	255	(mesityl)$_3$Al	260
^{i}Bu$_2$Al(bht)	196	^{ii}BuAl(bht)$_2$	109	Al(bht)$_3$	3
Cp$_3$Al	81.7	Cp$_2$MeAl	72.7	Cp$_2$Al(bht)	−15
CN = 4					
[Me$_3$Al]$_2$	153	Me$_3$Al(thf)	182	Et$_4$Al$^-$	154
[Et$_3$Al]$_2$	152	Et$_3$Al(py)	167	AlH$_4^-$	103
[Me$_2$AlH]$_3$	159	Me$_2$AlH(thf)	179	AlCl$_4^-$	102
[Me$_2$AlCl]$_2$	180	Me$_2$AlCl(pyrazole)	157	AlBr$_4^-$	80
[Me$_2$AlOMe]$_3$	152	Me$_2$Al(bht)(py)	134	AlI$_4^-$	−27
[Et$_2$AlOEt]$_2$	151	MeAl(bht)$_2$(py)	72	AlBr$_2$I$_2^-$	37
[Et$_2$AlNEt$_2$]$_2$	160	Al(Omesityl)$_3$(py)	50	Al(OH)$_4^-$	80
Me$_2$Al(dpa)	151	[Cl$_2$AlOEt]$_3$	98	[(^{t}BuO)$_3$Al]$_2$	48
Et$_2$Al(acac)	140	Cp*$_3$Al(thf)	151	[AlCl$_3$]$_2$	91
Cl$_2$Al(acac)	88	Me(Cp*)AlCl$_2$	8.5	[AlBr$_3$]$_2$	75
(Ph$_3$SiO)$_2$Al(acac)	53	[Cp*AlCl$_2$]$_2$	−53	[AlI$_3$]$_2$	−11
CN = 5					
Me$_2$Al(hacet)(py-Me)	118	[Me$_2$Al(lak)]$_2$	114	MeAl(hacet)$_2$	68
Me$_2$Al(amket)(py-Me)	120	[Et$_2$AlO(CH$_2$)$_2$NEt$_2$]$_2$	112	MeAl(amket)$_2$	70
MeAl(hacet)	64	[Et(EtO)Al(hacet)]$_2$	60	[Cl(EtO)Al(hacet)]$_2$	60
CN = 6					
Al(acac)$_3$	0.0	[Al(acac)$_2$(thf)$_2$]$^+$	4	[Al(OH)$_6^{3-}$]	0.0
Al(dpa)$_3$	26	[Al(H$_3$ppma)$_2$]$^{3+}$	−12.9	[AlF$_6$]$^{3-}$	−1.2
Al(maltolate)$_3$	38	[Al(MeCN)$_6$]$^{3+}$	−33	[Al(NCS)$_6$]$^{3-}$	−33.7

bht = deprotonated 2,6-di-*tert*-butyl-4-phenol; Cp = C$_5$H$_5$; Cp* = C$_5$Me$_5$; thf = tetrahydrofuran; py = pyridine; py-Me = γ-picoline; acac = acetylacetonate; hacet = deprotonated 2′-hydroxyacetophenone; lak = deprotonated ethyl lactate; amket = deprotonated 2′-aminoacetophenone; H$_3$ppma = tris[4-(phenylphosphinato)-3-methyl-3-azabutyl]amine; dpa = deprotonated 2,2-dipyridylamine.

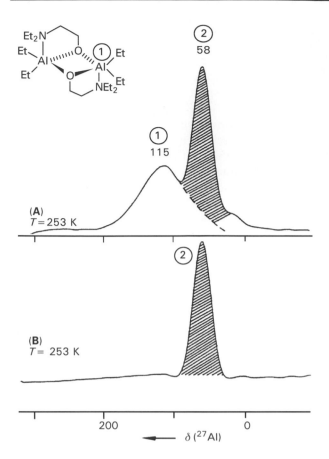

Figure 2 Background signal (2) in the ^{27}Al NMR spectrum of a multinuclear probe head. (A) ^{27}Al NMR spectrum of a 0.1 M solution of [Et$_2$AlO(CH$_2$)$_2$NEt$_2$]$_2$. (B) Spectrum obtained with the test-tube filled with solvent only. For resolution enhancement, the same Lorentz–Gauss transformation was used in both cases before Fourier transformation. Reprinted with permission of Wiley-VCH from Benn R and Rufinska A (1986) *Angewandte Chemie, International Edition in English* **25**: 861–881.

organoaluminium species in solution. Thus, the chemical shifts for the three-coordinate aluminium compounds range from 260 to 0 ppm, and the electronic nature of the ligand determines the extent of metal shielding. The majority of large chemical shift resonances of large chemical shift are generally exhibited by the coordinatively unsaturated three-coordinate R$_3$Al compounds (~260 ppm). Successive replacement of the alkyl substituent by a sterically hindered aryloxide ligand leads to a progressive decrease in chemical shift of the ^{27}Al resonance in the series of three-coordinate monoaryloxide (~190 ppm), bisaryloxide (~100 ppm) and trisaryloxide (~0 ppm) compounds. These phenomena can be explained in terms of a π-interaction between the aluminium empty p orbital and the aryloxide oxygen lone pairs. The increasing number of strong electronegative aryloxide ligands attached to aluminium

probably enhances the π-acceptor strength of aluminium and strengthens the Al–O π-bonding, which leads to decreasing chemical shift. In the corresponding four-coordinate Lewis acid–base adducts, R$_3$Al(L), the aluminium nucleus is significantly shielded due to the new dative bond formed and ^{27}Al chemical shifts are substantially smaller than that of the parent monomeric three-coordinate compound. A similar behaviour results when the alkyl groups in R$_3$Al(L) are successively replaced by aryloxide substituents, although in this case the observed decrease in chemical shift is not as great as for three-coordinate compounds. For the four-coordinate AlX$_4^-$ anionic species, where X = H, R, halogen, OH or OR, an appreciable increase in chemical shift is observed for the tetraalkylaluminium anions and a decrease is found for the AlI$_4^-$ anion. The δ ^{27}Al values of AlX$_4^-$ generally follow the electronegativity of X, and some discrepancies may be attributed to the anisotropic contribution of various ligands because no π-bonding is likely to take place in these anions. These highly symmetrical species can be easily identified by their chemical shift and by the very sharp resonance.

Recently, extensive X-ray work has revealed a large variety of solid-state structures of cyclopentadienylaluminium compounds. In combination with the X-ray data, ^{27}Al NMR spectroscopy in solution provides information about the contribution of π-bonding between aluminium and carbon centres. For example, inspection of selected data for the three- and four-coordinate cyclopentadienyl–aluminium complexes collated in **Table 3** shows that the shielding effect is most pronounced for situations where the aluminium centre, irrespective of its coordination number, bears electronegative substituents. This is consistent with the expectation that the electron-withdrawing group should enhance π-bonding in that Al–Cp interaction.

The δ ^{27}Al values for an extensive series of the four-coordinate R$_2$Al(μ–X)$_2$AlR$_2$ compounds with alkoxide or amide bridging ligands fall in a very narrow range of approximately 160–140 ppm. In the same range are observed chemical shifts of a substantially different class of compounds, the monomeric four-coordinate dialkylaluminium chelate complexes R$_2$Al(X,Y), where X,Y = O,O′–, O,N′– or N,N′– unsaturated bifunctional ligands with both unsaturated bifunctional atoms involved in a π conjugated bond system. However, a pronounced change in the shielding of aluminium resulting from the replacement of both R substituents in the R$_2$Al(O,O′)-type complexes by chloride or alkoxide ligands is reflected by increased shielding of the ^{27}Al resonance of approximately 60 ppm or 110 ppm, respectively. A

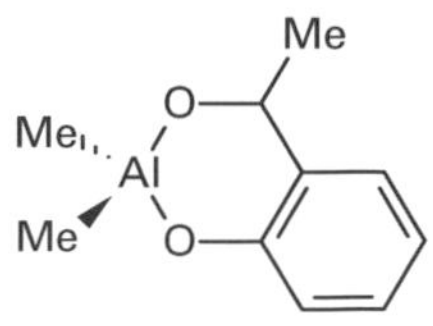

[6]
(δ 144 ppm)

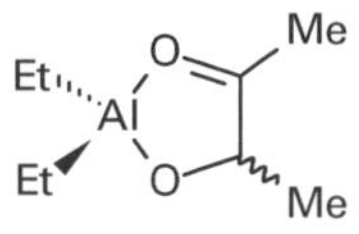

[7]
(δ 122 ppm)

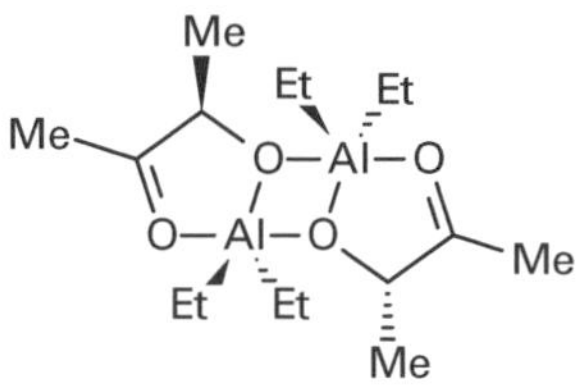

[8]
(δ 131 ppm)

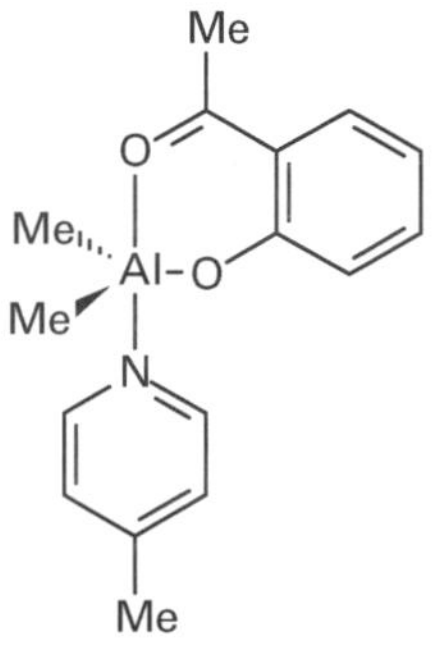

[9]
(δ 128 ppm)

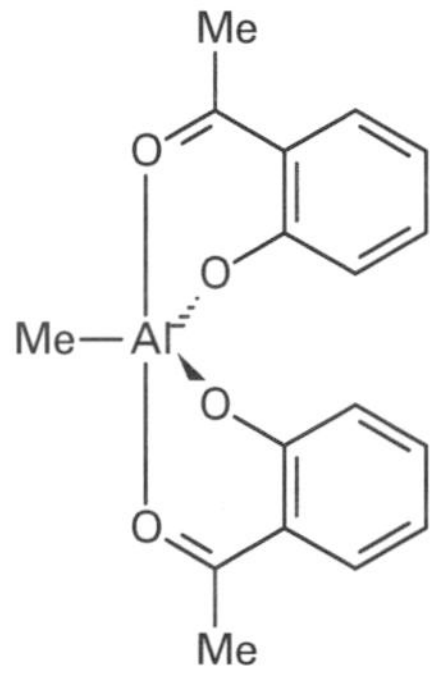

[10]
(δ 64 ppm)

unique shielding situation arises for the dialkylaluminium chelate derivatives of saturated ketone-functionalized alcohols. This group of compounds exists in solution as an equilibrium mixture of the four-coordinate monomer $R_2Al(O,O')$ and the five-coordinate dimer $[R_2Al(O,O')]_2$. The monomeric chelate complexes have the ^{27}Al shielding greater than those of related dialkylaluminium complexes derived from the unsaturated hydroxy carbonyl compound. This is illustrated by complexes $Me_2Al(hacet)$ [6] and $Et_2Al(acet)$ [7] (where hacet = deprotonated 2-hydroxyacetophenone, acet = deprotonated 3-hydroxy-2-butanone) and the corresponding chemical shift values. Furthermore, there is observed a decreased shielding of the ^{27}Al resonance with the association of monomeric species to the five-coordinate adduct which is opposite to the expected coordination number effects, e.g. [7] (δ 122 ppm) and $[Et_2Al(acet)]_2$ [8] (δ 131 ppm). The observed relatively high shielding of aluminium in the four-coordinate dialkylaluminium derivatives of saturated ketone-functionalized alcohols was reasonably explained by π-donation from the alkoxide-oxygen lone-pair orbitals to the σ^* antibonding orbital of the dative bond between the aluminium atom and the oxygen atom of the chelating carbonyl group. This type of interaction is rather unlikely for the related

aluminium compounds with unsaturated chelating ligands due to the involvement of oxygen lone pairs in a chelate conjugated bond system. Thus ^{27}Al NMR spectroscopy may be diagnostic of another type of π-interaction, i.e. π-bonding between the alkoxide oxygen lone pairs and Al–X σ^* antibonding orbitals in four-coordinate aluminium alkoxides.

Simple monomeric five-coordinate aluminium complexes are usually relatively unstable and/or highly reactive (when not stabilized by multidentate ligands) and ^{27}Al NMR spectroscopy becomes a powerful tool for the characterization of these species in solution. The available δ ^{27}Al data show that the aluminium chemical shifts in the series of the five-coordinate organoaluminium compounds varies with the number of alkyl groups attached to aluminium. The δ ^{27}Al values of the five-coordinate dialkylaluminium complexes span a relatively narrow range of approximately 135–100 ppm, irrespective of whether they occur as monomeric or dimeric moieties, e.g. $R_2Al(O,O')\cdot L$ or $[R_2Al(O,O')]_2$. The nature of chelate ligands bonded to aluminium does not affect significantly the ^{27}Al chemical shift. Furthermore, for the dialkylaluminium chelate derivatives of unsaturated hydroxy carbonyl compounds, $R_2Al(O,O')$, the observed shielding effect on the five-coordinate adduct formation, $R_2Al(O,O')\cdot L$, is approximately in the range of 40–15 ppm, e.g. the shielding effect for the chelate complex [6] and its adduct with γ-picoline [9] is 16 ppm. For the monoalkylaluminium five-coordinate compounds, an increase in shielding with respect to the dialkylaluminium derivatives is observed and their δ ^{27}Al fall in the range of 70–50 ppm (see for example MeAl(hacet)$_2$ [10]).

Al^{3+} species in solution

For years, ^{27}Al NMR spectroscopy has been extensively used to investigate the physical and chemical properties of aluminium salts in polar solvents with regard to solvation, ion pairing, hydrolysis and complexation. These studies involve mainly the octahedrally coordinated aluminium species since the solvates of Al^{3+} are octahedral, except where steric restrictions favour lower coordinated complexes. A correlation between the chemical shifts of aluminium hexasolvated complexes and the donor number (DN) (empirical scale of Lewis basicity) of the ligand has been observed, as might be expected for a metal ion with a large charge density whose chemical shift is dominated by the σ_{para} term. Al(III) solvation and ion pair $[Al(H_2O)_nL_{6-n}]^{3+}$ complexes can be readily studied. Quadrupole broadening usually precludes the observation of signals for geometrical isomers of these complexes. The δ ^{27}Al for certain Al^{3+} species is essentially independent of solution concentration and solvent composition. This indicates that the metal-ion electronic environment and subsequent chemical shift depend only on the composition of the principal solvation shell, and are not influenced by long-range outer-shell effects. For the octahedrally coordinated aluminium species $[Al(L^1)_{6-n}(L^2)_n]^{3+}$, when a L^1 ligand is gradually replaced by a L^2 ligand, the induced shift is usually additive. Very often the ligand exchange rate is slow enough to permit the observation of several solvated species and the kinetic and thermodynamic parameters may be determined.

^{27}Al NMR investigations play an important role in the control of the degree of hydrolysis and further of the polymerization of the aqueous Al(III) solution. Various aggregates give distinctive resonances of different AlO$_4$, AlO$_5$ and AlO$_6$ moieties, e.g. species such as the dimer $[Al_2(OH)_2(H_2O)_8]^{4+}$ and the polyoxocation $[AlO_4Al_{12}(OH)_{24}(H_2O)_{12}]^{7+}$ were identified. ^{27}Al NMR spectroscopy can be used to study the characterization of Al^{3+} species in aqueous solution over a wide range of concentration and pH. For example, a conceptual basic drawback in the interpretation of aluminium toxicological data lies to a great extent in the ill-defined nature of the aluminium species administered at physiological pH and in the absence of strongly complexing agents. Hence, there are numerous applications of ^{27}Al NMR measurements to probe biological processes, e.g. ^{27}Al NMR is a useful probe for the Al(III)–phosphate system and in this respect studies of aqueous solutions of Al(III) salts with biologically relevant phosphates were carried out.

Aluminium(I) compounds

Monovalent aluminium compounds have been synthesized and characterized during the last decade. The calculated δ ^{27}Al for the Al(I) compounds cover a relatively large range, over 1000 ppm, and indicate that the shift depends strongly on the nature of the bonded ligand. There are only a few experimental data. For example, the chemical shift for AlCl in toluene–ether solution is 35 ppm and for the tetramer [Cp*Al]$_4$ and the monomer Cp*Al (Cp* = C$_5$Me$_5$ in benzene the corresponding values are −80 ppm and −150 ppm, respectively. The observed line widths for the Al(I) compounds range from 100 to 1800 Hz.

Solid-state ^{27}Al MAS NMR

The increasing popularity of ^{27}Al NMR is strongly connected with high-resolution magic angle spinning (MAS) solid-state NMR, which allows investigation of polycrystalline and amorphous materials, which are of great practical importance. Understanding the structure of these materials is fundamental to many problems in materials science. High-resolution solid-state NMR is particularly valuable in quantifying the extent and nature of disorder. Another reason for the interest in solid-state studies is that the fluxional behaviour is often frozen out and consequently the NMR results are more amenable to correlation with X-ray studies, and these studies can act as a bridge between the crystal and solution structures. For example, ^{27}Al MAS NMR spectra have been extensively used to inspect the structural character of the aluminium species in zeolites. One of the most important applications yields the state of aluminium, shows the transformation of the zeolite symmetry, and yields access to the chemical status of the extra-framework aluminium species after stream treatment of zeolites. ^{27}Al MAS NMR is one of the main techniques for investigating cement, glasses, sol-gel precursors of the zeolites and other ceramic materials. Simple MAS spectra are usually limited by the low resolution generally caused by second-order broadening, and special techniques such as double rotation or two-dimensional MAS may be used to enhance the resolution.

^{71}Ga NMR spectroscopy

^{71}Ga NMR spectroscopic studies are much less common due to the characteristically broad resonances, associated with the large quadrupole moment, and low receptivity, and partly because Ga chemistry has attracted much less attention than that of B or Al. The chemical shifts are referred to $[Ga(H_2O)_6]^{3+}$. ^{71}Ga NMR data are rather scant and at present no meaningful conclusion with respect to the structure of gallium compounds in solution could be drawn from an interpretation of the chemical shift values. However, in view of the significant structural similarity of analogous gallium and aluminium compounds it is reasonable to expect that the general observation concerning the correlation between ^{27}Al shifts and the structure of aluminium species may be transferred to gallium. **Table 4** presents the ^{71}Ga chemical shifts for some gallium compounds.

^{71}Ga NMR spectroscopic studies have been used in the considerable research of the chemical nature of gallium(III) hydrolysis species. For example, it was found that the base hydrolysis of gallium(III) salt solutions led to the $[GaO_4Ga_{12}(OH)_{24}(H_2O)_{12}]^{7+}$ polyoxocation. However, much less of this species is formed than in the case of aluminium. Furthermore, ^{71}Ga NMR spectroscopy can be used to study the structure and stability of gallium complexes. The clinical use of ^{67}Ga as a radioactive tracer in tumour scanning and the inhibition of cellular incorporation *in vivo* by citrate, phosphate and other chelating agents served as the impetus for these studies. For example, in order to provide complexes with potentially greater *in vivo* stability with respect to acid-catalysed dissociation, model complexation studies with amino phosphonic acids have been carried out. It was found that the tripodal amino phosphinato ligand, tris[4-(phenylphosphinato)-3-methyl-3-aza-butyl]amine (H$_3$ppma), forms stable complexes of octahedral geometry at extremely low pH, a regime in which potentiometry is of no use for studying the solution chemistry. Other methods were sought to determine the stability constants, with NMR spectroscopy proving to be the only suitable tool. The

highly symmetric environment around the metal centre in $[Ga(H_3ppma)_2]^{3+}$ produces a low electric field gradient at the gallium nucleus, leading to an unusually narrow line (50 Hz) in the ^{71}Ga NMR spectrum. The narrow resonances made the complex amenable to stability constant studies via a combination of ^{71}Ga and ^{31}P NMR spectroscopies as illustrated in **Figure 3**. The chemical shifts observed for the gallium complexes are indicative of octahedral geometries. The gradual disappearance of the resonance for $[Ga(H_2O)_6]^{3+}$ (Ga) with the concomitant growth of the resonance for $[Ga(H_3ppma)_2]^{3+}$ (GaL$_2$) is evident as R is increased (concentration ratio $R = [H_3ppma]/[Ga^{3+}]$). The $[Ga(H_3ppma)(H_2O)_3]^{3+}$ species (GaL) is barely observable as an insignificant broad resonance and is only seen when R is small. The relative intensities of the gallium resonances correlate with their respective ^{31}P NMR spectra. The latter consist of three major resonances corresponding to the two diasteroisomers of GaL$_2$ and to the free ligand L, respectively. The formation constant of the GaL$_2$ complex was determined from the integration of the ^{71}Ga and ^{31}P NMR spectra.

The +1 oxidation state of gallium represents a relatively unexplored area. There are only a few available data. For example, the ^{71}Ga chemical shift for benzene solutions of GaAlX$_4$ compounds shows a considerable dependence on the type of AlX$_4^-$ anion and a decreased shielding of about 130 ppm was observed: from −680 ppm (AlCl$_4^-$) and −635 ppm (AlBr$_4^-$) to −552 ppm (AlI$_4^-$). The ^{71}Ga chemical shifts of GaAlX$_4$ were found to be solvent- and temperature-dependent. The observed line widths for the Ga(I) species are in the range of 80–1200 Hz.

^{115}In NMR spectroscopy

The first NMR measurements of In(III) salts in aqueous solution were reported in the 1950s. However, ^{115}In has a relatively large quadrupole moment which reduces its observability. ^{115}In NMR studies of indium complexes in solids or melts of indium metals, alloys and salts have been carried out. Relatively few experimental results are available in

Table 4 ^{71}Ga chemical shifts of selected gallium compounds

Compound	δ ^{71}Ga	Compound	δ ^{71}Ga	Compound	δ ^{71}Ga
[Me$_3$Ga]$_2$	720	Me$_3$Ga(OEt$_2$)	560	GaH$_4^-$	682
[Me$_2$GaCl]$_2$	355	Me$_3$Ga(NMe$_3$)	430	Ga(OH)$_4^-$	192
[MeGaCl$_2$]$_2$	295	Me$_3$Ga(SMe$_2$)	365	GaCl$_4^-$	250
[GaCl$_3$]$_2$	221	[Me$_2$GaOMe]$_2$	340	GaBr$_4^-$	69
[GaBr$_3$]$_2$	40	GaBr$_3$(OEt$_2$)	125	GaI$_4^-$	−27
[Ga(MeCN)$_6$]$^{3+}$	−76	[Ga(dmf)$_6$]$^{3+}$)	−25	[Ga(H$_2$O)$_6$]$^{3+}$	0

dmf = dimethylformamide.

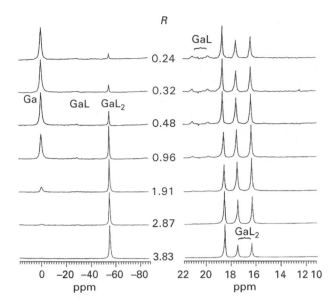

Figure 3 ^{71}Ga (65.7 MHz) (left) and ^{31}P (121.0 MHz) (right) NMR spectra for the stability constant study of the Ga(III)–H$_3$ppma system ($R = $ [H$_3$ppma]/[Ga^{3+}]). Reprinted with permission of the American Chemical Society from Lowe MP, Rettig SJ and Orvig C (1996) *Journal of the American Chemical Society* **118**: 10446–10456.

solution. The adopted reference of chemical shift is [In(H$_2$O)$_6$]$^{3+}$ and the reported line widths even for this symmetrical species are in the range of 1500–2000 Hz. For example, ^{115}In NMR studies of indium halide complex formation in an aqueous solvent mixture have been carried out. The chemical shifts were used along with the line width to elucidate the chemical equilibria and the exchange processes of indium halide complexes in HCl, HBr and HI aqueous solution and in water–acetone mixtures. The ^{115}In chemical shifts in the series of four-coordinate InX$_4^-$ ions are 420 ppm (InCl$_4^-$), 180 ppm (InBr$_4^-$) and -440 ppm (InI$_4^-$) and follow a trend similar to that found for the analogous Al and Ga species. Furthermore, δ ^{115}In for [In(H$_3$ppma)$_2$]$^{3+}$ in CD$_3$OD solution is -14.6 ppm. The relaxation rates of the ^{115}In nucleus were obtained for In(ClO$_4$)$_3$ in aqueous HClO$_4$, i.e. spin–lattice and spin–spin relaxation times of 0.33 ms and 0.24 ms, respectively.

^{205}Tl NMR spectroscopy

^{205}Tl is among the most sensitive NMR nuclides and was investigated quite early on. The first liquid- and solid-state spectra for thallium were measured in 1953. The resonance frequency of ^{205}Tl demanded a single-frequency tuned probe, except for new probe designs at lower fields. Thallium provides significant contrast from the lighter Group

13 elements both in terms of nuclear properties (see **Table 1**) and chemical behaviour. For thallium the +1 oxidation state is normally the more stable, an exception among the Group 13 elements, where the +3 oxidation state dominates. ^{205}Tl is the fourth most receptive spin-$\frac{1}{2}$ nuclide. The sensitivity of the NMR parameters, such as chemical shift, the spin–spin coupling constant and the relaxation time, to the changes in the chemical environment of the thallium nucleus is more pronounced than is that of the lighter Group 13 nuclei; this is quite reasonable in view of the greater polarizability of the sixth-row elements. Therefore, ^{205}Tl NMR spectroscopy is quite useful for investigating such phenomena as ion-pairing, solvent–solute interactions, phase changes, long-range substituent effects and binding constants.

The chemical shift ranges are approximately 3400 ppm for Tl(I) and 5000 ppm for Tl(III) compounds. The lower oxidation state ^{205}Tl is appreciably shielded. The ^{205}Tl chemical shift is highly sensitive to the anion, solvent, concentration, phase change, substituents and temperature effects. For example, the chemical shift of the ^{205}Tl(I) ion increased by about 2000 ppm as the solvent changed from water to liquid ammonia, and that of TlCl increased by 925 ppm from the melt to the solid. The anion- and concentration-dependent shifts are in the range of 10–100 ppm. The coupling constants involving the ^{205}Tl nucleus are usually large and vary significantly with changes in the physical and chemical environments. For example, $^1J_{TlC}$ is nearly 6000 Hz for CH$_3$TlX salts and $^2J_{TlH}$ ranges from -187 Hz to -1120 Hz for different organothallium(III) compounds. Only a few coupling constants have been observed for Tl(I) compounds due to the highly ionic nature of these complexes.

Tl(I) species in solution

^{205}Tl NMR has been used to determine ion-pair formation constants and the thermodynamic parameters of ion-pairing processes for Tl(I) salts in a number of solvents. The values of ion-pair formation constants have been found to be dependent on the dielectric constant of the solvent. A correlation between the ^{205}Tl chemical shift and the DN of the solvent has been indicated: the more basic the solvent, the lower the shielding of the Tl$^+$ ion. There have also been numerous studies of the preferential solvation of the Tl$^+$ ion in mixed solvent systems. The relative solvating abilities were analysed in terms of solvent donor–acceptor properties and the solvent DN was found to be most important in the preferential solvation.

An important recent application is the use of ^{205}Tl NMR measurements to probe biological processes in which Tl^+ replaces alkali cations; the Tl^+ ion has an ionic radius of 1.40 Å and in terms of size is close to K^+. ^{205}Tl NMR has been useful in the study of a wide range of the Tl(I) complexes with synthetic and biological ionophores. The mode of complexation may be indicated by the solvent effect. For example, the chemical shifts of the cryptate complexes of the Tl^+ ion are usually independent of solvent, while those of the various Tl(I)–crown complexes are very solvent dependent. The latter reflect the open axial position of the Tl(I)–crown, where the solvent and anion interaction with the complexed ion occurs. A series of the Tl(I)–antibiotics complexes has been characterized by the ^{205}Tl chemical shift and the spin–lattice relaxation. The trend in chemical shifts reflected the diversity of the ligand used to bind the Tl^+ ion. On the other hand, the magnitude of the chemical shift anisotropy was diagnostic of the symmetry of the Tl^+ binding site, and the magnitude of the dipolar contribution to the spin–lattice relaxation indicated the proximity of hydrogen atoms. Hence, ^{205}Tl NMR shows great promise as a probe of the nature of metal binding sites.

Much effort has been expended in studying the relaxation behaviour of the $^{205}Tl^+$ ion in solution and has found application in the study of enzyme structure. The spin–lattice relaxation time of the Tl(I) was found to be very sensitive to environmental effects. For example, the $^{205}Tl^+$ spin–lattice longitudinal relaxation times T_1 were found to vary by over one order of magnitude for the solvent series, ranging from 1.8 s in water to 0.08 s in n-butylamine. The relaxation time of the Tl(I) ion aqueous solution was independent of concentration, and spin–rotation was found to be the dominant relaxation mechanism. In nonaqueous solution ion pairing becomes important and the longitudinal relaxation time of the Tl^+ ion is dominated by the chemical shift anisotropy mechanism. Furthermore, in the case of Tl^+ complexes with antibiotics, the longitudinal relaxation time is determined by the contribution from spin rotation, dipolar, and chemical anisotropy shift mechanisms. A strong influence of dissolved dioxygen on the $^{205}Tl^+$ relaxation rate has also been found. Therefore, it is advisable to use a degassed solution before attempting relaxation measurements.

Organothallium(III) and miscellaneous compounds

The ^{205}Tl chemical shifts of organothallium(III) compounds cover a range of over 2000 ppm and they display solvent dependence, although not as striking as that for Tl(I). Solvent basicity and ion pairing contribute to the solvent dependence. The dependence of the ^{205}Tl chemical shift of Me_2TlX salts upon solvent, concentration, X anion and temperature has been studied. The latter factor was found to affect the chemical shifts the most. Substituent changes produce large chemical shifts, for example the difference in chemical shift between Me_3Tl and Et_3Tl is 130 ppm. The relaxation behaviour of ^{205}Tl(III) has been investigated for Me_2Tl^+ in solution as a function of temperature and magnetic field strength and the dominant relaxation mechanism was found to be chemical shift anisotropy.

List of symbols

I = spin; J = coupling constant; P_u = elements of the bond-order charge-density matrix (expresses population of the np orbitals and a p-electron 'imbalance' about the nucleus in question); r = radius of the valence p orbitals (r^{-3} is usually called the orbital expansion term); δ = chemical shift; ΔE = average electronic excitation energy; σ = nuclear screening constant; σ_{para} and σ_{dia} = local paramagnetic and diamagnetic, respectively, contributions to the nuclear screening constant.

See also: **NMR Relaxation Rates; NMR Spectrometers; Parameters in NMR Spectroscopy, Theory of; Relaxometers; Solid State NMR Using Quadrupolar Nuclei; Solid State NMR, Methods.**

Further reading

Akitt JW (1989) Multinuclear studies of aluminium compounds. *Progress in NMR spectroscopy* 21: 1–149.

Dechter JJ (1982) NMR of metal nuclides. Part I. The main group metals. *Progress in Inorganic Chemistry* 29: 351–369.

Harris RK and Mann BE (1978) Group III – aluminium, gallium, indium and thallium. In: *NMR and the Periodic Table*, Chapter 9A, pp 279–308. New York: Academic Press.

Hermanek S (1992) ^{11}B NMR spectra of boranes, main-group heteroboranes, and substituted derivatives. Factors influencing chemical shifts of skeletal atoms. *Chemical Reviews* 92: 325–362.

Hinton JF, Metz KR and Briggs RW (1988) Thallium NMR spectroscopy. *Progress in NMR Spectroscopy* 20: 497–513.

Nöth H and Wrackmeyer B (1978) Nuclear magnetic resonance spectroscopy of boron compounds. *NMR Basic Principles and Progress* 14: 1–461.

Siedle AR (1988) ^{11}B NMR spectroscopy. *Annual Reports on NMR Spectroscopy* 20: 205–314.

Wrackmeyer B (1988) Nuclear magnetic resonance spectroscopy of boron compounds containing two-, three- and four-coordinate boron. *Annual Reports on NMR Spectroscopy* 20: 61–203.

Heteronuclear NMR Applications (Ge, Sn, Pb)

Claudio Pettinari, Università di Camerino, Italy

MAGNETIC RESONANCE
Applications

Group 14 of the periodic table is surely one of the most interesting for the NMR spectroscopist. With the exception of germanium, all the other members have favourable NMR properties: tin, silicon and lead have at least one spin $\frac{1}{2}$ isotope of good sensitivity.

Silicon and tin have been studied extensively, and several applications have been found in chemistry. Only few reports are available on germanium, which has only one magnetic isotope, ^{73}Ge, characterized by a relatively large quadrupole moment; this has limited the number of compounds studied, and only a little more data is available on lead. There is a sizeable literature on tin, beginning with the first investigation in 1961. For example, there has been extensive compilation of solution and solid-state ^{119}Sn chemical shifts and coupling constants, mainly in organometallic chemistry. These data are often useful in the determination of molecular structures, not only of small molecules but also of polymeric and cluster compounds. In particular, solution data are the main tools for gaining information on the progress of a selected reaction, on the existence of donor–acceptors and/or solute–solvent interactions, of dissociation or autoassociation, and of intramolecular coordination.

Moreover in the last few years, one-dimensional (1D) and two-dimensional (2D) proton detected heteronuclear correlation NMR techniques involving ^{119}Sn and ^{207}Pb nuclei have been used for applications in organometallic NMR. In particular, the development of organometallic chemistry has been affected markedly by the ever-increasing capabilities of NMR instruments. Prominent NMR parameters such as chemical shifts, coupling constants and relaxation times have also been exploited to gain information on organo-E (E = Ge, Sn, Pb) compounds not readily available by other methods.

Properties of the nuclei

The nuclear properties are summarized in **Table 1**.

Germanium possesses only one magnetic isotope, ^{73}Ge, with spin $\frac{9}{2}$. This nucleus shows a combination of properties unfavourable to magnetic resonance studies, so that only a limited range of molecules can be studied: the natural abundance is 7.76%, the gyromagnetic ratio is very small ($\gamma = -0.9332 \times 10^7$ rad T^{-1} s^{-1}), and the nuclear quadrupole is moderately large. The sensitivity with respect to ^{1}H at natural abundance and constant field is 1.08×10^{-4}. Since quadrupolar relaxation dominates, dipole–dipole relaxation and nuclear Overhauser effects (NOE) are not important and unsymmetric germanium compounds often give very broad ^{73}Ge NMR signals.

Tin possesses three magnetically active tin isotopes with spin $I = \frac{1}{2}$ (^{115}Sn, ^{117}Sn, ^{119}Sn). Two of them, ^{117}Sn (7.51%) and ^{119}Sn (8.58%), have an appreciable natural abundance and receptivities ~20 and 25 times greater, respectively, than that of ^{13}C. The low natural abundance makes the ^{115}Sn resonance unfavourable. The basic resonance frequencies for ^{115}Sn, ^{117}Sn, and ^{119}Sn are 98.16, 106.90, and 111.87 MHz on a spectrometer that operates at a field strength of 7.05 T with ^{1}H NMR at 300 MHz.

In tin NMR experiments, the negative gyromagnetic ratio can lead to disadvantageous NOE behaviour, which can be suppressed by inverse gated decoupling or by using relaxation rate enhancement additives. This problem is not present in the NMR of lead compounds: in fact ^{207}Pb, the only lead isotope

Table 1 NMR properties^a of magnetic isotopes of germanium, tin and lead

Nucleus	Spin	NA(%)	R^H	R^C	γ (10^7 rad T^{-1} s^{-1})	Ξ (MHz)	Q (10^{-28} m^2)
^{73}Ge	$\frac{9}{2}$	7.76	1.08×10^{-4}	0.622	−0.9332	3.488 315	−0.17
^{115}Sn	$\frac{1}{2}$	0.35	1.24×10^{-4}	0.705	−8.792	32.718 780	–
^{117}Sn	$\frac{1}{2}$	7.51	3.49×10^{-3}	19.8	−9.578	35.632 295	–
^{119}Sn	$\frac{1}{2}$	8.58	4.51×10^{-3}	25.6	−10.021	37.290 662	–
^{207}Pb	$\frac{1}{2}$	22.6^b	2.1×10^{-3}	11.9	5.6264	20.920 597	–

^{a} NA is the natural abundance (%); R^H and R^C are the receptivities relative to that of ^{1}H and ^{13}C, respectively; γ is the gyromagnetic ratio; Ξ is the resonance frequency for the reference compound (Me)$_4$E (E = Ge, Sn or Pb) in a magnetic field for which (Me)$_4$Si has a proton resonance of 100 MHz; Q is the electric quadrupole moment.

^{b} Varies with sources.

with spin $\frac{1}{2}$ (the natural abundance is ~22% and the sensitivity to detection is 0.09% of that of the proton), has a positive gyromagnetic ratio. The maximum $^{207}Pb-\{^1H\}$ nuclear Overhauser effect is 350%.

Experimental aspects

Samples for solution studies are prepared in the same manner as for ^1H NMR spectroscopy. In most of the experiments, the standard reference compound is the tetramethyl-E (E = Ge, Sn, Pb) derivative, although sometimes $GeCl_4$ and $SnCl_4$ are used as references in ^{73}Ge and ^{119}Sn NMR spectra, respectively. The tetramethyl-E compound can be used externally; however, on modern NMR spectrometers it is generally preferable to use a known absolute frequency.

There are great difficulties in determining the chemical shifts of ^{73}Ge nuclei by direct observation owing to its low natural abundance and receptivity and large quadrupole moment. Coupling to ^{73}Ge is rarely reported; however, if appreciable $^1H-^{73}Ge$ coupling is present, the observation of ^{73}Ge resonances can be achieved by heteronuclear double-resonance experiments. Special pulse sequences such as INEPT (insensitive nuclei enhancement by polarization transfer) which effect nuclear spin polarization transfer from nuclei of high gyromagnetic ratio (γ), usually ^1H, to nuclei of small γ, and allow dramatic increases in signal-to-noise ratio and corresponding decreases of spectral accumulation time, have been used in the study of simple symmetrical molecules.

The ^{119}Sn nucleus is generally preferred in tin NMR experiments because of its greater natural receptivity. ^{119}Sn chemical information is obtained from 1D ^{119}Sn NMR and $^nJ(^{119}Sn, {}^1H)$, $^nJ(^{119}Sn, {}^{13}C)$ and $^nJ(^{119}Sn, {}^{117/119}Sn)$ coupling, or by using the $^1H-\{^{119}Sn\}$ heteronuclear double magnetic resonance (HDMR) method, polarization transfer techniques such as INEPT and DEPT (distortionless enhancement polarization transfer), or proton-detected heteronuclear $^1H-\{^{119}Sn\}$ correlation techniques such as HQMC (heteronuclear multiple-quantum coherence). HQMC NMR spectroscopy can be easily applied to the structure elucidation of organotin compounds, while in those containing at least a single Sn–N bond it is possible to measure very efficiently the $J(^{119}Sn, {}^{15}N)$ coupling constant from ^{15}N satellites in ^{119}Sn NMR spectra by application of the Hahn-echo extended (HEED) INEPT experiment.

The magnitude and sign of $^nJ(^{119}Sn,X)$ and $^nJ(^{207}Pb,X)$ coupling constants are fundamental requisites for interpreting the bonding and structure in tin and lead compounds. Relative signs of the coupling constants can be obtained efficiently from

2D heteronuclear shift correlation (HETCOR) experiments. This method is helpful also in determining small long-range coupling constants (e.g. $^4J(^{117/119}Sn, {}^1H)$), which are normally not resolved in 1D ^1H NMR spectra.

In order to obtain highly resolved, selective solid-state information, in recent years much work has been done on solid-state ^{119}Sn and ^{207}Pb NMR using magic angle spinning (MAS) and cross-polarization (CP) methods.

Chemical shifts

In agreement with Ramsey's theory, the variation of the 'paramagnetic term' σ_p is the dominating factor in the nuclear screening (Eqn [1]) for all three elements, i.e. germanium, tin and lead.

$$\sigma = \sigma_p + \sigma_d + \sigma_n \qquad [1]$$

In Equation [1] σ_d is the diamagnetic contribution to shielding; σ_n includes all diamagnetic and paramagnetic nonlocal contributions; and σ_p is a function of at least three factors: the effective nuclear charge, the p and d electron imbalance term, and the average excitation energies; σ_p and σ_d may be calculated in principle for a given system, but the theory has not yet been developed, so that this model may be applied with success only when calculating the difference in shielding in closely related compounds, whereas for heavy-atom substituents σ_p does not reflect the influence of nearest neighbours.

Germanium, tin and lead chemical shifts cover a very large range if inorganic and organometallic compounds are examined: more than 1200 ppm for ^{73}Ge, 5000 ppm for ^{119}Sn, and 12 000 ppm for ^{207}Pb. Tables 2, 3 and 4 exhibit selected ^{73}Ge, ^{119}Sn and ^{207}Pb data, all referred to $(Me)_4E$ (E = Ge, Sn or Pb), a negative sign indicating a higher shielding or a shift to lower frequencies.

Isotope effects

In the case of tin, the primary isotope effect (dependence of the screening upon the type of nuclide of tin investigated) should be considered. The chemical shift difference $[\delta(^{119}Sn) - \delta(^{117}Sn)]$ has been measured for a number of organotin compounds in non-coordinating solvents, and in most cases the primary isotope effect is negligibly low. Secondary isotope effects are more important: for example, ^{119}Sn chemical shifts have been measured for Me_2SnH_2, Me_2SnHD and Me_2SnD_2 and a shift of ~1 ppm to lower frequencies has been observed when one H is

Table 2 Selected ^{73}Ge NMR chemical shifts in ppm relative to Me$_4$Ge

Compound	δ(ppm)	Compound	δ(ppm)
Me$_4$Ge	0	Ge(CH=CH$_2$)$_4$	−58.7
Et$_4$Ge	17.3	Me$_3$Ge(CH=CH$_2$)	−15.6
Ph$_4$Ge	−31.6	Me$_2$Ge(CH=CH$_2$)$_2$	−30.6
GeH$_4$	−283.7	MeGe(CH=CH$_2$)$_3$	−44.9
EtGeH$_3$	−186.4	Ge(C≡CH)$_4$	−173
Et$_2$GeH$_2$	−88.4	MeGe(C≡CH)$_3$	−118
Et$_3$GeH	−15.7	Me$_2$Ge(C≡CH)$_2$	−77
GeCl$_4$	30.9	Me$_3$Ge(C≡CH)	−34
GeBr$_4$	−311.3	Ge(OMe)$_4$	−37.8
GeI$_4$	−1081.8	Ge(OEt)$_4$	−43.9
GeCl$_3$Br	−47.8	Ge(OPr)$_4$	−49.7
GeCl$_2$Br$_2$	−131.3	Ge(OCH$_2$CF$_3$)$_4$	−48.1
GeClBr$_3$	−219.4	Germacyclohexane	−131.2
GeCl$_4$(Bipy)	−313.7	1-Me-Germacyclohexane	−65.3
GeCl$_4$(phen)	−319.4	1-Ph-Germacyclohexane	−69.4
[Ge(NCS)$_6$]$^{2-}$	−442.5	Ge(NCS)$_4$(Bipy)	−351.8

increase in coordination number produces a significant increase in the shielding of Sn(IV) and Pb(IV).

Germanium(IV) compounds have been less studied in this regard; however, they seem to show the same behaviour. For example the GeCl$_4$-2,2′-bipyridine adduct [1] containing a hexacoordinate germanium has δ(^{73}Ge) ∼340 ppm upfield shifted with respect to the four-coordinate compound GeCl$_4$.

[1] δ (^{73}Ge): −313.7 ppm

replaced by one D. The largest isotopic effect has been observed with the [SnH$_3$]$^-$ anion, where the ^{119}Sn chemical shift moves by 9.84 ppm on perdeuteration. The isotopic effects in the case of germanium and lead compounds have not been widely studied: however, it has been found that they are generally greater for ^{207}Pb and less for ^{73}Ge with respect to ^{119}Sn.

Solvent and temperature dependence of tin, germanium and lead chemical shifts

Tin, germanium and lead chemical shifts are not very sensitive to temperature in the absence of chemical changes (e.g. autoassociation, dissociation, change in coordination number of the atom of the nucleus examined). Sometimes variations in conformational populations and electrostatic solvent–solute interactions can produce appreciable effects, and it has been found, for example, that shielding of the ^{119}Sn nucleus generally increases roughly linearly with increasing temperature. Analogously, change in solvent, in the absence of solvent–solute chemical interactions, autoassociation or dissociation affects the tin, germanium and lead chemical shifts only slightly with respect to their chemical shift ranges.

Coordination number dependence of tin, germanium and lead chemical shifts

Taking into account that relevant excitation energies are fairly constant, the paramagnetic term σ_p, which represents the imbalance of charge, could be related to local symmetry. It has been established that an

The knowledge of δ(^{119}Sn) chemical shifts is very important for predicting the geometry of coordination polyhedra of tin compounds in solution. In the case of tetravalent tin compounds, an increase in tin coordination number from four to five corresponds to an increase in tin shielding by ∼150–200 ppm. Similar increases are generally associated with six and seven coordination. The δ(^{119}Sn) for PhSn(edtc)$_2$Br [2], a hexacoordinate pseudooctahedral tin complex, is −704.1 ppm, whereas δ(^{119}Sn) for PhSn(edtc)$_3$ [3] in which the central Sn has coordination number seven and the shape is distorted pentagonal bipyramidal is −806.5 ppm.

The ^{119}Sn chemical shift also provides effective means of studying covalent solute–solvent interaction, autoassociation, dissociation, and intramolecular coordination. For example, the ^{119}Sn chemical shift is nearly concentration independent when tin(IV) and organotin compounds such as Me$_2$SnCl$_2$

[2] δ (^{119}Sn) : −704.1 ppm   [3] δ (^{119}Sn) : −806.5 ppm

Table 3 Selected ^{119}Sn NMR chemical shifts in ppm relative to Me_4Sn

Compound	δ(ppm)	Notes	Compound	δ(ppm)	Notes
Me_4Sn	0	Neat liquid	Me_3SnCl	+168.9	5% in CH_2Cl_2
$(CD_3)_4Sn$	+2.6	Neat liquid	Me_3SnBr	+128.0	Benzene
Et_4Sn	+1.4	20% in CCl_4	Me_3SnI	+38.6	3–20% toluene
Pr^i_4Sn	−43.9	Neat liquid	Ph_3SnCl	−44.7	$CDCl_3$
Ph_4Sn	−137 ± 2	30% in $C_2H_3Cl_3$	Ph_2SnCl_2	−32	CH_2Cl_2
Me_3SnEt	+3.0	CCl_4	Me_2SnCl_2	+137	30% in CH_2Cl_2
$(Me_3Sn)_4C$	+49.8	C_6D_6	$(Me_3Sn)_3P$	+36.3	65% in benzene
Me_3SnPh	−28.6	20% CH_2Cl_2	$[(Cl_3Sn)_5Pt]^{3-}$	−142.0	Acetone-d_6
$Me_3Sn(C{\equiv}CH)$	−68.1	CH_2Cl_2	Ph_3SnOH	−86	CH_2Cl_2
$Me_3SnC{\equiv}CH$-Ph	−69.0	$CDCl_3$	Me_3SnOH	+11.8	Saturated CH_2Cl_2
$Et_3SnCH{=}CH_2$	−42.0	50% CCl_4	$MeSnH_3$	−346	60% in toluene
Me_3SnPh	−30	Neat liquid	Me_2SnH_2	−229	Neat liquid
Me_2SnPh_2	−60	Neat liquid	Me_3SnH	−104.5	Benzene
$MeSnPh_3$	−98	CH_2Cl_2	Ph_3SnH	−148	Neat liquid
$SnCl_4$	−150	Neat liquid	$MeSnCl_3 \cdot 2DMSO$	−457	Saturated DMSO
$SnBr_4$	−638	Neat liquid	$Me_2SnCl_2 \cdot 2DMSO$	−246	Saturated DMSO
SnI_4	−1701	CS_2	$Me_3SnCl \cdot DMSO$	−86	Saturated CH_2Cl_2
$SnCl_3Br$	−265	1:1 M $SnBr_4/SnCl_4$	$SnCl_2$	−388	In 12 M HCl
$SnCl_2Br_2$	−387	1:1 M $SnBr_4/SnCl_4$	Cp_2Sn	−2199 ± 10	C_6H_{12}
$SnClBr_3$	−509	1:1 M $SnBr_4/SnCl_4$	Me_3SnOMe	+120.9	Benzene
$MeSnCl_3$	+18.7	Saturated benzene	$Me_2Sn(OMe)_2$	−126.3	Benzene
$PhSnCl_3$	−64	$CDCl_3$	$MeSn(OBu^i)_3$	−452.0	50% in mesitylene
$(Me_3Sn)_2NMe$	+81	20% C_6D_6	Me_3SnNMe_2	+75.5	10% in benzene

Table 4 Selected [207]Pb NMR chemical shifts in ppm relative to Me_4Pb

Compound	δ(ppm)	Notes	Compound	δ(ppm)	Notes
Me_4Pb	0	Containing 15% toluene	Me_3PbCl	+432	Saturated in CH_2Cl_2
Et_4Pb	+73.3	Neat liquid	Me_3PbBr	+367	Saturated in CH_2Cl_2
Vi_4Pb	−251	Neat liquid	Me_3PbI	+203.6	10% in CH_2Cl_2
$Me_3Pb(NMePh)$	+226	10% in C_6D_6	Me_3ViPb	−65.3	Neat liquid
$Me_3Pb(SnMe_3)$	−324	Saturated in C_6D_6	$Me_3(SMe)Pb$	+214	Neat liquid
$Ph_3PbGePh_3$	−271.5	$CDCl_3$	$Ph_3PbSnPh_3$	−256.5	$CDCl_3$
$Me_3(Pr^i)Pb$	62	Saturated $CDCl_3$	Me_2PbCl_2	−222	DMSO
$Et_3Pb(O_2CMe)$	317	$CDCl_3$	$(CH_2CMe_3)_3Pb$	+336	Saturated CH_2Cl_2
$Et_2Pb(O_2CMe)_2$	−441	$CDCl_3$	$[Pb(CH_2CMe_3)_3]_2$	−237	Saturated CH_2Cl_2
$Pb(O_2CMe)_4$	−1869	–	$Ph_3Pb(O_2CMe)$	−93	–
$PbCl_2$	−4750	Solid	$Pb(NO_3)_2$	−2961.2	1.0 M in H_2O
PbO_2	+4550 ± 30	Powder, 98% pure	$Pb(ClO_4)_2$	−2950	In H_2O
Pb_9^{4-}	−4152	Naked anion cluster	$Me_3PbPbMe_3$	−281	Saturated in C_6H_6
Pb_2Ph_6	−79.8	Saturated $CDCl_3$	$Pb(2$-naphthyl$)_6$	−74.6	$CDCl_3$

Vi = vinyl.

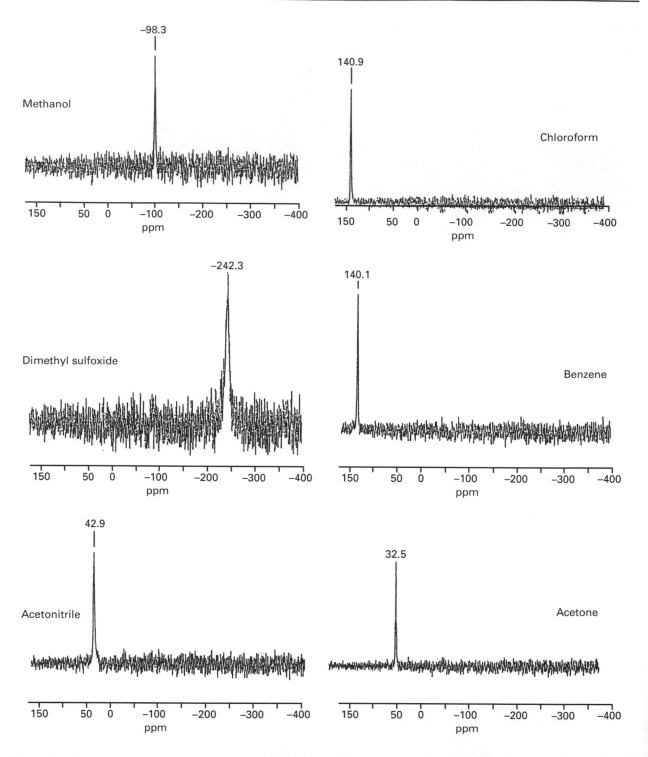

Figure 1 ^{119}Sn NMR spectra of Me_2SnCl_2 in chloroform, benzene, acetone, methanol, dimethyl sulfoxide and acetonitrile (concentration ~ 40 mg mL^{-1}).

are dissolved in noncoordinating solvents such as chloroform or benzene (**Figure 1**). On the other hand, when Me_2SnCl_2 is dissolved in various donor solvents such as acetonitrile, pyridine and dimethyl sulfoxide, the chemical shift shows dependence on the concentration.

The magnitude of the chemical shift can permit determination of association constants between the tin compound and the solvent. For example, formation of a 1:1 complex of trimethyltin chloride in dimethyl sulfoxide may be written as in Equation [2].

$$Bu^n_3SnCl + DMSO \underset{k_2}{\overset{k_1}{\rightleftharpoons}} Bu^n_3Sn(DMSO)Cl \quad [2]$$

The equilibrium constant can be calculated from the concentration dependence found for ^{119}Sn chemical shifts (Eqn [3]):

$$K = \frac{k_1}{k_2} = \frac{\delta_o(\delta_c - C\,\delta_o)}{\delta_o^2(1 - C) + C\,\delta_o^2 - \delta_o\,\delta_c} \quad [3]$$

where δ_o is the ^{119}Sn chemical shift for Bu^n_3SnCl, δ_c is the ^{119}Sn chemical shift for the $Bu^n_3Sn(DMSO)Cl$, and C is the molar fraction of Bu^n_3SnCl.

This type of behaviour is also displayed by lead and organolead compounds; the changes in $\delta(^{207}Pb)$ are generally larger than changes in $\delta(^{119}Sn)$: for example, the ^{207}Pb chemical shift for derivative $Pb(Ph)_2(O_2CMe)_2$ is -688 ppm in $CHCl_3$ and -859 ppm in pyridine.

Organotin(IV) thiolates show a low tendency to form complexes with molecules of coordinating solvent. However, it has been observed that although the $\delta(^{119}Sn)$ for the organotin(IV) thiolate compound $Bu^n_3Sn[S(1\text{-}C_{10}H_7)]$ [4] is concentration independent in noncoordinating solvent, the variation of $\delta(^{119}Sn)$ with temperature in pyridine(Py) is at first a sigmoidal curve which is typical of the equilibrium (Eqn [4]).

$$Bu^n_3Sn[S(1\text{-}C_{10}H_7)] + Py \rightleftharpoons$$
$$Bu^n_3 Sn[S(1\text{-}C_{10}H_7)](Py) \quad [4]$$

In several organometal derivatives, changes in the coordination number can arise from self-association, as in the case of organotin alkoxides. This self-association depends on steric and electronic effects and it can be studied with ^{119}Sn NMR. Several $R_{4-n}Sn(OR')_n$ compounds exhibit values of chemical shifts suggesting self-association as neat liquid or in noncoordinating solvents. For example, butyltin(IV) trialkoxides with an unbranched alkoxy

[4] δ (^{119}Sn) in $CDCl_3$: 89.1 ppm

group (such as $BuSn(OMe)_3$, [5], are six-coordinate at room temperature, the association presumably involving octahedral structures with both bridging and nonbridging alkoxy oxygen atom. Isopropoxy groups on tin, for example in [6], increase the steric effect and inhibit six-coordination. At room temperature, the chemical shift is typical of an essentially five-coordinate situation.

[5]

[6]

Effective nuclear charge and presence of low-lying excited states

Electronegative substituents generally lead to higher-frequency germanium, tin and lead chemical shifts. The effect of ligand electronegativity can be illustrated by the changes of ^{119}Sn chemical shifts along the series R_nSnCl_{4-n} (R = alkyl or aryl) as the value of the n changes from 2 to 0 (**Figure 2A**). However, the plot of the $\delta(^{119}Sn)$ for the compounds R_nSnCl_{4-n} (R = alkyl or aryl; n = 0, 1, 2, 3 or 4) in **Figure 2B**, indicates clearly that there are no obvious general relationships between $\delta(^{119}Sn)$ and the ligand electronegativity. The U-shape trend is usually attributed to change in the paramagnetic term due to p electron imbalance, generally maximum at $n = 2$.

The ^{73}Ge chemical shift of $GeCl_4$, $GeBr_4$, GeI_4 and the mixed tetrahalogermanes is not a linear function of halogen coordination but is consistent with the pattern of nuclear shielding established for halogens bonded to main-group elements: as expected, the nuclear shielding increases in the order Cl < Br < I.

The ^{119}Sn and ^{207}Pb chemical shifts follow the same trend shown by ^{73}Ge, but the larger lead and tin shift range allows a clearer distinction between closely related species (**Figure 3**).

In a series of methylvinylgermanes, the ^{73}Ge chemical shifts were compared with those of the corresponding methylvinylsilanes, and the correlation $\delta(^{73}Ge) = 1.96\,\delta(^{29}Si) + 2.37$ was found. It has been observed that the magnitude of the chemical shift cannot be simply attributed to the electron densities, but relative contributions of the diamagnetic and paramagnetic terms to the ^{73}Ge chemical shift should be considered.

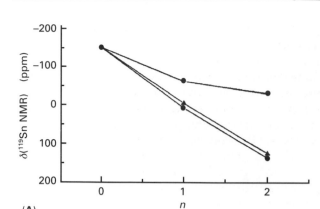

(A)

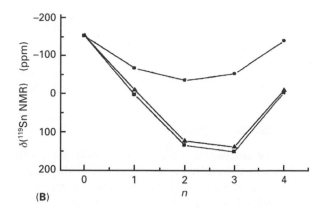

(B)

Figure 2 Plot of $\delta(^{119}Sn)$ against n for a series of monomeric tin(IV) compounds R_nSnCl_{4-n}. ■, Me; ●, Ph; ▲, Bu.

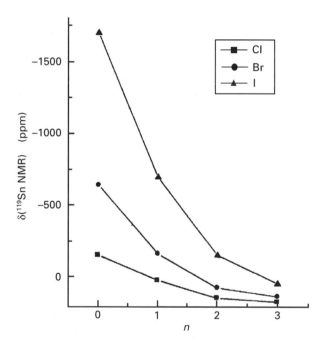

Figure 3 Plot of $\delta(^{119}Sn)$ against n for a series of monomeric methyltin(IV) halides. Me_nSnX_{4-n}. ■, Cl; ●, Br; — ▲, I.

[7] $\delta\,(^{119}Sn)$: +776 ppm   [8] $\delta\,(^{119}Sn)$: -2199 ppm

From a comparison between ethylvinylgermanes and ethynylstannanes, a similar linear correlation $\delta(^{73}Ge) = 0.481\,\delta(^{119}Sn) - 0.892$ was obtained.

Several studies have been made to evaluate the effects of the paramagnetic circulation of charge: in the monomeric bis[N,N-bis(trimethylsilyl)amino]-tin(II) compound [7], it has been found that circulation of charge from the trigonal plane into the underoccupied tin p_π orbitals deshields the tin atoms and produces a strong high-frequency shift (+776 ppm in C_6D_6) of the ^{119}Sn resonance. On the other hand, MO calculations indicate in monomeric tin(II) compounds such as stannocene, $(C_5H_5)_2Sn$ [8], a highly shielded tin atom (-2199 ppm in C_6H_{12}) as a consequence of an inefficient paramagnetic charge circulation due to the high LUMO energy of this molecule.

When tin is bonded to transition metals, large ^{119}Sn chemical shift ranges have been found that cannot be explained in terms of simple inductive and coordination number effects. The shielding could be interpreted in terms of d_π–d_π overlap and the effect on the average excitation energy.

Effects of interbond angles at the metal atoms

Changes in the bond angles in molecules in which the metal atom is a part of a ring system generally effect anomalous shifts of the resonances with respect to those found in analogous acyclic molecules. This is shown, for example, with tin–sulfur and lead–sulfur and with tin–carbon and lead–carbon bonds. A reduction of the cyclic interbond angles at the metal in five-membered rings generally decreases shielding. ^{119}Sn NMR data can be used to distinguish between six-membered and five-membered cyclic compounds: the $\delta(^{119}Sn)$ for compound [9] is -42.5 ppm, whereas that for the five-membered derivative [10] it is +53.5 ppm.

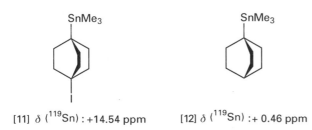

[9] δ (^{119}Sn) : -42.5 ppm   [10] δ (^{119}Sn) : +53.5 ppm

Other substituent effects

In the unassociated tin compounds $R_{4-n}SnX_n$ ($n = 0$–4), as the electron–releasing power of the organic group R increases, the tin atom is generally more shielded and the δ(^{119}Sn) moves to a higher field. This high-field shift, which cannot be explained in terms of p_π–d_π bonding between the π electron of the electron negative groups and the unfilled 5d orbital of the tin atom, is probably due to ring current effects by the π electron and/or a finite contribution to the shift from the diamagnetic term.

The values of the δ(^{119}Sn) chemical shift of diphenyltin(IV), dibenzyltin(IV) and di-*n*-butyltin(IV) compounds exhibit a good mutual correlation which can be expressed as in Equation [5].

$$\delta(^{119}\mathrm{SN})[\mathrm{R}_2^* \,\mathrm{Sn}] = a\delta(^{119}\mathrm{Sn})[\mathrm{R}_2^\# \,\mathrm{Sn}] + b \qquad [5]$$

where $[\mathrm{R}_2^*\mathrm{Sn}]$ and $[\mathrm{R}_2^\#\mathrm{Sn}]$ represent pairs of diorganotin(IV) compounds with different organic substituents, and a and b represent the slope and the intercepts of the linear correlation, respectively.

In some cases it has also been observed that a remote substituent in molecules in which direct π transmission mechanisms are prohibited can have a very strong influence on the nuclear shielding: for example, the ^{119}Sn substituent chemical shift (SCS) for derivative [11] (relative to parent molecule [12], which is due to both 'polar-field' and 'residual contribution' where the latter is probably due to 'through-bond' and/or 'through space electron delocalization', is very sensitive to polar substituent influences, like that derived from iodide groups.

SnMe₃   SnMe₃

[11] δ (^{119}Sn) : +14.54 ppm   [12] δ (^{119}Sn) : + 0.46 ppm

Coupling constants

The valence electron-mediated nuclear spin–spin couplings, $J(^{73}\mathrm{Ge},\mathrm{X})$, $J(^{117}\mathrm{Sn},\mathrm{X})$, $J(^{119}\mathrm{Sn},\mathrm{X})$ and $J(^{207}\mathrm{Pb},\mathrm{X})$, are dominated by the 'Fermi contact' term arising from the interaction of the nuclear spin magnetic dipole with the electron spin, which has a finite density at the nucleus. If the contact term is expressed with Pople and Santry's MO treatment and it is permissible to use the mean electronic excitation energy approximation, the coupling constants can be related to the hybridization (i.e. 's' electron density) and effective nuclear charge of the atoms participating in the bond. This implies that the parameters J(E,X) (E = Ge, Sn or Pb) are useful in the discussion of the structure and bonding.

The reduced couplings K(Ge,X) and K(Sn,X) are of opposite sign with respect to J(Ge,X) and J(Sn,X), respectively, owing to the negative magnetogyric ratios of ^{73}Ge and $^{115/117/119}$Sn.

In the case of X = ^{1}H, ^{19}F, ^{31}P, highly abundant spin-$\frac{1}{2}$ nuclei, the ^nJ(E,X) data are obtained conveniently from the X atom NMR spectra, whereas in the case of couplings with rare spin-$\frac{1}{2}$ nuclei (X = ^{13}C, ^{29}Si, ^{15}N), the ^nJ(E,X) can be obtained also from E NMR spectra by using Hahn-echo extended pulse sequences. In tin derivatives, in the case of couplings to quadrupolar X nuclei (X = ^{11}B, ^{14}N), it frequently happens that splitting due to ^{119}Sn–X coupling cannot be detected as a consequence of fast quadrupolar relaxation of X. The magnitude of J can be obtained from Equation [6].

$$J(\mathrm{Sn, X}) = \frac{\pi}{2}\left(\frac{3}{T_2^{\,\mathrm{SC}}(^{119}\mathrm{Sn})\; S_\mathrm{x}(S_\mathrm{x}+1)\; T_{(\mathrm{x})}^{\mathrm{Q}}}\right)^{1/2}$$

[6]

where S_x is the spin I of the quadrupolar nucleus, $T_2^{\,\mathrm{SC}}(^{119}\mathrm{Sn})$ is the transverse scalar relaxation time, and $T_\mathrm{X}^{\mathrm{Q}}$ is the quadrupolar relaxation time.

The most correct way to obtain the coupling sign of J(E,X), which is helpful in the discussion of the nature of chemical bonding, is based on the analysis of an energy level diagram that describes the interaction in a spin system. In a system containing at least three different magnetically active nuclei (E, A and X), two of them are defined as active spins, whereas the other is the passive one. If the relative signs of the coupling J(E,A) and J(E,X) and the absolute sign of one of these two are known, it is possible to determine the absolute sign of the other.

Spin couplings involving ^{73}Ge have been generally reported only in highly symmetrical compounds; **Table 5** gives a selection of the values the of coupling constants in germanium compounds of tetrahedral symmetry. The signs of these coupling constants are indicated only by comparison with those

Table 5 Some representative spin coupling involving ^{73}Ge

Compound	Coupling constant	(Hz)
GeH$_4$	$^1J(^{73}Ge,^1H)$:	−97.6
Me$_4$Ge	$^1J(^{73}Ge,^{13}C)$:	−18.7
GeF$_4$	$^1J(^{73}Ge,^{19}F)$:	−97.6
Me$_4$Ge	$^2J(^{73}Ge,^1H)$:	+3.0
(MeO)$_4$Ge	$^3J(^{73}Ge,^1H)$:	−1.9
(MeS)$_4$Ge	$^3J(^{73}Ge,^1H)$:	−2.5

of similar group 14 element compounds and not by double resonance experiments.

One-bond couplings

In the case of couplings with electropositive elements, the one-bond reduced coupling constants show the trend $^1K(Ge,X) < {}^1K(Sn,X) < {}^1K(Pb,X)$. In fact, the increase of $^1K(E,X)$ with increasing size of E corresponds to the change of valence s electron densities $\Psi_M(O)^2$. This relationship is not valid when X is a less electropositive or more polarizable atom. **Table 6** gives a selection of values of a one-bond coupling constant involving ^{119}Sn and ^{207}Pb.

$^1J(E,{}^1H)$

Very few one-bond coupling constants between ^{119}Sn or ^{207}Pb and H have been measured. This is in part due to the instability and the difficulty of preparing suitable species. Limited data are available for ^{119}Sn and indicate that the magnitude of the $^1J(^{119}Sn–H)$ in R$_{4-n}$SnH$_n$ increases with n when R = alkyl and decreases with n when R = phenyl, owing to greater electronegativity of phenyl with respect to alkyl groups. In SnH$_3^-$ the presence of an electron lone pair at tin leads to a very low s character for hybrid orbitals in Sn–H bonds, and hence to a very small magnitude of coupling.

$^1J(E,{}^{13}C)$

Owing to the large number of organolead and organotin species, there is a sizeable literature of data on $^1J(E,{}^{13}C)$ coupling constants (E = ^{119}Sn or ^{207}Pb).

The magnitude of $^1J(^{119}Sn,{}^{13}C)$ organotin compounds has been related quantitatively to the C–Sn–C bond angles. A simple linear relationship between $^1J(^{119}Sn,{}^{13}C)$ coupling constants of a group of dimethyltin(IV) compounds and their C–Sn–C angle (θ) has been found on the basis of the analysis of a large set of experimental data (Eqn [7]).

$$|{}^1J(^{119}Sn,{}^{13}C)| = (10.7 \pm 0.5)\theta - (778 \pm 64) \quad [7]$$

Table 6 Some representative one-bond spin $^1J(^{119}Sn,X)$ and $^1J(^{207}Pb,X)$

Compound	X	E	$^1J(E,X)$ values (Hz)
SnH$_4$	^1H	^{119}Sn	−1 930
MeSnH$_3$	^1H	^{119}Sn	−1 850
Me$_2$SnH$_2$	^1H	^{119}Sn	−1 797
Me$_3$SnH	^1H	^{119}Sn	−1 744
Ph$_3$SnH	^1H	^{119}Sn	−1 935.8
MeSnH$_2$Cl	^1H	^{119}Sn	−2 228
Me$_4$Sn	^{13}C	^{119}Sn	−340
Me$_3$SnCl	^{13}C	^{119}Sn	−380
Me$_2$SnCl$_2$	^{13}C	^{119}Sn	−566
Ph$_2$SnCl$_2$	^{13}C	^{119}Sn	−785
Me$_2$Sn(acac)$_2$	^{13}C	^{119}Sn	−966
Me$_3$SnLi·3THF	^{13}C	^{119}Sn	+155
(Me$_3$Sn)$_2$	^{13}C	^{119}Sn	−240
[SnF$_6$]$^{2-}$	^{19}F	^{119}Sn	1 550
[Sn(OH)F$_5$]$^{2-}$	^{19}F	^{119}Sn	1 776,127.8
cis-[Sn(OH)$_2$F$_4$]$^{2-}$	^{19}F	^{119}Sn	1 820,1 518
trans-[Sn(OH)$_2$F$_4$]$^{2-}$	^{19}F	^{119}Sn	1 956
(Me$_3$Sn)$_3$P	^{31}P	^{119}Sn	+832.5
SnCl$_4$(PEt$_3$)$_2$	^{31}P	^{119}Sn	2 383.1
Me$_3$SnPH$_2$	^{31}P	^{119}Sn	+463
Me$_3$SnSnMe$_3$	^{119}Sn	^{119}Sn	+4240
trans-Pt(SnCl$_3$)$_2$(PEt$_3$)$_2$	^{195}Pt	^{119}Sn	20 585
[Rh(SnCl$_3$)Cl$_5$]$^{3-}$	^{103}Rh	^{119}Sn	864
Me$_2$PbH$_2$	^1H	^{207}Pb	+2 454.6
Me$_3$PbH	^1H	^{207}Pb	+2 295
Ph$_4$Pb	^{13}C	^{207}Pb	+481
Me$_4$Pb	^{13}C	^{207}Pb	+320
Me$_3$PbBr	^{13}C	^{207}Pb	+246
Me$_3$PbPbMe$_3$	^{13}C	^{207}Pb	+28
Ph$_2$Pb(O$_2$CMe)$_2$	^{13}C	^{207}Pb	1 203
PhPb(O$_2$CMe)$_3$	^{13}C	^{207}Pb	2 100
Me$_3$PbSeMe	^{13}C	^{207}Pb	−1170
Me$_3$PbCH=CHMe	^{13}C	^{207}Pb	268
Me$_3$PbPbMe$_3$	^{207}Pb	^{207}Pb	+290
Me$_3$PbPbMe$_3$	^{119}Sn	^{207}Pb	−3 570
[Pt(PR$_3$)$_2$(PbPh$_3$)$_2$]	^{195}Pt	^{207}Pb	14 500–18 500
Pb$_9$Pt	^{195}Pt	^{207}Pb	4 122

A similar relation has been described for di-n-butyltin(IV) compounds (Eqn [8]).

$$|{}^1J(^{119}Sn,{}^{13}C)| = (9.99 \pm 0.73)\theta - (746 \pm 100) \quad [8]$$

and for tetra-, tri- and diphenyltin(IV) compounds (Eqn [9])

$$|{}^1J(^{119}Sn,{}^{13}C)| = (15.56 \pm 0.84)\theta - (1160 \pm 101) \quad [9]$$

Equation [7] is not valid for cyclic methyltin(IV) compounds.

The sign of $^1J(^{119}Sn,^{13}C)$ is generally negative for all tetraorganotin(IV) compounds, whereas $^1J(^{207}Pb,^{13}C)$ is positive for tetraorganolead(IV) derivatives.

In R_4Sn (R = alkyl) compounds, $^1J(^{119}Sn,^{13}C)$ decreases in magnitude with increasing chain length and branching. It has also been found that in the tetraorganotin derivatives the magnitude of $^1J(^{119}Sn,^{13}C)$ increases with change of the hybridization of the carbon atoms from sp^3 to sp.

The replacement of alkyl groups with electronegative substituents causes an increase in the magnitude of $^1J(^{119}Sn,^{13}C)$ as the result of a greater s character in the Sn–C hybrid orbital.

Very small values of $^1J(^{119}Sn,^{13}C)$ are found in cyclopentadienyltin(IV) compounds owing to the low carbon s character in the pentahapto link to tin, whereas the small values in allyl- and benzyltin(IV) compounds may be ascribed to σ–π conjugative connection of the polarized σ(Sn–C) bond with the adjacent π electron system.

As in the case of the $\delta(^{119}Sn)$ chemical shift, moderate linear correlations have been found between the values of triphenyltin(IV) compounds and those of analogous triorganotin compounds R_3SnX (R = Bu^n, Bz, Vi) which can be expressed as Equation [10]

$$^1J(^{119}Sn,^{13}C)(Ph_3Sn) = a\ ^1J(^{119}Sn,^{13}C)(R_3Sn) + b$$
[10]

a and *b* being slope and intercepts, respectively.

$^1J(E,X)$ (X = ^{15}N, ^{31}P, ^{19}F, ^{119}Sn, ^{207}Pb)

Several studies have been carried out on the bis(amino)stannylenes, where the coupling constants $^1J(^{119}Sn,^{15}N)$ are generally large and negative in accordance with a Ψ-pseudotrigonal-bipyramidal structure, whereas in stannylphosphane compounds $^1J(^{119}Sn,^{31}P)$ are generally positive and strongly dependent not only on changes in hybridization and substituent electronegativity, but also on s-overlap integral.

It has been found that the values of $^1K(^{119}Sn,^{31}P)$ for $(R_3Sn)_3P$ derivatives become less negative if the phosphorus lone electron pair is used in metal complexation.

$^1J(^{119}Sn,^{19}F)$ coupling constants are reported for hydroxo- and peroxotin(IV) fluoro complexes [13] and [14] and for $[SnF_6]^{2-}$ species. Their magnitude is strongly dependent on the stereochemistry.

[13]

$^1J(^{119}Sn,^{19}F)$:
1820 Hz (*trans*-OH)
1518 Hz (*trans*-F)

[14]

$^1J(^{119}Sn,^{19}F)$:
2064 Hz (*trans*-OOH)
1692 Hz (*trans*-F)

Several negative terms contribute to the extent of $^1K(E,^{19}F)$ so that no values are known for lead–fluorine and only a few tin–fluorine couplings have been reported.

$^1J(^{119}Sn,^{119}Sn)$ values are generally positive and show clear dependence upon tin hybridization. When very bulky substituents are on tin atoms, the tin–tin s-overlap integral is small, whereas the mutual polarizability term changes sign, so that very small or negative $^1J(^{119}Sn,^{119}Sn)$ values are found. Large values of $^1J(^{119}Sn,^{119}Sn)$ have been detected in tin–tin complexes containing electronegative substituents or characterized by greater coordination number.

$^1J(^{119}Sn,^{207}Pb)$ values are normally negative and often of very small magnitude owing to reduced lead–tin s-overlap.

$^1J(^{207}Pb,^{207}Pb)$ are positive, and values are also very small for the same reason: lead–lead compounds are in fact easily dissociated in solution owing small lead–lead overlap.

$^1J(E,$ transition metal)

Many tin–transition metal coupling constants are known owing to the ease of obtaining [(SnCl$_3$)–transition metal] derivatives. The $^1J(^{119}Sn,$ transition metal) values are generally large and are used to characterize complexes having tin–metal bonds also in solution. It has been observed that in square planar Pt(II) complexes [15] and [16], the magnitude of the $^1J(^{195}Pt,^{119}Sn)$ coupling constants depends on the magnitude of the *trans* effect of the respective substituent.

[15] $^1J(^{195}Pt,^{119}Sn)$: 16 321 Hz

[16] $^1J(^{195}Pt,^{119}Sn)$: 28 052 Hz

Geminal coupling constants, $^2J(E,Y)$

The geminal coupling constants $^2J(E-X-Y)$ (**Table 7**) are sensitive not only to the factors reported for one-bond couplings, but also to the nature of the intervening atom X, to the metal–X–Y bond angle, to the stereochemistry of the molecule and finally to the nature of other substituents.

$^2J(E-X-{}^1H)$

Several data have been reported, in particular for X = C. There is a linear relationship between $^2J(^{119}Sn,{}^1H)$ and $^2J(^{207}Pb,{}^1H)$ and also between $^2J(^{119}Sn,{}^1H)$ and $^1J(^{119}Sn,{}^{13}C)$ in related methyltin(IV) and methyllead(IV) derivatives. In several dimethyltin(IV) complexes the values of $^2J(^{119}Sn,{}^1H)$ can be used to calculate the bond angles C–Sn–C in solution (Eqn [11]).

$$(\text{angle C}-\text{Sn}-\text{C}) = 0.0161[\,^2J(^{119}Sn,{}^1H)_{Me}]^2 + 133.4 \qquad [11]$$

The sign of $^2J(^{119}Sn,{}^1H)$ is generally positive. Exceptionally the coupling can be negative, as in Me_3Sn^- and diorganostannylenes, whereas the $^2J(^{207}Pb,{}^1H)$ are negative and normally greater than $^2J(^{119}Sn,{}^1H)$ in related compounds. The magnitude of $^2J(E-C-{}^1H)$ increases with increasing coordination number on the metal. $^2J(^{207}Pb,{}^1H)$ values are generally less sensitive than $^2J(^{119}Sn,{}^1H)$ to structural geometrical differences. In several octahedral and square planar complexes, the magnitude of the $^2J(^{119}Sn-X-{}^1H)$ across a transition metal X can be used to distinguish the *trans* (compound [17]) and *cis* positions (compound [18]) of Sn and H.

[17] $^2J(^{119}Sn,{}^1H)$: 1740 Hz   [18] $^2J(^{119}Sn,{}^1H)$: ± 27 Hz

$^2J(E-X-{}^{13}C)$

The $^2J(^{119}Sn-X-{}^{13}C)$ couplings are generally small and positive in aliphatic and tetracoordinate tin derivatives, whereas they are large and negative in compounds containing an $Sn-C_{olefinic}-C$ bond and close to 0 in compounds containing an $Sn-C=C$ bond. In some cases the values are exceptionally large, as in $Sn(C\equiv CBu)_4$.

Other geminal couplings

The couplings best studied are those involving ^{119}Sn, ^{117}Sn, ^{19}F and ^{31}P. For example, in bis(triorganotin)oxides the geminal couplings $^2J(^{119}Sn,{}^{119}Sn)$ indicate a linear structure in solution. The $^2J(^{119}Sn-X-{}^{119}Sn)$, where the X intervening atom is a transition metal, show a strong dependence on the geometry and on the nature of other ligands on the metal or on tin atoms: for example, the $^2J(^{119}Sn-Pt-{}^{119}Sn)$ is 2601 Hz in *cis*-[PtCl$_2$(Cl$_3$Sn)$_2$] compound [19] and 36 286 Hz in *trans*-[PtCl$_2$(Cl$_3$Sn)$_2$] (compound [20]).

[19] $^2J(^{119}Sn,{}^{119}Sn)$: 2601 Hz [20] $^2J(^{119}Sn,{}^{119}Sn)$: 36286 Hz

A similar geometrical dependence has been shown by $^2J(^{119}Sn-Pt-{}^{31}P)$. Two-bond couplings involving ^{19}F are detected in $(CF_3)_4Ge$ (26.3 Hz), in $(CF_3)_4Sn$ (531 Hz) and also in the organolead derivatives $(Me)_3Pb(CF_3)$ (240 Hz) and $(Me)_2Pb(CF_3)_2$ (379 Hz), which show a strong dependence on the effective nuclear charge.

Table 7 Selected germinal coupling constants $^2J(E,Y)$

Compound	X	E	$^2J(E,Y)$ values (Hz)
MeSnH$_3$	^1H	^{119}Sn	+63.6
Me$_2$SnH$_2$	^1H	^{119}Sn	+58
Me$_2$SnH	^1H	^{119}Sn	+56.5
Me$_4$Sn	^1H	^{119}Sn	+53.9
Me$_2$SnCl	^1H	^{119}Sn	+58.2
Me$_2$SnCl$_2$	^1H	^{119}Sn	+68.9
MeSnCl$_3$	^1H	^{119}Sn	+96.9
Me$_2$SnPH$_2$	^1H	^{119}Sn	62.5
(Bu$_2$ClSn)$_2$O	^{119}Sn	^{119}Sn	74
(Me)$_3$SnCF$_2$H	^{19}F	^{119}Sn	265.6
Ph$_2$SnCl$_2$	^{13}C	^{119}Sn	63
Me$_2$SnPh	^{13}C	^{119}Sn	36.6
Et$_4$Sn	^{13}C	^{119}Sn	+23.5
Me$_4$Pb	^1H	^{207}Pb	−60.5
Me$_2$PbCl$_2$	^1H	^{207}Pb	−70
Me$_2$PbH$_2$	^1H	^{207}Pb	−76
Vl$_4$Pb	^1H	^{207}Pb	213.3
Ph$_4$Pb	^{13}C	^{207}Pb	−68
Me$_2$PbPbMe$_3$	^{13}C	^{207}Pb	+92

Table 8 Selected vicinal and long-range coupling constants $^nJ(E,Y)$

Compound	Y	E	n	$^nJ(E,Y)$ values (Hz)
Et_4Sn	^1H	^{119}Sn	3	−71.2
Vi_4Sn	^1H	^{119}Sn	3	183.1
$EtSnCl_3$	^1H	^{119}Sn	3	−242
Et_2SnCl_2	^1H	^{119}Sn	3	−130.8
$PhSnCl_3$	^1H	^{119}Sn	3	122
$(MeS)_4Sn$	^1H	^{119}Sn	3	−66
$(Me_2N)_4Sn$	^1H	^{119}Sn	3	−51.0
Me_3SnSMe	^1H	^{119}Sn	3	−37.5
$(Me_3Sn)_2C=CH_2$	^1H	^{119}Sn	3	124 (cis); 208 (trans)
$(CH_3)_2B–N(SnMe_3)_2$	^1H	^{119}Sn	3	33.0 (cis); 48.4 (trans)
$Me_2Sn(CF=CF_2)_2$	^{19}F	^{119}Sn	3	25 (cis); 29 (trans)
$Et_3Sn(CH_2)_2SnEt_3$	^{119}Sn	^{119}Sn	3	920
Bu_4Sn	^{13}C	^{119}Sn	3	52.0
$BuSnCl_3$	^{13}C	^{119}Sn	3	120.0
Ph_2SnCl_2	^{13}C	^{119}Sn	3	90
Ph_2SnCl_2	^{13}C	^{119}Sn	4	16
Me_3SnPh	^{13}C	^{119}Sn	4	10.8
Et_4Pb	^1H	^{207}Pb	3	+128.6
Ph_4Pb	^{13}C	^{207}Pb	3	80
Ph_4Pb	^{13}C	^{207}Pb	4	20
$(4-MeC_6H_4)_4Pb$	^1H	^{207}Pb	6	5.4
$Me_3PbPbMe_3$	^1H	^{207}Pb	3	+22.9
$(Me_3CCH_2)_4Pb$	^1H	^{207}Pb	4	5.3

Vicinal and long-range coupling constants, $^nJ(E,Y)$ ($n \geq 3$)

Many vicinal coupling constants (**Table 8**) are known, with different combinations of intervening atoms. In most cases the $^3J(^{119}Sn,X)$ data are dependent on the respective dihedral angles and follow a Karplus-type relationship. The sign of $^3J(^{119}Sn,^1H)$ and $^3J(^{119}Sn,^{13}C)$ is negative and the magnitudes of $^3J(^{119}Sn,^1H)$ and $^3J(^{119}Sn,^{13}C)$ are usually greater than that of $^2J(^{119}Sn,^1H)$ and $^2J(^{119}Sn,^{13}C)$, respectively.

In several olefinic compounds it has been observed that $^3J(^{119}Sn,^{13}C)_{trans}$ is greater than $^3J(^{119}Sn,^{13}C)_{cis}$. The magnitude of $^3J(^{119}Sn,^1H)$ and $^3J(^{119}Sn,^{13}C)$ is strongly dependent on the nature of the substituents, an increase being observed in the presence of electropositive substituents on the C=C double bond.

$^3J(^{119}Sn,^{119}Sn)$ often behaves in a similar manner to $^3J(^{119}Sn,^1H)$ and $^3J(^{119}Sn,^{13}C)$. For example, $^3J(^{119}Sn,^{119}Sn)$ is 491 Hz in cis- and 1012 Hz in trans-$Me_3SnCH=CHSnMe_3$, respectively. An exceptionally large value of $^3J(^{119}Sn,^{119}Sn)$ is found for the derivative [21] owing to an additive coupling pathway across the Pt–Pt bond, whereas very small $^3J(^{119}Sn,^{119}Sn)$ are observed across the C≡C bond.

[21] $^3J(^{119}Sn, ^{119}Sn)$: 25430 Hz

Several long-range coupling constants (**Table 8**) are reported for many different X nuclei, but they have not been systematically studied. Analogously to $^nJ(^1H,^1H)$ ($n \geq 3$), the $^nJ(E,Y)$ couplings are best transferred by π system or by σ–π interaction.

Relaxation behaviour

The dominant relaxation mechanism for ^{73}Ge is quadrupolar, and it is possible to obtain T_1 directly from the line width.

The tin and lead compounds so far described are generally small, highly symmetrical molecules in which the spin-rotation (SR) mechanism is important.

In the tetrahalogermanes GeX_4 (X = Cl, Br or I), T_1 is almost exclusively dominated by the quadrupole relaxation mechanism, while T_2 is dominated by the combination of the scalar coupling and the quadrupole relaxation mechanism in the high-temperature region. The contribution of scalar coupling to the relaxation mechanism decreases in the order

$$GeCl_4 > GeBr_4 > GeI_4$$

Relaxation studies on liquid tetrahalotin(IV) compounds, carried out to determine temperature and field dependences of relaxation, indicated that T_1 for $SnCl_4$, $SnMe_4$ and $SnMeCl_3$ is dominated by a spin-rotation mechanism, while both scalar (SC) and spin-rotation (SR) mechanisms contribute to T_1 for $SnBr_4$ and SnI_4. On the other hand, T_2 for $SnCl_4$, $SnBr_4$ and SnI_4 is scalar dominated. When the molecules have hydrogens directly linked to tin, $^{119}Sn/^1H$ dipolar relaxation is significant.

In the derivative $Bu_3SnSnBu_3$, the relaxation is about 75% T_1^{DD}.

In a 3 M solution of $Pb(ClO_4)_2$, the dominant T_1 relaxation is spin rotation. At higher concentration, CSA (chemical shift anisotropy) and dipolar contributions are also present and important. SC and SR mechanisms contribute to T_1 for neat $PbCl_4$. CSA is significant in $PbMe_3Cl$ in DMSO, which probably coordinates the lead atom.

Applications

Solid-state NMR

Solution and solid-state structures of metal and organometal derivatives are often not identical. High-resolution solid-state NMR experiments on metallic and organometallic compounds are important tools for bridging the information gap between X-ray diffraction results and solution-state NMR data. Until now, most of the high-resolution work in tin(IV), organotin(IV), lead and organolead(IV) chemistry has dealt with the combined use of cross-polarization and magic angle spinning (CP/MAS) techniques. The experimental conditions for ^{119}Sn and ^{207}Pb spectra are different from those usually employed for ^{13}C CP/MAS spectra of organic and organometallic compounds owing to smaller gyromagnetic ratios and larger shielding anisotropies. The usual CP/MAS situation for ^{119}Sn and ^{207}Pb is the presence of only a few resonances with large shielding anisotropies, so that it is very easy to determine shielding tensor components from one-dimensional CP/MAS spectra.

An example of application of CP/MAS is the $(Bu^i_3Sn)_2CO_3$ derivative [22]: the ^{119}Sn solution spectrum gives only one resonance at +101.7 ppm, in accordance with a monomeric tetracoordinated tin atom. The crystal structure of this compound clearly indicates a two-Sn environment: a polymeric chain of trigonal-bipyramidal $trans$-R_3SnO_2 units linked to each other via bridging carbonate anions and tetrahedral R_3SnO units bonded to the third oxygen atom of the CO_3^{2-} bridging group. In the ^{119}Sn CP/MAS spectrum, the presence of both tin fragments, a and b, is well resolved.

CP/MAS NMR spectra are also used to clarify the controversial view of the structures of R_2SnX_2 and R_3SnX species in the solid and solution state: for example, the tricyclohexyltin hydroxide, which is a linear polymer in the solid state ($\delta = -217$ ppm) is monomeric in solution ($\delta = +11.6$ ppm). The diorganotin derivatives, which are monomeric in both phases, exhibit very similar shifts both in ^{119}Sn CP/MAS and ^{119}Sn solution NMR spectra. Small chemical shifts are observed in diorganotin and triorganotin derivatives weakly associated in the solid state and not associated in solution.

The CP/MAS NMR technique is also useful for investigating dynamic processes in solids such as reorientational processes and structural phase transitions: an example of solid-state phase transition monitored by variable temperature ^{119}Sn CP/MAS is given by $Me_3Sn-C\equiv C-C\equiv C-SnMe_3$, for which a structural phase change occurs in the temperature range 232–248 K.

[22]

Other applications

The ^{73}Ge NMR chemical shifts of several methyl- and phenyl-substituted germacyclohexanes have been determined, and it has been shown that the chemical shifts are very sensitive to steric effects and can provide a useful tool for the conformational analysis of organogermanium compounds.

The progress of reactions involving tin and lead compounds can readily be monitored by ^{119}Sn and ^{207}Pb NMR. For example, the various products derived from the 1,1-organoboration of 1-alkynyltin compounds can be easily identified by ^{119}Sn NMR, compounds [23, 24, 25].

The content of complex reaction mixtures can be also analysed: for example, several species and intermediates are detectable in exchange reactions by ^{119}Sn NMR techniques, e.g. compounds in Equation [12].

1D and 2D ^{1}H–^{119}Sn HQMC experiments are useful in the investigation of weak coordination in functionalized triphenylvinyltin(IV) compounds of the type $Ph_3Sn-CH=CH-C(OH)RR'$.

The HQMC technique is widely employed in the study of the condensation product of salicylaldoxime (o-HO–N=CH–C_6H_4–OH) with di-n-butyl oxide, which in solution exists as a mixture of several coordination complexes such as [26]. In this case, two-dimensional ^{1}H–^{119}Sn HQMC spectroscopy yields essential data for the structure elucidation of

[12]

[23]   [24]

[25]

most of these complexes and also of the transient species, and allows several correlations between geometric parameters and $^nJ(^{119}Sn,^1H)$ to be established.

[26]

List of symbols

T_1 = spin–lattice relaxation times; T_2 = spin–spin relaxation time; T^Q = quadrupolar relaxation time; T^{SC} = transverse scalar relaxation time; δ = chemical shift; γ = gyromagnetic ratio; σ = shielding constant

(σ_d = diamagnetic, σ_n = nonlocal contribution, σ_p = paramagnetic).

See also: **NMR Principles; NMR Pulse Sequences; Parameters in NMR Spectroscopy, Theory of; ^{29}Si NMR; Solid State NMR, Methods; Solid State NMR Using Quadrupolar Nuclei; Structural Chemistry Using NMR Spectroscopy, Inorganic Molecules; Two-Dimensional NMR, Methods.**

Further reading

Dechter JJ (1982) NMR of metal nuclides. Part I. The main group metals. In: Lippard SJ (ed). *Progress in Inorganic Chemistry*, pp 285–385. New York: Wiley.

Hani R and Geanangel RA (1982) ^{119}Sn NMR in coordination chemistry. *Coordination Chemistry Reviews* **44**: 229–246.

Harris RK, Kennedy JD and McFarlane W (1978) Group IV–silicon, germanium, tin and lead. In: Harris RK and Mann BE (eds) *NMR and Periodic Table*, pp 340–371. New York: Academic Press.

Harrison PG (1989) Investigating tin compounds using spectroscopy. In: Harrison PG (ed) *Chemistry of Tin*, pp 60–117. Glasgow: Blackie.

Hunter BK and Reeves LW (1967) Chemical shift for compounds of the group IV elements silicon and tin. *Canadian Journal Chemistry* **46**: 1399–1414.

Kayser F, Biesemans M, Gielen M and Willem R (1996) Two dimensional ^1H–^{119}Sn proton detected correlation spectroscopy in coordination chemistry of hypervalent organotin compounds. In: Gielen M, Willem R and Wrackmeyer B (eds), *Advanced Applications of NMR to Organometallic Chemistry*, pp 45–86. Chichester: Wiley.

Kennedy JD and McFarlane W (1987) Silicon, germanium, tin and lead. In: Mason J (ed) *Multinuclear NMR*, pp 305–333. New York: Plenum Press.

Petrosyan VS (1977) NMR spectra and structures of organotin compounds. *Progress in NMR Spectroscopy* **11**: 115–148.

Sebald A (1996) Solid state NMR applications in organotin and organolead chemistry. In: Gielen M, Willem R and Wrackmeyer B (eds), *Advanced Applications of NMR to Organometallic Chemistry*, pp 123–157. Chichester: Wiley.

Smith PJ and Smith L (1973) Application of ^{119}Sn chemical shifts to structural tin chemistry. *Inorganica Chimica Acta Reviews* **7**: 11–33.

Smith PJ and Tupciauskas AP (1978) Chemical shifts of ^{119}Sn nuclei in organotin compounds. *Annual Reports on NMR Spectroscopy* **8**: 291–370.

Wrackmeyer B (1985) ^{119}Sn NMR parameters. *Annual Reports on NMR Spectroscopy* **16**: 73–185.

Wrackmeyer B (1996) Indirect nuclear ^{119}Sn–X spin–spin coupling. In: Gielen M, Willem R and Wrackmeyer B (eds), *Advanced Applications of NMR To Organometallic Chemistry*, pp 87–122. Chichester: Wiley.

Heteronuclear NMR Applications (La–Hg)

Trevor G Appleton, The University of Queensland,
Brisbane, Australia

Introduction

NMR spectroscopy for this group of metals is dominated by the three nuclei: ^{183}W, ^{195}Pt and ^{199}Hg. Each of these will be dealt with in turn, followed by brief mention of other nuclei with non-zero values of the nuclear spin quantum number, I.

Tungsten, ^{183}W

The only tungsten isotope with a non-zero value of I is ^{183}W, $I = \frac{1}{2}$, (14.4% natural abundance). Its receptivity relative to that of ^{1}H is low (1.06×10^{-5}). Its resonance frequency in a magnetic field of 2.35 T, where protons resonate at 100 MHz, is 4.17 MHz. This nucleus is radioactive, an α-particle emitter with a long half-life ($> 1 \times 10^{17}$ y). 'Satellite' peaks from coupling with ^{183}W are readily observed in the ^{31}P NMR spectra of tertiary phosphine complexes, the ^{1}H NMR spectra of hydrides, the ^{19}F NMR spectra of fluoro complexes and the ^{13}C NMR spectra of carbonyl complexes. The earliest measurements of tungsten chemical shifts were indirect measurements through INDOR (internuclear double resonance) experiments. Direct measurement is now commonplace, but large sensitivity enhancements are possible with two-dimensional indirect spectroscopy, using the highly receptive 100% abundant nuclei ^{31}P, ^{1}H or ^{19}F.

A solution of $Na_2[WO_4]$ (1 M in D_2O, pD 9) is now becoming accepted as the standard reference for ^{183}W NMR in preference to WF_6. Relaxation times for ^{183}W are frequently long. Since a major relaxation mechanism is through chemical shift anisotropy (CSA), shorter, more favourable relaxation times are obtained when the environment about the tungsten nucleus is less symmetric. Addition of a paramagnetic substance, such as $[Cr(acac)_3]$ can be used to shorten relaxation times which are inconveniently long. The tungsten chemical shift is very sensitive to the environment about the nucleus. The overall range is large (~6000 ppm). Shifts for some representative compounds are given in **Table 1**.

Variations in chemical shifts for heavy atom nuclei are dominated by changes in σ_p, the paramagnetic contribution to total nuclear screening. For a transition metal complex, a simplified expression for σ_p is

$$\sigma_p \propto -(r^{-3}) \sum_j^{occ} \sum_k^{unocc} (E_k - E_j)^{-1} C \qquad [1]$$

where r is an average distance of valence d-electrons from the nucleus, E_k and E_j are energies of unoccupied and occupied molecular orbitals, respectively, and C is a sum containing the coefficients of the metal s, p and d orbitals used in the summation to develop the various molecular orbitals. It will be noted that the tungsten nucleus in the series $[W(CO)_3(Cp)X]$ becomes progressively more shielded as the halide is changed from chloride to bromide to iodide, which is the 'normal halide dependence' order for transition metal complexes. This is opposite to the order expected from the effect of X^- on excitation energies, which must therefore be

Table 1 ^{183}W chemical shifts for some representative compounds

Compound	δ_W (ppm)
W(0)	
$W(CO)_6$	−3505
$W(CO)_5(PMePh_2)$	−3324
W(I)	
$[\{W(CO)_3(Cp)\}_2]$	−4040
W(II)	
$[W(CO)_3(Cp)Cl]$	−2406
$[W(CO)_3(Cp)Br]$	−2584
$[W(CO)_3(Cp)I]$	−2996
$[W(CO)_3(Cp)Me]$	−3549
$[W(CO)_3(Cp)H]$	−4017
W(IV)	
$[W(Cp)_2H_2]$	−4671
W(VI)	
WF_6	−1120
$[W_{10}O_{22}]^{4-}$	−43
$[WO_4]^{2-}$	0
$[W_6O_{19}]^{2-}$	+59
$[WCl_6]$	+2181

$Cp = \eta^5$-cyclopentadienyl.

outweighed by the effects of metal–ligand covalency on r and the coefficients C. For the W(VI) complexes WF_6 and WCl_6, however, the order expected from excitation energies is observed.

One area where ^{183}W NMR has made a major contribution is in the study of iso- and heteropolytungstate ions. Since the tungsten atoms are usually in highly distorted octahedral environments, the relaxation times are short enough to allow convenient direct observation. The sensitivity of δ_w to the tungsten environment, and the observation of W–O–W coupling patterns, facilitate structural assignments. Paramagnetic heteropolytungstate ions have been included in these studies.

Where similar series of tungsten and molybdenum compounds have been studied by ^{183}W and ^{95}Mo NMR, respectively, there are close parallels between the two series. ^{95}Mo ($I = \frac{5}{2}$) is quadrupolar, and lines are relatively broad in unsymmetrical environments. Magic-angle spinning (MAS) solid-state ^{183}W NMR spectra may be obtained, but long spectrometer times are usually required because of long spin–lattice relaxation times.

Platinum, ^{195}Pt

The only platinum nucleus with magnetic properties is ^{195}Pt, $I = \frac{1}{2}$, (33.7% abundance). The resonance frequency in a magnetic field of 2.35 T is approximately 21.4 MHz. 'Satellite' peaks from coupling with ^{195}Pt were observed in ^1H and ^{31}P NMR spectra in the 1960s, and much of the early work on ^{193}Pt detection used INDOR methods. Direct one-dimensional observation of ^{195}Pt NMR spectra is now routine. Because ^{195}Pt relaxation times are short for most compounds, there need be only a very short delay between pulses, allowing rapid accumulation. Two-dimensional inverse detection methods are also being increasingly used.

The most commonly used reference for ^{195}Pt NMR is an aqueous solution of $Na_2[PtCl_6]$. As a number of authors have pointed out, there are some disadvantages in using this substance. At high magnetic fields, peaks due to different combinations of chlorine isotopes may be resolved, and the resonance frequency is significantly dependent on temperature. This leads to some uncertainty (up to ± 5 ppm) in reported chemical shifts, but this is not very significant in the context of the overall chemical shift range for platinum (~15 000 ppm). However, this reference does have the great advantage of convenience, and is therefore unlikely to be supplanted. Other suggestions have been $[Pt(CN)_6]^{2-}$ (− 3863 ppm) and Ξ_{Pt}, a frequency of exactly 21.4 MHz in a magnetic field in which the resonance frequency of tetramethylsilane (TMS) is exactly 100 MHz (−4533 ppm). An aqueous solution of $K_2[PtCl_4]$ (δ_{Pt} −1624 ppm) is sometimes used as a secondary reference. Of greater concern than the small uncertainty in the reference frequency is the possibility, with a very wide chemical shift range for ^{195}Pt, of observing 'folded' peaks. If peaks are not carefully 'checked for folding', reported shifts may be in error by hundreds of ppm!

Chemical shifts for some representative platinum compounds are given in **Table 2**. The order of ligands in their effect on δ_{Pt} is similar to that on other well-studied transition metals (e.g. ^{59}Co, ^{103}Rh). In particular, ligands with heavy donor atoms tend to cause increased shielding of the metal nucleus so that, for example, the shielding increases from Cl^- to Br^- to I^-. As discussed above for W(II) compounds, this may be understood in terms of the heavier donor atoms having a greater effect on r and C in Equation [1].

While δ_{Pt} depends primarily on the donor atom set, the nature of the whole ligand does have an important effect (compare, for example, $[Pt(NH_3)_4]^{2+}$ and $[Pt(NO_2)_4]^{2-}$). Geometric isomers will also have different shifts, which are greatest when the different ligands bound to the metal are very different in their *trans* influence (compare the *cis* and *trans* pairs $[PtCl_2(NH_3)_2]$ and $[Pt(NH_3)_2(H_2O)_2]^{2+}$). There are also 'ring-size' effects, evident in the difference between the shifts for $[\{Pt(NH_3)_2(\mu\text{-}OH)\}_n]^{n-}$ with $n = 2$ (four-membered ring) and $n = 3$ (six-membered ring), and for *cis*-$[Pt(NH_3)_2(H_2O)_2]^{2+}$ and $[Pt(en)(H_2O)_2]^{2+}$ (five-membered chelate ring).

Chemical-shift anisotropy relaxation is important for platinum complexes, especially for square planar complexes, and at high magnetic field strengths. It is also enhanced when there is significant hydrogen-bonding between the solvent and solute (as with diammineplatinum compounds in water), and when platinum is coordinated to bulky ligands (both of which will decrease the rate of tumbling in solution). Fast CSA relaxation leads to broadening of peaks and loss of resolved splittings from couplings. It can also effectively decouple platinum from another nucleus which is being observed. Coupling constants involving ^{195}Pt [e.g. $^1J(Pt\text{–}^{31}P)$, $^1J(Pt\text{–}^{15}N)$, $^1J(Pt\text{–}^1H)$, $^1J(Pt\text{–}^{13}C)$, $^2J(Pt\text{–}CH_3)$, $^2J(Pt\text{–}CF_3)$] can provide useful structural information (e.g. through the *trans* influence of the *trans* ligand on these couplings) which may therefore be lost at high magnetic fields. In compounds containing platinum–platinum bonds, the magnitude of $J(Pt\text{–}Pt)$ does not appear to be related simply to bond length or bond strength, although $J(Pt\text{–}Pt)$ is affected in the same way as other coupling constants by the *trans* influence of the ligands *trans* to the metal–metal bond.

Table 2 ^{195}Pt chemical shifts for some representative compounds

Compound	δ_{Pt}
Pt (0)	
[Pt(PPh$_3$)$_3$]	−4583
[Pt(1,5-cyclooctadiene)$_2$]	−4636
[Pt(PMe$_2$Ph)$_4$]	−4728
[Pt(PCy$_3$)$_2$]	−6501
Pt(I) (Pt—Pt bond)	
[{PtCl$_2$(CO)}$_2$]$^{2-}$	−4162
Pt(II)	
[Pt(H$_2$O)$_4$]$^{2+}$	+31
[{Pt(NH$_3$)$_2$(μ-OH)}$_2$]$^{2-}$	−1153
[PtCl$_3$(H$_2$O)]$^-$	−1180
trans-[Pt(NH$_3$)$_2$(H$_2$O)$_2$]$^{2+}$	−1374
[{Pt(NH$_3$)$_2$(μ-OH)}$_3$]$^{3-}$	−1505
cis-[Pt(NH$_3$)$_2$(H$_2$O)$_2$]$^{2+}$	−1584
[PtCl$_4$]$^{2-}$	−1620
[PtCl$_3$(NMe$_3$)]$^-$	−1715
[Pt(en)(H$_2$O)$_2$]$^{2+}$	−1914
trans-[PtCl$_2$(NH$_3$)$_2$] (indmf)	−2101
cis-[PtCl$_2$(NH$_3$)$_2$] (indmf)	−2104
[Pt(NO$_2$)$_4$]$^{2-}$	−2166
[Pt(NH$_3$)$_4$]$^{2+}$	−2580
[PtBr$_4$]$^{2-}$	−2690
[PtCl$_3$(SMe$_2$)]$^-$	−2757
[PtCl$_3$(PMe$_3$)]$^-$	−3500
[Pt(CN)$_4$]$^{2-}$	−4746
Pt(III) (Pt—Pt bond)	
[{(H$_2$O)Pt(μ-SO$_4$)$_2$}$_2$]$^{2-}$	+1753
[{ClPt(μ-P(O)$_2$O(O)$_2$P)$_2$}$_2$]$^{4-}$	−4236
Pt(IV)	
[PtF$_6$]$^{2-}$	+7314
[Pt(OH)$_6$]$^{2-}$	+3277
[PtCl$_6$]$^{2-}$	0
fac-[PtMe$_3$(H$_2$O)$_3$]$^+$	−1794
[PtBr$_6$]$^{2-}$	−1860
[Pt(CN)$_6$]$^{2-}$	−3866

When platinum is bound by ligands with donor atoms having quadrupolar nuclei (e.g. ^{14}N, ^{75}As), NMR peaks are frequently broad, because the rate of quadrupole-induced relaxation is such that ^{195}Pt is only partially decoupled from the quadrupolar nucleus.

Solid state ^{195}Pt NMR spectra, including CP-MAS spectra, have been obtained. Much higher spinning rates are required for platinum(II) compounds than for most platinum(IV) complexes, because of their much greater shielding anisotropy.

Mercury, ^{199}Hg

Mercury has two isotopes with $I > 0$, ^{199}Hg ($I = \frac{1}{2}$; 16.8% abundance) and ^{201}Hg ($I = \frac{3}{2}$; 13.2% abundance). Because of the large quadrupole moment of ^{201}Hg, this nucleus is not readily observed, and only ^{199}Hg is of significance for NMR spectroscopy. Its resonance frequency in a magnetic field with strength 2.35 T is 17.9 MHz. There is general agreement on the use of neat dimethylmercury as reference, despite the high toxicity of this substance. An aqueous solution of Hg(ClO$_4$)$_2$ (−2253 relative to Me$_2$Hg) has been suggested as an alternative but its shift is dependent on concentration and temperature.

Common coordination behaviour for Hg(II) involves strong linear coordination by two ligands, often with additional weaker interactions from solvent molecules, counter-ions, etc. In some cases, (e.g. halides X$^-$) higher coordination numbers are possible (e.g. tetrahedral [HgX$_4$]$^{2-}$). There is considerable ligand lability, so that averaged signals are often observed. There is therefore often some uncertainty about the precise nature of the species being observed.

Chemical shift anisotropies are also large (except for highly symmetrical [HgX$_4$]$^{2-}$), leading to efficient relaxation, and some line broadening at high magnetic fields. Two-dimensional NMR spectra may be readily obtained.

Some ^{199}Hg chemical shift data are given in **Table 3** for some representative compounds whose structures appear to be fairly well defined in solution. As with typical transition metals, such as ^{195}Pt, and ^{113}Cd, the nuclear shielding increases in halide complexes as the halide ion becomes heavier ('normal halide dependence')

Coupling constants between ^{199}Hg and other nuclei [e.g ^{13}C, ^{31}P and ^{1}H(^2J, ^3J)] have been extensively studied. In a series of complexes (e.g. CH$_3$HgX) the coupling constants [^2J(Hg–CH$_3$) in this instance] may be used to establish a trans influence series which is similar to that obtained from coupling constants involving ^{195}Pt.

Table 3 ^{199}Hg shifts for some representative compounds

Compound	δ_{Hg} (ppm)
Hg(I)	
$[Hg_2]^{2+}(ClO_4)_2$ /H$_2$O(satd.)	−1614
Hg(II)	
Me$_2$Hg	0
MeHgCl	−813
MeHgBr	−915
[MeHg(NH$_3$)]BF$_4$	−943
MeHgI	−1097
[MeHg(OH$_2$)]$^+$	−1150
[HgCl$_4$]$^{2-}$	−1200
[HgBr$_4$]$^{2-}$	−1810
[Hg(H$_2$O)$_6$]ClO$_4$)$_2$	−2253
[HgI$_4$]$^{2-}$	−3510

Other elements, La–Hg

Lanthanum

Nuclei with $I > 0$ are ^{139}La ($I = \frac{7}{2}$; 99.91% abundance) and ^{138}La ($I = 5$; 0.09% abundance). Because of its low natural abundance, ^{138}La is of no practical importance for NMR spectroscopy. The common oxidation state, La(III), is diamagnetic but co-ordination complexes are usually labile, and quadrupole-induced relaxation causes broad line widths, except when the environment is highly symmetric, as in [LaX$_6$]$^{3-}$ complexes, which are presumably octahedral. Both the chemical shift and line width can provide information about the average environment of the La^{3+} ion. As expected for a d^0 ion, the metal nucleus is less shielded in [LaBr$_6$]$^{3-}$ (+1090 ppm from aqueous La^{3+}) than in [LaCl6]$^{3-}$ (+851 ppm).

Other lanthanides

Diamagnetic oxidation states tend to be unstable in aqueous solution, except for Yb(II) and Lu(III). Since ^{171}Yb (14.3% abundance) has $I = \frac{1}{2}$, there have been some solution studies on Yb(II) complexes. ^{169}Tm (100% abundance) also has $I = \frac{1}{2}$, but diamagnetic compounds are not stable in solution. ^{175}Lu ($I = \frac{7}{2}$, 97.4% abundance) has a large quadrupole moment, which limits its use. Solid state spectra have been obtained on paramagnetic compounds using nuclei with relatively low quadrupole moments: ^{141}Pr ($I = \frac{5}{2}$; 100% abundance), ^{151}Eu ($I = \frac{5}{2}$; 47.8% abundance), ^{159}Tb ($I = \frac{3}{2}$; 100% abundance), ^{165}Ho ($I = \frac{7}{2}$; 100% abundance) and with ^{169}Tm ($I = \frac{1}{2}$; 100% abundance).

Hafnium

^{171}Hf ($I = \frac{7}{2}$; 18.5% abundance) and ^{179}Hf ($I = \frac{9}{2}$; 13.8% abundance) have large quadrupole moments, causing severe line broadening.

Tantalum

^{181}Ta ($I = \frac{7}{2}$; 99.99% abundance) also possesses a large quadrupole moment so that, even in octahedral species such as [TaCl$_6$]$^-$, lines are broad.

Rhenium

^{185}Re ($I = \frac{5}{2}$; 37.1% abundance) and ^{187}Re ($I = \frac{5}{2}$; 62.9% abundance) also have large quadrupole moments, so that the line widths are large even for the symmetric species [ReO$_4$]$^-$.

Osmium

^{187}Os has $I = \frac{1}{2}$, but its low natural abundance (1.64%) and its very low receptivity make it difficult to observe. ^{189}Os ($I = \frac{3}{2}$; 16.1% abundance) has a large quadrupole moment, leading to very broad lines, except in highly symmetric environments. In some cases, ^{187}Os spectra may be obtained by indirect detection, using more sensitive nuclei such as ^{31}P. Coupling constants between ^{187}Os and other nuclei (e.g. ^{31}P, ^{13}C) may be easily observed, and provide structural information.

Iridium

^{191}Ir ($I = \frac{3}{2}$; 37.3% abundance) and ^{193}Ir ($I = \frac{3}{2}$; 62.7%) have large quadrupole moments, low receptivity and low resonance frequencies, which make observation difficult.

Gold

^{197}Au ($I = \frac{3}{2}$; 100% abundance) has a large quadrupole moment. With the preferred linear [Au(I)] and square planar [Au(III)] geometries, quadrupole-induced relaxation is very fast, and signals have not been observed in solution.

List of symbols

I = nuclear spin quantum number; J = coupling constant; σ_p = paramagnetic contribution to total nuclear screening.

See also: **Fourier Transformation and Sampling Theory; Heteronuclear NMR Applications (Sc–Zn); Heteronuclear NMR Applications (Y–Cd); NMR Relaxation Rates; NMR Spectrometers.**

Further reading

Appleton TG, Clark HC and Manzer LE (1973) The trans influence: its measurement and significance. *Coordination Chemistry Reviews* 10: 335–422.

Brevard C, Pregosin PS and Thouvenot R (1991) Group 6 chromium to tungsten. In: Pregosin PS (ed) *Transition Metal Nuclear Magnetic Resonance*, pp 59–89. Amsterdam: Elsevier.

Goodfellow RJ (1987) Group VIII transition metals. In: Mason J (ed) *Multinuclear NMR*, pp 521–561. New York: Plenum.

Goodfellow RJ (1987) Post-transition metals, copper to mercury. In: Mason J (ed) *Multinuclear NMR*, pp 563–589. New York: Plenum.

Granger P (1991) Groups 11 and 12 copper to mercury. In Pregosin PS (ed) *Transition Metal Nuclear Magnetic Resonance*, pp 265–346. Amsterdam: Elsevier.

Pregosin PS (1982) Platinum-195 nuclear magnetic resonance. *Coordination Chemistry Reviews* 44: 247–291.

Pregosin PS (1986) Platinum NMR spectroscopy. *Annual Reports on NMR Spectroscopy* 17: 285–349.

Pregosin PS (1991) Group 10 (nickel to platinum). In: Pregosin PS (ed) *Transition Metal Nuclear Magnetic Resonance*, pp 216–263. Amsterdam: Elsevier.

Rehder D (1987) Early transition metals, lanthanides and actinides. In: Mason J (ed) *Multinuclear NMR*, pp 479–519. New York: Plenum.

Rehder D (1991) Groups 3–5 scandium to tantalum. In: Pregosin PS (ed) *Transition Metal Nuclear Magnetic Resonance*, pp 1–58. Amsterdam: Elsevier.

Wrackmeyer B and Contreras R (1992) [199]Hg NMR parameters. *Annual Reports on NMR Spectroscopy* 24: 267–329.

Heteronuclear NMR Applications (O, S, Se and Te)

Ioannis P Gerothanassis, University of Ioannina, Greece

MAGNETIC RESONANCE

Applications

NMR spectroscopic properties and techniques for [17]O, [33]S, [77]Se and [123,125]Te

Nuclear properties

Both [17]O and [33]S are quadrupolar nuclei with very low natural abundance (**Table 1**). The NMR active isotopes of selenium and tellurium are spin $\frac{1}{2}$ nuclei. The receptivity of [77]Se is about three times larger than that of [13]C. The element tellurium has two active isotopes, [123,125]Te. [125]Te has been used in the majority of tellurium NMR investigations owing its significantly better receptivity.

Experimental techniques

The stringent requirements in ^{17}O and [33]S NMR studies of compounds at natural abundance are the high concentrations and extensive signal averaging needed. Recording of spectra can be greatly facilitated by the use of enriched samples. Synthesis with oxygen isotopes involves rather straightforward organic reactions. The quadrupolar moments of [17]O

Table 1 Nuclear properties of [17]O, [33]S, [77]Se, [123]Te and [125]Te

Isotope	Spin	Natural abundance (%)	NMR frequency[a] (MHz)	Receptivity	
				Relative[b]	Absolute[c]
[17]O	$\frac{5}{2}$	0.037	54.227	2.91×10^{-2}	1.08×10^{-5}
[33]S	$\frac{3}{2}$	0.76	30.678	2.26×10^{-3}	1.71×10^{-5}
[77]Se	$\frac{1}{2}$	7.58	76.270	6.93×10^{-3}	5.26×10^{-4}
[123]Te	$\frac{1}{2}$	0.87	104.831	1.80×10^{-2}	1.57×10^{-4}
[125]Te	$\frac{1}{2}$	6.99	126.387	3.15×10^{-2}	2.91×10^{-3}

[a] At 9.395 T (^{1}H frequency: 400 MHz).
[b] Relative to proton, at constant field, for equal number of nuclei.
[c] Product of relative receptivity and natural abundance.

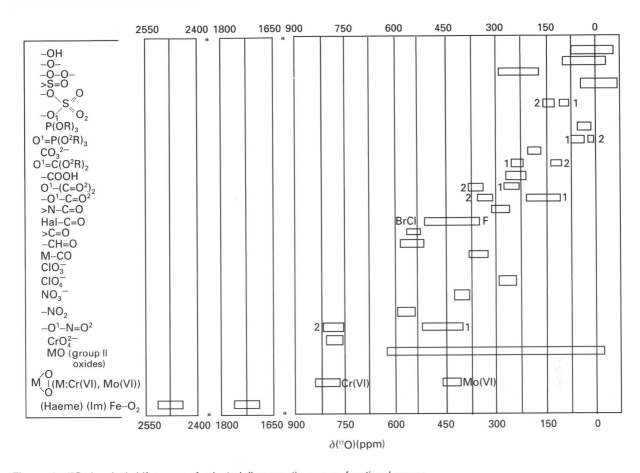

Figure 1 ^{17}O chemical shift ranges of selected diamagnetic oxygen functional groups.

and ^{33}S result in very effective relaxation and, thus, extensive line broadening with T_1 and T_2 relaxation times often in the range below 1 ms. A possible way to attenuate extensive line-broadening effects is to use higher temperatures and/or solvents of low viscosity. Observation of very broad (>1 kHz) resonances, however, is a potential problem due to baseline distortions. The most severe origin of baseline artifacts is the transient response of the NMR probe, often referred to as 'acoustic ringing', caused by the generation of ultrasonic waves from the action of the RF pulse. Several methods have been proposed to overcome this problem, with particular emphasis on multipulse sequences.

The longitudinal relaxation of ^{77}Se and ^{125}Te at high fields is governed by the spin–rotation (SR) mechanism with typical T_1 values in the range of 1–30 s. In the solid state the relaxation times are very long; consequently the use of MAS NMR with cross polarization from ^1H is necessary. The ^{77}Se and ^{125}Te shieldings indicate a strong temperature dependence (several ppm/K), and so extreme care is necessary to maintain constant temperature, particularly when broadband decoupling is employed.

As is common with many heteronuclei, a variety of reference compounds and referencing procedures (both internal and external) have been suggested. The most commonly used standards ($\delta = 0$, as external reference) are H_2O (a temperature dependence of the ^{17}O shielding has been reported) or 1,4-dioxane at 30°C for ^{17}O; $(NH_4)_2SO_4$ or Cs_2SO_4 in water for ^{33}S; Me_2Se in $CDCl_3$ for ^{77}Se and Me_2Te in $CDCl_3$ for ^{125}Te.

^{17}O NMR parameters and applications

^{17}O shieldings

The total range of ^{17}O shieldings of diamagnetic compounds is about 2600 ppm (**Figure 1**). For compounds containing oxygen bonded only to carbon and/or hydrogen there is a clear correlation between C–O π bond order and high frequency shift. This type of correlation arises from the multiple bond term in the Karplus–Pople equation. Bridging oxygens indicate a strong deshielding for π-accepting acyl substituents which introduce π character into the C–O bond of the bridging oxygen.

In nitrogen compounds, chemical shift values are also determined largely by π bonding effects. Bridging oxygen resonances are observed at higher frequencies in nitrites (R–O–N=O), where π bonding is appreciable. A lesser but nonetheless significant effect is observed for bridging oxygens in nitrates (R–O–NO$_2$). The most striking feature of the ^{17}O shifts observed for P–O bonds is the small chemical shift range (20–260 ppm) which cannot be related to P–O bond order in a straightforward fashion. However, ^{17}O resonances for non-equivalent oxygens in phosphates and polyphosphates could be resolved and could be utilized in stereochemical and binding site studies. For S–O bonding, similar trends to those of P–O bonds were observed. Modification of π-bond order in a X=O group due to substituents on X results in a decrease (or increase) in the net charge on X as the charge on O increases (or decreases). The chemical shift variations experienced by the nuclei X and O are then opposite and roughly proportional to the ratio $\langle r^{-3}\rangle 2P_X/\langle r^{-3}\rangle 2P_O$.

For a series of oxyanions and a series of C=O and N=O compounds a linear correlation between $\delta(^{17}\text{O})$ and the inverse of the lowest energy $n \rightarrow \pi^*$ transition was found. A linear correlation exists between acetone chemical shifts and the maximum of the $n \rightarrow \pi^*$ transition for acetone–water solutions or for acetone in different solvents. Similar trends have also been observed in a variety of substituted acetophenones; however, it was suggested that such trends may be fortuitous. In a series of aliphatic ethers a correlation between $\delta(^{17}\text{O})$ and the first ionization potential was suggested.

Compounds that contain oxygen–oxygen bonds and oxygen–fluorine bonds display large high frequency shifts. The ability, therefore, of ^{17}O NMR to unambiguously distinguish between non-equivalent oxygens has found application in several structural and mechanistic studies. Many stereochemical studies have dealt with polynuclear, anionic oxo complexes (polyoxoanions) of the early transition elements in their highest oxidation states. Despite their structural complexity, these polyoxoanions often yield well-resolved ^{17}O NMR spectra since large shielding ranges are observed and samples may be studied at elevated temperatures in nonviscous solvents. Furthermore, ^{17}O enrichment can easily be obtained.

Indirect spin coupling constants

Because of the breadth of ^{17}O resonances, only relatively large coupling constants can be measured and hence the majority of those recorded refer to couplings from directly bonded nuclei. $^1J(\text{O,H})$ in water and in alcohols has a value close to 80 Hz.

The $^1J(\text{P,O})$ couplings involving P=O groups cover a fairly wide range (145–210 Hz) and for a series of compounds of type OPX$_3$ (X=F, O in OMe, N in NMe$_2$ and C in Me) there is a consistent drop in coupling with decreasing electronegativity of X. The oxygens in P–OMe groups within phosphates give rise to a consistently smaller coupling constant, compared with P=O, of between 90 and 100 Hz. The reduced coupling constants decrease with increasing atomic number of the central atoms along the isoelectronic sequence VO$_4^{3-}$, CrO$_4^{2-}$ and MnO$_4^-$. An application of ^{17}O spin–spin couplings has been made to investigate H$_3$O$^+$ species.

Nuclear quadrupole coupling constants (χ)

Experimental χ values cover a range from –8 to 17 MHz. The experimental χ values for a series of compounds in the solid state with C=O, –O–, PO, NO, and SO bonds show the following trends:

$$O - O > N - O > NO_2 > C = O > P - OR \sim S = O >$$
$$16 - 17 \quad 16 \quad\quad 13 \quad\quad 11 - 9 \quad\quad 9.5 - 9$$

$$ROH > SO_2 > P = O \sim C = O > XO_n^{m-}$$
$$9 - 8 \quad 6.7 \quad\quad 4 \quad\quad\quad\quad 1$$

Strong hydrogen bonding reduces the χ values by up to 2.6 MHz. The χ values of the terminal CO groups in metallocarbonyls were found to be between 1–3 MHz although values below 0.1 MHz have also been observed. It was suggested that increased $d\pi$–$p\pi$ back bonding from the metal will increase electron density perpendicular to the C–O axis and so reduce χ values.

Dynamic and kinetic studies

Numerous dynamic and kinetic studies have been performed using ^{17}O NMR. Line shape analysis is not straightforward because the line width itself is temperature dependent. Nevertheless, it has been shown that dynamic ^{17}O NMR is a convenient alternative to ^{13}C NMR with, for example, methyl carbonyl compounds. A substantial number of publications have been concerned primarily with studies of the hydration (solvation) of ions in solution and the determination of rates and activation parameters for ligand substitution (a rapid development of high-pressure NMR has occurred in the last twenty years). There is an extensive literature of ^{17}O NMR spectroscopy used for the study of hyperfine interactions between the unpaired electrons in paramagnetic molecules and ions and the ^{17}O nucleus. The magnitudes and temperature dependences of line widths (or relaxation time measurements) of the 'bound' and 'free' resonances, and the chemical shift separations between those resonances, can in principle be used to derive information about the hyperfine splitting constant, the rate of the solvent interchange process and the solvation number. However, because the shifted resonance of the solvated water molecules is often very broad, it is usually not possible to derive the hydration number of the paramagnetic cation from straightforward area measurements. In these cases, the hydration numbers may be obtained by the use of theory developed by Swift and Connick.

Torsion angle and hydrogen bonding effects

^{17}O NMR spectroscopy is a powerful method for the detection of steric effects in molecules in which steric interactions are characterized by rotation of functional groups around single bonds to relieve van der Waals interactions or on rigid systems in which steric interactions are partially accommodated by bond angle and bond length distortions. Applications include aromatic and heteroaromatic nitro compounds, aryl ketones, aldehydes, 1,2-diketones, aromatic carboxylic acids, esters and amides.

^{17}O NMR appears to be especially promising for use in hydrogen bonding interaction studies because of the large chemical shift range of the oxygen nucleus. The dominance of intramolecular hydrogen bonding effects over substituent effects was clearly demonstrated in acetophenones, benzaldehydes and hydroxynaphthoquinones. Detailed studies of the influence of intermolecular solute–solvent hydrogen bonding in amides and peptides demonstrate that both long range dipole–dipole interactions and hydrogen bonds at the amide oxygen induce significant and specific shielding of the ^{17}O nucleus.

Water orientation in lyotropic phases and bilayers – hydration of proteins and DNA

^{17}O NMR studies have several advantages compared with ^1H and ^2D NMR for investigating water orientation in lyotropic phases and bilayers and the hydration of proteins and DNA since the relaxation effect is large and the intramolecular origin of the electric field gradient makes the quadrupolar interaction virtually independent of the molecular environment. Recently, it has been suggested that the relaxation dispersion in aqueous protein and DNA solutions in the MHz frequency range is not owing to a long-lived hydration layer at the biomolecular surface, but to a small number of crystallographically well-defined water molecules.

The solid state

Under conditions of magic-angle spinning (MAS) the powder band shape of the $(\frac{1}{2}, -\frac{1}{2})$ transition of quadrupolar nuclei with non-integral spin does not depend on first-order quadrupolar interactions but only upon second-order quadrupolar effects which do not eliminate the possibility of obtaining meaningful NMR spectra. In this case operation at very high magnetic fields will generally be most advantageous since the maximum second-order quadrupolar line width of the $(\frac{1}{2}, -\frac{1}{2})$ transition decreases linearly with the magnetic field. Typical results are those obtained for a range of oxides, oxyanions and high-temperature oxide superconductors.

^{33}S NMR parameters and applications

^{33}S shieldings

^{33}S shieldings cover a range of about 1000 ppm. Four regions (with several exceptions) can be identified: (a) singly bonded sulfur, (b) multiple bonded sulfur, (c) sulfur in delocalized π systems and (d) S=O bonds. The sulfur nucleus in thiols and thioethers is strongly shielded ($\delta = -300$ to -510 ppm). Oxidation to sulfoxides results in $300-400$ ppm deshielding ($\delta = +40$ to -100 ppm). Further oxidation to sulfones does not significantly change $\delta(^{33}S)$ values ($\delta = +40$ to -30 ppm). Sulfur in three- and four-membered ring compounds ([1]–[3]) is more shielded than in the corresponding acyclic and cyclic analogues of larger ring sizes (similar trends have been observed for ^1H, ^{13}C and ^{17}O nuclei).

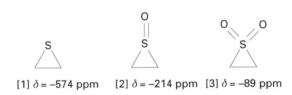

[1] $\delta = -574$ ppm   [2] $\delta = -214$ ppm   [3] $\delta = -89$ ppm

The resonance range of sulfones is about 60 ppm and it has been used to follow the analysis of sulfur-containing oils after oxidation of the sulfur species. The values of $\delta(^{33}S)$ in sulfonamides are quite similar to those of sulfones. Sulfoximines [R–(O=)S(=NR″)–R'] resonate at lower frequencies than the corresponding sulfones [R–S(=O)$_2$–R'], which may be attributed to the lower electronegativity of the doubly bonded NR group in the sulfoximine replacing the oxygen in sulfones. The opposite is seen when sulfimines [R–S(=NR″)–R'] are compared with the corresponding sulfoxides [R–S(=O)–R'].

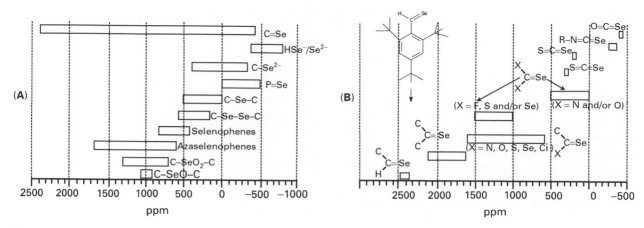

Figure 2   (A) ^{77}Se chemical shift ranges of selected selenium functionalities (excluding metal complexes). (B) ^{77}Se chemical shift ranges of compounds containing C=Se groups (excluding metal complexes). Reproduced by permission of John Wiley and Sons from Duddeck H (1996) Sulfur, selenium and tellurium NMR. In Grant DM and Harris RK (eds) *Encyclopedia of Nuclear Magnetic Resonance*, pp 4623–4635. Copyright 1996. John Wiley & Sons.

Sulfuric acid and sulfates resonate around $\delta = 0$. Significant deshieldings have been observed when oxygen is replaced gradually by sulfur in thiomolybdate anions. Thiotungstates show a similar trend. Sulfides cover a wide range, over 600 ppm (e.g. −680 and −42 ppm for Li_2S and BaS, respectively). This can be attributed to different bond characters (ionicities) in the crystal.

Substituent effects in diorganyl sulfones, 3- and *trans*-3,4-substituted thiolane 1,1-dioxides and 4-substituted 2-thiolene 1,1-dioxides have been investigated and correlations with ^{13}C and ^{17}O in analogous compounds as well as with Taft (Hammett) electronic parameters have been reported. Stereochemical effects on ^{33}S shielding have been studied such as the stereomeric sulfones [4] and [5].

[4] $\delta = -10$ ppm   [5] $\delta = -20$ ppm

^{33}S Coupling constants

There is a limited number of ^{33}S coupling constants in the literature due to severely broadened signals. $^{2}J(^{33}S-^{1}H)$ coupling constants of 3 and 4.5 Hz were measured for Me_2SO_2 and sulfolane and a $^{3}J(^{33}S-^{1}H)$ coupling constant of 6 Hz for 2,5-dihydrothiophene 1,1-dioxide. One-bond $^{33}S-^{19}F$ coupling of ∼252 Hz has been found for SF_6.

^{33}S Relaxation times

The line widths of ^{33}S cover a range of 10 to 300 Hz for $[MoS_nO_{4-n}]^{2-}$ and $[WS_nO_{4-n}]^{2-}$ ($n = 1-4$), >4 kHz

for sulfoxides and >9 kHz for sulfonium salts. The narrow ^{33}S signal in solid ZnS can be attributed to cubic symmetry and magnetic dilution of this compound.

^{77}Se NMR parameters and their applications

^{77}Se shieldings

The total range of ^{77}Se shieldings is about 3300 ppm. **Figure 2** illustrates the resonance ranges of various selenium functional groups. Electropositive substituents such as H_3Si and H_3Ge attached to selenium result in very strong shielding ($\delta = -666$ and −612 ppm, respectively). Selenide ions are even more shielded, e.g. K_2Se ($\delta = -670$ ppm) and $H_3Si-Se^-Li^+$ ($\delta = -736$ ppm). Increasing electronegativity of substituents directly attached to selenium in R–Se–X (R-organyl) results in deshielding. In selenium halides (SeX_2), strong shielding is observed with decreasing electronegativity of X in the series X=Cl ($\delta = 1758$ ppm), Br ($\delta = 1474$ ppm) and I ($\delta = 814$ ppm). Deshielding, but with an opposite trend to the halogen effect, is observed for the oxygenated species: $SeOF_2$ ($\delta = 1378$ ppm), $SeOCl_2$ ($\delta = 1479$ ppm) and $SeOBr_2$ ($\delta = 1559$ ppm).

Alkyl substituents in dialkyl and arylalkyl selenides have a significant deshielding effect on $\delta(^{77}Se)$, with β-effects being larger than α-effects. Analogous trends have been observed for phenylselenenylalkanes (PhSe–R), alkylselenols (RSeH), alkylselenolates (RSe–Na$^+$) and in dialkyl diselenides (RSeSeR). The γ-effects are substantially shielding. The large magnitudes of α-, β- and γ-effects have been discussed in terms of intramolecular polarizability effects. The

$\delta(^{77}Se)$ of selenoanisoles (Ph–Se–Me) substituted at the aromatic ring have been expressed in terms of dual substituent parameters (DSP).

Oxidation of Se(II) and Se(IV) generally results in significant deshielding: Me_2Se ($\delta = 0$ ppm), Me_2SeBr_2 ($\delta = 389$ ppm) and Me_2SeCl_2 ($\delta = 448$ ppm). The magnitudes of substituent effects of alkyl groups are approximately four times smaller in the selenoxides than in the corresponding selenides. Halogenated species show the opposite tendency.

Se(VI) species are shielded compared with the structurally related Se(IV) analogues: H_2SeO_4 ($\delta = 1001$ ppm) and H_2SeO_3 ($\delta = 1300$ ppm). The $\delta(^{77}Se)$ of hypervalent tetraarylselenium compounds are not significantly different compared with those of divalent diaryl selenides.

The C=Se functional group shows shieldings which cover the whole of the known range of $\delta(^{77}Se)$: selenoaldehyde ($\delta = 2398$ ppm) and O=C=Se ($\delta = -447$ ppm). Shieldings strongly depend on the nature of the atoms and substituents attached to the carbon of the C=Se group. Furthermore, correlations with n $\to$ $\pi*$ transitions and $\delta(^{17}O)$ in C=O analogues have been found.

Cyclic compounds with selenium in the ring(s) have been comprehensively investigated. Annulation of conjugated ring systems to selenophene causes considerable shieldings. Replacement of carbon by an electronegative element such as nitrogen atom results in a deshielding by up to 900 ppm. Even minor strain effects within the heterocycle can cause significant shielding. As expected, positively charged unsaturated ring systems cause extreme deshielding effects.

$\delta = 605$ ppm $\delta = 579$ ppm $\delta = 429$ ppm

$\delta = 1013$ ppm $\delta = 1511$ ppm $\delta = 2434$ ppm

P=Se bonds indicate negative $\delta(^{77}Se)$ values which was attributed to the dipolar mesomeric form II: $-P=Se$ (I) $\leftrightarrow$ $-P^+-Se^-$ (II). Shieldings depend strongly on the nature of the sustituent at phosphorus and the branching of carbon substituents.

There is a large number of $\delta(^{77}Se)$ of selenium-metal compounds which, according to Duddeck, can be classified into three types: (a) selenium with covalent bonds to main group metals. Deshielding is usually observed compared with the corresponding selenide anions. (b) Selenium σ-, π-, μ- or η-coordinated to transition metal atoms. The $\delta(^{77}Se)$ of these complexes cover the whole shift range. In most cases deshielding is observed compared with the free selenium-containing ligand. Strong shieldings have been observed in the case of the η⁴-cobalt complexes compared with the free heterocycles. (c) Selenium ligands with other atoms (usually phosphorus or nitrogen) directly coordinated to the metal atom. The $\delta(^{77}Se)$ of this type of metal complexes do not differ significantly from those of the free ligands.

Variable temperature NMR has been utilized in the conformational analysis of selenium-containing compounds such as the ring inversion process of phenylselenenylcyclohexane, 2,11-diselena[3.3] metacyclophane, [3.3]diselena- and [4.4]tetraselenacyclophanes and 1,4,5,7-tetrahydro-3H-2,6-benzo-diselenines.

^{77}Se NMR has been utilized as a tool for chiral recognition which occurs across several bonds such as in the derivatizing agent (4S,5S)-(–)-4-methyl-5-phenyloxazolidine-2-selone.

CP MAS ^{77}Se NMR in the solid has been utilized to investigate conformational properties in cyclic compounds and comparison with X-ray structures. $\delta(^{77}Se)$ were found to be strongly dependent on the crystallographic non-equivalences in the unit cell.

^{77}Se coupling constants

^{77}Se coupling constants are abundant in the literature since they can be investigated in a straightforward way from the spectra of the coupling partner such as ^1H, ^{13}C and ^{31}P nuclei. The J values have been reviewed extensively so in this article only a synopsis will be given.

$^nJ(^{77}Se, {}^1H)$ One-bond coupling constants are, presumably, positive with values between 65.4 Hz (H_2Se) to 5.3 Hz for Cp–W(CO$_3$)SeH. Geminal couplings are mostly positive and vary between 4 and 16 Hz for hydrogen atoms in aliphatic fragments, but can increase to 50 Hz if the fragment has further geminal heteratoms or if it is in a heteroatom ring. Two- and three-bond (vicinal) coupling constants show some stereochemical dependence.

[6] δ = 2858 ppm

[7] δ = 1418 ppm

[8] δ = −168 ppm

$^{n}J(^{77}Se, X$ where $X = {}^{13}C, {}^{15}N, {}^{19}F, {}^{29}Si, {}^{31}P, {}^{77}Se, {}^{125}Te$ and $^{129}Xe)$ One-bond couplings $^{1}J(^{77}Se, {}^{13}C)$ are negative and strongly depend on the hybridization of the carbon atom. Typical values are −45 to −100 Hz for Csp^3 — Se. Geminal couplings (< 15 Hz for aliphatic carbons) and vicinal couplings (3–12 Hz) are very little investigated and it is not clear whether a Karplus-type dependence exists.

There are very few reports of $^{77}Se–^{15}N$ couplings (60.1 Hz for CF$_3$Se–NH$_2$). One-bond $^{77}Se–^{19}F$ coupling constants considerably exceed 1000 Hz and strongly depend on the stereochemical position of the fluorine atoms (for SeF$_4$ the coupling to the axial fluorine atom is 284 Hz and that to the equatorial one 1206 Hz). Geminal couplings can be substantial (~120 Hz in fluorinated 1,3-diselenetanes) and four-bond coupling constants of ~12 Hz have been observed.

Few data of one bond $^{77}Se–^{29}Si$ have been published and are generally in the region of 110–49 Hz. One-bond $^{77}Se–^{31}P$ couplings are negative and show a strong dependence on both the bond order between phosphorus and selenium (−800 to −1200 Hz for P=Se double bonds and −200 to −620 Hz for single bonds) and the stereochemistry of the compounds investigated.

One-bond $^{77}Se–^{77}Se$ couplings have been observed in a number of diselenides (+22 to −67 Hz) and polyselenide anions with charged selenide atoms (250 ± 20 Hz). Two-bond couplings can have large values, particularly in cyclic sulfur–selenium compounds (95–114 Hz). Three-bond couplings (3–16 Hz) have been observed and four-bond couplings up to 16 Hz have been observed in sulfur–selenium cyclic compounds.

Couplings to metal nuclei Typical one-bond couplings to metal nuclei are ^{119}Sn, 500–1700 Hz; ^{207}Pb,

780–150 Hz; ^{103}Rh, 20–44 Hz; ^{113}Cd, 126–195 Hz; ^{183}W, 109–14 Hz; ^{195}Pt, 630–74 Hz (a strong dependence on the nature of the substituents and the stereochemistry has been emphasized) and ^{199}Hg, −1270 to −751 Hz.

^{125}Te NMR spectral parameters and their applications

^{125}Te shieldings

The total range of ^{125}Te shieldings is nearly 5000 ppm. The $\delta(^{125}Te)$ of phosphine tellurides, R$_3$PTe, lie between −513 (R = Me) and −1000 ppm (R = Pr^i) and have been interpreted in terms of the character of the P–Te bond.

The effects of replacing α-hydrogen atoms of dimethyl telluride and phenylmethyl tellurides or ditellurides are in close analogy to those in the corresponding selenides. The same holds for diorganyltellurium dichlorides and difluorides. Similarly, $\delta(^{125}Te)$ in ortho-substituted tellurophenetols can be correlated with analogous selenophenetols.

Telluroketones [6] indicate strong deshielding (δ = 2858 ppm). If a heteroatom is attached to the tellurocarbonyl group, such as in [7], then a shielding is observed. The unusually high shielding (δ = −168 ppm) for [8] has been interpreted in terms of the dipolar canonical form.

The (^{125}Te) of cyclic tellurium compounds indicate similar trends to those with the selenium analogues. Halogenation of the tellurium atom indicates strong deshielding. Furthermore, positively charged saturated and unsaturated ring systems cause extreme deshielding effects.

δ = 782 ppm

δ = 727 ppm

δ = 269 ppm

X = Br, δ = 940 ppm

δ = 257 ppm

δ = 1304 ppm

Variable temperature ^{125}Te NMR has been utilized to investigate ring inversion in phenyltellurenyl-substituted cyclohexane derivatives. ^{125}Te NMR was shown to be an excellent tool for chiral recognition. Thus, the ^{125}Te chemical shifts of the stereoisomeric bicyclic compounds [9] and [10] differ by 34 ppm. Interestingly long-range effects, through several bonds, of the menthyl residue in [11] on tellurium shieldings have been observed.

[9] $\delta = 175$ ppm   [10] $\delta = 209$ ppm   [11] $\delta = 989/983$ ppm

Very few CP MAS ^{125}Te NMR studies in the solid have been reported; however, they clearly demonstrate the great sensitivity of the method in investigating even minor distortions from octahedral crystallographic symmetry.

^{125}Te coupling constants

^{125}Te coupling constants are less abundant in the literature than those involving ^{77}Se and are of the opposite sign due to negative magnetogyric ratio. The ratio $J(^{125}\text{Te},X)/J(^{77}\text{Se},X)$ is 2–3 (in absolute value) and very probably reflects differences in the Fermi contact interaction. One-bond coupling constants $^1J(^{125}\text{Te},^1\text{H})$ are negative (–59 Hz for H_2Te), geminal couplings in dialkyltellurides are ~20 Hz and –90 Hz in tellurophene. Vicinal couplings are smaller (Et_2Te, –22.7 Hz).

One-bond $^1J(^{125}\text{Te},^{13}\text{C})$ couplings show a strong dependence on the hybridization state of the carbon atom (Me_2Te, 162 Hz; (H_2C=CH)$_2$Te, 285 Hz; Me–Te–C≡C–Bu, 531 Hz). One-bond $^1J(^{125}\text{Te},^{19}\text{F})$ values are very large (TeF$_6$, 3688 Hz; TeF$_7^-$, 2876 Hz; Ph$_2$TeF$_2$, 530 Hz). The $^1J(^{125}\text{Te},^{31}\text{P})$ couplings strongly depend on the bond order (R$_3$P=Te, 1548–1743 Hz; (Bu^t)$_2$P–Te–P(Bu^t)$_2$, 451 Hz).

$^1J(^{183}\text{W},^{125}\text{Te})$ and $^1J(^{195}\text{Pt},^{125}\text{Te})$ couplings have also been reported.

List of symbols

J = coupling constant; T_1, T_2 = relaxation constants; χ = nuclear quadrupole coupling constant.

See also: **High Pressure Studies Using NMR Spectroscopy; NMR in Anisotropic Systems, Theory; NMR of Solids; NMR Relaxation Rates; Solid State NMR, Methods; Solid State NMR Using Quadrupolar Nuclei.**

Further reading

Barbarella G (1995) Sulfur-33 NMR. *Progress in NMR Spectroscopy* 25: 1.

Boykin DW (ed) (1991) *^{17}O NMR Spectroscopy in Organic Chemistry*. Boston: CRC Press.

Duddeck H (1995) Selenium-77 nuclear magnetic resonance spectroscopy. *Progress in NMR Spectroscopy* 27: 1–323.

Duddeck H (1996) Sulfur, selenium & tellurium NMR. In: Grant DM and Harris RK (eds) *Encyclopedia of Nuclear Magnetic Resonance*, pp 4623–4635. Chichester: Wiley.

Gerothanassis IP (1997) Methods of avoiding the effects of acoustic ringing in pulsed Fourier transform nuclear magnetic resonance spectroscopy. *Progress in NMR Spectroscopy* 19: 267–329.

Gerothanassis IP (1996) Oxygen-17 NMR. In: Grant DM and Harris RK (eds) *Encyclopedia of Nuclear Magnetic Resonance*, pp 3430–3440. Chichester: Wiley.

Kintzinger JP (1981) Oxygen NMR, characteristic parameters and applications. In: Diehl P, Fluck E and Kosfeld R (eds) *NMR, Basic Principles and Progress*, Vol 17, pp 1–64. New York: Springer.

Pearson JG and Oldfield E (1996) Oxygen-17 NMR: applications in biochemistry. In: Grant DM and Harris RK (eds). *Encyclopedia of Nuclear Magnetic Resonance*, pp 3440–3443. Chichester: Wiley.

Heteronuclear NMR Applications (Sc–Zn)

Dieter Rehder, University of Hamburg, Germany

All of the 3d series metals have at least one magnetic nucleus and hence are accessible to NMR measurements. With the exception of ^{57}Fe, the nuclear spins are $>\frac{1}{2}$, i.e. the nuclei possess an electric nuclear quadrupole moment Q (an ellipsoidal distribution of the nuclear charge) which induces rapid 'quadrupolar' relaxation. Only one of these nuclei, viz. ^{51}V, belongs to the low-Q category and thus provides favourably sharp resonance signals even in compounds of low symmetry. Some of the nuclei have excellent relative receptivities r (^{45}Sc, ^{51}V, ^{55}Mn, ^{59}Co), comparable to that of ^{1}H; for others, like ^{57}Fe and ^{61}Ni, the very low r causes detection problems. When other spectroscopic methods for providing information on the metal and its coordination sphere are less meaningful, e.g. electron absorption spectroscopy in closed shell systems (d^0: Sc^{III}, Ti^{IV}, V^V, Cr^{VI}; d^{10}: Mn^{VII}, Mn^{-III}, Co^{-I}, Ni^0, Cu^I, Zn^{II}), metal NMR can be a valuable tool. The large shielding variations accompanying even minor changes in the coordination spheres of the metal nuclei in closed and open shell (d^4, d^6, d^8) compounds make direct NMR (i.e. probing of the metal nucleus itself) the method of choice when elucidating structural features, including their relation to the electron distribution in and the reactivity of a complex compound. Metal NMR is now being applied for the characterization of biological materials (note that V, Mn, Fe, Co, Ni, Cu and Zn are 'biometals') and their model compounds, and for catalyst systems based on the first transition series metals. Applications in organometallic and coordination chemistry and speciation in solution are other fields of interest, and the use of metal NMR parameters as a basis for establishing thermodynamic and kinetic parameters is an intriguing new development. Increasingly, static and non-static solid-state measurements are employed.

The main interest has been directed towards ^{51}V, ^{59}Co and ^{57}Fe (the latter nowadays commonly detected by indirect methods relying on scalar coupling to a sensitive nucleus); the large amount of data available on vanadium in all of its oxidation states has led to several data compilations (see Further reading).

Properties of the nuclei, sample preparation and spectrum measurements

Table 1 lists nuclear properties. The only spin-$\frac{1}{2}$ nucleus in the 3d series is ^{57}Fe with a rather low receptivity (4.2×10^{-3} relative to ^{13}C; 7.4×10^{-7} relative to ^{1}H), requiring special detection measures. Where applicable, ^{57}Fe NMR parameters are obtained by indirect methods, e.g. via ^{1}H or ^{31}P NMR in organoiron or phosphine–iron compounds. The rest of the nuclei possess an electric nuclear quadrupole moment and are therefore subject to effective quadrupole relaxation, i.e. short T_1 ($\approx T_2$) relaxation times and broad resonance lines. Three criteria have to be fulfilled if these nuclei are to be employed as NMR probes in a conventional solution NMR experiment, viz. (i) a reasonable receptivity, (ii) a reasonably small quadrupole moment, and (iii) a short molecular correlation time (τ_c)/small molecular electric field gradient (q). Except for ^{51}V, all quadrupolar nuclei belong to the medium quadrupole category, where criterion (iii) restricts conventional measurements to small, nonbulky molecules that have relatively high symmetries. ^{51}V, with a rather low quadrupole moment of -0.052×10^{-28} m^2, is susceptible to NMR in almost every environment, although there are examples, such as vanadate–protein complexes, where special detection modes (centred on the quadrupolar central transition) have to be employed to cope with the long τ_c. As to criterion (i), ^{45}Sc, ^{51}V, ^{55}Mn, ^{59}Co (and ^{63}Cu) are suitable nuclei. Resolution problems originating from broad resonance lines are often balanced by a very pronounced sensitivity of shielding towards even minor variations in the molecular structure.

Measurements in solution (including liquid-crystal solutions) are carried out using medium field strength spectrometers equipped with a multinuclear probe head. Some of the nuclei (^{45}Sc, ^{51}V, ^{55}Mn, ^{59}Co, ^{63}Cu) have measuring frequencies close to that of ^{13}C and can usually be detected using a broadband ^{13}C probe head. For nuclei exhibiting very large shielding ranges (mainly V, Cr, Mn, Fe, Co),

Table 1 NMR properties of 3d transition metal nuclei[a]

	ν^b (MHz)	Spin	Q^c	γ^d	Natural abundance (%)	r^e	Reference compound
^{45}Sc	24.330	7/2	−0.22	+6.5088	100	0.30	$[Sc(H_2O)_6]^{3+}$ [h]
^{47}Ti[f]	5.643	5/2	+0.29	−1.5106	7.28	1.5×10^{-4}	$TiCl_4$ [i]
^{49}Ti[f]	5.644	7/2	+0.24	−1.5110	5.51	1.5×10^{-4}	$TiCl_4$ [i]
^{50}V	9.988	6	+0.21	+2.6721	0.24	1.3×10^{-4}	$VOCl_3$ [j]
^{51}V*	26.350	7/2	−0.052	+7.0492	99.76	0.38	$VOCl_3$ [j]
^{53}Cr	5.636	3/2	−0.15[g]	−1.5007	9.55	8.6×10^{-3}	$[CrO_4]^{2-}$ [k]
^{55}Mn	24.840	5/2	+0.55	+6.6453	100	0.175	$[MnO_4]^{-}$ [l]
^{57}Fe	3.238	1/2	–	+0.8687	2.19	7.4×10^{-7}	$Fe(CO)_5$ [m]
^{59}Co	23.727	7/2	+0.42	+6.3015	100	0.28	$[Co(CN)_6]^{3-}$ [n]
^{61}Ni	8.936	3/2	+0.16	−2.3948	1.19	4.2×10^{-5}	$Ni(CO)_4$
^{63}Cu*	26.515	3/2	−0.21	+7.1088	69.09	0.064	$[Cu\{P(OMe)_3\}_4]^{+}$ [o]
^{65}Cu	28.404	3/2	−0.195	+7.6104	30.91	0.035	$[Cu\{P(OMe)_3\}_4]^{+}$ [o]
^{67}Zn	6.257	5/2	+0.15	+1.6778	4.11	1.2×10^{-4}	$[Zn(H_2O)_6]^{2+}$ [p]

[a] The nucleus usually employed in NMR experiments is denoted by an asterisk.
[b] Measuring frequency at 2.35 T (where ^{1}H of TMS resonates at 100 MHz).
[c] Electric nuclear quadrupole moment, in 10^{-28} m^2.
[d] Magnetogyric ratio, in 10^7 rad s^{-1} T^{-1}.
[e] Receptivity relative to ^{1}H = 1.
[f] $^{47/49}$Ti NMR signals appear 'paired', separated by 266 ppm.
[g] The quadrupole moment for ^{53}Cr is still under debate. Values for $|Q|$ as low as 0.04 have also been reported.
[h] Dilute, aqueous solution of $Sc(ClO_4)_3$.
[i] $[TiF_6]^{2-}$ is also used as a standard ($\delta = -1163$ ppm).
[j] An aqueous, alkaline solution of vanadate ($[VO_4]^{3-}$, $\delta = -576$ ppm) is sometimes employed as a standard in solution NMR, $Na_3[VO_4]$ in solid state NMR.
[k] Dilute, aqueous solution of $K_2[CrO_4]$.
[l] Saturated aqueous solution of $K[MnO_4]$.
[m] The following standards are frequently employed: $[Fe(CN)_6]^{4-}$ ($\delta = +2497$ ppm), (η^5-C_5H_5)$_2$Fe dissolved in CDCl$_3$ ($\delta = -954.2$ ppm).
[n] Alternative standards are: Co(acetylacetonate)$_3$ ($\delta = +12\,500$ ppm), $[Co(NH_3)_6]^{3+}$ (+8150 ppm), $[Co(ethylenediamine)_3]^{3+}$ (+7120 ppm).
[o] Me = CH_3. Alternatively (but not recommended) $[Cu(NCMe)_4][ClO_4/BF_4]$ in MeCN ($\delta = -83$ ppm) may be used.
[p] The standard commonly used is an aqueous solution of $Zn(ClO_4)_2$.

one has to take care that the signals appear in the correct window (i.e. are not folded). For the quadrupolar nuclei, it is advisable to set the pulse width to 60–90° and to apply a line broadening factor. Relaxation delays can usually be avoided. Except in special cases, ^{1}H broad-band decoupling is unnecessary. For very large molecules (metalloproteins), high magnetic field strengths are recommended to cope with second-order shift contributions to the $+\frac{1}{2} \rightarrow -\frac{1}{2}$ quadrupole component. See Individual nuclei and applications, vanadium, below).

Commercially available equipment is used for solid-state static, MAS, OFF-MAS and double rotation techniques, and precautions have to be taken to discriminate between resonance signals and rotation side-bands by measuring the spectra at different spinning frequencies. The information that can be extracted from spectra under the static versus non-static conditions (where quadrupole perturbations, dipole interaction and chemical shift anisotropy

effects are suppressed) will be exemplified for solid state Sc NMR spectra below. Note again that the position of the central $+\frac{1}{2} \rightarrow -\frac{1}{2}$ transition which, owing to large quadrupole coupling anisotropy often is the only accessible one, is not coincidental with the isotropic shift δ_{iso}.

Table 1 also contains chemical shift reference compounds (used as the zero point of the shift scales) and, in the table footnote, standards employed in practice, including their chemical shifts with respect to the reference if not identical. Chemical shifts depend on concentration, isotopic composition (in the first but also the second coordination sphere), nature of the solvent, ionic strength, counterion and other influence of the medium, and these dependences are proportional to the intrinsic shielding sensitivity of a nucleus. Where the shielding sensitivity is exceedingly large (^{59}Co), these effects are significant and have to be taken into account when preparing sample and standard solutions. Coupling constants

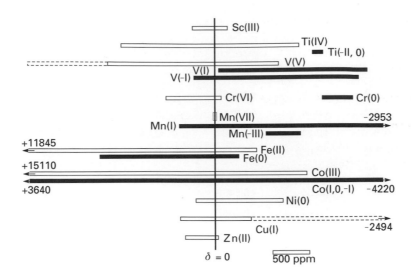

Figure 1 Chemical shift ranges for the 3d series metals. Open bars correspond to high, solid bars to low oxidation states. Chemical shift values δ have negative signs to the right of the vertical $\delta = 0$ line, which connects the references given in Table 1. Where the shift range extends beyond the scale of the figure, the limiting δ values are indicated.

are less dependent on medium effects, but relaxation times can also be very dependent on solution conditions. Other factors influencing shielding (and relaxation times) are temperature and pressure (see below). Chemical shifts should also be corrected with respect to field-lock shifts if different deuterium locks are used for the (external) standard/reference and the sample, and with respect to influences of the bulk susceptibility, if different sample shapes are compared (e.g. cylindrical versus spherical).

Chemical shifts (nuclear shielding)

Figure 1 is a summary of chemical shift ranges for the 3d series nuclei in their more common oxidation states in diamagnetic compounds, representing the present state of knowledge but also, to a certain extent, the intrinsic shielding sensitivities. Shift ranges for high and low oxidation states are represented by open and solid bars, respectively. Metals that form closed shell complexes (d^0: Sc^{III}; d^{10}: Cu^I, Zn^{II}) have relatively narrow shift ranges, while open shell systems (d^4: V^I; d^6: V^{-I}, Mn^I, Fe^{II}, Co^{III}; d^8: Fe^0, Co^I) give rise to extended shift ranges. Particularly noteworthy is the overall shift range of ~20000 ppm for ^{59}Co.

There is a tendency for deshielding of the nucleus (downfield shift with respect to the magnetic field, high-field shift with respect to the RF field) in compounds with high-valency metal nuclei, and for shielding in compounds containing the metal nuclei in a low oxidation state. However, as demonstrated by the large overlapping areas of open and solid bars for the same metal in **Figure 1**, no clear correlation between shielding and oxidation state exists, indicat-

ing that the nature of the coordination environment of the metal centre has to be considered. The factors influencing shielding are complex but can be described in a simplified manner by the expression $\sigma = \sigma(\text{dia}) + \text{const.}(\Delta E^{-1})_{av}\ (C^2 r^{-3})_{av}$, where σ is the overall shielding and $\sigma(\text{dia})$ the (essentially constant, since dominated by the core electrons) diamagnetic contribution. The second term, the paramagnetic contribution $\sigma(\text{para})$, the main factor influencing variations in shielding (and hence chemical shift), contains the mean HOMO-LUMO splitting ΔE, the LCAO coefficient C of relevant orbitals ($C = 1$ in the crystal field approximation), and a measure for the expansion of the orbitals, $\langle r^{-3} \rangle$. The main contributions to this term arise from the valence d orbitals, but in closed shell systems valence p contributions have to be taken into account and may become predominant. Based on this relationship, the following shielding trends can be predicted and have been verified by experiment.

(1) A strong ligand field decreases $\sigma(\text{para})$ via an increase of ΔE and thus gives rise to high overall shielding. In compounds with a low-valency metal centre, strong ligands are π acceptors such as CO, PR_3 and CNR; in compounds with a high-valency metal centre, 'hard' ligands such as oxo and nitrogen functional groups induce a large ΔE splitting. Consequently, the ^{51}V nucleus is shielded in $[VO_4]^{3-}$ ($\delta = -540$) as well as in $[V(CO)_6]^-$ (-1952), and deshielded in $[VS_4]^{3-}$ ($+1395$) as well as in $[V(NO)_2(\text{thf})_4]^+$ ($+269$ ppm).

(2) Mainly in open shell compounds, shielding is additionally influenced through the $(C^2 r^{-3})_{3d}$ term. This factor becomes small for ligands such

as H^-, R^-, I^-, SR_2 or, more generally, for ligands with a high polarizability/low electronegativity. In a series such as $[Co(NO)_2X]_2$, shielding increases in the sequence Cl ($\delta = +3640$), Br ($+3390$), I ($+2790$ ppm), known as the 'normal halogen dependence of chemical shielding'. Corresponding trends ('normal polarizability dependence') are commonly found in the series amine, phosphine, arsine and stibine, as well as in the series oxo, thio, seleno and telluro functional ligands. The inverse halogen (polarizability) dependence, i.e. a decrease of shielding (mainly as a consequence of a decrease in ΔE) usually prevails in closed shell d^0 systems; cf. Cp_2TiF_2 ($\delta = -1052$), Cp_2TiCl_2 (-772), Cp_2TiBr_2 (-668 ppm).

(3) Bulky ligands and strained ring structures (e.g. 3- and 4-membered chelate rings) induce deshielding in open shell and increase shielding in closed shell systems with respect to sterically normal conditions. This sensitivity of shielding to steric effects often allows for the distinction of conformers and diastereomeric pairs of enantiomers.

(4) Non-innocent ligands (typically catecholates, cat) can give rise to substantial deshielding of the metal nucleus, e.g. in VO(sb)cat (sb is a tridentate Schiff base dianion) $\delta = +480$ ppm instead of an expected $\delta \approx -540$ ppm for a vanadium compound with an 'innocent' N_xO_y donor set.

(5) Since the number of relevant transitions between highest occupied and lowest unoccupied molecular orbitals increases with a decrease in local symmetry, structural isomers (e.g. *cis/trans*, *fac/mer*) exhibit different shielding situations.

Secondary factors influencing shielding

Apart from effects arising directly from the coordinating functions and the arrangement of the ligands, various second-sphere and outer-sphere effects have to be considered, some of which will be noted here. (i) Ligand substituents, e.g. Z in phosphines PZ_3, very effectively influence the steric and electronic nature (donor/acceptor properties) of the ligand. $P(OR)_3$, for example, induces stronger shielding than $P(alkyl)_3$ and PMe_3 stronger shielding than $P(Bu^t)_3$. (ii) The shielding variations observed as the counterion or the solvent varies can indicate contact-ion pairing (e.g. in the series $[Co(en)_3]X_3$, X = Cl, Br, I: $\Delta\delta = 32$ ppm) and solvation, respectively. (iii) In the solid state, distinct crystallographic sites are represented by distinct resonance signals. The δ_{iso}(solid) are not usually equal to δ_{iso}(solution).

Isotopic substitution leads to increased shielding for the heavier isotope (e.g. ^{12}CO versus ^{13}CO; $C^{14}N^-$ versus $C^{15}N^-$). This isotope effect can be traced back to the same origin as the temperature and pressure dependence of shielding, i.e. they can be related to the vibrational fine structure of the highest electronic levels participating in electronic transitions: as the temperature decreases, vibronically excited levels become decreasingly populated; ΔE increases, and overall shielding decreases. An equivalent effect is responsible for increased shielding with increased pressure and an increase of shielding for a heavier with respect to a lighter isotopomer. **Table 2** provides selected examples and **Figure 2** illustrates isotope effects for $C_5H_5V(CO)_4$.

Anisotropy of chemical shifts

Under anisotropic conditions (solid state, mesophases), the chemical shift consists principally of three components (a parallel ($\delta_\parallel$) and two perpendicular ones (δ_1 and δ_2, where $\delta_\perp = \frac{1}{2}(\delta_1 + \delta_2)$)), the separation of which depends on the molecular anisotropy which, for quadrupolar nuclei, correlates with the extent of quadrupole perturbation and hence with the size of the nuclear quadrupole coupling constant e^2qQ/h. The extent of anisotropy is expressed by the molecular asymmetry parameter η, defined by $\eta = (\delta_2 - \delta_1)/\delta_\parallel - \delta_{iso})$, and the chemical shift anisot-ropy $\Delta\delta = |\delta_\parallel - \delta_\perp|$. Selected values for δ_{iso}, $\Delta\delta$ and η are listed in **Table 3** together with e^2qQ/h values. The chemical shift anisotropy

Table 2 Isotope, temperature and pressure effects on shielding

Isotope	Compound	Detected by	Isotope shift[a]	Temperature gradient[b]	Pressure gradient[c]
$^{1/2}H$	$[Sc(H_2O)_6]^{3+}$	^{45}Sc	-0.52		
	$C_5H_5V(CO)_4$	^{51}V	-0.715	0.61	
	$[(CpV(CO)_3H]^-$	^{51}V	-4.7		
	$[Co(NH_3)_6]^{3+}$	^{59}Co	-5.2	1.55	
$^{12/13}C$	$[V(CO)_6]^-$	^{51}V	-0.27	0.30	-0.032
	$[Co(CN)_6]^{3-}$	^{59}Co	-0.85	1.38	-0.018
$^{14/15}N$	$[Co(CN)_6]^{3-}$	^{59}Co	-0.197		
$^{16/18}O$	$[V(Co)_6]^-$	^{51}V	-0.10		
	$[MnO_4]^-$	^{55}Mn	-0.60		

[a] The one- and two-bond isotope shifts are indicated per isotopic substitution.

[b] The temperature gradients (tg) are given in ppm per degree, averaged over a temperature (T) range of ~230–320 K (^{51}V) and 290–350 K (^{59}Co); the (tg)/T relation is not strictly linear.

[c] kg cm^{-2}, for a pressure range of 1000 to 9000 kg cm^{-2} and at 291 K.

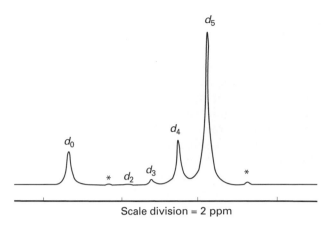

Scale division = 2 ppm

Figure 2 Two-bond deuterium isotope shifts on the ^{51}V chemical shifts in η^5-{$C_5{}^1H_{5-n}{}^2H_n$} $V(CO)_4$ (assigned d_n). The asterisks indicate the high-field components of the natural-abundance ^{13}C satellite doublets for d_0 and d_5. Reproduced with permission from Rehder D, Hoch M Jameson CJ (1990) Deuterium isotope effects and bonding in carbonylvanadium complexes. *Magnetic Resonance Chemistry* **28**: 138–144. © John Wiley & Sons Ltd.

produces an important relaxation mechanism in nonquadrupolar nuclei (^{57}Fe; see below).

Scalar coupling constants

Information on electron-mediated (scalar) spin–spin coupling J may be obtained from the NMR spectrum of the metal nucleus N_M or the ligand nucleus N_L.

When N_M is the NMR probe, two limiting cases have to be considered. One case is where coupling is fully effective and thus gives rise to a resolved first-order multiplet for the M resonance. This is the case if M is a spin-$\frac{1}{2}$ nucleus (^{57}Fe), or if the quadrupole moment of M is small (^{51}V) and L is a spin-$\frac{1}{2}$ nucleus or again a quadrupole nucleus with a small quadrupole moment (^{14}N). Examples are given in **Figure 3A** and **Figure 3B(ii)**. For the medium quadrupole nuclei, resolved J coupling is restricted to compounds where M is in a local symmetry providing small electrical field gradients at the nucleus and hence less effective quadrupole relaxation. The second case is where coupling is completely quenched. This is the case if fast relaxation leads to relaxation decoupling of N_M and N_L, or in the case of intermolecular exchange if it is accompanied by bond rupture. Fast relaxation is encountered with quadrupolar nuclei in all but highly symmetric environments, and with long molecular correlation times τ_c. Bulky and/or large molecules, and a high viscosity of the medium (polar solvents, low temperature, high concentration) will give rise to slow molecular tumbling and hence long τ_c.

If N_L is the NMR probe, we again have to deal with two limiting cases. One is when all of the coupling information is available. This is the case if M belongs to the low quadrupole category nuclei and/or M is at a highly symmetric site. The spectrum appears as a resolved nonbinomial multiplet of

Table 3 Solid-state chemical shifts δ_{iso}, chemical shift anisotropies $|\Delta\delta|$, asymmetry parameters η, and nuclear quadrupole coupling constants e^2qQ/h

Compound	Nucleus	$\delta_{iso}{}^a$	$\|\Delta\delta\|$	η	e^2qQ/h (MHz)
$Sc(O_2CMe)_3$	^{45}Sc	−40	–	0	5
α-$NaVO_3$	^{51}V	−578	365	0.70	2.95
$VOCl_3$	^{51}V	~0	Not reported	0.09	5.7
$[NH_4]_2[VO_2nta]^b$	^{51}V	−507	800	0.04	8.80
$CpV(CO)_4$	^{51}V	−1534	Not reported	0.11	2.79
$KMnO_4$	^{55}Mn	~0	Not reported	0.121	1.57
$Mn(CO)_5Cl$	^{55}Mn	~ −1000	1350	0.1	13.9
$Co(tpp)(im)_2{}^c$	^{59}Co	+8940	2497	0.17	5
$Co(acac)_3$	^{59}Co	+12505	1029	0.38	5.6
$FeCo_3(CO)_{12}H$	^{59}Co	−2833	1140	0.23	12.19
$Myoglobin.CNEt^d$	^{57}Fe	+9223	1288	–	–
CuI^e	^{63}Cu	~ −410	Not reported	Not reported	1.7
$Zn(O_2CMe)_2.2H_2O$	^{67}Zn	~0	Not reported	0.87	5.3

^{a} Referenced against the standards given in **Table 1**.
^{b} nta = nitrilotriacetate (3−).
^{c} tppp = tetraphenylporphyrin, im = imidazole.
^{d} From solution studies.
^{e} In $CuBr_xI_{1-x}$ crystals.

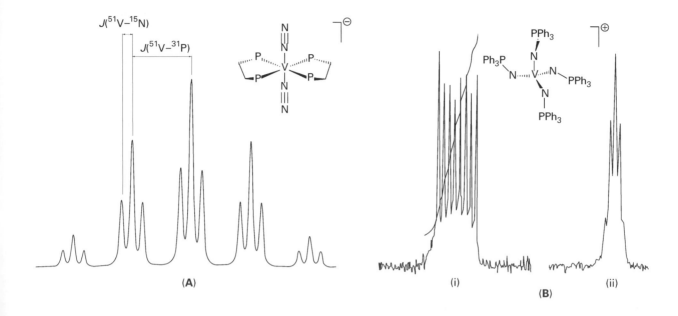

Figure 3 (A) One-bond scalar couplings in the ^{51}V NMR spectrum of *trans*-[V(^{15}N)$_2$(Me$_2$PCH$_2$CH$_2$PMe$_2$)$_2$]−: $^1J(^{51}V-^{31}P) = 314$ (quintet), $^1J(^{51}V-^{15}N) = 57$ Hz (triplet splitting). Reproduced with permission of the Royal Society of Chemistry from Rehder D, Woitha C, Priebsch W and Gailus H (1992) Structural characterization of a dinitrogenvanadium complex, a functional model for vanadiumnitrogenase. *Chemical Communications* 364–365. (B) Two-bond coupling $^2J(^{51}V-^{31}P) = 120$ Hz in the ^{31}P (i) and ^{51}V (ii) NMR spectra of [V(NPPh$_3$)$_4$]$^+$. Reprinted with permission from Aistars A, Doedens RJ and Dohesty NM (1994) Synthesis and characterization of vanadium(V) mono-, bis-, tris- and tetrakis(phosphoraniminato) complexes. *Inorganic Chemistry* **33**: 4360–4365. © 1994 American Chemical Society.

multiplicity $2I + 1$ (I is the nuclear spin of M; **Figure 3B(i)**) or as the envelope of such a multiplet, a typical plateau-like signal where the width at half-height divided by $(2I + 1)$ provides a good estimate for J. In the solid state $J(N_M N_L)$ multiplets can be distorted owing to the anisotropy of J and the quadrupole coupling interaction. An example is (Ph$_3$P)$_2$CuCl, where the ^{31}P quartet spacings are 1.43, 1.27 and 10.94 kHz. The second limiting case is again complete relaxation decoupling, resulting in a comparatively sharp resonance signal for N_L. More or less broad signals with a Lorentzian–Gaussian line shape are typical for intermediate situations.

Coupling constants $J(N_M N_L)$, in most cases one-bond couplings ^1J, are available for all of the 3d series nuclei except for ^{67}Zn. They are dominated by the Fermi contact term and the main factors in this term contributing to the size of J are the s electron densities $|S(0)|^2$ at the M and L nuclei, which are susceptible to the metal oxidation state and donor/acceptor properties of the ligands. For a direct comparison of coupling constants of different nuclei, it is advisable to employ the reduced coupling constant $K \propto J/\gamma_M\gamma_L$; γ is the magnetogyric ratio, see **Table 1** for the γ_M values. **Figure 4** summarizes $^1J/\gamma_M\gamma_L$ ranges for various M–L pairs in relation to the $|S(0)|^2$ terms.

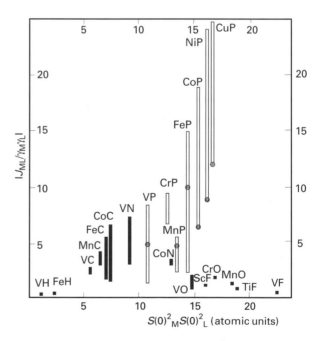

Figure 4 Reduced one-bond M–L coupling constants versus the product of the s electron densities at the respective nuclei (assuming effective zero charge). Open bars represent M–P (phosphine) coupling; the shaded circle represent phosphites [P(OR)$_3$]. $S(0)^2$ values have been taken from Pregosin PS and Kunz RW (1979) *NMR Basic Principles and Progress* **16**: 80–85.

Relaxation and line widths

The width of a resonance line, usually measured at half-height and quoted as $W_{1/2}$ (or $\Delta\nu_{1/2}$) in units of Hertz, is mainly determined by the spin–spin relaxation time T_2 ($W_{1/2} = (\pi T_2)^{-1}$) which, in nonviscous solutions and other than highly symmetric complexes, usually has about the same magnitude as the spin–lattice relaxation time T_1. In the solid state, T_1 and T_2 can deviate from each other considerably. The knowledge of T_1 values is a necessary precondition for optimizing NMR detection, especially for nuclei with long T_1, viz. ^{57}Fe. In the absence of quadrupolar ligand nuclei, the main relaxation mechanism in ^{57}Fe compounds with a nonisotropic coordination environment is the chemical shift anisotropy $\Delta\delta$: T_1 is inversely proportional to $(\Delta\delta)^2$, the square of the applied magnetic field B_0 and is a function of the correlation time τ_c. τ_c gains importance for large and/or bulky molecules; compare T_1 ($B_0 \approx 9$ T) = 4.2 s for ferrocene, and 0.02 s for myoglobin·CO.

Chemical shift anisotropy has also been noted for cobalt complexes as a B_0-dependent component to relaxation which, for quadrupolar nuclei, is otherwise dominated by the quadrupolar relaxation mechanism: $T_2^{-1} \propto f(I)(e^2qQ/h)^2\,(1+\eta^2/3)\tau_c$, where $f(I)$ is a function of the nuclear spin, and e^2qQ/h and η ($= 0$ under axial symmetry) are as defined above (cf. also **Table 2**). τ_c describes the intimacy of interaction between solvent and solute molecules and hence the ease by which a solute molecule tumbles in a solution. As noted previously, T_1 and T_2 are very similar. For example, in aqueous $ZnSO_4$, ^{67}Zn T_1 and T_2 are 9.3 and 9.8 ms; in aqueous $Co(acac)_3$ the ^{59}Co T_1 and T_2 are both 1.7 ms, and in neat $VOCl_3$ its ^{51}V T_1 and T_2 are both 17 ms while the ^{50}V T_1 and T_2 are both 5 ms. The latter example ($VOCl_3$) clearly demonstrates the importance of the size of the nuclear quadrupole moment Q (cf. **Table 1**), since all of the other parameters are identical. On the other hand, in a series of similar compounds containing the same metal nucleus (i.e. the same $f(I)$ and Q), T_2 reflects variations of τ_c and the electric field gradient q at the nucleus. For solutions, long T_2 (narrow resonance lines) can hence be expected for nuclei with a reasonably small Q (e.g. ^{51}V), for small, nonbulky molecules in nonviscous media (influence via τ_c), for molecules of cubic point symmetry and for a few other cases where q becomes small, e.g. octahedral complexes of C_{3v} symmetry, half-sandwich complexes having local C_{3v} or C_{4v} symmetry, or trigonal-pyramidal complexes of C_{3v} symmetry. The field gradient q also reflects the donor and acceptor power and related properties of the ligands.

Finally, it should be remembered that relaxation is also influenced by chemical exchange.

Individual nuclei and applications

Scandium and titanium

Scandium-45 is a relatively easily accessible nucleus owing to its 100% natural abundance and high magnetogyric ratio, but the comparatively large quadrupole Q has restricted its application. In the solid state, the large Q can provide valuable information on the quadrupolar interaction as exemplified for $ScCl_3 \cdot 6H_2O$ and $Sc(O_2CCH_3)_3$ in **Figure 5**. Relevant data can be extracted from static spectra, which depending on the strength of the applied magnetic field, reveal first-order or second-order quadrupole patterns (**Figure 5B**), while under MAS conditions (**Figure 5C**) the main spectral information is the exact isotropic chemical shift.

The concentration, counterion and pH dependences of $\delta(^{45}Sc)$ in aqueous solutions of scandium salts have revealed equilibria between outer-sphere versus inner-sphere complexes (contact ion pairs versus direct coordination of the anion), in addition to the equilibria between $[Sc(H_2O)_6]^{3+}$, which is present at low pH, its deprotonation ($[Sc(H_2O)_5(OH)]^{2+}$) and oligomerization products ($[Sc_2(H_2O)_x(OH)_2]^{4+}$, $[Sc_3(H_2O)_y(OH)_4]^{4+}$), present at pH values up to 4, and $[Sc(OH)_4]^-$, the species that is formed at high pH.

Owing to the very similar magnetogyric ratios, $^{47/49}Ti$, NMR spectra are twinned, the two resonances separated by only 266 ppm. The slightly smaller Q and larger spin of the ^{49}Ti nucleus make the low-field (high-frequency) twin the better-resolved one. As in the case of ^{45}Sc, the sizeable quadrupole moments restrict the observability of $^{47/49}Ti$ resonances to rather symmetric molecules. $[Ti(CO)_6]^{2-}$ marks the high-field, TiI_4 the low-field margin of the Ti chemical shift scale. Mono- and bis(cyclopentadienyl) complexes are in between. In the context of their application in Ziegler–Natta polymerization, a dependence of chemical shift upon the electronic nature of the substituents on cyclopentadienyl (Cp) is noteworthy, an effect which is also observed for $\delta(^{51}V)$ and $\delta(^{57}Fe)$ in Cp-vanadium and Cp-iron complexes.

Vanadium, chromium and manganese

The nucleus ^{51}V is the only one among the quadrupolar transition metals with a relatively low quadrupole moment *and* a high receptivity, and is hence a particularly well-suited NMR probe even in

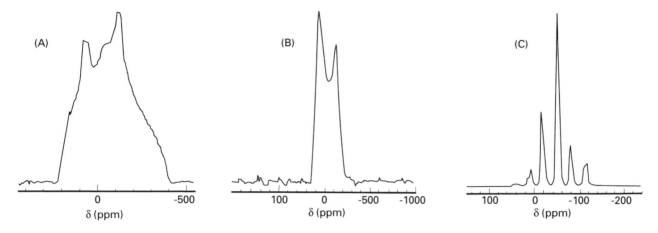

Figure 5 Solid state ^{45}Sc NMR spectra A: Static of ScCl$_3$.6H$_2$O at 8.45 T; B: static of Sc(O$_2$CCH$_3$)$_3$ at 3.52 T; C: MAS of Sc(O$_2$CCH$_3$)$_3$ at 8.45 T, showing the central line and spinning side-bands. Reproduced with permission of the Royal Society of Chemistry from Old-field E (1987) Solid-state scandium-45, yttrium-89 and lanthanum-139 nuclear magnetic resonance spectroscopy. *Chemical Communications*, 27–29.

compounds with no actual symmetry. Well-resolved coupling patterns, however, require compounds of medium to high local symmetry, as shown in **Figures 2** and **3** for complexes of idealized C$_{4v}$, D$_{4h}$ and T$_d$ symmetries. A large body of J and δ data has been accumulated (see Further reading for compilations), many of them in the context of investigations into biological/biomimetic vanadium systems, vanadium catalyst systems, correlations between NMR parameters and ligand properties, and vanadium speciation analysis.

More than a dozen species have been detected in aqueous vanadate solutions, and their interconversion and protonation/deprotonation equilibria have been studied by quantitative 1D-NMR, 2D-EXSY, and a combination of ^{51}V NMR with potentiometric measurements. The latter procedure has turned out to be a powerful tool for investigating the speciation in ternary systems containing, along with vanadates and protons, a biogenic ligand as a third component. **Figure 6** is an illustrative example for such a system. While the extreme narrowing conditions ($\omega\tau_c \ll 1$) are fulfilled in these systems, this is no longer so for large V^V–protein molecules with a tightly bound vanadium, placing the vanadium outside the extreme narrowing (though still within the motional narrowing) regime. Here, acquisition of spectra is restricted to the quadrupolar central $+\frac{1}{2}\rightarrow-\frac{1}{2}$ transition, which makes up only about 20% of the overall intensity (and is subject to a second-order shift contribution) but, contrasting the other three quadrupole components, does not suffer from severe relaxation broadening. The two slightly different metal ion binding sites in the C- and N-terminal lobes of transferrin have thus been characterized by ^{51}V NMR.

The main oxidation states investigated by ^{51}V NMR are the diamagnetic –I (carbonylvanadates), +I (half-sandwich complexes) and +V states (e.g. vanadates and other oxovanadium(V) compounds), but other oxidation states such as –III, low-spin +III and binuclear, and strongly antiferromagnetically coupled +IV have also been studied. Irrespective of the oxidation state, the concepts of electronic and steric ligand influences upon $\delta(M)$ outlined above are valid throughout. The steric effect has been employed to distinguish between diastereomers based on pairs of enantiomers in V^I and V^V complexes. Depending on the separation of the chiral centres, the diastereomer splittings amount to ~2 – 10 ppm.

Chemically there are many similarities between V, Cr and Mn in their low-valency diamagnetic states (d^6: V^{-I}, Cr^0, Mn^{+I}) and in their highest, closed shell (d^0) oxidation states. Where data are available (mainly for Cr-d^6 and Mn-d^6), dependences of chemical shifts $\delta(^{53}$Cr) and $\delta(^{55}$Mn) upon the donor/acceptor properties and sensitivities towards steric effects have been documented that parallel those reported for $\delta(^{51}$V) in V-d^6 and V-d^4 complexes.

The relative receptivity of the ^{55}Mn nucleus compares with that of ^{51}V; $Q(^{55}$Mn) is, however, an order of magnitude larger than $Q(^{51}$V), limiting the accessibility of ^{55}Mn as an NMR probe. The manganese shift range is flanked by [Mn(CO)$_3$(NCMe)$_3$]$^+$ at the low-field margin, and the formally Mn^{-III} complexes Mn(NO)$_3$L at the high-field margin. The only Mn^{VII} compound studied so far by NMR is KMnO$_4$, which gives a very sharp resonance line in solution in accordance with its T$_d$ symmetry but a very broad feature in the solid state as a consequence of effective

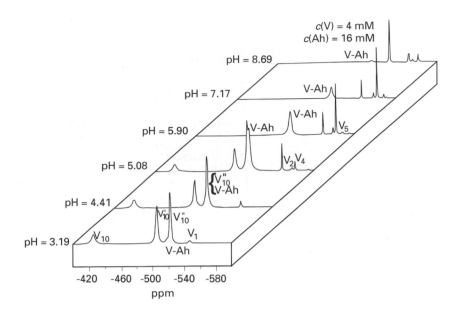

Figure 6 pH-dependent 131.5 MHz ^{51}V NMR spectra of the system vanadate/H+/alanylhistidine (Ah). The resonances correspond to decavanadate (V_{10}), vanadate-Ah complexes (V-Ah), monovanadate (V_1), divanadate (V_2), tetravanadate (V_4) and pentavanadate (V_5) in chemical equilibrium with each other. Reproduced with permission from Elvingsn K, Fritzsche M, Rehder D and Pettersson L (1994) Potentiometric and ^{51}V NMR study of aqueous equilibria in the H+-vanadate(V)-alanylhistidine system. *Acta Chimica Scandinavica* **48**: 881.

interaction between Q and a crystallographically imposed field gradient q. For $|Q(^{53}Cr)|$, reported values range between 0.04 and 0.15×10^{-28} m^2. NMR experience with this nucleus places it in the medium quadrupole category which, along with the low receptivity, makes ^{53}Cr a difficult nucleus to detect. The $\delta(^{53}Cr)$ shift range is flanked by $[CrO_3Br]^-$ (+670) and *cis*-$[Cr(CO)_2(PF_3)_4]$ (–1898 ppm).

Iron, cobalt and nickel

Iron-57 ($I = \frac{1}{2}$) is among the least NMR-sensitive nuclei of the periodic table. Isotopic enrichment or steady-state techniques (in the case of long T_1), high magnetic field strengths and large sample volumes may be applied in the case of direct detection, but 1D double resonance and 2D indirect techniques are nowadays much more commonly employed in spectrum acquisition wherever coupling to a sensitive ligand nucleus in the coordination sphere (such as ^{1}H, ^{13}C, ^{15}N, ^{31}P) is observable. The chemical shift range spanned by ^{57}Fe, ~12 000 ppm, is surmounted only by the ^{59}Co shift range of ~20 000 ppm. In the case of ^{59}Co, this extended shift range, reflecting a high intrinsic shielding sensitivity, somewhat neutralizes the disadvantage of the relatively large $Q(^{59}Co)$, leading to line widths commonly of several kHz. Nonetheless, abundant data on hexa- and tetracoordinated cobalt complexes in the oxidation states +III, +I and –I are in fact available. In contrast, ^{61}Ni, again a medium quadrupole category nucleus, but with a very low receptivity, causes severe detection problems. Only recently, through the use of special devices such as solenoidal coils, has a comprehensive series of NMR parameters been obtained on mixed phosphine/carbonylnickel(0) (d^8) complexes, which exhibit chemical shift dependences on the nature of the substituents Z in the phosphine PZ$_3$ comparable to what is known from open shell complexes of other transition metals.

Recent applications of ^{57}Fe and ^{59}Co NMR have been directed towards correlations between chemical shift and kinetic parameters on the one hand and the investigation of iron–porphyrin and cobalt–porphyrin systems on the other. In the half-sandwich complexes CpFe(CO)$_2$R ([1] in **Figure 7**) shielding decreases with increasing bulk of R ($\delta(^{57}Fe)$ values cover the range of +684 for R = Me to +805 for R = Bu^s), and $\delta(^{57}Fe)$ in turn correlates linearly with the rate constant k for the insertion of CO in the reaction

$$CpFe(CO)_2R + PPh_3 \rightarrow CpFeCO(PPh_3)COR$$

(increase in k as shielding decreases). The sensitivity of shielding to steric effects is also evident in the discrimination of conformers of the complexes (diene)Fe(CO)$_2$L [2]. The catalytic activity of the complexes Cp′Co(diene) ([3] in **Figure 7**) in the formation of substituted pyridines [4] by co-trimerization of

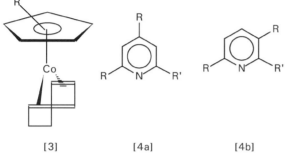

Figure 7 Applications of ^{57}Fe and ^{59}Co NMR. For compounds [1] $\delta(^{57}Fe)$ values correlate with the size of R and the rate of insertion of R into the Fe–CO bond. Compounds [2] are different conformers discernible by their $\delta(^{57}Fe)$ values. Compounds [3] catalyse the formation of [4]; the ratio of [4a] and [4b] depends on the $\delta(^{59}Co)$ values, which are related to the nature of R.

alkynes and isonitriles is closely related to $\delta(^{59}Co)$, which in turn is related to the nature of Cp′: an electron-rich Cp′, such as C_5Me_5 gives rise to high shielding of the ^{59}Co nucleus and induces the preferred formation of the symmetrically substituted pyridine [4a].

The presence in porphyrin and related systems of cobalt (cobalamines) and iron (heme proteins) has initiated metal NMR studies into native (myoglobin, cytochrome *c*) and model systems (substituted porphyrins) with the objective of elucidating the influence of the chemical nature of axial ligands (CO, isonitriles, N-donors) upon parameters such as δ and T_1 (and e^2qQ/h in the case of ^{59}Co), and hence of the functionally interesting metal centre. The differing binding strengths of isonitriles and carbon monoxide in myoglobins produce a change of the chemical shift anisotropy (CSA), viz. $\Delta\delta = 3600$ ppm for Mb·CO versus 1250 ppm for Mb·CNR, and thus clearly different CSA-induced T_1 values (17 ms for Mb·CO, 140 ms for Mb·CNR).

Copper and zinc

Cu^+ and Zn^{2+} are closed shell d^{10} systems and exhibit relatively low intrinsic shielding sensitivities. This fact, together with the unfavourable size of the quadrupole moment, the low receptivity in the case of ^{67}Zn, and the ease of chemical exchange in the case of

Cu complexes has restricted NMR access to these biologically important metal nuclei. The $\delta(^{63/65}Cu)$ range spans ~1000 ppm, with an extension from the high-field margin (–517 ppm, occupied by $[CuI_4]^{3-}$) to –2494 ppm for the cluster compound $[Cu_3Fe_3(CO)_{12}]^{3-}$. ^{67}Zn chemical shifts encompass about 450 ppm. Cu^I complexes with N-donor ligands such a nitriles and pyridines undergo exchange of the kind $[CuL_4]^+ \rightleftharpoons [CuL_3]^+ + L$, where the lower symmetry component considerably broadens the resonance line. Despite this complication, some insight has been obtained by ^{63}Cu NMR into the distribution of Cu^+ in mixed-solvent systems comprising nitriles and phosphites. A few series of tetrahedral Cu and Zn complexes, viz. $[Cu(phosphite)_4]^+$, $[Cu(ER_2)_4]^+$ (E = S, Se, Te) and $[Zn(ER_2)_4]^{2+}$ (E = S, Se) have been investigated. The latter are of interest in the context of biogenic {$ZnCys_4$} centres present in alcohol dehydrogenase and thioneins.

The $^{63/65}Cu$ probe is also increasingly used to elucidate solid-state features of copper-containing superconducting materials.

List of symbols

J = scalar coupling constant; q = electric field gradient; Q = nuclear quadrupole moment; r = receptivity; T_1 = spin–lattice relaxation time; T_2 = spin–spin relaxation time; $W_{1/2}$, $\Delta v_{1/2}$ = resonance line width [at half-height]; γ = magnetogyric ratio; δ_{iso} = isotropic chemical shift; η = molecular asymmetry parameter; σ = shielding; $\sigma(dia)$ = diamagnetic contribution to shielding; $\sigma(para)$ = paramagnetic contribution shielding; τ_c = molecular correlation time.

See also: **Chemical Exchange Effects in NMR; High Pressure Studies Using NMR Spectroscopy; NMR in Anisotropic Systems, Theory; NMR Relaxation Rates; Parameters in NMR Spectroscopy, Theory of; Solid State NMR Using Quadrupolar Nuclei; Solid State NMR, Methods.**

Further reading

Brevard C and Granger P (1981) *Handbook of High Resolution Multinuclear NMR*. New York: Wiley.

Gielen M, Willem R and Wrackmeyer B (1996) *Advanced Applications of NMR to Organometallic Chemistry*. Chichester: Wiley.

Howarth OW (1990) Vanadium-51 NMR. *Progress in Nuclear Magnetic Resonance Spectroscopy* 22: 453–485.

Juranic N (1989) Ligand field interpretation of metal NMR chemical shifts in octahedral d⁶ transition metal complexes. *Coordination Chemistry Reviews* 96: 253–290.

Laszlo P (1983) *NMR of Newly Accessible Nuclei*. New York: Academic Press.

Mason J (1987) *Multinuclear NMR*. New York: Plenum Press.

Pregosin PS (1991) *Transition Metal Nuclear Magnetic Resonance*. Amsterdam: Elsevier.

Rehder D (1984) NMR of the first transition series nuclei (Sc to Zn). *Magnetic Resonance Reviews* 9: 125–237.

Rehder D (1986) Applications of transition metal NMR spectroscopy in coordination chemistry. *Chimia* 6: 186–199.

Rehder D (1991) Metal NMR of organometallic (d-block) systems. *Coordination Chemistry Reviews* 110: 161–210.

Heteronuclear NMR Applications (Y–Cd)

Erkki Kolehmainen, Department of Chemistry, University of Jyväskylä, Jyväskylä, Finland

MAGNETIC RESONANCE

Applications

The elements yttrium (Y), zirconium (Zr), niobium (Nb), molybdenum (Mo), technetium (Tc), ruthenium (Ru), rhodium (Rh), palladium (Pd), silver (Ag) and cadmium (Cd) form a series of transition metals characterized by the electronic configurations $Kr4d^n5s^2$ where $n = 1-10$. This series of d-block metals is located in the 5th row of the periodic table, between the series from Sc to Zn (4th row) and from La to Hg (6th row). The NMR spectroscopy of transition metals is a rapidly increasing area of magnetic resonance with new experimental methods and new applications for both the solution and, especially, for solid state. There is an extensive literature on multinuclear NMR studies applied to the structures of polyoxometalates of many transition metal nuclei, especially Nb and Mo. The NMR data available for the series from Y to Cd depends strongly on the element. A literature search using the *Chemical Abstracts* data base reveals that in case of Y, Zr, Tc, Ru and Pd the number of reports is limited while for ^{113}Nb, ^{95}Mo, ^{103}Rh and ^{113}Cd there exists much experimental data. Cadmium-113 alone makes up about 20% of the total NMR papers dealing with these d-block metals. However, this situation might change significantly in the future. For example, with the advent of high temperature superconductors one might expect that ^{89}Y NMR could become a very common method in characterizing yttrium-containing superconducting materials.

Generally, in this article the main aim is to cover novel NMR techniques and promising new results, supplemented with characteristic reference data available in the excellent reviews and articles catalogued in the Further reading section at the end of this article. In selecting and arranging the reference material it has been the goal to demonstrate obvious trends in NMR data (mainly collected in tables of characteristic chemical shifts) that depend on the structural characteristics and measuring conditions: solution vs. solid state, the ligand in a metal complex and its substitution, coordination number, solvent, temperature, etc. From these tables the reader can get a hint as to where to hunt for a 'hidden' NMR signal. The text itself, dedicated to each of the NMR-nuclei concerned, serves as a general guide to select an appropriate measuring technique and to estimate if it is worth examining a given NMR-nucleus when compared with the other possible nuclei and techniques available.

Properties of the nuclei

Some important properties of the NMR active isotopes of Y, Zr, Nb, Mo, Tc, Ru, Rh, Pd, Ag and Cd are collected in **Table 1**. From the practical point of view the receptivity (R) relative to that of ^{13}C (=1) is one of the most important terms in estimating how easily a nucleus can be observed by NMR. In the case of nuclei with, $I = \frac{1}{2}$ such as ^{89}Y, ^{103}Rh, ^{107}Ag, ^{109}Ag, ^{111}Cd and ^{113}Cd, which are characterized by sharp NMR spectral lines, this parameter can be used as a direct basis in this estimation. However, a nuclear quadrupole moment can produce serious broadening of the spectral lines and hence decrease the signal-to-noise ratio in the NMR spectra. In the case of transition metal nuclei this signal broadening can be the most serious limiting factor from a practical NMR point of view. For example, ^{105}Pd should be a promising NMR nucleus based on its receptivity

Table 1 Istotope, spin (*I*), Natural abundance (%NA), frequency (*B*$_0$) relative to ^{1}H (at 100 MHz), receptivity (*R*) relative to ^{13}C (= 1), magnetogyric ratio (γ), magnetic moment (μ) and nuclear quadrupole moment (*Q*$_m$^{2})

Isotope	I	NA (%)	B$_0$	R	10^7 (γ rad T^{-1}s^{-1})	μ	10^{28}Q (m^2)
^{89}Y	$-\frac{1}{2}$	100	4.92	0.67	−1.31	−0.137	–
^{91}Zr	$\frac{5}{2}$	11.2	9.37	6.05	−2.50	−1.304	−0.21
^{93}Nb	$\frac{9}{2}$	100	24.55	2740	6.54	6.17	−0.32
^{95}Mo	$\frac{5}{2}$	15.9	6.52	2.88	1.74	−0.914	−0.015
^{97}Mo	$\frac{5}{2}$	9.55	6.65	1.84	−1.78	−0.934	0.17
^{99}Tc	$\frac{9}{2}$	0	22.5	–	6.05	5.66	−0.13
^{99}Ru	$\frac{5}{2}$	12.7	3.39	0.82	−1.23	−6.41	0.076
^{101}Ru	$\frac{5}{2}$	17.0	4.94	1.54	−1.38	−0.719	0.44
^{103}Rh	$-\frac{1}{2}$	100	3.17	0.18	−0.85	−0.088	–
^{105}Pd	$\frac{5}{2}$	22.3	4.58	1.43	−0.76	−0.642	0.8
^{107}Ag	$-\frac{1}{2}$	51.8	4.05	0.20	−1.08	−0.113	–
^{109}Ag	$-\frac{1}{2}$	48.2	4.65	0.28	−1.24	−0.131	–
^{111}Cd	$\frac{1}{2}$	12.8	21.2	7.01	−5.70	−0.595	–
^{113}Cd	$\frac{1}{2}$	12.2	22.2	7.59	−5.93	−0.622	–

and natural abundance but it suffers from a huge nuclear quadrupole moment such that in, for example, K_2PdCl_6 the line width at half-height is 25 kHz. Other limiting factors often met in the early days of NMR, viz. the limited spectral range of the spectrometer hardware, probe and amplifiers, are nowadays not usually problematic. A modern NMR spectrometer can be tuned and matched routinely to observe nuclei from ^{103}Rh to ^{31}P, ^{19}F and ^{1}H. This flexibility and the increased sensitivity of NMR spectrometers, e.g. inverse 2D detection techniques, are pushing the detection limits of transition metal nuclei to small concentrations and allowing complex structural characterizations.

Experimental methods

The variety of measuring techniques and applications in heteronuclear NMR is enormous. In the case of $I = \frac{1}{2}$ nuclei the same basic rules as for ^{1}H, ^{19}F and ^{13}C are true. However, one should remember that negative gyromagnetic ratios as in case of silver-107/109 and cadmium-111/113 isotopes prevent the signal enhancement by NOE effects. This is different to the case for ^{13}C NMR. For $I \geq 1$ nuclei the nuclear quadrupole moment opens an efficient relaxation channel and magnetization (free induction) decays often very rapidly. Starting the acquisition as soon as possible after the excitation pulse to avoid the loss of magnetization causes some distortions on the spectral base line (rolling) owing to the acoustic ringing of the probehead. This distortion can be eliminated by 'an antiringing' pulse sequence (at the expense of sensitivity) or by using base line correction routines that are now available as a standard in the spectrometer software.

Two-dimensional experimental techniques such as homo- and heteronuclear chemical shift correlations are frequently used in heteronuclear NMR applications. COSY (correlation spectroscopy) and INADEQUATE (incredible natural abundance double quantum transfer experiment) are suitable methods if the natural abundance is not very low, as in the case of ^{113}Cd. The most promising 2D techniques are, however, inverse detected heteronuclear chemical shift correlations such as ^{1}H,X HSQC (heteronuclear single quantum coherence), ^{1}H,X HMQC (heteronuclear multiple quantum coherence) and ^{1}H,X HMBC (heteronuclear multiple bond correlation) and, especially, ^{31}P,X HMQC and ^{31}P,X HMBC. If the scalar coupling constants between a proton (phosphorus or any nucleus of high natural abundance and high gyromagnetic ratio used for free induction decay (FID) detection) and the heteronucleus is small (<2 Hz) the delay in the evolution of these couplings increases to such a length that a significant loss of magnetization occurs and inverse heteronuclear multiple quantum (or bond) correlation becomes impractical. Using the proton for detection of heteronuclear chemical shift correlation can be done easily by a standard two-channel spectrometer equipped with a direct or inverse dual (proton and X) probehead. To observe heteronuclear chemical shift correlation by ^{31}P or some other heteronucleus the spectrometer should have two broad band channels and a triple resonance probehead to be able to perform proton decoupling during an experiment if necessary.

Low natural abundances, low gyromagnetic ratios and high quadrupole moments of several NMR-active isotopes of the series from yttrium to cadmium can cause difficulties in solution state NMR. However, these formidable obstacles in the solution state can be turned to advantages in solid state NMR. Low natural abundance (i.e. dilute spins) can help in solid state measurements because the problem raised by dipolar interactions between like spins is eliminated. Sensitivity problems found with low gyromagnetic ratios in the solution state can be overcome in the solid state by the cross-polarization (CP) technique by which the magnetization of the observed nucleus (S) can be increased by a factor, γ_H/γ_S. Further, a nuclear quadrupole moment can result in considerable sensitivity to local nuclear site symmetry, thus giving useful structural information.

The major technique to eliminate dipolar interactions in the solid state NMR is MAS (magic-angle spinning). Together with CP and high-power proton decoupling it has become the most often applied method in solid-state NMR. However, in this connection it is worth mentioning some other techniques such as double-rotation (DOR) and dynamic-angle spinning (DAS) which could remarkably increase the potential of the NMR of quadrupolar nuclei in the future. In DOR, MAS and DAS techniques the sample must be rotated at high speed, which can cause some technical problems. These are avoided by manipulation of nuclear spins by pulse sequences such as two-pulse free induction decay (TPF) and nutation spectroscopy. 2D NMR nutation spectroscopy of powder samples has been applied, for example, in ^{93}Nb NMR.

Yttrium-89

Although ^{89}Y is characterized by $I = \frac{1}{2}$, its long T_1 relaxation time, low magnetic moment and low measuring frequency have caused problems in ^{89}Y NMR. On the other hand, its 100% natural abundance and the increasing interest in organo-yttrium and solid state chemistry (high-temperature superconducting materials, e.g. $YBa_2Cu_3O_{7-x}$) have created strong interest in this NMR nucleus. Especially promising in this respect is the CP-MAS technique. At first, CP-MAS was applied in the case of the 'common' NMR nuclei such as ^{13}C, ^{27}Al, ^{29}Si and ^{31}P but has now expanded into 'more difficult' low γ-nuclei such as ^{89}Y. The first solid state ^{89}Y CP-MAS

NMR spectra of $Y(NO_3)_3 \cdot 6H_2O$, $Y_2(SO_4)_3 \cdot 8H_2O$, $YCl_3 \cdot 6H_2O$, $Y(OAc)_3 \cdot 4H_2O$ and $Y(acac)_3 \cdot 3H_2O$ were reported by Merwin and Sebald in 1990. Later, mono-, di- and polymetallic samples were characterized by ^{89}Y CP-MAS NMR, revealing a direct correlation between the number of ^{89}Y NMR signals and the unique solid state environments of the ^{89}Y nuclei.

In **Table 2** are collected some characteristic ^{89}Y NMR chemical shifts. Nowadays, a common reference compound for ^{89}Y NMR is aqueous $Y(NO_3)_3$ although aqueous YCl_3 also has been used. In fact, both salts form, in water, the same hexahydrate cation, $[Y(H_2O)_6]^{3+}$. Generally, the chemical shifts of the yttrium salts show similar concentration trends to those observed for scandium and lanthanum.

Zirconium-91

The number of papers dealing with ^{91}Zr NMR is still small. However, some ^{91}Zr NMR chemical shifts of organometallic compounds (mainly zirconocene derivatives) and inorganic ions are available and are collected in **Table 3**. The line widths, $LW_{1/2}$ of these compounds vary between 19–3100 Hz. The most often used standard is Cp_2ZrBr_2 because it shows a sharper resonance line than any other zirconium complex. Von Philipsborn and co-workers (Organisch-Chemisches Institut, Universität Zürich), have used zirconium-91 NMR chemical shifts and line widths as indicators of coordination geometry distortions in zirconocene complexes. Further, the experimental trends in ^{91}Zr NMR chemical shifts of these zirconocene complexes are well reproduced computationally by molecular orbital calculations using the

Table 2 Some characteristic ^{89}Y NMR chemical shifts from $[Y(H_2O)_6]^{3+}$ ($\delta = 0$ ppm)

Compound	Solvent/conditions	δ (^{89}Y)/(ppm)
$Y(NO_3)_3$	0.77 M in H_2O–OCMe$_2$ at 298 K	− 4.9
$Y(NO_3)_3 \cdot 6H_2O$	Solid	− 53.2
$YCl_3 \cdot 6H_2O$	Solid	+58
$Y(OAc)_3 \cdot 4H_2O$	Solid	+46.7
$Y(acac)_3 \cdot 3H_2O$	Solid	+21.5 and +27.9
$Y_2(SO_4)_3 \cdot 8H_2O$	Solid	− 40 and − 46.2
Y_2O_3	Solid	+270 and +315
$Y_2Sn_2O_7$	Solid	+150
$Cp'_3Y(THF)^a$	THF	− 371
$Cp'_2YCl(THF)$	THF	− 103
$[YCl_6]^{3-}$	Nitrobenzene	+240
$[Y(EDTA)]^-$	H_2O	+126

^{a}Cp$'$ = C_5H_4Me, THF = tetrahydrofuran.

Table 3 Some characteristic ^{91}Zr NMR chemical shifts from Cp$_2$ZrBr$_2$ ($\delta = 0$ ppm)

Compound	Solvent	δ (^{91}Zr) (ppm)
Zr(NEt$_2$)$_4$	–	+875
Zr(BH$_4$)$_4$	Benzene-d_6	+41
H$_2$ZrCl$_6$	Conc HCl	+601
[ZrF$_6$]$^{2-}$	D$_2$O	–191
Cp$_2$ZrCl$_2$^{a}	–	–122
Cp$_2$ZrBr$_2$	THF^b	0
Cp$_2$ZrI$_2$	–	+126
Cp$_2^*$ZrCl$_2$^{c}	THF^b	+82
Cp$_2$ZrCl(vinyl)	THF^b	+16
CpZr(η^3-allyl)(ip)^d	–	+132
Cp$_2$Zr (ip)	–	–325

^{a} Cp = C$_5$H$_5$.
^{b} THF = tetrahydrofuran.
^{c} Cp* = C$_5$Me$_5$.
^{d} ip = η^4-isoprene.

IGLO (individual gauge for localized orbitals) or GIAO (gauge including AOs) SCF methods employing large basis sets. In the solid state, Zr-metal, ZrH$_2$, ZrC, ZrCo, ZrO$_2$ and ZrO$_2$ doped with Y$_2$O$_3$ and MgO have been characterized by ^{91}Zr NMR. Solid state ^{91}Zr NMR line shapes have been found to be related to the Zr (or Nb) content in Zr$_{1-x}$Nb$_x$ alloys. These applications suggest an increasing interest in ^{91}Zr NMR in metallurgy and material sciences.

Niobium-93

There exist several reports on ^{93}Nb NMR, as expected based on a natural abundance of 100% and a receptivity of 2740 (**Table 1**). In fact, the receptivity of ^{93}Nb is one of the most promising among all the NMR-active nuclei. However, the influence of broad resonance lines again effectively diminishes its usefulness although its other NMR-properties are favourable. In **Table 4** are collected some characteristic ^{93}Nb NMR chemical shifts. In general the chemical shift trends of ^{93}Nb are similar to those observed for ^{51}V NMR. Recommended reference compounds for niobium-93 are NbOCl$_3$ in acetonitrile or the [NbF$_6$]$^-$ anion in concentrated aqueous HF, although [NbCl$_6$]$^-$ in acetonitrile has also been used. In addition to inorganic salts and their anions, and various carbonyl niobium complexes collected in **Table 4** there exist several cyclopentadienyl niobium complexes. The ^{93}Nb NMR chemical shifts of these orgnaometallic complexes are clearly shielded, varying between 1290 and 2258 ppm upfield from the reference. Their $LW_{1/2}$ vary from 20 Hz for

Table 4 Some characteristic ^{93}Nb NMR chemical shifts from [NbCl$_6$]$^-$ ($\delta = 0$ ppm)

Compound	Solvent/temperature (K)	δ(^{93}Nb)/ppm
Cu$_3$[NbS$_4$]	Solid	+380
Cu$_3$[NbSe$_4$]	Solid	+1110
Cu$_3$[NbTe$_4$]	Solid	+2590
[NbW$_5$O$_{19}$]$^{3-}$	–	–888
NbOCl$_3$	MeCN	–450
[NbOCl$_4$]$^-$	MeCN	–480
[NbOF$_4$]$^-$^{a}	MeCN	–1228
[NbOBr$_4$]$^-$	MeCN	–210
[NbSCl$_4$]$^-$	MeCN	+500
[NbSeCl$_4$]$^-$	MeCN	+970
[NbF$_6$]$^-$	MeCN	–1550
[NbCl$_6$]$^-$	MeCN	0
[NbBr$_6$]$^-$	MeCN	+735
[HNb(CO)$_5$]$^{2-}$	NH$_3$/223	–2122
[Nb(CO)$_6$]$^-$	NH$_3$/223	–2136
[Nb(CO)$_6$]$^-$	THF/298	–2121
[Nb(CO)$_5$NH$_3$]$^-$	NH$_3$/223	–1880

^{a} J(F, Nb) = 410 Hz.

CpNb(CO)$_4$ in THF–CH$_2$Cl$_2$ to 10 300 Hz in [Cp-Nb(AuPPh$_3$)(CO)$_3$]$^-$.

In addition to the previously mentioned 2D nutation NMR spectroscopy, an interesting new method in solids, first-satellite spectroscopy, applied to ^{93}Nb NMR of LiNbO$_3$ (and ^{27}Al NMR in Al$_2$O$_3$) has been reported by McDowell and his co-workers. This technique is experimentally easier and more sensitive than the other methods for studying wide, quadrupolar broadened spectra. Further, the experimental data are considerably simpler and easier to interpret than obtained by the other methods. Examples of the significance of ^{93}Nb NMR in materials science include novel solid state studies on ion ordering and ion shifts in relaxor ferroelectrics, structural distortions and intrinsic defects in LiNbO$_3$ and studies on superconducting Nb–Cu multilayers by the field cycling method.

Molybdenum-95 and -97

Molybdenum has two naturally abundant NMR-active isotopes (**Table 1**). The nuclear properties of the isotope ^{95}Mo are more favourable than those of ^{97}Mo. Malito has reviewed more than 150 original papers on ^{95}Mo NMR. The vast majority of those reports are solution state studies whilst very few of them deal with solid state $^{95/97}$Mo NMR. This scarcity of solid state $^{95/97}$Mo NMR data is owing to the relatively low NMR receptivity, long probe dead-times and acoustic ringing associated with the pulse-NMR observation of these low-frequency isotopes. Recently, however, it has been demonstrated that solid state MAS ^{95}Mo NMR studies of organometallic complexes of molybdenum are feasible, particularly if the compounds are known to give relatively narrow ^{95}Mo NMR spectral lines in solution.

As mentioned before (poly)oxometalates and related structures are traditional topics for multinuclear NMR studies. In the case of molybdenum, there exist systematic experimental NMR and *ab initio* molecular orbital studies on $[MoO_{4-n}X_n]^{2-}$ anions (X = S or Se and n = 0–4). It has been shown that the ^{95}Mo NMR chemical shifts of these anions can be reliably estimated by such calculations. The paramagnetic shielding term mainly determines the chemical shift variation, ligand induced changes in molybdenum valence d-orbitals correlate with the changes in the metal chemical shift and the excitation energy term, ΔE, in the Karplus–Pople equation is related to the character of the ligand.

Some characteristic ^{95}Mo NMR chemical shifts are collected in **Table 5**. A common reference for ^{95}Mo NMR is aqueous (0.2 M) Na_2MoO_4 (δ = 0 ppm). The chemical shift variation of ^{95}Mo is from +4000 to −2000 ppm and clearly depends on the oxidation state of the metal. ^{95}Mo NMR line widths can be very broad (>5 kHz). Generally, slower molecular motion (as a consequence of increased relative molecular mass) results in broader resonance lines. This fact can be a limiting factor in studies of molybdenum-containing metalloenzymes and other biopolymers by ^{95}Mo NMR. A very recent area of biochemical interest is the Mo(VI) complexes of carbohydrates such as alditols and corresponding acids. For example, it has been shown by ^{1}H, ^{13}C, ^{17}O and ^{95}Mo NMR that D-galacturonic and D-glucuronic acids in their aqueous solutions form complexes with molybdate anions. Although the acids predominately exist in the pyranose forms, their complexes involving the less stable α- and β-furanose anomers as well as the α-pyranose form was detected also. The 2:1 complexes with the α-pyranose forms, insofar as they involve metal binding to the ring oxygen atom, are considered to play an important role in the oxidation of the acids by Mo(VI).

Technetium-99

Technetium is an artificially produced element used for medical imaging. Reports dealing with ^{99}Tc NMR are very rare. In addition to unfavourable NMR properties (**Table 1**) a more complex situation

regarding practical measurements comes from the fact that ^{99}Tc is radioactive. Nevertheless, there are some ^{99}Tc NMR chemical shifts available, as collected in **Table 6**, referenced to the signal of TcO_4^- (δ = 0 ppm). ^{99}Tc NMR line widths can vary strongly from < 100 Hz to > 3 kHz.

Ruthenium-99 and -101

Ruthenium has seven stable isotopes, of which ^{99}Ru and ^{101}Ru are NMR active. Although their natural abundances and resonance frequencies are quite similar the smaller quadrupole moment of ruthenium-99 makes the former more favourable from a NMR spectroscopic point of view. The ^{99}Ru NMR chemical shift range is enormous (> 18 000 ppm). The shifts of ruthenium compounds and complexes follow roughly the same trends as those of iron and osmium. The reference compound in ^{99}Ru NMR is the $[Ru(CN)_6]^{4-}$ anion (δ = 0 ppm). Some characteristic ^{99}Ru NMR chemical shifts are collected in **Table 7**.

Ruthenium complexes with nitrogen-containing ligands show very interesting photochemical properties which can be monitored by ^{99}Ru NMR. In materials science, solid state ^{99}Ru and ^{99}Ru NMR has been recently applied in characterizing the packing of ruthenium metal. This result represents the first detection of Ru NMR in a paramagnetic solid.

Rhodium-103

Traditionally ^{103}Rh NMR spectroscopy has been regarded as problematic owing to the very low gyromagnetic ratio and long relaxation time of rhodium-103 although its natural abundance is 100% and $I = \frac{1}{2}$ (**Table 1**). However, by using modern broad band spectrometers rhodium-103 is an easy nucleus to observe. Taking into account the catalytic significance of rhodium compounds, ^{103}Rh NMR spectroscopy has turned out to be a very powerful method and there exist plenty of experimental data. Some characteristic ^{103}Rh NMR chemical shifts are collected in **Table 8**. In the case of ^{103}Rh NMR, several chemical shift references have been used but nowadays *mer*-$[RhCl_3((SCH_3)_2)_3]$ is generally accepted. Owing to the low gyromagnetic ratio of rhodium-103, its detection is now based on polarization transfer techniques such as INEPT (insensitive nuclei enhanced by polarization transfer), DEPT (distortionless enhancement by polarization transfer) or HMQC although earlier INDOR (internuclear double resonance) was also utilized. If a rhodium-bound proton is used for the detection of rhodium-103 an enhancement of $(\gamma_H/\gamma_{Rh})^{5/2} = 5635$ can be achieved in ^{1}H, ^{103}Rh HMQC. If the proton is not directly

Table 5 Some characteristic ^{95}Mo NMR chemical shifts from $[MoO_4]^{2-}$ (δ = 0 ppm)

Compound	Oxidation state	δ $(^{95}Mo)/(ppm)$	Compound	Oxidation state	δ $(^{95}Mo)/(ppm)$
Mo(CO)$_6$ (in DMF)	0	−1850	K$_4$[Mo$_2$Cl$_8$]	2	+3816
Mo(CO)$_6$ (in CD$_3$CN)	0	−1855	Mo$_2$[OCH(CH$_3$)$_2$]$_6$	3	+2444
Mo(CO)$_6$ (in (CH$_3$)$_2$CO)	0	−1856	Mo$_2$[OC(CH$_3$)$_3$]$_6$	3	+2645
Mo(CO)$_6$ (in CDCl$_3$)	0	−1857	Mo$_2$[N(CH$_3$)$_2$]$_6$ (in toluene)	3	+2430
Mo(CO)$_6$ (in C$_6$H$_6$)	0	−1858	Mo$_2$[N(CH$_3$)$_2$]$_6$ (in hexane)	3	+2420
Mo(CO)$_6$ (in CH$_2$Cl$_2$)	0	−1855 to −1858	(Cp)$_2$Mo$_2$[μ-S$_2$C$_2$(CF$_3$)$_2$]$_2$	3	+2301
Mo(CO)$_5$CH$_3$CN	0	−1440	K$_4$[Mo(CN)$_8$]	4	−1309
Mo(CO)$_5$Cl$^-$	0	−1513	[(MoO$_3$)$_2$(EDTA)$_2$]$^{4-}$	4	+63
Mo(CO)$_5$Br$^-$	0	−1540	[MoH$_2$(Cp)$_2$]	4	−2507
Mo(CO)$_5$I$^-$	0	−1660	(Cp*)$_2$Mo$_2$(μ-S)$_2$(μ-S$_2$)	4	+440
Mo(CO)$_4$(CH$_3$CN)$_2$	0	−1307	(Cp*)$_2$Mo$_2$(μ-S)$_2$(μ-Se$_2$)	4	+770
Mo(CO)$_3$(CH$_3$CN)$_3$	0	−1114	[MoOCl$_2$P(CH$_3$)$_3$]	4	+1890
Mo$_2$(CO)$_6$(Cp)$_2$	1	−1856	[MoOCl$_2$(PPh$_2$CH$_3$)$_3$]	4	+2180
Mo$_2$(CO)$_6$(Cp*)$_2$	1	−1701	[Mo$_2$O$_4$(EDTA)·H$_2$O]$^{2-}$	5	+609
Mo$_2$(CO)$_4$(Cp)$_2$	1	+182	[Mo$_2$O$_4$(EDTA)· 2H$_2$O]$^{2-}$	5	+982
Mo$_2$(CO)$_4$(Cp*)$_2$	1	+133	(Cp*)$_2$Mo$_2$O$_2$(μ-S)$_2$	5	−93
Mo$_2$(CH$_3$CO$_2$)$_4$	2	+3702	(Cp*)$_2$Mo$_2$S$_2$(μ-S)$_2$	5	+478
Mo$_2$(CH$_3$CH$_2$CH$_2$CO$_2$)$_4$	2	+3682	(Cp*)$_2$Mo$_2$O$_2$(μ-Se)$_2$	5	+131
Mo$_2$(CH$_3$CH$_2$CH$_2$CH$_2$CO$_2$)$_4$	2	+3661	Na$_2$MoO$_4$ (pH 7)	6	−11.71
Mo(η^6-C$_6$H$_6$)$_2$	2	−1362	Na$_2$MoO$_4$ (pH 9)	6	−10.33
Mo(η^6-C$_6$H$_6$CH$_3$)$_2$	2	−1270	MoO$_2$(NCS)$_4^{2-}$	6	−155
Mo(η^5-C$_7$H$_9$)(η^7-C$_7$H$_7$)	2	−469	MoNCl$_3$	6	+952
Mo(η^6-C$_7$H$_8$)$_2$	2	+358	Mo$_2$O$_5^{2-}$	6	+120
Mo(Cp)(CO)$_3$Cl	2	−836	Mo + D-glutaric acid	6	+45 to +115
Mo(Cp)(CO)$_3$Br	2	−956	Mo + threo-alditol	6	+22
Mo(Cp)(CO)$_3$I	2	−1248	Mo + erythro-alditol	6	+30 to +34

bound to rhodium and ^{1}H, ^{103}Rh HMBC must be used instead of HMQC, the delay required for polarization transfer [generally $J(^1$H,^{103}Rh$\leq$2 Hz)] is so long that an extensive loss of magnetization occurs. The magnitude of $^1J(^{31}$P, ^{103}Rh) is generally not less than 80 Hz and therefore ^{31}P is an excellent nucleus for indirect detection of rhodium NMR resonances although the enhancement by differences in gyromagnetic ratios is not as favourable as in the case of proton. **Figure 1** shows the ^{31}P, ^{103}Rh HMQC contour map obtained from a mixture of [Rh(H)(PPh$_3$)$_4$] (0.02 M), CySH (0.2 M) and Ph$_3$P (0.2 M) in 10% pyridine–toluene at −25°C. If the rhodium-103 nucleus is in an unsymmetric environment it can also be detected easily by direct observation but in a symmetric environment this can be rather difficult.

There exist many papers on ^{103}Rh NMR. For example, small rhodium particles have been characterized by the ^{103}Rh NMR line broadening and

Table 6 Some characteristic ^{99}Tc NMR chemical shifts from $[TcO_4]^-$ (δ = 0 ppm)

Compound	δ $(^{99}Tc)/(ppm)$
Tc(CO)$_3$ClPPh$_3$(trans)	−1481
Tc(CO)$_3$BrPPh$_3$(trans)	−1555
Tc(CO)$_3$BrPPh$_3$(cis)	−1460
Tc(CO)$_3$(CH$_3$CN)$_3$	−2853
Tc(CO)$_3$(CH$_3$CN)(PPh$_3$)$_2$	−3213
Tc(DMPE)$_3^{+a}$	−13

[a] 1,2-Bis(dimethylphosphino)ethane.

chemical shift changes which have been interpreted in terms of both the electron-orbital (chemical) and electron-spin (Knight) shifts. Further, alkylrhodoximes, [Rh(Hdmg)$_2$RL], where Hdmg is the monoanion of dimethylglyoxime, L is H$_2$O, pyridine or PPh$_3$ and R is alkyl or an halomethyl, have been extensively studied by ^{103}Rh NMR spectroscopy. The ^{103}Rh shielding decreases in the order R = Et > Me

Table 7 Some characteristic ^{99}Ru NMR chemical shifts from [Ru(CN)$_6$]$^{4-}$ (δ = 0 ppm)

Compound	δ (^{99}Ru)(ppm)
Ru(η^5-C$_5$H$_5$)	−1270
Ru$_3$(CO)$_{12}$	−1208
[Ru(CO)$_3$Cl$_2$]$_2$	+1204
[Ru(CO)$_3$Cl$_3$]$^-$	+1090
[Ru(CO)$_2$Cl$_4$]$^{2-}$	+2523
RuO$_4$	~+2000
[Ru(NH$_3$)$_6$]$^{2+}$	~+7750
[Ru(H$_2$O)$_6$]$^{2+}$	+16 050

Table 8 Some characteristic ^{103}Rh NMR chemical shifts to high frequency from Ξ (^{103}Rh) = 3.16 MHz

Compound	δ (^{103}Rh)/(ppm)
[(η^5-C$_5$H$_5$)Rh(η^4-C$_4$H$_4$)]	−2057
[(η^5-C$_5$H$_5$)Rh(η^4-C$_8$H$_8$)]	−348
[(η^5-C$_5$H$_5$)Rh(CO)$_2$]	−1322
[(η^5-C$_5$Me$_5$)RhMoO$_4$]$_4$	+4079
[Rh$_6$(CO)$_{16}$]	−426
[Rh$_6$(CO)$_{15}$C]$^{2-}$	−313
[Rh(CO)$_2$Cl$_2$]$^-$	+84
[Rh(acac)(CO)$_2$]	+309
[Rh(cod)Cl(PPh)$_2$]	+409
[Rh(cod)Cl]$_2$	+1112
[Rh(cod)(acac)]	+1306
[Rh(acac)(p-quinone)]	+2065
[RhCl$_6$]$^{3-}$	+8075
[Rh(H$_2$O)$_6$]$^{3-}$	+9931

> Bu > Pr^i > Bu^i > *neo*-pent > Bu^t. In the case of [(P$_2$)Rh(hfa-cac)], where P = bidentate chelating phosphane and hfacac = hexafluoroacetylacetonate, correlations between solid state structures, ^{103}Rh NMR chemical shifts and catalytic activities (CO$_2$ hydrogenation) have been investigated. ^{103}Rh NMR chemical shift correlations have also been clarified for rhodium(III) complexes with cyanide and sulfur-donor ligands. A very interesting report describes a correlation between the rhodium-103 shielding in rhodium enamide complexes and the stereoselectivity of dihydrogen (H$_2$) addition to diastereomeric olefin complexes.

Palladium-105

Palladium belongs to the same group of elements as nickel and platinum. The only report dealing with

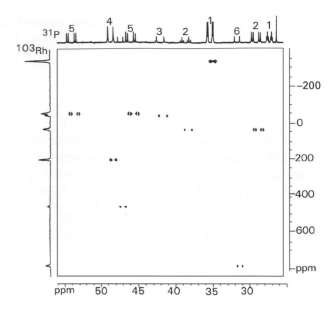

Figure 1 ^{31}P, ^{103}Rh HMQC spectrum obtained from a mixture of [Rh(H)(PPh$_3$)$_4$] (0.02 M), CySH (0.2 M) and Ph$_3$P (0.2 M) in 10% pyridine–toluene at −25°C. Reproduced with permission of John Wiley and Sons from Carlton L (1997) Rhodium-103 NMR of carboxylate and thiolate complexes by indirect detection using phosphorus. *Magnetic Resonance in Chemistry* **35**: 153–158. Copyright 1997, John Wiley & Sons.

^{105}Pd NMR is on K$_2$[PdCl$_6$] for which the $LW_{1/2}$ is 25 kHz.

Silver-107 and -109

From the NMR spectroscopic point of view, silver suffers from poor sensitivity owing to the low gyromagnetic ratio of both NMR-active isotopes. Further, long relaxation times make it necessary to incorporate long delays within the pulse sequence to avoid saturation effects during acquistion. Sensitivity enhancement by NOE effects is not possible owing to the negative γ value of both NMR-active silver nuclei. In spite of these unfavourable nuclear properties, there exist several papers on ^{109}Ag NMR. Although the natural abundance of the isotope-109 is less than that of 107, the larger γ value of 109 gives a better receptivity. As in the case of rhodium-103, direct observation of silver-109 is not useful and various polarization transfer techniques such as INEPT and HMQC are recommended. An example has been published by Berners-Price and co-workers where a retro-INEPT 2D ^{31}P-{^{109}Ag} pulse sequence (published by Bodenhausen and Ruben for ^{15}N-{^{1}H}) was utilized.

Remarkable progress in solid state ^{109}Ag NMR has taken place during the last few years. The first ^{109}Ag CP-MAS NMR spectrum was published by Merwin and Sebald in 1992. Some isotropic solid state ^{109}Ag

NMR chemical shifts are included in **Table 9** as well as a collection of characteristic solution-state data of silver-109.

Cadmium-111 and -113

Among the NMR-active isotopes included in this article the most exhaustively studied is certainly ^{113}Cd. The reasons for this are quite obvious: $I = \frac{1}{2}$ and a relatively good receptivity. Cadmium has also another NMR-active isotope (^{111}Cd) characterized with $I = \frac{1}{2}$ but with a somewhat smaller receptivity than ^{113}Cd. This is why ^{113}Cd is preferred over ^{111}Cd. According to a review by Granger, the number of publications dealing with Cd NMR from 1980 to 1987 is ~160. Since then the number of papers on Cd NMR has continuously increased. Therefore, including all the data available here is not possible. Nevertheless, owing to the significance of ^{113}Cd NMR in biological, inorganic, organic and organometallic chemistry extra attention is given to this NMR-active nucleus. At first sight it looks strange that ^{113}Cd NMR could possess any biological interest because cadmium is known to be a very toxic element for living organisms. However, the Cd^{2+} cation resembles closely the divalent cations of calcium and zinc which are involved in very many biological complexes. From an NMR spectroscopic point of view these two cations are very impractical. Replacing calcium and zinc by cadmium in these complexes is often easy and opens new versatile NMR spectroscopic possibilities in this area of research. ^{113}Cd NMR studies in biochemistry and bio-inorganic chemistry have been reported by Sadler and Viles on the binding sites of Cd^{2+} and Zn^{2+} in serum albumin and by Chung and co-workers on the binding of Cd^{2+} in soil fulvic acid.

The direct observation of ^{113}Cd is generally easy, although in some cases long relaxation times may cause prolonged acquisition times. This can be avoided by using a paramagnetic additive such as Gd^{3+} in the sample. If there exists a coupling between proton and cadmium, polarization transfer techniques such as HMQC are useful, as in the case of rhodium and silver. Owing to the relatively high natural abundance, homonuclear COSY can also be useful alternative in the case of ^{113}Cd. **Figure 2** shows the DQF (double quantum filtered) ^{113}Cd–^{113}Cd COSY spectrum of cadmium metallothionein 2 (MT-2).

Although the observation of ^{113}Cd NMR spectral lines is easy and obtainable by a plethora of various measuring techniques, there is an obvious danger of oversimplification of the chemical shift data. This is

Table 9 Some characteristic ^{109}Ag NMR chemical shifts from aqueous $AgNO_3$ ($\delta = 0$ ppm)

Compound	Solvent/temperature (K)	δ (^{109}Ag)/ (ppm)
$Ag(pyridine)_2^+Ac^-$	C_5H_5N	+350
$Ag(pyridine)_2^+NO_3^-$	C_5H_5N	+259
$Ag(C_5H_5)^+NO_3^-$	$H_2O/300$	+128
$Ag(CH_3CN)^+NO_3^-$	$H_2O/300$	+108
$Ag(DMSO)^+NO_3^-$	$H_2O/300$	+94
$Ag(C_6H_6)^+ClO_4^-$	–	+350
$Ag(NH_3)_2^+NO_3^-$	Conc. aq. NH_3/ 300	+593
$[(C_2H_5O)_3P]_4Ag^+SCN^-$	CH_2Cl_2	+1371
$[Ag(dppe)_2]^+NO_3^-$	$CDCl_3/300$	+1338
$Ag[CH_3CHC(OH)CO_2]$	Solid	+345.9,+320.2
		+219.7,+210.7
$Ag(CH_3CO_2)$	Solid	+401.2, +382.7
$Ag(p\text{-}CH_3C_6H_4SO_3)$	Solid	+44.1
$Ag(acac)$	Solid	+471.6
$AgN(SO_2Me)_2$	Solid	+210.3
$AgN(SO_2Me)_2 \cdot \frac{1}{4}H_2O$	Solid	+32.5

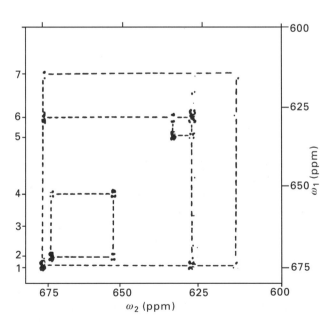

Figure 2 DQF ^{113}Cd–^{113}Cd COSY spectrum of MT-2. Reproduced with permission of the American Chemical Society from Frey MH, Wagner G, Vašák M, Sørensen OW, Neuhaus D, Wörgötter E, Kägi JHR, Ernst RR and Wüthrich K (1985) Polypeptide-metal cluster connectivities in metallothionein 2 by novel ^{1}H–^{113}Cd heteronuclear two-dimensional NMR experiments. *Journal of the American Chemical Society* **107**: 6847–6851. Copyright 1985, American Chemical Society.

because chemical exchange often occurs between different complexation or binding sites (including ligand and solvent) for cadmium and the observed Cd NMR line is a time average of the different chemical shifts from these different sites. Therefore it is recommended that ^{113}Cd NMR studies are carried out at low temperature to 'freeze out' these different species. **Figure 3** shows the ^{113}Cd NMR spectra of five different Cd(II) complexes of multiple piperazine–pyridine ligands measured in ethanol at −50°C.

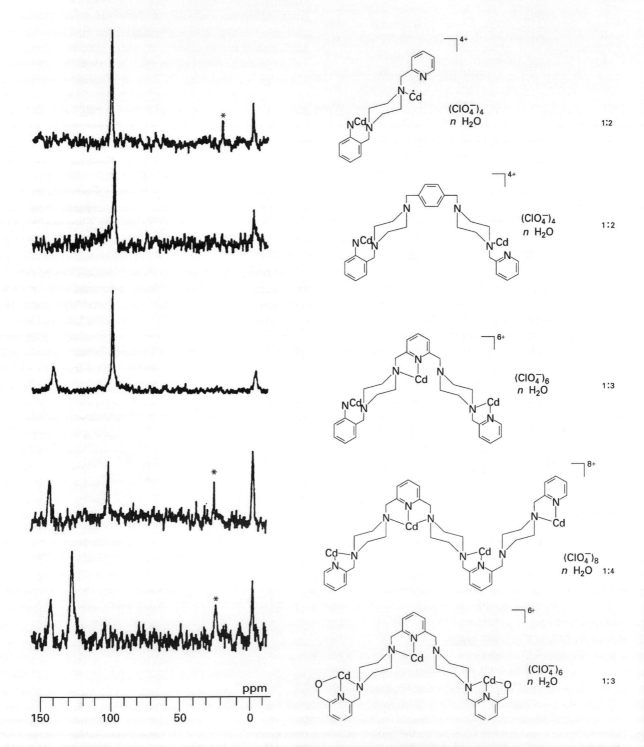

Figure 3 ^{113}Cd NMR spectra of the Cd(II) complexes of five different piperazine–pyridine ligands measured in ethanol at −50°C. The signal of the uncomplexed Cd^{2+} cation is marked by an asterisk. Reproduced with permission of VCH Verlagsgesellschaft from Ratilainen J, Airola K, Kolemainen E and Rissanen K (1997) Regioselective complexation of new multiple piperazine/pyridine ligands: differentiation by ^{113}Cd-NMR spectroscopy. *Chemische Berichte/Recueil* **130**: 1353–1359.

Cd^{2+} cations can form salts and complexes with numerous anions and molecules that bear free electron pairs. For example, in supercooled water at −80°C five different CdI$_n^{2-n}$ ($n = 0$–4) species are observed with the chemical shifts −86, +20, +43, +122 and +101 ppm, respectively. Generally, ^{113}Cd NMR chemical shifts are referenced to the signal of 0.1 M aqueous Cd(ClO$_4$)$_2$. For low-temperature measurements this reference is not suitable and 0.1 M Cd(ClO$_4$)$_2$ in ethanol can be used. In **Table 10** are collected some characteristic ^{113}Cd NMR chemical shifts. In addition to the monometal compounds and complexes, cadmium can form clusters with several cadmium and/or zinc atoms which also are characterized by ^{113}Cd NMR.

As noted before, the isotropic ^{113}Cd NMR chemical shift determined in solution is influenced by exchange processes, temperature, concentration and solvent effects which may limit the interpretation of the structural data. In the solid state, complete information concerning the shielding tensor can be obtained without the difficulties mentioned above. From single-crystal NMR experiments the orientation of the principal elements of the shielding tensor with respect of the coordinate system based on the molecule are obtained. Further, this knowledge provides information on the coordination of cadmium, for example in the active site of metalloproteins such as parvalbumin and concanavalin A. There exist many solid-state NMR studies on cadmium complexes with ligands that contain nitrogen, oxygen and sulfur donor groups. A novel application of solid state ^{113}Cd NMR is an investigation on host–guest systems. In these studies ^{113}Cd NMR has been used to differentiate the possible coordination environments.

Summary

In an article with a limited length and depth of coverage like this, dealing with 10 elements and 14 NMR-active nuclei, it is impossible to discuss all the important NMR parameters. Therefore, most attention has been paid to NMR chemical shifts while spin–spin coupling constants and relaxation times are only mentioned in the contexts where they should be taken into account to obtain useful NMR results. However, as well as direct NMR detection, one should remember, for example, the usefulness of the (inverse) two-dimensional pulse sequences and the MAS technique to utilize homo- and heteronuclear spin–spin coupling constants for polarization transfer between nuclei to enhance the sensitivity of the measurement. These approaches have increased the number of heteronuclear NMR applications dramatically and should play an important role in future studies.

Table 10 Some characteristic ^{113}Cd NMR chemical shifts from 0.1 M aqueous Cd(ClO$_4$)$_2$ ($\delta = 0$ ppm)

Compound	δ (^{113}Cd) (ppm)
[Cd(CH$_3$S)$_4$]$^{2-}$	+663
[Cd(C$_2$H$_5$S)$_4$]$^{2-}$	+648
[Cd(n–C$_3$H$_7$S)$_4$]$^{2-}$	+647
[Cd(n–C$_4$H$_9$S)$_4$]$^{2-}$	+644
[Cd(C$_6$H$_5$S)$_4$]$^{2-}$	+590
[Cd(C$_6$H$_5$CH$_2$S)$_4$]$^{2-}$	+646
[Cd(SCH$_2$CH$_2$S)$_2$]$^{2-}$	+829
[Cd(C$_6$H$_5$Se)$_4$]$^{2-}$	+541
[Cd(H$_2$O)$_6$]$^{2+}$	0.0
[Cd(NH$_3$S)$_6$]$^{2+}$	+287.4
[Cd(pyridine)]$^{2+}$	+10
[Cd(pyridine)$_2$]$^{2+}$	+40
[Cd(pyridine)$_3$]$^{2+}$	+70
[Cd(pyridine)$_4$]$^{2+}$	+95
Cd(CF$_3$SO$_3$)$_2$[P(n-C$_4$H$_9$)$_3$]	+167
Cd(CF$_3$SO$_3$)$_2$[P(n-C$_4$H$_9$)$_3$]$_2$	+379
Cd(CF$_3$SO$_3$)$_2$[P(n-C$_4$H$_9$)$_3$]$_3$	+505
Cd(CF$_3$CO$_2$)$_2$[P(n-C$_4$H$_9$)$_3$]	+194
Cd(CF$_3$CO$_2$)$_2$[P(n-C$_4$H$_9$)$_3$]$_2$	+431
Cd(CF$_3$CO$_2$)$_2$[P(n-C$_4$H$_9$)$_3$]$_3$	+485
Cd(ClO$_4$)$_2$[P(n-C$_4$H$_9$)$_3$]	+163
Cd(ClO$_4$)$_2$[P(n-C$_4$H$_9$)$_3$]$_2$	+354
Cd(ClO$_4$)$_2$[P(n-C$_4$H$_9$)$_3$]$_3$	+506
CdF$_2$(solid)	−233
CdCl$_2$(solid)	+211
CdBr$_2$(solid)	+41
CdI$_2$(solid)	−672
[Cd(S-2,4,6-Pr$^i_3$C$_2$H$_2$)$_2$(bpy)$_2$]	
Single crystal: σ_{11}	+814
σ_{22}	+630
σ_{33}	+32
[Cd(S-2,4,6-Pr$^i_3$C$_2$H$_2$)$_2$(phen)]	
Solid	+371.5
Solution	+450

List of symbols

I = nuclear spin quantum number; J = coupling constant; R = receptivity; T_1 = relaxation time; γ = gyromagnetic ratio; δ = chemical shift.

See also: ¹³C NMR, Methods; Heteronuclear NMR Applications (B, Al, Ga, In, Tl); Heteronuclear NMR Applications (La–Hg); Heteronuclear NMR Applications (Sc–Zn); Magnetic Resonance, Historical Perspective; MRI Theory; NMR Pulse Sequences; NMR of Solids; NMR Spectrometers; Parameters in NMR Spectroscopy, Theory of; Solid State NMR, Methods.

Further reading

Asaro F, Costa G, Dreos R, Pellitzer G and von Philipsborn W (1996) Steric effects in organometallic compounds. A ¹⁰³Rh NMR study of alkylrhodoximes. *Journal of Organometallic Chemistry* 513: 193–200.

Berners-Price SJ, Sadler PJ and Brevard C (1990) Tetrahedral, chelated, silver(I) diphosphine complexes. Rapid measurements of chemical shifts and couplings by two-dimensional ³¹P-{¹⁰⁹Ag} NMR spectroscopy. *Magnetic Resonance in Chemistry* 28: 145–148.

Bühl M, Hopp G, von Philipsborn W, Beck S, Prosenc M-H, Rief U and Brintzinger H-H (1996) Zirconium-91 chemical shifts and line widths as indicators of coordination geometry distortions in zirconocene complexes. *Organometallics* 15: 778–785.

Burgstaller A, Ebert H and Voitländer J (1993) NMR properties of hcp Ru metal-first detection of the ⁹⁹Ru and ¹⁰¹Ru NMR in a paramagnetic solid. *Hyperfine Interactions* 80: 1015–1018.

Carlton L (1997) Rhodium-103 NMR of carboxylate and thiolate complexes by indirect detection using phosphorus. *Magnetic Resonance in Chemistry* 35: 153–158.

Chung KH, Rhee SW, Shin HS and Moon CH (1996) Probe of cadmium(II) binding on soil fulvic acid investigated by ¹¹³Cd NMR spectroscopy. *Canadian Journal of Chemistry* 74: 1360–1365.

Davies JA and Dutremez S (1992) Solid state NMR studies of *d*-block and *p*-block metal nuclei: applications to organometallic and coordination chemistry. *Coordination Chemistry Reviews* 114: 201–247.

Malito J (1997) Molybdenum-95 NMR spectroscopy. *Annual Reports on NMR Spectroscopy* 33: 151–206.

McDowell A, Conradi MS and Haase J (1996) First-satellite spectroscopy, a new method for quadrupolar spins. *Journal of Magnetic Resonance, Series A* 119: 211–218.

Merwin LH and Sebald A (1990) The first ⁸⁹Y CP-MAS spectra. *Journal of Magnetic Resonance* 88: 167–171.

Pregosin PS (ed) (1991) *Transition Metal Nuclear Magnetic Resonance, Studies in Inorganic Chemistry*, Vol 13. New York: Elsevier.

Ratilainen J, Airola K, Kolehmainen E and Rissanen K (1997) Regioselective complexation of new multiple piperazine/pyridine ligands: differentiation by ¹¹³Cd NMR spectroscopy. *Chemische Berichte/Recueil* 130: 1353–1359.

Sadler PJ and Viles JH (1996) ¹H and ¹¹³Cd NMR investigations of Cd²⁺ and Zn²⁺ binding sites on serum albumin: competition with Ca²⁺, Ni²⁺, Cu²⁺ and Zn²⁺. *Inorganic Chemistry* 35: 4490–4496.

High Energy Ion Beam Analysis

Geoff W Grime, University of Oxford, UK

HIGH ENERGY SPECTROSCOPY
Methods & Instrumentation

The interactions between medium energy (1–5 MeV) atomic ions and the atoms of a sample produce a wide variety of reaction products; electromagnetic radiation from the electron shells or gamma rays from the nucleus, scattered particles from the primary beam which may be modified in direction and energy, or particles of a different species emitted from the nucleus. Any of these may be analysed spectroscopically using suitable detectors to provide information about the elemental and in some cases isotopic composition of the sample.

Analysis using MeV ions

Interactions of MeV ions in matter

MeV ions interact with matter through collisions with the electrons and nuclei of the atoms of the material. Ion–electron collisions give rise to reactions associated with the electron shells of atoms (photon emission, Auger electrons, conduction electrons or holes, secondary electron emission) and nuclear collisions give rise to reactions involving the nucleus (nuclear scattering, atomic displacement, sputtering, nuclear reactions resulting in gamma ray or particle emission). The products of these reactions may be detected and used for analytical purposes.

In nuclear scattering theory, the relative probability of a reaction taking place is expressed as a cross-section, σ. Each atom or nucleus is represented by a disk of area σ, and if a beam particle passes through this disk a reaction takes place resulting in the detection of a particle or photon in a detector with 1 steradian (sr) solid angle. Thus high yield reactions have a high cross-section. Values of cross-section are of the order of 10^{-24} cm², and this quantity is called 1 barn. Reaction cross-sections are usually quoted as barn sr^{-1}. The yield of a particular reaction may be calculated using the expression

$$Y = \Omega \sigma N n_{T} \qquad [1]$$

where Y is the number of reaction products (photons, particles) detected in a detector of solid angle Ω sr when a sample containing n_{T} target atoms per square centimetre is bombarded with N projectile ions. Here, the cross-section, σ, is assumed to be in square centimetres.

Cross-sections for ion–atom reactions vary over a wide range depending on the specific reaction, the ion and atom (or nuclear) species, the ion energy and the angle of detection. Clearly for analytical purposes, high cross-section reactions are preferable, but this also depends on the relative intensity of any competing reactions which produce a background signal. In many cases, ion beam analysis (IBA) reactions are free from background so that even low cross-section reactions may be used for analysis.

PIXE (proton induced X-ray emission)

In PIXE, the basic interaction between the beam and the sample which gives rise to the emission of characteristic X-rays is analogous to the processes involved in electron induced X-ray emission (EDX, or electron probe microanalysis, EPMA). In each case, the charged particle in the beam creates inner electron shell vacancies in the sample atoms, leaving them in an excited state which decays by electron transitions from higher shells, each of which causes the emission of an X-ray quantum whose energy is determined by the atomic number of the atom. This can be detected either using a wavelength dispersive crystal spectrometer (WDX) or an energy dispersive lithium drifted silicon detector (Si-Li) (EDX). The use of WDX detectors is not common in PIXE.

Because of the use of Si-Li detectors, EDX and PIXE share many characteristics, especially the advantage of rapid multielemental analysis, but the use of MeV protons as the incident particle gives one crucial advantage over the technologically simpler use of keV electrons in an electron microscope, and that is in reducing the background of broad spectrum non-characteristic X-rays. In EDX, the major source of background is created by the scattering and deceleration of the electrons of the primary beam as they collide with the electrons in the sample. Each collision gives rise to a large change in momentum of the primary electron and causes the emission of broad spectrum 'Bremsstrahlung' radiation which has a maximum energy equal to the beam energy. Bremsstrahlung is the major source of background in EDX and is a fundamental limitation to the minimum detectable limit (MDL).

In contrast, MeV ions lose negligible momentum in each collision with an electron, which means that the primary Bremsstrahlung is effectively absent, permitting the detection of weak characteristic X-ray lines and giving a reduction in MDL of 100 to 1000 times over EDX. PIXE is not totally free from background; the electrons created during the initial ionization generate a background at low energies (secondary electron Bremsstrahlung), which limits the MDL for elements in the range Na to Fe (depending on the primary beam energy), but a substantial improvement (two to three orders of magnitude) over EDX is achieved. **Figure 1** presents a comparison of EDX and PIXE on the same sample of brain tissue showing the improvement in sensitivity resulting from the lack of Bremsstrahlung.

Like EDX, the lightest element that may be detected is determined by the response of the detector to low energy X-rays, and with conventional beryllium window detectors, the lightest element that can be routinely detected is sodium. PIXE has a high cross-section (typically 10–1000 barn sr^{-1}) which makes it very well suited for use with microprobe systems.

Each element has a characteristic spectrum of X-ray emission lines with energies depending on the atomic number. These lines are categorized into groups depending on which electron shell was ionized in the collision. For example, the K series consists of lines denoted as $K_{\alpha 1}, K_{\alpha 2}, \ldots K_{\beta 1}, K_{\beta 2} \ldots$ resulting from transitions between the L subshells and the K subshells. The K_{α} and the K_{β} sublines have energies which are close in value, so that with a detector of moderate energy resolution such as Si-Li, only two peaks are observed. The L series, resulting from transitions between the M subshells and the L shells is more complex with three main groups (L_{α}, L_{β}, L_{γ}) and several less intense peaks. In heavy elements (above Ba), the M series (N shell to M shell)

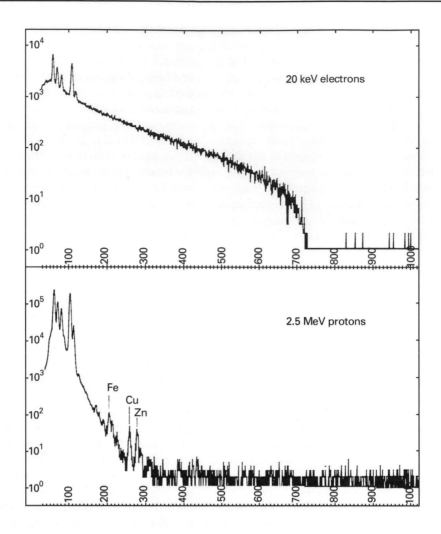

Figure 1 Energy dispersed X-ray spectra from the same sample excited by 20 keV electrons (top) and 2.5 MeV protons (bottom). The enhancement of the detection limit for the trace elements caused by the absence of primary Bremsstrahlung in the PIXE spectrum can be seen clearly. Reproduced with permission of Wiley from Johansson SAE, Campbell JL and Malmqvist KG (1995) *Particle-Induced X-ray Emission Spectrometry*. New York: Wiley.

may also be observed, but the M lines are not often used for quantification. **Table 1** shows the energy of the most intense K, L and M lines from selected elements, with those detectable using a typical Si-Li detector. The table shows that by using both the K series and the L series all elements heavier than Na may be detected using PIXE. It also highlights the possibility of interference between the series of different elements (e.g. As K_α has almost the same energy as Pb L_α). When all the sublines of each series of each element in the sample are taken into consideration, the probability of overlaps is very high, and spectrum processing software must be capable of handling this.

Figure 2 shows the PIXE spectrum from a single ambient aerosol particle, showing the major and trace elements detected.

Absorbers One characteristic of PIXE analysis is the wide dynamic range of the significant peaks. Intense peaks from major elements in the sample may contain millions of counts, while in the low background region of the spectrum, significant peaks containing less than 10 counts may be observed. For this reason, PIXE spectra are normally presented with a logarithmic vertical scale. Another consequence of this is that spurious responses from intense peaks in the detector can have a drastic effect in swamping the small peaks from trace elements. For this reason, the careful use of X-ray absorbers (filters) is crucial to obtaining the best performance with PIXE. For example, when analysing bone, an absorber (typically of Al or a polymer foil) may be fitted to reduce the intensity of the Ca peak to zero or a few parts per thousand to minimize the

Table 1 Energies (in keV) of the K_α, L_α, and M_α lines of selected elements. Energies in italic face may be detected using a typical Si-Li detector.

Element	K_α energy	L_α energy	M_α energy
Li	0.052		
O	0.523		
Na	1.041		
Si	1.740		
Ca	3.691	0.341	
Fe	6.403	0.704	
As	10.534	1.282	
Zr	15.774	2.042	0.350
Ag	22.162	2.984	0.568
Ba	32.191	4.467	0.972
W	59.310	8.396	1.774
Pb	74.957	10.549	2.342
Pu	103.653	14.297	3.350

background and spurious peaks due to the intense flux of Ca X-rays. This is crucial in improving the detection limits for the metals. Fitting an absorber means that X-ray lines of lower energy than the line being removed are lost completely, and it is becoming more common now to use two detectors simultaneously, one unfiltered for major elements and one with a suitable filter for the trace elements.

The art of PIXE analysis lies in the selection of the optimum filter.

Quantification of PIXE spectra Figure 2 shows that a PIXE spectrum consists of peaks which are approximately Gaussian sitting on top of a continuous background. The background-subtracted area of each peak is proportional to the concentration of the associated element, and a number of software packages are now available to perform the processing required to extract the areas and convert these to true concentrations.

For a sample consisting of a thin film (so that proton energy loss and X-ray absorption can be neglected), the yield of characteristic X-ray photons of energy E_x from element Z induced by particles of energy E is given by

$$N = C_z Q \Omega \varepsilon(E_x) Y(Z, E) \qquad [2]$$

where N is the total number of photons detected using a detector of solid angle Ω, Q is the total beam charge, C_z is the concentration of the element and $\varepsilon(E_x)$ is the dependence of detector efficiency on X-ray energy. $Y(Z, E)$ is the yield of the interaction expressed as counts per unit of concentration per unit of charge per unit of solid angle. This, in turn, is derived from the ionization cross-section (the probability of creating a vacancy) and the fluorescence yield (the number of photons emitted in each X-ray line for each vacancy).

In more realistic thick samples, the calculation must also take account of the energy loss of the particle as it penetrates the sample and also the absorption of the X-rays as they emerge from the sample. To do this, a knowledge of the bulk composition of the sample is required (together with any variation with depth) and both the stopping power of the ion in the matrix and the X-ray attenuation coefficient must be known. The calculation then involves a numerical integration of the total X-ray yield from each sublayer of the sample.

All of the physical parameters involved in this calculation (ionization cross-section, fluorescence yield, stopping power, X-ray attenuation and detector efficiency) have been measured and parameterized with sufficient accuracy that quantitative PIXE analysis may be carried out without routine reference to standards once the fixed system parameters have been determined. A small number of computer

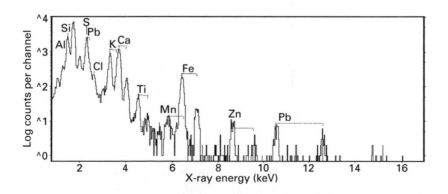

Figure 2 Proton-induced X-ray spectrum from a single ambient aerosol particle of 2 μm diameter. The spectrum was collected in 5 min using a beam of 3 MeV protons focused to a diameter of 1 μm at a current of 100 pA.

programs have been developed to carry out PIXE spectrum processing, and special mention must be made of GUPIX, developed at the University of Guelph, Ontario, which embodies the most detailed physical model of the PIXE process and also offers the capability to iterate the sample matrix composition (including user-defined 'invisible' elements such as oxygen) until the matrix composition matches the values obtained from the peak areas.

Accuracy and detection limit of PIXE The major sources of error in a PIXE measurement and the above calculation are as follows:

(1) *Inaccuracies in the basic assumptions about the target.* The calculation assumes that the sample is a flat homogenous slab of the given composition. Any departure from this model (surface roughness, compositional variation with depth, etc.) will introduce an error in the calculated relative intensity of the X-ray lines.

(2) *Inaccuracies in the physics database.* These may contribute errors to the relative X-ray intensities, but it is believed that the database is of sufficient accuracy that these are relatively insignificant compared with other sources of error.

(3) *Numerical and statistical errors in processing the spectrum.* The quality of the final fit depends on the fitting procedure and the amount of detail in the mathematical model. The precision of each peak determination also depends on the quality of the statistics in the experimental spectrum. Peaks with few counts will not be fitted so well as intense peaks.

(4) *Limitations in the model used for the detector response function and absorber transmission.* The response of the detector varies with X-ray energy (and also depends on any absorbers used). This can be simulated using a simple physical model of the detector and parameterized expressions for X-ray absorption coefficient. However, for the highest accuracy it is necessary to analyse standard samples and derive a correction factor for each X-ray line of interest.

(5) *Inaccuracies in determining the charge and detector solid angle.* These affect the absolute accuracy but not the relative concentrations. For this reason it is more accurate to use PIXE to determine element ratios rather than absolute values. The charge is difficult to measure accurately since the beam currents involved are generally quite small and also large numbers of secondary electrons are created in the beam impact on the sample, which, if they are not returned to the sample, will give an error in the measurement.

Beam current normalization is often carried out by using PIXE simultaneously with another technique (e.g. Rutherford backscattering (RBS), which also offers the possibility of determining the sample matrix independently).

When all these sources of error have been addressed, it should be possible to achieve a relative accuracy of the order of 5–10% and an absolute accuracy of 10–20%.

The detection limit depends on the height of the peak relative to the background (which now can mean both the continuum background and also any overlapping X-ray lines or detector artefacts). The detection limit is conventionally defined as 3 times the square root of the area of the background within one standard deviation of the peak centre (converted to concentration using the same factor as for the peak). This depends on all the parameters of the experiment, and **Figure 3** shows the theoretical detection limits for trace elements in a carbon matrix. This indicates that values of the order of 0.1–1 ppm may be expected under optimum conditions.

RBS (Rutherford backscattering)

Particle backscattering spectrometry, also known as Rutherford backscattering or RBS involves measuring the recoil energy of backscattered particles from the primary beam to obtain analytical information. Ions recoiling from direct elastic collisions with the nuclei of atoms in the sample lose energy according to the mass of the target atom; recoils from heavy nuclei have a higher energy than recoils from light nuclei. If the target atom is not at the surface, energy is also lost during the passage through the sample to and from the reaction site (**Figure 4**), so measuring the energy distribution of recoiling atoms gives

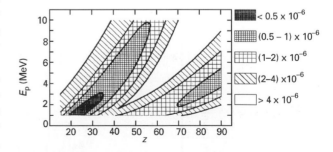

Figure 3 Minimum detectable limit of trace elements in a carbon matrix as a function of atomic number and proton energy. The calculation assumes a beam charge of 1 μC and curves are presented for K and L lines. Reproduced with permission of Wiley from Johansson SAE, Campbell JL and Malmqvist KG (1995) *Particle-Induced X-ray Emission Spectrometry*. New York: Wiley.

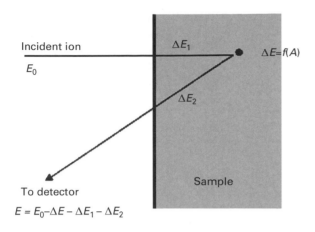

Figure 4 Typical geometry for Rutherford backscattering. The incident particle loses energy ΔE_1 going through the sample to the scattering site, loses energy ΔE (determined by the Rutherford formula) in the collision and loses energy ΔE_2 on the recoil path to the detector.

information on the major element composition and depth distribution in the sample. This technique is loosely complementary to PIXE in that the greatest mass resolution occurs for the light elements which are invisible using PIXE. The cross-sections are lower than for PIXE (typically 0.1–10 barn sr^{-1}) therefore the minimum detectable limit is not so low as for PIXE (typically 0.1%), but RBS can be used for determining major element composition and depth profiles. In certain favourable cases (e.g. heavy inclusions embedded in a light matrix) RBS can be used to carry out nondestructive three-dimensional mapping. Used simultaneously with PIXE, RBS provides, in principle, analysis of all elements above He in the periodic table.

Interpreting RBS spectra The classical treatment of the scattering of two charged particles yields the following expressions for the recoil energy and for the cross-section:

$$E = E_0 \frac{m_1^2}{(m_1 + m_2)^2}\left[\cos\theta + \left(\frac{m_2^2}{m_1^2} - \sin^2\theta\right)^{\frac{1}{2}}\right]^2 \quad [3]$$

$$\frac{\mathrm{d}\sigma}{\mathrm{d}\Omega} \approx 1.296\left(\frac{zZ}{E_0}\right)^2\left[\sin^{-4}\left(\frac{\theta}{2}\right) - 2\left(\frac{m_1}{m_2}\right) + \dots\right] \quad [4]$$

where E is the recoil energy of a particle of mass m_1, charge z and initial energy E_0 scattering through a total angle θ from a stationary nucleus of mass m_2 and charge Z. The form of these expressions is

shown in **Figure 5** for 3 MeV protons and alpha particles recoiling off nuclei up to mass 100. These curves show that the best mass resolution (change in recoil energy with mass) occurs for low masses. At high masses, the difference in recoil energy is too small to identify the mass of the scattering nucleus uniquely. The curves also show that the cross-section increases with mass. Alpha particles give both a higher cross-section and a better mass resolution, and these are normally used for RBS applications.

Because of the interplay of mass and depth information in RBS spectra, interpreting the data is not so straightforward as, for example, in PIXE spectroscopy where each peak is uniquely associated with a particular element. At present, RBS spectrum interpretation relies on simulating the spectrum from a proposed model of the sample using a suitable computer program, for example the commonly used program, RUMP. The parameters of the model can then be refined, either manually or using fitting routines, until a good match is obtained to the experimental data. Thus, some insight into the physics of the situation is required in order to decide on a suitable starting point for the modelling procedure. Some of the points to be considered are presented in **Figures 6A–E** which show simulated RBS spectra. In **Figure 6A** recoils from a very thin SiO$_2$ film give two narrow peaks, one at higher energy from Si and one at lower energy from O. The relative areas of the peaks are proportional to the ratio of Si to O. In **Figure 6B** a thicker film of SiO$_2$ gives two broader peaks, as recoils from deeper inside the sample occur at lower energies. The energy width of the peaks increases with increasing sample thickness. In the case of an infinitely thick sample, **Figure 6C**, the responses from Si and O merge and instead of seeing peaks, the presence of the two elements is seen as steps or edges

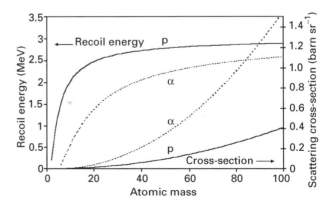

Figure 5 Recoil energy (left axis) and scattering cross-section (right axis) for 3 MeV protons (solid line) and alpha particles (dashed line) backscattering from nuclei of different masses. Scattering angle, θ, 150° (i.e. beam to detector angle 30°).

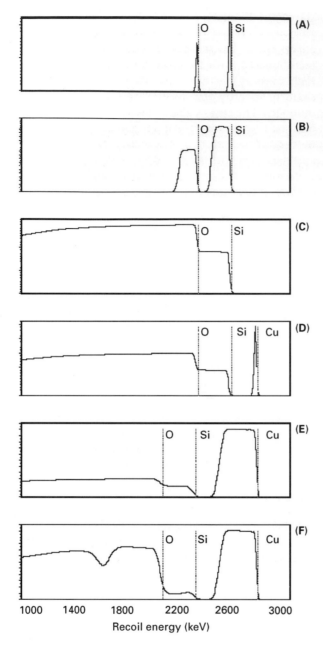

Figure 6 Simulated RBS spectra using 3 MeV protons scattered at an angle of 150° for the following samples: (A) 0.1 μm film of SiO_2, (B) 5 μm film of SiO_2, (C) Thick sample of SiO_2, (D) 0.01 μm Cu film on SiO_2, (E) 2.5 μm Cu film on SiO_2. Simulations (A)–(E) carried out using the Rutherford cross-section (Eqn [3]). (F) Simulation of 2.5 μm Cu film on SiO_2 (case (E)) using true light element cross-sections for Si and O.

corresponding to the recoil energy of that element at the surface. When a thin Cu film is added to the surface of the SiO_2, this is seen as a sharp peak at higher energy (**Figure 6D**), and if the Cu layer is made thicker (**Figure 6E**), the Cu peak becomes broader, but also the Si and O edges are shifted to lower energy because of the energy loss in the Cu film.

The use of RBS with protons on light nuclei is complicated by nuclear reactions (see below) and the cross-section may be much higher (or lower) depending on the energy and the scattering angle. This is illustrated in **Figure 6F**, which shows the same simulation as in **Figure 6E**, but now using true cross-sections rather than the theoretical Rutherford formula. It can be seen that the yield for O is significantly higher than the Rutherford case due to nuclear reactions, and this can be exploited to increase the sensitivity to O. The dip in the spectrum observed at about 1600 keV is due to a nuclear reaction in ^{16}O which reduces the cross-section over a limited energy range.

In spite of the relative complexity of interpreting the spectra, RBS with alpha particles is widely used for materials analysis for characterizing thin film structures or depth profiles of impurities or dopants. RBS with protons can be used simultaneously with PIXE to help to determine the bulk matrix composition and also the incident beam charge.

Nuclear reaction analysis (NRA)

If the incident ion penetrates the repulsive electrostatic field surrounding the nucleus of a target atom, it is possible for nuclear reactions to take place involving the weak and strong nucleonic forces. This can result in the formation of an excited unstable compound nucleus which may decay by the emission of particles or gamma rays to give a final nucleus which may be different from the original. Nuclear reactions in general have a much lower cross-section than atomic reactions such as PIXE (typically 10–100 mb sr^{-1}), but they often have resonant behaviour, so that provided the conditions are carefully chosen, NRA may give a useful yield and unique information. The cross-section and the types of reactions observed are strongly dependent on the isotope of the target atom and the beam energy and do not vary in a systematic way with atomic number, as do the yields and X-ray energies in PIXE. Nuclear reactions are often described using the notation A(a,b)B, where A and B are the initial and final states of the target nucleus, a is the projectile ion and b is the emerging reaction product (particle or gamma rays).

Nuclear reactions tend to be most important for light nuclei, as the Coulomb repulsion is less for nuclei with low Z. There are three type of nuclear reaction which are of interest for analytical purposes and these are described below. In general, because of the rapid and irregular variation of cross-section with energy, NRA techniques are difficult to quantify and in the majority of cases, quantitative data is obtained by comparison with known standards.

Resonant elastic scattering   At certain energies, the yield of RBS analysis may be strongly enhanced (or reduced) by interactions with the target nucleus. One example of this is resonant scattering of protons from ^{12}C, i.e. ^{12}C(p,p)^{12}C. At an energy of 1.76 ± 0.1 MeV, the cross-section for backscattered protons from ^{12}C increases by a factor of 60 relative to the classical Rutherford formula and this can be exploited to improve the sensitivity for carbon. This is shown in the simulations of **Figure 7**, for a 10 nm thick layer of carbon on a copper substrate.

Nuclear reactions resulting in gamma ray emission   Most nuclear reactions result in the emission of gamma rays of a well defined characteristic energy and these can be energy-analysed using a suitable detector and used for analysis. This technique is sometimes referred to as particle-induced gamma emission or PIGE. PIGE is useful for fluo-rine determination, for example, since the reaction ^{19}F(p, $\alpha\,\gamma$)^{16}O has a particularly high yield, though many other light elements up to Al can also be detected with a reasonable detection limit. **Figure 8** shows the gamma ray spectrum of a tourmaline, a mineral rich in light elements such as F, Li and B, bombarded with 3 MeV protons showing the characteristic lines from the light elements in the sample.

Nuclear reactions resulting in particle emission   Some nuclear reactions result in the emission of a particle (proton or alpha) with an energy higher than the primary beam energy so that they can be detected unambiguously in the detector used for RBS analysis. Relatively few reactions are useful for analytical purposes. One example is ^{7}Li (p,α) ^{4}He, which occurs when Li is bombarded with 2–3 MeV protons. Two alpha particles are emitted which have energies of 7–8 MeV and can be detected with no interference from the spectrum of backscattered protons.

Equipment for ion beam analysis

Accelerators

The major component of any nuclear microbeam facility is the particle accelerator. This must generate ions of the species and energy required to carry out the analysis (protons and alpha particles with energies of a few MeV), but the requirement of operating with a focusing system may also impose additional requirements on the energy spread and brightness of the beam.

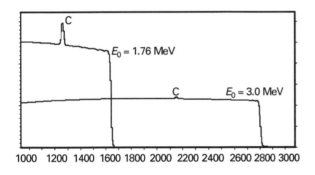

Figure 7   Simulated RBS spectra of a 10 nm layer of C on a Cu substrate using protons of 3 MeV and 1.76 MeV, which correspond to the (p,p) resonance in ^{12}C.

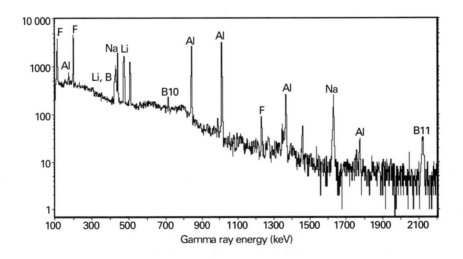

Figure 8   Gamma ray spectrum from a tourmaline bombarded with 3 MeV protons. Gamma ray lines corresponding to F, Li, B, Na and Al are marked. The unmarked lines are created by background sources within the fabric of the laboratory.

The accelerators most suitable for IBA are the electrostatic type, in which a static voltage equal to the beam energy required is generated, and the particles are accelerated in a single step (or two steps if the 'tandem' principle is used). The voltage may be generated either by an electrostatic system such as the popular 'van de Graaff' system, or by a voltage multiplier stack, in which an alternating voltage is rectified in such a way as to generate a DC voltage many times greater than the amplitude of the applied voltage. **Figure 9** shows schematically the principle of electrostatic accelerators.

Historically a nuclear accelerator formed part of a major facility and many IBA installations still compete for beamtime at a nuclear physics accelerator institute. In contrast, modern small accelerators represent a relatively modest investment in capital and support costs and increasingly now, new small accelerators are being purchased specifically for analytical applications. It is notable, for example, that the Louvre Museum in Paris now houses a modern accelerator facility (Accelerateur Grand Louvre pour Analyses Elémentaire – AGLAE) dedicated to the analysis of art objects and archaeological samples.

Detectors

The majority of detectors used in IBA are based on the reverse biased semiconductor junction. Ionizing radiation creates electron–hole pairs in the depletion region of the junction which drift under the applied electric field to the electrodes where they create a voltage pulse which is amplified. The material and dimensions of the detector depend on the radiation to be detected, and must be chosen so that the incident photon or particle has a high probability of interacting with an atom in the depletion region (**Table 2**). In order to reduce the noise due to thermal electrons, photon (X-ray and gamma) detectors must be operated with the detector crystal and preamplifier at a low temperature (usually with liquid nitrogen cooling, but newer semiconductor materials such as cadmium zinc telluride can give good performance with Peltier junction cooling devices). For particle detectors, other sources of noise dominate and these can be operated at room temperature.

Applications of nonspatially resolved IBA

PIXE and RBS are the major spectroscopic techniques used for materials analysis because of their high yield and relative ease of quantification. In many cases, however, the use of nonspatially resolved, or 'broad beam' PIXE has been displaced to some extent by inductively coupled plasma mass spectrometry (ICPMS), which offers the same advantages of rapid multielemental analysis with the additional advantage of very low detection limits and greater accessibility. However, PIXE still has advantages of speed and ease of sample preparation and is used routinely in a number of applications, especially ambient aerosol filter analysis. Broad beam RBS still has a wide application in characterizing thin films and diffusion profiles of impurities in solids, for example in measuring thickness of metallization layers on silicon or depth profiles following implantation of dopants into semiconductors. The technique may also be combined with channelling, in which the variation of yield as the beam is aligned

Table 2   Properties of semiconductor detectors used for spectroscopic analysis of charged particles, X-rays and gamma rays

Name ('common name')	Silicon surface barrier detector (surface barrier, SSB)	Lithium drifted silicon (Si-Li)	High purity germanium (HpGe)
Radiation	MeV charged particles	X-rays	Gamma rays
Physical form	Thin disk of Si typically 10–30 mm diameter with depletion layer formed beneath an insulating 'barrier' on the surface	Single crystal Si disk typically 10 mm diameter, 4 mm thick. Li diffused into Si to compensate defects and impurities to ensure efficient charge collection	High purity single crystal Ge cylinder typically 20 mm diameter, 80 mm long
Must be cooled?	No	Yes	Yes
Depletion layer thickness	Typically 100 μm	~4 mm	~80 mm
Entrance window	None	Varies: typically 10 μm Be. Also ultra-thin polymer windows to extend lower energy range	Thin (~1 mm) Al
Energy range	~10 keV to MeV. Upper limit depends on depletion depth	~1 keV–50 keV. Lower limit depends on window; upper limit depends on crystal thickness	~30 keV–10 MeV. Lower limit depends on window and electronic noise; upper limit depends on crystal dimensions

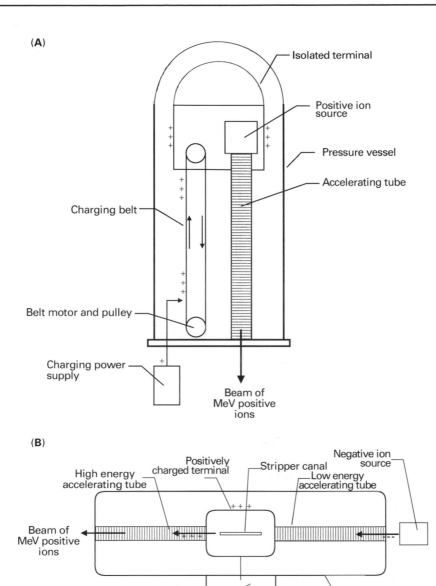

Figure 9 Schematic diagram of (A) single ended accelerator and (B) tandem accelerator. The terminal voltage may be generated either by a moving belt or chain (the van de Graaff or Pelletron principle) or by rectifying a high frequency alternating voltage (the Tandetron). In the single-ended accelerator a positive ion source is mounted in the terminal and the beam energy is equal to the terminal voltage. In the tandem accelerator an external *negative* ion source injects ions into the accelerator where they are accelerated towards the positive terminal. In the terminal they pass through a region of low pressure gas (the stripper) where collisions with the gas molecules remove some, or all, of the electrons. The resulting positive ion is further accelerated. The final energy is given by $V(1+q)$ where V is the terminal voltage and q is the charge state after stripping. The accelerator assembly is normally enclosed in a pressure vessel containing several atmospheres of an insulating gas such as SF_6 to avoid the possibility of electrical discharges.

with one of the major axes of a single crystal sample can be used to obtain information on lattice quality and the localization within the lattice of impurity atoms. A full discussion of channelling is beyond the scope of this article.

PIXE (and to a certain extent RBS) are now more commonly used in the spatially-resolved mode. This is because their high yield contributes to the attainment of high spatial resolution.

List of symbols

C_z = concentration of element; E = energy; m = particle mass; n_T = number of target atoms; N = number of projectile atoms, number of photons; q = terminal charge; Q = total beam charge; V = voltage; Y = number of reaction products; $Y(Z,E)$ = yield of interaction; z = charge; σ = cross-section; ε = detector efficiency; Ω = solid angle.

See also: **Atomic Absorption, Methods and Instrumentation; Atomic Absorption, Theory; Atomic Emission, Methods and Instrumentation; Atomic Fluorescence, Methods and Instrumentation; Fluorescence and Emission Spectroscopy, Theory; Geology and Mineralogy, Applications of Atomic Spectroscopy; Inductively Coupled Plasma Mass Spectrometry, Methods; Proton Microprobe (Method and Background); X-Ray Emission Spectroscopy, Applications; X-Ray Emission Spectroscopy, Methods; X-Ray Fluorescence Spectrometers; X-Ray Spectroscopy, Theory.**

Further reading

Breese MBH, Jamieson DN and King PJC (1996) *Materials Analysis Using a Nuclear Microprobe*. New York: Wiley.

Johansson SAE and Campbell JL (1988) *PIXE, A Novel Technique for Elemental Analysis*. Chichester: Wiley.

Johansson SAE, Campbell JL and Malmqvist KG (1995) *Particle-Induced X-ray Emission Spectrometry*. New York: Wiley.

Watt F and Grime GW (1987) *Principles and Applications of High Energy Ion Microbeams*. Bristol: Hilger.

High Pressure Studies Using NMR Spectroscopy

Jiri Jonas, University of Illinois at Urbana-Champaign, IL, USA

MAGNETIC RESONANCE
Applications

Introduction

The use of pressure as an experimental variable in NMR studies of chemical and biochemical systems leads to added complexity in instrumentation but the unique information gained from the combination of high pressure and NMR techniques justifies fully its use. There are several fundamental reasons for carrying out NMR experiments at high pressures.

First, to separate the effects of density and temperature on various dynamic processes, one has to perform the measurements as a function of pressure. Second, for liquids the use of pressure enables one to extend the measurement range well above the normal boiling point and thus to study supercritical fluids. Third, as noncovalent interactions play a primary role in stabilization of biochemical systems, the use of pressure allows one to change, in a controlled way, the intermolecular interactions without the major perturbations produced by changes in temperature and/or chemical composition. Fourth, pressure affects chemical equilibria and reaction rates. The following standard equations define the reaction volume, ΔV and the activation volume, $\Delta V^\ddagger$:

$$\Delta V = -\left[\frac{RT\,\partial \ln K}{\partial P}\right]_T, \quad \Delta V^\ddagger = -\left[\frac{RT\,\partial \ln k}{\partial P}\right]_T$$

where K is the equilibrium constant, and k is the reaction rate. With the knowledge of $\Delta V^\ddagger$ values one can draw conclusions about the nature of reaction and its mechanism. Fifth, the phase behaviour and dynamics of molecular solids and model membranes can be explored more completely by carrying out high pressure experiments. Sixth, the combination of advanced 2D, and 3D NMR techniques with high pressure capability represents a powerful new experimental tool in studies of protein folding.

The usual range of pressures used to investigate chemical and biochemical systems is from 0.1 MPa to 1 GPa (0.1 MPa = 1 bar; 1 GPa = 10 kbar); such pressures only change intermolecular distances and affect conformations but do not change covalent bond distances or bond angles.

Pressure effects on NMR spectra

This heading is related to the fact that the effects of pressure and/or density on chemical shifts and spin–spin-coupling in molecular systems are relatively small and difficult to interpret. In contrast, most of the high pressure NMR studies of chemical or biochemical systems deal with pressure or density effects on dynamical behaviour of the readily compressible gases, molecular liquids and aqueous solution of biomolecules. Increasing pressure changes both the density and shear viscosity of the liquids which in turn changes the relaxation times and chemical exchange. The majority of NMR studies deal with liquids of low viscosity for which $T_1 = T_2$ and the condition of extreme motional narrowing applies. In the case of rotational motions of a spherical top molecule the following expression can be used to calculate the rotational correlation time, τ_R, from proton relaxation times

$$\frac{1}{T_1} = \frac{3}{2}\hbar^2\gamma^4 \sum_{i,j} r_{ij}^{-6} \tau_R$$

where γ is the gyromagnetic ratio for hydrogen and r_{ij} is the distance between protons. However, for proton-containing molecules one has to separate the intramolecular and intermolecular contributions to the observed proton relaxation times. In the case of nuclei with spin $I > \frac{1}{2}$ the relaxation studies provide direct information about rotational motions because the electric quadrupole interactions provide the dominant relaxation mechanisms. For nuclei with spin $I = 1$ (e.g. ^{2}H, ^{14}N) the relaxation rate due to the quadrupolar interaction is given by the well-known expression

$$\frac{1}{T_1} = \frac{3}{8}(1 + \chi^2/3)(e^2qQ/\hbar)^2 \tau_R$$

where χ is the asymmetry parameter, $e^2qQ/\hbar$ is 2π times the quadrupole coupling constant in hertz, and τ_R is the rotational correlation time.

It is well established that the rotational correlation time, τ_R, for the isotopic rotation of a spherical top molecule can be expressed in terms of the macroscopic viscosity, η, using the Debye equation

$$\tau_R = \frac{4\pi a^2 \eta}{3k_B T}$$

where η is the viscosity, a is the radius of the molecule and the other symbols have their usual meaning.

In the case of the intermolecular contribution to relaxation times of proton-containing molecules and for the effect of pressure on self-diffusion one can use the Stokes–Einstein expression, which relates the shear viscosity, η, to the self-diffusion coefficient D:

$$D = \frac{k_B T}{6\pi a \eta}$$

where all symbols have their usual meaning.

For non-hydrogen bonded liquids the increase of pressure slows down both rotational and translational motions with the result that for low viscosity liquids one shortens the spin–lattice relaxation times and decreases the diffusion.

The slowing down of exchange owing to increased pressure in chemically exchanging systems also leads to changes in the NMR spectra. As discussed elsewhere in this encyclopedia, high resolution NMR spectroscopy is widely used to study chemical exchange processes. As an example of the effect of pressure on chemical exchange **Figure 1** shows the pressure dependence of hindered rotation about the C–N amide bond in N,N-dimethyltrichloracetamide (DMTCA). From the observed line shapes it is straightforward to calculate the observed reaction rate and from its pressure dependence one can calculate the volume of activation $\Delta V^{\ddagger}$. In the case of DMTCA the experimental rotation rate, k, decreases with increasing pressure and the correlation of the rates with measured viscosity (shear viscosity, η) shows that the hindered rotation in DMTCA falls in the strongly coupled diffusive regime.

There is yet another important origin of observed pressure changes in the NMR spectra of biomolecules in aqueous solutions. NMR studies of protein denaturation using temperature or chemical means are well established but pressure also leads to reversible denaturation as discussed in more detail in the section on applications. In the case of model membranes the use of pressure leads to very rich barotropic behaviour and high pressure NMR techniques can establish pressure–temperature phase diagrams for the membrane system studied.

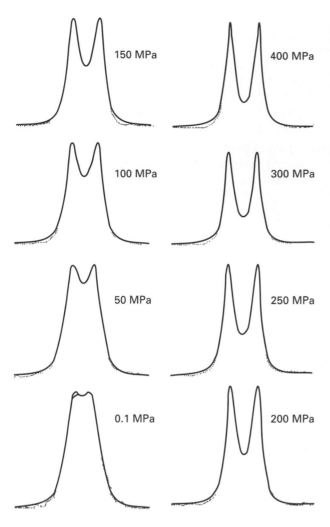

Figure 1 Pressure effects on the line shape of the N-methyl proton NMR spectrum of DMTCA in pentane at 282.3 K. The dotted lines denote the experimental line shapes, and the full lines denote the calculated line shapes.

In addition to gases and liquids the NMR spectra of molecular solids exhibit sensitivity to pressure. The relaxation behaviour is changed and, in addition, pressure also leads to changes in phase-transition temperatures.

The effect of pressure on the phase-transition temperature can be evaluated using the Clausius–Clapeyron equation:

$$\frac{\mathrm{d}T}{\mathrm{d}P} = T\left(\frac{\Delta V}{\Delta H}\right)$$

where T is the transition temperature, P is the pressure, ΔV is the molar volume change and ΔH is the change in molar enthalpy.

High pressure NMR instrumentation

Since high resolution NMR spectroscopy represents a spectroscopic technique of major importance for chemistry and biochemistry the overview of high pressure NMR instrumentation focuses on high resolution, high pressure NMR techniques.

Thanks to advances in superconducting magnet technology one can achieve a high homogeneity of the magnetic field over the sample volume even without sample spinning and, therefore, it is possible to construct high resolution NMR probes for work at high pressures. The high resolution, high pressure NMR equipment consists of three main parts: (a) pressure generating and pressure measuring systems; (b) nonmagnetic high pressure vessel; and (c) a NMR probe that includes a sample cell. All components necessary for building a high pressure setup, such as hand pumps, high pressure tubing, and valve intensifiers are currently available from commercial sources. **Figure 2** shows a schematic drawing of a high pressure generating system capable of producing a hydrostatic pressure up to 1 GPa. In addition to high pressure NMR instrumentation based on a high pressure vessel made from non-magnetic metallic alloy, one can also obtain high resolution NMR spectra using glass capillaries or special sample cells made from sapphire. **Table 1**, which gives characteristic performance features of high pressure NMR probes, also includes information on the diamond anvil cell (DAV) which is suitable for broad-line work up to pressures of 8 GPa. However, the extremely small sample size makes DAV probes difficult to operate and relatively insensitive.

A schematic drawing of a high pressure NMR vessel, and a sample cell used for high resolution NMR experiments at pressures up to 950 MPa, is shown in **Figure 3**. It is important to note that even at pressures of 1 GPa and sample diameter of 8 mm one can achieve a resolution of 3×10^{-9}. The natural abundance ^{13}C NMR spectrum of 2-ethylhexylcyclohexanecarboxylate recorded at 80°C and 500 MPa as shown in **Figure 4** serves as an illustrative example of the high quality NMR spectra which can be routinely obtained at high pressures.

Applications of high pressure NMR spectroscopy

Dynamic structure of liquids

Theoretical and experimental evidence indicate that many physical properties of liquids may be

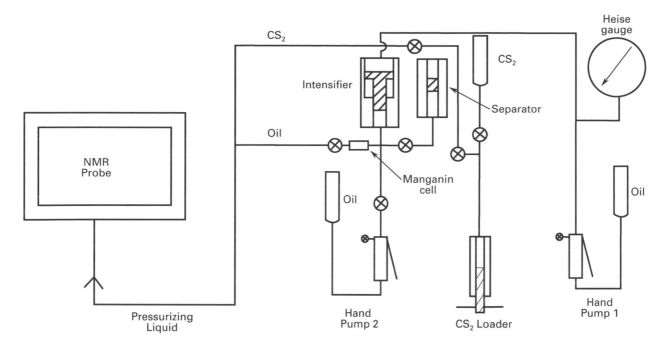

Figure 2 Schematic diagram of the high pressure generating equipment.

Table 1 Characteristic performance features of high pressure NMR probes

Probe type	Pressure range (MPa)	Temperature range (K)	Sample diameter (mm)	Resolution (× 10⁻⁹)	¹H Frequency (MHz)
Capillary cell	0.1–250	a	1–1.5	3	a
Sapphire cell	0.1–100	a	3.4	1.5	a
HP vessel	0.1–950	253–373	8	3	60–500
DAV[b]	0.1–8000	200–300	<<1	c	1–100

[a] Temperature and frequency range is determined by the commercial probe used.
[b] Diamond anvil cell.
[c] Broad line.

determined, to a large extent, by the size and shape of the constituent atoms or molecules. Owing to the close packing of molecules in liquids, even a small change in density can produce a considerable change in the molecular dynamics of the liquid; therefore, to test rigorously a theoretical model of a liquid, or a model of a specific dynamic process in a liquid, one must perform isochoric, isothermal, and isobaric experiments. **Figure 5**, which shows the temperature dependence of the self-diffusion coefficient, D, in liquid tetramethylsilane at constant density and constant pressure, illustrates the importance of separating the effects of density and temperature on molecular motions. Four main research directions can be identified in the high pressure NMR studies of liquids: (a) studies of self-diffusion; (b) investigations of reorientational motions; (c) studies of angular momentum behaviour; and (d) tests of

applicability of hydrodynamic equations at the molecular level.

The effect of pressure on the anisotropic reorientation of acetonitrile-d_3 in the liquid state can serve as in illustrative example of high pressure NMR studies of molecular reorientation in liquids. The deuteron and nitrogen spin–lattice relaxation times of acetonitrile-d_3 have been measured as a function of pressure up to 200 MPa at 23°C, using the NMR T_1 relaxation technique. Since the quadrupole coupling constants for nitrogen and deuterons in acetonitrile-d_3 are known, the experimental T_1 data are interpretable in terms of the rotational diffusion constants for motion perpendicular $D_\perp$ and parallel $D_\parallel$ to the symmetry axis.

From **Figure 6**, which plots the pressure dependence of $D_\perp$ and $D_\parallel$, it is readily apparent that $\Delta V^\ddagger(D_\perp) \gg \Delta V^\ddagger(D_\parallel)$. This experimental finding

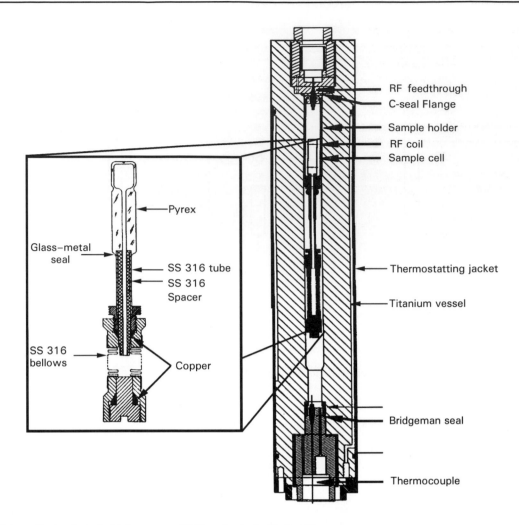

Figure 3 Schematic drawing of a high pressure NMR probe and sample cell.

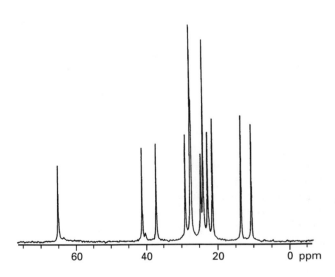

Figure 4 High resolution natural abundance ^{13}C NMR spectrum of liquid 2-ethylhexylcyclohexanecarboxylate at 80°C and 500 MPa.

reflects the difference in frictional torques connected with the reorientation about the individual molecular axis, or more generally stated, reflects the symmetry of intermolecular potentials. The low $\Delta V^{\ddagger}(D_{\parallel})$ value indicates that the intermolecular potential energy is largely independent of the angle of orientation about the main symmetry axis and thus the rotational frictional coefficient is very small.

Supercritical fluids

NMR techniques can be used to study supercritical fluids as the employed pressures range from 0.1 to 80 MPa. The reason for renewed interest in the properties of supercritical fluids can be traced to the great promise of supercritical fluid extraction techniques. High pressure NMR spectroscopy can not only be used to investigate transport and intermolecular interactions in compressed supercritical

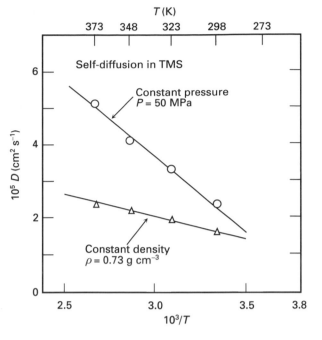

Figure 5 Temperature dependence of self-diffusion in liquid tetramethylsilane (TMS) at (○) constant pressure and (△) constant density.

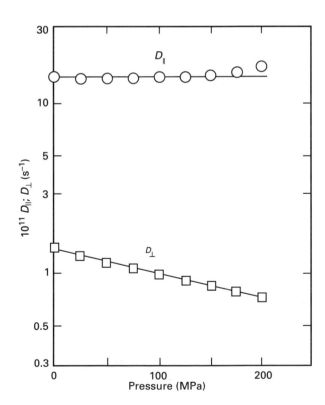

Figure 6 Pressure effffects on the rotational diffusion constants $D_\perp$ and $D_\parallel$ of acetonitrile-d_3 at 23°C.

fluids but an in situ NMR technique has also been developed for the determination of the solubility of solids in supercritical fluids.

The basic idea behind the determination of the solubility of solids in supercritical fluids is based on the radically different spin–spin relaxation rate between dissolved and solid material (T_2, solid $\ll T_2$, dissolved). Using the 90°–t–180° spin echo sequence with a pulse separation of $t \ll 1$ ms ensures that no contribution to the NMR echo signal can result from the quickly relaxing protons in the solid material. **Figure 7** gives the experimental solubilities for solid naphthalene in supercritical carbon dioxide as determined by the spin echo NMR technique. The three isotherms measured were selected near the upper supercritical end point (UCEP) for naphthalene in CO_2. In **Figure 7** the 58.5°C isotherm shows a large increase in solubility at about 23.5 Mpa; the slope of the isotherm is near zero. This indicates that one operates in close proximity to UCEP. This NMR technique can yield both solubility data and phase information when studying equilibria in supercritical fluid mixtures.

Phase transitions in molecular solids

Because molecular solids exhibit significant compressibility even in the pressure range 0.1 to 900 MPa one can use high pressure NMR relaxation techniques to investigate phase transitions. The high degree of molecular rational freedom in plastic crystals (orientationally disordered crystals)

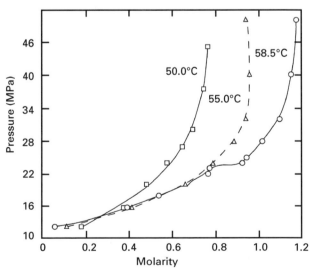

Figure 7 Experimental solubilities for solid naphthalene in supercritical CO_2 expressed in moles of napthalene dissolved per litre of solution.

provides a good illustrative example. Adamantane, $C_{10}H_{16}$, is a saturated hydrocarbon with an fcc crystal structure in its high temperature plastic phase (α-phase) whereas at lower temperatures it undergoes a phase transition to a tetragonal structure (β-phase) in which the rotation is quenched. Taking advantage of the abrupt change in T_1 or T_2 at the phase transition provides the means of detecting the precise phase-transition pressure. At the phase-transition pressure, the T_1 of adamantane decreases abruptly by a factor of 40 in passing from the orientationally disordered plastic α-phase to the β-phase. **Figure 8** shows the pressure dependence of the proton spin–lattice relaxation times, T_1, in solid adamantane in the α- and β-phases at different temperatures.

The relaxation time can be expressed in terms of the rotational correlation time τ, and the

$$\frac{1}{T_1} = C \left(\frac{\omega_0 \tau}{1 + \omega_0^2 \tau^2} + \frac{4\omega_0 \tau}{1 + 4\omega_0^2 \tau^2} \right) \quad [1]$$

second moments (M^2) where ω_0 is the Larmor frequency (Eqn [1]). The constant C is proportional to the proton second moment expressed in angular

frequency units. For the measured T_1 one can assume $\omega_0 \tau \gg 1$ for the β-phase rotation. Then the expressions for the relaxation times are as follows:

$$
\begin{aligned}
(1/T_1)_\alpha &= 5\omega_0\tau C_\alpha && \text{for } \omega_0\tau \ll 1 \\
(1/T_1)_\beta &= 2C_\beta/\omega_0\tau && \text{for } \omega_0\tau \gg 1 \quad [2]
\end{aligned}
$$

where $C_\alpha \omega_0 = 8.91 \times 10^9$ radian^2 s^{-2} and $2C_\beta/\omega_0 = 2.07 \times 10^{-7}$.

Assuming that rotational motion is a thermally activated process and that the second moments are independent of temperature and pressure within our experimental range, one can obtain the activation volume $\Delta V^\ddagger(T)$ at constant temperature from the pressure dependence of $\ln(T_1)$:

$$T_1(T, P) = A(T) \exp\{\pm[\Delta V^\ddagger(T)/RT]P\} \quad [3]$$

and the activation enthaply, $\Delta H^\ddagger(P)$, at constant pressure:

$$T_1(T, P) = B(P) \exp[\pm\Delta H^\ddagger(P)/RT] \quad [4]$$

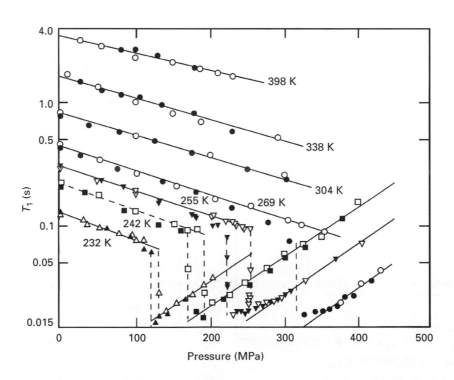

Figure 8 Pressure dependence of the proton spin–lattice relaxation times, T_1, of solid adamantane in its two phases at different temperatures. The open symbols denote values obtained in the compression run whereas the filled symbols give values for the decompression run.

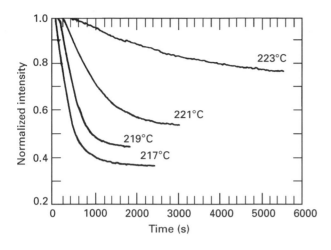

Figure 9 Normalized spin-echo amplitude as a function of time in polyethylene at 400 MPa and temperatures of 217 to 223°C.

where Equations [3] and [4] have a plus sign when $\omega_0\tau \gg 1$, and a minus sign when $\omega_0\tau \ll 1$.

The interesting feature of the experimental results given in **Figure 8** is the hysteresis of the observed transition pressure for the compression $\alpha \to \beta$ run and the decompression $\beta \to \alpha$ run. Analysis of this hysteresis can provide valuable information about the nature of order–disorder transitions.

Pressure effects on crystallization kinetics of polymers

NMR relaxation or line width measurements are sensitive to the degree of crystallinity in solid polymers, because the protons in the crystalline lattice experience strong dipole–dipole interactions which cause fast spin–spin relaxation and line broadening. As a result, line width and relaxation studies have been used to measure crystallinity in a broad range of polymers. The use of high pressure as an experimental variable in NMR studies of polymer crystallinity offers new details about polymer crystallization. The study of pressure and temperature effects on the

kinetics of crystallization of linear polyethylene can be used as an illustrative example of this specific application of high pressure NMR techniques. One can determine the amount of amorphous polyethylene present from the initial maximum by spin-echo measurements 90°–τ–180°–τ–echo with τ equal to 300 μs. **Figure 9**, which plots the amplitude of the echo signal against time, shows the crystallization data at 400 MPa for polyethylene. Before each measurement the sample was equilibrated at the desired temperature and a pressure 75 MPa below the crystallization pressure. The very fast relaxation of the crystalline material prevented it from contributing to the echo intensity. The isotherms as depicted in **Figure 9** show the characteristic crystallization behaviour: an induction time before crystallization, a primary crystallization, and a gradual decrease in crystallization rate as the final crystallinity is approached and secondary crystallization and perfection pressures occur. The quantitative analysis of the high pressure crystallization kinetics yields information about the formation of extended crystals, the value of the Avrami coefficient, the mechanism of extended chain crystallization, and the effect of pressure on the surface energies of the crystal nuclei. High pressure NMR experiments are interesting because of their direct sensitivity to polymer chain dynamics on the microscopic scale. In this way NMR complements techniques such as dilatometry which are sensitive to bulk properties.

Mechanistic studies of chemical reactions in solution

High pressure NMR studies of various chemical reactions provide information about the reaction mechanism. Almost all reactions exhibit a characteristic pressure dependence of the reaction rate which can be used to calculate the volume of activation, $\Delta V^{\ddagger}$. The $\Delta V^{\ddagger}$, extrapolated to ambient pressure, together with the partial molar volume for the reactants and products, or with the reaction volume

Table 2 High pressure kinetic studies of solvent exchange in inorganic systems

System	Solvent	Pressure (bar)	Temperature (K)	Nuclei
$[Cu(DMF)_6]^{2+}$, $[Cu(H2O)6]^{2+}$	DMF, H_2O	2000	217–300	^{17}O
$[Ln(PDTA)(H_2O)_2]^-$ (Ln = Tb, Dy, Er, Tm, Yb) $[Er(EDTA)(H_2O)_2]^-$	H2O	2000	272–292	^{17}O
$[Ln(DTPA-BMA)(H_2O)]$ (Ln = Nd, Eu, Tb, Dy, Ho)	H_2O	2000	na	^{17}O
$[M(en)_3](CF_3SO_3)_2$ (M = Co, Fe, Mn)	en[a]	2000	278–332	^{14}N
$[Cp^*M(H_2O)_3]^{2+}$ (M = Rh, Ir)	H_2O	2000	288–333	^{17}O

[a] Ethylenediamine.

of the overall reaction, allows one to create a volume profile that describes the volume changes along the reaction coordinate. The location of the transition state with respect to reactants and products determines the mechanism of the reaction investigated. Both organic and inorganic reactions can be studied. Specifically, solvent and ligand exchange, and ligand substitution reactions in inorganic systems have extensively been studied by high pressure NMR techniques. **Table 2** gives the activation parameters of solvent exchange studied in inorganic systems.

Biochemical applications

Model membranes

Phospholipid bilayers and monolayers have important roles in nature as components of cell membranes

Table 3 Examples of NMR studies of the pressure effects on model membranes[a]

System	Experiment	Result
DPPC[b]	Natural abundance $^{13}C, T_1, T_2$	Phase transitions
DPPC-d_{62} DPPC-d_{62}-TTC[c]	^2H line shapes	Phase diagram; order parameter; pressure reversal of the anaesthetic effect of tetracaine
DPPC-TTC[e]	^{31}P line shapes, T_1	Structure and dynamics of the head group; phase diagram
DPPC-d_2 (2,2); (9,9); (13,13)	^2H line shapes, T_1, T_2	Order parameters; chain motions
DPPC-d_{62}-cholesterol	^2H line shapes	Phase diagram
DPPC POPC[d]	$^1H T_1$ and $T_{1\rho}$	Lateral diffusion

[a] Pressure range from 0.1 to 500 MPa.
[b] DPPC = dipalmitoylphosphatidylcholine.
[c] TTC = tetracaine.
[d] POPC = palmitoyloleylphosphatidylcholine.

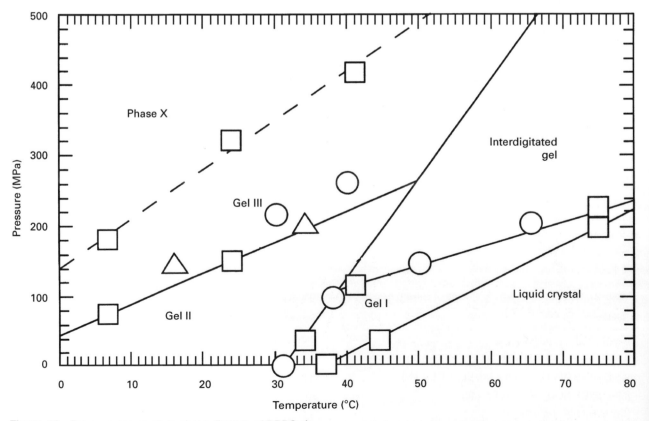

Figure 10 Pressure–temperature phase diagram of DPPC-d_{62}.

and lipoproteins. In cell membranes, phospholipid bilayers constitute the permeability barriers between aqueous compartments. Aside from their structural interfacial roles, aggregated phospholipids modulate the functions of associated proteins and enzymes by direct binding and by physical effects. Therefore, synthetic phospholipid bilayers, in multilamellar or small bilayer vesicle forms, have become models for the study of the structural and dynamic properties of natural phospholipid aggregates.

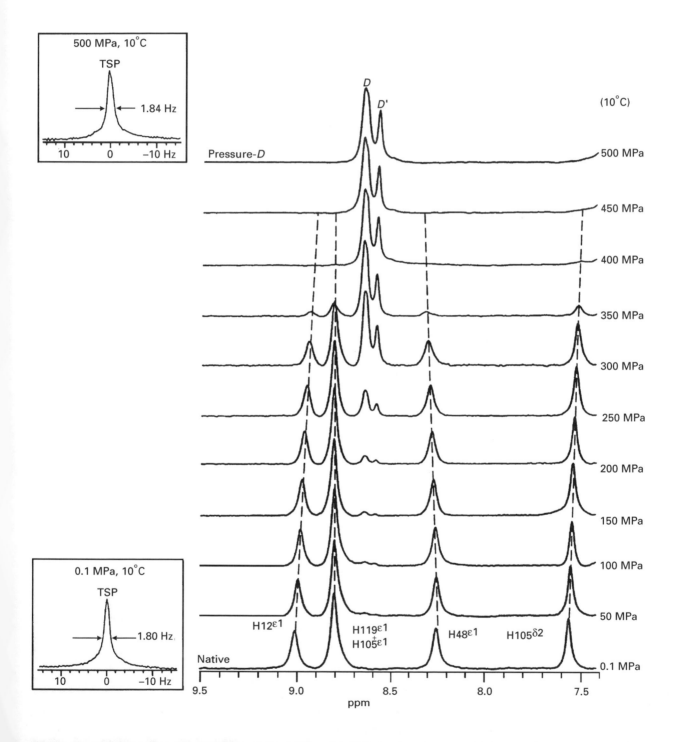

Figure 11 Histidine region of the ¹H NMR spectra of RNase A in D₂O at various pressures. The standard in the insets is sodium 3-(trimethylsilyl) tetradeuteriopropionate (TSP).

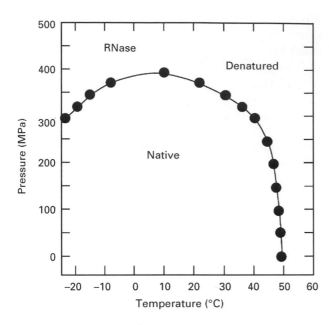

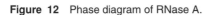

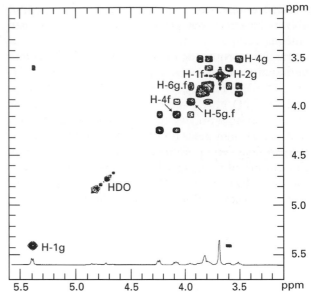

Figure 12   Phase diagram of RNase A.

Figure 13   2D ^{1}H Correlation spectroscopy spectrum with presaturation pulse for solvent suppression of 2 mM sucrose at 25°C and 475 MPa pressure recorded at 500 MHz.

Since the 1970s, NMR methods, in particular ^{2}H NMR, have been applied very effectively to investigate the biophysical properties of phospholipid bilayers. NMR experiments with high pressure and variable temperature measurements allow one to access diverse gel phases in phospholipid bilayers and to study the order and dynamics of phospholipids in bilayers as a function of volume changes.

Table 3 gives illustrative examples of high pressure NMR studies of model membranes and indicates the type of information that can be obtained from these studies. **Figure 10** shows the pressure–temperature phase diagram of deuterated dipalmitoylphosphatidylcholine,   DPPC-d_{62},   as determined by high pressure NMR techniques using direct measurement of the deuteron quadrupole splitting of the methyl group and first moment analysis.

Pressure denaturation of proteins

The relationship between the amino acid sequence and the structure of the native conformation of proteins represents the key unsolved problem of biochemistry and biology. Temperature and chemical perturbation represent the usual approach used in studies of protein stability and the folding problem. However, it is advantageous to use pressure to study protein solutions as pressure perturbs the protein environment in a controlled way by changing

only intermolecular interactions. Moreover, by taking advantage of the phase behaviour of water, high pressure can lower the freezing point of an aqueous solution and allow the study of cold denaturation of proteins. The reversible pressure denaturation of bovine ribonuclease A (RNase A) can be used as an illustrative example for pressure denaturation. In this specific protein the $\varepsilon1$ hydrogen signals of the four histidine residues are well resolved from other proton peaks in the 1D ^{1}H NMR spectrum of the native RNase A in D_2O. These peaks have been assigned and used to investigate the folding and unfolding of the protein. **Figure 11** shows the pressure effects on the proton NMR spectrum in the histidine region of RNase A at pH 2.0 and 10°C. As expected with increasing pressure, the intensity of the native histidine peaks decreases and at about 400 MPa their disappearance indicates that the protein is denatured. The appearance of peaks D and D' with increasing pressure signals the denaturation.

It is important to point out that during the pressure-assisted cold denaturation many proteins show significant residual secondary structure which parallels the structure of folding intermediates. It appears that pressure denaturation and pressure-assisted cold denaturation of proteins studied by high resolution NMR techniques provide novel information about the folding of proteins. In addition, these experiments allow the determination of phase diagrams for proteins. **Figure 12** shows the pressure–temperature phase diagram for RNase. Above the curve the protein is in the denatured state.

The most promising direction of the high pressure studies of proteins is the application of advanced 2D and 3D NMR techniques. **Figure 13** shows the high quality of 2D NMR spectra which can be routinely obtained at high pressures. Very often the 2D spectra obtained at high pressure offer new insights into the pressure-induced unfolding and pressure-induced dissociation of proteins when compared with results obtained by other high pressure spectroscopic techniques such as fluorescence methods.

List of symbols

a = molecular radius; D = Self-diffusion coefficient; I = nuclear spin quantum number; k = rate constant; K = equilibrium constant; P = pressure; r_{ij} = distance between protons; R = gas constant; t = pulse separation time; T = absolute temperature; T_1, T_2 = relaxation times; ΔV = reaction volume; $\Delta V^{\ddagger}$ = activation volume; γ = gyromagnetic ratio; η = shear viscosity; χ = asymmetry parameter; ω_0 = Larmor frequency.

See also: **Biofluids Studied By NMR; Diffusion Studied Using NMR Spectroscopy; High Pressure Studies Using NMR Spectroscopy; Labelling Studies in Biochemistry Using NMR; Membranes Studied By NMR Spectroscopy; NMR Relaxation Rates; Proteins Studied Using NMR Spectroscopy; Relaxometers; Two-Dimensional NMR, Methods.**

Further reading

Benedek GB (1963) *Magnetic Resonance at High Pressure.* New York: Interscience Publishers.

Jonas J (1973) NMR studies in liquids at high pressure. *Advances in Magnetic Resonance* 6: 73–139.

Jonas J (1975) Nuclear magnetic resonance at high pressure. *Annual Reviews in Physical Chemistry* 26: 167–190.

Jonas J (1978) Magnetic resonance spectroscopy at high pressure. In: Kelm H (ed) *Proceedings NATO ASI, Series C, High Pressure Chemistry*, pp 65–110. Dordrecht: Reidel.

Jonas J (1982) Nuclear magnetic resonance at high pressure. *Science* 216: 1179–1184.

Jonas J (1991) High pressure NMR studies of the dynamics in liquids and complex systems In: Jonas J (ed) *High Pressure NMR*, pp 85–128. Vol 24 in Monographs NMR Basic Principles and Progress. Heidelberg: Springer-Verlag.

Jonas J (1993) High pressure NMR studies of chemical and biochemical systems. *Proceedings NATO ASI, Series C, High Pressure Chemistry, Biochemistry and Materials Science*, 401: 393–441. Dordrecht: Reidel.

Jonas J and Jonas A (1994) High pressure NMR spectroscopy of proteins and membranes. *Annual Reviews in Biophysical and Biomolecular Structure* 23: 287–318.

Jonas J and Lamb DM (1987) Transport and intermolecular interactions in compressed supercritical fluids. In: Squires TE and Paulaitis ME (eds) *Supercritical Fluids Chemical and Engineering Principles and Applications*, pp 15–28. Vol 329 in ACS Symposium Series. Washington DC: American Chemical Society.

van Eldik R (1987) High pressure studies of inorganic reaction mechanisms. In: *Proceedings NATO ASI* 197: 333–356.

High Resolution Electron Energy Loss Spectroscopy, Applications

Horst Conrad, Fritz Haber Institute of the Max Planck Gesellschaft, Berlin, Germany
Martin E Kordesch, Ohio University, Athens, OH, USA

VIBRATIONAL, ROTATIONAL & RAMAN SPECTROSCOPIES
Applications

Introduction

High resolution electron energy loss spectroscopy, known as 'HREELS', is a method for measuring the vibrational spectra of molecules adsorbed on surfaces, and some properties of clean surfaces in ultrahigh vacuum. The technical aspects of the technique, theoretical basis and some examples of HREELS will be discussed.

HREELS is an 'electrons in–electrons out' spectroscopy that uses electrons typically in the 2–20 eV energy range as a probe, and detects electrons that have undergone both elastic and inelastic scattering from a surface or adsorbed layer. In practice, electrons are detected that have lost up to about 500 meV (≈ 4000 cm^{-1}, 8.06 cm^{-1} = 1 meV) relative to the primary beam energy in the scattering event. This range includes the most common vibrational modes of small molecules, and their overtones and combinations.

A major advantage of the HREELS method is the benefit derived from electron detection over optical detectors and methods, and a variety of scattering mechanisms. The major technical complexity of the HREELS method is the 'resolution' of 'the spectrometer', in this case the production of an intense, well-collimated beam of electrons with an energy width of less than 10 meV at the chosen primary energy (usually about 5 eV).

Detailed and comprehensive analysis of vibrational spectroscopy performed with electrons can be found in the classic work by Ibach and Mills. The details of the modern HREELS spectrometer are explored by the acknowledged master of the technique Harald Ibach in his monograph. A guide to surfaces and adsorbates studied with HREELS can be found in the proceedings of the 'Vibrations at Surfaces' Conferences, several of which are listed in the 'Further reading' section. A general introduction is often included in surface science or surface chemistry textbooks, for example the books by Lüth, Prutton and Somorjai. The World Wide Web is a resource for manufacturers of HREELS apparatus, e.g. Omicron Vakuumphysik GmbH (omicron-instruments.com) and LK Technologies, Inc. (lktech.com).

The ideal surfaces for study are smooth, metal single crystals that are clean and well-ordered on an atomic level. Such a specimen is 'highly reflecting' and maintains the signal in the specularly reflected electron beam. Practical surfaces range from polished single crystals, cleaved surfaces and semiconductor wafers to polycrystalline metals. The adsorbate can be a monoatomic or molecular species, such as atomic hydrogen, or large macromolecules and polymers. Insulators can be measured in special circumstances.

Many aspects of surface science and surface spectroscopy are concerned with the geometrical structure of surfaces, the composition of the surface and the identification of adatoms that may be present. Vibrational spectroscopy is a method for direct measurement of specific *chemical bonds* of adsorbed atoms and molecules, both between the adsorbate and the surface and the adatoms themselves. In the early days of HREELS, the 1970s, an added attraction for this type of spectroscopy was the ability to observe adsorbate–surface bonding modes (often < 125 meV = 1000 cm^{-1}), because the infrared spectrometers of the day used 'grating' spectrometers, and IR detectors that were useful only above 1600 cm^{-1}. The low cost and versatile Fourier transform infrared spectrometer (FTIR) and improved detector technology have eclipsed HREELS for routine surface chemical bond analysis. There are, however, some surface processes that can only be observed with electrons. Some diagnostic benefits that are related to the scattering mechanisms operative in HREELS continue to be useful for surface science. It should be noted that HREELS is usually performed on a 'known' adsorbate, with a focus on the details of a specific adsorption system. HREELS is seldom used for the identification of 'unknown' adsorbate species.

Basic concepts and requirements

Method

In HREELS experiments, electrons produced by a thermionic emission source are passed through an electrostatic monochromator, so that an electron beam of narrow angular width and narrow energy spread is produced that is then scattered from a surface. Electrons which scatter from the surface are analysed with an electrostatic analyser (similar or identical to the monochromator) and detected with an electron multiplier. The HREELS data is acquired as 'count rate' (= current exiting the analyser) vs the energy loss range scanned, from E_p to $(E_p - E_{scan})$, for example from 5 to 4.5 eV, that includes E_{loss}. In a typical HREELS spectrum, a 'no-loss' or 'elastic' peak will be observed in the specular scattering direction with energy E_p, and a FWHM of ΔE_p, and loss peaks will be observed in accordance with certain selection rules, at energies $E_{loss} = E_p - \hbar\omega$, where ω is the frequency of oscillation of the vibrational mode, and an energy width of $\Delta E_{loss} \geq \Delta E_p$. A general schematic is shown in **Figure 1**.

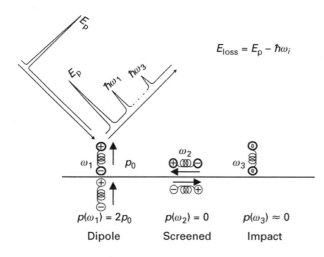

The 127° cylindrical capacitor analyser is currently the most advanced device for HREELS. The functional components of the HREELS spectrometer are shown as a block diagram in **Figure 2**; a representative schematic of a 127° cylinder spectrometer is also shown in **Figure 3**. There are various other HREELS instrument designs with individual merits.

HREELS operation requires specialized precautions and procedures. Many are determined by the trade-off between the spectrum intensity and resolution, the production of the low-energy electron beam, and mechanical restrictions. Practical size limits for HREELS analysers are about 100 mm radius. The thermionic electron source has an energy width of the order of 1 eV. Only a small fraction of the emitted current is passed by the monochromator. Electrons with 5 eV kinetic energy are easily deflected by stray electric and magnetic fields. All of the spectrometer parts are coated with graphite to create a uniform work function on parts in the electron path. The spectrometer itself is usually baked after the UHV chamber itself is baked, so that non-uniformities in the analyser due to unwanted adsorbed contaminants are removed. No magnetic parts are allowed in the vicinity of the spectrometer. Most HREELS chambers are doubly shielded against magnetic fields (< 1 mG cm^{-1} for the residual homogeneous DC field), and all of the electric potentials applied to the spectrometer are shielded and regulated to within a few percent of 1 mV. The specimen

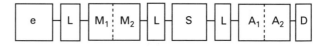

Figure 2 The necessary part of an HREELS spectrometer: e, electron source, L, lens to focus electrons onto the entrance slit of the monochromator, M, which can have one or two stages. Another lens, L, focuses the beam onto the sample, S. A lens, L, focuses electrons scattered from the surface onto the entrance slit of the analyser, A, which can also have two stages. The electrons are counted at the detector, D.

Figure 1 An overview of high resolution electron energy loss spectroscopy. Top left: the incident electron beam is shown as a narrow, intense peak on the intensity vs energy loss axes. The specularly reflected beam is shown with loss peaks due to adsorbed molecules, with modes $\hbar\omega$. Centre: The scattering mechanism is illustrated with the three diatomic molecules 'adsorbed' on the surface with perpendicular and parallel orientation relative to the surface. Mode ω_1 has a dynamic dipole moment p_0 which is perpendicular to the surface, and induces a second image dipole in the same direction, so that the electron scatters from a combined dipole moment of $2p_0$. This is the dipole scattering process. The mode ω_2 is parallel to the surface, and the induced image dipole cancels the molecular dynamic dipole moment. The mode is 'screened' and is not present in the spectrum if there is no impact contribution to the scattering. Mode ω_3 is shown with the dynamic dipole moment equal to zero (the orientation is not relevant). The mode will be observed as an 'impact' mode.

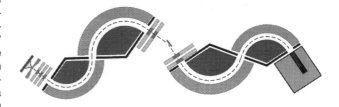

Figure 3 A typical 127° two-stage cylindrical capacitor analyser. The individual parts correspond to the components described in **Figure 2**. Usually, one of the arms of the spectrometer is made to rotate in-plane about the specimen axis to allow off-specular measurement.

must be introduced between the HREELS mono-chromator and analyser with no vibration, no magnetic parts, and no 'stray' potentials such as those that may arise from charging of ceramic insulators, etc. It is often beneficial to be able to cool the specimen to 100 K or lower, and sputter and heat the surface for cleaning.

The 'mechanics' of HREELS is now manageable using personal computer based spectrometer control. Typical spectral resolution of modern instruments is 1–5 meV. At the lower end of this range physical effects due to natural vibrational line widths are observable. In the early days of HREELS (which extend to the present, because many of the older spectrometers are still in use), single-pass cylindrical (127°) capacitor spectrometers, both fixed and with a movable analyser or monochromator, double-pass cylindrical (127°) capacitor analysers, spherical capacitor spectrometers and even some dual-use photoelectron spectrometers with enhanced resolution ranges were used, with spectral resolution in the 7–12 meV range.

Scattering mechanism

A non-mathematical description of the HREELS scattering mechanisms will be given briefly. In 'dipole scattering' an incident electron scatters from the long-range oscillating dipole field that arises when the adatom oscillates on the surface. In this mode, collecting only electrons scattered into the near-specular direction, HREELS information is essentially identical to data obtained from surface infrared vibrational spectroscopy. Electrons scatter from the long-range dipole fields of elementary excitations of the solid or adsorbate. A change in the dipole moment during the vibration is detected; consequently, there must be a non-zero dipole derivative with respect to time or a 'dynamic' dipole moment. **Figure 4** illustrates the scattering geometry for HREELS.

Energy and momentum conservation are simply expressed as: $E_{loss} = E_p \pm \hbar\omega_s$, where the momentum of the incident ('primary') electron is k_p, and k_s for the scattered electron, so that

$$E_{loss} = \frac{\hbar^2 k_s^2}{2m} = \frac{\hbar^2 k_p^2}{2m} \pm \hbar\omega_s(Q_\parallel)$$

where $Q_\parallel$ is the momentum transfer parallel to the surface, and the vector

$$k_{s\parallel} = k_{p\parallel} \pm Q_\parallel + G_\parallel$$

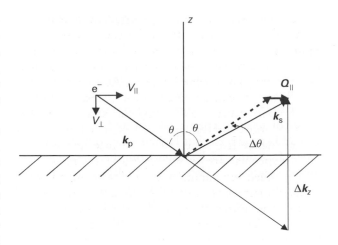

Figure 4 A schematic representation of specular reflection (θ) and momentum conservation in HREELS scattering events. The momentum parallel to the surface is conserved. The quantity $Q_\parallel$ is the parallel momentum transfer, and determines the angular deflection $\Delta\theta$. The incident electron is characterized by velocity and momentum components v and k_p, k_s is the scattered electron momentum, and Δk_z the change in momentum in the z direction.

where G is a reciprocal lattice vector (usually, $G_\parallel = (0,0)$, $G_\parallel \neq (0,0)$ is diffraction). The dipole excitations have a long-range potential, which ensures that the interaction time is large and that the probability for scattering is also high. Both losses and gains are possible.

In order to pass through the analyser entrance slit, the angular deflection (momentum transfer) caused by the collision must be small enough to keep the electron within a few degrees of the specular direction. An estimate for this deflection in dipole scattering is given by $\delta\theta = \hbar\omega_s/2E_p$, so that for 200 meV loss energy and 5 eV primary energy, the deflection is of the order 1°.

It is possible to detect electrons that are not scattered into the specular direction, either by rotation of the sample or moving the HREELS analyser or monochromator. The possibility of data collection 'off specular' sets HREELS apart from its optical counterparts. The most important 'off-specular' scattering mechanism is known as 'impact' scattering. This name is intended to denote a *short-range* inelastic scattering mechanism unrelated to the long-range dipole scattering mechanism. (The elastic 'impact' process is electron diffraction.) As a functional, but not entirely correct, definition, any intensity that remains in the 'off-specular' direction spectrum is termed 'impact' scattering. The intensity of 'impact modes' for a mode u is $I_{impact} = \frac{1}{2}(Q \cdot u)^2 I_0$, where u is the displacement of the atom from its equilibrium position, Q is the change in wavevector of the electron, and I_0 is the elastic scattering intensity. The wave vector transfer Q can be large, so that impact scattering is not restricted to the specular direction. Experimentally, the intensity of the 'dipole' loss peaks exhibits the same dependence on off-specular angle as the elastic peak intensity. As a consequence, impact modes can be much more intense relative to the dipolar contribution at high off-specular angles. Because most analysers are constructed for about 50° incidence from the sample surface normal, 25–30° off specular are the largest practical values.

A detailed examination of the electron–adatom–surface interaction must be divided into several parts: either direct scattering from the adatom in the specular direction, specular reflection at the surface followed by (forward) scattering from the adatom, or adatom scattering followed by specular reflection. Multiple combinations of these individual events can also result in an electron detected at the analyser. For each interaction, there are elastic and inelastic versions. A schematic representation is shown in **Figure 5**. To date, HREELS theory has dealt only with the large divisions of scattering events into elastic/inelastic and dipole/impact processes. Other phenomena have been observed, and can be regarded as significant scattering types observed in HREELS spectra, even if they are not entirely distinct from the dipole and impact mechanisms (dipole or impact scattering may still be *part* of the process). Molecules with large static dipole moments have a large scattering cross section, so that 'dipole' scattering from a partially covered surface may cause diffuse scattering of electrons, much like a rough surface causes diffuse optical reflection. There are many forms of resonance scattering: an electron can be scattered into a surface state or resonance, causing an increased residence time at the surface, which can influence spectral features. The incident electron can induce surface resonances, such as image potential states, which also capture the electron at the surface for increased residence time and possible non-specular scattering. Molecules can capture electrons in 'negative ion resonances', which also have consequences for peak intensities and multiple or combination scattering peaks. Ionic crystals and semiconductors can have surface phonon modes in the energy range which is accessible to HREELS. Some illustrative examples are given below.

Selection rules and symmetry

Selection rules for dipole scattering from surface–adatom symmetry configurations follow those for IR spectroscopy. Relevant discussion can be found in Ibach and Mills, and general references on symmetry, normal modes and selection rules in

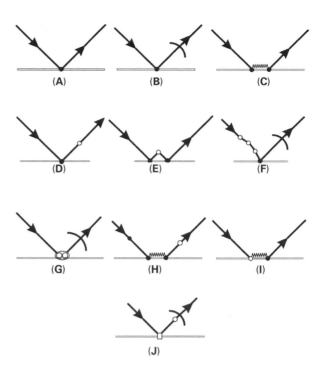

Figure 5 A schematic representation of scattering phenomena that can be observed in HREELS spectra. The diagram notation is loosely based on McRae and Ibach and Mills. The incident electron comes from the left. In each diagram, a time-reversed event is often possible, and these are not shown. Events that involve small-angle forward scattering preserve the specularly scattered intensity, and contribute to HREELS; large-angle scattering contributes less, and some of these diagrams are omitted. In (A) the no-loss, specular scattering from a surface is shown (a filled, black circle at the surface). The no-loss or elastic peak is a result of process (A). Process (B) represents dipolar elastic diffuse scattering (denoted with an arc on the exit beam), such as observed for no-loss peak intensity decreases upon adsorption (see **Figure 14**, **Figure 15B**). Diagram (C) shows a resonant process, where an electron is scattered elastically into and out of a surface resonance (wavy line) in the specular direction. An example is a clean-surface reflectivity measurement (see **Figure 17**) with a surface resonance and/or image potential states. In these processes, in order to scatter into a weakly bound surface resonance, energy conservation by an Umklapp process may occur. Diagram (D) shows the elastic, specular reflection at the surface followed by a loss (open circle) and small angle deflection. This is 'dipole scattering'; the reverse, loss then reflection, is equally important, but not shown. Most HREELS spectral features are due to processes (A) and (D). Diagram (E) shows an unlikely, high-angle scattering, inelastic process that is detected in the specular direction. The process (F) represents multiple, small-angle inelastic scattering followed by specular reflection, but at angles unrelated to the incident electron, which results in a 'diffuse' reflected beam, even though each inelastic scattering event was nearly specular (see **Figure 14**, HCN 'ice'). Process (G) represents a molecular resonance. The electron scatters out of this resoanance isotropically. The diagrams (H) and (I) show combinations of inelastic scattering and surface resonances with specular reflection. Process (J) represents diffuse (open square box) impact scattering followed by an inelastic loss event. Many more combinations of these mechanisms are observed in special cases.

spectroscopy. Briefly, for dipole scattering, those modes with a non-zero dynamic dipole moment are allowed. Because the electric field induced by the incident electron is perpendicular to the surface, only dipoles with a component in this direction can interact with the incident electron. This implies that the normal component (z) of the dynamic dipole moment must be totally symmetric with respect to all of the symmetry operations of a point group. Often, the rule is stated that fundamental modes observed in dipole scattering must belong to the A_1, A' and A totally symmetric representations. This is true for any surface that can effectively screen the parallel dipole moment of the adsorbate motion (most often, metals). The possibility remains that a mode exists that is normal to the surface, but has no dynamic dipole moment, or that a parallel mode that belongs to the correct representation (i.e. dipole allowed) could be observed due to charge transfer to orbitals that create a dynamic dipole moment normal to the surface.

The impact scattering selection rule is more complex, and can be found in Ibach and Mills.

Group frequencies for adsorbates are usually similar to those determined from IR spectroscopy for coordination compounds and from bulk measurements. Several reference works of this type are available.

Applications

Clean surfaces

In the 2–20 eV range, surface excitations related to the dielectric response function and free carrier density of the surface are observed. Clean-surface loss spectra can include features due to surface phonons, surface plasmons, interband transitions and surface optical phonons in ionic insulators. The probe depth for these phenomena in HREELS is about 10 nm. Transitions between surface states can also be observed in the loss spectrum. Some examples are given in the section related to hydrogen adsorption and surface states below.

Clean surfaces can be probed with electrons to determine vibrational, electronic and structural properties. Diffraction with electrons is well known; practical low-energy electron diffraction is performed with display analysers in the 50–500 eV range. Bulk electronic excitations such as plasmons are usually observed in the 10–50 eV range. While some HREELS spectrometers are capable of 200 eV primary beams for this kind of measurement, plasmons are more often investigated with analyser types designed for higher energy, such as those used for

Auger electron spectroscopy or X-ray and ultraviolet photoelectron spectroscopy.

Adsorbates on surfaces

Chemisorbed molecules Literally hundreds of adsorbate–surface systems have been investigated in the circa 20 years over which HREELS measurements have been made. A selection of work will be presented that represents the major principles of HREELS and what can be learned from this technique. It is not comprehensive, and an electronic literature search program is recommended for up-to-the-minute references to the 'state of the art'. Modes are often denoted by Greek letters, the most common are ν for stretching modes, δ for bending modes and ρ for wag, scissor or twisting modes. Aromatic molecules have several different labelling conventions.

In **Figure 6** the relative importance of charge transfer upon adsorption, bond order, the dipole moment perpendicular to the surface and screening is shown in a comparison of three isoelectronic adsorbates on Pd and Cu surfaces. In the earlier discussion (**Figure 1**), the adsorbate–surface modes were neglected for clarity. In **Figure 6**, NO, CO and CN HREELS spectra are shown; these adsorbates are present on the surface as NO^+, CO and CN^-. Each of these diatomic molecules will have a

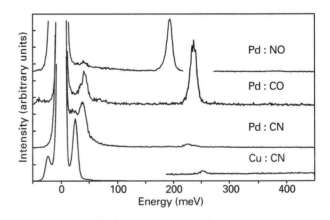

Figure 6 The HREELS spectra of saturation coverage of adsorbed NO, CO and CN on Pd(111) and Cu(111). In each case, a metal–adsorbate (M–R, R= NO,CO,CN) mode and a molecular stretching mode (ν(A–B)) is observed. For NO and CO, the relative bond order is observed in the position of the ν(N–O) and ν(C–O) stretching modes; the relative strength of the metal–adsorbate and molecular stretching dynamic dipole moments is seen in the ratio of the M–NO to ν(N–O) modes, 1:12; M–CO to ν(C–O), 1:2. For CN on Pd(111) and Cu(111), the ν(C–N) mode is parallel to the surface. The difference in charge transfer and bond strength is seen in the ratio M–CN to ν(C–N) which is 20:1 for Pd and 190:1 for Cu. The spectra are not normalized to each other, and are not intended for absolute comparison.

stretching mode and a molecule–surface vibrational mode known as a 'frustrated translation' perpendicular to the surface. For NO and CO, the axis passing through the molecule along its length is perpendicular to the surface; for CN on Pd, the C–N axis and the CN stretching motion are parallel to the surface and partially screened, on Cu the CN motion is again parallel to the surface and hence very weak, but the Cu–CN⁻ mode (near the elastic peak) is very intense due to the motion of the *entire* ion perpendicular to the surface. A gain is also observed. Adsorbed NO is also an ion, but the difference in charge transfer compared to adsorbed CN is significant, as is revealed by the difference in the adsorbate–metal modes.

Using a modern spectrometer with better resolution, **Figure 7** shows a spectrum of an ordered OH overlayer on Ag(110) generated by the reaction of water with pre-adsorbed oxygen (the oxygen spectrum is the lower curve in **Figure 7**). The OH radical is stabilized by the chemisorption bond to the silver surface and forms an ordered (1×2) layer. Note that the δ(O–H) mode is broadened relative to the elastic peak, and that phonons due to the silver surface are observed as both losses and gains (45, 83 cm^{-1}).

In **Figure 8**, the use of dipole selection rules to determine molecular bond orientation is shown. On Cu(111), for adsorbed CN, the CN axis is parallel to the surface. The C–N stretching mode is weak (screened), and measured to lie at 254 meV (2045 cm^{-1}); the Cu–CN mode is very strong. When CN is adsorbed on an oxygen pre-covered surface, the C–N stretching mode is observed at a higher frequency, 266 meV (2140 cm^{-1}), and several times more intense. No new modes are observed, so that

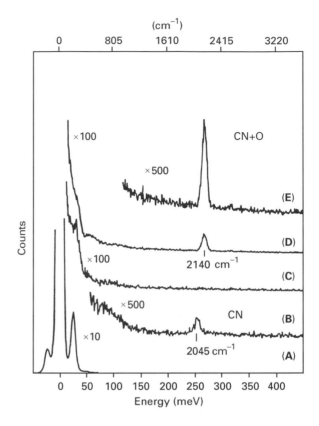

Figure 8   The reorientation of absorbed CN on oxygen pre-covered Cu(111). The bottom spectra (A) and (B) show 100 L (L = 1 × 10⁻⁶ torr s⁻¹) C₂N₂ dissociatively (as CN) absorbed on Cu(111) at 300 K. The loss and gain of the CN–Cu absorbate–metal frustrated translation is clearly evident at ±24 meV. Spectrum (C) shows the Cu(111) surface after 60 L oxygen exposure at 300 K. Spectra (D) and (E) show the surface after exposure to 20 L C₂N₂ after the oxygen exposure. The CN stretching mode is more intense, and shifts from 254 to 266 meV, and shows a 'dipolar' angular dependence. Reprinted from *Journal of Electron Spectroscopy and Related Phenomena*, **44**, M.E. Kordesch, HREELS of the adsorption and reaction of CN on clean and oxygen covered Cu (111), 154, 1987, with permission from Elsevier Science.

an O–CN bond is not formed, and the Cu–CN mode is reduced in intensity. These spectra are interpreted to indicate that the CN molecule initially bonds to the surface as an ion, with the CN axis parallel to the surface. In the presence of pre-adsorbed oxygen, the vertical bonding geometry, with the CN axis perpendicular to the surface, is observed. The shift in frequency is consistent with this observation, and the fact that oxygen adsorbs on Cu as O⁻ and thereby alters the CN bond geometry. Similar orientation effects are observed for molecules such as pyridine and benzene, thiophene, furan and pyrrole. A molecule may be observed to reorient as a function of surface coverage (i.e. a second layer), or as a function of the packing density of the chemisorbed layer (reorientation in the first layer).

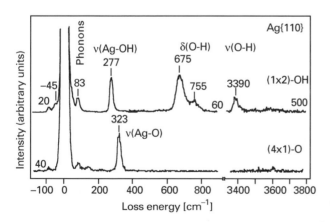

Figure 7   HREELS spectrum of OH and Ag(110) [top], and O [bottom]. A modern HREELS spectrometer was used with 2 meV elastic peak width. Note phonon modes and the width of the δ(O–H) relative to the ν(Ag–OH) mode.

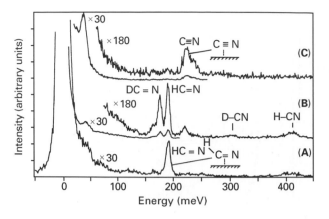

Figure 9 Deuteration and hydrogenation of CN to form DCN and HCN in situ during HREELS observation. (A) HREELS of 10 L HCN at 300 K. (B) 10 L NCCN at 300 K, after 30 min exposure to D_2 at 1×10^{-8} torr and hydrogen form the residual gas. (C) 10 L NCCN at 300 K. The absorbed CN comibnes to form DCN and HCN. Reprinted from *Surface Science*, **175**, M.E. Kordesch, The hydrogenation of CN on Pd(111) and Pd(100), L687–L692, 1986, with permission from Elsevier Science.

Another useful diagnostic method in HREELS (and vibrational spectroscopy in general) is the use of isotopic substitution to identify a particular vibrational mode by substitution of a different mass isotope and determining the shift of the mode frequency upon substitution. In **Figure 9**, the HCN molecule adsorbed on Pd(111) at 300 K is exposed to deuterium while the HREELS spectrum is monitored. New modes are observed; the C–H mode is replaced with a C–D mode at a new frequency related to the change in the mass of D relative to H. The altered mass of DCN relative to HCN also shifts the bending and molecule–substrate modes.

New HREELS spectrometers are able to distinguish shifts due to oxygen, carbon and nitrogen isotopes, among others, with no alteration in the adsorption geometry, simplifying the interpretation of the spectra. The shifts are not as simple as the H/D substitution, and must be calculated from first principles. In **Figure 10**, the reaction of isotopically labelled CO_2 with preadsorbed O to form adsorbed carbonate on Ag(111) is shown, with the relative

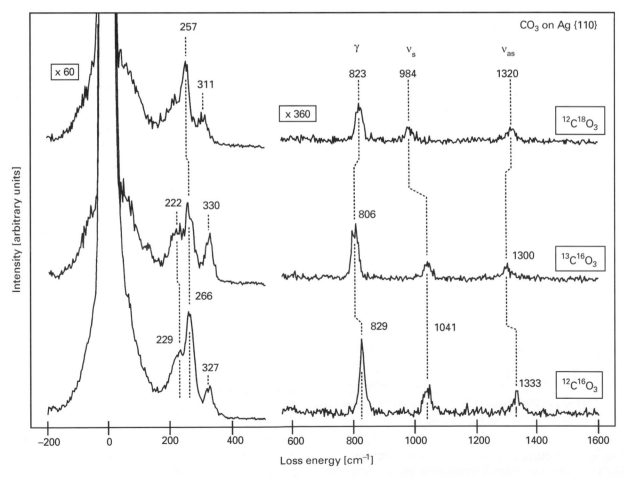

Figure 10 Carbonate formation on Ag(110). The new HREELS spectrometers can distinguish isotopically labelled molecular vibrations. The shifts are complex and must be calculated from first principles.

shifts of the peak positions due to the differences in isotope masses. The assignment of the carbonate modes is included in **Figure 11**. The frequencies are for $(CO_3)^-$ ions in solution. The delta-mode is not seen in specular HREELS. The extrinsic modes of CO_3 are at 229 and 266 cm^{-1} for the normal carbonate; the higher mode is assigned to the substrate stretch, the lower mode is a libration mode. The losses around 330 to 311 cm^{-1} are due to residual oxygen not bonded in a carbonate, shifted in accordance with the isotopic substitution, because the layers are prepared by pre-adsorption of ^{18}O or ^{16}O at 300 K and forming in the (4×1) reconstructed surface. During the reaction (exposure to CO_2, $^{13}C^{16}O_2$ or $^{12}C^{18}O_2$) only part of the oxygen reacts, the rest is compressed into the higher coverage reconstruction, the (2 × 1) surface, and survives. The LEED pattern is a superposition of (2 × 1)O and (1 × 2) CO_3.

Reactions can be observed in HREELS, most commonly the dissociation of large molecules into smaller fragments after heating. A correlation with temperature-programmed desorption is very useful. The reaction of small molecules to form larger compounds is also of interest. In **Figure 12**, the reaction of benzonitrile with oxygen on Cu(111) to form an 'N-bonded oxide' (or fulminate) is shown. The two reactants are adsorbed at low temperature, where no reaction occurs. The substrate is heated to successively higher temperatures and then observed in HREELS. The benzonitrile is adsorbed with the plane of the benzene ring parallel to the surface, and the benzene–CN bond is observed to remain intact in the spectrum during the reaction. At 170 K the new

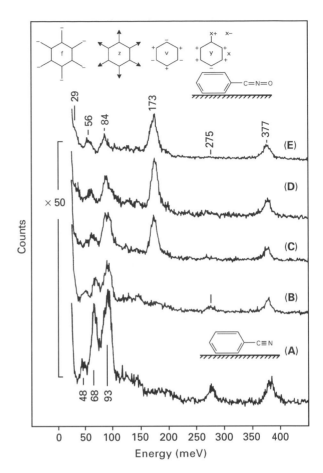

Figure 12 The reaction of benzonitrile and oxygen on Cu(111) to form an 'N-bonded oxide' or CN–O compound. Top left: Whiffen mode symbols for the monosubstituted benzenes: f: in-phase H motion relative to the ring plane (93 meV), Z: C–H stretching motion (377 meV), v: ring torsion (84–86 meV), y: a substituent-sensitive mode where the substituent (small x at the top of the benzene ring) moves out-of phase with the ring C atom opposite to the point of attachment (86–56 meV), x: the in-phase version of y (unresolved). (A) 100 L oxygen + 10 L Phenyl-CN (benzonitrile) at 100 K. (B), as (A) heated to 210 K, (C) heated to 250 K, (D) heated to 300 K. (E) 100 L oxygen + 10 L PhCN exposure at 300 K. The CN–O bond is observed at 173 meV. Reprinted from *Surface Science*, **211/212**, M.E. Kordesch, An HREELS characterization of a surface CN-oxygen compound formed on Cu(111), 1047, 1989, with permission from Elsevier Science.

CN–O stretching band is observed at 171 meV (1370 cm^{-1}). This example illustrates how HREELS is applied to specific chemical bonds.

The deuterium substitution reaction mentioned above raises the question of time-resolved HREELS. Some attempts have been made to use parallel detection methods in HREELS, so that time-resolved spectra can be acquired. The experimental realization of this method is complex; Wilson Ho has pioneered the implementation of time-resolved HREELS, and some results are given in his review.

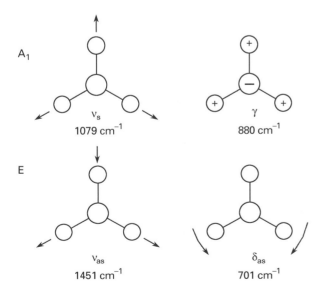

Figure 11 Normal modes for the carbonate ion.

FTIR instruments may prove more efficient in this area.

Large molecules pose a difficult interpretation task for HREELS. A molecule with many, closely spaced modes will produce a spectrum with unresolved vibrational modes. A large molecule may not have a single adsorption geometry, making the application of selection rules difficult. Some success has been achieved with polymers. An example is shown in **Figure 13**; the observation of Fomblin Y, a perfluoropolyether oil, on polycrystalline molybdenum. The polymer has a molecular mass of over 10 000, and contains several different monomer constituents. The fluorine groups, and some pendant group modes, can be observed. The individual fluorine groups, CF, CF_2, CF_3 are not usually resolved (157 meV, 1256 cm^{-1}), even in IR spectra, but several overtones of the C–F modes are observed. Useful information can still be obtained, even from very large molecules or polymers.

As the surface coverage increases, as in a condensed multilayer 'ice', the dipole selection rules are relaxed, because the screening of parallel modes is less effective in the outermost layers of the 'ice', because these layers are further from the surfaces. In **Figure 14**, a HREELS spectrum of HCN multilayer 'ice' condensed on Pd at 100 K is shown. The HCN molecule has three normal modes, ν(C–H) and ν(C–N) stretching modes along the molecule axis, and a doubly degenerate δ(H–CN) bending mode. There are at least 20 modes evident in this spectrum. At low temperature, HCN is thought to form hydrogen-bonded chains (a 'catamer'). The HCN molecule in the chain has a strong static dipole moment, and the motion of the chains parallel to the surface (a libration, or frustrated rotation) serves to induce a strong dynamic dipole moment in the direction perpendicular to the chains. The libration is observed along with the usual HCN modes.

The HCN 'ice' spectrum illustrates several aspects of HREELS. The strong dynamic dipole moment due to the libration mode causes multiple scattering from the molecules, so that there are several replica 'primary' peaks almost as intense as the elastic peak. Consequently, each of the usual HCN loss peaks generates a spectrum with the same spectral intensity relationships as the 'no-loss' peak. The multiple

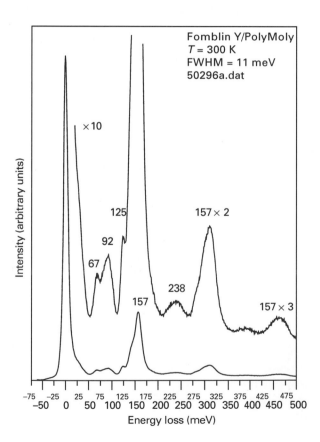

Figure 13 HREELS spectrum of Fomblin Y on polycrystalline molybdenum. The CF, CF_2, CF_3 modes are unresolved, at 157 meV. Overtones at 314 and 471 meV are observed. Other modes: δ(OCO) = 67 meV, ν(OCO) or ν(CC) = 92 meV, the mode at 125 meV is characteristic of Fomblin Y, a stretching mode in a monomer segment, ν(CF_3–C–CF_2).

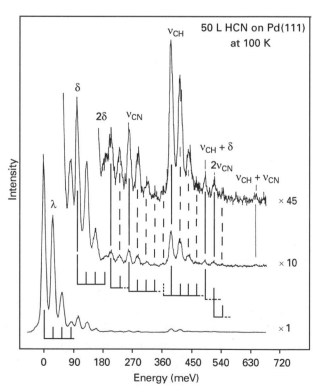

Figure 14 HREELS spectrum of 50 L HCN exposure at 100 K on Pd(111). The libration mode λ, the HCN bending mode δ, and the CN and CH stretching modes ν_{CH} and ν_{CN} are identified with some of their combinations. The vertical ladders show some of the Poisson series for the libration mode.

scattering intensities follow the Poisson relationship for multiple random events. The chain formation distorts the molecular potential, so that overtones and combinations are possible due to anharmonic coupling of the normal modes.

A common phenomenon observed upon adsorption of molecules is the decrease in the elastic peak intensity as a function of coverage. **Figures 15A, B** show the intensity and angular width of the elastic peak as a function of HCN coverage on Pd. The intensity correlates well with layer formation on the surface, much like RHEED oscillations. Even though the scattering mechanism is decidedly 'dipolar', the surface is made 'rough' by a partial coverage of HCN. As the surface layer is completed, the specular beam sharpens, but the phenomenon is repeated as the second layer adsorbs.

In dielectric layers, where a surface phonon mode may occur, or in ionic crystals, multiple scattering from the surface phonon mode can result in Poisson 'replicas' of the no-loss peak. These modes are referred to as 'Fuchs–Kliewer' modes; they are a general feature of HREELS spectra of ionic and polar materials, and metal oxides. Ordered overlays on surfaces can also exhibit collective modes, but at submonolayer coverages the HREELS loss peaks are due almost exclusively to single oscillations of the fundamentals. Substrate (silver) phonon modes are shown at 10 meV (83 cm^{-1}) in **Figure 7**.

Single crystals are understandably the most interesting substrates for surface science, because other spectral data and calculations are made easier by an ordered surface. For technology, however, such as lubrication studies, or adhesion, a polycrystalline surface is more relevant. Some success has been achieved with polycrystalline metal surfaces, but a case-by-case test for suitability is necessary. The spectrum in **Figure 13** (Fomblin Y perfluoropolyether) was nearly identical to that obtained on Mo(100), with the exception of a slightly broader elastic peak.

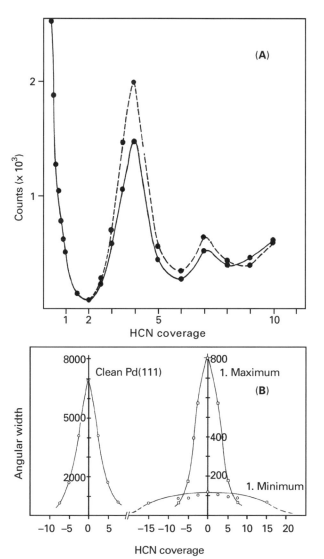

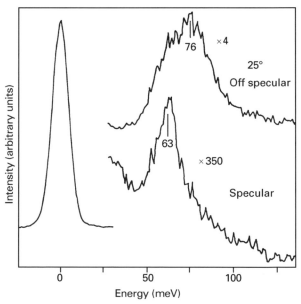

Figure 15 (A) The intensity of the no-loss peak as a function of HCN exposure in Langmuir at 100 K. The dotted line is with the spectrometer optimized to compensate for work function changes, the solid line is 'as measured'. (B) The angular width of the no-loss peak, measured for clean Pd(111), at the first minimum, 2 L, and at the first maximum in **Figure 9A**, about 4 L.

Figures 16 Specular and off-specular HREELS spectra of H adsorbed on Pd(100). The perpendicular mode is at 63 meV. The parallel mode is at 76 meV. Reprinted from *Surface Science*, **178**, M.E. Kordesch, Surface resonances in vibrational spectroscopy of hydrogen on transition metal surfaces: Pd(100) and Pd(111), 578–588, 1986, with permission from Elsevier Science.

Hydrogen Hydrogen is difficult to detect with many surface analytical methods. In HREELS, the adsorption of hydrogen and deuterium is relatively easy, both experimentally, and in 'first instance' interpretation of the spectral features. In the case of hydrocarbons and other molecules, the CH, NH or OH modes are often studied. The CH modes in aromatic molecules are observed to be partially 'impact' modes. **Figure 7** shows several OH modes; in particular it is possible to observe natural line width broadening for the δ(O–H) mode on Ag(110) with the new high resolution spectrometers.

On many transition metals, hydrogen adsorbs dissociatively, with atomic hydrogen adsorbed in high-symmetry sites. The modes perpendicular to the surface are usually dipole modes; parallel, frustrated translation, modes have some impact scattering character and can be observed in off-specular

spectra. **Figure 16** is an example of specular and off-specular HREELS spectra of H adsorbed on Pd(111).

Atomic hydrogen is small and adsorbs near to the surface. H modes can directly couple to surface states and resonances, such as image potential states, that may overlap in the near-surface region. An empirical verification of this phenomenon, observed on several metals (Pd, Pt, Rh, Ru) is the enhancement of surface resonances by adsorbed H, and the enhancement of H vibrational modes at primary energies which correspond to the population of the surface resonance with electrons from the incident HREELS beam. In **Figure 17**, the reflectivities of the Pd(111) and (100) surface with and without adsorbed hydrogen are shown. The relative intensities of the H frustrated translation and rotation (perpendicular and parallel) modes are shown in **Figure 18**.

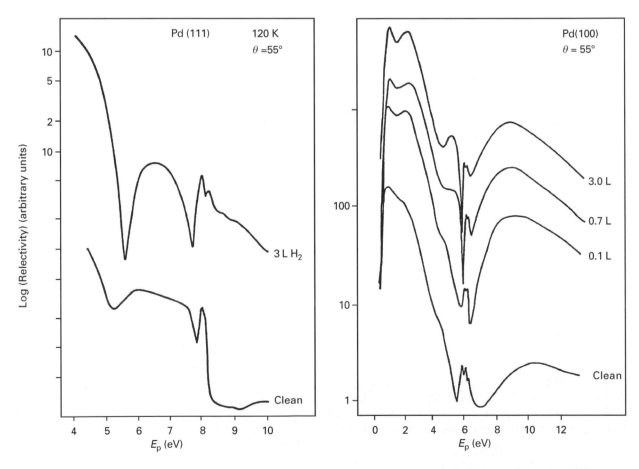

Figures 17 Surface reflectivity of Pd(111) and Pd(100) with and without adsorbed H. The reflectivity is on log scale. The addition of hydrogen shifts and intensifies the lowest energy surface resonance on Pd(111) (5.5 eV). The sharp drop in reflectivity at 8 eV corresponds to the emergence of a surface diffraction beam, and opens a new channel for electron interaction with the surface. The image potential states are just below this emergence threshold. On Pd(100) the curves are similar, but the energy scale is reduced due to the different crystal structure of the surface and different-sized surface Brillouin Zone. Reprinted from *Surface Science*, **178**, M.E. Kordesch, Surface resonances in vibrational spectroscopy of hydrogen on transition metal surfaces: Pd(100) and Pd(111), 578–588, 1986, with permission from Elsevier Science.

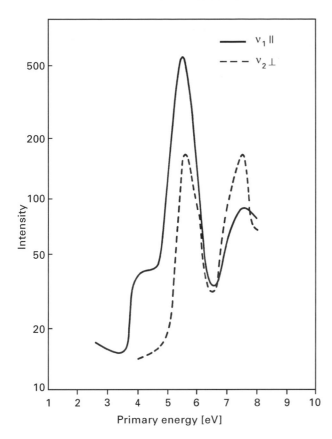

Figure 18 The intensity of the parallel and perpendicular H Vibrational modes at different primary energies on Pd(111). The maxima in the vibrational enhancement correspond to minima in the electron reflectivity curves. Reprinted from *Journal of Electron Spectroscopy and Related Phenomena*, **38**, M.E. Kordesch, Surface resonances on Pd(111)/H observed with HREELS, 289–298, 1986, with permission from Elsevier Science.

List of symbols

A = symmetry group designation; G = reciprocal lattice vector; I = intensity; k = crystal momentum; m = electron mass; p = dipole moment; ΔE = energy spread.

See also: **Hydrogen Bonding and Other Physicochemical Interactions Studied By IR and Raman Spectroscopy; IR Spectral Group Frequencies of Organic Compounds; IR Spectroscopy, Theory; Photoelectron Spectrometers; Polymer Applications of IR and Raman Spectroscopy; Surface Studies By IR Spectroscopy; Symmetry in Spectroscopy, Effects of.**

Further reading

Harris DC and Bertolucci MD (1989) *Symmetry and Spectroscopy: An Introduction to Vibrational and Electronic Spectroscopy*. New York: Dover Publications, Inc.

Herzberg G (1989) *Molecular Spectra and Molecular Structure: I. Spectra of Diatomic Molecules, II. Infrared and Raman Spectra of Polyatomic Molecules*. Malabar, Florida: Robert E. Krieger Publishing Co.

Ho W (1993) High resolution electron energy loss spectroscopy. In: Rossiter BW and Baetzold RC (eds) *Investigations of Surfaces and Interfaces – Part A*. Physical Methods of Chemistry Series, 2nd edn, Vol IXA, Ch 4, pp 209–320. New York: Wiley.

Ibach H and Mills DL (1982) *Electron Energy Loss Spectroscopy and Surface Vibrations*. New York: Academic Press.

Ibach H (1990) *Electron Energy Loss Spectrometers*, Springer Series in Optical Sciences, Vol 63. Berlin: Springer Verlag.

Kesmodel L (1995) High resolution electron energy loss spectroscopy. In: Hubbard AT (ed) *The CRC Handbook of Surface Imaging and Visualization*, pp 223–237. Boca Raton: CRC Press, Inc.

Lüth H (1993) *Surfaces and Interfaces of Solids*, 2nd edn. Berlin: Springer Verlag.

McRae EG (1979) Electronic surface resonances of crystals. *Reviews of Modern Physics* 51: 541–568.

Nakamoto K (1986) *Infrared and Raman Spectra of Inorganic and Coordination Compounds*, 4th edn. New York: Wiley.

Prutton M (1994) *Introduction to Surface Physics*. Oxford: Clarendon Press.

Somorjai GA (1993) *Introduction to Surface Chemistry and Catalysis*. New York: John Wiley and Sons.

Weidhem J, Mueller U and Dehnike K (1982) *Schwingungsspektroskopie*. New York: Thieme.

Yates JT Jr (1998) *Experimental Innovations in Surface Science*. New York: Springer Verlag.

Vibrations at Surfaces Conferences:

1st: Kernforschungsanlage Jülich Reports, Jülich, Germany, Conference 26 (1978).

2nd: Caudano R, Gilles J-M and Lucas AA (1982) *Vibrations at Surfaces*. New York: Plenum.

3rd: Brundle CR and Morawitz H (1982) *Vibrations at Surfaces 1982*. Amsterdam: Elsevier.

4th: King DA, Richardson NV and Holloway S (eds) (1986) *Vibrations at Surfaces 1985*. Amsterdam: Elsevier.

5th: Bradshaw AM and Conrad H (eds) (1987) *Vibrations at Surfaces 1987*. Amsterdam: Elsevier.

6th: Chabal YT, Hoffmann FM and Williams GP (eds) (1990) *Vibrations at Surfaces 1990*. Amsterdam: Elsevier.

7th: Proceedings in *Journal of Electron Spectroscopy and Related Phenomena* **64/65** (1993).

8th: Proceedings in *Surface Science* **368** (1996).

High Resolution IR Spectroscopy (Gas Phase) Instrumentation

Jyrki K Kauppinen and **Jari O Partanen**,
University of Turku, Finland

VIBRATIONAL, ROTATIONAL &
RAMAN SPECTROSCOPIES
Methods & Instrumentation

The definition of high resolution in IR spectroscopy has been changing with time. In this article, the limit of high resolution is defined as 0.1 cm^{-1}. This is the full width at half height (FWHH) of the line. In scientific language, the higher the resolution, the smaller the FWHH. In history, grating and even prism spectrometers, which have a lower resolution than 0.1 cm^{-1}, have also been regarded as high-resolution instruments, but the development of spectroscopic instrumentation has pushed the limit of high resolution towards smaller FWHH. Nowadays, high-resolution IR spectra are generally recorded by Fourier-transform IR (FT-IR) spectrometers, or by laser spectrometers.

Michelson, in the 1880s, invented his interferometer, which was used to study the speed of light and to fix the standard metre with the wavelength of a spectral line. The use of the Michelson interferometer in spectroscopic applications was started almost 60 years later.

The first FT-IR spectrum was computed in 1949 by Peter Fellgett, who used an interferometer to measure light from celestial bodies. Transforming an interferogram into a spectrum was, however, a laborious task, and FT-IR spectroscopy for years remained the field of only a few advanced research groups, who had large computers for time-consuming Fourier transforms. The first commercial FT-IR spectrometers appeared at the end of the 1960s, when microcomputers became available. The development of the Cooley–Tukey algorithm in 1965 for a quick performance of a Fourier transform (the fast Fourier transform, FFT) was an important step for FT-IR spectroscopy. The FT-IR spectrometers of the 1960s worked with a resolution of a few wavenumbers, but they were the high-resolution technology of their own time.

There are a few commercial high-resolution spectrometer models, and research laboratories have built their own instruments, which can achieve a resolution of 0.001 cm^{-1}, or even better.

The history of laser spectrometers is short, starting effectively at the beginning of the 1980s.

Basic definitions

IR radiation consists of electromagnetic waves, which oscillate with a frequency of 3×10^{11} to 4×10^{14} Hz. The corresponding wavelength range is 10^3 to 0.78 μm. The infrared waves travel with the group velocity of light $c_n = c_0/n$, where c_0 is the velocity of light in a vacuum (2.99792458×10^8 m s^{-1}) and n is the refractive index of the medium. IR radiation has shorter wavelength than microwaves but longer wavelength than visible light. The IR region may be subdivided into three: far-IR (10^3 to 30 μm), mid-IR (30–3 μm) and near-IR (3–0.78 μm).

Generally, in spectroscopy, the interaction between the matter and the electromagnetic waves takes place via absorption or emission. The matter can absorb the electromagnetic waves with wavelength λ in the transition $E'' \rightarrow E'$ if

$$hc/\lambda = E' - E'' \qquad [1]$$

where h is Planck's constant (6.626076×10^{-34} J s) and E'' and E' are lower and higher energy levels, respectively, of the atoms or the molecules. Equation [1] is a basic result of quantum physics. The emission is the inverse process, where the energy drop of the atoms or the molecules is emitted as an electromagnetic quantum hc/λ. In IR spectroscopy, generally the absorption is examined. When IR radiation is penetrating through the sample material (solid, liquid or gas) the absorption is described by Beer's law

$$I(\nu) = I_0(\nu) \exp[-\beta(\nu)Cd] \qquad [2]$$

where $I_0(\nu)$ and $I(\nu)$ are incident and transmitted intensities, respectively, $\nu = 1/\lambda$ is the wavenumber of radiation, $\beta(\nu)$ is the absorptivity, C is the concentration and d the thickness of the sample. For a solid sample $\beta(\nu)C = \alpha(\nu)$ is the absorption coefficient. The fraction $[I_0(\nu) - I(\nu)]/I_0(\nu)$ is called absorption, $I(\nu)/I_0(\nu)$ transmittance, and $-\ln[I(\nu)/I_0(\nu)] = \beta(\nu)Cd$ absorbance. Typical curves $I_0(\nu)$, $I(\nu)$, $I(\nu)/I_0(\nu)$ and $-\ln[I(\nu)/I_0(\nu)]$ are shown in **Figure 1**. The absorbance as a function of the wavenumber ν is called simply the (IR) absorption spectrum.

Fourier transform interferometers

Michelson interferometer

Figure 2 illustrates a basic optical layout of the original Michelson interferometer and **Figure 3** shows

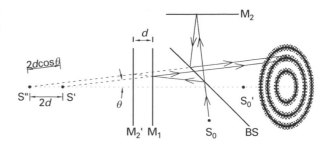

Figure 3 Michelson interferometer produces two coherent images S′ and S″ of the real source S_0. These coherent sources generate the interference fringe pattern.

schematically the ray travelling at the angle θ with the optical axis of the interferometer. If the beamsplitter divides the incident intensity I_0 into two equal parts, the output intensity is simply $I = 2I_0(1 + \cos\phi)$, with $\phi = (2\pi/\lambda)2nd\cos\theta = 2\pi\nu x$, where x is the optical path difference. If $\theta = 0$, $n = 1$, and S_0 is a monochromatic point source, then $x = 2d$, and

$$I = 2I_0[1 + \cos(2\pi\nu 2d)] \qquad [3]$$

Let us assume that S_0 has a wide-band continuous spectrum $E(\nu)$ as shown in **Figure 4**, and it is still a point source. Now the interference signal from the region between ν and $\nu + d\nu$ is given by $dF(x,\nu) = 2E(\nu)[1 + \cos(2\pi\nu x)]\,d\nu$, according to Equation [3]. At a given optical path difference x, the total signal is

$$F(x) = 2\int_0^\infty E(\nu)[1 + \cos(2\pi\nu x)]d\nu \qquad [4]$$

where $F(x)$ is called an interference record. By subtracting from $F(x)$ the constant term

$$\frac{1}{2}F(0) = 2\int_0^\infty E(\nu)d\nu \qquad [5]$$

an interferogram is obtained:

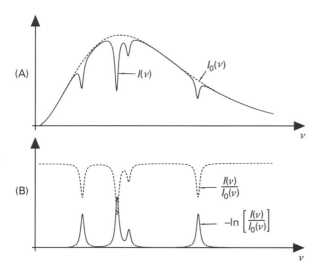

Figure 1 (A) Incident (or background) spectrum $I_0(\nu)$ and transmitted spectrum $I(\nu)$. (B) Transmittance spectrum $I(\nu)/I_0(\nu)$ and absorbance spectrum $-\ln[I(\nu)/I_0(\nu)]$.

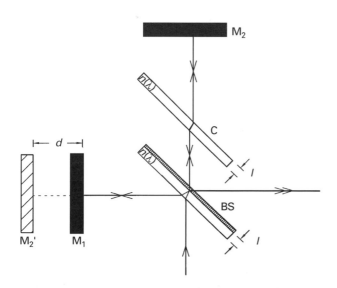

Figure 2 Optical arrangement of the Michelson interferometer. M_1 is a fixed and M_2 a moving mirror. BS is a beamsplitter and C a compensating plate. M_2' is the image of M_2 formed by BS.

$$I(x) = F(x) - \frac{1}{2}F(0) = 2\int_0^\infty E(\nu)\cos(2\pi\nu x)d\nu \quad [6]$$

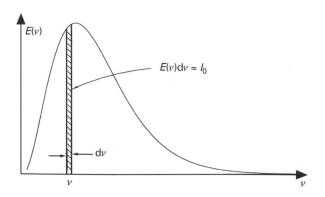

Figure 4 Wide band continuous spectrum $E(\nu)$ can be expressed as the sum of monochromatic infinitely narrow ($d\nu$) lines.

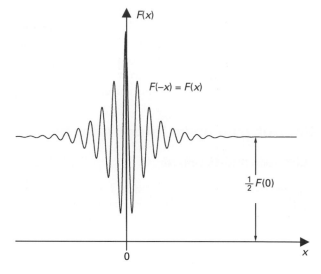

Figure 5 Typical interference record $F(x)$.

If it is assumed that $E(-\nu) = E(\nu)$, the computations become easier, because

$$I(x) = \int_{-\infty}^{\infty} E(\nu)\cos(2\pi\nu x)\mathrm{d}\nu$$

$$= \int_{-\infty}^{\infty} E(\nu)\exp(\mathrm{i}2\pi\nu x)\mathrm{d}\nu = \mathscr{F}\{E(\nu)\} \qquad [7]$$

Hence, I_x and $E(\nu)$ form a Fourier transform pair, and they can be written as follows

$$I(x) = \int_{-\infty}^{\infty} E(\nu)\exp\left(\mathrm{i}2\pi\nu x\right)\mathrm{d}\nu = \mathscr{F}\{E(\nu)\}$$
$$\qquad\qquad [8]$$
$$E(\nu) = \int_{-\infty}^{\infty} I(x)\exp\left(-\mathrm{i}2\pi\nu x\right)\mathrm{d}x = \mathscr{F}^{-1}\{I(x)\}$$

where $\mathscr{F}\{\ \}$ and $\mathscr{F}^{-1}\{\ \}$ are the Fourier transform and the inverse Fourier transform, respectively. Furthermore, one can notice that $\mathscr{F}\{\mathscr{F}^{-1}\{I(x)\}\} = I(x)$. A typical interference record is shown in **Figure 5**.

Because in practice it is impossible to make the inverse Fourier transform $\mathscr{F}^{-1}\{I(x)\}$ of the continuous $I(x)$, one has to use the discrete Fourier transforms. Thus the interferogram $I(x)$ is sampled at the discrete points $x_j = jh$, let us say, between $x = -L$ and $x = L$. The spectrum is now given by

$$E^h(\nu) = h\sum_{j=-N}^{N-1} I_j\exp(-\mathrm{i}2\pi\nu jh) \qquad [9]$$

In practice, the discrete Fourier transform is computed by the FFT in order to minimize the needed computation time. It can be shown that

$$E^h(\nu) = W^h(\nu) * E(\nu) \qquad [10]$$

where $*$ means convolution, and the instrumental profile due to the truncation of the interferogram at $\pm L$ and the discrete sampling is given by

$$W^h(\nu) = \sum_{k=-\infty}^{\infty} 2\operatorname{sinc}\left[2\pi\left(\nu + \frac{k}{h}\right)\right] \qquad [11]$$

And now the instrumental resolution (FWHH of the instrumental function) is given by $\Delta\nu = 1.207/(2L)$, where L is the maximum optical path difference (please note, a point source is assumed). Furthermore, the optimum sampling interval is given by $h = 1/(2\nu_{max})$ if $E(\nu) = 0$ with $|\nu| > \nu_{max}$. The optimum sampling interval guarantees there is no aliasing in the spectrum.

Collimated beam and extended source in Michelson interferometer

A practical Michelson interferometer (**Figure 6**) has an extended source, as shown in **Figure 7**, and a collimator mirror. The parabolic collimator mirror may be off-axis, as in **Figure 6**, or on-axis, as in **Figure 7**. When a denotes the area of the source, A the area of the collimator mirror, f the focal length

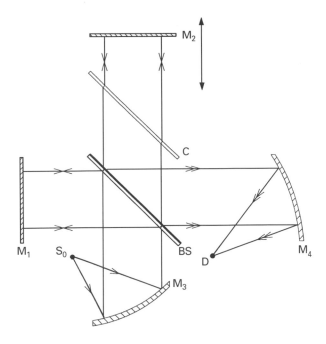

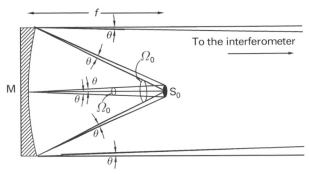

Figure 7 Extended IR radiation source S_0 at the focal point of the parabolic collimator mirror M.

Figure 6 Optical layout of a practical Michelson interferometer. S_0 is a black-body source and D an IR detector; M_3 and M_4 are parabolic off-axis mirrors. M_3 and M_4 may also be normal parabolic mirrors as in **Figure 7**.

of the collimator, $\Omega_0 = A/f^2$ the solid angle of the collimator, and $\Omega = a/f^2$ the solid angle of the source, then the signal is proportional to

$$aA/f^2 = \Omega A = a\Omega_0 \qquad [12]$$

On the other hand, the optical path difference is now $x \cos \alpha \approx x[1 - \Omega'/(2\pi)]$, where $x = 2nd$ and the angle α between the optical axis and the beam varies from 0 to θ, which is the maximum of α, while Ω' is changing from 0 to Ω. Thus the total interferogram is

$$I_\Omega(x) = \int_0^\Omega I\left[x\left(1 - \frac{\Omega'}{2\pi}\right)\right] d\Omega' =$$

$$\Omega \int_{-\infty}^{\infty} E(\nu)\mathrm{sinc}\left(\frac{\nu x\Omega}{2}\right)\exp\left[i2\pi\nu x\left(1 - \frac{\Omega}{4\pi}\right)\right]d\nu \quad [13]$$

In the case of a point source,

$$\lim_{\Omega \to 0} I_\Omega(x) = \int_{-\infty}^{\infty} E(\nu)\exp(i2\pi\nu x)\,d\nu$$

and thus the total instrumental function can be

given by

$$W_{L,\Omega}^h(\nu) = W_\Omega(\mp\nu_0, \nu)*$$

$$\left\{\sum_{k=-\infty}^{\infty} 2\ \mathrm{sinc}\left[2\pi\left(\nu + \frac{k}{h}\right)L\right]\right\} \qquad [14]$$

This has to be used in Equation [10] instead of $W_L^h(\nu)$. As one can see, the instrumental line shape function [i.e. the observed line shape if $E(\nu) = \delta(\nu + \nu_0) + \delta(\nu - \nu_0)$] is given by

$$W_{L,\Omega}(\nu) = W_\Omega(\nu_0, \nu) * 2L\,\mathrm{sinc}(2\pi\nu L) \qquad [15]$$

This situation is illustrated in **Figure 8** with three different Ω values. The optimum situation is the case where the aperture broadening $W_\Omega(\nu_0, \nu)$ and the truncation broadening $2L\mathrm{sinc}(2\pi\nu L)$ have approximately equal widths, i.e.

$$\frac{\nu_0\Omega}{2\pi} \approx \frac{1.21}{2L} \qquad [16]$$

In **Figure 8** the optimum situation is roughly in the second row. Furthermore, it can be seen that the position of the line at ν_0 is given by $\nu_{\mathrm{obs}} = [1 - \Omega/4\pi]\nu_0$ = constant $\times \nu_0$. Now, according to Equation [12], the signal is proportional to $aA/f^2 = \Omega A$, and for a given resolution, when Ω is fixed, the only parameter which has an effect on the signal, is the area A of the collimator mirror. The area a of the radiation source and the focal length f of the collimator mirror can be selected freely provided that $a/f^2 = \Omega$. A similar effect can occur due to the area of the detector. As mentioned earlier, the instrumental profile plays an important role in describing the behaviour of the

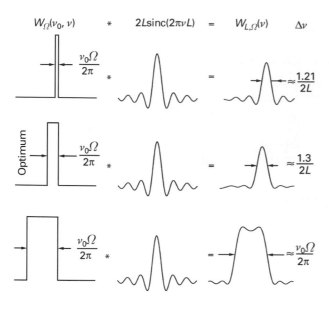

Figure 8 Total instrumental line shape function $W_{L,\Omega}(\nu)$ with three different Ω-values.

spectrometer. Especially in high-resolution gas-phase spectroscopy, where the instrumental profile $W_{L,\Omega}$ and the real (Lorentzian) line shape have approximately equal widths, the convolution by instrumental profile changes substantially the real line shape.

Other FT interferometers

The most difficult disadvantage of the Michelson interferometer is tilting of the moving plane mirror during scan. It is well known that the tilting angle of the moving mirror should be less than about $\lambda/(8D)$, where λ is the wavelength of the light under study and D is the diameter of the moving mirror. Thus the driving of the mirror is very difficult with high-resolution FT-IR spectrometers. One way to solve this problem is by using a dynamic alignment system, which measures the tilting angle and tries to keep it at zero during the run. However, as the resolution increases, the probability of malfunction of the dynamic alignment system during a very long run increases.

Another solution is to use cube-corner mirrors, which consist of three mutually orthogonal plane mirrors. The mirror system reflects any incident ray back in the opposite direction. The first very high resolution cube-corner interferometer that really worked was the Oulu interferometer in Finland. The Oulu interferometer is basically a Michelson interferometer, where the moving and the fixed mirror are cube-corners. If the corners are perfect, the tilt problem completely disappears. The only disadvantage of this type of interferometer is a shearing problem, i.e. the lateral shift of the moving cube-corner. This is the

case where the moving cube-corner does not follow the optical axis of the interferometer. The lower the resolution the more serious this disadvantage is.

A cube-corner interferometer is improved if two cube-corners are set to move back to back. Two cube-corners fixed back to back as a single moving part completely eliminates the tilting and the lateral shifts, if the cube-corners are perfect ($90° \times 90° \times 90°$), and the corners coincide exactly. Even further improvements are emerging at the University of Turku, Finland, namely a super high-resolution interferometer with a resolution of 4×10^{-4} cm^{-1} which has the moving cube-corner pair back-to-back and the beam focus at the beamsplitter (see **Figure 9**). This type of interferometer is almost ideal for very high resolution IR spectrometry.

An alternative interferometer to the cube-corner is the cat's-eye interferometer, where the cube-corners are replaced by cat's-eye retroreflectors. These are components that consist of one parabolic and one spherical mirror, and reflect an incident ray back in the opposite direction. Cat's-eye interferometers are also used in high-resolution spectroscopy.

Instrumental details of FT-IR spectrometers

Radiation source

Black-body radiators are used as broad band radiation sources in IR spectrometers. The only adjustable parameter of the source is the temperature; the higher the temperature, the higher the intensity. Typical temperatures are 1000–1800 K. The most commonly used sources in mid-IR are Globars (silicon carbide), Nernst glowers and nichrome coils. In the far-IR below 100 cm^{-1} a high-pressure mercury lamp is quite good. Nowadays, ceramic sources have become more popular. The radiation sources may need a cooling system, like water cooling.

Beamsplitter

The beamsplitter is one of the most expensive and sensitive components of an interferometer, and must be chosen carefully.

A pellicle beamsplitter is a high tensile strength elastic membrane, which is stretched like a drumhead over a flat frame. The membrane has a high refractive index and its absorption is negligible. The most common membrane material is Mylar. An ideal beamsplitter divides the incoming intensity into two equal parts. This is, however, achieved only in a definite wavenumber region, due to the interference

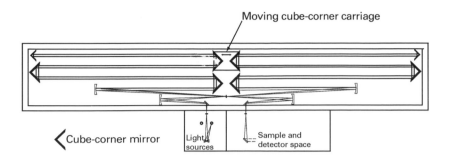

Figure 9 Layout of the Turku high-resolution interferometer.

phenomenon in a thin film. A Mylar beamsplitter works with wavenumbers less than 1000 cm^{-1}.

In the near-IR region a beam-splitting film should be very thin. In this case it is necessary to use a low-absorption dielectric coating, which is deposited on a suitable substrate plate. A thin Ge layer deposited on a KBr substrate, with a KBr compensating plate, is a good beamsplitter in the wavenumber region 1000–4000 cm^{-1}.

Optics

With the help of optical components the IR radiation is carried from the source through the whole spectrometer and finally to the detector. Mostly the needed accuracy is so-called geometrical accuracy. This means that the beam has to be guided to the appropriate places to within 0.1–1 mm. However, higher accuracy is needed in order to measure the wavelength of radiation in the interferometer. It is said that there is a need for interferometric accuracy, i.e., everything should be correct within a small fraction of wavelength, say 0.01–0.1 μm.

In IR spectroscopy there are only a few materials transparent enough; this means that lens optics is not usually used. Instead, the use of mirror optics is preferred and the transmission of any material is avoided. However, in certain circumstances, for convenience sake, attempts are made to use some materials, e.g. polyethylene and KBr lenses. Thus the weakest points in IR optics are, besides the beamsplitter, only the windows of the gas cell. Mirrors are often manufactured from Pyrex glass coated with metal. Gold coating has a good reflection coefficient especially in far-IR and aluminium is also quite good over the whole IR region. In IR spectrometers, plane mirrors, spherical mirrors, paraboloidal mirrors, off-axis paraboloidal mirrors, ellipsoidal mirrors, toroidal and off-axis ellipsoidal mirrors are used. In planning IR instruments one must bear in mind that a perfect imaging takes place only in plane mirrors. All others have some kind of aberration, like spherical aberration. These aberrations limit the performance of instruments. However, there are some special imaging situations where perfect imaging is possible, for example, a spherical mirror images perfectly only on the optical axis from the centre of curvature back to the centre of curvature. A paraboloidal mirror images from infinity to the focal point on the optical axis and the off-axis paraboloid images from infinity to the focal point, which is out of the optical axis. An ellipsoidal mirror images from one focus to another. Hence these mirrors should preferably be used, so that conditions are as near perfect as possible. This also minimizes aberrations. In IR spectroscopy it is very important that as much radiation power as possible is transmitted from the source through the whole optics to the detector.

Gas cell

The most common sample compartment in high-resolution FT-IR spectrometers is the White cell, where the beam passes a path of 1–1000 m. In a long-path gas cell, however, a small f/number (focal length of the mirror/diameter of the pupil) and a large size of the circular image may lead to a loss of effectivity. In the Oulu interferometer this problem was solved by an alternative multiple-path gas cell, where the images at the focal planes occur on a circle around the principal axis of the cell. The aberrations remain small in spite of a small f/number, because the successive images are relatively near to the principal axis of the optics.

Detectors

The detector of the FT-IR spectrometer is very important, because it determines the maximum possible signal-to-noise ratio (S/N; where S = signal and N = noise) and the minimum recording time of the spectrum. IR detectors can be divided into two categories: thermal detectors and quantum detectors. Thermal detectors sense the change of temperature of a detector material due to absorption of the IR radiation. The output signal is generated in the form of a thermal electromotive force (e.g. thermocouples), the change in resistance of a conductor (bolometers) or semiconductor (thermistor bolometers), or the change in pressure of a noble gas (pneumatic detectors). Thermal detectors respond to radiation over a wide range of wavenumbers, but they are usually slow with a time constant typically from 0.001 to 0.1 s. In pyroelectric bolometers, a heat-sensing ferroelectric material changes the degree of polarization when the temperature changes due to absorption. The voltage responsivity of a pyroelectric detector at frequency f is roughly proportional to $1/f$. The most commonly used material for the pyroelectric detector is deuterated triglycine sulfate.

In quantum detectors, IR radiation causes electrons to be excited to a higher energy level. In an n-type semiconductor, electrons in the valence band are unable to increase the conductivity. If IR radiation excites the electrons to the conduction band, they can act as current carriers. For good sensitivity, it is necessary to cool the detector. For certain detectors, such as PbS and PbSe, it may be sufficient to use a thermoelectric cooler to maintain their temperature just below ambient. Mercury cadmium telluride detectors must usually be maintained at liquid-nitrogen temperature (77 K), whereas others may require cooling to the temperature of liquid helium (4.2 K).

In the far-IR, where photon energies are very low, quantum detectors cannot be used at all. Two types of thermal detectors, each operating at liquid-helium temperatures, have been used. The first one is the germanium bolometer, often doped with a low level of copper, gallium or antimony. For increased responsivity, these detectors are cooled down to 1.5 K by pumping the detector cryostat. The second type is the InSb hot-electron detector.

The sensitivity of IR detectors is expressed by the noise equivalent power (NEP) of the detector. This NEP is the ratio of the root mean square (rms) noise voltage V_{rms} in the detector to the voltage responsivity, i.e. $NEP = V_{rms}/R$. If V_{rms} is in units of V Hz$^{-1/2}$ and R in units of V/W, then NEP is in units W Hz$^{-1/2}$. Usually the manufacturer gives a specific detectivity D^* as a function of wavelength or wavenumber. It is defined as $D^* = \sqrt{A_D}/NEP$, where A_D is the area of the detector. Some typical D^*-curves are shown in **Figure 10**.

Electronics

Nowadays, IR spectrometers are usually computer controlled. Thus the main component of the electronics is a microprocessor, which controls all the operations of the spectrometer, collects the data, and computes, saves, displays and plots final spectra. The computer is very important to perform any kind of data treatment needed in IR spectroscopy. The other electronic parts of FT-IR spectrometers are a preamplifier and a main amplifier of the IR detector, an

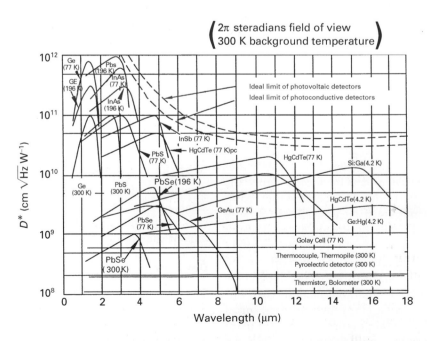

Figure 10 D^* curves for the most used IR detectors (Hamamatsu Photonics K. K., Solid State Division, 1990).

analogue-to-digital converter (AD converter), a set of bandpass electronic filters and power supplies. It is important that the numbers of bits and data rate of the AD converter are high enough.

Advantages of FT-IR spectrometers

Jacquinot advantage (throughput advantage)

In an interferometer the total radiation power on the detector is much higher than that in a grating instrument. In practice, the FT-IR spectrometer has an approximately 200 times larger S/N than the grating spectrometer with the same resolution and recording time. The reason is that at the optimum resolution the area of the circular aperture (the Jacquinot stop) in the FT-IR spectrometer is 200 times larger than the slit in the grating spectrometer.

Fellgett advantage (multiplex advantage)

In FT-IR spectrometers all the wavenumbers ν are recorded simultaneously. Let us assume that the wavenumber region under study consists of M resolution elements, i.e. $\nu_2 - \nu_1 = M\Delta\nu$. If the noise of a detector and an amplifier is greater than the noise of a radiation source, then the S/N is proportional to the square root of the total observation time T, i.e. $S/N \propto \sqrt{T}$. In fact $S \propto T$ and $N \propto \sqrt{T}$. In FT-IR spectrometers one observes all the times T, all the M spectral elements, and thus $(S/N)_{FT} = a\sqrt{T}$; in grating spectrometers the observation time of each spectral element is only T/M, and $(S/N)_G = a\sqrt{T/M}$, where in both equations a denotes the same constant. And finally $(S/N)_{FT}/(S/N)_G = \sqrt{M}$, which is the Fellgett multiplex advantage. The Jacquinot advantage and the Fellgett advantage together give $(S/N)_{FT}/(S/N)_G \approx 200\sqrt{M}$ in IR.

Other advantages

It is useful to define a K-factor as $K = \Delta\nu_{true}/\Delta\nu$, where $\Delta\nu$ is the FWHH of the instrument function $W_{\Omega,L}(\nu_0, \nu)$ and $\Delta\nu_{true}$ is the FWHH of the true line shape. The optimum instrumental resolution $\Delta\nu_0 = \Delta\nu_{true}/K_0 \approx 1.21/(2L_0)$ is the case where the maximum distortion of observed lines is roughly equal to the noise of the spectrum and the optimum K-value is given by $K_0 = 2\ln[(S/N)_0]/(1.21\pi) \approx \log_{10}[(S/N)_0]$. The greatest difference between FT and grating instruments is the instrumental profiles $2L\mathrm{sinc}(2\pi\nu L)$ (in FT) and $L'\mathrm{sinc}^2(\pi\nu L')$ (in grating). **Table 1** shows the optimum K-values, i.e., K_0, as a function of (S/N), when the instrumental profile is sinc and sinc^2 shape. **Table 1** shows that, for

Table 1 Optimum K-values as a function of the signal-to-noise ratio (S/N) are listed for sinc- and sinc^2-type instrumental functions

S/N	Sinc	Sinc^2
10	1.21	3.59
20	1.58	7.19
40	1.95	14.4
60	2.16	21.6
80	2.31	28.7
100	2.43	35.9
200	2.79	71.9
400	3.16	144
600	3.37	216
800	3.53	287
1 000	3.64	359
2 000	4.01	719
4 000	4.37	1 437
6 000	4.59	2 156
8 000	4.74	2 874
10 000	4.86	3 593

example, with $(S/N) = 1000$, the optimum value K_0 for grating spectrometers is 100 times higher than that for FT-IR spectrometers. In other words, a 100 times higher resolution is needed in the case of the grating spectrometer. This is usually impossible in practice. Now, it can be concluded that the optimum instrumental resolution of the FT-IR spectrometer is the lowest possible one or the FT-IR spectrometer measures a spectrum with the smallest possible distortions of observed lines. This is the resolution advantage of FT-IR spectrometers.

The fact that $\nu_{obs} = \text{constant} \times \nu_{true}$ is the linearity advantage of FT-IR spectrometers. It is roughly valid even though the interferometer does not work correctly, for example, in the case of phase errors in the interferogram. Furthermore, no ghost lines exist in the spectrum. However, in the grating spectra ghost lines are possible in the case of malfunction of the instrument.

Use of FT-IR spectrometers

Performance

The instrumental resolution of high-resolution FT-IR spectrometers is, in practice, of the order of 0.001 cm^{-1}. In addition to high resolution, a good absolute accuracy of the line positions in the spectrum is maintained. A wavenumber accuracy of 10^{-9}, or even better, has been achieved.

The main field of use of high-resolution FT-IR spectrometers is molecular spectroscopy. The spectrometers are used to study the geometric structures of molecules and their behaviour in rotations and vibrations. IR spectra are effective in revealing quantum mechanical phenomena and interactions in molecules.

An example of a typical rotation–vibration band, measured with a high-resolution FT-IR spectrometer, is shown in **Figure 11**. It is the bending band ν_2 of CO_2 near 667 cm^{-1}. This spectrum was recorded by the Oulu spectrometer with an instrumental resolution of 0.002 cm^{-1}. The relative errors of the wavenumber compared to model equations are of the order of 10^{-8}. The vibration energy levels of different normal modes are sometimes very close to each other. In this case these normal modes can be coupled so that the energy levels are shifted apart. Now the observed wavenumber ν_{obs} cannot be expressed in a closed form and the rotational lines of the spectrum are shifted. With the proper quantum

mechanical calculations it is possible to calculate the shifts. The high resolution and accuracy in wavenumber scale make an FT-IR spectrometer a very effective instrument to study these kinds of interactions in molecular spectra.

Calibration

In principle, the calibration of FT-IR spectrometers is performed with regard to intensity and wavenumber. However, the intensity calibration is not usually necessary because the absorbance scale is used, $-\ln(I/I_0)$, where the ratio I/I_0 automatically accomplishes the intensity calibration. The most important task in calibration is the wavenumber calibration. Usually the calibration takes place by simultaneous measurement of the known IR spectrum. For example, the user simply adds reference gas (known spectrum) into the sample gas under measurement. Using the known

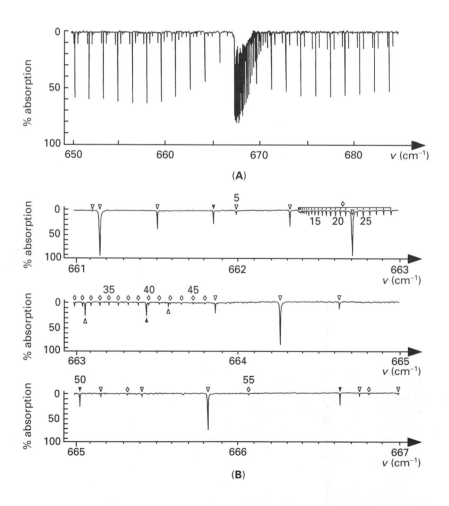

Figure 11 A part of the ν_2 bending band of CO_2 near 667 cm^{-1} measured by the Oulu FT-IR spectrometer with a resolution of 0.0045 cm^{-1} (A) and 0.002 cm^{-1} (B). In (B), ∇ indicates $^{12}C^{16}O_2$, $\blacktriangledown$ indicates $^{13}C^{16}O_2$ and $\diamond$ indicates $^{16}O^{12}C^{18}O$.

spectral lines of the reference gas it is possible to fix the wavenumber scale with the uncertainty, which is roughly the uncertainty of the reference spectral lines. The linearity advantage of FT-IR spectrometers makes it possible to reach a perfectly linear wavenumber scale. This is why only one reference line is needed, i.e. the reference laser line, to measure the optical path difference of the interferometer. The situation is completely different in grating, prism and laser spectrometers, where the wavenumber scale is more or less nonlinear with respect to the scanning parameter, like the rotation of grating or prism. Therefore in these spectrometers it is important to use many reference lines measured simultaneously with the sample spectrum in order to achieve a good wavenumber scale.

Data handling and computation

Nowadays, all modern IR spectrometers are equipped with a (micro) computer. Of course, the computer calculates absorbance $-\ln(I/I_0)$, displays the spectrum on the screen or plots the spectrum on paper, saves the spectra on disk, and so on. Modern FT-IR spectrometers are able to record very accurate spectra, where S/N is high and $|\mathrm{d}\nu/\nu|$ is very small, typically $S/N \approx 10^5$ and $|\mathrm{d}\nu/\nu| \approx 10^{-9}$. Also absorbance accuracy is quite high in FT-IR spectrometers. The spectra with high accuracy and high S/N include plenty of spectral information. Often spectral information is not in a proper form in the original recorded spectra. In order to derive all the possible information from high-quality IR spectra, good data-treatment algorithms are needed. If the spectrum consists of overlapping lines, resolution enhancement methods, like derivation, Fourier self-deconvolution, linear prediction (or LOMEP, line-shape optimized maximum entropy linear prediction) or band fitting and factor analysis, are very useful. Other possible data treatment methods, which the users may need, are, for example, smoothing, background elimination, interference fringe pattern elimination, peak-finding algorithm, maximum-entropy method, deconvolution and multicomponent analysis methods. In modern IR spectrometers good data treatment algorithms greatly increase the usefulness of IR spectroscopy in many applications.

Laser spectrometers

Laser spectrometers are not very useful in IR; they have more applications in the visible region, for example, LIDAR (light detection and ranging). The reason for this is the lack of good, cheap commercial lasers for IR. Wavenumber regions, due to tuning methods, are quite narrow, and the wavenumber calibration is a problem because of the lack of good calibration lines. In some cases it is also difficult to maintain a single-mode operation of the laser. The advantages of the laser spectrometer are a very high resolution, even sub-Doppler resolution, and a good S/N of the spectra.

There are two main techniques when using lasers as IR spectrometers. The first method is simply to tune the laser wavelength over a narrow wavelength region and simultaneously detect the laser signal $I(\nu)$ transmitted through the gas cell. The second possibility is to tune the wavelength of the absorption lines by applying the magnetic (Zeeman effect) or electric (Stark effect) field to the sample gas. When the wavelength of the tuned spectral lines matches the laser wavelength, absorption takes place. Thus absorption spectra are recorded as a function of the magnetic or electric field magnitude (strength).

A diode laser spectrometer is described as a typical example of this kind of spectrometer. A simple p–n junction laser is run by applying the forward bias voltage across the junction. The parallel polished crystal faces perpendicular to the junction form the mirrors acting as a laser resonator. The coherent radiation emits at the wavenumber $\nu_k = k/(2nL)$, where k is an integer, n is the refractive index and L is the length of the resonator. The wavenumbers ν_k are called the longitudinal modes of the laser. The number of modes is typically from three to ten. Operation in IR needs the cooling of the diode laser. Tuning can be accomplished by changing the term nL, which depends on the temperature. Temperature can be fine-controlled by changing current through the junction. It is possible to tune typically 200 cm^{-1} by changing the temperature and about 2 cm^{-1} by changing the current. Thus a spectrum has to be run in parts of 1–2 cm^{-1} by current tuning at a constant temperature. The temperatures of the parts are selected so that the parts together form a continuous spectrum over a desired region. Running the diode laser is very troublesome, because current tuning is needed at very many different temperatures, which makes calibration, for example, very difficult. In addition, a single-mode operation of the diode laser is needed, in other words, the integer k must be a constant. This can be performed by scanning a grating monochromator simultaneously with the laser tuning. The grating spectrometer follows the wavenumber $\nu_k = k/(2nL)$ of the mode. The change in wavenumber during tuning is measured by recording the interference fringes of a long Fabry–Perot etalon. The absolute value of the wavenumber in each part has to be fixed by calibration lines. In many applications the diode laser spectrometer solely is not sufficient. However, it is a very good instrument for looking at details over a wide region spectrum recorded by a FT-IR spectrometer. Thus the laser spectrometer is a supplementary instrument to the FT-IR spectrometer.

List of symbols

a = source area; A = collimator mirror area; A_D = detector area; c_0 = velocity of light in a vacuum; C = concentration; d = distance; D = mirror diameter; D^* = specific detectivity; E'' and E' = lower and higher energy levels; E = spectrum; f = focal length, or frequency; F = interference record; $\mathcal{F}\{\ \}$ = Fourier transform; $\mathcal{F}^{-1}\{\ \}$ = inverse Fourier transform; h = Planck's constant; $i = \sqrt{-1}$; I = radiation intensity; k = an integer; K = resolution enhancement factor; L = maximum optical path difference; M = number of resolution elements; n = refractive index; N = noise; S = signal; $(S/N)_{FT}$ = signal-to-noise-ratio of a FT-IR spectrometer; $(S/N)_G$ = signal-to-noise-ratio of a grating spectrometer; S_0 = radiation source; T = observation time; V_{rms} = root-mean-square voltage; W = instrumental function; α = angle between optical axis and beam; $\alpha(\nu)$ = absorption coefficient; $\beta(\nu)$ = absorptivity; δ = Dirac's delta function; λ = wavelength; ν = radiation wavenumber; Ω = solid angle.

See also: **Fourier Transformation and Sampling Theory; FT-Raman Spectroscopy, Applications; Gas Phase Applications of NMR Spectroscopy; High Resolution IR Spectroscopy (Gas Phase), Applications; Hydrogen Bonding and Other Physicochemical Interactions Studied By IR and Raman Spectroscopy; Laboratory Information Management Systems (LIMS); Laser Spectroscopy Theory; Light Sources and Optics; Vibrational, Rotational and Raman Spectroscopy, Historical Perspective.**

Further reading

Bell RJ (1972) *Introductory Fourier Transform Spectroscopy*. New York: Academic Press.

Bracewell R (1965) *The Fourier Transform and its Applications*. New York: McGraw-Hill.

Brigham EO (1974) *The Fast Fourier Transform*. Englewood Cliffs: Prentice-Hall.

Chamberlain J (1979) *The Principles of Interferometric Spectroscopy*. Chichester, UK: John Wiley.

Griffiths PR and de Haseth JA (1986) *Fourier Transform Infrared Spectrometry*. New York: John Wiley.

Herzberg G (1945) *Infrared and Raman Spectra of Polyatomic Molecules*. New York: Van Nostrand.

Mackenzie MW (ed) (1998) *Advances in Applied Fourier Transform Infrared Spectroscopy*. New York: John Wiley.

Marshall AG and Verdun FR (1990) *Fourier Transforms in NMR, Optical, and Mass Spectrometry, A Users Handbook*. New York: Elsevier.

High Resolution IR Spectroscopy (Gas Phase), Applications

E Canè and A Trombetti, Università di Bologna, Italy

> VIBRATIONAL, ROTATIONAL & RAMAN SPECTROSCOPIES
> Applications

The IR region of the electromagnetic spectrum extends from the visible ($\lambda \cong 800$ nm, $\tilde{\nu} \cong 12\ 500$ cm^{-1}) to the sub-mm waves ($\lambda < 1$ mm; $\tilde{\nu} > 10$ cm^{-1}) and the IR photon carries an energy in the 2.48×10^{-21} – 1.99×10^{-24} J range. This is also the energy exchanged by an atom or a molecule when IR radiation is emitted or absorbed. The interaction of the IR radiation with particles corresponds to transitions between rotation–vibration energy levels in molecules or between Ryberg levels in atoms.

High-resolution IR spectroscopy is confined to the study of atoms and molecules in the gas phase at low pressures, where the interaction energy between particles is orders of magnitude lower than in the condensed phases. The intermolecular interactions broaden the energy levels, and, consequently, the spectral lines, preventing the observation of finer details in the spectra. In the IR region the width of the spectral lines in the gas phase is caused mainly by the Doppler effect and by the broadening due to collisions. The concept of high resolution in IR spectroscopy has changed with the introduction of new experimental techniques such as FT-spectroscopy and laser techniques. Presently, an instrumental resolution of 2×10^{-3} cm^{-1} (measured as FWHM) or higher is achievable with commercial FT interferometers in the 10–4000 cm^{-1} range: this is comparable to the Doppler line width for a molecule with a mass of 50u at room temperature, calculated using the equation:

$$\Delta \tilde{\nu} = 2\tilde{\nu} \left(\frac{2T k_B \ln 2}{m c^2} \right)^{1/2} \quad [1]$$

where k_B is the Boltzmann constant, m is the particle mass, T is the absolute temperature and c is the speed of light. Thus the Doppler width can be assumed as the limit of high-resolution FTIR spectrometers that use a broadband radiation source of low spectral brightness. With the introduction of lasers in the infrared, sub-Doppler resolution can be obtained using the saturation lamb-dip or other specialized techniques. Sub-Doppler resolution can also be performed with molecular beams or in supersonic jet expansions, where molecules are rotationally and vibrationally cooled to a few K. Moreover, the enhancement of sensitivity of the new experimental techniques has made possible the detection and analysis of many unstable or transient species such as van der Waals complexes, free radicals and ions.

The main applications covered in this study are the accurate determination of rotation and rotation–vibration molecular energies; the determination of the molecular geometry of simple molecules; the evaluation of force field and of the vibration– and rotation–vibration interactions; the measurement of pressure broadening and pressure shift of the spectral lines; the determination of electric dipole moments via laser-Stark spectroscopy; the studies of intramolecular dynamics; the calculation of rate constants, equilibrium constants and other thermodynamic data; the evaluation of relaxation times.

Rotational and rovibrational energies

The results of the analysis of a high-resolution IR spectrum of a molecule is usually a large and accurate set of molecular constants. With these constants, rotational and rovibrational energy levels can be calculated within the range of the quantum numbers of the transitions from which they have been obtained. This is the primary application of high-resolution gas phase spectra in the IR region. The results of the analysis of the $\tilde{\nu}_1$, $2\tilde{\nu}_2$ and $2\tilde{\nu}_3 + \tilde{\nu}_6$ IR bands of COF$_2$

are presented as an example in **Table 1**. The wavenumbers of about 3200 spectral lines have been measured and assigned to specific rovibrational transitions, identified through the rotational quantum numbers of the lower and upper states of the transition. After the assignments a least-squares analysis was performed, where the parameters of **Table 1** and their standard deviations were determined.

The most important molecular constants are $\tilde{\nu}_0$, the energy of the vibrational levels above the rotationless ground state and the rotational constants B_x, B_y and B_z. The centrifugal distortion constants, D and d, H and h, are orders of magnitude smaller than B and account for the deviations from the rigid rotor energy levels. The constants W are necessary to reproduce the measured spectrum since the $\tilde{\nu}_1$ and $2\tilde{\nu}_2$ states are in Fermi resonance. The constants ξ take into account the Coriolis interaction between the $\tilde{\nu}_1$ and $2\tilde{\nu}_3 + \tilde{\nu}_6$. For the latter state no transitions have been observed and its constants have been obtained from the perturbation induced on the levels of $\tilde{\nu}_1$. The RMS in **Table 1** is the standard deviation normalized to a single rovibrational transition of the calculated and measured spectrum. This quantity should be very close to the precision of the wavenumber measurement if the theoretical model used for the analysis is satisfactory.

Determination of molecular geometry

Accurate molecular geometries can be obtained from the analysis of the high-resolution IR spectra. The analysis of rovibrational spectra provides, among other important data, the rotational constants of the molecules, which in turn allow the determination of the principal moments of inertia and, eventually, of the molecular geometry. Pure rotational spectroscopy provides the same information, usually for the ground state, except when the molecule under study has no permanent electric dipole moment. The principal moments of inertia are related to the bond lengths and bond angles of the molecule by equations that vary according to the geometry. Equilibrium geometry corresponds to the minimum of the potential energy surface of the molecule. Owing to the zero-point energy this minimum does not correspond to a stationary state of the molecule. The rovibrational analysis of fundamentals and overtones gives the rotational constants of the excited vibrational state, and of the lowest molecular state, which is above the minimum of the potential energy curve by the zero-point energy. The bond lengths obtained from the ground state rotational constants are approximately the mean values of the lengths in the ground state and

Table 1   Parameters for the $2\nu_2$, ν_1, and $2\nu_3+\nu_6$ states of COF_2. Reprinted with permission of Academic Press from D' Cunha *et al* (1997) *Journal of Molecular Spectroscopy* **186**: 363–373

	$2\nu_2$ state	ν_1 state	$2\nu_3+\nu_6$ state
ν_0	1 922.516 220(49)	1 935.940 117(47)	1 935.145(30)
B_x	0.390 996 08(127)	0.393 161 46(93)	0.394 881 1[a]
B_y	0.391 157 49(124)	0.390 005 56(107)	0.392 324 0(985)
B_z	0.194 993 62(21)	0.195 528 25(26)	0.196 974 7(632)
$D_J \times 10^6$	0.446 129(451)	0.437 270(163)	0.200 4(209)
$D_{JK} \times 10^6$	−0.766 714(945)	−0.742 981(653)	0.104 2(243)
$D_K \times 10^6$	0.353 319(598)	0.339 445(640)	4.007 2(161)
$d_1 \times 10^6$	−0.041 021(348)	−0.041 513(211)	−0.042 227[b]
$d_2 \times 10^6$	−0.008 716(227)	−0.008 832(183)	−0.010 031[b]
$H_J \times 10^{11}$	0.123 47(2150)	0.025 39[a]	0.089 36[b]
$H_{JK} \times 10^{11}$	−0.462 29(6740)	−0.155 58(2910)	−0.375 49[b]
$H_{KJ} \times 10^{11}$	0.556 43(7710)	0.217 96(3480)	0.480 50[b]
$H_K \times 10^{11}$	−0.238 54(2940)	−0.032 93[a]	−0.194 57[b]
$h_1 \times 10^{11}$	−0.009 40[c]	0.009 40[b]	0.009 40[b]
$h_2 \times 10^{11}$	0.003 66[b]	0.003 66[b]	0.003 66[b]
$h_3 \times 10^{11}$	0.008 25[c]	−0.008 25[b]	−0.008 25[b]
W_0		13.844 5[a]	
$W_J \times 10^4$		−0.074 8[a]	
$W_K \times 10^4$		0.042 0[a]	
$W_{XY} \times 10^4$		1.167 9(66)	
ξ			−0.006 711(95)
$\xi_J \times 10^5$			−0.126 49(615)
$\xi_K \times 10^5$			−0.127 92(544)
RMS dev	0.000 39	0.000 36	
Lines	1 529	1 675	

All values are in cm^{-1}.
[a] Constrained to the value obtained with a damping factor of 0.0001.
[b] Constrained to the ground state value.
[c] Sign changed to conform with III^l representation (see text).

are usually slightly larger than the equilibrium ones, and the geometry is called the r_0-structure. The equilibrium geometry, r_e-structure, can be determined from B_e if the rovibration interaction constants α_i, defined by the equation

$$B_{\nu_i} = B_e - \sum \alpha_i \left(\nu_i + \frac{1}{2} \right) \quad [2]$$

for all the normal vibrations of the molecule are known. Typical values of a r_0-structure are compared with a r_e-structure in **Table 2** for C_2H_2. Since for an asymmetric top molecule there are three principal moments of inertia a maximum of three molecular parameters, bond length and angles, can be obtained. Moreover, in planar molecules only two moments of inertia are independent and in linear molecules such as acetylene the two moments of inertia are equal. Thus, determination of geometry for polyatomic molecules seems to be impossible.

Isotopic substitution is the way to increase the number of independent moments of inertia so that the structure parameters of polyatomic molecules can also be inferred. In fact, molecules differing by one or more isotopes have the same r_e-structure, while the r_0-structure is slightly different since the zero-point energy is isotope dependent. For large polyatomic molecules the determination of the r_e-structure is nearly impossible. In benzene, C_6H_6, owing to the hexagonal symmetry, only the r_{C-H} and r_{C-C} interatomic distances are needed to determine the geometry. From the analysis of the high-resolution IR or UV spectrum only one independent rotational constant is obtained and an isotopically substituted benzene such as C_6D_6 has to be studied to obtain the r_0-structure. Since each isotopomer has 20 normal vibrations, and thus 20 α_i values, the analysis of 40 high-resolution bands is needed for the r_e-structure. To date, about 15 fundamental bands of C_6H_6 and C_6D_6 have been rotationally analysed.

Force field evaluation

Accurate *ab initio* theoretical calculations of harmonic and anharmonic force fields of molecules as large as benzene have recently been performed. By comparing the experimentally determined molecular constants with the calculated ones, the reliability of theoretical calculations can be assessed. All the molecular constants of COF_2 in **Table 1** can be evaluated from the theoretical anharmonic force field. For molecules such as ammonia the computed fundamental wavenumbers are within 3 cm^{-1} or better of the experiment. For anharmonic constants and centrifugal distortion constants, which are much smaller than the fundamental frequencies, the deviations are on average around 15%. The experimental determination of anharmonic constants x_{ij} that appear in the equation for the vibrational energy of a polyatomic molecule with the harmonic frequencies ω_i^0 and without degenerate vibrations:

$$G_0(v_1, v_2, v_3 \ldots) = \sum_i \omega_i^0 v_i + \sum_i \sum_{j \geq i} X_{ij}^0 v_i v_j \quad [3]$$

relies on the analysis of overtone and combination bands. In **Table 1** the data under the headings $\tilde{\nu}_1$, $2\tilde{\nu}_2$ and $2\tilde{\nu}_3 + \tilde{\nu}_6$ refer to a fundamental, an overtone and a combination band, respectively. The relations between $\tilde{\nu}_0$ of **Table 1** and ω_i^0, x_{ij}^0 are given by

for $\tilde{\nu}_1$ $\qquad\qquad \tilde{\nu}_0 = \omega_1^0 + x_{1,1}^0$ $\qquad$ [4]

for $2\tilde{\nu}_2$ $\qquad\qquad \tilde{\nu}_0 = 2\omega_2^0 + 4x_{2,2}^0$ $\qquad$ [5]

for $2\tilde{\nu}_3 + \tilde{\nu}_6$ $\quad \tilde{\nu}_0 = 2\omega_3^0 + \omega_6^0 + 4x_{3,3}^0 + 2x_{3,6}^0 + x_{6,6}^0$ [6]

The assessment of the reliability of theoretical calculations is particularly important when the experiment is unable to determine some of the molecular constants, which in turn can be calculated by *ab initio* methods. In benzene, for example, there are about 300 anharmonic constants and there is little hope that all of these could be obtained experimentally. The equilibrium structure r_e is also a very important result

Table 2 r_0- and r_e-structures of acetylene

	$r_0{}^a$ (pm)	r_e (pm)	$r_e{}^b$ (pm)
$r_{C\equiv C}$	120.85	120.241 (9)^c	120.33
r_{C-H}	105.72	106.25 (1)^c	106.05

^{a} From B_0 of C_2H_2 and C_2D_2.
^{b} Theoretical equilibrium structure.
^{c} Standard errors in parentheses refer to the last significant digits.

of the theoretical calculations of molecular force field, and is to be compared with the experimental structure (see **Table 2**).

Laser Stark IR spectroscopy

The laser Stark technique applied to IR spectroscopy is based on bringing in resonance, with a fixed frequency IR laser (such as CO_2, CO, N_2O lasers), the molecular transitions modulated via an electric field (Stark effect). From an analysis of the Stark spectra very accurate values of the electric dipole moments of molecules in the ground and excited vibrational states are obtained. **Table 3** presents some of the dipole moments that have been determined.

Pressure broadening and pressure shift

The pressure shift and pressure broadening of the spectral lines are primarily due to the collisions of the molecules. The study of these effects provides information on the mechanism of the molecular interactions which are mainly of dipole–dipole type. In addition, the experimental determination of the line broadening and line shift parameters is instrumental in the spectroscopic study of the atmosphere, where the absorption or emission of IR radiation takes place from regions where the pressure may change by several orders of magnitude. These effects are relatively small; for example, for the P and R branches of the ν_3 band of $^{12}C^{16}O_2$ at 2350 cm^{-1}, broadened by N_2 at 296 K, broadening coefficients in the 0.066–0.089 cm^{-1} atm^{-1} range and shift coefficients of -0.0016–0.0037 cm^{-1} atm^{-1} have been reported by Devi and co-workers.

Table 3 Electric dipole moments from laser Stark spectroscopy

Molecule	Vibrational state	μ (D)^{a,b}
PH_3	$v_2 = 1$	0.574 20(27)
	$v_4 = 1$	0.579 04(32)
HNO_3	$v_2 = 1$	2.12(4)
HCOOH	Ground state	1.395 7(9)
	$v_3 = 1$	1.421 6(10)
CH_3F	Ground state	1.857
	$v_3 = 1$	1.905
ND_3	Ground state	1.492(8)
	$v_2 = 1$	1.352(3)
AsH_3	Ground state	0.213(3)
	$v_2 = 1$	0.218(3)

^{a} 1 D = 3.335 64 × 10^{-30} cm
^{b} Standard errors in parentheses refer to the last significant digits.

High-resolution IR spectroscopy of weakly bound complexes

High-resolution IR spectroscopy has been used to elucidate the potential energy surfaces that govern the weak intermolecular forces between molecules, since it provides information on the vibrationally excited region of the potential energy surface of the ground electronic state. This region controls the long- and short-range collisional dynamics of the monomers. This application was delayed until the advent of adequately sensitive detection methods and of an appropriate light source. The experimental methods are based on either direct absorption techniques of light obtained from a laser source (from a tunable diode laser or from difference frequency mixing of a dye and Ar^+ lasers) by complexes in a cooled multiple pass gas cell or on FTIR techniques. A rotationally resolved spectrum can be recorded even though the line widths are Doppler limited or broadened by pressure or predissociation effects. From the analysis of these spectra thermodynamic properties, vibrational term values, rotational constants and bond dissociation energies may be extracted. Alternatively, supersonic molecular beam methods can be used with direct absorption laser spectroscopy or bolometric techniques. These experiments supply additional information on samples at rotational–vibrational temperatures ranging from 1–2 K for expansion from an ultracold pinhole, to 10–30 K for slit expansion. The spectroscopic techniques developed in the FAR-IR allow the investigation of the low-frequency intermolecular modes in complexes. Additional spectroscopic data on the large amplitude motions associated with these low frequency vibrations have been obtained from the analysis of the hot and/or combination bands observed in the mid/near-IR. Some of the complexes analysed so far are $(HF)_2$, $(DF)_2$, $(HCl)_2$, $(CO_2)_2$, $(HCN)_2$, $(HCCH)_2$, $(NH_3)_2$, H_2-inert gas, H_2–HF, HF–inert gas, HCl–inert gas, HF–N_2, HF–OCO, OC–HF, C_2H_2–HF, HCN–HF, HF–DF, $(DF)_3$. The HF and DF van der Waals complexes have attracted much attention because of their fundamental importance in the chemistry of hydrogen bonded species. Most of the spectra recorded in the near-IR are structured since the upper states are metastable with respect to vibrational (or rotational) predissociation. The main channel leading to the breakdown of the complex into its constituent subunits is the coupling of the intramolecular mode with the dissociating, low-frequency intermolecular modes. This coupling causes a redistribution of vibrational energy in the complex that may excite the intermolecular bond up to dissocia-

tion. The excited states have to last at least for the time needed to complete an end-over-end rotation in the complex to observe a rotationally resolved structure in the spectrum. From the experimental values of FWHM line width it is possible to calculate the lifetime of the metastable state, according to

$$\tau = 1/2\pi\Delta\nu_{FWHM} \qquad [7]$$

for a Lorentzian absorption line shape and a single exponential decay of the upper state population. However, another possible contribution to the homogeneous line width comes from intramolecular vibrational relaxation (IVR). Consequently, the collision-free line width depends in principle on the combined 'predissociation/relaxation' pathway. From the analysis of high-resolution spectra a very large range of dissociation/relaxation behaviour results [$\tau > 3 \times 10^{-4}$ s for ν_1 of Ar–HF (Huang and co-workers) to $\tau \geq 10$ ps for ν_2 of the linear $(HCN)_3$ (Jucks and Miller)] indicating strong sensitivity to small differences in the potential energy surfaces. In many dimers a strong dependence of the line widths from specific vibrational couplings has been observed as in $(HF)_2$ [$\Delta\nu_1 = 13.4$ MHz, $\Delta\nu_2 = 408$ MHz (Pine and co-workers and Huang and co-workers)], HCN–HF [$\Delta\nu_1 = 2722$ MHz, $\Delta\nu_2 = 11.8$ MHz, $\Delta\nu_3 = 558$ MHz (Wofford and co-workers)]. The analysis of spectroscopic data may give insight into the coupling between intermolecular bending and stretching modes, as indicated by the results of the Coriolis analysis in the Ar–HF dimer between the ν_2 perpendicular band and the $3\nu_3$ overtone of the intermolecular stretching vibration (Lovejoy and co-workers). Additional probes of intermolecular interactions are the intensity changes and the amount of the shift of the frequency of the monomer fundamentals in complexes from their unperturbed position. From IR experiments it is possible to infer the correct equilibrium structure of a complex, i.e. that corresponding to the absolute min-imum of the potential surface. Information about large amplitude motions from hot bands and combination bands analysis are also very useful to this end. Results indicate that for many complexes, i.e. CO_2, N_2O dimers, and HCN trimers, several structures, corresponding to a different minima in the potential surface, are possible and that the isomerization involves motion along intermolecular coordinates. In the case of Ar_nHF clusters a satisfactory agreement is observed between the isomer equilibrium structures predicted on the basis of pairwise additive potentials and the experimental observations (see **Figure 1**).

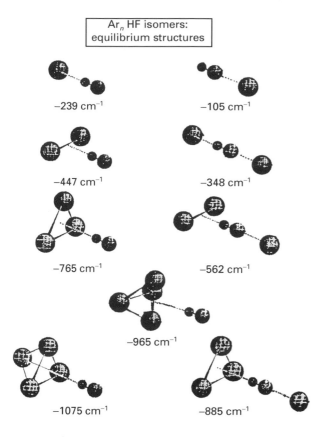

Figure 1 Equilibrium structures and energies for Ar_nHF ($n = 1-4$) predicted from pairwise additive potentials. In all cases, the lowest predicted energy structures (left) are in excellent agreement with experimental observation. Reprinted with permission of Annual Reviews from Nesbitt DJ (1994) High-resolution, direct infrared laser absorption spectroscopy in slit supersonic jet: intermolecular forces and unimolecular vibrational dynamics in clusters. *Annual Review of Physical Chemistry* **45**: 369–399.

High-resolution IR spectroscopy of free radicals and ions

IR laser spectroscopy has been employed to study free radicals and ions because of its sensitivity in the detection of compounds present in small concentrations. The spectroscopic characterization of radicals and ions is valuable in the fields of chemical kinetics, of astronomy and of plasma diagnosis. The experimental techniques used to produce unstable molecules and to record the spectra are here briefly summarized. Free radicals can be generated using the discharge-flow method, while ionic species can be obtained using a hollow-cathode discharge cell that, combined with magnetic-field modulation or amplitude-discharge modulation, allows the detection of the selected ionic species among neutral stable molecules. Alternatively, velocity modulation IR laser spectroscopy can be used to selectively detect

ionic species among neutral ones, exploiting the characteristic Doppler shift of charged molecules. HD^+, HeH^+, NeH^+, ArH^+, HCO^+ and HNN^+ are some of the ions studied with this technique in the infrared and whose fundamental frequencies and molecular constants have been determined. Both radicals and ions may be generated by photolysis or photoionization of a stable precursor seeded in a rare-gas free-jet expansion. Another powerful technique with which to study ions employs a fast ion beam extracted from a plasma ion source and velocity tuned. This is merged with CO_2 laser light to induce transitions from a vibrational bound level near the dissociation limit to a repulsive electronic state or to a predissociating bound level above the dissociation limit. Most of the systems studied are diatomic, such as HD^+, HeH^+, He_2^+, H_2^+ and D_2^+, and the information obtained concerns the description and spectroscopy of weakly bound states, to a infer deeper understanding of ion–neutral reaction dynamics and of the charge–dipole induced interaction. In addition, polyatomic unstable species have been the subject of detailed study with high-resolution IR spectroscopy, among those examined are CCH, C_3, C_4, C_9, C_{13}, NO_3, HO_2, HCCO and C_3H_3 radicals, and H_3^+, SiH_3^+, CH_3CNH^+, CH_2^+, NH_3^+ and H_2O^+ ions. The results of such experiments may be valuable in the field of chemical kinetics since they allow the determination of rate constants and an understanding of some aspects of the mechanism of the reaction. For example, the observation, in a hollow cathode discharge, of the ν_1 band of HOC^+ at 3268 cm^{-1} allowed the measurement of the abundance ratio $[HCO^+]:[HOC^+]$ in the laboratory from the relative intensity of the IR high-resolution lines of the two species, using the transition dipole moments for the band of HCO^+ (0.168 D) and HOC^+ (0.350 D), and assuming that the rotational temperature of both ions was equal to the cell temperature (223 K) (Amano). This information was particularly valuable in studying the formation and depletion of HOC^+ in the interstellar medium. A rate equation analysis was performed to rationalize the observed dependence of the relative intensities of HOC^+ and HCO^+ lines as a function of the pressures of H_2 and CO. The rate constants of some of the reactions considered in the rate equation analysis were then determined. The contribution that the analysis of the high-resolution IR spectra of such kind of molecules brings to astronomy is the determination with great accuracy of the molecular constants that allow one to search for the rotational lines in laboratory microwave spectra. These results may lead to the identification of unassigned interstellar lines. The nature and the amount of a species present in plasmas may be determined by high-resolution IR spectroscopy, as in the case of SiH_3^+, a transient molecular ion observed and analysed in a silane discharge plasma. The ν_2 and ν_4 bands, i.e. the out-of-plane bending and the in-plane degenerate bending vibration respectively, have been recorded and analysed (Davies and Smith), taking into account the Coriolis interaction between the two vibrationally excited states.

Molecular dynamics

Frequency domain spectroscopic techniques can be widely used to determine the homogeneous intramolecular vibrational redistribution (IVR) of energy, a phenomenon observed when molecules have enough energy to break bonds. An advantage of the frequency domain techniques is the ability to measure the rotational state-dependence of the IVR rates. Briefly, the background theory is that a single rotational state in an excited anharmonic normal-mode vibrational state (the bright state) carries all the oscillator strength and is coupled to near-resonant vibrational dark states through perturbation terms of the Hamiltonian. As a result of this coupling the bright state oscillator strength is redistributed among the dark states. The width of the intensity distribution is a measure of the time scale of energy localization in the bright state. Thus, one goal of IVR studies is the measurement of the relaxation rates that span a wide range (50 fs to 10 ns) and are dependent on molecular structure. The two major experimental methods used for IVR studies of isolated molecules in their ground electronic state can be divided into single photon and double resonance techniques. The general requirements of these experimental methods are both high resolution and sensitivity. Quantitative IVR rates of many molecules have been determined: they span from 10^8–10^{13} s^{-1}, with 3×10^9 s^{-1} being a typical value. These rates may be sufficiently slow to permit mode-selective chemistry in some systems. From a comparative analysis of the IVR rate values for the acetylenic C–H stretching for many molecules of the type $(CY_3)_3XCCH$ (see **Table 4**), it appears that the lifetime and density of states are not correlated, even if the state density must be large enough to provide a bath for energy redistribution. The IVR rate is controlled by the coupling of a few dark states that interact with the bright one through low order terms of the potential energy surface, which in turn must be efficiently coupled to other non-resonant states to allow fast energy flow. Other structural aspects of IVR such as the role of large amplitude motion in IVR dynamics, the change of

Table 4 Experimental IVR lifetimes and states densities for the acetylenic C–H stretch in molecules of the form $(CY_3)_3XCCH$. Adapted with permission of Annual Reviews from Lehmann KK, Scoles G and Pate BH (1994) Intramolecular dynamics from eigenstate-resolved infrared spectra. *Annual Review of Physical Chemistry* **45**: 241–274.

Molecule	Level of excitation	Density of states[a] (cm)	Lifetime (ps)
$(CH_3)_3CCCH$	$v = 1$	4.9×10^2	200
	$v = 2$	6.2×10^5	110
$(CD_3)_3CCCH$	$v = 1$	2.8×10^3	40
	$v = 2$	7.6×10^6	$< 40^b$
$(CH_3)_3SiCCH$	$v = 1$	1.0×10^4	2 000
	$v = 2$	2.9×10^7	4 000
$(CD_3)_3SiCCH$	$v = 1$	1.0×10^5	830
	$v = 2$	6.0×10^8	140
$(CH_3)_3SnCCH$	$v = 1$	1.0×10^6	6 000
	$v = 2$	1.0×10^9	$> 1\,000^c$
$(CF_3)_3CCCH$	$v = 1$	4.2×10^6	60
	$v = 2$	1.0×10^{11}	$< 40^b$

[a] The state densities (states cm^{-1}) are for vibrational bath states of A_1 symmetry, which can couple via anharmonic interactions.
[b] Upper limit of the lifetime. The overtone absorption was unobservable, despite observation of the fundamental.
[c] Lower limit of the lifetime. The Q branch structure was too sharp to obtain an estimate of the Lorentzian wing. Other inhomogeneities (either isotopes, torsional, or K structure) were also present.

IVR rates by structural modification, and the degree of dependence of the IVR rates on modes are currently being studied.

Thermodynamic and kinetic data from high-resolution spectroscopic IR studies

An example of this application is the FTIR and IR laser diode spectroscopy study of the reversible gas-phase reaction

$$HONO + NH_3 \rightleftarrows H_3N - HONO$$

A kinetic study has been carried out (Pagsberg *et al.*) by monitoring selected absorption signals of the rovibration transitions of the ν_3 and ν_5 IR bands of HONO as a function of time for different partial pressures of ammonia at a total pressure of 20 mbar. At 298 K the rate constant of the forward reaction and the equilibrium constant are $k_f = 2.2 \pm 0.2 \times 10^3$ M^{-1} s^{-1} and $K_c = 1.5 \pm 0.2 \times 10^5$ M^{-1}, respectively. From the observed temperature dependence of the equilibrium constant expressed in terms of the van't Hoff equation, the values of $\Delta_r H^\circ_{298} = -49.4 \pm 3.3$ kJ mol^{-1} and $\Delta_r S^\circ_{298} = -111.7 \pm 4.2$ J mol^{-1} K^{-1} have been derived.

List of symbols

B_e = equilibrium rotational constant; B_{v_i} = rotational constant in the v_i vibrational mode; B_x = rotational constant; D, d = quartic centrifugal distortion constants; $G_0(v_i)$ = vibrational term value when the energy levels are referred to the ground state as zero; H, h = sextic distorsion constants; k_B = Boltzmann constant; k_f = rate constant of the forward reaction; K_c = equilibrium constant; r_{C-H} = interatomic distance between C and H atoms; T = (thermodynamic) temperature; v_i = vibrational quantum number; W = Fermi resonance parameter; x^0_{ij} = vibrational anharmonic constants when the energy levels are referred to the ground state as zero; α_i = rovibration interaction constant; $\Delta_r H^\circ_{298}$ = standard reaction enthalpy; $\Delta_r S^\circ_{298}$ = standard reaction entropy; Δv = full line width at half intensity maximum; μ = dipole moment; $\tilde{v}(v_i)$ = transition wavenumber in vacuum; ξ = Coriolis constant; $\tilde{v}$ = wavenumber in vacuum; $v_i\tilde{v}_0$ = wavenumber of a rovibrational band origin; τ = relaxation time; ω^0_i = harmonic vibration wavenumber when the energy levels are referred to the ground state as zero.

See also: **High Resolution IR Spectroscopy (Gas Phase) Instrumentation; Interstellar Molecules, Spectroscopy of; IR Spectroscopy, Theory; Laser Magnetic Resonance; Microwave and Radiowave Spectroscopy, Applications; Rotational Spectroscopy, Theory.**

Further reading

Amano T (1988) High resolution infrared spectroscopy of molecular ions. *Philosophical Transactions of the Royal Society of London, Section A* **324**: 163–178.

Barrow RF and Crozet P (1997) Gas-phase molecular spectroscopy. *Annual Reports on the Progress of Chemistry, Section C, Physical Chemistry* **93**: 187–256.

Brian JH and Brown JM (1992) High resolution infrared spectroscopy. In: Andrews DL (ed) *Applied Laser Spectroscopy*, pp 185–225. New York: VCH.

Davies PB and Smith DM (1994) Diode laser spectroscopy and coupled analysis of the v_2 and v_4 fundamental bands of SiH$_3^+$. *Journal of Chemical Physics* **100**: 6166–6174.

Devi VM, Benner DC, Rinsland CP and Smith MAH (1992) Measurements of pressure broadening and pressure shifting by nitrogen in the 4.3 μm band of ^{12}C

^{16}O$_2$. *Journal of Quantum Spectroscopy and Radiation Transfer* **48**: 581–589.

Gudemann CS and Saykally RJ (1984) Velocity modulation infrared laser spectroscopy of molecular ions. *Annual Review of Physical Chemistry* **35**: 387–418.

Hirota E (1994) Rotational and vibrational spectra of free radicals and molecular ions. *Annual Reports on the1 Progress of Chemistry, Section C, Physical Chemistry* **91**: 3–36.

Hollas JM (1982) *High Resolution Spectroscopy*. London: Butterworth.

Huang ZS, Jucks KW and Miller RE (1986) The argon-hydrogen fluoride binary complex: an example of a long lived metastable system. *Journal of Chemical Physics* **85**: 6905–6909.

Huang ZS, Jucks KW and Miller RE (1986) The vibrational predissociation lifetime of the HF dimer upon exciting the "free-H" stretching vibration. *Journal of Chemical Physics* **85**: 3338–3341.

Jacox ME (1998) Vibrational and electronic energy levels of polyatomic transient molecules. *Journal of Physical and Chemical Reference Data* **27**: 115–412.

Jucks KW and Miller RE (1988) Near infrared spectroscopic observation of the linear and cyclic isomers of the hydrogen cyanide trimer. *Journal of Chemical Physics* **88**: 2196–2204.

Lehmann KK, Scoles G and Pate BH (1994) Intramolecular dynamics from eigenstate-resolved infrared spectra. *Annual Review of Physical Chemistry* **45**: 241–274.

Lovejoy CM, Schuder MD and Nesbitt DJ (1986) High resolution IR laser spectroscopy of the Van der Waals complexes in slit supersonic jets. *Journal of Chemical Physics* **85**: 4890–4902.

Nesbitt DJ (1994) High-resolution, direct infrared laser absorption spectroscopy in slit supersonic jet: intermolecular forces and unimolecular vibrational dynamics in clusters. *Annual Review of Physical Chemistry* **45**: 367–399.

Pagsberg P, Ratajczak E, Sillesen A and Lotaika Z (1994) Kinetics and thermochemistry of the reversible gas phase reaction HONO + NH$_3$ → H$_3$N-HONO studied by infrared diode laser spectroscopy. *Chemical Physics Letters* **227**: 6–12.

Pine AS, Lafferty WJ and Howard BJ (1984) Vibrational predissociation, tunneling, and rotational saturation in the HF and DF dimers. *Journal of Chemical Physics* **81**: 2939–2950.

Rao KN and Weber A (eds) (1992) *Spectroscopy of the Earth's Atmosphere and Interstellar Medium*. New York: Academic Press.

Weber WH, Tanaka K and Tanaka T (eds) (1987) Stark and Zeeman techniques in laser spectroscopy. *Journal of the Optical Society of America, Part B* **4**: 1141–1172.

Wofford BA, Bevan JW, Olson WB and Lafferty WT (1985) Rovibrational analysis of v_3 HCN–HF using Fourier transform infrared spectroscopy. *Journal of Chemical Physics* **83**: 6188–6192.

High Resolution Solid State NMR, 13C

Etsuko Katoh, National Institute of Agrobiological Resources, Tsukuba, Japan
Isao Ando, Tokyo Institute of Technology, Japan

Introduction

After the first NMR experiment for obtaining high-resolution spectra of solids was carried out with the high-speed magic-angle spinning (MAS) method in 1958, the cross polarization (CP) and multipulse methods were developed as important for high-resolution solid-state NMR experiments. On the basis of these methods a CP MAS technique that combines MAS and CP has been conventionally used to obtain high-resolution solid-state NMR spectra.

In the solution state, the observable NMR chemical shift is often the average value of some preferred conformations of a molecule because of its rapid motion, but in the solid state the chemical shift is often characteristic of a specified conformation because of the highly restricted molecular motion. Thus, the chemical shifts can be used as well as the spin–lattice relaxation time (T_1), the spin–spin relaxation time (T_2) and T_1 in the rotating frame ($T_{1\rho}$) to provide information on the dynamics. At present, ^{13}C CP MAS has been widely employed to characterize the structure and dynamics of solid samples such as polymers, biopolymers, catalysts etc. Here, we will introduce some applications of ^{13}C CP MAS NMR for the characterization of the structure and dynamics of polymers, including polypeptides and proteins and also briefly describe the applications of solid-state ^{13}C NMR spectroscopy to food materials and coals. There is also a large body of literature on high-resolution solid-state ^{13}C NMR spectroscopy of small molecules, although the methods used are largely the same as those given here.

Synthetic polymers

Polymer materials are almost always used as solids. It is well known that the properties of polymers depend on their structure and dynamics. Therefore, to design new polymer materials, it is necessary to understand more deeply the property–structure relationships. The X-ray diffraction method has provided the structure of polymers which have high crystallinity. However, most polymers have low crystallinity and so structural information about the non-crystalline region, which is the major component, cannot be obtained by X-ray studies. Therefore, X-ray diffraction is of limited use in the structural analysis for such systems. Chain segments in the non-crystalline region are sometimes in a mobile state and so the X-ray diffraction method provides no structural or dynamic information. On the other hand, solid-state NMR provides knowledge of the structure and dynamics of a sample irrespective of whether the region studied is crystalline or non-crystalline.

Polyethylene

Polyethylene (PE) is a typical plastic. The polymer chain exhibits a *trans* and two kinds of *gauche* conformations, between which the energy difference is small and so the polymer takes various configurations under different conditions. PE has two kinds of crystal forms, as determined by several spectroscopic methods such as NMR, X-ray diffraction, electron diffraction, IR and neutron diffraction methods. One form is the orthorhombic form which occurs under normal conditions and another is the monoclinic form which occurs under high pressure or drawn conditions. In both crystal forms the conformation is the all-*trans* zigzag conformation, but the chain arrangements are different from each other. The all-*trans* zigzag planes in the orthorhombic form are perpendicular to each other, and in the monoclinic form they are parallel. Further, there also exist the non-crystalline and interfacial phases. Their existence strongly affects the physical properties of the polymer. Therefore, it is very important to analyse the detailed structure of PE in the solid state to understand its physical properties.

Figure 1 shows ^{13}C CP MAS NMR spectra of single crystal PE (SC-PE), melt-quenched PE (MQ-PE) and drawn PE (DR-PE). Only one sharp peak (33.0 ppm) appears in the SC-PE spectrum. However, MQ-PE and DR-PE have two and three peaks, respectively. This means that MQ-PE and DR-PE have at least two or three kinds of magnetically inequivalent carbon atoms. These peaks have been assigned by T_1 measurements; peak a (31 ppm) comes from the non-crystalline region, which is a

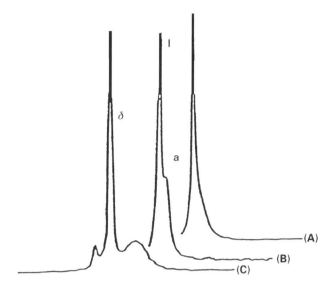

Figure 1 ^{13}C CP MAS NMR spectra of polyethylene. (A) SC-, (B) MQ- and (C) DR-PE. Parts (A) and (B) reproduced with permission of Elsevier Science from Ando I, Sorita T, Yamanobe T, Komoto T, Sato H, Deguchi K and Imanari M (1985) *Polymer* **26**: 11 864. Part (C) reproduced with permission of Elsevier Science from Van der Hart DL and Khoury F (1984) *Polymer* **25**: 1589.

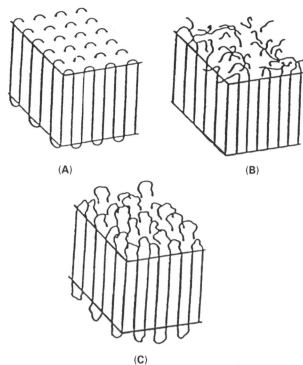

Figure 2   Schematic drawing of the conformation models of a polyethylene single crystal. (A) sharp-fold, (B) switch-board and (C) loose-loop model. Reproduced with permission of Elsevier Science from Ando I, Sorita T, Yamanobe T *et al* (1985) *Polymer* **26**: 1864.

mobile state, and peak I (33 ppm) from the crystalline region, which is in an immobile state. Further, the high-frequency peak of DR-PE (35.0 ppm) is assigned to the methylene carbons in the monoclinic crystal region. These assignments were justified by quantum chemical shielding calculations with the tight-binding (TB) sum-over-states (SOS) method. A peak from the interfacial region between the crystal and non-crystalline regions has also been reported.

The crystal structure of PE has been proposed from models such as sharp-fold, switch-board and loose-loops models as shown in **Figure 2A–C**. To obtain detailed information on the fold structure, SC- and MQ-PE have been studied by high-resolution solid-state ^{13}C NMR. This study reveals that SC- and MQ-PE take the adjacent re-entry type of macroconformation in addition to loose and long loops in the fold surface, as shown in **Figure 3**. The number of carbon atoms in the *trans* zigzag chain from one fold to the next is estimated to be ~100, which leads to an estimate of the stem length of about 125 Å. Its magnitude is consistent with the crystal thickness (120–150 Å) measured directly for PE single crystals by electron microscopy.

The physical properties of polymers in the solid state are strongly affected by temperature. Variable-temperature (VT) NMR techniques should provide useful information on the structural and dynamic aspects of polymers in the solid state. The advantage

of VT solid-state NMR techniques is that the temperature dependence of the conformation and molecular motion can be studied by observing the chemical shift. The structure of ultrahigh relative molecular weight PE (UHRMW-PE) has been investigated in the solid state by VT ^{13}C CP MAS NMR. The ^{13}C NMR spectrum of UHRMW-PE at room temperature has two peaks (the low and high frequency peaks are designated by A and I, respectively) as shown in **Figure 4**. Their chemical shifts agree with those for the melt-crystallized PE (MC-PE), and so peaks I and A have been assigned to the crystalline region with the *trans* zigzag conformation and to the non-crystalline region, respectively. When the temperature is decreased, peak A shifts to low frequency but peak I shifts to high frequency. The ^{13}C chemical shift behaviour for peak A of UHRMW-PE is similar to that of MQ-PE. At –108°C, the molecular motion is frozen and the chemical shift for peak A is about 32 ppm. Therefore, peak A is due to the methylene carbons in the *trans* conformation in the non-crystalline region since in the frozen state a methylene carbon in the *gauche* conformation should appear at about 27 ppm because of the γ-*gauche* effect. On the other hand, the ^{13}C chemical

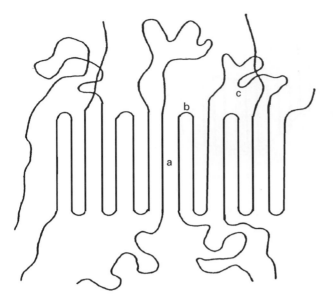

Figure 3 Schematic illustration of the folded chain conformation form of melt-quenched polyethylene. (a) *trans* zigzag, (b) sharp-fold and (c) loose and long loop. Reproduced with permission of Elsevier Science from Ando I, Sorita T, Yamanobe T, *et al.* (1985) *Polymer* **26**: 1864.

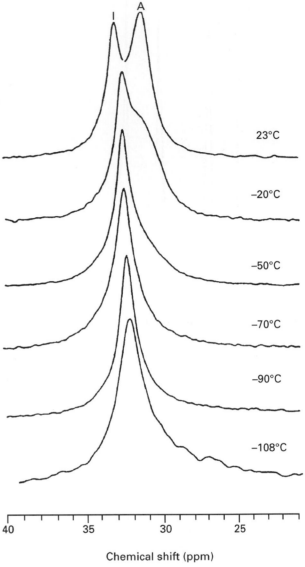

Figure 4 ^{13}C CP MAS NMR spectra of ultrahigh relative molecular mass polyethylene as a function of temperature. Reproduced with permission of John Wiley & Sons from Akiyama (1990) *Journal of Polymer Science, Part B: Polymer Physics* **28**: 587.

shift of peak I shifts to low frequency, which is opposite to the case of MQ-PE, with decreasing temperature. At –108°C, the chemical shift is about 32 ppm, and peaks A and I coalesce. It is difficult to conclude that the cause for the low frequency shift of peak I is a change of structure in going from the orthorhombic form to a different crystal structure. It is known that the ^{13}C shifts of methylene carbons in the orthorhombic, triclinic and monoclinic forms are about 33, 34 and 35 ppm, respectively. From these results, the ^{13}C shift of 32 ppm at –108°C indicates that the crystalline structure is different from other crystal structures found at room temperature. The other possibility for the low frequency shift is a distortion of the orthorhombic form.

The relaxation parameters, T_1, dipolar-dephasing relaxation times (T_{DD}), etc. can provide useful information on the dynamics of polymer motion in the solid state and to monitor solid-state polymer phase transitions.

The non-crystalline region of two kinds of PE samples, using single ^{13}C-labelled solution-crystallized PE (PE-SL) and single ^{13}C-labelled melt quenched PE (MQ-PE-SL), has been studied by VT ^{13}C CP MAS NMR. The dynamics of the non-crystalline region have been discussed by measuring ^{13}C T_1 and T_{DD} from –120 to 44°C. Each of these spectra consists of three peaks, corresponding to an orthorhombic crystalline peak at 33.0 ppm, a monoclinic crystalline peak at 34.4 ppm (a small shoulder on the left-hand side of the orthorhombic peak) and a non-crystalline peak which appears at 30.8–31.3 ppm. The ^{13}C T_1 data of these samples over a wide range of temperature, obtained using the inversion–recovery method with the pulse saturation transfer (PST) pulse sequence, are given. The PST pulse sequence enhances the intensity of mobile methylene carbons. The T_1 values for samples PE-SL and MQ-PE-SL are plotted against the inverse of the absolute temperature ($1/T$) in **Figure 5A**. It has been suggested previously that local molecular motion in the non-

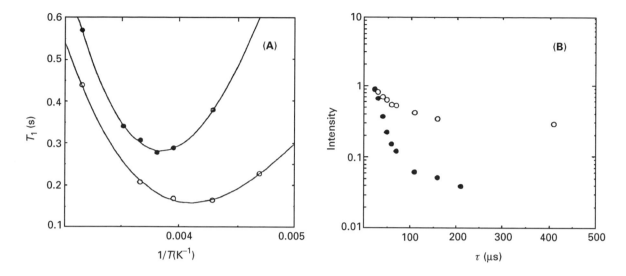

Figure 5 (A) Plots of non-crystalline ^{13}C spin–lattice relaxation times: T_1 of PE-SL samples (●) and MQ-PE-SL (O) versus the reciprocal absolute temperature. (B) Intensity of non-crystalline peaks of a sample of PE-SL and MQ-PE-SL versus delay time τ. The peak intensity was obtained from computer simulation of the ^{13}C partially relaxed dipolar-dephasing NMR spectra. Reproduced with permission of John Wiley & Sons from Chen Q, Yamada T, Kurosu H, Ando I, Shioino T and Doi Y (1992) *Journal of Polymer Science, Part B: Polymer Physics* **30**: 591.

crystalline region of PE is independent of the degree of crystallinity, higher-order structures or morphologies. However, these suggestions are not supported by the experimental results of PE-SL and MQ-PE-SL for two reasons: one is the difference in T_1 values and the other is the large difference in the temperature of the T_1 minimum of the non-crystalline region between PE-SL and MQ-PE-SL, as shown in **Figure 5A**. These facts show that local molecular motions in the non-crystalline regions of the PE-SL sample are more constrained than that of MQ-PE-SL. To study whether the dynamic behaviour of the two kinds of non-crystalline region is also different on the T_2 time-scale, T_{DD} values of samples of PE-SL and MQ-PE-SL were measured over a wide temperature range. The relative intensity of the non-crystalline peak obtained from computer simulation was plotted against the delay time τ in **Figure 5B**. It can be clearly seen that the non-crystalline peak of the sample of PE-SL relaxes more quickly than that of the MQ-PE-SL sample. The value of T_{DD} depends on molecular motion, carbon–proton dipolar interactions, MAS rate and spin diffusion. Fundamentally, it can be said that the dipolar dephasing time in the non-crystalline region becomes a measure of molecular motion because of the high mobility. Therefore, the longer T_{DD} value of MQ-PE-SL, compared with that of PE-SL obviously suggests that the carbon–proton dipolar interaction is partially averaged by molecular motion on the T_2 time-scale.

Poly(vinyl alcohol)

Poly(vinyl alcohol) (PVA) is used as a high-performance fibre and as a barrier material against oxygen. Detailed information about the structure and molecular motion is needed in order to develop and design this material. The ^{13}C CPMAS NMR spectra of PVA have been measured at room temperature. The methine carbon resonance split into three peaks in the CP MAS NMR spectrum. In the case of solution state NMR, the three methine carbon resonances and relative intensities correspond to the triad tacticity. However, the chemical shifts of the three methine carbons in the solid state are about 77, 71 and 65 ppm (**Figure 6**) and the relative intensities are not consistent with the triad tacticity in solution. The chemical shifts of two high-frequency peaks move significantly to higher frequencies in the solid state. These peaks, on the basis of the formation of intramolecular hydrogen bonds, have been assigned as follows. The highest frequency peak of the methine carbon is assigned to the *mm* triad with two intramolecular hydrogen bonds, the second highest one is assigned to the *mm* and *mr* triad with one intramolecular hydrogen bond and the other peak is assigned to the *mm*, *mr* and *rr* triads with no intramolecular hydrogen bonds.

Most recently, the ^{13}C CP MAS NMR spectra of PVA gels were measured to clarify the structure of their immobile component. In the ^{13}C CP MAS NMR

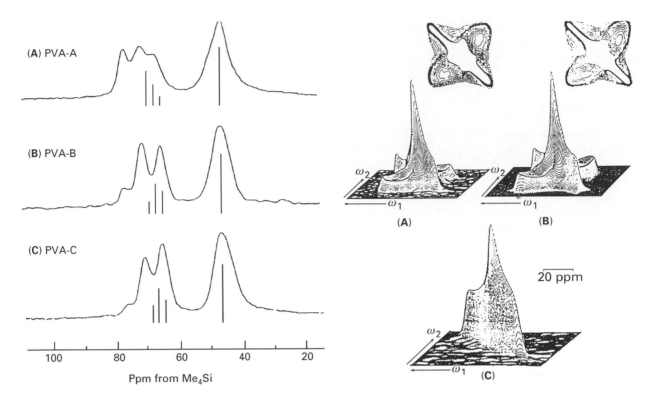

(A) PVA-A

(B) PVA-B

(C) PVA-C

100   80   60   40   20

Ppm from Me$_4$Si

Figure 6 ^{13}C CP-MAS spectra of poly(vinyl alcohol) at room temperature. (A) syndiotactic, (B) isotactic and (C) atactic. Line spectra represent the ^{13}C spectra in DMSO-d_6 solution at 353 K. Reproduced with permission of American Chemical Society from Terao T, Maeda S and Saika A (1983) *Macromolecules* **16**: 1535.

Figure 7 (A) Pure absorption-mode 2D exchange spectra of isotactic POM at $T = 360$ K and mixing time $t_m = 2$ s, (B) theoretical spectrum of (A) with $x = 2\omega$, $t_m = 2.5$ s and (C) $T = 252$ K and $t_m = 1$ s. Reproduced with permission of Academic Press from Hagemeyer A, Schmidt-Rohr K and Spiess HW (1989) *Advances in Magnetic Resonance* **13**: 85.

spectra, the three carbon peaks were clearly observed, and originate from the formation of strong intermolecular or intramolecular hydrogen bonds between OH groups such as PVA in the solid state as mentioned above. Further, the molecular motion of immobile and mobile regions of PVA gel through the observation of ^{13}C T_1 has been studied.

Polyoxymethylene

The molecular motion of polyoxymethylene (POM), which is the first member of the polyether series, has been investigated using the two-dimensional (2D) exchange experiment. The observed and calculated 2D exchange spectra of POM are shown in **Figures 7A** and **7B**, respectively. The spectrum observed at 252 K shows only diagonal signals and thus no molecular motion occurred during the mixing time. However, if molecular motion occurred during the mixing time, the spectrum would show steric effects as shown in **Figure 7C**. The calculations were performed by using the model of helical jump motions, assuming a one-dimensional random walk in continuous time and elementary processes of 200° jumps. Thus, at $T = 360$ K the subspectra can

be identified for the discrete steps a given CH$_2$ group of POM experiences, as the helix rotates.

Fluoropolymers

Fluoropolymers have excellent properties and have been widely used for various purposes under severe conditions. Although high-resolution solid-state NMR is a powerful tool with which to elucidate structures and dynamics of polymers in the solid state as mentioned above, the ^{13}C CP MAS experiment often fails to provide detailed information on the structures of fluoropolymers because their spectra gives rise to broad peaks. This is mainly due to two dominant factors, namely large dipolar coupling and the shielding anisotropy of fluorine nuclei. Although the high-speed magic-angle spinning (HS MAS) method leads to a significant reduction in such broadening for ^{19}F NMR in the solid state and has been used to characterize fluoropolymers, only a few ^{13}C solid-state NMR experiments have been reported. The simultaneous application of proton and fluorine decoupling drastically reduces line broadening caused by ^{1}H–^{13}C and ^{19}F–^{13}C dipolar interactions. The two kinds of poly(vinylidene fluoride), non-polar α phase and polar β phase, have been characterized by means of solid-state triple-resonance spectroscopy. There was a small difference (~ 4 ppm) in the chemical shift positions for these distinct crystalline modifications.

Blend polymers

Multicomponent organic polymer materials such as polymer blends are important in industry. Their macroscopic properties are determined not only by the molecular properties of the individual polymers, but also by the miscibility, morphology and structure of the interfaces between the polymers. Solid-state NMR is a useful tool to study the microscopic structure of heterogeneous polymer materials.

The miscibility of PMMA and poly(vinylidene fluoride) (PVDF) has been established by various methods. Both ^{19}F–^{13}C and ^{19}F–^{1}H CP and molecular miscibility in blends of PVDF and isotactic, syndiotactic and atactic PMMAs were investigated by triple resonance ^{1}H, ^{19}F, ^{13}C solid state CP MAS NMR. The fraction of nonmixed PMMA was determined for these blends, and it was found that this fraction was smaller for isotactic than for atactic and syndiotactic PMMAs.

A study of the polymer blends of poly(vinyl phenol) and poly(methyl acrylate) using 2D ^{13}C–^{1}H correlation NMR have been reported. From the spectrum, a direct interaction between the hydroxyl hydrogen of poly(vinyl phenol) and the carbonyl carbon of poly(methyl acrylate) can be deduced. The miscibility in annealed 50:50 blends of a random, copolyester Vectra-A, containing 73% *p*-hydroxybenzoic acid and 27% hydroxynaphthoic acid, and poly(ethylene terephthalate) (PET) has been discussed by using a standard 2D exchange experiment. The small cross peaks between the peak of aliphatic PET and that of quaternary Vectra-A carbons diagonal lines indicated intimate mixing.

Biopolymers

In the solution state, the NMR chemical shifts of biopolymers with internal rotations are often averaged values for all internal rotations because of rapid interconversion by rotation about peptide bonds. In the solid state, however, chemical shifts are often characteristic of specific conformations because of highly restricted rotation about the peptide bonds. The NMR chemical shift is affected by a change in the electronic state arising from the conformational change. NMR chemical shifts in the solid state, therefore, provide useful information about the electronic state and conformation of a biopolymer with fixed structure, especially membrane-bound

polymers. Several studies of typical synthetic polypeptides and membrane proteins using high-resolution ^{13}C NMR in the solid state are given here.

Synthetic polypeptides

Synthetic polypeptides consist of repeating sequences of certain amino acids and their structures are not as complicated as those of proteins. For this reason, synthetic polypeptides are sometimes used as model compounds for proteins. Their preferred conformations are classified as α helix, β sheet, ω helix and so on. High-resolution solid-state ^{13}C NMR spectroscopy has proved to be a very powerful tool for determining the structure of polypeptides in the crystalline state. For example, ^{13}C CP MAS NMR spectra of solid poly(L-alanine) ($[Ala]_n$) show the Cα, Cβ and C=O carbon signals to be well resolved between the α helix and β sheet forms. The chemical shifts of the Cα and C=O carbons of the α helix are displaced significantly to high frequency by 4.2 and 4.6 ppm, respectively, relative to those of the β sheet form, while the shift of the Cβ carbon of the α helix is displaced to low frequency by about 5 ppm with respect to that of the β sheet. For this reason, the value of the ^{13}C shift can be used to describe the local conformation. In addition, the ^{13}C shifts of randomly coiled $[Ala]_n$ in trifluoroacetic acid solution have values between those of the α helix and β sheet forms. The absolute ^{13}C shifts of the Cα and Cβ carbons are affected by the chemical structure of the individual amino acid residues and can be used effectively for conformational studies of particular amino acid residues in polypeptides and proteins. On the other hand, the C=O shifts do not seem to be affected by residue structure and can be used for diagnosing the main chain conformation.

Recently, the relation between the amide proton shift and the conformation of synthetic polypeptides has been studied in the solid state using ^{1}H combined rotation and multipulse spectroscopy (CRAMPS). It has been determined that the amide chemical shifts of a synthetic polypeptide are 8.0–8.1 ppm for the α helix and 8.6–9.1 ppm for the β sheet. It was shown that high-resolution solid-state ^{1}H NMR such as CRAMPS is also a useful tool for studying the molecular structure of polypeptides in addition to high-resolution solid-state ^{13}C NMR.

Poly(β-benzyl-L-aspartate) (PBLA) in the solid state adopts various conformations such as a right-handed α(α_R) helix, a left-handed α(α_L) helix, a left-handed ω(ω_L) helix and a β sheet form, depending on the temperature. VT ^{13}C CP MAS NMR spectra of PBLA are shown in **Figure 8**. From this, it is clear that the conformational changes such as α_R helix → ω_L helix → β sheet occur with increasing temperature.

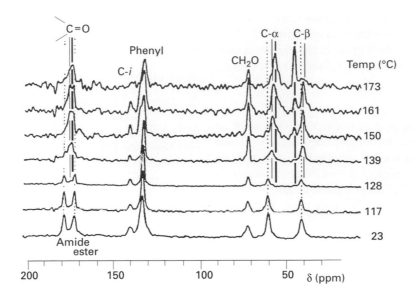

Figure 8 VT ¹³C CP MAS NMR spectrum with spinning sideband suppression of PBLA in the solid state at various temperatures: dashed line, α_R helix form, light solid line, ω_L helix form and bold solid line, β sheet form. Reproduced with permission of American Chemical Society from Akieda T, Mimura H, Kuroki S, Kurosu H and Ando I (1992) *Macromolecules* **25**: 5794.

Poly(γ-*n*-octadecyl-L-glutamate) (POLG) forms a crystalline phase composed of paraffin-like crystallites together with the α helical main chain, packing into a characteristic layer structure. The polymer forms thermotropic cholesteric liquid crystals by the melting of the side-chain crystallites. The conformational changes of POLG have been studied extensively by high-resolution solid-state NMR. **Figure 9** shows the ¹³C CP MAS NMR spectrum of POLG at room temperature together with the assignment of peaks. At room temperature, ¹³C shift values of the C=O and Cα carbons are 176.0 and 57.6 ppm, respectively. These values show that the main chain takes an α helix conformation. On the other hand, the interior CH₂ signal splits into two peaks. From reference data of *n*-alkanes and polyethylene, it is seen that the peak at 33.4 ppm is the carbon signal in the all-*trans* zig-zag conformation in a crystalline state and the peak at 30.6 ppm is the carbon signal in the non-crystalline phase or the liquid phase.

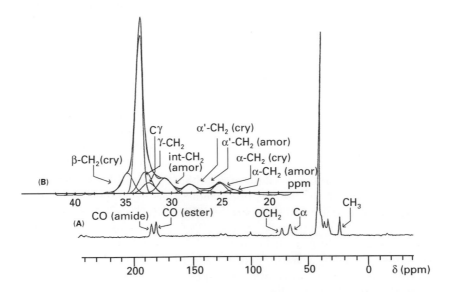

Figure 9 (A) ¹³C CP MAS NMR spectrum of POLG in the solid state at room temperature. (B) The methylene region is expanded and deconvoluted by computer-fitting with a Gaussian function. (cry and amor indicate crystalline and non-crystalline state, respectively.) Reproduced with permission of Elsevier Science from Katoh E, Kurosu H and Ando I (1994) *Journal of Molecular Structure* **318**: 123.

Figure 10 shows the VT ^{13}C CP MAS NMR spectra of POLG over the temperature range −40 to 230°C. The *n*-alkyl peaks change noticeably as the temperature is increased. The peak of the carbon signal in the non-crystalline state disappears above 50°C and the intensity of the interior CH$_2$ carbon in the non-crystalline state increases noticeably. This is due to the melting of side-chain crystallites. On the other hand, ^{13}C shift values of the C=O and Cα carbons are independent of temperature. It is shown that the main chain takes on a right-handed α helix conformation within the temperature range −40 to 230°C.

For samples which show a variation in molecular motion with temperature it is useful to measure the T_1 value using both the inversion–recovery method with the PST pulse sequence and Torchia's pulse sequence, depending on the type and rate of molecular motions. The molecular motions of POLG have been investigated within the temperature range −40 to 230°C, using both methods. Figure 11 shows the temperature dependence of T_1 for some typical carbons. According to the BPP theory, T_1 decreases, passes through a minimum (at a correlation time: τ_c) and

increases again when molecular motion increases, i.e., temperature is increased. The ^{13}C T_1 values of POLG are very short in spite of the fact that, as seen from the chemical shifts, the *n*-alkyl side-chains of POLG are in the crystalline state in the temperature range −40 to 40°C. If the *n*-alkyl side-chains are in the crystalline state, the ^{13}C T_1 values must be very long by analogy with the *n*-alkane data. However, the ^{13}C T_1 values for individual carbons of the *n*-alkyl side-chains are very short and very close to the T_1 values of *n*-alkanes in the rotator phase; in addition, the ^{13}C T_1 values for individual carbons of *n*-alkyl side-chains are almost the same as for *n*-alkanes in the rotator phase. From these results, it can be said that the *n*-alkyl side-chains of POLG are in a phase similar to the rotator phase, in which they undergo fast rotation such as libration, along the *trans* zigzag chain axis. The effect of the transitions at 45°, 60° and between 90° and 110°C on the individual carbon of the side-chains has also been investigated. The transition at 45°C comes from the melting of the side-chains, while other changes come from changes of the liquid crystalline phase, which has also been proved by other means.

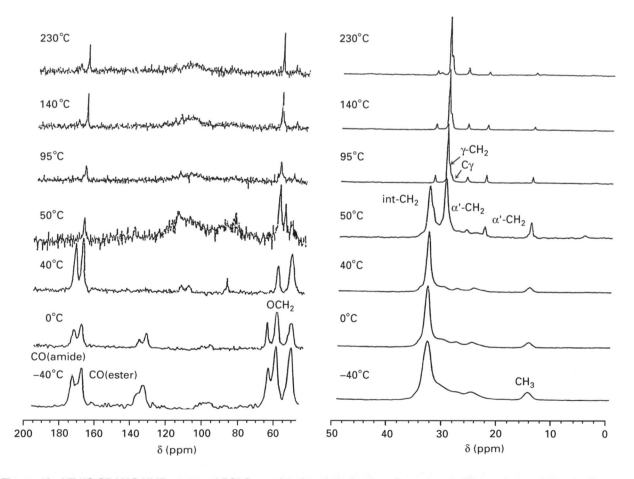

Figure 10 VT ^{13}C CP MAS NMR spectra of POLG as a function of temperature. Reproduced with permission of Elsevier Science from Katoh E, Kurosu H and Ando I (1994) *Journal of Molecular Structure* **318**: 123.

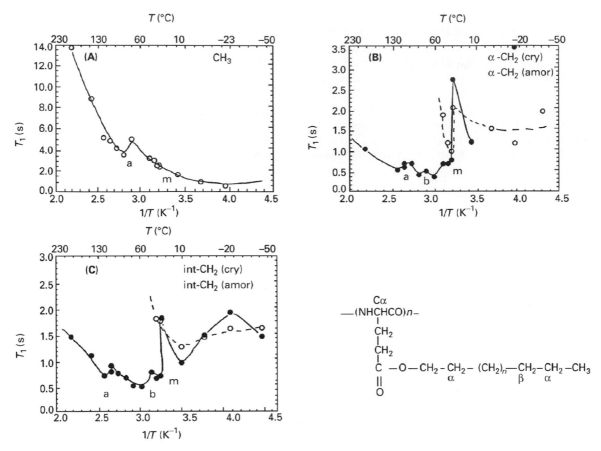

Figure 11   Temperature dependence of ^{13}C T_1 for individual carbons of POLG. (A) CH$_3$, (B) α-CH$_2$ and (C) int-CH$_2$ (a and b indicate transition temperatures. m indicates melting point.) For T_1 measurement in the temperature range -40 to 50°C Torchia's pulse sequence was used, and in the temperature range 50 to 230°C, the inversion–recovery pulse sequence was used. Reproduced with permission of Elsevier Science from Katoh E, Kurosu H, Kuroki S and Ando I (1994) *Journal of Molecular Structure* **326**: 145.

It has been demonstrated that the introduction of fluorine atoms into polymers induces new physical properties by changes in intra- and intermolecular interactions. Poly(γ-glutamate)s with *n*-fluoroalkyl side-chains have been studied using high-resolution solid-state NMR. From the experimental results, it was found that the conformation of the main chain depends on the carbon number of the *n*-fluoroalkyl side-chains. The main chain of poly(γ-alkyl-L-glutamate) with short *n*-fluoroalkyl side-chains takes the helix form and the poly(γ-alkyl-L-glutamate) with long *n*-fluoroalkyl side-chains takes the β sheet form. These conformational behaviours in the solid state are different from that of poly(γ-alkyl-L-glutamate) without *n*-fluoroalkyl side-chains.

Fibrous proteins

Since fibrous proteins generally have periodical amino acid sequences and higher-order structure, the clarification of their fine structure in the solid state becomes very important not only when discussing the physical and chemical properties, but when obtaining information about the molecular design of synthetic polypeptides. The conformation-dependent ^{13}C CP MAS NMR chemical shifts are particularly useful for the determination of the conformational features of fibrous proteins such as silk fibroin, collagen and wool keratin. For example, the conformational transition of *S*-carboxymethyl keratin that has low-sulfur fractions (SCMK low-sulfur) has been estimated. For SCMK low-sulfur heated at 200°C for 3 hr under vacuum, the ^{13}C MAS NMR spectrum shows that each signal becomes broader than those of other treated specimens. This indicates the existence of various conformations and/or different microenvironments in the heated SCMK low-sulfur. Thus, it can be said that the random coil form appears by heating. On the other hand, from the X-ray diffraction, the α helix form completely vanishes in SCMK low-sulfur under the same conditions. The difference between the results from X-ray diffraction and NMR spectroscopy suggests that only the packing of the ordered structure (α helix form) in SCMK

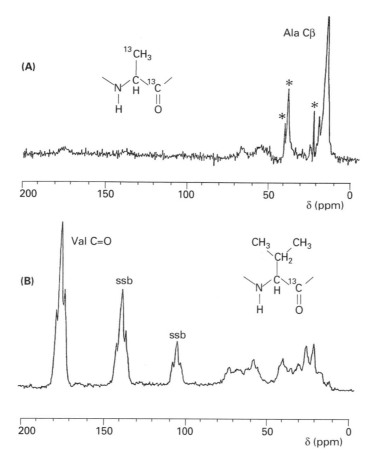

Figure 12 ¹³C CP MAS NMR spectra of (A) [3-¹³C] Ala bR and (B) [1-¹³C]Val bR in the purple membranes. (*) signals of ¹³C-labelled lipids. Reproduced with permission of Elsevier Science from Tuzi S, Naito A and Saito H (1994) *Biochemistry* **33**: 15 046.

low-sulfur is disrupted by heating, while the secondary structure is retained.

Membrane proteins

Membranes provide a vital interface between a biological organism and its environment. Furthermore, they provide a means for the formation of compartments within a cell. In every sense, they are vital to living systems. Indeed, since membranes and membrane proteins cannot easily be crystallized, NMR spectroscopy is a particularly powerful method for probing their structure and function. [1-¹³C]Ala, [3-¹³C]Ala and [1-¹³C]Val selected labelling of bacteriorhodopsin (bR) is a very convenient probe for conformation and dynamics studies. **Figure 12** shows the ¹³C CP MAS NMR spectra of [3-¹³C]Ala and [1-¹³C]Val bR. The peaks of [3-¹³C]Ala bR have been assigned to the α_{II} helix form at 16.3 ppm (60%), α_I helix form at 14.9 ppm (20%) and the loops at 17.2 ppm (20%), taking a variety of turn structures. Further, from the VT CP MAS ¹³C NMR spectrum of [3-¹³C]Ala labelled bR, well-resolved signals were

observed at both ambient temperature and −20°C but were broadened considerably at temperatures below −40°C. This situation was interpreted as showing that interconversion between several slightly different conformations is taking place. It was found that the exchange process was strongly influenced by the manner of organization of the lipid bilayers, and depended upon the presence or absence of cations responsible for electric shielding of the negative charge at the polar head groups. The secondary structures of [3-¹³C]Ala-labelled bR were not always identical at temperatures between ambient and lower temperatures, since the ¹³C shifts and relative peak intensities from purple membrane preparations containing these salts changed with temperature in the range −110 to 23°C. In particular, some residues involving Ala residues at the α_{II} helix and loop region were converted at temperatures below −60°C into a conformation of bR involving a α_I helix.

The rotational resonance (RR) method is one of several solid-state NMR approaches that have been developed recently for determining weak dipolar couplings under conditions of MAS. To measure heteronuclear dipolar interactions, such as between ^{13}C and ^{15}N, rotational echo double resonance (REDOR) and transferred-echo double resonance (TEDOR) NMR have been developed. The measurement of weak dipolar couplings in MAS experiments using any of these methods greatly enhances solid state NMR studies of biological systems since the dipolar couplings are directly related to internuclear distances which in turn provide constraints for the determination of molecular structure.

The interatomic distances in crystalline specimen of ^{13}C, ^{15}N doubly labelled simple peptides, N-acetyl-Pro-Gly-Phe, evaluated from REDOR data were compared with those from X-ray diffraction studies and found to justify such a novel approach. The REDOR-derived conformation of this peptide was β turn type I, consistent with the X-ray diffraction study. The maximum deviations of the distances determined by NMR and X-ray diffraction is 0.08 Å despite the complete neglect of the dipolar interactions with the labelled nuclei of neighbouring molecules and natural abundance nuclei. The precision and accuracy given by ^{13}C REDOR experiments are of the order of 0.05 Å. Distinction between the two types of β turn forms, including the β turn type II, found in the monoclinic crystal of this peptide (whose interatomic distances are different by about 0.57 Å) is made possible only by very accurate REDOR measurement.

More recently, it is the ability of solid-state NMR to give completely resolved spectra of immobile proteins that has enabled the structures of larger membrane proteins to be determined in the definitive environment of lipid bilayers. Further, high-resolution solid-state ^{13}C NMR spectra of the [3-^{13}C]Ala-labelled fragment of bR incorporated into dimyristoylphosphatidylcholine bilayers have been recorded. This approach has proved useful for the assignment of peaks as well as dynamic features of the transmembrane helices in lipid bilayers.

Coals and coal products

The complete molecular structure of coal is one of the unsolved secrets of nature. Of the various analytical methods available, NMR spectroscopy has proved to be of special importance in coal research. Solid-state NMR techniques allow the structure of coal and coal products to be investigated in a direct and nondestructive way.

The accuracy of aromaticity measurements on coals by CP MAS ^{13}C NMR, together with additional problems posed by high-field measurements and spectral editing, and some emerging techniques, have been discussed. It is suggested that a combination of low field, single-pulse excitation with long relaxation delays and the use of a suitable reagent to quench paramagnetic centres is the most satisfactory, albeit time consuming, recipe for obtaining reasonably reliable results on unknown samples.

One particular coal, Illinois No. 6, has been measured by 2D ^{13}C–^{1}H chemical shift correlation spectroscopy. A variant of dipolar shift correlation spectroscopy has been described and this is very convenient for determining the relative amounts of methylene, methine and methyl or non-protonated carbons in a complex mixture such as coal.

Foods

It has long been recognized that NMR can be of use to the food scientist and food processor. However, it is only within the last 15 years that there has been a consistent and widespread growth of the use of NMR. One reason for this is the development of new solid-state NMR methods.

Work on starch provides an excellent example of the information obtainable from solid-state NMR. Native starch granules have crystalline and non-crystalline regions. The crystalline regions are made up of an ordered arrangement of polymer chains in the form of double helices, whereas the amorphous regions contain single chains. The two polymorphs of starch, A (from cereals) and B (from tubers), have different packing arrangements of the double helices. The CP MAS NMR spectra of various native starches clearly demonstrate distinguishable crystalline and amorphous regions.

^{13}C CP MAS and single-pulse (SP) MAS spectra have been obtained for galactomannans and glucomannans as powders, hydrates or gels. Comparison of CP and SP experiments on galactomannan gels did not reveal any chemical shift changes, suggesting that the conformations of mobile and rigid segments are similar. CP MAS experiments have been performed on purified carrageenans and agaroses and on the seaweeds which are the source of these materials. Although the lines were broad, characterization of the intact algae was possible and various substituents were identified.

List of symbols

T_1 = spin–lattice relaxation time; T_2 = spin–spin relaxation time; $T_{1\rho}$ = spin–lattice relaxation time in the rotating frame; T_{DD} = dipolar-dephasing relaxation time.

See also: **Food Science, Applications of NMR Spectroscopy; High Resolution Solid State NMR, 1H, 19F; Membranes Studied By NMR Spectroscopy; Neutron Diffraction, Theory; NMR of Solids; NMR Pulse Sequences; NMR Spectrometers; Powder X-Ray Diffraction, Applications; Proteins Studied Using NMR Spectroscopy; Solid State NMR, Methods; Solid State NMR, Rotational Resonance.**

Further reading

Ando I and Asakura T (1998) *Solid State NMR of Polymers.* Amsterdam: Elsevier Science.

Ando I, Yamanobe T and Asakura T (1990) *Progress in NMR Spectroscopy* 22: 210.

Belton PS, Colquhoun IJ and Hills BP (1993) *Annual Reports in NMR Spectroscopy* 26: 1.

Dec SF, Wind RA and Maciel GE (1987) *Macromolecules* 20: 349.

Evans JS (ed) (1995) *Biomolecular NMR Spectroscopy.* New York: Oxford University Press.

Katoh E, Sugisawa H, Oshima A, Tabata Y, Seguchi T and Yamazaki T (1998) *Radiation Physics and Chemistry* (in the press).

Katoh E, Sugimoto H, Kita Y and Ando I (1995) *Journal of Molecular Structure* 355: 21.

Komoroski RA (ed) (1986) *High Resolution NMR of Synthetic Polymers in Bulks.* Florida: VCH.

Kurosu H, Ando S, Yoshimizu H and Ando I (1994) *Annual Reports in NMR Spectroscopy* 28: 189.

Harris RK and Jackson P (1991) *Chemical Reviews* 91: 1427.

Holstein P, Scheler U and Harris RK (1997) *Magnetic Resonance in Chemistry* 35: 647.

Meiler W and Meusinger R (1991) *Annual Reports in NMR Spectroscopy* 23: 375.

Saito H and Ando I (1989) *Annual Reports in NMR Spectroscopy* 21: 210.

Schmidt-Rohr K and Spiess HW (1994) *Multidimensional Solid State NMR and Polymers.* London: Academic Press.

High Resolution Solid State NMR, 1H, 19F

Anne S Ulrich, Friedrich-Schiller-University of Jena, Germany

MAGNETIC RESONANCE
Applications

Introduction

Hydrogen (^1H) and fluorine (^{19}F) possess the highest magnetic moments amongst all isotopes in the periodic table (besides the radioactive tritium, ^3H), which gives them the greatest NMR sensitivity and strongest dipolar couplings. Their spin of $I = \frac{1}{2}$ makes them well suited for structural investigations, and ^1H is indeed the most widely used nucleus in solution-state NMR. In the solid state, however, the very advantage of strong dipolar interactions, conveying long-range distance information, turns to a disadvantage when rapid spin diffusion occurs between nuclei in an extensively coupled network. The high abundance of ^1H in organic compounds thus leads to substantial line-broadening, and the same happens to ^{19}F in a perfluorinated or protonated environment. Several different line-narrowing approaches have been developed to achieve high resolution, namely magic-angle spinning (MAS), multiple-pulse sequences, a combination of both (CRAMPS), or the use of uniaxially oriented samples. Some important experiments will be illustrated here that are applicable to both ^1H and ^{19}F for determining chemical shift values, seg-

mental mobilities, and through-space correlations. Depending on the material, both nuclei have been fruitfully applied in the analysis of inorganic crystals organic solids, glasses, polymers, liquid crystals and biological systems. The extensive literature on broadline NMR studies of these systems will not be considered here.

The high-frequency channel of a standard solid-state NMR spectrometer tends to be optimized for ^1H, especially when ^1H-decoupling is required for observing ^{13}C and other low-band nuclei. For ^{19}F NMR experiments it is necessary to re-tune the transmitters, amplifiers and probes, and some extra hardware may need to be implemented when ^{19}F and ^1H are to be used simultaneously. The areas of ^1H and ^{19}F NMR have thus been pursued by different groups and with different objectives. If any general trends are to be identified, solid-state ^1H NMR appears to be moving towards higher resolution by working at high magnetic field strengths (currently up to 20 T), by applying high MAS speeds (beyond 35 kHz), by making use of multidimensional experiments (HETCOR, multiquantum), and with an awareness of the physical limitations to the inherent spectral line widths. Concerning the solid state ^{19}F NMR analysis of (per-)fluorinated materials, an observation of ^{19}F is often preferred over the otherwise very popular ^{13}C, because in these systems the ^{13}C resonances are strongly broadened and not so readily decoupled from the ^{19}F spins. With yet another motivation to study dilute ^{19}F-labels in a protonated environment, namely in organic and biological systems, many ^{19}F NMR experiments can be realized under efficient ^{1}H-decoupling that are analogous to ^{13}C NMR.

Properties of the nuclei

The nuclear spin properties of ^{1}H and ^{19}F are summarized and compared in **Table 1**. The high gyromagnetic ratios γ provide both isotopes with a much higher sensitivity than ^{2}H (1% of ^{1}H), ^{13}C (1.6%) or ^{15}N (0.1%), to name a few, and these factors are further enhanced by the respective natural abundances. The homo- and heteronuclear dipolar Hamiltonian also depends on γ, and it is given for a dilute pair of spins (I_1,I_2) by

$$\mathcal{H}_D = -\left(\frac{\mu_0}{4\pi}\right)\frac{\gamma_1\gamma_2\hbar^2}{r_{12}^3}\left(\frac{3\cos^2\theta - 1}{2}\right) \times (3I_{1z}I_{2z} - \mathbf{I}_1\mathbf{I}_2) \quad [1]$$

The coupling within a pair (or within a small group) of nuclei is well defined by their relative positions (distance r and angle θ), which makes their interaction inhomogeneous. The line shapes or dipolar splittings of dilute labels can thus be directly analysed to measure local structural parameters or motional averaging. In an extended network of abundant spins, on the other hand, the dipolar interactions are more complex than in Equation [1] and they become highly nonlinear. This leads to homogeneous line-broadening by spin diffusion, which is much harder to suppress experimentally.

The main difference between ^{1}H and ^{19}F lies in the ranges of their isotropic and anisotropic chemical shifts (anisotropic = dependent on the orientation of the molecule in space, i.e. on its alignment with respect to the magnetic field direction). Unlike ^{1}H, the ^{19}F atom is rather polarizable, which makes the chemical shift very sensitive to local charge densities and hydrogen-bonding, and which also leads to significant solvent and temperature effects. **Table 2** provides a comprehensive list of isotropic ^{19}F chemical shifts for inorganic and organic solid compounds, as compiled by Miller. Referencing in the solid state usually relies on comparison with a separate sample of TMS for ^{1}H, or CFCl$_3$ for ^{19}F, rather than using internal or external chemical shift standards. An

extensive list of chemical shielding anisotropies for ^{1}H and ^{19}F has been published by Duncan.

Methods for studying abundant and dilute spins

In the following overview of methods and applications in high-resolution solid-state ^{1}H and ^{19}F NMR it is instructive to consider different kinds of situations that the experiment may need to address. Depending on the abundance of spins, different line broadening mechanisms will be encountered, and different techniques are required to achieve high resolution. After a discussion of four such categories, some representative applications to various materials will be illustrated.

^{1}H NMR of abundant spins

In organic compounds, in polymers and in biological materials the ^{1}H-density is high and spin diffusion leads to rapid dephasing of the magnetization. The resulting homogenous line broadening can only be fully suppressed by MAS when the spinning frequency ν_r is much higher than the static line width (which is typically around 50 kHz in rigid solids). Since modern commercial NMR probes can reach speeds of up to 35 kHz with small rotors, this is not usually sufficient to achieve complete homonuclear dipolar decoupling in rigid solids. Only samples with a high intrinsic mobility, such as liquid crystals and rubber-like materials, will give narrow lines under high-speed MAS, with well resolved isotropic chemical shifts. Note that the chemical shielding anisotropy does not usually represent a problem for ^{1}H NMR, because it represents an inhomogenous interaction. It is efficiently narrowed by MAS already at low spinning speeds, resulting in a narrow isotropic line plus a set of side-bands that are separated by multiples of ν_r.

Table 1 Nuclear spin properties of ^{1}H and ^{19}F

Property	^{1}H	^{19}F
Natural abundance	99.98%	100%
Gyromagnetic ratio γ (10^7 rad s^{-1} T^{-1})	26.7522	25.1815
Relative sensitivity (at constant field)	1.00	0.83
Resonance frequency at 11.74 T	500 MHz	470.4 MHz
Chemical shift range (total) (organic compounds)	−20–30 ppm 0–12 ppm	−400–800 ppm −200–0 ppm
Shielding anisotropy (CSA)	30 ppm	200 ppm
Dominant relaxation mechanism	Dipolar	Dipolar, CSA

Table 2 Chemical shifts of fluorine in inorganic and organic solie samples. Reproduced with permission of Elsevier Science from Miller JM (1996) *Progress in NMR Spectroscopy*, Vol 28, pp 255–281. Amsterdam: Elsevier.

Sample	Chemical group	Chemical shift (relative to CFCl$_3$ in ppm)
Inorganic samples		
LiF	F$^-$	−204
LiF	F$^-$	−130[a]
NaF	F$^-$	−221
NaF	F$^-$	−221
NaF	F$^-$	−126[a]
NaF	F$^-$	−224
KF	F$^-$	−130
KF	F$^-$	−130.2
KF	F$^-$	−123
KF•2H$_2$O	F$^-$	−133
KF–CaF$_2$	F$^-$	−121
KF–alumina	F$^-$	−159
KF–alumina	F$^-$	−115
	AlF$_6^{3-}$	−156
KF–silica	SiF$_6^{2-}$	−129
	SiF$_4^-$	−116
	SiF$_4$/HF	−161
KAlF$_4$	AlF$_4^-$	−155
K$_3$AlF$_6$	AlF$_6^{3-}$	−155
K$_2$SiF$_6$	SiF$_6^{2-}$	−92
RbF	F$^-$	−90
RbF	F$^-$	−88
RbF•H$_2$O	F$^-$	−113
RbF−alumina	F$^-$	−109
	AlF$_6^{3-}$	−139
CsF	F$^-$	−8
CsF	F$^-$	−79[a]
CsF−CaF$_2$	F$^-$	−79
CsF•2H$_2$O	F$^-$	−97
CsF–alumina	F$^-$	−116
CsF–alumina	F$^-$	−88[a]
	AlF$_6^{3-}$	−111[a]
Cs$_2$SiF$_6$	SiF$_6^{2-}$	−92
MgF$_2$	F$^-$	−194.8
CaF$_2$	F$^-$	−104.8
CaF$_2$	F$^-$	−107.7
CaF$_2$	F$^-$	−106.4, −107.0
SrF$_2$	F$^-$	−84.1
CdF$_2$	F$^-$	−192.1
Hg$_2$F$_2$	F$^-$	−95.8
HgF$_2$	F$^-$	−196.4
SnF$_2$	F$^-$	−110.4
SnF$_4$	F$^-$	−146.9
(C$_4$H$_9$)$_3$SnF	F$^-$	−145
AlF$_3$	F$^-$	−174
α-PbF$_2$	F(1)	−20.5
	F(2)	−57.7
	Mobile F$^-$	−39.0
Na$_3$AlF$_6$	F$^-$	−189

Table 2 *Continued*

Sample	Chemical group	Chemical shift (relative to CFCl$_3$ in ppm)
K$_3$AlF$_6$	F$^-$	−190
Al$_2$(F,OH)$_2$SiO$_4$ (topaz)	F$^-$	−140
Na$_2$SiF$_6$	F$^-$	−152
Na$_2$SiF$_6$	F$^-$	−135
Na$_2$SiF$_6$	F$^-$	−151^a
K$_2$SiF$_6$	F$^-$	−135
NH$_4$F−silica	SiF$_6^{2-}$	−128
NH$_4$F−alumina	F$^-$	−178
	Al−F	−166
Et$_4$NF−alumina	F$^-$	−123
Et$_4$NF•2H$_2$O	F$^-$	−130
Et$_4$NF−cytosine	F$^-$	−148
Bu$_4$NF−alumina	F$^-$	−106
Bu$_4$NF•3H$_2$O	F$^-$	−109
Bu$_4$NF−4-cyanophenol	F$^-$	−144
Bu$_4$NF−uracil	F$^-$	−149
Bu$_4$NF−HOCPh$_3$	F$^-$	−147
HAP/SnF$_2$ dentifrice	FHAP	−103 to −99
	F$^-$	−119
FAP	Ca$_5$(PO$_4$)$_3$F	−98.9
FAP	Ca$_5$(PO$_4$)$_3$F	−101.0
FAP	Ca$_5$(PO$_4$)$_3$F	−99.0
Sb(III) subs., FAP	1F$^-$/Sb^{3+}	−97.5
	2F$^-$/Sb^{3+}	−95.4
	1F$^-$/Sb^{3+}	−89.9
Synthetic FAP	Ca$_5$(PO$_4$)$_3$F	−98.9
Heated FAP	Ca$_5$(PO$_4$)$_3$F	−100
Mineral	Ca$_5$(PO$_4$)$_3$F	−100.4
Synthetic FHAP	Ca$_5$(PO$_4$)$_3$F$_x$(OH)$_{1-x}$	−102.3
FHAP	Ca$_5$(PO$_4$)$_3$F$_x$(OH)$_{1-x}$	−99 to −103
	F$^-$(NSA)	−119
CaF$_2$−FAP	F$^-$(CaF$_2$)	−105
	FAP	−99
Gallophosphate cloverite	F$^-$ in DR cages 4Ga	−67.7
	3 Ga, 2Al	−72.8
	2 Ga, 2Al	−81.0
	1 Ga, 1Al	−89.2
	4Al	−95.5
	HO−Al−F	−145.6
	F in sodalite type cages	−178.2
Flurophosphate glass	F$^-$ coord. to Al/Ba	−99 to −90
Fluorinated AlPO$_4$	Al−F	−142
	F$^-$··CHAH$^+$	−123
Fluorinated AlPO$_4$	F$^-$ ion pair	−119 to −123
	Al−F	−135 to −139^a
Hexamethylenediamine	F$^-$ occluded in double 4-membered rings	−67.8
subs. Ga$_2$O$_3$−P$_2$O$_5$	Ga...F$^-$... Ga	−92.2, −113
Fluorinated SiO$_2$	F$^-$	−63 to −75
Templated with	Si−F	−155^a
R$_4$NF		
	F$^-$ Interaction with Ge/Ti/Al/Ga	−140
	F$^-$...NH$_4^+$	−128

Table 2 *Continued*

Sample	Chemical group	Chemical shift (relative to CFCl$_3$ in ppm)
Octadecasil	F$^-$ in 4-membered ring cages	−38
Fluoride subs, silicate	F$^-$	−64
F subs. LEV type zeolite	F$^-$ in framework	−82
	Si–F	−120
	Al–F	−150, −175
	SiF$_6^{2-}$	−129
Zn,Ti substituted layered silicate	F$^-$	−146.6, −169.1, −186.0
Ga layered silicate	F$^-$ in dioctahedral sheets	−108
F−hectorite	F$^-$ in trioctahedral Mg/Li-containing sheets	−182.8^a
F−montmorillonite	F$^-$ in trioctahedral Mg-containing sheets	−176.2^a
Barasym SMM 100	F$^-$ in dioctahderal Mg-containing sheets	−152.0^a
Synthetic dioctahedral	F$^-$ in dioctahedral Al-containing sheets	−131.9^a
2:1 layered aluminosilicate	F$^-$	−133.2^a
Tremolite	F$^-$	−171.7
Fluorscandium pargasite	F$^-$	−169.6
NaF−zeolite A(LTA)	F$^-$ in sodalite structure	−179
KF−montmorillonite K10	SiF$_6^{2-}$	−129
NH$_4$−montmorillonite K10	SiF$_6^{2-}$	−136
RbF−montmorillonite K10	SiF$_6^{2-}$	−123
	AlF$_4^-$	−141
CsF−montmorillonite K10	AlF$_6^{3-}$	−113
RbF−silica	SiF$_6^{2-}$	−122
Cd/Zn/CuF$_2^-$ montmorillonite K10	F$^-$	−151
CuF$_2$−silica	SiF$_6^{2-}$	−149
ZnF$_2$−silica	F$^-$/F−H$_3$O$^+$	−124
	SiF$_6^{2-}$	−149
CdF$_2$−silica	F$^-$/F−H$_3$O$^+$	−124
NaF$_4$−silica	F$^-$	−128
NaBF$_4$−silica	F$^-$	−128
[(PS)H$^+$] [OTeF$_5^-$]	F$^-$ triclinic, axial	−12.8
(PS)H$^+$ = protonated	F$^-$ triclinic, equatorial	−30.7
1.8-bis(dimethylamino)-	F$^-$ orthorhombic, axial	−17.2
naphthalene	F$^-$ orthorhombic, equatorial	−33.1
[N(n-Bu$_4^+$][OTeF$_5^-$)]	F$^-$ axial	−14.3
	F$^-$ equatorial	−33.6
[N(n-Bu)$_4^+$]	F$^-$ axial	−27.7, −33.1
[H(OTeF$_5$)$_2^-$]	F$^-$ equatorial	−42.4, −43.3
Organic samples		
(CF$_2$CFCl)$_n$	CFCl	−130.2
	CF$_2$	−106.2, −100.2
CH$_2$CF$_2$/CF$_3$CFCF$_2$/CF$_2$CFCl co- and ter-polymers	−CH$_2$CF$_2$CF(CF$_3$)CF$_2$CH$_2$−	−71.9, −72.1, −72.7 −72.9
	−CH$_2$CF$_2$CF(CF$_3$)CH$_2$CF$_2$−	−76.4, −76.6, −77.6 −77.9
	−CH$_2$CF$_2$CH$_2$−	−82.1
	−CF$_2$CH$_2$CF$_2$CH$_2$CF$_2$−	−91.4, −91.8, −90.8 −91.5, −89.7
	−CFClCH$_2$CF$_2$CH$_2$CF$_2$−	−90.9
	−CF$_2$CH$_2$CF$_2$CF(CF$_3$)CF$_2$−	−104.4, −104.7
	−CF$_2$CH$_2$CF$_2$CF$_2$CFCl−	−108.8, −109.1
	−CF(CF)$_3$CH$_2$CF$_2$CF$_2$CF(CF$_3$)−	−112.8
	−CF$_2$CH$_2$CF$_2$CF$_2$CF(CF$_3$)−	−111.3, −111.6

Table 2 *Continued*

Sample	Chemical group	Chemical shift (relative to $CFCl_3$ in ppm)
	$-CF_2CH_2CF_2CF_2CH_2-$	−115.2, −116.7
		−116.4
	$-CH_2CF_2CF_2CH_2CH_2-$	−116.8, −116.4
	$-CH_2CF_2CF_2CF(CF)_3CH_2-$	−119.3, −119.6
		−120.0, −120.2
	$-CH_2CF_2CF_2CFClCH_2-$	−119.9, −120.2
	$-(CF_2)_n-$	−123.2, −123.6
		−123.7, −123.8
	$-CF_2CF_2CFClCH_2CF_2-$	−122.3
	$-CF_2CFClCH_2-$	−127.7, −129.8
		−130.5
	$-CH_2CF_2CF_2CF_2CH_2-$	−127.3, −126.7
	$-CF_2CF(CF_3)CF_2CF_2CF(CF_3)-$	−128.4, −130.5
		−131.3
	$-CH_2CF(CF_3)CF_2CF_2CF(CF_3)-$	−135.7
	$-CF_2CF_2CF_2(CF_3)CH_2CF_2-$	−184.9, −185.0
		−185.4, −185.8
$(CF_2CF_2)_n$	$-CF_2CF_2CF_2-$	−123.2
$(-C(CF_3)=C(CF_3)-)_n$	CF_3-	−50.2
	CF_3-	−54.1
	CF_3-	−82.4
	CF_3CF_2-	−127.9
	$-CF_2CF_2CF_2-$	−123.7
Poly(vinyl difluoride)	Imperfections	−113
	C–F	−89
$C_8F_{17}SO_3Na$	CF_3-	−83.0
	CF_3CF_2-	−129.8
	$-CF_2CF_2CF_2-$	−124.7
	$-CF_2SO_3-$	−118.2
Perfluoronaphthalene	F_1	−146.9
($C_{10}F_8$)	F_2	−148.4
	F_3	−151.7
	F_4	−152.6
$C_6F_5NH_2$	$F_{2.6}$	−165.2
	$F_{3.5}$	−167.7
	F_4	−177.8
AgTFA	$CF_3CO_2^-$	−71.5
$CFCl_3$/PU foam	$CFCl_3$	−2.0
$CFCl_2CH_3$/PU foam	$CFCl_2CH_3$	−45.6
$CHCl_2CF_3$/ PU foam	$CHCl_2CF_3$	−79.6
C_6H_4F-co-polymer	p-F	−108.5
CF_4/plasma treated diamond	CF	−148
	CF_2	−106, −123
	CF_3	−78
C_6F_6 sorbed on polystyrene	Dual mode sorbed	−117, −161
	Mobile C_6F_6	−169
Bis(trifluoromethyl) aniline/polystyrene	Dual mode sorbed	−55.7
C_6F_6 sorbed on butyl rubber	Dual mode sorbed	−153.4
CF_2CFHCF_3 subs. adamantane	CF_3	−75
	CF_2	−122, −130
	CFH	−209

Table 2 *Continued*

Sample	Chemical group	Chemical shift (relative to CFCl₃ in ppm)
Fluorinated steroids	CF	−165.6, −165.2
		−171.4
	CHF	−187.9, −180.1
	CH₂F	−192.3

All shifts have been recalculated with respect to CFCl₃ as the reference.
[a] Possibly an inconsistent assignment. Use with caution.

As an alternative to averaging in real space, dipolar decoupling can also be achieved in spin space by using multiple-pulse sequences such as WAHUHA-4, MREV-8 or BR-24. The efficiencies and spectral sweep-widths of these sequences are limited by the cycle time, e.g. to average a dipolar width of 50 kHz the cycle would have to be short compared to 20 µs. This places considerable technical demands on the probe design, namely short ringdown times (<< 10 µs) and the capacity for short high-power pulses under high duty cycle. Homonuclear decoupling efficiency is further improved by Combined Rotation And Multiple-Pulse Spectroscopy (CRAMPS). Even though this kind of experiment was not considered in the past to be easily implemented, modern spectrometers can be more routinely adjusted with respect to accurate phase cycling and timing. CRAMPS measurements are usually carried out under quasi-static conditions (typically 2–3 kHz MAS), but recent experiments have also been designed for fast MAS (10–20 kHz) in combination with short windowless multiple-pulse sequences. To assign simple compounds, the line widths in a CRAMPS spectrum are often sufficient, even though they are still broader than in solution state ¹H NMR.

Figure 1 illustrates some sources of line-broadening that cannot be prevented even on a perfectly well adjusted spectrometer. The resolution of the glassy polymer polystyrene is about an order of magnitude worse than that of crystalline adipic acid, which is partially attributed to the dispersion of chemical shifts in the amorphous material. Additionally, it has recently been recognized that the anisotropic molecular magnetic susceptibility provides a dipolar broadening mechanism for aromatic groups, which limits their peak resolution to about 0.3 ppm. Further difficulties in a CRAMPS experiment will be encountered when a group undergoes motions with a correlation time comparable to the pulse width, to the multiple-pulse cycle time, or to the sample spinning speed, as this can lead to severe losses in signal intensity.

Multidimensional spectroscopy plays an essential role in solution state ¹H NMR, by providing correlations through bonds (COSY, TOCSY), through

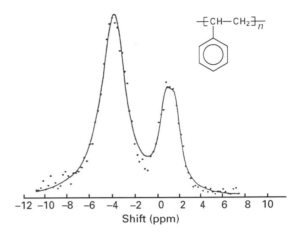

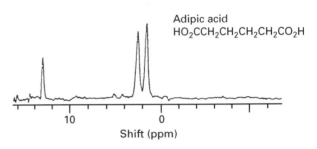

Figure 1 ¹H NMR CRAMPS spectra of amorphous polystyrene (top) and crystalline adipic acid (bottom), showing different degrees of resolution. Reproduced with permission of Elsevier Science from Gerstein BC (1998) ¹H NMR. In: Ando I and Asakura T (eds) *Solid State NMR of Polymers*, Vol 84, pp 166–189. Amsterdam: Elsevier Science.

space (NOESY), or to heteroatoms (HMQC, HSQC). Homonuclear 2D spectra of solids are only feasible when the sample possesses an intrinsically high mobility to give a sufficient resolution, which is the case for liquid crystals. Heteronuclear correlation (HETCOR) experiments, on the other hand, are a powerful tool to map even the very broad ¹H signals of a rigid molecule onto a second ¹³C dimension via the dipolar coupling through space. Since ¹³C possesses a chemical shift range of about 200 ppm, compared with 12 ppm for ¹H, the increase in resolution makes it possible to assign resonances in relatively large organic compounds and to study hydrogen-bonding in biological solids. During the evolution and mixing periods of a HETCOR experiment, the suppression of ^{1}H–^{1}H homonuclear and ^{1}H–^{13}C heteronuclear couplings can be achieved by multiple-pulse sequences and/or MAS. In the related wide-line separation experiment (WISE), where ^{1}H-line-widths are used as a measure of the respective segmental mobilities, the homonuclear interaction is allowed to be active during the evolution period. Whereas HETCOR is applicable to sites carrying any number of protons, the dipolar chemical shift correlation experiment (DIPSHIFT) has been designed to determine the accurate bond distance in a particular heteronuclear spin-pair, e.g. of ^{1}H–^{15}N in polypeptides.

Another sophisticated way of correlating spins with one another is possible by multiquantum spectroscopy, which can in principle resolve the connectivities of atoms that are present as an isolated cluster. Various multiple-pulse experiments have been designed to recouple isolated spins under fast MAS via multiquantum (or zero-quantum) coherences. When faced with an abundant network of ^{1}H, however, it is usually only feasible to monitor build-up rates qualitatively as a measure of the dipolar coupling strength. For example, when molecular motion reduces this interaction in polymers, the signal intensities after quantum production can be interpreted in terms of the respective local dynamics.

A different kind of ^{1}H NMR application in polymers is the use of long-range spin diffusion to study the heterogeneity of the sample. In this experiment a gradient of magnetization is set up between different regions that can be differentiated either by their mobilities (i.e. amorphous and crystalline) or by their chemical shifts. During a subsequent storage period the magnetization will diffuse over distances of up to 30 nm per second, such that domain sizes can be estimated by choosing suitable delay times. In these systems a CRAMPS experiment may be applied to resolve groups of signals, and multiple-pulse sequences have even been used as a chemical shift filter to 'catch' a resonance and keep it aligned with the static magnetic field direction. Strictly, however, the experiments mentioned in this paragraph are derived from broadline experiments and do not contribute themselves to an improvement in resolution.

^{19}F NMR of abundant spins

The situation with abundant ^{19}F spins, e.g. in fluoropolymers and inorganic fluorides, is similar to what has been outlined for ^{1}H. The same basic experiments apply, although the much broader range of ^{19}F isotropic chemical shifts (see **Table 2**) affords an intrinsically better resolution. ^{19}F MAS spectra of rigid solids will be accompanied by a higher number of spinning side-bands due to the increased chemical shielding anisotropy (CSA). Technically it becomes more demanding to excite uniformly the large spectral width of ^{19}F (90° pulse length of typically 1–2 µs), especially in multiple-pulse experiments which are quite sensitive to offset. To cover a broad spectral width in a CRAMPS experiment, it is essential to use the shortest possible cycle times. To that aim, the ring-down time of a probe can be reduced by Q-spoiling, which involves the addition of resistance to the RF circuit but has the disadvantage of requiring higher power levels for a given 90° pulse length. Alternatively, the signal can be built up in a point-wise fashion at the end of an incremented number of cycles. Since this single-point approach dispenses with the need for sampling during the pulse train, windowless or semi-windowless sequences can be used, although this method is associated with an obvious penalty in measurement time. In terms of hardware, a fast digitizer is required for ^{19}F NMR, and fluorine-containing materials should be avoided in the probe components. The relaxation mechanisms of ^{19}F are somewhat more complicated than those of ^{1}H, as they are affected not only by dipolar couplings but also by the chemical shielding anisotropy and hence magnetic field strength.

^{19}F NMR of dilute spins in the presence of protons

Dilute ^{19}F spins in a protonated environment do not suffer from direct homonuclear line broadening, but a new problem arises from the possibility of cross-relaxation to ^{1}H. In principle these heteronuclear interactions are inhomogenous, and MAS should be effective for narrowing. However, the situation becomes complicated when strong ^{1}H–^{1}H coupling leads to spin diffusion within the proton network, because this leads to a homogeneous dephasing of the ^{19}F magnetization. Therefore, to obtain well resolved ^{19}F NMR spectra strong ^{1}H-decoupling is essential (ideally up to 250 kHz). A double-resonance ^{19}F{^{1}H} probe should thus be capable of handling high power levels on both channels, and efficient filters are required to separate the two frequencies which are only 6% apart. With increasing magnetic field strengths the filtering becomes easier. Probe design may nevertheless require new approaches, such as transmission lines or cavity resonators, especially when a third channel is to be included for a low-frequency nucleus. As a consequence of high-power ^{1}H-decoupling, the ^{19}F resonances will be affected by the Bloch–Siegert shift, which is given by the ratio

$$\Delta = (\gamma_F \, B_{1H})^2/(\omega_F^2 - \omega_H^2) \qquad [2]$$

where $\gamma_F \, B_{1H}$ is the magnitude of the ^{1}H-decoupling field at the ^{19}F resonance frequency. On a 4.7 T spectrometer (200 MHz for ^{1}H) the shift is typically a few ppm towards lower frequency, but it becomes less significant at higher magnetic field strength.

Figure 2 illustrates the effects of ^{1}H-decoupling, of MAS, and of intrinsic molecular mobility on the ^{19}F NMR spectrum of the fluorinated sterol diflucortolon-21-valerate (DFC). While the static ^{19}F line shape of the crystalline material shows no recognizable features (**Figure 2A**), ^{1}H-decoupling reveals the two underlying CSA tensors (**Figure 2B**). With additional MAS at moderate speed (**Figure 2C**), the isotropic chemical shifts are resolved (as marked by an asterisk), and the principal values of the tensors can be calculated from the spinning sideband intensities using the Herzfeld–Berger approach. To demonstrate the molecular mobility in a liquid crystalline environment, in the next example (**Figure 2D**) the hydrophobic DFC was incorporated at a total concentration of 10% (w/w) into lipid bilayers composed of dimyristoylphosphatidylcholine (DMPC). Two narrow lines appear in this ^{19}F {^{1}H} MAS spectrum (**Figure 2D**), corresponding to the two ^{19}F-substituents on those sterol molecules that are partitioned into the bilayer. Since they undergo rapid long-axial rotation and restricted anisotropic tumbling, their total line width is less than the MAS spinning speed (5 kHz) and no sideband patterns are observed. Additionally, a residual contribution of crystalline material is visible in the same spectrum (**Figure 2D**), whose integrated intensity shows that DFC can only be dissolved up to 3.5% (w/w) in hydrated DMPC bilayers. Notably, the isotropic chemical shifts of DFC in the hydrophobic bilayer interior differ significantly from those in the crystalline environment, whereas they are rather similar (within 1 ppm) to the values obtained in CHCl$_3$ solution (**Figure 2E**).

An elegant way to distinguish ^{19}F-substituents based on their mobility or their proximity to protons is provided by cross-polarization (CP) from ^{1}H to ^{19}F, allowing, for example, the suppression of PTFE background signals. Having been originally developed under static conditions, this experiment can be used under MAS with some slight modifications. By choosing a short CP contact time, the ^{19}F-signals from immobile groups are selected since they are engaged in a rapid transfer of dipolar magnetization. Alternatively, the dipolar dephasing experiment can be used

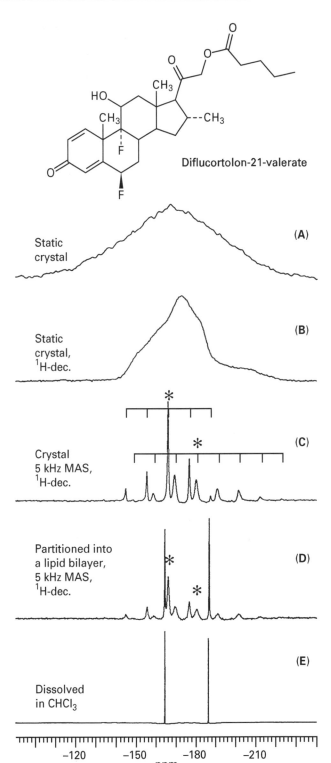

Figure 2 ^{19}F NMR spectra of diflucortolon-21-valerate (DFC) under different experimental conditions and in different environments, illustrating the effects of ^{1}H-decoupling, MAS and intrinsic molecular mobility.

to select resonances of ^{19}F-spins that are in comparatively mobile domains or remote from protons. CP MAS can also be combined with inversion recovery measurements, since the ^{19}F relaxation tends to be affected by cross-relaxation to ^{1}H. The T_1-values of ^{19}F are obtained either by direct polarization or by a suitable post-CP variable delay, whereas T_1-relaxation of the protons is revealed by inserting a pre-CP delay on ^{1}H during the ^{19}F {^{1}H}-experiment.

Homonuclear ^{19}F–^{19}F correlations provide information about spin diffusion through space, as, for example, in a 2D spin-exchange experiment. This approach requires sufficiently fast MAS and ^{1}H-decoupling to achieve complete resolution, since any magnetization exchange due to spectral overlap of the ^{19}F resonances has to be avoided. On the other hand, it may be desirable to stimulate a controlled magnetization exchange by introducing a certain spectral overlap under moderate MAS speeds. That is, when the spinning side-bands of one signal are allowed to overlap with the side-band pattern of a second signal, these two spins are recoupled by Rotational Resonance (RR). The RR experiment and related broadband recoupling schemes were originally developed to measure weak pairwise couplings between selective ^{13}C-labels and thus to determine their accurate distances (up to 7 Å). When applied to a fluorinated sterol, the RR effect was observed as a significant line-broadening rather than a dipolar fine-structure, which is consistent with rapid spin diffusion. Despite these difficulties, fluorinated substituents are desirable for RR and related experiments, because ^{19}F–^{19}F distances could theoretically be extended up to 16 Å. In fact, long-range heteronuclear distances (^{19}F to ^{13}C or ^{31}P) have been successfully determined by Rotational Echo Double Resonance (REDOR) and Transferred Echo Double Resonance (TEDOR) experiments. Owing to the fast spin diffusion of the ^{19}F magnetization it was usually necessary to observe the REDOR effect on the low-frequency channel under ^{19}F-dephasing. Overall, many of the experiments known for ^{13}C can be adapted for dilute ^{19}F spins provided that efficient ^{1}H-decoupling is available, and there is much potential for further development.

High-resolution ¹H and ¹⁹F NMR of oriented samples

Instead of averaging away the anisotropic nuclear spin interactions by MAS or multiple-pulse sequences, it is possible to take advantage of the anisotropy of these interactions, provided that macroscopically oriented samples are available. This kind of static solid-state NMR approach is entirely different from the experiments discussed above. It can lead to highly resolved ^{1}H and ^{19}F spectra with narrow lines, whose position carries information about the alignment of individual molecular segments. Oriented samples are typically prepared by stretching polymers or suitable proteins into fibres, or by aligning polymer films, liquid crystals or biological membranes on planar supports. For the most basic experimental set-up, the symmetry axis of the uniaxially oriented sample is aligned parallel to the static magnetic field direction. That way all groups of any one kind in the molecule will experience the same nuclear spin interaction. This interaction is characterized by a single value of the relevant anisotropy tensor, for example by the CSA tensor or by the inhomogenous dipolar coupling to another dilute spin (see Eqn [1]). The resulting ^{1}H or ^{19}F NMR signal will thus consist of a narrow line, whose resonance position reveals the geometric relationship between the respective tensor and the macroscopic axis of the oriented sample. In contrast, a non-oriented static sample gives rise to a broad powder line shape, and an MAS spectrum is simply reduced to the isotropic resonance. In practice, the effective line width of an oriented sample is limited by the quality of alignment, plus any residual broadening.

Applications of ¹H and ¹⁹F NMR

Inorganic solids

The classic inorganic compound investigated by solid state ^{1}H NMR is $CaSO_4 \cdot xH_2O$. As a powder it yields the well-known Pake pattern, and oriented single crystals have been used to analyse the angle-dependence of the homonuclear dipolar coupling. Today, most ^{1}H NMR applications employ fast MAS, preferentially at high field. Studies of hydroxyl groups in zeolites, in glasses, and on the surface of catalysts yield information about their local structure and acidity. Using ^{19}F MAS or CRAMPS experiments, the isotropic chemical shifts of many inorganic fluorides have been determined (see **Table 2**). Here, CaF_2 and its fluorohydroxyapatite derivatives $Ca_5(PO_4)_3 F_x(OH)_{1-x}$ are of particular interest due to their relevance in human dental enamel, in bone implants, as well as in fluorescent lights. In addition, the fluoride species formed upon adsorption onto alumina, silica, clays, zeolites or related materials have attracted attention in view of their role as catalysts or molecular sieves.

Organic molecules

Solid-state ^{1}H NMR constitutes a powerful approach to investigate the hydrogen-bonding and ionization states of small organic compounds, with similar applications in structural biology. High-resolution ¹H MAS and CRAMPS experiments have provided isotropic chemical shifts and CSA tensors for many simple molecules. In several instances a direct correlation with hydrogen-bonding lengths could be demonstrated, e.g. for amino acid carboxyl groups. Well resolved ¹H–¹³C HETCOR spectra have been obtained for various drugs (e.g. ibuprofen) and oligopeptides. With regard to ¹⁹F NMR, apart from the isotropic chemical shift values the ¹⁹F CSA tensors have been characterized for various aliphatic (e.g. silver trifluoroacetate, $C_8F_{17}SO_3Na$) and aromatic (e.g. pentafluoroaniline, perfluoronaphthalene) compounds by analysing the spinning side-band intensities. Pharamaceutically active sterols with one, two, and three ¹⁹F-substituents (cf. **Figure 2C**) were used to establish more advanced ¹⁹F{¹H} and triple-resonance ¹³C{¹⁹F}{¹H} NMR experiments. Since the 1D spectra were found to be rather sensitive to sample polymorphism, they can be used to detect different crystalline forms that may exhibit different pharmacological activities.

Polymers and rubbers

High-resolution ¹H NMR spectra of polymers are most readily obtained when the material possesses an intrinsically mobile backbone or side-chains (e.g. an elastomer), when measured at high temperature, or when a deuterated solvent has been imbibed [e.g. $CDCl_3$ in PMMA, or a swollen functionalized resin carrying a bound product]. For more rigid polymers, sufficiently narrow line widths have been achieved with the aid of fast MAS or CRAMPS to resolve aliphatic and aromatic resonances. Proton line widths and relaxation times have thus been used to describe segmental mobilities and crystallinity, to assess phase separation in co-polymers (e.g. PS–PVME), or to examine hydrogen-bonding and acidity (e.g in PVA or polysiloxanes). Unlike the case with protonated polymers, where ¹³C{¹H} NMR tends to be the method of choice, fluorinated polymers give strongly broadened ¹³C NMR spectra, which creates the need for direct ¹⁹F-observation. Owing to the wide chemical shift range of ¹⁹F NMR, simple broadline/relaxation NMR studies were quite popular in the past, but many high-resolution experiments have been demonstrated since. Studies of commercially important materials include PTFE, $(CF_2CF_2)_n$; PTrFE, $(CF_2CFH)_n$; PCTFE, $(CF_2CFCl)_n$; and PVDF, $(CF_2H_2)_n$; as well as co-polymers of the latter material with HFP, (CF_2CFCF_3); TFE, (CF_2CF_2); and CTFE, (CF_2CFCl). Some advanced triple-resonance REDOR experiments have been performed with ¹⁹F-labels in dendrimers and in

nanospheres, to visualize their packing density and the distribution of sequestered ligands, respectively.

Liquid crystals and lipid bilayers

The fast long-axial rotation ($\tau_c \approx 10^{-10}$ s) of liquid crystalline molecules leads to a partial averaging of the anisotropic interactions, thus improving the basic spectral resolution. For example, fast MAS of hydrated phospholipid dispersions yields well-resolved ¹H-NOESY spectra with distinct NOE cross peaks. Another important property of liquid crystals lies in the local alignment of the molecules about a director axis, which makes it useful to prepare macroscopically oriented samples. In a mechanically oriented sample, quasi-isotropic ¹H NMR spectra should be accessible by aligning the static sample axis along the magic angle. That way the intrinsic rotational diffusion of the molecules will average all anisotropic interactions to zero, although in practice any mis-alignment and slow collective motions of the molecules will reintroduce some line broadening. As an example, **Figure 3** illustrates the ¹H NMR spectra of an un-oriented dispersion of DMPC (**Figure 3A**) versus an oriented sample on glass plates (**Figure 3B**) aligned at the magic angle. Resolution can now be further enhanced by slowly spinning the oriented sample around the magic angle, which is called Magic-Angle Oriented Sample Spinning (MAOSS). The spectrum in **Figure 3A** shows narrow lipid signals (¹H line widths of 10–30 Hz) at a spinning speed of only 200 Hz, which compares favourably with the resolution achieved for lipid dispersions under high-speed MAS (14 kHz) or at a much higher magnetic field strength (17.6 T). It appears hopeful that even the signals of membrane-bound polypeptides (e.g. gramicidin, M13 coat protein) may be resolved using either of these MAS approaches.

Selective ¹⁹F labelling of lipids has contributed to a better understanding of the structural and motional properties of artificial model bilayers as well as biological membranes (e.g. of the bacterium *Acholeplasma laidlawii* when fed with ¹⁹F-labelled fatty acids). On the basis of early broadline NMR investigations, resolution has been improved by using oriented samples and multiple-pulse experiments to select certain anisotropic spin interactions. For example, the local order parameter of a particular acyl-chain segment can be calculated from the dipolar splitting of the corresponding CF_2-labelled group. **Figure 4** illustrates how the ¹⁹F–¹⁹F dipolar coupling of 4,4-DMPC-F_2 can be extracted, which is not resolved in the normal ¹⁹F NMR spectrum of an un-oriented lipid dispersion (**Figure 4A**), unless

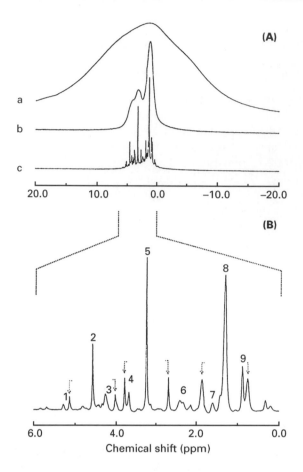

Figure 3 ¹H NMR spectra of DMPC bilayers as a non-oriented lipid dispersion (A), and as a macroscopically oriented sample that is aligned along the magic angle (B). The resolution is further improved by MAOSS at 200 Hz (see text). Reproduced with permission of Academic Press from Glaubitz C and Watts A (1998) *Journal of Magnetic Resonance* **130**: 305–316.

additional ¹H-decoupling is applied (**Figure 4B**). Alternatively, and without any need for ¹H-decoupling, the Carr–Purcell–Meiboom–Gill (CPMG) multiple-pulse sequence is able to suppress all unwanted line-broadening interactions such that only the desired homonuclear ¹⁹F coupling remains in the shape of a Pake pattern (**Figure 4C**). By extending the CPMG experiment to an oriented sample that is aligned with its symmetry axis perpendicular to the magnetic field (**Figure 4D**), an even better resolution (< 50 Hz line width) is achieved.

Peptides and proteins

Solid-state NMR lends itself to the structural analysis of proteins that do not crystallize, or when details about their functional mechanism or hydrogen-bonding pattern are to be investigated. For example, long-range distances have been measured by heteronuclear REDOR experiments on enzymes prepared with specifically ¹⁹F-labelled side-

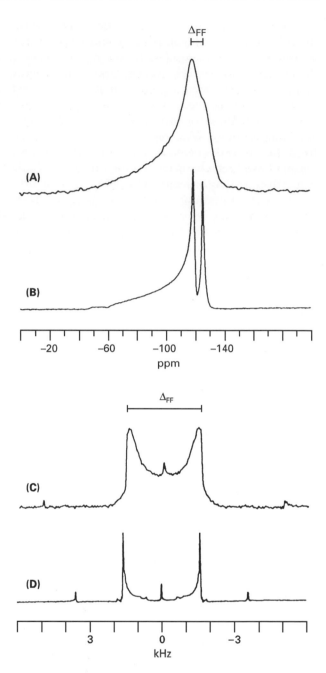

Figure 4 ¹⁹F NMR spectra of 4,4-DMPC-F_2 as a non-oriented lipid dispersion without ¹H-decoupling (A), and with ¹H-decoupling (B). The dipolar splitting Δ_{FF} is directly resolved without ¹H-decoupling in the CPMG experiment (C), and the lines are sharpened further when an oriented sample is prepared and aligned perpendicular to the magnetic field (D). Adapted with permission of Academic Press from Grage SL and Ulrich AS (1999) *Journal of Magnetic Resonance* **138**: 98–106.

chains in the frozen or lyophilized state. In this way it could be demonstrated that the substrate binding cleft of 5-enolpyruvylshikimate-3-phosphate synthase closes upon ligand binding, whereas the binding pocket of tryptophan synthase remains largely unperturbed.

Another, different kind of NMR approach has been developed to investigate the structure of membrane-associated peptides and proteins. This method relies on the use of oriented membrane samples, and it measures the ^{1}H–heteronuclear dipolar coupling and ^{15}N-CSA to determine the orientation of a labelled ^{1}H–^{15}N bond in the polypeptide backbone. To build a full molecular structure from such orientational constraints, usually a series of selectively labelled peptides have to be prepared and the alignment of each single backbone segment is then measured step-by-step. Recently, a large number (>100) of signals from a uniformly ^{15}N-labelled membrane protein were successfully resolved, using a 2D PISEMA experiment with homonuclear Lee–Goldburg decoupling. Once an assignment strategy for this novel approach is developed, solid-state NMR will provide a set of tools for a comprehensive structural analysis of membrane proteins.

List of symbols

$\hbar$ = Planck's constant/2π; I = nuclear spin quantum number; r = distance between spins; T_1 = relaxation time; Δ = Bloch–Siegert shift; γ = gyromagnetic ratio; μ_0 = permittivity of vaccum; ν_r = spinning frequency; τ_c = correlation time; θ = angle between internuclear vector and magnetic field direction; ω = Larmor frequency.

See also: ^{19}**F NMR Applications, Solution State; High Resolution Solid State NMR, ^{13}C; Liquid Crystals and Liquid Crystal Solutions Studied By NMR; Membranes Studied By NMR Spectroscopy; NMR of Solids; NMR in Anisotropic Systems, Theory; Structural Chemistry Using NMR Spectroscopy, Inorganic Molecules; Solid State NMR, Rotational Resonance; Solid State NMR, Methods; Structural Chemistry Using NMR Spectroscopy, Organic Molecules; Structural Chemistry Using NMR Spectroscopy, Peptides; Structural Chemistry Using NMR Spectroscopy, Pharmaceuticals.**

Further reading

Duncan TM (1990) *A Compilation of Chemical Shift Anisotropies.* Chicago: Farragut Press.

Gerstein BC (1998) ^{1}H NMR. In: Ando I and Asakura T (eds) *Solid State NMR of Polymers,* Vol 84, pp 166–189. Amsterdam: Elsevier Science.

Glaubitz C and Watts A (1998) Magic angle-oriented sample spinning (MAOSS): a new approach to biomembrane studies. *Journal of Magnetic Resonance* 130: 305–316.

Grage SL and Ulrich AS (1999) Structural parameters from ^{19}F homonuclear dipolar couplings, obtained by multipulse solid state NMR on static and oriented systems. *Journal of Magnetic Resonance* 138: 98–106.

Grant DM and Harris RK (eds) (1996) *Encyclopedia of Nuclear Magnetic Resonance,* Chapters on Bilayer Membranes: ^{1}H & ^{19}F NMR; CRAMPS; HETCOR in Organic Solids; Proton Chemical Shift Measurements in Solids. Chichester: Wiley.

Hafner S and Spiess HW (1998) Advanced solid-state NMR spectroscopy of strongly dipolar coupled spins under fast magic angle spinning. *Concepts in Magnetic Resonance* 10: 99–128.

Harris RK, Monti EA and Holstein P (1998) Fluoropolymers. In: Ando I and Asakura T (eds) *Solid State NMR of Polymers*, Vol 84, pp 667–712. Amsterdam: Elsevier Science.

Harris RK, Monti GA and Holstein P (1998) ^{19}F NMR. In: Ando I and Asakura T (eds) *Solid State NMR of Polymers*, Vol 84, pp 253–266. Amsterdam: Elsevier Science.

Harris RK and Jackson P (1991) High-resolution fluorine-19 magnetic resonance of solids. *Chemical Reviews* 91: 1427–1440.

Miller JM (1996) Fluorine-19 magic angle spinning NMR. *Progress in Nuclear Magnetic Resonance Spectroscopy* 28: 255–281.

Hole Burning Spectroscopy, Methods

Josef Friedrich, Technische Universität München, Germany

> **HIGH ENERGY SPECTROSCOPY**
> **Methods & Instrumentation**

Introduction

Much of our information on the microscopic features of matter in various aggregate states has been obtained from spectroscopy. How detailed such information can be depends on the resolution of the experiment. During the evolution of modern physics it often happened that theory predicted features that could not be verified experimentally because the resolution of the associated spectroscopic techniques was too low. As an example, consider the famous experiment by Pound and Rebka in 1960 by which the gravitational red shift of an electromagnetic wave could be verified. This red shift was predicted by Einstein in 1911 on the basis of general relativity, but the verification had to wait until there was a technique available with the necessary resolution. The required resolution is incredibly high, of the order of 10^{15}, but, nevertheless, could be achieved by making use of the Mössbauer effect. A resolution of 10^{15} means that the frequency of an electromagnetic wave of 10^{15} Hz can be determined with an accuracy of 1 Hz. The highest resolution that can be achieved in the spectroscopy of a particular system (e.g. of atoms, molecules, nuclei, etc.) is determined by the natural width of the spectral lines in these systems. According to the Heisenberg principle, this width is determined by the lifetime of the excited state. However, in almost all real systems this ultimate limit cannot be achieved in a straightforward way because the environment of the system broadens the respective levels. The Mössbauer effect is one technique where the natural line width can be exploited for high-resolution spectroscopy. Hole burning techniques also work at this ultimate limit of resolution, but the physical principle is totally different. Nevertheless, because of the ability to do spectroscopy at the limit determined by the natural line width, the hole burning technique is sometimes called the 'optical Mössbauer effect'. The following surveys how the hole burning technique evolved during the last few decades of the twentieth century, its physical basis, and the fields of spectroscopy and technology in which it is used.

Hole burning: a survey

Hole burning stands as a generalized synonym for all kinds of saturation spectroscopic techniques in inhomogeneously broadened bands. At the frequency where saturation is performed, a dip appears in the spectrum, the so-called hole. Generally speaking, the hole burning technique aims to unravel the features of the homogeneous line shape of the transition involved, which is obscured by the presence of strong inhomogeneous broadening. Hole burning was demonstrated for the first time in 1948 in NMR spectroscopy. Since then, it has been used in almost all fields of spectroscopy: NMR, ESR, dielectric, IR and optical spectroscopy. Hole burning experiments have been applied to almost all aggregate states: gases, liquids, crystals, glasses and polymers. More recently biological materials have also attracted increasing interest.

The power of the technique rests with a special variation of hole burning, so-called persistent spectral hole burning. Persistent spectral hole burning was discovered in 1974 by Gorokhovskii, Kharlamow and their respective co-workers. It is based on population storage in a long-lived intermediate, which is very often a photochemical state. In this case the method is called photochemical hole burning. It is the persistence of the holes that has made the hole burning method an important and attractive technique in a very broad field of spectroscopy techniques and applications. The application to optical data storage in the frequency domain, with the possibility of increasing the bit density beyond the diffraction limit, has attracted much interest. Information storage in the time domain is also a promising field of application. In addition, hole burning is a technique highly suited to storing multiple holograms in narrow frequency domains. Spectral holes can be exploited for frequency stabilization of laser radiation and for shaping of laser pulses as well as for ultra-narrow optical filters. In spectroscopy, the method has gained most attention in optical spectroscopy of the solid state. The following survey focuses on this field of study.

The homogeneous line shape function: the perfect crystal case

Consider an optical transition of a dye probe molecule doped a low concentration into a perfect crystal lattice (**Figure 1A**). Since all the probes have the same local environment, their absorption frequencies coincide and the line shape of the transition considered is representative of the line shape of a single probe molecule. Excitation of the probe molecule is accompanied by a charge redistribution in the excited state. This leads to a different equilibrium configuration of the lattice molecules in the excited state. As a consequence, there is a certain probability that the optical excitation will be accompanied by excitation of lattice motions that give rise to so-called phonon side bands. The intensity distribution is determined by the Franck–Condon principle. The relative intensity of the transition with no lattice phonon excitation, the so-called zero-phonon line, is given by the Debye–Waller factor α:

$$\alpha = \frac{\text{Intensity in zero-phonon line}}{\text{Total intensity in the band}}$$

As a rule, the zero-phonon line is much narrower than the accompanying phonon transitions. Consequently, its peak intensity is much higher than the peak intensity of the phonon wing (**Figure 1B**). The reason for the large line width of the transitions into phonon states is the short lifetime of these states, which is largely determined by vibrational relaxation.

In comparision with the phonon states, the lifetime T_1 of the purely electronic state can be orders of magnitudes longer. According to the Heisenberg law, the width $\Delta\omega_0$ of the zero-phonon line is given by

$$\Delta\omega_0 = \frac{1}{T_1} \qquad [1]$$

$\Delta\omega_0$ is called the natural width of the transition. As a rule, the transition attains this width only at very low temperatures, i.e. at around or below a few kelvin. At higher temperatures, the line width is determined by the thermal fluctuations of the lattice, which lead via some coupling mechanism to thermal fluctuations of the energy levels involved, i.e. of the electronic ground and excited states. Since these fluctuations disrupt the phase of the polarization, they are called dephasing processes. The decay time T_2^* of

the phase coherence is called the pure dephasing time. These dephasing processes provide an additional contribution to the line width, so that the total line width $\Delta\omega_h$ is determined by two contributions:

$$\Delta\omega_h = \frac{1}{T_1} + \frac{2}{T_2^*} \qquad [2]$$

$\Delta\omega_h$ is called the homogenous line width. As a rule, the associated line shape is Lorentzian. As the pure dephasing contribution $2/T_2^*$ dies out rapidly as the temperature approaches absolute zero, $\Delta\omega_h$ approaches its ultimate limit of $\Delta\omega_0$ at very low temperature. As an example, consider an organic dye molecule. T_1 is typically of the order of some tens of nanoseconds, so that $\Delta\omega_0$ is of the order of some tens

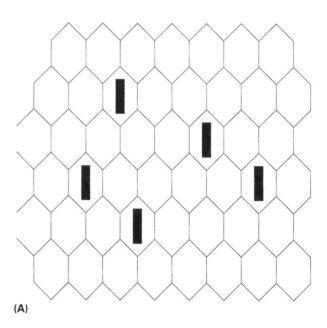

(A)

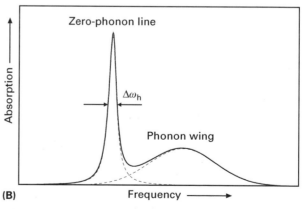

(B)

Figure 1 (A) Sketch of a perfect host lattice (honeycomb) doped with probe molecules (black bars). (B) Line shape of an ensemble of identical probe molecules in a perfect host lattice.

of MHz. In this case the ultimate limit of spectroscopic resolution is around 10^7.

Inhomogeneous broadening

A basic requirement for spectral hole burning to be observed is inhomogeneous line broadening. In NMR, inhomogeneous broadening originates mainly from field inhomogeneities. In the gas phase, inhomogeneous broadening comes from the velocity distribution of the molecules, which leads to a frequency distribution via the Doppler effect. In the solid state, inhomogeneous line broadening occurs because of structural disorder. For example, in a crystal there are lattice defects that cause statistically varying strain fields. These strain fields lead to statistically varying frequency shifts of the zero-phonon transitions of the probe molecules. As a consequence, the zero-phonon frequencies are spread out and the line experiences a broadening $\Delta\omega_i$. Whereas the homogenous line broadening $\Delta\omega_h$ is dynamic in nature, inhomogeneous broadening is, at sufficiently low temperature, essentially static in nature.

In a more general way this situation is sketched in **Figure 2A**. Each of the probe molecules sees a different environment and hence their absorption frequencies are different.

In organic crystals, the inhomogenous broadening is of the order of a wavenumber; in glasses it is of the order of several hundred wavenumbers. In this latter case, even the phonon side bands are largely buried beneath the inhomogeneous envelope (**Figure 2B**).

In contrast to the homogeneous line shape, which is Lorentzian, the inhomogeneous distribution is governed by the statistics of large numbers and hence is Gaussian. At sufficiently low temperatures, the ratio of the inhomogenous to the homogenous width may cover 5–6 orders of magnitude.

Hole burning

Suppose the dye probe molecule is photoreactive. Then, to understand the essential features of hole burning, it suffices to consider three states only: the ground state S_0, the first excited singlet state S_1, and the photoproduct state P_0 (**Figure 3**). For simplicity we assume that P_0 has an infinite lifetime. The absorption spectrum of P_0 is shifted as compared to that of S_0 because P_0 represents a structurally different molecule. Consider **Figure 4**: narrow-bandwidth laser light is tuned to some position within the inhomogeneous band. We denote the respective frequency by ν_L. Excitation leads to photochemistry; hence, population is transferred to P_0 and, concomitantly, S_0

is depleted. However, depletion in S_0 can only occur in a spectral range around ν_L roughly determined by the homogenous line width $\Delta\omega_h$. At temperatures of a few kelvin, $\Delta\omega_h$ is close to $\Delta\omega_0$ and hence is extremely narrow. As a consequence, the depletion dip is extremely narrow. The depletion dip is called the 'spectral hole'. If P_0 has an infinite lifetime, the lifetime of the hole is, of course, also infinite. In this case the hole is persistent. However, in the usual terminology holes are called persistent if their lifetime is much longer than the typical hole burning timescale determined by burning and reading the hole. A typical order of magnitude for this timescale is a few minutes. For vanishingly small depletion and sufficiently low power of the burning laser, the shape of the depletion hole is Lorentzian with a width close to $\Delta\omega_h$. This depletion hole is, for instance, detected by scanning the

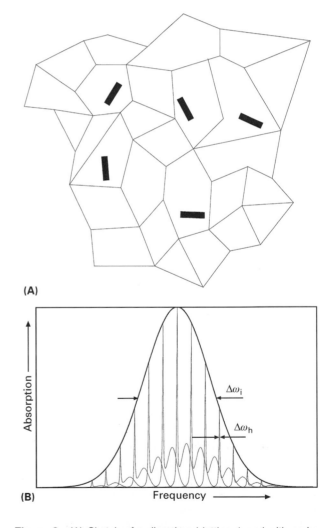

(A)

(B)

Figure 2 (A) Sketch of a disordered lattice doped with probe molecules (black bars). Note that each probe molecule has a different environment. (B) Inhomogeneous line broadening: the inhomogeneous line of width $\Delta\omega_i$ is built from an ensemble of homogeneous lines of width $\Delta\omega_h$.

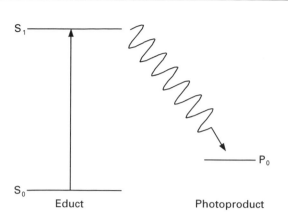

Figure 3 Three levels are necessary for persistent hole burning. P_0 has a long lifetime and acts as a storage state.

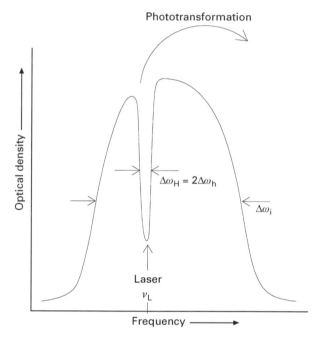

Figure 4 Hole burning: laser radiation with frequency ν_L creates a population dip and, concomitantly, a hole in the inhomogenous band.

laser over the depletion range. The resulting line shape of the recorded hole is a convolution of the depletion hole with the homogeneous line shape, i.e. a convolution of a Lorentzian with a Lorentzian that is itself a Lorentzian of twice the width:

$$\Delta\omega_H \Rightarrow 2\Delta\omega_h \qquad [3]$$

where $\Delta\omega_H$ is the hole width.

Figure 5 shows a hole in the inhomogenous absorption band of protoporphyrin IX-substituted myoglobin. The insert shows the hole on an enlarged scale.

Saturation broadening

Saturation spectroscopy of two-level systems results in a power-dependent line width of the transition concerned, a phenomenon termed power broadening. In a similar fashion, line broadening due to strong saturation can also occur in a three-level system with a long-lived storage state P_0. However, in this case, it is not the power that determines the broadening but rather the number of irradiating photons, which is proportional to the irradiated energy dose or to the laser fluence. Hence, we call this broadening 'fluence broadening'. It is easy to see why the irradiated energy and not the power is the important factor: the power needed to burn a hole with a certain depth can be made arbitrarily small by increasing the irradiation time accordingly. This has no effect on the hole area so long as P_0 has an infinite lifetime. However, if the irradiated energy increases, photochemistry in the centre of the line slows down because the number of absorbers decreases, but it is still going on in the wings. As a consequence, the centre of the line is more strongly depleted and the line flattens and becomes broader. Because of this saturation broadening, the homogeneous line width must be determined from a plot of the hole width as a function of laser fluence in the limit of vanishing fluence. Likewise, one can plot the hole width as a function of the hole depth (**Figure 6**): as the depth approaches zero, the hole width converges to $2\Delta\omega_h$.

Hole burning techniques

The deeper the holes, the better is the signal-to-noise ratio. However, deep holes are artificially broadened

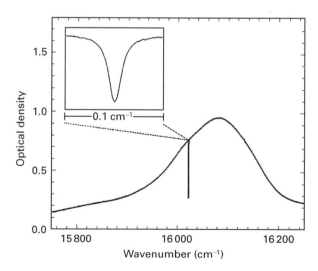

Figure 5 A spectral hole in the absorption spectrum of myoglobin complexed with protoporphyrin IX at $T = 1.5$ K.

owing to saturation. Hence, techniques have been developed for reading shallow spectral holes with a high signal-to-noise ratio.

The most straightforward way to read a hole is to scan the laser over the frequency range where burning has been performed. In the range of the hole, the number of absorbers is diminished and so there is an increased transmission. The transmission technique requires a baseline correction since the hole is detected on top of an inhomogeneous background absorption. For an optical density around 1, relative transmission changes of a few per cent can easily be detected in this way.

If the optical density at the burning frequency is well below 1, the fluorescence excitation mode is more appropriate for reading the hole. In this mode it is not the transmission but rather the fluorescence that is measured. If the laser is scanned across the spectral range of the hole, the fluorescence decreases. The relative change of the fluorescence as a function of frequency directly follows the shape of the hole. A baseline correction is required in this case, too.

There are some elaborate techniques that are extremely sensitive because they are based on zero-background signals. The most important of these are the dichroitic and the holographic techniques.

In the dichroitic technique, the hole is burnt with a polarized laser field, say in the z direction (**Figure 7**). The polarized laser radiation leads to an anisotropic distribution of the vector N, denoting the number density of absorbers oriented with their transition

dipole moment at an angle θ to the polarization of the burning field. Reading is performed with the sample between two crossed polarizers P_1 and P_2, inclined at say $-45°$ and $+45°$ to the z direction. If the sample is isotropic, it does not alter the polarization direction, so the reading laser field E_1 will be P_1-polarized and will be blocked by P_2. On the other hand, if at frequency $\nu - \nu_L$ the sample has a preferred transmission for z-polarized light, then the reading laser field E_1 will be rotated towards the z axis. As a consequence, the respective field vector E_1' will acquire a component E_2' along the P_2 direction and light will be transmitted. In this way the hole is measured against zero background intensity. Accordingly, the signal-to-noise ratio can be orders of magnitude better than with the transmission or fluorescence detection modes and tiny changes in optical density can be measured. However, two things have to be accounted for: the degree of anisotropy is not constant over the frequency range of the hole, and it also depends on the depth of the hole. This so-called polarization bleaching is very similar to saturation broadening. For strong polarization bleaching, the shape of the hole may be flattened or may even show a double peak structure. Apart from this effect, the hole shape may also suffer from interference with light which is not totally blocked by P_1 and P_2 as a result of small degrees of misalignment.

In the holographic technique two beams, 1 and 2, interfere at an angle of 2θ in the sample and write a spatial grating of the absorption coefficient and of the index of refraction (**Figure 8**). The modulation of this grating is frequency dependent via the line shape function. If one of the writing beams, say 2, is blocked, the other beam can serve as the reading beam. If it is tuned across the frequency range of the hole, the grating deflects the reading beam (1) into

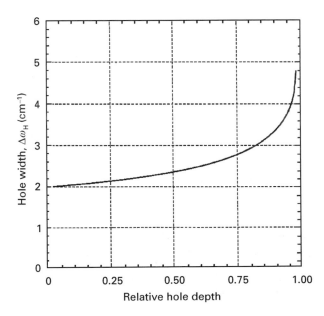

Figure 6 Saturation broadening due to laser fluence. Reproduced with permission from Dick B (1989) Habilitationsschrift, University of Göttingen.

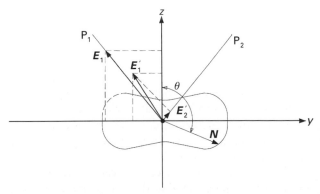

Figure 7 Exploiting the dichroism of holes for a low-noise reading process. N denotes the number density of absorbers with their dipole moments in the direction θ. P_1 is the polarizer, P_2 the analyser.

the direction of beam 2. The intensity of the diffracted light is proportional to the modulation depths of the absorption coefficient and of the index of refraction, and hence is strongest at the centre of the hole and vanishes far in the wings. Although the holographic technique reproduces the true line shape of the hole as long as the saturation of the hole is weak, it has characteristic properties that have to be taken into account. First, strong bleaching results in a strong deviation from a sinusoidal grating, which leads to higher diffraction orders. Second, strong bleaching also leads to a strong distortion of the Lorentzian line shape, which may eventually show a double peak structure.

Applications of hole burning

Spectroscopy

Hole burning has been used in a very broad field of scientific problems ranging from solid-state physics to biological physics. The most straightforward application is the measurement of the homogeneous line width. Apart from measuring vibrational relaxation in the excited electronic state in this way, most effort has been put into measuring the homogeneous line width of the purely electronic $S_1 \leftarrow S_0$ absorption of dopant molecules in glasses. The homogeneous line width in a glassy host material has quite specific properties compared to that in crystalline materials (**Figures 9A** and **B**): in the low-temperature regime, $T < 4$ K, the homogeneous width in a glass is an order of magnitude larger than in a crystal (**Figure 9A**). In a crystal, it is already close to its lifetime-limited value and hence is temperature independent (**Figure 9B**). In a glass, it depends on temperature down to the lowest temperatures measured in these types of experiments (~50 mK); however, its temperature dependence is weak. It is usually well described by a temperature power law $\sim T^\alpha$ with α between 1 and 2. In a crystal, the temperature dependence goes as T^7 or is very often activated. The reason for the specific features of the optical line width in glasses is the presence of specific degrees of freedom, so-called TLS (two-level system) modes, which reflect the disorder in the structure and which cannot be frozen out, not even at the lowest temperatures reached so far in solid-state materials.

The TLS degrees of freedom reflect the simplest approach to characterizing the complex energy landscape of disordered materials such as glasses and polymers. These materials are never in true thermodynamic equilibrium, and, hence, there are always some structural dynamics covering an extremely

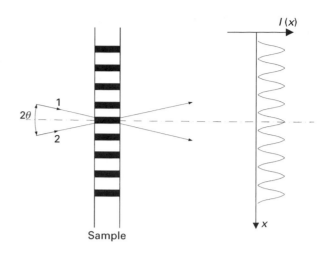

Figure 8 Holographic hole burning: burning is performed via two beams, 1 and 2, which interfere in the sample and create a spatial grating $I(x)$ of the absorption coefficient and the index of refraction.

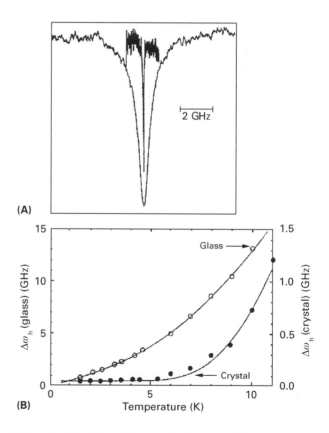

Figure 9 (A) Hole widths of chlorine-doped benzophenone glass (broad hole) and crystal (narrow hole). (B) The homogeneous line widths of chlorine-doped benzophenone glass and benzophenone crystal as a function of temperature. Note the different scales for glass and crystal.

broad range of relaxation rates of up to 15 orders of magnitude, from nanoseconds to months. These structural dynamics create locally fluctuating strain and/or electric fields that, in turn, lead to fluctuations of the electronic energy levels of the dye probe. These fluctuations are reflected in a time-dependent line broadening, called spectral diffusion (**Figure 10**). Measurement of the spectral diffusion broadening gives insight into the structural dynamics and the associated features of the energy landscape. This is especially interesting for biological materials.

Another major field in which the hole burning technique can be exploited concerns measuring the influence of external fields, such as electric, magnetic or pressure fields, on the electronic energy levels. For instance, measuring the Stark effect (electric field) even in good-quality crystals usually requires special modulation techniques since the spectral changes are rather small. In disordered materials, owing to the large inhomogeneous broadening, it is difficult to obtain good-quality data unless unusually large changes of the dipole moments are involved. In hole burning, moderate field strengths (10 kV cm^{-1}) and pressure levels (1 MPa) yield easily detectable line shifts and field-induced changes in the line widths. Moreover, the high resolution and the associated selectivity of these experiments enables one to investigate field effects as a function of frequency within the inhomogeneous band. The solvent shift can be varied simply by tuning the laser frequency and hence becomes a parameter of the experiment. From hole burning Stark effect experiments (**Figure 11**), information on the molecular dipole moments involved, on the respective polarizabilities, on matrix

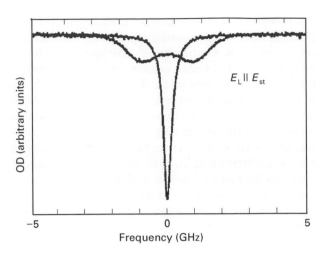

Figure 11 Deformation of a spectral hole in an external field. The external field E_{st} (about 12 kV cm^{-1}) is parallel to the laser field. The sample is protoporphyrin IX-substituted myoglobin in a glycerol/water glass. The double peak structure is indicative of a chromophore with a well-defined dipole moment.

fields and on associated symmetries can be obtained. High-quality quantum chemical calculations are usually desirable to support the interpretation of the experiments because there are always several parameters that are not very well known. From pressure experiments (**Figure 12**), information on local compressibilities, on molecule–matrix interactions and on the vacuum absorption frequency of the probe molecule can be obtained.

Hole burning has been used widely to investigate relaxation processes of coherent tunnelling in the solid state. An outstanding example in this context is the methyl group. Consider a dye probe molecule

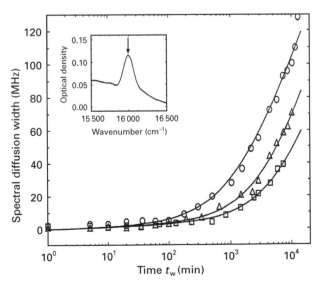

Figure 10 Spectral diffusion broadening as a function of waiting time t_w for different ageing times. The sample is protoporphyrin IX in a dimethylformamide/glycerol glass; temperature is 100 mK. Insert: Broad-band absorption spectrum. The arrow marks the wavenumber where hole burning was performed.

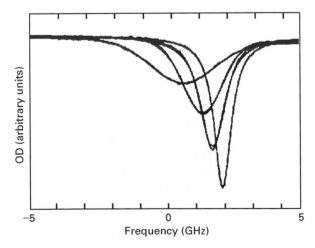

Figure 12 Shift and broadening of a spectral hole as a function of pressure. Maximum pressure is 0.6 MPa. The sample is resorufin in glycerol glass at liquid-helium temperature.

with one or several methyl groups attached, for instance dimethyl-s-tetrazine (**Figure 13**). Irrespective of how disordered the host matrix is, the methyl group always has at least a 3-fold rotational symmetry axis (along the C–C bond), which results in a 3-fold periodic potential $V = V_3 \cos 3\varphi$, where φ is the angle of rotation around the C–C bond and V_3 is the barrier height. The lowest states in this potential interact via the tunnelling interaction. As a result, the three states form a split pair, with A and E symmetry, where the E state is still doubly degenerate. Since the C_3 group is isomorphic to the group of even permutations of the protons, the wavefunctions of the methyl rotor have to be totally symmetric according to the Pauli principle. Relaxation between the tunnelling states requires an interaction, which breaks the permutation symmetry. These interactions necessarily contain the nuclear spin, and so are very small. Accordingly, the lifetimes of these states can be extremely long (e.g. months at 1 K) and thus population can be stored in these states. This is where hole burning comes in. Suppose laser light excites a molecule in its A-rotor (E-rotor) state from S_0 to S_1 (**Figure 14**). From S_1, intersystem crossing to the triplet state T_1 occurs. In the T_1 state the electron spin turns around the nuclear spin via the hyperfine interaction, the rotor converts from the A to the E state or from the E to the A state. Light irradiation thus creates a population redistribution among the rotational tunnelling states. The point is that this population redistribution is accompanied by a frequency transformation because the tunnelling splitting δ^* in S_1 may be different from that in S_0. Hence, a hole appears at the laser frequency, which is accompanied by two antiholes shifted symmetrically by the difference $|\delta - \delta^*|$ between the tunnelling splitting in S_0 and S_1 (**Figure 13**). The hole offers an easy way to measure the rotational tunnelling relaxation. All one has to measure is how the hole recovers or how the antiholes vanish. **Figure 15** shows the tunnelling relaxation rate as a function of temperature for the protonated and the deuterated rotor. The rate increases with T^7, indicating a Raman-type relaxation process. Deuteration speeds up the relaxation instead of slowing it down because of the nuclear quadrupole interaction associated with the deuterons and the fact that the methyl rotor in the example considered is almost free.

Summarizing the application of hole burning in spectroscopy, it should be noted that there is also a large field in which it is used in elucidating photochemical reactions. It is also increasingly used in studies of photosynthesis, to shed light on the energy and electron transfer processes, on the nature of the states involved, on electron phonon coupling and

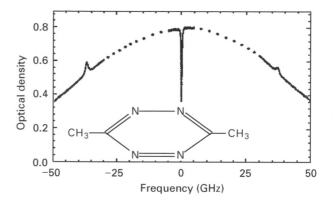

Figure 13 Hole burning via nuclear spin conversion in rotational tunnelling states of methyl groups. The sample is dimethyl-s-tetrazine in *n*-octane at 1.4 K. The central hole at the laser frequency is accompanied by two antiholes symmetrically shifted by 37 GHz.

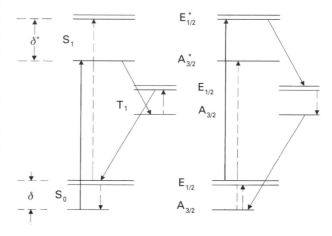

Figure 14 The level scheme of a methyl rotor attached to a chromophore. δ and δ^* are tunnelling splitting in S_0 and S_1, respectively. T_1 is the lowest triplet state. E and A label the symmetry species of the rotor states, the subscripts characterize the total nuclear spin of the methyl protons.

exciton coherence. Further, the hole burning technique offers great power to investigate the features of the energy landscape of proteins and the associated dynamics in conformation space.

Technology

In 1976 the use of hole burning was suggested for frequency-domain optical data storage. **Figure 16** sketches the principle. The storage medium could, for instance, be a dye-doped polymer film. In each spatial spot (**Figure 16A**) a number of about $\Delta\omega_i/\Delta\omega_H$ bits can be stored in the frequency domain (**Figure 16B**), with $\Delta\omega_i$ being the inhomogeneous width and $\Delta\omega_H$ the hole width. $\Delta\omega_i/\Delta\omega_H$ could be as large as 10^4–10^5. As a consequence, the capacity of

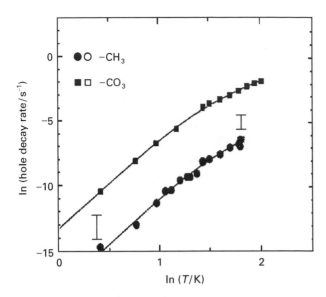

Figure 15 Rotational tunnelling relaxation rate as a function of temperature in a log–log representation. Fitted curves are based on Raman-type relaxation processes. Note that the tunnelling relaxation in the deuterated rotor is faster by two orders of magnitude. The sample is dimethyl-s-tetrazine in *n*-octane.

an optical storage device could be increased by several orders of magnitude beyond the diffraction limit, i.e. to about 10^{12} bits cm^{-2}. So far such a memory has not been realized because too many requirements have to be met simultaneously: the high bit density is achieved only at liquid helium temperatures; at

higher temperatures the information is, at least partly, lost; appropriate dye molecules should absorb in the frequency range where common laser diodes work; the respective Debye–Waller factors as well as the photochemical yields should be high so that information can be written and read in the nanosecond time regime; and repeated reading may deteriorate the stored information.

Solutions could be found for many of these requirements. For instance, the use of photon-gated hole burning ensures nondestructive read out. Many materials have been discovered in which this concept could be successfully demonstrated; however, attempts to unify all the requirements in one system have failed so far.

Instead of storing data bits in the frequency domain, they can be stored in the time domain, for instance as a pulse sequence E_S as shown in **Figure 17A**. Writing of the information is performed by interference of the signal E_S with a short reference pulse E_R in the photochemical storage material. If the spectral width of the reference pulse E_R is much larger than the width of the whole signal train E_S, which in turn is much larger than the homogeneous width of the respective optical transitions of the storage material, then the whole Fourier spectrum of the signal train can be stored as a grating in frequency space, i.e. as many holes in the inhomogeneous band whose width, of course, has to be even larger than the width of the reference pulse. **Figure 17B** demonstrates how

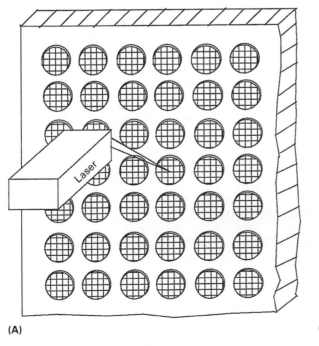

(A)

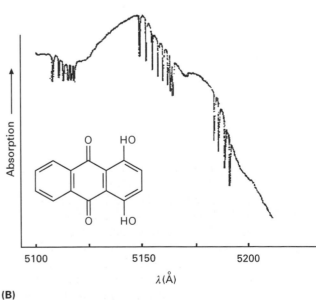

(B)

Figure 16 (A) Hole burning and data storage. (B) In each spatial spot many bits can be stored in the frequency or wavelength domain. The dye probe, in this case, is quimizarin whose structure is shown in the insert.

Writing

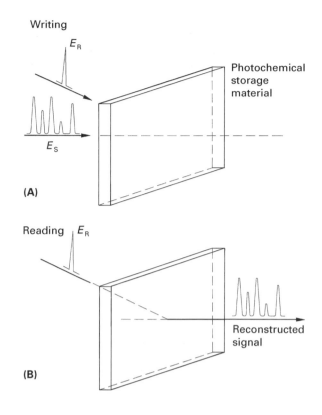

(A)

(B)

Figure 17 (A) Data storage in the time domain using the 'PASPE' technique. (B) Reading the information. The echo is stimulated by a single pulse.

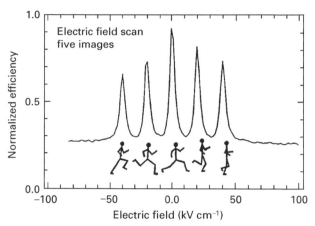

Figure 18 Five holograms stored as holes in the electric field domain. By scanning the field, the five holograms can be read out serially, resulting in a movie showing a running jogger. Reproduced by permission of the Society of Photo-Optical Instrumentation Engineers (SPIE) from Wild UP and Renn A (1988) *Proceedings of the SPIE*. **910**: 61.

the stored information is recovered. If a single pulse E_R hits the sample some time (hours, days, weeks) later, the sample regenerates the whole pulse sequence. Depending on how the reference pulse is applied in the writing and reading procedure, the time-reversed signal train can be generated as well. This is nothing other than a special form of the stimulated photon echo where the contrast of the stored information can be enhanced by accumulating the information over many writing cycles in the long-lived photochemical state. Accordingly, the technique has been called PASPE – photochemically accumulated stimulated photon echo.

As well as storing information bitwise, it is also possible to store information in an analogue fashion as a hologram of a picture. The concept is very similar to that shown in **Figures 17A** and **B**, but with E_R and E_S being the reference and the object waves, respectively. The reference wave can, for instance, be a plane wave. The object wave carries all the information of the picture. The two waves interfere for some time in the storage medium and create a hole at the laser frequency. In the frequency domain this hole is sharp, but in space it is distributed over a larger part of the storage material.

Again, one can store many holograms in the frequency range covered by the inhomogeneous band. The holograms can be read by scanning the reading laser beam over the frequency interval where the holograms are stored. If the reading laser beam is directed along the direction of the reference beam, the burnt-in hologram regenerates the original object wave. Likewise, reading can be performed in the electric field domain by making use of the Stark effect. In this case the holograms are stored with an electric field applied in a way that each hologram has its own specific Stark field. The frequency of the burning (and reading) laser is fixed. In reading the holograms, the Stark field is tuned. Whenever the field reaches one of the specific values used during the storage process, a holographic interference pattern with a spatially well-resolved modulation of the absorption coefficient and index of refraction appears that scatters the light from the reading laser such that the object wave is regenerated. **Figure 18** shows an example: five pictures are stored as holograms in the five holes shown. Since each hole has its specific Stark field, reading is performed by scanning the electric field. In this way, a movie is generated, showing a jogger, as indicated in the figure.

List of symbols

E = field vector $(E = |E|)$; E_R = reference pulse; E_S = signal pulse sequence; N = vector of number density of absorbers oriented at θ to the polarization of the burning field; T_1 = lifetime of purely electronic state; T_2^* = pure dephasing time; α = Debye–Waller factor, temperature-law exponent; δ = tunnelling splitting; θ = angle between transition dipole moment and burning field polarization, angle between holographic beams; ν = frequency; $\Delta\omega_0$ = natural width of zero-phonon line; $\Delta\omega_h$ = homogeneous width of zero-phonon line; $\Delta\omega_H$ = hole width; $\Delta\omega_i$ = inhomogeneous line width.

Further reading

Bloembergen N, Purcell EM and Pound RV (1948) Relaxation effects in nuclear magnetic resonance absorption. *Physical Review* 73:679–712.

Borczyskovski C, Oppenländer A, Trommsdorff HP and Vial J-C (1990) Optical measurements of methyl-group tunneling and nuclear-spin conversion. *Physical Review Letters* 65: 3277–3280.

De Caro C, Renn A and Wild UP (1991) Hole burning, Stark effect, and data storage: 2: holographic recording and detection of spectral holes. *Applied Optics* 30: 2890–2898.

Dick B (1989) Polarisationsspektroskopie permanenter und transienter spektraler Löcher, Habilitationsschrift, Universit of Göttingen.

Friedrich J and Haarer D (1984) Photochemical hole burning: A spectroscopic study of relaxation processes in polymers and glasses. *Angewandt Chemie, International Edition in English* 23: 113–140.

Gorokhovskii AA, Kaarli RK and Rebane LA (1974) Hole burning in the contour of a pure electronic line in a Shpol'skii system. *JETP Letters* 20: 216–218.

Kharlamow B, Personov RI and Bykovskaya LA (1974) Stable 'gap' in absorption spectra of solid solutions of organic molecules by laser irradiation. *Optics Communications* 12: 191–193.

Levenson MD, Macfarlane RM and Shelby RM (1980) Polarization-spectroscopy measurement of the homogeneous linewidth of an inhomogeneously broadened color-center band. *Physical Review B* 22: 4915–4920.

Meixner A, Renn A and Wild UP (1989) Spectral hole-burning and holography. I. Transmission and holographic detection of spectral holes. *Journal of Chemical Physics* 91: 6728–6736.

Moerner WE (ed) (1998) *Persistent Spectral Hole Burning: Science and Applications*. Berlin: Springer-Verlag.

Moerner WE (1995) Molecular electronics for frequency domain optical storage: persistent spectral hole-burning – a review. *Journal of Molecular Electronics* 1: 55–71.

Pinsker M and Friedrich J (1996) Hole burning spectroscopy and quantum phenomena in methyl groups. *Molecular Crystals and Liquid Crystals* 291: 97–102.

Rebane A (1998) Compression and recovery of temporal profiles of picosecond light signals by persistent spectral hole-burning holograms. *Optics Communications* 67: 301–304.

Thorn-Leeson D, Wiersma DA, Fritsch K and Friedrich J (1997) the energy landscape of myoglobin: An optical study. *Journal of Physical Chemistry* 101: 6331–6340.

Wegener H (1995) *Der Mössbauer-Effekt und seine Anwendung in Physik and Chemie*. Mannheim: Hochschultaschenbücher-Verlag.

Wild UP and Renn A (1988) Spectral hole burning and hologram storage. *Proceedings of the SPIE* 910: 61–65.

HPLC-IR, Applications

See **Chromatography-IR, Applications.**

HPLC-MS, Methods

See **Chromatography-MS, Methods; Hyphenated Techniques, Applications of in Mass Spectrometry.**

HPLC-NMR, Applications

See **Chromatography-NMR, Applications.**

HREELS, Applications

See **High Resolution Electron Energy Loss Spectroscopy, Applications.**

Hydrogen Bonding and Other Physicochemical Interactions Studied By IR and Raman Spectroscopy

AS Gilbert, Beckenham, Kent, UK

> **VIBRATIONAL, ROTATIONAL & RAMAN SPECTROSCOPIES**
> **Applications**

Vibrational spectra are sensitive to intermolecular interactions. This is most clearly evidenced by the differences seen in spectra of condensed phases compared to gases at low pressure which show not just the loss of rotational sidebands but also changes in profile, intensity and position. Hydrogen bonding (H-bonding) provides the most dramatic effects but even the weak van der Waals forces can significantly modify spectra. In liquids, collisions can distort molecular configuration leading to local breakdown of symmetry.

Hydrogen bonds

Introduction

The hydrogen bond is a weak, fairly directional, interaction represented here as a broken line in the scheme A–H---B. Atom A is electronegative, the A–H bond being therefore slightly ionic in character, and atom B possesses an area of basicity such as lone pairs on nitrogen, oxygen and halogen or π electron rings (here B is a group of atoms) in certain aromatic systems. The energies of interaction are of an order of magnitude less than covalent bonds but greater than the very simple non-directional non-covalent forces (van der Waals or dispersion forces) that occur between all molecules. The unit A is generally oxygen, nitrogen or halogen but S–H and in certain circumstances even C–H can engage in H-bonding.

Theoretical and experimental studies suggest that the hydrogen bond is not purely electrostatic but has some covalent character. It is observed from crystallographic and other investigations that the length of the A–H bond increases and the distance between A and B decreases with increasing strength of interaction. In the limit of the weakest H-bonds, A and B are separated by about the sum of their collision radii. This is slightly greater, by about 0.5 Å than the sum of the van der Waals radii.

Figure 1 shows IR spectra of isopropanol in solution (CCl_4) at various concentrations, the pathlength being adjusted to compensate as nearly as possible for the changes in molarity. It illustrates the dramatic effects of H-bonding (between the hydroxyl groups) on the O–H stretching mode. The free bond yields a relatively narrow band above 3600 cm^{-1} but the singly and doubly H-bonded groups give broad and intense absorptions near 3500 cm^{-1} and 3350 cm^{-1} respectively.

This behaviour of the stretching band v (A–H), of shift to lower frequency, increase in overall intensity and band broadening is highly characteristic and indicative of H-bonding. In extreme cases shifts of more than 2000 cm^{-1} are observed while breadth and intensity increases can be two orders of magnitude or so.

The band shift and intensity changes can be crudely rationalized on the basis of a decrease in force constant due to weakening (lengthening) of A–H and increased polarization on interaction with B.

Raman spectra show similar behaviour to IR in terms of shift and band breadth but the intensity increase is not observed. **Figure 2** illustrates part of the spectrum of a mixture of deuterium chloride and dimethyl ether in the gas phase. The stretching vibration of unassociated DCl can be seen at 2085 cm^{-1} along with rotational side bands, while H-bonded DCl yields a broad absorption underlying the sidebands centred near 1885 cm^{-1}.

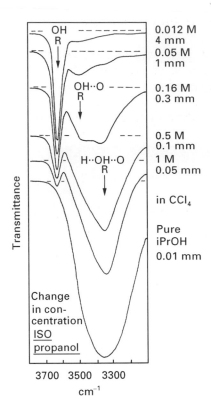

Figure 1 Effect of H-bonding on the O–H stretching vibration of isopropanol. Reproduced with permission from Colthup NB, Daly LH and Wiberley SE (1975) *Introduction to Infrared and Raman Spectroscopy*. New York, San Francisco and London: Academic Press.

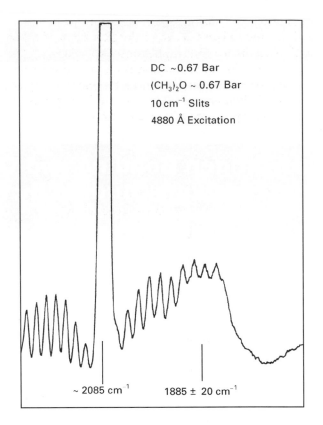

Figure 2 Part of the Raman spectrum of a gaseous mixture of $(CH_3)_2O$ and DCl. Reproduced with permission from Gilbert AS and Bernstein HJ (1974) Measurement of rotational temperatures by Raman spectroscopy: application of Raman spectroscopy to the acquisition of thermodynamic values in a chemical system. In: Lapp M and Penney CM (eds) *Laser Raman Gas Diagnostics*, pp 161–169. New York: Plenum Press.

The bending vibrations of A–H are affected much less than the stretching vibration. Some broadening and intensification takes place but the small shift is to higher wavenumbers.

The new vibration, ν(A–B), the overall stretching of the system and one which directly reflects H-bond strength, is found at low frequencies usually no higher than about 200 cm^{-1}. Probably the strongest H-bonded system in existence, the symmetrical bifluoride ion $(F–H–F)^{-1}$, has ν(F–F) near 600 cm^{-1}.

Finally, the vibrational modes of bonds directly attached to B are significantly affected. Carbonyl groups which have been extensively studied as H-bond acceptors show downward shifts in ν(C=O) of as much as 40 cm^{-1}.

Dependence on temperature and concentration

In gas and liquid phases the extent of H-bonding decreases with rise in temperature and quantitative band-intensity measurements can give values for various thermodynamic quantities.

In solution similar measurements on concentration dependencies can also provide information on energetics but in addition can be used to allow

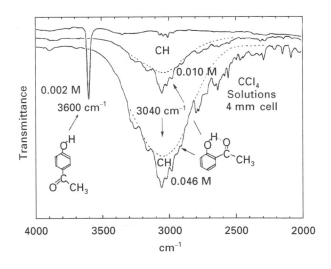

Figure 3 The IR spectra of hydroxyacetophenones in CCl$_4$ solution. Reproduced with permission from Colthup NB, Daly LH and Wiberley SE (1975) *Introduction to Infrared and Raman Spectroscopy*. New York, San Francisco and London: Academic Press.

qualitative conclusions about structure to be made. For instance, **Figure 3** shows the IR spectra of *ortho* and *para* hydroxyacetophenones in solution. The *para* isomer in dilute solution is almost entirely free but the *ortho* isomer by contrast exhibits complete internal H-bonding which, as would be expected, is unaffected by change in concentration.

Correlations with bond length

In crystals the stretching vibrating is observed to possess a sort of half-parabolic relationship with overall H-bond distance (i.e. length A–B). **Figure 4** shows data from a number of O–H---O systems. Note that data from other families (e.g. N–H---O)

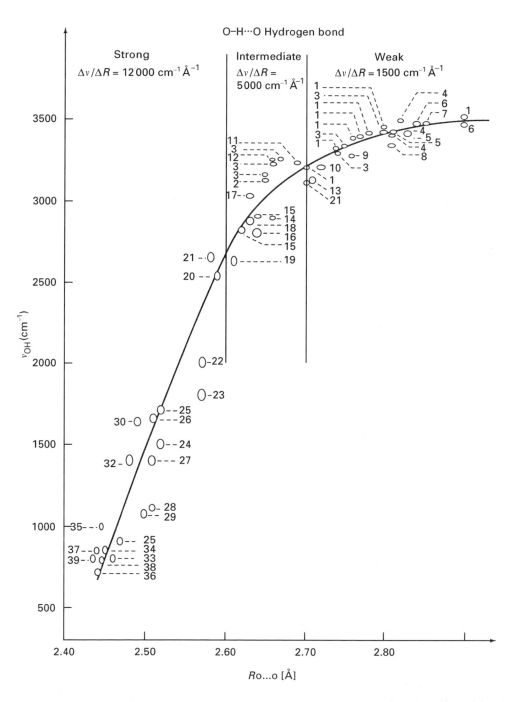

Figure 4 Correlation between the hydrogen-bonded O–H stretching frequency ν(O–H) and $R_{O...O}$ (the O$\cdots$O bond distance) for O–H---O H-bonds in crystals. The weakly bonded examples are mostly salt hydrates, the strongest examples (those with frequencies below 1000 cm^{-1}) are in the main organic singly-ionized diacid anions (oxygen analogues of bifluoride ion). Reproduced with permission of Springer-Verlag from Novak A (1974) Hydrogen bonding in solids: correlation of spectroscopic and crystallographic data. *Structure and Bonding* **18**: 177–216.

fall along separate curves. The sum of the van der Waals radii for two oxygens is about 2.8 Å.

Correlations with acidity/basicity and energy of interaction

The strength of H-bonding increases with greater proton-donating ability of the A–H moiety and increasing proton-accepting power of the base B. **Figure 5** shows the effect of differing basicity on isopropanol at constant concentration. Here the solvent itself plays the role of the base. In benzene (inert) the hydroxyl groups are mostly free (the weak sideband indicates a small amount of OH to OH bonding) but are almost entirely H-bonded to either tetrahydrofuran, pyridine or triethylamine. The shift $\Delta\nu$(A–H) is proportional to solvent pKa, the values being -2.1, 5.2 and 11 respectively.

The Badger–Bauer rule states that there is a linear relationship (in solution) between the shift $\Delta\nu$(A–H) and ΔH, the enthalpy of H-bond formation, but it was formulated originally from early and somewhat limited data. More comprehensive work has since shown that there are deviations from linearity, most obviously in the region of weaker bonding. Each individual system again has a separate relationship, each system here being a one-proton donor (e.g. phenol) with a range of organic bases or vice-versa in a particular solvent.

The solvents utilized must of course be inert; CCl_4 or benzene are commonly used. Detailed work suggests that family types must be very exclusive to have a smooth relationship, for example a range of nitrogen bases must be all of the same electronic configuration and must be free of the complications of steric interference.

When a single carbonylic base is used, $\Delta\nu$(C=O) is found also to yield largely linear relationships with ΔH. Other studies find quite reasonable linear relations between band shift and ΔG (the free energy difference derived from association constants) within specific families of donor and acceptor. While there is some theoretical justification for a linear relationship between shift and enthalpy (or more specifically with ΔE, the internal energy change, in the gas phase) the linear relation with ΔG is only empirical as the latter cannot be directly related with ΔH (or ΔE).

Estimation of thermodynamic quantities

IR absorbance spectroscopy obeys the Beer–Lambert law whereby band absorbance is linearly related to concentration of the molecule in solution. This allows the estimation of the association constant, and thereby ΔG, by measuring the decrease in the intensity of the free A–H band on formation of an H-bond along with knowledge of the total amounts of each substance present. The degree of association (whether dimer, trimer etc.) can be found by examining the way H-bond formation varies with change in concentration. The enthalpy can be estimated by measuring the association constant over a range of temperatures and employing the van't Hoff equation, subject to the usual caveats when using this relationship.

Strong H-bonds

On the basis of the simple electrostatic picture a much stronger H-bond should result if either A–H or B are ionized. This is indeed found from crystallographic data, the bifluoride ion being a well known example. Many very short and strongly bonded O–H–O systems exist; **Figure 6A, B** shows the IR spectra of sodium hydrogen diacetate, which contains $[CH_3C-(=O)O--H--O(O=)CCH_3]^-$, and its singly deuterated equivalent. The O–O distance is slightly less than 2.45 Å and the proton is located equidistant (or very nearly so) from the two oxygens. It should be noted that the spectra are displayed here

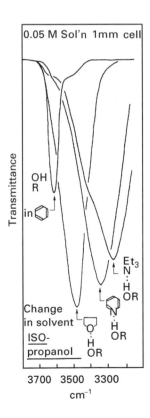

Figure 5 Effect of changing the solvent on the O–H stretching vibration of isopropanol. Reproduced with permission from Colthup NB, Daly LH and Wiberley SE (1975) *Introduction to Infrared and Raman Spectroscopy.* New York, San Francisco and London: Academic Press.

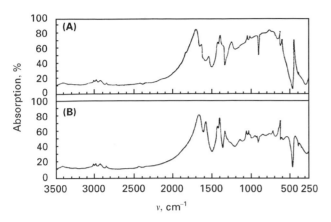

Figure 6 The IR spectra of (A) NaH(CH₃CO₂)₂ and (B) NaD-(CH₃CO₂)₂. Reproduced with permission from Hadzi D (1965) Infrared spectra of strongly hydrogen bonded systems. *Pure and Applied Chemistry* 11: 435–453.

in absorbance so that the bands go upwards. Both $\nu(\text{OH})$ and $\nu(\text{OD})$ yield very broad bands starting at around 2500 cm⁻¹ and extending to below 500 cm⁻¹. The band maximum is at about 750 cm⁻¹ in the first case and somewhere around 600 cm⁻¹ in the other. Both spectra display a sharp nick (window) at about 900 cm⁻¹ and such features often abound in spectra of these types of compounds to give very bizarre appearances.

Isotope shifts (H→D) in such compounds are often much lower than normal (which is slightly less than √2) though the jagged profiles of the bands can make estimation difficult. All the phenomena suggest that the potential surfaces for vibration are quite anharmonic.

Many medium-strength H-bonds also yield strange features in their stretching bands, carboxylic acids being a case in point.

Explanations for bandwidth increase

Surprisingly there is as yet no accepted explanation for the phenomenon of band broadening in H-bonding. It may be that a number of different causes are responsible. A major factor is undoubtedly the abnormality of the vibrational potential functions which appear both from experiment and theory to be often considerably more anharmonic than usual.

Various suggestions for the likely effects of anharmonicity include very short lifetime of the excited vibrational state (uncertainty broadening) and breakdown of the Born–Oppenheimer approximation (i.e. motion of electrons and nuclei are no longer independent, which might be viewed as leading to a multiplicity of potential functions). Special factors may operate in the liquid state; in water the combination of distribution of bond lengths and

high polarizability may be connected not only with broad bands but also with the presence of extensive continua that underly the spectrum. In solids, breakdown of selection rules for acoustic modes could contribute to breadth. Widespread interactions with other vibrational modes, including Fermi resonance, are responsible for the strange and broken profiles and can give the appearance of band plurality.

In some instances strong reflection from solids may artificially broaden IR absorbance spectra. This is certainly the case for the bifluoride ion.

Non-specific forces in the liquid state

Compared to the vapour, band shifts similar in sign though smaller in general magnitude to H-bonding, even in its absence, are observed in liquids and solutions. Thus stretching vibrations are slightly shifted to lower wavenumber; for instance $\nu(\text{C=O})$ from PhC(=O)CH₃ is 1709 cm⁻¹ in the vapour but 1697 cm⁻¹ in hexane and 1692 cm⁻¹ in CCl₄. As a number of different effects are likely to influence the electronic structure and bond dynamics, quantitative predictions are difficult. Attempts to relate shifts to bulk properties only, such as dielectric constant, while ignoring nearest-neighbour interactions have therefore met with little success.

While many of these interactions are essentially very weak they nevertheless can affect vibrational spectra quite significantly. Solute–solvent interactions though only causing minor changes in band profile or position can severely interfere with quantitative analysis. This is because the spectrum of the pure solvent will differ from when it is involved in solution, preventing accurate subtraction if solvent and solute bands overlie each other.

Solute–solvent interactions can occasionally be put to good use, as in the spectral resolution of enantiomeric pairs. By themselves monostereoisomers have identical spectra but when dissolved in an optically active solvent such as D- (or L-)2-octanol (available commercially) yield slightly different spectra. This is because of the asymmetry of the respective interactions which effectively create diastereoisometric pairs.

Collisional interactions

Collisions between molecules can cause distortion and alter the symmetry. Carbon disulfide is linear with a centre of symmetry (the two C=S bonds are equal in length) and thus the symmetric-stretching vibration at about 650 cm⁻¹ is forbidden in the IR. It is however observed as a weak band in IR spectra of liquid CS₂. Normally forbidden bands can also make appearances when charge-transfer complexes are formed, iodine–pyridine being an example.

Collisions are also responsible for inducing transient polarizations in non-polar molecules which result in a characteristic broad absorbance in the liquid in the range 40–100 cm^{-1}. Both geometric distortion and induced local dipole moments are involved.

Matrix isolation

It is possible to control and distinguish many interactions by use of the matrix-isolation technique. This involves highly diluting a species of interest with an 'inert' and IR-transparent gas, such as argon or nitrogen, and condensing the mixture at very low temperature (liquid helium) onto an IR-transparent window (or a Raman cell). By varying the dilution it is possible to isolate single molecules or controlled associations of molecules in 'cages' of the matrix gas. Reactive species can be trapped and studied at leisure. The very low temperature solid phase has the advantage of eliminating translations and collisions yet also yielding generally very sharp bands which can aid in distinguishing mixed species.

Figure 7 shows IR spectra of methanol in various phases including an argon matrix. The bands observed are all from O–H stretching and demonstrate the advantage of the matrix technique. The vapour-phase spectra are dominated by rotational side bands; only H-bonded multimer can be seen in pure liquid and solid, though solution shows dimer as well. Spectra from the matrix at two different dilutions can however distinguish dimer, trimer, tetramer and multimer.

Simple molecular aggregates are amenable to detailed *ab initio* molecular orbital calculations of vibrational modes, which can be compared to experiment to resolve structural dilemmas. For instance theory predicts that the linear-asymmetric H_2O dimer is more stable than the cyclic-symmetric alternative. Consistent with theory, the low-temperature matrix spectrum can only be fitted to the linear structure.

Inelastic neutron scattering

This technique complements IR and Raman spectroscopy in the study of H-bonding in the solid state.

List of symbols

$R = O \cdots O$ bond distance; ΔE = internal-energy change; ΔG = free-energy difference; ΔH = enthalpy of bond formation; $\Delta \nu$ = O–H stretching frequency.

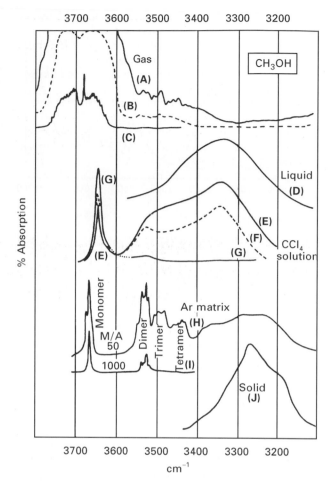

Figure 7 IR spectra of methanol in various phases. (A)–(C) gas; (D) pure liquid; (E)–(G) decreasing concentration in CCl_4 solution; (H)–(I) decreasing concentration in Ar matrix; (J) pure solid. Reproduced with permission from Hallam HE (1973) *Vibrational Spectroscopy of Trapped Species*. Wiley.

See also: **IR Spectrometers; IR Spectroscopy Sample Preparation Methods; IR Spectroscopy, Theory.**

Further reading

Abraham MH, Duce PP, Prior DV, Barratt DG, Morris JJ and Taylor PJ (1989) Hydrogen bonding: solute proton donor and proton acceptor scales for use in drug design. *Journal of the Chemical Society, Perkin Transactions II*, 1355–1375.

Bellamy LJ (1968) *Advances in Infrared Group Frequencies*. London: Methuen.

Clark RGH and Hester RE (eds) (1989) Spectroscopy of matrix isolated species. In: *Advances in Spectroscopy 17*. London and New York: Wiley.

Frohlich H (1993) Calculating the degree of association, equilibrium constant and bond energy for hydrogen bonding in benzyl alcohol and phenol. *Journal of Chemical Education* 70: A3–A6.

Hallam HE (1973) *Vibrational Spectroscopy of Trapped Species*. London and New York: Wiley.

Hadzi D (ed) (1997) *Theoretical Treatments of Hydrogen Bonding*. London and New York: Wiley.

Jeffrey GA (1997) *An Introduction to Hydrogen Bonding*. Oxford University Press.

Kamlet MJ, Solomonovici A and Taft RW (1979) Linear solvation energy relationships, 5: correlations between infrared Δv values and the β scale of hydrogen-bond acceptor basicities. *Journal of the American Chemical Society* 101: 3734–3739.

Pimentel GC and McClellan AL (1960) *The Hydrogen Bond*. San Francisco: Freeman.

Rao CNR, Dwivedi PC, Ratajczak H and Orville-Thomas WJ (1975) Relation between O–H stretching frequency and hydrogen bond energy: re-examination of the Badger–Bauer rule. *Journal of the Chemical Society, Faraday Transactions II* 71: 955–966.

Scheiner S (1997) *Hydrogen Bonding: A Theoretical Perspective*. Oxford University Press.

Tomkinson J (1992) The vibrations of hydrogen bonds. *Spectrochimica Acta* 48A: 329–348.

Hyphenated Techniques, Applications of in Mass Spectrometry

WMA Niessen, hyphen MassSpec Consultancy, Leiden, The Netherlands

MASS SPECTROMETRY
Applications

Introduction

One of the most fascinating fields of instrumental development in mass spectrometry (MS) is hyphenation: the on-line coupling of various techniques to MS. Apart from the obvious combinations of straightforward coupling of gas chromatography (GC) or liquid chromatography (LC) to MS, a wide variety of other combinations has been described. This contribution pays attention to the rationale of hyphenated techniques and briefly indicates a number of typical applications.

The key idea behind hyphenation is the significant gain in signal-to-noise ratio, and thus improvement in detection limit, that can be achieved by multidimensional methods. This concept was nicely pictured by Yost and co-workers, in the early 1980s (**Figure 1**). While in hyphenation the response or signal achieved decreases with the increasing number of couplings or dimensions, the (chemical) noise decreases even faster due to the increased selectivity, thus resulting in an improved signal-to-noise ratio. Another important issue is the ability to avoid sample losses and sample contamination in on-line rather than off-line combinations.

In principle, one can discriminate between various approaches to hyphenation, depending on the primary objective:

(a) On-line separation technique coupled to MS, e.g. GC-MS and LC-MS.

(b) On-line sample pretreatment in combination with a separation technique coupled to MS.

(c) On-line multidimensional separation techniques coupled to MS.

(d) On-line coupling of a separation technique to multiple detection strategies, one of which is MS.

Obviously, a combination of these strategies is also possible. Examples of these approaches are given below.

Tandem mass spectrometry

The first hyphenated approach to be considered is the on-line combination of MS and MS, i.e. tandem mass spectrometry (MS-MS). A variety of combinations of different mass analysers have been described, including quadrupole and magnetic-sector analysers as MS_1, and quadrupole, magnetic-sector, ion-trap and time-of-flight analysers as MS_2. Instruments like triple-quadrupoles are widely used for MS-MS, either as stand-alone systems with sample introduction via a solids insertion probe or flow-injection analysis, or in on-line combination with GC or LC. The work of Yost and co-workers and of Hunt and colleagues exemplify these methods.

In the most common mode, i.e. the product-ion mode, MS_1, is used to select a precursor ion with a particular m/z from the variety of ions generated in the ion source. The mass selected ions are dissociated via collisions with an inert gas in a collision cell, and

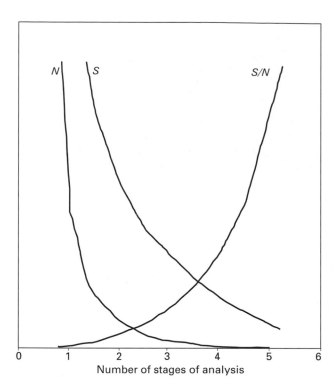

Figure 1 Hyphenation of analytical techniques. Dependence of the signal *S*, the noise *N*, and the signal-to-noise (*S/N*) ratio on the number of analytical stages.

the product-ions are subsequently mass-analysed by MS_2. In this setup, MS_1 can be considered as a separation technique, while MS_2 is operated as a conventional mass analyser and detector. Analogies between MS-MS and GC-MS have frequently been drawn (**Figure 2**). It has been argued whether a chromatographic separation is actually still required in a combination with MS-MS, given the excellent selectivity that can be achieved. However, it is currently generally agreed that some separation is required, at least in the analysis of samples of biological or environmental origin, in order to avoid rapid contamination of the ion source and in order to reduce and/or avoid analyte ion suppression effects, i.e. in electrospray ionization.

MS-MS is currently very widely used in combination with chromatographic separation methods, especially LC. The obvious reason for this is the frequent use of soft ionization techniques in LC-MS interfacing, i.e. electrospray and atmospheric-pressure chemical ionization. MS-MS allows additional structural information as well as the molecular mass information to be obtained. On-line LC-MS-MS is currently the method-of-choice in quantitative bioanalysis in (pre-)clinical pharmacological studies during drug development in pharmaceutical industries. In these studies, the instrument is operated in

selective-reaction monitoring (SRM) mode, selecting a particular precursor ion in MS_1 and selecting and detecting one or more product ions in MS_2. The excellent selectivity, the high level of confidence of identity, and the ability to use stable isotopically labelled internal standards are important arguments for the use of SRM in this type of routine applications.

On-line chromatography–mass spectrometry

The on-line combination of a chromatographic separation technique (GC, LC, but also thin-layer chromatography (TLC), capillary zone electrophoresis (CZE) and supercritical fluid chromatography (SFC)) with MS enables the mass spectrometric characterization of components in complex mixtures after separation with minimal or no sample loss. It is especially useful in the identification of minor or trace components that are difficult to collect by fractionation of the column effluent or would be easily lost. Furthermore, as already indicated above, on-line GC-MS and LC-MS-MS are important tools in quantitative bioanalysis as well. Obviously, fractionation in large series of samples for routine quantitative applications would be extremely time-consuming and ineffective.

GC-MS plays an important role in many application areas, including the characterization of components in petroleum and derived products and in essential oils, the identification and quantitation of compounds of environmental interest such as polychlorodibenzodioxins and related compounds, polycyclic aromatic hydrocarbons, pesticides and herbicides, and a variety of other microcontaminants. In addition, GC-MS is important in the analysis of compounds of pharmacological, forensic and/or toxicological interest, including drugs, anaesthetics, steroids, growth hormones and drugs of abuse.

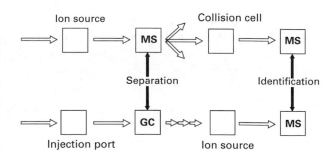

Figure 2 Schematic comparison of MS-MS and GC-MS.

LC-MS is applied in complementary fields, where the analytes are not amenable to GC-MS. LC-MS is for instance applied in the identification and quantitation of pesticides, herbicides, surfactants and (sulfonated) azo dyes in environmental samples, in the identification of drugs during drug development, their degradation products and metabolites, in the quantitative bioanalysis of drugs and related compounds in biological tissues and fluids, in the characterization of natural products, such as alkaloids, taxoids, toxins, as well as endogenous compounds like acylcarnitines, prostaglandins, bile acids. Furthermore, LC-MS plays an important role in biochemical and biotechnological applications of MS, via the electrospray MS and electrospray LC-MS analysis of biomacromolecules like peptides, proteins and DNA fragments.

Methods and some applications of GC-MS and LC-MS are discussed in a separate contribution.

The on-line combination of TLC and MS via FAB, liquid SIMS, or MALDI has also been frequently described. The TLC-MS combination, recently reviewed by Somsen and co-workers, is applied for a wide variety of compounds, including drugs and their metabolites, antibiotics, steroids, alkaloids, lipids, bile acids, porphyrins, dyes and peptides.

On-line sample pretreatment

With the advent of LC-MS technologies, the on-line combination with various sample pretreatment strategies received considerable attention. The rationale for on-line sample pretreatment is to avoid sample losses and sample contamination during the off-line transfer from one step of the analytical procedure to another. In addition, an on-line procedure greatly facilitates automation of the complete procedure, thereby speeding up the analysis. The most successful and most widely applied approach is on-line solid-phase extraction (SPE) in combination with LC-MS. A number of instruments have been developed and commercialized for this combination, e.g. the Varian AASP, the Gilson ASPEC, the Spark Holland Prospekt, and the Merck OSP-2. In all these automated systems, the sample constituents within a certain polarity range are trapped onto a short cartridge column (or Empore disk) containing reversed-phase LC packing material. The cartridge or disk may be washed with water to remove hydrophilic sample constituents. Subsequently, the analytes of interest are desorbed and transferred to an LC column for separation. These on-line SPE strategies serve for both sample pretreatment and analyte preconcentration. While on-line SPE-LC-MS was initially mainly

developed for quantitative bioanalysis, the most important current application is in environmental analysis, i.e. in the selective sample pretreatment and analyte preconcentration of pesticides, their degradation products as well as other microcontaminants from surface water as an on-line part of the SPE-LC-MS or SPE-LC-MS-MS analysis. The results of the group of Brinkman show that analyte preconcentration by factors up to 10^3 or 10^4 are feasible in pesticide analysis from surface water, enabling concentration detection limits well below 0.1 $\mu g L^{-1}$.

Consecutively, the same group demonstrated that SPE can also be applied in an on-line combination with GC-MS. An example is the SPE-GC-MS analysis of 10 mL of river Rhine water spiked at the 0.5 $\mu g L^{-1}$ level with 80 microcontaminants, such as chlorobenzenes, aromatic compounds, anilines, phenols and organonitrogen and organophosphorus pesticides.

In addition to SPE, other sample pretreatment methods have been combined with GC-MS (see Goosens and co-workers), LC-MS or to MS(-MS) directly, e.g. on-line membrane sample introduction, solid-phase micro-extraction (SPME), supercritical fluid extraction and membrane dialysis.

A variety of on-line sample pretreatment procedures have been described for the coupling to on-line capillary zone electrophoresis (CZE)-MS, including capillary isotachophoresis, electrodialysis, liquid–liquid electroextraction and SPE on Empore disks. The latter system was applied to the analysis of the neuroleptic drug haloperidol in patients' urine by on-line Empore-disk SPE-CZE-MS.

Multidimensional separation techniques

Multidimensional separation techniques have been developed with a number of objectives such as increased peak capacity and/or improved resolution for the separation of highly complex samples, shortened analysis time via heartcut and partial analysis of fractions from complex samples, and enhanced detection of trace components. Both GC-GC, LC-GC, and LC-LC methods as well as various other multidimensional combinations involving CZE or SFC have been described and coupled to mass spectrometry. In most cases, the multidimensional approach comprises a combination of two chromatographic columns, either containing two different stationary phases (both GC and LC), or developed with two different mobile phases (LC only). The sample is injected onto the first column. Part of the chromatogram is heartcut, either sampled onto a retention gap or short trapping column or transferred directly, and subsequently analysed on the second column, which is then interfaced to an MS. A typical setup for LC-LC-MS is shown in **Figure 3**.

Several examples of on-line multidimensional GC-MS were reviewed by Ragunathan and co-workers including the characterization of essential oils in order to determine the composition, to identify or quantify enantiomers, or to study the chemo-taxonomy of the oil, and the determination of coplanar polychlorobiphenyls congeners.

The on-line combination of LC-GC-MS is applied in a variety of analytical problems, such as the analysis of (trace amounts of) aromatic compounds in complex fossil fuel fractions or in vegetable oil, the identification of microcontaminants in surface water, and the identification of impurities in pharmaceutical products.

The on-line combination of LC-LC-MS has been investigated for a number of reasons. In addition to the general prospects of multidimensional separation techniques, especially enhanced selectivity, there was special interest in the ability to perform LC-LC with two different mobile-phase compositions. In this way, it should be possible to avoid problems with mobile-phase incompatibility due to the use of non-volatile mobile-phase constituents. A good example of this approach is the determination of enantiomers of β-blockers in plasma samples, described by Edholm and co-workers. Racemic mixtures of a β-blocker like metoprolol can be separated on a α_1-acid glycoprotein column. However, the chromatography requires the use of a 20 mM phosphate buffer (pH 7) in the mobile phase, which is not compatible with on-line LC-MS. Therefore, the chiral column was coupled via a set of two trapping columns to a common reversed-phase LC column. After separation, the two enantiomers were sepa-rately trapped onto two short columns and subsequently transferred to the second LC column for analysis with a mobile phase containing ammonium acetate, which is well compatible with the thermos-pray LC-MS interfacing applied in this study. Subsequently, it was demonstrated that by the use of MS-MS the second separation step could be avoided: the effluent from the trapping column can be directly introduced into the MS-MS system, operated in SRM mode. This phase-system switching approach is well suited for solving this type of mobile-phase incompatibility problem.

Multiple detection strategies

The increasingly more difficult analytical problems to be solved is an important impetus in all developments in hyphenated techniques, but especially in multidimensional detection strategies. Development of adequate separation techniques for the analysis of complex samples is often a difficult task, independent of the question whether one is interested in only one compound present at trace level, or in characterization and structure elucidation of most sample components. Multidimensional separation techniques were developed in order to achieve the separation of the components of very complex samples. MS is an extremely powerful technique for the identification and structure elucidation of unknowns. The sensitivity of MS as a fairly universal technique in this area is unsurpassed. However, in many instances the information obtained from MS is not sufficient to enable unambiguous identification of the unknown compounds. Furthermore, in certain application areas it is obligatory to identify the components with such a high level of confidence that independent results from a second technique in addition to MS are required, as for instance for impurity screening of pharmaceutical products. Improved levels of confidence in identification and structure elucidation can be achieved by the use of multiple detection strategies. Although detection with a variety of detectors can be achieved via multiple analysis of the sample on systems equipped with different detectors, there has been a growing interest and need for systems which enable the use of multiple detectors coupled to a single analytical separation system. This is especially important in two areas, i.e. in the analysis of very complex samples where multiple detection strategies coupled to one single data-acquisition and processing system can greatly facilitate the interpretation of the data, and in the high through-put analysis where minimization of the analysis time is of utmost importance.

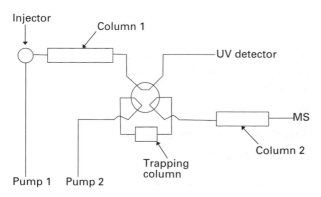

Figure 3 Schematic diagram for a coupled-column LC-LC-MS system.

Several dual or multiple detection systems involving MS have been described. In this respect, one should not forget that the mass spectrometer is a destructive detector. Therefore, the mass spectrometer should be the final detector in a series. Combination with another destructive detector, e.g. a flame ionization detector, is possible only after splitting the column effluent, while in-line coupling with a nondestructive detector, e.g. a UV absorbance detector, is only possible when this detector does not result in too much chromatographic peak broadening. In several cases, the second nondestructive detector is used in parallel rather than in series, e.g. because the sensitivities of both detectors are widely different, or in order to avoid the loss of chromatographic resolution due to excessive peak broadening by the first detector in the series. The most widely applied dual detector strategies are those where MS is combined with another spectrometric technique, in order to obtain complementary structural information.

Two types of spectrometric detectors have been widely used in an in-line combination with GC-MS, i.e. Fourier-transform infrared (FT-IR) and atomic emission spectrometry (AED). Both FT-IR and AED are used in parallel with the MS, i.e. after a split.

GC-FT-IR-MS systems are commercially available. When the measured IR spectrum of an unknown is not present in the IR spectral library, the FT-IR part of the system can help in the identification of an unknown by providing information on functional groups and the structural backbone. While homologues cannot readily be identified from IR spectra but easily discriminated by MS, the FT-IR part can be used as an important tool in discriminating between structural isomers where MS generally fails. Typical applications of GC-FT-IR-MS can be found in the analysis of essential oils, flavours and fragrances, and in the identification of isomeric reaction products or other types of isomeric compounds. The potential of a dual detection GC-FT-IR-MS system was also explored for the identification of illicit drugs, e.g. in tracing banned stimulants used by athletes.

The combination of GC-AED-MS to our knowledge is not (yet) commercially available. The element specificity of the AED can be of great help in identifying particular classes of compounds in complex samples, e.g. in environmental screening. The AED provides information on the presence of certain elements, e.g. halogens, nitrogen, or phosphorus, in the chromatographic peaks detected. This facilitates the search for relevant peaks as well as the interpretation of the mass spectra of particular compounds in a complex chromatogram. In addition, the AED can

provide a fairly accurate determination of the elemental composition, thereby often excluding the need for accurate mass determination by expensive magnetic sector instruments. Furthermore, GC-AED-MS can be applied in the analysis of organometallic compounds, e.g. organomercury and organotin compounds.

The availability of nondestructive flow-through detectors for LC, e.g. UV absorbance and fluorescence detectors, readily leads to the application of dual detection systems in LC-MS. While the additional information of single-wavelength UV detection is generally limited, the availability of efficient and sufficiently sensitive UV-photodiode array detectors enables the on-line acquisition of UV spectra as well as mass spectra (**Figure 4**). Although the structural information available from a UV spectrum is generally limited, especially in a relatively undefined solvent mixture during gradient elution, the on-line combination can successfully be used in facilitating identification when appropriate UV spectral libraries are available. The combination of retention time in a well defined LC system, the match of a UV spectrum, and the molecular mass (and possible structural information) from MS are especially useful for confirmation of identity, e.g. in multiresidue screening for pesticides in environmental samples or antibiotics in food products. Furthermore, an excellent feature of UV-PDA systems in combination with appropriate software is peak purity assessment, based on changes of the UV spectra over the chromatographic peak. This feature may be helpful in impurity profiling by LC-UV-PDA-MS.

Although the use of FT-IR in combination with LC is of growing importance, the on-line or in-line combination of FT-IR and MS is not yet frequently applied. However, the on-line combination of LC-MS and NMR is currently introduced as an important tool in identification of impurities, degradation products, and/or metabolites of pharmaceutical products. A recent example of on-line LC-NMR-MS is the characterization of ibuprofen metabolites in human urine. Simultaneous acquisition of electrospray MS and NMR spectra of synthetic drug products in an 'open-access' setting has also been reported.

Final considerations and perspectives

Hyphenation in relation to MS is an important research topic because in many instances the optimum tuning of the two or more parts of the setup requires

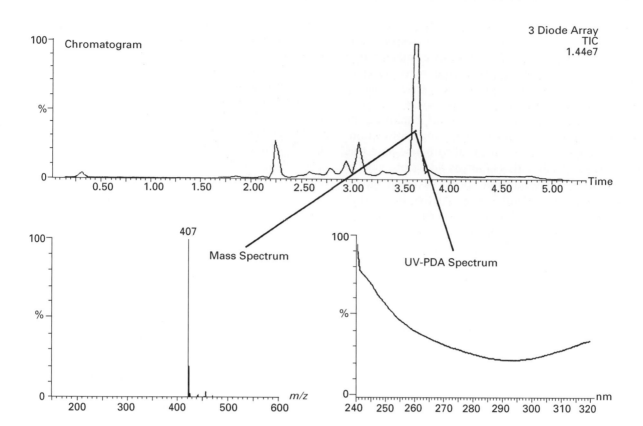

Figure 4 An example of typical results from an LC-UV-PDA-MS setup for dual detection of separated compounds.

considerable attention. Good knowledge of both parts is required in order to avoid the often inevitable compatibility problems and to obtain best possible results. However, hyphenated MS techniques have already widely proved their value in solving real analytical problems.

In principle, the use of advanced data-processing software could be considered as a hyphenated approach as well, but in general is not. However, one can question whether the use of principal component analysis and similar data-processing techniques does not have a similar impact on the analytical procedure and its result as the hardware hyphenation discussed above.

Software for both controlling the hyphenated combination of techniques and efficient processing of the data is of utmost importance to the breakthrough and general use of a hyphenated technique. In this respect, the development of automatic tuning and calibration software for GC-MS and more recently LC-MS, of computer spectral library searching after electron ionization, and of efficient quantitation software packages after both GC-MS and LC-MS can be considered as important steps in the development of hyphenated techniques. Control of the complete (often multivendor) instrumentation,

including the GC or LC chromatograph, the sample processor or autosampler, on-line or off-line second (scanning) detectors from within a single software platform, often the MS control and acquisition software, is also essential and obligatory for the success of the technique. Some instruments enable high levels of automation and control via the use of macro scripts or instrument control languages. Such approaches may be used for optimization and/or control of the various steps in the hyphenated technique, but also for artificial-intelligence type of applications where the software makes decisions concerning the analytical strategy to follow or the type of experiments to perform on the basis of the data acquired. The latter is, for instance, applied in data-dependent product-ion scanning during GC-MS-MS and LC-MS-MS.

Progress is also made in software for mass spectral interpretation, especially for the so-called deconvolution of (mixed) ion envelopes of multiply charged protein ions generated by electrospray ionization, and for the peptide sequencing by MS-MS techniques. In that respect, the development of tools to search extensive protein and DNA databases using peptide maps, peptide molecular masses, LC retention time data, and sequence tags is an enormous step forward in the routine application of hyphenated MS-MS.

Excellent examples of the hybrid of hyphenated hardware, automated sample processing via computer control of the instrumentation, and automated data processing can be found in approaches recently developed in relation to combinatorial chemistry. In order to screen a particular combinatorial library in a 96-well plate, high throughput electrospray LC-MS is applied. The data are postprocessed by means of a browser, which compares the molecular mass measured with expected values, taking into account the various cationized and anionized species that may be generated during electrospray ionization. By means of green and red colours, the software indicates on a screen representation of the 96-well plate which samples were found to be present and which were not. In addition to this, and using similar software strategies, the use of automated fraction collection after preparative LC under control of the data acquired by MS has been demonstrated. This hyphenated approach allows the purification of particular products from extensive combinatorial libraries based on the results of a first biological screening and prior to a more advanced biological screening, using the purified products.

See also: **Atmospheric Pressure ionization in Mass Spectrometry; Biochemical Applications of Mass Spectrometry; Chemical Structure Information from Mass Spectrometry; Chromatography-MS, Methods; Isotopic Labelling in Mass Spectrometry; Medical Applications of Mass Spectrometry; MS-MS and MS***^{n}**.**

Further reading

Brotherton HO and Yost RA (1984) Rapid screening and confirmation for drugs and metabolites in racing animals by tandem mass spectrometry. *American Journal of Veterinary Research* 45: 2436.

Edholm L-E, Lindberg C, Paulson J and Walhagen A (1998) Determination of drug enantiomers in biological samples by coupled column liquid chromatography and liquid chromatography–mass spectrometry. *Journal of Chromatography* 424: 61.

Hankemeier Th, Van Leeuwen SPJ, Vreuls RJJ and Brinkman UATh (1998) Use of a presolvent to include volatile organic analytes in the application range of on-line solid-phase extraction–gas chromatography–mass spectrometry. *Journal of Chromatography A* 811: 117.

Hogenboom AC, Speksnijder P, Vreeken RJ, Niessen WMA and Brinkman UATh (1997) Rapid target analysis of microcontaminants in water by on-line single-short-column liquid chromatography combined with atmospheric-pressure chemical ionization tandem mass spectrometry. *Journal of Chromatography A* 777: 81.

Hunt DF, Shabanowitz J, Harvey TM and Coates ML (1983) Analysis of organics in the environment by functional group using a triple quadrupole mass spectrometer. *Journal of Chromatography* 271: 93.

Kitson FG, Larsen BS and McEwen CN (1996) *Gas Chromatography and Mass Spectrometry. A Practical Guide.* London: Academic Press.

Niessen WMA (1998) *Liquid Chromatography–Mass Spectrometry,* 2nd edn. New York: Marcel Dekker.

Ragunathan N, Krock KA, Klawun C, Sasaki TA and Wilkins CL (1995) Multispectral detection for gas chromatography. *Journal of Chromatography A* 703: 335.

Slobodnik J, Hogenboom AC, Vreuls JJ *et al* (1996) Trace-level determination of pesticide residues using on-line solid-phase extraction–column liquid chromatography with atmospheric pressure ionization mass spectrometric and tandem mass spectrometric detection. *Journal of Chromatography A* 741: 59.

Somsen GW, Morden W and Wilson ID (1995) Planar chromatography coupled with spectroscopic techniques. *Journal of Chromatography A* 703: 613.

Vreuls JJ, Bulterman A-J, Ghijsen RT and Brinkman UATh (1992) On-line preconcentration of aqueous samples for gas chromatographic–mass spectrometric analysis. *Analyst* 117: 1701.

Walhagen A, Edholm L-E, Heeremans CEM *et al* (1989) Coupled-column chromatography–mass spectrometry. Thermospray liquid chromatographic–mass spectrometric and liquid chromatographic–tandem mass spectrometric analysis of metoprolol enantiomers in plasma using phase-system switching. *Journal of Chromatography* 474: 257.

Index

NOTE

Bold page number locators refer to complete articles on the various topics covered by this encyclopedia. Illustrations, including spectra, are indicated by *italic* page numbers.

Text and tables are located by page numbers in normal print.

Cross references, prefixed by *see* and *see also*, are also listed at the end of each article.

proton 2D rotating frame Overhauser effect
spectroscopy, polypeptide *1651*
proton affinities
1998 scale 1897–1899
definition 1893–1895
determination 1895–1897
ionization energy relationship 974
mass spectrometry **1893–1901**
thermospray ionization, mobile-phase
constituents 2357, 2356
proton bridging, stereochemistry 2212
proton decoupling
13C NMR 1098
in vivo NMR 852, 857
proton Fourier transform–pulsed gradient
spin-echo data, complex solutions *371*
proton induced X-ray emission (PIXE)
accuracy 754
analysis focusing 1901, 1902–1903,
1905
detector limits 754
principles *752, 753,* 751–753
quantification 753–754
spatial resolution 758–760
proton microprobes, high energy
spectroscopy **1901–1905**
proton NMR studies
see also in vitro NMR; *in vivo* NMR;
in vivo proton NMR
amniotic fluid 113–114
aqueous humour 115
bile 113
biofluids 102–107
resonance assignment *100, 98*
blood plasma *100,* 99–101
13C NMR comparison 149
cancer detection 101
carbohydrates *173, 174, 174, 175,*
173–177
2D homonuclear NMR studies *175,*
174–177
Gal*p*β1-4Glc*p 173, 174, 175, 173*
β1-4glucopyranose *174, 173*
homonuclear Hartmann–Hahn
spectroscopy 176
through-space dipolar interaction
methods 176–177
total correlation spectroscopy 176
castor bean stems *1535, 1534*
cells *188,* 185–188
cerebrospinal fluid 112
chemical shift 2235–2236
diabetes 108
digestive fluids 115
disadvantages 1097
drug interactions *2268,* 2267–2268
drug metabolism 376
electron paramagnetic resonance
comparison 437–438
erythrocytes 108–109
ethanol *1547,* 1546–1547
follicular fluid 113–114
high resolution solid-state NMR
813–825
kauradienoic acid spectrum *1557*
lipoprotein analysis 101
microscopy *1529, 1535, 1534*
organic molecules 2234–2238
pathological cyst fluid 115

proton NMR studies (*continued*)
perfused heart studies 1764, 1768
radiation damping 2150–2151
rat erythrocytes *188,* 187
renal failure 108
saliva 115
seminal fluid 112–113
single off-resonance peak in spectrum
1556
stereochemical assignments 2238
superconducting magnets *1578, 1579,*
1577–1578
synovial fluid 114
through-bond coupling 2236–2238
vitreous humour 115
whole blood 99–101, 108–109
proton nuclear Overhauser enhancement
spectroscopy, trypsin inhibitor *1649*
proton shielding, calculation 1748
proton–electron double resonance imaging
(PEDRI) 444
proton-bound molecule pairs 993
proton-decoupled 1D 13C NMR spectrum
1097
proton-observed carbon-edited techniques
(POCE) 860
protonated molecules (MH+)
CID mass spectrometry 2216
mass spectral stereochemical effects
2211
methane, mass spectrometry theory 1895
proton bridging 2212
stabilization 2212
protonation entropies, mass spectrometry
theory 1894, 1895
protons
13C NMR studies, irradiation 152, 151,
152
magnetic dipole moment, discovery 1232
membrane studies, NMR spectroscopy
1283, *1284,* 1282
nuclear properties 814
relaxation times, paramagnetic
compounds 1402
resonance frequencies, stray fields 1397
spin–lattice relaxation times, high
pressure NMR spectroscopy 766,
765–766
transfer
chemical ionization, ion trap MS
1008
ion–molecule *987,* 986–987
laser-induced optoacoustic
spectroscopy *1131,* 1131–1132
mass spectrometry theory 1894,
1895–1896
protons relaxation enhancement (PRE)
method, relaxation reagents *229,*
228–229
Prozac *see* fluoxetine
Prussian blue (PB;
iron(III)hexacyanoferrate(II)) 2171
PSD *see* phase sensitive detection;
polarization state detector; position-
sensitive detector; post-source decay
pseudo-double-beam instruments 59
pseudo-FID *see* pseudo-free induction
decay

pseudo-first-order-reactions, laser-induced
optoacoustic spectroscopy 1124
pseudo-free induction decay (pseudo-FID),
NMR RF gradients *1941,* 1940–1941
pseudo-Voigt functions *1869*
pseudocontact shifts *see* dipolar shifts
pseudoephedrines, vibrational circular
dichroism 2411
PSG *see* polarization state generator
195Pt-15N, spin–spin couplings 1513
195Pt, NMR spectroscopy 719–720
PTB *see* phosphotyrosine binding
PTFE *see* poly(tetrafluoroethylene)
PTS *see* photothermal spectroscopy
pullulan, Raman optical activity 1960
pulmonary arteries, MRI flow measurement
1366, 1370–1371
pulse effects, magnetic field gradients, high
resolution NMR spectroscopy
1224–1225
pulse radiolysis, UV-visible spectroscopy
218
pulse sandwiches
bilinear rotating decoupling *1566,*
1565–1566
NMR 1565–1567
pulse sequences
attached proton test *1560,* 1559
broad-band decoupling 1567
correlation spectroscopy *1563,*
1562–1563
frequency selective 1566–1567
gradient NMR 1565
heteronuclear single quantum coherence
1564
inversion–recovery NMR 1572
Jeener–Broekaert NMR 1572
NMR **1554–1567**
quantum mechanical methods 1561
rotational resonance *2140,* 2140
spectral editing 1558–1560
spin-echo *1560,* 1558
spin-echo NMR 1572
pulse techniques
13C NMR studies *152, 153, 154,*
152–154
electron paramagnetic resonance
196–197, 437–438, 463–464
nuclear quadrupolar resonance *1664,*
1663, *1669,* 1665–1668
pulsed characteristics, photoacoustic
spectroscopy 1817–1818, 1821
pulsed field gradient spin-echo (PFGSE),
NMR RF gradients *1940,*
1939–1940
pulsed field gradients (PFGs)
2D NMR 2377–2378
in vivo NMR 866
MRI 1376
solvent suppression *2147,*
2147–2148
pulsed gradient spin-echo (PGSE) 370–371,
370
eddy currents suppression 372
high perfomance instrumentation 372
NMR *1530, 1533,* 1533–1534
pulsed lasers 1115, 1162–1163, 1167–1168
laser-induced optoacoustic spectroscopy
1124

White, RL **288–293**

whole-body MR *see* magnetic resonance imaging, clinical applications

wide angle X-ray scattering (WAXS), fibres/films *531*, 530

wide-bore magnets
NMR spectroscopes 1577
pulsed gradient spin-echo measurements 372

wide-field geometry, electronic spectroscopy 77

wide-field illumination, Raman microspectroscopy *1951*, 1949

wideline methods, MRI of rigid solids 2012

wideline separation of interactions (WISE) *2134*, 2134

Wien's displacement law, blackbody radiation, thermal light sources *1162*, 1161

Wieser, ME **1072–1086**

wigglers
X-ray absorption fine structure experiments 2450
X-ray emissions theory 2495

Wigner matrices 1705

Wigner symbols 2346–2348
spherical tensors, rotational spectroscopy theory 2018–2020

Wigner–Eckart theorem 2339
nuclear quadrupolar resonance 1674
spherical tensors, rotational spectroscopy theory 2018

Wiley–McLaren time-of-flight focusing methods 2362, 2363

Wilkins, JPG **592–593**

Williams, AJ **1105–1113**

Williams, PM **2051–2059**

windows, IR spectroscopy samples *1059*, 1058

WISE *see* wideline separation of interactions

WKB scattering 960–961

Wlodarczak, G **1297–1307**

Wobrauschek, P **2478–2486**

Woizniak S **1594–1608**

Wolf Rayet stars 2201

wood
atomic spectroscopy **621–631**
preservatives 623, 624
atomic spectroscopy 622
penetration/retention 622–623

Woodward rules, spectral shifts 389

work allocation, laboratory information management 1108

workplace drugtesting 617, 615–616, 616–618

World Wide Web (WWW)
laboratory information management systems 1111–1112
S³LIMS 1112

WWW *see* World Wide Web

X

X-band imaging, electron paramagnetic resonance imaging *440*, *441*, *442*, 439–441

X-ray absorption fine structure (XAFS) spectroscopy
beam intensity 2449
core-hole lifetime broadening 2448
energy resolution 2448–2449
experiments 2447–2453, *2449*
harmonics 2449, 2450, 2452
tunability 2449

X-ray absorption spectrometers **2447–2454**, *2449*
detectors 2453
optics 2451–2453
sources 2449–2450

X-ray crystallography
NMR comparison 1209
proteins, NMR studies 1892–1893

X-ray detected magnetic circular dichroism (XMCD) 1218, 1222–1223

X-ray diffraction (XRD)
see also powder X-ray diffraction
anomalous scattering 1251–1253
atomic displacements *2112*, 2111–2113
atomic structure 2106–2112
bond distances 2108, 2109, 2107–2108
chiral molecules *2111*, 2110–2112
DNA, conformational transitions *534*, 533
electric field effects *1254*, 1254
electron density 2124–2115
fibres/films **529–539**
high pressure 930–931
inorganic compounds/minerals **924–933**
instrumentation *536*, 536
intermolecular contacts 2110, 2109–2110
ion implantation effects *1254*, 1254
magnetic scattering 1253
materials science applications **1248–1257**
muscle contraction 533–534
perturbations *1254*, 1254
polyethylene *536*, 536
poly(ethylene terephthalate) *535*, 534–536, 537–538
polymer applications *538*, 536–537, 537–538, 1860
principles 1250, *1250*, 1249–1250
semiconductors 1254
small molecule applications **2106–2115**
stress/strain relationships 1253–1254
structural databases 2110
structure determination *1252*, 1250–1252
synchrotron 928–933
time-resolved studies *1255*, 1256
topography *1255*, 1255
X-ray sources *1249*, 1248–1249

X-ray emission spectroscopy (XES)
applications **2455–2462**
lasers 1123
methods **2463–2467**

X-ray fluorescence spectroscopy (XFS/XRF) 2466
applications **2478–2486**
calibration 426
environmental/agricultural applications **422–429**
geological analysis 668
interference effects 424

X-ray fluorescence spectroscopy (XFS/XRF) (*continued*)
molecular orbitals 2455–2457
soil analysis 427
spectrometers 2447, **2467–2477**
sulphate ions *2457*

X-ray photoelectron spectroscopy (XPS) 1824, 1823, 2282, 2520
atomic charges 2457–2458

X-ray spectroscopy
dynamic structure factors 2082
theory **2487–2498**
see also photoelectron spectroscopy

X-ray synchrotron radiation 530–531
experimental techniques 531–533
powder X-ray diffraction 1867, 1872–1873

X-rays
absolute configuration determination 472
absorption
cross-sections
oxygen *2448*, 2447
platinum *2448*, 2447
proton induced X-ray emission 752–753
atomic inner shell transitions 2487–2492
characteristic 2487–2489
classification 2463
conversion efficiency 2493
critical frequency 2495
detection
Mössbauer spectroscopy *1316*, 1316–1318
X-ray fluorescence 2469–2472
dispersion, X-ray fluorescence 2470, 2469–2470
emission
by free electrons 2492–2495
characteristic *2464*, 2463–2465
chlorine compounds 2459, 2458–2459
complexation 2461–2462
continuous 2463, 2464–2465
dimethyl sulphide complexes 2460, *2460*
electron induced/proton induced comparison *752*, 751
intensities 2456
proton induced *752*, *753*, *754*, 751–754
rodano-groups *2461*, 2460–2461
sulphur *2455*, 2455–2456
generation
extreme nonlinear optics 1691, 1699–1700
inverse bremsstrahlung process 1700
L-emissions, 3d atomic orbitals 2458–2459
method of Bijvoet, molecular configuration 2403
microanalysis, X-ray emission spectroscopy 2466
muonic 2497
photoacoustic spectroscopy *1803*
powder X-ray diffractometers 1867
scattering cross-sections
oxygen *2448*, 2447
platinum *2448*, 2447